Achieving durable disease resistance in cereals

It is widely recognised that agriculture is a significant contributor to global warming and climate change. Agriculture needs to reduce its environmental impact and adapt to current climate change whilst still feeding a growing population, i.e. become more 'climate-smart'. Burleigh Dodds Science Publishing is playing its part in achieving this by bringing together key research on making the production of the world's most important crops and livestock products more sustainable.

Based on extensive research, our publications specifically target the challenge of climate-smart agriculture. In this way we are using 'smart publishing' to help achieve climate-smart agriculture.

Burleigh Dodds Science Publishing is an independent and innovative publisher delivering high quality customer-focused agricultural science content in both print and online formats for the academic and research communities. Our aim is to build a foundation of knowledge on which researchers can build to meet the challenge of climate-smart agriculture.

For more information about Burleigh Dodds Science Publishing simply call us on +44 (0) 1223 839365, email info@bdspublishing.com or alternatively please visit our website at www.bdspublishing.com.

Related titles:

Achieving sustainable cultivation of maize Volume 2: Cultivation techniques, pest and disease control
Print (ISBN 978-1-78676-012-8); Online (ISBN 978-1-78676-014-2, 978-1-78676-015-9)

Achieving sustainable cultivation of wheat Volume 1: Breeding, quality traits, pests and diseases
Print (ISBN 978-1-78676-016-6); Online (ISBN 978-1-78676-018-0, 978-1-78676-019-7)

Integrated disease management of wheat and barley
Print (ISBN 978-1-78676-216-0); Online (ISBN 978-1-78676-219-1, 978-1-78676-218-4)

Achieving sustainable cultivation of barley
Print (ISBN 978-1-78676-308-2); Online (ISBN 978-1-78676-310-5, 978-1-78676-311-2)

Chapters are available individually from our online bookshop:
https://shop.bdspublishing.com

climate•SMART•publishing
IN AGRICULTURAL SCIENCE

BURLEIGH DODDS SERIES IN AGRICULTURAL SCIENCE
NUMBER 106

Achieving durable disease resistance in cereals

Edited by Professor Richard Oliver, formerly Curtin University, Australia

burleigh dodds
SCIENCE PUBLISHING

Published by Burleigh Dodds Science Publishing Limited
82 High Street, Sawston, Cambridge CB22 3HJ, UK
www.bdspublishing.com

Burleigh Dodds Science Publishing, 1518 Walnut Street, Suite 900, Philadelphia, PA 19102-3406, USA

First published 2021 by Burleigh Dodds Science Publishing Limited

Library of Congress Control Number: 2021940786

British Library Cataloguing in Publication Data
A catalogue record for this book is available from the British Library

ISBN 978-1-78676-601-4 (Print)
ISBN 978-1-78676-604-5 (PDF)
ISBN 978-1-78676-603-8 (ePub)
ISSN 2059-6936 (print)
ISSN 2059-6944 (online)

DOI 10.19103/AS.2021.0092

Typeset by Deanta Global Publishing Services, Dublin, Ireland

Contents

Part 5 Fungal diseases of cereals: net blotch

Part 7 Barley yellow dwarf virus

Part 8 Fungal diseases of cereals: Regional strategies

Series list

Title | Series number

Achieving sustainable cultivation of maize - Vol 1 001
From improved varieties to local applications
Edited by: Dr Dave Watson, CGIAR Maize Research Program Manager, CIMMYT, Mexico

Achieving sustainable cultivation of maize - Vol 2 002
Cultivation techniques, pest and disease control
Edited by: Dr Dave Watson, CGIAR Maize Research Program Manager, CIMMYT, Mexico

Achieving sustainable cultivation of rice - Vol 1 003
Breeding for higher yield and quality
Edited by: Prof. Takuji Sasaki, Tokyo University of Agriculture, Japan

Achieving sustainable cultivation of rice - Vol 2 004
Cultivation, pest and disease management
Edited by: Prof. Takuji Sasaki, Tokyo University of Agriculture, Japan

Achieving sustainable cultivation of wheat - Vol 1 005
Breeding, quality traits, pests and diseases
Edited by: Prof. Peter Langridge, The University of Adelaide, Australia

Achieving sustainable cultivation of wheat - Vol 2 006
Cultivation techniques
Edited by: Prof. Peter Langridge, The University of Adelaide, Australia

Achieving sustainable cultivation of tomatoes 007
Edited by: Dr Autar Mattoo, USDA-ARS, USA; and Prof. Avtar Handa, Purdue University, USA

Achieving sustainable production of milk - Vol 1 008
Milk composition, genetics and breeding
Edited by: Dr Nico van Belzen, International Dairy Federation (IDF), Belgium

Achieving sustainable production of milk - Vol 2 009
Safety, quality and sustainability
Edited by: Dr Nico van Belzen, International Dairy Federation (IDF), Belgium

Achieving sustainable production of milk - Vol 3 010
Dairy herd management and welfare
Edited by: Prof. John Webster, University of Bristol, UK

Ensuring safety and quality in the production of beef - Vol 1 011
Safety
Edited by: Prof. Gary Acuff, Texas A&M University, USA; and Prof. James Dickson, Iowa State University, USA

Ensuring safety and quality in the production of beef - Vol 2 012
Quality
Edited by: Prof. Michael Dikeman, Kansas State University, USA

Achieving sustainable production of poultry meat - Vol 1 013
Safety, quality and sustainability
Edited by: Prof. Steven C. Ricke, University of Arkansas, USA

Achieving sustainable production of poultry meat - Vol 2 014
Breeding and nutrition
Edited by: Prof. Todd Applegate, University of Georgia, USA

Introduction

This volume offers a comprehensive review of research on achieving durable resistance to diseases in cereals. The book opens with a chapter on global patterns in cereals diseases and is then broken down into eight parts. Parts 1 through 6 cover various fungal diseases, such as leaf rust, Fusarium head blight, Septoria tritici blotch, Septoria nodorum blotch and spot blotch, net blotch, and finally, tan spot, blast and Ramularia. Chapters in Part 7 of the book focus on barley yellow dwarf virus (BYDV), specifically highlighting the advances in understanding the biology and epidemiology of BYDV and how identifying resistance genes can help breed durable disease-resistant cereals. The final part of the book discusses the regional strategies for breeding durable disease-resistant cereals, focusing on North America, Northwest Europe, North Africa and West Asia.

Chapter 1 sets the scene by highlighting the global patterns of cereal diseases and the impacts of breeding for host plant resistance. Plant breeding for host plant resistance to pathogens brings together different disciplinary domains, especially genetics and plant pathology. The strategies underpinning plant breeding have therefore, logically, been influenced by a number of paradigms that have dominated the field of disease management. The chapter provides a brief overview of these paradigms, where the authors link the implications for plant breeding. It also discusses the meaning of crop losses for plant breeding, then goes on to highlight three paradigms: pathogen elimination and total control, disease management and ecological disease management.

Part 1 Fungal diseases of cereals: rusts

The first chapter of Part 1 examines advances in understanding the biology and epidemiology of rust diseases in cereals. Chapter 2 provides an overview of the wheat rust pathogen lifecycle that has been critical to the design of effective disease management strategies and discusses recent integration of basic biological knowledge and genomic-led tools within an epidemiological framework. The chapter also includes a case study on the 'field pathogenomics' technique, illustrating the value of genomic-based tools in disease surveillance activities. Bringing together advances in understanding basic pathogen biology, developments in modelling for disease forecasting and identification, alongside genomic-led advances in surveillance and resistance gene cloning, holds great promise for curtailing the threat of these notorious pathogens.

Chapter 3 reviews advances in identifying stripe rust resistance genes in cereals. Stripe rust is one of the most serious wheat diseases of the world, usually resulting in massive loss of grain production. The most effective and environmentally friendly way to control the spread of stripe rust is to plant wheat varieties that carry stripe rust resistance genes. The identification and utilization of stripe rust resistance genes is very important for achieving this goal. The chapter summarises the hazards of stripe rust and the current progress in the discovery of stripe rust resistance genes. It also introduces the advanced methods to identify Yr genes. The chapter also shows the successful application of Yr genes in wheat breeding program in southwestern China, which is the largest epidemic area of stripe rust in the world. The further identification and applications of Yr genes are also discussed.

Part 2 Fungal diseases of cereals: Fusarium head blight

Part 2 opens with a chapter that reviews advances in understanding the epidemiology of *Fusarium* in cereals. Chapter 4 begins by discussing the life cycle of *Fusarium* head blight (FHB) and the factors that can affect the length of the life cycle, including environmental factors and disease management strategies. The chapter then goes on to address mycotoxin production and the influence of environmental factors on disease development, followed by a section on how management practices can influence the risk of disease. The chapter also examines the various methods that can be used to predict FHB epidemics, emphasises the importance of detecting and identifying plant pathogens for disease diagnosis and also provides a discussion on the importance of assessing FHB at the right time.

Moving on to Chapter 5, this chapter focuses on cereal-*Fusarium* interactions. All cereal crop species are vulnerable to root, stem-base and floral diseases caused by various *Fusarium* species. Most problematic is *Fusarium* head blight because grains become contaminated with harmful mycotoxins. Currently, *Fusarium* control using fungicides is only partially effective, whilst cultivars with acceptable resistance levels are scarce. The chapter reviews the recent advances in fundamental knowledge on the cell biology of interactions, the in-planta production of trichothecene mycotoxins, *Fusarium* genomes/pan-genomes, newly discovered *Fusarium* virulence factors, small, secreted effectors, plant defence components and resistance genes underlying major quantitative trait loci. These discoveries are discussed in the context of exploiting new intervention targets to achieve control through genetic modification, gene editing and HIGS. Finally, the chapter explores how to develop a more integrated approach that includes marker assisted selection in breeding programmes, removal of susceptibility loci and the inclusion of new approaches arising from plant defence, virulence and effector studies.

The final chapter of Part 2 of the book examines advances in genetic improvement of durable resistance to *Fusarium* head blight, specifically focusing on wheat. Wheat *Fusarium* head blight is a destructive disease in wheat worldwide. Wheat resistance to FHB is a complex with five types. Each type of resistance is controlled by multiple quantitative trait loci (QTLs) with most having minor effects and being affected by environments. Chapter 6 describes methodologies used for evaluating different types of resistance, consolidates the QTLs for type II and Type III resistance into 26 repeatable QTLs, discusses progresses made in genetics and breeding of wheat FHB resistance, and discusses possible new breeding strategies for FHB resistance improvement. The 26 repeatable QTL were located in ~100 Mb intervals based on IWGSC reference sequence map, which will be critical QTLs for functional marker development and for improvement of FHB resistance in breeding. Genomic selection (GS) together with marker-assisted selection (MAS) coupling with phenotypic selection will facilitate accumulation of multiple QTLs from different sources to create highly resistant cultivars.

Part 3 Fungal diseases of cereals: Septoria tritici blotch

The first chapter of Part 3 examines advances in understanding the epidemiology of Septoria tritici blotch in cereals. Chapter 7 begins by discussing the taxonomy of *Zymoseptoria tritici* and how Septoria tritici blotch is formed. The chapter then goes on to review the life cycle of *Zymoseptoria tritici* and the modes of nutrition at various stages in its growth cycle. Moving on from this, the chapter reviews the epidemiology of Septoria tritici blotch and how diagnosing the disease is important, especially at an early stage in its infection period. It concludes by providing an overview of the steps that need to be taken in order to progress in understanding the biology and epidemiology of Septoria tritici blotch.

Expanding on topics previously discussed in Chapter 7, Chapter 8 moves on to review plant-pathogen interactions in Septoria tritici blotch infection of cereals. The chapter begins by introducing the biology and epidemics of *Zymoseptoria tritici* on wheat, then moves on to discuss *Z. tritici* genomics, focusing specifically on genome characteristics, the *Z. tritici* pangenome, related *Zymoseptoria* species genomics and *Zymoseptoria* transposons. The chapter also discusses the molecular and biological tools available in *Z. tritici* as well as the molecular biology of *Z. tritici* infection, focusing on aspects such as cell biology, transcriptomics of *Z. tritici* infection and secondary metabolism. A section on Septoria tritici blotch disease management and control is also provided, before concluding with an overview of why analysing *Z. tritici* infections as a model to study plant-pathogen interactions is important.

The subject of Chapter 9 is advances in understanding breeding techniques for durable Septoria tritici blotch resistance in cereals. Septoria tritici blotch (STB), caused by the hemibiotrophic fungus *Zymoseptoria tritici*, is one of the most important foliar diseases of winter cereal crops. Recent advances are helping to understand the genetic basis and architecture of resistance to STB. To date, at least 22 genes for qualitative resistance and over 200 quantitative trait loci for quantitative resistance have been identified in cereals. This knowledge is enabling cereal breeding programs to develop varieties with more durable resistance to STB. The chapter reviews recent research on genetic resistance loci and breeding strategies based on both conventional and biotechnology-based breeding approaches (molecular marker/genomic-assisted breeding, genetic transformation, and gene-editing) to achieve achieving durable resistance to STB infection and minimise grain yield losses.

Part 4 Fungal diseases of cereals: Septoria nodorum blotch and spot blotch

Part 4 opens with a chapter that examines understanding the plant-pathogen interaction associated with Septoria nodorum blotch of wheat. Chapter 10 begins by reviewing the necrotrophic effector-host sensitivity gene interactions in the wheat-*P. nodorum* system. It then reviews the genetic relationship between NE-sensitivity gene interactions and the importance of these interactions in the field. Additional QTL associated with susceptibility/resistance to *P. nodorum* is also discussed, followed by a section on the impact of genome sequencing in characterizing NE-sensitivity gene interactions.

Chapter 11 analyses advances in genetic mapping of Septoria nodorum blotch resistance in wheat and applications in resistance breeding. Septoria nodorum blotch (SNB) caused by the necrotrophic fungus *Parastagonospora nodorum* is an important wheat disease in many high rainfall areas across the world. It reduces both yield and grain quality by causing symptoms on wheat leaves and glumes and can cause yield losses up to 30% under warm and humid conditions. The chapter gives an update on the recent progress in genetic mapping of SNB resistance in wheat, with focus on adult plant leaf blotch and glume blotch resistance with relevance to resistance breeding. This is followed by a case study on the investigation of the naturally occurring *P. nodorum* population in Norway and mapping of resistance loci in relevant wheat germplasm using multi-parent advanced generation inter-cross (MAGIC) populations and genome-wide association study (GWAS) panels as well as how this information can be used to improve resistance breeding and disease management. The chapter concludes by providing some future perspectives of SNB resistance breeding is provided.

The next chapter focuses on advances in breeding techniques for durable resistance to spot blotch in cereals. Chapter 12 starts by highlighting the spread, economic importance and the disease cycle of spot blotch. The chapter then goes on to examine the diversity of the pathogen and physiological specialization. This is then followed by a discussion on the identification of resistance sources, as well as the histological, biochemical and morphological components of resistance. The chapter also reviews molecular approaches for resistance breeding, specifically focusing on QTL and GWAS. Resistance genes and their possible deployment are also discussed, along with a section on low molecular weight toxins and their possible role in pathogenicity. The chapter also examines necrotrophic effector triggered susceptibility and associated genes and breeding for spot blotch resistance in wheat. Breeding for spot blotch resistance in barley is also discussed, before concluding with a discussion on farmers participatory research in the release of spot blotch resistant varieties.

Part 5 Fungal diseases of cereals: net blotch

Chapter 13 focuses on advances in understanding the epidemiology, molecular biology and control of net blotch and the net blotch barley interaction. Net blotches are the most widely distributed foliar diseases of barley worldwide, causing significant losses in grain yield. They occur as net form net blotch, caused by *Pyrenophora teres* f. *teres* and spot form net blotch caused by *P. teres* f. *maculata*. Both sexual and asexual reproduction play a role in the *P. teres* disease cycles leading to changes in genetic variation of populations. Breeding programs have to keep pace with pathogenic changes and ensure different sources of resistance are present in current barley cultivars. Knowledge of the genetic architecture and genes involved in virulence is thus vital to increase the durability of net blotch resistance in barley cultivars. The chapter explores the molecular biology, lifecycle and epidemiology of the net blotch fungi and discusses the key challenges we are facing in managing the net blotches using both fungicide resistance and breeding strategies to achieve durable disease resistance in barley.

The next chapter discusses understanding plant-pathogen interactions in net blotch infection of cereals. An economically important disease of barley that causes significant yield and quality losses is net blotch caused by the necrotrophic fungal pathogen *Pyrenophora teres*. To reduce the impact of net blotch the research community is engaging in basic and applied research to enhance genetic resistances, as it is the most economic and sustainable management strategy. Durable resistance against *P. teres* will be a major achievement towards the goal of developing widely adapted barley varieties that have high yields and quality across dynamic environments. Chapter 14 focuses on a thorough review of the latest knowledge of both host resistance/susceptibility and pathogen virulence/avirulence in this important pathosystem, and the implications this knowledge will have on deploying sustainable resistances to this destructive pathogen of barley.

The subject of Chapter 15 is breeding barley for durable resistance to net and spot forms of net blotch. It starts by reviewing how *Pyrenophora teres* f. *teres* can cause net form of net blotch. The chapter then goes on to examine the molecular markers that can be identified to provide resistances to net form net blotch. A section on the population dynamics of barley–*P. teres* f. *teres* interactions is also provided. The chapter also reviews how breeding crops with specific genes can help to create durable resistance to net form blotch. It moves on to discuss how *Pyrenophora teres* Drechs. *f. maculata* can cause spot form net blotch and how identifying specific molecular markers can help provide resistance to this form of net blotch. The chapter concludes by highlighting the importance of combining durable resistance to both forms of net blotch.

Part 6 Fungal diseases of cereals: tan spot, blast and Ramularia

The first chapter of Part 6 reviews tan spot disease under the lenses of plant pathologists. Chapter 16 begins by examining the global presence of tan spot. The chapter then goes on to review the fungal and pathogen forms of *Pyrenophora tritici-repentis* (Ptr). A case study on tan spot on the prairies is also provided, focusing specifically on its emergence and agronomic background, the different races of Ptr and the genetic diversity between them, focusing predominantly on ToxA. The chapter also discusses the development of the disease rating system, addressing how early screening for resistance is important for treating diseased crops. The chapter concludes by emphasising how much tan spot disease infection research has changed in the last 90 years.

Chapter 17 examines the development of an early warning system for wheat blast. Wheat blast is caused by the fungus *Pyricularia oryzae Triticum* pathotype (PoT). Significantly damaging wheat blast epidemics are sporadic and limited to tropical wheat growing areas in South America. Unexpectedly, wheat blast was reported in Bangladesh and Zambia in 2016 and 2020, respectively. The urgent need to deal with a poorly studied disease has mobilized the scientific community. Original research and reviews have been published in various venues. Nevertheless, disease control is still a difficult task. Much less research has, however, focused on crucially important and complex ecological interactions at the field, landscape, or regional levels. The chapter reviews aspects of the epidemiology of wheat blast, mainly those related to inoculum and its role for the epidemics. It then describes the models that have been developed by the authors as well as the decision support system.

Examples of the implementation of a warning system in Bangladesh and Brazil are also illustrated.

The subject of Chapter 18 is investigating the biology of rice blast disease and prospects for durable resistance. There are important biological processes involved in rice blast disease that are now well-studied during the early events in plant infection which include: the cell biology of appressorium formation, the biology of invasive growth and effector secretion, the two distinct mechanisms of effector secretion, the nature of the plant-pathogen interface, PAMP-triggered immunity modulation by secreted effectors and effector-triggered immunity and blast resistance. The devastating losses caused by the blast fungus have been documented in most grasses, but this chapter discusses the use of major resistance genes to rice blast and wheat blast disease as an emerging threat to global food security. The chapter also highlights an emerging approach to breed for durable resistance to plant pathogens using gene editing technologies with an example: CRISPR-Cas9 mutagenesis of dominant S-genes for disease control.

The final chapter of Part 6 examines Ramularia left spot in barley. Ramularia leaf spot is an emerging pathogen across barley growing regions of the world. Its rise from minor to major disease has been rapid over the last twenty years. Chapter 19 reviews the latest research into Ramularia biology and control and highlights the areas where recent advances have been made. The chapter begins by reviewing how the fungal pathogen Ramularia collo-cygni was first reported along with the symptoms that appear in barley crops. It also highlights where Ramularia leaf spot first emerged and how it was globally distributed when it was first discovered. The chapter also goes on to analyse the life cycle of the fungus as well as the impact of different environmental factors on disease expression. Sections on genetics of disease resistance, pathogen variability and control of Ramularia leaf spot by fungicides are also provided.

Part 7 Barley yellow dwarf virus

Part 7 opens with Chapter 20, which analyses the advances in understanding the biology and epidemiology of Barley yellow dwarf virus. A tri-trophic network of domesticated grasses (host), various aphids (vector) and barley yellow dwarf virus (pathogen) species has been spread by humans from Eurasia to the rest of the world. Understanding how climate, natural and agricultural landscapes challenge pathogens, vectors, and their natural enemies and shape their dynamics is the key to managing this pathosystem. The chapter provides an overview of this complex system and its evolution. The chapter includes a case study of biological control of aphids causing wheat BYDV in Brazil. The current challenge is to create tools that integrate knowledge of this complex pathosystem and facilitate monitoring and decision making for rational management to reduce the burden of disease.

Moving on from Chapter 20, Chapter 21 focuses on resistance breeding in barley against barley yellow dwarf virus. Barley yellow dwarf is one of the most widespread and damaging viral diseases of grasses and cereal crops worldwide. Due to an increasing risk of food losses e.g. in barley by Barley yellow dwarf virus as a consequence of climate change, associated by a strong demand to decrease the use of chemical insecticides, breeding for BYDV resistance is of prime importance today. The chapter describes the negative impact of BYDV on barley on multiple levels (anatomy, physiology and agronomic traits). It also demonstrates the benefits of BYDV resistance regarding a reduction in yield losses but also a decreased spread of BYDV in the field due to effects on the tritrophic interaction of virus, vector and plant. Until now, several genes and QTL are known that mediate tolerance or resistance against BYDV, respectively. The combination of genomic tools and phenotyping is the basis for the identification of these genes and recent developments facilitate to enhance this process.

Part 8 Fungal diseases of cereals: Regional strategies

The final part of the book begins with a chapter that focuses on the key challenges in breeding durable disease-resistant cereals in North America. Chapter 22 first describes the challenges of diverse climates, diseases, and market classes that face North American small-grain cereal breeders and producers. It discusses the challenges inherent in the complex systems of cereal breeding on the continent, and the changing resistance priorities brought about by shifting pathogen races and production practices. The remainder of the chapter is devoted (in rough order of priority) to the status and prospects for durable resistance to the main pests currently confronting the continent: Fusarium head blight, rusts, powdery mildew, leaf (and glume) blotches, viruses, Hessian fly, and bacterial leaf streak.

The next chapter reviews the achievements in breeding cereals with durable disease resistance in Northwest Europe. Chapter 23 begins by first identifying the main cereals produced in Northwest Europe, then goes on to highlight the additional sources of information about breeding for resistance, such as the AHDB Recommended List. The chapter also examines Plant Breeder's Rights, National List and Recommended List trials and how disease resistances can be rated on a scale of one to nine, one being most susceptible and nine being most immune. A section on the demand for disease resistance in cereals is also provided, emphasising why resistant varieties are needed and the minimum standards that need to be met. The chapter also highlights the use of a Variety Selection Tool. Sections on breeding elite cereal varieties with resistance to multiple diseases, durable and non-durable resistance are also included. This is then followed by separate sections on resistance to biotrophic fungal diseases, non-biotrophic fungal diseases, viruses and insects. The chapter also examines mitigating trade-offs of disease resistance before highlighting the future disease threats for cereals.

The final chapter of the book reviews the key challenges in breeding durable disease-resistant cereals in North Africa and West Asia. Chapter 24 reviews progress and challenges in breeding disease-resistant cereals in the West Asia and North Africa (WANA) region. It discusses challenges in ensuring durable resistance in the face of constantly-evolving pathogen threats. It includes case studies on progress in combatting rust diseases in Iran and combatting Septoria in North Africa.

Chapter 1

Global patterns of cereal diseases and the impacts of breeding for host plant resistance

Serge Savary and Laetitia Willocquet, Institut National de Recherche pour l'Agriculture, l'alimentation et l'Environnement (INRAE), France

1 Introduction

Kresek (Ou, 1987) occurs in rice stands a few days after crop establishment. Some leaves of the young seedlings show watery lesions, turn grey, and start folding around the midrib. Over a few days, these symptoms affect most or all the leaves of the same plant; dying leaves turn brown, droop, collapse, and while still attached to the plant, float on the surface of the irrigation water; soon, entire plants wither, turn greyish-brown, and the entire rice plot is covered by diseased and dying rice hills. Kresek can be seen wherever rice is cultivated in the Asian tropics, from India to the Philippines, and from Indonesia to Southern China. These symptoms are especially seen in the rainy season, in transplanted rice fields, when the weather is overcast, and rain is often accompanied by very strong wind, sometimes typhoons. For many generations of rice farmers and over millions of lowland hectares of cultivated rice, kresek has been destroying entire new crop stands within days, thwarting the growing season. Kresek is the most severe of the symptom forms of rice bacterial leaf blight caused by *Xanthomonas campestris* pv. *oryzae*.

Kresek is much less frequently observed today in Asia than three decades ago. This is because there has been much change in the way rice is being cultivated (Pandey et al., 2010; Savary et al., 2017; Becker and Angulo, 2019). First, transplanting of seedling in rice hills is being progressively replaced

http://dx.doi.org/10.19103/AS.2021.0092.01

by direct seeding (in various forms) as a crop establishment method: transplanting unavoidably leads to root wounds, which are entry points for the pathogen. Second, manual transplanting was frequently accompanied by the cutting of leaf tips, which increases root growth and favours plant establishment, but which is also a very effective way of spreading the disease in a crop stand. Third, irrigation water (a main source of inoculum in the tropics if contaminated) tends to be managed differently today where rice plots are larger, with lower water levels, and less chance for water to be wind-driven over soil levees between irrigated rice plots when strong winds occur. However, the main reason why kresek is much less frequently observed in Asia today is that the rice cultivars that are grown are, in most case, resistant to bacterial leaf blight.

For decades, bacterial leaf blight had been a major, massive, blockage to any agricultural development in tropical Asia. A major success of rice pathologists and breeders during the Green Revolution was to develop varieties that carried resistance genes to the disease (Bonman et al., 1992). Rice plant pathologists and breeders have essentially removed this obstacle, leading to a critical improvement of the food security status in this part of the world (Mew, 1991). Yet the disease has not disappeared – it is still a concern, especially in the western part of the Indo-Gangetic plains, where high-quality rice is cultivated, and it also remains a very important disease wherever the turn-over of rice varieties is slow, as in the North-East of India, Burma, and Cambodia. However, the main 'rice bowls' of Asia, in China, India, Vietnam, and the Philippines all grow rice cultivars with notable, usually very efficient, levels of resistance to bacterial blight.

As is often the case with major successes, the meaning of host plant resistance (HPR) in rice to bacterial leaf blight is now forgotten. The terribly severe epidemics of the past are seldom seen, and if so, at relatively small scales. As a characteristic of rice cultivars, resistance to bacterial leaf blight is now a routine: in India, for instance, it is illegal to release a rice variety that does not have an adequate level of resistance to the disease (Darshan Brar, Personal Communication). As a trait, resistance to this disease is so essential that it today is taken for granted.

Yet this success against bacterial blight of rice is not permanent. This is because the pathogen evolves as the crop, and its cultivation, does. Sources of HPR must be maintained, collections of cultivars need to be curated, the genetic diversity of the pathogen in the field and in landscapes must be monitored, and new material incorporating resistance genes must be shared amongst breeding networks (Garrett et al., 2017). All this plant material and the entire breeding programmes absolutely require field testing using standardised, shareable, transparent methods. The work of plant pathologists and plant breeders on bacterial blight well illustrates the challenges of disease resistance breeding, with permanent, very large, heavy and costly networks of field experiments, of storage systems, decades of work along the 'pipelines' of breeding programmes, and yet the necessity to rapidly adapt to new disease problems in changing agriculture.

2 The meaning of crop losses for plant breeding

Such huge investments in breeding programmes require a strategy, where hard choices often have to be made. These choices are first grounded on the needs of growing human populations and activities, which may be seen as a set of agricultural hotspots in the world. For annual crops such as cereals, one may consider the following hotspots: a series of main cereal (net) producers (the Midwest of the United States; Western Europe; Argentina and Brazil; Australia; Russia, Ukraine and Eastern Europe); a series of main cereal (net) consumers (sub-Saharan Africa; the Middle East and North Africa); and a series of main producer-consumers (Eastern Asia; South Asia; Southeast Asia).

Another set of criteria, which are specific to the breeding of resistant cultivated crop varieties, are the losses caused by pathogens and pests. Crop losses to pathogens (and to pests as well) constitute a primary criterion to consider, because they represent the size of breeding targets: the larger the losses, the greater the benefits that breeding could generate. Measurements of crop loss to pathogens and pests also constitute a permanent scale, a yardstick (Zadoks, 1985), against which the progress of plant breeding can be gauged: as progress takes place in disease management, and especially in breeding new cultivars with effective resistances, crop loss data provide quantitative, transparent, and verifiable information to judge the progress that has been made, and the progress which is still required.

An international survey on global crop losses caused by crop pathogens and pests was conducted in 2016–2017 (Savary et al., 2019). The survey concerned five major global food crops: rice, wheat, maize, potato, and soybean. Through a time-bound structured questionnaire, we were able to elicit 989 responses from 219 experts in 67 countries over the main producing regions of the world for the five crops. The survey also had a good coverage of the main food security hotspots: the United States Midwest and Canada (USM&C); South Brazil, Paraguay, Uruguay, and Argentina (SB&A); Northwest Europe (NWE); West Asia and North Africa (WANA); sub-Saharan Africa (SSA); mainland China (China); the Indo-Gangetic Plain (IGP); and Southeast Asia (SEA). Responses from the survey represented countries that account for a total of 83%, 94%, 79%, 69%, and 95% of the global production for wheat, rice, maize, potato, and soybean, respectively (based on the average production of the crops over 2010–2014; FAOSTAT, 2018). The survey provided responses on a total of 137 pathogens and pests on the five considered crops.

This survey led to produce yield loss (range) estimates at a global level and per hotspot for wheat (21.5% (range: 10.1%–28.1%)), rice (30.0% (range: 24.6%–40.9%)), maize (22.5% (range: 19.5%–41.1%)), potato (17.2% (range: 8.1%–21.0%)), and soybean (21.4% (range: 11.0–32.4%)). Table 1 provides a summary of these yield loss estimates (expressed as the percentage of global production), where we only show the five most damaging pathogens or pests per crop and the estimates of accumulated yield losses.

In wheat, global yield losses from individual pathogens or pests range from a maximum of 3.25% (leaf rust) to a minimum (fifteenth rank) of 0.24% (Rhizoctonia bare patch). In rice, the range of individual losses is from 6.78% (sheath blight) to 0.20% (sheath mite and kernel smut). In maize, the range is from 4.58% (Fusarium and Gibberella stalk rots) to 0.35% (corn stunt and tar spot). In potato, the range is from 5.98% (late blight) to 0.02% (aphids). And in soybean, the range is from 4.24% (cyst nematode) to 0.19% (frogeye leaf spot). The numbers in Table 1 may seem small for most plant-health specialists. This is because the discipline is mostly reporting yield loss at the plot or field scale. The figures of Table 1, however, are average yield losses for the global production of wheat, rice, maize, potato, and soybean. Converting the percentages of Table 1 in total yield biomass (Savary et al., 2019) may provide a better view of the challenge: wheat leaf rust causes yield losses in the range of 24 million tons each year (using averaged 2010–2014 data; FAOSTAT, 2018); losses to septoria tritici blotch – a major focus of disease management in Western Europe, for example – are about 18 million tons annually; and losses to spot blotch, a major disease in wheat in the Indo-Gangetic plains, 12 million tons. In rice, the estimated yield losses to sheath blight are about 58 million tons: this figure approaches the yearly total rice output of China as a whole, and this estimate is congruent with previous estimates involving surveys in farmers' fields (Savary et al., 2000a), crop loss experiments (Savary et al., 2000b), and yield loss simulation analyses (Willocquet et al., 2004). Other currently important pathogens and pests of rice include: the stem borers (several species, about 32 million tons), rice blast (37 million tons), and bacterial blight (still with 23 million tons losses, globally, today). Brown spot is another important disease of rice (32 million tons), especially in poorly endowed environments and soils, where water shortage is frequent, and heat stress regular. Maize includes a large diversity of yield reducers, with a strong contribution of animal pests: Fusarium and Gibberella stalk rots (53 million tons), the (emerging and very dangerous) fall armyworm (33 million tons), and Northern leaf blight (31 million tons). The case of soybean is also different, with a more narrow range of pathogen and pests, and an important role of nematodes: cyst nematode (18 million tons); white mould (17 million tons), and soybean rust (13 million tons).

This survey provides an overview of what the priorities of disease management, and therefore, of plant breeding, could globally be. It conforms

Table 1 Global yield loss reductions of five major food crops caused by crop pathogens or pests, expressed as percentages (adapted from Savary et al., 2019)

Wheat		Rice		Maize		Potato		Soybean	
Pathogen or pest	% losses	Pathogen or pest	% losses	Pathogen or pest	% losses	Pathogen or pest	% losses	Pathogen or pest	% losses
Leaf rust	3.25	Sheath blight	6.78	F&G stalk rots	4.58	Late blight	5.98	Cyst nematode	4.24
FHB-scab	2.85	Stem borers	5.57	Fall armyworm	2.85	Brown rot	3.10	White mould	3.90
Tritici blotch	2.44	Blast	4.33	Northern leaf blight	2.68	Early blight	2.59	Soybean rust	2.96
Stripe rust	2.08	Brown spot	3.77	F&G ear rots	2.38	Cyst nematode	1.52	Cercospora leaf blight	1.88
Spot blotch	1.67	Bacterial blight	2.72	Anthracnose stalk rot	1.65	Powdery scab	0.94	Brown Spot	1.32
Tan spot	1.64	Leaf folder	1.92	Southern rust	1.15	Verticillium wilt	0.68	Charcoal rot	1.31
Aphids	1.30	Brown plant hopper	1.31	Diabrotica	0.98	Apical leaf curl	0.67	Root knot nematodes	1.18
Powdery mildew	1.07	Bacterial panicle blight	0.87	Grey leaf spot	0.94	Potato leafhopper	0.65	Phytophthora root and stem rot	0.94
BYD	0.96	False smut	0.68	Common rust	0.75	Common scab	0.46	Sudden death	0.89
Stem rust	0.90	Sheath rot	0.40	Bacterial stalk Rot	0.64	Colorado potato beetle	0.24	Fusarium wilt and rot	0.67
Nodorum blotch	0.90	Ragged stunt	0.30	Southern leaf blight	0.55	Leaf miner	0.20	Rhizoctonia rot and blight	0.53
Crown and Root Rot	0.86	Stripe	0.27	White spot	0.42	Slugs	0.08	Soybean mosaic	0.43
Army worm	0.37	Rice tungro	0.27	Maize streak	0.37	Leaf worm	0.05	Target spot	0.23
Sharp Eye Spot	0.26	Kernel smut	0.20	Corn stunt	0.35	Canker and black scurf	0.03	Spider mites	0.20
Rhizoctonia Bare patch	0.24	Rice sheath mite	0.20	Tar Spot	0.35	Aphids	0.02	Frogeye leaf spot	0.19
Sum	21.47	Sum	30.03	Sum	22.51	Sum	17.22	Sum	21.40

to the available data that is available (see Savary et al., 2019 for a comprehensive list of references used to assess the results of this survey) from surveys in farmers' fields, crop loss experiments, and simulation outputs. The figures of Table 1 however invite three remarks:

- First, crop losses are not static; they do vary with shifts in crop health and changes in injury profiles, with changes in crop management, and with change in disease management methods (including resistant cultivars). The figures of Table 1 would therefore require to be regularly updated.
- Second, Table 1 only refers to yield, not to crop losses: the quality losses, such as those associated with mycotoxin contamination, are not included in this table. Taking into account the contamination by mycotoxins is a very real and pressing need, in view of the growing importance of the mycotoxin problems of the tropical world (see, for instance, maize in SSA: Nelson, 2020), and the temperate world alike (see, for instance, the mycotoxin contamination associated with Fusarium pathogens in wheat in Western Europe: Nielsen et al., 2011; Van Der Fels-Klerx et al., 2012).
- Third, Table 1 only reflects the current situation, represents only what appears today to be causing yield losses. Table 1 does not reflect the progress that has been made, and fails to spell out the efforts that are still required: bacterial leaf blight only 'ranks' 5 in the list of Table 1; it most certainly would rank much higher if we were to consider the potential losses this disease could cause. There certainly are many such examples in Table 1 (as well as in the more comprehensive list of Savary et al., 2019).

There are two other elements which are not, or very incompletely, reflected in Table 1, and which the 2016–2017 survey did highlight. One element is the analysis at the hotspot level (Savary et al., 2019): while there are 'global' pathogens and pests, clearly, the types and levels of injuries widely vary amongst food security hotspots. The analysis of crop loss levels at the food security hotspot level may enable identifying key, yet hotspot-specific breeding priorities, which would enable disease management and breeding programmes to strengthen their impacts. The second element generated by the survey is information on emerging diseases and pests. The survey indeed led to highlight the growing importance of emerging diseases and pests such as:

- in wheat: stem rust (sub-Saharan Africa); wheat blast (South Asia);
- in rice: false smut (South and Southeast Asia); stripe virus disease (East Asia); and
- in maize: false army worm (sub-Saharan Africa); maize lethal necrosis (sub-Saharan Africa).

Plant breeding for host plant resistance to pathogens brings together different disciplinary domains, especially genetics and plant pathology. The strategies underpinning plant breeding have therefore, logically, been influenced by a number of paradigms that have dominated the field of disease management. The following of this Introduction chapter is a very brief overview of these paradigms, where we link the implications for plant breeding. In doing so, we try to follow an approximate (1) linear and (2) chronological order. Yet we are aware that, depending on the considered crop and region, (1) there might have been occasional jumps and step-backs leading to non-linear paths, and (2) the chronology of events may have much varied.

3 Paradigm 1: pathogen elimination and total control

The idea of total annihilation of a plant pathogen was once widespread, being seen as economically justified and technically feasible. The idea was supported by the fantastic progress of synthetic chemistry in the first half of the twentieth century, which seemed to make total control over pathogens possible. The paradigm of total control of pathogen nowadays has lost its appeal, in part because of spectacular setbacks, because of scientific reasoning (see, e.g. Zadoks and Schein, 1979), and because of Rachel Carson's *Silent Spring*. It could be argued that qualitative host plant resistance, which involves major resistance genes that completely prevent pathogen infection and/or development, are instruments of total control, in the same way as a chemical would be. Similar to chemicals, many adaptations of pathogens (and pests) to major resistance genes have been documented, reinforcing the analogy.

Breeding for host plant resistance has long been dominated by qualitative, major genes, as targets. Such genes are comparatively easy to track in breeding programmes. Their efficiency has been demonstrated in many cases. In many cases, however, major, dominant genes have been overcome by the genetic flexibility of the targeted pathogen and the rapid selection of pathogens genotypes that had become adapted to the resistance gene. Pathogen adaptation to qualitative resistance has been documented in many cases, where the spatial and temporal breadth of deployment of a given resistance, combined with the evolutionary potential of the pathogen, was shown to be a major factor (McDonald, 2014; Mundt, 2014).

4 Paradigm 2: disease management

There is an important difference between a resistance gene (that is, the allele of a gene) for complete resistance and a chemical pesticide: genes (alleles) are not sprayed, more or less uniformly, on a cultivated landscape, as a pesticide would be; resistance genes are an integral part of genomes (where many interactions are at play). Further, a diversity of resistance genes exists in plants, spanning from complete (qualitative) resistance genes, to incomplete (partial, quantitative) resistance genes, of many kinds. This whole range of resistance genes displays 'shades of grey' (Poland et al., 2009) in their effects on the pathogen.

Since about the 1970s, 'disease management' has progressively replaced 'disease control' (Zadoks and Schein, 1979), reflecting the notions that pathogens are part of ecosystems, which themselves necessarily have been disturbed to give way to agriculture (Fresco and Kroonenberg, 1992), and that any action on an ecosystem is bound to bring about reactions and adaptations. Disease management implies the view that pathogen elimination is not agronomically needed, not environmentally sound, and not economically justified; it implies that plants, their pathogens, and farmers must co-exist. Partial host plant resistance is the key instrument for disease management: it does not prevent a pathogen to complete its life cycle, but prevents this life cycle to be rapid and to enable a large population of the pathogen to develop (Mundt, 2014; Willocquet et al., 2017).

Partial (or incomplete, or again, quantitative) host plant resistance usually is governed by multiple genes or quantitative trait loci (QTLs); therefore, it usually does not generate a strong selection pressure for the pathogen to adapt (McDonald, 2014); and, because it involves different genetic loci, is not easily overcome by the genetic flexibility of a pathogen to adapt (Poland et al., 2009; Willocquet et al., 2017). Quantitative HPR thus seems to be the best target for selection. Unfortunately, selection for quantitative resistance is much harder than for qualitative resistance: it involves different segments of genome, each associated with small contribution to overall resistance; these QTLs can be widely distributed in the plant genome and are sometimes associated with strongly unwanted traits; and sometimes QTLs are not available in the germplasm pools that breeders are commonly using. Importantly, too, is the fact that each QTL effect, being small, is hard to detect and to track in a selection scheme, even when molecular markers exist. Simply: quantitative HPR is often too cumbersome, too time-consuming, to be included in breeding programmes, where dozens of widely different traits are to be considered. A main area of possible advance in the future is simulation modelling, enabling *ex-ante* evaluation of the specific contribution of individual, selected, QTLs to overall resistance (Willocquet et al., 2017).

5 Paradigm 3: ecological disease management

Plant pathogens widely differ in the ways they disperse within and amongst agrosystems (e.g. Fitt et al., 1989; Xu and Ridout, 1998; Mundt et al., 2009), resulting in very different patterns of spatial disease distribution (e.g. Jeger, 1983; Gilligan and van den Bosch, 2008). Pathogens also differ widely in the ways they affect the physiology of growing plants and injure crop stands (e.g. Willocquet et al., 2004). Pathogens are also very different in their strategies of survival between cropping seasons (e.g. Zwankhuizen and Zadoks, 2002). These are important reasons why strategies for disease management need to consider the ecology of both the pathogens and crops. Depending on the pathogen and disease considered, therefore, the deployment of resistance genes needs to account for these epidemiological (pathogen) and physiological (disease) differences: making use of the proper type of resistance when it is required and within the spatial context where it will be the most effective will enable lasting and effective management.

6 Where to look for further information

A very recent and excellent text book on plant pathology, including aspects on breeding for host plant resistance is: Tronsmo, A. M., Collinge, D. B., Djurle, A., Munk, L., Yuen, J., Tronsmo, A., eds. (2020). *Plant Pathology and Plant Diseases*. Wallingford, UK, CAB International.

Savary et al. (2019) in the References section provides a detailed account of the importance of diseases on five staple crops globally and in food security hot-spots.

Poland et al. (2009) provides a thorough synthesis of quantitative host plant resistance: its expression, its molecular basis, and its underlying mechanisms.

McDonald, B. A. and Linde, C. (2002). Pathogen population genetics, evolutionary potential, and durable resistance. *Annu. Rev. Phytopathol.* 40, 349-379 provides a guideline to mobilise population genetics concepts in order to design strategies for durable host plant resistance.

7 References

Becker, M. and Angulo, C. (2019). The evolution of lowland rice-based production systems in Asia: historic trends, determinants of change, future perspective. *Adv. Agron.* 157: 293-327.

Bonman, J. M., Khush, G. S. and Nelson, R. J. (1992). Breeding rice for resistance to pests. *Annu. Rev. Phytopathol.* 30(1): 507-528.

FAOSTAT (Food and Agriculture Organization of the United Nations) (2018). Available at: http://www.fao.org/faostat/en.

Fitt, B. D. L., McCartney, H. A. and Walklate, P. J. (1989). The role of rain in dispersal of pathogen inoculum. *Annu. Rev. Phytopathol.* 27(1): 241-270.

Fresco, L. O. and Kroonenberg, S. B. (1992). Time and spatial scales in ecological sustainability. *Land Use Policy* 9(3): 155-168.

Garrett, K. A., Andersen, K. F., Asche, F., Bowden, R. L., Forbes, G. A., Kulakow, P. A. and Zhou, B. (2017). Resistance genes in global crop breeding networks. *Phytopathology* 107(10): 1268-1278.

Gilligan, C. A. and van den Bosch, F. (2008). Epidemiological models for invasion and persistence of pathogens. *Annu. Rev. Phytopathol.* 46: 385–418.

Jeger, M. J. (1983). Analysing epidemics in time and space. *Plant Pathol.* 32(1): 5–11.

McDonald, B. A. (2014). Using dynamic diversity to achieve durable disease resistance in agricultural ecosystems. *Trop. Plant Pathol.* 39(3): 191–196.

Mew, T. W. (1991). Disease management in rice. In: Pimentel, D. and Hansen, A. A. (Eds.). *CRC Handbook of Pest Management in Agriculture* (2nd edn.) (vol. III), Boca Raton, Ann Arbor, Boston, CRC Press Inc, 279–299.

Mundt, C. C. (2014). Durable resistance: a key to sustainable management of pathogens and pests. *Inf. Genet. Evol.* 27: 446–455.

Mundt, C. C., Sackett, K. E., Wallace, L. D., Cowger, C. and Dudley, J. P. (2009). Aerial dispersal and multiple-scale spread of epidemic disease. *EcoHealth* 6(4): 546–552.

Nelson, R. (2020). International plant pathology: past and future contributions to global food Ssecurity. *Phytopathology* 110(2): 245–253.

Nielsen, L. K., Jensen, J. D., Nielsen, G. C., Jensen, J. E., Spliid, N. H., Thomsen, I. K., Justesen, A. F., Collinge, D. B. and Jørgensen, L. N. (2011). Fusarium head blight of cereals in Denmark: species complex and related mycotoxins. *Phytopathology* 101(8): 960–969.

Ou, S. H. (1987). *Rice Diseases* (2nd edn.), Farnham House, Farnham Royal, Slough, England, CAB International.

Pandey, S., Byerlee, D., Dawe, D., Dobermann, A., Mohanty, S., Rozelle, S. and Hardy, B. (Eds.) (2010). *Rice in the Global Economy: Strategic Research and Policy Issues for Food Security*, Los Baños Philippines, International Rice Research Institute.

Poland, J. A., Balint-Kurti, P. J., Wisser, R. J., Pratt, R. C. and Nelson, R. J. (2009). Shades of gray: the world of quantitative disease resistance. *Trends Plant Sci.* 14(1): 21–29.

Savary, S., Willocquet, L., Elazegui, F. A., Teng, P. S., Du, P. V., Zhu, D., Tang, Q., Huang, S., Lin, X., Singh, H. M. and Srivastsava, R. K. (2000a). Rice pest constraints in tropical Asia: characterisation of injury profiles in relation to production situations. *Plant Dis.* 84(3): 341–356.

Savary, S., Willocquet, L., Elazegui, F. A., Castilla, N. P. and Teng, P. S. (2000b). Rice pest constraints in tropical Asia: quantification of yield losses due to rice pests in a range of production situations. *Plant Dis.* 84(3): 357–369.

Savary, S., McRoberts, N., Esker, P. D., Willocquet, L. and Teng, P. S. (2017). Production situations as drivers of crop health: evidence and implications. *Plant Pathol.* 66(6): 867–876.

Savary, S., Willocquet, L., Pethybridge, S. J., Esker, P., McRoberts, N. and Nelson, A. (2019) The global burden of pathogens and pests on major food crops. *Nat. Ecol. Evol.* 3(3): 430–439.

Van Der Fels-Klerx, H. J., Klemsdal, S., Hietaniemi, V., Lindblad, M., Ioannou-Kakouri, E. and Van Asselt, E. D. (2012). Mycotoxin contamination of cereal grain commodities in relation to climate in North West Europe. *Food Addit. Contam A* 29(10): 1581–1592.

Willocquet, L., Elazegui, F. A., Castilla, N., Fernandez, L., Fischer, K. S., Peng, S., Teng, P. S., Srivastava, R. K., Singh, H. M., Zhu, D. and Savary, S. (2004). Research priorities for rice pest management in tropical Asia: a simulation analysis of yield losses and management efficiencies. *Phytopathology* 94(7): 672–682.

Willocquet, L., Savary, S. and Yuen, J. (2017). Multiscale phenotyping and decision strategies in breeding for resistance. *Trends Plant Sci.* 22(5): 420–432.

Xu, X. M. and Ridout, M. S. (1998). Effects of initial epidemic conditions, sporulation rate, and spore dispersal gradient on the spatio-temporal dynamics of plant disease epidemics. *Phytopathology* 88(10): 1000–1012.

Zadoks, J. C. (1985). On the conceptual basis of crop loss assessment: the threshold theory. *Annu. Rev. Phytopathol.* 23(1): 455–473.

Zadoks, J. C. and Schein, R. D. (1979). *Epidemiology and Plant Disease Management*, New York, Oxford University Press.

Zwankhuizen, M. J. and Zadoks, J. C. (2002). *Phytophthora infestans*'s 10-year truce with Holland: a long-term analysis of potato late-blight epidemics in the Netherlands. *Plant Pathol.* 51(4): 413–423.

Part 1

Fungal diseases of cereals: rusts

Chapter 2

Advances in understanding the biology and epidemiology of rust diseases of cereals

Vanessa Bueno-Sancho, Clare M. Lewis and Diane G. O. Saunders, John Innes Centre, UK

1 Introduction

The rust fungi (order: Pucciniales) constitute the largest group of plant parasitic fungi, with more than 9000 species known worldwide (Perdomo-Sanchez and Piepenbring, 2008). They have been shown to infect a wide variety of hosts from gymnosperms to angiosperms. Within the Pucciniales, the largest genus by far is the *Puccinia* which contains up to 4000 characterised species including many economically important pathogens of staple crops such as cereals (Aime et al., 2017). In terms of scientific and economic importance the *Puccinia* were ranked third in a list of the most important plant pathogenic fungi (Dean et al., 2012). Economically this is largely due to the impact of the three wheat rust fungi – wheat stem (black) rust (*Puccinia graminis* f. sp. *tritici*), yellow (stripe) rust (*Puccinia striiformis* f. sp. *tritici*) and leaf (brown) rust (*Puccinia triticina*). The damage caused to wheat production by these three species alone is estimated annually to cost more than £4 billion (Figueroa et al., 2018). However, these pathogens are not just a contemporary threat to wheat production but have

http://dx.doi.org/10.19103/AS.2021.0092.02

threatened crop production throughout history, with the earliest archaeological evidence of stem rust spores being identified close to the area where cereal cultivation first began, on Neolithic potsherds from Jarmo, Iraq (Stewart and Robertson, 1968).

Beyond their economic importance, the complexities of the wheat rust fungi lifecycle, in particular, intrigued many of our scientific predecessors. After long-standing debate, the seminal experiments of De Bary in 1865 finally introduced the term heteroecism (De Bary, 1865–1866), and it is now known that wheat rust fungi require two hosts to complete their lifecycles. In the modern-day genomic era, scientific fascination with wheat rust fungi has turned to the dikaryotic nature of the asexual urediniospore. This spore type facilitates the rapid spread of the disease within cereal crops and is utilised as the most accessible stage for genome-sequencing projects. With two genetically distinct haploid nuclei, their separation has been key to building an understanding of the dikaryotic nuclear partnership. This chapter provides an overview of the wheat rust fungi lifecycle, discusses recent integration of basic biological knowledge and genomic-led tools and resources within an epidemiological framework. Such an approach will be crucial in making long-term predictions about invasion and persistence of wheat rust races in new areas (Gilligan and Van Den Bosch, 2008), which ultimately can be used to curtail the threat of these notorious wheat rust pathogens.

2 The wheat rust pathogen lifecycle

The three wheat rust fungi broadly follow the same macrocyclic lifecycle (Fig. 1), with a suite of five distinct spore types and two hosts, a 'primary' cereal host and a taxonomically unrelated 'alternate' host that includes *Berberis* and *Mahonia* spp. (for *P. graminis* and *P. striiformis*) or *Thalictrum flavum* (for *P. triticina*) (Roelfs et al., 1992, Jin, 2011). The asexual stage enables rapid proliferation of the pathogen through the production of large quantities of urediniospores in cycles approximately every 14 days on the 'primary' host that can be dispersed thousands of kilometres by wind currents (Brown and Hovmoller, 2002). This ability for long-distance dispersal helps these pathogens to spread widely on a global scale and leads to the emergence of epidemics in new locations (Nagarajan and Singh, 1990). For instance, it was recently shown that the Australian *P. graminis* f. sp. *tritici* population likely originated through an intercontinental high-altitude wind-borne transmission event of urediniospores that originated in South Africa (Visser et al., 2019). The asexual 'repeating-stage' continues *ad infinitum* during the spring and summer months until suboptimal disease conditions or a decrease in day length, cooler temperatures and host senescence occur (Singh et al., 2002). This signifies the onset of autumn and instigates a switch to the sexual stage of the lifecycle (Fig. 1).

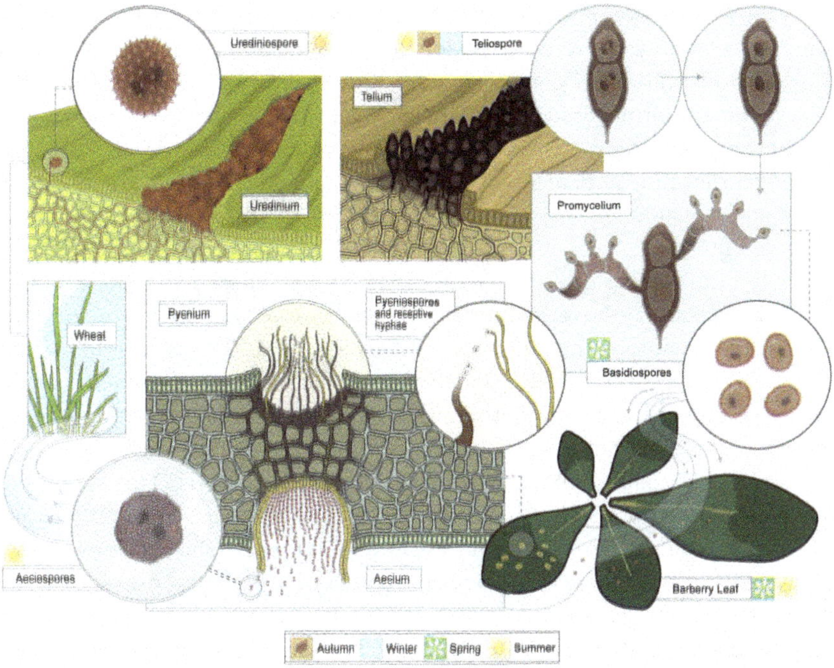

Figure 1 Illustration of the macrocyclic lifecycle of the wheat stem or yellow rust pathogens that broadly follow the same disease cycle.

The onset of the sexual cycle is denoted by telia formation on the infected 'primary' host, post-urediniospore infection. These telia undergo karyogamy to yield diploid teliospores, which have thick cell walls to avoid desiccation (Peterson, 1974). The teliospores enter a period of dormancy to overwinter on the 'primary' host, undergoing meiosis in the spring and forming structures called basidia which in-turn release the short-lived, airborne basidiospores. Only disseminated basidiospores that successfully adhere to the adaxial leaf surface of the alternate host will germinate and produce mycelia. Following mycelial growth, circular lesions form on the leaf surface, which within a few days differentiate to form flask-shaped pycnia. These pycnia have two mating types, either negative (−) or positive (+) and consist of two mating structures, aerial hyphae and pycniospores, that are immersed within a pycnial nectar. After compatible mating (where a hypha from a - pycnia fuses with a pycniospore from a + pycnia, or vice versa) aecia develop on the abaxial leaf surface, which undergo plasmogamy to release dikaryotic aeciospores (Rodriguez-Algaba et al., 2014). These aeciospores specifically infect the 'primary' host forming uredinia which release urediniospores and thus the asexual stage of the cereal rust life cycle again proliferates.

The mechanisms by which rust fungi proliferate are both a highly efficient diversion of host resources and a brutal assault on a susceptible host. The yield losses they cause are a direct result of the rust fungi absorbing nutrients that would, in a healthy plant, be used for grain development, and thus, causing wrinkled and nutritionally inferior grain (Roelfs, 1985). When uredia break through the epidermal tissue, the plant struggles to control transpiration and its metabolism is compromised. Additionally, the wounds created leave the plant prone to desiccation and/or infection by other plant pathogens and subsequent weakening of tillers leaving the host plant prone to 'lodging' (Schumann and Leonard, 2000).

In temperate regions, in the absence of the alternate host, explosive epidemics caused by the repeating stage of the asexual cycle are not uncommon. One uredinium alone can produce in excess of 100 000 urediniospores (Schumann and Leonard, 2000), and so when scaled up in multiparous uredinia, with a 14-day rotation, an outbreak can rapidly escalate to an epidemic. However, where the fungus survives year-round as uredinia/uredinospores on its gramineous host, genetic diversity is limited (Berlin, 2017), as was found with the now infamous Ug99 race of *P. graminis* f. sp. *tritici* which was first detected in Uganda in 1998 (Pretorius et al., 2000) and still predominates in East Africa some 15 years later (Newcomb et al., 2016). In the presence of the alternate host there is always the threat that new isolates will evolve via sexual recombination which have the potential to cause disease epidemics. Hence, regarding wheat stem rust this led historically to widespread legislation and exclusion campaigns to limit the prevalence of susceptible *Berberis* spp. across Europe and North America over the past century (Stakman, 1923).

Only recently has the alternate host of *P. striiformis* f. sp. *tritici* been defined and in the past decade sexual reproduction recorded on several *Berberis* species from China and Pakistan (Zhuang et al., 2019, Mehmood et al., 2019). Each cluster of aecia represents a high amount of diversity, with every single aecium containing a different genotype (Rodriguez-Algaba et al., 2017). Due to the important role of barberry in generating new races, identifying the locations in which the wheat rust pathogens are capable of carrying out sexual reproduction is essential (Jin, 2011). The occurrence of barberry does not always imply sexual reproduction, as in the United States, where *P. striiformis* f. sp. *tritici* does not reproduce sexually, unlike *P. graminis* f. sp. *tritici* (Wang et al., 2015).

Also in Western Europe, *P. striiformis* f. sp. *tritici* is found only in its asexual form and *P. graminis* f. sp. *tritici* which can undergo sexual reproduction in this region has been in abeyance for many decades. Hence, the prevalence of *Berberis* spp. has been increasing in Europe, with clear implications for cereal infections (Barnes et al., 2020). For instance, since repeal of the barberry exclusion laws in Sweden in 1994, the increase in barberry prevalence has been associated with significant enhancement of the genetic diversity among the oat stem rust fungus, which also utilises barberry as an alternate host (Berlin et al., 2013), while in some areas a highly diverse sexual population of wheat stem rust has recently emerged (Berlin, 2017). The knowledge of the wheat rust pathogen lifecycle will continue to be critical to the design of future disease management strategies.

3 Epidemiology of the wheat rust pathogens

Developing an epidemiological framework where the mechanisms that lead to disease epidemics are understood can be exceptionally valuable in targeting disease control measures. The generation of new wheat rust races through genetic mechanisms such as sexual recombination, somatic hybridisation and/or the introduction of new races through long-distance dispersal can lead to complete changes in the pathogen population structure and thereby particularly damaging epidemics (Li et al., 2019). For instance, prior to the turn of the last century, the *P. striiformis* f. sp. *tritici* population worldwide was divided into seven clear genetic groups, largely based on geographical origin, except in the Middle East where several genetic groups were found (Thach et al., 2016). In the 2000s, the situation changed as more aggressive races appeared that were better adapted to warmer climates and in certain regions replaced the older *P. striiformis* f. sp. *tritici* population (Milus et al., 2006, Milus et al., 2009). For instance, the *PstS1* race became dominant in Australia and the United States, in regions initially considered too warm for this pathogen to proliferate (Hovmoller et al., 2008).

In the past decade in Europe there has also been a complete replacement of the *P. striiformis* f. sp. *tritici* population, with a number of new *P. striiformis* f. sp. *tritici* races identified including Warrior (*PstS7*), Warrior (-) (*PstS10*), Kranich (*PstS8*) and Triticale aggressive being recorded since 2011 (Hovmoller et al., 2016, Hubbard et al., 2015, Bueno-Sancho et al., 2017). These races are thought to have originated in the sexually recombining populations in Asia and then spread to Europe due to long-distance dispersal (Hovmoller et al., 2016). This flux in the *P. striiformis* f. sp. *tritici* population that has accelerated in the past two decades is likely, at least in part, attributed to enhanced migration of *P. striiformis* f. sp. *tritici* races, which could be a consequence of a parallel increase in human movement (Wellings, 2007). Indeed, the low levels of population diversity found between countries and continental regions among the global *P. striiformis* f. sp. *tritici* population tends to support high levels of migration (Sharma-Poudyal et al., 2020).

Although the levels of genetic diversity of *P. striiformis* f. sp. *tritici* isolates between regions appear low, within countries individual isolates have been shown to be highly heterozygous, attributed to long-term asexual evolution (Schwessinger, 2017). Such high levels of heterozygosity have also been observed for *P. triticina*, which is described as highly clonal (McCallum et al.,

2016). *P. triticina* populations in the United States, Central Asia, Russia, the Middle East and Europe have all been reported to be closely related, which indicates frequent migration between these distinct geographical regions (Kolmer et al., 2019). Despite its clonality, variability in *P. triticina* virulence phenotypes (pathotypes) is often found, causing the emergence of new races (Kolmer and Ordonez, 2007). This is believed to occur due to high levels of mutation, genetic drift and subsequent host selection (Aoun et al., 2020). Furthermore, a high level of genetic variation associated with somatic hybridisation has been observed in *P. triticina* (Wu et al., 2020), which can act as an asexual mechanism to enhance its ability to overcome host resistance.

Out of the three wheat rusts, *P. triticina* is broadly the most widely distributed and present in all wheat production areas worldwide (Aoun et al., 2020). It is especially suited to humid and warm climates, such as Central Asia, South America, Great Plains in North America and the Mediterranean (Liu and Kolmer, 2013). Global warming has also increased its prevalence in areas previously too cold that now have milder summers. *P. striiformis* f. sp. *tritici* is also present in most wheat-growing areas, although it was traditionally found in cold and humid areas (De Vallavieille-Pope et al., 1995). This was partly due to urediniospore germination temperature optima (12°C–15°C), which was lower than that reported for *P. graminis* (15°C–24°C) and *P. triticina* (20°C) (Roelfs et al., 1992). However, since the 2000s, *P. striiformis* f. sp. *tritici* isolates have been found in warmer climates with optimum germination temperatures of 18°C (Chen, 2005).

For *P. graminis* f. sp. *tritici,* epidemics have changed dramatically since the appearance of the Ug99 race in Uganda in 1998 (Singh et al., 2011). Since then, three additional races (termed TKTTF, TTRTF and PTSK) believed to be variants of the Ug99 race have emerged and spread widely. The TKTTF race is reported to originate in the Middle East, spreading to Ethiopia and causing severe epidemics between 2013 and 2014 due to the wide-scale planting of the susceptible wheat variety Digalu (Olivera et al., 2015). This *P. graminis* f. sp. *tritici* race and its variants have also caused sporadic outbreaks in several European countries, including Germany, Denmark and the United Kingdom (Lewis et al., 2018). The TTRTF race was first detected in 2014 in Georgia (Olivera et al., 2019) and since, has spread to Kenya, Ethiopia and Egypt (Newcomb et al., 2016). Recent outbreaks reported in Sicily were also attributed to the *P. graminis* f. sp. *tritici* TTRTF race (Bhattacharya, 2017). A close relative to the TTRTF race was identified in Asia, Eastern Africa and the Middle East (race RRTTF) and more recently in Ecuador in 2016 (Barnes et al., 2018). Finally, the *P. graminis* f. sp. *tritici* race PTKS, previously found in Ethiopia, Kenya and Yemen, was found in South Africa in 2017 (Terefe et al., 2019). *P. graminis* f. sp. *tritici* is adapted to warmer conditions than the other two wheat rusts, and as climate changes this could facilitate the expansion in frequency of stem rust epidemics in regions that have previously been largely unsuitable for its proliferation (Saunders et al., 2019).

Annual movements of *P. graminis* f. sp. *tritici* spores have been reported to occur from northern Mexico to northern United States and Canada. This is believed to occur in the spring and has been named the 'American rust corridor' (Singh et al., 2013). Long-distance dispersal for *P. striiformis* f. sp. *tritici* has also been repeatedly reported in the United States, travelling more than 2400 km (Zadoks, 1961) and India, with distances of over 1000 km (Mehta, 1952). In Europe, *P. striiformis* f. sp. *tritici* tends to be carried on winds towards the north-east, along the coast. Transport of *P. triticina* inoculum across the Mediterranean, between the south of France and the north of Africa has also been reported. The similarities in *P. triticina* races found in Western Europe and North or even Central Africa suggest the existence of large-scale international exchange of inoculum (Hogg et al., 1969). The ability for long-distance dispersal of wheat rust races and their recent adaptation to new climatic zones continues to prove challenging when tackling epidemics.

4 Developing integrated disease forecasting for wheat rusts

To track the progression of wheat rust epidemics, monitoring programmes that aim to predict disease incidence have been established. One example is the multi-component and highly integrated disease monitoring and forecasting system developed in Ethiopia (Allen-Sader et al., 2019). Ethiopia is the largest wheat producer in sub-Saharan Africa and also a gateway for new *P. striiformis* f. sp. *tritici* isolates entering Africa, and hence the highest priority country in the region for wheat rust surveillance (Meyer et al., 2017). The Ethiopian Early Warning System uses field surveillance data, spore dispersal modelling and disease forecasting based on weather and data from wheat rust advisories and policymakers to provide information to farmers in a timely fashion, so they can take action before disease takes hold. This framework has been extremely valuable in reducing disease incidence and in advising policy makers and smallholder farmers about how to maximise the use of limited fungicide resources, as demonstrated during the 2017 and 2018 seasons (Allen-Sader et al., 2019).

Studying dispersal and spore deposition can help predict the severity and impact of spore transport under climate change (Prank et al., 2019). For instance, models for *P. striiformis* f. sp. *tritici* have shown that epidemics may well become more severe and appear earlier in the year as climate changes (Lyon and Broders, 2017). For *P. graminis* f. sp. *tritici*, climate change is expected to expand overwintering areas and increase the frequency of spore migration. However, the probability of infection following spore deposition may decrease due to drier conditions (Prank et al., 2019). For *P. triticina*, the resulting disease is expected to increase during winter and autumn seasons with a drop during summer (Launay et al., 2020). Weather models can also be extremely valuable for guiding fungicide spray regimes and have been integrated into neural networks to help design risk models to target specific regions for fungicide spray. Choosing the most impactful areas to spray lowers the level of fungicide application and the overall economic cost (El Jarroudi et al., 2020).

For *P. triticina* a platform called Injury Profile SIMulation (IPSIM) was developed to aid farmer decisions. This system uses various disease risk factors such as severity, soil information, weather conditions and cropping systems to generate a model for predicting disease risk (Robin et al., 2018). Other recent models have focussed on integrating satellite imagery with airborne inoculum and weather data to help generate model-based frameworks to control crop diseases (Newlands, 2018). Satellite imagery has indeed been used for the detection of wheat rust disease incidence since remote sensing technologies are able to monitor diseases more rapidly and accurately than humans (Chen et al., 2018). High-resolution multispectral imagery and hyperspectral unmanned aerial vehicle (UAV) images have been used to detect yellow rust and construct severity maps that assist with disease monitoring (Su et al., 2018, Zhang et al., 2019). This imagery has also been used for detection of leaf rust using variations of leaf area index at the canopy level (Azadbakht et al., 2019). The use of advanced machine-learning techniques has allowed the development of quick surveillance using UAV images and it is a matter of time until these methods become established in routine monitoring of wheat rust (Bohnenkamp et al., 2019).

5 Mechanisms of wheat rust control

Traditionally the most effective means of controlling wheat rust has been the use of wheat varieties resistant to infection, through integration of partial or race-specific resistance genes. This was exemplified during the Green Revolution in the mid-to-late twentieth century where breeding of durable stem rust resistance into high-yielding semi-dwarf wheat cultivars drove stem rust to a disease of near insignificance in many regions (Hodson, 2011). However, the emergence of the broadly virulent Ug99 race in 1998 in Uganda rendered up to 80% of the world's wheat varieties vulnerable once again to stem rust, re-invigorating the battle to identify and integrate resistance into elite wheat varieties (Bhavani et al., 2019). Plants have at their disposal a broad suite of innate and inducible resistance mechanisms which are primarily governed by the activity of resistance (R) proteins. These highly conserved proteins have evolved with functions not only in pathogen perception but also in the activation of host cellular processes to protect against the proliferation and elicitation of disease (Martin et al., 2003).

To date more than 300 plant *R* genes have been cloned, having roles in defence signalling against a range of plant pathogens including many governing resistances to the wheat rusts (Ellis et al., 2014). These cloned *R* genes have largely been elucidated through positional cloning or mutational genomics (Sanchez-Martin et al., 2016, Thind et al., 2017). However, these techniques are only possible where a single *R* gene exists in a genetic background susceptible to the pathogen of interest and such situations can be difficult and/or laborious to achieve. With the advance of novel genomic-based techniques such as SMRT RenSeq (Steuernagel et al., 2016) the number of annotated *R* genes is expected to increase at a much faster pace than before. This technique exploits the fact that wild relatives of crop plants often harbour multiple *R* genes and combines *R* gene sequence capture (RenSeq) with single-molecule real-time (SMRT) sequencing to rapidly clone *R* genes. In the past 4 years alone, SMRT RenSeq has been used to rapidly clone (amongst others) the stem rust *R* genes *Sr33*, *Sr22* and *Sr45* from hexaploid bread wheat (Steuernagel et al., 2016) and four stem rust *R* genes from *Aegilops tauschii* (Arora et al., 2019). However, despite the advances in *R* gene cloning, the underlying mechanisms of action remain largely elusive, which will be vital to guide rational engineering of disease resistance in future.

As a second line of defence, the routine application of fungicides can be used to effectively control the wheat rusts. This practice has been crucial in supporting high-intensity farming employed by modern agriculture, where timely and managed fungicide application is efficacious upon both crop yields and grain quality, and effectively sustains production (Loughman et al., 2005, Gianessi and Reigner, 2006). One effective group of systemic fungicides are the triazoles that were commercially introduced in the 1970s (Morton and Staub, 2008). This group of fungicides form the basis of cereal disease management strategies worldwide, particularly in Europe, North America, Australia and New Zealand.

In cereal disease management programmes products and application strategies vary between regions, although triazoles are often mixed with strobilurins which were introduced in 1992 and with the new generation pyrazole carboxamide (e.g. isopyrazam, bixafen) succinate dehydrogenase inhibitors (SDHIs), introduced over the past two decades (Ren et al., 2018). However, the reliance on fungicides in agricultural settings creates a selection pressure that can lead to the evolution of resistance that differs between fungicide classes and pathogen species. The wheat rusts have been classified as low-risk organisms (Oliver, 2014). However, reduced sensitivity of *P. striiformis* f. sp. *tritici* isolates to triazole fungicides has been reported in the United Kingdom, the United States and China (Bayles et al., 2000, Kang et al., 2019, Tian et al., 2019), highlighting the importance of monitoring programmes.

In addition to genetic and chemical control mechanisms, cultural practices can also play a vital role in disease control. For instance, the eradication of the alternate host for *P. graminis* f. sp. *tritici* has been exceptionally beneficial in reducing early season inoculum in regions where the pathogen is unable to overwinter in its urediniospore state and in curtailing the spawning of new races through sexual recombination. For instance, the large-scale removal of *Berberis* in the United States radically reduced the number of *P. graminis* f. sp. *tritici* races per year from 17 to 8 after eradication (Peterson, 2001). This illustrates how an integrated approach, utilising basic biological knowledge regarding the pathogen's lifecycle alongside chemical and genetic control mechanisms can act together to strengthen resilience to wheat rust in high-production agricultural settings.

6 Studying wheat rust pathogens in the genomic era

Like other obligate biotrophic pathogens, the wheat rusts are unable to be cultured in the absence of their host and are recalcitrance to genetic transformation. They also exhibit high degrees of host specificity precluding the use of more accessible model plant hosts as surrogates. However, the genomic era has been instrumental in accelerating molecular genetic research of such intractable pathosystems, providing novel insight into the biology of the wheat rust pathogens. Genomic-based approaches have generated new insight regarding proteins governing their pathogenicity, defined evolutionary mechanisms at the genomic level that have governed the generation of some of the most damaging wheat rust races such as the Ug99 race of *P. graminis* f. sp. *tritici* and markedly improved the resolution of population studies (Li et al., 2019, Figueroa et al., 2020), whilst also providing new knowledge regarding the hosts response to infection and accelerating the identification of host resistance genes (Chandra et al., 2017). Together, this has generated a foundation of genomic knowledge that can be utilised to address many long-standing fundamental questions about wheat rust biology.

Since assembly of the first rust pathogen genomes (Duplessis et al., 2011), a flurry of activity to define pathogenicity factors ensued with genome-wide analyses and transcriptomic studies used to develop rich catalogues of candidate secreted effector proteins for these agronomically important crop pathogens (Saunders et al., 2012). Effector proteins are secreted by pathogens during the infection process to promote successful colonisation. Certain effectors function in the plant apoplast and others are translocated into the host cell cytoplasm, interacting with host proteins to manipulate the physiological processes of the host for the benefit of the pathogen (Wang et al., 2017). On certain host genotypes some of these effectors, termed avirulence (Avr) proteins, are recognised by corresponding host immune receptors, commonly nucleotide-binding, leucine-rich repeat receptors (NLRs; (Wilmanski et al., 2008)).

Comparative genomic studies initially focussed on classifying effectors based on conserved features between pathosystems such as presence of a secretion signal (among other criteria). These potential-secreted proteins were then surveyed for modifications or deletions of candidate-secreted effector proteins in virulent strains as a mechanism of identifying Avr effectors that can lead to the evasion of recognition and the emergence of novel pathogen races (Wu et al., 2015). The identification of the first avirulence proteins in the wheat rust *P. graminis* f. sp. *tritici* (AvrSr50 and AvrSr35) was underpinned by comparative genomic studies with spontaneous and chemically induced mutants (Chen et al., 2017, Salcedo et al., 2017). The generation of genome assemblies of the corresponding wild-type strains was key in this process, although a multitude of inherent features continue to challenge the development of chromosome-level assemblies for the wheat rust pathogens.

Wheat rusts contain large genomes for basiodiomycete fungi (Raffaele and Kamoun, 2012), with *P. triticina* having the largest haploid genome size of the three (~135 Mb), followed by *P. striiformis* f. sp. *tritici* (≈117 Mb) and *P. graminis* f. sp. *tritici* (~ 90 Mb). All contain a high percentage of mobile elements, repeats and are highly heterozygous (Cuomo et al., 2017). With the most accessible stage of the lifecycle being the abundant dikaryotic urediniospores, this initially impeded genome assemblies using short-read Illumina sequencing, leading to the generation of many chimeric assemblies (Duplessis et al., 2011). However, recent improvements in long-read sequencing technology and genome assembly algorithms have aided the development of phased genomes where both haplotypes are represented. This has enabled comparative analysis between haplotypes to study evolutionary processes that have shaped rust genomic architecture. For instance, assemblies of complete karyon sequences of rust fungi facilitated the discovery of the origin of the Ug99 *P. graminis* f. sp. *tritici* race. Genome analysis showed that isolates belonging to this race had arisen by somatic hybridisation, since no recombination or chromosome rearrangement was observed, illustrating that it likely originated from a much older African lineage of *P. graminis* f. sp. *tritici* (Li et al., 2019).

High levels of diversity have also been reported, for instance, between haplotypes for *P. striiformis* f. sp. *tritici*, with 25% of the genes not having an allelic counterpart (Schwessinger et al., 2020). Furthermore, it has been observed that asexual isolates show much higher degrees of diversity between haplotypes than sexual isolates such as the *P. striiformis* f. sp. *tritici* Warrior race isolates that recently emerged in Europe and have large genomic structural variation (Figueroa et al., 2020). The ever-decreasing price of genome sequencing projects has been a key accelerant for the flurry of genomic-based projects that have significantly advanced our fundamental knowledge of the wheat rusts in the last decade (Kiran et al., 2016) and will undoubtedly continue to help build a better understanding of these agronomically important fungi.

7 Case study: field pathogenomics

Field pathogenomics was developed in 2013 as a rapid, high-resolution genotyping method to evaluate a reported change in the *P. striiformis* f. sp. *tritici* population in the United Kingdom that had been noted by breeders and agronomists (Hubbard et al., 2015). Traditionally, genotyping of *P. striiformis* f. sp. *tritici* isolates was reliant on PCR-based approaches applied largely to purified isolates with a handful of genetic markers such as polymorphic simple sequence repeat (SSR) loci to differentiate pathogen isolates (Thach et al., 2016). A fixed array of pre-defined single nucleotide polymorphisms (SNPs) did become available for studying *P. graminis* f. sp. *tritici* population dynamics (Pretorius et al., 2012) but its application was limited to a single lab and pathogen. The field pathogenomics methodology provided a *novel* and accessible genomics-based platform to quickly assess any pathogen population composition directly from field-collected infected plant samples as a complement to traditional pathology-based approaches (Fig. 2).

In traditional pathology, isolates are captured, purified, multiplied and inoculated onto a panel of differential wheat lines which carry known resistance genes whereby a virulence profile of the isolate is established. However, this lengthy process takes many months to complete and its labour-intensive nature severely restrains the number of isolates that can be analysed (Atkins and Clark, 2004). In contrast, the field pathogenomics method uses infected leaf samples directly captured in the field for RNA-seq analysis, quickly providing a genotypic profile of the pathogen population at unprecedented resolution. The close correlation between the genotype and pathology of *P. striiformis* f. sp. *tritici* isolates also enables genomics-based approaches to act as a proxy for inferring likely virulence profiles in the interim.

The application of field pathogenomics to monitor wheat rust populations has a number of advantages that include: (i) the ability to rapidly screen unprecedented numbers of isolates due to the lower labour input required, (ii) subverting isolate losses during their multiplication (a frequent problem for pathology studies) by ensuring genotype data is achieved from all samples as no multiplication is required (iii) helping with pre-selection of the most genotypically notable and diverse isolates to make the best use of subsequent labour-intensive pathology and (iv) ensuring any genotypic nuances are detected and fully evaluated. The latter became clear during the validation of the field pathogenomics method when a *P. striiformis* f. sp. *tritici* isolate with an identical virulence profile to a historic race collected from the wheat variety Solstice was found to be significantly genotypically distinct, a development

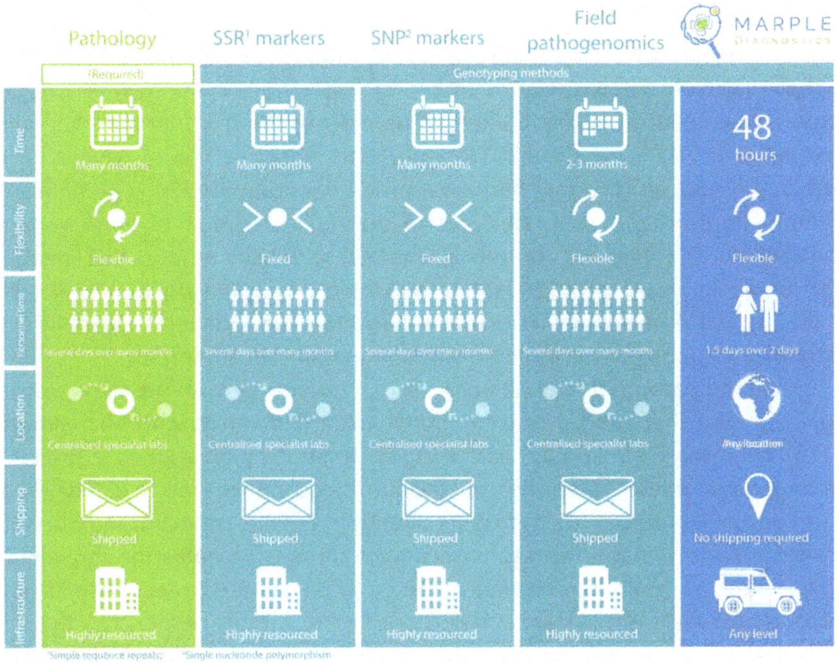

Figure 2 Infographic illustrating the key considerations of different approaches for genotypic characterisation of wheat rust pathogen isolates, alongside traditional pathology.

that would have been overlooked had phenotypic profiling been considered in isolation. Interestingly, this distinct group of isolates were genetically similar to the Warrior race and formed part of the influx of new races into Europe around 2011 (Hubbard et al., 2015). This also highlighted the immense value of aligning the field pathogenomics method and virulence profiling, illustrating how the two approaches can act hand-in-hand as valuable tools in the monitoring of pathogens that can rapidly shift population structures such as *P. striiformis* f. sp. *tritici*. Accordingly, these two methods have since worked in unison in the UK Cereal Pathogen Virulence Survey (UKCPVS) and been instrumental in monitoring the population structure of *P. striiformis* f. sp. *tritici* in the United Kingdom and beyond.

In the initial application of field pathogenomics in 2013, a comparison was carried out between isolates in field-collected samples with those from 20 years of prior diversity, revealing a complete shift in the *P. striiformis* f. sp. *tritici* population. These analyses showed that the older population of *P. striiformis* f. sp. *tritici* had been largely displaced with all 2013 isolates belonging to four exotic lineages that had recently entered the United Kingdom, including the infamous Warrior race (Hubbard et al., 2015). These new *P. striiformis* f.

sp. *tritici* isolates had much higher degrees of genetic diversity than the population that existed pre-2013 that was subsequently shown to be likely due to their origin among sexually recombining populations in the near-Himalayan region (Hovmøller et al., 2016). Further analysis of the *P. striiformis* f. sp. *tritici* population in 2014, 2015 and 2016 captured another new lineage shown to be part of the Kranich race that had previously been identified only in mainland Europe. Furthermore, this study revealed high degrees of host varietal specificity in the field and potential seasonal variability among the lineages with certain lineages detected later only in the growing season (Bueno-Sancho et al., 2017). This new knowledge regarding seasonal and varietal specificity is extremely valuable for informing disease management strategies.

The field pathogenomics strategy also generates gene expression profiles from the host that can be of exceptional value (Derevnina and Michelmore, 2015). For instance, being applied to study variations in host gene expression among wheat lines in response to *Fusarium* head blight (Fall et al., 2019). In addition, the speed and ability to work directly from field-collected samples that are stored and transported as inactivate biosamples obviates the movement of live samples, enabling the methodology to be utilised in response to new disease outbreaks, for instance, being applied to rapidly respond to the first outbreak of wheat blast in Bangladesh. Within just six weeks of receiving samples the causal agent was confirmed and through comparative genomics with other available genome sequences its likely origin and thereby route of entry determined (Islam et al., 2016).

To widen accessibility of such genomics-based approaches the field pathogenomics methodology recently formed the basis to develop a new method that substitutes reliance on large, immobile Illumina sequencing technology for the portable MinION nanopore sequencer, termed Mobile And Real-time PLant disEase (MARPLE) diagnostics (Radhakrishnan et al., 2019). Instead of carrying out full RNA-seq analysis, MARPLE diagnostics employs targeted re-sequencing of a small subset of highly variable genes that are sufficient to capture the diversity of the *P. striiformis* f. sp. *tritici* population. This new system is also highly suited to disease emergencies as it can generate strain-level data on *P. striiformis* f. sp. *tritici* within just 48 h of collecting a sample. The gene panel defined for targeted re-sequencing is also flexible and can easily accommodate additional genes as required. For instance, incorporating genes that encode proteins identified as conserved fungicide targets across fungal pathogens. In the future, this will allow real-time monitoring for mutations that could be linked to potential breakdowns in fungicide utility for *P. striiformis* f. sp. *tritici*.

However, focussing on just a handful of *P. striiformis* f. sp. *tritici* genes in MARPLE diagnostics also leads to a loss of information in comparison to the field pathogenomics method. For instance, at present, it cannot be used to study the host response. Furthermore, to design similar approaches for other pathosystems requires existing genomic data from a diverse set of isolates to be available to define variable genes for amplification. However, as the cost of genome sequencing continues to decrease, the high value of such genomic-based surveillance tools will undoubtedly lead to much wider adoption across many important pathosystems.

8 Conclusion and future trends

As the wheat rusts continue to threaten cereal production in all major wheat-growing areas worldwide, anticipating the severity of rust epidemics will require an integrated approach. Recent advances in genomic-based tools have markedly accelerated resistance gene cloning and the resolution of pathogen population studies. In addition, mathematical modelling offers an opportunity to understand under which conditions wheat rust diseases occur and also predict their spread. For instance, investigating urediniospore dispersal patterns can help understand migration between countries as has been the case for *P. graminis* f. sp. *tritici* (Meyer et al., 2017) and improve disease risk forecasting as was shown for *P. striiformis* f. sp. *tritici* (Newlands, 2018). However, the accessibility and importance of the wheat rust asexual lifecycle as the amplifier of disease epidemics has, to date, channelled research attention to the urediniospore stage.

Recently, a renewed emphasis on the sexual cycle has started to generate new knowledge that can help guide disease management in regions where the alternate hosts play a role in wheat rust epidemiology. Now the alternate host has been defined for all three wheat rust pathogens, and their crucial role has come back to the forefront. In future, including surveys of alternate hosts within surveillance programmes would enable the rapid identification of new races that may subsequently spread to cereals (Zhao et al., 2016). Furthermore, ensuring genomic-based tools for surveillance are simplified, as was the case with MARPLE diagnostics (Radhakrishnan et al., 2019), improves the accessibility of these latest technological advances ensuring they can be more widely utilised. As our understanding of basic pathogen biology continues to grow, informing modelling in disease forecasting and identification, alongside the integration of advances in resistance gene cloning, this holds great promise for ultimately curtailing the threat of these notorious wheat rust pathogens.

9 Where to look for further information

The following websites provide up-to-date information on the geographic distribution of wheat rust races and the utility of known rust resistance genes:

- Global Rust Reference Center, Aarhus University, Denmark: https://agro.au.dk/forskning/internationale-platforme/wheatrust/.
- Borlaug Global Rust Initiative, *R* gene list: https://www.globalrust.org/gene.

The following website provides information on the susceptibility of *Berberis* species to *P. graminis* f. sp. *tritici*, screened within the United States:

- United States Department of Agriculture, Animal and Plant Inspection Service: https://www.aphis.usda.gov/aphis/ourfocus/planthealth/plant-pest-and-disease-programs/pests-and-diseases/barberry/ct_barberry.

10 Acknowledgements

We thank Tobin Florio and Matt Heaton (JIC) for figure preparation. V.B.S. was supported by the UK Biological Sciences Research Council (BBSRC) Norwich Research Park Biosciences Doctoral Training Partnership (grant number BB/M011216/1), C.M.L. by funding from BBSRC (grant number BB/S003975/1). Additional funding was provided to D.G.O.S. by a European Research Council (no. 715638) grant, the BBSRC Institute Strategic Programmes BB/P012574/1 and BB/P016855/1 and the John Innes Foundation.

11 References

Aime, M. C., McTaggart, A. R., Mondo, S. J. and Duplessis, S. 2017. Phylogenetics and phylogenomics of rust fungi. *Advances in Genetics* 100, 267–307.

Allen-Sader, C., Thurston, W., Meyer, M., Nure, E., Bacha, N., Alemayehu, Y., Stutt, R. O. J. H., Safka, D., Craig, A. P., Derso, E., Burgin, L. E., Millington, S. C., Hort, M. C., Hodson, D. P. and Gilligan, C. A. 2019. An early warning system to predict and mitigate wheat rust diseases in Ethiopia. *Environmental Research Letters: ERL [Web Site]* 14(11), 115004.

Aoun, M., Kolmer, J. A., Breiland, M., Richards, J., Brueggeman, R. S., Szabo, L. J. and Acevedo, M. 2020. Genotyping-by-sequencing for the study of genetic diversity in *Puccinia triticina*. *Plant Disease* 104(3), 752–760.

Arora, S., Steuernagel, B., Gaurav, K., Chandramohan, S., Long, Y. M., Matny, O., Johnson, R., Enk, J., Periyannan, S., Singh, N., Asyraf Md Hatta, M., Athiyannan, N., Cheema, J., Yu, G., Kangara, N., Ghosh, S., Szabo, L. J., Poland, J., Bariana, H., Jones, J. D. G., Bentley, A. R., Ayliffe, M., Olson, E., Xu, S. S., Steffenson, B. J., Lagudah, E., Wulff, B. B. H. 2019. Resistance gene cloning from a wild crop relative by sequence capture and association genetics. *Nature Biotechnology* 37(2), 139–143.

Atkins, S. D. and Clark, I. M. 2004. Fungal molecular diagnostics: a mini review. *Journal of Applied Genetics* 45(1), 3–15.

Azadbakht, M., Ashourloo, D., Aghighi, H., Radiom, S. and Alimohammadi, A. 2019. Wheat leaf rust detection at canopy scale under different LAI levels using machine learning techniques. *Computers and Electronics in Agriculture* 156, 119–128.

Barnes, C. W., Ordonez, M. E., Hambleton, S. and Dadej, K. 2018. Detection of Puccinia graminis f. sp tritici race RRTTF in Ecuador in 2016. *Plant Disease* 102, 448–449.

Barnes, G., Saunders, D. G. O. and Williamson, T. 2020. Banishing barberry: the history of *Berberis vulgaris* prevalence and wheat stem rust incidence across Britain. *Plant Pathology* 69(7), 1193–1202.

Bayles, R. A., Stigwood, P. L. and Clarkson, J. D. S. 2000. Shifts in sensitivity of *Puccinia striiformis* to DMI fungicides in the UK. *Acta Phytopathologica et Entomologica Hungarica* 35, 381-382.

Berlin, A. 2017. *Stem Rust Attacks in Sweden Heralds the Return of a Previously Vanquished Foe*. Sweden: Swedish University of Agricultural Sciences, SLU News.

Berlin, A., Samils, B., Djurle, A., Wirsen, H., Szabo, L. and Yuen, J. 2013. Disease development and genotypic diversity of *Puccinia graminis* f. sp *avenae* in Swedish oat fields. *Plant Pathology* 62(1), 32-40.

Bhattacharya, S. 2017. Deadly new wheat disease threatens Europe's crops. *Nature* 542(7640), 145-146.

Bhavani, S., Hodson, D. P., Huerta-Espino, J., Randhawa, M. S. and Singh, R. P. 2019. Progress in breeding for resistance to Ug99 and other races of the stem rust fungus in CIMMYT wheat germplasm. *Frontiers of Agricultural Science and Engineering* 6(3), 210-224.

Bohnenkamp, D., Behmann, J. and Mahlein, A. K. 2019. In-field detection of yellow rust in wheat on the ground canopy and UAV scale. *Remote Sensing* 11(21), 2495.

Brown, J. K. and Hovmoller, M. S. 2002. Aerial dispersal of pathogens on the global and continental scales and its impact on plant disease. *Science* 297(5581), 537-541.

Bueno-Sancho, V., Persoons, A., Hubbard, A., Cabrera-Quio, L. E., Lewis, C. M., Corredor-Moreno, P., Bunting, D. C. E., Ali, S., Chng, S., Hodson, D. P., Madariaga Burrows, R., Bryson, R., Thomas, J., Holdgate, S. and Saunders, D. G. O. 2017. Pathogenomic analysis of wheat yellow rust lineages detects seasonal variation and host specificity. *Genome Biology and Evolution* 9(12), 3282-3296.

Chandra, S., Kazmi, A. Z., Ahmed, Z., Roychowdhury, G., Kumari, V., Kumar, M. and Mukhopadhyay, K. 2017. Genome-wide identification and characterization of NB-ARC resistant genes in wheat (*Triticum aestivum* L.) and their expression during leaf rust infection. *Plant Cell Reports* 36(7), 1097-1112.

Chen, D. M., Shi, Y. Y., Huang, W. J., Zhang, J. C. and Wu, K. H. 2018. Mapping wheat rust based on high spatial resolution satellite imagery. *Computers and Electronics in Agriculture* 152, 109-116.

Chen, J. P., Upadhyaya, N. M., Ortiz, D., Sperschneider, J., Li, F., Bouton, C., Breen, S., Dong, C., Xu, B., Zhang, X., Mago, R., Newell, K., Xia, X., Bernoux, M., Taylor, J. M., Steffenson, B., Jin, Y., Zhang, P., Kanyuka, K., Figueroa, M., Ellis, J. G., Park, R. F. and Dodds, P. N. 2017. Loss of *AvrSr50* by somatic exchange in stem rust leads to virulence for *Sr50* resistance in wheat. *Science* 358(6370), 1607-1610.

Chen, X. M. 2005. Epidemiology and control of stripe rust [*Puccinia striiformis* f. sp *tritici*] on wheat. *Canadian Journal of Plant Pathology* 27(3), 314-337.

Cuomo, C. A., Bakkeren, G., Khalil, H. B., Panwar, V., Joly, D., Linning, R., Sakthikumar, S., Song, X., Adiconis, X., Fan, L., Goldberg, J. M., Levin, J. Z., Young, S., Zeng, Q., Anikster, Y., Bruce, M., Wang, M., Yin, C., McCallum, B., Szabo, L. J., Hulbert, S., Chen, X. and Fellers, J. P. 2017. Comparative analysis highlights variable genome content of wheat rusts and divergence of the mating loci. *G3* 7(2), 361-376.

De Bary, A. 1865-1866. *Neue Untersuchungen über Uredineen, inbesondere die Entwicklung der Puccinia graminis und den Zusammenhang derselbe mit Aecidium berberidis*. Berlin: Monatsber. K. Preuss. Akad. Berlin Wiss.

De Vallavieille-Pope, C., Huber, L., Leconte, M. and Goyeau, H. 1995. Comparative effects of temperature and interrupted wet periods on germination, penetration, and infection of *Puccinia recondita* f. sp. *tritici* and *P. striiformis* on wheat seedlings. *Phytopathology* 85, 409–415.

Dean, R., Van Kan, J. A., Pretorius, Z. A., Hammond-Kosack, K. E., Di Pietro, A., Spanu, P. D., Rudd, J. J., Dickman, M., Kahmann, R., Ellis, J. and Foster, G. D. 2012. The Top 10 fungal pathogens in molecular plant pathology. *Molecular Plant Pathology* 13(4), 414–430.

Derevnina, L. and Michelmore, R. W. 2015. Wheat rusts never sleep but neither do sequencers: will pathogenomics transform the way plant diseases are managed? *Genome Biology* 16, 44.

Duplessis, S., Cuomo, C. A., Lin, Y. C., Aerts, A., Tisserant, E., Veneault-Fourrey, C., Joly, D. L., Hacquard, S., Amselem, J., Cantarel, B. L., Chiu, R., Coutinho, P. M., Feau, N., Field, M., Frey, P., Gelhaye, E., Goldberg, J., Grabherr, M. G., Kodira, C. D., Kohler, A., Kües, U., Lindquist, E. A., Lucas, S. M., Mago, R., Mauceli, E., Morin, E., Murat, C., Pangilinan, J. L., Park, R., Pearson, M., Quesneville, H., Rouhier, N., Sakthikumar, S., Salamov, A. A., Schmutz, J., Selles, B., Shapiro, H., Tanguay, P., Tuskan, G. A., Henrissat, B., Van de Peer, Y., Rouzé, P., Ellis, J. G., Dodds, P. N., Schein, J. E., Zhong, S., Hamelin, R. C., Grigoriev, I. V., Szabo, L. J. and Martin, F. 2011. Obligate biotrophy features unraveled by the genomic analysis of rust fungi. *Proceedings of the National Academy of Sciences of the United States of America* 108(22), 9166–9171.

El Jarroudi, M., Lahlali, R., El Jarroudi, H., Tychon, B., Belleflamme, A., Junk, J., Denis, A., El Jarroudi, M. and Kouadio, L. 2020. Employing weather-based disease and machine learning techniques for optimal control of Septoria leaf blotch and stripe rust in wheat. *Advanced Intelligent Systems for Sustainable Development (AI2SD'2019)*. Cham: Springer.

Ellis, J. G., Lagudah, E. S., Spielmeyer, W. and Dodds, P. N. 2014. The past, present and future of breeding rust resistant wheat. *Frontiers in Plant Science* 5, 641.

Fall, L. A., Salazar, M. M., Drnevich, J., Holmes, J. R., Tseng, M. C., Kolb, F. L. and Mideros, S. X. 2019. Field pathogenomics of *Fusarium* head blight reveals pathogen transcriptome differences due to host resistance. *Mycologia* 111(4), 563–573.

Figueroa, M., Dodds, P. N. and Henningsen, E. C. 2020. Evolution of virulence in rust fungi - multiple solutions to one problem. *Current Opinion in Plant Biology* 56, 20–27.

Figueroa, M., Hammond-Kosack, K. E. and Solomon, P. S. 2018. A review of wheat diseases-a field perspective. *Molecular Plant Pathology* 19(6), 1523–1536.

Gianessi, L. and Reigner, N. 2006. The importance of fungicides in U.S. crop production. *Outlooks on Pest Management* 17(5), 209–213.

Gilligan, C. A. and Van Den Bosch, F. 2008. Epidemiological models for invasion and persistence of pathogens. *Annual Review of Phytopathology* 46, 385–418.

Hodson, D. P. 2011. Shifting boundaries: challenges for rust monitoring. *Euphytica* 179(1), 93–104.

Hogg, W. H., Hounam, C. E., Mallik, A. K. and Zadoks, J. C. 1969. *Meterological Factors Affecting the Epidemiology of Wheat Rusts*. Geneva: World Meteorological Organization.

Hovmoller, M. S., Walter, S., Bayles, R. A., Hubbard, A., Flath, K., Sommerfeldt, N., Leconte, M., Czembor, P., Rodriguez-Algaba, J., Thach, T., Hansen, J. G., Lassen, P., Justesen, A. F., Ali, S. and De Vallavieille-Pope, C. 2016. Replacement of the European wheat yellow rust population by new races from the centre of diversity in the near-Himalayan region. *Plant Pathology* 65(3), 402–411.

Hovmoller, M. S., Yahyaoui, A. H., Milus, E. A. and Justesen, A. F. 2008. Rapid global spread of two aggressive strains of a wheat rust fungus. *Molecular Ecology* 17(17), 3818–3826.

Hubbard, A., Lewis, C. M., Yoshida, K., Ramirez-Gonzalez, R. H., De Vallavieille-Pope, C., Thomas, J., Kamoun, S., Bayles, R., Uauy, C. and Saunders, D. G. 2015. Field pathogenomics reveals the emergence of a diverse wheat yellow rust population. *Genome Biology* 16, 23.

Islam, M. T., Croll, D., Gladieux, P., Soanes, D. M., Persoons, A., Bhattacharjee, P., Hossain, M. S., Gupta, D. R., Rahman, M. M., Mahboob, M. G., Cook, N., Salam, M. U., Surovy, M. Z., Sancho, V. B., Maciel, J. L., NhaniJúnior, A., Castroagudín, V. L., Reges, J. T., Ceresini, P. C., Ravel, S., Kellner, R., Fournier, E., Tharreau, D., Lebrun, M. H., McDonald, B. A., Stitt, T., Swan, D., Talbot, N. J., Saunders, D. G., Win, J. and Kamoun, S. 2016. Emergence of wheat blast in Bangladesh was caused by a South American lineage of *Magnaporthe oryzae*. *BMC Biology* 14(1), 84.

Jin, Y. 2011. Role of *Berberis* spp. as alternate hosts in generating new races of *Puccinia graminis* and *P. striiformis*. *Euphytica* 179(1), 105–108.

Kang, Z. H., Li, X., Wan, A. M., Wang, M. N. and Chen, X. M. 2019. Differential sensitivity among *Puccinia striiformis* f. sp. *tritici* isolates to propiconazole and pyraclostrobin fungicides. *Canadian Journal of Plant Pathology* 41(3), 415–434.

Kiran, K., Rawal, H. C., Dubey, H., Jaswal, R., Devanna, B. N., Gupta, D. K., Bhardwaj, S. C., Prasad, P., Pal, D., Chhuneja, P., Balasubramanian, P., Kumar, J., Swami, M., Solanke, A. U., Gaikwad, K., Singh, N. K. and Sharma, T. R. 2016. Draft genome of the wheat rust pathogen (*Puccinia triticina*) unravels genome-wide structural variations during evolution. *Genome Biology and Evolution* 8(9), 2702–2721.

Kolmer, J. A. and Ordonez, M. E. 2007. Genetic differentiation of *Puccinia triticina* populations in central Asia and the Caucasus. *Phytopathology* 97(9), 1141 –1149.

Kolmer, J. A., Ordonez, M. E., German, S., Morgounov, A., Pretorius, Z., Visser, B., Goyeau, H., Anikster, Y. and Acevedo, M. 2019. Multilocus genotypes of the wheat leaf rust fungus *Puccinia triticina* in worldwide regions indicate past and current long-distance migration. *Phytopathology* 109(8), 1453–1463.

Launay, M., Zurfluh, O., Huard, F., Buis, S., Bourgeois, G., Caubel, J., Huber, L. and Bancal, M. 2020. Robustness of crop disease response to climate change signal under modeling uncertainties. *Agricultural Systems* 178, 102733.

Lewis, C. M., Persoons, A., Bebber, D. P., Kigathi, R. N., Maintz, J., Findlay, K., Bueno-Sancho, V., Corredor-Moreno, P., Harrington, S. A., Kangara, N., Berlin, A., García, R., Germán, S. E., Hanzalová, A., Hodson, D. P., Hovmøller, M. S., Huerta-Espino, J., Imtiaz, M., Mirza, J. I., Justesen, A. F., Niks, R. E., Omrani, A., Patpour, M., Pretorius, Z. A., Roohparvar, R., Sela, H., Singh, R. P., Steffenson, B., Visser, B., Fenwick, P. M., Thomas, J., Wulff, B. B. H. and Saunders, D. G. O. 2018. Potential for re-emergence of wheat stem rust in the United Kingdom. *Communications Biology* 1, 13.

Li, F., Upadhyaya, N. M., Sperschneider, J., Matny, O., Nguyen-Phuc, H., Mago, R., Raley, C., Miller, M. E., Silverstein, K. A. T., Henningsen, E., Hirsch, C. D., Visser, B., Pretorius, Z. A., Steffenson, B. J., Schwessinger, B., Dodds, P. N. and Figueroa, M. 2019. Emergence of the Ug99 lineage of the wheat stem rust pathogen through somatic hybridisation. *Nature Communications* 10(1), 5068.

Liu, M. and Kolmer, J. 2013. Population divergence in the wheat leaf rust fungus *Puccinia triticina* is correlated with wheat evolution. *Phytopathology* 103, 84–84.

Loughman, R., Jayasena, K. and Majewski, J. 2005. Yield loss and fungicide control of stem rust of wheat. *Australian Journal of Agricultural Research* 56(1), 91–96.

Lyon, B. and Broders, K. 2017. Impact of climate change and race evolution on the epidemiology and ecology of stripe rust in central and eastern USA and Canada. *Canadian Journal of Plant Pathology* 39(4), 385–392.

Martin, G. B., Bogdanove, A. J. and Sessa, G. 2003. Understanding the functions of plant disease resistance proteins. *Annual Review of Plant Biology* 54, 23–61.

McCallum, B. D., Seto-Goh, P. and Xue, A. 2016. Physiologic specialization of *Puccinia triticina*, the causal agent of wheat leaf rust, in Canada in 2010. *Canadian Journal of Plant Pathology* 38(4), 440–447.

Mehmood, S., Sajid, M., Zhao, J., Khan, T., Zhan, G. M., Huang, L. L. and Kang, Z. 2019. Identification of *Berberis* species collected from the Himalayan region of Pakistan susceptible to *Puccinia striiformis* f. sp. *tritici*. *Plant Disease* 103(3), 461–467.

Mehta, K. C. 1952. Further studies on cereal rusts in India. *Scientific Monograph Indian Council Agricultural Research* 2, 368.

Meyer, M., Cox, J. A., Hitchings, M. D. T., Burgin, L., Hort, M. C., Hodson, D. P. and Gilligan, C. A. 2017. Quantifying airborne dispersal routes of pathogens over continents to safeguard global wheat supply. *Nature Plants* 3(10), 780–786.

Milus, E. A., Kristensen, K. and Hovmoller, M. S. 2009. Evidence for increased aggressiveness in a recent widespread strain of *Puccinia striiformis* f. sp. *tritici* causing stripe rust of wheat. *Phytopathology* 99(1), 89–94.

Milus, E. A., Seyran, E. and Mcnew, R. 2006. Aggressiveness of *Puccinia striiformis* f. sp. *tritici* isolates in the south-central United States. *Plant Disease* 90(7), 847–852.

Morton, V. and Staub, T. 2008. A short history of fungicides. APSnet Features [Online]. Available at: https://www.apsnet.org/edcenter/apsnetfeatures/Pages/Fungicides.aspx.

Nagarajan, S. and Singh, D. V. 1990. Long-distance dispersion of rust pathogens. *Annual Review of Phytopathology* 28, 139–153.

Newcomb, M., Olivera, P. D., Rouse, M. N., Szabo, L. J., Johnson, J., Gale, S., Luster, D. G., Wanyera, R., Macharia, G., Bhavani, S., Hodson, D., Patpour, M., Hovmøller, M. S., Fetch, T. G. and Jin, Y. 2016. Kenyan isolates of *Puccinia graminis* f. sp. *tritici* from 2008 to 2014: virulence to *SrTmp* in the Ug99 race group and implications for breeding programs. *Phytopathology* 106(7), 729–736.

Newlands, N. K. 2018. Model-based forecasting of agricultural crop disease risk at the regional scale, integrating airborne inoculum, environmental, and Satellite-based monitoring data. *Frontiers in Environmental Science* 6.

Oliver, R. P. 2014. A reassessment of the risk of rust fungi developing resistance to fungicides. *Pest Management Science* 70(11), 1641–1645.

Olivera, P., Newcomb, M., Szabo, L. J., Rouse, M., Johnson, J., Gale, S., Luster, D. G., Hodson, D., Cox, J. A., Burgin, L., Hort, M., Gilligan, C. A., Patpour, M., Justesen, A. F., Hovmøller, M. S., Woldeab, G., Hailu, E., Hundie, B., Tadesse, K., Pumphrey, M., Singh, R. P. and Jin, Y. 2015. Phenotypic and genotypic characterization of race TKTTF of *Puccinia graminis* f. sp *tritici* that caused a wheat stem rust epidemic in southern Ethiopia in 2013–14. *Phytopathology* 105(7), 917–928.

Olivera, P. D., Sikharulidze, Z., Dumbadze, R., Szabo, L. J., Newcomb, M., Natsarishvili, K., Rouse, M. N., Luster, D. G. and Jin, Y. 2019. Presence of a sexual population of *Puccinia graminis* f. sp. *tritici* in Georgia provides a hotspot for genotypic and phenotypic diversity. *Phytopathology* 109(12), 2152–2160.

Perdomo-Sanchez, O. and Piepenbring, M. 2008. A new species of *Puccinia* (Pucciniales, Basidiomycota) and new records of rust fungi from Panama. *Mycological Progress* 7(3), 161–168.

Peterson, P. D. 2001. *Stem Rust of Wheat: From Ancient Enemy to Modern Foe*. St Paul, MN: Amer Phytopathological Society Press.

Peterson, R. H. 1974. The rust fungus life cycle. *The Botanical Review* 40(4), 453-513.

Prank, M., Kenaley, S. C., Bergstrom, G. C., Acevedo, M. and Mahowald, N. M. 2019. Climate change impacts the spread potential of wheat stem rust, a significant crop disease. *Environmental Research Letters* 14(12), 124053.

Pretorius, Z. A., Singh, R. P., Wagoire, W. W. and Payne, T. S. 2000. Detection of virulence to wheat stem rust resistance gene *Sr31* in *Puccinia graminis* f. sp. *tritici* in Uganda. *Plant Disease* 84(2), 203.

Pretorius, Z. A., Szabo, L. J., Boshoff, W. H. P., Herselman, L. and Visser, B. 2012. First report of a new TTKSF race of wheat stem rust (*Puccinia graminis* f. sp. *tritici*) in South Africa and Zimbabwe. *Plant Disease* 96(4), 590.

Radhakrishnan, G. V., Cook, N. M., Bueno-Sancho, V., Lewis, C. M., Persoons, A., Mitiku, A. D., Heaton, M., Davey, P. E., Abeyo, B., Alemayehu, Y., Badebo, A., Barnett, M., Bryant, R., Chatelain, J., Chen, X., Dong, S., Henriksson, T., Holdgate, S., Justesen, A. F., Kalous, J., Kang, Z., Laczny, S., Legoff, J. P., Lesch, D., Richards, T., Randhawa, H. S., Thach, T., Wang, M., Hovmøller, M. S., Hodson, D. P. and Saunders, D. G. O. 2019. MARPLE, a point-of-care, strain-level disease diagnostics and surveillance tool for complex fungal pathogens. *BMC Biology* 17(1), 65.

Raffaele, S. and Kamoun, S. 2012. Genome evolution in filamentous plant pathogens: why bigger can be better. *Nature Reviews Microbiology* 10(6), 417-430.

Ren, Z. L., Liu, H., Jiao, D., Hu, H. T., Wang, W., Gong, J. X., Wang, A. L., Cao, H. Q. and Lv, X. H. 2018. Design, synthesis, and antifungal activity of novel cinnamon-pyrazole carboxamide derivatives. *Drug Development Research* 79(6), 307-312.

Robin, M. H., Bancal, M. O., Cellier, V., Delos, M., Felix, I., Launay, M., Magnard, A., Olivier, A., Robert, C., Rolland, B., Sache, I. and Aubertot, J. N. 2018. IPSIM-Web, an online resource for promoting qualitative aggregative hierarchical network models to predict plant disease risk: application to brown rust on wheat. *Plant Disease* 102(3), 488-499.

Rodriguez-Algaba, J., Sorensen, C. K., Labouriau, R., Justesen, A. F. and Hovmoller, M. S. 2017. Genetic diversity within and among aecia of the wheat rust fungus *Puccinia striiformis* on the alternate host *Berberis vulgaris*. *Fungal Biology* 121(6-7), 541-549.

Rodriguez-Algaba, J., Walter, S., Sorensen, C. K., Hovmoller, M. S. and Justesen, A. F. 2014. Sexual structures and recombination of the wheat rust fungus *Puccinia striiformis* on *Berberis vulgaris*. *Fungal Genetics and Biology: FG and B* 70, 77-85.

Roelfs, A. P. 1985. Wheat and rye stem rust. In: Roelfs, A. P. A. B. and Bushnell W. R. (Eds), *The Cereal Rusts*. Orlando, FL: Academic Press, 3-37.

Roelfs, A. P., Singh, R. P. and Saari, E. E. 1992. *Rust Diseases of Wheat: Concepts and Methods of Disease Management*. Mexico, DF: CIMMYT.

Salcedo, A., Rutter, W., Wang, S. C., Akhunova, A., Bolus, S., Chao, S. M., Anderson, N., De Soto, M. F., Rouse, M., Szabo, L., Bowden, R. L., Dubcovsky, J. and Akhunov, E. 2017. Variation in the *AvrSr35* gene determines *Sr35* resistance against wheat stem rust race Ug99. *Science* 358(6370), 1604-1606.

Sanchez-Martin, J., Steuernagel, B., Ghosh, S., Herren, G., Hurni, S., Adamski, N., Vrána, J., Kubaláková, M., Krattinger, S. G., Wicker, T., Doležel, J., Keller, B. and Wulff, B. B. 2016. Rapid gene isolation in barley and wheat by mutant chromosome sequencing. *Genome Biology* 17(1), 221.

Saunders, D. G. O., Pretorius, Z. A. and Hovmoller, M. S. 2019. Tackling the re-emergence of wheat stem rust in Western Europe. *Communications Biology* 2, 51.

Saunders, D. G. O., Win, J., Cano, L. M., Szabo, L. J., Kamoun, S. and Raffaele, S. 2012. Using hierarchical clustering of secreted protein families to classify and rank candidate effectors of rust fungi. *PLoS ONE* 7(1), e29847.

Schumann, G. L. and Leonard, K. J. 2000. Stem rust of wheat (black rust). *The Plant Health Instructor*. doi: 10.1094/PHI-I-2000-0721-01.

Schwessinger, B. 2017. Fundamental wheat stripe rust research in the 21(st) century. *New Phytologist* 213(4), 1625–1631.

Schwessinger, B., Chen, Y. J., Tien, R., Vogt, J. K., Sperschneider, J., Nagar, R., McMullan, M., Sicheritz-Ponten, T., Sørensen, C. K., Hovmøller, M. S., Rathjen, J. P. and Justesen, A. F. 2020. Distinct life histories impact dikaryotic genome evolution in the rust fungus *Puccinia striiformis* causing stripe rust in wheat. *Genome Biology and Evolution* 12(5), 597–617.

Sharma-Poudyal, D., Bai, Q., Wan, A., Wang, M., See, D. and Chen, X. 2020. Molecular characterization of international collections of the wheat stripe rust pathogen *Puccinia striiformis* f. sp. *tritici* reveals high diversity and intercontinental migration. *Phytopathology* 110(4), 933–942.

Singh, A., Pandey, M. P., Singh, A. K., Knox, R. E., Ammar, K., Clarke, J. M., Clarke, F. R., Singh, R. P., Pozniak, C. J., Depauw, R. M., McCallum, B. D., Cuthbert, R. D., Randhawa, H. S. and Fetch, T. G. 2013. Identification and mapping of leaf, stem and stripe rust resistance quantitative trait loci and their interactions in durum wheat. *Molecular Breeding: New Strategies in Plant Improvement* 31(2), 405–418.

Singh, R. P., Hodson, D. P., Huerta-Espino, J., Jin, Y., Bhavani, S., Njau, P., Herrera-Foessel, S., Singh, P. K., Singh, S. and Govindan, V. 2011. The emergence of Ug99 races of the stem rust fungus is a threat to world wheat production. *Annual Review of Phytopathology* 49(49), 465–481.

Singh, R. P., Huerta-Espino, J. and Roelfs, A. P. 2002. The wheat rusts. In: Curtis, B. C., Rajaram, S. and Gomez Macpherson, H. (Eds), *Bread Wheat*. Rome: Food & Agriculture Organization of the United Nations.

Stakman, E. C. 1923. Barberry eradication prevents black rust in Western Europe. *United States Department of Agriculture, Department Circular* 269, 1–15.

Steuernagel, B., Periyannan, S. K., Hernandez-Pinzon, I., Witek, K., Rouse, M. N., Yu, G. T., Hatta, A., Ayliffe, M., Bariana, H., Jones, J. D., Lagudah, E. S. and Wulff, B. B. 2016. Rapid cloning of disease-resistance genes in plants using mutagenesis and sequence capture. *Nature Biotechnology* 34(6), 652–655.

Stewart, R. B. and Robertson, W. 1968. Fungal spores from prehistoric potsherds. *Mycologia* 60(3), 701–704.

Su, J. Y., Liu, C. J., Coombes, M., Hu, X. P., Wang, C. H., Xu, X. M., Li, Q., Guo, L. and Chen, W. 2018. Wheat yellow rust monitoring by learning from multispectral UAV aerial imagery. *Computers and Electronics in Agriculture* 155, 157–166.

Terefe, T., Pretorius, Z. A., Visser, B. and Boshoff, W. H. P. 2019. First report of *Puccinia graminis* f. sp. *tritici* race PTKSK, a variant of wheat stem rust race Ug99, in South Africa. *Plant Disease* 103(6), 1421–1421.

Thach, T., Ali, S., De Vallavieille-Pope, C., Justesen, A. F. and Hovmoller, M. S. 2016. Worldwide population structure of the wheat rust fungus *Puccinia striiformis* in the past. *Fungal Genetics and Biology: FG and B* 87, 1–8.

Thind, A. K., Wicker, T., Simkova, H., Fossati, D., Moullet, O., Brabant, C., Vrána, J., Doležel, J. and Krattinger, S. G. 2017. Rapid cloning of genes in hexaploid wheat using cultivar-specific long-range chromosome assembly. *Nature Biotechnology* 35(8), 793–796.

Tian, Y., Meng, Y., Zhao, X. C., Chen, X. M., Ma, H. B., Xu, S. D., Huang, L., Kang, Z. and Zhan, G. 2019. Trade-off between triadimefon sensitivity and pathogenicity in a selfed sexual population of *Puccinia striiformis* f. sp. *tritici*. *Frontiers in Microbiology* 10, 2729.

Visser, B., Meyer, M., Park, R. F., Gilligan, C. A., Burgin, L. E., Hort, M. C., Hodson, D. P. and Pretorius, Z. A. 2019. Microsatellite analysis and urediniospore dispersal simulations support the movement of *Puccinia graminis* f. sp. *tritici* from Southern Africa to Australia. *Phytopathology* 109(1), 133–144.

Wang, M. N., Wan, A. M. and Chen, X. M. 2015. Barberry as alternate host is important for *Puccinia graminis* f. sp *tritici* but not for *Puccinia striiformis* f. sp *tritici* in the US Pacific Northwest. *Plant Disease* 99(11), 1507–1516.

Wang, S., Boevink, P. C., Welsh, L., Zhang, R., Whisson, S. C. and Birch, P. R. J. 2017. Delivery of cytoplasmic and apoplastic effectors from *Phytophthora infestans* haustoria by distinct secretion pathways. *New Phytologist* 216(1), 205–215.

Wellings, C. R. 2007. *Puccinia striiformis* in Australia: a review of the incursion, evolution, and adaptation of stripe rust in the period 1979–2006. *Australian Journal of Agricultural Research* 58(6), 567–575.

Wilmanski, J. M., Petnicki-Ocwieja, T. and Kobayashi, K. S. 2008. NLR proteins: integral members of innate immunity and mediators of inflammatory diseases. *Journal of Leukocyte Biology* 83(1), 13–30.

Wu, J., Kou, Y., Bao, J., Li, Y., Tang, M., Zhu, X., Ponaya, A., Xiao, G., Li, J., Li, C., Song, M. Y., Cumagun, C. J., Deng, Q., Lu, G., Jeon, J. S., Naqvi, N. I. and Zhou, B. 2015. Comparative genomics identifies the *Magnaporthe oryzae* avirulence effector AvrPi9 that triggers Pi9-mediated blast resistance in rice. *New Phytologist* 206(4), 1463–1475.

Wu, J. Q., Dong, C., Song, L. and Park, R. F. 2020. Long-read-based de novo genome assembly and comparative genomics of the wheat leaf rust pathogen *Puccinia triticina* identifies candidates for three avirulence genes. *Frontiers in Genetics* 11, 521.

Zadoks, J. C. 1961. Yellow rust on wheat studies in epidemiology and physiologic specialization. *Tijdschrift over Plantenziekten* 67(3), 69–256.

Zhang, X., Han, L. X., Dong, Y. Y., Shi, Y., Huang, W. J., Han, L. H., González-Moreno, P., Ma, H., Ye, H. and Sobeih, T. 2019. A deep learning-based approach for automated yellow rust disease detection from high-resolution hyperspectral UAV images. *Remote Sensing* 11(13), 1554.

Zhao, J., Wang, M., Chen, X. and Kang, Z. 2016. Role of alternate hosts in epidemiology and pathogen variation of cereal rusts. *Annual Review of Phytopathology* 54, 207–228.

Zhuang, H., Zhao, J., Huang, L. L., Kang, Z. S. and Zhao, J. 2019. Identification of three *Berberis* species as potential alternate hosts for *Puccinia striiformis* f. sp. *tritici* in wheat-growing regions of Xinjiang, China. *Journal of Integrative Agriculture* 18(12), 2786–2792.

Chapter 3

Advances in identifying stripe rust resistance genes in cereals

Tianheng Ren, Zhi Li, Feiquan Tan, Cheng Jiang and Peigao Luo, Sichuan Agricultural University, China

1 Introduction

Wheat is one of the most important crops in the world. Approximately one-third of the world's population consumes wheat. Stripe rust, which is caused by *Puccinia striiformis* f. sp. *tritici* (Pst), is considered a devastating disease in cooler areas at higher latitudes or altitudes and is very detrimental to wheat production in the subtropical agricultural zone (Chen 2005, Ren et al. 2009, Chen et al 2014).

Although rust diseases have been documented for a long time (approximately 5000 years), the first detailed record of the occurrence of stripe rust was provided in 1827, when Schmidt discovered the pathogen that caused stripe rust and named it *Uredo Glumarum* (Stubbs 1985). However, stripe rust was not distinguished from two other rust diseases (leaf rust and stem rust) until 1896. Eriksson and Henning (1896) indicated that the cause of stripe rust was different from that of stem rust and leaf rust. They named the pathogen causing stripe rust *Puccinia glumarum*, which was renamed *Puccinia striiformis* by Hylander et al. in 1953.

The most important historical epidemiological areas of stripe rust include eastern Africa, eastern and western Asia and western Europe (Chen 2005). Stripe

http://dx.doi.org/10.19103/AS.2021.0092.04

rust usually results in the loss of approximately 5% of production. However, in some most prevalent areas, stripe rust can raise the loss to 20–30% of production (Chen 2014). In extreme situations, stripe rust can cause 100% loss in some individual regions. For instance, there were 14 serious outbreaks of stripe rust resulting in a 25% yield loss in the United States from 1958 to 2014 (Chen 2005, 2014). China is the largest producer of wheat and is considered the largest epidemic area of stripe rust in the world (Stubbs 1985). Stripe rust is epidemic almost every year in China, and violent epidemics occurred four times, in 1950, 1964, 1990 and 2002, causing yield losses of 6.0 million tons, 3.2 million tons, 1.8 million tons and 1.3 million tons, respectively (Wan et al. 2007).

However, many districts of northern Africa, central and southern Asia, eastern Europe, Australia, North America and South America have become new epidemic areas of stripe rust. For instance, in Central Asia and the Caucasus, four serious outbreaks of stripe rust resulted in 20–40% yield losses between 1999 and 2009. The same situation has occurred in Australia since stripe rust was first detected in 1979, resulting in 80% yield losses in 1984 and 1986 (Murray and Brennan 2009). All the available evidence has indicated that stripe rust is one of the most important wheat diseases and occurs over a wide range, covering almost the entire sown wheat area.

Chemical control is an important method for controlling the spread of stripe rust (Chen 2005, 2014). Effective systemic DMI, QoI and SDHI fungicides are mainly used for this purpose at present. However, the most effective and environmental-friendly way to control the spread of stripe rust is to plant wheat varieties that carry stripe rust resistance genes (Ren et al. 2009). It is worth noting that the pathogens and isolates of stripe rust mutate very quickly, and the occurrence and development of some pathotypes result from the interaction of pathotypes and wheat varieties under certain environmental conditions. Therefore, although many wheat cultivars have remained resistant to stripe rust for 5 years or more, some cultivars were infected with stripe rust before they were widely used. It is necessary to explore and utilize new stripe rust resistance genes to ensure wheat production safety and food security worldwide.

It has been more than 100 years since the British scientist Biffen (1905) proved that resistance to stripe rust is heritable and conforms to Mendel's law of inheritance. There have been great advances in identifying stripe rust genes. At the beginning of this work, scientists could only analyze the genetic laws of the phenotypes of stripe rust through conventional hybridization. However, with the rapid development of molecular biology and molecular genetics, advances in technology have allowed scientists to more efficiently and quickly study resistance gene loci. Many resistance genes have been identified and located on different chromosomes; moreover, several of them have been cloned.

2 Stripe rust resistance genes

2.1 The sources of Yr *genes and their application*

In 1962, the gene symbol *Yr* (yellow rust resistance) was first used to designate wheat stripe rust resistance genes by Lupton and Macer (1962). To date, scientists have identified more than 110 stripe rust resistance genes and more than 300 stripe rust resistance QTLs (Tables 1 and 2). These genes and QTLs are distributed throughout the wheat genome, and the B genome shows the greatest abundance of stripe rust resistance gene clusters. There have been more than 80 officially named *Yr* genes (*Yr1* to *Yr83*) (Table 1). Most of the *Yr* genes are dominant genes, while *Yr2*, *Yr6*, *Yr7*, *Yr19*, *Yr23* and *Yr51* are recessive.

According to their source, these *Yr* genes are mainly derived from common wheat (*Triticum aestivum* L.), while 25 are derived from related species of wheat, including *Triticum spelta* L. (*Yr5*) (Macer 1966), *Triticum durum* L. (*Yr7*, *Yr24*, *Yr53*, *Yr56*, *Yr64*, *Yr65*) (Macer 1966, McIntosh and Lagudah 2000, Xu et al. 2013, Cheng et al. 2014, Table 1), *Triticum vavilovii* L. (*Yr10*) (Macer 1975), *Triticum dicoccoides* L. (*Yr15*, *Yr35*, *Yr36*) (Gerechter-Amitai et al. 1989, Marais et al. 2005a, Uauy et al. 2005), *Triticum turgidum* (*Yr26*) (Ma et al. 2001), *Triticum cereale L.* (*Yr9*, *Yr83*) (Shi et al. 2001, Li, J. et al. 2020), *Thinopyrum* (*Yr50*, *Yr69*) (Liu et al. 2013, Hou et al. 2016) and *Aegilaps* (*Yr8*, *Yr17*, *Yr19*, *Yr28*, *Yr37*, *Yr38*, *Yr40*, *Yr42* and *Yr70*) (Riley et al. 1968, Bariana and McIntosh 1993, Chen and Line 1995, Singh et al. 2000, Marais et al. 2005b, 2006, Kuraparthy et al. 2007, Marais et al. 2009, Bansal et al. 2017). Recently, more resistance genes have been identified from many other wheat-related species, such as *Psathyrostachys huashanica* (Ma et al. 2016), *Leymus mollis* (Zhang et al. 2017a), *Leymus racemosus* (Rahmatov et al. 2017), *Haynaldia villosa* (Chen et al. 2002) and *Roegneria kamoji* (Zhao et al. 2017). It is important to introduce the resistance genes of these related species into wheat to improve wheat resistance ability and expand the diversity of resistance genes.

However, due to the changes in the prevalence of virulent pathotypes, the resistance of many *Yr* genes has completely broken down in the field. For instance, only *Yr5*, *Yr10*, *Yr12*, *Yr13*, *Yr14*, *Yr16*, *Yr18*, *Yr36*, *Yr41*, *Yr44*, *Yr46*, *Yr48*, *Yr50*, *Yr52*, *Yr59*, *Yr62*, *Yr69* and *Yr83* (Sharma-Poudyal et al. 2013, Hou et al. 2016, Jiang et al. 2020, Li, J. et al. 2020) remain resistant to the prevalent virulent pathotypes in China. Moreover, the carrier varieties of most resistance genes are often disliked by breeders due to poor agronomic traits; therefore, only a few resistance genes have been used in wheat breeding programs. For instance, Huang et al. (2017) indicated that the resistance ability of 79 wheat cultivars in the main wheat-producing areas in China to the prevalent pathotypes of stripe rust is low. Although *Yr18* shows high resistance to all prevalent pathotypes in China, only three cultivars harbor this effective resistance gene. Xue et al. (2014) also indicated that none of the 68 wheat cultivars that were planted over

Table 1 The symbols of stripe rust resistant genes from *Yr1* to *Yr83*

Symbol	Location	Origin	Lines/cultivars	Linked marker	Reference
Yr1	2AL	*Triticum aestivum L.*	Chinese166 Corin Dalee	*Xstm673acag*	Lupton and Macer (1962)
Yr2	7B	*Triticum aestivum L.*	Derius Flevina Hana HD2329	*Xwmc364*	Lupton and Macer (1962)
Yr3a	1B	*Triticum aestivum L.*	Bon Fermier, Nudif TP1	-	Lupton and Macer (1962)
Yr3b	1B	*Triticum aestivum L.*	Hybrid46	-	Lupton and Macer (1962)
Yr3c	1B	*Triticum aestivum L.*	Minister	-	Lupton and Macer (1962)
Yr4a	6B	*Triticum aestivum L.*	Vilnorin23, Mardler Argent	-	Lupton and Macer (1962)
Yr4b	6B	*Triticum aestivum L.*	Avalon, Opal, Startin	-	Lupton and Macer (1962)
Yr5	2BL	*Triticum spelta L.*	E5557, E8510	*STS7/8-Xbarc349*	Macer (1966)
Yr6	7BS	*Triticum aestivum L.*	Austerlitz, Fielder, Heines Kolben	*Xwmc76, Xwmc276*	Macer (1966)
Yr7	2BL	*Triticum durum L.*	Brock, Lee, Nudif Tp257	*Xgwm526*	Macer (1966)
Yr8	2M/2B	*Aegilaps comosa*	Hobbit Sib, Compair	*OP-D111265*	Riley et al. (1968)
Yr9	1RS/1BL	*Triticum cereale L.*	Aurora, Benno, Bezostaya	*AF1/AF4*	Macer (1975)
Yr10	1BS	*Triticum vavilovii L.*	Moro, PI178383	*Xpsp3000*	Macer (1975)
Yr11	-	*Triticum aestivum L.*	Joss Cambier	-	McIntosh (1988)
Yr12	-	*Triticum aestivum L.*	Mega	-	McIntosh (1988)
Yr13	-	*Triticum aestivum L.*	Hustler, Mardler, Marksman	-	McIntosh (1988)
Yr14	-	*Triticum aestivum L.*	Kador, Score,Wembley, Hobit	-	McIntosh (1988)
Yr15	1BS	*Triticum dicoccoides L.*	G-25	*Xgwm413, Xbarc8*	Gerechter-Amitai et al. (1989)
Yr16	2DS	*Triticum aestivum L.*	Bersee, Capelle Desprez	*Xpsr641, Xpsr681*	Worland and Law (1986)
Yr17	2AS	*Aegilops ventricosa*	RL6081, VPM1	*SC-Y15*	Bariana and McIntosh (1993)
Yr18	7DS	*Triticum aestivum L.*	Jupateco 73R,Lerma Rojo64	*Xcssfr1-7*	Singh (1992)
Yr19	5B	*Aegilops speltoids*	Compair	-	Chen and Line (1995)

Yr20	6D	*Triticum aestivum L.*	Fielder	–	Chen and Line (1995)
Yr21	1B	*Triticum aestivum L.*	Lemhi	*Ppslem*	Chen and Line (1995)
Yr22	4D	*Triticum aestivum L.*	Lee	–	Chen and Line (1995)
Yr23	6D	*Triticum aestivum L.*	Lee	–	Chen and Line (1995)
Yr24	1BS	*Triticum durum L.*	Meering, K773	*Xwe173*	McIntosh and Lagudah (2000)
Yr25	1D	*Triticum aestivum L.*	TP1295, TP981		Calonnec et al. (2002)
Yr26	6AS	*Triticum turgidum L.*	Yangmai-5, R55	*Xwe173, STSBO33*	Ma et al. (2001)
Yr27	2BS	*Triticum aestivum L.*	Ciano79, Selkirk	*Xcdo405, Xbcd152*	McDonald et al. (2004)
Yr28	4DS	*Aegilops tauschii*	Altar84/T, W7984	*Xwmg634*	Singh et al. (2000)
Yr29	1BL	*Triticum aestivum L.*	Lalbahadur, Pavon 76	*Xwmc44*	William et al. (2003)
Yr30	3BS	*Triticum aestivum L.*	Opata 85	*Xgwm389*	Singh et al. (2000)
Yr31	2BS	*Triticum aestivum L.*	Pastor	*Lr13, Lr23*	Wheat Catalogue
Yr32	2AL	*Triticum aestivum L.*	Anouska; Tres, Carstens V	*Xwmc198*	Eriksen et al. (2004)
Yr33	7DL	*Triticum aestivum L.*	Batavia	*Xwmc437, Xwmc111*	Wheat Catalogue
Yr34	5AL	*Triticum aestivum L.*	WAWHT2046	*B1, Xgwm410*	Bariana et al. (2006)
Yr35	6BS	*Triticum dicoccoides*	98M71	*Xcfd1*	Marais et al. (2005a)
Yr36	6BS	*Triticum dicoccoides*	RSL65	*Xucw113, Xucw127*	Uauy et al. (2005)
Yr37	2DL	*Aegilops Kotschyi*	Line S14	–	Marais et al. (2005b)
Yr38	6A	*Aegilops sharonensis*	Line 0352-4	–	Marais et al. (2006)
Yr39	7BL	*Triticum aestivum L.*	Alpowa	*Xwgp45, Xwgp36*	Lin and Chen (2007)
Yr40	5DS	*Aegilops geniculata*	TA 5602, TA5603		Kuraparthy et al. (2007)
Yr41	2BS	*Triticum aestivum L.*	Chuannong19,Chuannong27	*Xgwm410*, BE446068	Jiang et al. (2020)
Yr42	6AL	*Aegilops neglecta*	03M119-71A	–	Marais et al. (2009)
Yr43	2BL	*Triticum aestivum L.*	IDO377s	*Xgwm501, XBarc139*	Cheng and Chen (2010)

(Continued)

Table 1 (*Continued*)

Symbol	Location	Origin	Lines/cultivars	Linked marker	Reference
Yr44	2BL	*Triticum aestivum L.*	Zak	*Xgwm501*	Cheng and Chen (2010)
Yr45	3DL	*Triticum aestivum L.*	PI181434	*Xbarc6, Xwmc656*	Li, Q. et al. (2011)
Yr46	4DL	*Triticum aestivum L.*	RL6077	*Xgwm165, Xgwm192*	Herrera-Foessel et al. (2011)
Yr47	5BS	*Triticum aestivum L.*	AUS28183, AUS28187	*Xgwm234*	Bansal et al. (2011)
Yr48	5AL	*Triticum aestivum L.*	PI610750	*BE495011*	Lowe et al. (2011)
Yr49	3DS	*Triticum aestivum L.*	chuangmai18	-	McIntosh et al. (2016)
Yr50	4BL	*Thinopyrum* intermedium	CH223	*Xbarc1096, Xwmc47*	Liu et al. (2013)
Yr51	4AL	*Triticum aestivum L.*	AUS 27858	*Owm45F3R3, Sun104*	Randhawa et al. (2014)
Yr52	7BL	*Triticum aestivum L.*	PI183527	*Xwgp5258, Xbarc182*	Ren, R. S. et al. (2012)
Yr53	2BL	*Triticum durum L.*	PI 480148	*Xwmc441, XLRRrev, NLRRrev350*	Xu et al. (2013)
Yr54	2DL	*Triticum aestivum L.*	Quaiu 3	*wpt-667162, wpt-667054*	Basnet et al. (2014)
Yr55	2DL	*Triticum aestivum L.*	Frelon, AUS38882	*Xmag4089, Xmag3385*	Wheat Catalogue
Yr56	2AS	*Triticum durum L.*	Wollaroi	*Xsun167, Xsun168*	Wheat Catalogue
Yr57	3BS	*Triticum aestivum L.*	AUS27858	*BS00062676, Xgwm389*	Randhawa et al. (2015)
Yr58	3BS	*Triticum aestivum L.*	AUS19292	*3023704-sun476*	Chhetri et al. (2016)
Yr59	7BL	*Triticum aestivum L.*	PI178759	*Xbarc32*	Zhou et al. (2014a)
Yr60	4AL	*Triticum aestivum L.*	Almop (GID5934039)	*Xwmc219, Xwmc313*	Herrera-Foessel et al. (2015)
Yr61	7AS	*Triticum aestivum L.*	Pindong34	*Xwgp5467, Xwgp5765*	Zhou et al. (2014b)
Yr62	4BL	*Triticum aestivum L.*	PI192252	*Xgwm192*	Lu et al. (2014)
Yr63	7BS	-	AUS27955	*IWB33120, IWB52844*	McIntosh et al. (2016)
Yr64	1BS	*Triticum durum L.*	PI331260	*Xgwm413, Xgdm33*	Cheng and Chen (2014)

Yr65	1BS	*Triticum durum L.*	PI480016	*Xgwm18, Xgwm11*	Cheng and Chen (2014)
Yr66	3DS	-	MSP4543.1	*IWB47165, IWB18087*	Wheat Catalogue
Yr67	7BL	*Triticum aestivum L.*	AGG91585WHEA	*IWB62475, IWB71995*	Wheat Catalogue
Yr68	4BL	-	AGG91587WHEA1	*IWB74301-IWB28394*	McIntosh et al. (2016)
Yr69	2AS	*Thinopyrum ponticum*	CH7086	*X2AS33, Xmag3807*	Hou et al. (2016)
Yr70	5DS	*Aegilops* umbellulata	IL 393-4	*Xgwm190*	Bansal et al. (2017)
Yr71	3DL	*Triticum aestivum L.*	Sunco	*Xgwm114b*	Bariana et al. (2016)
Yr72	2BL	-	AUS27507	*Xsun481, IWB12294*	McIntosh et al. (2016)
Yr73	3DL	*Triticum aestivum L.*	Avocet R	-	Dracatos et al. (2016)
Yr74	5BL	*Triticum aestivum L.*	Avocet R	-	Dracatos et al. (2016)
Yr75	7AL	-	Nyabing-3	*Xcfa2016, IWB36240*	McIntosh et al. (2016)
Yr76	3AS	*Triticum aestivum L.*	Tyee	*Xwmc1, Xwmc532*	Xiang et al. (2016)
Yr77	6DS	-	PI 322118	*Xbarc54, Xcfd188*	McIntosh et al. (2017)
Yr78	6BS	*Triticum aestivum L.*	PI 519805	*IWA7257, Xwmc737*	Dong et al. (2017)
Yr79	7BL	*Triticum aestivum L.*	PI 182103	*Xbarc72, Xwmc335*	Feng et al. (2018)
Yr80	3BL	*Triticum aestivum L.*	Aus27284	*KASP_65624, KASP_53113*	Nsabiyera et al. (2018)
Yr81	6A	*Triticum aestivum L.*	Aus27430	*KASP_3077*	Gessese et al. (2019)
Yr82	3BL	*Triticum aestivum L.*	Aus27969	*sun KASP_300, KASP_8775*	Pakeerathan et al. (2019)
Yr83	6RL	*Triticum cereale L.*	T-701	-	Li, J. et al. (2020)

Table 2 Several major effective *QYr* for APR identified by PCR-based technology

QTLs	Location	Interval (markers)	PVE (%)	Reference
QPst.jic-1B	1BL	*Xgwm818*	18.0–22.0	Melichar et al. (2008)
QYr.cimmyt-1BL	1BL	*Xgwm259*	33.0–33.9	William et al. (2006)
QYrtm.pau-2A	2AS	*Xwmc407-Xwmc170*	14.0–27.0	Chhuneja et al. (2008)
QTL 2AS	2AS	*Xwmc407-Xgwm071d*	27.0	Christiansen et al. (2006)
QYr.inra-2AL	2AL	*Xgwm282a-Xgwm359*	20.0–40.0	Mallard et al. (2005)
QTL 2AL	2AL	*Xwmc198a-Xwmc170B*	40.0–41.0	Christiansen et al. (2006)
QYr.cim-2BS	2BS	*YrF-wmc474*	24.8	Pretorius et al. (2017)
QYR3	2BS	*Xcdo405-Xbcd152*	30.7	Boukhatem et al. (2002)
QYr.inra-2BS	2BS	*Xgpv3032-Xcfd50a*	22.0–70.0	Mallard et al. (2005)
QYr.sgi-2B.1	2BS	*Xgwm148-s12m60A*	17.0–46.0	Ramburan et al. (2004)
QYrlu.cau-2BS.1	2BS	*Xwmc154-Xgwm148*	36.6	Guo et al. (2008)
QYR1	2BL	*Xgwm47-Xgwm501*	45.8–46.0	Boukhatem et al. (2002)
QTL 2BL	2BL	*Xwmc149-Xwmc317a*	21.0	Christiansen et al. (2006)
QYr.inra-2BL	2BL	*Xbarc101-Xgwm120*	42.0–61.0	Mallard et al. (2005)
QYraq.cau-2BL	2BL	*Xwmc175-Xwmc332*	61.5	Guo et al. (2008)
QYr.inra-2DS	2DS	*Xgwm102-Xgwm539*	24.0–69.0	Mallard et al. (2005)
QYr.cimmyt-3BS	3BS	*Xgwm389*	16.0–28.0	Imtiaz et al. (2004)
QYr.cimmyt-4DS	4DS	*Xbcd265-Xmwg634*	9.0–31.0	Imtiaz et al. (2004)
QYR5	5AL	*Xfbb209-Xabg391*	15	Boukhatem et al. (2002)
QYrtm.pau-5A	5AL	*Xbarc151-Xcfd12*	24	Chhuneja et al. (2008)
QYr.inra-5BL.1	5BL	*Xgwm639a-Xgwm639c*	18.0–26.0	Mallard et al. (2005)
QYr.inra-5BL.2	5BL	*Xgwm234-XDuPW115a*	29.0–34.0	Mallard et al. (2005)

QYrex.wgp-6AS	6AS	*Xgwm344-Xwgp56*	24.5–30.9	Lin and Chen (2009)
QYrst.wgp-6BS.1	6BS	*Xbarc101-Xbarc136*	32.0–45.0	Santra et al. (2008)
QYrst.wgp-6BS.2	6BS	*Xgwm132-Xgdm113*	25.0–43.0	Santra et al. (2008)
QTL 6BL	6BL	*Xwmc397-Xwmc105b*	25.0–29.0	Christiansen et al. (2006)
QYr.msw-7B	7BL	*Xgwm611*	26.0	Imtiaz et al. (2004)
YrCK	7DS	*M59/P41-215-M49/P33-280*	18.0–30.0	Navabi et al. (2005)
QYr.cimmyt-7D	7DS	*Xbcd1438-Xwg834*	12.0–36.0	Singh et al. (2000)

68 000 ha contained *Yr5*, *Yr10*, *Yr15* and *Yr18*. Therefore, it is very important to identify new resistance sources and to successfully apply these resistance genes in wheat breeding.

2.2 Adult plant resistance genes

There are many criteria for the classification of wheat stripe rust resistance. For instance, resistant materials can be divided into those showing all-stage resistance (ASR) and adult plant resistance (APR). Currently, wheat breeders are paying more attention to APR materials (Singh et al. 2004, Wang et al. 2005). This is because APR materials are susceptible to stripe rust in the seedling stage but resistant to stripe rust in the adult plant stage. The resistance ability of APR materials increases as growth progresses (Roebbelen and Sharp 1978). During the filling stage, the leaves of the APR materials can also be infected by stripe rust. However, the pathogen spores of stripe rust would not diffuse heavily on the leaves and would make few losses of production.

Compared with ASR materials, the seedling stage of APR materials usually faces lower selection pressure from virulent pathotypes and is characterized by a prolonged incubation period of the pathogens, resulting in a decrease in spore production of 42–98% (Mares 1979, Singh et al. 2005). Thus, APR materials show broad-spectrum, persistent disease resistance to stripe rust.

At present, the generation of wheat cultivars with APR is considered the main breeding goal in major wheat-producing areas such as the United States, Europe and Australia (Singh et al. 2004). Several resistance genes conferring APR have also been identified. Among the 83 *Yr* genes with official names, 20 genes are APR genes (*Yr11*, *Yr12*, *Yr13*, *Yr14*, *Yr16*, *Yr29*, *Yr30*, *Yr34*, *Yr46*, *Yr48*, *Yr49*, *Yr54*, *Yr56*, *Yr58*, *Yr68*, *Yr71*, *Yr75*, *Yr77*, *Yr78*, *Yr80*).

Many APR traits are also controlled by quantitative trait loci (QTLs). These wheat materials usually carry more than one major-effect QTL and several minor-effect QTLs. It is worth noting that only when the major QTLs and the minor QTLs interact together will the plants show persistent adult resistance. The resistance provided by minor QTLs in this process is usually referred to as background resistance (Bennett 1984). Therefore, the resistance genes isolated from the major QTLs may confer lower resistance if they are present in wheat cultivars alone.

Another kind of APR trait exists that is very similar to that described above. The difference is that this kind of APR is controlled by several minor QTLs and confers high or intermediate resistance to stripe rust during the adult stage. Although spores may exist on the leaves of cultivars with this type of APR, the overall incidence remains very low, with little influence on the yield. Therefore, this kind of APR trait is referred to as 'slow resistance' (Bennett 1984).

With the rapid development of molecular marker technology, great progress has been made in the localization of APR QTLs. More than 100 APR

QTLs have been identified, which are distributed on 21 chromosomes of wheat. The QTL *QYr. inra-2BS*, which is located on chromosome 2BS, can explain from 22% to 70% of the observed APR phenotypic variation (Mallard et al. 2005) and is the most effective APR QTL. However, most APR QTLs explain less than 15% of APR phenotypic variation, and few QTLs are considered major APR QTLs (Table 2). Therefore, it is necessary to continue to identify more effective APR QTLs.

Most APR genes/QTLs have been identified by using SSR markers since the 2000s (Table 2). However, with the rapid development of sequencing technology, many sequencing-based technologies have been rapidly applied in APR gene/QTL identification. For instance, Liu et al. (2018) identified high-temperature APR QTLs in the wheat cultivar 'Madsen' with a 90K wheat SNP array. Yang et al. (2020) identified 17 APR loci by GWAS in 120 common wheat cultivars. Many other new APR loci have been identified by using SNP arrays, DArT, GWAS, BSA-seq and BSR-seq (Yang et al. 2017, Ren, Y. et al. 2017, Pretorius et al. 2017, Wu et al. 2017a,b, Zhang et al. 2019, Li et al. 2017, 2019, Feng et al. 2018, Zeng et al. 2019a,b, Yang et al. 2020). These results suggest that sequencing-based molecular markers could play more important roles in APR QTL/gene identification.

3 Traditional methods for identifying wheat resistance genes

3.1 Mendelian genetic analysis

Mendelian genetic analysis is a method whereby wheat resistance genes can be analyzed based on Mendel's theory of genetics. The steps of this method are usually as follows: the F_1 generation is obtained by the crossing of resistant plants (with *Yr* genes) and susceptible plants (without *Yr* genes). Then, a sufficient number of F_2 plants are reproduced by the selfing of F_1 plants. The relationship between *Yr* genes and resistance can then be verified according to whether the resistant/susceptible separation ratio of the F_2 generation conforms to the Mendelian inheritance law (Fig. 1).

Using this method, scientists usually observed resistance in F_1 plants first. If the F_1 plants exhibit resistance similar to the resistant parent, the *Yr* gene is presumed to be a dominant gene. Otherwise, the *Yr* gene is presumed to be recessive if the phenotype is similar to that of the susceptible parent. Then, scientists analyze the resistance separation ratio of the F_2 generation. If the resistance trait is controlled by a single dominant *Yr* gene, the separation ratio should be 3 resistant: 1 susceptible. Finally, the results should be examined in the F_3 generation or a backcross generation. Scientists generally prefer to adopt the backcrossing method for such assessments. For instance, if the *Yr* gene is presumed to be a dominant gene, the F_1 plants will be backcrossed with the susceptible parent, and the separation ratio in Ft should be 1 resistant:1 susceptible (Fig. 1).

Mendelian genetic analysis is a useful method for the discovery and identification of unknown *Yr* genes. The advantages of this method are that it is accurate and inexpensive. However, this method always takes a long time to produce results.

3.2 Genetic derivation

Genetic derivation is a method for analyzing *Yr* genes based on the interaction relationship between wheat and pathogens.

The main process of genetic derivation involves the inoculation of wheat materials harboring unknown resistance genes with Pst pathotypes with a virulence formula, followed by the determination of the existence of different *Yr* genes according to different resistance patterns. Finally, the genotypes of the tested wheat materials are confirmed according to the integrated resistance patterns of multiple pathotypes (Gao et al. 2019).

Genetic derivation avoids the disadvantages of developing hybrid combinations, which is time-consuming, usually requiring from 4 to 8 weeks. Therefore, this method has been widely used.

The gene derivation method can be used not only to identify recognized *Yr* genes in different wheat materials but also to identify whether unknown *Yr* genes are present in wheat materials. Dubin et al. (1989) summarized six basic principles of *Yr* gene derivation:

- i If the test materials and the controls exhibit the same Pst infection resistance patterns, it is assumed that the test materials and controls have the same *Yr* genes.
- ii If one recognized that *Yr* gene confers resistance to several Pst pathotypes but the test materials are susceptible to the same Pst pathotypes, it is assumed that the test materials do not contain the recognized *Yr* gene.
- iii If the test materials and the controls exhibit different resistance patterns under inoculation with the same Pst pathogen, it is assumed that the test materials and the controls may contain different *Yr* genes.
- iv If the response type of the test materials is weaker than that of the control, the test materials may harbor the same *Yr* gene as the controls. However, the test materials may also harbor another *Yr* gene. This unknown *Yr* gene has an epistatic effect on the recognized *Yr* gene of the controls and is associated with a weaker response pattern.
- v If a recognized *Yr* gene confers characteristic resistance in a suitable environment, it is assumed that this *Yr* gene is present in the test materials.
- vi The putative *Yr* gene of the test materials can be proven in the parent that is known to carry the gene.

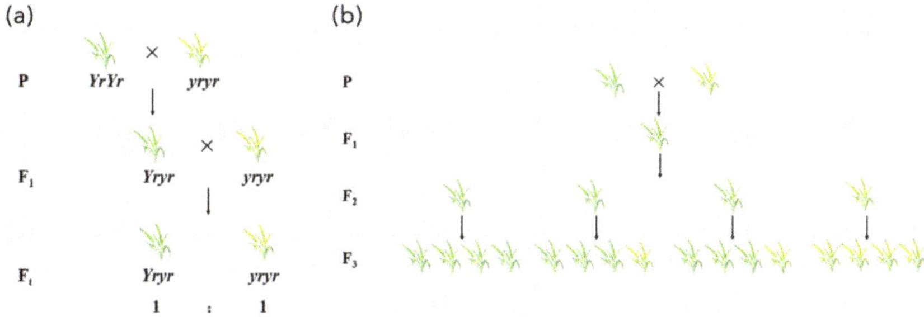

Figure 1 The process of Mendelian genetic analysis: (a) test cross method, (b) selfing method.

These principles indicate steps for standardizing the application of the genetic derivation method in wheat *Yr* gene identification research (Li, Q. et al. 2011, Randhawa et al. 2016). However, the genetic derivation method can only identify ASR genes and cannot identify APR genes directly. Therefore, if we want to identify APR genes via the genetic derivation method, we must first confirm the presence of the ASR genes in the materials, after which the APR data can be added to evaluate the resistance level at the adult stage (Amink et al. 2008). Moreover, *Yr* gene expression is sometimes influenced by the genetic background. Therefore, scientists usually combine genetic derivation methods and molecular marker methods for the identification of *Yr* genes.

4 Molecular marker methods for identifying wheat resistance genes

Due to rapid changes in the prevalence of virulent pathotypes and negative effects on agronomic traits, many *Yr* genes are not used in wheat breeding programs and disappear rapidly. Many recognized *Yr* genes are still difficult to use in wheat breeding programs. Therefore, the identification of additional broad-spectrum resistance genes with positive effects (or without negative effects) on agronomic traits is still a major challenge for wheat breeders.

Fortunately, with the great development of molecular biology technology, methods for identifying new *Yr* genes have rapidly improved. Many different kinds of PCR-based molecular markers have been developed and used in wheat *Yr* gene mapping. Moreover, several new technologies that could previously be used only in human or animal genomes or in plants with small genome sizes can now be used in wheat genomes for gene mapping. Due to the development of such technology, 40 of 83 formally defined *Yr* genes have been identified since 2010, and more *Yr* genes/QTLs with temporary names exhibiting ASR or APR for stripe rust have been identified. New technologies will continue to speed up the efficiency of *Yr* gene identification in the future.

4.1 PCR-based molecular markers

Molecular markers are used in detection methods focused at the molecular level. This approach has the characteristics of high efficiency and stability but also presents a very high requirement for the accuracy and stability of the markers. In recent years, with the improvement of molecular research methods, this strategy has been increasingly applied in research related to disease resistance breeding. The identification of resistance factors by combining the gene derivation and molecular marker methods can greatly shorten the research time and improve dependability and accuracy.

Restriction enzyme fragment length polymorphism (RFLP) markers were first used to map wheat resistance genes. Many RFLP markers, such as *Xcdo*, *Xbcd*, *Xwg* and *Xmwg*, were developed during the early application of gene mapping (Gill et al. 1991). By the end of the 20th century, molecular markers based on PCR technology were being developed, such as amplified fragment length polymorphisms (AFLPs) (Guyomarc'h et al. 2002), expressed sequence tags (ESTs) (Ishikawa et al. 2009) and simple sequence repeats (SSRs) (Röder et al. 1995), and these PCR-based markers were rapidly applied in gene mapping programs. SSR markers have been most widely used thus far due to their abundance and accuracy in gene mapping (Wang et al. 2008). The most commonly used SSR markers include the *Xgwm* (Röder et al. 1995), *Xwmc* (Gupta et al. 2002) and *Xbarc* (Song et al. 2005) markers, and many RFLP, AFLP and EST markers have been converted to SSR markers, such as *Xbcd*, *Xcdo* and *Xcfd*. Table 3 lists the most commonly used PCR-based molecular markers for gene mapping.

In addition, several new types of molecular markers have been developed and used for *Yr* gene identification and mapping.

The sequence-related amplified polymorphism (SRAP) strategy is a new technique involving molecular markers, with the advantages of simplicity, reliability and a moderate throughput ratio (Chen et al. 2012). The SRAP approach aims to amplify an open reading frame (ORF) by using special primers and makes it feasible to identify polymorphisms. Therefore, the SRAP method is very suitable for *Yr* gene mapping (Table 4) (Chen et al. 2012, Zhou et al. 2017).

The target region amplified polymorphism (TRAP) approach is a new molecular marker technique based on ESTs. The primers used for TRAP are designed on the basis of EST sequences, and specific regions of the genome can be amplified by PCR. Therefore, EST sequence information for the studied plants must be available when designing primers for TRAP. Due to the advantages of simple operation, good repeatability, high polymorphism and high efficiency, the TRAP approach has been successfully used in wheat *Yr* gene identification (Table 4) (Chen et al. 2012, Feng et al. 2014, Zhou et al. 2017).

According to the conserved domains of plant resistance genes, resistance gene analog polymorphism (RGAP) markers can be used to amplify structurally

Table 3 Common used PCR-based molecular markers for wheat gene mapping

Marker	Type	Source	Reference
Xcfd	AFLP	*Aegilops tauschii*	Guyomarc'h et al. (2002)
Xcfa	AFLP	*Triticum urartu*	Sourdille et al. (2003)
Xgwm	SSR	*Triticum aestivum L.*	Röder et al. (1995)
Xpsp	SSR	*Triticum aestivum L.*	Bryan et al. (1997)
Xgdm	SSR	*Aegilops tauschii*	Pestsova et al. (2000)
Xwmc	SSR	*Triticum aestivum L.*	Gupta et al. (2002)
Xbarc	SSR	*Triticum aestivum L*	Song et al. (2005)
Xgpw	SSR	*Aegilops speltoides*	Sourdille et al. (2004)
UGMS	SSR	cereal species	Parida et al. (2006)
Xksum	EST-SSR	*Triticum aestivum L*	Yu et al. (2004)
Xstm	EST-SSR	*Triticum aestivum L*	Hayden et al. (2006)
Xmag	EST-SSR	*Triticum aestivum L*	Xue et al. (2008)
Dup W	EST-SSR	*Triticum durum L.*	Eujayl et al. (2002)
Xswes	EST-SSR	*Triticum aestivum L*	Chen (2005)
Xcfe	EST-SSR	*Triticum aestivum L*	Zhang et al. (2005)
CWEM	EST-SSR	*Triticum aestivum L*	Peng and Lapitan (2005)
TNAC	EST	*Triticum aestivum L*	Ishikawa et al. (2009)

similar RGA sequences of plant DNA or cDNA sequences (Shi et al. 2001). Therefore, RGAP is another commonly used type of molecular marker for *Yr* gene identification (Shi et al. 2001, Sui et al. 2009, Li, Q. et al. 2011, Ren, R. S. et al. 2012, Xu et al. 2013, Zhou et al. 2017).

These three new techniques further strengthen the application of PCR-based molecular markers in *Yr* gene mapping. Many new *Yr* genes/QTLs have been mapped with these new techniques and are listed in Table 4.

4.2 SNP markers for stripe rust in wheat

Since 2012, the rapid development of wheat genomics has led to the development of single-nucleotide polymorphism (SNP) markers and diversity array technology (DArT). These kinds of markers cover the whole genome and present the advantages of high density, precision and throughput, which greatly accelerate the precise mapping and cloning of genes (Gupta et al. 2008).

Compared with PCR-based molecular markers, SNPs present technical advantages such as high stability, an abundance of loci, low costs, codominance, and easy-to-implement high-throughput analysis. Therefore, SNPs are widely used in the gene mapping of *Arabidopsis*, maize, rice and other plants.

Table 4 Stripe rust resistance genes identified by different methods

Gene/QTLs	Lines/cultivars	Methods	Reference
Yr6/Yr7	Heines Kolben/Lee	Chromosome analysis	Macer (1966)
Yr17	VPM1	Chromosome analysis	Bariana and McIntosh (1993)
Yr10	*Ae. tauschii*	RFLP	Spielmeyer et al. (2000)
Yr26	*Triticum turgidum* L	SSR	Ma et al. (2001)
Yr9	Weique	RGAP	Shi et al. (2001)
Yr27	Ciano 79	Chromosome analysis	McDonald et al. (2004)
Yr41	Chuan-nong19	SSR	Luo et al. (2008)
Yrcn17	Chuan-nong17	Chromosome analysis	Ren et al. (2009)
YrC591	C591	SSR/ AFLP	Li et al. (2009)
Yr45	PI 181434	RGAP/SSR	Li, Q. et al. (2011)
YrSph	D31	SSR/SRAP/TRAP	Chen et al. (2012)
Yr52	PI 183527	RGAP/SSR/BSA	Ren, R. S. et al. (2012)
Yr53	PI 480148	RGAP/SSR	Xu et al. (2013)
yrLM168	LM168-1	TRAP/RGAP/SSR	Feng et al. (2014)
Yr54	Quaiu 3	DArT/SSR	Basnet et al. (2014)
Yr 57	AUS27858	DArT/BSA	Randhawa et al. (2015)
Yr69		SSR	Hou et al. (2016)
Yr71	Sunco	90K Array/SSR	Bariana et al. (2016)
Yr76	Tyee	SSR	Xiang et al. (2016)
QYr.iiwbr-5B, *QYr.iiwbr-2D*	Cappelle-Desprez	DArT/SSR	Pawar et al. (2016)
Yr6	Cadenza	Exome capture sequencing	Gardiner et al. (2016)
YrHY	Hejiangyizai	SSR/SRAP/TRAP/RGAP	Zhou et al. (2017)

YrZM103	ZhengMai103	BSR	Zhang et al. (2017b)
Yr78	PI 519805	GWAS/ KASP array	Dong et al. (2017)
YrH62.A	03031-1-5 H62	90K array/BSA/SSR	Wu et al. (2018a)
Qyrmap.nwafu-2BS	Napo 63	90K array/BSA/SSR	Wu et al. (2017a)
Yr79	PI 182103	SSR/SNP	Feng et al. (2018)
QYr.nwafu-6BL	Friedrichswerther	90K array/BSA/SSR	Wu et al. (2018b)
QYrcem.nwafu-7BL	Centrum	35K array/660K array/BSA	Mu et al. (2018)
YrBai	Baidatou	DArT	Li et al. (2018)
Yr17	Mengmai58/Huaiyang1	BSR	Wang et al. (2018)
QYr.nwafu-3BS.2 QYr.nwafu-7BL	P9936	55K array	Huang et al. (2019)
QYr.hebau-5AL, QYr.hebau-4BS, QYr.hebau-6DS	SW8588/Thatcher.	55K array/SSR	Zhang et al. (2019)
QYr.sicau-1B.1	20828	55K Array/SSR	Ma et al. (2019)
QYr.nwafu-6BL.2	P10078 /Snb'S'	660K array/Pool analysis	Zeng et al. (2019a)
Yr81	Aus27430	90K array/SSR	Gessese et al. (2019)
Yr82	Aus27969	90K array/BSA	Pakeerathan et al. (2019)

Although the study of SNP markers in wheat began late due to the limitation of genome information, with the completion of a wheat draft genome and the development of SNP algorithms for polyploid genomes, wheat SNP research has achieved fruitful results in recent years (Jighly et al. 2015, Huang et al. 2019, Zhang et al. 2019).

The first 9K wheat SNP array was developed by the International Wheat SNP Working Group and Illumina, in which a transcript assembly and expressed sequence tag (ESTs) were applied in an Illumina Plant Microarray. Subsequently, a 90K array was developed based on the 9K array (Wang et al. 2014).

With the development of second-generation sequencing technology, it is possible to develop higher-density SNP arrays. For instance, The Cereal Genome Research Centre at the University of Bristol in the UK worked with Affymetrix to develop a high-density 820K SNP array (Axiom® HD Wheat Genotyping Array), which contained 819 571 SNP markers (Winfield et al. 2016). Subsequently, a commercial 35K array was developed by Affymetrix (Allen et al. 2016).

The Wheat660K array was designed by the Chinese Academy of Agricultural Sciences and synthesized by Affymetrix. This SNP array contains 630 518 SNP markers, and most of these markers have been compared to the common wheat genome published by IWGSC (Cui et al. 2017). Subsequently, a more practical Wheat55K array with 53 063 markers was developed based on the Wheat660K array (Ren et al. 2018b).

Compared to the 90K or 820K array, the 660K array is based on a design involving 106 437 genes from the wheat genome. For each wheat gene, there are 5–6 SNP markers in this array, and it is especially enriched in SNPs from the wheat D genome, which makes up for the lack of SNPs from the D genome in other SNP arrays. Sun et al. (2020) systematically evaluated seven arrays that are commonly used in wheat molecular breeding programs: the 9K, 15K, 90K, 55K, 660K, 35K and 820K arrays. In terms of the number of SNP markers and associated genes and the distribution and density of SNPs, the Wheat660K array contains the highest number of genome-specific SNPs with precise physical locations (99.05%), and the Wheat660K array shows the highest uniformity among these commonly used arrays. Moreover, 229 266 SNPs of the Wheat660K array are located in the promoter regions of 66 934 (63.52%) annotated genes, including almost all the annotated genes detected by other arrays. Therefore, the Wheat660K array shows high reliability and effectiveness and may be the best choice for use in wheat molecular marker-assisted selection (MAS) breeding in recent years.

These SNP arrays have been successfully used in *Yr* gene identification. For instance, Wu et al. (2018a) identified an APR gene, *YrH62*, located on chromosome 1B using the 90K array. Zeng et al. (2019a) identified a major QTL, *QYr.nwafu-6BL.2*, which is located on 6BL, by using the Wheat660K array. More new *Yr* genes have been identified by using the Wheat55K array since 2019 (Ma et al. 2019, Li et al. 2019, Huang et al. 2019). This is because all 53 063 markers of the Wheat55K Array were carefully selected from the Wheat660K array.

4.3 DArT molecular markers for stripe rust in wheat

DArT technology was developed by Jaccoud et al. in 2001 and is a molecular marker approach based on array hybridization. In 2006, Akbari et al. first introduced DArT technology to wheat; these researchers developed *wPt* series markers and used them to construct a wheat genetic map. With the rapid development of next-generation sequencing (NGS), Li et al. (2015) developed the DArT genotyping by sequencing (DArT-GBS) approach, which is characterized by higher accuracy and density.

Compared with SNP markers, DArT markers are less affected by the ploidy and genome size of species; therefore, DArT markers are very suitable for use in hexaploid wheat. Moreover, DArT markers present universality in Triticeae and may be suitable for the identification and mapping of some alien disease resistance genes.

DArT markers have been successfully used in wheat *Yr* gene mapping. For instance, Xing et al. (2012) reported the detection of two QTLs for APR to stripe rust on chromosomes 2AL and 2BL by using DArT markers. Jighly et al. (2015) reported 12 DArT and 29 SNP markers on chromosome arms 3AS, 3AL, 1AL, 2AL, 2BS, 2BL, 3BS, 3BL, 5BL, 6AL and 7DS significantly linked to *Yr* genes. Li et al. (2018) reported that *YrBai* was located in the 0.8 cM region on wheat chromosome 6D on the basis of DArT marker analysis. However, it is worth noting that DArT markers are usually combined with other molecular markers, such as SSRs or SNPs.

4.4 The genome-wide association study for stripe rust in wheat

The genome-wide association study (GWAS) approach was developed by Risch and Merikangas (1996) and was initially applied for the genetic study of complex traits in animals. This method was further applied for genetic analysis in humans and plants (Gupta et al. 2005). With the advance of sequencing techniques, GWAS has become an important way to study and analyze complex traits due to its advantages of high accuracy and reliability, time savings, efficiency and scientific feasibility (Gupta et al. 2005).

In recent years, the application of GWAS in wheat *Yr* gene mapping has been increasing, and great achievements have been made. For instance, Naruoka et al. (2015) used the 9K SNP array for GWAS and identified several new stripe rust resistance QTLs in winter wheat in the Pacific Northwest of the United States. Liu et al. (2016) applied the GWAS method and identified 11 APR loci with the 90K array in 232 durum wheat lines. Wu et al. (2020)

identified a candidate gene, TraesCS2B01G513100, for stripe rust resistance from 411 spring wheat materials by GWAS. These reports indicate that the GWAS approach has contributed greatly to the identification of *Yr* genes in wheat and may play a more important role in future research aimed at *Yr* gene identification.

5 Application of sequence-based technology in *Yr* gene mapping

5.1 Pooled DNA sequencing

Pooled DNA sequencing, which is based on bulked segregant analysis (BSA) and second-generation sequencing, is mainly used for gene mapping and cloning (Giovannoni et al. 1991, Michelmore et al. 1991).

The principle of BSA is based on the targeted character differences between two parents. In separate generations, a certain number of offspring with extreme phenotype differences are chosen, and DNA pools with different target characters are constructed. Then, the molecular markers associated with the characters are obtained by using PCR-based molecular markers or sequencing methods. Finally, the degree of linkage between the different markers or sequences identified is confirmed by genetic population verification (Fig. 2).

BSA has been widely used as a technology for rapidly mapping genes or identifying QTLs in the field of plant genetics research (Zou et al. 2016). BSA-seq (Pool-seq) was first used to identify mutant target genes in plants and is very effective for gene mapping and cloning in small species such as *Arabidopsis thaliana* (Schneeberger and Weigel 2011) and rice (*Oryza sativa*) (Takagi et al. 2013). Initially, BSA-seq was too difficult to apply in plant species with large complex genomes, especially in wheat, due to the high cost of sequencing. However, with sequence technology development, the price of genome sequencing is decreasing. BSA-seq has been further developed and rapidly used for *Yr* gene mapping and identification. Many new *Yr* genes, such as *Yr57* (Randhawa et al. 2015), *Yr82* (Pakeerathan et al. 2019) and *YrH62. A* (Wu et al. 2018a), have been successfully identified and mapped by BSA-seq.

5.2 Whole-exome capture sequencing of pools

The exome is the sum of all exon regions in the genome. High-throughput sequencing combined with exome capture can be used to directly explore the functional variation caused by the variation of protein-coding genes.

Exon capture sequencing can be divided into three steps: the capture and enrichment of regional exon sequences, DNA sequencing and bioinformatics analysis.

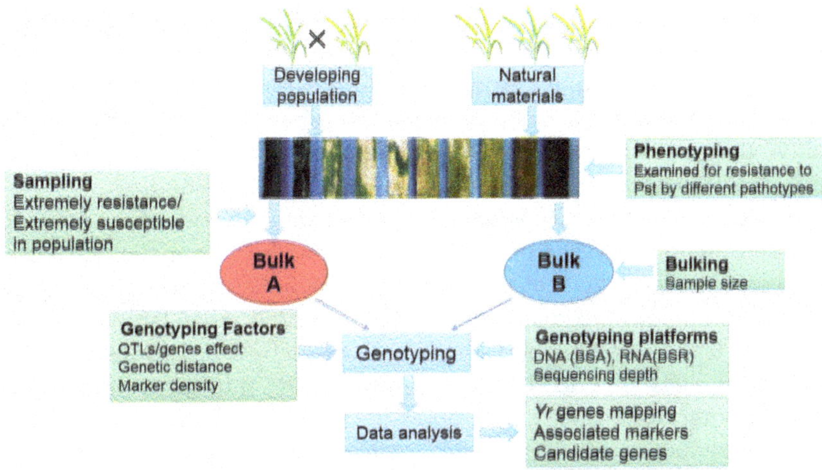

Figure 2 The pipeline of bulked sample analysis for *Yr* genes.

At present, NimbleGen and Agilent are the main platforms for exome capture. The NimbleGen platform is based on molecular hybridization and DNA microarray technology for capturing target regions. On the other hand, Agilent's exon capture system is a liquid-phase targeted sequence capture system based on oligonucleotide synthesis technology. Both platforms present the advantages of high specificity, high accuracy and good reproducibility.

However, due to the limitations of genome information and the high cost of exon capture sequencing, research on large-genome crops such as wheat started late. However, the development of this technology is progressing very quickly. For instance, Gardiner et al. successfully finely mapped *Yr6* by pools by whole-exome capture sequencing (2016).

5.3 Bulked segregant RNA-sequencing

Transcriptomics refers to all mRNAs and noncoding RNAs that are transcribed in a specific cell or tissue in a certain functional state. Transcriptional sequencing technology (RNA-seq) is a new generation of high-throughput sequencing technology for cDNA sequencing.

Compared with BSA-seq, the advantages of bulked segregant RNA-seq (BSR-seq) are obvious and are listed below (Liu et al. 2012): (i) BSR-seq can provide polymorphism information for the whole transcriptome as well as whole-genome expression information; (ii) because the mutant sites of target genes usually occur in the coding region, the genetic analysis of mutation sites can be realized at the transcriptome level, which greatly reduces the sequencing cost and improves the mapping efficiency; (iii) BSR-seq can exclude the interference caused by large genomes and abundant repeated sequences or homologous genes, which are highly abundant in wheat genome; (iv) polymorphism of transcriptome analyses can be performed more accurately with deeper sequencing; (v) various types of tissue samples can be used for BSR-seq, and the analysis is not limited to the expression of the target genes; and (vi) the genes mapped by BSR-seq can be mutated genes, quality trait genes or QTLs with a large PVE (phenotypic variation explained).

BSR-seq has been successfully applied in wheat *Yr* gene mapping. For instance, Ramirez-Gonzalez et al. (2015) obtained an F_2 generation derived from a cross between Avocet S and Avocet S+*Yr15*. Resistant and susceptible DNA pools were produced and then subjected to BSR-seq. A large number of differential SNPs were obtained. These SNPs could be further developed into Kompetitive allele-specific PCR (KASP) markers. The *Yr15* gene was mapped within an interval of 0.77 cM through the mapping of the F_2 generation population. Wang et al. (2018) also indicated that *Yrmm58* and *Yrhy1* on chromosome 2AS were *Yr17* on the basis of BSR-seq. More unknown *Yr* genes can be identified by BSR-seq technology. For instance, Zhang et al. (2017b) identified and mapped a new stripe rust resistance gene, *YrZM103*, on chromosome 7BL by BSR-seq.

For *Yr* gene identification, Mendelian genetic analysis and genetic derivation are usually carried out first to help scientists determine whether the studied materials harbor new disease resistance genes and to determine whether the genes are dominant or recessive. Based on the results of classic genetic analysis, multiple methods can be used to identify unknown *Yr* genes (Fig. 3). Initially, cytogenetics was generally used for *Yr* gene mapping (Macer 1966, Bariana and

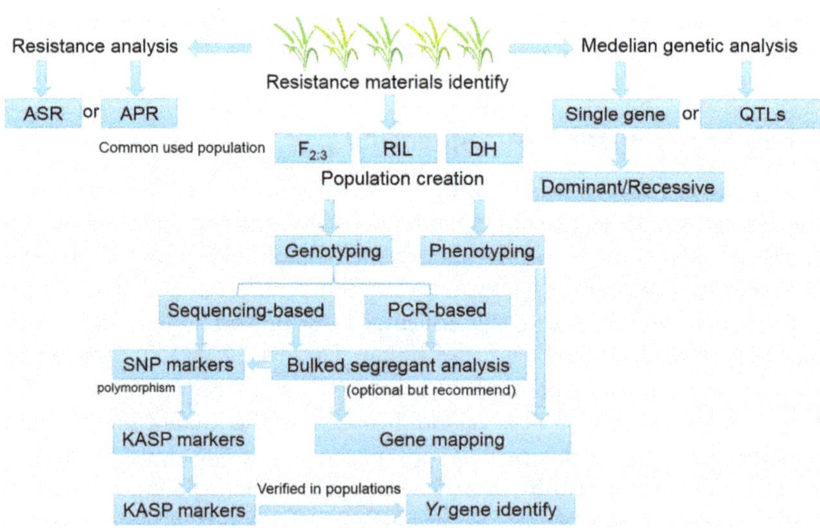

Figure 3 A common process of *Yr* genes identified by molecular markers.

McIntosh 1993, McDonald et al. 2004). With the advent of molecular markers, scientists came to preferentially use such markers for gene mapping. PCR-based molecular markers, especially SSRs and RGAPs, have played key roles in *Yr* gene identification since the late 1990s (Shi et al. 2001, Luo et al. 2005). However, cytogenetics still plays an important role in identifying resistance genes in wheat-related species (Ren et al. 2009, Li, J. et al. 2020). Along with the great improvement of sequence technology, since 2015, sequencing-based molecular marker approaches such as the application of SNP arrays (Gessese et al. 2019), GWAS (Dong et al. 2017), BSA-seq (Pakeerathan et al. 2019) and BSR-seq (Zhang et al. 2017b) have been used for rapid wheat *Yr* gene identification and mapping. The combined use of PCR-based and sequence-based methods accelerates the efficiency and accuracy of *Yr* gene mapping.

6 Case study: application of stripe resistance genes in southwestern China

China is the country with the highest wheat production. Stripe rust is the most important wheat disease in China and seriously affects the production of wheat. In recent years, Pst pathotypes CYR31 (239E175), CYR32 (111E239), CYR33 (111E222) and CYR34 (127E255) have been severely virulent in newly released wheat cultivars in the field (Wan et al. 2004, Ren, T. H. et al. 2012, 2017, 2018a).

Based on ecological conditions, wheat stripe rust in China can be divided into three major zones, namely, the autumn sources of inocula, the spring sources of inocula and spring epidemic areas (Chen et al., 2013). Sichuan Province is the most important source of spring stripe rust inocula in China (Chen et al., 2013). Therefore, Sichuan Province is not only the worst-hit area but also the main source of stripe rust infection in China. The special climatic conditions in Sichuan are conducive to the early occurrence of stripe rust, and the accumulated Pst pathogens spread to other regions. The occurrence of stripe rust in Sichuan has an important impact on wheat production in China. In the 1970s and 1980s, due to the application of Fan6 series cultivars harboring long-lasting resistance gene groups and the introduction of the *Yr9* and *Yr17* resistance genes, there were no violent occurrences of stripe rust in Sichuan Province for more than 20 years.

However, with the emergence of new pathotypes of Pst in the early 1990s, the resistance of these genes was lost successively, which led to the occurrence of stripe rust in a large area of wheat in Sichuan Province in successive years and an epidemic of stripe rust throughout the country, resulting in a massive loss of wheat production.

To solve this problem, breeders in Sichuan Province have tried many different approaches. For instance, breeders introduced additional *Yr* resistance genes into the wheat breeding program, such as *Yr26*. More resistance genes were introduced from several alien varieties, such as rye. Several new resistance genes were identified and continue to be applied in wheat breeding programs, such as *Yr41*. In this section, we will introduce several experiences of our breeding group regarding the identification of new resistance genes and their application in wheat breeding.

6.1 Exploration of new resistance genes from the Secale genus

Rye (*Secale cereal L.*) is the most important and valuable related species for the improvement of wheat genetics (Shi et al. 2001, Ren, T. H. et al. 2012, Li, J. et al. 2020). The *Yr9* gene from rye showing high resistance to stripe rust was transferred into wheat along with the corresponding chromosome translocation (Ren et al. 2009). However, since the 1990s, the resistance conferred by *Yr9* has broken down (Ren et al. 2009, 2012).

Rye is a cross-pollinated plant with high genetic diversity within the population of a variety. Therefore, for the further use of the 1RS.1BL translocation in wheat breeding programs, we suggest the introduction of a large amount of new genetic variation from many different rye sources into wheat (Ren et al. 2012).

In fact, we have already introduced many different rye variations into wheat, and many different homozygous translocation lines have been obtained. The methods that we used to transfer rye chromatin into the wheat genome are described below:

A rye variety was crossed as a pollen donor with common wheat cultivars/lines (we usually use wheat cultivars with a high yield that are susceptible to diseases as parents). The growth points of F1 hybrid plants were soaked in 0.05% colchicine + 3% dimethyl sulfoxide for 8 h and then washed overnight with running water to produce amphidiploids (C_1). Next, the C_1 plants were backcrossed twice using the same wheat parent to produce BC_2F_1 plants. Then, the seedling BC_2F_2 plants were identified by FISH (fluorescence in situ hybridization) and then transplanted in the field. During this generation, many wheat/rye monosomic addition plants, double monosomic addition plants and triple monosomic addition plants could be selected. Only the plants that contained rye chromosomes and exhibited resistance to stripe rust in the field were harvested and reproduced to obtain the next generation. Thereafter, only would the plants that exhibited resistance to stripe rust be harvested and reproduced by selfing until BC_2F_5 or BC_2F_6. Then, the seedling plants were evaluated again by FISH to identify plants showing homologous translocations, substitutions and additions.

All wheat parents used in the development process were susceptible. Therefore, the observed resistance obviously originated from rye.

In our group, variants from many different rye sources have been introduced into wheat, such as L155 rye, Baili rye, Weining rye, Aigan rye,

Kustro rye, Qinling rye, Chile rye, Anatolicum rye and Africanum rye. Nearly 1000 new translocation, substitution or addition lines were developed. Many forms of translocation were obtained, including all balanced translocations from 1R to 7R and many unbalanced translocations, such as 2RS.2BL, 3RS.3BL, 4RS.4DL, 3RS.6AL, 1RL.4AS, 1RS.5BL and 3RL.2AS. Most of these translocations conferred high resistance to stripe rust in the field.

However, most of the translocation lines used in actual breeding programs harbored the 1RS.1BL translocation. This was because 1RS.1BL does not only confer high resistance to diseases but is also associated with very favorable agronomic traits. Several new 1RS.1BL translocation lines developed by our group were also examined and approved as cultivars by the Sichuan Provincial Variety Examination and Committee, such as Chuan-nong12, Chuan-nong17 and Chuan-nong18, which are sown in large areas in southwestern China.

These results indicated that rye may contain abundant broad-spectrum resistance genes for stripe rust. Therefore, it is necessary to further develop rye genetic resources and explore resistance genes in the rye genome. However, only the translocation lines exhibited resistance to stripe rust, and these good agronomic traits could be used for wheat breeding programs directly.

6.2 Revaluation of the potential application of Yr14 in wheat breeding programs

Yr41 is an all-stage resistance gene for wheat stripe rust and was the first wheat stripe rust resistance gene independently discovered by Chinese scholars and officially named. Here, we would like to describe the discovery of *Yr41* and the great potential for the application of this resistance gene.

In the early 1990s, our breeding group extensively collected breeding resources for multiple disease resistance. An 'R' series of breeding materials that were resistant to stripe rust were efficiently and accurately identified by using an experimental design consisting of 'test materials + spreader materials', as shown in Fig. 4.

A new line, R88, was selected from the hybrid combination of Q1104A/R935, whose stripe rust resistance level ranged from high resistance to immunity. A new wheat cultivar, Chuan-nong19, was selected from R88 and released in 2003 in Sichuan Province due to its high resistance and very favorable agronomic traits.

Mendelian genetic analysis indicated that Chuan-nong19 contained a new dominant stripe rust resistance gene. Then, this resistance gene was preliminarily mapped on the chromosome arm of wheat 2BS by using SSR markers and was temporarily named *YrCN19* (Luo et al. 2005).

At that time, *Yr27*, *Yr31*, *YrSp* and *YrV23* were already reported to be located on 2BS; therefore, multiple methods, such as genetic derivation, pedigree

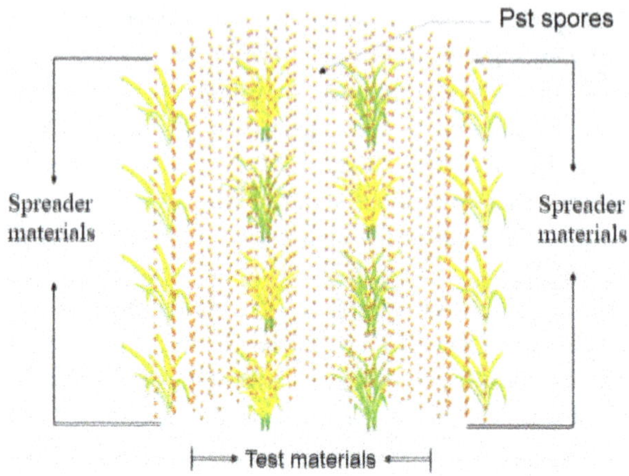

Figure 4 The experimental design of 'test materials + spreader materials' in field.

analysis and allelism analysis, were carried out, and the results indicated that *YrCN19* is a new stripe rust resistance locus on chromosome 2BS. Finally, this locus was named *Yr41* by the international wheat gene named committee.

Since the approval of Chuan-nong19 in 2003, the prevalence of Pst pathogens in China has changed many times. The appearance of CYR34 led to the breakdown of *Yr* genes widely used in China. However, *Yr41* maintained immunity or high resistance to all prevalent Pst pathogens in China. In wheat-producing areas where stripe rust is prevalent in China, *Yr41* has conferred persistent resistance for 20 years.

The cultivated life of any wheat cultivar is limited. If a resistance gene exists only in a single wheat cultivar and when that cultivar is eliminated due to its inability to adapt to new production conditions, even if the resistance gene continues to confer resistance, it will be eliminated along with the cultivar. Hence, some valuable strategies for the use of *Yr41* in wheat breeding have been reported (Luo et al. 2005, Jiang et al. 2020)

Our breeding group used molecular MAS to monitor and track the *Yr41* gene during the breeding process. At the same time, we selected a series of lines with very favorable agronomic traits with rich genetic diversity, which were referred to as 'agronomic parents with genetic diversity'. Based on the great resistance conferred by *Yr41*, a new method referred to as the 'Multiple Genetic Background conversion method' was developed. The key point of this method is to select a large number of plants showing the 'agronomic parents with genetic diversity' as receptors, transfer the resistance genes into different excellent genetic backgrounds and make them fully expressed (Fig. 5).

By using this method, the *Yr41* gene continued to be expressed in different wheat genetic backgrounds for more than 20 years, and several of these lines

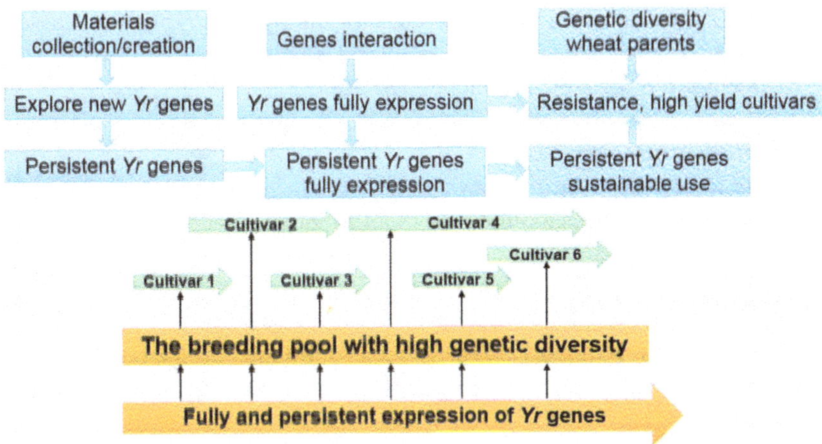

Figure 5 The process of multiple genetic background conversion method.

were released as commercial wheat cultivars, such as Chuan-nong21, Chuan-nong24, Chuan-nong26, Chuan-nong27, Chuan-nong29, Chuan-nong30, Chuan-nong32, Jianongmai809, Zhongkemai169, Chuanmai96, Chuanmai98, Yumai13 and Yumai14. These wheat cultivars have been used widely in southwestern China, and the sown area was calculated to be more than 6 000 000 ha in the past 18 years. This indicates that *Yr41* is effective and can play an important role in China, especially in the wheat resistance improvement breeding program in southwestern China.

7 Where to look for further information

In this section, we provide several useful websites for readers who are interested in wheat stripe rust resistance genes.

The Triticeae Multiomics Center: http://202.194.139.32.

This website provides wheat genome/gene/protein, wild emmer genome/gene/protein, Aegilops tauschii genome/gene/protein and barley genome/gene/protein information. This site also provides BLAST, JBrowse, primer design (including KASP/CAP primer design) and expression database searches.

Wheat-SnpHub-Portal: http://wheat.cau.edu.cn/Wheat_SnpHub_Portal/.

An easy-to-set web server framework for exploring large-scale genomic variation data in wheat (Wang et al. 2020).

GrainGenes: https://wheat.pw.usda.gov/GG3/.

GrainGenes is a digital platform that serves small grain research communities as a central data repository and as a facilitator of community activities, primarily related to wheat, barley, oat and rye.

HarvEST: https://harvest.ucr.edu/.

HarvEST provides information on ESTs for barley, *Brachypodium*, cassava, citrus, coffee, cowpea, *Musa* (banana/plantain), soybean, rice and wheat.

MASWheat: https://maswheat.ucdavis.edu/.

This website provides information on *Lr/Sr/Yr* genes and protocols, mainly for marker-assisted selection.

Stripe Rust: https://striperust.wsu.edu/.

This website provides a great deal of information about stripe rust.

8 Acknowledgements

We greatly thank for National Natural Science Foundation of China (#31801357) and Foundation of Sichuan Province Science and Technology Support Program (#2019YJ0510).

9 References

Akbari, M., Wenzl, P., Caig, V., Carling, J., Xia, L., Yang, S., Uszynski, G., Mohler, V., Lehmensiek, A., Kuchel, H., Hayden, M. J., Howes, N., Sharp, P., Vaughan, P., Rathmell, B., Huttner, E. and Kilian, A. (2006). Diversity arrays technology (DArT) for high-throughput profiling of the hexaploid wheat genome. *TAG. Theoretical and Applied Genetics. Theoretische und Angewandte Genetik* 113(8): 1409–1420.

Allen, A. M., Winfield, M. O., Burridge, A. J., Downie, R. C., Benbow, H. R., Barker, G. L., Wilkinson, P. A., Coghill, J., Waterfall, C., Davassi, A., Scopes, G., Pirani, A., Webster, T., Brew, F., Bloor, C., Griffiths, S., Bentley, A. R., Alda, M., Jack, P., Phillips, A. L. and Edwards, K. J. (2016). Characterization of a wheat breeders' array suitable for high-throughput SNP genotyping of global accessions of hexaploid bread wheat (*Triticum aestivum*). *Plant Biotechnology Journal* 15(3): 390–401.

Bansal, M., Kaur, S., Dhaliwal, H. S., Bains, N. S., Bariana, H. S., Chhuneja, P. and Bansal, U. K. (2017). Mapping of *Aegilops umbellulata*-derived leaf rust and stripe rust resistance loci in wheat. *Plant Pathology* 66(1): 38–44.

Bansal, U. K., Forrest, K. L., Hayden, M. J., Miah, H., Singh, D. and Bariana, H. S. (2011). Characterisation of a new stripe rust resistance gene *Yr47* and its genetic association with the leaf rust resistance gene *Lr52*. *TAG. Theoretical and Applied Genetics. Theoretische und Angewandte Genetik* 122(8): 1461–1466.

Bariana, H., Forrest, K., Qureshi, N., Miah, H., Hayden, M. and Bansal, U. (2016). Adult plant stripe rust resistance gene *Yr71* maps close to *Lr24* in chromosome 3d of common wheat. *Molecular Breeding* 36(7): 98.

Bariana, H. S. and McIntosh, R. A. (1993). Cytogenetic studies in wheat. XIV. Location of rust resistance genes in VPM1 and their genetic linkage with other disease resistance genes in chromosome 2a. *Genome* 36(3): 476–482.

Bariana, H. S., Parry, N., Barclay, I. R., Loughman, R., McLean, R. J., Shankar, M., Wilson, R. E., Willey, N. J. and Francki, M. (2006). Identification and characterization of stripe rust resistance gene *Yr34* in common wheat. *TAG. Theoretical and Applied Genetics. Theoretische und Angewandte Genetik* 112(6): 1143–1148.

Basnet, B. R., Singh, R. P., Ibrahim, A. M. H., Herrera-Foessel, S. A., Huerta-Espino, J., Lan, C. and Rudd, J. C. (2014). Characterization of *Yr54* and other genes associated with adult plant resistance to yellow rust and leaf rust in common wheat Quaiu 3. *Molecular Breeding* 33(2): 385-399.

Bennett, F. G. A. (1984). Resistance to powdery mildew in wheat: a review of its use in agriculture and lant breeding programs. *Plant Pathology* 33(3): 279-300.

Biffen, R. H. (1905). Mendel's laws of inheritance and wheat breeding. *Journal of Agricultural Science* 1(1): 4-48.

Boukhatem, N., Baret, P. V., Mingeot, D. and Jacquemin, J. M. (2002). Quantitative trait loci for resistance against yellow rust in two wheat-derived recombinant inbred line populations. *TAG. Theoretical and Applied Genetics. Theoretische und Angewandte Genetik* 104(1): 111-118.

Bryan, G. J., Collins, A. J., Stephenson, P., Orry, A., Smith, J. B. and Gale, M. D. (1997). Isolation and characterisation of microsatellites from hexaploid bread wheat. *Theoretical and Applied Genetics* 94(5): 557-563.

Calonnec, A., Johnson, R. and de Vallavieille-Pope, C. (2002). Genetic analyses of resistance of the wheat differential cultivars Carstens V and Spaldings Prolific to two races of *Puccinia striiformis*. *Plant Pathology* 51(6): 777-786.

Chen, H. M., Li, L. Z., Wei, X. Y., Li, S. S., Lei, T. D., Hu, H. Z., Wang, H. G. and Zhang, X. S. (2005). Development, chromosome location and genetic mapping of EST-SSR markers in wheat. *Chinese Science Bulletin* 50: 2328-2336.

Chen, Q., Conner, R. L., Li, H., Laroche, A., Graf, R. J. and Kuzyk, A. D. (2002). Expression of resistance to stripe rust, powdery mildew and the wheat curl mite in *Triticum aestivum* × *Haynaldia villosa* lines. *Canadian Journal of Plant Science* 82(2): 451-456.

Chen, S. S., Chen, G. Y., Chen, H., Wei, Y. M., Li, W., Liu, Y. X., et al. (2012). Mapping stripe rust resistance gene *Yrsph* derived from *Tritium sphaerococcum perc.* with SSR, SRAP, and Trap markers. *Euphytica* 185(1): 19-26.

Chen, W. Q., Wellings, C., Chen, X. M., Kang, Z. S. and Liu, T. G. (2014). Wheat stripe (yellow) rust caused by *Puccinia striiformis* f. sp. *tritici*. *Molecular Plant Pathology* 15(5): 433-446.

Chen, X. M. (2005). Epidemiology and control of stripe rust (*Puccinia striiformis* f. sp. *tritici*) on wheat. *Canadian Journal of Plant Pathology* 27(3): 314-337.

Chen, X. M. (2014). Integration of cultivar resistance and fungicide application for control of wheat stripe rust. *Canadian Journal of Plant Pathology* 36(3): 311-326.

Chen, X. M. and Line, R. F. (1995). Gene number and heritability of wheat cultivars with durable, high-temperature, adult-plant resistance and race-specific resistance to *Puccinia striiformis*. *Phytopathology* 85(5): 573-578.

Chen, W., Kang, Z., Ma, Z., Xu, S., Jin, S. and Jiang, Y. (2013). Integrated management of wheat stripe rust caused by Puccinia striiformis f. sp. tritici in China. *Scientia Agricultura Sinica* 46(20): 4254-4262.

Cheng, P. and Chen, X. M. (2010). Molecular mapping of a gene for stripe rust resistance in spring wheat cultivar IDO377s. *TAG. Theoretical and Applied Genetics. Theoretische und Angewandte Genetik* 121(1): 195-204.

Cheng, P., Xu, L. S., Wang, M. N., See, D. R. and Chen, X. M. (2014). Molecular mapping of genes *Yr64* and *Yr65* for stripe rust resistance in hexaploid derivatives of durum wheat accessions PI 331260 and PI 480016. *TAG. Theoretical and Applied Genetics. Theoretische und Angewandte Genetik* 127(10): 2267-2277.

Chhetri, M., Bariana, H., Kandiah, P. and Bansal, U. (2016). *Yr58*: a new stripe rust resistance gene and its interaction with *Yr46* for enhanced resistance. *Phytopathology* 106(12): 1530-1534.

Chhuneja, P., Kaur, S., Garg, T., Ghai, M., Kaur, S., Prashar, M., Bains, N. S., Goel, R. K., Keller, B., Dhaliwal, H. S. and Singh, K. (2008). Mapping of adult plant stripe rust resistance genes in diploid A genome wheat species and their transfer to bread wheat. *TAG. Theoretical and Applied Genetics. Theoretische und Angewandte Genetik* 116(3): 313-324.

Christiansen, M. J., Feenstra, B., Skovgaard, I. M. and Andersen, S. B. (2006). Genetic analysis of resistance to yellow rust in hexaploid wheat using a mixture model for multiple crosses. *TAG. Theoretical and Applied Genetics. Theoretische und Angewandte Genetik* 112(4): 581-591.

Cui, F., Zhang, N., Fan, X. L., Zhang, W., Zhao, C. H., Yang, L. J., Pan, R. Q., Chen, M., Han, J., Zhao, X. Q., Ji, J., Tong, Y. P., Zhang, H. X., Jia, J. Z., Zhao, G. Y. and Li, J. M. (2017). Utilization of a Wheat660K SNP array-derived high-density genetic map for high-resolution mapping of a major QTL for kernel number. *Scientific Reports* 7(1): 3788.

Dong, Z., Hegarty, J. M., Zhang, J., Zhang, W. and Dubcovsky, J. (2017). Validation and characterization of a QTL for adult plant resistance to stripe rust on wheat chromosome arm 6BS (*Yr78*). *Theoretical and Applied Genetics* 130(10): 2127 -2137.

Dracatos, P. M., Zhang, P., Park, R. F., McIntosh, R. A. and Wellings, C. R. (2016). Complementary resistance genes in wheat selection 'avocet R' confer resistance to stripe rust. *TAG. Theoretical and Applied Genetics. Theoretische und Angewandte Genetik* 129(1): 65-76.

Dubin, H. J., Johnson, R. and Stubbs, R. W. (1989). Postulated genes for resistance to stripe rust in selected CIMMYT and related wheats. *Plant Disease* 73(6): 472-475.

Eriksen, L., Afshari, F., Christiansen, M. J., McIntosh, R. A., Jahoor, A. and Wellings, C. R. (2004). *Yr32* for resistance to stripe (yellow) rust present in the wheat cultivar Carstens v. Theor. *Theoretical and Applied Genetics* 108(3): 567-575.

Eriksson, J. and Henning, E. (1896). *Die getreideroste. Ihre Geschichte und Natur sowie Massregein gegen dieselben*. P. A. Norstedt and Soner, Stockholm. 463.

Eujayl, I., Sorrells, M. E., Baum, M., Wolters, P. and Powell, W. (2002). Isolation of EST-derived microsatellite markers for genotyping the A and B genomes of wheat. *TAG. Theoretical and Applied Genetics. Theoretische und Angewandte Genetik* 104(2-3): 399-407.

Feng, J., Chen, G., Wei, Y., Liu, Y., Jiang, Q., Li, W., Pu, Z., Lan, X., Dai, S. and Zheng, Y. (2014). Identification and genetic mapping of a recessive gene for resistance to stripe rust in wheat line LM168-1. *Molecular Breeding* 33(3): 601-609.

Feng, J., Wang, M., See, D. R., Chao, S., Zheng, Y. L. and Chen, X. (2018). Characterization of novel gene *Yr79* and four additional QTL for all-stage and high-temperature adult-plant resistance to stripe rust in spring wheat PI 182103. *Phytopathology* 108(6): 737-747.

Gao, P., Zhou, Y., Gebrewahid, T. W., Zhang, P., Yan, X., Li, X., Yao, Z., Li, Z. and Liu, D. (2019). Identification of known leaf rust resistance genes in common wheat cultivars from Sichuan Province in China. *Crop Protection* 115: 122-129.

Gardiner, L. J., Bansept-Basler, P., Olohan, L., Joynson, R., Brenchley, R., Hall, N., O'Sullivan, D. M. and Hall, A. (2016). Mapping by sequencing in complex polyploid genomes using genic sequence capture: a case study to map yellow rust resistance in hexaploid wheat. *The Plant Journal: For Cell and Molecular Biology* 87(4): 403-419.

Gerechter-Amitai, Z. K., van Silfhout, C. H., Grama, A. and Kleitman, F. (1989). Yr15- a new gene for resistance to *Puccinia striiformis* in *Triticum dicoccoides sel*. G-25. *Euphytica* 43(1-2): 187-190.

Gessese, M., Bariana, H., Wong, D., Hayden, M. and Bansal, U. (2019). Molecular mapping of stripe rust resistance gene *Yr81* in a common wheat landrace AUS27430. *Plant Disease* 103(6): 1166-1171.

Gill, K. S., Lubbers, E. L., Gill, B. S., Raupp, W. J. and Cox, T. S. (1991). A genetic linkage map of *Triticum tauschii* (DD) and its relationship to the D genome of bread wheat (AABBDD). *Genome* 34(3): 362-374.

Giovannoni, J. J., Wing, R. A., Ganal, M. W. and Tanksley, S. D. (1991). Isolation of molecular markers from specific chromosomal intervals using DNA pools from existing mapping populations. *Nucleic Acids Research* 19(23): 6553-6558.

Guo, Q., Zhang, Z. J., Xu, Y. B., Li, G. H., Feng, J. and Zhou, Y. (2008). Quantitative trait loci for high-temperature adult-plant and slow-rusting resistance to *Puccinia striiformis* f. sp. *tritici* in wheat cultivars. *Phytopathology* 98(7): 803-809.

Gupta, K., Balyan, S., Edwards, J., Isaac, P., Korzun, V., Röder, M., Gautier, M. F., Joudrier, P., Schlatter, R., Dubcovsky, J., De La Pena, C., Khairallah, M., Penner, G., Hayden, J., Sharp, P., Keller, B., Wang, C., Hardouin, P., Jack, P. and Leroy, P. (2002). Genetic mapping of 66 new microsatellite (SSR) loci in bread wheat. *TAG. Theoretical and Applied Genetics. Theoretische und Angewandte Genetik* 105(2-3): 413 -422.

Gupta, P. K., Rustgi, S. and Kulwal, P. L. (2005). Linkage disequilibrium and association studies in higher plants: present status and future prospects. *Plant Molecular Biology* 57(4): 461-485.

Gupta, P. K., Rustgi, S. and Mir, R. R. (2008). Array-based high-throughput DNA markers for crop improvement. *Heredity* 101(1): 5-18.

Guyomarc'h, H., Sourdille, P., Charmet, G., Edwards, J. and Bernard, M. (2002). Characterisation of polymorphic microsatellite markers from *Aegilops tauschii* and transferability to the D-genome of bread wheat. *TAG. Theoretical and Applied Genetics. Theoretische und Angewandte Genetik* 104(6-7): 1164-1172.

Hayden, M. J., Stephenson, P., Logojan, A. M., Khatkar, D., Rogers, C., Elsden, J., Koebner, R. M., Snape, J. W. and Sharp, P. J. (2006). Development and genetic mapping of sequence-tagged microsatellites (STMs) in bread wheat (*Triticum aestivum* L.). *TAG. Theoretical and Applied Genetics. Theoretische und Angewandte Genetik* 113(7): 1271-1281.

Herrera-Foessel, S. A., Lagudah, E. S., Huerta-Espino, J., Hayden, M. J., Bariana, H. S., Singh, D. and Singh, R. P. (2011). New slow-rusting leaf rust and stripe rust resistance genes *Lr67*and *Yr46* in wheat are pleiotropic or closely linked. *Theoretical and Applied Genetics* 122(1): 239-249.

Herrera-Foessel, S. A., Singh, R. P., Lan, C. X., Huerta-Espino, J., Calvo-Salazar, V., Bansal, U. K., Bariana, H. S. and Lagudah, E. S. (2015). *Yr60*, a gene conferring moderate resistance to stripe rust in wheat. *Plant Disease* 99(4): 508-511.

Hou, L., Jia, J., Zhang, X., Li, X., Yang, Z., Ma, J., Guo, H., Zhan, H., Qiao, L. and Chang, Z. (2016). Molecular mapping of the stripe rust resistance gene *Yr69* on wheat chromosome 2AS. *Plant Disease* 100(8): 1717-1724.

Huang, L., Liu, T., Xiao, X., Qu, C., Liu, B., Gao, L., Luo, P. and Chen, W. (2017). Evaluation of stripe rust resistance and molecular detection of Yr Genes of 79 wheat varieties (Lines) in China. *Scientia Agricultura Sinica* 50(16): 3122-3134.

Huang, S., Wu, J., Wang, X., Mu, J., Xu, Z., Zeng, Q., Liu, S., Wang, Q., Kang, Z. and Han, D. (2019). Utilization of the genomewide wheat55K SNP array for genetic analysis of stripe rust resistance in common wheat line P9936. *Phytopathology* 109(5): 819-827.

Hylander, N., Jorstad, I. and Nannfeldt, J. A. (1953). Enumeratio uredionearum Scandinavicarum. *Opera Botanica* 1: 1-102.

Imtiaz, M., Ahmad, M., Cromey, M. G., Griffin, W. B. and Hampton, J. G. (2004). Detection of molecular markers linked to the durable adult plant stripe rust resistance gene *Yr18* in bread wheat (*Triticum aestivum* L.). *Plant Breeding* 123(5): 401-404.

Ishikawa, G., Nakamura, T., Ashida, T., Saito, M., Nasuda, S., Endo, T. R., Wu, J. and Matsumoto, T. (2009). Localization of anchor loci representing five hundred annotated rice genes to wheat chromosomes using PLUG markers. *TAG. Theoretical and Applied Genetics. Theoretische und Angewandte Genetik* 118(3): 499-514.

Jaccoud, D., Peng, K., Feinstein, D. and Kilian, A. (2001). Diversity arrays: a solid state technology for sequence information independent genotyping. *Nucleic Acids Research* 29(4): E25.

Jiang, C., Wang, X., Chen, W., Liu, T., Zhong, S., Huang, Q., et al. (2020). Resistance performance of wheat stripe rust resistance gene *Yr41* and its effect on yield parameters in F_2 populations under field conditions. *Crop Protection* 134: 105168.

Jighly, A., Oyiga, B. C., Makdis, F., Nazari, K., Youssef, O., Tadesse, W., Abdalla, O. and Ogbonnaya, F. C. (2015). Genome-wide DArT and SNP scan for QTL associated with resistance to stripe rust (*Puccinia striiformis* f. sp. *tritici*) in elite icarda wheat (*Triticum aestivum* L.) germplasm. *TAG. Theoretical and Applied Genetics. Theoretische und Angewandte Genetik* 128(7): 1277-1295.

Kuraparthy, V., Chhuneja, P., Dhaliwal, H. S., Kaur, S., Bowden, R. L. and Gill, B. S. (2007). Characterization and mapping of cryptic alien introgression from Aegilops geniculata with new leaf rust and stripe rust resistance genes *Lr57*and *Yr40* in wheat. *Theoretical and Applied Genetics* 114(8): 1379-1389.

Li, G., Tang, L., Yin, Y., Zhang, A., Yang, Z., Yang, E., Tang, Z., Fu, S. and Yang, Z. (2020). Molecular dissection of *Secale africanum* chromosome 6r^{afr} in wheat enabled localization of genes for resistance to powdery mildew and stripe rust. *BMC Plant Biology* 20(1): 134.

Li, H. H., Vikram, P., Singh, R. P., Kilian, A., Carling, J., Song, J., Burgueno-Ferreira, J. A., Bhavani, S., Huerta-Espino, J., Payne, T., Sehgal, D., Wenzl, P. and Singh, S. (2015). A high density GBS map of bread wheat and its application for dissecting complex disease resistance traits. *BMC Genomics* 16(1): 216.

Li, H. Y., Wang, Q. L., Xu, L. S., Mu, J., Wu, J., Zeng, Q., Yu, S., Huang, L., Han, D. and Kang, Z. (2017). Rapid identification of a major effect QTL conferring adult plant resistance to stripe rust in wheat cultivar Yaco S. *Euphytica* 213(6): 124.

Li, J., Dundas, I., Dong, C., Li, G., Trethowan, R., Yang, Z., Hoxha, S. and Zhang, P. (2020). Identification and characterization of a new stripe rust resistance gene *Yr83* on rye chromosome 6R in wheat. *TAG. Theoretical and Applied Genetics. Theoretische und Angewandte Genetik* 133(4): 1095-1107.

Li, M., Feng, J., Cao, S., Lin, R., Cheng, G., Yu, Y., Chen, W. and Xu, S. (2011). Postulation of seedlings resistance genes to yellow rust in commercial wheat cultivars from Yunnan Province in China. *Agricultural Sciences in China* 10(11): 1723-1731.

Li, Q., Chen, X. M., Wang, M. N. and Jing, J. X. (2011). *Yr45*, a new wheat gene for stripe rust resistance on the long arm of chromosome 3d. *TAG. Theoretical and Applied Genetics. Theoretische und Angewandte Genetik* 122(1): 189-197.

Li, Q., Guo, J., Chao, K., Yang, J., Yue, W., Ma, D. and Wang, B. (2018). High-density mapping of an adult-plant stripe rust resistance gene *YrBai* in wheat landrace Baidatou using the whole genome DArTseq and SNP analysis. *Frontiers in Plant Science* 9: 1120.

Li, Y., Niu, Y. C. and Chen, X. M. (2009). Mapping a stripe rust resistance gene *Yrc591* in wheat variety C591 with SSR and AFLP markers. *TAG. Theoretical and Applied Genetics. Theoretische und Angewandte Genetik* 118(2): 339–346.

Lin, F. and Chen, X. M. (2007). Genetics and molecular mapping of genes for race-specific all-stage resistance and non-race-specific high-temperature adult-plant resistance to stripe rust in spring wheat cultivar Alpowa. *TAG. Theoretical and Applied Genetics. Theoretische und Angewandte Genetik* 114(7): 1277–1287.

Lin, F. and Chen, X. M. (2009). Quantitative trait loci for non-race-specific, high-temperature adult-plant resistance to stripe rust in wheat cultivar Express. *TAG. Theoretical and Applied Genetics. Theoretische und Angewandte Genetik* 118(4): 631–642.

Liu, J., Chang, Z., Zhang, X., Yang, Z., Li, X., Jia, J., Zhan, H., Guo, H. and Wang, J. (2013). Putative *Thinopyrum intermedium*-derived stripe rust resistance gene Yr50 maps on wheat chromosome arm 4BL. *TAG. Theoretical and Applied Genetics. Theoretische und Angewandte Genetik* 126(1): 265–274.

Liu, L., Wang, M. N., Feng, J. Y., See, D. R., Chao, S. M. and Chen, X. M. (2018). Combination of all-stage and high-temperature adult-plant resistance QTL confers high-level, durable resistance to stripe rust in winter wheat cultivar Madsen. *TAG. Theoretical and Applied Genetics. Theoretische und Angewandte Genetik* 131(9): 1835–1849.

Liu, S., Yeh, C. T., Tang, H. M., Nettleton, D. and Sclmable, P. S. (2012). Gene mapping via bulked segregant RNA-Seq(BSR-Seq). *PLoS ONE* 7(5): e36406.

Liu, W. Z., Maccaferri, M., Bulli, P., Rynearson, S., Tuberosa, R., Chen, X. M. and Pumphrey, M. (2016). Genome-wide association mapping for seedling and field resistance to Puccinia striiformis f. sp. tritici in elite durum wheat. *Theoretical and Applied Genetics* 130(4): 1–19.

Lowe, I., Jankuloski, L., Chao, S., Chen, X., See, D. and Dubcovsky, J. (2011). Mapping and validation of QTL which confer partial resistance to broadly virulent post-2000 North American races of stripe rust in hexaploid wheat. *TAG. Theoretical and Applied Genetics. Theoretische und Angewandte Genetik* 123(1): 143–157.

Lu, Y., Wang, M., Chen, X., See, D., Chao, S. and Jing, J. (2014). Mapping of *Yr62* and a small-effect QTL for high-temperature adult-plant resistance to stripe rust in spring wheat PI 192252. *TAG. Theoretical and Applied Genetics. Theoretische und Angewandte Genetik* 127(6): 1449–1459.

Luo, P. G., Hu, X. Y., Ren, Z. L.*, Zhang, H. Y., Shu, K. and Yang, Z. J. (2008). Allelic analysis of stripe rust resistance genes on wheat chromosome 2BS. *Genome* 51: 922–927.

Luo, P. G., Ren, Z. L., Zhang, H. Q. and Zhang, H. Y. (2005). Identification, chromosome location, and diagnostic markers for a new gene (*YrCN19*) for resistance to wheat stripe rust. *Phytopathology* 95(11): 1266–1270.

Lupton, F. G. H. and Macer, R. C. F. (1962). Inheritance of resistance to yellow rust (*Pucciniaglumarum* Erikss., and Henn.) in seven varieties of wheat. *Transactions of the British Mycological Society* 45(1): 21–45.

Ma, D. F., Fang, Z. W., Yin, J. L., Chao, K. X., Jing, J. X., Li, Q. and Wang, B. T. (2016). Molecular mapping of stripe rust resistance gene *YrHu* derived from *Psathyrostachys huashanica*. *Molecular Breeding* 36(6): 64.

Ma, J., Qin, N., Cai, B., Chen, G., Ding, P., Zhang, H., Yang, C., Huang, L., Mu, Y., Tang, H., Liu, Y., Wang, J., Qi, P., Jiang, Q., Zheng, Y., Liu, C., Lan, X. and Wei, Y. (2019). Identification and validation of a novel major QTL for all-stage stripe rust resistance on 1BL in the winter wheat line 20828. *TAG. Theoretical and Applied Genetics. Theoretische und Angewandte Genetik* 132(5): 1363-1373.

Ma, J., Zhou, R., Dong, Y., Wang, L., Wang, X. and Jia, J. (2001). Molecular mappingand detection of the yellow rust resistance gene *Yr26* in wheat transferred from *Triticum turgidum* L. using microsatellite markers. *Euphytica* 120(2): 219-226.

Macer, R. C. F. (1966). The formal and monosomic genetic analysis of stripe rust (*Pucciniastriiformis*) resistance in wheat. In Proceedings of the 2nd International Wheat Genetics Symposium. 19-24 August 1963, Lund, Sweden. J. MacKey (Ed). Hereditas 2(Suppl.). 1966, 127-142.

Macer, R. C. F. (1975). Plant pathology in a changing world. *Transactions–British Mycological Society* 65: 351-374.

Mallard, S., Gaudet, D., Aldeia, A., Abelard, C., Besnard, A. L., Sourdille, P. and Dedryver, F. (2005). Genetic analysis of durable resistance to yellow rust in bread wheat. *TAG. Theoretical and Applied Genetics. Theoretische und Angewandte Genetik* 110(8): 1401-1409.

Marais, F., Marais, A., McCallum, B. and Pretorius, Z. (2009). Transfer of leaf rust and stripe rust resistance genes *Lr62* and *Yr42* from *Aegilops neglecta* Req. ex *Bertol*. to Common Wheat. *Crop Science* 49(3): 871-879.

Marais, G. F., McCallum, B. and Marais, A. S. (2006). Leaf rust and stripe rust resistance genes derived from *Aegilops sharonensis*. *Euphytica* 149(3): 373-380.

Marais, G. F., Pretorius, Z. A., Wellings, C. R., McCallum, B. and Marais, A. S. (2005a). Leaf rust and stripe rust resistance genes transferred to common wheat from *Triticum dicoccoides*. *Euphytica* 143(1-2): 115-123.

Marais, G. F., McCallum, B., Snyman, J. E., Pretorius, Z. A. and Marais, A. S. (2005b). Leaf rust and stripe rust resistance genes *Lr54* and *Yr37* transferred to wheat from *Aegilops kotschyi*. *Plant Breeding* 124(6): 538-541.

Mares, D. J. (1979). Microscopic study of the development of yellow rust (*Puccinia striiformis*) in a wheat cultivar showing adult plant resistance. *Physiological Plant Pathology* 15(3): 289-296.

McDonald, D. B., McIntosh, R. A., Wellings, C. R., Singh, R. P. and Nelson, J. C. (2004). Cytogenetical studies in wheat. XIX. Location and linkage studies on gene *Yr27* for resistance to stripe (yellow) rust. *Euphytica* 136(3): 239-248.

McIntosh, R. A. (1988). Catalogue of gene symbols for wheat. In Proceedings of the 7th International Wheat Genetics Symposium. 14-19 July 1988, Cambridge, UK. T. E. Miller and R. M. D. Koebner (Ed). Institute of Plant Science Research, Cambridge, UK. Vol. 2. pp. 1225-1323.

McIntosh, R. A., Dubcovsky, J., Rogers, J., Morris, C., Appels, R. and Xia, X. C. (2016). Catalogue of gene symbols for wheat: 2016 supplement. Available at: http://www.shigen.nig.ac.jp/wheat/komugi/genes/macgene/supplement2015-2016.pdf.2016.

McIntosh, R. A., Dubcovsky, J., Rogers, J., Morris, C., Appels, R. and Xia, X. C. (2017). Catalogue of gene symbols for wheat: 2017 supplement. Available at: http://www.shigen.nig.ac.jp/wheat/komugi/genes/macgene/supplement2017.pdf.2017.

McIntosh, R. A. and Lagudah, E. S. (2000). Cytogenetical studies in wheat. XVIII. Gene *Yr24* for resistance to stripe rust. *Plant Breeding* 119(1): 81-83.

Melichar, J. P. E., Berry, S., Newell, C., MacCormack, R. and Boyd, L. A. (2008). QTL identification and microphenotype characterization of the developmentally regulated yellow rust resistance in the UK wheat cultivar Guardian. *TAG. Theoretical and Applied Genetics. Theoretische und Angewandte Genetik* 117(3): 391-399.

Michelmore, R. W., Paran, I. and Kesseli, R. V. (1991). Identification of markers linked to disease-resistance genes by bulked segregant analysis: a rapid method to detect markers in specific genomic regions by using segregating populations. *Proceedings of the National Academy of Sciences of the United States of America* 88(21): 9828-9832.

Mu, J., Huang, S., Liu, S., Zeng, Q. and Han, D. (2018). Genetic architecture of wheat stripe rust resistance revealed by combining QTL mapping using SNP-based genetic maps and bulked segregant analysis. *Theoretical and Applied Genetics* 132(2): 443-455.

Murray, G. M. and Brennan, J. P. (2009). The current and potential costs from diseases of wheat in Australia. Australian Grains Research and Development Corporation Report Available at: http://www.grdc.com.au (vol. 69).

Naruoka, Y., Garland-Campbell, K. A. and Carter, A. H. (2015). Genome-wide association mapping for stripe rust (Puccinia striiformis F. sp. tritici) in US Pacific Northwest winter wheat (*Triticum aestivum* L.). *TAG. Theoretical and Applied Genetics. Theoretische und Angewandte Genetik* 128(6): 1083-1101.

Navabi, A., Tewari, J. P., Singh, R. P., McCallum, B., Laroche, A. and Briggs, K. G. (2005). Inheritance and QTL analysis of durable resistance to stripe and leaf rusts in an Australian cultivar, *Triticum aestivum* 'Cook'. *Genome* 48(1): 97-107.

Nsabiyera, V., Bariana, H. S., Qureshi, N., Wong, D., Hayden, M. J. and Bansal, U. K. (2018). Characterisation and mapping of adult plant stripe rust resistance in wheat accession Aus27284. *TAG. Theoretical and Applied Genetics. Theoretische und Angewandte Genetik* 131(7): 1459-1467.

Pakeerathan, K., Bariana, H., Qureshi, N., Wong, D. and Bansal, U. (2019). Identification of a new source of stripe rust resistance *Yr82* in wheat. *Theoretical and Applied Genetics* 132(11): 3169-3176.

Pathan, A. K., Wellings, C. R., Barianam H. S. and Park, R. F. (2008). Evaluation of seedling and adult plant resistance in European wheat cultivars to Australian isolates of *Puccinia striiformis* f. sp. *tritici*. *Euphytica* 163(2): 283-301.

Pawar, S. K., Sharma, D., Duhan, J. S., Saharan, M. S., Tiwari, R. and Sharma, I. (2016). Mapping of stripe rust resistance QTL in Cappelle-Desprez × PBW343 RIL population effective in northern wheat belt of India. *3 Biotech* 6(1): 76.

Parida, S. K., Kumar, A. R. K., Dalal, V., et al. (2006). Unigene derived microsatellite markers for the cereal genomes. *Theoretical and Applied Genetics* 112: 808-817.

Peng, J. H. and Lapitan, N. L. V. (2005). Characterization of EST-derived microsatellites in the wheat genome and development of eSSR markers. *Functional and Integrative Genomics* 5(2): 80-96.

Pestsova, E., Ganal, M. W. and Röder, M. S. (2000). Isolation and mapping of microsatellite markers specific for the D genome of bread wheat. *Genome* 43(4): 689-697.

Pretorius, Z. A., Lan, C. X., Prins, R., Knight, V., McLaren, N. W., Singh, R. P., Bender, C. M. and Kloppers, F. J. (2017). Application of remote sensing to identify adult plant resistance loci to stripe rust in two bread wheat mapping populations. *Precision Agriculture* 18(4): 411-428.

Rahmatov, M., Hovmoller, M. S., Nazari, K., et al. (2017). Seedling and adult plant stripe rust resistance in diverse wheat-alien introgression lines. *Crop Science* 57(4): 1–11.

Ramburan, V. P., Pretorius, Z. A., Louw, J. H., Boyd, L. A., Smith, P. H., Boshoff, W. H. and Prins, R. (2004). A genetic analysis of adult plant resistance to stripe rust in the wheat cultivar Kariega. *TAG. Theoretical and Applied Genetics. Theoretische und Angewandte Genetik* 108(7): 1426–1433.

Ramirez-Gonzalez, R. H., Segovia, V., Bird, N., Fenwick, P., Holdgate, S., Berry, S., Jack, P., Caccamo, M. and Uauy, C. (2015). RNA-Seq bulked segregant analysis enables the identification of high-resolution genetic markers for breeding in hexaploid wheat. *Plant Biotechnology Journal* 13(5): 613–624.

Randhawa, M., Bansal, U., Lillemo, M., Miah, H. and Bariana, H. (2016). Postulation of rust resistance genes in Nordic spring wheat genotypes and identification of widely effective sources of resistance against the Australian rust flora. *Journal of Applied Genetics* 57(4): 453–465.

Randhawa, M., Bansal, U., Valarik, M., Klocová, B., Doležel, J. and Bariana, H. (2014). Molecular mapping of stripe rust resistance gene *Yr51* in chromosome 4AL ofwheat. *TAG. Theoretical and Applied Genetics. Theoretische und Angewandte Genetik* 127(2): 317–324.

Randhawa, M. S., Bariana, H. S., Mago, R. and Bansal, U. K. (2015). Mapping of a new stripe rust resistance locus *Yr57* on chromosome 3BS of wheat. *Molecular Breeding* 35(2): 65.

Ren, R. S., Wang, M. N., Chen, X. M. and Zhang, Z. J. (2012). Characterization and molecular mapping of *Yr52* for high-temperature adult-plant resistance to stripe rust in spring wheat germplasm PI 183527. *TAG. Theoretical and Applied Genetics. Theoretische und Angewandte Genetik* 125(5): 847–857.

Ren, T., Hu, Y., Tang, Y., Li, C., Yan, B., Ren, Z., Tan, F., Tang, Z., Fu, S. and Li, Z. (2018b). Utilization of a Wheat55K SNP array for mapping of major QTL for temporal expression of the tiller number. *Frontiers in Plant Science* 9: 333.

Ren, T. H., Chen, F., Yan, B. J., Zhang, H. Q. and Ren, Z. L. (2012). Genetic diversity of wheat-rye 1BL.1RS translocation lines derived from different wheat and rye sources. *Euphytica* 183(2): 133–146.

Ren, T. H., Ren, Z. L., Yang, M. Y., Yan, B. J., Tan, F. Q., Fu, S. L., Tang, Z. and Li, Z. (2018a). Novel source of 1RS from Baili rye conferred high resistance to diseases and enhanced yield traits to common wheat. *Molecular Breeding* 38(8): 101.

Ren, T. H., Tang, Z. X., Fu, S. L., Yan, B. J., Tan, F. Q., Ren, Z. and Li, Z. (2017). Molecular cytogenetic characterization of novel wheat-rye T1RS.1BL translocation lines with high resistance to diseases and great agronomic traits. *Frontiers in Plant Science* 8: 799.

Ren, T. H., Yang, Z. J., Yan, B. J., Zhang, H. Q., Fu, S. L. and Ren, Z. L. (2009). Development and characterization of a new 1BL.1RS translocation line with resistance to stripe rust and powdery mildew of wheat. *Euphytica* 169(2): 207–213.

Ren, Y., Singh, R. P., Basnet, B. R., Lan, C. X., Huerta-Espino, J., Lagudah, E. S. and Ponce-Molina, L. J. (2017). Identification and mapping of adult plant resistance loci to leaf rust and stripe rust in common wheat cultivar Kundan. *Plant Disease* 101(3): 456–463.

Riley, R., Chapman, V. and Johnson, R. (1968). The incorporation of alien disease resistance in wheat by genetic interference with the regulation of meiotic chromosome synapsis. *Genetical Research* 12(2): 199–219.

Riseh, N. and Merikangas, K. (1996). The future of genetic studies of complex human diseases. *Science* 273(3): 350–354.

Röder, M. S., Plaschke, J., König, S. U., Börner, A., Sorrells, M. E., Tanksley, S. D. and Ganal, M. W. (1995). Abundance, variability and chromosomal location of microsatellites in wheat. *Molecular and General Genetics: MGG* 246(3): 327–333.

Roebbelen, G. and Sharp, E. L. (1978). Mode of inheritance, interaction and application of genes conditioning resistance to yellow rust. *Forschritte Der Pflanzenzuechtung* 19(1): 1-1.

Santra, D. K., Chen, X. M., Santra, M., et al. (2008). Identification and mapping QTL for high-temperature adult-plant resistance to stripe rust in winter wheat (Triticum aestivum L.) cultivar Stephens. *Theoretical and Applied Genetics* 117: 793 -802.

Schneeberger, K. and Weigel, D. (2011). Fast-forward genetics enabled by new sequencing technologies. *Trends in Plant Science* 16(5): 282–288.

Sharma-Poudyal, D., Chen, X. M., Wan, A. M., Zhan, G. M., Kang, Z. S., Cao, S. Q., Jin, S. L., Morgounov, A., Akin, B., Mert, Z., Shah, S. J. A., Bux, H., Ashraf, M., Sharma, R. C., Madariaga, R., Puri, K. D., Wellings, C., Xi, K. Q., Wanyera, R., Manninger, K., Ganzález, M. I., Koyda, M., Sanin, S. and Patzek, L. J. (2013). Virulence characterization of international collections of the wheat stripe rust pathogen, *Puccinia striiformis* f.sp *tritici*. *Plant Disease* 97(3): 379–386.

Shi, Z. X., Chen, X. M., Line, R. F., Leung, H. and Wellings, C. R. (2001). Development of resistance gene analog polymorphism markers for the *Yr9* gene resistance to wheat stripe rust. *Genome* 44(4): 509–516.

Singh, R. P. (1992). Genetic association of leaf rust resistance gene *Lr34* with adult-plant resistance to stripe rust in bread wheat. *Phytopathology* 82(8): 835–838.

Singh, R. P., Huerta-Espino, J. and William, M. H. (2005). Genetics and breeding for durable resistance to leaf and stripe rust in wheat. *Turkish Journal of Agriculture and Forestry* 29(2): 121–127.

Singh, R. P., Nelson, J. C. and Sorrells, M. E. (2000). Mapping *Yr28* and other genes for resistance to stripe rust in wheat. *Crop Science* 40(4): 1148–1155.

Singh, R. P., William, H. M., Huerta-Espino, J., et al. (2004). Wheat rust in Asia: meeting the challenges with old and new technologies//New Directions for a Diverse Planet. Proceedings of the 4th International Crop Science Congress, Brisbane, Australia, 2004.

Spielmeyer, W., Huang, L., Bariana, H., Laroche, A., Gill, B. S. and Lagudah, E. S. (2000). NBS-LRR sequence family is associated with leaf and stripe rust resistance on the end of homoeologous chromosome group 1S of wheat. *Theoretical and Applied Genetics* 101(7): 1139–1144.

Song, Q. J., Shi, J. R., Singh, S., Fickus, E. W., Costa, J. M., Lewis, J., Gill, B. S., Ward, R. and Cregan, P. B. (2005). Development and mapping of microsatellite (SSR) markers in wheat. *TAG. Theoretical and Applied Genetics. Theoretische und Angewandte Genetik* 110(3): 550–560.

Sourdille, P., Cadalen, T., Guyomarc'h, H., Snape, J. W., Perretant, M. R., Charmet, G., Boeuf, C., Bernard, S. and Bernard, M. (2003). An update of the Courtot×Chinese spring intervarietal molecular marker linkage map for the QTL detection of agronomic traits in wheat. *TAG. Theoretical and Applied Genetics. Theoretische und Angewandte Genetik* 106(3): 530–538.

Sourdille, P., Singh, S., Cadalen, T., Brown-Guedira, G. L., Gay, G., Qi, L., Gill, B. S., Dufour, P., Murigneux, A. and Bernard, M. (2004). Microsatellite-based deletion bin system for the establishment of genetic physical map relationships in wheat (*Triticum aestivum* L.). *Functional and Integrative Genomics* 4(1): 12–25.

Sui, X. X., Wang, M. N. and Chen, X. M. (2009). Molecular mapping of a stripe rust resistance gene in spring wheat cultivar Zak. *Phytopathology* 99(10): 1209 -1215.

Sun, C. W., Dong, Z. D., Zhao, L., Ren, Y., Zhang, N. and Chen, F. (2020). The Wheat 660K SNP array demonstrates great potential for marker-assisted selection in polyploid wheat. *Plant Biotechnology Journal* 18: 1354–1360.

Stubbs, R. W. (1985). Stripe rust. In: Roelfs, A. P. and Bushnell, W. R. (Eds) *The Cereal Rusts II*. Academic Press, Orlando. pp. 61-101.

Takagi, H., Uemura, A., Yaegashi, H., Tamiru, M., Abe, A., Mitsuoka, C., Utsushi, H., Natsume, S., Kanzaki, H., Matsumura, H., Saitoh, H., Yoshida, K., Cano, L. M., Kamoun, S. and Terauchi, R. (2013). MutMap-Gap: whole-genome resequencing of mutant F_2 progeny bulk combined with de novo assembly of gap regions identifies the rice blast resistance gene Pii. *New Phytologist* 200(1): 276–283.

Uauy, C., Brevis, J. C., Chen, X., Khan, I., Jackson, L., Chicaiza, O., Distelfeld, A., Fahima, T. and Dubcovsky, J. (2005). High-temperature adult-plant (HTAP) stripe rust resistance gene *Yr36* from *Triticum turgidum* ssp. *dicoccoides* is closely linked to the grain protein content locus Gpc-B1. *TAG. Theoretical and Applied Genetics. Theoretische und Angewandte Genetik* 112(1): 97–105.

Wan, A., Zhao, Z., Chen, X., He, Z., Jin, S., Jia, Q., Yao, G., Yang, J., Wang, B., Li, G., Bi, Y. and Yuan, Z. (2004). Wheat stripe rust epidemic and virulence of *Puccinia striiformis* f. sp. *tritici* in China in 2002. *Plant Disease* 88(8): 896–904.

Wan, A. M., Chen, X. M. and He, Z. H. (2007). Wheat stripe rust in china. *Australian Journal of Agricultural Research* 58(6): 605–619.

Wang, C., Zhang, Y., Han, D., Kang, Z., Li, G., Cao, A. and Chen, P. (2008). SSR and STS markers for wheat stripe rust resistance gene *Yr26*. *Euphytica* 159(3): 359–366.

Wang, S., Wong, D., Forrest, K., Allen, A., Chao, S., Huang, B. E., Maccaferri, M., Salvi, S., Milner, S. G., Cattivelli, L., Mastrangelo, A. M., Whan, A., Stephen, S., Barker, G., Wieseke, Plieske, J., International Wheat Genome Sequencing Consortium, Lillemo, M., Mather, D., Appels, R., Dolferus, R., Brown-Guedira, G., Korol, A., Akhunova, A. R., Feuillet, C., Salse, J., Morgante, M., Pozniak, C., Luo, M. C., Dvorak, J., Morell, M., Dubcovsky, J., Ganal, M., Tuberosa, R., Lawley, C., Mikoulitch, I., Cavanagh, C., Edwards, K. J., Hayden, M. and Akhunov, E. (2014). Characterization of polyploid wheat genomic diversity using a high-density 90,000 single nucleotide polymorphism array. *Plant Biotechnology Journal* 12(6):787–796.

Wang, W., Wang, Z., Li, X., Ni, Z., Hu, Z., Xin, M., Peng, H., Yao, Y., Sun, Q. and Guo, W. (2020). SnpHub: an easy-to-set-up web server framework for exploring large-scale genomic variation data in the post-genomic era with applications in wheat. *GigaScience* 9(6): giaa060.

Wang, Y., Zhang, H., Xie, J., Guo, B., Chen, Y., Zhang, H., Lu, P., Wu, Q., Li, M., Zhang, D., Guo, G., Yang, J., Zhang, P., Zhang, Y., Wang, X., Zhao, H., Cao, T. and Liu, Z. (2018). Mapping stripe rust resistance genes by BSR-Seq: *YrMM58* and *YrHY1* on chromosome 2AS in Chinese wheat lines Mengmai 58 and Huaiyang 1 are *Yr17*. *The Crop Journal* 6(1): 91–98.

Wang, Z. L., Li, L. H., He, Z. H., Duan, X. Y., Zhou, Y. L., Chen, X. M., Lillemo, M., Singh, R. P., Wang, H. and Xia, X. C. (2005). Seeding and adult-plant resistance to powdery mildew in Chinese bread wheat cultivars and lines. *Plant Disease* 89(5): 457–463.

William, H. M., Singh, R. P., Huerta-Espino, J., Palacios, G. and Suenaga, K. (2006). Characterization of genetic loci conferring adult plant resistance to leaf rust and stripe rust in spring wheat. *Genome* 49(8): 977–990.

William, M., Singh, R. P., Huerta-Espino, J., Islas, S. O. and Hoisington, D. (2003). Molecular marker mapping of leaf rust resistance gene *Lr46* and its association with stripe rust resistance gene *Yr29* in wheat. *Phytopathology* 93(2): 153–159.

Winfield, M. O., Allen, A. M., Burridge, A. J., Barker, G. L. A., Benbow, H. R., Wilkinson, P. A., Coghill, J., Waterfall, C., Davassi, A., Scopes, G., Pirani, A., Webster, T., Brew, F., Bloor, C., King, J., West, C., Griffiths, S., King, I., Bentley, A. R. and Edwards, K. J. (2016). High-density SNP genotyping array for hexaploid wheat and its secondary and tertiary gene pool. *Plant Biotechnology Journal* 14(5): 1195–1206.

Worland, A. J. and Law, C. N. (1986). Genetic analysis of chromosome 2d of wheat. I. The location of genes affecting height, day-length insensitivity, hybrid dwarfism and yellow rust resistance. *Z. Pflanzenzücht* 96: 331–345.

Wu, J., Wang, Q., Liu, S., Huang, S., Mu, J., Zeng, Q., Huang, L., Han, D. and Kang, Z. (2017a). Saturation mapping of a major effect QTL for stripe rust resistance on wheat chromosome 2b in cultivar Napo 63 using SNP genotyping arrays. *Frontiers in Plant Science* 8: 653.

Wu, J., Wang, Q., Kang, Z., Liu, S., Li, H., Mu, J., Dai, M., Han, D., Zeng, Q. and Chen, X. (2017b). Development and validation of KASP-SNP markers for QTL underlying resistance to stripe rust in common wheat cultivar P10057. *Plant Disease* 101(12): 2079–2087.

Wu, J., Wang, Q., Xu, L., Chen, X., Li, B., Mu, J., Zeng, Q., Huang, L., Han, D. and Kang, Z. (2018a). Combining SNP genotyping array with bulked segregant analysis to map a gene controlling adult-plant resistance to stripe rust in wheat line 03031-1-5. *Phytopathology* 108(1): 103–113.

Wu, J., Liu, S., Wang, Q., Zeng, Q., Mu, J., Huang, S., Yu, S., Han, D. and Kang, Z. (2018b). Rapid identification of an adult plant stripe rust resistance gene in hexaploid wheat by high-throughput snp array genotyping of pooled extremes. *TAG. Theoretical and Applied Genetics. Theoretische und Angewandte Genetik* 131(1): 43–58.

Wu, J., Yu, R., Wang, H., Zhou, C., Huang, S., Jiao, H., Yu, S., Nie, X., Wang, Q., Liu, S., Weining, S., Singh, R. P., Bhavani, S., Kang, Z., Han, D. and Zeng, Q. (2020). A large-scale genomic association analysis identifies the candidate causal genes conferring stripe rust resistance under multiple field environments. *Plant Biotechnology Journal* 19(1): 177–191. doi:10.1111/pbi.13452.

Xiang, C., Feng, J. Y., Wang, M. N., Chen, X. M., See, D. R., Wan, A. M. and Wang, T. (2016). Molecular mapping of stripe rust resistance gene *Yr76* in winter club wheat cultivar tyee. *Phytopathology* 106(10): 1186–1193.

Xing, L., Li, Z. and Liu, D. (2012). QTL mapping of adult-plant resistance genes to stripe rust and leaf rust in CIMMYT wheat PBW343 and Muu. *Journal of Agricultural University of Hebei* 35(5): 45–50.

Xu, L. S., Wang, M. N., Cheng, P., Kang, Z. S., Hulbert, S. H. and Chen, X. M. (2013). Molecular mapping of *Yr53*, a new gene for stripe rust resistance in durum wheat accession PI 480148 and its transfer to common wheat. *TAG. Theoretical and Applied Genetics. Theoretische und Angewandte Genetik* 126(2): 523–533.

Xue, S., Zhang, Z., Lin, F., Kong, Z., Cao, Y., Li, C., Yi, H., Mei, M., Zhu, H., Wu, J., Xu, H., Zhao, D., Tian, D., Zhang, C. and Ma, Z. (2008). A high-density intervarietal map of the wheat genome enriched with markers derived from expressed sequence tags. *TAG. Theoretical and Applied Genetics. Theoretische und Angewandte Genetik* 117(2): 181-189.

Xue, W., Xu, X., Mu, J., Wang, Q., Wu, J., Huang, L., Kang, Z. and Han, D. (2014). Evaluation of stripe rust resistance and genes in Chinese elite wheat varieties. *Journal of Triticeae Crops*34(8): 1054-1060.

Yang, F., Liu, J., Guo, Y., He, Z., Rasheed, A., Wu, L., Cao, S., Nan, H. and Xia, X. (2020). Genome-wide association mapping of adult-plant resistance to stripe rust in common wheat (*Triticum aestivum*). *Plant Disease* 104(8): 2174-2180. doi:10.1094/PDIS-10-19-2116-RE.

Yang, L. J., Zhang, X. J., Zhang, X., Wang, J., Luo, M., Yang, M., Wang, H., Xiang, L., Zeng, F., Yu, D., Fu, D. and Rosewarne, G. M. (2017). Identification and evaluation of resistance to powdery mildew and yellow rust in a wheat mapping population. *PLoS ONE* 12(5): e0177905.

Yu, J. K., Dake, T. M., Singh, S., Benscher, D., Li, W., Gill, B. and Sorrells, M. E. (2004). Development and mapping of EST-derived simple sequence repeat markers for hexaploid wheat. *Genome* 47(5): 805-818.

Zeng, Q., Wu, J., Liu, S., Huang, S., Wang, Q., Mu, J., Yu, S., Han, D. and Kang, Z. (2019a). A major QTL co-localized on chromosome 6BL and its epistatic interaction for enhanced wheat stripe rust resistance. *Theoretical and Applied Genetics* 132(5): 1409 -1424.

Zeng, Q., Wu, J., Huang, S., Yuan, F., Liu, S., Wang, Q., Mu, J., Yu, S., Chen, L., Han, D. and Kang, Z. (2019b). SNP-based linkage mapping for validation of adult plant stripe rust resistance QTL in common wheat cultivar Chakwal 86. *The Crop Journal* 7(2): 176-186.

Zhang, A., Li, W., Wang, C., Yang, X., Chen, C., Zhu, C., Peng, N., Tian, Z., Wang, Y., Zhang, H., Liu, X. and Ji, W. (2017a). Molecular cytogenetics identification of a wheat-*Leymus mollis* double disomic addition line with stripe rust resistance. *Genome* 60(5): 375-383.

Zhang, H., Xie, J., Chen, Y., Liu, X., Wang, Y., Yan, S., Yang, Z., Zhao, H., Wang, X., Jia, L., Cao, T. and Liu, Z. (2017b). Mapping stripe rust resistance gene *Yrzm103* in wheat cultivar Zhengmai 103 by BSR-seq. *Acta Agronomica Sinica* 43(11): 1643.

Zhang, L. Y., Bernard, M., Leroy, P., Feuillet, C. and Sourdille, P. (2005). High transferability of bread wheat EST-derived SSRs to other cereals. *TAG. Theoretical and Applied Genetics. Theoretische und Angewandte Genetik* 111(4): 677-687.

Zhang, P., Li, X., Gebrewahid, T. W., Liu, H., Xia, X., He, Z., et al. (2019). QTL mapping of adult-plant resistance to leaf and stripe rust in wheat cross Sw 8588/thatcher using the Wheat55K SNP array. *Plant Disease* 103(12): 3041-3049.

Zhao, F. Q., Li, Q., Chen, G. Y., Li, J., Kang, H. Y., Wang, Y., Fan, X., Sha, L., Zhou, Y. and Zhang, H. (2017). Stripe rust resistance in *Roegneria kamoji* (*Poaceae: Triticeae*) and its genetic analysis. *Journal of Phytopathology* 165(3): 157-161.

Zhou, L., Liu, T., Cheng, Y., Ye, X., Li, W., Pu, Z., Jiang, Q., Liu, Y., Wei, Y., Deng, M., Zheng, Y. and Chen, G. (2017). Molecular mapping of a stripe rust resistance gene in Chinese wheat landrace "Hejiangyizai" using SSR, RGAP, TRAP, and SRAP markers. *Crop Protection* 94: 178-184.

Zhou, X. L., Han, D. J., Chen, X. M., Gou, H. L., Guo, S. J., Rong, L., Wang, Q. L., Huang, L. L. and Kang, Z. S. (2014a). Characterization and molecular mapping of stripe rust resistance gene *Yr61* in winter wheat cultivar Pindong 34. *TAG. Theoretical and Applied Genetics. Theoretische und Angewandte Genetik* 127(11): 2349–2358.

Zhou, X. L., Wang, M. N., Chen, X. M., Lu, Y., Kang, Z. S. and Jing, J. X. (2014b). Identification of *Yr59* conferring high-temperature adult-plant resistance to stripe rust in wheat germplasm PI 178759. *TAG. Theoretical and Applied Genetics. Theoretische und Angewandte Genetik* 127(4): 935–945.

Zou, C., Wang, P. and Xu, Y. (2016). Bulked sample analysis in genetics, genomics and crop improvement. *Plant Biotechnology Journal* 14(10): 1941–1955.

Part 2

Fungal diseases of cereals: Fusarium head blight

Chapter 4

Advances in understanding the epidemiology of *Fusarium* in cereals

Stephen N. Wegulo, University of Nebraska-Lincoln, USA

1 Introduction

Cereal grains are a major component of the human diet, contributing more than half of the global daily calorie intake (Awika, 2011; McKevith, 2004). The four major cereal crops grown in the world are maize, rice, wheat, and barley. Sorghum, millet, oats, rye, and triticale are also grown, but to a lesser extent. In addition to human food, cereals supply the majority of nutrients for livestock, thus contributing indirectly to the human diet. Additional products of economic importance derived from cereals include industrial ethanol used to make fuels and sweeteners for human consumption (Awika, 2011).

Cereal crops are vulnerable to attack by many plant diseases, which cause yield and quality loss. Among them are diseases caused by *Fusarium* pathogens, which result in major losses that threaten global food and economic security. Many *Fusarium* species are causal agents of cereal diseases, but a few cause diseases of major economic importance, including Fusarium head blight (FHB) of wheat and barley, ear and stalk rots of maize, and crown and root rots of wheat and barley. In addition to yield reduction, *Fusarium* pathogens also

http://dx.doi.org/10.19103/AS.2021.0092.06

produce mycotoxins, which reduce grain quality and are harmful to humans and animals (Desjardins, 2006).

This chapter reviews the epidemiology of *Fusarium* in cereals with emphasis on *F. graminearum* (teleomorph: *Gibberella zeae*) as a representative species because it is a predominant pathogen of cereals that causes several different diseases including head blights, crown and root rots, seedling blights, and ear and stalk rots. FHB, a disease of major economic importance worldwide that affects all small grain cereals, but especially wheat and barley, is used to illustrate *F. graminearum* epidemiology. *F. graminearum* is a member of the *Fusarium graminearum* species complex (FGSC), which consists of 16 different species ((Aoki et al., 2012), several of which are causal agents of FHB. Using published records and climate envelope modeling with BIOCLIM, a species distribution modeling system, Backhouse (2014) recorded *F. graminearum* from all continents except Antarctica. It was predicted to be present in most major rain-fed wheat-growing regions of the world except the hottest parts of south Asia.

Although *F. graminearum* is found mainly on wheat, corn, and barley, it can also occur on other annual and perennial hosts. In agar culture, *F. graminearum* appears gray, pink, brown, or burgundy. The fungus can grow on agar media in which it forms red pigments (Leslie and Summerell, 2006). On potato dextrose agar, *F. graminearum* grew rapidly between 12°C and 28°C with maximal growth at 28°C (Bai and Shaner, 1996). *F. graminearum* produces asexual spores known as macroconidia primarily in fruiting structures known as sporodochia (singular: sporodochium) (Leslie and Summerell, 2006). Macroconidia of *F. graminearum* are slender, thick-walled, curved to straight, tapered at both ends, with 5–7 septa. Macroconidia are 35–62 × 2.5–5.0 μm in size (Booth, 1971). Globose chlamydospores (thick-walled asexual resting spores) are produced mainly in macroconidia, but may also form in mycelia. The sexual stage of *F. graminearum* (*G. zeae*) produces perithecia (sexual fruiting structures) superficially on cereal host residue. Perithecia are dark purple or black and form abundantly on crop residue, especially maize residue, as well as in culture (Sutton, 1982; Trail and Common, 2000). Within the perithecia, ascospores (sexual spores) develop within sacs known as asci (singular: ascus). Each ascus contains eight ascospores that have 0–3 septa. Ascospores measure 17–25.5 × 3–5 μm (Booth, 1971).

FHB was first described in England by W. G. Smith in 1884 and several years later it was reported in the USA by F. D. Chester in 1890 and J. C. Arthur in 1891 (Parry et al., 1995; Stack, 2003). Since then several major epidemics have occurred worldwide. In the USA, there were five major FHB epidemics from the late 1910s to early 1930s (Stack, 2003). The disease re-emerged in the 1980s and 1990s and since then epidemics of varying intensity have occurred, more frequently in the eastern half of the USA (McMullen et al., 1997; McMullen et al., 2012). The re-emergence of FHB in the USA is attributed in part to a combination of favorable weather conditions (wet, humid, and warm weather) before, during and after anthesis, and an abundant supply of inoculum due to reduced tillage practices that leave crop residue on the soil surface (Shaner, 2003).

A typical symptom of FHB is premature whitening of the spike (Dill-Macky, 2010). One or more spikelets or the entire spike appears prematurely whitened or bleached. Initial symptoms generally are visible near the middle of the spike on the first florets to flower, but can occur anywhere on the spike. Premature bleaching progresses with time to most or all of the spikelets, causing the entire spike to be bleached. The bleached spikes are readily visible in a green field. Infection of the peduncle (stem immediately below the spike) may also occur, causing a brown or purple discoloration. If the peduncle is infected early, the entire head becomes sterile. The fungus sporulates on infected spikelets and glumes during prolonged wet weather. This results in pink to salmon-orange spore masses. Bleached spikelets are sterile or contain kernels that are shriveled and/or appear chalky white or pink. These kernels are often referred to as *Fusarium*-damaged kernels (FDK), scabby kernels, or 'tombstones' (McMullen et al., 1997). Kernels that are apparently healthy may also be infected, especially if infection occurred late in kernel development. *F. graminearum* produces trichothecene myctoxins, the major one of which is deoxinivalenol (DON), also known as vomitoxin. The mycotoxins accumulate in grain, lowering its quality and causing it to be harmful to humans and animals.

FHB can cause substantial economic losses. In the USA in 1917, losses caused by the disease were estimated at 288000 metric tons in 31 out of 40 states surveyed (Atanasoff, 1920). Dickinson and Mains (1929) reported that an epidemic that occurred in 1919 in the USA resulted in a loss of 2.18 million metric tons of wheat. Mains et al. (1929) estimated that an epidemic in 1928 caused a 15% yield loss in wheat in Indiana. Significant additional losses occurred in the 1930s (Dickinson and Mains, 1942). Major economic losses occurred again in the 1980s (McMullen et al., 1997), with total losses in the USA's wheat production totaling 2.72 million metric tons in 1982 alone (Boosalis et al., 1983). Johnson et al. (2003) estimated that FHB caused direct losses in wheat and barley totaling more than $1.3 billion in the USA during the period from 1991 to 1997. They estimated the total economic impact in rural communities and businesses related to grain production and marketing to be 3–4 times this amount. Nganje et al. (2004) estimated the cumulative direct economic losses due to FHB from 1998 to 2001 in nine states in the northern Great Plains and the central part of the USA to be $871 million. Secondary economic losses were estimated at $1.8 billion, resulting in a total economic impact of $2.7 billion in the 4-year period.

Direct losses due to FHB are manifested as reduced yield due to FDK or spikelet sterility, low test weight of harvested grain, costs incurred to remove FDK from grain, and price discounts at grain elevators (points of sale and storage) due to DON contamination. In addition, germination of *Fusarium*-infected grain can be reduced to unacceptable levels (Argyris et al., 2003) and if the grain is used as seed for the next season's crop, seedling blights and root and crown rots can significantly reduce stand establishment and subsequently yield. In years in which severe and widespread FHB epidemics occur, there can be scarcity of seeds needed for planting the next season's crop. Indirect losses result from poor-quality food products (e.g. bread and beer) made from *Fusarium*-damaged and DON-contaminated grain, reduced productivity of livestock due to toxicosis or feed refusal, costs incurred in the treatment of toxicosis in humans and animals, and costs associated with fungicide seed treatment to control seedling and root and crown rot diseases. Other indirect losses result from loss of profits or failure of businesses related to grain production and marketing, since the economic linkages and activities associated with crop production (e.g. tillage, planting, and harvesting) are covered for the most part through revenues that producers receive from crop sales (Nganje et al., 2001). Worldwide, losses due to FHB can be staggering in years when environmental conditions favor disease development.

2 Life cycle

FHB is primarily a monocyclic disease (Fernando et al., 1997). Only one cycle of infective inoculum is produced per growing season. After initial infection of spikes, symptoms progress on individual spikes, but there is little or no secondary infection of the same spike or other spikes. The life cycle of *F. graminearum* consists of the following phases: survival during intercrop periods, dispersal during the crop-growing season and infection of the crop culminating in disease. Several factors can influence the life cycle, including environmental factors (moisture, temperature and relative humidity (RH)) and disease management strategies and tactics (e.g. crop rotation, tillage type, host resistance, and chemical and biological disease control).

2.1 Survival

F. graminearum survives as a necrotroph on living plants or as a saprophyte on crop residues in the soil or on the soil surface. It survives as mycelium, perithecia, or chlamydospores (Khonga and Sutton, 1988; Pereyra and Dill-Macky, 2004). In a field experiment, Khonga and Sutton (1988) inoculated maize and wheat plants with *F. graminearum* to produce infested residues (stems, ears, spikelets, and grains) which they then placed on, above, or in the soil during the autumn.

They retrieved samples over a period of three years and assessed them for production of perithecia and macroconidia. Both types of inoculum were produced on the residues placed on or above the soil for up to three years, but not on buried residue. Most of the inoculum was produced in the first two years, and macroconidia were produced mostly during the first year. Summerell and Burgess (1988) found *F. graminearum* to survive for at least 104 weeks on wheat residue left on the soil surface or incorporated into the soil, but not on buried residue.

Pereyra and Dill-Macky (2008) examined the presence of *Fusarium* spp. in residues of various crops common in wheat and barley cropping systems. The residues were collected from no-tillage and reduced-tillage field plots and included residues of wheat, barley, maize, sunflower, pasture, and gramineous weed species. *F. graminearum* was recovered from most of the residues, but most frequently from wheat and barley residues. Populations of the pathogen declined over time in all residues examined, and it colonized wheat and barley residues longer in no-till than in reduced tillage production systems. In a separate study, Pereyra and Dill-Macky (2004) found that survival of *F. graminearum* on wheat residue was related to the rate of residue decomposition. Survival on node tissues decreased over time and the rate of colonization of node tissues on the soil surface was twice the rate of colonization of buried node tissues. Inch and Gilbert (2003) investigated the survival of *F. graminearum* in *Fusarium*-damaged wheat kernels (FDK) under field conditions. The pathogen was isolated at a frequency from 85% to 99% after 24 months from FDK left on the soil surface or buried at 5 cm or 10 cm and perithecia formed on the FDK in all treatments.

In addition to cereals, *F. graminearum* has been isolated from many wild or weedy gramineous and non-gramineous hosts (Lofgren et al., 2018; Mourelos et al., 2014; Turkington et al., 2011). Under controlled environment conditions, Fulcher et al. (2019) compared the ability of *F. graminearum* to overwinter on the senesced stems of spring wheat (*Triticum aestivum* L.) and five grasses common to cereal production sites in the state of New York, USA. The grasses were *Bromus inermis,* Leyss (smooth brome), *Dactylis glomerata* L. (orchardgrass), *Lolium perenne* L. (perennial ryegrass), *Phalaris arundinacea* L. (reed canarygrass), and *Elymus repens* (L.) Gould (quackgrass). The artificially infested stems of both the spring wheat and the grasses retained *F. graminearum* over two years, but the retention was higher in the first year (25–100%) than in the second year (0–40%). Ascospores and macroconidia of *F. graminearum* were produced to varying levels on all six species. The survival and reproduction of *F. graminearum* on wild grasses suggests that these grasses can serve as inoculum reservoirs that can contribute to the initiation and severity of cereal disease epidemics incited by the pathogen.

2.2 Inoculum development and release

Crop residue is the primary source of *F. graminearum* inoculum (Atanasoff, 1920; Andersen, 1948). Infections of cereals by the pathogen are incited by ascospores or macroconidia produced in perithecia and sporodochia, respectively, on crop residue. Perithecia begin forming on cereal crop residue in the autumn, but ascospores in the perithecia mature the following spring or summer (White, 2016). Under controlled environment conditions, Dufault et al. (2006) found that moisture levels of −0.45 MPa and −1.30 MPa and a temperature range from 16°C to 24°C promoted the production and development of perithecia on inoculated maize stalks incubated in growth chambers set at a16-hour daily photoperiod. In contrast, moisture levels of −2.36 MPa and −4.02 MPa and temperatures of 12°C and 28°C limited or slowed down perithecial production and development. Manstretta and Rossi (2016) found that perithecia were produced on maize stalks at temperatures ranging from 5°C to 30°C with an optimum temperature of 21.7°C, but matured only at 20°C and 25°C. Perithecium production occurred when relative humidity (RH) was 75% or higher, but maturation occurred only when RH was 85% or higher. Both perithecium production and maturation increased with RH. Macroconidia are produced copiously in slimy masses borne on sporodochia on crop residue or on the surface of infected plants during wet weather (Trail, 2009). Temperature requirements for macroconidia production are similar to those for perithecia and ascospore formation with a temperature range of 16-36°C and an optimum temperature of 32°C (Andersen, 1948). In in vitro experiments conducted by Rossi et al. (2002), *F. graminearum* macroconidia were produced over a temperature range of 10–35°C with an optimum temperature of 32°C.

Ascopores of *F. graminearum* are discharged forcibly from perithecia (Trail et al., 2005, Schmale et al., 2005a). Trail et al. (2005) found that ascospores of *F. graminearum* were discharged with a launch speed of 34.5 m s^{-1}, which required a pressure of 1.54 MPa and an acceleration of 870 000 g. David et al. (2016) investigated the effects of temperature and RH on ascospore release and discharge distances of *F. graminearum* in controlled environment chambers. The number of discharged ascospores increased with RH, with the highest number discharged at 15°C and 95% RH compared to 25°C and lower RH levels of 60% and 75%. Ascospores were discharged to distances ranging from 0.5 mm to longer than 12 mm, peaked in distance traveled at 4-5 mm, and traveled farther with increasing RH and at 25°C compared to 15°C. Schmale et al. (2005a) observed ascospore discharge distances of <1 mm to nearly 10 mm with an average of 4.6 mm from 6-day-old perithecia of *F. graminearum*. Macroconidia are liberated by rain splash or wind-driven rain (Sutton, 1982).

2.3 Dispersal

Ascospores are the major source of airborne inoculum (Andersen, 1948; Manstretta et al., 2015; Markell and Francl, 2003). Their forcible discharge from perithecia and the rain splash liberation of macroconidia from spordochia enable the spores to surpass the thin laminar boundary layer of air surrounding crop residue or living plants and become airborne. The ascospores can then be transported over short distances within a field by air currents and wind or over long distances by wind when they escape the crop canopy. Macroconidia usually are transported over short distances by rain splash (Sutton, 1982), but can also become airborne and be transported over long distances (Markell and Francl, 2003). The curved nature of macroconidia and their production in slimy masses borne on sporodochia are considered to be adaptations for rain splash dispersal (Deacon, 2006; Trail, 2009). Paul et al. (2004) investigated rain splash dispersal of *F. graminearum* in wheat fields with maize residue or *F. graminearum*-infested maize kernels on the soil surface. Both macroconidia and ascospores were recovered from splash samples collected at 0, 30, and 100 cm (height of wheat spikes) above the soil surface. Spore flux density was higher at 0 cm and 30 cm compared to 100 cm and in wheat fields with *F. graminearum*-infested maize kernels than those with maize residue.

Prussin et al. (2014) monitored the long-distance transport of *F. graminearum* from a field-scale source of inoculum (0.372 ha of an inoculated wheat field). They recovered a clonal isolate of the pathogen at distances of up to 750 m from the source. Using remote piloted vehicles fitted with spore sampling devices, Maldonado-Ramirez et al. (2005) collected, over a period of 4 years, 12 858 viable spores of *F. graminearum* at 60 m above the ground during 158 flights conducted during the 4 years over agricultural fields during May and June in the state of New York, USA. In Virginia, USA, Schmale III et al. (2012) used autonomous unmanned aerial vehicles to collect 11 isolates of *F. graminearum* at heights from 40 m to 320 m above ground level. All 11 isolates caused FHB on wheat and produced trichothecene mycotoxins. The abundance of viable spores of *F. graminearum* in the atmosphere as demonstrated in these studies suggests that long-distance transport of the pathogen can play an important role in the development of disease epidemics on a regional scale.

To initiate disease, airborne spores must be deposited on the plant surface followed by infection. Schmale III et al. (2005b,c) assessed the spatial patterns of viable spore deposition of *F. graminearum* in rotational wheat and maize fields lacking cereal residue. During 54 day and night sampling periods in 68 field environments over 2 years in wheat fields, 93% of the spore deposition patterns were random and 7% were aggregated. During 27 day and night sampling periods over 2 years in maize fields, 92% of spore deposition patterns at the silk and tassel levels were random and the remaining 8% were aggregated. In both wheat and maize fields, all the aggregated patterns were observed at night, and this was attributed to the stable atmosphere during the night, which may be favorable to the settling of uneven clouds of spores. Manstretta et al. (2015) also found random deposition patterns of ascospores of *F. graminearum* within a wheat canopy when they placed spore traps at heights of 10, 30, 60, and 90 cm.

Schmale III et al. (2006) examined the temporal patterns of spore deposition of *F. graminearum* in rotational wheat fields without cereal residue and found that out of a total of 73 day and night sampling periods over three years, the majority (94%, 86%, and 82%) of spore deposition occurred at night. They attributed the high spore deposition during the night to the rapid cooling of the earth's surface compared to the atmosphere, a slowdown in turbulent mixing and a downward transfer of heat, resulting in an inversion layer and a subsequent gravitational settling of the spores from the upper atmosphere. Conversely, the low daytime spore deposition was explained by vertical mixing by turbulent air currents, which may have transported the spores high into the atmosphere. Temporal patterns of spore deposition were identical over all landscape environments sampled, which suggested the deposition of spores over kilometer distances from a uniformly mixed atmospheric inoculum source.

2.4 Infection and symptom spread

In wheat, FHB infections occur after spike emergence. Spikes are most susceptible to infection from anthesis through the soft dough stage partly because the anthers contain stimulants for spore germination and pathogen growth (Schroeder and Christensen, 1963; Strange and Smith, 1971). During warm temperatures (from 25°C to 30°C) and wet conditions, blight symptoms develop within 2–4 days after infection. Primary inoculum consists mainly of ascospores produced in perithecia, which form on crop residues in the autumn and spring in the presence of moisture and as temperatures rise in the spring. Ascospores or macroconidia are released from crop residues and spread by wind or rain splash. When they land on spikes during wet, warm weather, they germinate to form mycelium which extends to points of entry to the spike's interior: the stomates, anthers, and crevices between the palea and lemma (Lewandowski and Bushnell, 2001).

Pritsch et al. (2000) found that following spray inoculation of wheat spikes with macroconidia of *F. gramimearum*, the macroconidia germinated within 6–12 hours after inoculation (hai) on the inoculated (abaxial) surface of glumes. Fungal mycelium penetrated the glumes through stomatal pores into the substomatal chambers. There was no evidence of direct penetration of epidermal cells by hyphae growing on the glume's surface. The mycelium exhibited subcuticular growth along stomatal rows followed by colonization of glume parenchyma cells and sporulation within 48–76 hours. Intracellular growth of hyphae was observed within 48–76 hai in the epidermis and parenchyma tissue. Colonization was similar between the resistant cultivar Sumai 3 and the susceptible cultivar Wheaton.

Brown et al. (2010) studied the pathways of spikelet to spikelet colonization in wheat ears of the FHB-susceptible cultivar Bobwhite following point inoculation of the two middle spikelets. At 2 days post inoculation (dpi), the first microscopic sign of fungal colonization was observed in the palea and was limited to subcuticular and intercellular hyphal growth. By 5 dpi, the hyphae had colonized all cell types of the palea and the lower non-lignified parenchyma had lost cell integrity, and by 12 dpi, the parenchyma layer had collapsed. Intercellular hyphae colonized the rachilla cortex by 5 dpi, then intracellular lateral hyphal growth penetrated the sclerenchyma, ruptured the epidermis, and grew on the surface. By 12 dpi, the cortex and phloem collapsed. The rachis was initially colonized through vertical intercellular and lateral intracellular hyphal growth. Intercellular hyphae grew from the rachilla of the inoculated spikelet and entered the node cortex and colonized its vasculature by 5 dpi. The hyphae then progressed vertically down the node to the rachis. Hyphae also grew laterally and intracellularly from the rachis node to the rachis through the pit fields of the sclerenchyma cells that separate the two tissues. Once in the rachis, hyphal growth became vertical and colonized the rachis above and below the inoculated spikelet through intercellular growth in the cortex and intracellular growth in the vasculature. This vertical hyphal growth in the rachis was predominant at 5 dpi. Uninoculated spikelets above and below the inoculated spikelet were colonized via intercellular lateral growth of hyphae that progressed beyond the sclerenchyma cells separating the rachis and the rachis node. Colonization was rapid between 2 dpi and 5 dpi, by which time hyphal infection was observed in the third rachis node below the inoculated spikelet, but symptoms (bleaching and necrosis) were confined to the inoculated spikelet. Beyond 5 dpi, symptom development progressed and by 12 dpi the entire spike was bleached.

Boenisch and Schäfer (2011) investigated, following inoculation with macroconidia, the infection strategy of *F. graminearum* on detached floral organs (caryopses, paleas, lemmas and glumes) of susceptible and moderately resistant wheat cultivars. They identified three stages of infection. In stage I (initial colonization), germination of macroconidia occurred within 6–12 hours post inoculation (hpi) and hyphal growth spread on the surface of the organs. Short infection hyphae developed, indicating direct penetration. Homogeneous hyphal networks formed until 1–2, 4–5, and 6–7 days post inoculation (dpi) on caryopses, paleas, and lemmas and glumes, respectively. After these periods, hyphal morphology became heterogeneous. No symptoms were observed during stage I of infection.

In stage II (main infection), runner hyphae branched at high frequency and formed foot structures, lobate appresoria and infection cushions surrounded by necrotic plant cells. The lesions appeared starting from 3, 5, and 7 dpi on caryopses, paleas, and lemmas and glumes, respectively. During this stage, a specific *TRI5* (trichothecene mycotoxin) induction was observed in infection cushions but not in runner hyphae. Direct penetration of epidermal cells by infection structures was observed. Necrosis and chlorosis developed rapidly during this stage. In stage III (final infection), the entire tissue was necrotic and aerial hyphae and sporodochia were formed. All three infection stages were similar between a resistant cultivar (Sumai 3) and a susceptible cultivar (Nandu).

On intact plants following point inoculation of spikelets, all three stages of infection were observed. Stage I: macroconidia germinated within 6–12 hpi. Stage II: infection cushions developed on caryopses, paleas, and lemmas between 3 dpi and 4 dpi. Stage III: aerial hyphae were formed starting from 4 dpi and tissues were completely necrotic between 5 dpi and 7 dpi. Colonization of glumes on intact plants did not occur within 7 dpi. All three stages of infection in intact plants were similar between a moderately resistant cultivar (Amaretto) and the susceptible cultivar Nandu.

3 Mycotoxin production

During their interaction with cereal plant hosts, *F. graminearum* and other *Fusarium* species in the FGSC produce mycotoxins, which contaminate grain and are harmful to humans and animals (Desjardins, 2006). The most abundant and frequently detected mycotoxin in cereal grains is deoxynivalenol (DON), a Type B trichothecene produced by *F. graminearum* and other members of the FGSC (Foroud et al., 2019). Recently, *F. graminearum* isolates collected during a survey of the pathogen in North America produced a new Type A trichothecene mycotoxin, NX-2 and its deacylated form, NX-3 (Varga et al., 2015). To assess the geographic distribution of NX-2 isolates, Kelly et al. (2016) used a PCR-RFLP assay targeting variation in the *TRI1* gene to survey 2515 *F. graminearum* isolates from 19 countries. They found that isolates with the NX-2 allele were limited to southern Canada and the northern USA where they were observed at a low frequency.

The acetylated derivatives of DON, 3-ADON, and 15-ADON, have been frequently detected in DON-contaminated grain (Mirocha et al., 2003). Isolates of *F. graminearum* that produce only 3-ADON have the 3-ADON chemotype; those that produce only 15-ADON have the 15-ADON chemotype (Miller et al., 1991). Trichothecene mycotoxins play a role in *F. graminearum* aggressiveness or virulence on wheat. Desjardins et al. (1996) showed that a trichothecene non-producing mutant of *F. graminearum* was less virulent than a wild-type, trichothecene-producing strain. Hernandez Nopsa et al. (2014) and Panthi et al. (2014) found isolates of *F. graminearum* with the 15-ADON chemotype collected from wheat grain elevators (points of sale and storage of grain) and fields in Nebraska, USA, to vary considerably in the amount of DON they produced in vitro and in grain on wheat spikes. In greenhouse studies, these isolates differed in aggressiveness on wheat spikes. Those that produced higher DON concentrations were more aggressive than those that produced lower concentrations of the mycotoxin. Using isolates of *F. graminearum* and *F. boothii* (a member of the FGSC) more recently collected from Nebraska wheat fields, Valverde-Bogantes et al. (2020) similarly demonstrated in a greenhouse study that in each species, the most aggressive isolate produced the most DON and the least aggressive isolate produced the least DON. Puri and Zhong (2010) demonstrated that *F. graminearum* isolates with the 3-ADON chemotype were more aggressive and produced more DON in grain than isolates with the 15-ADON chemotype. These observations suggest that regardless of chemotype, *F. graminearum* isolates in a given population differ in aggressiveness and the ability to produce DON. The presence of the NIV chemotype in Louisiana and Arkansas, USA (Gale et al., 2011; Horevaj et al., 2011) and the recently discovered NX producers (Varga et al., 2015) shows that *F. graminearum* populations in North America are complex and diverse.

4 Influence of environmental factors on disease development

The environmental factors that mostly influence the development of FHB of small grain cereals are temperature, moisture, and RH. These factors are in turn influenced by climate change, which can cause excessive rainfall or drought (Miller, 2008). Andersen (1948) conducted elaborate experiments on the effects of temperature and moisture on FHB development in spring wheat. Plants inoculated with macroconidia of *F. graminearum* at the past anthesis growth stages (BBCH 73–85, Lancashire et al., 1991; Zadoks et al., 1974) were incubated at temperatures in the range from 15°C to 30°C and were concomitantly exposed to continuous wetness for periods of 24–60 hours. The optimum temperature for FHB development was found to be 25°C regardless of the wetness duration. Higher disease severity (percent of spikelets killed) resulted from exposure to longer periods of wetness.

In a study to determine the effects of post-anthesis moisture on DON accumulation in soft red winter wheat under field conditions, Cowger et al. (2009) inoculated FHB-moderately resistant or susceptible cultivars with macroconidia of *F. graminearum* at mid anthesis (50% of spikes with extruded anthers) and then subjected them to different durations of mist ranging from 0 to 30 days. Misting between 10 days and 20 days significantly increased FHB and DON compared to no misting in both moderately resistant and susceptible cultivars. Climatic data analyzed in relation to DON accumulation in winter wheat over a 7-year period revealed a strong positive correlation between the number of days with a mean RH above 80% and mean annual DON content in grain (Chandelier et al., 2011).

The relationships between environmental factors and biological variables (FHB intensity, fungal biomass, and DON) associated with FHB in wheat were analyzed using measurements obtained from 150 location-years originating in three European countries (Kriss et al., 2012). Results showed that from 18 days before anthesis to harvest, moisture-related variables (average RH, hours of RH above 80%) were highly correlated with biological variables. There were significant semi-partial correlations between RH variables and DON in harvested grain. In a similar study conducted in the USA with longer window lengths (time of planting to crop maturity), it was determined that moisture- or wetness-related variables (daily average RH, total daily precipitation) were positively correlated with FHB intensity (Kriss et al., 2010).

5 Influence of management practices on disease risk

Krupinsky et al. (2002) reviewed the influence of management practices on the incidence and severity of plant diseases including FHB. Management practices can significantly influence the risk level of FHB and other cereal diseases caused by *F. graminearum*. These practices include tillage, crop rotation, choice of cultivar planted, irrigation management, and chemical and biological control. They influence disease risk by increasing or decreasing the amount of inoculum available to initiate and sustain disease or an epidemic during favorable environmental conditions. The greater the amount of inoculum, the higher the disease risk. In a greenhouse experiment, Stein et al. (2009) inoculated spring wheat plants at anthesis with increasing concentrations of *F. graminearum* macroconidia. They found that FHB incidence and severity sharply increased with increasing inoculum concentration. In field experiments, Keller et al. (2011) monitored the spread of a released clone of *F. graminearum* from small plots with increasing amounts of infested maize stalks. The intensity of recovery of the clone and the distance over which it was recovered were greater from plots with larger amounts of infested residue, indicating that larger amounts of inoculum posed a higher risk for disease development and spread.

5.1 Tillage and crop sequence

F. graminearum survives on residues of maize, wheat, barley, and other cereals (Sutton, 1982; Parry et al., 1995). Field studies have demonstrated that tillage to bury host crop residues and crop rotation with non-hosts reduce FHB intensity and DON accumulation in cereal grains. Dill-Macky and Jones (2000) found that FHB intensity was highest when wheat followed maize and least when wheat followed soybeans (*Glycine max*). They also demonstrated that FHB intensity in moldboard-plowed plots was lower than that in chisel-plowed or no-till plots. In the same study, DON concentration in a soybean-wheat rotation was 25% lower than in a wheat-wheat rotation and 49% lower than in a corn-wheat rotation. By serving as a source of primary *F. graminearum* inoculum, cereal crop residues contribute to atmospheric inoculum, which is transported over long distances and can contribute to the development of FHB epidemics on a regional scale. Therefore, cereal crop residues play a major role in the initiation and development of both local and regional FHB epidemics (Keller et al., 2014).

5.2 Choice of cultivar

Under favorable environmental conditions during the growing season, FHB intensity and DON accumulation in grain will be higher in susceptible compared to moderately resistant cultivars (Wegulo et al., 2011; Wiersma et al., 1996). If minimum or reduced tillage is practiced, the crop residue from a susceptible cultivar will have a greater amount of fungal biomass and therefore will provide a higher concentration of primary inoculum during the following growing season than the crop residue from a moderately resistant cultivar. Therefore, choice of cultivar can influence the severity of an epidemic during both the current and the following growing season. Planting a susceptible cultivar increases disease risk whereas planting a moderately resistant cultivar reduces disease risk.

Some agronomic traits in small grain crops are associated with passive resistance to FHB. Mesterházy (1995) found that under natural conditions, short phenotypes of winter wheat were more severely affected by FHB than tall phenotypes. This can be explained by a difference in the microclimate around the wheat spikes between the short and tall phenotypes. In tall phenotypes, exposure of the spikes to wind and air currents is greater, which results in faster drying following night dew or rain events, and this can significantly reduce disease progression.

Cultivars that are prone to lodging are at a higher risk for FHB and DON. Langseth and Stabbetorp (1996) showed that DON content in grain from the lodged sections of fields (1369 ppb) was nearly twice that from the non-lodged sections (745 ppb). Nakajima et al. (2008) investigated the effects of lodging on DON and NIV accumulation in wheat, barley, and rice naturally or artificially infected with *F. graminearum*. They found the concentrations of DON and NIV to be significantly higher in grain from lodged compared to non-lodged plants. These studies indicate that lodging can significantly increase DON concentration in small grain cereals. The increase in DON due to lodging can be attributed to a moist and more humid environment near the soil surface compared to higher locations in the plant canopy where air circulation limits the duration of free moisture and reduces RH. In addition to moisture and humidity, spikes on lodged plants are closer to pathogen inoculum on the soil surface, especially in no-till fields that have abundant crop residue. This can lead to more infections per spike and therefore higher FHB severity and DON accumulation in grain.

Cleistogamous (closed flowering) cultivars of barley were shown to be more resistant to FHB than chasmogamous (open-flowering) cultivars (Yoshida et al., 2005, 2007, 2008). Where cleistogamous barley cultivars are available and adapted with desirable agronomic traits, planting them can help to reduce FHB and DON. Although wheat is mostly chasmogamous and extrudes anthers at flowering, Kubo et al. (2010) hypothesized that the risk of FHB infections is lower in cleistogamous than in chasmogamous wheat cultivars. They produced recombinant inbred lines (RILs) and showed that there was less initial FHB infection in cleistogamous compared to chasmogamous RILs.

Because the majority of FHB infections occur during flowering (Andersen, 1948), if several cultivars of a small grain cereal crop with different flowering dates are planted, disease risk is reduced because some may escape infections depending on the timing of favorable environmental conditions for infection. Disease risk is lower for cultivars with shorter flowering periods than those with longer flowering periods (Krupinsky et al., 2002; Wegulo et al., 2015).

5.3 Irrigation management

Sufficient moisture is a requirement for the normal development of small grain cereals and other crops. In growing seasons with insufficient rainfall, some growers use irrigation as a source of moisture. However, excessive irrigation before, during, and after anthesis in wheat or barley can significantly increase the risk of FHB development and DON accumulation in grain. Research has shown that increasing the amount of moisture following anthesis increases FHB intensity, FDK and DON. In field experiments conducted under irrigated and rain-fed environments in a wet and a dry growing season, Bolanos-Carriel et al. (2020) found that in a moderately resistant wheat cultivar, FHB index, FDK, and DON were higher and yield was lower under irrigated conditions and in the wet season compared to rain-fed conditions and the dry season. Cowger et al. (2009) found that in soft red winter wheat, 10 or 20 days of post-anthesis mist significantly increased FHB intensity, FDK, DON, and FDK compared to no mist. Some investigators have reported a reduction in DON due to extended post-anthesis irrigation (Culler et al., 2007; Gautam and Dill-Macky, 2012, Lemmens et al., 2004), indicating a benefit from post-anthesis moisture. However, this benefit of DON reduction is minimal when compared to the loss in yield and grain quality that results from excessive moisture before, during, and after anthesis. Avoiding irrigation during the flowering and grain filling periods reduces disease risk. When dry conditions dictate that the cereal crop must be irrigated during these periods, managing irrigation to allow the crop canopy to dry between irrigations will reduce disease risk.

5.4 Chemical and biological control

Fungicide application is one of the management tactics for FHB. Demethylation inhibitor (DMI) fungicides are the most effective and most widely used (McMullen et al., 2012; Paul et al., 2018). Applying DMI fungicides significantly reduces FHB, DON, FDK, and yield loss (Bolanos-Carriel et al., 2020; Paul et al., 2018; Wegulo et al., 2011). Because applying a fungicide reduces fungal biomass by killing the pathogen, less inoculum will form on the residue, which reduces disease risk during the following growing season. A similar effect will be achieved with biological control agents (BCAs). Optimal timing for fungicide and BCA application is from early anthesis up to 6 days later (Bolanos-Carriel et al., 2020). The BCA *Clonostachys rosea*, in addition to reducing FHB intensity, FDK, and DON, significantly reduces perithecial production on crop residues in the field (Xue et al., 2009; Xue et al., 2014).

6 Predicting *Fusarium* head blight epidemics

Forecasting systems use input variables such as weather and disease resistance levels of wheat varieties to calculate the risk of disease outbreaks. Hence, growers can deploy management tactics such as fungicide application in a timely manner and only when necessary. De Wolf et al. (2003) used non-parametric correlation analysis and stepwise logistic regression analysis of environmental data collected at 50 location-years in four states representing three wheat production regions in the USA to identify combinations of weather variables and their occurrence and duration relative to anthesis that accurately predicted FHB outbreaks in the field. The most useful predictor variables were the duration (hours) of rainfall 7 days before anthesis, the duration (hours) when the temperature was between 15°C and 30°C 7 days before anthesis, and the duration (hours) when the temperature was between 15°C and 30°C and RH was equal to or greater than 90%. These variables were incorporated into models that predict FHB epidemics in the USA with over 75% accuracy (Fusarium Risk Tool, http://www.wheatscab.psu.edu/).

Shah et al. (2013) developed 15 logistic regression models for predicting major FHB epidemics using weather-based variables in pre- and post-anthesis time windows. Except for one rainfall-based predictor, the best weather-based predictors identified and used in these models were all derivatives of RH or temperature. The models had indicator variables for cultivar resistance, presence or absence of maize residue and wheat type (winter or spring). Their predictive performance was improved by adding up to four weather-based predictors, and they had an average test misclassification rate of 19% lower than that of the models deployed in the Fusarium risk tool at that time.

Shah et al. (2014) investigated the utility of boosted regression trees (BRTs) for predictive modeling of FHB epidemics in the USA and compared the performance of BRTs to that of the 15 previously developed logistic regression models mentioned in the preceding paragraph. They found that the misclassification rate of BRTs was 31% lower than the average misclassification rate of the 15 logistic regression models. The strongest predictors of FHB using BRTs were mean daily RH, mean daily temperature and the number of hours during which RH was 90% or greater and temperature was between 9°C and 30°C simultaneously. The predicted risk of major FHB epidemics significantly increased when mean daily RH exceeded 70%, a threshold that is lower than in most previously developed plant disease prediction models. The investigators noted, however, that whereas the predictive performance and sensitivity of BRTs were better than those of logistic regression models, fitting BRTs is computationally much more expensive than fitting logistic regression models, which presents a challenge in deploying the BRTs in a real-time, rapid-update online system such as the Fusarium risk tool which is currently deployed in the USA.

The DONcast® model (http://www.weatherinnovations.com/doncast.cfm) is used in Canada to predict DON accumulation in wheat at full spike emergence, enabling growers to make timely fungicide application decisions. The model uses weather variables but also incorporates cultivar, crop rotation, tillage type, and spike emergence date. In Switzerland, the FusaProg information system (Musa et al., 2007) assesses the risk of *Fusarium* infection and DON contamination in wheat using weather variables, crop rotation, tillage type, and cultivar. Use of these models can save growers time and money by preventing unnecessary application of fungicides and they significantly reduce losses due to FHB and DON by enabling growers to make informed and timely fungicide application decisions in FHB-favorable growing seasons.

7 Detection and identification

Accurate detection and identification of plant pathogens is critical for the correct diagnosis of diseases, which enables the deployment of timely, effective management strategies and tactics. The FGSC contains at least 16 species that cause diseases on cereal and other crops. Traditional methods of identification rely on growing the pathogen in culture and examining several characteristics including spore morphology, spore size, and pigment production (Leslie and Summerell, 2006). However, because of the close phylogenetic relatedness and high similarity among them (Aoki et al., 2012), it is extremely difficult to distinguish between members of the FGSC using traditional methods.

Molecular assays have been developed that accurately identify species in the FGSC based on genetic differences. Polymerase chain reaction (PCR), in which a nucleic acid sequence is amplified exponentially in vitro, was first published by Mullis and Faloona (1987) and has been used widely in the detection and identification of microorganisms (Kuzdraliński et al., 2017; Niessen and Vogel, 2010). Ward et al. (2008) developed, through a multilocus genotyping (MLGT) assay, a multiplex PCR that enabled the simultaneous identification of large numbers of species and trichothecene chemotypes of FHB-causing *Fusarium* isolates collected from cereal grain producers in Canada. This was achieved by developing 41 oligonucleotide probes targeting species or trichothecene chemotype-specific variation within six genes, based on differences in nucleotide sequences identified among 53 B-trichothecene-producing isolates. Recently, Hafez et al. (2020), designed, based on the DNA sequence of the translational elongation factor 1α (*TEF1α*) gene, a PCR restriction fragment length polymorphism (RFLP) assay that can accurately and rapidly differentiate *F. graminearum* sensu stricto from all other *Fusarium* spp.

Because PCR requires expensive lab equipment, trained personnel, and agarose gel electrophoresis (AGE) to separate products and visualize them, it is time-consuming and costly (Niessen, 2015). An alternative method for the detection and identification of *Fusarium* spp. that is rapid and less costly has been developed. Loop-mediated isothermal amplification (LAMP) was described by Notomi et al. (2000). It amplifies DNA rapidly, efficiently, and with high specificity under constant temperature. Unlike PCR, it can be deployed on-site, such as in a crop field, to provide rapid diagnosis. Niessen and Vogel (2010) developed a LAMP assay for the specific detection of *F. graminearum* based on the *gooA* (galactose oxidase) gene of the pathogen. The assay was optimized to be rapid, sensitive, and specific and detected the presence of less than 2 pg of purified target DNA per reaction within half an hour. Using the assay, it was possible to detect *F. graminearum* from fungal cultures, barley grains, and bulk samples of wheat grains. Several studies have utilized LAMP to detect FHB-causing *Fusarium* spp. (Abd-Elsalam et al., 2011; Denschlag et al., 2014; Gupta et al., 2020; Komura et al., 2018).

8 Disease assessment

Assessment of FHB is usually done approximately three weeks after anthesis, but this period can be shorter if disease progression is fast due to favorable environmental conditions. FHB can be quantified visually as disease incidence (percentage of symptomatic spikes), disease severity (percentage of symptomatic spikelets on a spike), or index, which is an estimate of the overall FHB intensity in a field. Two types of index have been used in quantifying FHB. The first one is a product of severity on symptomatic spikes and incidence

(Shaner, 2003). The second one is defined as the percentage of symptomatic spikelets per spike on all spikes sampled including non-symptomatic spikes, that is, field or plot level severity (Paul et al., 2005b). Incidence and severity are usually determined using an arbitrary number of spikes per field plot. These spikes can be randomly selected within each plot or from an arbitrary number of spots in a plot (Bolanos-Carriel et al., 2020; Wegulo et al., 2011). Using 10 data sets of FHB incidence and severity collected from wheat fields, Paul et al. (2005a) found a highly significant and consistent relationship between incidence and severity. Estimation of severity was more precise at lower than at higher incidence values. Incidence is much easier to measure than severity, which is time-consuming and labor-intensive to measure. During surveys of many fields, incidence can be used to estimate severity, thereby saving time and resources. Index and yield, yield and FDK, and yield and DON, are negatively correlated whereas index and FDK, index and DON, and FDK and DON are positively correlated. Bolanos-Carriel et al. (2020) and Wegulo et al. (2011) showed that these correlations were stronger in susceptible compared to moderately resistant winter wheat cultivars and in FHB-favorable compared to unfavorable conditions during the growing season. In FHB-favorable conditions, index and DON, and index and FDK were strongly and positively correlated, which confirmed the usefulness of index as a measurement of FHB (Bolanos-Carriel et al. (2020).

The most commonly used method to quantify *Fusarium*-damaged kernels (FDK) is to visually separate FDK from healthy kernels in grain samples and then calculate the percentage of FDK. FDK has also been estimated using near-infrared spectroscopy. Dowell et al. (1999) and Peiris et al. (2010) used a single kernel near-infrared (SKNIR) system to sort FDK from visually sound kernels. The SKNIR system classified FDK and visually sound kernels with a near 100% accuracy. Wegulo and Dowell (2008) evaluated the ability of the SKNIR system to detect FDK by comparing FDK sorted by the system to FDK sorted visually by two human raters. Visual sorting was strongly correlated with sorting by the SKNIR system ($0.89 \leq r \leq 0.91$), but the SKNIR system had a wider range of FDK detection and was more consistent. Compared to the SKNIR system, visual raters overestimated FDK in samples with a low percentage of *Fusarium*-damaged grain and underestimated FDK in samples with a high percentage of *Fusarium*-damaged grain.

Various chromatographic methods have been used to measure DON. They include liquid chromatography (LC) with mass spectrometric detection, LC-diode array detection, gas chromatography with electron capture detection (GC-ECD), and gas chromatography with mass spectrometry (GC-MS) (Neuhof et al., 2009; Tacke and Casper, 1996). Enzyme-linked immunosorbent assay (ELISA) kits are also used to measure DON (Han et al., 2020). However, caution should be exercised when interpreting results from ELISA kits, because they have a linear range of quantitation that may be too narrow and values outside this range are extrapolations based on linear regression (Panthi et al., 2014). LC, GC, and ELISA require grain to be ground to flour. Non-destructive methods that leave kernels intact have been developed to estimate DON in grain. They include near-infrared spectroscopy (Dowell et al., 1999; Peiris et al., 2010; Peiris et al., 2017) and hyperspectral imaging (Barbedo et al., 2017; Alisaac et al., 2019). These methods only predict, but do not measure the actual amount of DON present. They are suitable in situations where investigators such as plant breeders want to know whether wheat kernels of a specific breeding line have high or low DON, and want to preserve the wheat kernels for advancing lines with low DON in the breeding program. For accurate measurements of DON, chromatographic methods are recommended.

9 Future trends in research

To enhance our knowledge of the epidemiology and management of *Fusarium* diseases in cereals, research is needed to improve the accuracy of disease and mycotoxin prediction systems that are currently deployed in various parts of the world. A new area of research that has gained interest during the last decade is the aerobiology of *F. graminearum*. Isolates of the pathogen recovered from the atmosphere at heights of 40 m to over 300 m have caused FHB and produced mycotoxins in wheat. Knowledge on how widespread the pathogen is in the atmosphere, how long it can survive there, and the extent to which atmospheric inoculum contributes to local and regional epidemics will be invaluable. Similarly, it will be useful to know the amount and extent of inoculum produced on non-cultivated hosts like wild grasses and how much this inoculum contributes to disease outbreaks in cereal crops. New detection methods that are quick, efficient and deployable on-site will enable rapid and accurate diagnosis, which will lead to timely deployment of disease management measures.

10 Where to look for further information

The following books provide a good overview of the topic covered in this chapter.

- Fusarium Head Blight of Wheat and Barley, edited by Kurt J. Leonard and William R. Bushnell. The American Phytopathological Society (2003).
- Fusarium Head Blight in Latin America, edited by Teresa M. Alconada Magliano and Sofía Noemí Chulze. Springer (2013).
- Fusarium Mycotoxins: Chemistry, Genetics, and Biology by Anne E. Desjardins. The American Phytopathological Society (2006).

11 References

Abd-Elsalam, K., Bahkali, A., Moslem, M., Amin, O. E. and Niessen, L. (2011). An optimized protocol for DNA extraction from wheat seeds and Loop-Mediated Isothermal amplification (LAMP) to detect Fusarium graminearum contamination of wheat grain, *Int. J. Mol. Sci.* 12(6): 3459–3472.

Alisaac, E., Behmann, J., Rathgeb, A., Karlovsky, P., Heinz-Wilhelm Dehne, H. W. and Mahlein, A. K. (2019). Assessment of Fusarium infection and mycotoxin contamination of wheat kernels and flour using hyperspectral imaging, *Toxins* 11(10): 556, https://doi.org/10.3390/toxins11100556.

Andersen, A. L. (1948). The development of *Gibberella zeae* head blight of wheat, *Phytopathology* 38: 599–611.

Aoki, T., Ward, T. J., Kistler, H. C. and O'Donnell, K. (2012). Systematics, phylogeny and trichothecene mycotoxin potential of Fusarium head blight cereal pathogens, *Mycotoxins* 62(2): 91–102.

Argyris, J., van Sanford, D. and TeKrony, D. (2003). *Fusarium graminearum* infection during wheat seed development and its effect on seed quality, *Crop Sci.* 43(5): 1782–1788.

Atanasoff, D. (1920). Fusarium blight (scab) of wheat and other cereals, *J. Agric. Res.* 20: 1–32.

Awika, J. M. (2011). Major cereal grains production and use around the world. In: *Advances in Cereal Science: Implications to Food Processing and Health Promotion* (Vol. 1089), Awika, J. M., Piironen, V. and Bean, S. (Eds), American Chemical Society, Washington, DC, pp. 1–13.

Backhouse, D. (2014). Global distribution of *Fusarium graminearum*, *F. asiaticum*, and *F. boothii* from wheat in relation to climate, *Eur. J. Plant Pathol.* 139(1): 161–173.

Bai, G. H. and Shaner, G. (1996). Variation in *Fusarium graminearum* and cultivar resistance to wheat scab, *Plant Dis.* 80(9): 975–979.

Barbedo, J. G. A., Tibola, C. S. and Lima, M. I. P. (2017). Deoxynivalenol screening in wheat kernels using hyperspectral imaging, *Biosyst. Eng.* 155: 24–32.

Boenisch, M. J. and Schäfer, W. (2011). *Fusarium graminearum* forms mycotoxin producing infection structures on wheat, *BMC Plant Biol.* 11: 110.

Bolanos-Carriel, C., Wegulo, S. N., Baeziger, P. S., Funnell-Harris, D., Hallen-Adams, H. E. and Eskridge, K. M. (2020). Effects of fungicide chemical class, fungicide application timing, and environment on Fusarium head blight in winter wheat, *Eur. J. Plant Pathol.* 158(3): 667–679, https://doi.org/10.1007/s10658-020-02109-3.

Boosalis, M. G., Doupnik, B. L., Wysong, D. and Watkins, J. (1983). The wheat scab epidemic of 1982. *University of Nebraska Farm, Ranch, and Home Quarterly* 29: 7–9.

Booth, C. (1971). *The Genus Fusarium*, Commonwealth Mycological Institute, Kew, Surrey.

Brown, N. A., Urban, M., van de Meene, A. M. L. and Hammond-Kosack, K. E. (2010). The infection biology of *Fusarium graminearum*: defining the pathways of spikelet to spikelet colonisation in wheat ears, *Fungal Biol.* 114(7): 555–571.

Chandelier, A., Nimal, C., André, F., Planchon, V. and Oger, R. (2011). *Fusarium* species and DON contamination associated with head blight in winter wheat over a 7-year period (2003–2009) in Belgium, *Eur. J. Plant Pathol.* 130(3): 403–414.

Cowger, C., Patton-Özkurt, J., Brown-Guedira, G. and Perugini, L. (2009). Post-anthesis moisture increased Fusarium head blight and deoxynivalenol levels in North Carolina winter wheat, *Phytopathology* 99(4): 320–327.

Culler, M. D., Miller-Garvin, J. E. and Dill-Macky, R. (2007). Effect of extended irrigation and host resistance on deoxynivalenol accumulation in *Fusarium*-infected wheat, *Plant Dis.* 91(11): 1464–1472.

David, R. F., Marr, L. C. and Schmale III, D. G. (2016). Ascospore release and discharge distances of *Fusarium graminearum* under controlled temperature and relative humidity, *Eur. J. Plant Pathol.* 146(1): 59–69.

De Wolf, E. D., Madden, L. V. and Lipps, P. E. (2003). Risk assessment models for wheat Fusarium head blight epidemics based on within-season weather data, *Phytopathology* 93(4): 428–435.

Deacon, J. W. (2006). *Fungal Biology* (4th edn.), Blackwell Publishing, Oxford.

Denschlag, C., Rieder, J., Vogel, R. F. and Niessen, L. (2014). Real-time loop-mediated isothermal amplification (LAMP) assay for group specific detection of important trichothecene producing *Fusarium* species in wheat, *Int. J. Food Microbiol.* 177: 117–127.

Desjardins, A. E. (2006). *Fusarium Mycotoxins. Chemistry, Genetics, and Biology*, The American Phytopathological Society, St. Paul.

Desjardins, A. E., Proctor, R. H., Bai, G., McCormick, S. P., Shaner, G., Buechley, G. and Hohn, T. M. (1996). Reduced virulence of trichothecene-nonproducing mutants of *Gibberella zeae* in wheat field tests, *Mol. Plant Microbe Interact.* 9(9): 775–781.

Dickinson, J. G. and Mains, E. B. (1929). Scab of wheat and barley and its control, *U.S Dep. Agric. Farmers Bull.* 1599: 1–17.

Dickinson, J. G. and Mains, E. B. (1942). Scab of wheat and barley and its control, *U.S Dep. Agric. Farmers Bull.* 1599(Revised): 1–22.

Dill-Macky, R. (2010). Fusarium head blight (scab). In: *Compendium of Wheat Diseases and Pests*, Bockus, W. W., Bowden, R. L., Hunger, R. M., Morrill, W. L., Murray, T. D. and Smiley, R. W. (Eds), The American Phytopathological Society, St. Paul, pp. 34–36.

Dill-Macky, R. and Jones, R. K. (2000). The effect of previous crop residues and tillage on Fusarium head blight of wheat, *Plant Dis.* 84(1): 71–76.

Dowell, F. E., Ram, M. S. and Seitz, L. M. (1999). Predicting scab, vomitoxin, and ergosterol in single wheat kernels using near-infrared spectroscopy, *Cereal Chem.* 76(4): 573–576.

Dufault, N. S., De Wolf, E. D., Lipps, P. E. and Madden, L. V. (2006). Role of temperature and moisture in the production and maturation of *Gibberella zeae* perithecia, *Plant Dis.* 90(5): 637–644.

Fernando, W. G. D., Paulitz, T. C., Seaman, W. L., Dutilleul, P. and Miller, J. D. (1997). Head blight gradients caused by Gibberella zeae from area sources of inoculum in wheat field plots, *Phytopathology* 87(4): 414–421.

Foroud, N. A., Baines, D., Gagkaeva, T. Y., Thakor, N., Badea, A., Steiner, B., Bürstmayr, M. and Bürstmayr, H. (2019). Trichothecenes in cereal grains–an update, *Toxins* 11(11): 634.

Fulcher, M. R., Garcia, J. P., Damann, K. C. M. and Bergstrom, G. C. (2019). Variable interactions between non-cereal grasses and *Fusarium graminearum*, *Can. J. Plant Pathol.* 41(3): 450–456.

Gale, L. R., Harrison, S. A., Ward, T. J., O'Donnell, K., Milus, E. A., Gale, S. W. and Kistler, H. C. (2011). Nivalenol-type populations of *Fusarium graminearum* and *F. asiaticum* are prevalent on wheat in southern Louisiana, *Phytopathology* 101(1): 124–134.

Gautam, P. and Dill-Macky, R. (2012). Free water can leach mycotoxins from *Fusarium*-infected wheat heads, *J. Phytopathol.* 160(9): 484–490.

Gupta, S., Saharan, M. S., Gurjar, M. S., Singh, J., Bashyal, B. M. and Aggarwal, R. (2020). Molecular detection of *Fusarium graminearum* causing head blight of wheat by loop mediated isothermal amplification (LAMP) assay, *Indian Phytopathol.* 73(4): 667–672, https://doi.org/10.1007/s42360-020-00261-8.

Hafez, M., Abdelmagid, A., Adam, L. R. and Daayf, F. (2020). Specific detection and identification of Fusarium graminearum sensu stricto using a PCR-RFLP tool and specific primers targeting the translational elongation factor 1α gene, *Plant Dis.* 104(4): 1076–1086.

Han, L., Li, Y.-T., Jiang, J.-Q., Li, R.-F., Fan, G.-Y., Lv, J.-M., Zhou, Y., Wen-Ju Zhang, W.-J.. and Wang, Z.-L. (2020). Development of a direct competitive ELISA kit for detecting deoxynivalenol contamination in wheat, *Molecules* 25(1): 50, https://doi.org/10.3390/molecules25010050.

Hernandez Nopsa, J. F., Wegulo, S. N., Panthi, A., Hallen-Adams, H. E., Harris, S. D. and Baenziger, P. S. (2014).Characterization of Nebraska isolates of *Fusarium graminearum* causing head blight of wheat, *Crop Sci.* 54(1): 310–317.

Horevaj, P., Gale, L. R. and Milus, E. A. (2011). Resistance in winter wheat lines to initial infection and spread within spikes by deoxynivalenol and nivalenol chemotypes of *Fusarium graminearum*, *Plant Dis.* 95(1): 31–37.

Inch, S. A. and Gilbert, J. (2003). Survival of Gibberella zeae in *Fusarium*-damaged wheat kernels, *Plant Dis.* 87(3): 282–287.

Johnson, D. D., Flaskerud, G. K., Taylor, R. D. and Satyanarayana, V. (2003). Quantifying economic impacts of Fusarium head blight in wheat. In: *Fusarium Head Blight of Wheat and Barley*, Leonard, K. J. and Bushnell, W. R. (Eds), The American Phytopathological Society, St. Paul, pp. 461–483.

Keller, M. D., Bergstrom, G. C. and Shields, E. J. (2014). The aerobiology of *Fusarium graminearum*, *Aerobiologia* 30(2): 123–136.

Keller, M. D., Thomason, W. E. and Schmale III, D. G. (2011). The spread of a released clone of *Gibberella zeae* from different amounts of infested corn residue, *Plant Dis.* 95(11): 1458–1464.

Kelly, A., Proctor, R. H., Belzile, F., Chulze, S. N., Clear, R. M., Cowger, C., Elmer, W., Lee, T., Obanor, F., Waalwijk, C. and Ward, T. J. (2016). The geographic distribution and complex evolutionary history of the NX-2 trichothecene chemotype from *Fusarium graminearum*, *Fungal Genet. Biol.* 95: 39–48.

Khonga, E. B. and Sutton, J. C. (1988). Inoculum production and survival of Gibberella zeae in maize and wheat residues, *Can. J. Plant Pathol.* 10(3): 232–239.

Komura, R., Kawakami, T., Nakajima, K., Suzuki, H. and Nakashima, C. (2018). Simultaneous detection of benzimidazole-resistant strains of Fusarium head blight using the loop-mediated isothermal amplification-fluorescent loop primer method, *J. Gen. Plant Pathol.* 84(4): 247-253.

Kriss, A. B., Paul, P. A. and Madden, L. V. (2010). Relationship between yearly fluctuations in Fusarium head blight intensity and environmental variables: a window-pane analysis, *Phytopathology* 100(8): 784–797.

Kriss, A. B., Paul, P. A., Xu, X., Nicholson, P., Doohan, F. M., Hornok, L., Reitini, A., Edwards, S. G. and Madden, L. V. (2012). Quantification of the relationship between the environment and Fusarium head blight, Fusarium pathogen density, and mycotoxins in winter wheat in Europe, *Eur. J. Plant Pathol.* 133(4): 975–993.

Krupinsky, J. M., Bailey, K. L., McMullen, M. P., Gossen, B. D. and Turkington, T. K. (2002). Managing plant disease risk in diversified cropping systems, *Agron. J.* 94(2): 198–209.

Kubo, K., Kawada, N., Fujita, M., Hatta, K., Oda, S. and Nakajima, T. (2010). Effect of cleistogamy on Fusarium head blight resistance in wheat, *Breed. Sci.* 60(4): 405–411.

Kuzdraliński, A., Kot, A., Szczerba, H., Nowak, M. and Marta Muszyńska, M. (2017). A review of conventional PCR assays for the detection of selected phytopathogens of wheat, *J. Mol. Microbiol. Biotechnol.* 27(3): 175–189.

Lancashire, P. D., Bleiholder, H., Boom, T. V. D., Langelüddeke, P., Stauss, R., Weber, E. and Witzenberger, A. (1991). A uniform decimal code for growth stages of crops and weeds, *Ann. Appl. Biol.* 119(3): 561-601.

Langseth, W. and Stabbetorp, H. (1996). The effect of lodging and time of harvest on deoxynivalenol contamination in barley and oats, *J. Phytopathol.* 144(5): 241-245.

Lemmens, M., Buerstmayr, H., Krska, R., Schuhmacher, R., Grausgruber, H. and Ruckenbauer, P. (2004). The effect of inoculation treatment and long-term application of moisture on Fusarium head blight symptoms and deoxynivalenol contamination in wheat grains, *Eur. J. Plant Pathol.* 110(3): 299–308.

Leslie, J. F. and Summerell, B. A. (2006). *The Fusarium Laboratory Manual*, Blackwell Publishing, Oxford.

Lewandowski, S. and Bushnell, W. R. (2001). Development of *Fusarium graminearum* on floret surfaces of field-grown barley. In: S. M. Canty, J. Lewis, L. Siler, and R. W. Ward (Eds.), 2001 National Fusarium Head Blight Forum Proceedings, East Lansing, Michigan, USA. U.S. Wheat & Barley Scab Initiative, p. 128.

Lofgren, L. A., LeBlanc, N. R., Certano, A. K., Nachtigall, J., LaBine, K. M., Riddle, J., Broz, K., Dong, Y., Bethan, B., Kafer, C. W. and Kistler, H. C. (2018). *Fusarium graminearum*: pathogen or endophyte of North American grasses?, *New Phytol.* 217(3): 1203-1212.

Mains, E. B., Vestal, C. M. and Curits, P. B. (1929). Scab of small grains and feeding trouble in Indiana in 1928, *Proc. Indiana Acad. Sci.* 39: 101–110.

Maldonado-Ramirez, S. L., Schmale III, D. G., Shields, E. J. and Bergstrom, G. C. (2005). The relative abundance of viable spores of *Gibberella zeae* in the planetary boundary layer suggests the role of long-distance transport in regional epidemics of Fusarium head blight, *Agr. Forest Meteorol.* 132(1–2): 20–27.

Manstretta, V., Gourdain, E. and Rossi, V. (2015). Deposition patterns of *Fusarium graminearum* ascospores and conidia within a wheat canopy, *Eur. J. Plant Pathol.* 143(4): 873–880.

Manstretta, V. and Rossi, V. (2016). Effects of temperature and moisture on development of *Fusarium graminearum* perithecia in maize stalk residues, *Appl. Environ. Microbiol.* 82(1): 184–191.

Markell, S. G. and Francl, L. J. (2003).Fusarium head blight inoculum: species prevalence and *Gibberella zeae* spore type, *Plant Dis.* 87(7): 814–820.

McKevith, B. (2004). Nutritional aspects of cereals. *Br. Nutr. Bull.* 29(2): 111-142.

McMullen, M., Bergstrom, G., De Wolf, E., Dill-Macky, R., Hershman, D., Shaner, G. and Van Sanford, D. (2012). A unified effort to fight an enemy of wheat and barley: fusarium head blight, *Plant Dis.* 96(12): 1712–1728.

McMullen, M., Jones, R. and Gallenberg, D. (1997). Scab of wheat and barley: a re-emerging disease of devastating impact, *Plant Dis.* 81(12): 1340-1348.

Mesterházy, A. (1995). Types and components of resistance to Fusarium head blight of wheat, *Plant Breed.* 114(5): 377-386.

Miller, J. D. (2008). Mycotoxins in small grains and maize; old problems, new challenges, *Food Addit. Contam.* 25(2): 219-230.

Miller, J. D., Greenhalgh, R., Wang, Y. and Lu, M. (1991). Trichothecene chemotypes of three *Fusarium* species, *Mycologia* 83(2): 121-130.

Mirocha, C. J., Xie, W. and Filho, E. R. (2003). Chemistry and detection of Fusarium mycotoxins. In: *Fusarium Head Blight of Wheat and Barley*, Leonard, K. J. and Bushnell, W. R. (Eds), APS Press, St. Paul, pp. 144-164.

Mourelos, C. A., Malbran, I., Balatti, P. A., Ghiringhelli, P. D. and Lori, G. A. (2014). Gramineous and non-gramineous weed species as alternative hosts of *Fusarium graminearum*, causal agent of Fusarium head blight of wheat, in Argentina, *Crop Prot.* 65: 100-104.

Mullis, K. B. and Faloona, F. A. (1987). Specific synthesis of DNA in vitro via a polymerase-catalyzed chain reaction, *Meth. Enzymol.* 155: 335-350.

Musa, T., Hecker, A., Vogelgsang, S. and Forrer, H. R. (2007). Forecasting of Fusarium head blight and deoxynivalenol content in winter wheat with FusaProg, *OEPP EPPO Bull.* 37(2): 283-289.

Nakajima, T., Yoshida, M. and Tomimura, K. (2008). Effect of lodging on the level of mycotoxins in wheat, barley, and rice infected with the *Fusarium graminearum* species complex, *J. Gen. Plant Pathol.* 74(4): 289-295.

Neuhof, T., Ganzauer, N., Koch, M. and Nehls, I. (2009). A comparison of chromatographic methods for the determination of deoxynivalenol in wheat, *Chromatographia* 69(11-12): 1457-1462.

Nganje, W. E., Bangsund, D. A., Leistritz, F. L., Wilson, W. W. and Tiapo, N. M. (2004). Regional economic impacts of Fusarium head blight in wheat and barley, *Rev. Agric. Econ.* 26(3): 332-347.

Nganje, W. E., Johnson, D. D., Wilson, W. W., Leistritz, F. L., Bangsund, D. A. and Tiapo, N. M. (2001). *Economic Impacts of Fusarium Head Blight in Wheat and Barely.* Agribusiness and Applied Economics Report No. 464, Department of Agribusiness and Applied Economics, Agricultural Experiment Station North Dakota State University, Fargo.

Niessen, L. (2015). Current state and future perspectives of loop-mediated isothermal amplification (LAMP)-based diagnosis of filamentous fungi and yeasts, *Appl. Microbiol. Biotechnol.* 99(2): 553-574.

Niessen, L. and Vogel, R. F. (2010). Detection of Fusarium graminearum DNA using a loop-mediated isothermal amplification (LAMP) assay, *Int. J. Food Microbiol.* 140(2-3): 183-191.

Notomi, T., Okayama, H., Masubuchi, H., Yonekawa, T., Watanabe, K., Amino, N. and Hase, T. (2000). Loop-mediated isothermal amplification of DNA, *Nucleic Acids Res.* 28(12): E63.

Panthi, A., Hallen-Adams, H., Wegulo, S. N., Hernandez Nopsa, J. and Baenziger, P. S. (2014). Chemotype and aggressiveness of isolates of *Fusarium graminearum* causing head blight of wheat in Nebraska, *Can. J. Plant Pathol.* 36(4): 447-455.

Parry, D. W., Jenkinson, P. and McLeod, L. (1995). *Fusarium* ear blight (scab) in small grain cereals–a review, *Plant Pathol.* 44(2): 207-238.

Paul, P. A., Bradley, C. A., Madden, L. V., Lana, F. D., Bergstrom, G. C., Dill-Macky, R., Esker, P. D., Wise, K. A., McMullen, M., Grybauskas, A., Kirk, W. W., Milus, E. and Ruden, K. (2018). Meta-analysis of the effects of QoI and DMI fungicide combinations on Fusarium head blight and deoxynivalenol in wheat, *Plant Dis.* 102(12): 2602-2615.

Paul, P. A., El-Allaf, S. M., Lipps, P. E. and Madden, L. V. (2004). Rain splash dispersal of *Gibberella zeae* within wheat canopies in Ohio, *Phytopathology* 94(12): 1342-1349.

Paul, P. A., El-Allaf, S. M., Lipps, P. E. and Madden, L. V. (2005a). Relationships between incidence and severity of Fusarium head blight on winter wheat in Ohio, *Phytopathology* 95(9): 1049–1060.

Paul, P. A., Lipps, P. E. and Madden, L. V. (2005b). Relationship between visual estimates of Fusarium head blight intensity and deoxynivalenol accumulation in harvested wheat grain: a meta-analysis, *Phytopathology* 95(10): 1225–1236.

Peiris, K. H. S., Dong, Y., Davis, M. A., Bockus, W. W. and Dowell, F. E. (2017). Estimation of the deoxynivalenol and moisture contents of bulk wheat grain samples by FT-NIR spectroscopy, *Cereal Chem.* 94(4): 677–682.

Peiris, K. H. S., Pumphrey, M. O., Dong, Y., Maghirang, E. B., Berzonsky, W. and Dowell, F. E. (2010). Near-infrared spectroscopic method for identification of Fusarium head blight damage and prediction of deoxinivalenol in single wheat kernels, *Cereal Chem.* 87(6): 511–517.

Pereyra, S. A. and Dill-Macky, R. (2004). Survival and inoculum production of *Gibberella zeae* in wheat residue, *Plant Dis.* 88(7): 724–730.

Pereyra, S. A. and Dill-Macky, R. (2008). Colonization of the residues of diverse plant species by *Gibberella zeae* and their contribution to Fusarium head blight inoculum, *Plant Dis.* 92(5): 800–807.

Pritsch, C., Muehlbauer, G. J., Bushnell, W. R., Somers, D. A. and Vance, C. P. (2000). Fungal development and induction of defense response genes during early infection of wheat spikes by *Fusarium graminearum*, *Mol. Plant Microbe Interact.* 13(2): 159–169.

Prussin, A. J., II, Li, Q., Malla, R., Ross, S. D. and Schmale III, D. G. (2014). Monitoring the long-distance transport of *Fusarium graminearum* from field-scale sources of inoculum, *Plant Dis.* 98(4): 504–511.

Puri, K. D. and Zhong, S. (2010). The 3ADON population of *Fusarium graminearum* found in North Dakota is more aggressive and produces a higher level of DON than the prevalent 15ADON population in spring wheat, *Phytopathology* 100(10): 1007–1014.

Rossi, V., Pattori, E., Ravanetti, A. and Giosuè, S. (2002). Effect of constant and fluctuating temperature regimes on sporulation of four fungi causing head blight of wheat, *J. Plant Pathol.* 84(2): 95–105.

Schmale III, D. G., Arntsen, Q. A. and Bergstrom, G. C. (2005a). The forcible discharge of ascospores of *Gibberella zeae*, *Can. J. Plant Pathol.* 27(3): 376–382.

Schmale III, D. G., Shah, D. A. and Bergstrom, G. C. (2005b). Spatial patterns of viable spore deposition of *Gibberella zeae* in wheat fields, *Phytopathology* 95(5): 472–479.

Schmale III, D. G., Shah, D. A. and Bergstrom, G. C. (2005c). Spatial patterns of viable spore deposition of the corn ear rot pathogen, *Gibberella zeae*, in first-year corn fields, *Can. J. Plant Pathol.* 27: 225–233.

Schmale III, D. G., Ross, S. D., Fetters, T. L., Tallapragada, P., Wood-Jones, A. K. and Dingus, B. (2012). Isolates of *Fusarium graminearum* collected 40–320 meters above ground level cause Fusarium head blight in wheat and produce trichothecene mycotoxins, *Aerobiologia* 28(1): 1–11.

Schmale III, D. G., Shields, E. J. and Bergstrom, G. C. (2006). Night-time spore deposition of the fusarium head blight pathogen, *Gibberella zeae*, in rotational wheat fields, *Can. J. Plant Pathol.* 28: 100–108.

Schroeder, H. W. and Christensen, J. J. (1963). Factors affecting resistance of wheat to scab caused by *Gibberella zeae*, *Phytopathology* 53: 831–838.

Shah, D. A., De Wolf, E. D., Paul, P. A. and Madden, L. V. (2014). Predicting Fusarium head blight epidemics with boosted regression trees, *Phytopathology* 104(7): 702–714.

Shah, D. A., Molineros, J. E., Paul, P. A., Willyerd, K. T., Madden, L. V. and De Wolf, E. D. (2013). Predicting Fusarium head blight epidemics with weather-driven pre- and post-anthesis logistic regression models, *Phytopathology* 103(9): 906–919.

Shaner, G. (2003). Epidemiology of Fusarium head blight of small grain cereals in North America. In: *Fusarium Head Blight of Wheat and Barley*, Leonard, K. J. and Bushnell, W. R. (Eds), The American Phytopathological Society, St. Paul, pp. 84–119.

Stack, R. W. (2003). History of Fusarium head blight with emphasis on North America. In: *Fusarium Head Blight of Wheat and Barley*, Leonard, K. J. and Bushnell, W. R. (Eds), APS Press, St. Paul, pp. 1–34.

Stein, J. M., Osborne, L. E., Bondalapati, K. D., Glover, K. D. and Nelson, C. A. (2009). Fusarium head blight severity and deoxynivalenol concentration in wheat in response to *Gibberella zeae* inoculum concentration, *Phytopathology* 99(6): 759–764.

Strange, R. N. and Smith, H. (1971). A fungal growth stimulant in anthers which predisposes wheat to attack by *Fusarium graminearum*, *Physiol. Plant Pathol.* 1(2): 141–150.

Summerell, B. A. and Burgess, L. W. (1988). Stubble management practices and the survival of Fusarium graminearum Group 1 in wheat stubble residues, *Austral. Plant Pathol.* 17(4): 88–93.

Sutton, J. C. (1982). Epidemiology of wheat head blight and maize ear rot caused by *Fusarium graminearum*, *Can. J. Plant Pathol.* 4(2): 195–209.

Tacke, B. K. and Casper, H. H. (1996). Determination of deoxynivalenol in wheat, barley, and malt by column cleanup and gas chromatography with electron capture detection, *J. Assoc. Off. Anal. Chem.* 79(2): 472–475.

Trail, F. (2009). For blighted waves for grain: *fusarium graminearum* in the postgenomics era, *Plant Physiol.* 149(1): 103–110.

Trail, F. and Common, R. (2000). Perithecial development by Gibberella zeae: a lightmicroscopy study, *Mycologia* 92(1): 130–138.

Trail, F., Gaffoor, I. and Vogel, S. (2005). Ejection mechanics and trajectory of the ascospores of Gibberella zeae (anamorph Fusarium graminearum), *Fungal Genet. Biol.* 42(6): 528–533.

Turkington, T. K., Clear, R. M., Demeke, T., Lange, R., Xi, K. and Kumar, K. (2011). Isolation of *Fusarium graminearum* from cereal, grass and corn residues from Alberta, 2001-2003, *Can. J. Plant Pathol.* 33(2): 179–186.

Valverde-Bogantes, E., Bolanos-Carriel, C., Hallen-Adams, H. E., McMaster, N., Schmale III, D. G. and Wegulo, S. N. (2020). Aggressiveness and deoxynivalenol production of Nebraska isolates of *Fusarium boothii* and F. graminearum, *Plant Health Prog.* 21(2): 97–102.

Varga, E., Wiesenberger, G., Hametner, C., Ward, T. J., Dong, Y., Schöfbeck, D., McCormick, S., Broz, K., Stückler, R., Schuhmacher, R., Krska, R., Kistler, H. C., Berthiller, F. and Adam, G. (2015). New tricks of an old enemy: isolates of *Fusarium graminearum* produce a type A trichothecene mycotoxin, *Environ. Microbiol.* 17(8): 2588 -2600.

Ward, T. J., Clear, R. M., Rooney, A. P., O'Donnell, K., Gaba, D., Patrick, S., Starkey, D. E., Gilbert, J., Geiser, D. M. and Nowicki, T. W. (2008). An adaptive evolutionary shift in Fusarium head blight pathogen populations is driving the rapid spread of more toxigenic *Fusarium graminearum* in North America, *Fungal Genet. Biol.* 45(4): 473–484.

Wegulo, S. N., Baenziger, P. S., Hernandez Nopsa, J., Bockus, W. W. and Hallen-Adams, H. (2015). Management of Fusarium head blight of wheat and barley, *Crop Prot.* 73: 100–107.

Wegulo, S. N., Bockus, W. W., Hernandez Nopsa, J. H., De Wolf, E. D., Eskridge, K. M., Peiris, K. H. S. and Dowell, F. E. (2011). Effects of integrating cultivar resistance and fungicide application on Fusarium head blight and deoxynivalenol in winter wheat, *Plant Dis.* 95(5): 554–560.

Wegulo, S. N. and Dowell, F. E. (2008). Near-infrared versus visual sorting of Fusarium-damaged kernels in winter wheat, *Can. J. Plant Sci.* 88(6): 1087–1089.

White, D. G. (2016). Gibberella stalk rot. In: *Compendium of Corn Diseases* (4th edn.), Munkvold, G. P. and White, D. G. (Eds), APS Press, St. Paul, pp. 74–76.

Wiersma, J. V., Peters, E. L., Hanson, M. A., Bouvette, R. J. and Busch, R. H. (1996). Fusarium head blight in hard red spring wheat: cultivar responses to natural epidemics, *Agron. J.* 88(2): 223-230.

Xue, A. G., Chen, Y. H., Santanna, S. M. R., Voldeng, H. D., Fedak, G.,Savard, M. E., Längle, T., Zhang, J. X. and Harman, G. E. (2014). Efficacy of CLO-1 biofungicide in suppressing perithecial production by *Gibberella zeae* on cropresidues, *Can. J. Plant Pathol.* 36(2): 161-169.

Xue, A. G., Voldeng, H. D., Savard, M. E., Fedak, G., Tian, X. and Hsiang, T. (2009). Biological control of Fusarium head blight of wheat with Clonostachys rosea strain ACM941, *Can. J. Plant Pathol.* 31(2): 169-179.

Yoshida, M., Kawada, N. and Nakajima, T. (2007). Effect of infection timing on Fusarium head blight and mycotoxin accumulation in open and closed-flowering barley, *Phytopathology* 97(9): 1054-1062.

Yoshida, M., Kawada, N. and Tohnooka, T. (2005). Effect of row type, flowering type and several other spike characters on resistance to Fusarium head blight in barley, *Euphytica* 141(3): 217-227.

Yoshida, M., Nakajima, T., Arai, M., Suzuki, F. and Tomimura, K. (2008). Effect of the timing of fungicide application on Fusarium head blight and mycotoxin accumulation in closed-flowering barley, *Plant Dis.* 92(8): 1164–1170.

Zadoks, J. C., Chang, T. T. and Konzak, C. F. (1974). A decimal code for the growth stages of cereals, *Weed Res.* 14(6): 415-421.

Chapter 5

Cereal-*Fusarium* interactions: Improved fundamental insights into *Fusarium* pathogenomics and cereal host resistance reveals new ways to achieve durable disease control

Claire Kanja, Ana K. Machado Wood, Laura Baggaley, Catherine Walker and Kim E. Hammond-Kosack, Rothamsted Research, UK

1 Introduction

Ensuring global food security for the ever-growing world population is a major concern. Fungal pathogens destroy a substantial amount of all food crops each year (~15%). Particularly serious threats to most cereal crops are several Ascomycete species in the genus *Fusarium* that cause a wide spectrum of seedling, root, stem-base and/or floral diseases (Table 1). Fusarium head blight (FHB) disease, also known as ear blight or head scab disease, is the number-one floral disease of small grain cereal crops globally (Baldwin et al., 2018) and has the potential to devastate crops just weeks before harvest (Dean et al.,

http://dx.doi.org/10.19103/AS.2021.0092.07

Table 1 Fusarium species infecting different cereal species

Host	Main Fusarium species	Disease name (tissue)	Further information
Durum wheat	*F. avenaceum* *F. graminearum* *F. poae* *F. culmorum*	Fusarium head blight (wheat spikes)	Alternative names - Fusarium ear blight or wheat scab Goswami and Kistler (2005) Beccari et al. (2018) Haile et al. (2019)
Bread wheat	*F. graminearum* (worldwide) *F. culmorum* (Europe) *F. asiaticum* (Asia) *F. meridionale* (South America) *F. boothii* *F. sporotrichioides* (Russia)	Fusarium head blight (wheat spikes)	Parry et al. (1995) Goswami et al. (2005) McMullen et al. (2012) Przemieniecki et al. (2014) Van Der Lee et al. (2015)
Wheat	*F. pseudograminearum* *F. graminearum* *F. culmorum*	Fusarium crown/foot rot (stem base)	Backhouse et al. (2004) Hogg et al. (2007)
Barley	*F. graminearum* *F. culmorum*	Fusarium head blight (barley spikes)	McMullen et al. (2012) Schwarz et al. (1997)
Oats	*F. graminearum* *F. poae* *F. langsethiae*	Fusarium head blight (oat panicle)	Tekauz et al. (2004) Edwards (2009)
Rice	*F. anthophilum* *F. graminearum* *F. asiaticum*	Fusarium head blight (rice panicles)	Nyvall et al. (1999) Goswami et al. (2005)
Rye	*F. graminearum* *F. culmorum*	Fusarium head blight (rye spikes)	Miedaner (1997)
Triticale	*F. graminearum* *F. culmorum*	Fusarium head blight (triticale spikes)	Miedaner (1997) Veitch et al. (2008)
Maize	*F. graminearum* *F. moniliforme* *F. subglutinans* (S Korea) *F. temperatum* (S Korea) *F. brachygibbosum* (China)	Fusarium stalk rot	Nelson (1992) Shin et al. (2014) Shan et al. (2017)
Maize	*F. verticillioides* *F. subglutinans* *F. proliferatum* *F. graminearum* *F. culmorum*	Fusarium ear rot	Nelson (1992) Munkvold (2003)

2012; McMullen et al., 2012; Parry et al., 1995). It typically takes 14–21 days from initial infection to complete spike colonisation and disease formation. FHB epidemics can cause 50–70% loss of marketable grain. This disease, first described in the late nineteenth century, became serious in the United States in the 1910s before declining but has become particularly prevalent globally since the 1990s (Goswami and Kistler, 2004; McMullen et al., 2012; Parry

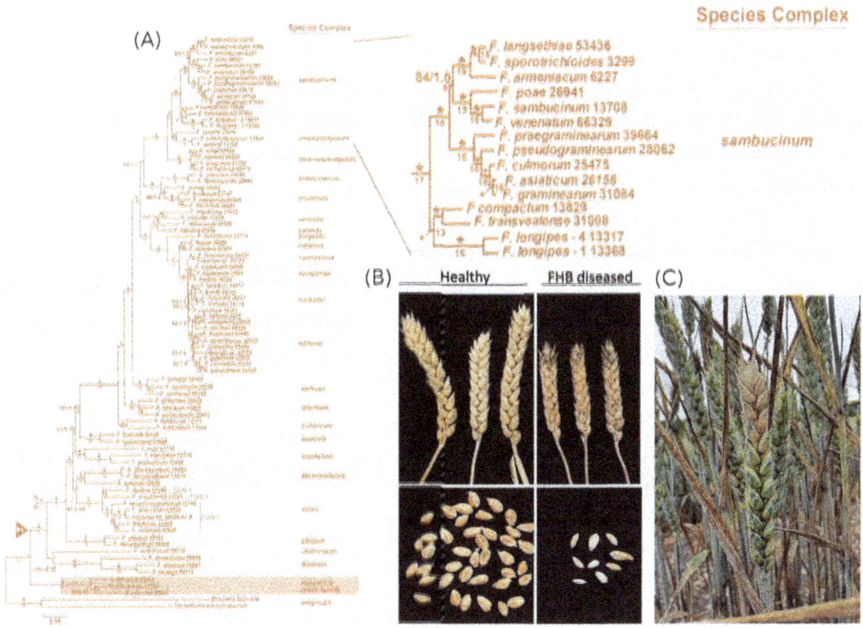

Figure 1 Phylogenetic tree of Fusarium species and their species complexes and images of FHB disease in the field, at harvest and the effects on grain. (A) Partitioned maximum likelihood bootstrapped (ML-BS) phylogram showing the 23 species complexes of Fusarium, taken from Geiser et al. (2020). The Fusarium species that cause Fusarium head blight in cereals are members of the *Fusarium sambucinum* species complex (FSSC) (enlarged region of Panel A). Based on full-length exonic nucleotide (nt) sequences of 19 protein-coding genes inferred using IQ-TREE (Nguyen et al., 2015; Geiser et al., 2020). Nodes supported by 100% ML bootstrap (5000 replicates; BS) and 1.0 Bayesian posterior probability (BPP), including F1 and F2, are indicated with asterisks; ML-BS/BPP values are shown for nodes receiving less than 100%/1.0 support. Numerals situated adjacent to nodes represent gene concordance factors (gCFs) calculated by IQ-TREE 2 for that node, expressed as the number of loci that resolve it out of 19. (B) Healthy wheat spikes and grain (left) and wheat spikes and grain following infection with *Fusarium graminearum* (right). (C) Premature bleaching of FHB infected wheat in the field. These images came from a 2020 Cadenza wheat field trial spray inoculated with *F. graminearum* Rothamsted field isolate FG602.10 at early anthesis (spore dose spore dose 2 × 10^4/ml). Post inoculation, the floral tissue was bagged overnight to ensure a high level of humidity and good levels of initial infection. In this field trial there was a high level of natural yellow (stripe) rust infection on leaves which lead to premature leaf senescence visible in Panel C.

et al., 1995). Affected wheat grains are small, discoloured and shrivelled due to the loss of albumin and glutenin proteins in the endosperm (Boyacioğlu and Hettiarachchy, 1995; Goswami and Kistler, 2004). Brown lesions with clear ring-like centres can often be seen on the external surfaces of the glumes, and subsequently, premature total bleaching of the spike occurs (Fig. 1). FHB disease not only reduces grain yield and grain quality but also poses a considerable risk to human, animal and ecosystem health, owing to the ability of Fusaria hyphae to contaminate grains with various mycotoxins. The common mycotoxins associated with Fusarium floral diseases in wheat, barley and maize crops are deoxynivalenol (DON), nivalenol (NIV), zearalenone (ZEA) and fumonisins (FUM) (Bennett and Klich, 2003). DON is the most frequently detected mycotoxin in harvested grains. The direct effect of DON is to inhibit protein synthesis by binding to the ribosome. DON contamination causes vomiting, feed refusal, anorexia and immune dysregulation effects as well as growth, reproductive and teratogenic effects in farmed animals and poultry (Pestka, 2010). To minimise human and animal exposure to DON, regulatory organisations in numerous countries have established maximum permissible levels for DON in cereals and their products (Bianchini et al., 2015; European Commission, 2006; Pinotti et al., 2016). In the European Union (EU), 56% of cereal grain samples contained DON above the recommended threshold for human consumption in 2019 (BIOMIN, 2019). Due to climate change as well as changes in agricultural practices, both locally and globally, DON contamination has become a more pressing and challenging societal issue due to the increased frequency and severity of FHB epidemics (Backhouse, 2014; McMullen et al., 2012; Nganje et al., 2004; Vaughan et al., 2016). FHB disease can also cause other less well-known negative impacts on plant breeding, soil health, grain shipments as well as directly altering agricultural and processing practices (Table 2). New and diverse approaches to control FHB disease are urgently required, especially those delivered through germplasm improvements.

Epidemiologically, FHB disease is considered monocyclic. Floral infections only occur during the short period of crop anthesis (~10 days) and negligible spike-to-spike infections occur between tillers. In minimum cultivation systems, high levels of inoculum (both conidia and ascospores) are present in infected crop debris left on the soil surface. Slow to degrade infected maize stubble is a particularly problematic local inoculum source to subsequent cereal crops. In addition, traditional non-cereal break crops, for example, soybean, sugar beet and other vegetable species, often succumb to Fusaria root and stem-base diseases caused by the same species capable of causing FHB in cereal crops. This renders the break crop year less effective and limits farmers' rotation options. In some global regions, long-distance air movement of FHB ascospore inoculum is also problematic (Del Ponte et al., 2005). For the different cereal crops, the dominant Fusarium species causing FHB or other floral diseases can vary (Fig. 1 and Table 1). Globally, *F. graminearum* is the most problematic species.

Unlike numerous plant-pathogen interactions, Fusarium-plant interactions do not conform to the classical gene-for-gene relationship. Natural resistance sources explored in each cereal species are based upon numerous major and

Table 2 Well-known and less well-known direct effects and consequences of Fusarium diseases in cereal crops and harvested grains

Direct effects	Subsequent consequences	References and URLs
Well-documented effects		
Harvested grains contaminated with a range of mycotoxins	Official testing of grain shipments to ensure maximum permitted thresholds for different end uses are not exceeded. High testing costs, delays and additional transport costs and/or blending when use downgraded from human consumption to animal feedstuffs or other uses Not detected highly/moderately contaminated grain supplies when processed cause vomiting and feed refusal in humans and farmed animals – cattle, pigs and poultry	Bennett and Klich (2003) Ferrigo et al. (2016) https://www.biomin.net/solutions/mycotoxin-survey https://scabusa.org/index.php/ https://ahdb.org.uk/knowledge-library/fusarium-and-microdochium-in-cereals
Harvested grains contaminated with Fusarium hydrophobins	Incorrect fermentation process leads to a highly detrimental and costly excess foaming phenomenon called 'gushing' in beer production	Sarlin et al. (2012)
Harvested grains have reduced gluten content	Lower milling quality flour which can no longer be used for making bread or other bakery products	Boyacioğlu and Hettiarachchy (1995); Goswami and Kistler (2004)
Fusarium-infected maize stubble remains at the soil surface in fields for 2–3 seasons resulting in excess perithecia formation and greater/longer releases of ascospores	In field maize, crops residues cause the infection of subsequent wheat and barley crops leading to local increases in FHB disease incidence and severity Reduced crop rotation options in maize growing regions	McMullen et al. (2012); Goswami and Kistler (2004)
During bulk grain shipment at below 14% moisture, Fusarium fungal growth and/or mycotoxin production can recommence from infected grains leading to increased DON and/or hydrophobin content	Re-sampling and testing of bulk grain shipments at destination ports adds costs and potentially alters/downgrades use options	Magan et al. (2010); Schmidt-Heydt et al. (2011)

(*Continued*)

Table 2 (*Continued*)

Direct effects	Subsequent consequences	References and URLs
Contamination of distillers' dried grains (DDGS), a fuel ethanol co-product, arising from the processing of predominantly maize cobs infected with DON and other mycotoxins	Sampling and testing of DDGS for mycotoxins increases costs and potentially alters use options, that is, restricts use as an animal feedstuff	Pinotti et al. (2016) Khatibi et al. (2014)
Less well-known effects		
FHB causing Fusarium species can also infect and cause serious diseases in sugar beet, soybean and vegetable crops	More limited rotation options for farmers and growers	Estrada et al. (2010) Arias et al. (2013)
Pre-infections by Barley Yellow Dwarf Virus (BYDV) increase FHB disease and DON mycotoxin content	Further reductions to grain yield, quality and safety More limited use options for farmers	Liu and Buchenauer (2005)
Maize stalks severely infected with Fusarium rot disease may contain significant mycotoxin levels	Maize stalks unacceptable for use as silage More limited use options for farmers	Goswami and Kistler (2004) Mirocha and Christensen (1974)
Commercial production of hybrid wheat and barley seed is highly susceptible to FHB disease because at anthesis the individual flowers remain open for longer	Hybrid cereal seed production regions limited to geographical regions with a drier climate where FHB disease rarely occurs	Personal communications from several commercial hybrid cereal seed producers

Pre-harvest heavy rain events leach out the water-soluble DON from the FHB infected floral tissues into rainwater and hence the soil below	Soil microbial communities, soil macrofauna and soil animals exposed to DON mycotoxin contamination which may adversely alter ecosystem composition and health	Gautam and Dill-Macky (2012)
Pre-harvest heavy rain events lead to a redistribution of the water-soluble DON from the bran to the white milled flour	Mycotoxin distribution within mill fractions from grain may vary. Greater milling precision may be required to ensure legislative limits are met	Edwards et al. (2018)
Delayed harvest coincident with rainfall leads to a significant increase in Fusarium biomass and accumulation of mycotoxins	Barley grain destined for malting should be harvested with minimum delay once ripe	AHDB SAFEMalt project recommendations – https://croprotect.com/diseases/fusarium-ear-blight
In FHB infected fields, the settings on the combine harvester equipment are altered to leave the more lightweight Fusarium mycotoxin contaminated grains in the field thereby lowering DON content in the final harvested grain	Post-harvest more wildlife (avian and mammal) are exposed to contaminated grains remaining in the field which may adversely alter ecosystem composition and health	Personal communications from several UK agronomists and farmers
Mycotoxin adsorbing clays added to compound animal feeds processed from light to moderately FHB infected cereal grains	This treatment successfully reduces animal exposure to harmful mycotoxins but adds extra costs to animal feed production and may also affect the uptake of minerals by livestock	Elliott et al. (2020)
Heavily DON mycotoxin contaminated FHB grains burnt as an alternative fuel source in coal-fired power stations	One of the few use options available in countries where blending grain samples from different sources to achieve legal DON limits is not permitted	Personal communications from several UK sources

minor quantitative trait loci (QTLs) (Venske et al., 2019). Also the genes so far identified underlying these QTLs are not the classical resistance gene type, that is, intracellular NBS-LRR proteins or extracellular LRR receptors with or without a cytoplasmic kinase domain (see Section 6). This lack of a classical gene-for-gene relationship in Fusarium-cereal interactions has two potential underlying causes. Firstly, the inhibitory effect of the DON mycotoxins on protein translation would dramatically restrict the success of any transcriptionally induced plant defence responses. Secondly, the infection sites are typically various young floral tissues, which are nutrient-rich sink tissues primarily focussed on the processes of fertilisation and supporting grain development and not plant defence. Elsewhere in this chapter (see Section 6), natural sources of FHB resistance, their QTL locations and underlying resistance gene types are described and discussed in detail.

The other major approach to disease control is the use of fungicides. Effective chemical control is difficult because of an early symptomless disease phase (described in Section 2), which means the floral applications (T3 sprays) must occur before symptoms appear to protect the entire newly emerged flowering tiller populations. For this prophylactic treatment to be effective, advanced FHB disease forecasting systems are increasingly used to pinpoint vulnerable cereal crops, whilst reducing costly and unnecessary sprays (Haran et al., 2006). Also, the globally dominant Fusarium species, *F. graminearum*, is inherently resistant to azole fungicides due to the presence of multiple copies of the gene coding for the target enzyme CYP51 (Fan et al., 2013).

This chapter reviews the advances in Fusarium-cereal research over the last 10 years. The primary focus is wheat, barley and maize crops and mycotoxin-producing Fusarium species. In Section 2 we explore the cell biology of the infection process, the role(s) of the DON mycotoxin, the significance of plasmodesmatal colonisation and discuss the current infection model and some of the key points of pathogen vulnerability. We then explore in Sections 3 and 4 the *F. graminearum* genome and predicted gene repertoire, highlighting the features also found in other pathogenic Fusaria which are distinct from other cereal-infecting fungi. Also included in these sections are descriptions and discussion of the various *in vitro* and *in planta* transcriptome studies which have considerably increased our knowledge of the spatial and temporal patterns of gene expression in different plant species/tissue types which have been successfully linked to detailed cell biology analyses. Section 5 details the numerous large-scale and more hypothesis-driven reverse genetics experiments that have resulted in ~10% of the total gene repertoire of *F. graminearum* being tested primarily for roles in FHB disease formation. In this section, special emphasis has been given to host perception, signalling, transcription factors, *in planta*-induced secondary metabolite clusters (SMCs) and secreted effectors. In Section 6, the complex repertoire of Fusarium-induced responses during floral defence is described and discussed, along with the gene types functionally linked to QTL-mediated resistance. Section 7 provides a synthesis of the new genetic modification (GM), gene editing (GE) and host-induced gene silencing (HIGS) approaches to disease control, devised and tested based upon the improved fundamental insights that have arisen from studying Fusarium genomes, virulence requirements and the mechanisms underlying compatible and/or incompatible interactions. In the final section, a list of pressing unanswered questions is presented to provoke further discussions and experimentation.

2 Infection biology, trichothecene mycotoxins and plasmodesmata function

2.1 Entry and colonisation

Spike infection of small grain cereals is initiated when airborne sexual ascospores or asexual conidia are deposited by wind or rain splash onto flowering florets and spikelets (Parry et al., 1995). The most well-studied interaction is between *F. graminearum* and wheat. On the outer wheat glumes, *F. graminearum* spores germinate, forming specialised unbranched runner hyphae (RH). From these RH, multicellular, complex appressoria develop known as infection cushions (ICs), which generate multiple penetration sites to aid fungal entry into the plant cuticle (Fig. 2) (Boenisch and Schäfer, 2011; Brown et al., 2010; Mentges et al., 2020). These ICs have been shown to secrete an arsenal of proteins and metabolites, including the mycotoxin DON, enabling invasion of wheat epidermal cells. In contrast, few genes are specifically upregulated in RH, with the exception of genes involved in the biosynthesis of the secondary metabolite (SM) aurofusarin which has broad antibiotic activity (Mentges et al., 2020).

An alternative infection route is when surface growing hyphae from the *F. graminearum* germlings gain entry via natural openings such as stomata or cracked anthers, then grow down the anther filament into the host or grow down the inside of the glumes, lemma and palea and directly infect at the floret base (Pritsch et al., 2000). Under moist conditions, abundant aerial mycelia are visible to the naked eye extending from infected anthers. Post-initial spikelet colonisation, further internal infection of the wheat spike is achieved by the intercellular growth of hyphae from the infected spikelet, through the rachilla into the rachis node, and subsequently into other spikelets, both up and down the entire spike (Fig. 2) (Brown et al., 2010; Brown et al., 2011). The establishment of this asymptomatic phase in the rachis and rachis node is characterised by the advancement of the fungal infection front localised solely to the apoplast of the cortex where parenchyma cells remain intact. Behind the advancing infection, as fungal biomass increases, the development of the

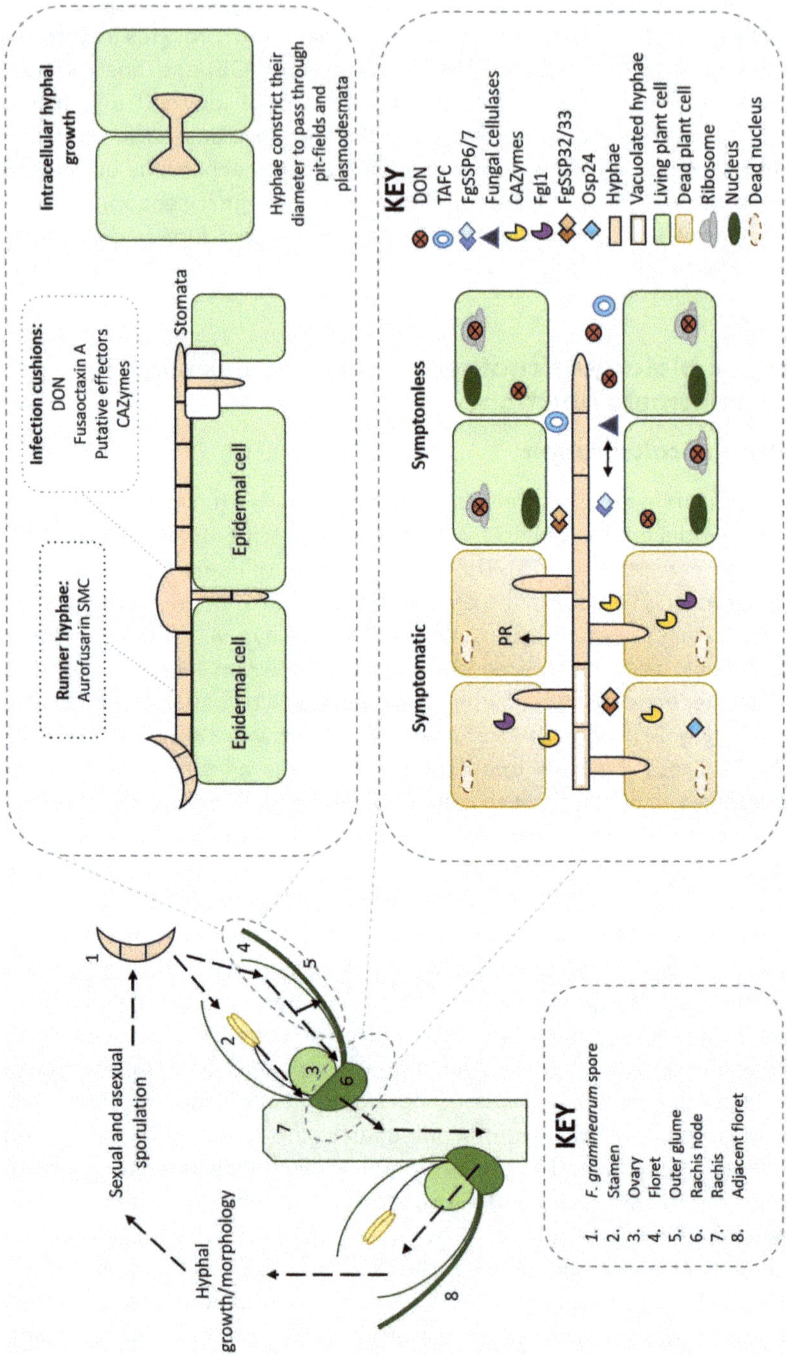
Intracellular hyphal growth
Hyphae constrict their diameter to pass through pit-fields and plasmodesmata
Infection cushions:
DON
Fusaoctaxin A
Putative effectors
CAZymes
Runner hyphae:
Aurofusarin SMC
Stomata
Epidermal cell
Epidermal cell
KEY
DON
TAFC
FgSSP6/7
Fungal cellulases
CAZymes
Fgl1
FgSSP32/33
Osp24
Hyphae
Vacuolated hyphae
Living plant cell
Dead plant cell
Ribosome
Nucleus
Dead nucleus
Symptomless
Symptomatic
PR
Sexual and asexual sporulation
Hyphal growth/morphology
KEY
1. F. graminearum spore
2. Stamen
3. Ovary
4. Floret
5. Outer glume
6. Rachis node
7. Rachis
8. Adjacent floret

Figure 2 A working model for the *Fusarium graminearum*-wheat interaction in spike tissue. This updated model depicts the *F. graminearum* life cycle, compares the symptomless and symptomatic stages of infection, and demonstrates the transition in a compatible interaction from intercellular to intracellular hyphal growth as infection progresses. A) Fusarium spore germinates at the tip of a wheat floret (1). Hyphae (dashed arrow) grow on the stamen, filament and inner sides of the palea and lemma (2, 4 and 5) before invading the ovary leading to the abortion of ovary development (3). Hyphal growth then colonise into adjacent florets within the same spikelet followed by infection of the rachis node (6). Hyphae transition in the rachis, infecting the cortical cells surrounding the vascular system, and subsequently the xylem and phloem tissues collectively allowing fungal growth up and down the wheat spike (7) and into neighbouring spikelets (8). Hyphae grow on crop stubble and produce asexual spores and sexual spores in perithecia. B) Spores germinate on the outer wheat glumes forming runner hyphae or specialised infection cushions enabling penetration of the plant cuticle. Hyphae may also use natural openings such as stomata to gain entry to the host. Once within a plant cell, hyphae can constrict through plasmodesmata and pit-fields enabling entry into neighbouring cells. C) During infection, a range of secreted molecules are produced in the symptomless and symptomatic phases. Abbreviations : CAZyme, carbohydrate-active enzyme; DON, deoxynivalenol; FgSSP, *F. graminearum* small secreted protein; PR, plant response; SMC, secondary metabolite gene cluster; TAFC, triacetyl fusarinine C.

first macroscopically visible disease symptoms coincides with intracellular colonisation of wheat cells, cell death and the loss of cellular contents. At this later stage of infection, inter- and intra-cellular hyphae become abundant, and once the spike is completely colonised, fungal mycelium accumulate below the rachis surface, rupturing the epidermis and giving rise to aerial mycelium, conidiophores and conidia. In the older regions of the colony, hyphal contents are lost and/or recycled and are considered unlikely to be acquiring nutrients externally (Brown et al., 2010; Brown et al., 2011). *F. graminearum* infections of wheat spikes can therefore be considered to exist along a continuum - from symptomless biotrophic infection to symptomatic necrotrophic infection - the extremes of which are phenotypically, spatially and temporally distinct.

Compared with FHB, the symptoms of stem-base disease called Fusarium crown rot (FCR) develop far more slowly, usually taking 4–8 weeks to become visible. Three distinct FCR infection phases have been identified: initial spore germination followed by the formation of a superficial hyphal mat, colonisation of the adaxial epidermis of the outer leaf sheath and mycelial growth into the crown, concomitant with a drop in fungal biomass and finally extensive colonisation of the internal crown tissue (Stephens et al., 2008). Fusarium root rot (FRR) disease is more rapid, taking between 1 and 2 weeks to become visible and then subsequently extends vertically into the stem base, leaf sheaths and stem apex (Wang et al., 2015). Three distinct FRR infection phases of young wheat seedlings have been identified: initial spore germination followed by the formation of a superficial hyphal mat and simple infection hyphae, then the formation of IC, intra- and extracellular colonisation of root cortex concomitant, with a rise in fungal biomass coupled to the onset of spore formation (sporodochia, macroconidia and chlamydospores), and finally vertical hyphal colonisation through the pith parenchyma, the bundle sheath and the entire vascular tissues (Wang et al., 2015).

In barley, *F. graminearum* infections of the spike usually occur post-anthesis. Hyphae colonise the brush hairs of the extruding seed tip before gaining entry into the developing caryopsis (Leonard and Bushnell, 2003; Skadsen and Hohn, 2004). In contrast to wheat infection, wild-type *F. graminearum* infections of barley are confined to the initially infected spikelets and florets and no colonisation of the rachis node and rachis occurs (Langevin et al., 2004; Lewandowski et al., 2006). Therefore, barley spikes often exhibit a patchier distribution of diseased spikelets than observed in wheat spikes.

In maize, floral infections begin when spores land and germinate on the maize silks and hyphae progress, both internally and externally along the silks to the top of the ear. Silk cell walls thicken upon infection indicating a host defence response (Miller et al., 2007). Alternatively, Fusarium stem and ear infections occur via the deep wounds arising from chewing insects, particularly problematic is the European corn borer (*Ostrinia nubilalis*) (Sutton, 1982). Within the maize ears, the immature kernels and rachis become rapidly infected and Fusarium hyphae colonise the parenchyma and vasculature (Miller et al., 2007). When maize stalk infections occur, there is a short asymptomatic phase, where most hyphae grow longitudinally between the cell walls, before transitioning from intercellular to intracellular hyphal growth to gain entry and fully occupy host cells within 72 h (Zhang et al., 2016).

Aside from causing diseases of cereal crops, Fusarium-host interactions have also been extensively investigated using two model plant species. For *Arabidopsis thaliana*, both floral and leaf infection assays have been established to utilise the wide diversity of ecotypes, single-gene mutants and other genetic and genomic resources available in this small, rapid life-cycle species (Brewer and Hammond-Kosack, 2015; Chen et al., 2006; Urban et al., 2002). For those preferring to solely study Fusarium interactions in a cereal species, the *Brachypodium distachyon* pathosystem has been developed (Dinolfo et al., 2021; Draper et al., 2001; Peraldi et al., 2011). Brachypodium's small, well-annotated genome (Huo et al., 2009), rapid life-cycle and relatively small size makes it a favourable experimental system (Draper et al., 2001). Infection development in Brachypodium spikelets is similar to that in wheat (Su et al., 2018) and therefore provides an alternative for the wheat spikelet infection assay.

A high throughput coleoptile tissue infection bioassay has been developed to explore and intercompare some of the fine details of the colonisation processes in wheat-*F. graminearum* interactions (Zhang et al., 2012). Although rapid, taking just 10 days from seed sowing to final disease assessments, and clearly exhibiting some cellular colonisation hallmarks present in floral infection, considerable transcriptional differences between coleoptile and floral infection have been noted (Zhang et al., 2012). Potentially, this coleoptile bioassay could be a space-saving way to evaluate strain differences in virulence/aggressiveness by measuring lesion size, compared to the typical wheat spike bioassay. However, of the seven *F. graminearum* single-gene deletion mutants identified by Zhang et al. (2012) shown to be reduced in coleoptile infection, only four showed reduced infection of spikes. Only through further experimentation will the niche or wider research potential of the coleoptile bioassay become clearer.

2.2 Trichothecene mycotoxin production during infection

Most of the genes required for trichothecene (*Tri*) mycotoxin biosynthesis and regulation have now been identified and extensively characterised. In multiple

Fusarium species, the genes involved in *Tri*/DON biosynthesis, termed the *Tri* genes, have been shown to reside at one major locus and two minor loci, as shown in Fig. 3A. Most *Tri* genes reside within the ~25 Kb *Tri5* cluster on chromosome 2, and this includes genes coding for the global transcriptional regulators *Tri6* and *Tri10* which flank the pathway committing *Tri5* gene which codes for trichodiene synthase, the first enzyme specifically required for *Tri* biosynthesis. Whereas other upregulated genes required for the later stages of the isoprenoid pathways, from which *Tri* biosynthesis pathway emanates, are located in other genome locations.

During most plant infections or when grown under specific nutrient conditions *in vitro*, many *Tri* genes are rapidly induced in complex spatial and temporal patterns (Fig. 3B). The genes within the *Tri* cluster are induced in a tissue-specific manner and, while strong induction of the *Tri* genes is observed during infection of the wheat and barley spikes, no such induction is observed during the infection of the wheat coleoptile. Besides regulation by *Tri6* and *Tri10*, *Tri* biosynthesis is regulated by other global regulators which control responses to carbon, nitrogen, light and pH (Gardiner et al., 2009; Hou et al., 2015; Merhej et al., 2011 Nakajima et al., 2020). Also, specific polyamines, low molecular weight polycations aliphatic compounds, including putrescine and agmatine, act as potent inducers of DON production (Gardiner et al., 2010). However, a recent study indicates that the constant lowering of pH by agmatine addition under *in vitro* conditions is the more likely underlying induction mechanism (Nakajima et al., 2020). The host polyamine biosynthetic pathway, induced in the early stages of *F. graminearum* infection, is hypothesised to hijack this defence response to elevate the production of DON which acts as a virulence factor (Gardiner et al., 2010). DON production can also be induced by reactive oxygen species (ROS), including elevated hydrogen peroxide levels (Ponts et al., 2009), which occur during plant defence induction in most tissue types. Therefore in situations where DON production is not induced, that is, the coleoptile, or not required, that is, the roots, other repressive mechanisms are likely to operate.

Various *Tri* producing chemotypes exist in different Fusarium species and within local/national populations. These chemotypes are caused by the variant alleles of the *Tri8* gene possessed by each Fusarium strain (Alexander et al., 2011). The chemotypes so far identified include the 3A-DON and 15A-DON chemotypes, the NIV chemotypes and more recently recognised rarer 3ANX and NX chemotypes (Crippin et al., 2019). Most of these chemotypes are pathogenic towards various cereal and/or non-cereal species.

Within the wheat floral infection cycle, there are two well-studied phases where the production of *Tri* mycotoxins has been described in detail, namely early during the formation of ICs harbouring specialised toxisome structures, and later during the colonisation of the rachis node and rachis

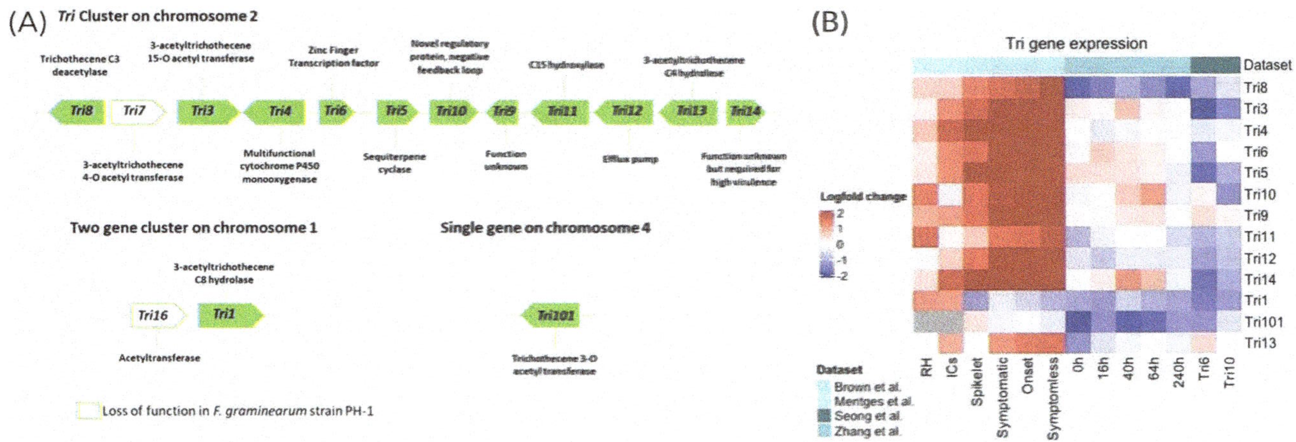

Figure 3 Organisation of the *Tri* gene clusters and *in planta* expression patterns of *Tri* genes during the different infection phases in various wheat tissue types involving either the wild type *Fusarium graminearum* strain or isogenic strains harbouring a deletion of either the *Tri6* or *Tri10* genes. (A) Genomic organisation in *F. graminearum* of the three *Tri* gene clusters and the known function of each *Tri* gene product. (B) Expression patterns of individual *Tri* genes (logfold change compared to control conditions) belonging to the three *F. graminearum* PH-1 *Tri* clusters under different conditions obtained from several transcriptome experiments: i) The Mentges et al. (2020) experiment exploring gene expression in runner hyphae (RH) and infection cushions (ICs) relative to mycelia during wheat infection, ii) the Brown et al. (2017) experiment exploring gene expression within the spikelet and during the symptomatic, onset and symptomless phases of rachis infection in wheat relative to expression in complete media, iii) the Zhang et al. (2012) experiment exploring gene expression during infection of wheat coleoptiles demonstrating the specificity of the *Tri* gene induction to the wheat spikelet and iv) the Seong et al. (2009) experiment which explores gene expression in the delta *Tri6* and delta *Tri10* mutants *in vitro*. In Table 5, additional details on these four transcriptomic data sets are given.

internode tissues (Fig. 2). Despite the specific induction of the *Tri* genes responsible for DON production in ICs, DON biosynthesis is not required for either IC development or plant cell wall penetration (Mentges et al., 2020). DON production during this phase of infection is instead proposed to function as a post-penetration virulence factor. This hypothesis is supported by the observation that DON induces programmed cell death (PCD) in plant tissues which may facilitate the release of host nutrients for fungal sequestration, aiding these initial phases of infection (Blumke et al., 2015; Desmond et al., 2008; Diamond et al., 2013). In contrast, the requirement for DON during the colonisation of the rachis is well established. Genetic disruption of *Tri5* produces a fungal mutant which is unable to grow beyond an inoculated wheat spikelet and colonise the rachis node and rachilla (Bai et al., 2001; Jansen et al., 2005; Proctor et al., 1995b). Infection with *Δtri5* mutants leads to enhanced plant defence responses in wheat spikes characterised by cell wall thickening impeding rachis colonisation, primarily callose deposition, which is accompanied by the production of jasmonate-related defence compounds (Bonnighausen et al., 2019; Jansen et al., 2005). The *Δtri5* mutant causes the formation on the outer glumes of a characteristic restricted eye-shaped lesion with a black-purple border surrounding a bleached centre (Cuzick et al., 2008). In addition, fungal biomass, floret necrosis and bleaching are reduced upon infection with the *Δtri5* mutant (Boddu et al., 2007; Jansen et al., 2005). DON is also hypothesised to function in a concentration-dependent manner whereby at lower concentrations DON suppresses host defences until the accumulation of DON to higher titres occurs, for example, during the symptomatic phase, and instead induces PCD and nutrient release to aid the necrotrophic growth of *F. graminearum* (Fig. 2) (Brown et al., 2017; Desmond et al., 2008; Diamond et al., 2013).

During the infection of other plant species and tissue types, differences have been noted in *Tri* gene induction and/or a DON requirement for virulence as described in Fig. 3 and Table 3. In most floral tissues post initial infection, the *Δtri5* mutant strain tends to colonise more slowly and causes less disease than the wild-type progenitor strain. Whereas in stem-base tissues and young seedlings, the wild-type strain and the *Δtri5* mutant strain cause similar levels of disease development and necrosis (Table 3). Recently, DON production by Fusarium was shown to be detrimental to the early stages of wheat root colonisation (Winter et al., 2019).

Collectively, these analyses of *Tri* gene induction, combined with determining the different contributions that DON provides to Fusarium virulence, indicates that control strategies that specifically target DON reduction/elimination during the early colonisation stages of FHB formation in wheat spikes will probably be highly productive.

Table 3 The requirement for deoxynivalenol (DON) as a virulence factor in different plant species and various tissue types. Assessed by comparing the disease-causing abilities of isogenic wild type and delta *tri5* gene mutant *F. graminearum* strains

	Plant tissue colonised				
Plant species*	Initial floral infection	Floral rachis/ other floral organs	Stem base/ crown/stalk	Seedling and roots	Coleoptile
Wheat	WT [1-7]	RV [1-7]	WT [8]	RV [1]	WT [9]
Barley	WT [3, 4]	NA			
Maize	WT [4]	RV [4]	WT [10]	WT [1]	
Brachypodium	RV [11]	RV [11]			
Arabidopsis	WT [6]	WT [6]			

*The DON requirement for virulence has also been reported for oats, rye and soybean in a limited number of bioassays. The experiments done and the results obtained are described in the PHI-base database (www.PHI-base.org) by the search for *Tri5* gene (Urban et al. (2020).
WT - wild-type virulence, RV reduced virulence, NA - not possible to assess, Blank - not reported
1. Proctor et al. (1995b)
2. Bai et al. (2001)
3. Jansen et al. (2005)
4. Maier et al. (2006)
5. Boddu et al. (2007)
6. Cuzick et al. (2008)
7. Mentges et al. (2020)
8. Mudge et al. (2006)
9. Zhang et al. (2012)
10. Quesada-Ocampo et al. (2016)
11. Pasquet et al. (2014)

2.3 The role of the plasmodesmata (PD) in hyphal colonisation

The transition from asymptomatic to symptomatic infection in wheat corresponds with a switch from intercellular to intracellular hyphal growth (Fig. 2). To achieve this transition *F. graminearum* hyphae utilise thin areas of the cell wall that are abundant in 'pits', collectively known as pit fields. In healthy plants, pit fields permit fluid movement and molecular exchange via plasmodesmata (PD) - including RNAs and proteins. PD can be defined as plant-specific channels between adjacent cells, connecting the cytoplasm, plasma membrane and endoplasmic reticulum via a structure called the desmotubule (Burch-Smith and Zambryski, 2012; Robards, 1971; Tilney et al., 1991). The capacity of these small structures is huge, with thousands being present at a single cellular interface (Cheval et al., 2020), acting as both local and systemic communication channels, making them ideal targets for pathogen manipulation.

The permeability of the PD structures can be controlled via two main mechanisms: callose deposition and degradation as well as PD-associated proteins, restricting or permitting the passage of certain molecules at any one time. Partially resistant wheat cultivars, such as the Chinese cultivar Sumai-3 that possess type II resistance as well as DON modification mechanisms, exhibit increased callose deposits, where the rachilla meets the rachis, when inoculated with *F. graminearum*, in comparison to more susceptible cultivars (Ribichich et al., 2000). This suggests *F. graminearum* requires PD utilisation and an absence of callose deposition for successful infection. Interestingly, the SM, Fusaoctaxin A required for *F. graminearum* virulence (Section 5), has recently been shown to be essential for the inhibition of callose deposition in wheat (Jia et al., 2019).

Microscopic studies of the *Magnaporthe oryzae*-rice leaf interaction have revealed constriction of invasive hyphae (IH) during PD passage but no evidence of permanent physiological PD remodelling or damage (Kankanala et al., 2007). The diameter of the hyphae decreases to approximately 0.5 μm during this process, which is still 10 times larger than the maximum PD size of 50 nm (Kankanala et al., 2007). Detailed microscopic studies have demonstrated *F. graminearum* operates a similar mechanism to that shown by *M. oryzae*, whereby intracellular hyphae constrict to pass through the PD before hyphal diameter enlargement in the adjacent cell (Brown et al., 2010) (Fig. 2). Some of the molecular mechanisms responsible for this highly successful and repeatedly re-iterated PD traversing intracellular colonisation behaviour have been investigated in *M. oryzae*. A mitogen-activated protein kinase (MAPK), Pmk1, has been shown to be required to prevent excessive callose deposition at PD (Sakulkoo et al., 2018). In *F. graminearum*, the corresponding MAP1/PMK1 protein is required for both initial infection and several aspects of post-invasive growth (see Sections 5 and 6) (Urban et al., 2003). Additionally, unpublished findings suggest fungal effectors can localise to the PD suggesting these channels are manipulated during infection. Research on this topic is ongoing and involves several pathosystems including Fusarium and wheat.

2.4 Fusarium infection model and possible targets for intervention

The infection biology of the *F. graminearum*-wheat interaction described above is summarised in Fig. 2. This provides an updated working infection model, building upon the previous spatial and temporal models proposed and refined by Brown et al. (2011, 2017).

This updated model highlights several critical points during the infection process suitable for targeted interventions. Strategies may include disrupting infection cushion formation, reducing early DON production, minimising host cell sensing, altering fungal secretion patterns during the apoplastic symptomless infection phase, minimising PD colonisation and /or disrupting the autophagy process within the older Fusarium hyphae to minimise nutrient redistribution to the extending hyphal fronts. As discussed later in this chapter (see Section

6), host resistance to *F. graminearum* is usually only partial and is often directed at DON detoxification and/or the formation of defensive barriers. Therefore, the identification of host cellular intervention targets required for successful pathogenesis will also be crucial to control Fusarium. For example, PD are critical for intracellular hyphal spread and host colonisation. Alteration of PD permeability in the host could lead to decreased rates of tissue colonisation. However, many of the genes responsible for PD permeability control have multiple functions, and therefore this sort of host manipulation, unless fine-tuned, could be detrimental to normal plant growth and/or development. It is also likely that specific combinations of host and pathogen intervention targets, underpinned by new fundamental insights into the cell biology of compatible interactions, will need to be used to prevent, slow down or halt Fusarium infections.

3 *Fusarium* genomes

F. graminearum was the third plant pathogenic filamentous fungus for which an assembled and highly annotated whole genome sequence became available. This was published in 2007 (Cuomo et al., 2007), after the genome sequences of the rice blast fungus *Magnaporthe grisea* (Dean et al., 2005) and the corn smut fungus *Ustilago maydis* (Kämper et al., 2006). The subsequent advent of high-throughput next-generation sequencing (NGS) technologies combined with improved assembly and bioinformatics pipelines accelerated the publication of genomes, and especially the assembly of the repeat-rich and/or hypervariable regions. Thereafter, the quality and completeness of published genomes have improved and this helped to reveal the complexity, extent and location of the hypervariable regions which had previously been difficult to identify. In addition, the ever-increasing availability of RNA seq-based transcriptomic datasets led to several revisions of the gene calls and the release of new gene annotations. In fungal plant pathogens, the hypervariable genome regions between strains are often the location of genes associated with pathogenicity, including effectors, other secreted proteins, *in planta* expressed genes and SMCs (Cuomo et al., 2007; Fouché et al., 2018; Jones et al., 2018).

The first *F. graminearum* strain sequenced was PH-1 (NRRL 32084) from North America. This was chosen as the reference strain due to its high and stable virulence, high DON production, good growth in culture and sporulation stability. Today, the genome databases at the National Center for Biotechnology Information (NCBI) and EBI-EMBL ENSEMBL contain > 14 *F. graminearum* genomes. Besides *F. graminearum*, four other Fusarium species belonging to the *F. sambucinum* species complex (FSSC) capable of causing FHB disease on wheat, barley and /or maize (Fig. 1) now have genome sequencing information available at different levels of annotation and assembly qualities (Table 4).

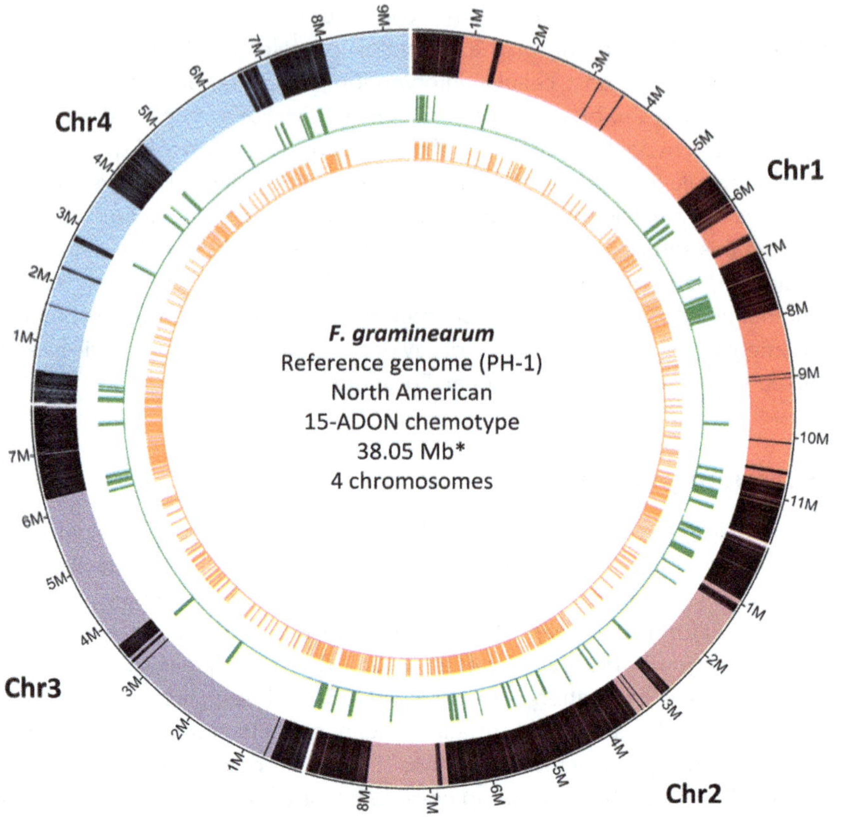

KEY
Black lines – high recombination regions
Green lines – secondary metabolite clusters
Orange lines – genes coding for secreted proteins

Figure 4 Circos plot of the fully sequenced, assembled and annotated *F. graminearum* genome of the global reference strain PH-1. The four chromosomes are presented in four different colours (outer layer).

The global reference *F. graminearum* PH-1 genome is 38.06 Mb in size, maps to four large chromosomes and contains minimal repetitive sequences, no active transposons but possesses a surprisingly large number of hypervariable regions (Cuomo et al., 2007). By comparing two strains, gene diversity, measured by single-nucleotide polymorphism density, was found to be richest in the telomere proximal regions and in several discrete internal chromosome regions (Fig. 4). In a genetic map, these hypervariable regions exhibited elevated levels of recombination (Cuomo et al., 2007; King et al., 2015). The internal hypervariable regions were hypothesised to represent ancestral telomeric fusion

Table 4 FGSC genome accessions available in various genome databases

Species	Strain code	Assembly	Scaffolds	Source	Country of origin	Host	Reference/submitter
F. graminearum	PH-1[1]	GCA_000240135.3	31	NCBI	USA	maize	Cuomo et al. (2007)
F. graminearum	PH-1[1]	GCA_900044135.1	6	NCBI	USA	maize	King et al. (2017)
F. graminearum	CS3005	GCA_000599445.1	199	NCBI	Australia	wheat	Gardiner et al. (2014)
F. graminearum	MDC_Fg1	GCA_900492705.1	96	NCBI	France	wheat	Alouane et al. (2018)
F. graminearum	ITEM 124	GCA_002352725.1	67	NCBI	Italy	rice	University of Pisa[2]
F. graminearum	FG078	GCA_006942295.1	335	NCBI	France	wheat	Laboratory of Microbiology, Parasitology and Hygiene[3]
F. graminearum	TaB10	GCA_012959185.1	54	NCBI	Serbia	apple	University of Massachusetts Amherst[4]
F. graminearum	MDC_Fg13	GCA_901446245.1	283	NCBI	France	wheat	UMR 1095 INRA/UCA, Genetics, Diversity and Ecophysiology of Cereals[5]
F. graminearum	FN009	GCA_900476405.1	715	NCBI	France	wheat	Laboratoire Agronomie et Environnement[6]
F. graminearum	NRRL28336	GCA_001717905.1	303	NCBI	USA	wheat	Walkowiak et al. (2016)
F. graminearum	DAOM180378	GCA_001717915.1	520	NCBI	Canada	maize	Walkowiak et al. (2016)
F. graminearum	DAOM241165	GCA_000966645.1	486	NCBI	Canada	wheat	Walkowiak et al. (2015)
F. graminearum	DAOM233423	GCA_000966635.1	869	NCBI	USA	wheat	Walkowiak et al. (2015)
F. graminearum	CML 3066	PRJEB12819	5	EBI	Brazil	wheat	Wood et al. (2020b)
F. asiaticum	NRRL6101	GCA_001717845.1	265	NCBI	Japan	barley	Walkowiak et al. (2016)
F. asiaticum	NRRL28720	GCA_001717835.1	353	NCBI	Nepal	maize	Walkowiak et al. (2016)
F. austroamericanum	Faus151	GCA_009617505.1	619	NCBI	Brazil	wheat	Tralamazza et al. (2019)

(Continued)

Table 4 (*Continued*)

Species	Strain code	Assembly	Scaffolds	Source	Country of origin	Host	Reference/submitter
F. austroamericanum	Faus154	GCA_009617525.1	534	NCBI	Brazil	wheat	Tralamazza et al. (2019)
F. cortaderiae	Fcor153	GCA_009617495.1	573	NCBI	Brazil	wheat	Tralamazza et al. (2019)
F. meridionale	Fmer152	GCA_009617515	694	NCBI	Brazil	wheat	Tralamazza et al. (2019)
F. meridionale	NRRL28721	GCA_001717825.1	676	NCBI	Nepal	maize	Walkowiak et al. (2016)
F. meridionale	NRRL28723	GCA_001717855	287	NCBI	Nepal	maize	Walkowiak et al. (2016)

1. Global reference strain (NRRL 31084).
2. Unpublished. Release date: 25 Sep 2017.
3. Unpublished. Release date: 12 Jul 2019.
4. Unpublished. Release date: 04 May 2020.
5. Unpublished. Release date: 13 May 2019.
6. Unpublished. Release date: 18 Jun 2018.

events that led to the formation of larger chromosomes (Cuomo et al., 2007; Paper et al., 2007). How and why these regions have maintained high genetic diversity remains unknown. In comparison to other *Fusarium* spp. genomes, the *F. graminearum* genome is considerably smaller and more compact, with *F. oxysporum* and *F. verticillioides*, for instance, possessing 59.9 Mb and 41.7 Mb genomes mapping to 15 and 11 chromosomes, respectively (Ma et al., 2010). The compactness of the *F. graminearum* genome is, in part, due to the lack of high-identity duplicated sequences in comparison to other filamentous fungi – a reduction that can be partly accounted for by the homothallic nature of this species, which rarely outcrosses, thereby limiting repeat acquisition (Goswami and Kistler, 2004). The lack of repetitive sequences is also attributable to the presence of an active repeat-induced point (RIP) mutation system – a fungi-specific system which selectively mutates duplicated sequences, thought to provide defence against the spread of transposable elements at the expense of gene family expansion (Cuomo et al., 2007; Galagan and Selker, 2004; Rep and Kistler, 2010). The most recent fully annotated and assembled *F. graminearum* genome is provided by King et al. (2017) and describes multiple new genes and gene models and brings the annotated gene total to 14145. Also the complete set of centromeric and telomeric sequences is now present (Fig. 4).

The availability of numerous annotated genomes has led to the development of bespoke bioinformatic pipelines for the prediction of candidate *F. graminearum* secreted proteins, SMCs and many other protein classes. Secretome and effectorome predictions in fungal genomes have received considerable attention, specifically because most fungal effectors are likely to be small secreted proteins (≤ 300 amino acids) and rich in cysteine residues (Lo Presti et al., 2015). The latest secretome analysis predicts that *F. graminearum* PH-1 contains nearly 900 secreted proteins and over 180 putative effectors. Interestingly, most of these genes are concentrated in the hypervariable regions in the genome (Fig. 4). Many of the predicted secreted proteins lack any annotation or known putative domains. Of the secreted proteins with predicted function, plant cell wall degrading enzymes (PCWDEs) and other enzymes are present in greater abundance in *F. graminearum* than in many other cereal- and non-cereal-infecting fungal pathogens (Brown et al., 2012).

An extensive repertoire of > 70 small gene clusters responsible for the production of various known toxins and other SMs are located across the *F. graminearum* genome often in the hypervariable regions (Fig. 4). The number of SM clusters (SMC) varies among strains of the same species and between species within the *Fusarium sambucinum* species complex (FSSC). The known and/or suspected functions of some of these SMC in disease formation and/or plant defence suppression are described in Section 5.

Due to the lack of a classical 'gene-for-gene' relationship between FHB causing Fusarium species and their small grain cereal hosts, the primary focus of inter-strain and inter-species genome comparisons is to develop pan-genomes and thereby define the 'core' and 'variable' gene repertoire for each species as well as for the wider FSSC. This approach should help to refine the invariant, stable core gene sets of which a subset is hypothesised to be integral for Fusarium virulence.

4 Transcriptome analysis of compatible interactions in multiple cereal species

Transcriptome studies permit the identification of genes differentially or specifically expressed during *in planta* infection and/or fungal development time series compared to vegetative *in vitro* growth on different media. Transcriptomics has become a popular experimental approach often used to refine gene models in genome analyses, provide evidence for alternative splicing events and is used to draw up lists of candidate virulence genes for functional evaluation. For *F. graminearum,* the initial transcriptomic studies were done using a high-throughput DNA-based microarray technology. The *F. graminearum* Affymetrix array was based on the Version 3 gene call (Guldener et al., 2006). However, since 2015, *F. graminearum* transcriptome analyses have increasingly been done using NGS technologies such as RNA seq and ChIP-seq, which permit the simultaneous analysis of fungal and host gene expression patterns.

F. graminearum transcriptome experiments range from *in vitro* growth under mycotoxin and non-mycotoxin inducing conditions, sexual and asexual spore germination time series, perithecia formation, infection time series of wheat or barley spikes, maize silks, young wheat seedlings, wheat and maize stem bases, comparison of infections of resistant and susceptible cultivars under non-field or field conditions to comparisons of WT strains and single-gene deletion strains under various *in vitro* and/or *in planta* conditions. The most well-used and cited transcriptome data sets available are summarised in Table 5.

The first *in planta F. graminearum* transcriptome study which explored barley spike colonisation revealed considerable upregulation of genes involved in virulence such as *Tri* biosynthesis and fungal CWDEs compared to growth in media, but the nutrient starvation conditions of minimal media did not promote *in vitro* expression of pathogenicity genes as previously thought (Solomon et al., 2003; Talbot et al., 1997).

Transcriptome studies using *Tri6* and *Tr10* deletion *F. graminearum* mutants (Seong et al., 2009) revealed the importance of TRI6 in the regulation of the isoprenoid pathway, acetyl-CoA biosynthesis and in regulating *F. graminearum* genes involved in virulence (Lu and Faris, 2019; Seong et al., 2009). Whereas, later studies indicate TRI6 regulates the expression of several SMC, both directly and indirectly (Shostak et al., 2020).

Table 5 The range of *in planta* and *in vitro* Genechip and RNAseq transcriptome experiments available for *F. graminearum*

Exp. type	*Fusarium graminearum* strain(s)/origin	Experiment	Transcriptomics technique	Reference
In vitro/in planta	PH-1[1] (NRRL31084) /USA	*In planta* - FHB[2] infected barley (spikes) (5 TP[3]). *In vitro* - CM[4], MMC[5] and MMN[6]	*Fg*GeneChip[7]	Guldener et al. (2006)
In vitro	PH-1	Vegetative growth and perithecium stages (3 TP) in CA[8]	*Fg*GeneChip	Qi et al. (2006)
In vitro	PH-1 PH-1Δ*Cch1* (FGSG_01364)	Vegetative growth (1 TP) and perithecium stages (2 TP) in CA	*Fg*GeneChip	Hallen and Trail (2008)
In vitro	PH-1	Conidia germination stages (3 TP) in CM	*Fg*GeneChip	Seong et al. (2008)
In planta	PH-1	Crown rot infected wheat (stem base) (3 TP)	*Fg*GeneChip	Stephens et al. (2008)
In vitro	Z3639 (NRRL 29169) Δ*tri5* (FGSG_03537)/USA	Mycelium in response to Trichodiene treatment (1 TP)	*Fg*GeneChip	Seong et al. (2009)
In planta	PH-1 PH-1Δ*Tri6* (FGSG_16251) PH-1Δ*Tri10* (FGSG_03538)	FHB infected wheat (spikes) (1 TP)	*Fg*GeneChip	Seong et al. (2009)
In vitro	CS3005/Australia	Vegetative growth in non-inducing and inducing DON media (1 TP)	*Fg*GeneChip	Gardiner et al. (2009)
In planta	PH-1	Crown rot infected wheat (stem base) (4 TP)	*Fg*GeneChip	Guenther et al. (2009)
In vitro/in planta	PH-1 PH-1Δ*FgStuA* (FGSG_10129)	*In vitro* - sporulation and SM[9] induction media. *In planta* - wheat (spikes) (1 TP)	*Fg*GeneChip	Lysøe et al. (2011)
In planta	PH-1	FHB infected wheat (spikes) (6 TP)	*Fg*GeneChip	Lysøe et al. (2011)
In vitro	PH-1 PH-1Δ*Fgp1* (FGSG_12164)	Vegetative growth in putrescine medium (1 TP)	*Fg*GeneChip	Jonkers et al. (2012)
In planta	PH-1 expressing a reporter AmCyan construct	Infected wheat coleoptile (4 TP)	*Fg*GeneChip	Zhang et al. (2012)

(Continued)

Table 5 (*Continued*)

Exp. type	*Fusarium graminearum* strain(s)/origin	Experiment	Transcriptomics technique	Reference
In vitro	PH-1 FG1 (3-ADON) FG2 (15-ADON)	Toxin inducing liquid media - individual strains and combination of two strains (1TP)	RNAseq	Walkowiak et al. (2015)
In planta	DAOM18037/Canada	Infection in wheat, barley and maize (3 TP)	Custom oligomer microarrays[10]	Harris et al. (2016)
In planta	PH-1	Sequentially infected wheat rachis segments (1 TP)	*Fg*GeneChip	Brown et al. (2017)
In vitro/in planta	PH-1	*In vitro* - mycelium in DON inducing media. *In planta* - infected wheat with pathogenic and saprophytic growth (1 TP)	RNAseq (*Fg*[11] transcripts analysis only)	Boedi et al. (2016)
In vitro/in planta	3ADON-type and 15ADON-type populations (10 strains each)/USA	*In vitro* - spores harvested from MBA[12] *In planta* - FHB infected wheat spikes (3 TP)	RNAseq (*Fg* transcript analysis only)	Puri et al. (2016)
In planta	Butte 86ADA-11/USA	FHB infected wheat spikelets and rachis (1 TP)	RNAseq (*Fg* and wheat transcript analysis)	Hofstad et al. (2016)
In planta	DAOM233423/USA	FHB infected wheat spikes - four cultivars[13] (2 TP)	RNAseq (*Fg* and wheat transcript analysis)	Pan et al. (2018)
In vitro/in planta	Field isolates/USA	*In vitro* - axenic culture. *In planta* - naturally infected wheat cultivars[15] from four locations (1 TP)	RNAseq (*Fg* and wheat transcript analysis)	Fall et al. (2019)
In vitro/in planta	PH-1 PH-1*Δ16221_3* (FGSG_16221)	*In vitro* - hyphae grown in YPD[14]. *In planta* - sequentially infected wheat rachis segments (2 TP)	RNAseq (*Fg* transcript analysis only)	Dilks et al. (2019)

In vitro/in planta	Fg-8/1/Germany	*In vitro* - mycelium grown in CM. *In planta* - RH and IC[16] from infected wheat spikes (1 TP)	RNAseq (*Fg* transcript analysis only)	Mentges et al. (2020)
In vitro	NRRL29169 NRRL29169Δ*tri6* NRRL29169Δ*tri6*/*Tri6*	Growth in preferred nutrient (PN) and non-preferred (NPN) media (1TP)	RNAseq	Shostak et al. (2020)

1. *Fusarium graminearum* reference strain (Cuomo et al., 2007).
2. FHB - Fusarium head blight.
3. TP - Time-point(s).
4. CM - Complete medium.
5. MMC - Minimal medium lacking any carbon source.
6. MMN - Minimal medium lacking any nitrogen source.
7. *F. graminearum* Affymetrix GeneChip.
8. CA - Carrot agar medium.
9. SM - Secondary metabolites.
10. Array data has been deposited at NCBI (GEO accession #GSE37886) - Harris et al., 2016.
11. *Fg* - *Fusarium graminearum*.
12. MBA - Mung Bean Agar medium.
13. Four wheat cultivars were used in this study: Wuhan 1 (Type II FHB resistant), Nyubai (Type I FHB resistant), HC374 (a FHB-resistant double haploid line derived from crossing Wuhan 1 with Nyubai) and Shaw (a FHB-susceptible cultivar).
14. YPD - Yeast peptone dextrose.
15. Three wheat cultivars with distinct levels of FHB resistance: resistant (R: IL11-28222), intermediate (I:IL10-19464) and susceptible (S: Pioneer 25R47).
16. RH and IC - Runner hyphae and infection cushions.

Multiple host interaction studies have revealed that the expression of SMs and transport genes are upregulated in a host-specific manner (Harris et al., 2016). In maize, 69 *F. graminearum* genes were found to be specifically upregulated *in planta* when compared to barley or wheat infection datasets. These comparative studies also revealed that the transcriptome of *F. graminearum* on wheat was more similar to that on barley than that on maize (Harris et al., 2016).

An *F. graminearum* transcriptome study explored the spatial-temporal variation during Fusarium infection in wheat spikes and reported the differential expression of putative effectors among the three distinct phases: symptomless, onset and symptomatic (Brown et al., 2017). In addition, very few differentially expressed *F. graminearum* genes were found in common between infected spikelet and rachis samples in two independent studies (Brown et al., 2017; Hofstad et al., 2016), suggesting that the strategies used by the pathogen when colonising different tissues might be dissimilar.

Recent RNA sequencing studies have compared *F. graminearum* differentially expressed genes (DEGs) when infecting wheat cultivars with varying degrees of susceptibility. In one study, genes involved in *Tri* biosynthesis are differentially upregulated only during the infection of resistant cultivars (Hofstad et al., 2016) (see Section 6). In another, the fungal DEGs on all cultivars, regardless of FHB susceptibility, are enriched in secreted proteins (Fall et al., 2019). These studies highlight the complexity of the interplay between the fungal transcriptome and host genotype.

Fungal transcriptome data for *F. graminearum* during *in planta* colonisation can indicate genes required for pathogenicity and therefore possible targets for chemical or genetic control. Whereas the corresponding transcriptome of an infected host can highlight genes associated with susceptibility or resistance. Whilst transcription factors such as WRKYs and bZIPs as well as membrane activity associated with host susceptibility have been found using large-scale microarray experiments (Erayman et al., 2015), RNA seq experiments have narrowed the field of candidates for genes associated with the currently unknown FHB resistance QTL located on the arm of chromosome 2DL (Biselli et al., 2018).

5 Virulence determinants, secondary metabolite clusters and effectors

In this section we discuss the role(s) of Fusarium virulence determinants primarily required to cause FHB disease. The multiple factors produced by *F. graminearum* that are deemed 'essential-for-life', whilst being required for pathogenicity, are not considered virulence determinants (Korves and Colosimo, 2009), but may serve as fungicide targets or become the targets of transgene-based control options. Also, many of these newly recognised Fusarium virulence factors, the core members of SMCs and effectors described below could potentially become new fungicide targets and/or targets for RNA interference (RNAi) technologies.

Through single- and higher-order gene deletion experiments, ~10% of the predicted *F. graminearum* gene repertoire (i.e. 1357/14145) has now been tested for function in 1669 interactions and published through peer review (www.PHI-base.org, 4.10 (Nov 2020 release), Urban et al., 2020). Other unpublished gene deletion studies are tracked in the FgMutant database (https://scabusa.org/FgMutantDb, Baldwin et al., 2018). From these reverse genetics experiments numerous intracellular and membrane spanning/ membrane associated virulence factors belonging to multiple protein classes have been identified. Many of these proteins have conserved virulence functions in other cereal and non-cereal infecting fungi. Several genome-wide functional gene analysis studies of whole gene families and intracellular signalling cascades have been completed. These studies have revealed specific G-protein coupled receptors (GPCRs), G-protein signalling, protein kinase signalling cascades, phosphatases, subtilases and transcription factors which are required for virulence, DON mycotoxin production, sexual and asexual spore formation and/or growth under nutrient-rich or -poor conditions (Table 6). These systematic gene deletion analyses have identified ~130 *F. graminearum* virulence determinants of the 603 currently reported, with the remainder being identified through more targeted studies (Urban et al., 2020). Collectively, these

Table 6 The outcomes of eight large-scale gene deletion studies in *F. graminearum* PH-1

Gene family	No. putative genes in PH-1	No. genes deleted*	Deletions resulting in reduced virulence	Reference
Polyketide synthases	15	15	0 (0%)	Gaffoor et al. (2005)
Protein kinases	116	96	42 (44%)	Wang et al. (2011)
Transcription factors	709	657	26 (4%)	Son et al. (2011)
Phosphatases	82	71	25 (35%)	Yun et al. (2015)
Autophagy associated genes (ATG)	28	28	27 (96%)	Lv et al. (2017)
G-protein coupled receptors (GPCRs)	123/105	33/105	2/5 (6%/5%)	Dilks et al. (2019)/Jiang et al. (2019)
Secretory orphan† genes	50	50	3 (6%)	Jiang et al. (2020)
Subtilases	32	32	1 (3%)	Xu et al. (2020)

* Non-lethal gene deletions.

† Genes encoding proteins with no homologues in available genomes (2010) of closely related Fusarium species.

studies have revealed a hierarchy in the role of different gene families for fungal virulence. At one end of the scale, no individual polyketide synthase is essential for *F. graminearum* virulence on wheat, whereas 96% of autophagy-associated genes are required (Gaffoor et al., 2005; Lv et al., 2017) (Table 6). The key features emerging from the gene deletion studies published prior to 2012 have already been previously reviewed (Kazan et al., 2012; Rauwane et al., 2020; Urban and Hammond-Kosack, 2013). Therefore, in this section we will highlight the more recently identified virulence factors and emerging underlying virulence mechanisms, and comment upon how some of these new fundamental insights could be used to devise new approaches to disease control.

5.1 Host plant perception and intercellular signalling

Fungal GPCRs communicate environmental changes to intracellular G-proteins within the fungal cell which then direct the appropriate biological response. The conserved GPCR FgSTE2 is involved in host-directed chemotropism, mediated by the detection of secreted wheat peroxidases (Sridhar et al., 2020). Interestingly, the deletion of *FgSte2* does not affect growth in media but decreases virulence on wheat coleoptiles highlighting the association between *F. graminearum* responding to external stimuli and virulence. The first crystal structure of a fungal GPCR was published in 2020 for the yeast protein ScSte2 (Velazhahan et al., 2020). This opens the possibility for the design of novel drugs targeting Ste2 and other fungal GPCRs. Recently, a new class of GPCRs (X-GPCRs) (Brown et al., 2018) has been associated with *F. graminearum* virulence (Dilks et al., 2019; Jiang et al., 2019). A direct interaction *in vitro* was shown between two Class X-GPCRs (FGRRES_07792 and FGRRES_16221) and their intracellular G-proteins but this interaction is yet to be confirmed *in planta*. The absence of FGRRES_16221 leads to the dysregulation of other virulence factors and the dramatic enhancement in wheat defence responses which restricts disease progression (Dilks et al., 2019). Whereas the deletion of the GPCR FgGIV2 leads to the restriction of hyphal colonisation to the apoplast (Jiang et al., 2019). Understanding the various mechanisms underpinning fungal/host plant perception should reveal future targets for pathogen control strategies.

5.2 Response to plant hormones and other defence signalling molecules

Upon perception of phytohormone induction and/or plant molecules orchestrating defensive responses within a host, *F. graminearum* upregulates a series of virulence factors involved in the transport and degradation of these molecules. Salicylic acid (SA) is a key induced component associated with wheat resistance to fungal pathogens including *F. graminearum* (Sorahinobar et al., 2016). The adenosine triphosphate-binding cassette transporter FgABCC9 is upregulated *in planta* in response to SA accumulation. FgABCC9 facilitates the export of SA from fungal cells, and *ΔFgabcc9* mutants exhibit increased sensitivity to SA, lower levels of DON accumulation and an overall reduction in virulence on wheat (Qi et al., 2018). *F. graminearum* can utilise SA as a carbon source (Qi et al., 2012) and the ability to metabolise SA also contributes to fungal virulence. The salicylate hydroxylase FgNahG catalyses the conversion of SA to catechol (Rabe et al., 2013), thus reducing SA accumulation in wheat spikes and perturbing FHB resistance mechanisms mediated by SA (see Section 6).

In a complex interplay with other phytohormones, ethylene (ET) plays an important role in plant defence (Li et al., 2019b) with ET signalling mediating FHB resistance in wheat (Foroud et al., 2019) (see Section 6). *F. graminearum* induces wheat spike bleaching brought on by DON production. This has been identified as premature senescence, a process linked to the ET signalling pathway (Bai and Shaner, 2004; Wang et al., 2018). Early studies suggested that *F. graminearum* exploits plant defence signalling by producing its own ET (Chen et al., 2009). However, no genes associated with ET production have so far been reported. Conversely, *F. graminearum* appears to manipulate *in planta* ET production via the degradation of the hormone's precursor 1-aminocyclopropane carboxylic acid (ACC) using the ACC deaminase enzyme FgACD2 (Svoboda et al., 2019). This enzyme is, however, dispensable for full virulence.

5.3 Transcriptional regulation

The orchestration of gene expression by transcription factors, themselves frequently under the control of different cellular signalling pathways, is an important mediator of cellular functions and the pathogenicity of fungi (Shelest, 2008). Their role in modulating pathogenicity-related gene expression, such as cell wall degrading enzymes and SMCs has previously been well documented and reviewed (Keller, 2019; Kubicek et al., 2014 Rauwane et al., 2020). For *F. graminearum*, key transcription factors required for full pathogenicity, primarily although not exclusively due to their regulation of DON biosynthesis, include FgStuA, TRI6, FgMED-1, the MADS-box transcription factor FgMcm1 and the velvet complex (FgVeA/FgVelB), which includes the transcription factor FgLaeA (Fan et al., 2019; Jiang et al., 2011; Jiang et al., 2012; Kim et al., 2013; Proctor et al., 1995a; Seong et al., 2009; Shostak et al., 2020; Yang et al., 2015).

FgStuA is conserved among Ascomycetes and functions as a global regulator of sporulation, mycotoxin biosynthesis and the induction of several groups of cell wall-associated proteins. The *Tri* biosynthesis pathway is not expressed in FgStuA mutants and as a consequence virulence is highly reduced (Lysøe et al., 2011). The velvet complex (FgVelB/FgVeA/FgLaeA) is also highly conserved throughout the Ascomycetes and, in *F. graminearum*, is required for the biosynthesis of DON and many other SMs as well as for the regulation of many aspects of fungal development. However, the molecular mechanisms of SMC regulation by the velvet complex remain largely unknown (Bayram et al., 2008; Chen et al., 2019). Recent studies in *Fusarium fujikuori* suggest that the velvet protein LaeA, with its methyltransferase domain, may regulate SM biosynthesis through epigenetic modifications such as histone acetylation (Niehaus et al., 2018). TRI6 (Fig. 3) has a far more restricted taxonomic distribution and is a global regulator of many SMC genes in Fusarium and Trichoderma species. Many SMC, such as the *Tri* cluster (C23, Table 7), are regulated directly by TRI6, whereas the regulation of the Gramillin SMC (cluster C02, Table 7) occurs indirectly through the interaction between TRI6 and the cluster-specific transcription factor GRA2 (Seong et al., 2009; Shostak et al., 2020). Recently, FgCrz1A, a calcineurin-responsive transcription factor, has been shown to be required for virulence and DON biosynthesis in wheat spikes and maize silks, to control responses to various stresses and for different aspects of *F. graminearum* development (Chen et al., 2019a). FgCrz1A was also shown to act as a downstream regulator of Ca^{2+} signalling in *F. graminearum* which possibly accounts for the plethora of affected responses when this transcription factor is absent.

The genome-wide *F. graminearum* transcription factor study identified six genes predicted to code for DNA binding proteins, which when deleted reduced virulence but did not compromise DON production (Son et al., 2011). Of these six genes FGSG_10179, FGSG_08737, FGSG_08182, FGSG_08028, FGSG_01293 and FGSG_00719, the latter four have been designated as uncharacterised Zn2-C6 fungal-specific transcription factors. This study reveals the possibility that other transcriptionally controlled gene regulatory network(s) mediate the *in planta* Fusarium penetration and colonisation process alongside those which regulate DON mycotoxin production.

Although transcription factors per se are unlikely to be exploited directly as drug targets, gene sequences coding for transcription factors have already been tested in plant biotechnology approaches to control cereal infecting Fusarium species using the technique called HIGS (see Section 7).

5.4 Transport

Multiple SNARE proteins, required for intracellular vesicle trafficking and membrane fusion, are involved in the production of DON mycotoxin (Adnan et al., 2020a,b; Han et al., 2017). The SNARE protein FgSec22 localises to the endoplasmic reticulum of fungal cells. *Fgsec22* deletion mutants show a significant downregulation of genes in the *Tri* cluster compared to wild type, as well as reduced vegetative growth and disease symptom development in wheat spikes (Adnan et al., 2020a). The sub-apically localised t-SNARE protein

FgSso2 is required for the export of DON out of the fungal cell and is required for pathogenicity (Hong et al., 2010; O'Mara et al., 2020). Upon deletion of *Sso2*, mutants exhibit downregulation in DON biosynthesis genes and an upregulation in genes associated with cellular repair mechanisms, possibly due to DON accumulation within the fungal cell (O'Mara et al., 2020). Interestingly the sub-apical membrane fusion mediated by FgSso2 may be an important mechanism contributing to *F. graminearum* virulence independent of its role in DON export. *Δsso2* deletion mutants display reduced virulence during barley colonisation, where DON is not required for colonisation throughout the spike (Hong et al., 2010; Maier et al., 2006). Targeting the disruption of specific transport mechanisms and especially *Tri* transportation is an attractive option for disease control because of the range of physiological and molecular feedback loops potentially triggered that could stress Fusarium cells and thereby alter *in planta* communication events. In addition, targeting transport mechanisms could hinder the hyphae's ability to suppress or detoxify host defences and/or perturb the essential autophagy process.

5.5 Secondary metabolites

The *F. graminearum* genome is predicted to contain between 67 and 86 SM gene clusters that are each responsible for the production of specific characterised and uncharacterised metabolites (Bakker et al., 2018; Brown and Proctor, 2016; King et al., 2015; Sieber et al., 2014). Several are known to function as harmful mycotoxins, whilst others play roles in host infection, inter-species competition and/or defence against mycophagy (Hoogendoorn et al., 2018). The signature genes in SMCs are mainly terpene synthases (TPS), polyketide synthases (PKS), non-ribosomal peptide synthetases (NRPS) and dimethylallyltryptophan synthases (DMATS) (Sieber et al., 2014).

One of the most well-known SMCs in *F. graminearum* is the *Tri* gene cluster (C23) (Fig. 3). Most of the genes that participate in trichothecene biosynthesis are found in a 10-member gene cluster of approximately 25 Kb in length. The importance of these *Tri* genes and trichothecenes during Fusarium-wheat infection is explained in detail in Section 2.2. Other well-known and well-characterised metabolites involved in *F. graminearum* virulence are the siderophores. Siderophores are low molecular weight organic compounds with strong iron-chelating activity biosynthesised by NRPSs (Lee et al., 2005; Oide et al., 2006). Siderophores are usually involved in reproductive and pathogenic developmental processes and can provide a protective antioxidant mechanism against ROS. *F. graminearum* harbours three *NRPS* genes, each residing in a different SMC, involved in the production of siderophores: *NPS1, NPS2 and NPS6* in SMCs, C63, C33 and C21, respectively (Table 7). Deletion of the *NPS2* gene, responsible for the biosynthesis of the intracellular

Table 7 *In planta* expression of the 86 designated *Fusarium graminearum* secondary metabolites clusters

Cluster number[1]	Cluster name/ metabolite produced[2]	Core enzyme[3]/ number of TF[4]	Expression[5]
C01		/ 1 TF	PI[6]
C02	Gramillin	NRPS8 / 2 TF	PI/PR[7]
C03		/ 1 TF	PI/PR
C04		/ 1 TF	NI[8]
C05		/ no TF	NI
C06		NRPS16, NRPS19 / no TF	PR[9]
C07		/ 3 TF	PI/PR
C08		PKS11 / 1 TF	NI
C09		/ no TF	PR
C10		/ no TF	PI/PR
C11		/ 1 TF	PI
C12		/ 4 TF	PR
C13	Aurofusarin	NRPS4, PKS12 / 2 TF	PI/PR
C14		/ 1 TF	NI
C15	Zearalenone	NRPS15, PKS13, PKS4 / 1 TF	NI
C16		PKS15 / no TF	PI
C17		/ no TF	PI
C18	Orcinol/orsellinic acid	PKS14 / no TF	PI/PR
C19		/ 4 TF	PI/PR
C20		/ 2 TF	NI
C21	Triacetylfusarinin	NRPS6 / no TF	PI
C22		/ no TF	PI/PR
C23	Trichothecene	/ 2 TF (*Tri6* and *Tri10*)	PI
C24		/ no TF	PI
C25		/ 1 TF	PI
C26		PKS8 / no TF	PI
C27		/ no TF	PI/PR
C28	Carotenoid	/ no TF	PI/PR
C29		/ 1 TF	PI/PR
C30		/ 1 TF	NI
C31/C32		PKS2 / 2 TF	NI
C33	Ferricrocin/koraiol	NRPS2 / 1 TF	NI
C34		PKS5 / no TF	NI
C35		/ 1 TF	PI
C36		/ 2TF	PI/PR

Cluster number[1]	Cluster name/ metabolite produced[2]	Core enzyme[3]/ number of TF[4]	Expression[5]
C37		/ 2 TF	PI/PR
C38		NRPS10 / 1 TF	PI/PR
C39		/ 2 TF	PI/PR
C40		/ 3 TF	PI/PR
C41	FusarinC	/ 1 TF	PR
C42		PKS10 / no TF	NI
C43		/ no TF	PI/PR
C44		PKS7 / 3 TF	PI/PR
C45		/ no TF	NI
C46		/ 1 TF	PR
C47	FusaristatinA	PKS6 and NRPS7 / no TF	PR
C48	guaia-6,10-diene	/ 2 TF	NI
C49	Butenolide	/ 1 TF	PI
C50		/ 1 TF	PI/PR
C51		/ 4 TF	PI/PR
C52		/ 4 TF	PI/PR
C53		PKS3 / 1 TF	NI
C54		/ no TF	PI/PR
C55		/ 2 TF	PI/PR
C56		/ no TF	PI/PR
C57		/ 1 TF	NI
C58		/ no TF	NI
C59	Culmorin	/ no TF	PI/PR
C60	Fusarielin	PKS9 / 1 TF	PI/PR
C61		PKS1, NRPS18 / no TF	PR
C62		/ no TF	PR
C63	Malonichrome	NRPS1 / no TF	PI
C64	Fusaoctaxin A	NRPS5, NRPS9 / no TF	PI
C65		/ 1 TF	NI
C66	Chrysogine	NRPS14 / 1 TF	PI/PR
C67		/ 1 TF	NI
C68		NRPS3 / 1 TF	PI/PR
C69		/ 1 TF	NI
C70*		/ no TF	-
C71*		/ 1 TF	-
C72*		/ 1 TF	-

(*Continued*)

Table 7 (*Continued*)

Cluster number[1]	Cluster name/ metabolite produced[2]	Core enzyme[3]/ number of TF[4]	Expression[5]
C73*		/ 1 TF	-
C74*		PKS52 / no TF	-
C75*		/ no TF	-
C76*		/ no TF	-
C77*		/ no TF	-
C78*		/ no TF	-
C79*		/ no TF	-
C80*		/ no TF	-
C81*		/ no TF	-
C82		/ 2 TF	PI/PR
C83		/ no TF	PI
C84		/ no TF	NI
C85		/ no TF	PI
C86		/ no TF	PI/PR

1. Cluster number according to Sieber et al. (2014), King et al. (2018).
2. Cluster name according to Sieber et al. (2014); Bahadoor et al. (2018), Jiang et al. (2019), Sørensen et al. (2014) Burkhardt et al. (2016) and Wollenberg et al. (2017).
3. Core enzyme names according to Hansen et al. (2015).
4. TF - predicted transcription factors according to Sieber et al. (2014), Shostak et al. (2020).
5. Expression analysis based on three transcriptome datasets: Boedi et al. (2016), Brown et al. (2017) and Mentges et al. (2020). All three datasets have analysed the expression of Fg genes both *in vitro* and *in planta*.
6. PI (plant induced) - expression of core genes was upregulated *in planta* in all three datasets.
7. PI/PR (plant induced/plant repressed) - expression of core genes was either up- or downregulated *in planta* depending on each dataset.
8. NI (non-induced) - expression of core genes not reported to change significantly in both *in vitro* and *in planta* conditions.
9. PR (plant repressed) - expression of core genes was downregulated *in planta* in all three datasets.
*Predicted secondary metabolites clusters absent in the reference strain PH-1. No transcriptome data is available for these clusters.

siderophore ferricrocin, results in the inability of the fungus to produce sexual spores. However, the *sid1* mutant which lacks the siderophore biosynthetic gene *SID1* in the ferricrocin cluster (C33) was unable to grow beyond the single inoculated spikelet in infected wheat. Deletion of the *NPS6* gene, required for the biosynthesis of the extracellular siderophore triacetyl fusarinine C (TAFC) (Fig. 5), did not affect sexual reproduction but caused enhanced sensitivity to iron starvation and oxidative stress leading to reduced virulence (Oide et al., 2006). *NPS1* is responsible for the biosynthesis of a second extracellular siderophore malonichrome which is not required for virulence. However, triple *nps1nps2nps6* deletion strains, lacking all three siderophores, are even more impaired in virulence than the double *nps2nps6* strains (Oide et al., 2015).

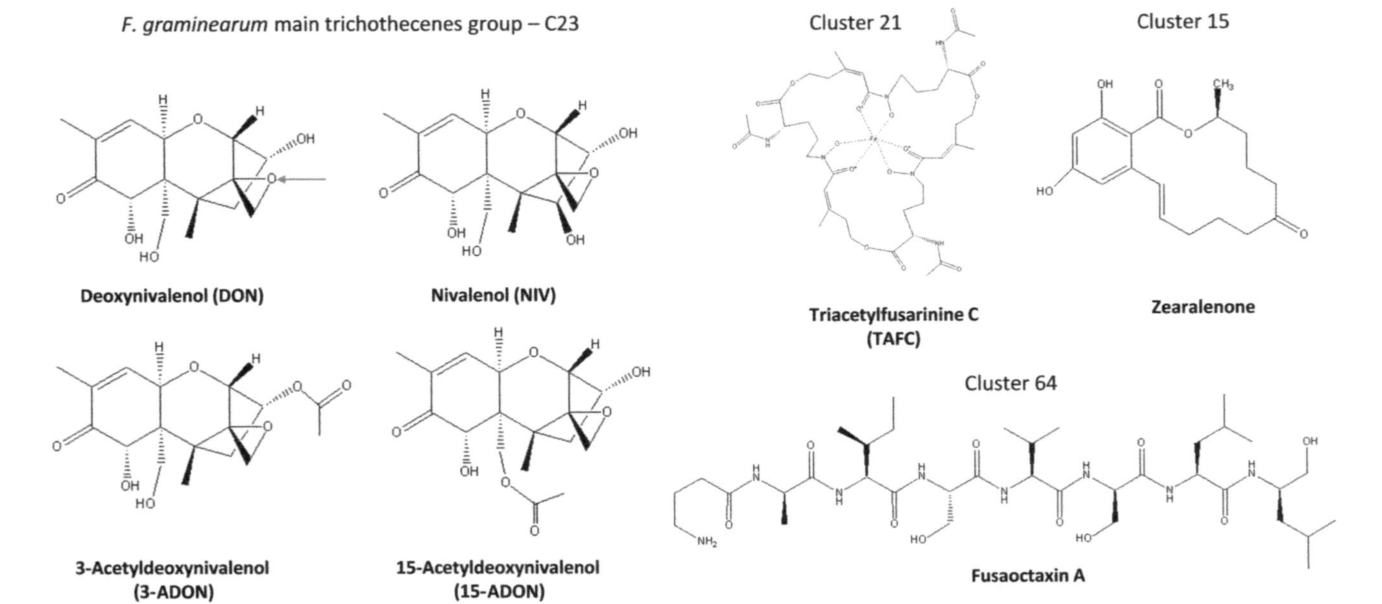

Figure 5 Chemical structures of secondary metabolites often secreted during cereal infection by various Fusarium species. Clusters are numbered according to Table 7. Trichothecenes, TAFC and Fusaocataxin A are important for *F. graminearum* virulence in wheat. Toxicity of trichothecenes relies upon the epoxide moiety (red arrow). The wheat *Fhb7* resistance gene codes for a glutathione S-transferase (GST) which catalyses conjugation of a glutathione (GSH) unit onto the toxic epoxide moiety of the DON mycotoxin and thereby detoxifies this Fusarium virulence promoting mycotoxin (described further in Section 6).

More recently, another SM gene cluster was shown to be important for *F. graminearum* virulence. Cluster C64 is *in planta* inducible and contains two NPSs (NPS5 and NPS9) (Sieber et al., 2014). The SMC product was identified as fusaoctaxin A, a linear, C-terminally reduced and D-amino acid residue-rich octapeptide required for *in planta* cell-to-cell invasion (Fig. 5) (Jia et al., 2019). Interestingly, the deletion of two genes within this cluster caused reduced *F. graminearum* virulence in wheat spikes and seedlings, but not in maize stalks (Jia et al., 2019). A contrasting study revealed that the *F. graminearum NRP9* deletion mutant is not able to synthesise fusaoctaxin A, yet shows an increase in virulence on wheat spikes compared to the wild-type strain. Therefore, the various role(s) of this octapeptide in pathogenicity remains to be fully elucidated (Westphal et al., 2019).

Identification of the function(s) of individual SM along with the various intracellular and extracellular host components targeted could potentially reveal a wealth of new intervention options to be pursued via GE approaches and/or via the screening of cereal landrace collections or TILLING populations for variants less exploitable by secreted Fusarium SMs.

5.6 Effectors

Throughout infection of the wheat spike, *F. graminearum* hyphae are predicted to secrete a substantial repertoire of small, secreted and cysteine-rich proteins in a phase-specific manner (Brown et al., 2017). Some of these proteins are predicted to be classical effectors - defined as small proteins (or non-proteinaceous molecules) which function to facilitate host colonisation often by manipulating the host immune system (Lo Presti et al., 2015). Many effectors characterised in other fungal species function in a classical gene-for-gene manner whereby Avr effectors interact either directly or indirectly with host resistance/guardee proteins or susceptibility targets to bring about either immunity or susceptibility, respectively, in a strain-dependent manner. By contrast, no Avr effectors have yet been identified in *F. graminearum* which may be due to (1) the major requirement for DON and other secreted metabolites for Fusarium virulence, (2) tissue-based floral resistance to Fusarium in all cereal species is controlled by either major or minor QTLs with predominantly additive effects and/or (3) the continually advancing biotrophic phase of the infection is relatively short-lived and appears to be sustained by a far longer necrotrophic phase within which fungal autophagy contributes to nutrient recycling from the older to the younger hyphae within the *in planta* mycelial colony. Instead several non-Avr effectors have been characterised in *F. graminearum* and are described in Table 8 and the remainder of this section. These putative effectors are hypothesised to aid infection by either (1) suppressing host defences to facilitate the symptomless phase of infection or (2) activating host defences

Table 8 The various non-Avr effectors produced by *Fusarium graminearum*

Effector/ FGSG locus ID	Mature protein size (aa)	Known or hypothesised function(s)	Domains	Single gene deletion strain phenotype/ BSMV-VOX phenotype	Localisation*	Plant species used in the bioassay	References
FGL1 05906	335	Secreted lipase involved in lipid degradation and inhibition of callose formation	Lipase	Reduced virulence, no growth of mutant beyond inoculated spikelet/N/A	apoplastic	Maize Wheat Soybean	Voigt et al. (2005) Blümke et al. (2014)
FgSSP32 15251	47	Induces low molecular weight compound production in *N. benthamiana*	N/A	Wild-type/reduced virulence	apoplastic	Wheat Nicotiana	Walker (2020)
FgSSP33 15437	53	Induces low molecular weight compound production in *N. benthamiana*	N/A	Wild-type/reduced virulence	apoplastic	Wheat Nicotiana	Walker (2020)
FgSSP6 10212	121	Hypothesised to enhance enzymatic activity of CWDEs	Cerato-platanin	Wild-type/enhanced susceptibility	apoplastic	Wheat Nicotiana	Machado (2017)
FgSSP7 11205	122	Hypothesised to enhance enzymatic activity of CWDEs	*Cerato-platanin*	*Wild-type/enhanced* susceptibility	*apoplastic*	Wheat Nicotiana	Machado (2017)
FgRALF 15123	59	A Group III rapid alkalinisation factor with a potential accessory role in *F. graminearum* virulence*	'YISY' motif required for alkalinising activity in plants	Wild-type/enhanced susceptibility	apoplastic	Wheat Arabidopsis	Wood et al. (2020a)
Osp24 11564	113	Binds to and destabilises TaSnRK1α a SNF1-related protein kinase required for metabolism and FHB resistance	N/A	Reduced virulence, mutant is defective in wheat rachis colonisation/N/A	cytoplasmic	Wheat	Jiang et al. (2020)
Fg12 11190	112	Possesses ribonuclease activity that induces plant cell death	Ribonuclease	Reduced virulence on soybeans hypocotyls/N/A	Unknown	Soybean	Yang et al. (2020)

CWDEs - cell wall degrading enzymes.
*The *F. oxysporum* f.sp. *lycopersici* homolog, the Group I RAFF F. RALF, induces alkalisation in host tissue and is required for virulence (Masachis et al., 2016).

to induce localised cell death leading to the release of nutrients for fungal sequestration during the symptomatic phase of infection. However, our understanding of the precise interactions between these secreted proteins and host factors remains limited.

One of the first effectors identified in *F. graminearum* and the most well characterised is the secreted lipase FGL1. FGL1 is required for full virulence of *F. graminearum* on wheat spikes and acts to suppress wheat callose synthesis in rachis node tissues through the release of polyunsaturated free fatty acids. FGL1 mutants are unable to overcome this enhanced physical barrier which is macroscopically visible and are therefore unable to grow beyond the initial infected spikelet (Blümke et al., 2014; Voigt et al., 2005). The abundance of FGL1 lipase is regulated by the Map1 protein kinase signalling cascade (Salomon et al., 2012).

Two cerato-platanin (CP) effectors, FgSSP6 and FgSSP7, have been shown to contribute to *F. graminearum*-wheat floral infections. The CP protein family, which has a proven role in facilitating the loosening of cell wall matrices (Baccelli, 2015), has been implicated in the pathogenicity of several other filamentous fungi including *Botrytis cinerea* and *Sclerotinia sclerotium* (Frias et al., 2013; Yang et al., 2018). While *FgSSP6* and *FgSSP7* gene deletion strains retain wild-type pathogenicity, overexpression of either protein in wheat prior to inoculation using the BSMV-VOX system (Fig. 6) enhances susceptibility to *F. graminearum* (Machado, 2017). As accessory virulence genes, *FgSSP6* and *FgSSP7* are hypothesised to act synergistically with fungal cellulases by making the physical barrier of the cell walls more accessible to enzymatic degradation. Hence overexpression of these SSPs ahead of the Fusarium infection may enhance infection by 'opening up' cell wall matrices to the enzymatic activity of the subsequently secreted FgCWDEs (Machado, 2017).

FgRALF, an effector belonging to the rapid alkalinisation factor (RALF) protein family, has been characterised in *F. graminearum*. While RALF proteins are predominantly found in plant species, having a role in development, *ralf* genes have also been identified in many fungal species including the Fusaria. *F. oxysporum* f.sp. *lycopersici* mutants lacking the *ralf* gene exhibit significantly attenuated virulence on tomato roots (Masachis et al., 2016). In contrast, deletion of *FgRALF* in *F. graminearum* does not lead to attenuated virulence. However, overexpression of FgRALF during wheat spike infection causes a slightly enhanced susceptibility towards *F. graminearum* suggesting that FgRALF may play an accessory role in virulence via an unknown mechanism (Wood et al., 2020a).

Large-scale gene deletions of putative effectors identified the gene *Osp24* as encoding an effector required for host-specific *F. graminearum* virulence on wheat. Once secreted, Osp24 enters the host cell and directly interacts with the wheat SNF1-related protein kinase TaSnRK1α, accelerating its degradation (Jiang et al., 2020). Osp24 competes with the DON-induced wheat protein *Triticum aestivum* Fusarium Resistance Orphan Gene (TaFROG) which, upon direct interaction, acts to stabilise TaSnRK1α and enhance FHB resistance (Perochon et al., 2015). Although originally considered to be an *F. graminearum* orphan gene, BLAST analyses reveal the *Osp24* gene is present in other Fusarium species with either a pathogenic or non-pathogenic lifestyle as well as several insect-infecting Ascomycetes (E value cut off $1e^{-30}$).

Recently a member of a widely conserved ribonuclease family has been identified as a secreted effector in *F. graminearum*. Fg12 was isolated through a proteomics approach and has been shown to induce cell-death in tobacco and tomato leaves, dependent on its RNAse activity (Yang et al., 2020). A reduction in virulence on soybean hypocotyls was induced upon deletion of *Fg12*; however, its role in wheat pathogenicity has yet to be determined. The ortholog of Fg12 in *Zymoseptoria tritici*, Zt6, induced a similar cell-death phenotype upon transient expression in tobacco; however, Zt6 was not required for virulence on wheat (Kettles et al., 2017).

Two additional effectors, FgSSP32 and FgSSP33, have been identified which are specific to the *Fusarium* genus. Both proteins are secreted during the necrotrophic phase of fungal infection, and heterologous expression of these proteins in the model species *Nicotiana benthamiana* leads to the induction of vascular-associated cell death as well as the induction of a low molecular weight plant metabolite Walker (2020). Once again, deletion of these genes in *F. graminearum* did not reduce virulence on wheat spikes or coleoptiles, whereas pre-expression of the effectors in wheat spikes via the BSMV-VOX system leads to a reduction in virulence Walker (2020).

With the exception of FGL1 and Osp24, the other five effectors described above are not individually required for virulence in the *F. graminearum*-wheat interaction. This may be a result of (1) genetic redundancy, where multiple effectors have the same function, or (2) buffering, where the host compensates or interferes in signalling by using alternative pathways (Hillmer et al., 2017; Tyler, 2017). In order to overcome this problem, *in planta* overexpression of recombinant *F. graminearum* effectors in wheat has been achieved using viral vectors to enable dissection of the subtler *in planta* contribution(s) of individual effectors. The most widely used viral vector for expression in wheat and barley is the Barley Stripe Mosaic Virus (BSMV); a commonly used vector for virus-induced gene silencing (VIGS) studies (Lee et al., 2012). This BSMV-virus-mediated overexpression (BSMV-VOX) system (Fig. 6) has been used to characterise effectors such as the *Pyrenophora tritici-repentis* effector ToxA (Manning et al., 2010) and more recently to characterise the aforementioned FgRALF, FgSSP6, FgSSP7, FgSSP32 and FgSSP33 (Machado, 2017, Walker, 2020).

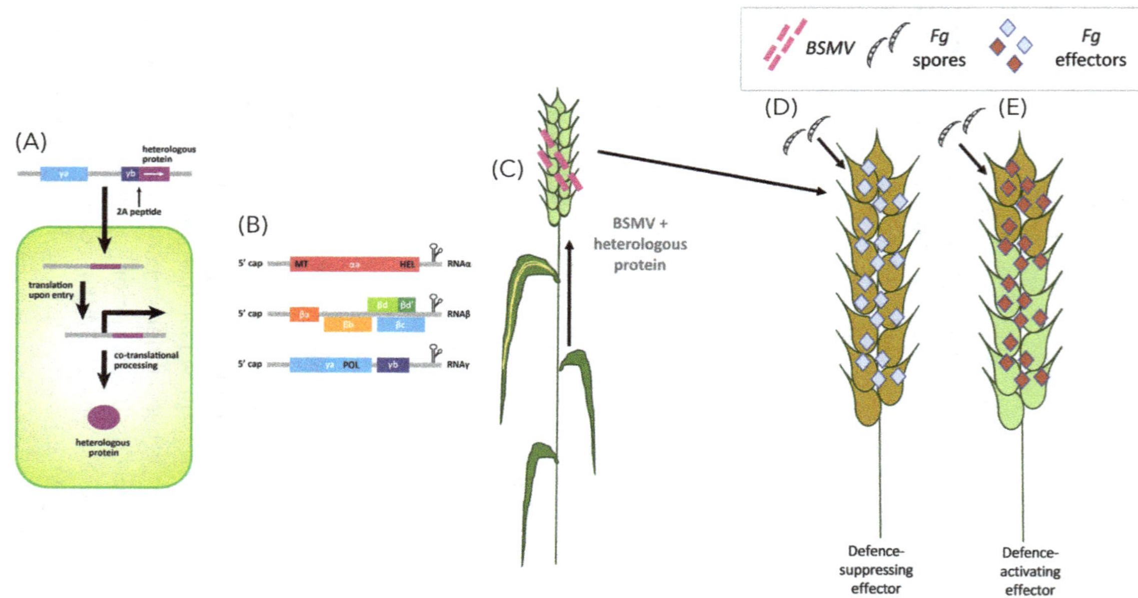

Figure 6 The Barley Stripe Mosaic Virus – virus-mediated overexpression (BSMV-VOX) technology. Panels (A) and (B) of this figure are adapted from Lee et al. (2012). (A) The details of virus-mediated overexpression (VOX) system. The heterologous protein coding sequence is inserted in the γ genome of BSMV, upstream of the in frame stop codon in the γb ORF. A gene for the autoproteolytic peptide 2A is also inserted between the 3′ terminus of the γb ORF and the gene of interest for processing the fusion protein during translation, thus releasing the heterologous protein of interest. (B) The BSMV genome is composed of three RNAs that are capped at the 5′ end and form a tRNA-like hairpin secondary structure at the 3′ terminus. RNAα encodes the αa replicase protein containing methyl transferase and helicase domains. RNAβ encodes coat and movement proteins whilst RNAγ encodes the polymerase (POL) component of replicase, and the cysteine-rich γb protein involved in viral pathogenicity. (C) Following inoculation of wheat leaves with BSMV containing the heterologous protein, the virus travels to the wheat spike whilst expressing the protein of interest. The wheat spikes are then inoculated with *F. graminearum* spores at anthesis and go on to cause FHB disease. (D) Overexpression of defence-suppressing effector proteins causes increased pathogenicity of *F. graminearum* and a greater number of wheat spikelet exhibiting FHB disease symptoms. Overexpression of defence-activating effectors (E) reduce the virulence of *F. graminearum* and fewer infected spikelets are recorded.

Although effectors are often considered to be species or even lineage-specific, several conserved effector families present in phytopathogenic fungi are notably absent from the *F. graminearum* genome. The LysM effector family, for instance, is defined by the presence of multiple, typically three, extracellular carbohydrate-binding LysM domains which bind to and sequester fungal chitin oligosaccharides, conserved fungal cell wall components that elicit host immune responses (De Jonge and Thomma, 2009). Over 300 putative LysM containing effectors have been identified in 62 published fungal genomes. These include Vd2LysM, Ecp6 and FvLcp1 from *Verticillium dahlia, Cladosporium fulvum* and *Fusarium verticillioides,* respectively, which each contain a triple LysM domain and have been shown to bind chitin, suppress various induced plant defence responses and/or contribute to virulence (Bolton et al., 2008; Kombrink et al., 2017; Zhang et al., 2019).

The necrosis and ET-inducing protein 1 (NEP1)-like proteins (NLPs) are also widely distributed across the bacterial and fungal kingdoms (Gijzen and Nürnberger, 2006). NLPs act as cytotoxic proteins solely in dicotyledonous plant species, triggering leaf necrosis and immune defence responses. The number of NLP family members varies significantly between species - the oomycete *Phytophthora sojae*, for instance, has 33 members (Dong et al., 2012). In contrast, the NLP protein family is absent in *F. graminearum* but present in other Fusarium species including *F. pseudograminearum, F. culmorum, F. verticillioides, F. fujikuroi, F. solani* and *F. oxysporum*.

Homologues of other highly conserved fungal effectors have been identified in *F. graminearum* during various bioinformatic analyses, yet functional characterisation of these proteins is still required. The *Colletotrichum tofieldiae* effector NIS1 is well represented in both Ascomycetes and Basidiomycetes species. NIS1 suppresses PAMP-triggered immunity by targeting the BAK1 immune kinases to impair intracellular plant signalling (Irieda et al., 2019; Yoshino et al., 2012). Within the *Fusarium* genus, NIS1 has been identified in the soybean root pathogen *Fusarium virguliforme* (Chang et al., 2016) and in *F. graminearum*. Whilst the former is responsible for the Sudden Death Syndrome (SDS) symptoms on infected soybean leaves, the role of NIS1 in *F. graminearum* virulence is yet to be reported.

By using effectors to screen for the presence of resistance (*R*) genes or susceptibility loci in germplasms of crop cultivars or a sexually compatible species, new disease-resistant cultivars can be produced (Vleeshouwers and Oliver, 2014). However, without the presence of the classical *R* - *Avr* gene combinations identified in any cereal host - Fusarium species combination tested, high-throughput screening for R protein function within any cereal species is not possible. Instead, the exploitation of the new knowledge gained on effector biology and function during FHB disease formation will most likely focus instead on devising methods to achieve direct effector inhibition and/or alter effector targets within the host to develop resistant cultivars. So far only two *F. graminearum* effectors, FGL1 and Osp24, are required for full virulence on wheat, and only Osp24 has a proteinaceous host-target identified (Jiang et al., 2020). Therefore it is vital to further explore and characterise the vast repertoire of putative core Fusarium candidate effectors as well as identify their primary plant targets to find more avenues for possible FHB control. With the increased availability over the coming years of fully sequenced and assembled wheat, barley, rice and maize genomes, some of these host vulnerability targets could simply be removed or replaced via marker-assisted breeding with alternative wheat sequences shown under laboratory conditions not to bind or only weakly bind Fusarium effectors that are required for full virulence. In addition, these Fusarium effectors could potentially become new fungicide targets and/or targets for RNAi technologies.

6 Plant defence pathways and resistance QTLs

Natural plant defences against cereal-infecting Fusarium species have been extensively explored using four main experimental approaches. Mutant analyses in wheat, barley and/or the model hosts Arabidopsis and Brachypodium have focussed in particular on the plant defence signalling components already well characterised in other plant-pathogen interactions, the roles of various phytohormones and the mechanisms underlying DON mycotoxin detoxification (reviewed by Brewer and Hammond-Kosack (2015) and Peraldi et al. (2011)). Numerous transcriptome analyses have explored FHB-resistant and FHB-susceptible cereal genotypes harbouring different QTL resistance loci in isogenic backgrounds, conferring either resistance to initial infection (type I), a reduced rate of disease spread in the rachis from the initially infected spikelet (type II), or a reduction in DON levels (type III) as well as the host plants responses to infections in different tissue types, infections with different *F. graminearum* mutants compromised in virulence or following exogenous application of DON mycotoxin. These transcriptome studies have greatly increased knowledge on the vast repertoire of differentially altered in abundance gene transcripts in the different plant species as well as in distinct tissue types (reviewed by Kazan and Gardiner, 2018). In addition, some of these transcriptome studies have been extended in scope by including similarly detailed metabolomics and/or proteomics analyses. Two less often used experimental approaches have been the analysis of selected genes via overexpression in stably transformed cereal or Arabidopsis plants and/or via transient VIGS of cereal host genes in floral tissues ahead of Fusarium infection using predominantly the BSMV vector system (Brauer et al., 2020; Lee et al., 2012) (Fig. 6).

Overall, the new fundamental knowledge that has emerged from these studies is complex and clearly indicates that different plant species and tissue types activate different responses to Fusarium infections or to just the DON mycotoxin (Brewer and Hammond-Kosack, 2015; Harris et al., 2016; Kazan and Gardiner, 2018; Walter et al., 2010). In floral tissues, the current evidence indicates that both initial colonisation and subsequent disease development are limited by SA and NPR1 mediated defence signalling, whilst jasmonic acid (JA) and ET mediated signalling facilitate early infection. However, roles for both JA and ET in resistance to late stages of infection have also been noted. Comparisons between FHB-resistant and FHB-susceptible cereal genotypes harbouring different QTL resistance loci clearly reveal that the ability to mount an earlier and stronger defence response seems to be a factor contributing to FHB resistance (Kazan and Gardiner, 2018). Upregulation of the phenylpropanoid pathway, lignin biosynthesis and many of the classic PR genes induced in other host-pathogen interactions have been evident in multiple FHB-cereal analyses. However, there is also emerging evidence for differential QTL specific transcriptome, proteome and metabolite responses as well as the induction of specific NBS-LRR, receptor kinases, WRKY transcriptions factors associated with some of the QTL-mediated resistance responses (Kazan and Gardiner, 2018).

Interestingly, there is considerable evidence for the suppression of pathogen virulence mechanisms by the host. For example, the increase in abundance of transcripts for numerous host genes encoding specific ABC transporters and specific UDP-glucosyl transferases in situations where Fusarium hyphae are producing DON or where DON alone has been applied. This minimisation of the effects of fungal toxins and enzymes is likely to be a factor contributing to disease resistance (Erayman et al., 2015). Also, one comparative barley study identified hundreds of *F. graminearum* inducible long-non-coding RNAs (lncRNAs) with possible roles in the regulation of barley defences (Huang et al., 2016).

Several overexpression studies in wheat or barley conferring increased FHB resistance have been reported. Defensive roles have been revealed for NPR1 (Makandar et al., 2006), the lipid transfer protein TaLTP5 (Zhu et al., 2012), the genes *PFT* and *WFhb1-1* which reside within the Fhb1 QTL region on chromosome 3B (see below), and the DON responsive gene *Ta-NACL-D1* (Perochon et al., 2019). These proteins and their effects on FHB disease development are presented and discussed in Section 7.

Host-induced gene silencing experiments done through either stable transformation or the more versatile genotype independent BSMV-VIGS method have been particularly informative in dissecting out several new facets of the defence responses in floral tissues. In barley, the silencing of HvCERK1, a putative chitin-binding receptor, increased the susceptibility to *F. graminearum*, suggesting that chitin perception is potentially required for effective defence (Karre et al., 2017). Silencing of TaLRRK-6D, a gene highly induced early during wheat infection, has formally linked for the first time a specific leucine rich-repeat receptor-like kinase (TaLRRK-6D) to basal defence against FHB disease in both resistant and susceptible genotypes. Also VIGS of *TaLRRK-6D* suppressed the expression of genes involved in SA signalling (Thapa et al., 2018). A metabolo-transcriptome study identified an inducible laccase TaLAC4 and an inducible plant-specific NAC transcription factor TaNAC032 as candidate genes within the QTL-Fhb1 derived from Sumai 3. VIGS analysis revealed that both genes independently had a QTL-dependent role in pathogen-induced lignification of secondary cell walls in the rachis and were required to prevent Fusarium hyphal growth from the spikelet into the rachis (Soni et al., 2020; Soni et al., 2021). VIGS experiments to explore the molecular basis of DON tolerance have shown that inhibiting the expression of the transporter TaABCC3.1 enhanced DON toxin sensitivity (Walter et al., 2015). Whilst both TaFROG and its interacting partner, the sucrose non-fermenting-1 (SNF1)-related protein kinase 1 catalytic subunit α (SnRK1α), are required to activate specific intracellular signalling events to enhance wheat's tolerance to DON (Perochon et al., 2019).

A lesser used but effective experimental approach to explore defence requirements in susceptible genotypes (wheat, barley and Brachypodium) has been the exogenous application of metabolites and phytohormones to alter the interaction outcome. Both floral and root infections have been explored. Exogenous application of JA, methyl-JA and ET increased floral susceptibility to FHB whilst increasing resistance to root rot (Haidoulis and Nicholson, 2020; Qi et al., 2016; Sun et al., 2016). SA and methyl-SA applications lead to enhanced floral resistance but increased susceptibility to root rot (Haidoulis and Nicholson, 2020; Sorahinobar et al., 2016). Application of the growth-associated phytohormone auxin increased resistance to both floral and root diseases whilst cytokinin applications increased both tissues' susceptibility (Haidoulis and Nicholson, 2020; Petti et al., 2012). Finally, exogenous application of brassinosteroids increased FHB resistance in barley (Ali et al., 2013). Collectively, these studies indicate that the interaction outcome can be modified via altering the hormone balance around the time of anthesis. But in a few studies, the negative effects of hormone applications on fungal growth *in vitro* and/or causing adverse effects on grain development have also been noted.

Three successful map-based cloning approaches to isolate the genes conferring QTL-mediated resistance have been reported. In none of these studies has an NBS-LRR type of resistance protein sequence been found. The gene responsible for conferring type II resistance residing within the wheat Fhb1 locus on chromosome 3B has been identified in two different Chinese sources (the Sumai 3 derivative Ning7840 and the landrace Wangshuibai). Several lines of evidence presented by two independent research groups indicate that resistance is provided by the histidine-rich calcium-binding protein *TaHRC*, which may be nuclear localised (Li et al., 2019a; Su et al., 2019). The wild-type *TaHRC* allele confers FHB susceptibility whereas the mutant allele with a large 5′ deletion in the open reading frame results in FHB resistance. However, how this large deletion and the altered protein sequence/structure confers resistance remains to be elucidated and is currently a subject of considerable debate. Recently, a second gene associated with resistance at the Fhb1 QTL locus has been reported (Paudel et al., 2020). The gene designated *WFhb1-1* codes for a 127 amino acid protein which has been shown in *in vitro* tests to weakly inhibit Fusarium growth. In *WFhb1-1* gene-edited lines of the cultivar Bobwhite, where the allele had been modified at the N terminus to confer resistance, significant upregulation of *WFhb1-1* was noted, suggesting that *TaHRC* might negatively regulate *WFhb1-1* expression (Paudel et al., 2020; Su et al., 2019).

By breeding from the wheat grass species *Thinopyrum ponticum*, which confers both type II and type III resistance, the resistance gene *Fhb7* has recently been identified (Guo et al., 2015; Wang et al., 2020a). *Fhb7* codes for a glutathione S-transferase (GST) which catalyses conjugation of a glutathione (GSH) unit onto the toxic epoxide moiety of the DON mycotoxin and thereby detoxifies this Fusarium virulence promoting mycotoxin (see Section 2 and Fig. 5). Interestingly, in a taxonomically related fungal species to the Fusaria, the non-pathogenic but endophytic species *Epichloë aotearoae* known to infect *Thinopyrum elongatum*, a GST sequence homologous to *Fhb7* was identified with 97% sequence identity. This suggests the possibility of horizontal gene transfer (HGT) of the GST from *Epichloë aotearoae* to *T. elongatum*.

In durum wheat, the gene underlying the QTL *QFhb.mgb-2A* conferring type II resistance has been identified as a wall-associated receptor-like kinase (*WAK2*) (Gadaleta et al., 2019). This finding suggests that the predicted plasma membrane spanning WAK2 protein somehow senses changes in the extracellular *in planta* environment (i.e. the plant cell wall and/or the apoplast) and thereby successfully activates intracellular defence signalling. These authors also propose that TaWAK2 negatively modulates the expression of a plant pectin methylesterase leading to a more rigid cell wall that prevents Fusarium penetration.

7 Crop improvements using genetic modification (GM), gene editing (GE) and host-induced gene silencing (HIGS) approaches

Genetic control of Fusarium diseases via breeding for resistance has been intensively explored since the 1990s, but so far, no wheat, barley, triticale, oat, rye, maize, rice or sorghum cultivar has been developed and released with complete resistance or immunity to floral, crown rot, stalk rot, root rot and/or seedling blight diseases (Dweba et al., 2017; Rawat et al., 2016). As discussed in Section 6, most genetic studies have focussed on identifying and exploiting newly recognised QTLs responsible for mediating one or more of the different types of FHB resistance. Through various breeding strategies, several promising moderately Fusarium-resistant, higher-yielding cultivars have gradually been released in different global regions and have been well adopted by farmers and growers to lower the impact of Fusarium diseases. However, other seed-based solutions and strategies to achieve higher levels of Fusarium control in all cereal crop species are still urgently needed. As outlined above the detailed fundamental studies on the Fusarium infection biology, the Fusarium genomes, the compatible and incompatible Fusarium-cereal interactions as well as host defences and mycotoxin detoxification mechanisms now provide a wealth of new insights from which novel targets and intervention strategies can be devised and exploited. In this section, we will focus on novel disease control approaches already arising via GE or GM and speculate on the new innovative ways that could be devised in the future.

GM in wheat for FHB resistance initially focussed on overexpression of genes encoding plant defence proteins. This included defence-signalling components, such as *Arabidopsis thaliana* NPR1 protein (Makandar et al., 2006), or cell-wall-associated defence proteins such as the wheat lipid transfer protein LTP5 (Zhu et al., 2012). Expression of each protein in wheat led to type II FHB resistance (Table 9). An alternative approach was to overexpress proteins that could detoxify the harmful DON mycotoxin produced by *F. graminearum* which is essential for floral virulence (as described above in Sections 2 and 5 and Figs. 3 and 5). Several studies focussed on expressing enzymes that could modify the DON structure to a non-toxic compound. The UDP-glucosyl transferase enzyme significantly detoxifies DON into the non-toxic DON-3-O-glucoside (D3G) and is also able to convert NIV into the non-toxic nivalenol-3-O-beta-D-glucoside (Li et al., 2017). This enzyme was first isolated and characterised in Arabidopsis (Poppenberger et al., 2003) and other functional homologues were subsequently identified in Brachypodium, rice and barley (Pasquet et al., 2016; Schweiger et al., 2010) (Table 9). Transgenic wheat plants expressing DON detoxification enzymes had elevated type II FHB resistance. Unfortunately, many aspects of this GM approach still need to be further studied. For instance, there is still insufficient information detailing the underlying detoxification mechanism(s) and whether detoxification can be reversed during some types of grain processing activities or following ingestion.

Since 2010, HIGS has been reported as an entirely new way to control several problematic pathogenic species that cause plant disease including cereal infecting Fusaria (Koch et al., 2013; Nowara et al., 2010) (Fig. 7). In HIGS, transgenic plants induce silencing of target gene transcripts in plant pathogens via RNAi immediately upon attempted infection. RNAi-mediated gene silencing is a post-transcriptional gene silencing (PTGS) mechanism and can be triggered by expressing a double-stranded RNA (dsRNA) construct complementary to the gene to be silenced. In transgenic plants, dsRNA can be produced from a DNA sequence containing inverted repeat sequences through RNA transcription. In transcribed RNA, these repeats fold back on themselves

Table 9 Examples of gene editing (GE) and genetic modification (GM) approaches demonstrated to control Fusarium disease control in wheat or barley. Four different strategies (A through D) have been tested

Gene	FHB resistance phenotype[1]/ Cereal species	Reference
Overexpression of plant-defence-related genes		
Arabidopsis thaliana NPR1	Type II resistance/wheat spikes	Makandar et al. (2006)
Wheat *β-1,3-glucanase, α-1-purothionin* and *thaumatin-like protein 1*	Type II, type III and/or type IV resistance/wheat spikes	Mackintosh et al. (2007)
Wheat *TaLTP5*	Type II FHB resistance to/wheat spikes	Zhu et al. (2012)
Common bean PvPGIP2	Type II and type III resistance/ wheat spikes	Tundo et al. (2016)
Rye NPR1	Type II resistance/wheat spikes	Yu et al. (2017)
Arabidopsis thaliana AtLTP4.4	Type II resistance/wheat spikes	McLaughlin et al. (2021)
Wheat Ta-NACL-D1	Type II and type IV resistance/ wheat spikes	Perochon et al. (2019)
Overexpression of plant DON detoxification genes		
Brachypodium UDP-glucosyltransferase	Type II and type III resistance/ wheat spikes	Pasquet et al. (2016)
Barley UDP-glucosyltransferase	Type II and type III resistance/ wheat spikes	Li et al. (2017)
Barley UDP-glucosyltransferase *HvUGT13248*	Type II and type III resistance/ wheat spikes, and resistance to Fusarium crown rot/wheat roots	Mandala et al. (2019)
Host-induced gene silencing (HIGS) of fungal genes		
F. graminearum Cyp51 A, B and C	Resistance to *F. graminearum* infection/barley leaves	Koch et al. (2013)
F. graminearum chitin synthase 3b	Type II resistance/wheat spikes and wheat coleoptiles	Cheng et al. (2015)
F. culmorum β-1, 3-glucan synthase FcGls1	Type II resistance/wheat spikes	Chen et al. (2016)
F. graminearum stacked genes FgSGE1-STE12-PP1	Type II and type III resistance/ wheat spikes	Wang et al. (2020b)
Gene editing (CRISPR/Cas9) of plant susceptibility factor genes		
Wheat TaHRC	Type II resistance/wheat spikes	Su et al. (2018)
Wheat TaNFXL1	Type II resistance/wheat spikes	Brauer et al. (2020)

[1]Resistance types: Type I - resistance to initial infection; Type II - reduced rate of disease spread in the rachis from the initially infected spikelet; Type III - reduction in DON levels, Type IV - reduction in Fusarium damaged kernels (FDK); Type V - lower yield losses.

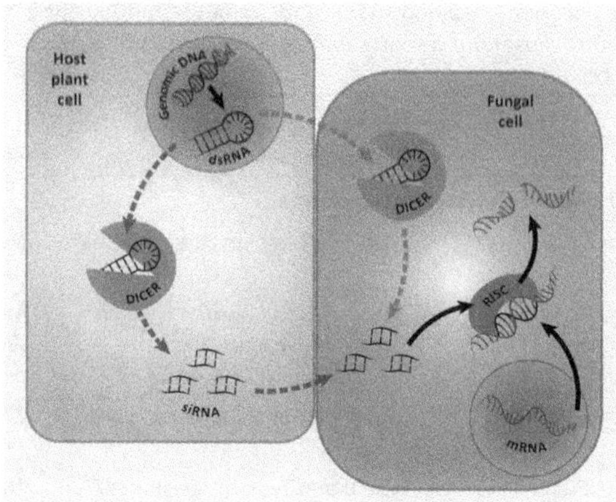

Figure 7 Host-induced gene silencing. Each HIGS construct targets a pathogen gene required for virulence and/or mycotoxin formation. The construct encodes an inverted sequence that forms a hairpin dsRNA following transcription and is introduced into the host plant either by transient or stable transformation. The dsRNA is processed to form small interfering RNA (siRNA) either before or after delivery to the pathogen cell using the plant's innate RNAi machinery. Once inside the pathogen cells the siRNA silences the targeted genes by interfering with mRNA transcripts. Adapted from Kanja and Hammond-Kosack (2020).

to form a double-stranded hairpin (hpRNA) (Baulcombe, 2015; Machado et al., 2018). These long hpRNA/dsRNA are then cleaved by a dicer protein into short ~25 bp RNA duplexes known as small interfering RNA (siRNAs). An RNA-induced silencing complex (RISC) recognises and binds to one strand of siRNA (the guide), whereas the other strand (the passenger) is degraded. An Argonaute protein forms the catalytic centre of the RISC. RISC degrades target mRNAs that are nearly perfectly complementary to the loaded guide strand of siRNA (Koch and Kogel, 2014; Wang and Jin, 2017). During HIGS, either host-produced long dsRNAs or siRNAs are transferred from the plant cells to the fungal cells and targets silencing the specific fungal genes in a homology-dependent manner. If the longer dsRNA is transferred, then the fungal dicer proteins must be involved in the production of siRNAs. HIGS has been shown to reduce *F. graminearum* infection in Arabidopsis, barley leaves and wheat floral tissues by silencing genes important for cell survival, for example, the enzyme chitin synthase 3b (Chs3b) required for cell wall formation, or enzymes required for the biosynthesis of sterols which are critical components of cell membranes, for example, the three cytochrome P450 paralogues CYP51 A, B and C (Cheng et al., 2015; Koch et al., 2013) of which CYP51C is a key virulence determinant and CYP51B is essential for ascospore formation (Fan et al., 2013). Although HIGS may be effective in controlling fungal diseases, this is a GM approach and remains limited in commercial use due to the strict legislation of genetically modified organisms (GMOs) in food and feed production in many countries (Rampersad, 2020). An alternative non-transgenic RNAi approach is spray-induced gene silencing (SIGS), which exploits the RNAi mechanism through the exogenous application of long dsRNA and siRNAs (Machado et al., 2018).

New GE techniques such as 'clustered regulatory interspaced short palindromic repeats/CRISPR-associated protein 9 (CRISPR/Cas9)' allows precise targeting and modification of genes of interest (GOI) and have been proposed or shown to be an efficient route to achieve FHB resistance (Brinegar et al., 2017). For example, recently the reintroduction into susceptible cultivars by GE of missing NAC domains in transcription factors controlling lignification and cell wall reinforcement responses has been proposed as a way to enhance type II FHB resistance in wheat (Soni et al., 2021). But to date for wheat, the GE approach has focussed primarily on the elimination of known susceptibility genes to increase resistance to FHB and other plant pathogens. A reduction in the expression of the wheat *NFXL1* gene, which codes for a transcription factor that represses *Tri*-induced defence responses, increased type II resistance to *F. graminearum* (Brauer et al., 2020) (Table 9). A previous study indicated that the NFXL1 ortholog represses *Tri*-induced defence responses as well as resistance mechanisms against bacterial infections in Arabidopsis (Asano et al., 2008). Although only a few studies have so far explored the use of CRISPR/Cas9 mediated GE to achieve FHB resistance, this approach appears to be a very promising strategy to control FHB. Other FHB susceptibility loci have already been formally recognised through classical genetic studies, for example, those closely linked to the semi-dwarfing genes *Rht-B1b* (*Rht1*) and *Rht-D1b* (*Rht2*) (He et al., 2016; Srinivasachary et al., 2009) and another on chromosome 4D (Hales et al., 2020). In some countries, such as the United States and Australia, GE crops that do not contain any foreign DNA, such as removed susceptibility loci, are not considered GMOs under the regulatory legislation and can be deployed freely (Hamburger, 2019). However, in the EU, the world's largest wheat producer (FAO, 2019), a technology-driven approach to regulation is adopted, where GMs achieved through means other than conventional breeding are subjected to strict limitations to commercialisation regardless of the final crop product/new trait (Brinegar et al., 2017; Callaway, 2018).

7.1 The challenges of exploiting virulence determinant for Fusarium control

Single virulence determinants are both exciting and challenging targets for Fusarium control. This is because it is already well known that the emergence of sequence variants can rapidly render resistance to traditional control chemistries that target single sites thereby reducing efficacy by causing shifts to fungicide resistance within pathogen populations (Fisher et al., 2012, 2018).

For effectors, the high level of genetic redundancies, rapid rates of evolution and lack of conservation amongst Fusarium strains collectively means most individual effectors are unsuitable control targets. Instead, focussing on more broad-spectrum targets may negate the issue of genetic redundancy and the rapid development of resistance, but at the same time this approach may increase the risk of detrimental off-target effects on beneficial organisms (Panwar et al., 2016). For example, the well-studied fungal intracellular MAPK signalling cascades, which control multiple disease establishment and propagation events (Jenczmionka et al., 2003; Urban et al., 2003; Wang et al., 2011), could be a promising target for disease control. However, the widespread presence of highly conserved MAPKs in multiple fungal taxa would mean a high likelihood of off-target effects on beneficial fungi including mycorrhiza or biocontrol agents that could negatively affect crop health and productivity.

Specifically, targeting DON mycotoxin production could reduce both mycotoxin contamination of grain and reduce the overall virulence of several Fusarium species on wheat in most tissues (see Table 3). In addition, in scenarios where DON levels have been reduced experimentally, the rapid localised elevation of plant defence responses to restrict infections has been observed and would be additionally beneficial. However, targeting DON mycotoxin production comes with caveats. The gene profile of the *Tri* biosynthesis (*Tri*) cluster varies between different isolates of *F. graminearum* (Amarasinghe and Fernando, 2016). Also, no single gene deletion, including the global regulator *Tri6*, has resulted in a complete loss of pathogenicity (Rampersad, 2020). Therefore, currently identifying targets for broad-spectrum control within the *Tri* cluster remains challenging. Alternatively, targeting the entire cluster for silencing and/or removal of the function of multiple enzymatic steps could avoid issues with isolate variation. However, to our knowledge, this has not yet been suggested or investigated.

7.2 Possible new targets for exploration

From the wealth of new fundamental biological and molecular insights on Fusarium-cereal interactions that have emerged in recent years, two topics merit a special mention as possible new targets for intervention, namely, plasmodesmatal colonisation and fungal autophagy. Repeated colonisation of plasmodesmata is crucial for successful colonisation of cereal host tissue and clearly requires a sophisticated dialogue to develop and be maintained between the advancing Fusarium hyphae and the soon to be invaded plant cell. Understanding this small part of the infection process may bring about some interesting insights as to how to control Fusarium infections in the future. Secondly, of all the virulence mechanisms explored by large-scale single deletion experiments (see Table 6), the process of autophagy has generated the highest proportion of mutants with reduced virulence/loss of pathogenicity phenotypes, that is, 96%. This highlights that the autophagy occurring in the older parts of the mycelial colony is essential for providing the building blocks of life that are trafficked and continually needed to feed the advancing Fusarium hyphae front. By finding ways to prematurely trigger autophagy in younger hyphae and/or prevent/disrupt the autophagy occurring in the more mature hyphae should lead to fungal starvation, reduced *in planta* growth rates and as a consequence provide more time for plant cells to establish effective defence responses.

Another possible target for control of FHB is the pattern-triggered immunity (PTI) pathway. PTI is activated in wheat by *F. graminearum* colonisation, and induction of this pathway by either the bacterial PAMP flagellin or by the ectopic expression of the transcription factor WRKY29 enhances FHB resistance in both Arabidopsis and wheat (Sarowar et al., 2019). Whilst investigation of this enhanced resistance against various chemotypes and other FHB causing species is required, the pleiotropic effects of prolonged induction of PTI in this way, for example, on final fertile tillers numbers, grain numbers and grain size, also need to be studied.

7.3 How to achieve and predict long-term durability?

One approach to achieving long-term durability is to strictly combine control options targeted upon reducing the effectiveness of specific pathogen virulence proteins/metabolites with control options focussed on either enhancing host resistance or lowering host disease susceptibility (Hammond-Kosack, 2014). This could be achieved by the targeted integration of GM solutions focussed on reducing Fusarium virulence to chromosomal regions already harbouring major QTLs conferring different types of disease resistance. Subsequently in cereal breeding programmes, marker-assisted selection would be used to ensure these stacked traits which control Fusarium by different mechanisms are never decoupled.

To predict potential long-term durability, disease-resistant germplasm developed through GE, GM or HIGS needs to be tested during the early development stages in multiple geographical locations to ensure these new type(s) of resistance are not rapidly overcome by other species in the Fusarium complex which are usually present at a very low incidence. In addition, by surveying and sequencing the genomes of the Fusarium species and strains that are still able to infect the new GE, GM and HIGS resistant germplasm, this would rapidly identify the gene sequence variants located in the variable portion of the Fusarium genome and possibly even in the core genome which allow the fungus to overcome the novel disease control approaches. If the outcomes detected indicate a high likelihood that the new technologies could be rapidly overcome, then specific limitations on the use of the new resistant germplasm within an integrated, multi-component control strategy, that is, with chemistries and/or biocontrol organisms, would need to be considered and implemented to ensure durability.

As the severity of FHB epidemics is strongly influenced by meteorological factors, climate change is predicted to acutely intensify the risk of these epidemics, further compromising small grain security and safety (Vaughan et al., 2016). Therefore, the development of durable and sustainable disease control measures is integral to stable food production systems (Wegulo et al., 2015). Greater understanding of the molecular interactions underpinning *F. graminearum* infection as well as during mixed Fusarium species infections under both current and future predicted environmental scenarios will be integral to the establishment of novel disease management strategies that also have a high likelihood for long-term durability.

8 Unanswered questions, future outlook and concluding remarks

Fusarium species are one of the most successful genera of pathogenic fungi (Geiser et al., 2020) inciting a broad spectrum of seedling, root, stem-base and floral diseases of cereal crops and contaminate harvested cereal grains and other harvested tissues with various mycotoxins. Collectively, these and other disease problems in cereal crops continue to destabilise global food security. In this chapter, we have described and discussed the increased understanding of the many mechanisms successfully deployed by Fusarium fungi to cause disease as well as the increased knowledge on host resistance mechanisms and the plant targets exploited to enhance Fusarium susceptibility. In order to achieve durable disease control in cereal crops, many unanswered questions still need to be addressed (Box 1). The need to continue investigations into both sides of the cereal-Fusarium interaction, as well as compare with other cereal-fungal pathosystems, for example, wheat blast, remains critically important.

Box 1 Questions that remain unanswered.

- How does Fusarium perceive its host?
- What signals (if any) induce the changes from symptomless to symptomatic Fusarium infections?
- What mechanisms allow Fusarium to grow rapidly cell-to-cell via the plasmodesmata? Do effectors and other metabolites manipulate these interactions?

- Why is the DON mycotoxin only required for the infection of some tissue types and plant host species?
- How is the secretion of Fusarium effectors regulated, is it tissue and host specific?
- What mechanism(s) deliver Fusarium effectors and metabolites into host plant cells both at and in advance of the invading hyphal front?
- Why and how are the various discrete high-recombination chromosomal regions retained in the *F. graminearum* genome?
- What are the evolutionary implications of the lack of accessory chromosomes in most cereal infecting Fusarium species?
- What differences at the transcriptomic, post-transcriptomic, proteomic and metabolomic levels allow the Fusarium species to infect various tissues, and how does this relate to the different infection strategies?
- How can machine learning and artificial intelligence approaches be used to gain new knowledge on fungal virulence and host resistance from the existing large repertoire of Fusarium-cereal data sets?
- Is there an interplay between effectors and the DON mycotoxin in suppressing the plant's defence responses in the different tissue types and hosts?
- Under natural field conditions, what combinations of defence mechanisms and/or lack of susceptibility factors in cereals make leaf tissue totally resistant to Fusarium infections whilst the floral tissue remains fully susceptible?
- Are there additional secondary metabolites synthesised by other gene clusters induced during cereal colonisation and what role(s) does each fulfil for the individual species within the *Fusarium sambucinum* species complex and beyond?
- How can the recently devised GM and GE approaches be best used in combination with current strategies to achieve durable disease control?
- Could a more interconnected approach from a local to an international level be adopted to control Fusarium diseases and mycotoxin contamination globally?

Increasingly, *in planta* transcriptome analyses are being used to identify the differentially expressed Fusarium and cereal host genes in numerous types of wild-type and 'experimentally engineered' compatible and incompatible interactions in different biological niches. There is a need for these experiments to become even more sophisticated involving not only carefully devised spatial and temporal analyses, but also to include complementary detailed cell biology, *in vitro*, proteomic, metabolomic and pangenome studies. In addition, these complex datasets when published each need to be deposited in formats that will permit easy interoperability with other datasets and data types so that machine learning and artificial intelligence approaches can be adopted to drive the pace of discovery. Similarly, the wealth of data and insights arising from Fusarium-cereal crop studies which focus on whole plant, whole crop or landscape scales of investigation need to be accessible in formats that permit relatively seamless interoperability with the wealth of data arising from cell biology and the various 'omics' based investigations. With this greater connectivity in knowledge graphs derived from experimentation at multiple scales, greater insights can be gained, hitherto unsuspected targets for disease control pinpointed and tested *in silico* and the challenges and possibility of unintentional consequences also explored. In Table10, a list of additional information sources and datasets on *Fusarium graminearum* and FHB disease are provided.

Over the next 10 years, fundamental knowledge on crop plant-Fusarium communication events will improve dramatically because of increasing amounts of cereal and Fusarium genomic and pangenomic information becoming available and through the application of new technologies at higher throughput and precision, for example, GE, laser capture microscopy and proximity labelling Kanja and Hammond-Kosack, (2020). This should permit the identification and characterisation of additional cereal resistance genes and the first susceptibility loci as well as various bonafide Fusarium effectors, their primary plant targets and numerous additional virulence genes, interfering iRNAs and metabolites essential for sensing and communicating between plant cells and Fusarium cells. This will then permit a greater comparison between compatible and incompatible Fusarium and non-Fusarium pathosystems, the various host tissue niches occupied and the core and variable components within the cereal host and Fusarium genomes. As a result, new solutions to Fusarium control can be devised which are either cereal species neutral or more likely targeted to specific tissue type and cereal host species. The latter is probably the more desirable outcome, to ensure that a more diverse pipeline of intervention strategies is devised which, if correctly deployed, both spatially and temporally, should lead to a higher likelihood of durable disease control of Fusarium and other co-occurring problematic phytopathogenic species.

Finally, within most cereal-Fusarium pathosystems, the production of harmful mycotoxins by Fusarium occurs very early in the infection process, prior to macroscopic disease symptom development. Therefore as future control technologies and strategies are devised and refined, to achieve the greatest societal, economic and ecological impacts, there is a constant need to minimise initial infections and thereby reduce the risk of mycotoxins entering the food and feed chains.

Table 10 Sources of additional information for *Fusarium graminearum* and FHB disease

Resource	URL	Description	Available data
JGI-MycoCosm fungal Portal	https://mycocosm.jgi.doe.gov/mycocosm/home	Provides data access, visualisation and analysis tools for comparative genomics across fungal genomes	Genome assemblies Genome annotations Secondary metabolite clusters
FungiDB	https://fungidb.org/fungidb/app	Bioinformatics data resource for fungi and oomycetes	Integrated datasets Genome summaries Transcriptomics Taxonomy Protein features Genome location synteny
PHI-base	http://www.phi-base.org/ https://github.com/PHI-base	Manually curated database of experimentally verified pathogenicity genes for over 250 pathogens. *Fusarium graminearum* has > 600 entries	Gene-ID Pathogen ID/Strain Host ID/Strain Experimental technique and phenotype
FgMutantDb	https://docs.google.com/spreadsheets/d/1ZJXGqvqKi2jVUCkQmxeGvfgf7BKLX1l1-AR4WqerG0Q/edit#gid=362556506	Community-generated database of gene deletion mutants for *Fusarium graminearum* – includes non-peer-reviewed results	Gene details Experimental references Gene-ID cross referencing
Ensembl Fungi	https://fungi.ensembl.org/index.html	Ensembl Fungi is a browser for fungal genomes. The majority of these are taken from the databases of the International Nucleotide Sequence Database Collaboration (the European Nucleotide Archive at the EBI, GenBank at the NCBI, and the DNA Database of Japan)	Genome summaries Gene and proteins details Direct links to the phenotypes reported in the PHI-base database

(Continued)

Table 10 (*Continued*)

Resource	URL	Description	Available data
AHDB – Agriculture and Horticulture Development Board	https://ahdb.org.uk/knowledge-library/fusarium-and-microdochium-in-cereals	Farmer/grower-funded board to provide practical knowledge exchange to industry stakeholders	Provides practical information on *Fusarium sp.* infection and life cycles as well as risk factors and management strategies for multiple cereal hosts
The US Wheat and Barley scab initiative	https://scabusa.org/index.php/	A collaboration between public and private sector researchers and growers to develop effective control measures. To enhance food safety and supply by reducing the impact of Fusarium head blight (scab) on wheat and barley	Provides resources for growers on topics such as grain sampling for DON, FHB forecasting and guides to integrated control strategies (via scabsmart (https://scabsmart.org/))
World Mycotoxin Survey	https://www.biomin.net/solutions/mycotoxin-survey	Detailed annual report of mycotoxin levels and contamination predictions for upcoming maize and wheat harvests for regions and sub-regions	Risk level indicators for farm animal species of six major mycotoxins including DON
Fusarium ear blight	https://en.wikipedia.org/wiki/Fusarium_ear_blight	Provides general information on multiple topics including other control measures	Multiple useful links

9 Acknowledgements

C. K. and C. W. are funded by the UK Biotechnology and Biological Sciences Research Council (BBSRC) and the Lawes Trust, Doctoral Training Programme grant (BB/M008770/1). The CAPES Foundation of Brazil is thanked for A. K. M. W's PhD scholarship (BEX 1266-13-6). L. B. is funded by the UK BBSRC-funded South West Biosciences Doctoral Training Partnership (BBSRC SWBio DTP) (BB/M009122/1). K. H.-K. and A. K. M. W. received BBSRC grant-aided support as part of the Institute Strategic Programme grant Designing Future Wheat Grant (BB/P016855/1) and previously received support from the bilateral BBSRC-EMBRAPA grants (BB/N004493/1, BB/N018095/1). We thank Dr Michael Hammond-Kosack for taking the wheat spike and grain images presented in Fig. 1. We also thank Drs Kostya Kanyuka and Martin Urban (Rothamsted Research, UK) and Neil Brown (University of Bath, UK) for providing critical feedback to improve specific sections of this chapter.

10 References

Adnan, M., Fang, W., Sun, P., Zheng, Y., Abubakar, Y. S., Zhang, J., Lou, Y., Zheng, W. and Lu, G. D. 2020a. R-SNARE FgSec22 is essential for growth, pathogenicity and DON production of *Fusarium graminearum*. *Current Genetics* 66(2), 421–435.

Adnan, M., Islam, W., Noman, A., Hussain, A., Anwar, M., Khan, M. U., Akram, W., Ashraf, M. F. and Raza, M. F. 2020b. Q-SNARE protein FgSyn8 plays an important role in growth, DON production and pathogenicity of *Fusarium graminearum*. *Microbial Pathogenesis* 140, 103948.

Alexander, N. J., Mccormick, S. P., Waalwijk, C., Van Der Lee, T. and Proctor, R. H. 2011. The genetic basis for 3-ADON and 15-ADON trichothecene chemotypes in *Fusarium*. *Fungal Genetics and Biology* 48(5), 485–495.

Alouane, T., Rimbert, H., Fabre, F., Cambon, F., Langin, T. and Bonhomme, L. 2018. Genome sequence of *Fusarium graminearum* strain MDC_Fg1, isolated from bread wheat grown in France. *Microbiology Resource Announcements* 7(19), e01260-18.

Ali, S. S., Kumar, G. B., Khan, M. and Doohan, F. M. 2013. Brassinosteroid enhances resistance to fusarium diseases of barley. *Phytopathology* 103(12), 1260–1267.

Amarasinghe, C. C. and Fernando, W. G. 2016. Comparative analysis of deoxynivalenol biosynthesis related gene expression among different chemotypes of *Fusarium graminearum* in spring wheat. *Frontiers in Microbiology* 7, 1229.

Arias, M. M. D., Leandro, L. F. and Munkvold, G. P. 2013. Aggressiveness of *Fusarium* species and impact of root infection on growth and yield of soybeans. *Phytopathology* 103(8), 822–832.

Asano, T., Yasuda, M., Nakashita, H., Kimura, M., Yamaguchi, K. and Nishiuchi, T. 2008. The AtNFXL1 gene functions as a signaling component of the type A trichothecene-dependent response. *Plant Signaling and Behavior* 3(11), 991–992.

Baccelli, I. 2015. Cerato-platanin family proteins: one function for multiple biological roles? *Frontiers in Plant Science* 5, 769.

Backhouse, D. 2014. Global distribution of *Fusarium graminearum*, *F. asiaticum* and *F. boothii* from wheat in relation to climate. *European Journal of Plant Pathology* 139(1), 161–173.

Backhouse, D., Abubakar, A. A., Burgess, L. W., Dennisc, J. I., Hollaway, G. J., Wildermuth, G. B., Wallwork, H. and Henry, F. J. 2004. Survey of *Fusarium* species associated with crown rot of wheat and barley in eastern Australia. *Australasian Plant Pathology* 33(2), 255-261.

Bahadoor, A., Brauer, E. K., Bosnich, W., Schneiderman, D., Johnston, A., Aubin, Y., Blackwell, B., Melanson, J. E. and Harris, L. J. 2018. Gramillin A and B: cyclic lipopeptides identified as the nonribosomal biosynthetic products of *Fusarium graminearum*. *Journal of the American Chemical Society* 140(48), 16783-16791.

Bai, G. and Shaner, G. 2004. Management and resistance in wheat and barley to Fusarium head blight. *Annual Review of Phytopathology* 42, 135-161.

Bai, G.-H., Plattner, R., Desjardins, A., Kolb, F. and McIntosh, R. 2001. Resistance to Fusarium head blight and deoxynivalenol accumulation in wheat. *Plant Breeding* 120(1), 1-6.

Bakker, M. G., Brown, D. W., Kelly, A. C., Kim, H.-S., Kurtzman, C. P., McCormick, S. P., O'Donnell, K. L., Proctor, R. H., Vaughan, M. M. and Ward, T. J. 2018. Fusarium mycotoxins: a trans-disciplinary overview. *Canadian Journal of Plant Pathology* 40(2), 161-171.

Baldwin, T. T., Basenko, E., Harb, O., Brown, N. A., Urban, M., Hammond-Kosack, K. E. and Bregitzer, P. P. 2018. Sharing mutants and experimental information prepublication using FgMutantDb (https://scabusa.org/FgMutantDb). *Fungal Genetics and Biology* 115, 90-93.

Baulcombe, D. C. 2015. VIGS, HIGS and FIGS: small RNA silencing in the interactions of viruses or filamentous organisms with their plant hosts. *Current Opinion in Plant Biology* 26, 141-146.

Bayram, O., Krappmann, S., Ni, M., Bok, J. W., Helmstaedt, K., Valerius, O., Braus-Stromeyer, S., Kwon, N. J., Keller, N. P., Yu, J. H. and Braus, G. H. 2008. VelB/VeA/LaeA complex coordinates light signal with fungal development and secondary metabolism. *Science* 320(5882), 1504-1506.

Beccari, G., Colasante, V., Tini, F., Senatore, M. T., Prodi, A., Sulyok, M. and Covarelli, L. 2018. Causal agents of Fusarium head blight of durum wheat (*Triticum durum* Desf.) in central Italy and their in vitro biosynthesis of secondary metabolites. *Food Microbiology* 70, 17-27.

Bennett, J. W. and Klich, M. 2003. Mycotoxins. *Clinical Microbiology Reviews* 16, 497-516.

Bianchini, A., Horsley, R., Jack, M. M., Kobielush, B., Ryu, D., Tittlemier, S., Wilson, W. W., Abbas, H. K., Abel, S., Harrison, G., Miller, J. D., Shier, W. T. and Weaver, G. 2015. DON occurrence in grains: a North American perspective. *Cereal Foods World* 60(1), 32-56.

BIOMIN 2019. Biomin Mycotoxin Survey [Online]. Available at: https://www.biomin.net/en/biomin-mycotoxin-survey/ [Accessed December 2020].

Biselli, C., Bagnaresi, P., Faccioli, P., Hu, X., Balcerzak, M., Mattera, M. G., Yan, Z., Ouellet, T., Cattivelli, L. and Valè, G. 2018. Comparative transcriptome profiles of near-isogenic hexaploid wheat lines differing for effective alleles at the 2DL FHB resistance QTL. *Frontiers in Plant Science* 9, 37.

Blümke, A., Falter, C., Herrfurth, C., Sode, B., Bode, R., Schäfer, W., Feussner, I. and Voigt, C. A. 2014. Secreted fungal effector lipase releases free fatty acids to inhibit innate immunity-related callose formation during wheat head infection. *Plant Physiology* 165(1), 346-358.

Blumke, A., Sode, B., Ellinger, D. and Voigt, C. A. 2015. Reduced susceptibility to Fusarium head blight in *Brachypodium distachyon* through priming with the Fusarium mycotoxin deoxynivalenol. *Molecular Plant Pathology* 16(5), 472–483.

Boddu, J., Cho, S. and Muehlbauer, G. J. 2007. Transcriptome analysis of trichothecene-induced gene expression in barley. *Molecular Plant-Microbe Interactions* 20(11), 1364–1375.

Boedi, S., Berger, H., Sieber, C., Munsterkotter, M., Maloku, I., Warth, B., Sulyok, M., Lemmens, M., Schuhmacher, R., Guldener, U. and Strauss, J. 2016. Comparison of *Fusarium graminearum* transcriptomes on living or dead wheat differentiates substrate-responsive and defense-responsive genes. *Frontiers in Microbiology* 7, 1113.

Boenisch, M. J. and Schäfer, W. 2011. *Fusarium graminearum* forms mycotoxin producing infection structures on wheat. *BMC Plant Biology* 11, 110.

Bolton, M. D., Van Esse, H. P., Vossen, J. H., De Jonge, R., Stergiopoulos, I., Stulemeijer, I. J., Van Den Berg, G. C., Borras-Hidalgo, O., Dekker, H. L., De Koster, C. G., De Wit, P. J., Joosten, M. H. and Thomma, B. P. 2008. The novel *Cladosporium fulvum* lysin motif effector Ecp6 is a virulence factor with orthologues in other fungal species. *Molecular Microbiology* 69(1), 119–136.

Bonnighausen, J., Schauer, N., Schafer, W. and Bormann, J. 2019. Metabolic profiling of wheat rachis node infection by *Fusarium graminearum* – decoding deoxynivalenol-dependent susceptibility. *New Phytologist* 221(1), 459–469.

Boyacioğlu, D. and Hettiarachchy, N. S. 1995. Changes in some biochemical components of wheat grain that was infected with *Fusarium graminearum*. *Journal of Cereal Science* 21(1), 57–62.

Brauer, E. K., Balcerzak, M., Rocheleau, H., Leung, W., Schernthaner, J., Subramaniam, R. and Ouellet, T. 2020. Genome editing of a deoxynivalenol-induced transcription factor confers resistance to *Fusarium graminearum* in wheat. *Molecular Plant-Microbe Interactions* 33(3), 553–560.

Brewer, H. C. and Hammond-Kosack, K. E. 2015. Host to a Stranger: Arabidopsis and Fusarium Ear Blight. *Trends in Plant Science* 20(10), 651–663.

Brinegar, K., K Yetisen, A., Choi, S., Vallillo, E., Ruiz-Esparza, G. U., Prabhakar, A. M., Khademhosseini, A. and Yun, S. H. 2017. The commercialization of genome-editing technologies. *Critical Reviews in Biotechnology* 37(7), 924–932.

Brown, D. W. and Proctor, R. H. 2016. Insights into natural products biosynthesis from analysis of 490 polyketide synthases from Fusarium. *Fungal Genetics and Biology* 89, 37–51.

Brown, N. A., Antoniw, J. and Hammond-Kosack, K. E. 2012. The predicted secretome of the plant pathogenic fungus *Fusarium graminearum*: a refined comparative analysis. *PLoS ONE* 7(4), e33731.

Brown, N. A., Bass, C., Baldwin, T. K., Chen, H., Massot, F., Carion, P. W., Urban, M., Van De Meene, A. M. and Hammond-Kosack, K. E. 2011. Characterisation of the *Fusarium graminearum*-wheat floral interaction. *Journal of Pathogens*, 626345.

Brown, N. A., Evans, J., Mead, A. and Hammond-Kosack, K. E. 2017. A spatial temporal analysis of the *Fusarium graminearum* transcriptome during symptomless and symptomatic wheat infection. *Molecular Plant Pathology* 18(9), 1295–1312.

Brown, N. A., Schrevens, S., Van Dijck, P. and Goldman, G. H. 2018. Fungal G-protein-coupled receptors: mediators of pathogenesis and targets for disease control. *Nature Microbiology* 3(4), 402–414.

Brown, N. A., Urban, M., Van De Meene, A. M. and Hammond-Kosack, K. E. 2010. The infection biology of *Fusarium graminearum*: defining the pathways of spikelet to spikelet colonisation in wheat ears. *Fungal Biology* 114(7), 555-571.

Burch-Smith, T. M. and Zambryski, P. C. 2012. Plasmodesmata paradigm shift: regulation from Without Versus within. In: Merchant, S. S. (Ed) *Annual Review of Plant Biology* (vol. 63), Palo Alto: Annual Reviews, 297-260.

Burkhardt, I., Siemon, T., Henrot, M., Studt, L., Rösler, S., Tudzynski, B., Christmann, M. and Dickschat, J. S. 2016. Mechanistic characterisation of two sesquiterpene cyclases from the plant pathogenic fungus *Fusarium fujikuroi*. *Angewandte Chemie* 55(30), 8748-8751.

Callaway, E. 2018. CRISPR plants now subject to tough GM laws in European Union. *Nature* 560(7716), 16.

Chang, H. X., Domier, L. L., Radwan, O., Yendrek, C. R., Hudson, M. E. and Hartman, G. L. 2016. Identification of multiple phytotoxins produced by *Fusarium virguliforme* including a phytotoxic effector (FvNIS1) associated with Sudden Death Syndrome foliar symptoms. *Molecular Plant-Microbe Interactions* 29(2), 96-108.

Chen, L., Tong, Q., Zhang, C. and Ding, K. 2019a. The transcription factor FgCrz1A is essential for fungal development, virulence, deoxynivalenol biosynthesis and stress responses in *Fusarium graminearum*. *Current Genetics* 65(1), 153-166.

Chen, S., Tao, L., Zeng, L., Vega-Sanchez, M. E., Umemura, K. and Wang, G. L. 2006. A highly efficient transient protoplast system for analyzing defence gene expression and protein-protein interactions in rice. *Molecular Plant Pathology* 7(5), 417-427.

Chen, W., Kastner, C., Nowara, D., Oliveira-Garcia, E., Rutten, T., Zhao, Y., Deising, H. B., Kumlehn, J. and Schweizer, P. 2016. Host-induced silencing of *Fusarium culmorum* genes protects wheat from infection. *Journal of Experimental Botany* 67(17), 4979-4991.

Chen, Y., Kistler, H. C. and Ma, Z. 2019. *Fusarium graminearum* trichothecene mycotoxins: Biosynthesis, regulation, and management. *Ann. Rev.* Phytopathol. 57, 15-39.

Chen, X., Steed, A., Travella, S., Keller, B. and Nicholson, P. 2009. *Fusarium graminearum* exploits ethylene signalling to colonize dicotyledonous and monocotyledonous plants. *New Phytologist* 182(4), 975-983.

Cheng, W., Song, X. S., Li, H. P., Cao, L. H., Sun, K., Qiu, X. L., Xu, Y. B., Yang, P., Huang, T., Zhang, J. B., Qu, B. and Liao, Y. C. 2015. Host-induced gene silencing of an essential chitin synthase gene confers durable resistance to Fusarium head blight and seedling blight in wheat. *Plant Biotechnology Journal* 13(9), 1335-1345.

Cheval, C., Samwald, S., Johnston, M. G., De Keijzer, J., Breakspear, A., Liu, X., Bellandi, A., Kadota, Y., Zipfel, C. and Faulkner, C. 2020. Chitin perception in plasmodesmata characterizes submembrane immune-signaling specificity in plants. *Proceedings of the National Academy of Sciences of the United States of America* 117(17), 9621-9629.

Crippin, T., Renaud, J. B., Sumarah, M. W. and Miller, J. D. 2019. Comparing genotype and chemotype of *Fusarium graminearum* from cereals in Ontario, Canada. *PLoS ONE* 14(5), e0216735.

Cuomo, C. A., Güldener, U., Xu, J.-R., Trail, F., Turgeon, B. G., Di Pietro, A., Walton, J. D., Ma, L.-J., Baker, S. E., Rep, M., Adam, G., Antoniw, J., Baldwin, B., Calvo, S., Chang, Y.-L., DeCaprio, D., Gale, L., Gnerre, S., Goswami, R. S., Hammond-Kosack, K. E., Harris, L. J., Hilburn, K., Kennell, J., Kroken, S., Magnuson, J. K., Mannhaupt, G., Mauceli,

E., Mewes, H.-W., Mitterbauer, R., Muehlbauer, G., Münsterkötter, M., Nelson, D., O'Donnell, K., Ouellet, T., Qi, W., Quesneville, H., Roncero, M. I. G., Seong, K.-Y., Tetko, I. V., Urban, M., Waalwijk, C., Ward, T., Yao, J., Birren, B. W. and Kistler, H. C. 2007. The *Fusarium graminearum* genome reveals a link between localized polymorphism and pathogen specialization. *Science* 317, 1400–1402.

Cuzick, A., Urban, M. and Hammond-Kosack, K. E. 2008. *Fusarium graminearum* gene deletion mutants *map1* and *tri5* reveal similarities and differences in the pathogenicity requirements to cause disease on Arabidopsis and wheat floral tissue. *New Phytologist* 177, 990–1000.

De Jonge, R. and Thomma, B. P. 2009. Fungal LysM effectors: extinguishers of host immunity? *Trends in Microbiology* 17(4), 151–157.

Dean, R., Van Kan, J. A., Pretorius, Z. A., Hammond-Kosack, K. E., Di Pietro, A., Spanu, P. D., Rudd, J. J., Dickman, M., Kahmann, R., Ellis, J. and Foster, G. D. 2012. The Top 10 fungal pathogens in molecular plant pathology. *Molecular Plant Pathology* 13(4), 414–430.

Dean, R. A., Talbot, N. J., Ebbole, D. J., Farman, M. L., Mitchell, T. K., Orbach, M. J., Thon, M., Kulkarni, R., Xu, J. R., Pan, H., Read, N. D., Lee, Y. H., Carbone, I., Brown, D., Oh, Y. Y., Donofrio, N., Jeong, J. S., Soanes, D. M., Djonovic, S., Kolomiets, E., Rehmeyer, C., Li, W., Harding, M., Kim, S., Lebrun, M. H., Bohnert, H., Coughlan, S., Butler, J., Calvo, S., Ma, L. J., Nicol, R., Purcell, S., Nusbaum, C., Galagan, J. E. and Birren, B. W. 2005. The genome sequence of the rice blast fungus *Magnaporthe grisea*. *Nature* 434(7036), 980–986.

Del Ponte, E. M., Fernandes, J. M. C. and Pierobom, C. R. 2005. Factors affecting density of airborne *Gibberella zeae* inoculum. *Fitopatologia Brasileira* 30(1), 55–60.

Desmond, O. J., Manners, J. M., Stephens, A. E., Maclean, D. J., Schenk, P. M., Gardiner, D. M., Munn, A. L. and Kazan, K. 2008. The Fusarium mycotoxin deoxynivalenol elicits hydrogen peroxide production, programmed cell death and defence responses in wheat. *Molecular Plant Pathology* 9(4), 435–445.

Diamond, M., Reape, T. J., Rocha, O., Doyle, S. M., Kacprzyk, J., Doohan, F. M. and McCabe, P. F. 2013. The Fusarium mycotoxin deoxynivalenol can inhibit plant apoptosis-like programmed cell death. *PLoS ONE* 8(7), e69542.

Dilks, T., Halsey, K., De Vos, R. P., Hammond-Kosack, K. E. and Brown, N. A. 2019. Non-canonical fungal G-protein coupled receptors promote Fusarium head blight on wheat. *PLoS Pathogens* 15(4), e1007666.

Dinolfo, M. I., Martínez, M., Nogueira, M. S., Nicholson, P. and Stenglein, S. A. 2021. Evaluation of interaction between *Brachypodium distachyon* roots and Fusarium species. *European Journal of Plant Pathology* 159(2), 269–278.

Dong, S., Kong, G., Qutob, D., Yu, X., Tang, J., Kang, J., Dai, T., Wang, H., Gijzen, M. and Wang, Y. 2012. The NLP toxin family in *Phytophthora sojae* includes rapidly evolving groups that lack necrosis-inducing activity. *Molecular Plant-Microbe Interactions* 25(7), 896–909.

Draper, J., Mur, L. A. J., Jenkins, G., Ghosh-Biswas, G. C., Bablak, P., Hasterok, R. and Routledge, A. P. M. 2001. *Brachypodium distachyon*. A new model system for functional genomics in grasses. *Plant Physiology* 127(4), 1539–1555.

Dweba, C. C., Figlan, S., Shimelis, H. A., Motaung, T. E., Sydenham, S., Mwadzingeni, L. and Tsilo, T. J. 2017. Fusarium head blight of wheat: pathogenesis and control strategies. *Crop Protection* 91, 114–122.

Edwards, S. G. 2009. Fusarium mycotoxin content of UK organic and conventional oats. *Food Additives and Contaminants. Part A, Chemistry, Analysis, Control, Exposure and Risk Assessment* 26(7), 1063-1069.

Edwards, S. G., Kharbikar, L. L., Dickin, E. T., MacDonald, S. and Scudamore, K. A. 2018. Impact of pre-harvest rainfall on the distribution of fusarium mycotoxins in wheat mill fractions. *Food Control* 89, 150-156.

Elliott, C. T., Connolly, L. and Kolawole, O. 2020. Potential adverse effects on animal health and performance caused by the addition of mineral adsorbents to feeds to reduce mycotoxin exposure. *Mycotoxin Research* 36(1), 115-126.

Erayman, M., Turktas, M., Akdogan, G., Gurkok, T., Inal, B., Ishakoglu, E., Ilhan, E. and Unver, T. 2015. Transcriptome analysis of wheat inoculated with *Fusarium graminearum*. *Frontiers in Plant Science* 6, 867.

Estrada, R., Gudmestad, N. C., Rivera, V. V. and Secor, G. A. 2010. *Fusarium graminearum* as a dry rot pathogen of potato in the USA: prevalence, comparison of host isolate aggressiveness and factors affecting aetiology. *Plant Pathology* 59(6), 1114-1120.

European Commission. 2006. Commission Regulation (EC) No 1881/2006 of 19 December 2006 setting maximum levels for certain contaminants in foodstuffs. Off. J. Eur. Union, 364.

Fall, L. A., Salazar, M. M., Drnevich, J., Holmes, J. R., Tseng, M. C., Kolb, F. L. and Mideros, S. X. 2019. Field pathogenomics of Fusarium head blight reveals pathogen transcriptome differences due to host resistance. *Mycologia* 111(4), 563-573.

Fan, G., Zhang, K., Zhang, J., Yang, J., Yang, X., Hu, Y., Huang, J., Zhu, Y., Yu, W., Hu, H., Wang, B., Shim, W. and Lu, G. D. 2019. The transcription factor FgMed1 is involved in early conidiogenesis and DON biosynthesis in the plant pathogenic fungus *Fusarium graminearum*. *Applied Microbiology and Biotechnology* 103(14), 5851-5865.

Fan, J., Urban, M., Parker, J. E., Brewer, H. C., Kelly, S. L., Hammond-Kosack, K. E., Fraaije, B. A., Liu, X. and Cools, H. J. 2013. Characterization of the sterol 14α-demethylases of *Fusarium graminearum* identifies a novel genus-specific CYP51 function. *New Phytologist* 198(3), 821-835.

FAO. 2019. *Food Outlook–Biannual Report on Global Food Markets*, Rome, Italy.

Ferrigo, D., Raiola, A. and Causin, R. 2016. Fusarium toxins in cereals: occurrence, legislation, factors promoting the appearance and their management. *Molecules* 21(5), 627.

Fisher, M. C., Henk, D. A., Briggs, C. J., Brownstein, J. S., Madoff, L. C., McCraw, S. L. and Gurr, S. J. 2012. Emerging fungal threats to animal, plant and ecosystem health. *Nature* 484, 186-194.

Fisher, M. C., Hawkins, N. J., Sanglard, D. and Gurr, S. J. 2018. Worldwide emergence of resistance to antifungal drugs challenges human health and food security. *Science*, 360, 739-742.

Foroud, N. A., Pordel, R., Goyal, R. K., Ryabova, D., Eranthodi, A., Chatterton, S. and Kovalchuk, I. 2019. Chemical activation of the ethylene signaling pathway promotes *Fusarium graminearum* resistance in detached wheat heads. *Phytopathology* 109(5), 796-803.

Fouché, S., Plissonneau, C. and Croll, D. 2018. The birth and death of effectors in rapidly evolving filamentous pathogen genomes. *Current Opinion in Microbiology* 46, 34-42.

Frias, M., Brito, N. and Gonzalez, C. 2013. The *Botrytis cinerea* cerato-platanin BcSpl1 is a potent inducer of systemic acquired resistance (SAR) in tobacco and generates a wave of salicylic acid expanding from the site of application. *Molecular Plant Pathology* 14(2), 191-196.

Gadaleta, A., Colasuonno, P., Giove, S. L., Blanco, A. and Giancaspro, A. 2019. Map-based cloning of QFhb. mgb-2A identifies a WAK2 gene responsible for Fusarium head blight resistance in wheat. *Scientific Reports* 9(1), 6929.

Gaffoor, I., Brown, D. W., Plattner, R., Proctor, R. H., Qi, W. and Trail, F. 2005. Functional analysis of the polyketide synthase genes in the filamentous fungus *Gibberella zeae* (anamorph *Fusarium graminearum*). *Eukaryotic Cell* 4(11), 1926–1933.

Galagan, J. E. and Selker, E. U. 2004. RIP: the evolutionary cost of genome defense. *Trends in Genetics* 20(9), 417–423.

Gardiner, D. M., Kazan, K., Praud, S., Torney, F. J., Rusu, A. and Manners, J. M. 2010. Early activation of wheat polyamine biosynthesis during Fusarium head blight implicates putrescine as an inducer of trichothecene mycotoxin production. *BMC Plant Biol*, 10, 289.

Gardiner, D. M., Osborne, S., Kazan, K. and Manners, J. M. 2009. Low pH regulates the production of deoxynivalenol by *Fusarium graminearum*. *Microbiology* 155(9), 3149–3156.

Gardiner, D. M., Stiller, J. and Kazan, K. 2014. Genome sequence of *Fusarium graminearum* isolate CS3005. *Genome Announcements* 2(2), e00227-14.

Gautam, P. and Dill-Macky, R. 2012. Free water can leach mycotoxins from Fusarium-infected wheat heads. *Journal of Phytopathology* 160(9), 484–490.

Geiser, D. M., Al-Hatmi, A., Aoki, T., Arie, T., Balmas, V., Barnes, I., Bergstrom, G. C., Bhattacharyya, M. K. K., Blomquist, C. L., Bowden, R. and Brankovics, B. 2020. Phylogenomic analysis of a 55.1 kb 19-gene dataset resolves a monophyletic Fusarium that includes the *Fusarium solani* species Complex. *Phytopathology*, https://doi.org/10.1094/PHYTO-08-20-0330-LE.

Gijzen, M. and Nürnberger, T. 2006. Nep1-like proteins from plant pathogens: recruitment and diversification of the NPP1 domain across taxa. *Phytochemistry* 67(16), 1800–1807.

Goswami, R. S. and Kistler, H. C. 2004. Heading for disaster: *Fusarium graminearum* on cereal crops. *Molecular Plant Pathology* 5(6), 515–525.

Goswami, R. S. and Kistler, H. C. 2005. Pathogenicity and in planta mycotoxin accumulation among members of the *Fusarium graminearum* species complex on wheat and rice. *Phytopathology* 95, 1397–1404.

Guenther, J. C., Hallen-Adams, H. E., Bücking, H., Shachar-Hill, Y. and Trail, F. 2009. Triacylglyceride metabolism by *Fusarium graminearum* during colonization and sexual development on wheat. *Molecular Plant-Microbe Interactions* 22(12), 1492–1503.

Guldener, U., Seong, K. Y., Boddu, J., Cho, S., Trail, F., Xu, J. R., Adam, G., Mewes, H. W., Muehlbauer, G. J. and Kistler, H. C. 2006. Development of a *Fusarium graminearum* Affymetrix GeneChip for profiling fungal gene expression *in vitro* and *in planta*. *Fungal Genetics and Biology* 43(5), 316–325.

Guo, J., Zhang, X., Hou, Y., Cai, J., Shen, X., Zhou, T., Xu, H., Ohm, H. W., Wang, H., Li, A., Han, F., Wang, H. and Kong, L. 2015. High-density mapping of the major FHB resistance gene *Fhb7* derived from *Thinopyrum ponticum* and its pyramiding with *Fhb1* by marker-assisted selection. *Theoretical and Applied Genetics* 128(11), 2301–2316.

Haidoulis, J. F. and Nicholson, P. 2020. Different effects of phytohormones on Fusarium head blight and Fusarium root rot resistance in *Brachypodium distachyon*. *Journal of Plant Interactions* 15(1), 335–344.

Haile, J. K., N'Diaye, A., Walkowiak, S., Nilsen, K. T., Clarke, J. M., Kutcher, H. R., Steiner, B., Buerstmayr, H. and Pozniak, C. J. 2019. Fusarium head blight in durum wheat: recent status, breeding directions, and future research prospects. *Phytopathology* 109(10), 1664–1675.

Hales, B., Steed, A., Giovannelli, V., Burt, C., Lemmens, M., Molnar-Lang, M. and Nicholson, P. 2020. Type II Fusarium head blight susceptibility conferred by a region on wheat chromosome 4D. *Journal of Experimental Botany* 71(16), 4703–4714.

Hallen, H. E. and Trail, F. 2008. The L-type calcium ion channel cch1 affects ascospore discharge and mycelial growth in the filamentous fungus *Gibberella zeae* (anamorph *Fusarium graminearum*). *Eukaryotic Cell* 7(2), 415–424.

Hamburger, D. 2019. Comparative analysis: the regulation of plants derived from genome editing in Argentina, Australia, Canada, the European Union, Japan and the United States. *Regulation of Genome Editing in Plant Biotechnology.* Springer.

Hammond-Kosack, K. E. 2014. Biotechnology: Plant protection. In: Van Alfen, N. (Ed-in-chief) *Encyclopedia of Agriculture and Food Systems* (vol. 2). San Diego: Elsevier, 134–152.

Han, J., Pluhackova, K. and Böckmann, R. A. 2017. The multifaceted role of SNARE proteins in membrane fusion. *Frontiers in Physiology* 8, 5.

Hansen, F. T., Gardiner, D. M., Lysøe, E., Fuertes, P. R., Tudzynski, B., Wiemann, P., Sondergaard, T. E., Giese, H., Brodersen, D. E. and Sørensen, J. L. 2015. An update to polyketide synthase and non-ribosomal synthetase genes and nomenclature in Fusarium. *Fungal Genetics and Biology* 75, 20–29.

Haran, M., Molineros, J. and Patil, G. 2006. Large scale plant disease forecasting: case study of fusarium head blight. *Center for Statistical Ecology and Environmental Statistics*, 9–10, https://citeseerx.ist.psu.edu/viewdoc/download?doi=10.1.1.115.2033&rep=rep1&type=pdf#page=15.

Harris, L. J., Balcerzak, M., Johnston, A., Schneiderman, D. and Ouellet, T. 2016. Host-preferential Fusarium graminearum gene expression during infection of wheat, barley, and maize. *Fungal Biology* 120(1), 111–123.

He, X., Singh, P. K., Dreisigacker, S., Singh, S., Lillemo, M. and Duveiller, E. 2016. Dwarfing genes *Rht-B1b* and *Rht-D1b* are associated with both Type I FHB susceptibility and low anther extrusion in two bread wheat populations. *PLoS ONE* 11(9), e0162499.

Hillmer, R. A., Tsuda, K., Rallapalli, G., Asai, S., Truman, W., Papke, M. D., Sakakibara, H., Jones, J. D. G., Myers, C. L. and Katagiri, F. 2017. The highly buffered Arabidopsis immune signaling network conceals the functions of its components. *PLoS Genetics* 13(5), e1006639.

Hofstad, A. N., Nussbaumer, T., Akhunov, E., Shin, S., Kugler, K. G., Kistler, H. C., Mayer, K. F. X. and Muehlbauer, G. J. 2016. Examining the transcriptional response in wheat *Fhb1* near-isogenic lines to *Fusarium graminearum* infection and deoxynivalenol treatment. *The Plant Genome* 9(1), 1–15.

Hogg, A. C., Johnston, R. H. and Dyer, A. T. 2007. Applying real-time quantitative PCR to Fusarium crown rot of wheat. *Plant Disease* 91(8), 1021–1028.

Hong, S. Y., So, J., Lee, J., Min, K., Son, H., Park, C., Yun, S. H. and Lee, Y. W. 2010. Functional analyses of two syntaxin-like SNARE genes, GzSYN1 and GzSYN2, in the ascomycete *Gibberella zeae*. *Fungal Genetics and Biology* 47(4), 364–372.

Hoogendoorn, K., Barra, L., Waalwijk, C., Dickschat, J. S., Van Der Lee, T. A. J. and Medema, M. H. 2018. Evolution and diversity of biosynthetic gene clusters in Fusarium. *Frontiers in Microbiology* 9, 1158.

Hou, R., Jiang, C., Zheng, Q., Wang, C. and Xu, J. R. 2015. The AreA transcription factor mediates the regulation of deoxynivalenol (DON) synthesis by ammonium and cyclic adenosine monophosphate (cAMP) signalling in *Fusarium graminearum*. *Molecular Plant Pathology* 16(9), 987–999.

Huang, Y., Li, L., Smith, K. P. and Muehlbauer, G. J. 2016. Differential transcriptomic responses to *Fusarium graminearum* infection in two barley quantitative trait loci associated with Fusarium head blight resistance. *BMC Genomics* 17, 387.

Huo, N. X., Vogel, J. P., Lazo, G. R., You, F. M., Ma, Y., McMahon, S., Dvorak, J., Anderson, O. D., Luo, M. C. and Gu, Y. Q. 2009. Structural characterization of Brachypodium genome and its syntenic relationship with rice and wheat. *Plant Molecular Biology* 70(1–2), 47–61.

Irieda, H., Inoue, Y., Mori, M., Yamada, K., Oshikawa, Y., Saitoh, H., Uemura, A., Terauchi, R., Kitakura, S., Kosaka, A., Singkaravanit-Ogawa, S. and Takano, Y. 2019. Conserved fungal effector suppresses PAMP-triggered immunity by targeting plant immune kinases. *Proceedings of the National Academy of Sciences of the United States of America* 116(2), 496–505.

Jansen, C., Von Wettstein, D., Schäfer, W., Kogel, K. H., Felk, A. and Maier, F. J. 2005. Infection patterns in barley and wheat spikes inoculated with wild-type and trichodiene synthase gene disrupted *Fusarium graminearum*. *Proceedings of the National Academy of Sciences of the United States of America* 102(46), 16892–16897.

Jenczmionka, N. J., Maier, F. J., Lösch, A. P. and Schäfer, W. 2003. Mating, conidiation and pathogenicity of *Fusarium graminearum*, the main causal agent of the head-blight disease of wheat, are regulated by the MAP kinase gpmk1. *Current Genetics* 43(2), 87–95.

Jia, L. J., Tang, H. Y., Wang, W. Q., Yuan, T. L., Wei, W. Q., Pang, B., Gong, X. M., Wang, S. F., Li, Y. J., Zhang, D., Liu, W. and Tang, W. H. 2019. A linear nonribosomal octapeptide from *Fusarium graminearum* facilitates cell-to-cell invasion of wheat. *Nature Communications* 10(1), 922.

Jiang, C., Cao, S., Wang, Z., Xu, H., Liang, J., Liu, H., Wang, G., Ding, M., Wang, Q., Gong, C., Feng, C., Hao, C. and Xu, J. R. 2019. An expanded subfamily of G-protein-coupled receptor genes in *Fusarium graminearum* required for wheat infection. *Nature Microbiology* 4(9), 1582–1591.

Jiang, C., Hei, R., Yang, Y., Zhang, S., Wang, Q., Wang, W., Zhang, Q., Yan, M., Zhu, G., Huang, P., Liu, H. and Xu, J. R. 2020. An orphan protein of *Fusarium graminearum* modulates host immunity by mediating proteasomal degradation of TaSnRK1α. *Nature Communications* 11(1), 4382.

Jiang, J., Liu, X., Yin, Y. and Ma, Z. 2011. Involvement of a velvet protein FgVeA in the regulation of asexual development, lipid and secondary metabolisms and virulence in *Fusarium graminearum*. *PLoS ONE* 6(11), e28291.

Jiang, J., Yun, Y., Liu, Y. and Ma, Z. 2012. FgVELB is associated with vegetative differentiation, secondary metabolism and virulence in *Fusarium graminearum*. *Fungal Genetics and Biology* 49(8), 653–662.

Jones, D. A., Bertazzoni, S., Turo, C. J., Syme, R. A. and Hane, J. K. 2018. Bioinformatic prediction of plant-pathogenicity effector proteins of fungi. *Current Opinion in Microbiology* 46, 43–49.

Jonkers, W., Dong, Y., Broz, K. and Kistler, H. C. 2012. The Wor1-like protein Fgp1 regulates pathogenicity, toxin synthesis and reproduction in the phytopathogenic fungus *Fusarium graminearum*. *PLoS Pathogens* 8(5), e1002724.

Kämper, J., Kahmann, R., Bölker, M., Ma, L. J., Brefort, T., Saville, B. J., Banuett, F., Kronstad, J. W., Gold, S. E., Müller, O., Perlin, M. H., Wösten, H. A., De Vries, R., Ruiz-Herrera, J., Reynaga-Peña, C. G., Snetselaar, K., Mccann, M., Pérez-Martín, J., Feldbrügge, M., Basse, C. W., Steinberg, G., Ibeas, J. I., Holloman, W., Guzman, P., Farman, M., Stajich, J. E., Sentandreu, R., González-Prieto, J. M., Kennell, J. C., Molina, L., Schirawski, J., Mendoza-Mendoza, A., Greilinger, D., Münch, K., Rössel, N., Scherer, M., Vranes, M., Ladendorf, O., Vincon, V., Fuchs, U., Sandrock, B., Meng, S., Ho, E. C., Cahill, M. J., Boyce, K. J., Klose, J., Klosterman, S. J., Deelstra, H. J., Ortiz-Castellanos, L., Li, W., Sanchez-Alonso, P., Schreier, P. H., Häuser-Hahn, I., Vaupel, M., Koopmann, E., Friedrich, G., Voss, H., Schlüter, T., Margolis, J., Platt, D., Swimmer, C., Gnirke, A., Chen, F., Vysotskaia, V., Mannhaupt, G., Güldener, U., Münsterkötter, M., Haase, D., Oesterheld, M., Mewes, H. W., Mauceli, E. W., Decaprio, D., Wade, C. M., Butler, J., Young, S., Jaffe, D. B., Calvo, S., Nusbaum, C., Galagan, J. and Birren, B. W. 2006. Insights from the genome of the biotrophic fungal plant pathogen *Ustilago maydis*. *Nature* 444(7115), 97–101.

Kanja, C. and Hammond-Kosack, K. E. 2020. Proteinaceous effector discovery and characterization in filamentous plant pathogens. *Molecular Plant Pathology* 21(10), 1353–1376.

Kankanala, P., Czymmek, K. and Valent, B. 2007. Roles for rice membrane dynamics and plasmodesmata during biotrophic invasion by the blast fungus. *Plant Cell* 19(2), 706–724.

Karre, S., Kumar, A., Dhokane, D. and Kushalappa, A. C. 2017. Metabolo-transcriptome profiling of barley reveals induction of chitin elicitor receptor kinase gene (HvCERK1) conferring resistance against *Fusarium graminearum*. *Plant Molecular Biology* 93(3), 247–267.

Kazan, K. and Gardiner, D. M. 2018. Transcriptomics of cereal-*Fusarium graminearum* interactions: what we have learned so far. *Molecular Plant Pathology* 19(3), 764–778.

Kazan, K., Gardiner, D. M. and Manners, J. M. 2012. On the trail of a cereal killer: recent advances in *Fusarium graminearum* pathogenomics and host resistance. *Molecular Plant Pathology* 13(4), 399–413.

Keller, N. P. 2019. Fungal secondary metabolism: regulation, function and drug discovery. *Nature Reviews. Microbiology* 17(3), 167–180.

Kettles, G. J., Bayon, C., Canning, G., Rudd, J. J. and Kanyuka, K. 2017. Apoplastic recognition of multiple candidate effectors from the wheat pathogen *Zymoseptoria tritici* in the nonhost plant *Nicotiana benthamiana*. *New Phytologist* 213(1), 338–350.

Khatibi, P. A., McMaster, N. J., Musser, R. and Schmale, D. G. 2014. Survey of mycotoxins in corn distillers' dried grains with solubles from seventy-eight ethanol plants in twelve states in the US in 2011. *Toxins* 6(4), 1155–1168.

Kim, H. K., Lee, S., Jo, S. M., Mccormick, S. P., Butchko, R. A., Proctor, R. H. and Yun, S. H. 2013. Functional roles of FgLaeA in controlling secondary metabolism, sexual development, and virulence in *Fusarium graminearum*. *PLoS ONE* 8(7), e68441.

King, R., Brown, N. A., Urban, M. and Hammond-Kosack, K. E. 2018. Inter-genome comparison of the Quorn fungus *Fusarium venenatum* and the closely related plant infecting pathogen *Fusarium graminearum*. *BMC Genomics* 19(1), 269.

King, R., Urban, M. and Hammond-Kosack, K. E. 2017. Annotation of *Fusarium graminearum* (PH-1) Version 5.0. *Genome Announcements* 5(2), e01479-16.

King, R., Urban, M., Hammond-Kosack, M. C., Hassani-Pak, K. and Hammond-Kosack, K. E. 2015. The completed genome sequence of the pathogenic ascomycete fungus *Fusarium graminearum*. *BMC Genomics* 16, 544.

Koch, A. and Kogel, K. H. 2014. New wind in the sails: improving the agronomic value of crop plants through RNAi-mediated gene silencing. *Plant Biotechnology Journal* 12(7), 821–831.

Koch, A., Kumar, N., Weber, L., Keller, H., Imani, J. and Kogel, K.-H. 2013. Host-induced gene silencing of cytochrome P450 lanosterol C14 -demethylase-encoding genes confers strong resistance to Fusarium species. *Proceedings of the National Academy of Sciences* 110(48), 19324–19329.

Kombrink, A., Rovenich, H., Shi-Kunne, X., Rojas-Padilla, E., Van Den Berg, G. C., Domazakis, E., De Jonge, R., Valkenburg, D. J., Sánchez-Vallet, A., Seidl, M. F. and Thomma, B. P. 2017. *Verticillium dahliae* LysM effectors differentially contribute to virulence on plant hosts. *Molecular Plant Pathology* 18(4), 596–608.

Korves, T. and Colosimo, M. E. 2009. Controlled vocabularies for microbial virulence factors. *Trends in Microbiology* 17(7), 279–285.

Kubicek, C. P., Starr, T. L. and Glass, N. L. 2014. Plant cell wall-degrading enzymes and their secretion in plant-pathogenic fungi. *Annual Review of Phytopathology* 52, 427–451.

Langevin, F., Eudes, F. and Comeau, A. 2004. Effect of trichothecenes produced by *Fusarium graminearum* during Fusarium head blight development in six cereal species. *European Journal of Plant Pathology* 110(7), 735–746.

Lee, B. N., Kroken, S., Chou, D. Y., Robbertse, B., Yoder, O. C. and Turgeon, B. G. 2005. Functional analysis of all nonribosomal peptide synthetases in *Cochliobolus heterostrophus* reveals a factor, NPS6, involved in virulence and resistance to oxidative stress. *Eukaryotic Cell* 4(3), 545–555.

Lee, W. S., Hammond-Kosack, K. E. and Kanyuka, K. 2012. Barley stripe mosaic virus-mediated tools for investigating gene function in cereal plants and their pathogens: virus-induced gene silencing, host-mediated gene silencing, and virus-mediated overexpression of heterologous protein. *Plant Physiology* 160(2), 582–590.

Leonard, K. J. and Bushnell, W. R. 2003. *Fusarium Head Blight of Wheat and Barley*. American Phytopathological Society. St Paul, USA: APS Press.

Lewandowski, S. M., Bushnell, W. R. and Evans, C. K. 2006. Distribution of mycelial colonies and lesions in field-grown barley inoculated with *Fusarium graminearum*. *Phytopathology* 96(6), 567–581.

Li, G., Zhou, J., Jia, H., Gao, Z., Fan, M., Luo, Y., Zhao, P., Xue, S., Li, N., Yuan, Y., Ma, S., Kong, Z., Jia, L., An, X., Jiang, G., Liu, W., Cao, W., Zhang, R., Fan, J., Xu, X., Liu, Y., Kong, Q., Zheng, S., Wang, Y., Qin, B., Cao, S., Ding, Y., Shi, J., Yan, H., Wang, X., Ran, C. and Ma, Z. 2019a. Mutation of a histidine-rich calcium-binding-protein gene in wheat confers resistance to Fusarium head blight. *Nature Genetics* 51(7), 1106–1112.

Li, N., Han, X., Feng, D., Yuan, D. and Huang, L. J. 2019b. Signaling crosstalk between salicylic acid and ethylene/jasmonate in plant defense: do we understand what they are whispering? *International Journal of Molecular Sciences* 20(3), 671.

Li, X., Michlmayr, H., Schweiger, W., Malachova, A., Shin, S., Huang, Y., Dong, Y., Wiesenberger, G., Mccormick, S., Lemmens, M., Fruhmann, P., Hametner, C., Berthiller, F., Adam, G. and Muehlbauer, G. J. 2017. A barley UDP-glucosyltransferase inactivates nivalenol and provides Fusarium Head Blight resistance in transgenic wheat. *Journal of Experimental Botany* 68(9), 2187–2197.

Liu, Y. and Buchenauer, H. 2005. Interactions between Barley yellow dwarf virus and Fusarium spp. affecting development of Fusarium head blight of wheat. *European Journal of Plant Pathology* 113(3), 283–295.

Lo Presti, L., Lanver, D., Schweizer, G., Tanaka, S., Liang, L., Tollot, M., Zuccaro, A., Reissmann, S. and Kahmann, R. 2015. Fungal effectors and plant susceptibility. *Annual Review of Plant Biology* 66, 513-545.

Lu, S. and Faris, J. D. 2019. *Fusarium graminearum* KP4-like proteins possess root growth-inhibiting activity against wheat and potentially contribute to fungal virulence in seedling rot. *Fungal Genetics and Biology* 123, 1-13.

Lv, W., Wang, C., Yang, N., Que, Y., Talbot, N. J. and Wang, Z. 2017. Genome-wide functional analysis reveals that autophagy is necessary for growth, sporulation, deoxynivalenol production and virulence in *Fusarium graminearum*. *Scientific Reports* 7(1), 11062.

Lysøe, E., Pasquali, M., Breakspear, A. and Kistler, H. C. 2011. The transcription factor FgStuAp influences spore development, pathogenicity, and secondary metabolism in *Fusarium graminearum*. *Molecular Plant-Microbe Interactions* 24(1), 54-67.

Ma, L. J., Van Der Does, H. C., Borkovich, K. A., Coleman, J. J., Daboussi, M. J., Di Pietro, A., Dufresne, M., Freitag, M., Grabherr, M., Henrissat, B., Houterman, P. M., Kang, S., Shim, W. B., Woloshuk, C., Xie, X., Xu, J. R., Antoniw, J., Baker, S. E., Bluhm, B. H., Breakspear, A., Brown, D. W., Butchko, R. A. E., Chapman, S., Coulson, R., Coutinho, P. M., Danchin, E. G. J., Diener, A., Gale, L. R., Gardiner, D. M., Goff, S., Hammond-Kosack, K. E., Hilburn, K., Houterman, P. M., Hua-Van, A., Jonkers, W., Kazan, K., Kodira, C. D., Koehrsen, M., Kumar, L., Lee, Y. H., Li, L., Manners, J. M., Miranda-Saavedra, D., Mukherjee, M., Park, G., Park, J., Park, S. Y., Proctor, R. H., Regev, A., Ruiz-Roldan, M. C., Sain, D., Sakthikumar, S., Sykes, S., Schwartz, D. C., Turgeon, B. G., Wapinski, I., Yoder, O., Young, S., Zeng, Q., Zhou, S., Galagan, J., Cuomo, C. A., Kistler H. C., and Rep. M. 2010. Comparative genomics reveals mobile pathogenicity chromosomes in *Fusarium oxysporum*. *Nature* 464, 367-373.

Machado, A. K., Brown, N. A., Urban, M., Kanyuka, K. and Hammond-Kosack, K. E. 2018. RNAi as an emerging approach to control Fusarium head blight disease and mycotoxin contamination in cereals. *Pest Management Science* 74(4), 790-799.

Machado, A. K. D. F. 2017. An exploration of the function of specific components of the predicted secretome of *Fusarium graminearum* during wheat infection. PhD thesis, University of Exeter.

Mackintosh, C. A., Lewis, J., Radmer, L. E., Shin, S., Heinen, S. J., Smith, L. A., Wyckoff, M. N., Dill-Macky, R., Evans, C. K., Kravchenko, S., Baldridge, G. D., Zeyen, R. J. and Muehlbauer, G. J. 2007. Overexpression of defense response genes in transgenic wheat enhances resistance to Fusarium head blight. *Plant Cell Reports* 26(4), 479-488.

Magan, N., Aldred, D., Mylona, K. and Lambert, R. J. 2010. Limiting mycotoxins in stored wheat. *Food Additives and Contaminants. Part A, Chemistry, Analysis, Control, Exposure and Risk Assessment* 27(5), 644-650.

Maier, F. J., Miedaner, T., Hadeler, B., Felk, A., Salomon, S., Lemmens, M., Kassner, H. and Schaefer, W. 2006. Involvement of trichothecenes in fusarioses of wheat, barley and maize evaluated by gene disruption of the trichodiene synthase (*Tri5*) gene in three field isolates of different chemotype and virulence. *Molecular Plant Pathology* 7(6), 449-461.

Makandar, R., Essig, J. S., Schapaugh, M. A., Trick, H. N. and Shah, J. 2006. Genetically engineered resistance to Fusarium head blight in wheat by expression of Arabidopsis NPR1. *Molecular Plant-Microbe Interactions* 19(2), 123-129.

Mandala, G., Tundo, S., Francesconi, S., Gevi, F., Zolla, L., Ceoloni, C. and D'Ovidio, R. 2019. Deoxynivalenol detoxification in transgenic wheat confers resistance to Fusarium Head Blight and Crown Rot diseases. *Molecular Plant-Microbe Interactions* 32(5), 583-592.

Manning, V. A., Chu, A. L., Scofield, S. R. and Ciuffetti, L. M. 2010. Intracellular expression of a host-selective toxin, ToxA, in diverse plants phenocopies silencing of a ToxA-interacting protein, ToxABP1. *New Phytologist* 187(4), 1034-1047.

Masachis, S., Segorbe, D., Turrà, D., Leon-Ruiz, M., Fürst, U., El Ghalid, M., Leonard, G., López-Berges, M. S., Richards, T. A., Felix, G. and Di Pietro, A. 2016. A fungal pathogen secretes plant alkalinizing peptides to increase infection. *Nature Microbiology* 1(6), 16043.

McLaughlin, J. E., Al Darwish, N. I., Garcia-Sanchez, J., Tyagi, N., Trick, H. N., Mccormick, S., Dill-Macky, R. and Tumer, N. E. 2021. A lipid transfer protein has antifungal and antioxidant activity and suppresses Fusarium head blight disease and DON accumulation in transgenic wheat. *Phytopathology®* 111(4), 671-683.

McMullen, M., Bergstrom, G., De Wolf, E., Dill-Macky, R., Hershman, D., Shaner, G. and Van Sanford, D. 2012. A unified effort to fight an enemy of wheat and barley: Fusarium head blight. *Plant Disease* 96(12), 1712-1728.

Mentges, M., Glasenapp, A., Boenisch, M., Malz, S., Henrissat, B., Frandsen, R. J. N., Guldener, U., Munsterkotter, M., Bormann, J., Lebrun, M. H., Schafer, W. and Martinez-Rocha, A. L. 2020. Infection cushions of *Fusarium graminearum* are fungal arsenals for wheat infection. *Molecular Plant Pathology* 21(8), 1070-1087.

Merhej, J., Richard-Forget, F. and Barreau, C. 2011. The pH regulatory factor Pac1 regulates *Tri* gene expression and trichothecene production in *Fusarium graminearum*. *Fungal Genetics and Biology* 48(3), 275-284.

Miedaner, T. 1997. Breeding wheat and rye for resistance to Fusarium diseases. *Plant Breeding* 116(3), 201-220.

Miller, S. S., Reid, L. M. and Harris, L. J. 2007. Colonization of maize silks by *Fusarium graminearum*, the causative organism of Gibberella ear rot. *Canadian Journal of Botany* 85(4), 369-376.

Mirocha, C. J. and Christensen, C. M. 1974. Fungus metabolites toxic to animals. *Annual Review of Phytopathology* 12(1), 303-330.

Mudge, A. M., Dill-Macky, R., Dong, Y., Gardiner, D. M., White, R. G. and Manners, J. M. 2006. A role for the mycotoxin deoxynivalenol in stem colonisation during crown rot disease of wheat caused by *Fusarium graminearum* and *Fusarium pseudograminearum*. *Physiological and Molecular Plant Pathology* 69 (1-3), 73-85.

Munkvold, G. P. 2003. Epidemiology of Fusarium diseases and their mycotoxins in maize ears. *European Journal of Plant Pathology* 109(7), 705-713.

Nakajima, Y., Akasaka, M., Shiobara, T., Kitou, Y., Maeda, K., Kanamaru, K., Ohsato, S., Kobayashi, T., Nishiuchi, T. and Kimura, M. 2020. Impact of nitrogen metabolism-associated culture pH changes on regulation of Fusarium trichothecene biosynthesis: revision of roles of polyamine agmatine and transcription factor area. *Current Genetics* 66(6), 1179-1190.

Nelson, P. E. 1992. Taxonomy and biology of *Fusarium moniliforme*. *Mycopathologia* 117(1-2), 29-36.

Nganje, W. E., Bangsund, D. A., Leistritz, F. L., Wilson, W. W. and Tiapo, N. M. 2004. Regional economic impacts of Fusarium head blight in wheat and barley. *Review of Agricultural Economics* 26(3), 332-347.

Nguyen, L. T., Schmidt, H. A., Von Haeseler, A. and Minh, B. Q. 2015. IQ-TREE: a fast and effective stochastic algorithm for estimating maximum-likelihood phylogenies. *Molecular Biology and Evolution* 32(1), 268–274.

Niehaus, E. M., Rindermann, L., Janevska, S., Münsterkötter, M., Güldener, U. and Tudzynski, B. 2018. Analysis of the global regulator Lae1 uncovers a connection between Lae1 and the histone acetyltransferase HAT1 in *Fusarium fujikuroi*. *Applied Microbiology and Biotechnology* 102(1), 279–295.

Nowara, D., Gay, A., Lacomme, C., Shaw, J., Ridout, C., Douchkov, D., Hensel, G., Kumlehn, J. and Schweizer, P. 2010. HIGS: host-induced gene silencing in the obligate biotrophic fungal pathogen *Blumeria graminis*. *The Plant Cell* 22(9), 3130–3141.

Nyvall, R. F., Percich, J. A. and Mirocha, C. J. 1999. Fusarium head blight of cultivated and natural wild rice (*Zizania palustris*) in Minnesota caused by *Fusarium graminearum* and associated *Fusarium* spp. *Plant Disease* 83(2), 159–164.

O'Mara, S. P., Broz, K., Boenisch, M., Zhong, Z., Dong, Y. and Kistler, H. C. 2020. The *Fusarium graminearum* t-SNARE Sso2 is involved in growth, defense, and DON accumulation and virulence. *Molecular Plant-Microbe Interactions* 33(7), 888–901.

Oide, S., Berthiller, F., Wiesenberger, G., Adam, G. and Turgeon, B. G. 2015. Individual and combined roles of malonichrome, ferricrocin, and TAFC siderophores in *Fusarium graminearum* pathogenic and sexual development. *Frontiers in Microbiology* 5, 759.

Oide, S., Moeder, W., Krasnoff, S., Gibson, D., Haas, H., Yoshioka, K. and Turgeon, B. G. 2006. NPS6, encoding a nonribosomal peptide synthetase involved in siderophore-mediated iron metabolism, is a conserved virulence determinant of plant pathogenic ascomycetes. *The Plant Cell* 18(10), 2836–2853.

Pan, Y., Liu, Z., Rocheleau, H., Fauteux, F., Wang, Y., McCartney, C. and Ouellet, T. 2018. Transcriptome dynamics associated with resistance and susceptibility against Fusarium head blight in four wheat genotypes. *BMC Genomics* 19(1), 642.

Panwar, V. P., McCallum, B., Jordan, M., Loewen, M., Fobert, P., McCartney, C. and Bakkeren, G. 2016. RNA silencing approaches for identifying pathogenicity and virulence elements towards engineering crop resistance to plant pathogenic fungi. *CAB Reviews: Perspectives in Agriculture, Veterinary Science, Nutrition and Natural Resources* 11(27), 1–13.

Paper, J. M., Scott-Craig, J. S., Adhikari, N. D., Cuomo, C. A. and Walton, J. D. 2007. Comparative proteomics of extracellular proteins *in vitro* and *in planta* from the pathogenic fungus *Fusarium graminearum*. *Proteomics* 7(17), 3171–3183.

Parry, D. W., Jenkinson, P. and McLeod, L. 1995. Fusarium ear blight (scab) in small grain cereals–a review. *Plant Pathology* 44(2), 207–238.

Pasquet, J. C., Changenet, V., Macadre, C., Boex-Fontvieille, E., Soulhat, C., Bouchabke-Coussa, O., Dalmais, M., Atanasova-Penichon, V., Bendahmane, A., Saindrenan, P. and Dufresne, M. 2016. A brachypodium UDP-glycosyltransferase confers root tolerance to deoxynivalenol and resistance to fusarium infection. *Plant Physiology* 172, 559–574.

Pasquet, J.-C., Chaouch, S., Macadré, C., Balzergue, S., Huguet, S., Martin-Magniette, M.-L., Bellvert, F., Deguercy, X., Thareau, V., Heintz, D., Saindrenan, P. and Dufresne, M. 2014. Differential gene expression and metabolomic analyses of *Brachypodium distachyon* infected by deoxynivalenol producing and non-producing strains of *Fusarium graminearum*. *BMC Genomics* 15(1), 629.

Paudel, B., Zhuang, Y., Galla, A., Dahal, S., Qiu, Y., Ma, A., Raihan, T. and Yen, Y. 2020. WFhb1-1 plays an important role in resistance against Fusarium head blight in wheat. *Scientific Reports* 10(1), 7794.

Peraldi, A., Beccari, G., Steed, A. and Nicholson, P. 2011. *Brachypodium distachyon*: a new pathosystem to study Fusarium head blight and other Fusarium diseases of wheat. *BMC Plant Biology* 11, 100.

Perochon, A., Jianguang, J., Kahla, A., Arunachalam, C., Scofield, S. R., Bowden, S., Wallington, E. and Doohan, F. M. 2015. TaFROG encodes a Pooideae orphan protein that interacts with SnRK1 and enhances resistance to the mycotoxigenic fungus *Fusarium graminearum*. *Plant Physiology* 169(4), 2895–2906.

Perochon, A., Váry, Z., Malla, K. B., Halford, N. G., Paul, M. J. and Doohan, F. M. 2019. The wheat SnRK1α family and its contribution to Fusarium toxin tolerance. *Plant Science* 288, 110217.

Pestka, J. J. 2010. Deoxynivalenol: mechanisms of action, human exposure, and toxicological relevance. *Archives of Toxicology* 84(9), 663–679.

Petti, C., Reiber, K., Ali, S. S., Berney, M. and Doohan, F. M. 2012. Auxin as a player in the biocontrol of Fusarium head blight disease of barley and its potential as a disease control agent. *BMC Plant Biology* 12, 224.

Pinotti, L., Ottoboni, M., Giromini, C., Dell'orto, V. and Cheli, F. 2016. Mycotoxin contamination in the EU feed supply chain: a focus on cereal byproducts. *Toxins* 8(2), 45.

Ponts, N., Couedelo, L., Pinson-Gadais, L., Verdal-Bonnin, M. N., Barreau, C. and Richard-Forget, F. 2009. Fusarium response to oxidative stress by H_2O_2 is trichothecene chemotype-dependent. *FEMS Microbiology Letters* 293(2), 255–262.

Poppenberger, B., Berthiller, F., Lucyshyn, D., Sieberer, T., Schuhmacher, R., Krska, R., Kuchler, K., Glossl, J., Luschnig, C. and Adam, G. 2003. Detoxification of the Fusarium mycotoxin deoxynivalenol by a UDP-glucosyltransferase from *Arabidopsis thaliana*. *Journal of Biological Chemistry* 278(48), 47905–47914.

Pritsch, C., Muehlbauer, G. J., Bushnell, W. R., Somers, D. A. and Vance, C. P. 2000. Fungal development and induction of defense response genes during early infection of wheat spikes by *Fusarium graminearum*. *Molecular Plant-Microbe Interactions* 13(2), 159–169.

Proctor, R. H., Hohn, T. M., McCormick, S. P. and Desjardins, A. E. 1995a. *Tri6* encodes an unusual zinc finger protein involved in regulation of trichothecene biosynthesis in *Fusarium sporotrichioides*. *Applied and Environmental Microbiology* 61(5), 1923–1930.

Proctor, R. H., Hohn, T. M. and McCormick, S. P. 1995b. Reduced virulence of Gibberella zeae caused by disruption of a trichothecene toxin biosynthetic gene. *Molecular Plant-Microbe Interactions* 8(4), 593–601.

Przemieniecki, S. W., Kurowski, T. P. and Korzekwa, K. 2014. Chemotypes and geographic distribution of the *Fusarium graminearum* species complex. *Environmental Biotechnology* 10(2), 45–59.

Puri, K. D., Yan, C., Leng, Y. and Zhong, S. 2016. RNA-Seq revealed differences in transcriptomes between 3ADON and 15ADON populations of *Fusarium graminearum in vitro* and *in planta*. *PLoS ONE* 11(10), e0163803.

Qi, P.-F., Balcerzak, M., Rocheleau, H., Leung, W., Wei, Y.-M., Zheng, Y.-L. and Ouellet, T. 2016. Jasmonic acid and abscisic acid play important roles in host-pathogen interaction between *Fusarium graminearum* and wheat during the early stages of Fusarium head blight. *Physiological and Molecular Plant Pathology* 93, 39–48.

Qi, P. F., Johnston, A., Balcerzak, M., Rocheleau, H., Harris, L. J., Long, X. Y., Wei, Y. M., Zheng, Y. L. and Ouellet, T. 2012. Effect of salicylic acid on *Fusarium graminearum*, the major causal agent of Fusarium head blight in wheat. *Fungal Biology* 116(3), 413–426.

Qi, P. F., Zhang, Y. Z., Liu, C. H., Zhu, J., Chen, Q., Guo, Z. R., Wang, Y., Xu, B. J., Zheng, T., Jiang, Y. F., Wang, J. P., Zhou, C. Y., Feng, X., Kong, L., Lan, X. J., Jiang, Q. T., Wei, Y. M. and Zheng, Y. L. 2018. *Fusarium graminearum* ATP-binding cassette transporter gene FgABCC9 is required for its transportation of salicylic acid, fungicide resistance, mycelial growth and pathogenicity towards wheat. *International Journal of Molecular Sciences* 19(8), 2351.

Qi, W., Kwon, C. and Trail, F. 2006. Microarray analysis of transcript accumulation during perithecium development in the filamentous fungus *Gibberella zeae* (anamorph *Fusarium graminearum*). *Molecular Genetics and Genomics* 276(1), 87–100.

Quesada-Ocampo, L., Al-Haddad, J., Scruggs, A. C., Buell, C. R. and Trail, F. 2016. Susceptibility of maize to stalk rot caused by *Fusarium graminearum* deoxynivalenol and zearalenone mutants. *Phytopathology* 106(8), 920–927.

Rabe, F., Ajami-Rashidi, Z., Doehlemann, G., Kahmann, R. and Djamei, A. 2013. Degradation of the plant defence hormone salicylic acid by the biotrophic fungus *Ustilago maydis*. *Molecular Microbiology* 89(1), 179–188.

Rampersad, S. N. 2020. Pathogenomics and management of Fusarium diseases in plants. *Pathogens* 9(5), 340.

Rauwane, M. E., Ogugua, U. V., Kalu, C. M., Ledwaba, L. K., Woldesemayat, A. A. and Ntushelo, K. 2020. Pathogenicity and virulence factors of *Fusarium graminearum* including factors discovered using Next Generation Sequencing technologies and proteomics. *Microorganisms* 8, 305.

Rawat, N., Pumphrey, M. O., Liu, S., Zhang, X., Tiwari, V. K., Ando, K., Trick, H. N., Bockus, W. W., Akhunov, E., Anderson, J. A. and Gill, B. S. 2016. Wheat *Fhb1* encodes a chimeric lectin with agglutinin domains and a pore-forming toxin-like domain conferring resistance to Fusarium head blight. *Nature Genetics* 48(12), 1576–1580.

Rep, M. and Kistler, H. C. 2010. The genomic organization of plant pathogenicity in Fusarium species. *Current Opinion in Plant Biology* 13(4), 420–426.

Ribichich, K. F., Lopez, S. E. and Vegetti, A. C. 2000. Histopathological spikelet changes produced by *Fusarium graminearum* in susceptible and resistant wheat cultivars. *Plant Disease* 84(7), 794–802.

Robards, A. W. 1971. The ultrastructure of plasmodesmata. *Protoplasma* 72(2–3), 315–323.

Sakulkoo, W., Oses-Ruiz, M., Garcia, E. O., Soanes, D. M., Littlejohn, G. R., Hacker, C., Correia, A., Valent, B. and Talbot, N. J. 2018. A single fungal MAP kinase controls plant cell-to-cell invasion by the rice blast fungus. *Science* 359(6382), 1399–1403.

Salomon, S., Gácser, A., Frerichmann, S., Kröger, C., Schäfer, W. and Voigt, C. A. 2012. The secreted lipase FGL1 is sufficient to restore the initial infection step to the apathogenic *Fusarium graminearum* MAP kinase disruption mutant Δgpmk1. *European Journal of Plant Pathology* 134(1), 23–37.

Sarlin, T., Kivioja, T., Kalkkinen, N., Linder, M. B. and Nakari-Setälä, T. 2012. Identification and characterization of gushing-active hydrophobins from *Fusarium graminearum* and related species. *Journal of Basic Microbiology* 52(2), 184–194.

Sarowar, S., Alam, S. T., Makandar, R., Lee, H., Trick, H. N., Dong, Y. and Shah, J. 2019. Targeting the pattern-triggered immunity pathway to enhance resistance to *Fusarium graminearum*. *Molecular Plant Pathology* 20(5), 626-640.

Schmidt-Heydt, M., Parra, R., Geisen, R. and Magan, N. 2011. Modelling the relationship between environmental factors, transcriptional genes and deoxynivalenol mycotoxin production by strains of two Fusarium species. *Journal of the Royal Society. Interface* 8, 117-126.

Schwarz, P., Casper, H., Barr, J. and Musial, M. 1997. Impact of Fusarium head blight on the malting and brewing quality of barley. *Cereal Research Communications* 25(3), 813-814.

Schweiger, W., Boddu, J., Shin, S., Poppenberger, B., Berthiller, F., Lemmens, M., Muehlbauer, G. J. and Adam, G. 2010. Validation of a candidate deoxynivalenol-inactivating UDP-glucosyltransferase from barley by heterologous expression in yeast. *Molecular Plant-Microbe Interactions* 23(7), 977-986.

Seong, K. Y., Pasquali, M., Zhou, X., Song, J., Hilburn, K., Mccormick, S., Dong, Y., Xu, J. R. and Kistler, H. C. 2009. Global gene regulation by *Fusarium transcription* factors Tri6 and Tri10 reveals adaptations for toxin biosynthesis. *Molecular Microbiology* 72(2), 354-367.

Seong, K. Y., Zhao, X., Xu, J. R., Güldener, U. and Kistler, H. C. 2008. Conidial germination in the filamentous fungus *Fusarium graminearum*. *Fungal Genetics and Biology* 45(4), 389-399.

Shan, L. Y., Cui, W. Y., Zhang, D. D., Zhang, J., Ma, N. N., Bao, Y. M., Dai, X. F. and Guo, W. 2017. First report of *Fusarium brachygibbosum* causing maize stalk rot in China. *Plant Disease* 101(5), 837.

Shelest, E. 2008. Transcription factors in fungi. *FEMS Microbiology Letters* 286(2), 145-151.

Shin, J. H., Han, J. H., Lee, J. K. and Kim, K. S. 2014. Characterization of the maize stalk rot pathogens *Fusarium subglutinans* and *F. temperatum* and the effect of fungicides on their mycelial growth and colony formation. *The Plant Pathology Journal* 30(4), 397-406.

Shostak, K., Bonner, C., Sproule, A., Thapa, I., Shields, S. W. J., Blackwell, B., Vierula, J., Overy, D. and Subramaniam, R. 2020. Activation of biosynthetic gene clusters by the global transcriptional regulator TRI6 in *Fusarium graminearum*. *Molecular Microbiology* 114(4), 664-680.

Sieber, C. M., Lee, W., Wong, P., Münsterkötter, M., Mewes, H. W., Schmeitzl, C., Varga, E., Berthiller, F., Adam, G. and Güldener, U. 2014. The *Fusarium graminearum* genome reveals more secondary metabolite gene clusters and hints of horizontal gene transfer. *PLoS ONE* 9(10), e110311.

Skadsen, R. W. and Hohn, T. M. 2004. Use of *Fusarium graminearum* transformed with gfp to follow infection patterns in barley and Arabidopsis. *Physiological and Molecular Plant Pathology* 64(1), 45-53.

Solomon, P. S., Tan, K. C. and Oliver, R. P. 2003. The nutrient supply of pathogenic fungi; a fertile field for study. *Molecular Plant Pathology* 4(3), 203-210.

Son, H., Seo, Y. S., Min, K., Park, A. R., Lee, J., Jin, J. M., Lin, Y., Cao, P., Hong, S. Y., Kim, E. K., Lee, S. H., Cho, A., Lee, S., Kim, M. G., Kim, Y., Kim, J. E., Kim, J. C., Choi, G. J., Yun, S. H., Lim, J. Y., Kim, M., Lee, Y. H., Choi, Y. D. and Lee, Y. W. 2011. A phenome-based functional analysis of transcription factors in the cereal head blight fungus, *Fusarium graminearum*. *PLoS Pathogens* 7(10), e1002310.

Soni, N., Altartouri, B., Hegde, N., Duggavathi, R., Nazarian-Firouzabadi, F. and Kushalappa, A. C. 2021. TaNAC032 transcription factor regulates lignin-biosynthetic genes to combat Fusarium head blight in wheat. *Plant Science* 304, 110820.

Soni, N., Hegde, N., Dhariwal, A. and Kushalappa, A. C. 2020. Role of laccase gene in wheat NILs differing at QTL-Fhb1 for resistance against Fusarium head blight. *Plant Science* 298, 110574.

Sorahinobar, M., Niknam, V., Ebrahimzadeh, H., Soltanloo, H., Behmanesh, M. and Enferadi, S. T. 2016. Central role of salicylic acid in resistance of wheat against *Fusarium graminearum*. *Journal of Plant Growth Regulation* 35(2), 477–491.

Sørensen, J. L., Sondergaard, T. E., Covarelli, L., Fuertes, P. R., Hansen, F. T., Frandsen, R. J. N., Saei, W., Lukassen, M. B., Wimmer, R., Nielsen, K. F., Gardiner, D. M. and Giese, H. 2014. Identification of the biosynthetic gene clusters for the lipopeptides fusaristatin A and W493 B in *Fusarium graminearum* and *F. pseudograminearum*. *Journal of Natural Products* 77(12), 2619–2625.

Sridhar, P. S., Trofimova, D., Subramaniam, R., Fundora, D. G.-P., Foroud, N. A., Allingham, J. S. and Loewen, M. C. 2020. Ste2 receptor-mediated chemotropism of *Fusarium graminearum* contributes to its pathogenicity against wheat. *Scientific Reports* 10, 1.

Srinivasachary, Gosman, N., Steed, A., Hollins, T., Bayles, R., Jennings, P. and Nicholson, P. 2009. Semi-dwarfing *Rht-B1* and *Rht-D1* loci of wheat differ significantly in their influence on resistance to Fusarium head blight. *Theoretical and Applied Genetics*, 118, 695–702.

Stephens, A. E., Gardiner, D. M., White, R. G., Munn, A. L. and Manners, J. M. 2008. Phases of infection and gene expression of *Fusarium graminearum* during crown rot disease of wheat. *Molecular Plant-Microbe Interactions* 21(12), 1571–1581.

Su, P., Guo, X., Fan, Y., Wang, L., Yu, G., Ge, W., Zhao, L., Ma, X., Wu, J., Li, A., Wang, H. and Kong, L. 2018. Application of Brachypodium genotypes to the analysis of type II resistance to Fusarium head blight (FHB). *Plant Science* 272, 255–266.

Su, Z., Bernardo, A., Tian, B., Chen, H., Wang, S., Ma, H., Cai, S., Liu, D., Zhang, D., Li, T., Trick, H., St Amand, P., Yu, J., Zhang, Z. and Bai, G. 2019. A deletion mutation in TaHRC confers Fhb1 resistance to Fusarium head blight in wheat. *Nature Genetics* 51(7), 1099–1105.

Sun, Y., Xiao, J., Jia, X., Ke, P., He, L., Cao, A., Wang, H., Wu, Y., Gao, X. and Wang, X. 2016. The role of wheat jasmonic acid and ethylene pathways in response to *Fusarium graminearum* infection. *Plant Growth Regulation* 80(1), 69–77.

Sutton, J. C. 1982. Epidemiology of wheat head blight and maize ear rot caused by *Fusarium graminearum*. *Canadian Journal of Plant Pathology* 4(2), 195–209.

Svoboda, T., Parich, A., Güldener, U., Schöfbeck, D., Twaruschek, K., Václavíková, M., Hellinger, R., Wiesenberger, G., Schuhmacher, R. and Adam, G. 2019. Biochemical characterization of the *Fusarium graminearum* candidate ACC-deaminases and virulence testing of knockout mutant strains. *Frontiers in Plant Science* 10, 1072.

Talbot, N. J., McCafferty, H. R. K., Ma, M., Moore, K. and Hamer, J. E. 1997. Nitrogen starvation of the rice blast fungus *Magnaporthe grisea* may act as an environmental cue for disease symptom expression. *Physiological and Molecular Plant Pathology* 50(3), 179–195.

Tekauz, A., McCallum, B., Ames, N. and Fetch, J. M. 2004. Fusarium head blight of oat–current status in western Canada. *Canadian Journal of Plant Pathology* 26(4), 473–479.

Thapa, G., Gunupuru, L. R., Hehir, J. G., Kahla, A., Mullins, E. and Doohan, F. M. 2018. A pathogen-responsive leucine rich receptor like kinase contributes to Fusarium resistance in cereals. *Frontiers in Plant Science* 9, 867.

Tilney, L. G., Cooke, T. J., Connelly, P. S. and Tilney, M. S. 1991. The structure of plasmodesmata as revealed by plasmolysis, detergent extraction and protease digestion. *Journal of Cell Biology* 112(4), 739–747.

Tralamazza, S. M., Rocha, L. O., Oggenfuss, U., Corrêa, B. and Croll, D. 2019. Complex evolutionary origins of specialized metabolite gene cluster diversity among the plant pathogenic fungi of the *Fusarium graminearum* species complex. *Genome Biology and Evolution* 11(11), 3106–3122.

Tundo, S., Janni, M., Moscetti, I., Mandala, G., Savatin, D., Blechl, A., Favaron, F. and D'Ovidio, R. 2016. PvPGIP2 accumulation in specific floral tissues but not in the endosperm limits *Fusarium graminearum* infection in wheat. *Molecular Plant-Microbe Interactions* 29(10), 815–821.

Tyler, B. M. 2017. The fog of war: how network buffering protects plants' defense secrets from pathogens. *PLoS Genetics* 13(5), e1006713.

Urban, M., Cuzick, A., Seager, J., Wood, V., Rutherford, K., Venkatesh, S. Y., De Silva, N., Martinez, M. C., Pedro, H., Yates, A. D., Hassani-Pak, K. and Hammond-Kosack, K. E. 2020. PHI-base: the pathogen-host interactions database. *Nucleic Acids Research* 48(D1), D613–D620.

Urban, M., Daniels, S., Mott, E. and Hammond-Kosack, K. 2002. Arabidopsis is susceptible to the cereal ear blight fungal pathogens *Fusarium graminearum* and *Fusarium culmorum*. *The Plant Journal* 32(6), 961–973.

Urban, M. and Hammond-Kosack, K. E. 2013. *Molecular Genetics and Genomic Approaches to Explore Fusarium Infection of Wheat Floral Tissue. Fusarium Genomics and Molecular and Cellular Biology*. Proctor, R. H. and Brown, D. (Eds), Caister Academic Press, Norfolk (UK), 43–79. ISBN:978-1-908230-25-6, http://www.horizonpress.com/fusarium.

Urban, M., Mott, E., Farley, T. and Hammond-Kosack, K. 2003. The *Fusarium graminearum* MAP1 gene is essential for pathogenicity and development of perithecia. *Molecular Plant Pathology* 4(5), 347–359.

Van Der Lee, T., Zhang, H., Van Diepeningen, A. and Waalwijk, C. 2015. Biogeography of *Fusarium graminearum* species complex and chemotypes: a review. *Food Additives and Contaminants. Part A, Chemistry, Analysis, Control, Exposure and Risk Assessment* 32(4), 453–460.

Vaughan, M., Backhouse, D. and Ponte, E. M. D. 2016. Climate change impacts on the ecology of *Fusarium graminearum* species complex and susceptibility of wheat to Fusarium head blight: a review. *World Mycotoxin Journal* 9(5), 685–700.

Veitch, R. S., Caldwell, C. D., Martin, R. A., Lada, R., Salmon, D., Anderson, D. M. and MacDonald, D. 2008. Susceptibility of winter and spring triticales to Fusarium head blight and deoxynivalenol accumulation. *Canadian Journal of Plant Science* 88(4), 783–788.

Velazhahan, V., Ma, N., Pándy-Szekeres, G., Kooistra, A. J., Lee, Y., Gloriam, D. E., Vaidehi, N. and Tate, C. G. 2020. Structure of the class D GPCR Ste2 dimer coupled to two G proteins. *Nature* 589, 148–153.

Venske, E., Dos Santos, R. S., Farias, D. D. R., Rother, V., Da Maia, L. C., Pegoraro, C. and Costa De Oliveira, A. 2019. Meta-analysis of the QTLome of Fusarium head blight resistance in bread wheat: refining the current puzzle. *Frontiers in Plant Science* 10, 727.

Vleeshouwers, V. G. and Oliver, R. P. 2014. Effectors as tools in disease resistance breeding against biotrophic, hemibiotrophic, and necrotrophic plant pathogens. *Molecular Plant-Microbe Interactions* 27(3), 196–206.

Voigt, C. A., Schafer, W. and Salomon, S. 2005. A secreted lipase of *Fusarium graminearum* is a virulence factor required for infection of cereals. *The Plant Journal* 42(3), 364–375.

Walker, C. H. 2020. Effector discovery and characterisation in the *Fusarium graminearum*-wheat floral interaction. PhD thesis, University of Nottingham.

Walkowiak, S., Bonner, C. T., Wang, L., Blackwell, B., Rowland, O. and Subramaniam, R. 2015. Intraspecies interaction of *Fusarium graminearum* contributes to reduced toxin production and virulence. *Molecular Plant-Microbe Interactions* 28(11), 1256–1267.

Walkowiak, S., Rowland, O., Rodrigue, N. and Subramaniam, R. 2016. Whole genome sequencing and comparative genomics of closely related Fusarium Head Blight fungi: *Fusarium graminearum*, *F. meridionale* and *F. asiaticum*. *BMC Genomics* 17(1), 1014.

Walter, S., Kahla, A., Arunachalam, C., Perochon, A., Khan, M. R., Scofield, S. R. and Doohan, F. M. 2015. A wheat ABC transporter contributes to both grain formation and mycotoxin tolerance. *Journal of Experimental Botany* 66(9), 2583–2593.

Walter, S., Nicholson, P. and Doohan, F. M. 2010. Action and reaction of host and pathogen during Fusarium head blight disease. *New Phytologist* 185(1), 54–66.

Wang, C., Zhang, S., Hou, R., Zhao, Z., Zheng, Q., Xu, Q., Zheng, D., Wang, G., Liu, H., Gao, X., Ma, J. W., Kistler, H. C., Kang, Z. and Xu, J. R. 2011. Functional analysis of the kinome of the wheat scab fungus *Fusarium graminearum*. *PLoS Pathogens* 7(12), e1002460.

Wang, H., Sun, S., Ge, W., Zhao, L., Hou, B., Wang, K., Lyu, Z., Chen, L., Xu, S., Guo, J. and Li, M. 2020a. Horizontal gene transfer of *Fhb7* from fungus underlies Fusarium head blight resistance in wheat. *Science* 368, 6493.

Wang, M., Wu, L., Mei, Y., Zhao, Y., Ma, Z., Zhang, X. and Chen, Y. 2020b. Host-induced gene silencing of multiple genes of *Fusarium graminearum* enhances resistance to Fusarium head blight in wheat. *Plant Biotechnology Journal* 18(12), 2373–2375.

Wang, L., Li, Q., Liu, Z., Surendra, A., Pan, Y., Li, Y., Zaharia, L. I., Ouellet, T. and Fobert, P. R. 2018. Integrated transcriptome and hormone profiling highlight the role of multiple phytohormone pathways in wheat resistance against fusarium head blight. *PLoS ONE* 13(11), e0207036.

Wang, M. and Jin, H. 2017. Spray-Induced Gene Silencing: a powerful innovative strategy for crop protection. *Trends in Microbiology* 25(1), 4–6.

Wang, Q., Vera Buxa, S., Furch, A., Friedt, W. and Gottwald, S. 2015. Insights into *Triticum aestivum* seedling root rot caused by *Fusarium graminearum*. *Molecular Plant-Microbe Interactions* 28(12), 1288–1303.

Wegulo, S. N., Baenziger, P. S., Nopsa, J. H., Bockus, W. W. and Hallen-Adams, H. 2015. Management of Fusarium head blight of wheat and barley. *Crop Protection* 73, 100–107.

Westphal, K. R., Nielsen, K. A. H., Wollenberg, R. D., Møllehøj, M. B., Bachleitner, S., Studt, L., Lysøe, E., Giese, H., Wimmer, R., Sørensen, J. L. and Sondergaard, T. E. 2019. Fusaoctaxin A, an example of a two-step mechanism for non-ribosomal peptide assembly and maturation in fungi. *Toxins* 11(5), 277.

Winter, M., Samuels, P. L., Dong, Y. and Dill-Macky, R. 2019. Trichothecene production is detrimental to early root colonization by *Fusarium culmorum* and *F. graminearum* in fusarium crown and root rot of wheat. *Plant Pathology* 68(1), 185–195.

Wollenberg, R. D., Saei, W., Westphal, K. R., Klitgaard, C. S., Nielsen, K. L., Lysøe, E., Gardiner, D. M., Wimmer, R., Sondergaard, T. E. and Sørensen, J. L. 2017. Chrysogine biosynthesis is mediated by a two-module nonribosomal peptide synthetase. *Journal of Natural Products* 80(7), 2131–2135.

Wood, A. K. M., Walker, C., Lee, W. S., Urban, M. and Hammond-Kosack, K. E. 2020a. Functional evaluation of a homologue of plant rapid alkalinisation factor (RALF) peptides in *Fusarium graminearum*. *Fungal Biology* 124(9), 753–765.

Wood, A. M., King, R., Urban, M., Nicolli, C., Del Ponte, E. and Hammond-Kosack, K. E. 2020b. Genome sequence of *Fusarium graminearum* strain CML3066 isolated from a wheat spike in Southern Brazil. *Microbiology Resource Announcements* 9(999): e00157-00120. DOI: 10.1128/MRA.00157-20.

Xu, L., Wang, H., Zhang, C., Wang, J., Chen, A., Chen, Y. and Ma, Z. 2020. System-wide characterization of subtilases reveals that subtilisin-like protease FgPrb1 of *Fusarium graminearum* regulates fungal development and virulence. *Fungal Genetics and Biology: FG and B* 144, 103449.

Yang, B., Wang, Y., Tian, M., Dai, K., Zheng, W., Liu, Z., Yang, S., Liu, X., Shi, D., Zhang, H., Wang, Y., Ye, W. and Wang, Y. 2020. Fg12 ribonuclease secretion contributes to *Fusarium graminearum* virulence and induces plant cell death. *Journal of Integrative Plant Biology* 63(2), 365–377.

Yang, C., Liu, H., Li, G., Liu, M., Yun, Y., Wang, C., Ma, Z. and Xu, J. R. 2015. The MADS-box transcription factor FgMcm1 regulates cell identity and fungal development in *Fusarium graminearum*. *Environmental Microbiology* 17(8), 2762–2776.

Yang, Y., Yang, X., Dong, Y. and Qiu, D. 2018. The *Botrytis cinerea* xylanase BcXyl1 modulates plant immunity. *Frontiers in Microbiology* 9, 2535.

Yoshino, K., Irieda, H., Sugimoto, F., Yoshioka, H., Okuno, T. and Takano, Y. 2012. Cell death of *Nicotiana benthamiana* is induced by secreted protein NIS1 of *Colletotrichum orbiculare* and is suppressed by a homologue of CgDN3. *Molecular Plant-Microbe Interactions* 25(5), 625–636.

Yu, G., Zhang, X., Yao, J., Zhou, M. and Ma, H. 2017. Resistance against Fusarium Head Blight in transgenic wheat plants expressing the Sc NPR 1 gene. *Journal of Phytopathology* 165(4), 223–231.

Yun, Y., Liu, Z., Yin, Y., Jiang, J., Chen, Y., Xu, J. R. and Ma, Z. 2015. Functional analysis of the *Fusarium graminearum* phosphatome. *New Phytologist* 207(1), 119–134.

Zhang, H., Kim, M. S., Huang, J., Yan, H., Yang, T., Song, L., Yu, W. and Shim, W. B. 2019. FvLcp1, a type-D fungal LysM protein with Chitin-binding domains, is a secreted protein involved in host recognition and fumonisin production in *Fusarium verticillioides*-maize kernel interaction. *bioRxiv*, 789925.

Zhang, X. W., Jia, L. J., Zhang, Y., Jiang, G., Li, X., Zhang, D. and Tang, W. H. 2012. *In planta* stage-specific fungal gene profiling elucidates the molecular strategies of *Fusarium graminearum* growing inside wheat coleoptiles. *The Plant Cell* 24(12), 5159–5176.

Zhang, Y., He, J., Jia, L. J., Yuan, T. L., Zhang, D., Guo, Y., Wang, Y. and Tang, W. H. 2016. Cellular tracking and gene profiling of *Fusarium graminearum* during maize stalk rot disease development elucidates its strategies in confronting phosphorus limitation in the host apoplast. *PLoS Pathogens* 12(3), e1005485.

Zhu, X., Li, Z., Xu, H., Zhou, M., Du, L. and Zhang, Z. 2012. Overexpression of wheat lipid transfer protein gene TaLTP5 increases resistances to *Cochliobolus sativus* and *Fusarium graminearum* in transgenic wheat. *Functional and Integrative Genomics* 12(3), 481–488.

Chapter 6

Advances in genetic improvement of durable resistance to *Fusarium* head blight in wheat

Guihua Bai, USDA-ARS, USA

1 Introduction

Wheat *Fusarium* head blight (FHB) is a fungal disease that is incited by several *Fusarium* species (Bai and Shaner 2004). In North America and many other countries, *Fusarium graminearum* Schwabe is the predominant pathogen for FHB, but *F. culmorum* and other closely related *Fusaria* are predominant in cooler climates (Buerstmayr et al. 2009). During wheat anthesis, airborne spores land and germinate on the extruded anthers or other floret tissues of wheat to start initial infection when moisture is available. The fungal mycelium then ramifies throughout the floral bracts, developing caryopsis and rachis to spread to entire wheat spikes in susceptible genotypes (Bai and Shaner 1994, 2004). The FHB symptoms start with dark brown, water-soaked spots on the infected spikelets (Fig. 1a). The infection then progresses internally from the initially

http://dx.doi.org/10.19103/AS.2021.0092.08
Published by Burleigh Dodds Science Publishing Limited, 2022.

Figure 1 Symptoms of *Fusarium* head blight on wheat spikes showing (a) initial infection symptoms in a spikelet, (b) spread of dark brown symptoms from initially infected spikelet to rachis, (c) spread of disease symptoms from the initial infection site to rest of the spike and (d) pink mold growing on the infected spikelets.

infected spikelet to its neighboring spikelets through the rachis to develop brownish chlorotic symptoms (Fig. 1b). The symptoms continuously spread upward and downward and eventually to entire spike in a highly susceptible genotype (Fig. 1c). Under warm and humid conditions, some infected spikes show pink color due to pink mold growing on the infected spikelets (Fig. 1d); therefore, wheat FHB is also called 'red-mold disease' in Japan and China (Ma et al. 2020). Later in the season, perithecia of *G. zeae* may appear as raised purplish-black spots (Wiese 1987).

FHB occurs on wheat spikes and directly affects kernel filling and eventually grain quality. Infected florets are usually poorly filled, and many florets fail to produce harvestable grain if severe infection occurs in early grain filling stage, which can cause significant losses in grain yield. FHB can infect wheat, barley and several other cereal crops and has the ability to completely destroy a potentially high-yielding crop within a few weeks of harvest, in particular when frequent rainfalls, high humidity or heavy dews occur from the flowering stage up through the soft dough stage of kernel-filling and high temperatures appear in later kernel-filling stage. Meantime, grains harvested from infected plants can contain a high proportion of discolored and shriveled kernels (Fig. 2), called 'tombstones' or *Fusarium* damaged kernels (FDK), which can dramatically lower test weight. The tombstones are usually contaminated with a high level of mycotoxins produced by the pathogen during infection, in particular, deoxynivalenol (DON), which is harmful to the health of human

Figure 2 Wheat kernels harvested from infected wheat spikes showing healthy kernels from a resistant wheat genotype (a) and more *Fusarium*-damaged kernel (FDK) in a moderately susceptible genotype (b). Both are harvested from the same field. Arrows point to FDK.

beings and livestock (Ferrigo et al. 2016; Foroud et al. 2019b). Wheat grain with DON contamination or low test-weight usually experiences difficulties in marketing, exporting and processing and has discounted price on sale.

FHB was first documented as 'wheat scab' in 1884 (Smith 1884). Earliest epidemics were recorded in the U.S. soft winter wheat states of Indiana and Ohio in the late nineteenth century (Chester 1890; Arthur 1891). In the early twentieth century, FHB epidemics had spread to other major wheat-growing areas of the United States (Dickson 1942) and Japan (Nisikado 1959). Since the middle of the twentieth century, FHB epidemics have been reported in most of the wheat production areas worldwide and become a major threat to wheat production in many countries (Bai and Shaner 1994; Dubin et al. 1997; Goswami and Kistler 2004; McMullen et al. 2012; Ma et al. 2020). In the United States, the economic losses in wheat alone caused by FHB were estimated to be at least $17 billion (Nganje et al. 2004; Wilson et al. 2017). In China, FHB epidemics have frequently occurred with at least 10% acreage affected in each epidemic since 1950s and caused ~20–50% yield losses in severe epidemics (Bai and Shaner 2004). Since 2012, FHB has spread to Huang-Huai-River Valleys where FHB has not been a major wheat disease before. The 2012 epidemic affected about 10 million hectares of wheat and resulted in more than 2 million tons of wheat yield losses (Liu et al. 2016; Ma et al. 2020).

FHB epidemics mainly depend on the availability of sufficient initial inocula and high humidities and temperatures before and at anthesis for spore formation and germination in wheat spike tissues, respectively. Global warming and adaptation of conservation tillage in modern agriculture have contributed to recent frequent and severe FHB epidemics worldwide. In the U.S. Great Plains and Huang-Huai-River Valleys in China, increased maize acreages might be another factor responsible for widespread epidemics in recent years by providing abundant pathogen inocula. Therefore, using cultural practices to minimize the sources of inoculum, such as deep plowing to bury plant residuals that carry abundant *Fusarium* inoculum and crop rotation with non-host crops to break disease cycle, is not feasible for disease control in modern agriculture. Fungicide application can reduce FHB damage, but it remains challenging because of the short time window for application, environmental contamination and increased cost of wheat production (McMullen et al. 2012). In addition, fungicide application on FHB susceptible cultivars is not effective. Growing resistant cultivars is a prerequisite for FHB control (McMullen et al. 2012; Bai et al. 2018; Ma et al. 2020).

However, most cultivars currently used in production worldwide are FHB susceptible (Bai and Shaner 2004; Bai et al. 2018). Breeding for wheat resistance to FHB remains a big challenge because (1) wheat resistance to FHB is a complex trait consisting of several types of resistance, (2) FHB resistance has to be evaluated and selected in an FHB nursery that is difficult to create, and (3) most of QTLs/genes identified to date show minor and inconsistent effects. This chapter will provide a review on how to evaluate different types of resistance of breeding materials, what might be more stable QTLs which need attention in breeding and how to use modern genomic technologies to improve FHB resistance in wheat.

2 *Fusarium* head blight resistance types

Several types of FHB resistance have been reported in wheat and wheat relatives based on their reactions to Fusarium infection or the methods used to evaluate their resistance. Based on host responses to pathogen infection, plant resistance to FHB can be either active or passive. Active resistance refers to that wheat gene products provide plant defense against pathogen infection through physiological or biochemical mechanisms, whereas passive resistance refers to plant traits that can indirectly lower disease severity. Many of these passive mechanisms may not be breeders' interest in modern breeding. For example, tall plants may have better resistance than shorter plants (Mao et al. 2010), but tall plants usually have lodging issue that is not desirable for modern wheat production. Therefore, breeders usually work on the types of resistance that can be phenotypically measured and has no or minimum effect on yield and quality traits. Based on wheat response to *Fusarium* infection, five types of resistance have been proposed. Schroeder and Christensen (1963) first observed two types of resistance: Type I as resistance to *Fusarium* initial infection and Type II as resistance to spread of FHB symptoms within a spike. In field conditions, a genotype that shows fewer infected spikes in a plot or fewer initially infected spikelets in a spike (Fig. 3a) is likely to have Type I resistance. However, this type

Published by Burleigh Dodds Science Publishing Limited, 2022.

Figure 3 *Fusarium*-infected wheat spikes from a field nursery to show Type I and Type II resistance. (a) A spike with both types of resistance has one initially infected site and limited spread of disease symptoms to a few neighboring spikelets; (b) disease symptoms spread to the entire spike of a susceptible genotype that lacks Type II resistance even if only one spikelet is initially infected in the spike; and (c) a spike with Type II resistance, but lack of Type I resistance, has multiple independent infected spikelets.

of resistance can be significantly affected by inoculum availability and weather conditions at wheat anthesis (He et al. 2016a, 2016b; Mao et al. 2010). When a high volume of inocula is available and weather conditions are inducive for initial infection, this type of resistance may disappear and infection in a 'resistant genotype' can be as severe as in the susceptible genotypes if it does not carry other types of resistance. Type I resistance, although important and reported in some cultivars, remains largely elusive because (1) germplasm with FHB immunity (complete Type I resistance) has not been found; (2) evaluation of Type I resistance is usually compromised by Type II resistance; (3) Type I resistance relies more on environmental conditions which affect natural *Fusarium* inoculum formation and dispersion, as well as spore germination in plant tissues; (4) artificial inoculation is difficult to provide consistent infection among experiments and high inoculum concentration may cause more initially infected spikelets in a spike; (5) Type I resistance is present at a low frequency in wheat germplasm. In addition, Type I resistance is difficult to be detected if plants have no Type II resistance, because one infected floret in a spike can cause an entire spike to be blighted if the plant lacks Type II resistance (Fig. 3b).

Different from Type I resistance, Type II resistance measures the resistance to disease spread within a spike after successful establishment of initial infection in a spikelet. In a genotype with high Type II resistance, FHB symptoms can be confined to the initially inoculated spikelet and do not spread to neighboring spikelets in the highly resistant genotypes (Fig. 3c) or spread to a few neighboring spikelets in the moderately resistant genotypes (Fig. 3a), whereas in the highly susceptible genotypes, the FHB symptoms continue to spread to uninoculated neighboring spikelets until the entire spike is bleached (Fig. 3b). In the moderately resistant or moderately susceptible genotypes, the number of spikelets to which FHB can spread varies with the resistance levels of a host genotype and environmental conditions where the plants are grown. In a moderately resistant genotype, most spikelets in the inoculated spike remain uninfected (Fig. 3a), whereas a moderately susceptible genotype may have most symptomatic spikelets in the inoculated spike. In highly susceptible genotypes, the inoculated spike can be bleached within 7–10 days after initial infection (Fig. 3b). Type II is the most stable type of resistance among the five types and is easier to screen compared to other resistance types; thus, it is the major type of resistance that has been selected in breeding programs (Bai and Shaner 2004; Bai et al. 2018). Most resistant cultivars currently available have Type II resistance. Also, Type II resistance is the foundation for Types III, IV and V resistance. Plants with high Type II resistance usually have low DON (Type III) and FDK (Type IV) and less yield loss (Type V) and thus it is the key FHB resistance component and has been extensively studied.

FHB infection not only reduces grain yield, but also grain quality because the *Fusarium* produces mycotoxin, mainly DON, during and after infection. The mycotoxins not only damage grain proteins to lower grain end-use quality and also are toxic to human and animals (Bai et al. 2018; Ma et al. 2020). Miller et al. (1985) observed that some cultivars accumulated less DON than others. They compared DON content between resistant and susceptible cultivars after adding DON to their embryo callus cultures and found that Frontana degraded more DON (18%) than susceptible cultivar Casavant (5%), which led to the hypothesis of Type III resistance as the ability to prevent synthesis and/or degrade DON (Miller and Arnison 1986). Most studies showed that plants with low FHB severity or disease index (Types I and II resistance) usually have lower DON content (Type III resistance). Liu et al. (2009) summarized the QTLs for different FHB resistance components and found that only 25 out of 209 QTLs were for Type III, and none of them was independent from Type I or Type II. High DON content in susceptible genotypes is most likely due to more mycelium grown in the infected grain. Many factors affect the DON content detected in harvested grain, including the plant growth stage when the infection starts (longer the fungal growth period in the infected grain, more DON produced in the grain), harvesting methods that may blow away lightweight grains with high DON content, DON evaluation methods or amount of DON degraded by plants. Type III resistance we measured is based on the final DON content in the harvested grain, thus, Type III can be described as low DON content resulting from either DON degradation or low DON production. Many studies demonstrated low DON production (low fungal biomass) in most resistant genotypes (Bai and Shaner 2004; Bai et al. 2018); however, DON degradation is also possible in some genotypes. Recent cloning of *Fhb7* showed that the *Fhb7* gene can directly degrade DON and other toxins produced by Fusarium species (Wang et al. 2020), which provided direct evidence to support DON degradation theory. Lemmens et al. (2005) reported a QTL in *Fhb1* region that can convert DON to non-toxic DON-3-O-glucoside and hypothesized that *Fhb1* might either encode a DON-glucosyltransferase or regulate its expression. However, later cloning of *Fhb1* did not find a connection between the *TaHRC* gene and DON degradation (Su et al. 2019; Li et al. 2019). Since significant impact of DON contamination on wheat quality, grain sale price and food safety, low mycotoxin accumulation in wheat and barley grain has become a major breeding objective in many breeding programs worldwide.

Mesterhazy (1995) further added two types of resistance to the list: resistance to kernel infection (hereafter designated as Type IV) and FHB tolerance (designated as Type V). However, resistance to kernel infection is usually measured by percentage of *Fusarium*-damaged kernels (FDK) in harvested kernels (Fig. 2), which is highly correlated with Type II resistance; thus, it may not be an independent type of resistance (Bai et al. 2001). QTL analyses found that only one-tenth of the QTLs that have been reported for different types of FHB resistance were for Type IV, and they were all associated with either Type I or Type II resistance (Liu et al. 2009). For Type V resistance, yield reduction can be caused by many factors that are not easily separated from FHB, and a robust evaluation method for Type V resistance is not available; therefore, germplasm with Type V resistance has not been documented. Gong et al. (2020) examined both Types IV and V resistance using *Fhb1* near isogenic lines and suggested that Type V and Type IV can be merged as Type IV and can be evaluated using kernel weight reduction.

3 Inoculation methods

FHB resistance is a quantitative trait, and accurate evaluation of resistant breeding materials is a big challenge for breeders. Inoculation methods, inoculum concentrations, plant growth stages and environmental conditions all affect the expression of wheat FHB resistance (Bai and Shaner 1994; Bai et al. 2018). In addition, disease scoring methods also impact the final disease ratings.

Among the inoculation methods used for FHB resistance evaluation, single floret inoculation has been commonly accepted as the standard method for evaluation of the Type II resistance (Bai and Shaner 1994). Because Type II resistance measures the degree of FHB symptom spread within a spike, single floret inoculation in a greenhouse excludes the possibility of multiple infection sites in a spike caused by natural infection and final disease severity solely results from the disease spread from the inoculated floret (Type II) without interference by Type I. To perform the single floret inoculation, a 10-μL conidial suspension containing ~1000 spores is injected into a single floret in the central spikelet of a spike at anthesis using a syringe (Fig. 4a). After initial infection in the inoculated spikelet, disease symptoms spread upward and downward along the rachis to uninoculated spikelets in the inoculated spike. The number of spikelets that the fungus spreads to depends on the resistance levels of tested genotypes. Disease can be rated when the susceptible check is completely blighted.

To evaluate Type I resistance in a greenhouse, a conidia spore suspension can be sprayed over spikes using a plastic spray bottle. Each flowering spike can be sprayed with ~2 mL of conidial spores at 10,000 spores per mL. All inoculated plants are placed inside a sealed plastic chamber to keep moisture after drying for half an hour. The plant canopy and inside chamber can be slightly misted with clean water using a spray bottle before closing the chamber. About

Figure 4 Inoculation of *Fusarium* pathogen in greenhouses and fields. (a) Single point inoculation to inject a *Fusarium* spore suspension into a single floret using a syringe; (b) infected corn kernel covered with Fusarium perithecia on soil surface; (c) a sprinkler system used in FHB screening nursery to maintain high moisture from flowering up through the soft dough stage of kernel development.

48–60 hours after incubation at 20–22°C, the inoculated plants are moved to a greenhouse bench for disease development after dark brown water sock symptoms appear on the glum of inoculated floret (Fig. 1a). The greenhouse is usually set at 22°C for 12 hours of daytime with supplemental light and at 18°C for 12 hours at dark. FHB symptoms may start to spread from the inoculated spikelet to neighboring spikelets in a spike 2–3 days after moving out of the moist chamber and may reach 100% PSS about 8–10 days after inoculation in highly susceptible genotypes. Type I resistance can be scored 5–6 days after spraying inoculation before symptoms spread to uninoculated spikelets.

In fields, FHB inoculation methods include single floret inoculation (Zhang et al. 2010; Dhariwal et al. 2018), spraying conidial spore suspension over flowering wheat spikes (Holzapfel et al. 2008; Sari et al. 2018) and grain-spawn inoculation at booting stage (Petersen et al. 2016; Lemes da Silva et al. 2019). The single floret inoculation is the same as that described for greenhouse experiments. The inoculated spikes can be bagged with a sealed small plastic sandwich bag to maintain moisture when misting system is not available in a field. At least ten random plants should be inoculated in each plot. The inoculated spikes can be labeled with inoculation day using a piece of colored masking tape. About 15–25 days after inoculation, depending on temperatures from inoculation to disease rating, the inoculated plants can be rated for FHB severity when most plants in the susceptible controls are completely blighted.

Spraying inoculation in a field is usually conducted at the flowering stage by spraying a conidial spore suspension over flowering spikes. This is usually done in late afternoon so dew formed during night can help spore germination on spike tissues. Significant variation in flowering times in a field can be observed among genotypes. Flowering times can last several days from starting to finishing within a plot (genotype) and two to four weeks among genotypes in some experiments. Inoculation days should be used to determine optimum times for disease notes.

For grain-spawn inoculation in fields, *Fusarium*-infected corn (or wheat) kernels (30kg/ha) are uniformly scattered on soil surface about four weeks before wheat heading and then put the same amount two weeks after first application. If weather is dry after inoculation, soil surface needs to be watered to help development of perithecia (Fig. 4b). Starting from early flowering stage, the FHB nursery needs to be misted till all genotypes reach the soft dough stage of kernel development to ensure appropriate moisture for *Fusarium* germination and establishment in wheat spike tissues. To create a low-cost misting nursery, a sprinkler system (Fig. 4c) can be set up to cover the entire nursery uniformly and a timer can be used to control daily misting schedule automatically.

Among the three inoculation methods, single floret inoculation can give the best estimation of Type II resistance if natural infection is low in field. The other two methods evaluate combined resistance of both Types I and II because one spike can have multiple florets infected simultaneously (Fig. 3c) under field conditions. Single floret and spraying inoculation can ensure infection start early, while infection timing of grain-pawn inoculation depends on weather conditions for the development of *Fusarium* fruit body and release of ascospores to plant tissues. However, spraying spore suspensions over a same plot multiple times may over-inoculate early flowering plants. Also, it is difficult to determine the optimum spore concentration that can accurately distinguish resistant from susceptible genotypes. Grain-spawn inoculation coupling with a misting system provides natural infection and sufficient inocula to ensure successful FHB infection in most years, therefore, is a preferred method for field inoculation.

Type II resistance evaluated by single floret inoculation may be cofounded by natural infection in an FHB field nursery with a misting system. Although all three inoculation methods can be used to evaluate the five types of resistance, Type I resistance usually cannot be clearly separated from Type II because of the high possibility of initial infections occuring in multiple spikelets of a spike in a field experiment. Therefore, the final FHB PSS in a field usually reflect a combination of both resistance types with Type II resistance as the predominant type.

4 *Fusarium* head blight disease rating methods and the relationships among different types of resistance

FHB disease rating is a complicated process. Different rating methods are used for different types of resistance. To measure Type I resistance, the percentage of initially infected spikelets in a spike can be recorded before FHB symptoms spread to the neighboring spikelets. Among the three inoculation methods, spraying inoculation may provide a better estimate of Type I if inoculum concentration is appropriate. To minimize interference from Type II resistance, the percentage of initially infected spikelets needs to be rated before spread of the FHB symptoms to neighboring spikelets (Fig. 3c), which is 5–7 days after inoculation in greenhouses. In fields, the percentage of initially infected spikes (FHB incidence) in a plot is also used to estimate Type I resistance if disease pressure is not very high (Bai and Shaner 1994). Accurate scoring of Type I resistance is difficult in fields, which may be affected by inoculation methods (spraying vs. grain-spawn inoculation) and inoculation times, flowering times, inoculum concentrations, and weather conditions after inoculation. In breeding programs, selecting plants with both types of resistance is sufficient for the improvement of FHB resistance.

Several rating systems have been used for type II resistance. Wang et al. (1982) proposed a 1–4 rating scale based on whether or not the fungus spread

Published by Burleigh Dodds Science Publishing Limited, 2022.

in the rachis from the site of initial infection, with 1 for symptoms appeared only in the inoculated spikelet, 2 for fungal spread to rachis but not to uninoculated spikelets, 3 for fungal spread to one uninoculated neighboring spikelet via the rachis and 4 for spread of infection to two or more uninoculated spikelets in the inoculated spike. On this scale, plants with grades 1–1.9 are resistant, 2–2.9 are moderately resistant and 3–4 are susceptible. However, under favorable conditions, even highly resistant wheat accessions can have FHB symptoms spread to rachis (Fig. 1b); thus, this scale rates some highly resistant genotypes as moderately resistant or moderately susceptible, whereas some moderately resistant and moderately susceptible genotypes are rated highly susceptible. However, moderately resistant and moderately susceptible genotypes are important sources of FHB resistance QTLs in modern breeding. Later, Xu and Fan (1985) slightly modified this scale by redefining grade 4 as one uninoculated spikelet to a half of the spike showing FHB symptoms, and grade 5 as more than a half of the spike showing FHB symptoms. This modification separates moderately resistant genotypes from highly resistant and susceptible genotypes.

Both versions of this rating system need to track the initial infection point to determine if and how far the FHB infection spreads from the inoculation point, which may not be easy in field conditions due to multiple infection sites from natural infection (Fig. 2a). Percentage of symptomatic spikelets in a spike (PSS) reflects overall disease severity in each spike without the need to track the initial infection site(s). PSS can accurately measure FHB severity in each spike and distinguish subtle phenotypic differences among genotypes. PSS was first used in a genetic study in 1989 in China (Bai et al. 1989), since then it has been used in most genetic researches worldwide and for evaluation of breeding materials in many wheat breeding programs (Bai et al. 2018).

The time for FHB note is important for FHB rating accuracy. In a greenhouse experiment, the optimum time to FHB note may vary with the resistance levels of materials to be evaluated. To separate highly susceptible plants from resistant plants, FHB notes can be taken at 10–12 days after inoculation when most of susceptible control reached 100% PSS. To separate highly resistant plants from moderate resistant plants, FHB symptoms can be rated at 18–21 days after inoculation depending on temperatures in the greenhouses. In any case, to evaluate Type II resistance accurately and generate repeatable data across different seasons and environments, using proper controls as references for different levels of resistance are critical.

In the greenhouse experiments, FHB is rated by counting the numbers of symptomatic and total spikelets in a spike. In fields, however, hundreds of lines may need to be scored in a short time window and counting all spikelets in all plants in a short time window may not be feasible. Therefore, several modified scoring methods have been used to simplify the spikelet counting process for phenotyping breeding materials. A 0–9 scale based on PSS has been used in some breeding programs, in which 0 refers to no infected spikelet in a spike, 1 refers to 10% infected spikelets per spike, and so on.

To determine optimum time for taking FHB note in fields, several factors need to be considered. Heading or flowering dates are used to estimate the time length after inoculation by either spraying or grain-spawn inoculation. Usually, early flowering plants have earlier infection, but we still need to check the disease levels in the known control genotypes that flowered at similar times. In elite breeding line performance trials, several notes can be taken for each tested line, with the first note starting when symptoms start spread in the susceptible controls, then take second and third note of these lines every two or three days. AUDPC can be calculated to measure the FHB progress or the final reading can be used as the disease rating for the line. In a genetics study and breeding line screening nursery, this may not be feasible because of too many tested lines in each experiment. In this case, one or two notes of final FHB severity for each line may be sufficient because AUDPC is highly correlated with final PSS. If highly susceptible controls reach 90–100% PSS, then tested lines with a similar flowering time can be rated for PSS in the same day. In a misting grain-spawn or spraying inoculated FHB nursery, disease incidence is usually high, with almost all spikes infected even in some plants of resistant cultivars, therefore, FHB rating is mainly to score FHB severity. In each plot, variation in plant flowering times makes visual disease rating more difficult. Using an average score from two or more people for each plot and rechecking the plot two days after first note may be able to correct obvious rating errors and improve rating accuracy. In any case, disease has to be scored before the spike tissues turn yellow, in which case the infected spikes may not be able to be distinguished from ripen spikes. Because all FHB inoculation methods are affected by environmental conditions, control cultivars with known levels of resistance have to be used in each phenotyping experiment to get consistent and comparable data among different environments and years. The resistance level of a cultivar should be determined by comparing performance of test lines with known control cultivars in each experiment.

In recent years, DON content in wheat kernels has drawn increased attention because economic penalties and food safety concerns directly associate with high DON levels in wheat grains (Foroud et al. 2019b). Therefore, Type III (low DON content) and Type IV (low FDK) resistance have attracted more interest from wheat researchers and breeders. DON content in harvested grain mainly depends on percentage of FDK because DON mainly resides in FDK (Mesterhazy et al. 1999; Bai and Shaner 2004). A wheat grain sample with more FDK usually has higher DON content, but, FDK in a genotype can be affected by many factors including FHB incidence (Type I) and severity (Type II), threshing methods that determine how much FDK can be kept in harvested grain and DON measuring methods. If the infection occurs in early wheat kernel developmental stage, grain might develop poorly, and tiny FDK that contain large amount of DON may be blown out during threshing, thus the DON content is far underestimated and does not reflect actual DON content produced in the tested lines. Measurements of DON content can be carried out by Gas Chromatography-Mass Spectrometry (GC-MS) (Mirocha et al. 1998) and enzyme-linked immunosorbent assays (ELISAs) (Casale et al. 1988). Recently near-infrared (NIR) transmittance kernel sorting technology has been developed for quick estimation of FDK and DON (Kautzman et al. 2015). Both ELISA- and NIR-based technologies can be used for quick screening of breeding materials for DON content; however, GC-MS or other chromatographic techniques may provide more accurate measurement of DON of wheat samples (Agriopoulou et al. 2020).

FDK is measured by percentage of FDK in a sample. Several methods can be used to evaluate FDK: (1) direct counting of FDK and healthy kernels to calculate FDK% in each sample, (2) visually estimating FDK% by comparing the testing samples with a series of controls with known percentages of FDK. Direct counting is the most accurate, but time-consuming and may not be feasible for evaluating thousands of samples in a short time period in a breeding program. Visual FDK estimation is widely used for quick screening of a large number of samples, but visual FDK estimation may need a skilled person to generate more reliable data. NIR-based seed sorting device is also used for FDK measurement in some of the U.S. breeding programs (Peiris et al. 2010). More recently, a grain analyzer has been optimized for FDK evaluation (https://www.vibeia.com/qm3-grain-analyzer). This grain analyzer analyzes wheat seed samples through image acquisition, processing and visualization and uses machine learning to classify sound kernels and FDK. This type of machines can analyze large number of samples daily without need of skilled people and provide more consistent results. However, FDK is highly correlated with FHB severity, and may not be an independent type of resistance (Bai et al. 2001; Foroud et al. 2019b). Selecting for Type II resistance remains the major strategy for breeders to reduce DON (Type IV) and FDK (Type III) in wheat breeding programs.

5 Preparation of *Fusarium* inoculum

To evaluate FHB resistance, a mixture of field isolates is preferred for inoculation (Bai et al. 2018; Ma et al. 2020), which not only reduces risk of selecting a low virulent isolate but also avoids an extra step for single spore isolation. To prepare a mixture of *Fusarium* isolates, FHB-infected seeds from a field are surface sterilized for 1 minute in 5% chlorine bleach, rinsed three times using sterile water and then placed on a potato dextrose agar medium (PDA) in a petri dish. *Fusarium*, bacteria and other microbes in the kernel can all grow in the PDA, but *Fusarium* usually grows much faster than other microbes, thus continuous transfer of an uncontaminated PDA disk with clean mycelium onto fresh PDA plates can eliminate contamination and a pure *Fusarium* culture can be obtained. When mycelium covers the entire petri dish, the petri dish can be sealed and stored in a refrigerator at 4°C for future use. The *Fusarium* culture needs to be refreshed to a new PDA medium annually to maintain viability of the culture.

Conidial suspensions are usually used for artificial inoculation (Bai et al. 2018). Several methods can be used to produce conidial inoculum. The easiest method is to produce conidial spores in mung bean broth, which is easy and safe with a low-cost for massive conidial production. To prepare 1-L of this medium, 40 g of intact green mung bean is placed in a 1-L Erlenmeyer flask containing 1-L boiling water. After the mung bean in the flask is boiled for 10 minutes on a hot plate, mung beans are removed through three layers of cheesecloth and the remaining liquid is aliquoted into ten 200-mL Erlenmeyer flasks. Mung bean needs to be kept intact during boiling process because starch leaking from mung bean to the broth will promote fast mycelium growth and reduce conidia production. All the flasks with mung bean broth need to be autoclaved and inoculated with a 3-mm^2 disc from a PDA culture of *Fusarium graminearum* after the broth is cooled down to a room temperature. The inoculated flasks are shacked at 250 times per minute on a shaker at 24°C for four days, and then the aqueous culture is filtered through a three-layer cheese cloth to remove mycelium. Conidiospores in the suspension were counted under microscope with a hemocytometer and normalized to 10^5-mL^{-1} using autoclaved water. The normalized spore suspension is stored in a refrigerator. The containers with the spore suspension are shaken thoroughly before each use to ensure consistent spore concentrations in each inoculation. It is recommended to make aliquots for different inoculations when first time to use it.

To produce grain spawn, corn or wheat kernels are socked in water for 24 hours and autoclaved after dumping excess water. After the autoclaved grains cool down to room temperature in large flasks, a 5-mL conidial suspension or several disks of PDA with mycelium can be transferred into each flask. All the flasks are shaken daily for two weeks to mix the fungal inoculum with the kernels. When white mold grows uniformly on most of the kernels in the flasks, the kernels are removed from the flasks and dried on a greenhouse bench. The dried grains are ready for field inoculation.

6 Sources of *Fusarium* head blight resistance

Early search for FHB resistance sources has been focused on Asian landraces from these areas where FHB epidemics had been frequent and severe, in particular in China and Japan, which resulted in the identification of many

FHB-resistant and moderately resistant wheat landraces and other germplasm (Bai and Shaner 2004; Ma et al. 2020). Among these resistant sources identified during that period, Chinese landraces such as Wangshuibai, Haiyangzhou and Huangcandou and Japanese cultivars such as Shinchunaga, Nobeokabozu Komugi and Nyu Bai are still considered the most resistant germplasms today (Bai et al. 2018; Ma et al. 2020). Because these landraces usually carry many undesirable agronomic traits, progress in using these resistant sources in breeding has been limited. Fortunately, a Chinese breeding program in Jiangsu province selected several FHB-resistant cultivars from the cross between a Chinese FHB-resistant landrace Taiwan Wheat and an Italy wheat cultivar Funo in 1970s. Among the several selected progeny lines, Sumai 3 showed the best FHB resistance and agronomic performance, and thus, Sumai 3 and its derivatives have been widely used as the major source of FHB resistance in wheat breeding programs in China and other areas (Bai et al. 2018; Ma et al. 2020). After several decades of breeding efforts, many highly resistant breeding lines and cultivars have been released in the Middle and Low Valleys of Yangtze River in China and the hard spring wheat and soft winter wheat regions in the United States (Kolb et al. 2001; Rudd et al. 2001; Liu et al. 2009; Anderson 2012; Anderson et al. 2012b; Bai et al. 2018; Ma et al. 2020).

Extensive search for new FHB-resistant sources in wheat and its relatives did not find any FHB immune material. However, several alien species have been identified to carry FHB resistance genes, including *Aegilops. ventricosa*, *Ae. speltoides*, *Thinopyrum ponticum*, *Th. elongatum*, *Th. intermedium*, *Dasypyrum villosum*, *Leymus racemosus* and *Elymus tsukushiensis* (Ma et al. 2020). The alien chromosome segments with the FHB resistance genes have been successfully transferred into common wheat using chromosome engineering technologies. Among them, three genes have been officially named: *Fhb3* from alien species *Leymus racemosus* (Qi et al. 2008), *Fhb6* from *Elymus tsukushiensis* (Cainong et al. 2015) and *Fhb7* from *Thinopyrum ponticum* (Guo et al. 2015). More recently, *Fhb7* has been successfully cloned, which demonstrated a high level of FHB resistance in wheat backgrounds (Wang et al. 2020) and is attracting more and more breeders' attention worldwide.

In general, most cultivars that are used in current wheat production are still susceptible to FHB, but significant differences in FHB resistance have been observed among the cultivars (Jin et al. 2013). Some locally adapted cultivars have moderate FHB resistance, good adaptation and other desirable agronomic traits, thus, resistance genes in these cultivars can be transferred to elite breeding lines without significant penalty of yield and other agronomic traits (Zhang et al. 2020). Some examples of these cultivars are soft winter wheat Ernie, Freedom, Truman and Roane (Liu et al. 2007; Bai et al. 2018) and hard winter wheat Heyne, Everest and Overland from the United States (Jin et al. 2013, 2014); Arina, Renan and Fundulea 201R from Europe (Gervais et al.

2003; Shen et al. 2003a; Paillard et al. 2004) and Yangmai 158 and Zhenmai 9023 from China (Zhang et al. 2020). These moderately FHB-resistant cultivars do not carry major resistance genes from Sumai 3, and the minor QTLs in these cultivars can be pyramided together with major resistance QTLs from exotic sources or alien segments to enhance the levels of FHB resistance in new wheat cultivars

7 Quantitative trait loci for wheat resistance to *Fusarium* head blight

FHB resistance is a quantitative trait, and each FHB-resistant wheat genotype may carry several QTLs (Bai and Shaner 2004; Zhang et al. 2020). Different resistant varieties may possess different numbers and combinations of FHB resistance QTLs. To date, more than 500 QTLs have been reported for different types of resistance from different sources worldwide and these QTLs are distributed on all 21 wheat chromosomes (Martin et al. 2009; Bai et al. 2018; Ma et al. 2020). Many of these QTLs are overlapped with each other on similar chromosome regions and show minor effects (Liu et al. 2009; Venske et al. 2019). Some of the QTLs showed a major effect in one background or environment, but minor effects in the others. Thus, it is difficult to confirm their allelic relationships using the allelism test as used for wheat resistance genes to rust diseases. To effectively use these QTLs, it is important to identify unique and repeatable QTLs that showed relatively stable, large effects in multiple studies. Using meta-analysis, several studies consolidated the list to less than 100 meta-QTLs based on their linkage map locations (Cai et al. 2019; Liu et al. 2009; Venske et al. 2019). However, QTL positions in these linkage maps were calculated from recombination frequency in each population, and thus, in some populations, a QTL interval of several centimorgans may be equal to a physical distance of several-hundreds of million base pairs (Mb) in a chromosome. Therefore, establishing these important QTL regions based on their physical distance may facilitate further use of these QTLs in genetic improvement of wheat FHB resistance.

To date, seven QTLs have been officially named, with four (*Fhb1*, *Fhb2*, *Fhb4* and *Fhb5*) from common wheat and three (*Fhb3*, *Fhb6* and *Fhb7*) from alien species. Among the four wheat QTLs, *Fhb1* on the chromosome arm 3BS and *Fhb2* on chromosome 6B were originally identified from Sumai 3 (Cuthbert et al. 2006, 2007), whereas *Fhb4* on chromosomes 4B (Xue et al. 2010a) and *Fhb5* on 5A were identified in Wangshuibai (Xue et al. 2011). Among them, *Fhb1* was the first mapped QTL for FHB resistance using DNA markers (Bai et al. 1999; Waldron et al. 1999). Later, the QTL was designated as *Fhb1* (Cuthbert et al. 2006). This QTL shows a large effect on Type II resistance and is the only QTL that has proven to be stable across multiple genetic backgrounds and environments (Bai et al. 2018). Currently used *Fhb1* in most breeding programs can be traced back to the Chinese landrace Taiwan Wheat (Su et al. 2019).

Owing to great importance of *Fhb1*, significant international efforts have been made to clone *Fhb1*. Schweiger et al. (2016) sequenced the 1 Mb chromosome region harboring *Fhb1* in a Sumai 3's derivative CM-8203 and delimited *Fhb1* to a 703 kb suppressed recombination region. Rawat et al. (2016) further reported a chimeric lectin with agglutinin domains and a pore-forming toxin-like domain (*PFT*) as an *Fhb1* causal gene. They reported that the wheat genotypes with the functional *PFT* are *Fhb1* resistant, whereas the genotypes without *PFT* or a malfunctioned *PFT* are *Fhb1* susceptible after screened a small set of germplasm lines from Asia. However, further screening of the *PFT* gene in a large collection of wheat accessions worldwide found that many highly susceptible genotypes also carried the functional *PFT*, suggesting that *PFT* alone may not be sufficient for *Fhb1* resistance (Jia et al. 2018; Su et al. 2019). Recently, two independent research teams in the United States (Su et al. 2019) and China (Li et al. 2019) simultaneously reported a histidine-rich calcium-binding protein (*TaHRC*) as the causal gene for *Fhb1*. The identical *TaHRC* gene was isolated from two different sources, one from a Sumai 3 derivative (Ning7840) and the other from a Chinese landrace (Wangshuibai). Results from both the teams confirmed that the wild type allele of *TaHRC* confers FHB susceptibility and the mutant allele with a large deletion in the start codon of the open reading frame results in FHB resistance (Li et al. 2019; Su et al. 2019). However, how this deletion in *TaHRC* regulates the resistance remains disputative. Su et al. (2019) used gene editing to knock out the functional *TaHRC* in a highly susceptible cultivar Bobwhite and showed that the transgenic plants significantly increased FHB resistance, whereas overexpression of the cDNA of *TaHRC* in Bobwhite did not alter the FHB susceptibility in the transgenic plants, demonstrating that the loss of function of the gene is responsible for FHB resistance. However, Li et al. (2019) reported that overexpression of *TaHRC* in Yangmai 158 (a moderately resistant cultivar) and ND183 (a susceptible line) driven by the native promoter increased FHB resistance. They further hypothesized that the large deletion in *TaHRC* creates a new upstream translation start codon encoding a functional TaHRC-R protein that is 14 amino acids longer than TaHRC-S protein and concluded that the gain-of-function of *TaHRC* is responsible for FHB resistance. The discrepancy between the two studies could be due to a dominant negative effect as explained by Lagudah and Krattinger (2019). Additional experiments need to be done to settle the discrepancy.

Among the three other named QTLs from wheat, *Fhb5* has been reported in many studies from many germplasm lines worldwide. This QTL has been located in a ~0.3 cM interval between Xgwm304 and Xgwm415 on chromosome 5A (Xue et al. 2011). More recently, haplotype analysis further mapped *Fhb5* in a 46–111 Mb region using a group of *Fhb5* near-isogenic lines (Brar et al.

2019). The QTL region has been reported in many germplasm lines worldwide (Table 1). Although *Fhb5* was reported to show a similar effect as *Fhb1* in some backgrounds (Brar et al. 2019), its average effect was much smaller than *Fhb1*, with a mean value of phenotypic variation explained (PVE) by the QTL of 10.9% (Table 1). *Fhb5* contributes both Type I (Lin et al. 2006) and Type II resistance (Liu et al. 2009) in field conditions and is present in many modern breeding materials; therefore, it is a useful QTL for breeding.

Fhb2 on wheat chromosome 6B explained a wide range of the phenotypic variation from 4.4–24%, mainly for type II resistance (Table 1). *Fhb2* was originally identified from Sumai 3 (Waldron et al. 1999). Cuthbert et al. (2007) mapped it as a major gene for Type II resistance. Later it has been reported mainly near the marker Xgwm644 in several other Chinese germplasm with minor-to-medium effects in most of the studies (Liu et al. 2009). *Fhb4* was originally reported as a major QTL for Type I resistance between markers Xgwm149 and Xgwm192 on chromosome 4BL of Wangshuibai (Lin et al. 2006; Xue et al. 2010a), but QTLs identified in the chromosome region (450-550 Mb) from other studies mainly showed minor-to-moderate effects on Type II resistance (Liu et al. 2005; Bonin and Kolb 2009). This QTL was not only reported in Chinese sources, also in U.S. wheat such as Ernie (Liu et al. 2007) and IL-95-1653 (Bonin and Kolb 2009), in European wheat such as Pirat (Holzapfel et al. 2008) and Arina (Miedaner et al. 2012) and in Brazilian wheat such as Frontana (Liu et al. 2009; Ágnes et al. 2014).

Besides QTL identified from wheat, three officially named QTLs (*Fhb3*, *Fhb6* and *Fhb7*) from alien species have been transferred into wheat. *Fhb7* has been recently cloned and shows a similar major effect on FHB resistance as *Fhb1* (Wang et al. 2020). *Fhb7* is originally from wheat grass, *Thinopyrum elongatum*. Using map-based cloning and the newly assembled E genome reference, Wang et al. (2020) confirmed that a glutathione S-transferase (GST) is responsible for FHB resistance of *Fhb7* by catalyzing the conjugation of a glutathione (GSH) unit onto their toxic epoxide moiety to detoxify DON in infected grains. More interestingly, the homolog of GST sequence for *Fhb7* was found in *Epichloë aotearoae*, an endophytic fungal species known to infect *T. elongatum,* with 97% sequence identity, suggesting possible horizontal gene transfer (HGT) of GST from *Epichloë aotearoae* to *T. elongatum*. *Fhb7* has resistance to a wide spectrum of *Fusarium* species and showed resistance not only to FHB, but also to crown rot caused by a different *Fusarium* species. Cloning of *Fhb7* provides a new type of resistance gene that directly detoxifies DON in wheat grain, which is different from *Fhb1*. *Fhb7* showed a major effect on Type II and Type III resistance in various Chinese cultivars without significant yield penalty, which demonstrated that *Fhb7* from *T. elongatum* can be directly used in wheat breeding. Cloning of *Fhb7* highlights the potential impacts and applications of alien sources for improvement of wheat resistance in breeding.

Table 1 Repeatable QTLs and their chromosome regions, phenotypic variation explained (PVE%) by the quantitative trait locus (QTL), germplasm accessions carrying the QTL and the references that reported these QTLs

Chr. Arm	QTL overlapping region (Mb)	No. of studies	No. of QTL donors	Mean PVE (%)	PVE range (%)	Resistance sources and references
1BS	10.0–100.0	9	9	8.6	2.7–16.0	Arina (Buerstmayr and Buerstmayr 2015; Draeger et al. 2007); DT707 (Sari et al. 2018); Everest (Lemes da Silva et al. 2019); F201R (Shen et al. 2003a); Lynx (Schmolke et al. 2005); Naxos (Lu et al. 2013); RL4137 (Srinivasachary et al. 2008a); TMP16315 (Dhariwal et al. 2018).
1BL	500.0–600.0	12	11	9	3.0–17.0	Wangshuibai (Lin et al. 2004; Zhang et al. 2010); Wesley-Fhb1 (Eckard et al. 2015a); Biscay (Holzapfel et al. 2008); Romanus (Holzapfel et al. 2008); Cansas (Klahr et al. 2007); HFZ (Li et al. 2012); SYN1 (Zhu et al. 2016b), IAS20*5/H567.71 (He et al. 2019); CM-82036 (Buerstmayr et al. 2002); Rialto (Srinivasachary et al. 2008b); Frontana (Ágnes et al. 2014).
2AS	30.0–130.0	7	7	9.7	7.0–14.0	Arina (Buerstmayr and Buerstmayr 2015); *T. macha* (Buerstmayr et al. 2011); Joppa (Zhao et al. 2018a); NC-Neuse (Petersen et al. 2016); Wangshuibai (Ma et al. 2006c); Rubens (Holzapfel et al. 2008); Spark (Srinivasachary et al. 2008b).
2BS	110.0–200.0	7	7	12.9	4.9–21.1	Renan (Gervais et al. 2003); 02-5B-318 (Giancaspro et al. 2016); IL94-1653 (Bonin and Kolb 2009); Nc-neuse (Petersen et al. 2017); Nanda2419 (Ma et al. 2008c); Bess (Petersen et al. 2017); NW87 (Xue et al. 2010b).
2DS	10.0–60.0	16	15	13.1	3.2–30.7	Kenyon (McCartney et al. 2016); YZ1 (Chao et al. 2014); Sumai 3 (Basnet et al. 2012); Naxos (He et al. 2016b; Lu et al. 2013); Frontana (Ágnes et al. 2014); Wangshuibai (Jia et al. 2005); Gamenya (Handa et al. 2008); Wangshuibai (Zhang et al. 2010); Ning 894037 (Shen et al. 2003b); Art (Clinesmith et al. 2019); Jagger (Cai and Bai 2014); Truman (Islam et al. 2016); Wuhan-1 (Somers et al. 2003); Biscay (Holzapfel et al. 2008); Maxine (Tamburic-Ilincic and Rosa 2019).

(*Continued*)

Table 1 (*Continued*)

Chr. Arm	QTL overlapping region (Mb)	No. of studies	No. of QTL donors	Mean PVE (%)	PVE range (%)	Resistance sources and references
2DL	540–640	16	15	14.6	3.1–25.0	CJ 9306 (Jiang et al. 2007a,b); Sumai 3 (Handa et al. 2008dar); Art (Clinesmith et al. 2019); Wuhan 1 (Zhu et al. 2020); SHA3/CBRD (Lu et al. 2013); SYN1 (Zhu et al. 2016b); Line 685 (Lu et al. 2011); Frontana (Ágnes et al. 2014); Soru#1 (He et al. 2016a); Solitär (Herter et al. 2018); Arina (Paillard et al. 2004); Sumai 3 (Suzuki et al. 2012); Nasma (He et al. 2019); C615 (Yi et al. 2018); VA00W-38 (Liu et al. 2012); Naxos (Lu et al. 2013).
3AS	10.0–70.0	14	12	17.8	4.0–41.0	Ernie (Li et al. 2010); Rialto (Srinivasachary et al. 2008b); Soru#1 (He et al. 2016a); TMP16315 (Dhariwal et al. 2018); Wangshuibai (Jia et al. 2005); Langdon (Otto et al. 2002; Chen et al. 2007; Zhu et al. 2016a); DT707 (Sari et al. 2018); Heyne (Zhang et al. 2012a); Israel A (Buerstmayr et al. 2013); PI 41025 (Zhang et al. 2014); HCD (Cai and Bai 2014); Spark (Srinivasachary et al. 2008b).
3BS	1.0–20	40	23	25.8	5.0–56.8	ND2603 (Anderson et al. 2001); CM-8206 (Buerstmayr et al. 2002, 2003); Wuhan-1 (Somers et al. 2003); Sumai 3 (Liu and Anderson 2003a,b; Suzuki et al. 2012; Basnet et al. 2012; Anderson et al. 2001; Zhou et al. 2010; Zhang and Mergoum 2007; Jayatilake et al. 2011), Wangshuibai (Ma et al. 2008; Lin et al. 2004; Mardi et al. 2005; Yu et al. 2008b; Jia et al. 2005; Ma et al. 2006c; Naji et al. 2008; Zhang et al. 2010), NW87 (Xue et al. 2010b); W14 (Chen et al. 2006); Huapei 57-2 (Bourdoncle and Ohm 2003); DH181 (Yang et al. 2005b); CS-SM3-7ADS (Ma et al. 2006b); CJ9306 (Jiang et al. 2007a,b); Baishanyehuang (Zhang et al. 2012b); TD161 (Buerstmayr et al. 2012); Zhoumai-27 (Shah et al. 2017); DBC-480 (Prat et al. 2017); Line 685 (Lu et al. 2011); HFZ (Li et al. 2012); Ning7840 (Yu et al. 2008a; Zhou et al. 2002; Bernardo et al. 2012); HCD (Cai and Bai 2014); ND2710 (Zhao et al. 2018b); Ning 894037 (Shen et al. 2003b); Catbird (Cativelli et al. 2013); Wesley-Fhb1-BC5 (Eckard et al. 2015).
3BL	450–550	8	8	8.1	2.1–12.4	Sumai 3 (Zhang and Mergoum 2007); Apache (Holzapfel et al. 2008); Baishanyehuang (Zhang et al. 2012b); Wangshuibai (Yu et al. 2008b); Truman (Islam et al. 2016); HCD (Cai and Bai 2014); Zhoumai-27 (Shah et al. 2017); Arina (Buerstmayr and Buerstmayr 2015)

4AL	600–740	17	14	9.7	5.0–21.1	CS-SM3-7ADS (Ma et al. 2006b); Nc-neuse (Petersen et al. 2016; Petersen et al. 2017); Frontana (Ágnes et al. 2014; Szabó-Hevér et al. 2012); FL62R1 (Zhang et al. 2018); Kenyon (McCartney et al. 2016); Soru#1 (He et al. 2016a); DH181 (Yang et al. 2005b); FTHP Redeemer (Tamburic-Ilincic and Rosa 2019); Arina (Buerstmayr and Buerstmayr 2015; Paillard et al. 2004); Pirat (Holzapfel et al. 2008); Apache (Holzapfel et al. 2008); SHA3/CBRD (Lu et al. 2013); LSP2 (Ren et al. 2016); Heyne (Zhang et al. 2012a).
4BS	30–60	9	10	17	5.0–64.2	Td161 (Buerstmayr et al. 2012); DH181 (Yang et al. 2005); Wesley-Fhb1-BC56 (Eckard et al. 2015); Kukeiharu 14 (Suzuki et al. 2012); Everest (Lemes da Silva et al. 2019); DBC-480 (Prat et al. 2017); Naxos (Lu et al. 2013); Nasma (He et al. 2019); Vienna (Tamburic-Ilincic and Rosa 2017); NX188 (Chao et al. 2014).
4BL	450–550	11	9	11.6	4.7–18.5	DT735 (Ruan et al. 2012); Frontana (Ágnes et al. 2014; Szabó-Hevér et al. 2012); Wangshuibai (Jia et al. 2005; Lin et al. 2006); Art (Clinesmith et al. 2019); IL94-1653 (Bonin and Kolb 2009); Ernie (Liu et al. 2007); Pirat (Holzapfel et al. 2008); Arina (Miedaner et al. 2012); NW66 (Xue et al. 2010b).
4DS	5.0–30	9	10	23.9	5.4–53.1	Apache (Holzapfel et al. 2008); History (Holzapfel et al. 2008); Romanus (Holzapfel et al. 2008); Spark (Srinivasachary et al. 2008b); Avle (Lu et al. 2011); Soissons (Srinivasachary et al. 2009); Arina (Draeger et al. 2007); Soru#1 (He et al. 2016a); History (Miedaner et al. 2012); IAS20*5/H567.71 (He et al. 2019); Nc-neuse (Petersen et al. 2017).
5AS	20–110	21	18	10.9	3.5–39.0	Spark (Srinivasachary et al. 2008b), Frontana (Ágnes et al. 2014), Haiyanzhong (Cai et al. 2016; Li et al. 2011), Naxos (Lu et al. 2013), Luke (Ren et al. 2019), Strongfield (Somers et al. 2006), Ernie (Liu et al. 2007), Remus (Steiner et al. 2019), W14 (Chen et al. 2006), NW66 (Xue et al. 2010b), CM-82036 (Buerstmayr et al. 2002, 2003) Baishanyehuang (Zhang et al. 2012b), CJ 9306 (Jiang et al. 2007a,b), Wangshuibai (Lin et al. 2006; Yu et al. 2008b), Overland-Fhb1 (Eckard, et al. 2015), PI277012 (Chu et al. 2011), Sumai 3 (Suzuki et al. 2012),

(*Continued*)

Table 1 (*Continued*)

Chr. Arm	QTL overlapping region (Mb)	No. of studies	No. of QTL donors	Mean PVE (%)	PVE range (%)	Resistance sources and references
5ALC	400–470	9	9	11.1	6.3–20.0	PI277012 (Chu et al. 2011); FL62R1 (Zhang et al. 2018); Everest (Lemes da Silva et al. 2019); BGRC3487 (Ruan et al. 2012); DT707 (Sari et al. 2018); Line 685 (Lu et al. 2011); Ben (Pirseyedi et al .2019); C615 (Yi et al. 2018); Wangshuibai (Ma et al. 2006c).
5AL	600–700	13	12	16.1	4.0–32.0	Renan (Gervais et al. 2003); Arina (Paillard et al. 2004; Buerstmayr and Buerstmayr, 2015); TMP16315 (Dhariwal et al. 2018); Joppa (Zhao et al. 2018a); PI 41025 (Zhang et al. 2014); Zhoumai-27 (Shah et al. 2017); SHA3/CBRD (Lu et al. 2013); *T. macha* (Buerstmayr et al. 2011); Ultima (Dhariwal et al. 2018); TC 67 (Malihipour et al. 2017); Apache (Holzapfel et al. 2008); Pirat (Holzapfel et al. 2008).
5BL	500–550	14	13	11.2	4.8–23.5	Arina (Paillard et al. 2004; Miedaner et al. 2012); DH181 (Yang et al. 2005b); Ernie (Li et al. 2010); Frontana (Ágnes et al. 2014); Ben (Pirseyedi et al. 2019); IAS20*5/H567.71 (He et al. 2019); Naxos (Lu et al. 2013); Solitär (Herter et al. 2018); VA00W-38 (Liu et al. 2012); NC-Neuse (Petersen et al. 2016); Kenyon (McCartney et al. 2016); LSP2 (Ren et al. 2016); BGRC3487 (Ruan et al. 2012).
5DL	430–490	7	7	9.3	3.5–15.0	Chokwang (Yang et al. 2005a); Naxos (Lu et al. 2013); Catbird (Cativelli et al. 2013); Soru#1 (He et al. 2016a); YZ1 (Chao et al. 2014); Renan (Gervais et al. 2003); Wangshuibai (Yu et al. 2008b).
6AL	520–620	7	6	10.7	5.9–17.4	AC Brio (Malihipour et al. 2017); NC-Neuse (Petersen et al. 2016; 2017); Remus (Steiner et al. 2004); Arina (Paillard et al. 2004); Ernie (Li et al. 2010); Frontana (Ágnes et al. 2014).
6BS	100–200	12	10	11.2	4.4–24.0	Wangshuibai (Zhang et al. 2010; Ma et al. 2008; Lin et al. 2004; Jia et al. 2005); Patton (Bonin and Kolb 2009); ND2603 (Anderson et al. 2001); Sumai 3 (Anderson et al. 2001); Ning8026 (Häberle et al. 2009); Haiyanzhong (Li et al. 2011); Td161 (Buerstmayr et al. 2012); DT707 (Sari et al. 2018); BW278 (Cuthbert et al. 2007); Ning 894037 (Shen et al. 2003b).

6BL	560–660	8	6	12.9	4.9–22.0	Frontana (Szabó-Hevér et al. 2012; Ágnes et al. 2014); Arina (Draeger et al. 2007; Buerstmayr and Buerstmayr 2015); Israel A (Buerstmayr et al. 2013); Wangshuibai (Lin et al. 2004); Sumai 3 (Basnet et al. 2012); Apache (Holzapfel et al. 2008).
7AS	20–120	8	7	12.3	7.4–24.0	RL4137 (Srinivasachary et al. 2008a); Frontana (Mardi et al. 2006); Wangshuibai (Zhang et al. 2010); 10Ae564 (Zhao et al. 2018a); DT735 (Ruan et al. 2012); VA00W-38 (Liu et al. 2012); Sumai 3 (Jayatilake et al. 2011).
7AL	600–700	14	12	10.3	3.2–19.0	Spark (Srinivasachary et al. 2008b); Wangshuibai (Zhang et al. 2004; Jia et al. 2005; Yu et al. 2008b); NK93604 (Semagn et al. 2007); DT707 (Sari et al. 2018); PI478742 (Kumar et al. 2007); Romanus (Holzapfel et al. 2008); SHA3/CBRD (Lu et al. 2013); SYN1 (Zhu et al. 2016b); Arina (Skinnes et al. 2009); HFZ (Li et al. 2012); Stoa (Zhang and Mergoum 2007); Vienna (Tamburic-Ilincic and Rosa 2017).
7BS	100–200	9	10	12.9	3.0–27.0	FTd (Buerstmayr et al. 2012); HTd (Buerstmayr et al. 2012); Ben (Pirseyedi et al. 2019); Frontana (Szabó-Hevér et al. 2012); Cansas (Klahr et al. 2007); Dream (Schmolke et al. 2005); Rubens (Holzapfel et al. 2008); Ning8026 (Häberle et al. 2009); DT735 (Ruan et al. 2012); Wheaton (Eckard et al. 2015).
7BL	600–740	6	7	8.3	2.8–11.4	Frontana (Ágnes et al. 2014); C615 (Yi et al. 2018); Riband (Draeger et al. 2007); DH181 (Yang et al. 2005b); History (Holzapfel et al. 2008); Romanus (Holzapfel et al. 2008); Wangshuibai (Jia et al. 2005).
7DS	90–190	6	5	13.2	3.4–21.7	Wesley-Fhb1 (Eckard et al. 2015); Kenyon (McCartney et al. 2016); Haiyanzhong (Li et al. 2011; Cai et al. 2016); LSP2 (Ren et al. 2016); Catbird (Cativelli et al. 2013).

Fhb3 is from an alien segment of *Leymus racemosus* that was transferred through whole-arm Robertsonian translocation 7AL.7Lr#1S (Qi et al. 2008). To study its effect on type II resistance in hard winter wheat backgrounds, an *Fhb3*-resistant line in Jagger background (JaggerFhb3) was crossed to JaggerFhb1 and OverlandFhb1, respectively. JaggerFhb1 and OverlandFhb1 are *Fhb1* backcrossing lines using Jagger and Overland as the recurrent parents, respectively. The F_4 recombinant inbred lines from JaggerFhb3 × JaggerFhb1 and JaggerFhb3 × OverlandFhb1 were grouped into four possible genotypic groups (*Fhb1/Fhb3*, *Fhb1/fhb3*, *fhb1/Fhb3* and *fhb1/fhb3*) based on the diagnostic marker data for the two genes. Comparisons of the FHB phenotypic data among the four groups found no significant difference in PSS between the haplotype groups *Fhb1/Fhb3* and *Fhb1/fhb3*, and between the groups *fhb1/Fhb3* and *fhb1/fhb3* (Fatima 2016), suggesting that *Fhb3* did not increase FHB resistance in the two wheat backgrounds with or without *Fhb1* gene. Therefore, pyramiding *Fhb3* with *Fhb1* does not enhance the FHB resistance of *Fhb1*.

Fhb6 was originally identified from the short arm of chromosome $1E^{ts}$#1S of *E. tsukushiensis*. It shows high Type II resistance and has been transferred into wheat chromosome 1AS (Cainong et al. 2015). Because *Fhb6* is relatively new in wheat, it has not been extensively used in breeding. More research is needed to determine its resistance performance in different wheat genetic backgrounds.

In addition to the seven named QTLs, many other unnamed QTLs have also been frequently detected in different studies. Although effects of these genes may not be as large as *Fhb1* and *Fhb7*, most of these QTLs were identified from locally adapted germplasm or cultivars; therefore, they can be more easily used in breeding than those from exotic sources. Many of these QTLs were mapped in large genetic intervals, often with inconsistent chromosome locations and varied QTL effects among these studies due to variations in FHB evaluation methods, testing environments, different types of resistance evaluated and types of markers used among different studies. Therefore, these QTLs have not been well utilized in breeding. However, some of these QTLs have been frequently reported in the similar chromosome regions and often showed medium-to-large effects on resistance in some studies. These QTL regions likely carry repeatable QTLs and further characterization of these QTL regions will facilitate deployment of these QTLs in breeding. Using sequence of flanking markers and IWGSC reference genome sequence map (IWGSC, 2018), 26 QTL regions can be located in different chromosomes (Table 1). Most of these QTLs are defined in ~100 Mb regions based on the physical positions of their flanking markers and each region has QTLs reported from at least six different studies with at least three studies that reported a PVE values >10% for the QTL. These repeatable QTL regions mainly showed resistance to Type

II, some also showed resistance to Types I and III, and thus, can be very useful resources for genetic improvement of wheat FHB resistance.

In chromosome group I, two repeatable QTL regions are identified in each arm of chromosome 1B. One QTL region was located in the 10-100 Mb region at the terminal end of 1BS. This region explained 8.6% (2.7-16%) of the phenotypic variation in different field and greenhouse experiments. It was detected in eight wheat accessions mainly from Europe and America. Another QTL was located in 500-600 Mb in 1BL and explained 9.04% (3-17%) of the phenotypic variation for Type II and Type III resistance in different field and greenhouse experiments. This region was detected in 11 wheat accessions including Chinese landraces and cultivars from Europe and America (Table 1).

Four repeatable QTL regions were located in chromosome group II with one each in 2AS, 2BS, 2DS and 2DL. On chromosome 2D, the QTL in the 10-60 Mb region of 2DS has been reported in 15 germplasm lines from China, United States and Europe. This QTL showed a major effect in several populations with a mean PVE of 13.1% (3.2-30.7%) for Type II and Type III resistance. Two important plant developmental genes *Rht8* and *Ppd-D1* were mapped within this region. It remains unknown if it is due to pleiotropic effects of *Rht8* or *Ppd-D1* or an FHB resistance gene tightly linked to the two plant-development-related genes in the region. Another 2D QTL is in the 540-640 Mb region of 2DL. This QTL region showed a slightly higher effect and had a mean PVE of 14.6% (3.1-24.3%) with PVE values >10% in 11 out of 16 studies. The wheat accessions carry this QTL region include Wuhan1, Sumai 3 and CJ9306 from China, Soru#1 and SHA3/CBRD from CIMMYT, Kenyon from Canada, Arina from Europe, Art from the United States and Frontana from Brazil (Table 1). This QTL was located near Xgwm539 (Liu et al. 2009; Buerstmayr et al. 2009) and conferred both Type II and Type III resistance showing an additive effect with *Fhb1* in different backgrounds (Clark et al. 2016; Jiang et al. 2007a,b).

The 2AS QTL region was identified from seven wheat or durum cultivars/landraces from different continents. This QTL is in the 30-130 Mb region and showed a relatively small effect with a PVE of 9.7% (7-14%). The 2BS QTL was located in the 110-200 Mb region on 2BS and showed a relatively larger effect, with a mean PVE of 12.9% (4.9-17.9%), than that of 2AS QTL. This QTL region has been reported in seven accessions mainly from the United States (Table 1).

Two repeatable QTL regions were identified on the group III chromosomes with one each from 3A and 3B. The 3B QTL is *Fhb1*. This QTL showed the largest effect among all QTLs reported to date with a mean PVE of 26.4% (Table 1). This QTL has been reported in at least 23 accessions from 40 studies with the mean PVE value >10% in 36 studies. Its donor parents are either Chinese landraces or Sumai 3 and its derivatives. The 3B QTL has been reported in Sumai 3 and Wangshuibai in multiple studies using different populations. The 3A QTL region had a mean PVE of 17.8% (Table 1) and was also a major QTL for Type II resistance. QTLs in this region have been reported in at least 14 studies and the closest marker to the QTL is Xgwm2. The QTL spans the region from 10 to 70 Mb on 3AS. FHB resistance allele has been identified in Wangshuibai and HCD from China, Ernie and Heyne from the United States, Tmp16315 from Canada, Frontana from Brazil and Soru#1 from Mexico (Table 1). This QTL was also reported in a durum wheat Langdon with the wild type allele for susceptibility (Chen et al. 2007).

Four QTL regions were detected in chromosome group 4, with one each from 4AL and 4DS and two from 4B (Table 1). The first QTL on 4B was located in the 450–550 Mb region on 4BL. This QTL matches with the chromosome location of *Fhb4*. This region was originally detected in Chinese landrace Wangshuibai and was later reported in several wheat cultivars or breeding lines from the United States, Europe and Brazil. The second QTL region on 4B was located between 30 Mb and 60 Mb on 4BS where *Rht-B1* for reduced plant height gene is located. This QTL showed a large effect on FHB resistance with a mean PVE of 17% from 13 populations, and the tall allele usually contributed resistance. This QTL has been reported in many different genetic backgrounds, including both bread wheat and durum wheat from Canada, Europe, China and the United States. On chromosome 4DS, an FHB resistance QTL in 5–30 Mb region was co-located with another plant height reducing gene *Rht-D1*. This QTL has been frequently reported in European wheat cultivars such as Spark, Avle, Apache and Arina and in wheat from the United States (Nc-neuse), CIMMYT (Soru#1) and Brazil (IAS20/H567.71). This QTL region also showed a large effect on FHB resistance, in particular, for these accessions from Europe, with a mean PVE of 17.2% from 11 populations. Similar to *Rht-B1*, the tall allele usually contributed resistance. However, Srinivasachary et al. (2009) tested NILs with contrasting alleles at *Rht-B1* and *Rht-D1* and found that the two plant height genes mainly affected Type I resistance. For Type II resistance *Rht-D1* had no effect, whereas short allele of *Rht-B1* contributed to resistance. It remains to be investigated if the resistance of the two QTLs is due to pleiotropic effects of the two plant height genes or closely linked genes for FHB resistance.

The 4AL QTL region was located in the terminal end (600–740 Mb) of chromosome 4AL (Table 1). This QTL was identified from 17 studies and explained 9.7% of the phenotypic variation (5–21%) for Type II and Type III resistance. The donor germplasm lines are mainly from the United States, Canada, CYMMIT, Brazil and Europe.

On chromosome 5A, FHB resistance QTLs were frequently reported on three regions with one each on the distal ends of each arm (20–110 Mb on 5AS and 600–700 Mb on 5AL) and the third one near centromere region (400–470 Mb) on 5AL (Table 1). The 5AS QTL overlapped with *Fhb5* described earlier. The 5AL terminal QTL showed a much larger effect than *Fhb5* in the 13 germplasm lines evaluated. The major effects of the QTL were reported in PI 277012 with a PVE of 32%, and two triticale lines with PVE values of 29% and 30%, respectively. This QTL is near the Q gene and is mainly present in European wheat cultivars such as Renan, Apache, Pirat, and Arina, and some durum and Triticale accessions (Gervais et al. 2003; Holzapfel et al. 2008; Buerstmayr et al. 2009). The QTL near centromere of 5AL was reported in at least nine studies and also showed a slightly larger effect than *Fhb5* with a mean PVE of 12.0%. This QTL was detected in wheat cultivars from the United States, Canada and China and also in durum wheat.

Two other QTLs on the group 5 chromosomes were located on chromosome arms 5BL and 5DL. The 5BL QTL was located in the 500–550 Mb region and reported from 14 studies. This QTL overall explained 11.2% of the phenotypic variations with the highest PVE (23.5%) identified in Ernie (Li et al. 2010). It has been detected in many wheat cultivars including Frontana, Ben, Naxos, Solitar, VA00W-38, NC-Neuse and so forth from different regions and also in durum wheat (Table 1). The 5DL QTL was located in the 430–490 Mb region on 5DL. This QTL explained 9.3% of the phenotypic variations with the highest PVE (12.9%) identified in Y21. This QTL can be found in seven germplasm accessions from different regions worldwide (Table 1).

In chromosome group 6, three QTL have been frequently reported with one in 6AL and two on 6B (Table 1). The 6A QTL was located in the 520–620 Mb region and explained 10.7% of the phenotypic variation for Type II resistance. It was detected in six cultivars from different countries. Two QTL regions on 6B were identified in the 100–200 Mb region on 6BS and the 560–660 Mb region on 6BL, respectively. The 6BS QTL is *Fhb2* discussed earlier. The 6BL QTL had a mean PVE of 12.9% (4.95–22%) and was reported from the eight studies. This QTL is mainly detected in Chinese and European germplasm.

In chromosome group 7, five QTLs have been frequently reported (Table 1). One QTL was located in the 20–120 Mb region on chromosome arm 7AS. This QTL region showed a major effect on Type II resistance in most reported populations, with a mean PVE of 12.3% (7.4–24%). The second QTL on 7A was located in the 600–700 Mb region on 7AL. This QTL had a mean PVE of 10.3%, ranging from 3.2 to 19% among different populations and was mainly reported in European germplasm and Chinese landraces. Two QTLs were identified on 7B with one in the 100–200 Mb region on 7BS and another in the 600–740 Mb region on 7BL. The QTL on 7BS showed a moderate effect on Type II resistance with a mean PVE of 12.9%, ranging from 3 to 27%. This QTL was detected in both durum and wheat. The 7BL QTL showed a smaller effect than the 7AS QTL with a mean PVE of 8.3%, ranging from 2.8 to 11.4%. This QTL was mainly detected in several European and Chinese wheat. The last QTL was detected in the 90–190 Mb region on chromosome arm 7DS. This QTL was only reported in six studies but showed a major effect in most of the populations with a mean PVE of 13.1%, ranging from 3.4 to 21.7% (Table 1). These QTLs were reported in several populations from different studies; therefore, they are valuable QTLs for breeding. However, functional markers linked to these QTLs need to be developed to improve selection efficiency in breeding programs.

8 Resistance mechanisms

Understanding resistance mechanisms will facilitate effective utilization of resistant sources in genetic improvement of wheat for better FHB resistance. However, information on how a plant fights against the pathogen remains limited. Both active and passive resistance mechanisms have been proposed in wheat. In active resistance, resistant wheat plants may either produce physical barriers (such as a thickened cell wall) or accumulate toxic phenolic compounds or triticens to prevent or delay the mycelial growth after initial infection (Ribichich et al. 2000; Zhang et al. 2013), whereas in passive resistance, plants indirectly lower disease levels by using their phenotypic traits such as heading date or plant height to escape or delay infection or provide unfavorable conditions against pathogen infection.

Fusarium graminearum may induce defense-responsive genes to encode antifungal proteins that differentially expressed between resistant and susceptible genotypes (Pritsch et al. 2000; Zhang et al. 2013). Some studies observed that *Fusarium* infection induced biosynthesis of more jasmonic acid (JA), ethylene (ET) and higher expression of lipoxygenase (LOX2) and chalcone synthase in resistant wheat plants than in susceptible plants (Li and Yen 2008; Chen et al. 2009; Ding et al. 2011; Foroud et al. 2019a). In addition, many other biochemical compounds such as choline, betaine, trans-aconitate, sucrose, glutamine and glutamate alanine have also been associated with fungal hyphal growth (Browne and Brindle 2007). Another active resistance mechanism is that plant genes can enhance wheat resistance by detoxifying mycotoxins, in particular, DON (Ma et al. 2020; Wang et al. 2020). This detoxification theory has been supported by several evidence: (1) knocking out a DON-producing gene in *Fusarium* prevented disease spread within a wheat spike (Bai et al. 2002), which suggests DON may be a virulence factor and degradation of DON in a plant may enhance plant resistance; (2) transgenic plants with barley detoxification gene in wheat improved wheat resistance (Li et al. 2015, 2017); and (3) recent cloning of *Fhb7* provided direct evidence that *Fhb7* gene directly detoxified DON to improve wheat FHB resistance (Wang et al. 2020). In addition, many genes for plant defense, phytohormone-related defense signaling and phenylpropanoid pathways have been associated with wheat active FHB resistance, and a recent review by Ma et al. (2020) provides a detailed description on these resistance mechanisms.

Although active resistance mechanisms are important to prevent FHB damage on wheat plant, passive resistance factors such as morphological traits that may also play significant roles in avoiding, escaping or reducing initial infection (Type I) to provide protection of wheat plants from FHB damage. Morphological traits including plant height, heading and flowering dates, anther and spike traits have been frequently reported to be associated with FHB infection (Hilton et al. 1999; Jones et al. 2018). In fields, tall plants usually have less severe FHB than short plants (Klahr et al. 2007; Mao et al. 2010; Lu et al. 2013). Plant height may affect FHB resistance by influencing microclimate for pathogen landing on spike tissue and initiating infection. Relative humidity, leaf wetness duration and temperatures on a spike might be different between tall and short plants (Hilton et al. 1999; Buerstmayr and Buerstmayr 2015; Jones et al. 2018). As described earlier, FHB resistance QTLs have been frequently reported in the *Rht-B1* region on 4BS, *Rht-D1* regions on 4DS and *Rht8* region on 2DS (Table 1). However, more resistance in short plants than taller plants was also observed when testing for Type II or Type III resistance (Srinivasachary et al. 2009; Yan et al. 2011; Saville et al. 2012; He et al. 2016b). Therefore, it remains unclear if these plant height genes are directly involved in the regulation of resistance or tightly linked to FHB resistance. Significant associations between flowering time and FHB ratings have been observed in some studies, but the trend was not consistent. Early flowering plants showed more disease in some studies but less disease in the others, which might vary among years and locations. As successful FHB infection depends on the availability of pathogen inocula and favorable weather during flowering. Flowering time mainly affects FHB by avoiding peak inoculum and favorable weather conditions, in particular, temperature and moisture, for infection initiation. For example, if early flowering plants meet cool and/or dry weather, the plants will have lower infection than later plants that meet warmer and wet weather at anthesis. Therefore, the mechanisms for wheat resistance to FHB are complicated. *Fhb7* regulates FHB resistance by detoxifying DON, but *Fhb1* may regulate FHB resistance by disrupting a functional susceptible gene. Thus, different QTLs may use different approaches to regulate plant resistance to FHB.

9 Breeding strategies

Traditionally, to breed for FHB-resistant cultivars, breeders usually cross a resistant parent to a parent with desirable agronomic and adaptation traits to generate a segregating population and then phenotypically select for FHB-resistant plants in the segregating progeny in an FHB nursery. To become a cultivar, the selected elite breeding lines need to have not only FHB resistance, but also desired agronomic traits and resistance to other diseases. Because most of the highly FHB-resistant parents are landraces or other germplasm lines with many undesired traits, it is difficult to simultaneously combine multiple FHB resistance QTLs with genes for resistance to other diseases and for desired agronomic traits using traditional phenotypic selection. Also, the expression of the resistance QTLs can be significantly affected by environments, which requires breeding selection to be made in an FHB nursery where consistent FHB selection pressure needs to be maintained throughout a breeding cycle. All these factors limited the progress in integration of the resistance QTLs into new cultivars. Fortunately, modern breeding technologies are available to simultaneously select multiple QTLs without the need for phenotyping, which can be used to pyramid different resistance QTLs together with the other genes controlling desirable agronomic traits in new cultivars as well as to create new sources of FHB resistance.

9.1 Resistance sources used for breeding

Several sources of resistance genes/QTLs can be used to improve wheat resistance to FHB. These sources include (1) genes from exotic sources with a major effect on FHB resistance such as *Fhb1*; (2) repeatable resistance genes/QTLs with moderate and minor effects on FHB resistance from locally adapted cultivars; (3) FHB resistance genes from alien species such as *Fhb7*; (4) novel sources of FHB resistance genes created by mutagenesis, gene editing or other new biotechnologies. *Fhb1* showed a large effect on FHB resistance in multiple backgrounds (Table 1) and reduced 20-50% FHB severity in locally adapted backgrounds. However, *Fhb1* showed only partial resistance and a plant with *Fhb1* alone may be moderately susceptible. A highly resistant line with *Fhb1* needs to carry not only *Fhb1*, also several other minor QTLs with additive effects (Cai et al. 2016; Li et al. 2011). Currently, *Fhb1* has not been widely deployed in most major wheat-growing areas worldwide. Since it has been transferred to many locally adapted lines or cultivars, many of these lines/cultivars carrying *Fhb1* can be directly used as the *Fhb1* donor parents without poor linkage drag. Also, diagnostic markers have been developed for *Fhb1* (Su et al. 2018), which can accurately track *Fhb1* movement in breeding lines. Thus, deployment of *Fhb1* has the potential to quickly enhance FHB resistance in new wheat cultivars. Most recent cloning of *Fhb7* provides another major gene for FHB resistance. Preliminary data showed that transferring *Fhb7* to adapted Chinese cultivars has no grain yield penalty; thus, this gene is ready to be used in breeding to reduce FHB infection and DON content in new wheat cultivars.

Besides *Fhb1* and *Fhb7*, many wheat cultivars from different countries carry native resistance QTLs (Table 1). These QTLs may not have a large effect as *Fhb1* and *Fhb7*, but they may significantly enhance plant resistance when they are pyramided with these major genes (Cai et al. 2019). Also, accumulation of several such QTLs with medium and minor effects can also achieve high levels of resistance (Cai et al. 2016). Most of the QTLs listed in Table 1 are present in many germplasm accessions that are distributed worldwide. All these QTLs are repeatable QTL with noticeable effects and many of them are present in locally adapted cultivars, and thus, they are important sources of FHB resistance QTLs for breeding. Several moderately FHB-resistant cultivars without *Fhb1* have been identified in the U.S. soft winter wheat (SWW), including Truman, Massy, Ernie, Roane and Freedom (Rudd et al. 2001; Sneller et al. 2012; Jin et al. 2013; Liu et al. 2013), and in the hard winter wheat (HWW), including Everest, Overland, Lyman, Heyne and Hondo (Bockus et al. 2009; Zhang et al. 2012a; Jin et al. 2013). These are locally adapted cultivars that meet growers' requirements for yield, quality and resistance to other diseases besides FHB. Therefore, FHB resistance QTLs in these wheat cultivars can be readily pyramided with the major QTLs to improve the level of FHB resistance in new wheat cultivars. Also, these cultivars may carry several different QTLs, combining these QTLs together from different cultivars may produce highly resistant cultivars by selecting transgressive segregants. One example is an HWW cultivar Everest from Kansas. It has moderate Type I FHB resistance under field conditions, but none of its parents in its pedigree (HBK1064-3/Betty 'S'//VBF0589-1/ IL89-6483) have any noticeable resistance. Many QTLs for FHB resistance show additive effects (Bai et al. 2001; Jiang et al. 2007a,b), and pyramiding these additive QTLs from different sources may achieve a high level of resistance.

Several resistance genes from alien species have been identified to show resistance in wheat. *Fhb6* and *Fhb7* showed a major effect on FHB resistance and they are good sources of resistance for breeding. Meantime, further screening of wheat wide relatives may identify other resistance genes for FHB resistance. As the development of new biotechnologies, resistance sources can be created through unconventional methods. For example, *Fhb1* cloning demonstrated that wild-type allele of *Fhb1* is responsible for susceptibility, which suggests that highly susceptible varieties may carry many susceptibility loci, and genetic modification of these susceptibility alleles may provide new sources of FHB resistance genes.

9.2 Marker-assisted selection

As described previously, individual QTL may only provide partial resistance and some of these QTLs may not show distinguishable phenotypic effects in many testing environments, thus, MAS may provide a powerful tool for transferring these QTLs to new cultivars. MAS uses molecular markers that are tightly linked to target genes/QTLs to indirectly select the genes/QTLs and track their movement from generation to generation in a breeding cycle. The selection can be made in early seedling stage by sampling a small piece of leaf tissue without damaging the plants. The non-destructive selection for disease resistance QTLs at the disease-free conditions makes MAS attractive for QTL pyramiding in a single cultivar by simultaneously selecting multiple QTLs for both disease resistance and agronomic traits. Thus, MAS can significantly improve efficiency of the selection for FHB-resistant cultivars. A successful example is to transfer *Fhb1* to locally adapted wheat cultivars. Previous fine mapping identified Xumn10 as a tightly linked marker for MAS of *Fhb1* (Liu et al. 2008). This marker has been successfully used as the near diagnostic marker of *Fhb1* in U.S. wheat breeding programs for about a decade, which led to successful release of many cultivars with significantly improved FHB resistance and locally adapted germplasm lines in the United States. Using MAS, several hard spring wheat cultivars with *Fhb1*, including Sabin from Minnesota, and Alsen, ND2710 and Glenn from North Dakota, have been released for production (Mergoum et al. 2007; Anderson 2012; Anderson et al. 2012). *Fhb1* was also transferred into many U.S. soft winter wheat cultivars such as Pioneer Brands 25R18, 25R42 and 25R51 by MAS (Brown-Guedira et al. 2008). These cultivars with *Fhb1* have played a significant role in combating FHB in U.S. wheat production.

Most recently, *Fhb1* has been cloned. Based on candidate gene sequence polymorphism between NILs contrasting in *Fhb1* alleles, Su et al. (2018) developed *Fhb1* diagnostic markers for MAS. These markers were validated in more than 1000 germplasm lines worldwide and have been used for routine screening of *Fhb1* in all USDA Genotyping Laboratories and many breeding programs worldwide. The development of the *Fhb1* diagnostic markers will facilitate quick deployment of *Fhb1* in commercial cultivars to minimize FHB damage in wheat production. More recently, *Fhb7* has been cloned and the gene marker can be diagnostic after further fine-tuning.

For MAS, availability of diagnostic markers for target QTLs is critical to MAS efficiency. Although linked markers for many FHB resistance QTL have been reported, currently only the markers for *Fhb1* are diagnostic. This may be due to (1) only a few genes have been cloned and gene sequence are not available for the development of gene-specific markers; (2) many QTLs were located in a large interval of several hundred mega-bases (Mb) on a chromosome, thus, frequent crossovers between the markers and the target genes result in high false-positive rates in MAS; (3) majority of the identified QTLs show minor effects and their expressions are affected by environments. Thus, fine-mapping is not easy to narrow down these QTLs to a small non-recombination region for functional markers development. Therefore, many reported markers are still population specific and have high false-positive rates if they are used to screen a diverse breeding population. Further fine mapping and eventual cloning some of these QTLs will improve marker quality and MAS efficiency.

9.3 Marker-assisted backcrossing (MABC)

To date, all reported QTLs show only partial resistance to FHB, including *Fhb1* and *Fhb7* that have a major effect on FHB resistance. These QTLs may show additive effects in many cases when they are combined with other QTLs. Therefore, transferring these exotic genes into locally adapted moderately susceptible or moderately resistant cultivars may significantly improve the levels of FHB resistance in commercial wheat cultivars (Kolb et al. 2001). However, direct transfer of *Fhb1* from Chinese sources to U.S. hard winter wheat breeding programs has not been successful because the donor parents usually have many undesired agronomic traits. Backcrossing several times of adapted recurrent parents to the selected progeny with the donor gene can quickly remove donor parent's backgrounds. However, phenotypic selection of FHB-resistant progeny for backcrossing in each generation is difficult and very time consuming with a high false-positive rate. Marker-assisted backcrossing (MABC) may significantly reduce selection time and cost with improved selection efficiency. In the past decade, the Central Small Grain Genotyping Laboratory in U.S. Department of Agriculture has transferred *Fhb1* to 18 locally adapted hard winter wheat cultivars or elite breeding lines released from the hard winter wheat programs in the U.S. Great Plains using MABC and demonstrated that selected *Fhb1* lines reduced FHB severity by 20–50% in different genetic backgrounds compared to their recurrent parents (Bernardo et al. 2014; Jin et al. 2013). Significant variations in FHB severity were observed among the selected *Fhb1* lines within a population and across populations. The differences among *Fhb1* lines might be due to segregation of other minor QTLs in the selected *Fhb1* lines. These minor QTLs in the backgrounds usually are unknown and markers for these QTLs are not available; therefore, phenotypic selection is necessary to select highly resistant lines from these marker-selected lines. Most of the selected *Fhb1* lines showed moderate-to-high FHB resistance. These selected highly resistant lines have been transferred to the hard winter wheat breeding programs as new *Fhb1* donor parents. Some of them may be selected for commercial production in severe epidemic years after further yield trails. Using MABC, selected *Fhb1* lines showed expected type II resistance after it was successfully transferred into durum wheat (Prat et al. 2017), which provides an important source of resistance for durum breeding. These *Fhb1* backcrossing lines with different genetic backgrounds will be important resources for breeders to quickly deploy *Fhb1* in new wheat and durum cultivars.

The major purposes of MABC are to accurately transfer target genes from the donor and quickly recover as much recurrent parent's genetic backgrounds as possible in the selected plants. In traditional backcrossing breeding, several backcrosses are needed to recover a high proportion of genetic background of the recurrent parent, which needs a long breeding cycle because each backcross takes one growing season for phenotyping and another season for making backcross. MABC can eliminate tedious and costly phenotyping process and directly track the target genes/QTLs using tightly linked markers. Furthermore, marker-assisted background selection can maximize the recovery rate of the recurrent parent's genome and shorten the breeding cycle by minimizing the number of backcrosses. Marker-assisted background selection needs to use markers that uniformly distribute across the genome to monitor how much of the donor's genome has been removed in the selected progeny. Since each chromosome may have only a few crossovers in a plant in each generation, 2–3 markers in each chromosome arm may be sufficient for background selection. Therefore, a low-cost per sample, next-generation-sequencing (NGS)-based marker system can serve the purpose. Recently, Bernardo et al. (2019) developed a new NGS-based high-throughput marker platform called Multiplex Restriction Amplicon Sequencing (MRASeq). This platform uses two-step PCR for library construction and has advantages of even distribution over the wheat genome, low cost per sample, and high throughput with at least several hundreds of polymorphic markers per bi-parental population. This platform has no patent requirement; thus, it is an ideal marker platform for background selection in MABC.

9.4 Genomic selection

To date, functional markers have been only available for a few FHB resistance genes. As described earlier, most FHB resistance QTLs show minor effects, and thus, they are hardly distinguishable using a conventional phenotyping method. Also it is difficult to develop functional markers for these QTLs. To achieve high resistance in a line requires simultaneous selection of several QTLs. Many reported markers for these QTLs are not diagnostic for these QTLs when they are used to screen diverse germplasm, which significantly discounts the MAS efficiency. Genomic selection (GS) can use information from a large number of genome-wide, high-density markers to capture diversity in that genome and to estimate breeding values without need of previous knowledge on the locations of the target genes. Therefore, GS is a useful approach for improving quantitative traits such as FHB resistance. Genomic prediction that combines marker data with phenotypic and pedigree data (when available) can increase the prediction accuracy of the breeding values, and thus, has the advantages of incorporating thousands of markers simultaneously and capturing variation of usually undetected minor QTLs (Heffner et al. 2009). Currently, genome-wide high-density markers can be generated in a short time at affordable costs (Bernardo et al. 2019), which makes it possible for using GS in breeding to improve FHB resistance.

Rutkoski et al. (2012) used both MAS and GS to predict FHB resistance traits and DON content in wheat using different GS models and found that GS had higher prediction accuracies than MAS, which is supported by Arruda et al. (2015, 2016). These studies indicate that GS is appropriate for improving FHB resistance in breeding populations. Jiang et al. (2015) revealed that marker density used for GS had only a marginal impact on the prediction accuracy. Even a small subset of specifically selected markers can provide more accurate predictions than a large number of random markers (Hoffstetter et al. 2016). A selected subset of 300 markers on the reduced set of training lines led to high prediction accuracies (0.7–0.8). Therefore, the development of SNP arrays or multiplex PCR systems that can genotype several hundred known SNPs at a low cost per sample may facilitate the application of GS in breeding for FHB resistance.

9.5 Genetic engineering

Wheat-expressing resistance genes from different sources can significantly increase wheat FHB resistance, thus direct transferring of the FHB resistance genes from different species into wheat plants using transgenic approach may provide a new tool to breeding for FHB resistance and also may create new sources of resistance. Many genes from different species have been reported to have antifungal or detoxification functions and some of these can serve as transgenes to improve wheat resistance. Li et al. (2015, 2017) transferred a barley UDP-glucosyltransferase into susceptible wheat and demonstrated that the transgenic plants had much higher type II resistance and lower toxin content than controls. Mackintosh et al. (2007) demonstrated that transgenic plants overexpressing a β-1,3-glucanase showed enhanced Types II, III and IV resistance. Recent cloning of *Fhb7* identified a detoxification gene (Wang et al. 2020) and transgenic plants carrying *Fhb7* showed significantly improved FHB resistance and lower DON content. Significant attempts have been made to transfer fungal detoxification genes into wheat for decades (Ma et al. 2020); however, transgenic plants showing consistent resistance against FHB are still not available from these efforts.

As previously described, the wild-type allele of *Fhb1* causal gene *TaHRC* encodes susceptible phenotype and a loss-of-function mutation results in resistance. Su et al. (2019) demonstrated that most wheat cultivars carry susceptible alleles at *Fhb1* and knocking out a susceptibility allele can improve resistance. *TaHRC* is a conserved gene present not only in wheat, but also in durum, barley and other cereal crops. In addition, *Fhb1* may not be the only case of susceptibility genes and some other studies also suggested the presence of other FHB susceptibility genes in wheat genome (Ma et al. 2006a; Handa et al. 2008; Basnet et al. 2012; Fabre et al. 2020; Hales et al. 2020). Disabling plant disease susceptibility genes have already been demonstrated to provide durable disease resistance in crops (Pavan et al. 2010), which opens a new avenue for FHB control and complements to introgression of gain-of-function forms of resistance genes. Therefore, mutation breeding that knock out these susceptible genes may create novel FHB resistant sources. One of the successful examples is the development of FHB-tolerant cultivar Emai 6 in China. Emai 6 was selected from 60Co-γ ray-treated Nanda2419 seeds. Continuous selection from Emai 6 produced another moderately FHB-resistant cultivar Emai 9. Recently, gene editing has been widely used to create genetic diversity and new sources of genes for breeding. Using gene editing to knock out target susceptible genes provides another avenue to the improvement of FHB resistance. Knocking out *TaHRC* in susceptible wheat Bobwhite reduced 30–40% FHB (Su et al. 2019). Therefore, gene editing shows promise to improve wheat FHB resistance by manipulating FHB-susceptible genes in commercial cultivars.

10 Conclusion and future trends

Wheat resistance to FHB is a complex quantitative trait with five types. Each type of resistance is controlled by multiple QTLs with most having minor effects and are significantly affected by environments. Among these resistance types, Type II and Type III resistance are more stable and directly affect grain yield and quality as well as human and animal health, therefore, are the focuses of the presentation in this chapter. To date, hundreds of QTLs for the two types of resistance have been reported. However, most of these QTLs can be located to 26 chromosome regions of ~100 Mb intervals based on IWGSC reference sequence map. These critical QTLs can be considered major targets for the improvement of FHB resistance in breeding. To date, only *Fhb1* and *Fhb7* have shown stable, high levels of FHB resistance in different genetic backgrounds. Recent cloning of *Fhb1* and *Fhb7* has made diagnostic markers available for marker-assisted breeding. Due to the nature of multiple genes controlling FHB resistance, further improvement of FHB resistance in new cultivars needs pyramiding multiple resistance QTLs with various effects together with genes for other agronomic traits. To achieve this objective, using modern breeding techniques will be especially critical to success. MABC coupling with marker-assisted background selection can provide a quick solution to improve FHB resistance of commercial wheat cultivars by direct transfer of *Fhb1* and *Fhb7* to these cultivars to quickly relieve FHB damage. GS together with MAS and phenotypic selection may be able to combine the minor QTLs from different locally adapted sources and major genes from exotic sources to create highly resistant cultivars for long-term FHB improvement. In addition, manipulation of FHB susceptibility genes in wheat provides the opportunity for creating new sources of resistance by mutation breeding and gene editing.

11 Where to look for further information

A website for up to date FHB research in the United States is: http://scabusa.org/. This website is an official website of the US Wheat and Barley Scab Initiative, an official USDA-funded organization that coordinates all aspects of wheat and barley FHB research in the United States, and contains all proceedings for annual FHB forums and all research projects funded by USWBSI since 1998, protocols for different FHB related researches and other FHB related databases and information.

12 Acknowledgements

This project was partly funded by the U.S. Wheat and Barley Scab Initiative and the National Research Initiative Competitive Grants 2017-67007-25939 from the U.S. Department of Agriculture National Institute of Food and Agriculture. Mention of trade names or commercial products in this publication is solely for the purpose of providing specific information and does not imply recommendation or endorsement by the U.S. Department of Agriculture. The U.S. Department of Agriculture is an equal opportunity provider and employer.

13 References

Ágnes, S., Szabolcs, L., Mónika, V., László, P., János, P., Csaba, L. and Ákos, M. (2014). Differential influence of QTL linked to Fusarium head blight, Fusarium-damaged kernel, deoxynivalenol contents and associated morphological traits in a Frontana-derived wheat population. *Euphytica* 200(1): 9-26. doi:10.1007/s10681-014-1124-2.

Agriopoulou, S., Stamatelopoulou, E. and Varzakas, T. (2020). Advances in analysis and detection of major mycotoxins in foods. *Foods* 9: 518. doi:10.3390/foods9040518.

Anderson, J. A. (2012). Overview of breeding for FHB resistance in wheat-where we've come from and where we are. In: *Proceedings of the 2012 National Fusarium Head Blight Forum* Canty, S., Clark, A., Anderson-Scully, A. and Van Sanford, D. (Eds). U.S. Wheat and Barley Scab Initiative, Orlando, FL. p. 45.

Anderson, J. A., Stack, R. W., Liu, S., Waldron, B. L., Fjeld, A. D., Coyne, C., Moreno-Sevilla, B., Fetch, J. M., Song, Q. J., Cregan, P. B. and Frohberg, R. C. (2001). DNA markers for Fusarium head blight resistance QTLs in two wheat populations. *Theor. Appl. Genet.* 102(8): 1164-1168. doi:10.1007/s001220000509.

Anderson, J. A., Wiersma, J. J., Linkert, G. L., Kolmer, J. A., Jin, Y., Dill-Macky, R., Wiersma, J. V. and Hareland, G. A. (2012). Registration of 'Sabin' wheat. *J. Plant Reg.* 6(2): 174-179.

Arruda, M. P., Brown, P. J., Lipka, A. E., Krill, A. M. and Kolb, F. L. (2015). Genomic selection for predicting Fusarium head blight resistance in a wheat breeding program. *Plant Gen.* 8: 3. doi:10.3835/plantgenome2015.01.0003.

Arruda, M. P., Lipka, A. E., Brown, P. J., Krill, A. M., Thurber, C., Brown-Guedira, G., Dong, Y., Foresman, B. J. and Kolb, F. L. (2016). Comparing genomic selection and marker-assisted selection for Fusarium head blight resistance in wheat (*Triticum aestivum* L.). *Mol. Breed.* 36(7): 84. doi:10.1007/s11032-016-0508-5.

Arthur, J. C. (1891). Wheat scab. *Ind. Agric. Exp. Stat Bull.* 36: 129-138.

Bai, G., Kolb, F. L., Shaner, G. E. and Domier, L. L. (1999). Amplified fragment length polymorphism markers linked to a major quantitative trait locus controlling scab resistance in wheat. *Phytopathology* 89(4): 343-348.

Bai, G. and Shaner, G. (2004). Management and resistance in wheat and barley to Fusarium head blight. *Annu. Rev. Phytopathol.* 42: 135–161.

Bai, G., Su, Z. and Cai, J. (2018). Wheat resistance to Fusarium head blight. *Can. J. Plant Pathol.* 40(3): 336–346. doi: 10.1080/07060661.2018.1476411.

Bai, G. H., Desjardins, A. E. and Plattner, R. D. (2002). Deoxynivalenol-nonproducing *Fusarium graminearum* causes initial infection, but does not cause disease spread in wheat spikes. *Mycopathologia* 153(2): 91–98.

Bai, G. H., Plattner, R., Desjardins, A. and Kolb, F. (2001). Resistance to Fusarium head blight and deoxynivalenol accumulation in wheat. *Plant Breed.* 120(1): 1–6.

Bai, G. H. and Shaner, G. E. (1994). Scab of wheat: prospects and control. *Plant Dis.* 78(8): 760–766.

Bai, G. H., Xiao, Q. P. and Mei, J. F. (1989). Genetic analysis of resistance to scab in six wheat cultivars. *Acta. Agric. Shanghai* 5: 17–23.

Basnet, B. R., Glover, K. D., Ibrahim, A. M. H., Yen, Y. and Chao, S. (2012). A QTL on chromosome 2DS of 'Sumai 3' increases susceptibility to Fusarium head blight in wheat. *Euphytica* 186(1): 91–101. doi:10.1007/s10681-011-0495-x.

Bernardo, A., Bai, G. H., Yu, J. B., Kolb, F. L., Bockus, W. and Dong, Y. H. (2014). Registration of near-Isogenic winter wheat germplasm contrasting in *Fhb1* for Fusarium head blight resistance. *J. Plant Reg.* 8(1): 106–108. doi:10.3198/jpr2013.05.0021crgs.

Bernardo, A., St Amand, P., Le, H., Su, Z. and Bai, G. (2019). Multiplex Restriction Amplicon Sequencing (MRASeq), a novel next generation sequencing based marker platform for high-throughput genotyping. *Plant Biotech. J.* 18: 254–265. doi:10.1111/pbi.13192.

Bernardo, A. N., Ma, H., Zhang, D. and Bai, G. (2012). Single nucleotide polymorphism in wheat chromosome region harboring Fhb1 for Fusarium head blight resistance. *Mol. Breed.* 29(2): 477–488. doi:10.1007/s11032-011-9565-y.

Bockus, W. W., Fritz, A. K. and Martin, T. J. (2009). Reaction of the 2008 Kansas Intrastate Nursery to Fusarium head blight. *Plant Dis. Man. Rep.* 3: CF009.

Bonin, C. M. and Kolb, F. L. (2009). Resistance to Fusarium head blight and kernel damage in a winter wheat recombinant inbred line population. *Crop Sci.* 49(4): 1304–1312.

Bourdoncle, W. and Ohm, H. W. (2003). Quantitative trait loci for resistance to Fusarium head blight in recombinant inbred wheat lines from the cross Huapei 57-2 / Patterson. *Euphytica* 131(1): 131–136. doi:10.1023/A:1023056207513.

Brar, G. S., Brûlé-Babel, A. L., Ruan, Y., Henriquez, M. A., Pozniak, C. J., Kutcher, H. R. and Hucl, P. J. (2019). Genetic factors affecting Fusarium head blight resistance improvement from introgression of exotic Sumai 3 alleles (including *Fhb1*, *Fhb2*, and *Fhb5*) in hard red spring wheat. *BMC Plant Biol.* 19(1): 179. doi:10.1186/s12870-019-1782-2.

Browne, R. A. and Brindle, K. M. (2007). H NMR-based metabolite profiling as a potential selection tool for breeding passive resistance against Fusarium head blight (FHB) in wheat. *Mol. Plant Pathol.* 8(4): 401–410.

Brown-Guedira, G., Griffey, C., Kolb, F., McKendry, A., Murphy, J. and Sanford, D. (2008,). Breeding FHB-resistant soft winter wheat: progress and prospects. *Cereal Res. Comm.* 36(Supplement 6): 31–35.

Buerstmayr, H., Ban, T. and Anderson, J. A. (2009). QTL mapping and marker-assisted selection for Fusarium head blight resistance in wheat: a review. *Plant Breed.* 128(1): 1–26.

Buerstmayr, H., Lemmens, M., Hartl, L., Doldi, L., Steiner, B., Stierschneider, M. and Ruckenbauer, P. (2002). Molecular mapping of QTLs for *Fusarium* head blight resistance in spring wheat. I. Resistance to fungal spread (Type II resistance). *Theor. Appl. Genet.* 104(1): 84-91. doi:10.1007/s001220200009.

Buerstmayr, H., Steiner, B., Hartl, L., Griesser, M., Angerer, N., Lengauer, D., Miedaner, T., Schneider, B. and Lemmens, M. (2003). Molecular mapping of QTLs for Fusarium head blight resistance in spring wheat. II. Resistance to fungal penetration and spread. *Theor. Appl. Genet.* 107(3): 503-508. doi:10.1007/s00122-003-1272-6.

Buerstmayr, M., Alimari, A., Steiner, B. and Buerstmayr, H. (2013). Genetic mapping of QTL for resistance to Fusarium head blight spread (type 2 resistance) in a *Triticum dicoccoides* × *Triticum durum* backcross-derived population. *Theor. Appl. Genet.* 126(11): 2825-2834. doi:10.1007/s00122-013-2174-x.

Buerstmayr, M. and Buerstmayr, H. (2015). Comparative mapping of quantitative trait loci for Fusarium head blight resistance and anther retention in the winter wheat population Capo × Arina. *Theor. Appl. Genet.* 128(8): 1519-1530. doi:10.1007/s00122-015-2527-8.

Buerstmayr, M., Huber, K., Heckmann, J., Steiner, B., Nelson, J. C. and Buerstmayr, H. (2012). Mapping of QTL for Fusarium head blight resistance and morphological and developmental traits in three backcross populations derived from *Triticum dicoccum* × *Triticum durum*. *Theor. Appl. Genet.* 125(8): 1751-1765. doi:10.1007/s00122-012-1951-2.

Buerstmayr, M., Lemmens, M., Steiner, B. and Buerstmayr, H. (2011). Advanced backcross QTL mapping of resistance to Fusarium head blight and plant morphological traits in a *Triticum macha* × *T. aestivum* population. *Theor. Appl. Genet.* 123(2): 293-306. doi:10.1007/s00122-011-1584-x.

Cai, J. and Bai, G. (2014). Quantitative trait loci for Fusarium head blight resistance in Huangcandou × 'Jagger' wheat population. *Crop Sci.* 54(6): 2520-2528.

Cai, J., Wang, S., Li, T., Zhang, G. and Bai, G. (2016). Multiple minor QTLs are responsible for Fusarium head blight resistance in Chinese wheat landrace Haiyanzhong. *PLoS ONE* 11(9): e0163292. doi:10.1371/journal.pone.0163292.

Cai, J., Wang, S., Su, Z., Li, T., Zhang, X. and Bai, G. (2019). Meta-analysis of QTL for Fusarium head blight resistance in Chinese wheat landraces. *Crop J.* 7(6): 784-798.

Cainong, J. C., Bockus, W. W., Feng, Y., Chen, P., Qi, L., Sehgal, S. K., Danilova, T. V., Koo, D. H., Friebe, B. and Gill, B. S. (2015). Chromosome engineering, mapping, and transferring of resistance to Fusarium head blight disease from *Elymus tsukushiensis* into wheat. *Theor. Appl. Genet.* 128(6): 1019-1027.

Casale, W. L., Pestka, J. J. and Hart, L. P. (1988). Enzyme-linked immunosorbent assay employing monoclonal antibody specific for deoxynivalenol (vomitoxin) and several analogues. *J. Agric. Food Chem.* 36(3): 663-668.

Cativelli, M., Lewis, S. and Appendino, M. L. (2013). A Fusarium head blight resistance quantitative trait locus on chromosome 7d of the spring wheat cultivar catbird. *Crop Sci.* 53(4): 1464-1471.

Chao, L., Song, Y., Gao, L., Yao, Q., Zhou, R., Xu, R. and Jia, J. (2014). Integration of QTL detection and marker assisted selection for improving resistance to Fusarium head blight and important agronomic traits in wheat. *Crop J.* 2: 72-78.

Chen, J., Griffey, C. A., Saghai Maroof, M. A., Stromberg, E. L., Biyashev, R. M., Zhao, W., Chappell, M. R., Pridgen, T. H., Dong, Y. and Zeng, Z. (2006). Validation of two major quantitative trait loci for Fusarium head blight resistance in Chinese wheat line W14. *Plant Breed.* 125(1): 99–101.

Chen, X., Faris, J. D., Hu, J., Stack, R. W., Adhikari, T., Elias, E. M., Kianian, S. F. and Cai, X. (2007). Saturation and comparative mapping of a major Fusarium head blight resistance QTL in tetraploid wheat. *Mol. Breed.* 19(2): 113–124. doi:10.1007/s11032-006-9049-7.

Chen, X., Steed, A., Travella, S., Keller, B. and Nicholson, P. (2009). Fusarium graminearum exploits ethylene signalling to colonize dicotyledonous and monocotyledonous plants. *New Phytol.* 182(4): 975–983.

Chester, F. D. (1890). The scab of the wheat. *Del. Agric. Exp. Sta. Rep.* 3: 89–90.

Chu, C., Niu, Z., Zhong, S., Chao, S., Friesen, T. L., Halley, S., Elias, E. M., Dong, Y., Faris, J. D. and Xu, S. S. (2011). Identification and molecular mapping of two QTLs with major effects for resistance to Fusarium head blight in wheat. *Theor. Appl. Genet.* 123(7): 1107–1119. doi:10.1007/s00122-011-1652-2.

Clark, A. J., Sarti-Dvorjak, D., Brown-Guedira, G., Dong, Y., Baik, B. K. and Van Sanford, D. A. (2016). Identifying rare FHB resistant segregants in intransigent backcross and F_2 winter wheat populations. *Front. Microbiol.* 7: 277.

Clinesmith, M. A., Fritz, A. K., Lemes da Silva, C., Bockus, W. W., Poland, J. A., Dowell, F. E. and Peiris, K. S. H. (2019). QTL mapping of Fusarium Head Blight resistance in winter wheat cultivars art and Everest. *Crop Sci.* 59: 911–924.

Cuthbert, P. A., Somers, D. J. and Brulé-Babel, A. (2007). Mapping of *Fhb2* on chromosome 6BS: a gene controlling Fusarium head blight field resistance in bread wheat (*Triticum aestivum* L.). *Theor. Appl. Genet.* 114(3): 429–437. doi:10.1007/s00122-006-0439-3.

Cuthbert, P. A., Somers, D. J., Thomas, J., Cloutier, S. and Brule-Babel, A. (2006). Fine mapping *Fhb1*, a major gene controlling Fusarium head blight resistance in bread wheat (*Triticum aestivum* L.). *Theor. Appl. Genet.* 112(8): 1465–1472.

Dhariwal, R., Fedak, G., Dion, Y., Pozniak, C., Laroche, A., Eudes, F. and Randhawa, H. S. (2018). High density single nucleotide polymorphism (SNP) mapping and quantitative trait loci (QTL) analysis in a Biparental spring *Triticale* Population Localized Major and Minor Effect Fusarium Head blight resistance and associated traits QTL. *Genes* 9(1): 19.

Dickson, J. G. (1942). Scab of wheat and barley and its control. *USDA Farmer Bull.* 1599: 1–22.

Ding, L., Xu, H., Yi, H., Yang, L., Kong, Z., Zhang, L., Xue, S., Jia, H. and Ma, Z. (2011). Resistance to hemi-biotrophic *F. graminearum* infection is associated with coordinated and ordered expression of diverse defense signaling pathways. *PLoS ONE* 6(4): e19008.

Draeger, R., Gosman, N., Steed, A., Chandler, E., Thomsett, M., Srinivasachary, S., Schondelmaier, J., Buerstmayr, H., Lemmens, M., Schmolke, M., Mesterhazy, A. and Nicholson, P. (2007). Identification of QTLs for resistance to Fusarium head blight, DON accumulation and associated traits in the winter wheat variety Arina. *Theor. Appl. Genet.* 115(5): 617–625. doi:10.1007/s00122-007-0592-3.

Dubin, J., Gilchrist, L., Reeves, J. and McNab, A. (Eds.) (1997). Fusarium head scab: global status and future prospects. In *Proceedings of a Workshop*, El Batan (Mexico). Mexico, DF (Mexico). CIMMYT, 130 pages.

Eckard, J. T., Glover, K. D., Mergoum, M., Anderson, J. A. and Gonzalez-Hernandez, J. L. (2015). Multiple Fusarium head blight resistance loci mapped and pyramided onto elite spring wheat *Fhb1* backgrounds using an IBD-based linkage approach. *Euphytica* 204(1): 63–79. doi:10.1007/s10681-014-1333-8.

Eckard, J. T., Gonzalez-Hernandez, J. L., Caffe, M., Berzonsky, W., Bockus, W. W., Marais, G. F. and Baenziger, P. S. (2015a). Native Fusarium head blight resistance from winter wheat cultivars Lyman, Overland, Ernie, and Freedom mapped and pyramided onto 'Wesley'-*Fhb1* backgrounds. *Mol. Breed.* 35(1): 6. doi:10.1007/s11032-015-0200-1.

Fabre, F., Rocher, F., Alouane, T., Langin, T. and Bonhomme, L. (2020). Searching for FHB resistances in bread wheat: susceptibility at the crossroad. *Front. Plant Sci.* 11: 731. doi:10.3389/fpls.2020.00731.

Fatima, N. (2016). Identification and deployment of QTLs for Fusarium head blight resistance in U.S. winter wheat. Ph.D. Thesis, Kansas State University.

Ferrigo, D., Raiola, A. and Causin, R. (2016). Fusarium toxins in cereals: occurrence, legislation, factors promoting the appearance and their management. *Molecules* 21(5): 627.

Foroud, N., Pordel, R., Goyal, R. K., Ryabova, D., Eranthodi, A., Chatterton, S. and Kovalchuk, I. (2019a). *Fusarium graminearum* chemical activation of the ethylene signaling pathway promotes resistance in detached wheat heads. *Phytopathology* 109: 796–803.

Foroud, N. A., Baines, D., Gagkaeva, T. Y., Thakor, N., Badea, A., Steiner, B., Bürstmayr, M. and Bürstmayr, H. (2019b). Trichothecenes in cereal grains–an update. *Toxins* 11(11): 634.

Gervais, L., Dedryver, F., Morlais, J. Y., Bodusseau, V., Negre, S., Bilous, M., Groos, C. and Trottet, M. (2003). Mapping of quantitative trait loci for field resistance to Fusarium head blight in an European winter wheat. *Theor. Appl. Genet.* 106(6): 961-970. doi:10.1007/s00122-002-1160-5.

Giancaspro, A., Giove, S. L., Zito, D., Blanco, A. and Gadaleta, A. (2016). Mapping QTLs for *Fusarium* head blight resistance in an interspecific wheat population. *Front. Plant Sci.* 7: 1381. doi:10.3389/fpls.2016.01381.

Gong, X., He, X., Zhang, Y., Li, L., Sun, Z., Bai, G., Singh, P. K. and Li, T. (2020). Development of an evaluation system for *Fusarium* resistance in wheat grains and its application in assessment of the corresponding effects of *Fhb1*. *Plant Dis.* 104(8): 2210–2216 doi:10.1094/PDIS-12-19-2584-RE.

Goswami, R. S. and Kistler, H. C. (2004). Heading for disaster: *Fusarium graminearum* on cereal crops. *Mol. Plant Pathol.* 5(6): 515–525.

Guo, J., Zhang, X., Hou, Y., Cai, J., Shen, X., Zhou, T., Xu, H., Ohm, H. W., Wang, H., Li, A., Han, F., Wang, H. and Kong, L. (2015). High-density mapping of the major FHB resistance gene *Fhb7* derived from *Thinopyrum ponticum* and its pyramiding with *Fhb1* by marker-assisted selection. *Theor. Appl. Genet.* 128(11): 2301–2316.

Häberle, J., Schweizer, G., Schondelmaier, J., Zimmermann, G. and Hartl, L. (2009). Mapping of QTL for resistance against Fusarium head blight in the winter wheat population Pelikan//Bussard/Ning8026. *Plant Breed.* 128(1): 27–35. doi:10.1111/j.1439-0523.2008.01540.x.

Hales, B., Steed, A., Giovannelli, V., Burt, C., Lemmens, M., Molnár-Láng, M. and Nicholson, P. (2020). Type II Fusarium head blight susceptibility factor identified in wheat. *bioRxiv*. Preprint. doi:10.1101/2020.02.06.937425.

Handa, H., Namiki, N., Xu, D. and Ban, T. (2008). Dissecting of the FHB resistance QTL on the short arm of wheat chromosome 2d using a comparative genomic approach: from QTL to candidate gene. *Mol. Breed.* 22(1): 71–84. doi:10.1007/s11032-008-9157-7.

He, X., Dreisigacker, S., Singh, R. P. and Singh, P. K. (2019). Genetics for low correlation between Fusarium head blight disease and deoxynivalenol (DON) content in a bread wheat mapping population. *Theor. Appl. Genet.* 132(8): 2401–2411. doi:10.1007/s00122-019-03362-9.

He, X., Lillemo, M., Shi, J., Wu, J., Bjørnstad, Å., Belova, T., Dreisigacker, S., Duveiller, E. and Singh, P. (2016a). QTL characterization of Fusarium head blight resistance in CIMMYT bread wheat line Soru#1. *PLoS ONE* 11(6): e0158052. doi:10.1371/journal.pone.0158052.

He, X., Singh, P. K., Dreisigacker, S., Singh, S., Lillemo, M. and Duveiller, E. (2016b). Dwarfing genes *Rht-B1b* and *Rht-D1b* are associated with both type I FHB susceptibility and low anther extrusion in two bread wheat populations. *PLoS ONE* 11(9): e0162499.

Heffner, E. L., Sorrells, M. E. and Jannink, J. (2009). Genomic selection for crop improvement. *Crop Sci.* 49(1): 1–12 doi: 10.2135/cropsci2008.08.0512.

Herter, C. P., Ebmeyer, E., Kollers, S., Korzun, V., Leiser, W. L., Würschum, T. and Miedaner, T. (2018). Rht24 reduces height in the winter wheat population 'Solitär × Bussard' without adverse effects on Fusarium head blight infection. *Theor. Appl. Genet.* 131(6): 1263–1272. doi:10.1007/s00122-018-3076-8.

Hilton, A. J., Jenkinson, P., Hollins, T. W. and Parry, D. W. (1999). Relationship between cultivar height and severity of Fusarium ear blight in wheat. *Plant Pathol.* 48(2): 202–208.

Hoffstetter, A., Cabrera, A., Huang, M. and Sneller, C. (2016). Optimizing training population data and validation of genomic selection for economic traits in soft winter wheat. *G3 (Bethesda)* 6(9): 2919–2928. doi:10.1534/g3.116.032532.

Holzapfel, J., Voss, H. H., Miedaner, T., Korzun, V., Häberle, J., Schweizer, G., Mohler, V., Zimmermann, G. and Hartl, L. (2008). Inheritance of resistance to Fusarium head blight in three European winter wheat populations. *Theor. Appl. Genet.* 117(7): 1119–1128 doi:10.1007/s00122-008-0850-z.

International Wheat Genome Sequencing Consortium (IWGSC) (2018). Shifting the limits in wheat research and breeding using a fully annotated reference genome. *Science* 361(6403): eaar7191. doi:10.1126/science.aar7191.

Islam, M. S., Brown-Guedira, G., Van Sanford, D., Ohm, H., Dong, Y. and McKendry, A. L. (2016). Novel QTL associated with the Fusarium head blight resistance in Truman soft red winter wheat. *Euphytica* 207(3): 571–592 doi:10.1007/s10681-015-1550-9

Jayatilake, D. V., Bai, G. H. and Dong, Y. H. (2011). A novel quantitative trait locus for Fusarium head blight resistance in chromosome 7a of wheat. *Theor. Appl. Genet.* 122(6): 1189–1198. doi:10.1007/s00122-010-1523-2.

Jia, G., Chen, P., Qin, G., Bai, G., Wang, X., Wang, S., Zhou, B., Zhang, S. and Liu, D. (2005). QTLs for Fusarium head blight response in a wheat DH population of Wangshuibai/Alondra's'. *Euphytica* 146(3): 183–191. doi:10.1007/s10681-005-9001-7.

Jia, H. Y., Zhou, J. Y., Xue, S. L., Li, G. Q., Yan, H. S., Ran, C., Zhang, Y., Shi, J., Jia, L., Wang, X., Luo, J. and Ma, Z. (2018). A journey to understand wheat Fusarium head blight resistance in the Chinese wheat Wangshuibai. *Crop J.* 6(1): 48–59.

Jiang, G. L., Dong, Y., Shi, J. and Ward, R. W. (2007a). QTL analysis of resistance to Fusarium head blight in the novel wheat germplasm CJ 9306. II. Resistance to deoxynivalenol accumulation and grain yield loss. *Theor. Appl. Genet.* 115(8): 1043–1052.

Jiang, G. L., Shi, J. and Ward, R. W. (2007b). QTL analysis of resistance to Fusarium head blight in the novel wheat germplasm CJ 9306. I. Resistance to fungal spread. *Theor. Appl. Genet.* 116(1): 3–13. doi:10.1007/s00122-007-0641-y.

Jiang, Y., Zhao, Y., Rodemann, B., Plieske, J., Kollers, S., Korzun, V., Ebmeyer, E., Argillier, O., Hinze, M., Ling, J., Röder, M. S., Ganal, M. W., Mette, M. F. and Reif, J. C. (2015). Potential and limits to unravel the genetic architecture and predict the variation of Fusarium head blight resistance in European winter wheat (*Triticum aestivum* L.). *Heredity* 114(3): 318–326. doi:10.1038/hdy.2014.104.

Jin, F., Bai, G. H., Zhang, D. D., Dong, Y. H., Ma, L. J., Bockus, W. and Dowell, F. (2014). Fusarium-damaged kernels and deoxynivalenol in Fusarium-infected U.S. winter wheat. *Phytopathology* 104(5): 472–478.

Jin, F., Zhang, D., Bockus, W., Baenziger, P. S., Carver, B. and Bai, G. (2013). Fusarium head blight resistance in U.S. winter wheat cultivars and elite breeding lines. *Crop Sci.* 53(5): 2006–2013.

Jones, S., Farooqi, A., Foulkes, J., Sparkes, D. L., Linforth, R. and Ray, R. V. (2018). Canopy and ear traits associated with avoidance of Fusarium head blight in wheat. *Front. Plant Sci.* 9: 1021. doi: 10.3389/fpls.2018.01021.

Kautzman, M. E., Wickstrom, M. L. and Scott, T. A. (2015). The use of near infrared transmittance kernel sorting technology to salvage high quality grain from grain downgraded due to *Fusarium* damage. *Anim. Nutr.* 1(1): 41–46.

Klahr, A., Zimmermann, G., Wenzel, G. and Mohler, V. (2007). Effects of environment, disease progress, plant height and heading date on the detection of QTLs for resistance to Fusarium head blight in a European winter wheat cross. *Euphytica* 154(1–2): 17–28. doi:10.1007/s10681-006-9264-7.

Kolb, F. L., Bai, G. H., Muehlbauer, G. J., Anderson, J. A., Smith, K. P. and Fedak, G. (2001). Host plant resistance genes for Fusarium head blight: mapping and manipulation with molecular markers. *Crop Sci.* 41(3): 611–619.

Kumar, S., Stack, R. W., Friesen, T. L. and Faris, J. D. (2007). Identification of a novel Fusarium head blight resistance quantitative trait locus on chromosome 7a in tetraploid wheat. *Phytopathology* 97(5): 592–597. doi:10.1094/PHYTO-97-5-0592.

Lagudah, E. S. and Krattinger, S. G. (2019). A new player contributing to durable Fusarium resistance. *Nat. Genet.* 51(7): 1070–1071.

Lemes da Silva, C., Fritz, A., Clinesmith, M., Poland, J., Dowell, F. and Peiris, K. (2019). QTL mapping of Fusarium head blight resistance and deoxynivalenol accumulation in the Kansas wheat variety 'Everest'. *Mol. Breed.* 39(3): 35. doi:10.1007/s11032-019-0937-z.

Lemmens, M., Scholz, U., Berthiller, F., Dall'Asta, C., Koutnik, A., Schuhmacher, R., Adam, G., Buerstmayr, H., Mesterházy, A., Krska, R. and Ruckenbauer, P. (2005). The ability to detoxify the mycotoxin deoxynivalenol colocalizes with a major quantitative trait locus for Fusarium head blight resistance in wheat. *Mol. Plant Microbe. Interact.* 18(12): 1318–1324. doi:10.1094/MPMI-18-1318.

Li, G. and Yen, Y. (2008). Jasmonate and ethylene signaling pathway may mediate Fusarium head blight resistance in wheat. *Crop Sci.* 48(5): 1888–1896.

Li, G., Zhou, J., Jia, H., Gao, Z., Fan, M., Luo, Y., Zhao, P., Xue, S., Li, N., Yuan, Y., Ma, S., Kong, Z., Jia, L., An, X., Jiang, G., Liu, W., Cao, W., Zhang, R., Fan, J., Xu, X., Liu, Y., Kong, Q., Zheng, S., Wang, Y., Qin, B., Cao, S., Ding, Y., Shi, J., Yan, H., Wang, X., Ran, C. and Ma, Z. (2019). Mutation of a histidine-rich calcium-binding-protein gene in wheat confers resistance to Fusarium head blight. *Nat. Genet.* 51(7): 1106–1112.

Li, H. B., Xie, G. Q., Ma, J., Liu, G. R., Wen, S. M., Ban, T., Chakraborty, S. and Liu, C. J. (2010). Genetic relationships between resistances to Fusarium head blight and crown rot in bread wheat (*Triticum aestivum* L.). *Theor. Appl. Genet.* 121(5): 941–950. doi:10.1007/s00122-010-1363-0.

Li, T., Bai, G., Wu, S. and Gu, S. (2011). Quantitative trait loci for resistance to Fusarium head blight in a Chinese wheat landrace Haiyanzhong. *Theor. Appl. Genet.* 122(8): 1497–1502. doi:10.1007/s00122-011-1549-0.

Li, T., Bai, G., Wu, S. and Gu, S. (2012). Quantitative trait loci for resistance to Fusarium head blight in the Chinese wheat landrace Huangfangzhu. *Euphytica* 185(1): 93-102. doi:10.1007/s10681-012-0631-2.

Li, X., Michlmayr, H., Schweiger, W., Malachova, A., Shin, S., Huang, Y., Dong, Y., Wiesenberger, G., McCormick, S., Lemmens, M., Fruhmann, P., Hametner, C., Berthiller, F., Adam, G. and Muehlbauer, G. J. (2017). A barley UDP-glucosyltransferase inactivates nivalenol and provides Fusarium head blight resistance in transgenic wheat. *J. Exp. Bot.* 68(9): 2187–2197.

Li, X., Shin, S., Heinen, S., Dill-Macky, R., Berthiller, F., Nersesian, N., Clemente, T., McCormick, S. and Muehlbauer, G. J. (2015). Transgenic wheat expressing a barley UDP-glucosyltransferase detoxifies deoxynivalenol and provides high levels of resistance to *Fusarium graminearum*. *Mol. Plant Microbe. Interact.* 28(11): 1237–1246.

Lin, F., Kong, Z. X., Zhu, H. L., Xue, S. L., Wu, J. Z., Tian, D. G., Wei, J. B., Zhang, C. Q. and Ma, Z. Q. (2004). Mapping QTL associated with resistance to Fusarium head blight in the Nanda2419 × Wangshuibai population. I. Type II resistance. *Theor. Appl. Genet.* 109(7): 1504–1511. doi:10.1007/s00122-004-1772-z.

Lin, F., Xue, S. L., Zhang, Z. Z., Zhang, C. Q., Kong, Z. X., Yao, G. Q., Tian, D. G., Zhu, H. L., Li, C. J., Cao, Y., Wei, J. B., Luo, Q. Y. and Ma, Z. Q. (2006). Mapping QTL associated with resistance to Fusarium head blight in the Nanda2419 × Wangshuibai population. II: Type I resistance. *Theor. Appl. Genet.* 112(3): 528–535. doi:10.1007/s00122-005-0156-3.

Liu, S., Abate, Z. A., Lu, H., Musket, T., Davis, G. L. and McKendry, A. L. (2007). QTL associated with Fusarium head blight resistance in the soft red winter wheat Ernie. *Theor. Appl. Genet.* 115(3): 417–427. doi:10.1007/s00122-007-0577-2.

Liu, S., Abate, Z. A. and McKendry, A. L. (2005). Inheritance of Fusarium head blight resistance in the soft red winter wheat Ernie. *Theor. Appl. Genet.* 110(3): 454–461.

Liu, S. and Anderson, J. A. (2003a). Marker assisted evaluation of Fusarium head blight resistant wheat germplasm. *Crop Sci.* 43(3): 760–766, doi:10.2135/cropsci2003.7600

Liu, S. and Anderson, J. A. (2003b). Targeted molecular mapping of a major wheat QTL for Fusarium head blight resistance using wheat ESTs and synteny with rice. *Genome* 46(5): 817–823. doi:10.1139/g03-066.

Liu, S., Christopher, M. D., Griffey, C. A., Hall, M. D., Gundrum, P. G. and Brooks, W. S. (2012). Molecular characterization of resistance to Fusarium head blight in U.S. soft red winter wheat breeding line VA00W-38. *Crop Sci.* 52(5): 2283–2292.

Liu, S., Griffey, C. A., Hall, M. D., McKendry, A. L., Chen, J., Brooks, W. S., Brown-Guedira, G., Van Sanford, D. and Schmale, D. G. (2013). Molecular characterization of field resistance to Fusarium head blight in two U.S. soft red winter wheat cultivars. *Theor. Appl. Genet.* 126(10): 2485–2498.

Liu, S., Hall, M. D., Griffey, C. A. and McKendry, A. L. (2009). Meta-analysis of QTL associated with Fusarium head blight resistance in wheat. *Crop Sci.* 49(6): 1955–1968.

Liu, S., Pumphrey, M. O., Gill, B. S., Trick, H. N., Zhang, J. X., Dolezel, J., Chalhoub, B. and Anderson, J. (2008). Toward positional cloning of fhb1, a major QTL for Fusarium head blight resistance in wheat. *Cereal Res. Comm. Suppl.* 36(Supplement 6): 195–201.

Lu, Q., Lillemo, M., Skinnes, H., He, X., Shi, J., Ji, F., Dong, Y. and Bjørnstad, A. (2013). Anther extrusion and plant height are associated with Type I resistance to Fusarium head blight in bread wheat line 'Shanghai-3/Catbird'. *Theor. Appl. Genet.* 126(2): 317–334. doi:10.1007/s00122-012-1981-9.

Lu, Q., Szabo-Hever, A., Bjørnstad, Å., Lillemo, M., Semagn, K., Mesterhazy, A., Ji, F., Shi, J. and Skinnes, H. (2011). Two major resistance quantitative trait loci are required to counteract the increased susceptibility to Fusarium head blight of the *Rht-D1b* dwarfng gene in wheat. *Crop Sci.* 51(6): 2430–2438. doi:10.2135/cropsci2010.12.0671.

Ma, H. X., Bai, G. H., Gill, B. S. and Hart, L. P. (2006a). Deletion of a chromosome arm altered wheat resistance to Fusarium head blight and deoxynivalenol accumulation in Chinese spring. *Plant Dis.* 90(12): 1545–1549. doi:10.1094/PD-90-1545.

Ma, H. X., Bai, G. H., Zhang, X. and Lu, W. Z. (2006b). Main effects, epistasis, and environmental interactions of quantitative trait loci for Fusarium head blight resistance in a recombinant inbred population. *Phytopathology* 96(5): 534–541. doi:10.1094/PHYTO-96-0534.

Ma, H. X., Zhang, K. M., Gao, L., Bai, G. H., Chen, H. G., Cai, Z. X. and Lu, W. Z. (2006c). Quantitative trait loci for resistance to Fusarium head blight and deoxynivalenol accumulation in Wangshuibai wheat under field conditions. *Plant Pathol.* 55(6): 739–745. doi:10.1111/j.1365-3059.2006.01447.x.

Ma, Z., Xie, Q., Li, G., Jia, H., Zhou, J., Kong, Z., Li, N. and Yuan, Y. (2020). Germplasms, genetics and genomics for better control of disastrous wheat Fusarium head blight. *Theor. Appl. Genet.* 133(5): 1541–1568 doi:10.1007/s00122-019-03525-8.

Ma, Z., Xue, S., Lin, F., Yang, S., Li, G., Tang, M., Kong, Z., Cao, Y., Zhao, D., Jia, H., Zhang, Z. and Zhang, L. (2008). Mapping and validation of scab resistance QTLs in the Nanda2419 × Wangshuibai population. *Cereal Res. Com.* 36(Supplement 6): 245–251 doi:10.1556/CRC.36.2008.Suppl.B.22.

Mackintosh, C. A., Lewis, J., Radmer, L. E., Shin, S., Heinen, S. J., Smith, L. A., Wyckoff, M. N., Dill-Macky, R., Evans, C. K., Kravchenko, S., Baldridge, G. D., Zeyen, R. J. and Muehlbauer, G. J. (2007). Overexpression of defense response genes in transgenic wheat enhances resistance to Fusarium head blight. *Plant Cell Rep.* 26(4): 479–488.

Malihipour, A., Gilbert, J., Fedak, G., Brûlé-Babel, A. and Cao, W. (2017). Mapping the A genome for QTL conditioning resistance to Fusarium head blight in a wheat population with *Triticum timopheevii* background. *Plant Dis.* 101(1): 11–19. doi:10.1094/PDIS-02-16-0144-RE.

Mao, S. L., Wei, Y. M., Cao, W., Lan, X. J., Yu, M., Chen, Z., Chen, G. and Zheng, Y. (2010). Confirmation of the relationship between plant height and Fusarium head blight resistance in wheat (*Triticum aestivum* L.) by QTL meta-analysis. *Euphytica* 174(3): 343–356.

Mardi, M., Buerstmayr, H., Ghareyazie, B., Lemmens, M., Mohammadi, S. A., Nolz, R. and Ruckenbauer, P. (2005). QTL analysis of resistance to Fusarium head blight in wheat using a 'Wangshuibai'-derived population. *Plant Breed.* 124(4): 329–333.

Mardi, M., Pazouki, L., Delavar, H., Kazemi, M. B., Ghareyazie, B., Steiner, B., Nolz, R., Lemmens, M. and Buerstmayr, H. (2006). QTL analysis of resistance to Fusarium head blight in wheat using a 'Frontana'-derived population. *Plant Breed.* 125(4): 313–317. doi:10.1111/j.1439-0523.2006.01228.x.

Martin, L., Chris-Carolin, S., and Thomas, M. (2009),. Revealing the genetic architecture of FHB resistance in hexaploid wheat (*Triticum aestivum* L.) by QTL meta-analysis. *Mol. Breed.* 23(3): 473–488.

McCartney, C. A., Brûlé-Babel, A. L., Fedak, G., Martin, R. A., McCallum, B. D., Gilbert, J., Hiebert, C. W. and Pozniak, C. J. (2016). Fusarium head blight resistance QTL in the spring wheat cross Kenyon/86ISMN 2137. *Front. Microbiol.* 7: 1542. doi:10.3389/fmicb.2016.01542.

McMullen, M., Bergstrom, G., De Wolf, E., Dill-Macky, R., Hershman, D., Shaner, G. and Van Sanford, D. (2012). A unified effort to fight an enemy of wheat and barley: Fusarium head blight. *Plant Dis.* 96(12): 1712–1728.

Mergoum, M., Frohberg, R. C. and Stack, R. W. (2007). Breeding hard red spring wheat for Fusarium head blight resistance, successes and challenges. In: *Wheat Production in Stressed Environments. Proceedings of the 7th International Wheat Conference* Buck, H. T., Nisi, J. E., Salomo´n, N. (Eds), November 27–December 2, 2005. Mar del Plata, Argentina. Springer, Netherlands. pp. 161–167.

Mesterhazy, A. (1995). Types and components of resistance to Fusarium head blight of wheat. *Plant Breed.* 114(5): 377–386.

Mesterhazy, Á, Bartok, T., Mirocha, C. G. and Komoroczy, R. (1999). Nature of wheat resistance to Fusarium head blight and the role of deoxynivalenol for breeding. *Plant Breed.* 118(2): 97–110.

Miedaner, T., Risser, P., Paillard, S., Schnurbusch, T., Keller, B., Hartl, L., Holzapfel, J., Korzun, V., Ebmeyer, E. and Utz, H. F. (2012). Broad-spectrum resistance loci for three quantitatively inherited diseases in two winter wheat populations. *Mol. Breed.* 29(3): 731–742 doi:10.1007/s11032-011-9586-6.

Miller, J. D. and Arnison, P. G. (1986). Degradation of deoxynivalenol by suspension cultures of the Fusarium Head blight resistant wheat cultivar Frontana. *Can. J. Plant Pathol.* 8(2): 147–150. doi:10.1080/07060668609501818.

Miller, J. D., Young, J. C. and Sampson, D. R. (1985). Deoxynivalenol and Fusarium head blight resistance in spring cereals. *Phytopathol. Z.* 113(4): 359–367.

Mirocha, C. J., Kolaczkowski, E., Xie, W., Yu, H. and Jelen, H. (1998). Analysis of deoxynivalenol and its derivatives (batch and single kernel) using gas chromatography/mass spectrometry. *J. Agric. Food. Chem.* 46: 1414–1418.

Naji, A. M., Moghaddam, M., Ghaffari, M. R., PourIrandoost, H., Farsad, L. K., Pirseyedi, S. M., Mohammadi, S. A., Ghareyazie, B. and Mardi, M. (2008). Validation of EST-derived STS markers localized on *Qfhs.ndsu-3BS* for Fusarium head blight resistance in wheat using a Wangshuibai derived population. *J. Genet. Gen* 35: 625–629.

Nganje, W. E., Kaitibie, S., Wilson, W. W., Leistritz, F. L. and Bangsund, D. A. (2004) Economic impacts of Fusarium head blight in wheat and Barley: 1993–2001. *Agribus Appl Econ Rep.* No. 538: 1–53.

Nisikado, Y. (1959). Studies on the wheat scab, caused by *Gibberella zeae* (Schw.) Petch, and its control. Report of the Ohara Institute for Agricultural Biology, Okayama University, pp. 141–165.

Otto, C. D., Kianian, S. F., Elias, E. M., Stack, R. W. and Joppa, L. R. (2002). Genetic dissection of a major Fusarium head blight QTL in tetraploid wheat. *Plant Mol. Biol.* 48(5–6): 625–632. doi:10.1023/A:1014821929830.

Paillard, S., Schnurbusch, T., Tiwari, R., Messmer, M., Winzeler, M., Keller, B. and Schachermayr, G. (2004). QTL analysis of resistance to Fusarium head blight in Swiss winter wheat (Triticum aestivum L.). *Theor. Appl. Genet.* 109(2): 323–332. doi:10.1007/s00122-004-1628-6.

Pavan, S., Jacobsen, E., Visser, R. G. F. and Bai, Y. (2010). Loss of susceptibility as a novel breeding strategy for durable and broad-spectrum resistance. *Mol. Breed.* 25(1): 1–12. doi:10.1007/s11032-009-9323-6.

Peiris, K. H. S., Pumphrey, M. O., Dong, Y., Maghirang, E. B., Berzonsky, W. and Dowell, F. E. (2010). Near-infrared spectroscopic method for identification of Fusarium head blight damage and prediction of deoxynivalenol in single wheat kernels. *Cereal Chem.* 87(6): 511–517.

Petersen, S., Lyerly, J. H., Maloney, P. V., Brown-Guedira, G., Cowger, C., Costa, J. M., Dong, Y. and Murphy, J. P. (2016). Mapping of Fusarium head blight resistance quantitative trait loci in winter wheat cultivar NC-Neuse. *Crop Sci.* 56(4): 1473–1483. doi: 10.2135/cropsci2015.05.0312.

Petersen, S., Lyerly, J. H., McKendry, A. L., Islam, M. S., Brown-Guedira, G., Cowger, C., Dong, Y. and Murphy, J. P. (2017). Validation of Fusarium head blight resistance QTL in US winter wheat. *Crop Sci.* 57(1): 1–12. doi: 10.2135/cropsci2015.07.0415.

Pirseyedi, S. M., Kumar, A., Ghavami, F., Hegstad, J. B., Mergoum, M., Mazaheri, M., Kianian, S. F. and Elias, E. M. (2019). Mapping QTL for Fusarium head blight resistance in a Tunisian-derived durum wheat population. *Cereal Res. Com.* 47(1): 78–87 doi:10.1556/0806.46.2018.053.

Prat, N., Guilbert, C., Prah, U., Wachter, E., Steiner, B., Langin, T., Robert, O. and Buerstmayr, H. (2017). QTL mapping of Fusarium head blight resistance in three related durum wheat populations. *Theor. Appl. Genet.* 130(1): 13–27. doi:10.1007/s00122-016-2785-0.

Pritsch, C., Muehlbauer, G. J., Bushnell, W. R., Somers, D. A. and Vancel, C. P. (2000). Fungal development and induction of defense response genes during early infection of wheat spikes by *Fusarium graminearum*. *Mol. Plant Microbe. Interact.* 13(2): 159–169.

Qi, L. L., Pumphrey, M. O., Friebe, B., Chen, P. D. and Gill, B. S. (2008). Molecular cytogenetic characterization of alien introgressions with gene *Fhb3* for resistance to Fusarium head blight disease of wheat. *Theor. Appl. Genet.* 117(7): 1155–1166.

Rawat, N., Pumphrey, M. O., Liu, S., Zhang, X., Tiwari, V. K., Ando, K., Trick, H. N., Bockus, W. W., Akhunov, E., Anderson, J. A. and Gill, B. S. (2016). Wheat *Fhb1* encodes a chimeric lectin with agglutinin domains and a pore-forming toxin-like domain conferring resistance to Fusarium head blight. *Nat. Genet.* 48(12): 1576–1580.

Ren, J., Wang, Z., Du, Z., Che, M., Zhang, Y., Quan, W., Wang, Y., Jiang, X. and Zhang, Z. (2019). Detection and validation of a novel major QTL for resistance to Fusarium head blight from *Triticum aestivum* in the terminal region of chromosome 7DL. *Theor. Appl. Genet.* 132(1): 241–255. doi:10.1007/s00122-018-3213-4.

Ren, R., Foulkes, J., Mayes, S., Yang, X. and Ray, R. V. (2016). Identification of novel quantitative trait loci for resistance to Fusarium seedling blight caused by *Microdochium majus* and *M. nivale* in wheat. *Field Crops Res.* 191: 1–12. doi:10.1016/j.fcr.2016.03.011.

Ribichich, K. F., Lopez, S. E. and Vegetti, A. C. (2000). Histopathological spikelet changes produced by *Fusarium graminearum* in susceptible and resistant wheat cultivars. *Plant Dis.* 84(7): 794–802.

Ruan, Y., Comeau, A., Langevin, F., Hucl, P., Clarke, J. M., Brule-Babel, A. and Pozniak, C. J. (2012). Identification of novel QTL for resistance to Fusarium head blight in a tetraploid wheat population. *Genome* 55(12): 853–864. doi:10.1139/gen-2012-0110.

Rudd, J. C., Horsley, R. D., McKendry, A. L. and Elias, E. M. (2001). Host plant resistance genes for fusarium head blight: sources, mechanisms, and utility in conventional breeding systems. *Crop Sci.* 41(3): 620–627.

Rutkoski, J., Benson, J., Jia, Y., Brown-Guedira, G., Jannink, J. and Sorrells, M. (2012). Evaluation of genomic prediction methods for Fusarium head blight resistance in wheat. *Plant Gen.* 5(2): 51–61. doi:10.3835/plantgenome2012.02.0001.

Sari, E., Berraies, S., Knox, R. E., Singh, A. K., Ruan, Y., Cuthbert, R. D., Pozniak, C. J., Henriquez, M. A., Kumar, S., Burt, A. J., N'Diaye, A., Konkin, D. J., Cabral, A. L., Campbell, H. L., Wiebe, K., Condie, J., Lokuruge, P., Meyer, B., Fedak, G., Clarke, F. R., Clarke, J. M., Somers, D. J. and Fobert, P. R. (2018). High density genetic mapping of Fusarium head blight resistance QTL in tetraploid wheat. *PLoS ONE* 13(10): e0204362. doi:10.1371/journal.pone.0204362.

Saville, R. J., Gosman, N., Burt, C. J., Makepeace, J., Steed, A., Corbitt, M., Chandler, E., Brown, J. K., Boulton, M. I. and Nicholson, P. (2012). The 'Green Revolution' dwarfing genes play a role in disease resistance in *Triticum aestivum* and *Hordeum vulgare*. *J. Exp. Bot.* 63(3): 1271–1283. doi:10.1093/jxb/err350.

Schmolke, M., Zimmermann, G., Buerstmayr, H., Schweizer, G., Miedaner, T., Korzun, V., Ebmeyer, E. and Hartl, L. (2005). Molecular mapping of Fusarium head blight resistance in the winter wheat population Dream/Lynx. *Theor. Appl. Genet.* 111(4): 747–756. doi:10.1007/s00122-005-2060-2.

Schroeder, H. and Christensen, J. (1963). Factors affecting resistance of wheat to scab caused by *Gibberella zeae*. *Phytopathology* 53: 831–838.

Schweiger, W., Steiner, B., Vautrin, S., Nussbaumer, T., Siegwart, G., Zamini, M., Jungreithmeier, F., Gratl, V., Lemmens, M., Mayer, K. F., Bérgès, H., Adam, G. and Buerstmayr, H. (2016). Suppressed recombination and unique candidate genes in the divergent haplotype encoding *Fhb1*, a major Fusarium head blight resistance locus in wheat. *Theor. Appl. Genet.* 129(8): 1607–1623.

Semagn, K., Skinnes, H., Bjørnstad, Å., Marøy, A. G. and Tarkegne, Y. (2007). Quantitative trait loci controlling Fusarium head blight resistance and low deoxynivalenol content in hexaploid wheat population from Arina and NK93604. *Crop Sci.* 47(1): 294–303. doi:10.2135/cropsci2006.02.0095.

Shah, L., Si, H., Wang, S., Zhu, Y., Jiang, H., Cao, J., Ali, A. and Ma, C. (2017). Quantitative trait loci associated to Fusarium head blight resistance in wheat populations containing Annong-1124 and Zhoumai-27. *Physiol. Mol. Plant Pathol.* 100: 67–74.

Shen, X., Ittu, M. and Ohm, H. W. (2003a). Quantitative trait loci conditioning resistance to Fusarium head blight in wheat line F201R. *Crop Sci.* 43(3): 850-857. doi:10.2135/cropsci2003.8500.

Shen, X., Zhou, M., Lu, W. and Ohm, H. (2003b). Detection of Fusarium head blight resistance QTL in a wheat population using bulked segregant analysis. *Theor. Appl. Genet.* 106(6): 1041-1047. doi:10.1007/s00122-002-1133-8.

Skinnes, H., Semagn, K., Tarkegne, Y., Marøy, A. G. and Bjørnstad, Å. (2009). The inheritance of anther extrusion in hexaploid wheat and its relationship to Fusarium head blight resistance and deoxynivalenol content. *Plant Breed.* 129(2): 149–155. doi:10.1111/j.1439-0523.2009.01731.x.

Smith, W. G. (1884). *Diseases of Field and Garden Crops*. MacMillan and Co., London.

Sneller, C., Paul, P., Herald, L. and Sugerman, B. (2012). Report on the 2011-2012 Northern uniform winter wheat scab nurseries (NUWWSN and PNUWWSN). In *Proceedings of the 2012 National Fusarium Head Blight Forum*. Edited by Canty, S., Clark, A., Anderson-Scully, A. and Van Sanford, D. US Wheat and Barley Scab Initiative. pp. 100-104.

Somers, D. J., Fedak, G., Clarke, J. and Cao, W. (2006). Mapping of FHB resistance QTLs in tetraploid wheat. *Genome* 49(12): 1586–1593. doi:10.1139/g06-127.

Somers, D. J., Fedak, G. and Savard, M. (2003). Molecular mapping of novel genes controlling Fusarium head blight resistance and deoxynivalenol accumulation in spring wheat. *Genome* 46(4): 555–564. doi:10.1139/g03-033.

Srinivasachary, S., Gosman, N., Steed, A., Faure, S., Bayles, R., Jennings, P. and Nicholson, P. (2008a). Mapping of QTL associated with Fusarium head blight in spring wheat RL4137. *Czech J. Genet. Plant Breed.* 44(4): 147–159.

Srinivasachary, S., Gosman, N., Steed, A., Hollins, T. W., Bayles, R., Jennings, P. and Nicholson, P. (2009). Semi-dwarfing *Rht-B1* and *Rht-D1* loci of wheat differ significantly in their influence on resistance to Fusarium head blight. *Theor. Appl. Genet.* 118(4): 695-702. doi:10.1007/s00122-008-0930-0.

Srinivasachary, S., Gosman, N., Steed, A., Simmonds, J., Leverington-Waite, M., Wang, Y., Snape, J. and Nicholson, P. (2008b). Susceptibility to Fusarium head blight is associated with the Rht-D1b semi-dwarfing allele in wheat. *Theor. Appl. Genet.* 116(8): 1145-1153. doi:10.1007/s00122-008-0742-2.

Steiner, B., Buerstmayr, M., Wagner, C., Danler, A., Eshonkulov, B., Ehn, M. and Buerstmayr, H. (2019). Fine-mapping of the Fusarium head blight resistance QTL Qfhs.ifa-5A identifies two resistance QTL associated with anther extrusion. *Theor. Appl. Genet.* 132(7): 2039-2053. doi:10.1007/s00122-019-03336-x.

Steiner, B., Lemmens, M., Griesser, M., Scholz, U., Schondelmaier, J. and Buerstmayr, H. (2004). Molecular mapping of resistance to Fusarium head blight in the spring wheat cultivar Frontana. *Theor. Appl. Genet.* 109(1): 215–224. doi:10.1007/s00122-004-1620-1.

Su, Z., Bernardo, A., Tian, B., Chen, H., Wang, S., Ma, H., Cai, S., Liu, D., Zhang, D., Li, T., Trick, H., St Amand, P., Yu, J., Zhang, Z. and Bai, G. (2019). A deletion mutation in TaHRC confers *Fhb1* resistance to Fusarium head blight in wheat. *Nat. Gen.* 51(7): 1099–1105. doi:10.1038/s41588-019-0425-8.

Suzuki, T., Sato, M. and Takeuchi, T. (2012). Evaluation of the effects of five QTL regions on Fusarium head blight resistance and agronomic traits in spring wheat (I L.). *Breed. Sci.* 62(1): 11–17 doi:10.1270/jsbbs.62.11.

Szabó-Hevér, Á, Lehoczki-Krsjak, S., Tóth, B., Purnhauser, L., Buerstmayr, H., Steiner, B. and Mesterházy, Á. (2012). Identification and validation of Fusarium head blight and Fusarium-damaged kernel QTL in a Frontana/Remus DH mapping population. *Can. J. Plant Pathol.* 34(2): 224-238. doi:10.1080/07060661.2012.676571.

Tamburic-Ilincic, L. and Rosa, S. B. (2017). Alleles on the two dwarfing loci on 4B and 4D are main drivers of FHB-related traits in the Canadian winter wheat population "Vienna" × *"25R47". Plant Breed.* 136: 799-808. doi:10.1111/pbr.12527.

Tamburic-Ilincic, L. and Rosa, S. B. (2019). QTL mapping of Fusarium head blight and Septoria tritici blotch in an elite hard red winter wheat population. *Mol. Breed.* 39(7): 94 doi:10.1007/s11032-019-0999-y.

Venske, E., dos Santos, R. S., Farias, D. D. R., Rother, V., da Maia, L. C., Pegoraro, C. and Costa de Oliveira, A. (2019). Meta-Analysis of the QTLome of Fusarium head blight resistance in bread wheat: refining the current puzzle. *Front. Plant Sci.* 10: 727. doi:10.3389/fpls.2019.00727.

Waldron, B. L., Moreno-Sevilla, B., Anderson, J. A., Stack, R. W. and Frohberg, R. C. (1999). RFLP mapping of QTL for Fusarium head blight resistance in wheat. *Crop Sci.* 39(3): 805–811.

Wang, H., Sun, S., Ge, W., Zhao, L., Hou, B., Wang, K., Lyu, Z., Chen, L., Xu, S., Guo, J., Li, M., Su, P., Li, X., Wang, G., Bo, C., Fang, X., Zhuang, W., Cheng, X., Wu, J., Dong, L., Chen, W., Li, W., Xiao, G., Zhao, J., Hao, Y., Xu, Y., Gao, Y., Liu, W., Liu, H., Yin, H., Li, J., Li, X., Zhao, Y., Wang, X., Ni, F., Ma, X., Li, A., Xu, S. S., Bai, G., Nevo, E., Gao, C., Ohm, H. and Kong, L. (2020). *Fhb7* Horizontal gene transfer from fungus underlies head blight resistance in wheat. *Science* 368: eaba5435, doi:10.1126/science.aba5435.

Wang, Y. Z., Yang, X. N. and Xiao, Q. P. (1982). The improvement of identification technique of scab resistance of wheat and the development of resistant sources. *Sci. Agric. Sin.* 5: 67–77.

Wiese, M. V. (1987). Scab (head blight). In: Wise, M. V. (Ed.) *Compendium of Wheat Disease* (2nd edn.). APS Press, St. Paul, MN. 16–18.

Wilson, W. W., McKee, G., Nganje, W., Dahl, B. and Bangsund, D. (2017). Economic impact of USWBSI's Scab Initiative to reduce FHB. *Agribus. Appl. Econ.* 774. doi:10.22004/ag.econ.264672.

Xu, Y. G. and Fan, Z. D. (1985). Evaluation and differentiation of pathogenicity of *Fusarium graminearum* on wheat cultivars. *Acta Phytopathol. Sin.* 12: 53–57.

Xue, S., Li, G., Jia, H., Xu, F., Lin, F., Tang, M., Wang, Y., An, X., Xu, H., Zhang, L., Kong, Z. and Ma, Z. (2010a). Fine mapping *Fhb4*, a major QTL conditioning resistance to Fusarium infection in bread wheat (*Triticum aestivum* L.). *Theor. Appl. Genet.* 121(1): 147–156.

Xue, S., Li, G., Jia, H., Lin, F., Cao, Y., Xu, F., Tang, M., Wang, Y., Wu, X., Zhang, Z., Zhang, L., Kong, Z. and Ma, Z. (2010b). Marker-assisted development and evaluation of near-isogenic lines for scab resistance QTLs of wheat. *Mol. Breed.* 25(3): 397–405. doi:10.1007/s11032-009-9339-y.

Xue, S., Xu, F., Tang, M., Zhou, Y., Li, G., An, X., Lin, F., Xu, H., Jia, H., Zhang, L., Kong, Z. and Ma, Z. (2011). Precise mapping *Fhb5*, a major QTL conditioning resistance to Fusarium infection in bread wheat (*Triticum aestivum* L). *Theor. Appl. Genet.* 123(6): 1055–1063.

Yan, W., Li, H. B., Cai, S. B., Ma, H. X., Rebetzke, G. J. and Liu, C. J. (2011). Effects of plant height on type I and type II resistance to Fusarium head blight in wheat. *Plant Pathol.* 60(3): 506–512.

Yang, J., Bai, G. and Shaner, G. E. (2005a). Novel quantitative trait loci (QTL) for Fusarium head blight resistance in wheat cultivar Chokwang. *Theor. Appl. Genet.* 111(8): 1571–1579. doi:10.1007/s00122-005-0087-z.

Yang, Z., Gilbert, J., Fedak, G. and Somers, D. J. (2005b). Genetic characterization of QTL associated with resistance to Fusarium head blight in a doubled-haploid spring wheat population. *Genome* 48(2): 187–196. doi:10.1139/g04-104.

Yi, X., Cheng, J., Jiang, Z., Hu, W., Bie, T., Gao, D., Li, D., Wu, R., Li, Y., Chen, S., Cheng, X., Liu, J., Zhang, Y. and Cheng, S. (2018). Genetic analysis of Fusarium head blight resistance in CIMMYT bread wheat line C615 using traditional and conditional QTL mapping. *Front. Plant Sci.* 9: 573. doi:10.3389/fpls.2018.00573.

Yu, G. H., Ma, H. X., Bai, G. H. and Tang, K. X. (2008a). Single-strand conformational polymorphism markers associated with a major QTL for Fusarium head blight resistance in wheat. *Mol. Biol.* 42: 504–513.

Yu, J. B., Bai, G. H., Zhou, W. C., Dong, Y. H. and Kolb, F. L. (2008b). Quantitative trait loci for Fusarium head blight resistance in a recombinant inbred population of Wangshuibai/Wheaton. *Phytopathology* 98(1): 87–94. doi:10.1094/PHYTO-98-1-0087.

Zhang, G. and Mergoum, M. (2007). Molecular mapping of kernel shattering and its association with Fusarium head blight resistance in a Sumai3 derived population. *Theor. Appl. Genet.* 115(6): 757–766. doi:10.1007/s00122-007-0606-1.

Zhang, M., Zhang, R., Yang, J. and Luo, P. (2010). Identification of a new QTL for Fusarium head blight resistance in the wheat genotype "Wang shui-bai". *Mol. Biol. Rep.* 37(2): 1031–1035. doi:10.1007/s11033-009-9809-7.

Zhang, P., Guo, C., Liu, Z., Bernardo, A., Ma, H., Jiang, P., Song, G. and Bai, G. (2020). Quantitative trait loci for Fusarium head blight resistance in wheat cultivars Yangmai 158 and Zhengmai 9023. *Crop J.* doi:10.1016/j.cj.2020.05.007.

Zhang, Q., Axtman, J. E., Faris, J. D., Chao, S., Zhang, Z., Friesen, T. L., Zhong, S., Cai, X., Elias, E. M. and Xu, S. S. (2014). Identification and molecular mapping of quantitative trait loci for Fusarium head blight resistance in emmer and durum wheat using a single nucleotide polymorphism-based linkage map. *Mol. Breed.* 34(4): 1677–1687. doi:10.1007/s11032-014-0180-6.

Zhang, W., Francis, T., Gao, P., Boyle, K., Jiang, F., Eudes, F., Cuthbert, R., Sharpe, A. and Fobert, P. R. (2018). Genetic characterization of type II Fusarium head blight resistance derived from transgressive segregation in a cross between Eastern and Western Canadian spring wheat. *Mol. Breed.* 38(1): 13. doi:10.1007/s11032-017-0761-2.

Zhang, X., Bai, G., Bockus, W., Ji, X. and Pan, H. (2012a). Quantitative trait loci for Fusarium head blight resistance in U.S. hard winter wheat cultivar Heyne. *Crop Sci.* 52(3): 1187–1194. doi:10.2135/cropsci2011.08.0418.

Zhang, X., Pan, H. and Bai, G. (2012b). Quantitative trait loci responsible for Fusarium head blight resistance in Chinese landrace Baishanyuehuang. *Theor. Appl. Genet.* 125(3): 495–502. doi:10.1007/s00122-012-1848-0.

Zhang, X., Fu, J., Hiromasa, Y., Pan, H. and Bai, G. (2013). Differentially expressed proteins associated with Fusarium head blight resistance in wheat. *PLoS ONE* 8(12): e82079.

Zhang, X., Zhou, M., Ren, L., Bai, G., Ma, H., Scholten, O. E., Guo, P. and Lu, W. (2004). Molecular characterization of Fusarium head blight resistance from wheat variety Wangshuibai. *Euphytica* 139(1): 59–64. doi:10.1007/s10681-004-2298-9

Zhao, M., Leng, Y., Chao, S., Xu, S. S. and Zhong, S. (2018a). Molecular mapping of QTL for Fusarium head blight resistance introgressed into durum wheat. *Theor. Appl. Genet.* 131(9): 1939–1951. doi:10.1007/s00122-018-3124-4.

Zhao, M., Wang, G., Leng, Y., Wan, H., Xi, P., Grosz, M. D., Mergoum, M. and Zhong, S. (2018b). Molecular mapping of Fusarium Head Blight resistance in the spring wheat Line ND2710. *Phytopathology* 108(8): 972–979. doi:10.1094/PHYTO-12-17-0392-R.

Zhou, M. P., Hayden, M. J., Zhang, Z. Y., Lu, W. Z. and Ma, H. X. (2010). Saturation and mapping of a major Fusarium head blight resistance QTL on chromosome 3BS of Sumai 3 wheat. *J. Appl. Genet.* 51(1): 19–25. doi:10.1007/BF03195706.

Zhou, W., Kolb, F. L., Bai, G., Shaner, G. and Domier, L. L. (2002). Genetic analysis of scab resistance QTL in wheat with microsatellite and AFLP markers. *Genome* 45(4): 719–727. doi:10.1139/g02-034.

Zhu, X., Zhong, S., Chao, S., Gu, Y. Q., Kianian, S. F., Elias, E. and Cai, X. (2016a). Toward a better understanding of the genomic region harboring Fusarium head blight resistance QTL Qfhs.ndsu-3AS in durum wheat. *Theor. Appl. Genet.* 129(1): 31–43. doi:10.1007/s00122-015-2606-x.

Zhu, Z., Bonnett, D., Ellis, M., He, X., Heslot, N., Dreisigacker, S., Gao, C. and Singh, P. (2016b). Characterization of Fusarium head blight resistance in a CIMMYT synthetic-derived bread wheat line. *Euphytica* 208(2): 367–375. doi:10.1007/s10681-015-1612-z.

Zhu, Z., Chen, L., Zhang, W., Yang, L., Zhu, W., Li, J., Liu, Y., Tong, H., Fu, L., Liu, J., Rasheed, A., Xia, X., He, Z., Hao, Y. and Gao, C. (2020). Genome-wide association analysis of Fusarium head blight resistance in Chinese elite wheat lines. *Front. Plant Sci.* 11: 206. doi:10.3389/fpls.2020.00206.

Part 3

Fungal diseases of cereals: Septoria tritici blotch

Chapter 7

Advances in understanding the epidemiology of Septoria tritici blotch in cereals

Stephen B. Goodwin, USDA-ARS, USA

1 Introduction

The Septoria complex is an economically important disease that occurs in most wheat-growing regions every year. Losses can range up to 30-50% (Eyal et al., 1987) and can be higher in disease-favorable years. Three primary pathogens are associated with the Septoria complex: *Zymoseptoria tritici*, the cause of Septoria tritici blotch (STB); *Parastagonospora nodorum*, causing Stagonospora nodorum leaf and glume blotch (a.k.a. Septoria nodorum blotch); and *Parastagonospora avenae* f. sp. *tritici*, causing Stagonospora avenae blotch. Among these, STB is the most common and causes the largest problem in most parts of the world; *Parastagonospora nodorum* also is very important and will be discussed in other chapters. In addition to yield losses that reduce the quantity of grain, STB lowers test weight (pounds per bushel) and can negatively affect the quality of the grain produced (Eyal et al., 1987). Losses to STB are minor if the disease is confined to the lower leaves but increase dramatically if the flag and flag minus one leaf are affected.

http://dx.doi.org/10.19103/AS.2021.0092.09
Published by Burleigh Dodds Science Publishing Limited, 2022.

In Europe, average annual losses to STB with moderately resistant wheat cultivars and appropriate fungicide applications average 5-10% (Fones and Gurr, 2015), and in the United Kingdom are closer to 20% when susceptible cultivars are grown without chemical protection. By taking a midpoint of the average annual losses reported by Fones and Gurr (2015) for STB in France, Germany and the United Kingdom, we can calculate a conservative economic loss of 1.08 billion euros, or approximately 1.32 billion dollars, in lost revenue due to STB every year for just those three countries. Similar estimates are not available for the USA but the loss numbers can be much higher in some areas where there has been less emphasis on breeding resistant cultivars, and low profit margins can preclude fungicide use. The value of the United States wheat crop was estimated to be $9.3 billion during 2020 (Crop Values 2020 Summary (February 2021), USDA-National Agricultural Statistical Service). If we assume a modest 5% loss due to STB, the economic damage caused by this disease likely approximated $465 million in 2020 and similar losses to United States wheat growers probably occur every year.

In addition to the USA and Europe, STB causes major losses in other wheat-growing regions throughout the world (Goodwin, 2018), particularly in North Africa, Ethiopia and southern South America where it is often the most economically damaging disease of wheat (reviewed in Goodwin, 2018). The only major wheat-growing areas where STB appears to be rare or absent are in parts of Asia and Australia. The disease occurs only during very wet years in India, Pakistan and Afghanistan and otherwise is extremely rare, most likely because the climate generally is too hot and dry (Saari and Wilcoxson, 1974); presumably it persists during most years in the lower leaves without causing much disease or being noticed, but then explodes during epidemics when environmental conditions are favorable. A big mystery is why the disease seems to be rare or absent in China (Goodwin, 2018), one of the biggest wheat-producing countries in the world. China has a temperate climate that should be favorable for STB development and Chinese wheat cultivars seem to be susceptible, yet the disease has been documented there only rarely, even though it is a major problem in some adjoining countries such as Kazakhstan and Russia.

The huge economic losses inflicted by *Z. tritici* on wheat production have stimulated a large global research effort. This research has significantly expanded our understanding of the biology, genetics and interaction of this pathogen with its wheat host. Availability of a completely sequenced genome (Goodwin et al., 2011) has greatly facilitated these efforts, although much recent progress has come from microscopy and other approaches that do not rely on molecular technologies. Of course, there remains much to be learned, but some of the recent changes in our understanding of the biology of this organism and in the epidemiology of the disease it causes will be discussed below.

Published by Burleigh Dodds Science Publishing Limited, 2022.

2 What's in a name: taxonomy of *Z. tritici*

Taxonomically the fungus *Z. tritici* is in the phylum Ascomycota, class Dothideomycetes, order Mycosphaerellales, family Mycosphaerellaceae. This family includes a huge number of plant pathogens plus fungi with saprobic or other lifestyles. Some related genera with large numbers of plant pathogens include *Cercospora*, a huge genus containing pathogens of diverse plants mostly characterized by production of the light-activated phytotoxin cercosporin, an example being *C. zeae-maydis* on maize; *Dothistroma*, such as the pine pathogen *D. septosporum*, which produces the toxin dothistromin; *Mycosphaerella*, one of which (*M. fragariae*) causes common spot on strawberry; *Pseudocercospora*, including the cause of black Sigatoka (or leaf streak) on banana, *P. fijiensis*; *Ramularia*, species of which cause leaf-spot diseases on a wide diversity of dicots and monocots including *R. collo-cygni* on barley; and *Septoria*, another large genus of leaf-spotting fungi including *S. lycopersici*, which almost anyone with a home garden will have seen as brownish/black spots on the leaves of their tomato plants. Therefore, evolutionarily the genus *Zymoseptoria* is in very good company within a family containing thousands of other plant pathogens.

The specific name of the fungus that causes the wheat disease Septoria tritici blotch has changed multiple times during the past 50 years. Prior to the advent of molecular methods for identifying species, mycologists used two sets of names: one for the anamorphs (asexual stages, with spores generated through mitosis) and another for teleomorphs (sexual stages, with spores the result of meiosis). During those dark times, there were often two legitimate names for the same species. The only way to link them would be to see both types of spores (sexual and asexual) produced by a single culture or to start a culture from one type of spore and see it produce the other. Some fungi then had three names: one for the asexual stage, another for the sexual stage and a third to unite both types (the holomorph, for a fungus, when both anamorph and teleomorph are known, usually the same as the teleomorph name). In plants, this would be the equivalent of having one set of legitimate, scientific names based only on leaf shape and structure, and another, completely independent nomenclature based on the flowers. Having multiple names for the same fungus was very confusing.

Mycologists clung to this archaic system long after sufficient evidence was available to unite the names into a single fungus, mainly because many scientists had devoted their careers to analyzing only one form or the other, and also because the names had meanings. For example, anamorphs often were named because they produced a certain spore type, and by seeing the name you could know what the fungus would look like; that information would be lost if anamorph names were subsumed into the teleomorphs. To those of us who were not trained as mycologists, this system seemed overly complicated and unnecessary once it became easy to associate anamorphs and teleomorphs based on DNA sequence data, but it was debated ferociously for many years. Fortunately, reason prevailed eventually and a proposal was put forward with rules to join anamorph and teleomorph into a single name for each fungus (Hawksworth et al., 2011). With DNA sequence data, the name of a fungus now can be placed in its phylogenetic framework and a single name assigned that hopefully will not change in the future. The changing nomenclatures caused confusion initially, but for anyone new to science they only learned the most recent name and probably never gave it a second thought.

This historical mess has impacted the name of the fungus that causes STB and has left the name of the disease as a relic of those times. Originally this fungus was known only in its asexual stage and was given the name *Septoria tritici* (Dezmazieres, 1842), indicating a fungus with septoria type of spores growing in leaves of wheat (called 'Triticorum cultorum' in the original description). This appellation persisted for the next 130 years and is how the name of the disease caused by this fungus came to be called Septoria tritici blotch. Septorias cause leaf spot diseases of many crops and share a number of characteristics (Verkley et al., 2013), one of which typically is long, thin, multicellular spores. No teleomorphs are known for most species and for many years they were assumed to be strictly asexual (Verkley et al., 2013), including *S. tritici*. This changed in 1972 when Sanderson (1972) discovered the teleomorph for this fungus in New Zealand and subsequently gave it the name *Mycosphaerella graminicola* (Sanderson, 1976). The genus *Mycosphaerella* is characterized by producing sexual structures (pseudothecia) that are buried within the fungal body or its host, containing a cavity with a pore or opening through which the spores are ejected. So by 1976 the fungus that causes Septoria tritici blotch had two valid scientific names: *Septoria tritici* for the asexual stage (and the name of the disease); and *Mycosphaerella graminicola* for the sexual stage. By the 1980s most scientists were switching to the teleomorph name as the holomorph for the fungus and most publications from then until 2011 referred to it as *M. graminicola*. However, changing the name of the disease after more than 100 years of widespread use was deemed likely to cause too much confusion among growers and crop scientists (and even many plant pathologists!), so it remained the same even though neither Septoria nor tritici were part of the scientific binomial commonly used by most academics to describe the causal organism.

Another problem with this situation was that the genus *Mycosphaerella* as defined was very broad and covered a large group of often unrelated organisms (Crous, 2009). Detailed analyses of multi-gene phylogenies and life-history traits revealed that the *Mycosphaerella* species pathogenic on grass hosts formed a separate evolutionary group that had a characteristic dimorphic lifestyle; in addition to the usual mycelial stage, these fungi also could produce yeast-like growth under particular conditions (Quaedvlieg et al., 2011). Due to these differences, and taxonomic rules that prevented other, earlier names from being used for this group, the new genus name *Zymoseptoria* was created for these related fungi growing as pathogens on grass hosts with septoria-like anamorphs and both mycelial and yeast-like growth forms (Quaedvlieg et al., 2011). The 'Zymo' part of the name refers to the yeast-like growth stage (it actually means fermentation but this is associated with yeast). Fortunately, the authors were able to retain the specific name *tritici* from the original by Dezmazieres way back in 1842, ultimately giving us *Zymoseptoria tritici* for the name of the pathogen that causes Septoria tritici blotch of wheat. With the new molecular-assisted taxonomies this name is likely to remain unchanged, although the need to associate this with previous names used in the literature may continue to cause some confusion in the future. There does not appear to be any interest in changing the name of the disease, so it likely will continue to be abbreviated STB rather than ZTB.

3 Life cycle of *Z. tritici*

The life cycle of *Z. tritici* was always thought to be relatively simple, but some wrinkles to the previous dogma have been added recently. This fungus is not an obligate pathogen and can survive without a living host, although its competitive abilities as a saprotroph are thought to be weak and it probably does not grow much outside of crop debris. During the host-free period (summer for winter wheat, winter for spring wheat) the fungus persists on host debris from the previous crop on or in the soil. In theory, it could survive on volunteer wheat or alternative hosts but its high host specificity and lack of volunteers after wheat crops make this unlikely. Ascospores seem to be released when leaves begin to dry after becoming wet and are ejected enough to be picked up by wind and carried to a wheat plant, where they germinate and grow toward stomata to penetrate the leaf. After colonizing the substomatal cavities (Duncan and Howard, 2000), the pathogen produces asexual spore structures called pycnidia. Under humid conditions, these exude masses of the long, thin, multicellular, septoria type of asexual spores (or conidia) in a gelatinous matrix that tends to bend in response to gravity. Under a microscope, these look like what you would see if you were to squeeze a tube of toothpaste straight up until the stream of paste bends down under its own weight. The matrix holds the spores together until they can be dispersed by rain splash to new leaves, where they can germinate to start another cycle of asexual reproduction. The extracellular matrix also may protect the spores from UV radiation and desiccation while outside of the host, although this does not appear to have been tested explicitly.

Sexual structures (pseudothecia) can form if individuals of opposite mating types meet in the same leaf. This species is strictly heterothallic, with mating types Mat-1 and Mat-2; one of each mating type is required for sexual reproduction. Like other Ascomycetes, genes for the two mating types of *Z. tritici* are idiomorphs - not just alleles (different forms of the same gene) but completely different genes at the same genetic locus, which can vary quite a bit in size. So far no homothallic (self-fertile, needing only one individual to produce sexual structures) isolates are known, although they do occur among other fungi in the same class (Yun et al., 1999, 2013).

Pseudothecia of *Z. tritici* are produced in dead or dying wheat leaves, typically, like the asexual pycnidia, in substomatal cavities. Within these structures, fusion of the two mating strains occurs followed by meiosis to generate four haploid products. Each of these products then undergoes one cycle of mitosis to form eight ascospores within a sac-like structure called an ascus. Depending on the growth stage of the host and environmental conditions, ascospores may be released immediately to reinfect the same crop, or they could remain dormant in the leaves to carry the fungus through the host-free period over summer or winter. Or both - these results are not mutually exclusive. Ascospores continue to be released following periods of high humidity (Cordo et al., 2017), implying that the same pseudothecia retain the capacity to produce spores over a long period of time. These pseudothecia continue to release ascospores, eventually infecting a new crop to begin the life cycle again.

Two recent discoveries have complicated this simplistic view of the *Z. tritici* life cycle. The first was the discovery that *Z. tritici* can produce chlamydospore-like cells (Francisco et al., 2019). Chlamydospores are known in other fungi, where they typically function as resting structures that facilitate survival during unfavorable environmental conditions, such as heat, cold or desiccation. They were not believed to be produced by *Z. tritici* or any of its relatives. Therefore, nobody was looking for them or expecting to find them. The chlamydospores of *Z. tritici* were more resistant to drought, cold or heat stresses relative to blastospores or pseudohyphae, and could germinate both on agar media or on the surface of wheat leaves (Francisco et al., 2019). Therefore, presumably they could survive and give rise to new infections even after extremely unfavorable environmental conditions, which would give this fungus a great survival advantage in nature by providing a long-lived source of inoculum. The discovery of these structures after so many years of believing they did not exist is a testament to perseverance and to reporting what was seen regardless of dogma. Part of the reason they were discovered is that multiple isolates of the pathogen were analyzed, one of which produced many more chlamydospores than the others and brought this phenomenon to the attention of scientists.

The second wrinkle to the previous ideas about the life cycle of this pathogen was the finding that it can ramify epiphytically (Fones et al., 2017) as blastospores (Francisco et al., 2019) for a period of time before penetration. Previously it was believed that the ascospores or pycnidiospores germinated on the leaf and that the germ tubes grew toward the stomata and penetrated, period. Now it is known that, in fact, they can proliferate on the leaf surface, in the form of yeast-like blastospores (Francisco et al., 2019) for up to 10 days (Fones et al., 2017) before penetration. Although the importance of this phenomenon in nature is not known, such epiphytic proliferation could provide additional propagules for rain splash to new plants, thus facilitating the spread of epidemics.

4 Mode of nutrition

Although *Z. tritici* is commonly thought to be a hemibiotroph, its exact modes of nutrition at various stages in its growth cycle remain murky. Biotrophic fungi obtain nutrients from living hosts. After penetrating a plant's cuticle and cell wall, these fungi establish an intimate relationship with host cells through specialized feeding structures called haustoria, fingerlike projections that press against but not through the host cell membranes. The host cells remain alive, but the haustoria allow the pathogen to extract nutrients for itself. The main body of the pathogen usually remains above the host epidermis. Typical biotrophic pathogens of wheat include those that cause powdery mildew (*Blumeria graminis* forma specialis *tritici*) and rust (several species in the Basidiomycetes genus *Puccinia*).

In contrast, necrotrophic fungi kill plant cells and then consume the released nutrients. They typically do not produce any kind of specialized feeding structures and the fungal bodies often are mostly or entirely within the killed host tissue. Killing of host cells usually is facilitated by producing some sort of toxin, which may be host specific (often limited only to certain genotypes of a host) or can be more general, affecting the host broadly, and sometimes multiple host species depending on the fungus. Some of the toxins function by stimulating apoptosis in the host to induce cell death (Laluk and Mengiste, 2010). Some necrotrophic pathogens of wheat include those that cause the diseases tan spot (*Pyrenophora tritici-repentis*), Stagonospora nodorum leaf and glume blotch (*Parastagonospora nodorum*), and sharp eyespot (*Rhizoctonia cerealis*) (Oliver et al., 2016; Zhu et al., 2017). The pathogen that causes Fusarium head blight or scab, *Fusarium graminearum*, previously was considered to be a necrotroph but now is thought to have a brief biotrophic phase before rapidly switching to necrotrophy, so currently is believed to be a facultative hemibiotroph (Trail, 2009; Zhang et al., 2012).

In contrast, *Z. tritici* has always been considered to be a hemibiotroph, although this has been questioned recently (Sánchez-Vallet et al., 2015). After entering the host through the stomata, *Z. tritici* colonizes the substomatal cavity and gradually spreads outwards (Duncan and Howard, 2000) but does not increase much in biomass. During this phase, it grows between the cells, but they are not penetrated nor are haustoria produced (Kema et al., 1996). Typically this phase does not induce any symptoms on the host. Because the cells are not killed this phase cannot be considered necrotrophic, yet without penetrating the cell walls or producing haustoria it is also different from what is done by other fungi that are biotrophic. How it obtains nutrients during this phase of growth is not known. Due to a reduced number of genes for carbohydrate degradation but a similar number of genes for proteases as endophytes it was proposed that *Z. tritici* may use protein catabolism during this phase of growth (Goodwin et al., 2011), while others have proposed that it may survive off of stored lipids (Rudd et al., 2015). Much more research is needed to fully understand how *Z. tritici* obtains nutrients during this phase of growth.

Eventually, host cells around the invasion point rapidly die and the fungus switches to necrotrophic growth. Exactly how or why this switch occurs is not known, although recent research suggests it might be triggered through quorum sensing (Francisco et al., 2019) and involve the production of toxins (Kettles and Kanyuka, 2016). Following this switch, the fungus grows rapidly (Keon et al., 2007) and begins to produce spore structures in the substomatal cavities. At this point, the fungus is feeding off of dead host cells so it is definitely a necrotroph. Because the mode of nutrition during the early, symptomless phase is not known, some have suggested that *Z. tritici* should be considered a latent necrotroph rather than a hemibiotroph (Sánchez-Vallet et al., 2015).

Another way to identify potential nutritional modes and lifestyles of fungi is to analyze the gene contents of sequenced genomes. This could be used to predict major lifestyle characteristics (e.g. plant pathogen versus saprotroph) (Haridas et al., 2020), or to examine specific modes of nutrition for plant pathogens (Hane et al., 2020). Hane et al. (2020) consider the commonly used trichotomy of pathogenic fungal nutritional modes into biotrophs, necrotrophs and hemibiotrophs to be too simplistic to adequately describe the full range of the observed diversity. Based solely on an analysis of carbohydrate-active enzymes (CAZymes) within genome sequences, they divide fungal nutritional modes into five main classes, two of which have two subclasses giving seven in total. These categories are (1) Saprotroph (non-pathogens); (2) Monomertroph (consisting of biotrophs plus symbionts); (3) Mesotroph - extracellular and (4) Mesotroph - intracellular (encompassing most of the former hemibiotrophs); (5) Vasculartroph (some hemibiotrophs plus wilts, anthracnoses and rots); and (6) Narrow or (7) Broad-host-range polymertrophs (for the necrotrophs) (Hane et al., 2020). Based on the CAZyme composition of its genome, Hane et al. (2020) predicted that *Z. tritici* should have a major class of Mesotroph, with subclasses in Monomertroph, and Mesotroph - extracellular. In other words,

Published by Burleigh Dodds Science Publishing Limited, 2022.

a hemibiotroph. How this will play out in the future remains to be seen, but it definitely highlights the odd biology and difficult classification of *Z. tritici*.

5 Epidemiology

Epidemics of STB are initiated soon after seeds germinate in fall or spring by ascospores produced on crop residues remaining from the previous growing season, and possibly also by pycnidiospores. Ascospores have an advantage because they are forcibly ejected from the asci. How far the ascospores of *Z. tritici* are ejected does not seem to have been measured but it must be up to approximately 8 mm at least and could be much farther. One way to collect ascospores is to place leaf pieces on the tops of inverted Petri plates and wait for them to be ejected onto the agar surface above. If the inside distance between top and bottom of the plates is 12 mm, the layer of agar is 3-4 mm deep, and the leaf pieces are 1-3 mm from the inverted tops depending on their degree of curl, then the ascospores must be ejected at least 5-8 mm to reach the agar surface. Placing agar at increasing heights over the leaf pieces would be a good way to measure the ejection height of ascospores produced by *Z. tritici*. The vertical distance spores can be discharged in some other Ascomycetes ranges from 10 mm to 450 mm (Ingold, 1933), and they leave the asci at a very high speed (Trail, 2007). Although the speed and height of ascospore discharge are not known for *Z. tritici*, they should be sufficient for the spores to clear the substrate and be picked up by the prevailing air currents.

The ejected ascospores will be carried by the wind until they are deposited onto a hard surface. If that surface is a wheat leaf, the spores can germinate and send out one or two germ tubes to infect and form primary lesions on the new seedlings. These lesions mature to produce pycnidia that contain the asexual pycnidiospores. During the next few weeks, it is not clear how much of the spread of new disease is due to rain splash of pycnidiospores from existing lesions versus ascospores that continue to be produced from stubble of the previous season's crop. What is known is that, for fall-planted wheat in the northern hemisphere, lesions can be found on developing wheat seedlings by late November or December (personal observations).

Once the weather warms up in spring, the overwintered lesions can produce pycnidiospores to spread the epidemic to new, uninfected wheat seedlings. Unlike the forcibly ejected ascospores, the asexual pycnidiospores ooze from the pycnidia in gelatinous masses that are ideally suited for dispersal by rain splash. These spores help the fungus spread from lower to higher leaves as the crop grows (Shaw, 1987) and also can be spread to new plants. Dispersal distance is not known for certain but decreases exponentially above 5-10 cm (Shaw, 1987). Horizontal dispersal distances could be much higher, especially during storms. Pycnidia can form in new lesions after about three weeks leading to multiple cycles of infection during the course of an epidemic. Ascospores also can be released during the growing season but the relative contributions of both spore types to disease spread are not known for certain.

As leaves senesce at the end of the growing season, pycnidia and especially pseudothecia can survive to produce inoculum for the next season's epidemic. An unknown is what effect chlamydospores might have on epidemics. They are not involved in dispersal, except possibly by flowing water or in soil or plant material moved through human activities, and probably do not affect the pace or development of epidemics. However, they could provide a long-lived source of inoculum to survive droughts or other long host-free periods, such as during crop rotations. It is interesting that for some countries where STB generally is rare or non-existent, it can reappear quickly once a more mesic, favorable environment returns, such as in India and Australia (reviewed in Goodwin, 2018). For example, during the Millennium drought from 2001 to 2009 in Australia, STB essentially disappeared, yet quickly became common again once the rains returned (McDonald et al., 2016). How the pathogen survived this long dry period is not known, but chlamydospores (Francisco et al., 2019) would certainly provide the potential for long-term survival. It would be interesting to test populations of *Z. tritici* from India, Australia and other locations with hot, dry climates plagued by periodic droughts to test whether they produce more chlamydospores compared to those from locations that do not experience such extreme meteorological conditions.

Other unknowns about their role in epidemiology include epiphytic growth and light. The pathogen certainly can ramify on the leaf surface during protected conditions in a laboratory (Fones et al., 2017), but whether this has any practical effects on the speed or development of epidemics is not known. In many areas, storms are followed by rapid drying and a return to high levels of solar radiation. Under these conditions, epiphytic growth would be extremely difficult for the pathogen and its role probably is negligible. However, during prolonged periods of damp, cloudy weather it certainly might be possible for the fungus to produce more inoculum that could spread to additional leaves even before it has penetrated at the original location. The fungus also senses and responds to light (McCorison and Goodwin, 2020), so different intensities of solar radiation possibly could explain regional differences in epidemic development.

A final factor affecting the development of epidemics could be the recent discovery that the fungus can still penetrate the host, grow and even mate during incompatible interactions (Kema et al., 2018). Usually, we assume that a resistance gene will stop incompatible individuals of the pathogen from infecting. They die and that is the end of their story. However, Kema et al. (2018) showed conclusively that, not only can a fungal individual penetrate an incompatible host, but it can mate and contribute its genes as ascospores produced during the next generation, which could be during the next infection cycle but also as part of the same epidemic. This is very concerning, because it will maintain a high effective population size for the pathogen, with its concomitant generation of new genetic variation through reassortment and mutation, even on cultivars containing resistance genes. Incompatible individuals only function as a male during mating (Kema et al., 2018) so variation in mitochondrial DNA will still be reduced, but this should allow a very high level of genetic variation in nuclear genes to be maintained within the population, despite the best efforts at control through resistance. The true effect of this biological phenomenon on the epidemiology of *Z. tritici* remains to be tested, but could be very high where resistant cultivars are deployed widely,

6 Diagnostics

Diagnosing STB by itself is usually straightforward and presents few problems. The lesions usually are somewhat vein delimited so tend to be long and thin with black pycnidia and/or pseudothecia spread out according to the placement of stomata on the leaf since these structures are formed in the substomatal cavities. Unfortunately, disease epidemics rarely occur in isolation, and correctly diagnosing STB can become problematic when multiple pathogens are present on the same leaf. The most commonly confused disease is Stagonospora nodorum blotch (SNB), caused by *Parastagonospora nodorum*. Typically SNB lesions tend to be more lens shaped rather than delimited by the veins as for STB, but during severe epidemics, the lesions can merge making it difficult to discern their shape. Tan spot, caused by *Pyrenophora tritici-repentis*, also can look similar, but those symptoms usually are not vein delimited and typically have a yellow halo around the lesion, which is characteristic for that disease. Latent infections also cannot be identified and senescent leaves make it difficult to discern lesion boundaries to identify the cause of their demise.

To alleviate these problems molecular diagnostics can be applied. Most of these involve some sort of polymerase chain reaction (PCR)-based system (for example Fraaije et al., 1999) and can be excellent when multiple pathogens are present in the same field, which is a common occurrence in northwestern Europe (Justesen et al., 2021) and probably most other locations worldwide. Performing PCR requires some specialized equipment and technical knowledge. A simpler approach that may allow for greater dissemination of diagnostic capabilities is Loop-Mediated Isothermal Amplification (LAMP). This approach involves using multiple sets of primers that generate a loop structure that allows for very fast, efficient amplification without the need for the dissociation and re-annealing steps required for PCR (Iwamoto et al., 2003). Therefore, LAMP can be done at a single temperature - typically 65°C - and it is not as sensitive to contamination so the DNA template does not need to be as pure as with PCR. These advantages make LAMP easily portable to other locations and it can be deployed without the need for extensive training or expensive equipment. Unfortunately, so far there does not appear to be a LAMP protocol available for *Z. tritici*, but this approach holds much promise going forward.

7 Conclusion and future trends

Great progress has been made in our understanding of the biology of *Z. tritici* and the epidemiology of STB during the past few years, but of course much remains to be learned. We still do not know much about the pathogenic lifestyle of this fungus and how it survives during its latent, asymptomatic phase, what triggers the switch to necrotrophy and all of the stages involved. The roles of chlamydospores and epiphytic growth on leaf surfaces in maintaining or spreading epidemics are not known, nor do we understand the long-term effects of matings on incompatible hosts and what the concomitant genetic diversity means for disease management. STB clearly co-occurs with other pathogens but how they influence each other is not known. Availability of a LAMP protocol or other easily used diagnostic tool would help shed light on the occurrence of this disease worldwide and could help identify latent or asymptomatic infections to guide disease-management decisions. With the recent rapid pace of progress, many of these questions are likely to be addressed in the future.

8 Where to look for further information

Global interest in STB is high due to its prominence as one of the most important diseases of wheat worldwide. To address the need for rapid exchange of information, a series of conferences devoted to the cereal leaf blights but with a major focus on STB of wheat have been held almost every year in different international settings. The International Symposium on Cereal Leaf Blights (ISCLB) was last held in Dublin, Ireland from May 22 to 24, 2019. The next meeting was supposed to have been in Hammamet, Tunisia during May of 2020 and was moved to May 19-21, 2021 because of Covid-19. The second date also was scrapped for the same reason and the meetings now are scheduled to resume some time during 2022, Covid willing. Once they resume, those meetings gather the top scientists from around the world to present their latest research results. Previous incarnations of those meetings occurred every 5 years, but the intervals between meetings were shortened to 3 years and then to every year due to rapid progress and high interest. After they are held in Tunisia (hopefully during 2022 but by 2023 at the latest) then the meetings are scheduled to be moved to Denmark, with future locations and dates determined as needed by the organizing committee. Highlights of

Published by Burleigh Dodds Science Publishing Limited, 2022.

the 2019 meetings were published as a research topic in Frontiers (https://www.frontiersin.org/research-topics/10489/highlights-of-isclb2019---international-symposium-on-cereal-leaf-blights-2019).

Other sources of general information about STB can be found through the American Phytopathological Society (https://www.apsnet.org/edcenter/disandpath/fungalasco/pdlessons/Pages/Septoria.aspx) and at CIMMYT, the International Maize and Wheat Improvement Center. CIMMYT has hosted the Septoria meetings several times during the past and has published volumes of meeting proceedings that serve as a nice summary of the global research effort, the last of which was from 1999 (van Ginkel, M., A. McNab, and J. Krupinsky, eds, 1999. Septoria and Stagonospora Diseases of Cereals: A Compilation of Global Research. Proceedings of the Fifth International Septoria Workshop, September 1999, CIMMYT, Mexico). Many additional sources of information can be found online, including:

- In Victoria, Australia: https://agriculture.vic.gov.au/biosecurity/plant-diseases/grain-pulses-and-cereal-diseases/septoria-tritici-blotch-of-wheat.
- In private industry: https://www.pioneer.com/us/agronomy/Septoria-Tritii-Blotch.html.
- In California, USA: http://ipm.ucanr.edu/PMG/r730100111.html.
- In North Dakota, USA: https://www.ag.ndsu.edu/publications/crops/fungal-leaf-spot-diseases-of-wheat-tan-spot-septoria-stagonospora-nodorum-blotch-and-septoria-tritici-blotch.
- In the Pacific Northwest of the USA: https://pnwhandbooks.org/plantdisease/host-disease/wheat-triticum-aestivum-septoria-tritici-blotch.
- By the Grains Research & Development Corporation, Australia: https://grdc.com.au/__data/assets/pdf_file/0023/159341/final-septoria-tritici-blotch-fact-sheet-hr-pdf.pdf.pdf.
- From Kansas State University, USA: https://bookstore.ksre.ksu.edu/pubs/EP133.pdf.
- In the UK: https://ahdb.org.uk/knowledge-library/septoria-tritici-in-winter-wheat.

However, the best way to keep abreast of the latest research on STB will be to follow the ISCLB meetings once they resume their normal schedule.

9 References

Cordo, C. A., Mónaco, C. I., Altamirano, R., Perelló, A. E., Larrán, S., Kripelz, N. I. and Simón, M. R. 2017. Weather conditions associated with the release and dispersal of *Zymoseptoria tritici* spores in the Argentine Pampas region. *International Journal of Agronomy* 2017: 1–13. https://doi.org/10.1155/2017/1468580.

Crous, P. W. 2009. Taxonomy and phylogeny of the genus Mycosphaerella and its anamorphs. *Fungal Diversity* 38: 1–24.

Dezmazieres, J. 1842. Neuvième notice sur quelques plantes cryptogames, la plupart inèdites, récemment découvertes en France, et qui vont paraître en nature dans le collection publiée par l'auteur. Ann des Sciences Naturelles. *Botanique Série* 2(t.17): 91–118.

Duncan, K. E. and Howard, R. J. 2000. Cytological analysis of wheat infection by the leaf blotch pathogen *Mycosphaerella graminicola*. *Mycological Research* 104(9): 1074–1082.

Eyal, Z., Scharen, A. L., Prescott, J. M. and van Ginkel, M. 1987. *The Septoria Diseases of Wheat: Concepts and Methods of Disease Management*. Mexico: CIMMYT. 46 pp.

Fones, H. N., Eyles, C. J., Kay, W., Cowper, J. and Gurr, S. J. 2017. A role for random, humidity-dependent epiphytic growth prior to invasion of wheat by *Zymoseptoria tritici*. *Fungal Genetics and Biology: FG and B* 106: 51–60.

Fones, H. and Gurr, S. 2015. The impact of Septoria tritici Blotch disease on wheat: an EU perspective. *Fungal Genetics and Biology: FG and B* 79: 3–7. https://doi.org/10.1016/j.fgb.2015.04.004.

Fraaije, B. A., Lovell, D. J., Rohel, E. A. and Hollomon, D. W. 1999. Rapid detection and diagnosis of *Septoria tritici* epidemics in wheat using a polymerase chain reaction/PicoGreen assay. *Journal of Applied Microbiology* 86(4): 701–708.

Francisco, C. S., Ma, X., Zwyssig, M. M., McDonald, B. A. and Palma-Guerrero, J. 2019. Morphological changes in response to environmental stresses in the fungal plant pathogen *Zymoseptoria tritici*. *Scientific Reports* 9(1): 9642. https://doi.org/10.1038/s41598-019-45994-3.

Goodwin, S. B. 2018. Diseases affecting wheat: *Septoria tritici* blotch. pp. 47–68. In: *Integrated Disease Management of Wheat and Barley* Oliver, R. P. (Ed). Cambridge, UK: Burleigh Dodds Science Publishing.

Goodwin, S. B., Ben M'Barek, S. B., Dhillon, B., Wittenberg, A. H. J., Crane, C. F., Hane, J. K., Foster, A. J., Van der Lee, T. A. J., Grimwood, J., Aerts, A., Antoniw, J., Bailey, A., Bluhm, B., Bowler, J., Bristow, J., van der Burgt, A., Canto-Canche, B., Churchill, A. C. L., Conde-Ferràez, L., Cools, H. J., Coutinho, P. M., Csukai, M., Dehal, P., De Wit, P., Donzelli, B., van de Geest, H. C., van Ham, R. C. H. J., Hammond-Kosack, K. E., Henrissat, B., Kilian, A., Kobayashi, A. K., Koopmann, E., Kourmpetis, Y., Kuzniar, A., Lindquist, E., Lombard, V., Maliepaard, C., Martins, N., Mehrabi, R., Nap, J. P. H., Ponomarenko, A., Rudd, J. J., Salamov, A., Schmutz, J., Schouten, H. J., Shapiro, H., Stergiopoulos, I., Torriani, S. F. F., Tu, H., de Vries, R. P., Waalwijk, C., Ware, S. B., Wiebenga, A., Zwiers, L. H., Oliver, R. P., Grigoriev, I. V. and Kema, G. H. J. 2011. Finished genome of the fungal wheat pathogen *Mycosphaerella graminicola* reveals dispensome structure, chromosome plasticity, and stealth pathogenesis. *PLoS Genetics* 7(6): e1002070. https://doi.org/10.1371/journal.pgen.1002070.

Hane, J. K., Paxman, J., Jones, D. A. B., Oliver, R. P. and de Wit, P. 2020. "CATAStrophy," a genome-informed trophic classification of filamentous plant pathogens – how many different types of filamentous plant pathogens are there? *Frontiers in Microbiology* 10: 3088. https://doi.org/10.3389/fmicb.2019.03088.

Haridas, S., Albert, R., Binder, M., Bloem, J., LaButti, K., Salamov, A., Andreopoulos, B., Baker, S. E., Barry, K., Bills, G., Bluhm, B. H., Cannon, C., Castanera, R., Culley, D. E., Daum, C., Ezra, D., González, J. B., Henrissat, B., Kuo, A., Liang, C., Lipzen, A.,

Lutzoni, F., Magnuson, J., Mondo, S. J., Nolan, M., Ohm, R. A., Pangilinan, J., Park, H. J., Ramírez, L., Alfaro, M., Sun, H., Tritt, A., Yoshinaga, Y., Zwiers, L. H., Turgeon, B. G., Goodwin, S. B., Spatafora, J. W., Crous, P. W. and Grigoriev, I. V. 2020. 101 Dothideomycetes genomes: a test case for predicting lifestyles and emergence of pathogens. *Studies in Mycology* 96: 141-153.

Hawksworth, D. L., Crous, P. W., Redhead, S. A., Reynolds, D. R., Samson, R. A., Seifert, K. A., Taylor, J. W., Wingfield, M. J., Abaci, O., Aime, C., Asan, A., Bai, F. Y., de Beer, Z. W., Begerow, D., Berikten, D., Boekhout, T., Buchanan, P. K., Burgess, T., Buzina, W., Cai, L., Cannon, P. F., Crane, J. L., Damm, U., Daniel, H. M., van Diepeningen, A. D., Druzhinina, I., Dyer, P. S., Eberhardt, U., Fell, J. W., Frisvad, J. C., Geiser, D. M., Geml, J., Glienke, C., Gräfenhan, T., Groenewald, J. Z., Groenewald, M., de Gruyter, J., Guého-Kellermann, E., Guo, L. D., Hibbett, D. S., Hong, S. B., de Hoog, G. S., Houbraken, J., Huhndorf, S. M., Hyde, K. D., Ismail, A., Johnston, P. R., Kadaifciler, D. G., Kirk, P. M., Kõljalg, U., Kurtzman, C. P., Lagneau, P. E., Lévesque, C. A., Liu, X., Lombard, L., Meyer, W., Miller, A., Minter, D. W., Najafzadeh, M. J., Norvell, L., Ozerskaya, S. M., Öziç, R., Pennycook, S. R., Peterson, S. W., Pettersson, O. V., Quaedvlieg, W., Robert, V. A., Ruibal, C., Schnürer, J., Schroers, H. J., Shivas, R., Slippers, B., Spierenburg, H., Takashima, M., Taşkın, E., Thines, M., Thrane, U., Uztan, A. H., van Raak, M., Varga, J., Vasco, A., Verkley, G., Videira, S. I. R., de Vries, R. P., Weir, B. S., Yilmaz, N., Yurkov, A. and Zhang, N. 2011. The Amsterdam Declaration on fungal nomenclature. *IMA Fungus* 2(1): 105-112.

Ingold, C. T. 1933. Spore discharge in the Ascomycetes I. Pyrenomycetes. *New Phytologist* 32(3): 175-196.

Iwamoto, T., Sonobe, T. and Hayashi, K. 2003. Loop-mediated isothermal amplification for direct detection of *Mycobacterium tuberculosis* complex, *M. avium*, and M. intracellulare in sputum samples. *Journal of Clinical Microbiology* 41(6): 2616-2622. https://doi.org/10.1128/JCM.41.6.2616-2622.2003.

Justesen, A. F., Corsi, B., Ficke, A., Hartl, L., Holdgate, S., Jørgensen, L. N., Lillemo, M., Lin, M., Mackay, I. J., Mohler, V., Stadlmeier, M., Tan, K.-C., Turner, J., Oliver, R. P. and Cockram, J. 2021. Hidden in plain sight: a molecular field survey of three wheat leaf blotch fungal diseases in North-Western Europe shows co-infection is widespread. *European Journal of Plant Pathology* 160(4): 949-962. https://doi.org/10.1007/s10658-021-02298-5.

Kema, G. H. J., Gohari, A. M., Aouini, L., Gibriel, H. A. Y., Ware, S. B., van den Bosch, F., Manning-Smith, R., Alonso-Chavez, V., Helps, J., Ben M'Barek, S., Mehrabi, R., Diaz-Trujillo, C., Zamani, E., Schouten, H. J., van der Lee, T. A. J., Waalwijk, C., de Waard, M. A., de Wit, P. J. G. M., Verstappen, E. C. P., Thomma, B. P. H. J., Meijer, H. J. G. and Seidl, M. F. 2018. Stress and sexual reproduction affect the dynamics of the wheat pathogen effector AvrStb6 and strobilurin resistance. *Nature Genetics* 50(3): 375-380. https://doi.org/10.1038/s41588-018-0052-9.

Kema, G. H. J., Yu, D. Z., Rijkenberg, H. J., Shaw, M. W. and Baayen, R. P. 1996. Histology of the pathogenesis of *Mycosphaerella graminicola* in wheat. *Phytopathology* 86(7): 777-786. Https://doi.org/10.1094/Phyto-86-777.

Keon, J. P. R., Antoniw, J. F., Carzaniga, R., Deller, S., Ward, J. L., Baker, J. M., Beale, M. H., Hammond-Kosack, K. E. and Rudd, J. J. 2007. Transcriptional adaptation of *Mycosphaerella graminicola* to programmed cell death (PCD) of its susceptible wheat host. *Molecular Plant-Microbe Interactions* 20(2): 178-193. https://doi.org/10.1094/MPMI-20-2-0178.

Kettles, G. J. and Kanyuka, K. 2016. Dissecting the molecular interactions between wheat and the fungal pathogen *Zymoseptoria tritici*. *Frontiers in Plant Science* 7: 508. https://doi.org/10.3389/fpls.2016.00508.

Laluk, K. and Mengiste, T. 2010. Necrotroph attacks on plants: wanton destruction or covert extortion? *Arabidopsis Book* 8: e0136. https://doi.org/10.1199/tab.0136.

McCorison, C. B. and Goodwin, S. B. 2020. The wheat pathogen *Zymoseptoria tritici* senses and responds to different wavelengths of light. *BMC Genomics* 21(1): 513 https://doi.org/10.1186/s12864-020-06899-y.

McDonald, M. C., McGinness, L., Hane, J. K., Williams, A. H., Milgate, A. and Solomon, P. S. 2016. Utilizing gene tree variation to identify candidate effector genes in *Zymoseptoria tritici*. *G3* 6(4): 779–791.

Oliver, R. P., Tan, K.-C. and Moffat, C. S. 2016. Necrotrophic pathogens of wheat. pp. 273–278. In: *Encyclopedia of Food Grains* (2nd edn.) Wrigley, C., Corke, H., Seetharaman, K. and Faubion, J. (Eds). Oxford: Academic Press.

Quaedvlieg, W., Kema, G. H., Groenewald, J. Z., Verkley, G. J., Seifbarghi, S., Razavi, M., Mirzadi Gohari, A., Mehrabi, R. and Crous, P. W. 2011 Zymoseptoria gen. nov.: a new genus to accommodate Septoria-like species occurring on graminicolous hosts. *Persoonia* 26: 57–69. Https://doi.org/10.3767/003158511X571841.

Rudd, J. J., Kanyuka, K., Hassani-Pak, K., Derbyshire, M., Andongabo, A., Devonshire, J., Lysenko, A., Saqi, M., Desai, N. M., Powers, S. J., Hooper, J., Ambroso, L., Bharti, A., Farmer, A., Hammond-Kosack, K. E., Dietrich, R. A. and Courbot, M. 2015. Transcriptome and metabolite profiling of the infection cycle of *Zymoseptoria tritici* on wheat reveals a biphasic interaction with plant immunity involving differential pathogen chromosomal contributions and a variation on the hemibiotrophic lifestyle definition. *Plant Physiology* 167(3): 1158–1185.

Saari, E. E. and Wilcoxson, R. D. 1974. Plant disease situation of high-yielding dwarf wheats in Asia and Africa. *Annual Review of Phytopathology* 12(1): 49–68.

Sánchez-Vallet, A., McDonald, M. C., Solomon, P. S. and McDonald, B. A. 2015. Is *Zymoseptoria tritici* a hemibiotroph? *Fungal Genetics and Biology: FG and B* 79: 29–32.

Sanderson, F. R. 1972. A *Mycosphaerella* species as the ascogenous state of *Septoria tritici* Rob. ex Desm. *New Zealand Journal of Botany* 10(4): 707–709.

Sanderson, F. R. 1976. *Mycosphaerella graminicola* (Fuckel) Sanderson comb. nov., the ascogenous state of *Septoria tritici* Rob. apud Desm. *New Zealand Journal of Botany* 14(4): 359–360.

Shaw, M. W. 1987. Assessment of upward movement of rain splash using a fluorescent tracer method and its application to the epidemiology of cereal pathogens. *Plant Pathology* 36(2): 201–213.

Trail, F. 2007. Fungal cannons: explosive spore discharge in the Ascomycota. *FEMS Microbiology Letters* 276(1): 12–18.

Trail, F. 2009. For blighted waves of grain: *Fusarium graminearum* in the postgenomics era. *Plant Physiology* 149(1): 103–110. https://doi.org/10.1104/pp.108.129684.

Verkley, G. J. M., Quaedvlieg, W., Shin, H. D. and Crous, P. W. 2013. A new approach to species delimitation in *Septoria*. *Studies in Mycology* 75(1): 213–305.

Yun, S. H., Berbee, M. L., Yoder, O. C. and Turgeon, B. G. 1999. Evolution of the fungal self-fertile reproductive life style from self-sterile ancestors. *Proceedings of the National Academy of Sciences of the United States of America* 96(10): 5592–5597.

Yun, S. H., Yoder, O. C. and Turgeon, B. G. 2013. Structure and function of the mating-type locus in the homothallic ascomycete, *Didymella zeae-maydis*. *Journal of Microbiology* 51(6): 814–820. https://doi.org/10.1007/s12275-013-3465-2.

Zhang, X. W., Jia, L. J., Zhang, Y., Jiang, G., Li, X., Zhang, D. and Tang, W. H. 2012. In planta stage-specific fungal gene profiling elucidates the molecular strategies of *Fusarium graminearum* growing inside wheat coleoptiles. *The Plant Cell* 24(12): 5159–5176.

Zhu, X., Lu, C., Du, L., Ye, X., Liu, X., Coule, A. and Zhang, Z. 2017. The wheat NB-LRR gene *TaRCR1* is required for host defence response to the necrotrophic fungal pathogen *Rhizoctonia cerealis*. *Plant Biotechnology Journal* 15(6): 674–687.

Published by Burleigh Dodds Science Publishing Limited, 2022.

Chapter 8

Understanding plant-pathogen interactions in *Septoria tritici* blotch infection of cereals

Y. Petit-Houdenot and M.-H. Lebrun, UMR Bioger, Université Paris Saclay, INRAE, AgroParistech, France; and G. Scalliet, Syngenta Crop Protection AG, Switzerland

1 Introduction

The wheat pathogen *Zymoseptoria tritici* (Capnodiales; Dothideomycetes) (syn. *Mycosphaerella graminicola*; Quaedvlieg et al., 2011) is haploid and heterothallic (Mat1-1 and Mat1-2; Waalwijk et al., 2002). In addition, the *Zymoseptoria* genus also contains four other species. *Z. pseudotritici* and *Z. ardabiliae* were isolated from wild grasses in Iran and are pathogenic on wheat. *Z. passerinii* is a pathogen of barley, while Z. *brevis* was isolated from grasses. The closest genus to *Zymoseptoria* is *Ramularia* which includes many plant pathogens, including the barley pathogen *Ramularia collo-cygni* (Quaedvlieg et al., 2011; Stukenbrock et al., 2011, 2012a,b; Feurtey et al., 2020). *Z. tritici* is a major fungal pathogen of bread and durum wheat causing *Septoria tritici* blotch disease (STB; Fones and Gurr, 2015). STB is an important disease in most wheat-producing regions (Boukef et al., 2013; Pakholkova et al., 2016; Hartmann et al., 2018; Hassine et al., 2019), including Europe, Russia, Ukraine, Mediterranean countries, North America (USA, Canada, Mexico), South America (Argentina, Chile), Africa (Mekonnen et al., 2020) and Australia (McDonald et al., 2019).

http://dx.doi.org/10.19103/AS.2021.0092.10

For unknown reasons, STB has not been reported in China, even though the environmental conditions in some parts of the country should be favorable for STB epidemics. Yield losses associated with STB in Europe range from 5% to 10%, despite the use of resistant varieties and fungicide treatments (Fones and Gurr, 2015). Yield losses higher than 50% have been reported in wheat fields with susceptible cultivars not treated with fungicides (Berraies et al., 2014; Tadesse et al., 2019; Hailemariam et al., 2020; Jalli et al., 2020).

Both sexual and asexual stages of *Z. tritici* are commonly observed in infected wheat fields during polycyclic STB epidemics (Suffert et al., 2011). Sexual reproduction takes place during the intercrop season on wheat debris, where pseudothecia release wind-dispersed ascospores (sexual spores) which act as the primary inoculum in autumn/winter on wheat seedlings (Zhan et al., 2001; Eriksen and Munk, 2003; Suffert et al., 2011; Morais et al., 2016). The density of leaf lesions is a crucial determinant of sexual reproduction at the plant and field scales (Suffert and Thompson, 2018) because physical contact between two compatible strains is required (Waalwijk et al., 2002). More severe epidemics are usually associated with higher levels of ascospore production on wheat debris from the previous year (Cowger et al., 2002; Suffert et al., 2011; Morais et al., 2016; Chaloner et al., 2019). Wheat residue management has long been known to be a factor controlling STB epidemics (Suffert et al., 2011), but it was shown that the amount of primary airborne inoculum available in a wheat crop is not actually a limiting factor for the onset of epidemics in the wheat production areas (Morais et al., 2016). The lifestyle of the fungus during the cropping season is characterized by a symptomless, relatively long, latent period (Suffert and Thompson, 2018; Chaloner et al., 2019). The rate of development of the epidemic, which progresses vertically in the canopy (Lovell et al., 2004), is determined by the number of asexual multiplication cycles completed by the pathogen (usually 5–7), which depends principally on temperature and rain events (Chaloner et al., 2019). Once a plant is infected, pycnidia release splash-dispersed pycnidiospore (asexual spores) that are responsible for secondary infections (Shaw, 1987; Karisto et al., 2021). Because of its tremendous impact on agriculture, *Z. tritici* biology, in particular asexual infection at the leaf scale, has been intensively studied (reviews: Suffert et al., 2011; Dean et al., 2012; Perez-Nadales et al., 2014; Rudd, 2015; McDonald et al., 2015a; Kettles and Kanyuka, 2016; Duba et al., 2018; Brennan et al., 2019). The description of the infection cycle at the cytological level has also been well-documented (see Section 4.1 for references). Infection begins with the deposition of spores on the leaf surface, their germination and the development of epiphytic hyphae on the leaf surface. These hyphae penetrate the leaf mostly through stomata. After penetration, *Z. tritici* infection is characterized by two distinct stages. An initial, relatively long, asymptomatic phase (10–12 days), is followed by a second phase characterized by the formation of large necrotic lesions associated with plant cell death (Keon et al., 2007). During this necrotrophic phase, *Z. tritici* produces a large number of pycnidia within necrotic lesions that release pycnidiospores expelled through stomata openings. Two types of pycnidiospores have been described. Macro-pycnidiospores (pycnidiospores *sensu stricto*) are composed of 4–8 cells and they are the main spores involved in disease dynamics during the crop season (reviewed in Suffert et al., 2011). Micro-pycnidiospores were sometimes reported in empty pycnidia present on wheat residues in dry conditions (Djerbi et al., 1974). These small spores were suspected to result from pycnidiospore germination inside the pycnidium, and they bear some resemblance to the yeast-like cells, called blastospores, produced in axenic cultures that divide by budding (Steinberg, 2015). Blastospores were also produced at an early stage of infection by germinating pycnidiospores (Francisco et al., 2019). In addition, chlamydospores, produced in response to nutritional and temperature stresses, were able to survive extreme cold, heat and drought treatments (Francisco et al., 2019).

2 *Z. tritici* genomics

2.1 **Z. tritici** *genome characteristics and annotation*

The first *Z. tritici* genome sequence was obtained from the bread wheat European isolate IPO323, using Sanger sequencing (Goodwin et al., 2011). This 39.7 Mb sequence is complete from telomere to telomere and includes centromeres (Goodwin et al., 2011; Schotanus et al., 2015). The *Z. tritici* genome is composed of 13 core chromosomes (CCs: 1.2–6.1 Mb), which are conserved across isolates, and up to 8 accessory chromosomes (ACs: 0.4–0.8 Mb), which vary in number and size among *Z. tritici* isolates (Goodwin et al., 2011; McDonald et al., 2015b; Badet et al., 2020). The first IPO323 genome annotation was obtained in 2011, using Fgenesh and Genewise *ab initio* gene predictors supported by EST and protein sequences (10 933 genes; Goodwin et al., 2011; available at NBCI, Mycocosm and Ensembl). A second IPO323 genome annotation was obtained in 2015, using GeneMark-ES, GeneMark-HMM and Augustus *ab initio* gene predictors supported by RNAseq and evolutionary evidences (11 839 genes; Grandaubert et al., 2015; available at FungiDB and Ensembl). Finally, a third IPO323 genome annotation was obtained in 2016, using the MAKER-HMM *ab initio* gene predictor with RNAseq evidence (13 583 genes; RRES available at Ensembl). These three annotations only share 4000 identical gene models, suggesting there are major differences in gene structures across predictions. A refined genome annotation with curated gene models is therefore needed. Specific annotations were performed for genes encoding plant cell wall-degrading enzymes (Brunner et al., 2013), secreted proteins/effectors (Morais do Amaral et al., 2012; Mirzadi Gohari et al., 2015) and genes involved in secondary metabolism (Cairns and Meyer, 2017).

2.2 **Z. tritici** *pangenome*

The genomes of 18 additional *Z. tritici* isolates from different geographic origins were characterized using long-read sequencing and gene predictions with *ab initio* gene predictors and RNAseq evidence (Badet et al., 2020). These genomes varied in size from 37.1 Mb to 41.7 Mb, and they were organized in 13 CCs and 4–8 ACs. All ACs were similar to those identified in IPO323 (no new ACs were found). The individual gene repertoires of these 18 isolates and IPO323 varied from 11 657 to 12 787 genes. Comparison of these repertoires highlighted 15 474 orthologous gene families with 15 451 single gene orthogroups (pangenome). A sub-set of these orthogroups is conserved across isolates (9193 conserved genes, 60%) while accessory orthogroups (40%) were either confined to a subset of isolates (4690 genes, 30%) or isolate specific (1592 genes, 10%). This is the largest accessory genome reported in fungi to date. Conserved genes were enriched in genes encoding proteins with housekeeping functions and carbohydrate-active enzymes (92% of CAZymes). This enrichment might reflect the conservation of basic cellular functions needed for both growth/survival and plant infection. Secondary metabolism gene clusters were distributed between conserved (70%) and accessory (30%) genes, in similar proportions to the pangenome. However, accessory genes were enriched in genes encoding putative effectors (60%), defined as small secreted proteins, that could account for differences in the infection behavior among isolates.

2.3 Related **Zymoseptoria** *species genomics*

The genomes of four related *Zymoseptoria* species (*Z. ardibilae*, *Z. brevis*, *Z. passerinii*, *Z. pseudotritici*) were characterized using long-read sequencing and gene prediction with *ab initio* gene predictors and RNAseq evidence (Feurtey et al., 2020). These genomes varied in size from 38.1 Mb to 41.6 Mb, and they were organized in 13 CCs largely similar to those of *Z. tritici*. The number of ACs varied from 4 to 5. Large regions of these ACs were similar to ACs from *Z. tritici*. The gene repertoires varied from 10 528 (*Z. passerinii*) to 11 661 (*Z. pseudotritici*). Comparison of these gene repertoires highlighted 18 237 orthogroups within the *Zymoseptoria* genus including *Z. tritici*, among which 44 % were conserved (8310 core genes), and 56% were accessory either at the species (46%) or isolate level (10%). The number of core genes conserved across these five species (8310) is close to the number of core genes identified in *Z. tritici* (9193 genes, Badet et al., 2020). In addition, 25%

of *Z. tritici* isolate-specific genes were identified in other species, suggesting a specific loss of these genes that are conserved at the genus level, in some *Z. tritici* isolates. Overall, a comparison of the genomes of these five species from the *Zymoseptoria* genus showed that they were highly related in both genome organization and gene repertoires. This similarity could result from a certain level of interspecific gene flow, as suggested by the detection of introgressions between these species (Feurtey et al., 2019).

2.4 Zymoseptoria *transposons*

A large proportion of the *Z. tritici* genome is composed of transposable elements (TEs). IPO323 genome TE content was estimated between 17% and 20% (Dhillon et al., 2014; Grandaubert et al., 2015, respectively), while the TE content of the global pangenome was estimated between 17% and 24% (19 isolates; 304 TE families; Badet et al., 2020). These variations in TE content across isolates explain 70% of their genome size variations. Related *Zymoseptoria* species have a similar TE content (13–25%), leading to 497 TE families in all *Zymoseptoria* species, among which 76 were conserved across species (Grandaubert et al., 2015; Feurtey et al., 2020). The most abundant TEs in *Z. tritici* are Class I retroelements (70%) from LTR-Gypsy and LTR-Copia super-families, followed by Class II DNA transposons (22%, Dhillon et al., 2014; Grandaubert et al., 2015; Badet et al., 2020). The distribution of TEs in the genome is not random. A small proportion of TEs were inserted in CDS (1-2%) or in promoter/terminator regions (4–6%, <1 Kb from CDS; Fouché et al., 2020; Badet et al., 2020). These insertions have a significant effect on the transcription of *Z. tritici* genes, including silencing of adjacent genes under specific conditions (Krishnan et al., 2018a; Fouché et al., 2020; Meile et al., 2020). The physical vicinity of TEs to genes might also influence their evolutionary dynamics by inducing mutations through the cell defense mechanism called repeat induced point mutations (RIP, Singh et al., 2021). In addition, a significant proportion of *Z. tritici* TEs (70%) was expressed, and specific TE families display a high level of expression during infection and under stress conditions (Fouché et al., 2020; Badet et al., 2020).

2.5 Zymoseptoria *core and accessory chromosomes*

All the information gathered on the *Z. tritici* genome allowed a detailed comparison of CCs and ACs. Compared to CCs, ACs displayed a lower gene density (2-fold, Goodwin et al., 2011; Kellner et al., 2014; Feurtey et al., 2020) and a highly reduced number of genes encoding secreted proteins (30-fold, Goodwin et al., 2011; Morais do Amaral et al., 2012). Compared to CC genes, AC genes displayed lower levels of expression *in vitro* (5–10 fold; Kellner et al.,

2014; Rudd et al., 2015; Palma-Guerrero et al., 2017; Feurtey et al., 2020). Only a small number of AC genes were differentially expressed during infection (79 genes; Rudd et al., 2015). ACs had a higher content of TEs than CCs (2-fold, Goodwin et al., 2011; Grandaubert et al., 2015; Badet et al., 2020). The role of ACs in the biology of *Z. tritici* is still a debate. Since there are no field isolates without ACs (McDonald et al., 2015b; Badet et al., 2020), and because many ACs genes are conserved in related *Zymoseptoria* species (Feurtey et al., 2020), they likely carry genes involved in functions that are important for the *Z. tritici* life cycle. Recently, forward genetics studies demonstrated that the experimental loss of some ACs from the IPO323 isolate leads to mutants with a quantitative increase in pathogenicity (2-fold) specifically on one of the four wheat cultivars tested (Habig et al., 2017). This result suggested that ACs could carry genes involved in adaptation to the host plant.

3 Molecular and biological tools available in *Z. tritici*

A number of molecular genetics tools are available for *Z. tritici* (Talbot, 2015). Genetic transformation is performed using either protoplast PEG-mediated transformation (Payne et al., 1998) or *Agrobacterium tumefaciens*-mediated-transformation (ATMT) (Zwiers and De Waard, 2001). ATMT is the method of choice since protoplast formation is now difficult to obtain due to changes in commercial fungal cell-wall degrading enzymes and possible differences among isolates in their sensitivity to these enzymes (Khan, 2017). ATMT vectors with different resistance markers (hygromycin, geneticin, sulfonylurea, basta) are available (Sidhu et al., 2015b). Targeted gene replacement is relatively easy in *Z. tritici* (5–40%; Bowler et al., 2010). *Ku70* deletion mutant strains deficient for the non-homologous end-Joining (NHEJ) DNA repair pathway displayed a strong increase in targeted gene replacement frequency (up to 100%; Bowler et al., 2010; Sidhu et al., 2015a,b). Vectors for GFP/RFP transcriptional and translational fusions are also available (Kilaru et al., 2015a; Schuster et al., 2015; Kilaru et al., 2017). Expression vectors with *Z. tritici* promoters were constructed either for over-expression (Sidhu et al., 2015a; Mehrabi et al., 2015) or for controlled expression (Marchegiani et al., 2015; Kilaru et al., 2015b; Sidhu et al., 2015a; Steinhauer et al., 2019; Fantozzi et al., 2020a), including vectors for infection-specific expression (Fantozzi et al., 2021a). Promoter replacement was achieved to identify essential genes (Marchegiani et al., 2015). Yeast-two-hybrid libraries (Ma et al., 2015) and an ORFeome library (Chaudhari et al., 2019) are available. All these molecular tools allow the genetic characterization of any gene of interest in *Z. tritici*. However, molecular tools such as *CAS9* gene editing, temperature-sensitive mutants, collections of genome-wide deletion and overexpression transformants and transposon-based insertional mutagenesis are still missing in *Z. tritici*. A first step toward genome-wide insertional mutagenesis was recently obtained by introducing the *Fusarium oxysporum TC1-mariner* transposable element *impala* (Dufresne and Daboussi, 2010) into *Z. tritici* (Pitarch, A., Delude, C., Petit, Y., Scalliet, G., and Lebrun, M. H., pers. com.). Transposon excision and re-insertion events were identified showing that *impala* is active in *Z. tritici* and that *impala* has an insertion preference for genes located on core chromosomes (84%), near transcriptional start sites (TSS, 50%). Collections of *impala* insertion mutants are currently screened to identify pathogenicity mutants. *Z. tritici* pathogenicity has been well-studied in laboratory conditions for decades, allowing the production of typical disease symptoms (chlorosis, necrosis), and the formation of pycnidia and pycnidiospores on experimentally infected leaves (Suffert et al., 2013; Rudd et al., 2015). Improved infection protocols have been developed to better quantify infection efficiency (Fones et al., 2015; Kay et al., 2019), as well as image analysis of infected leaves (Stewart and McDonald, 2014; Stewart et al., 2017). Controlled crosses between strains of opposite mating types have been obtained using infected plants left in outdoor conditions (Kema et al., 2008; Kema et al., 2018; Suffert et al., 2016; Suffert et al., 2019). Such crosses can also be performed under controlled greenhouse conditions (pers. comm. G. Scalliet and G. Kema). Protocols for isolation of *Z. tritici* from infected wheat leaves, *in vitro* culture, nucleic acid extraction, molecular characterization and pathogenicity assays in controlled conditions, were recently summarized by Fagundes et al. (2020).

4 Molecular biology of infection

4.1 Cell biology of wheat infection by Z. tritici

The *Z. tritici* infection process has been intensively studied using cytological methods (Cohen and Eyal, 1993; Kema, 1996; Duncan and Howard, 2000; El Chartouni et al., 2012; Yang et al., 2013; Steinberg, 2015; Fones et al., 2017; Haueisen et al., 2018; Francisco et al., 2020; Fantozzi et al., 2021; reviewed in Brennan et al., 2019). *Z. tritici* infection has been subdivided into six phases by Fantozzi et al. (2021): (1) spore deposition/adhesion on the leaf surface (surface resting), (2) spore germination and leaf surface colonization by epiphytic hyphae (surface exploration), (3) stomata penetration, (4) mesophyll/apoplast colonization, (5) initiation of pycnidia formation in the sub-stomatal cavity (fruiting body initiation), (6) pycnidia development and sporulation (fruiting body maturation). The infection of wheat by *Z. tritici* begins with the deposition and adhesion of spores on the leaf surface and the induction of their germination by an as-yet-unknown signal (Duncan and Howard, 2000). Germinating hyphae develop into epiphytic hyphae that grow abundantly on the leaf surface up to 18 dpi (Fones et al., 2017; Fantozzi et al., 2021). These epiphytic hyphae penetrate leaves mostly through stomata from 2 dpi to 13 dpi (Fantozzi et al., 2021). However, penetration events mostly occur around 3–6 dpi depending on strains, cultivars and inoculation methods (reviewed in Brennan et al., 2019), but this estimate awaits more quantitative observations. Alternative penetration events through wounds or junctions between epidermal cells have been reported, but they have not been quantified accurately (Kema, 1996; Duncan and Howard, 2000; El Chartouni et al., 2012). *Z. tritici* does not form single-celled appressoria. Nevertheless, appressoria-like structures (hyphal swellings) were frequently described on the borders of stomatal openings during penetration (Kema, 1996; Duncan and Howard, 2000). The growth of epiphytic hyphae toward stomata occurs most likely through a stochastic process (Fones et al., 2017). However, it has been suggested that thigmotropic signals could direct hyphae toward stomata (Duncan and Howard, 2000). After penetration, infectious hyphae invade mesophyll intercellular spaces (apoplast) without causing visible symptoms (asymptomatic phase) up to 10–12 dpi. The duration of the asymptomatic phase varies according to strains, cultivars and leaf age (Suffert et al., 2013; Pnini-Cohen et al., 2000; Yang et al., 2013; Hehir et al., 2018; Haueisen et al., 2018; Rahman et al., 2020). Infectious hyphae were observed in the apoplast at a relatively long distance from the sub-stomatal cavity colonized by penetrating hyphae (Fantozzi et al., 2021). No specialized feeding structures such as haustoria were observed, nor intracellular hyphae. During this relatively long asymptomatic mesophyll/apoplast colonization, fungal biomass and transcripts are in low abundance (Kema, 1996; Pnini-Cohen et al., 2000; Yang et al., 2013; Rudd et al., 2015; Rahman et al., 2020), and plant defense responses are almost not induced (Yang et al., 2013; Orton et al., 2017; Rudd et al., 2015). Overall, during this asymptomatic phase, *Z. tritici* colonizes only slightly the sub-stomatal cavities, and instead, the fungus largely explores the mesophyll intercellular space, behaving almost like an endophyte. At 10–14 days after infection, *Z. tritici* switches to a necrotrophic behavior that starts by an exponential increase of fungal biomass and the development of chlorotic and necrotic leaf symptoms (Kema, 1996; Pnini-Cohen et al., 2000; Rudd et al., 2015; Rahman et al., 2020). During this infection phase, the sub-stomatal cavity is colonized by hyphae that initiate the formation of pycnidia (Kema, 1996; Duncan and Howard, 2000; Francisco et al., 2020; Fantozzi et al., 2021). During their maturation, pycnidia enlarge, melanize and produce a large number of pycnidiospores embedded in a hydrophilic jelly of unknown composition that is extruded through the stomatal opening (Duncan and Howard, 2000).

4.2 Transcriptomics of Z. tritici *infection*

After early transcriptomics experiments using microarrays (Keon et al., 2007; Shetty et al., 2007; Kema et al., 2008), several NGS RNAseq studies surveyed

Published by Burleigh Dodds Science Publishing Limited, 2022.

the genome-wide expression patterns of *Z. tritici* during infection (Brunner et al., 2013; Yang et al., 2013; Kellner et al., 2014; Rudd et al., 2015; Palma-Guerrero et al., 2016, 2017; Haueisen et al., 2018; Ma et al., 2018). These high-quality transcriptomic analyses of the *Z. tritici* infection process have highlighted fungal and plant genes that are specifically expressed at different stages of infection, giving more insights into the *Z. tritici* infection process.

4.2.1 The fungal side

Using RNAseq, Brunner et al. (2013) analyzed three infection stages (7 dpi, 13 dpi and 56 dpi) of young wheat leaves by the Swiss isolate 3D7 (compatible interaction), for expression of genes encoding plant cell-wall degrading enzymes (PCWDEs). Yang et al. (2013) studied expression of genes from both the host and the pathogen (compatible and incompatible interactions) during the asymptomatic phase (4 dpi and 10 dpi) and the switch to necrotrophy (14 dpi), but this experiment suffered from a low read coverage (less than 10% of *Z. tritici* predicted genes), and a lack of comparison with the expression pattern in axenic culture. Kellner et al. (2014) compared expression of *Z. tritici* genes from isolate IPO323 in three different conditions: axenic culture, infected young wheat leaves at 4 dpi (compatible interaction) and infected young *Brachypodium* leaves at 4 dpi (incompatible non-host interaction), but this study did not include an analysis of infection kinetics. Haueisen et al. (2018) compared expression of genes from three *Z. tritici* isolates infecting young leaves from wheat (all compatible interactions) at four similar infection stages (stage A: 3-4-6 dpi, stage B: 8–11 dpi, stage C: 13 dpi, stage D: 20–24 dpi), but a comparison with expression patterns from axenic cultures was lacking. To date, the most comprehensive RNAseq transcriptomic studies of *Z. tritici* infection were performed by Rudd et al. (2015), Palma-Guerrero et al. (2016, 2017). Rudd et al. (2015) analyzed young wheat leaves infected by IPO323 at 1, 4, 9, 14 and 21 dpi and two axenic cultures. Metabolomics analyses of the same samples were also performed (Rudd et al., 2015). Palma-Guerrero et al. (2016, 2017) analyzed young wheat leaves infected by four Swiss *Z. tritici* isolates (3D7, 3D1, 1A5 and 1E4) at 3, 7, 11, 14, 21 and 56 dpi (all compatible interactions). Most results presented in Sections 4.2.1.1 to 4.2.1.3 are coming from these three transcriptomic experiments (Rudd et al., 2015; Palma-Guerrero et al. 2016, 2017).

4.2.1.1 Early infection

During the early stages of infection (1–4 dpi, surface resting and exploration stages, early penetration events) 700–1000 fungal genes were specifically expressed compared to axenic culture. They encode mostly proteins involved in lipid degradation/metabolism and response to stress. A set of 155 genes encoding early effectors were up-regulated, including LysM proteins that protect the fungus from plant recognition of fungal chitin (Lee et al., 2014, Sánchez-Vallet et al., 2020). The most highly expressed genes at that early stage of infection, included chloroperoxidases, hydrophobins, lipases, fatty acid hydrolases and alcohol/aldehyde dehydrogenases. In addition, a strong expression of genes encoding multicopper oxidases, cytochrome P450s and oxygenases, was observed (4 dpi; Kellner et al., 2014), suggesting an active detoxification process. Overall, the transcriptomic signatures of early infection stages suggest that *Z. tritici* mostly relies on intracellular lipids from spores as a carbon source, circumvents plant defenses by detoxifying plant reactive oxygen and toxic compounds, and avoids its recognition by the plant by masking its cell wall using hydrophobins and LysM effectors.

4.2.1.2 Switch to necrotrophy

The switch to necrotrophy (9, 11, and 12 dpi, late mesophyll/apoplast colonization, fruiting body initiation stages), was characterized by the specific expression of 1000–2000 fungal genes compared to axenic cultures. These genes mostly encode proteins involved in nitrogen metabolism, nutrient uptake and growth. In addition, 235 secreted proteins without known functions, and known effectors (ECP2, Trp18, Trp21, LysM), were specifically expressed at this stage of infection. Some of these effectors were among the most highly expressed genes at that stage of infection (30% of the Top50). A strong up-regulation of genes encoding proteases, PCWDEs and lipases, was also observed, in contrast to earlier stages of infection when they were either silent or poorly expressed. In addition, genes encoding enzymes involved in secondary metabolite biosynthesis were up-regulated at this stage of infection. Overall, these transcriptomic signatures of the switch to necrotrophy suggest that the fungus is actively growing, using plant carbon and nitrogen sources, likely degraded by fungal proteases and PCWDEs, and manipulating the plant using both protein effectors and secondary metabolites. Similar to the early infection stage, the fungus continues to circumvent plant defenses by detoxifying plant reactive oxygen and toxic compounds, degrading plant defense proteins using proteases, and preventing its recognition by the plant by masking its cell wall using LysM effectors.

4.2.1.3 Necrotrophy

Early necrotrophy (14 dpi, fruiting body initiation and early maturation stages) was characterized by the over-expression of 1000 genes compared to axenic culture conditions. They mostly encode secreted proteins either with no known functions (200 putative effectors) or with degradative activities (proteases, lipases, PCWDEs). PCWDEs were highly expressed at this stage of infection, despite their low number in the *Z. tritici* genome (Goodwin et al., 2011; Morais do Amaral et al., 2012). This necrotrophic-specific expression of proteases and PCWDEs was also observed by Haueisen et al. (2018). This suggests that *Z. tritici* utilize carbohydrates obtained by the degradation of plant cell walls as a carbon source during necrotrophy. Specific sets of genes encoding secreted effectors and enzymes involved in secondary metabolism (polyketide synthases, ABC-transporters, PKS-non-ribosomal peptide synthetase hybrid) were also expressed at this stage of infection. Overall, the transcriptomic signatures of early necrotrophy suggest that *Z. tritici* use plant carbon and nitrogen sources, likely released by fungal proteases and PCWDEs, and manipulates/damages the plant using both protein effectors and secondary metabolites that differ from those highly expressed during the transition to necrotrophy.

Overall, the *Z. tritici* infection transcriptome is characterized by waves of expression of genes encoding effectors (secreted proteins, secondary metabolite biosynthesis enzymes), proteases, lipases and PCWDEs that are specific to each stage of infection (early, mesophyll colonization, switch to necrotrophy, necrotrophy). This behavior is typical of hemibiotrophic fungal plant pathogens that display similar waves of expression of stage-specific effectors and a massive expression of genes encoding proteases and PCWDEs only during the switch to necrotrophy (O'Connell et al., 2012).

4.2.2 The plant side

To date, the most comprehensive RNAseq transcriptomic studies of the wheat side of *Z. tritici* infection were performed by Rudd et al. (2015), and Ma et al. (2018). Rudd et al. (2015) focused on wheat responses at 5-time points (1, 4, 9, 14 and 21 dpi) after infection of young wheat leaves with IPO323 (compatible), mirroring the fungal side RNAseq analysis performed by these authors. Ma et al. (2018) focused on wheat responses at 4-time points (7, 12, 14, 28 dpi) after infection of leaves with 3D7, 3D1, 1A5 and 1E4 *Z. tritici* isolates (compatible), corresponding to the fungal side RNAseq analyses of Palma-Guerrero et al. (2016, 2017). Rudd et al. (2015) showed that the asymptomatic phase of infection (4–11 dpi) is characterized by the absence or very weak defense response from the host plant, suggesting that *Z. tritici* is able to suppress its recognition by the plant in a compatible interaction such as in the highly susceptible cultivar Obelisk. However, when Ma et al., (2018) looked at transcriptional responses of the moderately susceptible wheat cultivar Drifter infected by different Swiss isolates at 7 dpi, they observed an up-regulation of wheat defense genes, including receptor-like kinases, pathogenesis-related proteins and the MLO gene. These apparent discrepancies could reflect differences in the abilities of wheat cultivars to react to *Z. tritici* infection, although the moderate resistance of Drifter was not sufficient to stop infection. Transcriptomics analysis of the necrotrophic phase was characterized by a large transcriptional reprogramming of wheat plants. Rudd et al. (2015) observed that plant defense-related genes were up-regulated at this stage of infection culminating with hypersensitive response-like cell death as already suggested by Keon et al. (2007) for the same wheat cultivar Obelisk. Ma et al. (2018) observed that in infected leaves of the Drifter wheat cultivar, plant defense genes were greatly downregulated at the necrotrophic stage (14 dpi). However, at this stage of infection, a sharp decrease in wheat RNAseq reads was observed, suggesting a lack of transcriptional response due to extensive wheat cell collapse associated with the necrotrophic phase.

4.3 Z. tritici *secondary metabolism and infection*

Until now, there are no chemically characterized *Z. tritici* secondary metabolites apart from the melanin precursor di-hydroxy-napthalene (DHN; Sterigopoulos et al., 2013; Lendenmann et al., 2014; Muria-Gonzalez et al., 2015; Derbyshire et al., 2018). However, the genome of *Z. tritici* carries a significant number of genes involved in secondary metabolism (Goodwin et al., 2011; Ohm et al., 2012; Cairns and Meyer, 2017; Gluck-Thaler et al., 2020). Using Anti-Smash and SMURF, Cairns and Meyer (2017) predicted 32 putative biosynthetic gene clusters (BGCs) involved in secondary metabolism. These BGCs could possibly biosynthesize 11 polyketides (PKS and PKS-NRPS), 10 peptides (NRPS) and 5 terpenes (TSD). A significant number of these BCGs were located in subtelomeric regions (37%). All BCGs were located on core chromosomes. Most *Z. tritici* BCGs (80%) were expressed either in axenic culture or during infection with different stage-specific expression profiles (Rudd et al., 2015; Palma-Guerrero et al., 2017; Cairns and Meyer, 2017). In particular, two PKS and two NRPS BGCs were specifically expressed either at the onset or during the necrotrophic phase, suggesting that the corresponding secondary metabolites are involved in necrotic symptoms development. Even though BGCs were expressed, their corresponding secondary metabolites have not been characterized so far. Nevertheless, some *Z. tritici* BGCs are sufficiently related to known fungal BGCs to deduce which metabolite is likely biosynthesized. For example, the gene encoding the NPRS involved in the biosynthesis of the siderophore fusarine-C (NRPS1) was deleted by targeted gene replacement (Derbyshire et al., 2018). The corresponding null mutant was hypersensitive to iron and pathogenic on wheat, showing that *Z. tritici* does not require such an iron acquisition pathway to infect wheat. Similarly, the gene encoding the PKS involved in the biosynthesis of melanin precursor DHN (PKS1) was deleted by targeted gene replacement (Derbyshire et al., 2018). The corresponding null mutant was still pathogenic on wheat, showing that *Z. tritici* does not require DHN melanin to infect wheat, even though pycnidia from this mutant were completely unpigmented. In addition, deletion of phosphopantetheinyl transferase encoding gene (PPT1) leads to a strong defect in pathogenicity. PPT1 provides an essential co-factor for all PKS and NRPS, including the NRPS involved lysine biosynthesis, the NRPS involved in siderophore biosynthesis (NRPS1) and the PKS involved in the biosynthesis of the melanin precursor DHN (PKS1, Derbyshire et al., 2018). Siderophore and DHN melanin are not essential for pathogenicity, while lysine biosynthesis is. Consequently, these authors concluded that the pathogenicity defect of their PPT1 deletion mutant is caused by its lysine auxotrophy. However, adding lysine to spores did not rescue the pathogenicity defect of PPT1 mutants and the lysine auxotroph mutant AAR1. This is in contrast with similar experiments performed on other plant pathogenic fungi, in which the pathogenicity defects of PPT1 null mutants and lysine auxotrophs were partially rescued by exogenous lysine or wounding (Horbach et al., 2009; Zainudin et al., 2015). These first results await additional experiments, but it is tempting to speculate that, as in other fungi, the pathogenicity defect of the *Z. tritici PPT1* mutant is due to a lack of production of yet unknown secondary metabolites, independently of its lysine auxotrophy, and DHN melanin/siderophore deficiencies. Overall, *Z. tritici* has the potential to produce at least 20 secondary metabolites that are currently unknown, some of which are likely produced during wheat leaf infection, since the genes encoding the corresponding biosynthetic enzymes are expressed during infection (Rudd et al., 2015; Palma-Guerrero et al., 2017; Cairns and Meyer, 2017).

4.4 Z. tritici *secretome and infection*

Z. tritici genome has about 970 genes encoding putative secreted proteins according to bioinformatic criteria (N-terminal signal peptide), among which 492 belong to the manually curated (refined) secretome published by Goodwin et al. (2011) and Morais do Amaral et al. (2012). One striking feature of the *Z. tritici* genome is that none of the genes encoding proteins from the refined secretome were located on dispensable chromosomes. Among the refined secretome, 321 proteins (65%) have a known functional domain, corresponding mostly to enzymes active on plant polymers (proteases, lipases and PCWDEs) and to fungal cell wall-associated proteins (LysM, hydrophobins). Another set of 171 secreted proteins encode putative effectors, including cysteine-rich small proteins without known function (n: 94). Only a small number of these secreted proteins have been functionally characterized. The most studied effectors of *Z. tritici* are the LysM effectors, homologs of the *Cladosporium fulvum* Ecp6 effector involved in protecting fungal cell walls from plant chitinases. Three LysM effector-encoding genes are present in the genome of *Z. tritici* (*Mg1LysM*, *Mg3LysM* and *MgxLysM*; Marshall et al., 2011). Mg1LysM and Mg3LysM both interact with chitin. The crystal structure or Mg1LysM in complex with chitin was obtained showing that chitin-dependent dimerization of Mg1LysM is crucial to mask chitin from the fungal cell wall (Sánchez-Vallet et al., 2020). Only the *ΔMg3LysM* mutant is unable to colonize wheat leaves and is impaired in sporulation (Marshall et al., 2011). Silencing of the wheat chitin receptors CERK1- and CEBiP-encoding genes fully restored the pathogenicity defect of the *ΔMg3LysM* mutant (Lee et al., 2014), demonstrating that the main role of fungal LysM effector is to prevent the recognition of the fungus by these plant immunity receptors. Other well-studied *Z. tritici* effectors are the necrosis-inducing proteins expressed during the switch to necrotrophy, namely ZtNIP1 and ZtNIP2. These effectors were identified using a proteomics strategy and were able to induce chlorotic and/or necrotic lesions on several wheat cultivars (Ben M'Barek et al., 2015). *Z. tritici* has a single necrosis-inducing like protein (NLP1), expressed during the switch to necrotrophy that induces necrosis in *Arabidopsis thaliana,* but not wheat (Motteram et al., 2009). The screen for *Z. tritici* effectors inducing necrosis in *Nicotiana benthamiana* has revealed 14 proteins out of 63 candidates that likely induce resistance in this non-host plant (Kettles et al., 2017). However, these effectors were not toxic to wheat. Few other *Z. tritici* secreted proteins have been characterized for their role in pathogenicity. A secreted ribonuclease (Zt6) acting on rDNA from both plants and microbes, but not from *Z. tritici*, is essential for pathogenicity (Kettles et al., 2018). A secreted glycosyltransferase (ZtGT2) is involved in hyphal growth and pathogenicity (King et al., 2017). A secreted cysteine-rich protein interacting with a wheat E3 ubiquitin ligase is involved in infection (Karki et al., 2021). The secretion of proteins by *Z. tritici* has also been studied. Fungal extracellular vesicles (EVs) containing secreted proteins have been isolated from *Z. tritici* axenic culture filtrates. Proteomics analysis identified 240 proteins similar to those of known fungal EVs (Hill and Solomon, 2020). Whether those EVs and their secreted proteins content play a role in *Z. tritici* pathogenicity remains to be determined. The deletion of the *ZtALG2* gene encoding a protein involved in protein glycosylation led to a mutant defective in the secretion of many proteins (Motteram et al., 2011). This mutant also displayed a strong defect in pathogenicity suggesting,that the glycosylation of secreted proteins is indeed important for infection, as observed for LysM effectors.

4.5 Other mechanisms involved in **Z. tritici** *pathogenicity*

Fungal small RNAs (sRNAs) have been hypothesized to play a role in infection, but their suppression by the deletion of genes encoding enzymes involved in their production had no effect on pathogenicity (Kettles et al., 2019; Ma et al., 2020). *Z. tritici* is able to switch between a filamentous and a yeast-like morphology (Steinberg, 2015), and several signaling pathways are involved in this dimorphic switch. ZtWOR1 encodes a general transcription factor (TF) required for the formation of yeast-like cells that is critical for pathogenicity (Mirzadi Gohari et al., 2014). Zt107320, a TF homologous to *M. oryzae* COD1, is quantitatively involved in the switch from yeast to hyphae and in pycnidia formation (Habig et al., 2020). The mitogen-activated protein-kinases (MAPK) such as ZtHog1 and ZtSlt2 (Mehrabi et al., 2006a,b), some key components of the cAMP pathway (Mehrabi et al., 2009) and cyclin (Choi and Goodwin, 2011) are involved in the switch from yeast to hyphae and in infection. The ZtALG2 gene required for protein glycosylation was unable to switch from yeast to hyphae and was non-pathogenic (Motteram et al., 2011). More recently, a T-DNA insertion mutagenesis identified 11 genes involved in the dimorphic switch, 2 of which encode a two-component response regulator SSK1p and a phosphoribosylamine-glycine ligase (Yemelin et al., 2017). However, these mutants were not affected in pathogenicity. Still, these results suggest that some of the genes involved in the dimorphic switch are necessary for pathogenicity. Vegetative hyphal fusions are also important during the late stages of infection, since the deletion of the ZtSof1 gene abolishes anastomosis and leads to an absence of pycnidia (Francisco et al., 2020). Few other genes involved in infection have been characterized. Deletion of the ABC transporter ZtATR4 leads to a quantitative defect in pathogenicity (Sterigopoulos et al., 2003). Deletion of the TF-encoding gene ZtVf1 leads to a quantitative defect in pathogenicity due to a reduced penetration frequency and pycnidia formation (Mohammadi et al., 2017). ZtRlm1, a TF of the MADS-box family that is critical for aerial mycelium formation, hyphal growth and melanization, is also quantitatively involved in pycnidia formation (Mohammadi et al., 2020). Finally, homologs of *Aspergillus nidulans* conidiation genes were studied in *Z. tritici* (Tiley et al., 2018). Out of five genes encoding TFs known to be involved in sporulation in fungi, *ZtFlbC* and *ZtBrlA2* were shown to be involved in pycnidiospore formation, whereas *ZtStuA* deletion resulted in non-pigmented mutants unable to differentiate pycnidia and pycnidiospores (Tiley et al., 2018; Derbyshire et al., 2018).

5 *S. tritici* blotch disease management

Fungicide applications are currently the main strategy used to control STB. This is particularly true in Europe, where most fungicide applications on cereals are dedicated to control STB. However, a major problem for chemical control is the emergence of fungicide-resistant *Z. tritici* isolates that rapidly become dominant in populations. The deployment of cultivars resistant to STB is the most economical method for managing STB. However, almost all highly resistant cultivars have not remained effective for a long time, due to the emergence of virulent *Z. tritici* isolates. Cultural practices and biological control can reduce the incidence and severity of STB, but these methods are not very efficient.

5.1 S. tritici *blotch control using fungicides*

STB epidemics are mainly controlled by fungicide applications. In Europe, about 70% of the wheat fungicide market is targeting the control of *Z. tritici* (Fones and Gurr, 2015; Torriani et al., 2015). In practice, up to four fungicide applications are performed in the United Kingdom and Ireland, where the climate is strongly conducive for STB (Bouma, 2005; Chaloner et al., 2019). The earliest foliar treatments are performed at GS30, at the start of stem elongation (see https://www.far.org.nz/assets/files/uploads/Iss_02_Cereal_Growth_Stages_June_09.pdf) in order to limit primary STB infections of young leaves. In most cases, the first foliar treatment is performed at the first node stage (GS31-32). Predictive tools that consider meteorological events and wheat varietal resistance help growers in deciding the timing of the first foliar application (http://www.barometre-maladies.arvalis-infos.fr/bletendre/). Additional treatments are focused on protecting the upper leaves from STB between the start of flag leaf emergence and ear emergence (GS39-49; Carmona et al., 2020). Since the introduction of synthetic fungicides to protect cereals from fungal diseases in the 1940s, four groups of single-site fungicides were used for STB control (Morton and Staub, 2008). The benzimidazoles, targeting fungal tubulin, were introduced in the late-1970s. This fungicide class is currently rarely used for STB control since the resistance-conferring beta-tubulin E198A-target mutation is widespread in *Z. tritici* populations (Lucas et al., 2015; Garnault et al., 2019). The DMI or azole fungicides, inhibitors of lanosterol 14α-demethylase in the fungal ergosterol biosynthesis pathway, were introduced in the 1970s and are still widely used for STB control (Clarke, 2006). Resistance toward DMIs occurs either through mutations of the *CYP51* gene encoding the target and/or by mutations of its promoter leading to its overexpression (Leroux and Walker, 2011; Cools et al., 2012; Cools and Fraaije, 2013). A third resistance mechanism is based on an increased efflux through the overexpression of the MgMfs1 transporter (Omrane et al., 2015; Omrane et al., 2017; Maë et al. 2020). The increased occurrence of less sensitive phenotypes from year to year correlates with a gradual erosion of DMI fungicide activity in the field (Blake et al., 2018), due to accumulation of non-synonymous substitutions in *CYP51* (Brunner et al., 2008). Azole-resistant populations also emerged in Australia sometime after 2011 (Milgate et al., 2016) with similar alleles to those found in European populations (McDonald et al., 2019). At a global scale, fungicide applications (mostly DMIs) have been a major driver of *Z. tritici* evolution, both through *de novo* mutations and gene flow (Hartmann et al., 2020). Nevertheless, the increase in DMI resistance frequency in populations is slow and azole fungicides are still effective, despite the presence of resistant alleles (Garnault et al., 2019). This could be due to the spectrum of resistance of each allele that confers resistance only to specific azole fungicides (incomplete cross-resistance, Fraaije et al., 2007; Leroux et al., 2007). In addition, primary CYP51 azole-resistant alleles display quantitative enzymatic defects, suggesting that *Z. tritici* populations slowly evolved complex CYP51 alleles (multiple mutations) to compensate for these defects (Cools et al., 2011; Cools and Fraaije, 2013). The quinone outside inhibitors (QoIs or strobilurins fungicides), targeting respiration complex III and displaying a high activity against STB, were introduced in the 1990s (Morton and Staub, 2008). However, resistance, caused by a mutation of the mitochondrial *cytb* gene, resulting in a G143A protein substitution, led to a major decrease of efficacy shortly after their introduction (Grasso et al., 2006; Sierotzki et al., 2007; Gisi et al., 2002; Fraaije et al., 2005; Torriani et al., 2009). Transmission of the mitochondrial *cytb* G148A resistant allele was facilitated during sexual reproduction on treated wheat plants (Kema et al., 2018). QoIs resistance is now fixed within most *Z. tritici* populations (Cheval et al., 2017). The succinate dehydrogenase inhibitors (SDHis), targeting complex II of the mitochondrial respiratory chain, were the last class of single-site inhibitors introduced for STB control in the late 2000s (Sierotzki and Scalliet, 2013). Resistance to SDHIs has arisen through mutations in genes encoding subunits of the ubiquinone binding pocket of the enzyme (Fraaije et al., 2012; Scalliet et al., 2012; Dooley et al., 2016), conferring different resistance levels and cross-resistance profiles across the SDHIs (Fraaije et al., 2012; Scalliet et al., 2012). Numerous field mutations were observed since 2012 (Garnault et al., 2020; https://www.frac.info/frac-teams/working-groups/sdhi-fungicides/), while the H152R mutation conferring a high level of resistance, was only reported after 2015 (Dooley et al., 2016; Rehfus et al., 2018). Subsequently, this H152R mutation is only present at a low frequency, likely as a result of a fitness penalty (Scalliet et al., 2012). Interestingly, a dispensable paralog of the C-subunit encoding gene was discovered that is affecting a particular subclass of SDHIs (Yamashita and Fraaije, 2018; Steinhauer et al., 2019). This is the first example of a pre-existing resistance in *Z. tritici* populations toward agrochemicals. The last group of inhibitors used for STB control is multisite inhibitors (e.g. folpet, chlorothalonil and mancozeb). These are thiol-reactive molecules acting on multiple targets (Medina-Cleghorn et al., 2015) and no resistance has been reported in *Z. tritici* so far (Birr et al., 2021). Recently, two novel complex III inhibitors, fenpicoxamid and metyltetraprole, have been reported. Fenpicoxamid is a semi-synthetic derivative of UK-2A, which is a natural molecule produced by *Streptomyces* sp. (Owen et al., 2017; Shibata et al., 1998). Fenpicoxamid is converted to UK-2A by *Z. tritici* and plant cells enabling the UK-2A molecule, structurally related to antimycin A, to specifically bind and inhibit the cytochrome bc1 complex at the quinone inside (Qi) site (Machida et al., 1999; Owen et al., 2017; Young et al., 2018). UK2A

Published by Burleigh Dodds Science Publishing Limited, 2022.

binds to a different target site of the cytochrome bc1 complex, and it does not display cross-resistance with QoI-resistant isolates. Metyltetraprole is a member of a new generation of QoI fungicides compatible with existing QoI resistance mutations that inhibit mitochondrial electron transport at the Qo site of the cytochrome bc1complex (Arakawa et al., 2018; Suemoto et al., 2019; Matsuzaki et al., 2020). Qi mutations, causing fenpicoxamid/UK-2A resistance, were generated in *Z. tritici* (Fouché et al., 2021) and *Saccharomyces cerevisiae* (Young et al., 2018). Novel molecules for the control of STB were recently reported. A novel C_{18} mono-alkyl lipophilic cation, displaying a high efficacy against STB, acts on mitochondrial respiration by inhibiting oxidative phosphorylation and inducing reactive oxygen species production at the level of respiratory complex I (Steinberg et al., 2020). Semi-synthetic rhamnolipids (RLs) were active on *Z. tritici* (Platel et al., 2020). Semi-synthetic sesamol esters displayed high activity against *Z. tritici* (Damiens et al., 2021). Effusol, a 9,10-dehydrophenanthrene natural product from rhizomes of the plant *Juncus maritimus*, is active against *Z. tritici* (Sahli et al., 2018). Cyclic lipopeptides from *Bacillus subtilis* such as mycosubtilin was active on *Z. tritici in vitro* and *in planta* (Mejri et al., 2018). In addition, Kerdraon et al. (2019) showed that *Z. tritici* alters microbial community networks during its crop residue pathogen stage. However, no significant direct interactions between microbes and *Z. tritici* were found. A deeper understanding of microbial networks will support the discovery of species which might be developed as biocontrol agents. Overall, because of the rapid emergence and spread of *Z. tritici* isolates resistant to most fungicides, there is a need for the discovery of the novel mode of actions and molecules (Torriani et al., 2015), and the development of application methods that lower the negative impact of fungicide resistance (timing of application, application rate, fungicides mixtures or alternation; van den Bosch et al., 2014).

5.2 S. tritici *blotch control using wheat resistant cultivars*

STB epidemics are also controlled by resistant wheat cultivars, carrying major resistance genes (*Stb*) and/or quantitative polygenic resistance genes. To date, 22 *Stb* genes have been identified in cultivated, landraces, wild and synthetic hexaploid wheat (Brown et al., 2015; Arraiano and Brown, 2017; Yang et al., 2018). All *Stb* genes have been overcome by virulent *Z. tritici* isolates (Cowger et al., 2000; Krenz et al., 2008; Brown et al., 2015; McDonald and Mundt, 2016). For example, *Stb1* and *Stb4* were efficient for over a decade, but they became ineffective due to the emergence of virulent isolates in *Z. tritici* populations which became widespread (Ponomarenko et al., 2011). To date, two wheat *Stb* resistance genes have been characterized. *Stb6* encodes a wheat receptor-like kinase conferring resistance against *Z. tritici* isolates carrying the *AvrStb6* avirulence gene (Saintenac et al., 2018). *Stb16q* encodes a wheat cysteine-rich receptor-like kinase conferring resistance against avirulent *Z. tritici* isolates (Saintenac et al., 2021). *AvrStb6* encodes a secreted protein with an unknown function that is expressed during infection (Zhong et al., 2017; Kema et al., 2018). Virulent isolates carry alleles of *AvrStb6* that are highly divergent from those of avirulent isolates (Zhong et al., 2017; Brunner and McDonald, 2018; Stephens et al., 2021). Another avirulence gene of *Z. tritici*, *Avr3D1*, also encodes a secreted protein with unknown function that is recognized by resistant wheat cultivars likely carrying the *Stb7* resistance gene (Meile et al., 2018). However, this recognition only induces partial resistance. Several sources of partial resistances have been identified as QTLs, although few turned out to be monogenic. For example, *Stb17* is involved in a quantitative resistance trait expressed at the adult plant stage (Tabib Ghaffary et al., 2011, 2012). To date, 167 QTLs of resistance have been detected in bi-parental mapping populations (Brown et al., 2015). A total of 105 resistance QTLs have been integrated into 27 meta-QTLs (Goudemand et al., 2013). These resistance QTLs were associated with either a reduction of leaf surface covered with necrosis or pycnidia or a reduction of disease progression (area under the disease progress curve, AUDPC). Two minor QTLs increasing the latent period have also been detected (Tabib Ghaffary et al., 2011; Hehir et al., 2018). A recent GWAS study of 175 winter wheat landraces and cultivars, identified 10 novel resistance QTLs (Odilbekov et al., 2019). Other sources of quantitative resistances have been discovered using molecular strategies, such as wheat small secreted proteins (SSPs) interacting with *Z. tritici* SSPs (Zhou et al., 2020), and taxonomically restricted wheat genes encoding proteins interacting with *Z. tritici* SSPs (Brennan et al., 2020). Unexpected responses to STB were identified in the wheat cultivar Stigg, displaying a partial resistance associated with a long asymptomatic phase (Benbow et al., 2020), and a weak defense response likely delaying the switch to necrotrophy (Benbow et al., 2020). In addition, the impact of STB can be reduced by traits contributing to disease escape, which limits the spread of fungal inoculum within crops (Arraiano et al., 2009). For example, wheat cultivars which are taller and later-heading will reduce the spread of spores to the upper leaves, and slow down the epidemic. However, disease escape traits can be undesirable in terms of agronomic properties and yield. Wheat genes suspected to enhance susceptibility to STB are thought to have been introduced inadvertently into UK wheat cultivars between the 1950s and 1980s, as a consequence of breeding programs aimed at increasing yield, rust and eyespot resistance (Arraiano and Brown, 2017). Indeed, in the United Kingdom, STB was not a major wheat disease from the 1900s to 1980s (Bearchell et al., 2005; Shaw et al., 2008). It re-occurred in the 1980s on newly released cultivars such as Norman and Longbow that turned out to be highly susceptible to STB (Arraiano and Brown, 2017). Overall, breeding for STB-resistance over the last 30 years in Europe presumably happened by phenotypic selection of field resistance rather than marker-assisted breeding, leading to the introduction of some *Stb* genes and the accumulation of resistance QTLs (Miedaner et al., 2013; Torriani et al., 2015; Mikaberidze and McDonald, 2020). Even if defeated, *Stb* resistance genes could be used either for pyramiding or for mixture/alternation deployment strategies (Kristoffersen et al., 2020). QTLs involved in partial resistance still needs to be better characterized at the genetic/molecular level, but their accumulation using marker-assisted breeding is feasible. Improvement of *Stb* genes by molecular techniques is also a promising area, as it may lead to novel resistance alleles recognizing virulent isolates, as proposed for other plant-microbe interactions (Bisht et al., 2019).

6 Conclusion

Z. tritici is an important fungal wheat pathogen both at the economic level and as a model to study plant-fungal interactions, their evolutionary dynamics and the molecular mechanisms involved. As such, it is considered among the top 10 fungal plant pathogens (Dean et al., 2012). *Z. tritici* has been the subject of intense genetics, molecular and genomics analyses (see Sections 2- 4). In particular, it has the largest number of ACs reported so far in fungi. These ACs have specific features compared to other chromosomes, such as a lower number of genes and a higher content of TEs. In addition, genes from ACs are less expressed than those from CCs including during infection. The role of ACs in the biology of *Z. tritici* is still open to debate, but there are no field isolates without ACs. In addition, many ACs genes are conserved in related *Zymoseptoria* species (Feurtey et al., 2020). Therefore, they likely carry non-essential genes important for the *Z. tritici* life cycle, including genes involved in adaptation to the host plant (Habig et al., 2017).

The molecular mechanisms underlying *Z. tritici* infection process are still poorly understood. Nevertheless, some important features of *Z. tritici* infection have been uncovered. Massive transcriptomic analyses have laid the foundations for studying the molecular mechanisms involved in this particular infection process. The early stages of *Z. tritici* infection are now better described. Spores germinate on the leaf surface and develop immediately into epiphytic hyphae that grow abundantly on the leaf surface up to 18 dpi (Fones et al., 2017; Fantozzi et al., 2021). This epiphytic behavior may have important consequences for the infection process and STB control. Indeed, it allows the penetration into wheat leaves through stomata over a long period (2 dpi to 13 dpi, Fantozzi et al., 2021). Since most penetration events are relatively late (4–6 dpi), the destruction of epiphytic hyphae that are easily accessible to fungicides could hamper infection. The penetration of *Z. tritici* through stomata is well-documented, but the mechanisms involved are poorly understood. *Z. tritici* does not differentiate single-cell appressoria. However, hyphal swellings were frequently described on the borders of stomata openings during penetration (Kema, 1996; Duncan and Howard, 2000). These swellings could help the fungus to redirect hyphal growth from the leaf surface plane to the interior of the stomata, as most fungal appressoria-like structures do (Demoor et al., 2019). These structures that have so far received little attention could be essential for the success of penetration. Transcriptomics of early infection stages (1-4 dpi, before penetration) are characterized by expression of fungal genes encoding proteins involved in detoxifying processes and stage-specific effectors and proteases, suggesting that *Z. tritici* is actively modifying its environment and manipulating its host. After penetration, infectious hyphae invade the mesophyll intercellular space (apoplast) without causing visible symptoms (asymptomatic phase) nor inducing plant defenses responses, up to 10 dpi. These infectious hyphae grow a long distance away from the sub-stomatal cavity colonized by penetrating hyphae (Fantozzi et al., 2021). This large exploration of the apoplast by *Z. tritici* hyphae creates a web of fungal cells in close contact with plant cells, as observed for cereal endophytes. This stage of infection is also not yet sufficiently documented, although transcriptomics identified stage-specific effectors and enzymes from secondary metabolism, suggesting that *Z. tritici* is actively manipulating its host plant. The switch to the necrotrophic stage is associated with an exponential fungal growth, expression of a large number of stage-specific effectors and enzymes involved in plant cell wall degradation, as well as stage-specific enzymes involved in secondary metabolism, although the triggers of this switch are not known. The role of these late effectors and secondary metabolites is still unknown, but some are likely involved in the development of chlorotic and necrotic symptoms. After this switch, *Z. tritici* recolonizes sub-stomatal cavities to differentiate pycnidia. Understanding this fungal development process (Tiley et al., 2018; Francisco et al., 2020) will be key for controlling this disease. Overall, *Z. tritici* infection is characterized by waves of expression of stage-specific effectors (proteins, secondary metabolites) and degradative enzymes (proteases, lipases and PCWDEs). These waves of fungal gene expression were also observed in hemibiotrophic fungal plant pathogens, including the massive expression of genes encoding proteases and PCWDEs during the switch to necrotrophy (O'Connell et al., 2012). These advances in understanding *Z. tritici* infection strongly suggest that it behaves as a hemibiotrophic fungus, acting as an endophyte during its asymptomatic phase. Even though the trigger that induces the switch to necrotrophy is unknown, the behavior of *Z. tritici* during this stage of infection relies on the massive expression of effectors and degradative enzymes, as observed in hemibiotrophic fungi. The regulatory networks involved in the control of these infection stage-specific waves of expression are unknown, but their study will likely bring important information on this infection process.

Z. tritici has not been always a major pathogen of wheat across the last century in Europe. Long-term trials in the United Kingdom (Broadbalk experiment from 1843 to 2003 in the United Kingdom in Bearchell et al., 2005; Shaw et al., 2008, and from 2015 to 2017 at other UK sites in Justesen et al., 2021) have shown that *Parastagonospora nodorum* and *Z. tritici* had shared and distinct occurrences on wheat over a long period of time. Both are leaf wheat pathogens from the Dothideomycetes, producing pycnidia on infected wheat leaves, although using very different infection strategies. Nevertheless, they frequently occur on the same infected wheat leaves (Justesen et al., 2021). PCR quantification of each fungal species in samples from these long-term trials have clearly shown that both fungi were detected in similar abundance in wheat leaves from 1840 to 1860, followed by a near absence of both pathogens until 1910. From 1910 to 1980, *P. nodorum* was the dominant species on wheat leaves, although *Z. tritici* was detected in low abundance. Since the 1980s, *P. nodorum* incidence has rapidly decreased, and it is now completely absent from most bread wheat fields in the United Kingdom and most European countries. Still, it is responsible for significant epidemics on *durum* wheat in Southern Europe (Iori et al., 2015), and on bread wheat in Northern Europe (Blixt et al., 2010; Justesen et al., 2021), Northern America and Australia (Downie et al., 2020). Since the 1980s, *Z. tritici* has emerged as the major pathogen of wheat in the United Kingdom and Europe, replacing *P. nodorum*. The biological causes underlying this intriguing switch in dominance between two fungal species having the same ecological niche (wheat) are still open to debate. Originally, it was proposed that the decrease in *P. nodorum* observed in the 1980s was due to the increase in SO_2 pollution, suggesting that *Z. tritici* was more resistant to this pollution. Subsequently, other parameters have been suggested to be involved, including the use of wheat cultivars that are highly susceptible to *Z. tritici* in the 1980s (Arraiano and Brown, 2017), the massive use of fungicides since the 1970s, and the development of resistance to these chemicals in *Z. tritici* (Lucas et al., 2015), but not in *P. nodorum*, which is still sensitive to many fungicides (Blixt et al., 2009). Despite being responsible for epidemics on wheat worldwide, *Z. tritici* is absent from some regions such as China, in which *P. nodorum* epidemics are frequent, and it is not the main pathogen of wheat in some regions such as Australia and North America, where *P. nodorum* dominates (Downie et al., 2020). These fluctuations in *Z. tritici* occurrence over time and space, are puzzling questions to tackle, but the answers could help to control this disease. In regions where *Z. tritici* is responsible for sustained and massive epidemics, this disease is still difficult to control. Fungicides have been the main control method in Europe since the 1980s, but the rapid emergence and spread of *Z. tritici* fungicide-resistant isolates have dramatically reduced the efficiency of some fungicides (Lucas et al., 2015). For example, strobilurins resistant *Z. tritici* isolates carrying a mutation of the mitochondrial *Cytb* gene have emerged

Published by Burleigh Dodds Science Publishing Limited, 2022.

rapidly (in a few years) after the release of these fungicides in the 1990s (Morton and Staub, 2008; Lucas et al., 2015). Because this resistance was rapidly fixed in most *Z. tritici* populations, the use of these fungicides was discontinued for controlling *Z. tritici*, but not other wheat fungal diseases. As a consequence, there is strong demand for the discovery of novel fungicides controlling *Z. tritici* populations resistant to previously developed molecules (Torriani et al., 2015). In addition, methods that counteract/lower the emergence of fungicide resistance are more than ever needed (van den Bosch et al., 2014). The same scenario has been observed for *Stb*-resistant wheat cultivars, and all *Stb* resistance genes have been defeated by virulent isolates (Brown et al., 2015), likely carrying a mutation in a single gene (avirulence gene, Zhong et al., 2017; Brunner and McDonald, 2018). As a consequence, there is a need for identifying new major resistance genes in wheat. Even if defeated, *Stb* resistance genes could be used either for pyramiding or for mixture/alternation deployment strategies, to counteract/lower the incidence of virulent isolates (Pilet-Nayel et al., 2017; Kristoffersen et al., 2020). QTLs involved in partial resistance are thought to be more durable, and their use in breeding programs is now more feasible, but their erosion due to constantly evolving *Z. tritici* populations is still possible, as it has been observed in other fungal plant interactions (Mundt, 2014).

7 Acknowledgements

Richard O'Connell is greatly acknowledged for his comments and proofreading the manuscript. We would also like to thank Richard Oliver, Cyrille Saintenac, Frederic Suffert, and Anne-Sophie Walker for sharing articles/ideas and for comments on the manuscript.

8 Where to look for further information

A large community of scientists is working on *Z. tritici*. A list of the laboratories involved is available as well as introductions to the different aspects of *Z. tritici* research, on the website: http://www.septoria-tritici-blotch.net/.

The *Z. tritici* research community is very active and can be followed up on Twitter: https://twitter.com/Zymoseptoria.

Regular meetings are organized (ISCLB). The last ISCLB was organized in Dublin in 2019. The program and abstracts are available at: https://researchrepository.ucd.ie/handle/10197/11177.

Many protocols for *Z. tritici* are available, and the best introductions are:

Fungal Genetics and Biology special issue on *Z. tritici* (2015) vol. 79.

Fagundes, W. C., Haueisen, J. and Stukenbrock, E. H. (2020), Dissecting the Biology of the Fungal Wheat Pathogen *Zymoseptoria tritici*: A Laboratory Workflow, *Current Protocols in Microbiology*, 59, e128.

https://www.protocols.io/workspaces/zymoseptoria-community-protocols.

Z. tritici genomics websites displaying different annotations of IPO323 genomic sequence.

https://mycocosm.jgi.doe.gov/Mycgr3/Mycgr3.home.html.

https://fungi.ensembl.org/Zymoseptoria_tritici/Info/Index.

https://fungidb.org/fungidb/app/record/dataset/DS_36dcafc6dd.

Genomic websites for the other *Z. tritici* stains and other *Zymoseptoria* species are available at Ensembl and Mycocosm.

Updates on fungicide resistance surveys of *Z. tritici* population and recommendations on using fungicides for its control are displayed at: FRAC: https://www.frac.info/frac-teams/working-groups/.

Impact of STB, disease management practices and cultivar resistance are displayed at:

ARVALIS (France): https://www.arvalis-infos.fr/septoriose-observer-pour-decider-d-une-intervention-@/view-18536-arvarticle.html.

AHDB (UK): https://ahdb.org.uk/knowledge-library/septoria-tritici-in-winter-wheat.

TEAGASC (Ireland): https://www.teagasc.ie/crops/crops/cereal-crops/spring-cereals/disease/.

GRDC (Australia): https://grdc.com.au/resources-and-publications/grdc-update-papers/tab-content/grdc-update-papers/2021/08/cereal-disease-update-2021.

FAR (New Zealand): https://www.far.org.nz/articles/tags/2/cereal/15/wheat.

International research centers working on wheat and *Z. tritici*:

CIMMYT: https://www.cimmyt.org/ and https://repository.cimmyt.org/.

ICARDA: https://www.icarda.org/.

9 References

Arakawa, A., Kasai, Y., Yamazaki, K. and Iwahashi, F. (2018). Features of interactions responsible for antifungal activity against resistant type cytochrome bc1: a data-driven analysis based on the binding free energy at the atomic level, *PLoS ONE* 13(11): e0207673. https://doi.org/10.1371/journal.pone.0207673.

Arraiano, L. S. and Brown, J. K. (2017). Sources of resistance and susceptibility to *Septoria tritici* blotch of wheat, *Mol. Plant Pathol.* 18(2), 276–292.

Arraiano, L. S., Balaam, N., Fenwick, P. M., Chapman, C., Feuerhelm, D., Howell, P., Smith, S. J., Widdowson, J. P. and Brown, J. K. M. (2009). Contributions of disease resistance and escape to the control of *Septoria tritici* blotch of wheat, *Plant Pathol.* 58(5), 910–922.

Badet, T., Oggenfuss, U., Abraham, L., McDonald, B. A. and Croll, D. (2020). A 19-isolate reference-quality global pangenome for the fungal wheat pathogen *Zymoseptoria tritici*, *BMC Biol.* 18(1), 12.

Bearchell, S. J., Fraaije, B. A., Shaw, M. W. and Fitt, B. D. (2005). Wheat archive links long-term fungal pathogen population dynamics to air pollution, *Proc. Natl Acad. Sci. U. S. A.* 102(15), 5438-5442.

Benbow, H. R., Brennan, C. J., Zhou, B., Christodoulou, T., Berry, S., Uauy, C., Mullins, E. and Doohan, F. M. (2020). Insights into the resistance of a synthetically-derived wheat to *Septoria tritici* blotch disease: less is more, *BMC Plant Biol.* 20(1), 407.

Ben M'Barek, S., Cordewener, J. H. G., Tabib Ghaffary, S. M., van der Lee, T. A. J., Liu, Z., Mirzadi Gohari, A., Mehrabi, R., America, A. H. P., Robert, O., Friesen, T. L., Hamza, S., Stergiopoulos, I., de Wit, P. J. and Kema, G. H. (2015). FPLC and liquid-chromatography mass spectrometry identify candidate necrosis-inducing proteins from culture filtrates of the fungal wheat pathogen *Zymoseptoria tritici*, *Fungal Genet. Biol.* 79, 54-62.

Berraies, S., Salah Gharbi, M., Rezgui, S. and Yahyaoui, A. (2014). Estimating grain yield losses caused by *Septoria* leaf blotch on durum wheat in Tunisia, *Chil. J. Agric. Res.* 74(4), 432-437.

Birr, T., Hasler, M., Verreet, J.-A. and Klink, H. (2021). Temporal changes in sensitivity of *Zymoseptoria tritici* field populations to different fungicidal modes of action, *Agriculture* 11(3), 269.

Bisht, D. S., Bhatia, V. and Bhattacharya, R. (2019). Improving plant-resistance to insect-pests and pathogens: the new opportunities through targeted genome editing, *Semin. Cell Dev. Biol.* 96, 65-76.

Blake, J. J., Gosling, P., Fraaije, B. A., Burnett, F. J., Knight, S. M., Kildea, S. and Paveley, N. D. (2018). Changes in field dose-response curves for demethylation inhibitor (DMI) and quinone outside inhibitor (QoI) fungicides against *Zymoseptoria tritici*, related to laboratory sensitivity phenotyping and genotyping assays, *Pest Manag. Sci.* 74(2), 302-313.

Blixt, E., Djurle, A., Yuen, J. and Olson, Å. (2009). Fungicide sensitivity in Swedish isolates of *Phaeosphaeria nodorum*, *Plant Pathol.* 58(4), 655-664.

Blixt, E., Olson, Å., Lindahl, B., Djurle, A. and Yuen, J. (2010). Spatiotemporal variation in the fungal community associated with wheat leaves showing symptoms similar to Stagonospora nodorum blotch, *Eur. J. Plant Pathol.* 126(3), 373-386.

Boukef, S., Yahyaoui and Rezgui, S. (2013). Geographical distribution of a specific mitochondrial haplotype of *Zymoseptoria tritici*, *Phytopathol. Mediterr.* 52(3), 466-471.

Bouma, E. (2005). Development of comparable agro-climatic zones for the international exchange of data on the efficacy and crop safety of plant protection products, *EPPO Bulletin* 35(2), 233-238.

Bowler, J., Scott, E., Tailor, R., Scalliet, G., Ray, J. and Csukai, M. (2010). New capabilities for Mycosphaerella graminicola research, *Mol. Plant Pathol.* 11(5), 691-704.

Brennan, C. J., Benbow, H. R., Mullins, E. and Doohan, F. M. (2019). A review of the known unknowns in the early stages of *Septoria tritici* blotch disease of wheat, *Plant Pathol.* 68(8), 1427-1438.

Brennan, C. J., Zhou, B., Benbow, H. R., Ajaz, S., Karki, S. J., Hehir, J. G., O'Driscoll, A., Feechan, A., Mullins, E. and Doohan, F. M. (2020). Taxonomically restricted wheat genes interact with small secreted fungal proteins and enhance resistance to *Septoria tritici* Blotch disease, *Front. Plant Sci.* 11, 433.

Brown, J. K. M., Chartrain, L., Lasserre-Zuber, P. and Saintenac, C. (2015). Genetics of resistance to *Zymoseptoria tritici* and applications to wheat breeding, *Fungal Genet. Biol.* 79, 33-41.

Brunner, P. C. and McDonald, B. A. (2018). Evolutionary analyses of the avirulence effector AvrStb6 in global populations of *Zymoseptoria tritici* identify candidate amino acids involved in recognition, *Mol. Plant Pathol.* 19(8), 1836–1846.

Brunner, P. C., Stefanato, F. L. and McDonald, B. A. (2008). Evolution of the CYP51 gene in *Mycosphaerella graminicola*: evidence for intragenic recombination and selective replacement, *Mol. Plant Pathol.* 9(3), 305–316.

Brunner, P. C., Torriani, S. F. F., Croll, D., Stukenbrock, E. H. and McDonald, B. A. (2013). Coevolution and life cycle specialization of plant cell wall degrading enzymes in a hemibiotrophic pathogen, *Mol. Biol. Evol.* 30(6), 1337–1347.

Cairns, T. and Meyer, V. (2017). *In silico* prediction and characterization of secondary metabolite biosynthetic gene clusters in the wheat pathogen *Zymoseptoria tritici*, *BMC Genomics* 18(1), 631.

Carmona, M., Sautua, F., Pérez-Hérnandez, O. and Reis, E. M. (2020). Role of fungicide applications on the integrated management of wheat stripe rust, *Front. Plant Sci.* 11, 733.

Chaloner, T. M., Fones, H. N., Varma, V., Bebber, D. P. and Gurr, S. J. (2019). A new mechanistic model of weather-dependent *Septoria tritici* blotch disease risk, *Philos. Trans. R. Soc. Lond. B Biol. Sci.* 374(1775), 20180266.

Chaudhari, Y., Cairns, T. C., Sidhu, Y., Attah, V., Thomas, G., Csukai, M., Talbot, N. J., Studholme, D. J. and Haynes, K. (2019). The *Zymoseptoria tritici* ORFeome: a functional genomics community resource, *Mol. Plant Microbe Interact.* 32(12), 1564–1570.

Cheval, P., Siah, A., Bomble, M., Popper, A. D., Reignault, P. and Halama, P. (2017). Evolution of QoI resistance of the wheat pathogen *Zymoseptoria tritici* in Northern France, *Crop Prot.* 92, 131–133.

Choi, Y. E. and Goodwin, S. B. (2011). Gene encoding a c-type cyclin in *Mycosphaerella graminicola* is involved in aerial mycelium formation, filamentous growth, hyphal swelling, melanin biosynthesis, stress response, and pathogenicity, *Mol. Plant Microbe Interact.* 24(4), 469–477.

Clarke, W. S. (2006). *Septoria tritici* and azole performance, *Aspects Appl. Biol.* 78, 127–132.

Cohen, L. and Eyal, Z. (1993). The histology of processes associated with the infection of resistant and susceptible wheat cultivars with *Septoria tritici*, *Plant Pathol.* 42(5), 737–743.

Cools, H. J. and Fraaije, B. A. (2013). Update on mechanisms of azole resistance in *Mycosphaerella graminicola* and implications for future control, *Pest Manag. Sci.* 69(2), 150–155.

Cools, H. J., Mullins, J. G. L., Fraaije, B. A., Parker, J. E., Kelly, D. E., Lucas, J. A. and Kelly, S. L. (2011). Impact of recently emerged sterol 14{alpha}-demethylase (CYP51) variants of Mycosphaerella graminicola on azole fungicide sensitivity, *Appl. Environ. Microbiol.* 77(11), 3830–3837.

Cools, H. J., Bayon, C., Atkins, S., Lucas, J. A. and Fraaije, B. A. (2012). Overexpression of the sterol 14α-demethylase gene (MgCYP51) in *Mycosphaerella graminicola* isolates confers a novel azole fungicide sensitivity phenotype, *Pest Manag. Sci.* 68(7), 1034–1040.

Cowger, C., Hoffer, M. E. and Mundt, C. C. (2000). Specific adaptation by *Mycosphaerella graminicola* to a resistant wheat cultivar, *Plant Pathol.* 49(4), 445–451.

Cowger, C., McDonald, B. A. and Mundt, C. C. (2002). Frequency of sexual reproduction by Mycosphaerella graminicola on partially resistant wheat cultivars, *Phytopathology* 92(11), 1175–1181.

Damiens, A., Alebrahim, M. T., Léonard, E., Fayeulle, A., Furman, C., Hilbert, J. L., Siah, A. and Billamboz, M. (2021). Sesamol-based terpenoids as promising bio-sourced crop protection compounds against the wheat pathogen *Zymoseptoria tritici*, *Pest Manag. Sci.* 77(5), 2403–2414.

Dean, R., Van Kan, J. A., Pretorius, Z. A., Hammond-Kosack, K. E., Di Pietro, A., Spanu, P. D., Rudd, J. J., Dickman, M., Kahmann, R., Ellis, J. and Foster, G. D. (2012). The Top 10 fungal pathogens in molecular plant pathology, *Mol. Plant Pathol.* 13(4), 414–430.

Demoor, A., Silar, P. and Brun, S. (2019). Appressorium: the breakthrough in *Dikarya*, *J. Fungi (Basel)* 5(3), 72.

Derbyshire, M. C., Gohari, A. M., Mehrabi, R., Kilaru, S., Steinberg, G., Ali, S., Bailey, A., Hammond-Kosack, K., Kema, G. H. J. and Rudd, J. J. (2018). Phosphopantetheinyl transferase (Ppt)-mediated biosynthesis of lysine, but not siderophores or DHN melanin, is required for virulence of *Zymoseptoria tritici* on wheat, *Sci. Rep.* 8(1), 17069.

Dhillon, B., Gill, N., Hamelin, R. C. and Goodwin, S. B. (2014). The landscape of transposable elements in the finished genome of the fungal wheat pathogen *Mycosphaerella graminicola*, *BMC Genomics* 15, 1132.

Djerbi, M., Kerlan, C. and Bompeix, G. (1974). Observations sur la morphogenèse et la cytologie des fructifications de *Septoria tritici*, *Ann. Inst. Natl Rech. Agronomique Tunis.* 47, 3–53.

Dooley, H., Shaw, M. W., Spink, J. and Kildea, S. (2016). Effect of azole fungicide mixtures, alternations and dose on azole sensitivity in the wheat pathogen *Zymoseptoria tritici*, *Plant Pathol.* 65(1), 124–136.

Downie, R. C., Lin, M., Corsi, B., Ficke, A., Lillemo, M., Oliver, R. P., Phan, H., Tan, K. C. and Cockram, J. (2020). Septoria nodorum blotch of wheat: disease management and resistance breeding in the face of shifting disease dynamics and a changing environment, *Phytopathol* 10, 1094.

Duba, A., Goriewa-Duba, K. and Wachowska, U. (2018). A review of the Interactions between wheat and wheat pathogens: *Zymoseptoria tritici*, Fusarium spp. and Parastagonospora nodorum, *Int. J. Mol. Sci.* 19(4): 1138. https://doi.org/10.3390/ijms19041138.

Dufresne, M. and Daboussi, M. J. (2010). Development of *impala*-based transposon systems for gene tagging in filamentous fungi, *Methods Mol. Biol., (Clifton, N.J.)* 638, 41–54.

Duncan, K. E. and Howard, R. J. (2000). Cytological analysis of wheat infection by the leaf blotch pathogen *Mycosphaerella graminicola*, *Mycol. Res.* 104(9), 1074–1082.

El Chartouni, L., Randoux, B., Duyme, F., Renard-Merlier, D., Tisserant, B., Bourdon, N., Pillon, V., Sanssené, J., Durand, R., Reignault, P. and Halama, P. (2012). Correlation of cytological and biochemical parameters with resistance and tolerance to *Mycosphaerella graminicola* in wheat, *Plant Biol. (Stuttg.)* 14(Suppl. 1), 11–21.

Eriksen, L. and Munk, L. (2003). The occurrence of *Mycosphaerella graminicola* and its anamorph *Septoria tritici* in winter wheat during the growing season, *Eur. J. Plant Pathol.* 109(3), 253–259.

Fagundes, W. C., Haueisen, J. and Stukenbrock, E. H. (2020). Dissecting the biology of the fungal wheat pathogen *Zymoseptoria tritici*: a laboratory workflow, *Curr. Protoc. Microbiol.* 59(1), e128.

Fantozzi, E., Kilaru, S., Cannon, S., Schuster, M., Gurr, S. J. and Steinberg, G. (2021a). Conditional promoters to investigate gene function during wheat infection by *Zymoseptoria tritici*, *Fungal Genet. Biol.* 146, 103487.

Fantozzi, E., Kilaru, S., Gurr, S. J. and Steinberg, G. (2021b). Asynchronous development of *Zymoseptoria tritici* infection in wheat, *Fungal Genet. Biol.* 146, 103504.

Feurtey, A., Stevens, D. M., Stephan, W. and Stukenbrock, E. H. (2019). Interspecific gene exchange introduces high genetic variability in crop pathogen, *Genome Biol. Evol.* 11(11), 3095-3105.

Feurtey, A., Lorrain, C., Croll, D., Eschenbrenner, C., Freitag, M., Habig, M., Haueisen, J., Möller, M., Schotanus, K. and Stukenbrock, E. H. (2020). Genome compartmentalization predates species divergence in the plant pathogen genus *Zymoseptoria*, *BMC Genomics* 21(1), 588.

Fones, H. N. and Gurr, S. J. (2015). The impact of *Septoria tritici* Blotch disease on wheat: an EU perspective, *Fungal Genet. Biol.* 79, 3-7.

Fones, H. N., Steinberg, G. and Gurr, S. J. (2015). Measurement of virulence in *Zymoseptoria tritici* through low inoculum-density assays, *Fungal Genet. Biol.* 79, 89-93.

Fones, H. N., Eyles, C. J., Kay, W., Cowper, J. and Gurr, S. J. (2017). A role for random, humidity-dependent epiphytic growth prior to invasion of wheat by *Zymoseptoria tritici*, *Fungal Genet. Biol.* 106, 51-60.

Fouché, G., Michel, T., Lalève, A, Wang, N.X., Young, D. H., Meunier, B., Debieu, D., Fillinger, S and Walker, A. S. (2021). Directed evolution predicts cytochrome b G37V target site modification as probable adaptive mechanism towards the QiI fungicide fenpicoxamid in *Zymoseptoria tritici*, *Environ. Microbiol.* (accepted).

Fouché, S., Badet, T., Oggenfuss, U., Plissonneau, C., Francisco, C. S. and Croll, D. (2020). Stress-driven transposable element de-repression dynamics and virulence evolution in a fungal pathogen, *Mol. Biol. Evol.* 37(1), 221-239.

Fraaije, B. A., Cools, H. J., Fountaine, J., Lovell, D. J., Motteram, J., West, J. S. and Lucas, J. A. (2005). Role of ascospores in further spread of QoI-Resistant cytochrome b alleles (G143A) in field populations of Mycosphaerella graminicola, Phytopathol. 95, 933-941.

Fraaije, B. A., Cools, H. J., Kim, S. H., Motteram, J., Clark, W. S. and Lucas, J. A. (2007). A novel substitution I381V in the sterol 14alpha-demethylase (CYP51) of *Mycosphaerella graminicola* is differentially selected by azole fungicides, *Mol. Plant Pathol.* 8(3), 245-254.

Fraaije, B. A., Bayon, C., Atkins, S., Cools, H. J., Lucas, J. A. and Fraaije, M. W. (2012). Risk assessment studies on succinate dehydrogenase inhibitors, the new weapons in the battle to control *Septoria* leaf blotch in wheat, *Mol. Plant Pathol.* 13(3), 263-275.

Francisco, C. S., Ma, X., Zwyssig, M. M., McDonald, B. A. and Palma-Guerrero, J. (2019). Morphological changes in response to environmental stresses in the fungal plant pathogen *Zymoseptoria tritici*, *Sci. Rep.* 9(1), 9642.

Francisco, C. S., Zwyssig, M. M. and Palma-Guerrero, J. (2020). The role of vegetative cell fusions in the development and asexual reproduction of the wheat fungal pathogen *Zymoseptoria tritici*, *BMC Biol.* 18(1), 99.

Garnault, M., Duplaix, C., Leroux, P., Couleaud, G., Carpentier, F., David, O. and Walker, A. S. (2019). Spatiotemporal dynamics of fungicide resistance in the wheat pathogen *Zymoseptoria tritici* in France, *Pest Manag. Sci.* 75(7), 1794-1807.

Garnault, M., Duplaix, C., Leroux, P., Couleaud, G., David, O., Walker, A. S. and Carpentier, F. (2020). Large-scale study validates that regional fungicide applications are major determinants of resistance evolution in the wheat pathogen *Zymoseptoria tritici* in France, *New Phytol.* 229(6), 3508-3521.

Gisi, U., Sierotzki, H., Cook, A. and McCaffery, A. (2002). Mechanisms influencing the evolution of resistance to Qo inhibitor fungicides, *Pest Manag. Sci.* 58(9), 859–867.

Gluck-Thaler, E., Haridas, S., Binder, M., Grigoriev, I. V., Crous, P. W., Spatafora, J. W., Bushley, K. and Slot, J. C. (2020). The architecture of metabolism maximizes biosynthetic diversity in the largest class of fungi, *Mol. Biol. Evol.* 37(10), 2838–2856.

Goodwin, S. B., Ben M'Barek, S. B., Dhillon, B., Wittenberg, A. H. J., Crane, C. F., Hane, J. K., Foster, A. J., Van der Lee, T. A. J., Grimwood, J., Aerts, A., Antoniw, J., Bailey, A., Bluhm, B., Bowler, J., Bristow, J., van der Burgt, A., Canto-Canché, B., Churchill, A. C., Conde-Ferràez, L., Cools, H. J., Coutinho, P. M., Csukai, M., Dehal, P., De Wit, P., Donzelli, B., van de Geest, H. C., van Ham, R. C., Hammond-Kosack, K. E., Henrissat, B., Kilian, A., Kobayashi, A. K., Koopmann, E., Kourmpetis, Y., Kuzniar, A., Lindquist, E., Lombard, V., Maliepaard, C., Martins, N., Mehrabi, R., Nap, J. P., Ponomarenko, A., Rudd, J. J., Salamov, A., Schmutz, J., Schouten, H. J., Shapiro, H., Stergiopoulos, I., Torriani, S. F., Tu, H., de Vries, R. P., Waalwijk, C., Ware, S. B., Wiebenga, A., Zwiers, L. H., Oliver, R. P., Grigoriev, I. V. and Kema, G. H. (2011). Finished genome of the fungal wheat pathogen *Mycosphaerella graminicola* reveals dispensome structure, chromosome plasticity, and stealth pathogenesis, *PLoS Genet.* 7(6), e1002070.

Goudemand, E., Laurent, V., Duchalais, L., Tabib Ghaffary, S. M., Kema, G. H. J., Lonnet, P., Margale, E. and Robert, O. (2013). Association mapping and meta-analysis: two complementary approaches for the detection of reliable *Septoria tritici* blotch quantitative resistance in bread wheat (*Triticum aestivum* L.), *Mol. Breed.* 32(3), 563–584.

Grandaubert, J., Bhattacharyya, A. and Stukenbrock, E. H. (2015). RNA-seq-based gene annotation and comparative genomics of four fungal grass pathogens in the genus *Zymoseptoria* identify novel orphan genes and species-specific invasions of transposable elements, *G3 (Bethesda)* 5(7), 1323–1333.

Grasso, V., Palermo, S., Sierotzki, H., Garibaldi, A. and Gisi, U. (2006). Cytochrome b gene structure and consequences for resistance to Qo inhibitor fungicides in plant pathogens, *Pest Manag. Sci.* 62(6), 465–472.

Habig, M., Quade, J. and Stukenbrock, E. H. (2017). Forward genetics approach reveals host genotype-dependent importance of accessory chromosomes in the fungal wheat pathogen *Zymoseptoria tritici*, *mBio* 8(6): e01919–17.

Habig, M., Bahena-Garrido, S. M., Barkmann, F., Haueisen, J. and Stukenbrock, E. H. (2020). The transcription factor Zt107320 affects the dimorphic switch, growth and virulence of the fungal wheat pathogen *Zymoseptoria tritici*, *Mol. Plant Pathol.* 21(1), 124–138.

Hailemariam, B. N., Kidane, Y. G. and Ayalew, A. (2020). Epidemiological factors of *Septoria tritici* blotch (*Zymoseptoria tritici*) in durum wheat (*Triticum turgidum*) in the highlands of Wollo, Ethiopia, *Ecol. Process.* 9(1), 61.

Hartmann, F. E., McDonald, B. A. and Croll, D. (2018). Genome-wide evidence for divergent selection between populations of a major agricultural pathogen, *Mol. Ecol.* 27(12), 2725–2741.

Hartmann, F. E., Vonlanthen, T., Singh, N. K., McDonald, M. C., Milgate, A. and Croll, D. (2020). The complex genomic basis of rapid convergent adaptation to pesticides across continents in a fungal plant pathogen, *Mol. Ecol.*, 1–16. https://doi.org/10.1111/mec.15737.

Hassine, M., Siah, A., Hellin, P., Cadalen, T., Halama, P., Hilbert, J. L., Hamada, W., Baraket, M., Yahyaoui, A., Legrève, A. and Duvivier, M. (2019). Sexual reproduction of *Zymoseptoria tritici* on durum wheat in Tunisia revealed by presence of airborne inoculum, fruiting bodies and high levels of genetic diversity, *Fungal Biol.* 123(10), 763–772.

Haueisen, J., Möller, M., Eschenbrenner, C. J., Grandaubert, J., Seybold, H., Adamiak, H. and Stukenbrock, E. H. (2018). Highly flexible infection programs in a specialized wheat pathogen, *Ecol. Evol.* 9(1), 275–294.

Hehir, J. G., Connolly, C., O'Driscoll, A., Lynch, J. P., Spink, J., Brown, J. K. M., Doohan, F. and Mullins, E. (2018). Temporal and spatial field evaluations highlight the importance of the pre-symptomatic phase in supporting strong partial resistance in *Triticum aestivum* against *Zymoseptoria tritici*, *Plant Pathol.* 67(3), 573–583.

Hill, E. H. and Solomon, P. S. (2020). Extracellular vesicles from the apoplastic fungal wheat pathogen *Zymoseptoria tritici*, *Fungal Biol. Biotechnol.* 7, 13.

Horbach, R., Graf, A., Weihmann, F., Antelo, L., Mathea, S., Liermann, J. C., Opatz, T., Thines, E., Aguirre, J. and Deising, H. B. (2009). Sfp-type 4'-phosphopantetheinyl transferase is indispensable for fungal pathogenicity, *Plant Cell* 21(10), 3379–3396.

Iori, A., Scala, V., Cesare, D., Pinzari, F., D'Egidio, M. G., Fanelli, C., Fabbri, A. A., Reverberi, M. and Serranti, S. (2015). Hyperspectral and molecular analysis of *Stagonospora nodorum* blotch disease in durum wheat, *Eur. J. Plant Pathol.* 141(4), 689–702.

Jalli, M., Kaseva, J., Andersson, B., Ficke, A., Nistrup-Jørgensen, L., Ronis, A., Kaukoranta, T., Ørum, J. and Djurle, A. (2020) Yield increases due to fungicide control of leaf blotch diseases in wheat and barley as a basis for IPM decision-making in the Nordic-Baltic region, *Eur. J. Plant Pathol.* 158(2), 315–333.

Justesen, A. F., Corsi, B., Ficke, A., Hartl, L., Holdgate, S., Jorgensen, L. N., Lillemo, M., Lin, M., MacKay, I. J., Mohler, V., Stadlmeier, M., Tan, K. C., Turner, J., Oliver, R. P. and Cockram, J. (2021). Hidden in plain sight: a molecular field survey of three wheat leaf blotch fungal diseases in North-Western Europe shows co-infection is widespread, *Eur. J. Plant Pathol.*

Karisto, P., Suffert, F. and Mikaberidze, A. (2021). Measuring splash-dispersal of a major wheat pathogen in the field, BioRxiv, 2021.03.23.436423.

Karki, S. J., Reilly, A., Zhou, B., Mascarello, M., Burke, J., Doohan, F., Douchkov, D., Schweizer, P. and Feechan, A. (2021). A small secreted protein from *Zymoseptoria tritici* interacts with a wheat E3 ubiquitin to promote disease, *J. Exp. Bot.* 72(2), 733–746.

Kay, W. T., Fones, H. N. and Gurr, S. J. (2019). Rapid loss of virulence during submergence of *Z. tritici* asexual spores, *Fungal Genet. Biol.* 128, 14–19.

Kellner, R., Bhattacharyya, A., Poppe, S., Hsu, T. Y., Brem, R. B. and Stukenbrock, E. H. (2014). Expression profiling of the wheat pathogen *Zymoseptoria tritici* reveals genomic patterns of transcription and host-specific regulatory programs, *Genome Biol. Evol.* 6(6), 1353–1365.

Kema, G. (1996). Histology of the pathogenesis of Mycosphaerella graminicola in wheat, Phytopathol 86, 777.

Kema, G. H., van der Lee, T. A., Mendes, O., Verstappen, E. C., Lankhorst, R. K., Sandbrink, H., van der Burgt, A., Zwiers, L. H., Csukai, M. and Waalwijk, C. (2008). Large-scale gene discovery in the *Septoria tritici* blotch fungus *Mycosphaerella graminicola* with a focus on in planta expression, *Mol. Plant Microbe Interact.* 21(9), 1249–1260.

Kema, G. H. J., Mirzadi Gohari, A., Aouini, L., Gibriel, H. A. Y., Ware, S. B., van den Bosch, F., Manning-Smith, R., Alonso-Chavez, V., Helps, J., Ben M'Barek, S., Mehrabi, R., Diaz-Trujillo, C., Zamani, E., Schouten, H. J., van der Lee, T. A. J., Waalwijk, C., de Waard, M. A., de Wit, P. J. G. M., Verstappen, E. C. P., Thomma, B. P. H. J., Meijer, H. J. G. and

Seidl, M. F. (2018). Stress and sexual reproduction affect the dynamics of the wheat pathogen effector *AvrStb6* and strobilurin resistance, *Nat. Genet.* 50(3), 375–380.

Keon, J., Antoniw, J., Carzaniga, R., Deller, S., Ward, J. L., Baker, J. M., Beale, M. H., Hammond-Kosack, K. and Rudd, J. J. (2007). Transcriptional adaptation of *Mycosphaerella graminicola* to programmed cell death (PCD) of its susceptible wheat host, *Mol. Plant Microbe Interact.* 20(2), 178–193.

Kerdraon, L., Barret, M., Laval, V. and Suffert, F. (2019). Differential dynamics of microbial community networks help identify microorganisms interacting with residue-borne pathogens: the case of *Zymoseptoria tritici* in wheat, *Microbiome* 7(1), 125.

Kettles, G. J. and Kanyuka, K. (2016). Dissecting the molecular interactions between wheat and the fungal pathogen *Zymoseptoria tritici*, *Front. Plant Sci.* 7, 508.

Kettles, G. J., Bayon, C., Canning, G., Rudd, J. J. and Kanyuka, K. (2017). Apoplastic recognition of multiple candidate effectors from the wheat pathogen *Zymoseptoria tritici* in the nonhost plant *Nicotiana benthamiana*, *New Phytol.* 213(1), 338–350.

Kettles, G. J., Bayon, C., Sparks, C. A., Canning, G., Kanyuka, K. and Rudd, J. J. (2018). Characterization of an antimicrobial and phytotoxic ribonuclease secreted by the fungal wheat pathogen *Zymoseptoria tritici*, *New Phytol.* 217(1), 320–331.

Kettles, G. J., Hofinger, B. J., Hu, P., Bayon, C., Rudd, J. J., Balmer, D., Courbot, M., Hammond-Kosack, K. E., Scalliet, G. and Kanyuka, K. (2019). sRNA Profiling combined with gene function analysis reveals a lack of evidence for cross-kingdom RNAi in the wheat - *Zymoseptoria tritici* pathosystem, *Front. Plant Sci.* 10, 892.

Kilaru, S., Schuster, M., Studholme, D., Soanes, D., Lin, C., Talbot, N. J. and Steinberg, G. (2015a). A codon-optimized green fluorescent protein for live cell imaging in *Zymoseptoria tritici*, *Fungal Genet. Biol.* 79, 125–131.

Kilaru, S., Ma, W., Schuster, M., Courbot, M. and Steinberg, G. (2015b). Conditional promoters for analysis of essential genes in *Zymoseptoria tritici*, *Fungal Genet. Biol.* 79, 166–173.

Kilaru, S., Schuster, M., Ma, W. and Steinberg, G. (2017). Fluorescent markers of various organelles in the wheat pathogen *Zymoseptoria tritici*, *Fungal Genet. Biol.* 105, 16–27.

King, R., Urban, M., Lauder, R. P., Hawkins, N., Evans, M., Plummer, A., Halsey, K., Lovegrove, A., Hammond-Kosack, K. and Rudd, J. J. (2017). A conserved fungal glycosyltransferase facilitates pathogenesis of plants by enabling hyphal growth on solid surfaces, *PLoS Pathog.* 13(10), e1006672.

Khan, H. (2017). Protoplast isolation from *Zymoseptoria tritici*, In protocols.io. DOI:10.17504/protocols.io.k2mcyc6.

Krenz, J. E., Sackett, K. E. and Mundt, C. C. (2008). Specificity of incomplete resistance to *Mycosphaerella graminicola* in wheat, *Phytopathology* 98(5), 555–561.

Krishnan, P., Ma, X., McDonald, B. A. and Brunner, P. C. (2018a). Widespread signatures of selection for secreted peptidases in a fungal plant pathogen, *BMC Evol. Biol.* 18(1), 7.

Krishnan, P., Meile, L., Plissonneau, C., Ma, X., Hartmann, F. E., Croll, D., McDonald, B. A. and Sánchez-Vallet, A. (2018b). Transposable element insertions shape gene regulation and melanin production in a fungal pathogen of wheat, *BMC Biol.* 16(1), 78.

Kristoffersen, R., Jørgensen, L. N., Eriksen, L. B., Nielsen, G. C. and Kiær, L. P. (2020). Control of Septoria tritici blotch by winter wheat cultivar mixtures: meta-analysis of 19 years of cultivar trials, *Field Crops Res.* 249, 107696.

Lee, W. S., Rudd, J. J., Hammond-Kosack, K. E. and Kanyuka, K. (2014). *Mycosphaerella graminicola* LysM effector-mediated stealth pathogenesis subverts recognition through both CERK1 and CEBiP omologuhes in wheat, *Mol. Plant Microbe Interact.* 27(3), 236–243.

Lendenmann, M. H., Croll, D., Stewart, E. L. and McDonald, B. A. (2014). Quantitative trait locus mapping of melanization in the plant pathogenic fungus *Zymoseptoria tritici*, *G3 (Bethesda)* 4(12), 2519–2533.

Leroux, P. and Walker, A. S. (2011). Multiple mechanisms account for resistance to sterol 14α-demethylation inhibitors in field isolates of *Mycosphaerella graminicola*, *Pest Manag. Sci.* 67(1), 44–59.

Leroux, P., Albertini, C., Gautier, A., Gredt, M. and Walker, A. S. (2007). Mutations in the CYP51 gene correlated with changes in sensitivity to sterol 14 alpha-demethylation inhibitors in field isolates of *Mycosphaerella* Graminicola, *Pest Manag. Sci.* 63(7), 688–698.

Lovell, D. J., Parker, S. R., Hunter, T., Welham, S. J. and Nichols, A. R. (2004). Position of inoculum in the canopy affects the risk of *Septoria tritici* blotch epidemics in winter wheat, *Plant Pathol.* 53(1), 11–21.

Lucas, J. A., Hawkins, N. J. and Fraaije, B. A. (2015). The evolution of fungicide resistance, *Adv. Appl. Microbiol.* 90, 29–92.

Ma, W., Kilaru, S., Collins, C., Courbot, M. and Steinberg, G. (2015). Libraries for two-hybrid screening of yeast and hyphal growth forms in *Zymoseptoria tritici*, *Fungal Genet. Biol.* 79, 94–101.

Ma, X., Keller, B., McDonald, B. A., Palma-Guerrero, J. and Wicker, T. (2018). Comparative transcriptomics reveals how wheat responds to infection by *Zymoseptoria tritici*, *Mol. Plant Microbe Interact.* 31(4), 420–431.

Ma, X., Wiedmer, J. and Palma-Guerrero, J. (2020). Small RNA bidirectional crosstalk during the interaction between wheat and *Zymoseptoria tritici*, *Front. Plant Sci.* 10, 1669.

Machida, K., Takimoto, H., Miyoshi, H. and Taniguchi, M. (1999). UK-2A,B,C and D, novel antifungal antibiotics from *Streptomyces* sp.517.02. V. Inhibition mechanism of bovine heart mitochondrial cytochrome bc1 by the novel antibiotic UK-2A, *J. Antibiot. (Tokyo)* 52(8), 748–753.

Mäe, A., Fillinger, S., Sooväli, P. and Heick, T. M. (2020). Fungicide sensitivity shifting of *Zymoseptoria tritici* in the Finnish-Baltic region and a novel Insertion in the *MFS1* promoter, *Front. Plant Sci.* 11, 385.

Marchegiani, E., Sidhu, Y., Haynes, K. and Lebrun, M. H. (2015). Conditional gene expression and promoter replacement in *Zymoseptoria tritici* using fungal nitrate reductase promoters, *Fungal Genet. Biol.* 79, 174–179.

Marshall, R., Kombrink, A., Motteram, J., Loza-Reyes, E., Lucas, J., Hammond-Kosack, K. E., Thomma, B. P. H. J. and Rudd, J. J. (2011). Analysis of two in planta expressed LysM effector homologs from the fungus *Mycosphaerella graminicola* reveals novel functional properties and varying contributions to virulence on wheat, *Plant Physiol.* 156(2), 756–769.

Matsuzaki, Y., Kiguchi, S., Suemoto, H. and Iwahashi, F. (2020). Antifungal activity of metyltetraprole against the existing QoI-resistant isolates of various plant pathogenic fungi: metyltetraprole against QoI-R isolates, *Pest Manag. Sci.* 76(5), 1743–1750.

McDonald, B. A. and Mundt, C. C. (2016). How knowledge of pathogen population biology informs management of Septoria tritici Blotch, *Phytopathology* 106(9), 948–955.

McDonald, M. C., McDonald, B. A. and Solomon, P. S. (2015a). Recent advances in the *Zymoseptoria tritici*-wheat interaction: insights from pathogenomics, *Front. Plant Sci.* 6, 102.

McDonald, M. C., Williams, A. H., Milgate, A., Pattemore, J. A., Solomon, P. S. and Hane, J. K. (2015b). Next-generation re-sequencing as a tool for rapid bioinformatic screening of presence and absence of genes and accessory chromosomes across isolates of *Zymoseptoria tritici*, *Fungal Genet. Biol.* 79, 71-75.

McDonald, M. C., Renkin, M., Spackman, M., Orchard, B., Croll, D., Solomon, P. S. and Milgate, A. (2019). Rapid parallel evolution of azole fungicide resistance in Australian populations of the wheat pathogen *Zymoseptoria tritici*, *Appl. Environ. Microbiol.* 85(4), e01908-18.

Medina-Cleghorn, D., Bateman, L. A., Ford, B., Heslin, A., Fisher, K. J., Dalvie, E. D. and Nomura, D. K. (2015). Mapping proteome-wide targets of environmental chemicals using reactivity-based chemoproteomic platforms, *Chem. Biol.* 22(10), 1394-1405.

Mehrabi, R., Zwiers, L. H., de Waard, M. A. and Kema, G. H. J. (2006a). MgHog1 regulates dimorphism and pathogenicity in the fungal wheat pathogen *Mycosphaerella graminicola*, *Mol. Plant Microbe Interact.* 19(11), 1262-1269.

Mehrabi, R., van der Lee, T., Waalwijk, C. and Kema, H. J. (2006b). MgSlt2, a cellular integrity MAP kinase gene of the fungal wheat pathogen *Mycosphaerella graminicola*, is dispensable for penetration but essential for invasive growth, *Mol. Plant Microbe Interact.* 19(4), 389-398.

Mehrabi, R., M'Barek, S. B., Lee, T. A. J. van der, Waalwijk, C., Wit, P. J. G. M. de and Kema, G. H. J. (2009). Gα and Gβ proteins regulate the cyclic AMP pathway that is required for development and pathogenicity of the phytopathogen *Mycosphaerella graminicola*, *Eukaryot. Cell* 8, 1001-1013.

Mehrabi, R., Mirzadi Gohari, A., da Silva, G. F., Steinberg, G., Kema, G. H. and de Wit, P. J. (2015). Flexible gateway constructs for functional analyses of genes in plant pathogenic fungi, *Fungal Genet. Biol.* 79, 186-192.

Meile, L., Croll, D., Brunner, P. C., Plissonneau, C., Hartmann, F. E., McDonald, B. A. and Sánchez-Vallet, A. (2018). A fungal avirulence factor encoded in a highly plastic genomic region triggers partial resistance to *Septoria tritici* blotch, *New Phytol.* 219(3), 1048-1061.

Meile, L., Peter, J., Puccetti, G., Alassimone, J., McDonald, B. A. and Sánchez-Vallet, A. (2020). Chromatin dynamics contribute to the spatiotemporal expression pattern of virulence genes in a fungal plant pathogen, *mBio* 11(5), e02343-20.

Mejri, S., Siah, A., Coutte, F., Magnin-Robert, M., Randoux, B., Tisserant, B., Krier, F., Jacques, P., Reignault, P. and Halama, P. (2018). Biocontrol of the wheat pathogen *Zymoseptoria tritici* using cyclic lipopeptides from *Bacillus subtilis*, *Environ. Sci. Pollut. Res. Int.* 25(30), 29822-29833.

Mekonnen, T., Haileselassie, T., Goodwin, S. B. and Tesfayea, K. (2020). Genetic diversity and population structure of *Zymoseptoria tritici* in Ethiopia as revealed by microsatellite markers, *Fungal Genet. Biol.* 141, 103413.

Miedaner, T., Zhao, Y., Gowda, M., Longin, C. F., Korzun, V., Ebmeyer, E., Kazman, E. and Reif, J. C. (2013). Genetic architecture of resistance to *Septoria tritici* blotch in European wheat, *BMC Genomics* 14(1), 858.

Mikaberidze, A. and McDonald, B. A. (2020). A tradeoff between tolerance and resistance to a major fungal pathogen in elite wheat cultivars, *New Phytol.* 226(3), 879-890.

Milgate, A., Adorada, D., Orchard, B. and Pattemore, J. (2016). First report of resistance to DMI fungicides in Australian populations of the wheat pathogen *Zymoseptoria tritici*, *Plant Dis.* 100(2), 522–522.

Mirzadi Gohari, A. M., Mehrabi, R., Robert, O., Ince, I. A., Boeren, S., Schuster, M., Steinberg, G., de Wit, P. J. G. M. and Kema, G. H. J. (2014). Molecular characterization and functional analyses of ZtWor1, a transcriptional regulator of the fungal wheat pathogen *Zymoseptoria tritici*, *Mol. Plant Pathol.* 15(4), 394–405.

Mirzadi Gohari, A. M., Ware, S. B., Wittenberg, A. H. J., Mehrabi, R., M'Barek, S. B., Verstappen, E. C. P., Lee, T. A. J. van der, Robert, O., Schouten, H. J., Wit, P. P. J. G. M. de and Kema, G. H. (2015). Effector discovery in the fungal wheat pathogen *Zymoseptoria tritici*, *Mol. Plant Pathol.* 16(9), 931–945.

Mohammadi, N., Mehrabi, R., Mirzadi Gohari, A., Mohammadi Goltapeh, E., Safaie, N. and Kema, G. H. J. (2017). The ZtVf1 transcription factor regulates development and virulence in the foliar wheat pathogen *Zymoseptoria tritici*, *Fungal Genet. Biol.* 109, 26–35.

Mohammadi, N., Mehrabi, R., Mirzadi Gohari, A., Roostaei, M., Mohammadi Goltapeh, E., Safaie, N. and Kema, G. H. J. (2020). MADS-Box transcription factor ZtRlm1 is responsible for virulence and development of the fungal wheat pathogen *Zymoseptoria tritici*, *Front. Microbiol.* 11, 1976.

Morais, D., Sache, I., Suffert, F. and Laval, V. (2016). Is the onset of *Septoria tritici* blotch epidemics related to the local pool of ascospores?, *Plant Pathol.* 65(2), 250–260.

Morais do Amaral, A. M. do, Antoniw, J., Rudd, J. J. and Hammond-Kosack, K. E. (2012). Defining the predicted protein secretome of the fungal wheat leaf pathogen *Mycosphaerella graminicola*, *PLoS ONE* 7(12), e49904.

Morton, V. and Staub, T. (2008). A short history of fungicides. Online, *APSnet* features. Available at: https://www.apsnet.org/edcenter/apsnetfeatures/Pages/Fungicides.aspx.

Motteram, J., Küfner, I., Deller, S., Brunner, F., Hammond-Kosack, K. E., Nürnberger, T. and Rudd, J. J. (2009). Molecular characterization and functional analysis of MgNLP, the sole NPP1 domain-containing protein, from the fungal wheat leaf pathogen Mycosphaerella graminicola, *Mol. Plant Microbe Interact.* 22(7), 790–799.

Motteram, J., Lovegrove, A., Pirie, E., Marsh, J., Devonshire, J., van de Meene, A., Hammond-Kosack, K. and Rudd, J. J. (2011). Aberrant protein N-glycosylation impacts upon infection-related growth transitions of the haploid plant-pathogenic fungus *Mycosphaerella graminicola*, *Mol. Microbiol.* 81(2), 415–433.

Mundt, C. C. (2014). Durable resistance: a key to sustainable management of pathogens and pests, *Infect. Genet. Evol.* 27, 446–455.

Muria-Gonzalez, M. J., Chooi, Y. H., Breen, S. and Solomon, P. S. (2015). The past, present and future of secondary metabolite research in the Dothideomycetes, *Mol. Plant Pathol.* 16(1), 92–107.

O'Connell, R. J., Thon, M. R., Hacquard, S., Amyotte, S. G., Kleemann, J., Torres, M. F., Damm, U., Buiate, E. A., Epstein, L., Alkan, N., Altmüller, J., Alvarado-Balderrama, L., Bauser, C. A., Becker, C., Birren, B. W., Chen, Z., Choi, J., Crouch, J. A., Duvick, J. P., Farman, M. A., Gan, P., Heiman, D., Henrissat, B., Howard, R. J., Kabbage, M., Koch, C., Kracher, B., Kubo, Y., Law, A. D., Lebrun, M. H., Lee, Y. H., Miyara, I., Moore, N., Neumann, U., Nordström, K., Panaccione, D. G., Panstruga, R., Place, M., Proctor, R. H., Prusky, D., Rech, G., Reinhardt, R., Rollins, J. A., Rounsley, S., Schardl, C. L., Schwartz, D. C., Shenoy, N., Shirasu, K., Sikhakolli, U. R., Stüber, K., Sukno, S.

A., Sweigard, J. A., Takano, Y., Takahara, H., Trail, F., van der Does, H. C., Voll, L. M., Will, I., Young, S., Zeng, Q., Zhang, J., Zhou, S., Dickman, M. B., Schulze-Lefert, P., Ver Loren van Themaat, E., Ma, L. J. and Vaillancourt, L. J. (2012). Lifestyle transitions in plant pathogenic *Colletotrichum* fungi deciphered by genome and transcriptome analyses, *Nat. Genet.* 44(9), 1060–1065.

Odilbekov, F., Armonienė, R., Koc, A., Svensson, J. and Chawade, A. (2019). GWAS-assisted genomic prediction to predict resistance to *Septoria tritici* Blotch in Nordic winter wheat at seedling stage, *Front. Genet.* 10, 1224.

Ohm, R. A., Feau, N., Henrissat, B., Schoch, C. L., Horwitz, B. A., Barry, K. W., Condon, B. J., Copeland, A. C., Dhillon, B., Glaser, F., Hesse, C. N., Kosti, I., LaButti, K., Lindquist, E. A., Lucas, S., Salamov, A. A., Bradshaw, R. E., Ciuffetti, L., Hamelin, R. C., Kema, G. H., Lawrence, C., Scott, J. A., Spatafora, J. W., Turgeon, B. G., de Wit, P. J., Zhong, S., Goodwin, S. B. and Grigoriev, I. V. (2012). Diverse lifestyles and strategies of plant pathogenesis encoded in the genomes of eighteen Dothideomycetes fungi, *PLoS Pathog.* 8(12), e1003037.

Omrane, S., Sghyer, H., Audéon, C., Lanen, C., Duplaix, C., Walker, A. S. and Fillinger, S. (2015). Fungicide efflux and the MgMFS1 transporter contribute to the multidrug resistance phenotype in *Zymoseptoria tritici* field isolates, *Environ. Microbiol.* 17(8), 2805–2823.

Omrane, S., Audéon, C., Ignace, A., Duplaix, C., Aouini, L., Kema, G., Walker, A. S. and Fillinger, S. (2017). Plasticity of the MFS1 promoter leads to multidrug resistance in the wheat pathogen *Zymoseptoria tritici*, *mSphere* 2(5), e00393-17.

Orton, E. S., Rudd, J. J. and Brown, J. K. (2017). Early molecular signatures of responses of wheat to *Zymoseptoria tritici* in compatible and incompatible interactions, *Plant Pathol.* 66(3), 450–459.

Owen, W. J., Yao, C., Myung, K., Kemmitt, G., Leader, A., Meyer, K. G., Bowling, A. J., Slanec, T. and Kramer, V. J. (2017). Biological characterization of fenpicoxamid, a new fungicide with utility in cereals and other crops, *Pest Manag. Sci.* 73(10), 2005–2016.

Pakholkova, E. V., Sal'nikova, N. N. and Kurkova, N. A. (2016). Genetic structure of regional populations of *Mycosphaerella graminicola* (*Septoria tritici*), *Sel'skokhozyaistvennaya Biologiya* (*Agricultural Biology*) 51(5), 722–730.

Palma-Guerrero, J., Torriani, S. F. F., Zala, M., Carter, D., Courbot, M., Rudd, J. J., McDonald, B. A. and Croll, D. (2016). Comparative transcriptomic analyses of *Zymoseptoria tritici* strains show complex lifestyle transitions and intraspecific variability in transcription profiles, *Mol. Plant Pathol.* 17(6), 845–859.

Palma-Guerrero, J., Ma, X., Torriani, S. F., Zala, M., Francisco, C. S., Hartmann, F. E., Croll, D. and McDonald, B. A. (2017). Comparative transcriptome analyses in *Zymoseptoria tritici* reveal significant differences in gene expression among strains during plant infection, *Mol. Plant Microbe Interact.* 30(3), 231–244.

Payne, A. C., Grosjean-Cournoyer, M. C. and Hollomon, D. W. (1998). Transformation of the phytopathogen *Mycosphaerella graminicola* to carbendazim and hygromycin B resistance, *Curr. Genet.* 34(2), 100–104.

Perez-Nadales, E., Nogueira, M. F. A., Baldin, C., Castanheira, S., El Ghalid, M., Grund, E., Lengeler, K., Marchegiani, E., Mehrotra, P. V., Moretti, M., Naik, V., Oses-Ruiz, M., Oskarsson, T., Schäfer, K., Wasserstrom, L., Brakhage, A. A., Gow, N. A., Kahmann, R., Lebrun, M. H., Perez-Martin, J., Di Pietro, A., Talbot, N. J., Toquin, V., Walther, A. and Wendland, J. (2014). Fungal model systems and the elucidation of pathogenicity determinants, *Fungal Genet. Biol.* 70, 42–67.

Pilet-Nayel, M. L., Moury, B., Caffier, V., Montarry, J., Kerlan, M. C., Fournet, S., Durel, C. E. and Delourme, R. (2017). Quantitative resistance to plant pathogens in pyramiding strategies for durable crop protection, *Front. Plant Sci.* 8, 1838.

Platel, R., Chaveriat, L., Le Guenic, S., Pipeleers, R., Magnin-Robert, M., Randoux, B., Trapet, P., Lequart, V., Joly, N., Halama, P., Martin, P., Höfte, M., Reignault, P. and Siah, A. (2020). Importance of the C12 carbon chain in the biological activity of rhamnolipids conferring protection in wheat against *Zymoseptoria tritici*, *Molecules* 26(1), 40.

Pnini-Cohen, S., Zilberstein, A., Schuster, S., Sharon, A. and Eyal, Z. (2000). Elucidation of *Septoria tritici* × Wheat interactions using GUS-expressing isolates, *Phytopathology* 90(3), 297–304.

Ponomarenko, A., Goodwin, S. B. and Kema, G. H. K. (2011). *Septoria tritici* blotch (STB) of wheat, *Plant Health Instructor*, DOI:10.1094/PHI-I-2011-0407-01.

Quaedvlieg, W., Kema, G. H., Groenewald, J. Z., Verkley, G. J., Seifbarghi, S., Razavi, M., Mirzadi Gohari, A., Mehrabi, R. and Crous, P. W. (2011). *Zymoseptoria* gen. nov.: a new genus to accommodate *Septoria*-like species occurring on graminicolous hosts, *Persoonia* 26, 57–69.

Rahman, A., Doohan, F. and Mullins, E. (2020). Quantification of in planta *Zymoseptoria tritici* progression through different infection phases and related association with components of aggressiveness, *Phytopathology* 110(6), 1208–1215.

Rehfus, A., Strobel, D., Bryson, R. and Stammler, G. (2018). Mutations in Sdh genes in field isolates of *Zymoseptoria tritici* and impact on the sensitivity to various succinate dehydrogenase inhibitors, *Plant Pathol.* 67(1), 175–180.

Rudd, J. J. (2015). Previous bottlenecks and future solutions to dissecting the *Zymoseptoria tritici*-wheat host-pathogen interaction, *Fungal Genet. Biol.* 79, 24–28.

Rudd, J. J., Kanyuka, K., Hassani-Pak, K., Derbyshire, M., Andongabo, A., Devonshire, J., Lysenko, A., Saqi, M., Desai, N. M., Powers, S. J., Hooper, J., Ambroso, L., Bharti, A., Farmer, A., Hammond-Kosack, K. E., Dietrich, R. A. and Courbot, M. (2015). Transcriptome and metabolite profiling of the infection cycle of *Zymoseptoria tritici* on wheat reveals a biphasic interaction with plant immunity involving differential pathogen chromosomal contributions and a variation on the hemibiotrophic lifestyle definition, *Plant Physiol.* 167(3), 1158–1185.

Sahli, R., Rivière, C., Siah, A., Smaoui, A., Samaillie, J., Hennebelle, T., Roumy, V., Ksouri, R., Halama, P. and Sahpaz, S. (2018). Biocontrol activity of effusol from the extremophile plant, *Juncus maritimus*, against the wheat pathogen *Zymoseptoria tritici*, *Environ. Sci. Pollut. Res. Int.* 25(30), 29775–29783.

Saintenac, C., Lee, W. S., Cambon, F., Rudd, J. J., King, R. C., Marande, W., Powers, S. J., Bergès, H., Phillips, A. L., Uauy, C., Hammond-Kosack, K. E., Langin, T. and Kanyuka, K. (2018). Wheat receptor-kinase-like protein Stb6 controls gene-for-gene resistance to fungal pathogen *Zymoseptoria tritici*, *Nat. Genet.* 50(3), 368–374.

Saintenac, C., Cambon, F., Aouini, L., Verstappen, E., Ghaffary, S. M. T., Poucet, T., Marande, W., Berges, H., Xu, S., Jaouannet, M., Favery, B., Alassimone, J., Sánchez-Vallet, A., Faris, J., Kema, G., Robert, O. and Langin, T. (2021). A wheat cysteine-rich receptor-like kinase confers broad-spectrum resistance against *Septoria tritici* blotch, *Nat. Commun.* 12(1), 433.

Sánchez-Vallet, A., Tian, H., Rodriguez-Moreno, L., Valkenburg, D. J., Saleem-Batcha, R., Wawra, S., Kombrink, A., Verhage, L., de Jonge, R., van Esse, H. P., Zuccaro, A., Croll, D., Mesters, J. R. and Thomma, B. P. H. J. (2020). A secreted LysM effector protects

Published by Burleigh Dodds Science Publishing Limited, 2022.

fungal hyphae through chitin-dependent homodimer polymerization, *PLoS Pathog.* 16(6), e1008652.

Scalliet, G., Bowler, J., Luksch, T., Kirchhofer-Allan, L., Steinhauer, D., Ward, K., Niklaus, M., Verras, A., Csukai, M., Daina, A. and Fonné-Pfister, R. (2012). Mutagenesis and functional studies with succinate dehydrogenase inhibitors in the wheat pathogen *Mycosphaerella graminicola*, *PLoS ONE* 7(4), e35429.

Schotanus, K., Soyer, J. L., Connolly, L. R., Grandaubert, J., Happel, P., Smith, K. M., Freitag, M. and Stukenbrock, E. H. (2015). Histone modifications rather than the novel regional centromeres of *Zymoseptoria tritici* distinguish core and accessory chromosomes, *Epigenetics Chromatin* 8, 41.

Schuster, M., Kilaru, S., Guo, M., Sommerauer, M., Lin, C. and Steinberg, G. (2015). Red fluorescent proteins for imaging *Zymoseptoria tritici* during invasion of wheat, *Fungal Genet. Biol.* 79, 132–140.

Shaw, M. W. (1987). Assessment of upward movement of rain splash using a fluorescent tracer method and its application to the epidemiology of cereal pathogens, *Plant Pathol.* 36(2), 201–213.

Shaw, M. W., Bearchell, S. J., Fitt, B. D. L. and Fraaije, B. A. (2008). Long-term relationships between environment and abundance in wheat of *Phaeosphaeria nodorum* and *Mycosphaerella graminicola*, *New Phytol.* 177(1), 229–238.

Shetty, N. P., Mehrabi, R., Lütken, H., Haldrup, A., Kema, G. H. J., Collinge, D. B. and Jørgensen, H. J. L. (2007). Role of hydrogen peroxide during the interaction between the hemibiotrophic fungal pathogen *Septoria tritici* and wheat, *New Phytol.* 174(3), 637–647.

Shibata, K., Hanafi, M., Fujii, J., Sakanaka, O., Iinuma, K., Ueki, M. and Taniguchi, M. (1998). UK-2A, B, C and D, novel antifungal antibiotics from *Streptomyces* sp. 517-02. III. Absolute configuration of an antifungal antibiotic, UK-2A, and consideration of its conformation, *J. Antibiot. (Tokyo)* 51(12), 1113–1116.

Sidhu, Y. S., Chaudhari, Y. K., Usher, J., Cairns, T. C., Csukai, M. and Haynes, K. (2015a). A suite of Gateway® compatible ternary expression vectors for functional analysis in *Zymoseptoria tritici*, *Fungal Genet. Biol.* 79, 180–185.

Sidhu, Y. S., Cairns, T. C., Chaudhari, Y. K., Usher, J., Talbot, N. J., Studholme, D. J., Csukai, M. and Haynes, K. (2015b). Exploitation of sulfonylurea resistance marker and non-homologous end joining mutants for functional analysis in *Zymoseptoria tritici*, *Fungal Genet. Biol.* 79, 102–109.

Singh, N. K., Badet, T., Abraham, L. and Croll, D. (2021). Rapid sequence evolution driven by transposable elements at a virulence locus in a fungal wheat pathogen, *BMC Genomics* 22(1), 393.

Sierotzki, H. and Scalliet, G. (2013). A review of current knowledge of resistance aspects for the next-generation succinate dehydrogenase inhibitor fungicides, *Phytopathology* 103(9), 880–887.

Sierotzki, H., Frey, R., Wullschleger, J., Palermo, S., Karlin, S., Godwin, J. and Gisi, U. (2007). Cytochrome b gene sequence and structure of *Pyrenophora teres* and *P. tritici-repentis* and implications for QoI resistance, *Pest Manag. Sci.* 63(3), 225–233.

Steinberg, G. (2015). Cell biology of *Zymoseptoria tritici*: pathogen cell organization and wheat infection, *Fungal Genet. Biol.* 79, 17–23.

Steinberg, G., Schuster, M., Gurr, S. J., Schrader, T. A., Schrader, M., Wood, M., Early, A. and Kilaru, S. (2020). A lipophilic cation protects crops against fungal pathogens by multiple modes of action, *Nat. Commun.* 11(1), 1608.

Steinhauer, D., Salat, M., Frey, R., Mosbach, A., Luksch, T., Balmer, D., Hansen, R., Widdison, S., Logan, G., Dietrich, R. A., Kema, G. H. J., Bieri, S., Sierotzki, H., Torriani, S. F. F. and Scalliet, G. (2019). A dispensable paralog of succinate dehydrogenase subunit C mediates standing resistance towards a subclass of SDHI fungicides in *Zymoseptoria tritici*, *PLoS Pathog.* 15(12), e1007780.

Stephens, C., Ölmez, F., Blyth, H., McDonald, M., Bansal, A., Turgay, E. B., Hahn, F., Saintenac, C., Nekrasov, V., Solomon, P., Milgate, A., Fraaije, B., Rudd, J. and Kanyuka, K. (2021). Remarkable recent changes in the genetic diversity of the avirulence gene *AvrStb6* in global populations of the wheat pathogen *Zymoseptoria tritici*, *Mol. Plant Pathol.*, Advance online publication.

Stergiopoulos, I., Zwiers, L. H. and De Waard, M. A. (2003). The ABC transporter MgAtr4 is a virulence factor of Mycosphaerella graminicola that affects colonization of substomatal cavities in wheat leaves, *Mol. Plant Microbe Interact.* 16(8), 689–698.

Stergiopoulos, I., Collemare, J., Mehrabi, R. and De Wit, P. J. (2013). Phytotoxic secondary metabolites and peptides produced by plant pathogenic Dothideomycete fungi, *FEMS Microbiol. Rev.* 37(1), 67–93.

Stewart, E. L. and McDonald, B. A. (2014). Measuring quantitative virulence in the wheat pathogen Zymoseptoria tritici using high-throughput automated image analysis, *Phytopathology* 104(9), 985–992.

Stewart, E. L., Croll, D., Lendenmann, M. H., Sanchez-Vallet, A., Hartmann, F. E., Palma-Guerrero, J., Ma, X. and McDonald, B. A. (2017). Quantitative trait locus mapping reveals complex genetic architecture of quantitative virulence in the wheat pathogen *Zymoseptoria tritici*, *Mol. Plant Pathol.* 19(1), 201–216.

Stukenbrock, E. H., Bataillon, T., Dutheil, J. Y., Hansen, T. T., Li, R., Zala, M., McDonald, B. A., Wang, J. and Schierup, M. H. (2011). The making of a new pathogen: insights from comparative population genomics of the domesticated wheat pathogen *Mycosphaerella graminicola* and its wild sister species, *Genome Res.* 21(12), 2157–2166.

Stukenbrock, E. H., Christiansen, F. B., Hansen, T. T., Dutheil, J. Y. and Schierup, M. H. (2012a). Fusion of two divergent fungal individuals led to the recent emergence of a unique widespread pathogen species, *Proc. Natl Acad. Sci. U. S. A.* 109(27), 10954–10959.

Stukenbrock, E. H., Quaedvlieg, W., Javan-Nikhah, M., Zala, M., Crous, P. W. and McDonald, B. A. (2012b). *Zymoseptoria ardabiliae* and *Z. pseudotritici*, two progenitor species of the *Septoria tritici* leaf blotch fungus Z. tritici (synonym: mycosphaerella graminicola), *Mycologia* 104(6), 1397–1407.

Suemoto, H., Matsuzaki, Y. and Iwahashi, F. (2019). Metyltetraprole, a novel putative complex III inhibitor, targets known QoI-resistant strains of *Zymoseptoria tritici* and Pyrenophora teres, *Pest Manag. Sci.* 75(4), 1181–1189.

Suffert, F., Sache, I. and Lannou, C. (2011). Early stages of Septoria tritici blotch epidemics of winter wheat: build-up, overseasoning, and release of primary inoculum, *Plant Pathol.* 60(2), 166–177.

Suffert, F., Sache, I. and Lannou, C. (2013). Assessment of quantitative traits of aggressiveness in *Mycosphaerella graminicola* on adult wheat plants, *Plant Pathol.* 62(6), 1330– 1341.

Suffert, F. and Thompson, R. N. (2018). Some reasons why the latent period should not always be considered constant over the course of a plant disease epidemic, *Plant Pathol.* 67(9), 1831–1840.

Suffert, F., Delestre, G., Carpentier, F., Gazeau, G., Walker, A. S., Gélisse, S. and Duplaix, C. (2016). Fashionably late partners have more fruitful encounters: impact of the timing of co-infection and pathogenicity on sexual reproduction in *Zymoseptoria tritici*, *Fungal Genet. Biol.* 92, 40–49.

Suffert, F., Delestre, G. and Gélisse, S. (2019). Sexual reproduction in the fungal foliar pathogen *Zymoseptoria tritici* is driven by antagonistic density dependence mechanisms, *Microb. Ecol.* 77(1), 110–123.

Tabib Ghaffary, S. M., Robert, O., Laurent, V., Lonnet, P., Margale, E., van der Lee, T. A. J., Visser, R. G. F. and Kema, G. H. J. (2011). Genetic analysis of resistance to *Septoria tritici* blotch in the French winter wheat cultivars Balance and Apache, *Theor. Appl. Genet.* 123(5), 741-754.

Tabib Ghaffary, S. M., Faris, J. D., Friesen, T. L., Visser, R. G. F., van der Lee, T. A. J., Robert, O. and Kema, G. H. J. (2012). New broad-spectrum resistance to *Septoria tritici* blotch derived from synthetic hexaploid wheat, *Theor. Appl. Genet.* 124(1), 125–142.

Tadesse, Y., Chala, A. and Kassa, B. (2019). Survey of *Septoria tritici* Blotch (*Septoria tritici*) of bread wheat (*Triticum aestivum* L.) in the Central Highlands of Ethiopia, *Am. J. Biosci. Bioeng.* 6, 36.

Talbot, N. J. (2015). Taming a wild beast: developing molecular tools and new methods to understand the biology of *Zymoseptoria tritici*, *Fungal Genet. Biol.* 79, 193–195.

Tiley, A. M. M., Foster, G. D. and Bailey, A. M. (2018). Exploring the genetic regulation of asexual sporulation in *Zymoseptoria tritici*, *Front. Microbiol.* 9, 1859.

Torriani, S. F., Brunner, P. C., McDonald, B. A. and Sierotzki, H. (2009). QoI resistance emerged independently at least 4 times in European populations of *Mycosphaerella graminicola*, *Pest Manag. Sci.* 65(2), 155–162.

Torriani, S. F. F., Melichar, J. P. E., Mills, C., Pain, N., Sierotzki, H. and Courbot, M. (2015). *Zymoseptoria tritici*: a major threat to wheat production, integrated approaches to control, *Fungal Genet. Biol.* 79, 8–12.

van den Bosch, F., Oliver, R., van den Berg, F. and Paveley, N. (2014). Governing principles can guide fungicide-resistance management tactics, *Annu. Rev. Phytopathol.* 52, 175–195.

Waalwijk, C., Mendes, O., Verstappen, E. C. P., de Waard, M. A. and Kema, G. H. J. (2002). Isolation and characterization of the mating-type idiomorphs from the wheat *Septoria* leaf blotch fungus *Mycosphaerella graminicola*, *Fungal Genet. Biol.* 35(3), 277-286.

Yamashita, M. and Fraaije, B. (2018). Non-target site SDHI resistance is present as standing genetic variation in field populations of *Zymoseptoria tritici*, *Pest Manag. Sci.* 74(3), 672-681.

Yang, F., Li, W. and Jørgensen, H. J. L. (2013). Transcriptional reprogramming of wheat and the hemibiotrophic pathogen *Septoria tritici* during two phases of the compatible interaction, *PLoS ONE* 8(11), e81606.

Yang, N., McDonald, M. C., Solomon, P. S. and Milgate, A. W. (2018). Genetic mapping of Stb19, a new resistance gene to *Zymoseptoria tritici* in wheat, *Theor. Appl. Genet.* 131(12), 2765–2773.

Yemelin, A., Brauchler, A., Jacob, S., Laufer, J., Heck, L., Foster, A. J., Antelo, L., Andresen, K. and Thines, E. (2017). Identification of factors involved in dimorphism and pathogenicity of *Zymoseptoria tritici*, *PLoS ONE* 12(8), e0183065.

Young, D. H., Wang, N. X., Meyer, S. T. and Avila-Adame, C. (2018). Characterization of the mechanism of action of the fungicide fenpicoxamid and its metabolite UK-2A, *Pest Manag. Sci.* 74(2), 489–498.

Zainudin, N. A., Condon, B., De Bruyne, L., Van Poucke, C., Bi, Q., Li, W., Höfte, M. and Turgeon, B. G. (2015). Virulence, host-selective toxin production, and development of three cochliobolus phytopathogens lacking the sfp-type 4′-phosphopantetheinyl transferase ppt1, *Mol. Plant Microbe Interact.* 28(10), 1130–1141.

Zhan, J., Mundt, C. C. and McDonald, B. A. (2001). Using restriction fragment length polymorphisms to assess temporal variation and estimate the number of ascospores that initiate epidemics in field populations of *Mycosphaerella graminicola*, *Phytopathology* 91(10), 1011–1017.

Zhong, Z., Marcel, T. C., Hartmann, F. E., Ma, X., Plissonneau, C., Zala, M., Ducasse, A., Confais, J., Compain, J., Lapalu, N., Amselem, J., McDonald, B. A., Croll, D. and Palma-Guerrero, J. (2017). A small secreted protein in *Zymoseptoria tritici* is responsible for avirulence on wheat cultivars carrying the Stb6 resistance gene, *New Phytol.* 214(2), 619–631.

Zhou, B., Benbow, H. R., Brennan, C. J., Arunachalam, C., Karki, S. J., Mullins, E., Feechan, A., Burke, J. I. and Doohan, F. M. (2020). Wheat encodes small, secreted proteins that contribute to resistance to *Septoria tritici* Blotch, *Front. Genet.* 11, 469.

Zwiers, L. H. and De Waard, M. A. (2001). Efficient *Agrobacterium tumefaciens*-mediated gene disruption in the phytopathogen *Mycosphaerella graminicola*, *Curr. Genet.* 39(5–6), 388–393.

Chapter 9

Advances in breeding techniques for durable Septoria tritici blotch (STB) resistance in cereals

Harsh Raman, NSW Department of Primary Industries, Australia

1 Introduction

The human population across the globe is expected to exceed 9 billion by 2050 (UN Department of Economic and Social Affairs, 2015). To meet increasing demand, food production will need to increase by more than 50% compared to the current level (Tadesse et al., 2019). Better management of diseases of major cereal food crops will be crucial to achieving this target. These staple crops were domesticated more than 10000 years ago by our ancestors and remain as the main sources of dietary protein and starch, accounting for more than 20% of energy intake (Fu et al., 2009).

http://dx.doi.org/10.19103/AS.2021.0092.12

Septoria tritici blotch (STB) is caused by an ascomycete hemibiotrophic fungus, *Zymoseptoria tritici* (previously teleomorph: *Mycosphaerella graminicola* (Fuckel) J. Schröter in Cohn anamorph: *Septoria tritici* Roberge ex Desmaz). STB is one of the major foliar diseases that can reduce grain yield by up to 50% (Karlstedt et al., 2019, Eyal and Levy, 1987, Goodwin, 2007, Eyal et al., 1987, Savary et al., 2019, Fones and Gurr, 2015). *Z. tritici* is a strictly apoplastic pathogen with a latent necrotrophic lifestyle (Sánchez-Vallet et al., 2015) and infects cereal crops such as common wheat (*Triticum aestivum*), durum wheat (*Triticum durum* Desf) and spelt wheat (*Triticum aestivum* L. subsp. *spelta* (L.) Thell) (Gharbi et al., 2000, Goodwin, 2007, Simon et al., 2010, Seifbarghi et al., 2009). STB symptoms appear on leaves as chlorotic and necrotic blotches that lead to the production of asexual spores which spread the disease during a growing season (Kema et al., 1996, Duncan and Howard, 2000). STB primarily reduces the photosynthetic capacity of leaves, leading to significant yield losses (Fones and Gurr, 2015).

STB can be controlled by various agronomic practices such as crop rotation, stubble management, eradication of alternative host species, seed dressing and foliar application of fungicide. In Europe, approximately 70% of the fungicide use is currently targeted to control STB in wheat (Torriani et al., 2015). However, with the intensive application of fungicides, *Z. tritici* has developed resistance to the main fungicide groups, including multi-site inhibitors, sterol 14α-demethylation inhibitors, succinate dehydrogenase inhibitors and quinone-outside inhibitors (Jørgensen et al., 2018, Chartrain et al., 2004a, Fraaije et al., 2005, Heick et al., 2017a,b, Cools and Fraaije, 2008, Mohd-Assaad et al., 2016, McDonald et al., 2019). This problem highlights the need for alternative strategies for controlling yield losses.

Deployment of genetic resistance is considered one of the most effective and environmentally safe strategies for reducing yield losses in cereal crops. There have been significant efforts to evaluate germplasm under field and/or glasshouse conditions to identify donor sources of resistance to STB. Several sources of resistance have been identified among accessions of diploid species such as *T. monococcum* (Jing et al., 2008), modern wheat and durum varieties, landraces and synthetic wheat derived from *Triticum dicocoides* and *Aegilops tauschii* (Ouaja et al., 2020, Adhikari et al., 2003, Adhikari et al., 2004a, Arraiano et al., 2001b, Kosellek et al., 2013, Vagndorf et al., 2017, May and Lagudah, 1992, Ghaffary et al., 2012). The presence of STB resistance due to the *Stb6* gene in these varieties can be traced back to traditional landraces of the mid-Neolithic era (Chartrain et al., 2005b, Dutta et al., 2021). Some of these sources of resistance have been exploited globally in resistance breeding programmes and subsequently deployed in commercial wheat and durum varieties using conventional breeding methods.

There has been great interest in the last three decades in investigating the range of genes conferring resistance to STB. This research is particularly important for the *Z. tritici* pathogen that is characterised by a 'boom-bust' cycle (in which the pathogen rapidly adapts to resistant varieties), affecting sustainable production of cereals especially in Australia, Europe and North America. There is great concern that single resistance genes in resistant varieties may cease to be effective against avirulent populations of *Z. tritici*. There is therefore a need to revisit the existing tools and technologies available for durable disease resistance breeding.

Tremendous progress has been made recently in cereal genetics and genomics research in areas such as:

- innovation in high-throughput phenotypic and genotyping platforms;
- understanding of genome-phenotype associations; and
- development of genomic resources and analytic pipelines for identification of candidate genes associated with resistance to diseases.

This chapter reviews recent research on genetic resistance loci and breeding strategies based on both conventional and biotechnology-based breeding approaches (molecular marker/genomic-assisted breeding, genetic transformation and gene editing) to achieve durable resistance to STB infection and minimise grain yield losses.

2 Challenges in achieving durable resistance: qualitative and quantitative resistance

Johnson (1984) defined durable resistance as resistance that remains effective over a long time and over a large area where conditions are favourable to the disease. However, the significance of 'durability' varies with different stakeholders (e.g. grower or researcher). Resistance which 'holds' under field conditions and does not 'significantly' affect return on the investment on a longer-term basis is good enough for most farmers. Although a significant advance in research, developing more complete 'immunity' by pyramiding (also called 'stacking') multiple resistance genes may not be ideal for fungal population fitness, as it may lead to the evolution of new virulent races of *Z. tritici*, a pathogen known for its high genetic diversity and gene flow (McDonald and Linde, 2002, McDonald, 1999, Zhan et al., 2003, Rudd, 2015).

Research has shown that not all sources of resistance are equally effective in controlling STB over time and thus differ in durability. STB resistance in the wheat variety 'Gene', for example, became ineffective in five years after its release due to evolution of virulent genotypes of *Z. tritici* in the United States (Cowger et al., 2000), whereas resistance in 'Kavkaz-K4500', 'Veranopolis' and

'Bulgaria-88' remained effective for many years (Eyal, 1999, Dubin and Rajaram, 1996). To achieve durable resistance in commercial varieties, comprehensive understanding of the genetics of host resistance (e.g. inheritance, allelic effects, allelic interactions and stability of resistance across environments) and of pathogen characteristics (e.g. race structure, mutation rate, mode of reproduction and diversity in avirulence genes/effectors) is essential.

Host-pathogen interaction and genetic inheritance studies have revealed two types of resistance to *Z. tritici* (Adhikari et al., 2003, 2004a,b,c, 2015, Goodwin et al., 2008, Mergoum et al., 2013, Arraiano and Brown, 2006, Arraiano et al., 2001b, 2007, Chartrain et al., 2004b, 2005b,c, Brading et al., 2002, McCartney et al., 2002, 2003, Rillo and Caldwell, 1966, Rosielle, 1972, Shaner and Finney, 1982, Somasco et al., 1996, Wilson, 1985). These are:

- qualitative resistance; and
- quantitative resistance.

These are discussed in more detail below.

Qualitative resistance refers to a typical race-specific resistance (also called innate immunity, vertical or hypersensitivity/basal resistance). It is controlled by a single *R* gene with large allelic effect, providing 'complete' resistance to avirulent isolates in a host (Niks et al., 2011, Van der Plank, 1968). Qualitative resistance component traits include the extent of chlorosis, pycnidial density and pycnidial size. They exhibit distinct phenotypes (resistance and susceptible individuals) and show a bimodal distribution for a single gene. Generally, qualitative resistance follows the classical 'gene-for-gene' relationship (Flor, 1971). In this relationship, the host resistance (*R*) protein interacts with the cognate *avirulence* (*Avr*) gene (effector protein) of the fungal pathogen to initiate a defence response (Dangl and Jones, 2001). In the wheat-*Z. tritici* pathosystem, the infection process and fungal effectors involved in both the long biotrophic phase of symptomless colonisation and the necrotrophic phase are only partly understood (Marshall et al., 2011, Kettles et al., 2018, Motteram et al., 2009).

The *AcrStb6* gene that secretes the protein for avirulence in wheat cultivars carrying the *Stb6* resistance gene has been cloned and functionally characterised (Zhong et al., 2017, Brading et al., 2002, Kema et al., 2000). *Stb6* for genetic resistance in wheat has also been cloned, encoding a wall-associated receptor kinase (Saintenac et al., 2018). STB-mediated resistance does not relate to the typical hypersensitive response. Both *AvrStb6* and *Stb6* show a gene-for-gene interaction, similar to other *R* genes that encode intracellular nucleotide-binding leucine-rich-repeat proteins (NBS-LRRs) that recognise pathogen-secreted isolate-specific *Avr* effectors delivered to the host cytoplasm (Jones and Dangl, 2006, Dodds and Rathjen, 2010). Race-specific resistance can be easily evaluated under controlled conditions using leaf assays (Arraiano et al., 2001a).

In contrast to qualitative resistance, quantitative resistance relates to non-race-specific interaction (also called horizontal, partial or polygenic/background resistance). Quantitative resistance is controlled by a large number of genes and their networks, contributing to small and cumulative allelic effects (Van der Plank, 1968, Parlevliet, 1979, Kump et al., 2011, Corwin et al., 2016, Niks et al., 2015). Quantitative resistance is generally assessed under field conditions by assessing the extent of disease severity at the later stages of plant development. Unlike qualitative resistance, the molecular basis of quantitative resistance is not well understood but may involve a common set of *R* genes. For example, Meile et al. (2018) identified gene-for-gene interaction underlying partial resistance to *Z. tritici*. Adult plant resistance (APR) to *Z. tritici*, which is expressed at the adult plant development stage but not at the seedling stage, has also been reported in synthetic wheat, for example, *Stb17* (Ghaffary et al., 2012). This type of resistance is described as 'quantitative (minor gene)' resistance (Mundt, 2018), although APR could be due to a single *R* gene (Krattinger et al., 2009).

3 Molecular marker technologies for genetic mapping of Septoria tritici blotch resistance genes

Genetic mapping of traits has been carried out in plants for almost 100 years (Sax, 1923, Sprague, 1941). Developments in modern molecular technologies have enabled researchers to identify genetic control, chromosomal location and allelic effects of loci involved in a trait of interest. Figure 1 shows the basic processes for genetic mapping and marker-assisted selection (MAS) for breeding disease-resistant varieties. There are two main marker technologies:

- hybridisation-based markers; and
- Polymerase chain reaction-based markers.

Hybridisation-based markers include:

- restriction fragment length polymorphism (RFLP);
- single feature polymrphism (SFP); and
- diversity arrays markers (DArTs) (Tamburic-Ilincic and Rosa, 2019, Adhikari et al., 2015, Mergoum et al., 2013).

PCR-based markers include:

- random amplified polymorphic DNA (RAPD) (Hawkins et al., 2005, Somers et al., 2001, Adhikari et al., 2004c);

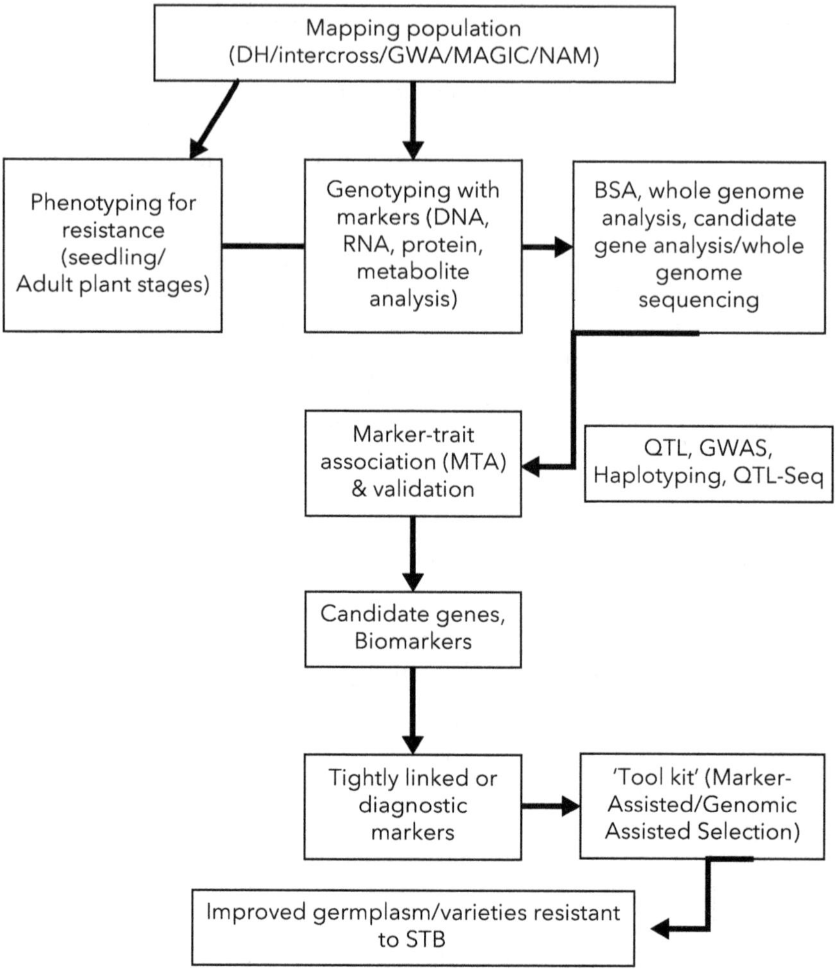

Figure 1 Generalised scheme for marker-assisted selection (MAS) approach in plant breeding programmes. Based on phenotypic expression of response to resistance (resistance and susceptibility), populations (doubled haploid: DH/intercross: $F_{2:3}$/ recombinant inbred lines: RIL; backcross, nested association mapping: NAM; multiparent advanced generation intercross: MAGIC) and genetic diversity/reference, germplasm panels (genome-wide association: GWA) can be phenotyped for resistance and genotyped with molecular markers. Linkage and whole genome-wide analyses such as QTL (quantitative trait loci) and GWAS (genome-wide association studies) can be used to identify marker-trait associations (MTAs). The MTAs (genomic regions) that show significant association between DNA/protein/metabolite markers and resistance to STB can be verified/validated in different backgrounds/GWAS panels to test their usefulness in the breeding programmes. Using reference genomes and pan-genomes, candidate genomic regions can be delimited to a single locus and further verified using large populations (intercross and GWAS panels). Tightly linked markers/genome-based markers can be deployed for marker-assisted selection (MAS) to monitor gene introgression for increasing selection efficiency for trait(s) of interest, that is, resistance to STB.

- amplified fragment length polymorphism (AFLP) (Honsdorf et al., 2010, Medini et al., 2014, Adhikari et al., 2004c);
- simple sequence repeat (SSR) (Lowe et al., 2004, Ghaffary et al., 2012, Mergoum et al., 2013);
- sequence-related amplified polymorphism (SRAP) (Sun et al., 2007, Li et al., 2007); and
- cleaved amplified polymorphic sequence (CAPS) markers.

All marker systems have been used for genetic analysis of traits. Besides genomic markers, markers derived from cDNA (RFLP, AFLP, RNA differential display and SSR) expressed sequence tags, and gene sequences/unigenes have also been used in genetic analysis (Gupta and Rustgi, 2004).

Recent innovations in the development of next generation sequencing (NGS) and availability of bioinformatics tools have made it possible to develop new alternative marker approaches. Examples include single nucleotide polymorphism (SNP) arrays at different densities such as Axiom (35K, 660K, 820K) and Illumina Infinium arrays (90K) and Affymetrix GeneChips which are being used for genetic analysis (Chen et al., 2013, Cavanagh et al., 2013, Maccaferri, 2014, Maccaferri et al., 2015, van Poecke et al., 2013). PCR-based NGS approaches include:

- restriction site associated DNA (RAD) (Miller et al., 2007, Baird et al., 2008, Chen et al., 2017, Stewart et al., 2018);
- intron polymorphisms;
- single nucleotide polymorphism (SNP) (Karlstedt et al., 2019, Muqaddasi et al., 2019, Ando et al., 2018, Miedaner et al., 2013);
- genotyping by sequencing (GBS) (Elshire et al., 2011); and
- DArTseq (Bemister et al., 2019, Raman et al., 2014).

These have been employed for a range of applications, including identification of marker-trait associations (MTAs). DNA and RNA sequencing/resequencing approaches are also increasingly being used for mapping genes of interest (Liu et al., 2012, Edae and Rouse, 2019, Huang et al., 2009).

Whole genome sequencing (WGS)-based approaches are being routinely utilised for mapping QTLs using pooled samples from contrasting phenotypes. WGS-based approaches include:

- MutMap (Abe et al., 2012);
- QTL-seq (Takagi et al., 2013);
- target enriched X-QTL (Guo et al., 2015);
- exome QTL sequencing (Hisano et al., 2017); and
- Indel-seq (Singh et al., 2017).

Table 1 Molecular mapping of qualitative genes for resistance to Septoria tritici blotch (Raman and Milgate, 2012, Goodwin, 2007, Brown et al., 2015)

Gene	Chromosomal location/arm	Marker Type	Resistance Source	Reference
Stb1	5BL	SSR	BULGARIA 88	(Adhikari et al., 2004c)
Stb2	1BS	SSR	VERANOPOLIS	(Adhikari et al., 2004b, Liu et al., 2013)
Stb3	7AS	SSR	ISRAEL 493	(Adhikari et al., 2004b, Goodwin and Thompson, 2011)
Stb4	7DS	SSR	TADINIA	(Adhikari et al., 2004a)
Stb5	7DS	SSR	SYNTHETIC 6X	(Arraiano et al., 2001b)
Stb6	3AS	SSR	FLAME, HEREWARD	(Brading et al., 2002)
Stb7	4AL	SSR	ST6	(McCartney et al., 2003)
Stb8	7BL	SSR	SYNTHETIC W7984	(Adhikari et al., 2003)
Stb9	2BL	SSR	COURTOT, TONIC	(Chartrain et al., 2009)
Stb10	1D (centromeric region)	SSR	KAVKAZ-K4500	(Chartrain et al., 2005a)
Stb11	1BS	SSR	TE9111	(Chartrain et al., 2005c)
Stb12	4AL	SSR	KAVKAZ-K4500	(Chartrain et al., 2005a)
Stb13	7BL	SSR	SALAMOUNI	(Cowling et al., 2004, Cowling, 2006)
Stb14	3BS	SSR	SALAMOUNI	(Cowling et al., 2004, Cowling, 2006)
Stb15	6AS	SSR/RFLP	ARINA, RIBAND	(Arraiano et al., 2007)
Stb16q	3DL	SSR	SH M3	(Ghaffary et al., 2012)
Stb17	5AL	SSR	SH M3	(Ghaffary et al., 2012)
Stb18	6DS	SSR	BALANCE	(Ghaffary et al., 2011)
Stb19	1DS	KASP	SUMMIT	(Yang et al., 2018)
StbWW	1BS	SSR	WW1842, WW2449, WW2451	(Raman et al., 2009)
StbSm3	3AS	SSR	SALAMOUNI	(Cuthbert, 2011)
TmStb1	$7A^mS$	SSR	MDR043 (*T. monococcum*)	(Jing et al., 2008)

More recently, *R* gene enrichment sequencing (AgRenSeq) has been used to identify rust resistance genes by exploiting pan-genome variation in wild diploid wheat (Arora et al., 2019). This approach has a potential to identify novel genes associated with STB resistance from diverse sources of germplasm, including wild species. Untapped natural variation may improve grain yield by reducing losses due to diseases in cereal crops.

These molecular marker systems and genetic mapping approaches have helped to identify loci associated with qualitative and quantitative resistance in biparental, multi-parental and diverse panels of cereals. Genetic loci associated with STB resistance have been dissected primarily by the use of QTL (quantitative trait loci) and genome-wide association studies (GWAS) approaches in wheat (Kollers et al., 2013, Miedaner et al., 2013, Gurung, 2014, Ando et al., 2018, Muqaddasi et al., 2019, Vagndorf et al., 2017) and in durum populations (Ballini et al., 2020).

To date, at least 22 *R* genes for resistance to STB has been described in cereals (Table 1). Some of the 'uncharacterised' *R* genes may relate to known STB resistance genes. For example, *StbWW* present in Australian wheat lines, WW1842, WW2449 and WW2451, may relate to *STB11* locus on chromosome 1B (Raman et al., 2009). In addition to qualitative genes with major allelic effects, more than 167 QTLs associated with resistance have also been reported (http://www.shigen.nig.ac.jp/wheat/komugi/genes/symbolClassList.jsp). Information on loci for qualitative and quantitative resistance to STB is provided in a number of reviews (Raman and Milgate, 2012, Goodwin, 2007, Brown et al., 2015). An updated list of QTL for STB resistance is shown in Table 2.

It is unclear whether some of the loci involved in qualitative and quantitative types of resistance are unique, as shown by their allelic effects and genomic locations. Many loci for qualitative and quantitative resistance map to similar marker intervals on chromosomes 1B, 2B, 3A, 3B, 4A, 5B, 6A, 6D, 7A, 7B and 7D (Brown et al., 2015). Further research is required to determine whether genes for STB resistance expressed at the seedling (for qualitative resistance) and adult plant stages under field conditions (qualitative and quantitative resistance) are the same and indeed map to the same positions in reference genome assemblies.

4 Ways of increasing durability of resistance: planting of varietal mixtures

Different approaches have been suggested to increase the durability of resistance. These approaches include:

- planting of varietal mixtures;
- deployment of qualitative and quantitative genes;

Table 2 QTL for resistance to STB in cereals (since Brown et al., 2015). DH: doubled haploid; GWAS: genome-wide association panel; RIL: recombinant inbred lines

Germplasm/ population	QTL name	Linkage group/ Chromosome	Associated molecular marker (interval) in bracket)	Seedling stage inoculum	Attribute analysed for resistance	P VALUE (−Log 10*p*)	Phenotypic/ genotypic* variation (%)	Donor Source of resistance	Co-localisation with known locus/gene	Reference
LIWILLA/BEGRA (DH)	*QStb.ihar -3A.2*	3A	wPt1688-gwm666.1	isolate IPO323 (2008)	Necrosis	9.11	50.00	Liwilla	*Stb6/QStb. risø-3A.1/ QStb.ihar-3 A.2*	Radecka-Janusik and Czembor (2014)
	QStb.ihar -3A.2	3A	wPt1688-gwm666.1	isolate IPO323 (2010)	Necrosis	7.23	42.60	Liwilla		Radecka-Janusik and Czembor (2014)
	QStb.ihar -3A.2	3A	wPt1688-gwm666.1	isolate IPO323 (2010)	Pycnidia	9.69	50.30	Liwilla		Radecka-Janusik and Czembor (2014)
	QStb.ihar -3A.2	3A	wPt1688-gwm666.1	isolate IPO323 (2010)	Necrosis	7.15	23.90	Liwilla		Radecka-Janusik and Czembor (2014)
	QStb.ihar -3A.2	3A	wPt1688-gwm666.1	isolate IPO323 (2010)	Pycnidia	7.92	43.10	Liwilla		Radecka-Janusik and Czembor (2014)
	QStb.ihar -6A	6A	wPt9679-wPt7027	isolate IPO323 (2010)		3.26	9.50	Liwilla	*Stb15*	Radecka-Janusik and Czembor (2014)
	QStb.ihar -7A.2	7A	wPt0114-wPt6184	IPO92006 (2006)	Necrosis	2.86	14.00	Liwilla	*Stb3*	Radecka-Janusik and Czembor (2014)
	QStb.ihar -7A.2	7A	wPt0040-wPt0114	Isolate St1-2 (2006)	Necrosis	4.36	18.00	Liwilla		Radecka-Janusik and Czembor (2014)
	QStb.ihar -7A.2	7A	wPt5153-wPt0429	St1-2 (2007)	Necrosis	4.98	25.40	Liwilla		Radecka-Janusik and Czembor (2014)

QStb.ihar-7A.2	7A	wPt5153-wPt0429	St1-2(2009)	Necrosis	7.24	30.30	Liwilla		Radecka-Janusik and Czembor (2014)
QStb.ihar-7A.2	7A	wPt0429-wPt5590	St1-2 (2009)	Pycnidia	6.57	28.90	Liwilla		Radecka-Janusik and Czembor (2014)
QStb.ihar-7A.2	7A	wPt0040-wPt0114	St1-2 (2010)	Pycnidia	2.76	13.50	Liwilla		Radecka-Janusik and Czembor (2014)
QStb.ihar-1B	1B	wPt8779-wPt7188	IPO92006 isolate (2006),	Necrosis	6.47	40.00	Liwilla	*StbWW, Stb2. Stb11*	Radecka-Janusik and Czembor (2014)
QStb.ihar-1B	1B	wPt8779-wPt7188	IPO92006 (2006)	Necrosis	7.17	39.10	Liwilla		Radecka-Janusik and Czembor (2014)
QStb.ihar-1B	1B	wPt8779-wPt7188	IPO92006 (2009)	Necrosis	5.1	19.80	Liwilla		Radecka-Janusik and Czembor (2014)
QStb.ihar-1B	1B	wPt8779-wPt7188	IPO92006 (2009)	Pycnidia	4.15	16.40	Liwilla		Radecka-Janusik and Czembor (2014)
QStb.ihar-1B	1B	wPt8779-wPt7188	IPO92006 (2010)	Pycnidia	3.43	15.80	Liwilla		Radecka-Janusik and Czembor (2014)
QStb.ihar-2B.2	2B	wPt8460-wPt7859	St1-2 isolate (2009)	Necrosis	3.8	14.40	Liwilla	*QStb.risø-2B*	Radecka-Janusik and Czembor (2014)
QStb.ihar-3B	3B	wPt1171-wPt1804	IPO92006 (2009)	Pycnidia	3.1	17.80	Liwilla	*QStb.risø-3B*	Radecka-Janusik and Czembor (2014)
QStb.ihar-3B	3B	wPt7186-wPt9912	St1-2 (2009)	Pycnidia	3.62	13.10	Liwilla		Radecka-Janusik and Czembor (2014)

(*Continued*)

Table 2 (*Continued*)

Germplasm/ population	QTL name	Linkage group/ Chromosome	Associated molecular marker (interval) in bracket)	Seedling stage inoculum	Attribute analysed for resistance	P VALUE (−Log 10*p*)	Phenotypic/ genotypic* variation (%)	Donor Source of resistance	Co-localisation with known locus/gene	Reference
	QStb.ihar -3B	3B	wPt7186-wPt9912	St1-2 (2010)	Necrosis	3.32	16.40	Liwilla		Radecka-Janusik and Czembor (2014)
	QStb.ihar -5D	5D	gwm205.1-barc110	86036 isolate (2008)	Necrosis	3.63	35.10	Begra		Radecka-Janusik and Czembor (2014)
	QStb.ihar -5D	5D	barc110-gwm174	St1-2 (2010)	Pycnidia	4.14	26.30	Begra		Radecka-Janusik and Czembor (2014)
Synthetic wheat-LARGO/ SPRING wheat line-ND495 (RIL)	-	Unknown	wPt_664276	Ma04-9-4 isolate	Necrosis and pycnidia	2.40	0.09			Adhikari et al. (2015)
	-	Unknown	wPt_666095	Ma04-9-4 isolate		2.24	0.08			Adhikari et al. (2015)
	Ma04-9-4 isolate	Unknown	wPt_4555			2.205	0.08			Adhikari et al. (2015)
	Ma04-9-4 isolate	2A-a	wPt_3896-wpt_4197/ wPt_4555			2.182	0.08			Adhikari et al. (2015)
GWAS panel of wheat	*QStb. NS-1B*	1BL	TA001769-0538	mixed 8 isolates	Leaf area with lesions (%)	0.000147	9.00			Vagndorf et al. (2017)
	QStb. NS-2A	7AL	BS00024643_51	mixed 8 isolates	Leaf area with lesions (%)	0.000948	7.00			Vagndorf et al. (2017)

	QStb. NS-5D	5DL	RAC875_c8100_163	mixed 8 isolates	Leaf area with lesions (%)	0.000339	9.00	Vagndorf et al. (2017)
	QStb. NS-7A	7AL	Excalibur_rep_ c102327_102	mixed 8 isolates	Leaf area with lesions (%)	0.000237	8.00	Vagndorf et al. (2017)
GWAS panel of wheat		1A	wPt4765	FALP0112	Necrosis (%)	1.1 E-02-0.0075	0.0789-0.0864	Gerard et al. (2017)
		1A	wPt7030	FALP0112	Necrosis (%)	0.01-6.7E-03	0.0806-0.0890	Gerard et al. (2017)
		2B	wPt0694	FALP0112	Necrosis (%)	0.012-4.7 E-02	0.0396-0.0787	Gerard et al. (2017)
		3A	wPt6854	FALP0112	Necrosis (%)	0.034-3.9E-02	0.0640-0.0672	Gerard et al. (2017)
		3B	wPt7341	FALP0112	Necrosis (%)	0.049	0.0460-0.064	Gerard et al. (2017)
		3B	wPt8855	FALP0112	Necrosis (%)	0.031-4.1E-02	0.0424-0.0648	Gerard et al. (2017)
		3B	wPt2748	FALP0112	Necrosis (%)	0.031-4.1E-02	0.0424-0.0648	Gerard et al. (2017)
		6B	wPt6116	FALP0112	Necrosis (%)	0.042 to	0.0493-0.1070	Gerard et al. (2017)
		1B	wPt8168	FALP0212	Necrosis (%)	2.2E-02-0.024	0.0542-0.0579	Gerard et al. (2017)
		1B	wPt6240	FALP0212	Necrosis (%)	4.3E-02-0.0019	0.0589-0.0972	Gerard et al. (2017)
		2A	wPt7306	FALP0212	Necrosis (%)	4.1E-02-0.005	0.0593-0.0895	Gerard et al. (2017)
		2D	wPt4144	FALP0212	Necrosis (%)	1.8E-02-0.049	0.0422-0.0797	Gerard et al. (2017)

(Continued)

Table 2 (*Continued*)

Germplasm/ population	QTL name	Linkage group/ Chromosome	Associated molecular marker (interval) in bracket)	Seedling stage inoculum	Attribute analysed for resistance	P VALUE (−Log 10*p*)	Phenotypic/ genotypic* variation (%)	Donor Source of resistance	Co-localisation with known locus/gene	Reference
		4B/7A	wPt8650	FALP0212	Necrosis (%)	3.2E-02-0.042	0.0337-0.0649			Gerard et al. (2017)
		5A	wPt3924	FALP0212	Necrosis (%)	7.4E-02-0.003	0.0428-0.0855			Gerard et al. (2017)
		1A	wPt8455	FALP0112	pycnidial coverage	0.0048-6.9E-03	0.101-0.1370			Gerard et al. (2017)
		1A	wPt4029	FALP0112	pycnidial coverage	0.0026-9.4E-03	0.106-0.1280			Gerard et al. (2017)
		1B	wPt8267	FALP0112	pycnidial coverage	0.016-4.2E-02	0.0518-0.0883			Gerard et al. (2017)
		1B	wPt7094	FALP0112	pycnidial coverage	0.034-3.9E-02	0.0559-0.0622			Gerard et al. (2017)
		1B	wPt8168	FALP0112	pycnidial coverage	0.031-6.6E-03	0.066-0.1810			Gerard et al. (2017)
		2B	wPt2106	FALP0112	pycnidial coverage	2.5E-02-0.059	0.0638-0.0838			Gerard et al. (2017)
		2B	wPt0049	FALP0112	pycnidial coverage	2.1E-02-0.05	0.0457-0.0542			Gerard et al. (2017)
		4B/7A	wPt8650	FALP0112	pycnidial coverage	3.6E-02-0.0075	0.0589-0.0923			Gerard et al. (2017)
		5B	wPt3457	FALP0112	pycnidial coverage	1.1E-02-0.016	0.0829-0.0871			Gerard et al. (2017)
		7A	wPt7763	FALP0112	pycnidial coverage	0.029-3.9E-02	0.1040-0.074			Gerard et al. (2017)
		1A	wPt8455	FALP0212	pycnidial coverage	1.1E-02-0.0043	0.0959-0.13			Gerard et al. (2017)
		1A	wPt4029	FALP0212	pycnidial coverage	1.0E-02-0.0067	0.0925-0.104			Gerard et al. (2017)

		1B	**wPt8267**	FALP0212	pycnidial coverage	3.3E-02-0.049	0.0503-0.0669	Gerard et al. (2017)
		1B	**wPt8168**	FALP0212	pycnidial coverage	2.7E-02-0.0012	0.0845-0.148	Gerard et al. (2017)
		1B	**wPt3451**	FALP0212	pycnidial coverage	1.4E-02-0.047	0.0513-0.0815	Gerard et al. (2017)
		1B	**wPt6240**	FALP0212	pycnidial coverage	4.2E-02-0.004	0.0525-0.119	Gerard et al. (2017)
		3D	wPt2757	FALP0212	pycnidial coverage	3.1E-02-0.0033	0.0651-0.115	Gerard et al. (2017)
		4B	**wPt7062**	FALP0212	pycnidial coverage	4.9E-02-0.022	0.0416-0.0743	Gerard et al. (2017)
		4B/7A	**wPt8650**	FALP0212	pycnidial coverage	2.1E-02-0.047	0.0436-0.0822	Gerard et al. (2017)
		6A	**wPt0832**	FALP0212	pycnidial coverage	0.013-3.7E-02	0.0575-0.101	Gerard et al. (2017)
		6A/6B	**wPt7599**	FALP0212	pycnidial coverage	2.7E-02-0.00065	0.0637-0.154	Gerard et al. (2017)
		7A	wPt3883	FALP0212	pycnidial coverage	0.021-4.2E-02	0.0511-0.084	Gerard et al. (2017)
		7A	wPt7734	FALP0212	pycnidial coverage	0.022-3.6E-02	0.0594-0.0831	Gerard et al. (2017)
		7D	wPt4555	FALP0212	pycnidial coverage	2.3E-02-0.0026	0.0726-0.115	Gerard et al. (2017)
Durum wheat GWAS panel	*qSTB.1*	1A	IACX3496/Kukri_c10239_2186		Severity			Kidane et al. (2017)
	qSTB.2	2B	Jagger_c9472_305		Severity			Kidane et al. (2017)
	qSTB.3	3A	wsnp_Ex_c3478_6369892/wsnp_RFL_Contig4404_5157920		Septoria progress coefficient			Kidane et al. (2017)

(Continued)

Table 2 (*Continued*)

Germplasm/ population	QTL name	Linkage group/ Chromosome	Associated molecular marker (interval) in bracket)	Seedling stage inoculum	Attribute analysed for resistance	P VALUE (−Log 10*p*)	Phenotypic/ genotypic* variation (%)	Donor Source of resistance	Co-localisation with known locus/gene	Reference
	qSTB.4	4A	Excalibur_ c113341_139		Septoria progress coefficient					Kidane et al. (2017)
	qSTB.5	5A	T . durum_ contig10210_425		Septoria progress coefficient					Kidane et al. (2017)
wheat GWAS panel	*STB.1*	5A	RAC875_ c16943_404			<0.001	4.20			Wurschum et al. (2017)
	STB.2	6A	Excalibur_ c12085_276			<0.001	29.90			Wurschum et al. (2017)
	STB.3	7A	BS00033442_51			<0.001	7.50			Wurschum et al. (2017)
	STB.4	7A	Kukri_c51068_528			<0.001	4.10			Wurschum et al. (2017)
	STB.5	7A	Ku_c29856_132			<0.001	6.20			Wurschum et al. (2017)
	STB.6	3B	tplb0059m03_622			<0.001	5.50			Wurschum et al. (2017)
	STB.7	3B	BobWhite_ c25670_425			<0.001	2.70			Wurschum et al. (2017)
	STB.8	6B	RAC875_ c26887_562			<0.001	4.00			Wurschum et al. (2017)
Wheat GWAS panel	*MQTL4*	2A	IWA8491		Severity (%)		0.2-6.3			Ando et al. (2018)
	QTL3	2B	IWA6263		Severity (%)		1.6-7.1			Ando et al. (2018)

	QStb. Isa_af-3B, MQTL14	3B	IWA4653	Severity (%)		0-5.6	Ando et al. (2018)
	MQTL14, QStb.cim -3BL	3B	IWA3901	Severity (%)		0-5.7	Ando et al. (2018)
	Stb15	6A	IWA3207	Severity (%)		0.2-6.0	Ando et al. (2018)
	Qstb.psr -6B-1, Qstb.6B, QStb.riso -6B.2	6B	IWA8011	Severity (%)		1.8-6.8	Ando et al. (2018)
	QStb.riso -7B	7B	IWA814	Severity (%)		1.2-5.7	Ando et al. (2018)
	QStb.ipk -7B	7B	IWA3513	Severity (%)		1.5-5.7	Ando et al. (2018)
Wheat GWAS panel		1A	wsnp_Ex_rep_ c109742_92411838	Severity (%)	3.03	0.87*	Muqaddasi et al. (2019)
		1B	IAAV3905		3.01	6.24	Muqaddasi et al. (2019)
		2B	AX_94734086		3.42	6.64	Muqaddasi et al. (2019)
		2D	Ppd_D1		5.98	14.93	Muqaddasi et al. (2019)
		4A	wsnp_JD_ c27162_22206547		3.31	5.80	Muqaddasi et al. (2019)
		5A	15 collated markers (AX_94955360, AX_95628896)		3.38-4.06	2.11-2.95	Muqaddasi et al. (2019)
		5A	AX_95073216, Tdurum_ contig56267_180, BobWhite_ c40643_370		3.38-3.84	2.10-2.75	Muqaddasi et al. (2019)

(*Continued*)

Table 2 (*Continued*)

Germplasm/ population	QTL name	Linkage group/ Chromosome	Associated molecular marker (interval) in bracket)	Seedling stage inoculum	Attribute analysed for resistance	P VALUE (−Log 10*p*)	Phenotypic/ genotypic* variation (%)	Donor Source of resistance	Co-localisation with known locus/gene	Reference
		5A	AX_95681646, BS00022378_51			3.16–3.31	−0.28			Muqaddasi et al. (2019)
		6D	Excalibur_rep_ c106283_82			3.04	2.13			Muqaddasi et al. (2019)
		7A	BS00052668_51, wsnp_CAP11_ c651_429263			3.03	2.59			Muqaddasi et al. (2019)
		7A	AX_94466147			3.03	2.59			Muqaddasi et al. (2019)
Triticum aestivum subsp. *spelta* (HTRI1410) × Alcedo, *Triticum aestivum* subsp. *spelta* (HTRI1410) × *Triticum aestivum* subsp. *spelta* (HTRI1410) × Solitar (DH)	*QTL_5A*	5A	RAC875_ c64905_274, BS00094342_51, Ku_c4925_624	IPO98028, IPO323, IPO98001	Chlorosis (%), Necrosis	6.83	17.50	HTRI1410		Karlstedt et al. (2019)
	QTL_4B	4B	Kukri_c32958_390, Excalibur_rep_ c102761_90, Tdurum_ contig42141_2028, wsnp_Ra_ c60161_61164325			3.25	8.80			Karlstedt et al. (2019)
	QTL_7B_1	7B	Excalibur_ c2217_731, BobWhite_ c3541_152			3.12	8.50	HTRI1410		Karlstedt et al. (2019)

	QTL_7B_2	7B	*T.durum_*contig9966_724, wsnp_BF291608B_Ta_2_2, Excalibur_c6871_217			6.11	15.30		Karlstedt et al. (2019)
Multiparent advanced generation intercross (MAGIC) Multiparental population	*QStb.lfl-1A*	1A		Mixed spore from 12 isolates	Leaf area with lesions (%)	1.2	6.10		Stadlmeier et al. (2019)
	QStb.lfl-1B	1B			Leaf area with lesions (%)	2.2	5.90		Stadlmeier et al. (2019)
	QStb.lfl-2B.1	2B			Leaf area with lesions (%)	1.1	5.30		Stadlmeier et al. (2019)
	QStb.lfl-2B.2	2B			Leaf area with lesions (%)	1.2	4.70		Stadlmeier et al. (2019)
	QStb.lfl-2B.3	2B			Leaf area with lesions (%)	3.5	6.30	Ambition, 'Firl3565	Stadlmeier et al. (2019)
	QStb.lfl-2D	2D			Leaf area with lesions (%)	5.4	8.50		Stadlmeier et al. (2019)
	QStb.lfl-4B	4B			Leaf area with lesions (%)	1.5	4.30		Stadlmeier et al. (2019)

(*Continued*)

Table 2 (*Continued*)

Germplasm/ population	QTL name	Linkage group/ Chromosome	Associated molecular marker (interval) in bracket)	Seedling stage inoculum	Attribute analysed for resistance	P VALUE (−Log 10*p*)	Phenotypic/ genotypic* variation (%)	Donor Source of resistance	Co-localisation with known locus/gene	Reference
Maxine/FTHP Redeemer (DH)		2D	DArT	MG-9626, MG-2		7.58	24.30	Redeemer		Tamburic-Ilincic and Rosa (2019)
		4B	DArT	MG-9626, MG-2, natural infection		3.23	12.00	Redeemer		
		7A	DArT	MG-9626, MG-2, natural infection		3.37	12.60	Maxine		
Durum wheat GWAS panel	*STB_ Nec3A*	3A		P1a/M1a	Necrosis (%)	13	0.45		*Stb6*	Ballini et al. (2020)
	STB_ Nec5A	5A			Necrosis (%)	5	0.11			Ballini et al. (2020)
	STB_ Pyc3A	3A			Pycnidia density %	6.7	0.11			Ballini et al. (2020)
Winter wheat GWAS panel		1A	12369332	Natural infection	Pycnidia grey value	6.64	0.083		*Qstb.1A*	Yates et al (2019)
		1A	472140874	Natural infection	Pycnidia grey value	5.53	0.063			Yates et al (2019)
		2A	635581134	Natural infection	Pycnidia grey value	6.12	0.078		*MQTL15*	Yates et al (2019)
		2B	648912764	Natural infection	Pycnidia density (%)	5.93	0.078			Yates et al (2019)
		2B	781584349	Natural infection	Pycnidia density (%)	5.59	0.085			Yates et al (2019)

2B	245888219	Natural infection	Pycnidia area	5.5	0.059	*MQTL17*	Yates et al (2019)
3B	30319666	Natural infection	Pycnidia grey value	5.6	0.075	*QStb.3B*	Yates et al (2019)
4A	648893451	Natural infection	Pycnidia density (%)	6.37	0.021		Yates et al (2019)
4D	485524935	Natural infection	Pycnidia grey value	8.71	0.106		Yates et al (2019)
5A	644275*	Natural infection	leaf lesion (%)	5.51	0.065	*MQTL19*	Yates et al (2019)
5A	20816445*	Natural infection	leaf lesion (%)	5.95	0.068	*MQTL19*	Yates et al (2019)
5A	685438196	Natural infection	Pycnidia grey value	5.67	0.075	*QTL9*	Yates et al (2019)
5B	671238689*	Natural infection	leaf lesion (%)	5.93	0.089		Yates et al (2019)
5B	442374822	Natural infection	Pycnidia grey value	6.4	0.096	*Stb1*	Yates et al (2019)
5D	1612786	Natural infection	leaf lesion (%)	9.16	0.103		Yates et al (2019)
5D	52615122	Natural infection	Pycnidia density (%)	6.13	0.044		Yates et al (2019)
5D	403075850	Natural infection	Pycnidia density (%)	6.19	0.066		Yates et al (2019)
6A	411095227	Natural infection	Pycnidia density (%)	6.24	0.019	*Stb15*	Yates et al (2019)

(*Continued*)

Table 2 (*Continued*)

Germplasm/ population	QTL name	Linkage group/ Chromosome	Associated molecular marker (interval) in bracket)	Seedling stage inoculum	Attribute analysed for resistance	P VALUE (−Log 10*p*)	Phenotypic/ genotypic* variation (%)	Donor Source of resistance	Co-localisation with known locus/gene	Reference
		6A	421527350	Natural infection	Pycnidia density (%)	6.47	0.091		*Stb15*	Yates et al (2019)
		6B	175719208	Natural infection	Pycnidia density (%)	5.79	0.073		*QStb.6B*	Yates et al (2019)
		6B	453931609	Natural infection	Pycnidia density (%)	6.03	0.062		*QStb.6B*	Yates et al (2019)
		6B	462130236	Natural infection	Pycnidia density (%)	6.99	0.052		*QStb.6B*	Yates et al (2019)
		6B	472609890	Natural infection	Pycnidia density (%)	7.1	0.093		*QStb.6B*	Yates et al (2019)
		6D	394936301	Natural infection	Pycnidia density (%)	5.7	0.074			Yates et al (2019)
		7B	265441725	Natural infection	Pycnidia density (%)	5.46	0.034			Yates et al (2019)
		7B	65661891	Natural infection	Pycnidia grey value	6.34	0.078			Yates et al (2019)
Winter wheat GWAs panel		1D	D_F5XZDLF 01A85DT_301	Natural infection	Disease score (0 to 5)	3.64 to 3.859848 96920862	0.0664374 879971906 to 0.07041356 20775426			Alemu et al (2021)

	1D	IAAV7510	Natural infection	Disease score (0 to 5)	3.0 to 3.40	0.0664374879971906 to 0.072		Alemu et al (2021)
	2A	BobWhite_c26374_339	Natural infection	Disease score (0 to 5)	3.03414194967821 to 3.27	0.0661754954822826 to 0.073		Alemu et al (2021)
	3B	RAC875_rep_c116515_181	Natural infection	Disease score (0 to 5)	3.914415962634 to 4.06475617423615			Alemu et al (2021)
	3B	Excalibur_rep_c106935_390	Natural infection	Disease score (0 to 5)	3.45311481364713 to 3.83	0.0673710341233633 to 0.084		Alemu et al (2021)
	3B	BS00005265_51/tplb0024c09_1335	Natural infection	Disease score (0 to 5)	3.29782438817223 to 3.41	0.067 to 0.076		Alemu et al (2021)
	3B	RAC875_c62223_86	Natural infection	Disease score (0 to 5)	3.25 to 3.25603830293077	0.067 to 0.72		Alemu et al (2021)
	3B	tplb0024c09_1335	Natural infection	Disease score (0 to 5)	3.17 to 3.18	0.067 to 0.070		Alemu et al (2021)
	4A	wsnp_Ex_c33012_41567026/RAC875_c11702_1015	Natural infection	Disease score (0 to 5)	3.17 to 4.48343862085241	0.070 to 0.099	*Stb7*, *Stb12*	Alemu et al (2021)
	4A	Excalibur_c56798_364	Natural infection	Disease score (0 to 5)	3.01	0.066		Alemu et al (2021)

(*Continued*)

Table 2 (*Continued*)

Germplasm/ population	QTL name	Linkage group/ Chromosome	Associated molecular marker (interval) in bracket)	Seedling stage inoculum	Attribute analysed for resistance	P VALUE (−Log 10*p*)	Phenotypic/ genotypic* variation (%)	Donor Source of resistance	Co-localisation with known locus/gene	Reference
		4B	Tdurum_ contig76559_124	Natural infection	Disease score (0 to 5)	3.479297 84758385 to 3.64	0.068 to 0.072			Alemu et al (2021)
		4B	BS00011510_51	Natural infection	Disease score (0 to 5)	to 3.51	0.066 to 0.078			Alemu et al (2021)
		4B	Kukri_c32958_390	Natural infection	Disease score (0 to 5)	3.08 to 3.47	0.066 to 0.078			Alemu et al (2021)
		6A	wsnp_Ku_ c38451_47086066	Natural infection	Disease score (0 to 5)	3.16 to 3.7267679 5898511	to 0.085		*Stb15*	Alemu et al (2021)
		6B	Kukri_c61725_545	Natural infection	Disease score (0 to 5)	3.12 to 3.13	0.066 to 0.069			Alemu et al (2021)
		6B	RAC875_ c23654_214	Natural infection	Disease score (0 to 5)	3.12	0.069			Alemu et al (2021)
		7A	Excalibur_c662_229	Natural infection	Disease score (0 to 5)	3.07	0.067			Alemu et al (2021)
Spring wheat GWAS panel	*APS_ MCH17_1*	4A	GENE-2768_150		Mixture of 13 Isolates	3.27	3.47			Louriki et al (2021)
	APS_ MCH17_2	4A	tplb0051b16_1324		Mixture of 13 Isolates	3	3.13			Louriki et al (2021)

APS_MCH17_3	4B	Ra_c27465_564	Mixture of 13 Isolates	3.01	3.17	Louriki et al (2021)
APS_MCH17_4	5A	wsnp_Ku_rep_c72362_72059764	Mixture of 13 Isolates	3.25	3.34	Louriki et al (2021)
APS_SAT17_1	1A	RAC875_c95364_259	Mixture of 13 Isolates	3.25	3.67	Louriki et al (2021)
APS_SAT17_2	1A	Kukri_c13107_100	Mixture of 13 Isolates	3.39	3.8	Louriki et al (2021)
APS_SAT17_3	1A	wsnp_Ku_c15214_23756800	Mixture of 13 Isolates	4.35	5.08	Louriki et al (2021)
APS_SAT17_4	1A	wsnp_Ku_c1642_3232242	Mixture of 13 Isolates	3.65	4.15	Louriki et al (2021)
APS_SAT17_5	1A	Ku_c1313_1673	Mixture of 13 Isolates	3.25	3.63	Louriki et al (2021)
APS_SAT17_6	1A	wsnp_JD_c6544_7697578	Mixture of 13 Isolates	4.16	4.85	Louriki et al (2021)
APS_SAT17_7	1A	RAC875_c80876_67	Mixture of 13 Isolates	4.3	5.03	Louriki et al (2021)
APS_SAT17_8	1A	Excalibur_c44668_382	Mixture of 13 Isolates	3.81	4.41	Louriki et al (2021)
APS_SAT17_9	1A	RAC875_c45115_509	Mixture of 13 Isolates	3.47	4.02	Louriki et al (2021)
APS_SAT17_10	1A	Kukri_c18017_1696	Mixture of 13 Isolates	4	4.69	Louriki et al (2021)
APS_SAT17_11	2A	RFL_Contig5277_1141	Mixture of 13 Isolates	3.4	3.8	Louriki et al (2021)
APS_SAT17_12	4A	RAC875_c83401_115	Mixture of 13 Isolates	3.75	4.46	Louriki et al (2021)

(Continued)

Table 2 *(Continued)*

Germplasm/ population	QTL name	Linkage group/ Chromosome	Associated molecular marker (interval) in bracket)	Seedling stage inoculum	Attribute analysed for resistance	P VALUE (−Log 10*p*)	Phenotypic/ genotypic* variation (%)	Donor Source of resistance	Co-localisation with known locus/gene	Reference
	APS_ SAT17_13	6B	GENE-3758_627		Mixture of 13 Isolates	4.23	4.85			Louriki et al (2021)
	APS_ SAT17_14	6B	BS00037462_51		Mixture of 13 Isolates	4.24	4.87			Louriki et al (2021)
	APS_ SAT17_15	6B	BobWhite_ c13202_399		Mixture of 13 Isolates	3.48	3.93			Louriki et al (2021)
	APS_ SAT17_16	7A	Ku_c15750_761		Mixture of 13 Isolates	3.06	3.33			Louriki et al (2021)
	APS_ SAT18_1	7B	BobWhite_rep_ c66957_84		Mixture of 13 Isolates	3.62	4.11			Louriki et al (2021)
	APS_ SAT18_2	7B	RFL_ Contig2540_306		Mixture of 13 Isolates	3.41	3.88			Louriki et al (2021)
	APS_ SAT18_3	7B	BS00095819_51		Mixture of 13 Isolates	3.42	3.88			Louriki et al (2021)
	SRT_ 71-R3_1	1A	wsnp_BE586140A_ Ta_2_1		isolate 71-R3	4.31	5.47			Louriki et al (2021)
	SRT_ 71-R3_2	2B	Excalibur_ c5438_274 2		isolate 71-R3	3.2	3.85			Louriki et al (2021)
	SRT_ 71-R3_3	2B	BobWhite_ c10520_1517		isolate 71-R3	5.19	6.66			Louriki et al (2021)
	SRT_ 71-R3_4	2B	GENE-1056_55 2B 145 7		isolate 71-R3	3.1	3.64			Louriki et al (2021)
	SRT_ 71-R3_5	4A	tplb0051b16_1324 4A 4		isolate 71-R3	3.49	4.21			Louriki et al (2021)

SRT_ 71-R3_6	4B	Tdurum_ contig60051_83	isolate 71-R3	3.63	4.45	Louriki et al (2021)
SRT_ SAT2_1 R	1A	AC875_ c20883_801 1A 1	isolate SAT2	3.5	4.16	Louriki et al (2021)
SRT_ SAT2_2 T	1A	durum_ contig44888_837	isolate SAT2	3.36	4.03	Louriki et al (2021)
SRT_ SAT2_3 R	2A	AC875_ c17787_274 2A 9	isolate SAT2	3.12	3.59	Louriki et al (2021)
SRT_ SAT2_4 R	2A	a_c22880_760 2A 104 4	isolate SAT2	3.69	4.5	Louriki et al (2021)
SRT_ SAT2_5 I	3B	AAV6442 3B 30 16.44 3	isolate SAT2	3.99	5.06	Louriki et al (2021)
SRT_ SAT2_6 T	4A	durum_ contig31218_279	isolate SAT2	3.18	3.83	Louriki et al (2021)
SRT_ SAT2_7 w	4A	snp_Ex_ c5072_9006666	isolate SAT2	3.46	4.18	Louriki et al (2021)
SRT_ SAT2_8 R	4B	AC875_ c24515_602 4B 8	isolate SAT2	3.29	3.91	Louriki et al (2021)

- pyramiding of qualitative genes and quantitative genes;
- stacking of several resistance genes with different recognition spectra and environmental optima into a single background; and
- integrated disease management (Pilet-Nayel et al., 2017, Mundt, 2002, Bailey-Serres et al., 2019, McDonald and Mundt, 2016, Latz et al., 2020).

This section discusses the planting of varietal mixtures.

Plant breeding programmes routinely evaluate germplasm across multiple environments to identify the best 'performers' for making genetic gains. Whilst there have been significant improvements in grain yield and other agronomic traits, intensive selection within elite germplasm has also led to a reduction in genetic diversity (Marshall, 1977, O'Driscoll et al., 2014). This has also meant a lack of diversity in wheat germplasm providing adequate resistance to *Z. tritici*.

Cultivation of mixed varieties (varietal mixtures) with different spectra of resistance has been employed to increase the durability and control yield losses due to STB infection (O'Driscoll et al., 2014, Mundt, 2002). Heterogeneous varietal mixtures are likely to restrict pathogen evolution as compared to monoculture/homogeneous/homozygous varieties (Simmonds, 1979, Sharma, 2001). However, deployment of mixed varieties is not a preferred strategy for broadacre growers who want varieties with specific high-end product quality traits (e.g. high protein content with polyphenol oxidase activity and premium noodle quality). These varieties cannot be mixed with low quality lines, as this practice may compromise quality attributes and market value. In addition, differences in heading date and maturity time may complicate harvest operations.

To address this problem, multilines which differ in their resistance genes but have similar maturity can be deployed for disease management. Wolfe et al. (1992) reported that a multiline strategy is more successful than using the same resistance alleles in a pure monoculture or combining them in a single host genotype. Ben M'Barek et al. (2020) also showed that introducing as little as 25% of a disease-resistant cultivar into a pure stand of a susceptible cultivar provides a substantial reduction of almost 50% in STB severity compared to the susceptible stand alone. For small land holders and mixed farming systems (e.g. systems including pigs), cultivation of mixed varieties of the same or different species may be a useful approach for disease management. Mixed grain-based products such as flour/pasta/bread/biscuit may also to appeal to health-conscious consumers.

5 Ways of increasing durability of resistance: deployment of qualitative resistance genes

Monogenic sources for genetic resistance are often deployed in commercial varieties, as they provide 'complete' resistance against pathogen(s). For such resistance, host-pathogen interaction response (resistance/susceptibility) is easier to evaluate under glasshouse and field conditions based on phenotypic expression at seedling stage. However, monogenic resistance sources only provide short-term protection to diseases as the pathogen evolves to escape recognition of the *R-Avr* gene interaction. Monogenic resistance is not durable, as seen in several 'boom and bust' plant pathosystems, including for *Z. tritici* (Rouxel and Balesdent, 2017, Krenz et al., 2008, Cowger et al., 2000, Li et al., 2006).

Brown et al. (2015) reported that the majority of *R* genes for STB resistance are now ineffective in European wheat varieties and have been rapidly overcome by *Z. tritici* races, due to mutation in the corresponding *Avr* genes (Cowger et al., 2000). The 'breakdown' of resistance is a major setback to plant breeding programmes, as it not only affects a current variety, but also affects their breeding pipeline, for example, advanced breeding lines producing potential new varieties. Breeding programmes often rely on a limited number of donor sources, especially for resistance to diseases and abiotic stresses, which are not as genetically diverse compared to other productivity traits.

Some monogenic sources are more 'durable' in conferring resistance compared to others. They may have effective background resistance gene(s) in 'matching' *Avr* gene(s) prevalent in *Z. tritici* populations. For example, *Stb16q* has more durable resistance compared to wheat cultivar 'Gene' (Ghaffary et al., 2012). Like other 'boom and bust' pathogens, *Z. tritici* co-adapts rapidly to resistant hosts. Kildea et al.(2020) found high levels of STB disease on the 'Cellule' wheat variety following inoculation with Cellule isolates, confirming the presence of *Stb16q* attacking isolates in the Irish *Z. tritici* population.

Despite the vulnerability of qualitative resistance genes to erosion of effectiveness, plant breeders still look for new sources of resistance to the pathogen within primary, secondary and tertiary gene pools. It is, however, challenging to incorporate genes from 'wild' sources into adapted varieties due to their unadapted genetic background, suppressed recombination and associated 'linkage drag'. However, unwanted genes can be eliminated via intensive backcrossing, use of molecular markers and speed breeding approaches (Watson et al., 2018), recovering desirable background. Despite this intensive effort, the effectiveness of disease resistance genes can never be assured, as durability in resistance can only be demonstrated under natural field conditions over time across different environments.

Combining major *R* genes in the same variety (pyramiding) is one of the options to increase durability of resistance. This strategy has proven useful in controlling fungal pathogens, especially those having an asexual mode of reproduction. However, sexual and asexual spores of *Z. tritici* cause infection under field conditions (Palmer and Skinner, 2002). Multigenic sources of STB resistance has been identified in some wheat varieties such as Kavkaz-K4500

(*Stb10* and *Stb12*) and Salamouni (*Stb13*, *Stb14* and *StbSm3*). Although these sources may not provide durable resistance across multiple environments over a long period, *R* genes can be deployed at various levels, such as specific adaptation and spatial or temporal scales (Day-Rubenstein et al., 2005). This strategy may increase the durability of resistance but it needs to be proven in growing environments where *R* genes have not been previously deployed in commercial varieties (Mundt, 2015). The allele frequency of *Avr* genes corresponding to multiple *R* gene may not change so rapidly as compared to when genes are deployed individually. Multiple alleles for genetic resistance can be easily combined in advanced breeding lines/varieties using conventional breeding methods. However, it is difficult to evaluate the phenotypic expression of different alleles conferring resistance using conventional phenotypic assays due to masking/residual effects. This limitation can be overcome through MAS that makes it possible to monitor the introgression of resistance alleles derived from different loci.

6 Ways of increasing durability of resistance: pyramiding of qualitative and quantitative resistance genes

Achieving durable resistance based on qualitative genes exclusively may not be possible for sexually reproducing pathogens, as pathogen races keep on evolving and changing their mode of action. However, genetic loci involved in stable quantitative resistance (basal/APR) with 'moderate' effects can be stacked in elite genetic backgrounds having *R* genes. There are many examples in the literature suggesting that quantitative resistance is more durable as compared to qualitative resistance, as observed in several pathosystems (Parlevliet, 2002, Turkensteen, 1993, Schurnbusch et al., 2004, Kolmer et al., 2008, Chen et al., 2003, Raman et al., 2018, Meile et al., 2018, Delourme et al., 2006).

It has been hypothesised that the durability of resistance could be due to decreased selective pressures for pathogens to overcome host quantitative resistance (Fukuoka et al., 2009). Several studies have shown that the life of qualitative resistance genes can be increased through combination with quantitative resistance. Palloix et al. (2009) showed that *R*-gene-mediated resistance is overcome at high frequency when introgressed in a susceptible background, whereas it is durable when combined with quantitative resistance. Brun et al. (2010) recommended combining quantitative resistance with *R*-gene-mediated resistance to enhance disease control and crop production against *Leptosphaeria maculans* which causes blackleg disease in oilseed rape. This study showed that qualitative resistance by *Rlm6* in a susceptible background remained effective for three cropping seasons, whereas the resistance was still effective after five years of selection by combining *Rlm6* with quantitative resistance. Later on, Delourme et al. (2014) showed that both *Rlm6*-mediated resistance and quantitative resistance to *L. maculans* became ineffective after the eighth cycle of selection.

Another approach is pyramiding of quantitative resistance genes to achieve more durable resistance. Quantitative resistance is being recognised as a 'silver' bullet to achieve more durable resistance compared to qualitative resistance mediated by effector triggered immunity (Jones and Dangl, 2006). Bhavani et al. (2019) reported that combination of multiple quantitative resistance genes proved to be highly effective in the rust resistance breeding programme at CIMMYT. Several studies have identified QTL for resistance by analysing germplasm (mapping populations, breeding lines, landraces and synthetic accession derived from related species) in wheat and durum wheat (see reviews by Brown et al., 2015, Raman and Milgate, 2012). These studies revealed low to moderate heritability for resistance and a relatively large number of QTLs (some of them may be in linkage drag/pleiotropy) with low genotypic variance. This suggests that identification of stable quantitative resistance across environments is still challenging due to genotype x environment interactions (Karlstedt et al., 2019). Once robust and stable associations across multi-environments are identified, those genomic regions can be pyramided together in the same background using marker-trait selection strategies, including genomic selection (GS). This approach has made it possible to improve the rate of genetic gains by improving selection efficiency for quantitative traits in cereal breeding programmes.

Besides quantitative resistance, single genes can also provide durable resistance to several pathogens (Fu et al., 2009, Krattinger et al., 2009, Moore et al., 2015). Examples include the wheat *Lr34* gene for resistance to leaf rust (showing complete segregation with the APR locus), *Yr18* against stripe rust (caused by *Puccinia striiformis* West. f. sp. *tritici*) and *Pm38* against powdery mildew (PM) (caused by *Blumeria graminis* (DC) EO Speer f. sp. *tritici*). These have provided broad-spectrum durable resistance in wheat over the last 90 years (Dyck, 1977, 2011, Spielmeyer et al., 2005). *Lr34* has also been associated with tolerance to barley yellow dwarf virus (Singh, 1993). The wheat *Lr34* gene also provides resistance to other fungal diseases in rice, barley and sorghum (Krattinger et al., 2016, Risk et al., 2013, Schnippenkoetter et al., 2017). This type of broad-spectrum resistance has also been referred as multiple disease resistance (Wiesner-Hanks and Nelson, 2016).

The effectiveness of quantitative resistance has been challenged by increased virulence. Le Guen et al. (2007) reported that pathogen populations adapt to polygenic resistance mechanisms, bypassing all or part of the resistance QTL. Despite this, erosion of quantitative resistance is likely to be slower compared to qualitative resistance. To date, none of the QTL for STB resistance has been cloned in cereals. Cloning for quantitative disease resistance has been hampered due to their small and unstable allelic effects in growing environments, and genotype x environment interactions. Further research is required to:

- identify stable QTL across multiple environments; and
- associate nucleotide sequence variants that are responsible for resistance.

This will make it possible to identify the nature of disease resistance genes and the molecular basis of stable QTL for STB resistance.

7 Ways to increase durability of resistance: manipulation of plant architecture genes and deployment of multiple resistance loci

Genetic studies based on QTL analyses have revealed some interesting findings. For example, some genomic regions associated with STB resistance have also been associated with phenological traits such as plant height and heading date in wheat (Simón et al., 2004, Zwart et al., 2010). Tamburic-Ilincic and Rosa (2019) also identified QTL for STB resistance on chromosomes 2D, 4B and 7A near the *Rht8* and *Rht-b1* genes for plant height. The authors found it possible to select short and early maturing lines with good resistance to Fusarium head blight and STB. However, it is not clear whether multi-trait associations are due to a genetic linkage or pleiotropy.

There are STB escape traits such as early vigour, growth rate, plant height, leaf length, leaf spacing, prostrate leaves, leaf insertion angle, flag leaf emergence and heading time (Arraiano et al., 2009, Brown et al., 2015). These can be combined with resistance traits to develop improved wheat varieties to further reduce disease progression and maintain yields (Benbow et al., 2020). Although 'disease escape' traits may be useful in reducing overall STB infection, some of these traits (e.g. late-heading and late maturity) are not desirable in environments prone to terminal drought and heat stress. In contrast, Arama et al. (1999) found no association between phenological traits and resistance to STB. Mapping of STB loci in structured and GWAS populations having similar phenologies would substantiate the role of plant development genes in STB resistance.

Another approach is deployment of multiple resistance loci. Loci for broad-spectrum resistance have been identified that confer resistance to STB, Fusarium head blight and *Parastangnospora nodorum* in wheat (Zwart et al., 2010, Miedaner et al., 2012). Zwart et al. (2010) reported QTL for resistance to STB and yellow leaf spot inherited from one parent, while QTLs for leaf rust and stem rust were inherited from the other parent (in repulsion). Fine mapping of QTLs is required to verify whether QTLs are indeed in replusion. Ballini et al. (2020) conducted genome-wide association mapping in *T. turgidum*, using a composite cross population of durum wheat, and identified QTLs for leaf rust and STB. Stadlmeier et al. (2019) mapped a winter wheat MAGIC (multiparent advanced generation intercross) population of bread wheat and identified QTL for resistance to PM, STB and tan spot (TS). Seven QTL for STB resistance were identified on 1A, 1B, 2B, 2D and 4D. This study also identified three QTLs for disease resistance: one QTL for resistance to STB and TS on chromosome 2B and two QTLs on chromosome 1AS and 7AL. Based on results, Stadlmeier et al. (2019) concluded that parallel marker-assisted breeding for resistance to PM, STB and TS appears to be feasible. Multiple disease resistance loci are excellent candidates for MAS and may increase durability of resistance through basal resistance mechanisms.

8 Marker-assisted selection (MAS) and genomic selection (GS) for Septoria tritici blotch resistance breeding

Cereal breeders have long been selecting resistant varieties against *Z. tritici* populations prevalent under field conditions. It is comparatively easy to select for qualitative resistance in field and glasshouse conditions whereas it is difficult to select for quantitative resistance. Selection can be very subjective due to factors such as infection from other pathogens and genotype x environment interactions. Selection for quantitative resistance is complicated by low heritability and genotype x environment interactions.

MAS has been widely used to monitor introgression of desirable alleles at the various stages in a breeding cycle. Despite genetic mapping of qualitative and quantitative resistance loci for STB resistance (Tables 1 and 2), there are very few examples of MAS in the published literature (Goodwin et al., 2008). This could be due to lack of tightly linked and validated markers of effective resistance genes/loci for MAS. MTAs for both qualitative and quantitative resistance need to be validated in a range of germplasm relevant to breeding programmes to select candidate resistance loci. The most promising QTL/race-specific genes can then be selected in forward and backcrossing programmes (such as marker-assisted backcrossing, marker-assisted gene stacking, marker-assisted recurrent selection and genome selection) using MAS strategies.

Several public and private cereal breeding programmes use molecular markers to select traits such as adaptation, grain yield and quality. Markers linked with STB resistance must be compatible with other marker systems. The development and validation of molecular markers suitable for high-throughput analysis will increase their adoption in breeding programmes. The majority of marker technologies used for mapping STB resistance (such as SNP arrays, AFLP, SSR, GBS, DArT and DArTseq) are not suitable for routine high-throughput MAS in cereal breeding programmes. CAPS markers are not well suited for routine MAS, as they require an additional step of digestion of PCR amplicons with restriction enzyme(s) before they are analysed for allelic variation. For routine

MAS, only a limited number of cost-effective and 'user-friendly' molecular marker approaches are suitable for high-throughput analysis. SNP and Indels/ presence–absence variants associated with STB resistance can be assayed via Illumina Golden Gate assay (Illumina), Kompetitive Allele Specific PCR (KASP) markers (KBioscience/LGC) and DArTTAg formats (Diversity Arrays Technology).

Makhoul et al. (2020) have described a way to convert effective SNP into KASP markers in wheat. A typical KASP marker assay involves two phases:

- the development of KASP assay; and
- its routine application in MAS.

Approximately 100 bp polymorphic sequences are interrogated for their uniqueness in the reference genome. Locus-specific (two pairs of) primers are then designed to target a region of interest. When marker sequences are short (such as in GBS pipelines), the corresponding sequence can be interrogated from reference genomes and then utilised for the development of KASP assays, as shown by Yang et al. (2018). Temperature Switch PCR and multiplex PCR assays can also be used for SNP genotyping (Hayden et al., 2009, Tabone et al., 2009).

Effective MAS also requires tightly linked markers associated with a trait of interest. Several genomic regions associated with qualitative and quantitative resistance have been identified so far, related to large chromosomal intervals (Tables 1 and 2). These need to be fine mapped to ensure introgression of favourable alleles (but not of alleles associated with undesirable traits) in elite lines. Molecular markers based on causal genes (also referred as perfect markers) are preferred by commercial plant breeding programmes, due to their discrimination ability between desirable and undesirable individuals. This is not always the case as shown for the *TaALMT1* gene involved in aluminium tolerance in wheat (Sasaki et al., 2004, Raman et al., 2005, 2009).

Genomes of *T. urartu* (Ling et al., 2013), *Ae. tauschii* (Jia et al., 2013) and *T. aestivum* (Brenchley et al., 2012, Raats et al., 2013, Walkowiak et al., 2020) have recently been sequenced. However, a single reference genome cannot uncover the entire gene content of a species due to extensive structural variants (Hurgobin et al., 2018, Bayer et al., 2020). To address this issue, pan-genomes of several cultivars, including Chinese spring wheat, have been constructed that display species-wide SNPs, small indels and structural variations which occur in pre- and post-domestication bottlenecks (Montenegro et al., 2017). These genomic resources provide a wealth of sequence variation and will assist in developing high yielding and disease-resistant varieties of cereals through molecular breeding (genomic-assisted breeding and gene engineering) approaches. Physical positions of significant MTAs and haplotypes associated with STB resistance can be determined on the reference sequenced/pan-genomes to identify candidate genes and their genetic architecture and gene functions. Diagnostic markers for resistance could provide better tools to speed up the development of new varieties and simplify the time-consuming and laborious germplasm evaluation pipeline. Functional analysis will also provide insights into mechanisms underlying disease resistance linking *R* protein(s) and QTL involved in STB resistance. At this stage, only one gene for STB resistance, *Stb6*, has been cloned in wheat. In the near future several genes for qualitative and quantitative resistance will be cloned to provide gene-based markers for precision breeding. While we wait for cloning of other *R* and *QR* genes, linked markers to STB resistance (located within 5cM) could be used to enhance selection efficiency.

Developments in omics approaches have provided insights into host-pathogen (*Z. tritici*) interactions (Seybold et al., 2020) and provide additional markers for monitoring STB resistance. Benbow et al. (2020) has performed transcriptomic analysis of STB resistant and STB susceptible cultivars and identified a Pathogen-associated molecular patterns (PAMPs)-triggered response in the STB susceptible cultivar 'Longbow', in contrast to few PAMP-related up-regulated genes in the resistant cultivar 'Stigg'. Zhou et al. (2020) have described two novel wheat genes: *TaSSP6* and *TaSSP7* that encode for small, secreted proteins and contribute to resistance to STB disease. A virus-induced gene silencing system was used to study the function of *TaSSP6* and *TaSSP7*. Silencing of *TaSSP7* caused a significant decrease in transcript abundance of TaSSP7 by a factor of 24 (construct BSMV:*TaSSP7*-V1) and a factor of 23 (construct BSMV:*TaSSP7*-V2). This study also showed that wheat proteins interact with small, secreted fungal proteins and this interaction may contribute to resistance. Cloned effector genes of *Z. tritici* (Zhong et al., 2017) and leaf infiltration assays using transient expression assay of effector proteins can provide insights into the process of host-pathogen interactions. Detailed knowledge of host-pathogen interaction genes and their durability across diverse environments will facilitate the development of STB management strategies.

Understanding genes underlying host interaction can be accelerated by developing reliable phenotyping tools for different components of resistance. High-throughput automated phenotyping methods are becoming available to better understand the genetic architecture of STB resistance in wheat. Odilbekov et al. (2018) reported that proxy measurements for photosystem II, chlorophyll content, carotenoid and anthocyanin levels and leaf surface temperature can be used to determine variation in STB. Yates et al. (2019) evaluated STB resistance of 335 European winter wheat cultivars using precision phenotyping methods and identified 26 genomic regions associated with quantitative resistance. Genetic analysis based on 5.4 million phenotypic datapoints and 13 648 SNP markers revealed distinct loci-affecting pathogen reproduction and leaf damage caused by naturally occurring population of *Z. tritici*. This study suggests that identification of alleles associated with different measures of STB resistance (e.g. leaf damage, pycnidial density and pycnidial size) will make it possible to combine different forms of quantitative resistance to improve durability of STB resistance.

In addition to the approaches described above, genetic gains for quantitative resistance can be made via GS. This approach predicts the genetic merit of the individuals based on genome-wide marker effects and calculates genomic estimated breeding values for selection (Meuwissen and Goddard, 2010). Typically, GS requires two types of populations:

- a training/reference population (with genotypic and phenotypic data); and
- a breeding/selection population (with genotypic data).

Phenotyping is not essential in the latter population. Experimental data on STB and other diseases suggest that GS has potential in breeding for improved quantitative resistance (Miedaner et al., 2013, Mirdita et al., 2015).

Muqaddasi et al. (2019) dissected the genetic basis of STB infection in 371 European wheat varieties. They explored the potential of genomic prediction in accounting for additive effects and additive plus epistatic interactions among loci for improving STB resistance. This study used GWAS analysis based on high-density SNP marker and phenotypic data. Significant marker (SNP)–trait (STB resistance) associations were identified on chromosomes 1A, 1B, 2B, 2D, 4A, 5A, 6D and 7A (plus unassigned groups). There are several associations with qualitative *R* genes such as *Stb2*, *Stb11*, *Stb9* and *Stb12*, as well as novel associations with chromosomes 6AL and 6DL. Genomic prediction values were estimated which can further be validated for their robustness and durability in wheat breeding programmes.

It is unclear from the published literature how many breeding programmes are implementing marker and genomic-assisted selection strategies for accelerating genetic gains in improving STB resistance. As described earlier, the majority of *R* genes for STB resistance are ineffective in some environments, suggesting the rate of marker implementation for STB is low, especially in Europe. Cereal breeding programmes using GS are likely to combine qualitative and quantitative loci to increase the durability of resistance.

9 The use of genetic transformation and gene editing techniques

Besides naturally occurring variation, novel genes conferring resistance to diseases can be deployed to enhance durability. Genetic engineering technologies facilitate direct transfer of genes of interest without any crossability barriers such as sexual incompatibility that occurs frequently in wide crosses. Methods for routine genetic transformation and gene editing with high frequency of plant regeneration are becoming available in major food crops (Hayta et al., 2019, Kong et al., 2020, Wang et al., 2017, Hua et al., 2019). Debernardi et al. (2020) have reported that expression of GROWTH-REGULATING FACTOR 4 (GRF4) and its cofactor GRF-INTERACTING FACTOR 1 increase the regeneration efficiency of wheat, triticale, rice and dicot species.

Different strategies for creating broad-spectrum disease resistance can be optimised and deployed (Keller et al., 1999). New genes (including from diverse organisms), to which resistance genes with different recognition spectra have not been exposed previously, can be assembled in cassettes (Michelmore et al., 2017). Multiple gene stacks can be combined at a single locus as shown in potato (Ghislain et al., 2019, Wulff and Moscou, 2014). Disease susceptibility genes can be edited to correct genetic defects. Spray-on gene editing technologies based on nanoparticles and RNA virus transfection (Doyle et al., 2019, Torti et al., 2021) are particularly promising strategies to reduce yield losses due to devastating disease epidemics. In a recent study, Campanaro et al., (2020) showed that virus induced gene silencing of TaWRKY10 transcription factor leads to increased resistance to *Z. tritici*. TaWRKY10 gene silenced wheat plants showed an up-regulated jasmonic acid response. Authors postulated that manipulating jasmonic acid level may provide enhance STB resistance in wheat.

10 Conclusion and future trends

A comprehensive knowledge of the genetic architecture of STB resistance and stability of resistance genes across growing environments is crucial for deploying integrated strategies to effectively control yield losses. Over the decades, several loci for qualitative and quantitative resistance to STB have been identified especially in common wheat. Whilst the majority of *R* genes are ineffective in Europe, they can be effectively deployed in other countries at local and regional scales. A vast reservoir of genetic resources including landrace and wild species accessions are available in cereals crops. Untapped sources provide us an opportunity to mine and deploy the best allele combinations for resistance to STB and other traits of interest to produce new varieties. This can be achieved via a thorough understanding of genetic interactions (additive and epistasis) that may affect STB expression. It is, however, important to avoid incorporating susceptibility genes to other diseases.

High-throughput screening methods are becoming available for characterisation of STB resistance (Odilbekov et al., 2018, Yates et al., 2019). New phenotypic methods based on machine learning and high-throughput analysis will make it possible to:

- identify different components of genetic resistance to STB at seedling and adult plant stages;

- enrich our knowledge of the genetic architecture of resistance; and
- combine multiple alleles for stable resistance in desired backgrounds.

Genome-phenome association studies based on ultrahigh-density markers in large and diverse panels of germplasm accessions would provide a suite of molecular tools for the selection of qualitative and quantitative resistance genes to achieve higher genetic gains in breeding programmes.

Given that *Z. tritici* shows a degree of evolutionary 'plasticity' which may allow it to keep pace with innovations in disease control with relative ease (Fones and Gurr, 2015), breeders need to stay ahead by developing strategies that minimise yield losses. New methodologies aimed at identification of quantitative resistance should be developed so that populations can be evaluated under controlled conditions. This will increase heritability (reliability) by reducing genotype x environment interaction. Durability of genetic resistance to STB can be improved via strategies such as spatial and temporal gene deployment. Better understanding of G X E X M (management) and phenotypic trait expression will enable marker-assisted selection for range of traits including STB resistance.

Multifaceted strategies based on agronomic measures, application of effective fungicides, systemic acquired resistance and biocontrol agents should be integrated to increase the durability of genetic resistance. This is especially crucial for environments where host resistance and chemical interventions are not effective in controlling disease losses.

Despite of great strides made in cereal genetics and the development of improved varieties, our knowledge of the molecular basis of both qualitative and quantitative resistance to STB remains limited. Innovative developments in genomics and phenomics, sensor technologies, statistical genetics and bioinformatics will make it possible to pinpoint candidate genes underlying resistance to STB and their molecular basis. Whole genome analysis revealing structural variation (e.g. copy number variation, homoeologous recombination) will provide insights into complexity, especially of quantitative resistance. This will accelerate knowledge-driven breeding for resistance. The function of these resistance genes can be established via gene expression tools (RNA, gene editing, genetic transformation and genetic mutants).

11 References

Abe, A., Kosugi, S., Yoshida, K., Natsume, S., Takagi, H., Kanzaki, H., Matsumura, H., Yoshida, K., Mitsuoka, C., Tamiru, M., Innan, H., Cano, L., Kamoun, S. and Terauchi, R. 2012. Genome sequencing reveals agronomically important loci in rice using MutMap. *Nature Biotechnology* 30(2), 174–178.

Adhikari, T. B., Anderson, J. M. and Goodwin, S. B. 2003. Identification and molecular mapping of a gene in wheat conferring resistance to *Mycosphaerella graminicola*. *Phytopathology* 93(9), 1158–1164.

Adhikari, T. B., Cavaletto, J. R., Dubcovsky, J., Gieco, J. O., Schlatter, A. R. and Goodwin, S. B. 2004a. Molecular mapping of the *Stb4* gene for resistance to Septoria tritici blotch in wheat. *Phytopathology* 94(11), 1198–1206.

Adhikari, T. B., Wallwork, H. and Goodwin, S. B. 2004b. Microsatellite markers linked to the *Stb2* and *Stb3* genes for resistance to Septoria tritici blotch in wheat. *Crop Science* 44(4), 1403–1411.

Adhikari, T. B., Yang, X., Cavaletto, J. R., Hu, X., Buechley, G., Ohm, H. W., Shaner, G. and Goodwin, S. B. 2004c. Molecular mapping of *Stb1*, a potentially durable gene for resistance to Septoria tritici blotch in wheat. *Theoretical and Applied Genetics*. 109(5), 944–953.

Adhikari, T. B., Mamidi, S., Gurung, S. and Bonman, J. M. 2015. Mapping of new quantitative trait loci (QTL) for resistance to Septoria tritici blotch in spring wheat (*Triticum aestivum* L.). *Euphytica* 205(3), 699–706.

Alemu, A., Brazauskas, G., Gaikpa, D. S., Henriksson, T., Islamov, B., Jørgensen, L. N., Koppel, M., Koppel, R., Liatukas, Ž., Svensson, J. T. and Chawade, A. 2021. Genome-wide association analysis and genomic prediction for adult-plant resistance to Septoria tritici blotch and powdery mildew in winter wheat. *Frontiers in Genetics*, 12. https://doi.org/10.3389/fgene.2021.661742.

Ando, K., Rynearson, S., Muleta, K. T., Gedamu, J., Girma, B., Bosque-Perez, N. A., Chen, M. S. and Pumphrey, M. O. 2018. Genome-wide associations for multiple pest resistances in a Northwestern United States elite spring wheat panel. *PLoS ONE* 13(2), e0191305.

Arama, P. F., Parlevliet, J. E. and Van Silfhout, C. H. 1999. Heading date and resistance to Septoria tritici blotch in wheat not genetically associated. *Euphytica* 106(1), 63–68.

Arora, S., Steuernagel, B., Gaurav, K., Chandramohan, S., Long, Y., Matny, O., Johnson, R., Enk, J., Periyannan, S., Singh, N., Asyraf Md Hatta, M., Athiyannan, N., Cheema, J., Yu, G., Kangara, N., Ghosh, S., Szabo, L. J., Poland,J., Bariana, H., Jones, J. D. G., Bentley, A. R., Ayliffe, M., Olson, E., Xu, S. S., Steffenson, B. J., Lagudah, E. and Wulff, B. B. H. 2019. Resistance gene cloning from a wild crop relative by sequence capture and association genetics. *Nat. Biotechnol.* 37, 139–143.

Arraiano, L. S., Balaam, N., Fenwick, P. M., Chapman, C., Feuerhelm, D., Howell, P., Smith, S. J., Widdowson, J. P. and Brown, J. K. M. 2009. Contributions of disease resistance and escape to the control of septoria tritici blotch of wheat. *Plant Pathology* 58, 910–922.

Arraiano, L. S., Brading, P. A. and Brown, J. K. M. 2001a. A detached seedling leaf technique to study resistance to *Mycosphaerella graminicola* (anamorph *Septoria tritici*) in wheat. *Plant Pathology* 50(3), 339–346.

Arraiano, L. S., Worland, A. J., Ellerbrook, C. and Brown, J. K. M. 2001b. Chromosomal location of a gene for resistance to Septoria tritici blotch (Mycosphaerella graminicola) in the hexaploid wheat 'synthetic 6x'. *Theoretical and Applied Genetics* 103(5), 758–764.

Arraiano, L. S. and Brown, J. K. M. 2006. Identification of isolate-specific and partial resistance to Septoria tritici blotch in 238 European wheat cultivars and breeding lines. *Plant Pathology* 55(6), 726–738.

Arraiano, L. S., Chartrain, L., Bossolini, E., Slatter, H. N., Keller, B. and Brown, J. K. M. 2007. A gene in European wheat cultivars for resistance to an African isolate of *Mycosphaerella graminicola*. *Plant Pathology* 56(1), 73–78.

Bailey-Serres, J., Parker, J. E., Ainsworth, E. A., Oldroyd, G. E. D. and Schroeder, J. I. 2019. Genetic strategies for improving crop yields. *Nature* 575(7781), 109–118.

Baird, N. A., Etter, P. D., Atwood, T. S., Currey, M. C., Shiver, A. L., Lewis, Z. A., Selker, E. U., Cresko, W. A. and Johnson, E. A. 2008. Rapid SNP discovery and genetic mapping using sequenced RAD markers. *PLoS ONE* 3(10), e3376.

Ballini, E., Tavaud, M., Ducasse, A., Sanchez, D., Paux, E., Kitt, J., Charmet, G., Audigeos, D., Roumet, P., David, J. and Morel, J.-B. 2020. Genome wide association mapping for resistance to multiple fungal pathogens in a panel issued from a broad composite cross-population of tetraploid wheat *Triticum turgidum*. *Euphytica* 216(6), 92.

Bayer, P. E., Golicz, A. A., Scheben, A., Batley, J. and Edwards, D. 2020. Plant pan-genomes are the new reference. *Nature Plants* 6(8), 914–920.

Bemister, D. H., Semagn, K., Iqbal, M., Randhawa, H., Strelkov, S. E. and Spaner, D. M. 2019. Mapping QTL associated with stripe rust, leaf rust, and leaf spotting in a Canadian spring wheat population. *Crop Science* 59(2), 650–658.

Ben M'Barek, S., Karisto, P., Abdedayem, W., Laribi, M., Fakhfakh, M., Kouki, H., Mikaberidze, A. and Yahyaoui, A. 2020. Improved control of Septoria tritici blotch in durum wheat using cultivar mixtures. *Plant Pathology* 69(9), 1655–1665.

Benbow, H. R., Brennan, C. J., Zhou, B., Christodoulou, T., Berry, S., Uauy, C., Mullins, E. and Doohan, F. M. 2020. Insights into the resistance of a synthetically-derived wheat to Septoria tritici blotch disease: less is more. *BMC Plant Biology* 20(1), 407.

Bhavani, S., Hodson, D. P., Huerta-Espino, J., Randhawa, M. S. and Singh, R. P. 2019. Progress in breeding for resistance to Ug99 and other races of the stem rust fungus in CIMMYT wheat germplasm. *Frontiers of Agricultural Science and Engineering* 6(3), 210–224. https://doi.org/10.15302/J-FASE-2019268.

Brading, P. A., Verstappen, E. C. P., Kema, G. H. J. and Brown, J. K. M. 2002. A gene-for-gene relationship between wheat and *Mycosphaerella graminicola*, the Septoria tritici blotch pathogen. *Phytopathology* 92(4), 439–445.

Brenchley, R., Spannagl, M., Pfeifer, M., Barker, G. L., D'amore, R., Allen, A. M., Mckenzie, N., Kramer, M., Kerhornou, A., Bolser, D., Kay, S., Waite, D., Trick, M., Bancroft, I., Gu, Y., Huo, N., Luo, M. C., Sehgal, S., Gill, B., Kianian, S., Anderson, O., Kersey, P., Dvorak, J., Mccombie, W. R., Hall, A., Mayer, K. F. X., Edwards, K. J., Bevan, M. W. and Hall, N. 2012. Analysis of the bread wheat genome using whole-genome shotgun sequencing. *Nature* 491(7426), 705–710. https://doi.org/10.1038/nature11650.

Brown, J. K. M., Chartrain, L., Lasserre-Zuber, P. and Saintenac, C. 2015. Genetics of resistance to *Zymoseptoria tritici* and applications to wheat breeding. *Fungal Genetics and Biology* 79, 33–41.

Brun, H., Chevre, A. M., Fitt, B. D. L., Powers, S., Besnard, A. L., Ermel, M., Huteau, V., Marquer, B., Eber, F., Renard, M. and Andrivon, D. 2010. Quantitative resistance increases the durability of qualitative resistance to *Leptosphaeria maculans* in *Brassica napus*. *New Phytologist* 185(1), 285–299.

Campanaro, A., Srivastava, A. K., Zhang, C., Lee, J., Millyard, L., Gatehouse, A. M. R., Byrne, E. and Sadanandom, A. 2021. TaWRKY10 transcription factor is a novel jasmonic acid signalling regulator involved in immunity against Septoria tritici blotch disease in wheat. *Plant Pathology*, 70, 1397–1408.

Cavanagh, C. R., Chao, S., Wang, S., Huang, B. E., Stephen, S., Kiani, S., Forrest, K., Saintenac, C., Brown-Guedira, G. L., Akhunova, A., See, D., Bai, G., Pumphrey, M., Tomar, L., Wong, D., Kong, S., Reynolds, M., Da Silva, M. L., Bockelman, H., Talbert, L., Anderson, J. A., Dreisigacker, S., Baenziger, S., Carter, A., Korzun, V., Morrell, P. L., Dubcovsky, J., Morell, M. K., Sorrells, M. E., Hayden, M. J. and Akhunov, E. 2013. Genome-wide comparative diversity uncovers multiple targets of selection for improvement in hexaploid wheat landraces and cultivars. *Proceedings of the National Academy of Sciences of the United States of America* 110(20), 8057-8062.

Chartrain, L., Berry, S. T. and Brown, J. K. M. 2005a. Resistance of wheat line Kavkaz-K4500 L 6(A).4 to septoria tritici blotch controlled by isolate-specific resistance genes. *Phytopathology* 95, 664-671.

Chartrain, L., Brading, P. A. and Brown, J. K. M. 2005b. Presence of the *Stb6* gene for resistance to Septoria tritici blotch (*Mycosphaerella graminicola*) in cultivars used in wheat-breeding programmes worldwide. *Plant Pathology* 54(2), 134-143.

Chartrain, L., Brading, P. A., Makepeace, J. C. and Brown, J. K. M. 2004a. Sources of resistance to Septoria tritici blotch and implications for wheat breeding. *Plant Pathology* 53(4), 454-460.

Chartrain, L., Brading, P. A., Widdowson, J. P. and Brown, J. K. M. 2004b. Partial resistance to Septoria tritici blotch (*Mycosphaeria graminicola*) in wheat cultivars Arina and Riband. *Phytopathology* 94(5), 497-504.

Chartrain, L., Joaquim, P., Berry, S. T., Arraiano, L. S., Azanza, F. and Brown, J. K. M. 2005c. Genetics of resistance to Septoria tritici blotch in the Portuguese wheat breeding line TE 9111. *Theoretical and Applied Genetics*. 110(6), 1138-1144.

Chartrain, L., Sourdille, P., Bernard, M. and Brown, J. K. M. 2009. Identification and location of *Stb9*, a gene for resistance to Septoria tritici blotch in wheat cultivars Courtot and Tonic. *Plant Pathology* 58(3), 547-555.

Chen, H., He, H., Zhou, F., Yu, H. and Deng, X. W. 2013. Development of genomics-based genotyping platforms and their applications in rice breeding. *Current Opinion in Plant Biology* 16(2), 247-254. https://doi.org/10.1016/j.pbi.2013.04.002.

Chen, H., Wang, S., Xing, Y., Xu, C., Hayes, P. M. and Zhang, Q. 2003. Comparative analysis of genomic locations and race specificities of loci for quantitative resistance to *Pyricularia grisea* in rice and barley. *Proceedings of the National Academy of Sciences of the United States of America* 100(5), 2544-2549.

Chen, J., Wang, B., Zhang, Y., Yue, X., Li, Z. and Liu, K. 2017. High-density ddRAD linkage and yield-related QTL mapping delimits a chromosomal region responsible for oil content in rapeseed (*Brassica napus* L.). *Breeding Science* 67(3), 296-306.

Cools, H. J. and Fraaije, B. A. 2008. Are azole fungicides losing ground against Septoria wheat disease? Resistance mechanisms in *Mycosphaerella graminicola*. *Pest Management Science* 64(7), 681-684. https://doi.org/10.1002/ps.1568.

Corwin, J. A., Copeland, D., Feusier, J., Subedy, A., Eshbaugh, R., Palmer, C., Maloof, J. and Kliebenstein, D. J. 2016. The quantitative basis of the Arabidopsis innate immune system to endemic pathogens depends on pathogen genetics. *PLoS Genetics* 12(2), e1005789.

Cowger, C., Hoffer, M. E. and Mundt, C. C. 2000. Specific adaptation by *Mycosphaerella graminicola* to a vertically resistant wheat cultivar. *Plant Pathology* 49(4), 445-451.

Cowling, S. G. 2006. Identification and mapping of host resistance genes to Septoria tritici blotch of wheat. PhD Thesis, University of Manitoba, Winnipeg, MB, Canada.

Cowling, S. G., Brule-Babel, A., Somers, D. and Lamari, L. 2004. Identification and mapping of host resistance genes to Septoria tritici blotch of wheat. *Phytopathology* 94.

Cuthbert, R. 2011. Molecular mapping of Septoria tritici blotch in hexaploid wheat (*Triticum aestivum* L.). Ph.D. Thesis, University of Manitoba, Canada.

Dangle. J. and Jones, J. 2001. Plant pathogens and integrated defence responses to infection. *Nature* 411: 826–833.

Day-Rubenstein, K. A. H., Shoemaker, Paul W., Robbin, A., Sullivan, J. and Frisvold, G. B. 2005. Crop genetic resources: an economic appraisal. Economic Information Bulletin 59388, United States Department of Agriculture, Economic Research Service.

Debernardi, J. M., Tricoli, D. M., Ercoli, M. F., Hayta, S., Ronald, P., Palatnik, J. F. and Dubcovsky, J. 2020. A GRF–GIF chimeric protein improves the regeneration efficiency of transgenic plants. *Nature Biotechnology* 38(11), 1274–1279.

Delourme, R., Bousset, L., Ermel, M., Duffe, P., Besnard, A. L., Marquer, B., Fudal, I., Linglin, J., Chadoeuf, J. and Brun, H. 2014. Quantitative resistance affects the speed of frequency increase but not the diversity of the virulence alleles overcoming a major resistance gene to *Leptosphaeria maculans* in oilseed rape. *Infection, Genetics and Evolution* 27, 490–499.

Delourme, R., Chevre, A. M., Brun, H., Rouxel, T., Balesdent, M. H., Dias, J. S., Salisbury, P., Renard, M. and Rimmer, S. R. 2006. Major gene and polygenic resistance to *Leptosphaeria maculans* in oilseed rape (*Brassica napus*). *European Journal of Plant Pathology* 114(1), 41–52.

Dodds, P. N. and Rathjen, J. P. 2010. Plant immunity: towards an integrated view of plant-pathogen interactions. *Nature Reviews Genetics* 11(8), 539–548.

Doyle, C., Higginbottom, K., Swift, T. A., Winfield, M., Bellas, C., Benito-Alifonso, D., Fletcher, T., Galan, M. C., Edwards, K. and Whitney, H. M. 2019. A simple method for spray-on gene editing in planta. *bioRxiv*, 805036. https://doi.org/10.1101/805036.

Dubin, H. J. and Rajaram, S. 1996. Breeding disease-resistant wheats for tropical highlands and lowlands. *Annual Review of Phytopathology* 34, 503–526.

Duncan, K. E. and Howard, R. J. 2000. Cytological analysis of wheat infection by the leaf blotch pathogen *Mxycosphaerella graminicola*. *Mycological Research* 104(9), 1074–1082.

Dutta, A., Croll, D., Mcdonald, B. A. and Krattinger, S. G. 2021. Genome-wide association study for Septoria tritici blotch resistance reveals the occurrence and distribution of *Stb6* in a historic Swiss landrace collection. *Euphytica* 217, 108.

Dyck, P. L. 1977. Genetics of leaf rust reaction in three introductions of common wheat. *Canadian Journal of Genetics and Cytology* 19(4), 711–716. https://doi.org/10.1139/g77-077.

Dyck, P. L. 2011. The association of a gene for leaf rust resistance with the chromosome 7d suppressor of stem rust resistance in common wheat. *Genome* 29(3), 467–469.

Edae, E. A. and Rouse, M. N. 2019. Bulked segregant analysis RNA-seq (BSR-Seq) validated a stem resistance locus in Aegilops umbellulata, a wild relative of wheat. *PLoS ONE* 14(9), e0215492.

Elshire, R. J., Glaubitz, J. C., Sun, Q., Poland, J. A., Kawamoto, K., Buckler, E. S. and Mitchell, S. E. 2011. A robust, simple Genotyping-by-Sequencing (GBS) approach for high diversity species. *PLoS ONE* 6(5), e19379.

Eyal, Z. (Eds.) 1999. *Breeding for Resistance to Septoria and Stagnospora Diseases of Wheat*, CABI Publishing, Cambridge, pp. 332–344.

Eyal, Z. and Levy, E. 1987. Variations in pathogenicity patterns of *Mycosphaerella graminicola* within *Triticum spp.* in Israel. *Euphytica* 36(1), 237–250.

Eyal, Z., Scharen, A. L., Precott, M. and Van Ginkel, M. 1987. *The Septoria Diseases of Wheat: Concepts and Methods of Disease Management: CIMMYT, Mexico*, CIMMYT, Mexico.

Flor, H. H. 1971. Current status of the gene-for-gene concept. *Annual Review of Phytopathology* 9(1), 275–296.

Fones, H. and Gurr, S. 2015. The impact of Septoria tritici blotch disease on wheat: an EU perspective. *Fungal Genetics and Biology : FG and B* 79, 3–7.

Fraaije, B. A., Cools, H. J., Fountaine, J., Lovell, D. J., Motteram, J., West, J. S. and Lucas, J. A. 2005. Role of ascospores in further spread of QoI-resistant cytochrome-b alleles (G143A) in field populations of *Mycosphaerella graminicola*. *Phytopathology* 95(8), 933–941.

Fu, D., Uauy, C., Distelfeld, A., Blechl, A., Epstein, L., Chen, X., Sela, H., Fahima, T. and Dubcovsky, J. 2009. A kinase-START gene confers temperature-dependent resistance to wheat stripe rust. *Science* 323(5919), 1357–1360.

Fukuoka, S., Saka, N., Koga, H., Ono, K., Shimizu, T., Ebana, K., Hayashi, N., Takahashi, A., Hirochika, H., Okuno, K. and Yano, M. 2009. Loss of function of a proline-containing protein confers durable disease resistance in rice. *Science* 325(5943), 998–1001.

Gerard, G. S., Kobiljski, B., Lohwasser, U., Börner, A. and Simón, M. R. 2018. Genetic architecture of adult plant resistance to leaf rust in a wheat association mapping panel. *Plant Pathology* 67, 584–594.

Ghaffary, S. M., Robert, O., Laurent, V., Lonnet, P., Margalé, E., Van Der Lee, T. A., Visser, R. G. and Kema, G. H. 2011. Genetic analysis of resistance to Septoria tritici blotch in the French winter wheat cultivars Balance and Apache. *Theoretical and Applied Genetics.* 123(5), 741–754.

Ghaffary, S. M. T., Faris, J. D., Friesen, T. L., Visser, R. G. F., Van Der Lee, T. A. J., Robert, O. and Kema, G. H. J. 2012. New broad-spectrum resistance to Septoria tritici blotch derived from synthetic hexaploid wheat. *Theoretical and Applied Genetics* 124, 125–142.

Gharbi, M. S., Deghais, M. and Amar, F. B. 2000. Breeding for resistance to Septoria tritici in durum wheat. In: Royo, C., Nachit, M. M., Di Fonzo, N. and Araus, J. L. (Eds), *Durum Wheat Improvement in the Mediterranean Region: New Challenges. Proceedings of the a Seminar* (vol. 2001), Zaragoza, Spain.

Ghislain, M., Byarugaba, A. A., Magembe, E., Njoroge, A., Rivera, C., Román, M. L., Tovar, J. C., Gamboa, S., Forbes, G. A., Kreuze, J. F., Barekye, A. and Kiggundu, A. 2019. Stacking three late blight resistance genes from wild species directly into African highland potato varieties confers complete field resistance to local blight races. *Plant Biotechnology Journal* 17(6), 1119–1129.

Goodwin, S. B. 2007. Back to basics and beyond: increasing the level of resistance to Septoria tritici blotch in wheat. *Australasian Plant Pathology* 36(6), 532-538.

Goodwin, S. B., Cavaletto, J. R., Lowe, L., Thompson, I., Xu, S. X., Adhikari, T. B. and Dubcovsky, J. 2008. Validation of a new map location for the *Stb3* gene for resistance to Septoria tritici blotch in wheat. *7th International Symposium on Mycospaherella and Stagnospora Diseases of Cereals.* Available at: http://www.path.ethz.ch/news/conferences/Mycosphaerella_Ascona_2007/00022_poster abstract.pds.

Goodwin, S. B. and Thompson, I. 2011 (Special Issue). Development of isogenic lines for resistance to Septoria tritici Blotch in wheat. *Czech Journal of Genetics and Plant Breeding* 47(Special Issue), S98–S101.

Guo, J., Fan, J., Hauser, B. A. and Rhee, S. Y. 2015. Target enrichment improves mapping of complex traits by deep sequencing. *G3* 6(1), 67–77.

Gupta, P. K. and Rustgi, S. 2004. Molecular markers from the transcribed/expressed region of the genome in higher plants. *Functional and Integrative Genomics* 4(3), 139–162.

Gurung, S., Mamidi, S., Bonman, J. M., Xiong, M., Brown-Guedira, G. and Adhikari, T. B. 2014. Genome-wide association study reveals novel quantitative trait loci associated with resistance to multiple leaf spot diseases of spring wheat. *PLoS ONE* 9(9), e108179.

Hawkins, G. P., Zhang, X., Thiagarjah, M. R., Corrigan, L. M. and Stringam, G. R. 2005. Identification of RAPD markers linked to pod length in *Brassica napus* L. canola. *Canadian Journal of Plant Science* 85(4), 803–808.

Hayden, M. J., Tabone, T. and Mather, D. E. 2009. Development and assessment of simple PCR markers for SNP genotyping in barley. *Theoretical and Applied Genetics.* 119(5), 939–951.

Hayta, S., Smedley, M. A., Demir, S. U., Blundell, R., Hinchliffe, A., Atkinson, N. and Harwood, W. A. 2019. An efficient and reproducible Agrobacterium-mediated transformation method for hexaploid wheat (*Triticum aestivum* L.). *Plant Methods* 15, 121.

Heick, T. M., Justesen, A. F. and Jørgensen, L. N. 2017a. Anti-resistance strategies for fungicides against wheat pathogen *Zymoseptoria tritici* with focus on DMI fungicides. *Crop Protection* 99, 108–117.

Heick, T. M., Justesen, A. F. and Jørgensen, L. N. 2017b. Resistance of wheat pathogen *Zymoseptoria tritici* to DMI and QoI fungicides in the Nordic-Baltic region—a status. *European Journal of Plant Pathology* 149(3), 669–682.

Hisano, H., Sakamoto, K., Takagi, H., Terauchi, R. and Sato, K. 2017. Exome QTL-seq maps monogenic locus and QTLs in barley. *BMC Genomics* 18(1), 125.

Honsdorf, N., Becker, H. C. and Ecke, W. 2010. Association mapping for phenological, morphological, and quality traits in canola quality winter rapeseed (*Brassica napus* L.). *Genome* 53(11), 899–907.

Hua, K., Zhang, J., Botella, J. R., Ma, C., Kong, F., Liu, B. and Zhu, J. K. 2019. Perspectives on the application of genome-editing technologies in crop breeding. *Molecular Plant* 12(8), 1047–1059.

Huang, X., Feng, Q., Qian, Q., Zhao, Q., Wang, L., Wang, A., Guan, J., Fan, D., Weng, Q., Huang, T., Dong, G., Sang, T. and Han, B. 2009. High-throughput genotyping by whole-genome resequencing. *Genome Research* 19(6), 1068–1076.

Hurgobin, B., Golicz, A. A., Bayer, P. E., Chan, C.-K. K., Tirnaz, S., Dolatabadian, A., Schiessl, S. V., Samans, B., Montenegro, J. D., Parkin, I. A. P., Pires, J. C., Chalhoub, B., King, G. J., Snowdon, R., Batley, J. and Edwards, D. 2018. Homoeologous exchange is a major cause of gene presence/absence variation in the amphidiploid *Brassica napus*. *Plant Biotechnology Journal* 16(7), 1265–1274.

Jia, J., Zhao, S., Kong, X., Li, Y., Zhao, G., He, W., Appels, R., Pfeifer, M., Tao, Y., Zhang, X., Jing, R., Zhang, C., Ma, Y., Gao, L., Gao, C., Spannagl, M., Mayer, K. F., Li, D., Pan, S., Zheng, F., Hu, Q., Xia, X., Li, J., Liang, Q., Chen, J., Wicker, T., Gou, C., Kuang, H., He, G., Luo, Y., Keller, B., Xia, Q., Lu, P., Wang, J., Zou, H., Zhang, R., Xu, J., Gao, J.,

Middleton, C., Quan, Z., Liu, G., Wang, J., International Wheat Genome Sequencing Consortium, Yang, H., Liu, X., He, Z., Mao, L. and Wang, J. 2013. Aegilops tauschii draft genome sequence reveals a gene repertoire for wheat adaptation. *Nature* 496(7443), 91-95.

Jing, H. C., Lovell, D., Gutteridge, R., Jenk, D., Kornyukhin, D., Mitrofanova, O. P., Kema, G. H. J. and Hammond-Kosack, K. E. 2008. Phenotypic and genetic analysis of the *Triticum monococcum-Mycosphaerella graminicola* interaction. *New Phytologist* 179(4), 1121-1132.

Johnson, R. 1984. A critical analysis of durable resistance. *Annual Review of Phytopathology* 22(1), 309-330.

Jones, J. D. and Dangl, J. L. 2006. The plant immune system. *Nature* 444(7117), 323-329.

Jørgensen, L. N., Oliver, R. P. and Heick, T. M. 2018. Occurence and avoidance of fungicide resistance in cereal diseases. In: Oliver, R. P. (Ed), *Integrated Disease Management of Wheat and Barley. Burleigh Dodds Series in Agricultural Science*, Burleigh Dodds Science Publishing, Cambridge, UK, pp. 235-259.

Karlstedt, F., Kopahnke, D., Perovic, D., Jacobi, A., Pillen, K. and Ordon, F. 2019. Mapping of quantitative trait loci (QTL) for resistance against *Zymoseptoria tritici* in the winter spelt wheat accession HTRI1410 (*Triticum aestivum* subsp. *spelta*). *Euphytica* 215(6), 108.

Keller, H., Pamboukdjian, N., Ponchet, M., Poupet, A., Delon, R., Verrier, J. L., Roby, D. and Ricci, P. 1999. Pathogen-induced elicitin production in transgenic tobacco generates a hypersensitive response and nonspecific disease resistance. *The Plant Cell* 11(2), 223-235.

Kema, G. H. J., Verstappen, E. C. and Waalwijk, C. 2000. Avirulence in the wheat Septoria tritici leaf blotch fungus *Mycosphaerella graminicola* is controlled by a single locus. *Molecular Plant-Microbe Interactions* 13(12), 1375- 1379.

Kema, G. H. J., Yu, D. Z., Rijkenberg, F. H. J., Shaw, M. W. and Baayen, R. P. 1996. Histology of the pathogenesis of *Mycosphaerella graminicola* in wheat. *Phytopathology* 86(7), 777-786.

Kettles, G. J., Bayon, C., Sparks, C. A., Canning, G., Kanyuka, K. and Rudd, J. J. 2018. Characterization of an antimicrobial and phytotoxic ribonuclease secreted by the fungal wheat pathogen *Zymoseptoria tritici*. *New Phytologist* 217(1), 320-331.

Kidane, Y. G., Hailemariam, B. N., Mengistu, D. K., Fadda, C., Pè, M. E. and Dell'acqua, M. 2017. Genome-wide association study of septoria tritici blotch resistance in Ethiopian durum wheat landraces. *Front Plant Sci* 8, 1586.

Kildea, S., Byrne, J. J., Cucak, M. and Hutton, F. 2020. First report of virulence to the Septoria tritici blotch resistance gene *Stb16q* in the Irish *Zymoseptoria tritici* population. *New Disease Reports* 41(1), 13. http://doi.org/10.5197/j.2044-0588.2020.041.013.

Kollers, S., Rodemann, B., Ling, J., Korzun, V., Ebmeyer, E., Argillier, O., Hinze, M., Plieske, J., Kulosa, D., Ganal, M. W. and Roder, M. S. 2013. Genetic architecture of resistance to Septoria tritici blotch (*Mycosphaerella graminicola*) in European winter wheat. *Molecular Breeding* 32(2), 411-423.

Kolmer, J. A., Singh, R. P., Garvin, D. F., Viccars, L., William, H. M., Huerta-Espino, J., Ogbonnaya, F. C., Raman, H., Orford, S., Bariana, H. S. and Lagudah, E. S. 2008. Analysis of the *Lr34/Yr18* rust resistance region in wheat germplasm. *Crop Science* 48(5), 1841-1852.

Kong, J., Martin-Ortigosa, S., Finer, J., Orchard, N., Gunadi, A., Batts, L. A., Thakare, D., Rush, B., Schmitz, O., Stuiver, M., Olhoft, P. and Pacheco-Villalobos, D. 2020.

Overexpression of the transcription factor GROWTH-REGULATING FACTOR5 improves transformation of dicot and monocot species. *Frontiers in Plant Science* 11, 572319.

Kosellek, C., Pillen, K., Nelson, J. C., Weber, W. E. and Saal, B. 2013. Inheritance of field resistance to Septoria tritici blotch in the wheat doubled-haploid population Solitar x Mazurka. *Euphytica* 194(2), 161–176.

Krattinger, S. G., Lagudah, E. S., Spielmeyer, W., Singh, R. P., Huerta-Espino, J., Mcfadden, H., Bossolini, E. S., Selter, L. L. and Keller, B. 2009. A putative ABC transporter confers durable resistance to multiple fungal pathogens in wheat. *Science* 323(5919), 1360–1363.

Krattinger, S. G., Sucher, J., Selter, L. L., Chauhan, H., Zhou, B., Tang, M., Upadhyaya, N. M., Mieulet, D., Guiderdoni, E., Weidenbach, D., Schaffrath, U., Lagudah, E. S. and Keller, B. 2016. The wheat durable, multipathogen resistance gene *Lr34* confers partial blast resistance in rice. *Plant Biotechnology Journal* 14(5), 1261–1268. https://doi.org/10.1111/pbi.12491.

Krenz, J. E., Sackett, K. E. and Mundt, C. C. 2008. Specificity of incomplete resistance to *Mycosphaerella graminicola* in wheat. *Phytopathology* 98(5), 555–561.

Kump, K. L., Bradbury, P. J., Wisser, R. J., Buckler, E. S., Belcher, A. R., Oropeza-Rosas, M. A., Zwonitzer, J. C., Kresovich, S., Mcmullen, M. D., Ware, D., Balint-Kurti, P. J. and Holland, J. B. 2011. Genome-wide association study of quantitative resistance to southern leaf blight in the maize nested association mapping population. *Nature Genetics* 43(2), 163–168.

Latz, M. A. C., Jensen, B., Collinge, D. B. and Lyngs Jørgensen, H. J. 2020. Identification of two endophytic fungi that control Septoria tritici blotch in the field, using a structured screening approach. *Biological Control* 141, 104128.

Le Guen, V., Garcia, D., Mattos, C. R. R., Doaré, F., Lespinasse, D. and Seguin, M. 2007. Bypassing of a polygenic *Microcyclus ulei* resistance in rubber tree, analysed by QTL detection. *New Phytologist* 173(2), 335–345.

Li, H., Barbetti, M. J. and Sivasithamparam, K. 2006. Concomitant inoculation of an avirulent strain of Leptosphaeria maculans prevents break-down of a single dominant gene-based resistance in *Brassica napus* cv. Surpass 400 by a virulent strain. *Field Crops Research* 95(2-3), 206–211.

Li, Y. Y., Shen, J., Wang, T., Chen, Q., Zhang, X., Fu, T., Meng, J., Tu, J. and Ma, C. 2007. QTL analysis of yield-related traits and their association with functional markers in *Brassica napus* L. *Australian Journal of Agricultural Research* 58(8), 759-766.

Ling, H. Q., Zhao, S., Liu, D., Wang, J., Sun, H., Zhang, C., Fan, H., Li, D., Dong, L., Tao, Y., Gao, C., Wu, H., Li, Y., Cui, Y., Guo, X., Zheng, S., Wang, B., Yu, K., Liang, Q., Yang, W., Lou, X., Chen, J., Feng, M., Jian, J., Zhang, X., Luo, G., Jiang, Y., Liu, J., Wang, Z., Sha, Y., Zhang, B., Wu, H., Tang, D., Shen, Q., Xue, P., Zou, S., Wang, X., Liu, X., Wang, F., Yang, Y., An, X., Dong, Z., Zhang, K., Zhang, X., Luo, M. C., Dvorak, J., Tong, Y., Wang, J., Yang, H., Li, Z., Wang, D., Zhang, A. and Wang, J. 2013. Draft genome of the wheat A-genome progenitor *Triticum urartu*. *Nature* 496(7443), 87–90.

Liu, S., Yeh, C. T., Tang, H. M., Nettleton, D. and Schnable, P. S. 2012. Gene mapping via bulked segregant RNA-seq (BSR-Seq). *PLoS ONE* 7(5), e36406.

Liu, Y., Zhang, L., Thompson, I. A., Goodwin, S. B. and Ohm, H. W. 2013. Molecular mapping re-locates the *Stb2* gene for resistance to Septoria tritici blotch derived from cultivar Veranopolis on wheat chromosome 1BS. *Euphytica* 190(1), 145–156.

Louriki, S., Rehman, S., El Hanafi, S., Bouhouch, Y., Al-Jaboobi, M., Amri, A., Douira, A. and Tadesse, W. 2021. Identification of Resistance Sources and Genome-Wide Association Mapping of Septoria Tritici Blotch Resistance in Spring Bread Wheat Germplasm of ICARDA. *Frontiers in Plant Science*, 12. https://doi.org/10.3389/fpls.2021.600176.

Lowe, A. J., Moule, C., Trick, M. and Edwards, K. J. 2004. Efficient large-scale development of microsatellites for marker and mapping applications in Brassica crop species. *Theoretical and Applied Genetics*.108(6), 1103-1112.

Maccaferri, M., Cane', M. A., Sanguineti, M. C., Salvi, S., Colalongo, M. C., Massi, A., Clarke, F., Knox, R., Pozniak, C. J., Clarke, J. M., Fahima, T., Dubcovsky, J., Xu, S., Ammar, K., Karsai, I., Vida, G. and Tuberosa, R. 2014. A consensus framework map of durum wheat (*Triticum durum* Desf.) suitable for linkage disequilibrium analysis and genome-wide association mapping. *BMC Genomics* 15, 873.

Maccaferri, M., Ricci, A., Salvi, S., Milner, S. G., Noli, E., Martelli, P. L., Casadio, R., Akhunov, E., Scalabrin, S., Vendramin, V., Ammar, K., Blanco, A., Desiderio, F., Distelfeld, A., Dubcovsky, J., Fahima, T., Faris, J., Korol, A., Massi, A., Mastrangelo, A. M., Morgante, M., Pozniak, C., N'diaye, A., Xu, S. and Tuberosa, R. 2015. A high-density, SNP-based consensus map of tetraploid wheat as a bridge to integrate durum and bread wheat genomics and breeding. *Plant Biotechnology Journal* 13(5), 648-663.

Makhoul, M., Rambla, C., Voss-Fels, K. P., Hickey, L. T., Snowdon, R. J. and Obermeier, C. 2020. Overcoming polyploidy pitfalls: a user guide for effective SNP conversion into KASP markers in wheat. *Theoretical and Applied Genetics*. 133(8), 2413-2430.

Marshall, D. R. 1977. The advantages and hazards of genetic homogeneity. *Annals of the New York Academy of Sciences* 287(1 The Genetic B), 1-20.

Marshall, R., Kombrink, A., Motteram, J., Loza-Reyes, E., Lucas, J., Hammond-Kosack, K., Thomma, B. and Rudd, J. 2011. Analysis of two in planta expressed LysM effector homologues from the fungus *Mycosphaerella graminicola* reveals novel functional properties and varying contributions to virulence on wheat. *Plant Physiology*, 111, 176347.

May, C. E. and Lagudah, E. S. 1992. Inheritance in hexaploid wheat of Septoria tritic blotch resistance and other characteristics derived from *Trititcum tauschii*. *Australian Journal of Agricultural Research* 43(3), 433-442.

McCartney, C. A., Brûlé-Babel, A. L. and Lamari, L. 2002 Inheritance of Race-Specific Resistance to *Mycosphaerella graminicola* in Wheat. *Phytopathology* 92(2), 138-144.

McCartney, C. A., Brule-Babel, A. L., Lamari, L. and Somers, D. J. 2003. Chromosomal location of a race-specific resistance gene to Mycosphaerella graminicola in the spring wheat ST6. Theor Appl Genet 107, 1181-1186.

McDonald, B. A. 1999. The population genetics of plant pathogens and resistance breeding strategies. Vortrage für Pflanzenzüchtung. *Proc German Soc of Plant Breeders and Plant Pathologists*, pp. 235-244.

McDonald, B. A. and Linde, C. 2002. The population genetics of plant pathogens and breeding strategies for durable resistance. *Euphytica* 124(2), 163-180.

McDonald, B. A. and Mundt, C. C. 2016. How knowledge of pathogen population biology informs management of Septoria tritici blotch. *Phytopathology* 106(9), 948-955.

McDonald, M. C., Renkin, M., Spackman, M., Orchard, B., Croll, D., Solomon, P. S. and Milgate, A. 2019. Rapid parallel evolution of azole fungicide resistance in Australian populations of the wheat pathogen *Zymoseptoria tritici*. *Applied and Environmental Microbiology* 85(4), e01908-18. https://doi.org/10.1128/AEM.01908-18.

Medini, M., Ferjaoui, S., Bahri, B., Mhri, W., Hattab, S. and Hamza, S. 2014. Bulk segregant analysis and marker-trait association reveal common AFLP markers for resistance to Septoria leaf blotch in Tunisian old durum wheat. *Biotechnologie Agronomie Societe et Environnement* 18, 3-10.

Meile, L., Croll, D., Brunner, P. C., Plissonneau, C., Hartmann, F. E., McDonald, B. A. and Sanchez-Vallet, A. 2018. A fungal avirulence factor encoded in a highly plastic genomic region triggers partial resistance to Septoria tritici blotch. *New Phytologist* 219(3), 1048-1061.

Mergoum, M., Harilal, V. E., Singh, P. K., Adhikari, T. B., Kumar, A., Ghavami, F., Elias, E., Alamri, M. S. and Kianian, S. F. 2013. Genetic analysis and mapping of seedling resistance to Septoria tritici Blotch in 'Steele-ND'/'ND 735' bread wheat population. *Cereal Research Communications* 41(2), 199-210.

Meuwissen, T. and Goddard, M. 2010. Accurate prediction of genetic values for complex traits by whole-genome resequencing. *Genetics* 185(2), 623-631.

Michelmore, R., Coaker, G., Bart, R., Beattie, G., Bent, A., Bruce, T., Cameron, D., Dangl, J., Dinesh-Kumar, S., Edwards, R., Eves-Van Den Akker, S., Gassmann, W., Greenberg, J. T., Hanley-Bowdoin, L., Harrison, R. J., Harvey, J., He, P., Huffaker, A., Hulbert, S., Innes, R., Jones, J. D. G., Kaloshian, I., Kamoun, S., Katagiri, F., Leach, J., Ma, W., Mcdowell, J., Medford, J., Meyers, B., Nelson, R., Oliver, R., Qi, Y., Saunders, D., Shaw, M., Smart, C., Subudhi, P., Torrance, L., Tyler, B., Valent, B. and Walsh, J. 2017. Foundational and translational research opportunities to improve plant health. *Molecular Plant-Microbe Interactions : MPMI* 30(7), 515-516.

Miedaner, T., Risser, P., Paillard, S., Schnurbusch, T., Keller, B., Hartl, L., Holzapfel, J., Korzun, V., Ebmeyer, E. and Utz, H. F. 2012. Broad-spectrum resistance loci for three quantitatively inherited diseases in two winter wheat populations. *Molecular Breeding* 29(3), 731-742.

Miedaner, T., Zhao, Y. S., Gowda, M., Longin, C. F. H., Korzun, V., Ebmeyer, E., Kazman, E. and Reif, J. C. 2013. Genetic architecture of resistance to Septoria tritici blotch in European wheat. *BMC Genomics* 14, 858.

Miller, M. R., Atwood, T. S., Eames, B. F., Eberhart, J. K., Yan, Y. L., Postlethwait, J. H. and Johnson, E. A. 2007. RAD marker microarrays enable rapid mapping of zebrafish mutations. *Genome Biology* 8(6), R105.

Mirdita, V., Liu, G. Z., Zhao, Y. S., Miedaner, T., Longin, C. F. H., Gowda, M., Mette, M. F. and Reif, J. C. 2015. Genetic architecture is more complex for resistance to Septoria tritici blotch than to Fusarium head blight in Central European winter wheat. *BMC Genomics* 16, 430.

Mohd-Assaad, N., McDonald, B. A. and Croll, D. 2016. Multilocus resistance evolution to azole fungicides in fungal plant pathogen populations. *Molecular Ecology* 25(24), 6124-6142.

Montenegro, J. D., Golicz, A. A., Bayer, P. E., Hurgobin, B., Lee, H., Chan, C.-K. K., Visendi, P., Lai, K., Doležel, J., Batley, J. and Edwards, D. 2017. The pangenome of hexaploid bread wheat. *The Plant Journal: For Cell and Molecular Biology* 90(5), 1007-1013.

Moore, J. W., Herrera-Foessel, S., Lan, C., Schnippenkoetter, W., Ayliffe, M., Huerta-Espino, J., Lillemo, M., Viccars, L., Milne, R., Periyannan, S., Kong, X., Spielmeyer, W., Talbot, M., Bariana, H., Patrick, J. W., Dodds, P., Singh, R. and Lagudah, E. 2015. A recently evolved hexose transporter variant confers resistance to multiple pathogens in wheat. *Nature Genetics* 47(12), 1494-1498.

Motteram, J., Küfner, I., Deller, S., Brunner, F., Hammond-Kosack, K. E., Nürnberger, T. and Rudd, J. J. 2009. Molecular characterization and functional analysis of MgNLP, the sole NPP1 domain-containing protein, from the fungal wheat leaf pathogen *Mycosphaerella graminicola*. *Molecular Plant-Microbe Interactions* 22(7), 790-799.

Mundt, C. C. 2002. Use of multiline cultivars and cultivar mixtures for disease management. *Annual Review of Phytopathology* 40, 381-410.

Mundt, C. C. 2015. Durable resistance: a key to sustainable management of pathogens and pests. *Infection, Genetics and Evolution* 27, 446-455.

Mundt, C. C. 2018. Pyramiding for resistance durability: theory and practice. *Phytopathology* 108(7), 792-802.

Muqaddasi, Q. H., Zhao, Y., Rodemann, B., Plieske, J., Ganal, M. W. and Röder, M. S. 2019. Genome-wide association mapping and prediction of adult stage Septoria tritici blotch infection in European winter wheat via high-density marker arrays. *The Plant Genome* 12(1), 30951099.

Niks, R. E., Parlevliet, J. E., Lindhout, P. and Bai, Y. 2011. *Breeding Crops with Resistance to Diseases and Pests*, Wageningen Academic Publishers, Wageningen.

Niks, R. E., Qi, X. and Marcel, T. C. 2015. Quantitative resistance to biotrophic filamentous plant pathogens: concepts, misconceptions, and mechanisms. *Annual Review of Phytopathology* 53, 445-470.

Odilbekov, F., Armonienė, R., Henriksson, T. and Chawade, A. 2018. Proximal phenotyping and machine learning methods to identify septoria tritici blotch disease symptoms in wheat. *Frontiers in Plant Science* 9. https://doi.org/10.3389/fpls.2018.00685.

O'Driscoll, A., Kildea, S., Doohan, F., Spink, J. and Mullins, E. 2014. The wheat-Septoria conflict: a new front opening up? *Trends in Plant Science* 19(9), 602-610. https://doi.org/10.1016/j.tplants.2014.04.011.

Ouaja, M., Aouini, L., Bahri, B., Ferjaoui, S., Medini, M., Marcel, T. C. and Hamza, S. 2020. Identification of valuable sources of resistance to *Zymoseptoria tritici* in the Tunisian durum wheat landraces. *European Journal of Plant Pathology* 156(2), 647-661.

Palloix, A., Ayme, V. and Moury, B. 2009. Durability of plant major resistance genes to pathogens depends on the genetic background, experimental evidence and consequences for breeding strategies. *New Phytologist* 183(1), 190-199.

Palmer, C. L. and Skinner, W. 2002. *Mycosphaerella graminicola*: latent infection, crop devastation and genomics. *Molecular Plant Pathology* 3(2), 63-70.

Parlevliet, J. E. 1979. Components of resistance that reduce the rate of epidemic development. *Annual Review of Phytopathology* 17(1), 203-222.

Parlevliet, J. E. 2002. Durability of resistance against fungal, bacterial and viral pathogens; present situation. *Euphytica* 124(2), 147-156.

Pilet-Nayel, M. L., Moury, B., Caffier, V., Montarry, J., Kerlan, M. C., Fournet, S., Durel, C. E. and Delourme, R. 2017. Quantitative resistance to plant pathogens in pyramiding strategies for durable crop protection. *Frontiers in Plant Science* 8, 1838.

Raats, D., Frenkel, Z., Krugman, T., Dodek, I., Sela, H., Simková, H., Magni, F., Cattonaro, F., Vautrin, S., Bergès, H., Wicker, T., Keller, B., Leroy, P., Philippe, R., Paux, E., Doležel, J., Feuillet, C., Korol, A. and Fahima, T. 2013. The physical map of wheat chromosome 1BS provides insights into its gene space organization and evolution. *Genome Biology* 14(12), R138. https://doi.org/10.1186/gb-2013-14-12-r138.

Radecka-Janusk M. and Czember, P. 2014. Genetic mapping of quantitative trait loci (QTL) for resistance to *Septoria tritici* blotch in winter wheat. *Euphytica* 200: 109-125.

Raman, H. and Milgate, A. 2012. Molecular breeding for resistance to Septoria tritici blotch in wheat. *Cereal Research Communications* 40(4), 451–466. https://doi.org/10.1556/crc.40.2012.4.1.

Raman, H., Raman, R., Diffey, S., Qiu, Y., Mcvittie, B., Barbulescu, D. M., Salisbury, P. A., Marcroft, S. and Delourme, R. 2018. Stable quantitative resistance loci to blackleg disease in canola (*Brassica napus* L.) over continents. *Frontiers in Plant Science* 9, 1622. https://doi.org/10.3389/fpls.2018.01622.

Raman, H., Raman, R., Kilian, A., Detering, F., Carling, J., Coombes, N., Diffey, S., Kadkol, G., Edwards, D., Mccully, M., Ruperao, P., Parkin, I. A. P., Batley, J., Luckett, D. J. and Wratten, N. 2014. Genome-wide delineation of natural variation for pod shatter resistance in *Brassica napus*. *PLoS ONE* 9(7), e101673. https://doi.org/10.1371/journal.pone.0101673.

Raman, H., Raman, R., Pirathiban, R., Mcvittie, B., Sharma, N., Liu, S., Qiu, Y., Zhu, A., Killian, A., Cullis, B., Farquhar, G. D., Williams, H. S., White, R., Tabah, D., Easton, A. and Zhang, Y. 2021. Multi-environment QTL analysis delineates a major locus associated with homoeologous exchanges for water-use efficiency and seed yield in allopolyploid *Brassica napus*. *bioRxiv*, 2021.07.08.451711.

Raman, H., Zhang, K., Cakir, M., Appels, R., Garvin, D. F., Maron, L. G., Kochian, L. V., Moroni, J. S., Raman, R., Imtiaz, M., Drake-Brockman, F., Waters, I., Martin, P., Sasaki, T., Yamamoto, Y., Matsumoto, H., Hebb, D. M., Delhaize, E. and Ryan, P. R. 2005. Molecular characterization and mapping of *ALMT1*, the aluminium-tolerance gene of bread wheat (*Triticum aestivum* L.). *Genome* 48, 781–791.

Raman, R., Milgate, A. W., Imtiaz, M., Tan, M.-K., Raman, H., Lisle, C., Coombes, N. and Martin, P. 2009. Molecular mapping and physical location of major gene conferring seedling resistance to Septoria tritici blotch in wheat. *Molecular Breeding* 24(2), 153–164.

Rillo, A. O. and Caldwell, R. M. 1966. Inheritance of resistance to Septoria tritici in *Triticum aestivum* subsp. *vulgare* 'Bulgaria 88'. *Phytopathology* 56, 897. [Abstract].

Risk, J. M., Selter, L. L., Chauhan, H., Krattinger, S. G., Kumlehn, J., Hensel, G., Viccars, L. A., Richardson, T. M., Buesing, G., Troller, A., Lagudah, E. S. and Keller, B. 2013. The wheat *Lr34* gene provides resistance against multiple fungal pathogens in barley. *Plant Biotechnology Journal* 11(7), 847–854.

Rosielle, A. A. 1972. Sources of resistance in wheat to speckled leaf blotch caused by Septoria tritici. *Euphytica* 21(1), 152–161.

Rouxel, T. and Balesdent, M. H. 2017. Life, death and rebirth of avirulence effectors in a fungal pathogen of Brassica crops, *Leptosphaeria maculans*. *New Phytologist* 214(2), 526–532.

Rudd, J. J. 2015. Previous bottlenecks and future solutions to dissecting the *Zymoseptoria tritici*-wheat host-pathogen interaction. *Fungal Genetics and Biology* 79, 24–28.

Saintenac, C., Cambon, F., Aouini, L., Verstappen, E., Ghaffary, S. M. T., Poucet, T., Marande, W., Berges, H., Xu, S., Jaouannet, M., Favery, B., Alassimone, J., Sánchez-Vallet, A., Faris, J., Kema, G., Robert, O. and Langin, T. 2021. A wheat cysteine-rich receptor-like kinase confers broad-spectrum resistance against Septoria tritici blotch. *Nat Commun*, 12, 433.

Saintenac, C., Lee, W. S., Cambon, F., Rudd, J. J., King, R. C., Marande, W., Powers, S. J., Bergès, H., Phillips, A. L., Uauy, C., Hammond-Kosack, K. E., Langin, T. and Kanyuka, K. 2018. Wheat receptor-kinase-like protein *Stb6* controls gene-for-gene resistance to fungal pathogen *Zymoseptoria tritici*. *Nature Genetics* 50(3), 368–374.

Sánchez-Vallet, A., McDonald, M. C., Solomon, P. S. and McDonald, B. A. 2015. Is *Zymoseptoria tritici* a hemibiotroph? *Fungal Genetics and Biology: FG and B* 79, 29-32.

Sasaki, T., Yamamoto, Y., Ezaki, B., Katsuhara, M., Ahn, S. J., Ryan, P. R., Delhaize, E. and Matsumoto, H. (2004). A wheat gene encoding an aluminum-activated malate transporter. *Plant Journal* 37, 645–653.

Savary, S., Willocquet, L., Pethybridge, S. J., Esker, P., Mcroberts, N. and Nelson, A. 2019. The global burden of pathogens and pests on major food crops. *Nature Ecology and Evolution* 3(3), 430–439.

Sax, K. 1923. The association of size differences with seed-coat pattern and pigmentation in *Phaseolus vulgaris*. *Genetics* 8(6), 552-560.

Schnippenkoetter, W., Lo, C., Liu, G., Dibley, K., Chan, W. L., White, J., Milne, R., Zwart, A., Kwong, E., Keller, B., Godwin, I., Krattinger, S. G. and Lagudah, E. 2017. The wheat *Lr34* multipathogen resistance gene confers resistance to anthracnose and rust in sorghum. *Plant Biotechnology Journal* 15(11), 1387–1396.

Schurnbusch, T., Paillard, S., Schori, A., Messner, M., Schachermayr, G., Winzeler, M. and Keller, B. 2004. Dissection of quantitative and durable leaf rust resistance in Swiss winter wheat reveals a major resistance QTL in the *Lr34* chromosomal region. *Theoretical and Applied Genetics.* 108(3), 477–484.

Seifbarghi, S., Razavi, M., Aminian, H., Zare, R. and Etebarian, H.-R. 2009. Studies on the host range of Septoria species on cereals and some wild grasses in Iran. *Phytopathologia Mediterranea* 48, 422–429.

Seybold, H., Demetrowitsch, T. J., Hassani, M. A., Szymczak, S., Reim, E., Haueisen, J., Lübbers, L., Rühlemann, M., Franke, A., Schwarz, K. and Stukenbrock, E. H. 2020. A fungal pathogen induces systemic susceptibility and systemic shifts in wheat metabolome and microbiome composition. *Nature Communications* 11(1), 1910.

Shaner, G. and Finney, R. E. 1982. Resistance in soft red winter wheat to *Mycosphaerella graminicola*. *Phytopathology* 72(1), 154–158.

Sharma, J. R. 2001. *Principles and Practices of Plant Breeding*, McGraw-Hill Publishing Company Limited, Tata, New Delhi.

Simmonds, N. W. 1979. *Principles of Crop Improvement*, Longman, London and New York.

Simón, M. R., Khlestkina, E. K., Castillo, N. S. and Borner, A. 2010. Mapping quantitative resistance to Septoria tritici blotch in spelt wheat. *European Journal of Plant Pathology* 128(3), 317–324.

Simón, M. R., Worland, A. J. and Struik, P. C. 2004. Influence of plant height and heading date on the expression of the resistance to Septoria tritici blotch in Near isogenic lines of wheat. *Crop Science* 44(6), 2078–2085.

Singh, R. P. 1993. Genetic association of gene *Bdv1* for tolerance to Barley Yellow Dwarf Virus with genes *Lr34* and *Yr18* for adult plant resistance to rusts in bread wheat. *Plant Diseases* 77(1103-1106). .

Singh, V. K., Khan, A. W., Saxena, R. K., Sinha, P., Kale, S. M., Parupalli, S., Kumar, V., Chitikineni, A., Vechalapu, S., Sameer Kumar, C. V., Sharma, M., Ghanta, A., Yamini, K. N., Muniswamy, S. and Varshney, R. K. 2017. Indel-seq: a fast-forward genetics approach for identification of trait-associated putative candidate genomic regions and its application in pigeonpea (*Cajanus cajan*). *Plant Biotechnology Journal* 15(7), 906-914.

Somasco, O. A., Qualset, C. O. and Gilchrist, D. G. 1996. Single gene resistance to Septoria tritici blotch in the spring wheat cultivar 'Tadinia'. *Plant Breeding* 115(4), 261-267.

Somers, D. J., Rakow, G., Prabhu, V. K. and Friesen, K. R. 2001. Identification of a major gene and RAPD markers for yellow seed coat colour in *Brassica napus*. *Genome* 44(6), 1077-1082.

Spielmeyer, W., Mcintosh, R. A., Kolmer, J. and Lagudah, E. S. 2005. Powdery mildew resistance and Lr34/Yr18 genes for durable resistance to leaf and stripe rust cosegregate at a locus on the short arm of chromosome 7d of wheat. *Theoretical and Applied Genetics*. 111(4), 731-735.

Sprague, G. F. 1941. The location of dominant favorable genes in maize by means of an inversion. *Genetics* 26, 143-149.

Stadlmeier, M., Jorgensen, L. N., Corsi, B., Cockram, J., Hartl, L. and Mohler, V. 2019. Genetic dissection of resistance to the three fungal plant pathogens Blumeria graminis, *Zymoseptoria tritici*, and Pyrenophora tritici-repentis using a multiparental winter wheat population. *G3* 9(5), 1745-1757.

Stewart, E. L., Croll, D., Lendenmann, M. H., Sanchez-Vallet, A., Hartmann, F. E., Palma-Guerrero, J., Ma, X. and McDonald, B. A. 2018. Quantitative trait locus mapping reveals complex genetic architecture of quantitative virulence in the wheat pathogen *Zymoseptoria tritici*. *Molecular Plant Pathology* 19(1), 201-216.

Sun, Z., Wang, Z., Tu, J., Zhang, J., Yu, F., McVetty, P. B. and Li, G. 2007. An ultradense genetic recombination map for *Brassica napus* consisting of 13551 SRAP markers. *Theoretical and Applied Genetics*. 114(8), 1305-1317.

Tabone, T., Mather, D. E. and Hayden, M. J. 2009. Temperature Switch PCR (TSP): robust assay design for reliable amplification and genotyping of SNPs. *BMC Genomics* 10, 580.

Tadesse, W., Sanchez-Garcia, M., Assefa, S. G., Amri, A., Bishaw, Z., Ogbonnaya, F. C. and Baum, M. 2019. Genetic gains in wheat breeding and its role in feeding the world. *Crop Breeding. Genetics and Genomics* 1, e190005.

Takagi, H., Abe, A., Yoshida, K., Kosugi, S., Natsume, S., Mitsuoka, C., Uemura, A., Utsushi, H., Tamiru, M., Takuno, S., Innan, H., Cano, L. M., Kamoun, S. and Terauchi, R. 2013. QTL-seq: rapid mapping of quantitative trait loci in rice by whole genome resequencing of DNA from two bulked populations. *Plant Journal: For Cell and Molecular Biology* 74(1), 174-183.

Tamburic-Ilincic, L. and Rosa, S. B. 2019. QTL mapping of Fusarium head blight and Septoria tritici blotch in an elite hard red winter wheat population. *Molecular Breeding* 39(7), 94.

Torriani, S. F., Melichar, J. P., Mills, C., Pain, N., Sierotzki, H. and Courbot, M. 2015. *Zymoseptoria tritici*: a major threat to wheat production, integrated approaches to control. *Fungal Genetics and Biology* 79, 8-12.

Torti, S., Schlesier, R., Thümmler, A., Bartels, D., Römer, P., Koch, B., Werner, S., Panwar, V., Kanyuka, K., Wirén, N. V., Jones, J. D. G., Hause, G., Giritch, A. and Gleba, Y. 2021. Transient reprogramming of crop plants for agronomic performance. *Nature Plants* 7(2), 159-171.

Turkensteen, L. 1993. Durable resistance of potatoes against Phytophthora infestans. In: Jacobs, T. and Parlevliet, J. E. (Eds), *Durability of Disease Resistance*, Academic Press Publishers, Dordrecht, the Netherlands, pp. 115-124.

Vagndorf, N., Nielsen, N. H., Edriss, V., Andersen, J. R., Orabi, J., Jorgensen, L. N. and Jahoor, A. 2017. Genomewide association study reveals novel quantitative trait loci associated with resistance towards Septoria tritici blotch in north European winter wheat. *Plant Breeding* 136(4), 474–482.

Van Der Plank, J. E. 1968. *Disease Resistance in Plants* (2nd edn.). Academic Press, New York and London.

Van Poecke, R. M. P., Maccaferri, M., Tang, J., Truong, H. T., Janssen, A., Van Orsouw, N. J., Salvi, S., Sanguineti, M. C., Tuberosa, R. and Van Der Vossen, E. A. G. 2013. Sequence-based SNP genotyping in durum wheat. *Plant Biotechnology Journal* 11(7), 809–817.

Walkowiak, S., Gao, L., Monat, C., Haberer, G., Kassa, M. T., Brinton, J., Ramirez-Gonzalez, R. H., Kolodziej, M. C., Delorean, E., Thambugala, D., Klymiuk, V., Byrns, B., Gundlach, H., Bandi, V., Siri, J. N., Nilsen, K., Aquino, C., Himmelbach, A., Copetti, D., Ban, T., Venturini, L., Bevan, M., Clavijo, B., Koo, D. H., Ens, J., Wiebe, K., N'diaye, A., Fritz, A. K., Gutwin, C., Fiebig, A., Fosker, C., Fu, B. X., Accinelli, G. G., Gardner, K. A., Fradgley, N., Gutierrez-Gonzalez, J., Halstead-Nussloch, G., Hatakeyama, M., Koh, C. S., Deek, J., Costamagna, A. C., Fobert, P., Heavens, D., Kanamori, H., Kawaura, K., Kobayashi, F., Krasileva, K., Kuo, T., Mckenzie, N., Murata, K., Nabeka, Y., Paape, T., Padmarasu, S., Percival-Alwyn, L., Kagale, S., Scholz, U., Sese, J., Juliana, P., Singh, R., Shimizu-Inatsugi, R., Swarbreck, D., Cockram, J., Budak, H., Tameshige, T., Tanaka, T., Tsuji, H., Wright, J., Wu, J., Steuernagel, B., Small, I., Cloutier, S., Keeble-Gagnère, G., Muehlbauer, G., Tibbets, J., Nasuda, S., Melonek, J., Hucl, P. J., Sharpe, A. G., Clark, M., Legg, E., Bharti, A., Langridge, P., Hall, A., Uauy, C., Mascher, M., Krattinger, S. G., Handa, H., Shimizu, K. K., Distelfeld, A., Chalmers, K., Keller, B., Mayer, K. F. X., Poland, J., Stein, N., Mccartney, C. A., Spannagl, M., Wicker, T. and Pozniak, C. J. 2020. Multiple wheat genomes reveal global variation in modern breeding. *Nature* 588(7837), 277–283.

Wang, K., Liu, H., Du, L. and Ye, X. 2017. Generation of marker-free transgenic hexaploid wheat via an Agrobacterium-mediated co-transformation strategy in commercial Chinese wheat varieties. *Plant Biotechnology Journal* 15(5), 614–623.

Watson, A., Ghosh, S., Williams, M. J., Cuddy, W. S., Simmonds, J., Rey, M.-D. Asyraf Md Hatta, M., Hinchliffe, A., Steed, A., Reynolds, D., Adamski, N. M., Breakspear, A., Korolev, A., Rayner, T., Dixon, L. E., Riaz, A., Martin, W., Ryan, M., Edwards, D., Batley, J., Raman, H., Carter, J., Rogers, C., Domoney, C., Moore, G., Harwood, W., Nicholson, P., Dieters, M. J., Delacy, I. H., Zhou, J., Uauy, C., Boden, S. A., Park, R. F., Wulff, B. B. H. and Hickey, L. T. 2018. Speed breeding is a powerful tool to accelerate crop research and breeding. *Nature Plants* 4, 23–29.

Wiesner-Hanks, T. and Nelson, R. 2016. Multiple disease resistance in plants. *Annual Review of Phytopathology* 54, 229–252.

Wilson, R. E. 1985. Inheritance of resistance to Septoria tritici in wheat. In: Scharen, A. L. (Ed), *Septoria of Cereals. Proceedings of the Second International Septoria Workshop*. Bozeman, Montana.

Wolfe, M. S., Brändle, U., Koller, B., Limpert, E., Mcdermott, J. M., Müller, K. and Schaffner, D. 1992. Barley mildew in Europe: population biology and host resistance. *Euphytica* 63(1–2), 125–139.

Wulff, B. B. and Moscou, M. J. 2014. Strategies for transferring resistance into wheat: from wide crosses to GM cassettes. *Frontiers in Plant Science* 5, 692.

Wurschum, T., Leiser, W. L. and Longin, C. F. H. 2017. Molecular genetic characterization and association mapping in spelt wheat. *Plant Breeding* 136, 214–223.

Yang, N., Mcdonald, M. C., Solomon, P. S. and Milgate, A. W. 2018. Genetic mapping of *Stb19*, a new resistance gene to *Zymoseptoria tritici* in wheat. *Theoretical and Applied Genetics.* 131(12), 2765–2773.

Yates, S., Mikaberidze, A., Krattinger, S. G., Abrouk, M., Hund, A., Yu, K., Studer, B., Fouche, S., Meile, L., Pereira, D., Karisto, P. and Mcdonald, B. A. 2019. Precision phenotyping reveals novel loci for quantitative resistance to septoria tritici blotch. *Plant Phenomics*, 3285904.

Zhan, J., Pettway, R. E. and Mcdonald, B. A. 2003. The global genetic structure of the wheat pathogen Mycosphaerella graminicola is characterized by high nuclear diversity, low mitochondrial diversity, regular recombination, and gene flow. *Fungal Genetics and Biology* 38(3), 286–297.

Zhong, Z., Marcel, T. C., Hartmann, F. E., Ma, X., Plissonneau, C., Zala, M., Ducasse, A., Confais, J., Compain, J., Lapalu, N., Amselem, J., Mcdonald, B. A., Croll, D. and Palma-Guerrero, J. 2017. A small secreted protein in *Zymoseptoria tritici* is responsible for avirulence on wheat cultivars carrying the *Stb6* resistance gene. *New Phytologist* 214(2), 619–631.

Zhou, B., Benbow, H. R., Brennan, C. J., Arunachalam, C., Karki, S. J., Mullins, E., Feechan, A., Burke, J. I. and Doohan, F. M. 2020. Wheat encodes small, secreted proteins that contribute to resistance to Septoria tritici blotch. *Frontiers in Genetics* 11, 469.

Zwart, R. S., Thompson, J. P., Milgate, A. W., Bansal, U. K., Williamson, P. M., Raman, H. and Bariana, H. S. 2010. QTL mapping of multiple foliar disease and root-lesion nematode resistances in wheat. *Molecular Breeding* 26(1), 107–124.

Part 4

Fungal diseases of cereals: Septoria nodorum blotch and spot blotch

Chapter 10

Understanding the plant-pathogen interaction associated with Septoria nodorum blotch of wheat

Gayan K. Kariyawasam, North Dakota State University, USA; and Timothy L. Friesen, Edward T. Schafer Agricultural Research Center, USDA-ARS, USA

1 Introduction

Wheat (*Triticum aestivum* L. and *T. turgidum* L.) is one of the major food crops of the world, accounting for 20% of the calories consumed by humans (Brenchley et al. 2012). World population is projected to reach ~9.7 billion by 2050 (Searchinger et al. 2019), and the demand for wheat is expected to increase annually by 1.6% (Singh et al. 2016). Improvement in wheat production is heavily impacted by plant diseases including rusts, leaf spots, root diseases and viral diseases (Langridge et al. 2001; Figueroa et al. 2018).

Septoria nodorum blotch (SNB) is a destructive foliar and glume disease found in major wheat-growing regions of the world including North America, Europe and Australia (Singh et al. 2016; Solomon et al. 2006). Even though in Western Europe SNB has generally been replaced by septoria tritici blotch (STB), in Australia it was estimated that SNB caused annual yield losses worth AUD $108 million (Murray and Brennan 2009), and continues to be a problem in wheat-producing regions of Northern Europe, Southern Brazil and the United States (Mehra et al. 2019). SNB is caused by the necrotrophic fungal

http://dx.doi.org/10.19103/AS.2021.0092.14

pathogen *Parastagonospora nodorum* ((Berk.) Quaedvlieg, Verkley & Crous.) (teleomorph: *Phaeosphaeria nodorum* (Hedjar.) syn. *Leptosphaeria nodorum* (Müll.), syn. *Septoria nodorum* (Berk.), syn. *Stagonospora nodorum* (Berk.)) (Oliver et al. 2012; Quaedvlieg et al. 2013). *P. nodorum* is an ascomycete fungus placed in the order Pleosporales within the Dothideomycete class (Solomon et al. 2006; Friesen et al. 2008b). Upon infection *P. nodorum* causes lens-shaped necrotic lesions surrounded by a chlorotic halo on leaves of susceptible wheat genotypes, whereas small restricted lesions are observed on resistant wheat genotypes (Friesen and Faris 2010). *P. nodorum* also causes dark brown-purple blotches on the glumes, a condition known as 'glume blotch' (Bergstrom 2010). Under favorable conditions, the disease can cause yield losses up to 50% (Eyal et al. 1987). Selecting for genetic resistance against SNB has been a problem for breeders mostly due to the lack of understanding of the molecular basis of the host-pathogen interaction (Friesen and Faris 2010).

Wheat-biotroph interactions often follow a classical gene-for-gene model, which was first described by Flor (1956) using the flax-rust system. In the classical gene-for-gene model, the pathogen-derived avirulence factors are recognized directly or indirectly by the products of the corresponding host resistance gene, resulting in a cascade of events that leads to a resistance response (Fig. 1). Necrotrophic fungal pathogens can be categorized as either necrotrophic generalists that have multiple hosts or necrotrophic specialists that have one or a few related host species (Andrew et al. 2012). Necrotrophic generalists often use cell wall degrading enzymes (CWDE) and non-specific toxins to penetrate the host and kill host tissue to acquire nutrients (van Kan 2006). In addition to

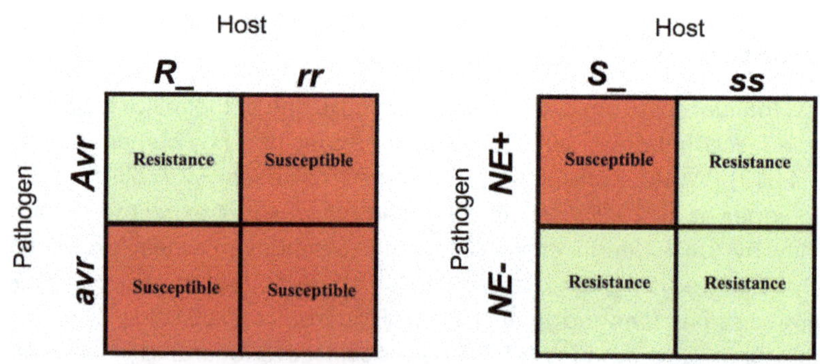

Figure 1 Left: Classical gene-for-gene interaction found in plant - biotrophic pathogen interactions where the direct or indirect recognition of an avirulent effector (*Avr*) gene product by a host resistance (*R*) gene product results in host resistance through effector-triggered immunity (ETI). Right: Inverse gene-for-gene interaction, typically found in plant - necrotrophic pathogen interactions where the direct or indirect interaction between the necrotrophic effector gene (*NE*) product and the corresponding sensitivity (*S*) gene product results in host susceptibility through necrotrophic effector-triggered susceptibility (NETS).

the release of CWDE, wheat-necrotrophic specialists such as *P. nodorum* and *Pyrenophora tritici-repentis* follow an inverse gene-for-gene model (Friesen and Faris 2010; Oliver et al. 2012; Faris et al. 2013). The inverse gene-for-gene model is a mirror image of the classical gene-for-gene model where direct or indirect recognition of pathogen-derived necrotrophic effectors (NE) (also known as host selective toxins) by the products of host dominant susceptibility/sensitivity genes leads to a susceptible reaction (Wolpert et al. 2002) (Fig. 1).

Recognition of a pathogen-produced effector by a corresponding plant resistance gene (R-genes), which are often nucleotide-binding, leucine-rich-repeat (NLR) genes, initiates the plant immune response known as effector-triggered immunity (ETI) (Jones and Dangl 2006). ETI includes the accumulation of reactive oxygen species, production of antimicrobial compounds, upregulation of pathogenicity-related genes, DNA laddering and electrolytic leakage, all contributing to programmed cell death (PCD) (Jones and Dangl 2006; Dodds and Rathjen 2010). Additionally, pathogen-associated molecular pattern (PAMP)-triggered immunity (PTI) can also happen, where perception of PAMPs such as fungal chitin by the pattern recognition receptors (PRR) leads to an early partial resistance response (Jones and Dangl 2006; Teixeira et al. 2019). PTI and ETI responses often work in synergy to prevent the progression of biotrophs by self-destruction of the cells surrounding the infection point (Cook et al. 2015). Necrotrophic pathogens, however, have targeted the ETI and PTI defense response pathways to induce PCD to obtain nutrients. This response is known as necrotrophic effector-triggered susceptibility (NETS) (Lorang et al. 2007; Faris et al. 2010; Liu et al. 2012; Shi et al. 2016a) and has been shown to be present in several necrotrophic specialist fungal pathogens.

Like other necrotrophic fungi, *P. nodorum* is known to produce an array of CWDEs, such as polygalacturonases, xylanases, glycanases, glucanases, proteases, esterases and cellulases, that play a major role in the infection of wheat tissue (Magro 1984; Lehtinen 1993; Carlile et al. 2000; Lalaoui et al. 2000). These CWDEs breakdown the components of the cell wall but have not been shown to be host specific in their activity (Friesen and Faris 2010). Bousquet and Skajennikoff (1974) were the first to report a toxic compound known as ochracine that was secreted by *P. nodorum* and inhibited the growth of roots. Since ochracine was not found in necrotic lesions caused by the fungus, it was not considered a major virulence factor of the disease. Multiple studies have shown that *P. nodorum* produces phytotoxic compounds such as septorine that hinder the growth of susceptible wheat cultivars (Devys et al. 1978; Bousquet et al. 1980; Devys et al. 1994). However, none of these compounds were shown to be host specific.

Keller et al. (1994) was the first to show host selective toxicity by showing that crude extracts of *P. nodorum* had varying degrees of embryotoxic effect on wheat genotypes showing susceptible and resistant responses for glume blotch. Wicki et al. (1999) used the same crude extract method to screen populations derived from eight different wheat parental lines for resistance against glume blotch. Therefore, it was possible that these crude extracts contained one or multiple NEs that selectively interacted with the more susceptible host genotypes.

2 Necrotrophic effector-host sensitivity gene interactions in the wheat-*P. nodorum* system

Liu et al. (2004a) was the first to report a host selective toxin, referred to from here on as necrotropic effectors (NE), and its corresponding wheat-sensitivity gene involved in SNB. Since this 2004 report, nine NE-host susceptibility gene interactions have been reported (Liu et al. 2004a; Friesen et al. 2006, 2007, 2008b, 2012; Abeysekara et al. 2009; Gao et al. 2015; Shi et al. 2016a). These interactions will be discussed in detail in this section.

2.1 *SnTox1-Snn1*

The *SnTox1-Snn1* interaction was the first NE-sensitivity gene interaction to be identified and was later molecularly fully characterized in the wheat-*P. nodorum* system (Liu et al. 2004a; Liu et al. 2012; Shi et al. 2016a). Liu et al. (2004a) infiltrated the International Triticeae Mapping Initiative (ITMI) population with partially purified SnTox1 from culture filtrates of the isolate Sn2000. Subsequent quantitative trait loci (QTL) analysis showed that a host gene, *Snn1*, residing on chromosome arm 1BS, was responsible for the necrosis induced by SnTox1 (Table 1, Fig. 2.). SnTox1 was further characterized as a protein with a size between 10 and 30 kDa. Based on the wheat genetic background, the importance of the *SnTox1-Snn1* interaction has varied where it explained as low as 0% and as high as 58% of the variation in disease (Liu et al. 2004b; Friesen et al. 2007; Chu et al. 2010; Phan et al. 2016).

SnTox1, the gene that encodes SnTox1, was the third NE-encoding gene to be cloned in *P. nodorum*. Liu et al. (2012) screened the SN15 reference sequence of *P. nodorum* for candidate genes based on previously known characteristics of SnTox1 and previously cloned genes *SnToxA* and *SnTox3*. *SNOG_20078* was chosen as the top candidate based on effector criteria and was therefore expressed in *Pichia pastoris*. Culture filtrates were used to validate *SNOG_20078* using infiltration results from the ITMI population, Chinese Spring (*Snn1*), CS-1BS-18 (*snn1*) and CSems237 (*snn1*) (Liu et al. 2012). Furthermore, transfer of a functional copy of *SNOG_20078* converted the avirulent isolate Sn79-1087 into a virulent isolate and deletion of the *SNOG_20078* in Sn2000 converted it to an avirulent isolate on *Snn1* differential lines (Liu et al. 2012). *SNOG_20078* was located on *P. nodorum* chromosome 6 and was designated *SnTox1* (Table 2). *SnTox1* encoded for a protein with 117 amino acids where the

Table 1 NE-wheat susceptibility gene interactions of *Parastagonospora nodorum*-wheat pathosystem

Necrotrophic effector	Host susceptibility gene	Host chromosome arm location	Markers	Reference
SnToxA	*Tsn1*	5BL	*Xfcp623*	Faris et al. (2010)
SnTox1	*Snn1*	1BS	*B500093078_5, XBE498831, Xfcp624, Xfcp618, X3476283*	Liu et al. (2004b), Liu et al. (2012), Cockram et al. (2015), Liu et al. (2016), Shi et al. (2016a)
SnTox2	*Snn2*	2DS	*Xcfd51, XTC253803*	Friesen et al. (2007)
SnTox3	*Snn3-B, Snn3-D*	5BS, 5DS	*Snn3-B1:Xfcp654, Xfcp665, Xmag705* *Snn3-D1:Xcfd18, Xhbg337*	Friesen et al. (2007), Liu et al. (2009), Zhang et al. (2011)
SnTox4	*Snn4*	1AS	*XBG262267, XBG262975, Xcfd58*	Abeysekara et al. (2009)
SnTox5	*Snn5*	4BL	*Xwmc349, Xcfd22*	Friesen et al. (2012)
SnTox6	*Snn6*	6AL	*XBE403326, XBE424987*	Gao et al. (2015)
SnTox7	*Snn7*	2DL	*Xcfd44, Xgwm349*	Shi et al. (2015)

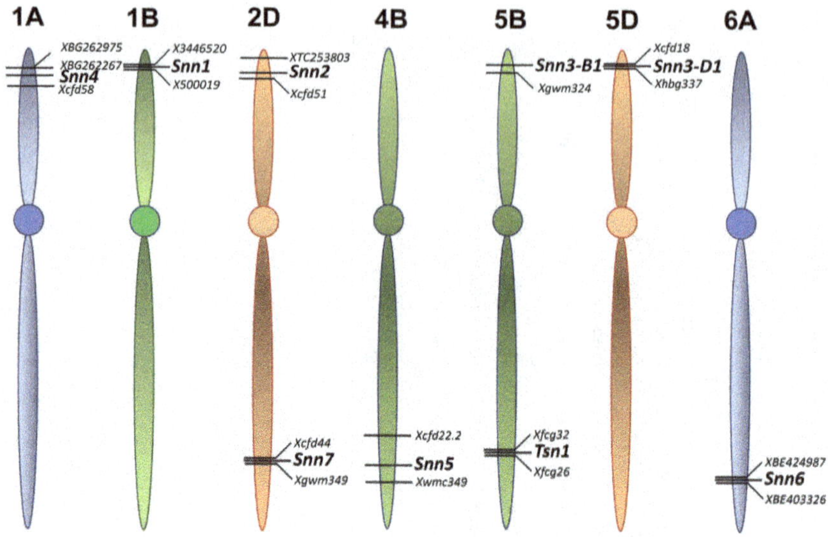

Figure 2 Chromosomal locations of *Parastagonospora nodorum* necrotrophic effector-sensitivity genes and flanking molecular markers in wheat. Short arms of chromosomes are located above the centromere and long arms are located below. See Table 1 for marker citations.

Table 2 Chromosomal locations of necrotrophic effector genes of *Parastagonospora nodorum* based on the newly published reference quality genome of Sn4 (Richards et al. 2018)

Necrotrophic effector gene	Pathogen chromosome	Physical location of the gene	
		Start	End
SnToxA	8	444359 bp	445675 bp
SnTox1	6	1558994 bp	1559174 bp
SnTox3	11	3133 bp	4616 bp

first 17 amino acids were predicted as a signal peptide. SnTox1 was a cysteine-rich protein with 16 cysteines present in the mature 100 amino acid protein. These cysteines provide stability to the protein against physical and chemical degradation. Mature SnTox1 was 10.33 kDa, which was within the size range of 10–30 kDa as predicted by Liu et al. (2004a).

SnTox1 plays a major and unique role in penetration during the initial stages of the infection process (Liu et al. 2016). SnTox1 fused with green fluorescent protein (SnTox1+GFP) showed that during the penetration process, SnTox1 localized to the outside of the fungal cell wall and to the penetration structure. SnTox1 that resides outside of the fungal cell wall protects the fungal mycelium from wheat chitinase degradation due to the predicted chitin-binding domain (CBD) (Fig. 3). Furthermore, transfer of a functional copy of *SnTox1* to other

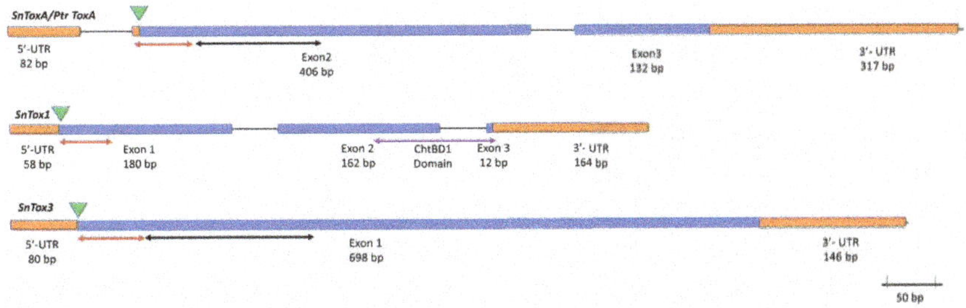

Figure 3 Structure of *Parastagonospora nodorum* necrotrophic effector genes. 5′ and 3′ UTR regions are indicated by orange boxes and exons are indicated by blue boxes. Black lines between the boxes represent introns. Green triangle represents the start codon. Red, black and purple double-headed arrows represent the signal peptide, pro-sequence and the predicted chitin binding domain, respectively. Exon1 of *SnToxA/Ptr ToxA* resides in the 5′-UTR region of the gene.

fungal pathogens such as *Cercospora beticola*, *Pyrenophora teres* f. *teres* and *Neurospora crassa* showed that SnTox1 also provided protection of these fungi from the same wheat chitinases (Liu et al. 2016).

Transcription analysis showed that *in planta*, expression of *SnTox1* peaked at 72 h post inoculation, correlating with the onset of necrosis at the same time point (Liu et al. 2012). Unique to this NE, the spraying of purified SnTox1 on to the leaves was enough to cause wide-spread necrosis (Liu et al. 2016). Infiltration of leaves of Chinese Spring (*Snn1*) and CSems237 (*snn1*) with SnTox1 followed by 3′-3′-diaminobenzidine (DAB) staining showed that the production of H_2O_2 was induced providing evidence that the SnTox1-Snn1 interaction induced a host-initiated oxidative burst *in planta*. Liu et al. (2012) infiltrated the same two lines with SnTox1 and analyzed the transcription of 28 wheat genes in an RT-PCR study to show that pathogenesis-related genes, PR-1-A1, a thaumatin-like protein gene and a chitinase gene were significantly upregulated in Chinese Spring (*Snn1*) compared to CSems237 (*snn1*). All three genes showed at least two-fold upregulation at 36 h post inoculation. Furthermore, at the same time point, DNA laddering was observed, indicating PCD. Production of reactive oxygen species (ROS), upregulation of expression of pathogenesis-related genes and PCD are all hallmarks of ETI induced upon the recognition of avirulence effectors by R-genes. These results indicated that SnTox1 interacted with Snn1 and hijacked the host defense system to elicit PCD allowing *P. nodorum* to acquire nutrients. Therefore, SnTox1 plays multiple roles in colonization of the pathogen in the host (Liu et al. 2012).

Infiltration of SnTox1-His and subsequent labeling with Alexa Fluor 488 showed that SnTox1 localized around the epidermal cells in the initial stages of infection, followed by localization surrounding the mesophyll cells in later stages in SnTox1-sensitive host genotypes (Liu et al. 2016). However, SnTox1 was not shown to be taken into the cells at any stage of infection (Liu et al.

2016). Therefore, it was predicted that the target for SnTox1 was outside the host cell, which was later shown by Shi et al. (2016a).

Snn1 mapped to wheat chromosome arm 1BS, using the ITMI population, with the candidate region for *Snn1* lying between the chromosomal deletion bins 1BS-4 and 1BS-18, accounting for a genetic distance of 27 cM (Liu et al. 2004a). Liu et al. (2004b) showed that the Chinese Spring-*T. dicoccoides* (CS-DIC) disomic chromosome substitution line CS-DIC 1B was insensitive to SnTox1 and less susceptible to the SnTox1 producing *P. nodorum* isolate Sn2000. In a follow-up study, Reddy et al. (2008) used an F_2 population derived from the cross CS (*Snn1*) × CS-DIC 1B (*snn1*) to delineate *Snn1* to a 0.46 cM region corresponding to 279 kb. However, efforts to locate a strong candidate gene failed since the region was gene rich and harbored multiple NLR genes (Reddy et al. 2008).

To clone *Snn1*, an F_2 population consisting of 8 500 progenies was developed from a cross between CS and CS with a pair of 1B chromosomes from the wheat variety 'Hope' (CS-Hope 1B) (Shi et al. 2016a). Additional genetic markers were derived from the end sequences of the 1B bacterial artificial chromosome (BAC) and used on the CS×CS-Hope 1B population to further delineate *Snn1* to a 0.16 cM region (Shi et al. 2016a). Linkage mapping showed that only one candidate gene, a member of the wall-associated kinase (WAK) genes, *TaWAK*, co-segregated with *Snn1* (Shi et al. 2016a). Introduction of mutations to *TaWAK* using the chemical mutagen ethyl methane sulfonate (EMS) converted SnTox1-sensitive CS to insensitive. *TaWAK* was further validated as *Snn1* by transforming the SnTox1-insensitive wheat genotype 'Bobwhite' with *TaWAK* cDNA coupled with the maize ubiquitin promoter. Transgenic plants containing and transcribing *TaWAK* were sensitive to SnTox1, whereas plants that were insensitive lacked the gene. Therefore, *TaWAK* was designated as *Snn1* (Shi et al. 2016a).

Snn1 encoded for a protein with four conserved domains. The extracellular domains, including the wall-associated receptor kinase galacturonan-binding domain (GUB_WAK) and the epidermal growth factor-calcium-binding domain (EGF_CA) were connected by a transmembrane domain to the intracellular serine/threonine protein kinase domain (S/TPK) (Fig. 4) (Shi et al. 2016a). Mutations induced in the GUB_WAK, EGF_CA or the S/TPK domains made CS insensitive to SnTox1 indicating that each of these domains was critical for a sensitive response. Yeast two-hybrid (Y2H) analysis performed using SnTox1 and Snn1(full or partial) provided evidence that SnTox1 directly bound to a region of 140 amino acids between the extracellular domains GUB_WAK and EGF_CA. This direct binding to the predicted extracellular domain provided further evidence that the SnTox1-Snn1 interaction takes place extracellularly as predicted by Liu et al. (2016).

Shi et al. (2016a) showed that *TaMAPK3* was upregulated within 15 min of the SnTox1-Snn1 interaction. Early upregulation of *MAPK* genes is a hallmark

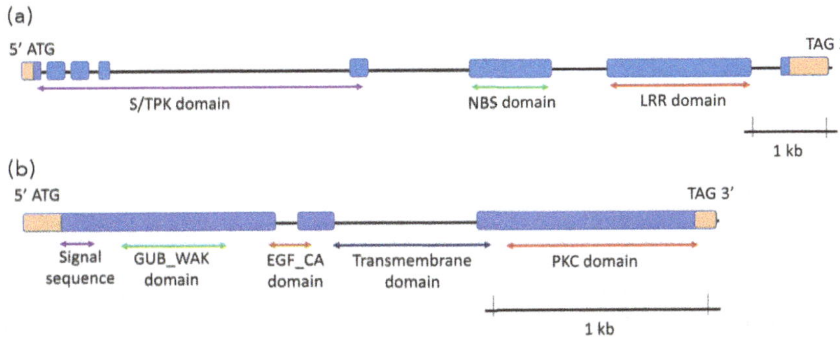

Figure 4 The structure of cloned necrotrophic effector-sensitivity genes in wheat. (a) Structure of the *Tsn1* gene that confers sensitivity to SnToxA. Exons and UTR regions are represented by blue and orange boxes, respectively. Colored double-headed arrows indicate the three regions encoding for domains: serine/threonine protein kinase domain, nucleotide binding domain and leucine-rich repeat domain (Faris et al. 2010). A kilobase scale is indicated at the bottom. (b) Gene structure of the *Snn1* gene that confers sensitivity to SnTox1. Exons and UTR regions are represented by blue and orange boxes, respectively. Double-headed colored arrows represent domains: wall-associated receptor kinase galacturonan binding domain, epidermal growth factor-calcium binding domain, serine/threonine protein kinase domain and the transmembrane domain (Shi et al. 2016a).

of PTI. Given the data, *Snn1* was considered a PRR class gene and this was the first report of a fungal NE manipulating the host PTI system. Transcription analysis showed that SnTox1 did not induce or upregulate the expression of *Snn1*. Instead, upon SnTox1-Snn1 interaction, transcription of *Snn1* was downregulated over time and the specific reason for the downregulation is still unknown (Shi et al. 2016a). Furthermore, transcription analysis revealed that *Snn1* was expressed in leaves but not in roots, and expression was influenced by light. CS plants subjected to continuous darkness showed continuous low expression of *Snn1* compared to plants subjected to a 12-h light/dark cycle showing a gradual increase in expression of *Snn1* during the dark hours, prior to peak at subjective dawn, followed by a gradual decrease during the day time. Therefore, the lack of necrotic reaction in dark conditions after inoculation with the Sn79+Tox1A1 and infiltration with the SnTox1 (Liu et al. 2012) could be due to the lack of *Snn1* expression.

2.2 SnToxA-Tsn1

The *ToxA-Tsn1* interaction has been the most intensively studied NE-susceptibility gene interaction in any wheat-necrotrophic pathosystem. *ToxA* is unique since it has been identified in three necrotrophic fungal pathogens of wheat including *P. tritici-repentis*, *P. nodorum* and *Bipolaris sorokiniana* (Tomás and Bockus 1987; Friesen et al. 2006; McDonald et al. 2018; Friesen et al. 2018).

ToxA was the first proteinaceous NE to be identified in any necrotrophic fungal pathosystem, and it was originally purified from *P. tritici-repentis*, the causal agent of tan spot disease of wheat (Tomás and Bockus 1987; Balance et al. 1989). Since ToxA was identified in *P. tritici-repentis*, it was originally designated as Ptr ToxA (Ciuffetti et al. 1997, 1998). The *ToxA* locus consisted of a promoter region of 278 bp, a predicted open-reading frame of 534 bp including two introns (Fig. 3) (Ciuffetti et al. 1997).

Ptr ToxA contained a 23 amino acid signal peptide and a 37 amino acid pro-domain that played a role in protein folding (Tuori et al. 2000). Both the signal sequence and the pro-domain were cleaved prior to secretion, resulting in a mature protein of 13.2 kDa consisting of 178 amino acids (Ciuffetti et al. 1997; Balance et al. 1996). Further analysis showed that Ptr ToxA contained a conserved RGD (Arg-Gly-Asp) motif that was known to be involved in protein-protein interactions (Meinhardt et al. 2002; Sarma et al. 2005). Studies showed that the RGD motif was essential for transportation of Ptr ToxA into the cytoplasm of the mesophyll cells in susceptible wheat genotypes prior to subcellular localization to the chloroplast (Manning and Ciuffetti 2005; Manning et al. 2007). Like the SnTox1-Snn1 interaction, perception of Ptr ToxA triggers transcriptional reprogramming of multiple genes including downregulation of genes encoding enzymes important in detoxifying ROS associated with the chloroplast (Pandelova et al. 2009). Therefore, accumulation of ROS in the chloroplasts was observed and such accumulation disrupted the protein homeostasis in both photosystems, which eventually leads to PCD (Manning et al. 2009; Pandelova et al. 2009). In addition, Adhikari et al. (2009) showed the upregulation of defense-related genes such as *WAKs*, *RLKs* and *MAPKs* which are also involved in PCD.

Manning et al. (2007) screened a yeast two-hybrid library and found that ToxA directly interacts with the protein designated ToxA-Binding Protein 1 (ToxABP1), a conserved chloroplast protein. Furthermore, Manning et al. (2007) revealed that ToxA binds to ToxABP1 via amino acid threonine 137 of ToxA. In a similar yeast two-hybrid study it was found that ToxA also interacted with plastocyanin, an important component of the electron transport system of photosynthesis (Tai et al. 2007). In addition, Tai et al. (2007) showed that the RGD motif was required for the recognition of the host target, and that the region of ToxA that spans from E145 to D149 was required for the protein-protein interaction and subsequent PCD (Tai et al. 2007). Another yeast two-hybrid experiment and co-immunoprecipitation assay showed that ToxA also interacted with the PR-1-type pathogenesis-related protein PR-1-5 (Lu et al. 2014). Site-specific mutational analysis showed that the N102 residue of ToxA and the N141 residue of the PR-1-5 were essential for the interaction. ToxA-N102A mutants failed to induce necrosis, whereas PR-1-5-N141A mutants failed to promote necrosis when co-infiltrated with ToxA into ToxA-sensitive genotypes (Lu et al. 2014). Finally, differential expression of PR-1-5 in ToxA-sensitive and insensitive wheat genotypes treated with ToxA, also showed that ToxA may be targeting PR-1-5 to induce necrosis (Lu et al. 2014).

Following the genome sequencing of *P. nodorum*, Friesen et al. (2006) found that the *P. nodorum* gene SNU16571.1 showed 99.7% homology to *Ptr ToxA*. The gene was designated *SnToxA*. Evaluation of the gene sequence of *SnToxA* and *Ptr ToxA*, in 95 *P. nodorum* isolates and 54 *P. tritici-repentis* isolates, respectively, revealed eleven haplotypes in the *P. nodorum* population compared to a single haplotype in the *P. tritici-repentis* population. High levels of sequence diversity in *SnToxA* suggested that the gene had been present in *P. nodorum* for a longer time and was likely transferred horizontally from *P. nodorum* to *P. tritici-repentis*. Stukenbrock and McDonald (2007) analyzed the gene sequence of *ToxA* in a worldwide collection of *P. nodorum* isolates and suggested that *P. nodorum* probably acquired the gene from another organism. Since the highest number of haplotypes of *SnToxA* was identified in South Africa, Stukenbrock and McDonald (2007) speculated that *SnToxA* was horizontally transferred to *P. nodorum* in South Africa.

Friesen et al. (2006) showed that in addition to the *ToxA* gene, an 11 kb region flanking *ToxA* was also found in both *P. nodorum* and *P. tritici-repentis*, and the flanking region consisted of a transposase-like gene that could have facilitated the movement. Recently, a 12 kb region with 91.9% homology to the horizontally transferred region that contained the *ToxA* gene was also found in *B. sorokiniana* (McDonald et al. 2018). Even though several variations were observed among the three 12 kb flanking regions, one of the two haplotypes of *ToxA* identified in *B. sorokiniana* (here after referred to as *BsToxA*) was identical to the haplotype identified in *P. tritici-repentis*. Therefore, it was possible that *ToxA* was recently transferred from *P. tritici-repentis* to *B. sorokiniana* (McDonald et al. 2018). However, in-depth analysis of the *ToxA* gene and its flanking regions in *B. sorokiniana* enabled the identification of the terminal inverted repeats (TIR) that demarcated the putative horizontally transferred gene fragment, and it spanned a more than 14 kb region (McDonald et al. 2019). Discovery of TIRs expanded the horizontally transferred gene fragment to 63 kb and 370 kb in *P. tritici-repentis* and *P. nodorum*, respectively (McDonald et al. 2019). McDonald et al. (2019) proposed a second model suggesting *B. sorokiniana* as the donor of *ToxA* to *P. tritici-repentis* since *B. sorokiniana* seems to have the more intact version of the horizontally transferred fragment. Very recently a large collection of *P. nodorum* isolates collected from Iran was analyzed for genetic diversity, and the data suggested that *P. nodorum* originated in the Fertile Crescent and the original pathogen populations already carried NE genes including *SnToxA* (Ghaderi et al. 2020). Therefore, Ghaderi et al. (2020) concluded that their findings supported the hypothesis that *P. nodorum* was the original source of *ToxA* that is now found in *P. tritici-repentis* and *B. sorokiniana*.

Even though the SnToxA-Tsn1 interaction plays a major role in disease, not all *P. nodorum*, *P. tritici-repentis* or *B. sorokiniana* isolates contain the *ToxA* gene. Friesen et al. (2006) screened 755 isolates of *P. nodorum* and only 24% of them contained the gene. In a very recent study, Richards et al. (2019) evaluated 197 isolates of *P. nodorum* collected from various wheat-growing regions of the United States and found *SnToxA* was most prevalent in isolates collected from regions where cultivated wheat genotypes harbored *Tsn1*. Richards et al. (2019) also showed that isolates collected from regions where *Tsn1* is less common in local wheat lines, typically lacked the *SnToxA* gene.

SnToxA, BsToxA and Ptr ToxA induced a similar necrotic response and isolates of each pathogen carrying the *ToxA* gene have increased virulence on wheat lines harboring *Tsn1* (Faris et al. 1996; Liu et al. 2006; Friesen et al. 2018; McDonald et al. 2018). Faris et al. (1996) was the first to genetically map *Tsn1* to the long arm of chromosome 5B. Tetraploid and hexaploid QTL mapping studies have shown that the *ToxA-Tsn1* interaction can account for 25%–95% of the variation in disease caused by *P. nodorum* (Friesen et al. 2006; Liu et al. 2006; Friesen et al. 2008b; Faris and Friesen 2009; Chu et al. 2010; Faris et al. 2011; Virdi et al. 2016). *SnToxA* mutants of *P. nodorum* created by site-directed gene knockouts caused less disease on wheat genotypes with *Tsn1* compared to their corresponding wild-type strains, also suggesting that the SnToxA-Tsn1 interaction plays a major role in disease (Friesen et al. 2006).

Faris et al. (1996) mapped *Tsn1* to a 22.2 cM region on the long arm of chromosome 5B. In a high-resolution mapping study, Haen et al. (2004) introduced more markers into the region and delineated the *Tsn1* locus to a 0.8 cM region. Lu et al. (2006) developed and assembled BAC contigs to the *Tsn1* locus. Using BAC sequences, Lu et al. (2006) was able to develop two simple sequence repeat (SSR) markers that flanked *Tsn1* at 0.3 and 0.5 cM.

Chromosome walking of BAC contigs and subsequent development of markers delineated the candidate region for *Tsn1* to 120 kb, which harbored only four genes (Faris et al. 2010). ToxA insensitive mutants were generated using EMS and sequence analysis suggested that *Tsn1* was a gene with serine/threonine protein kinase (S/TPK), nucleotide-binding site (NBS) and leucine-rich repeat (LRR) domains (S/TPK-NBS-LRR-like gene), all of which have been identified in classical resistance genes (Faris et al. 2010). Tsn1 consisted of 1 490 amino acids. The first five exons encoded for an S/TPK domain that consisted of three conserved motifs including an ATP-binding site, a substrate-binding site and an activation loop (Faris et al. 2010). The sixth exon encoded for an NBS domain and the seventh exon encoded for the 24 LRRs. No known motifs were observed in the portion of the protein encoded by the eighth exon (Fig. 4).

Unlike the SnTox1-Snn1 interaction, yeast two-hybrid experiments could not show that ToxA and Tsn1 interacted directly (Faris et al. 2010). Therefore, ToxA likely targets Tsn1 indirectly since Tsn1 is essential for ToxA to be taken up into the cell. As mentioned above, research has revealed that ToxA directly interacts with plastocyanin and ToxABP1 (Tai et al. 2007; Manning et al. 2007). Yeast two-hybrid results also have not been able to show a direct Tsn1 interaction with either the plastocyanin or ToxABP1 (Faris et al. 2010). Therefore, the involvement of Tsn1 in triggering necrosis is yet to be explained. Furthermore, Manning and Ciuffetti (2005) first showed that the ToxA-Tsn1 interaction depends on light. Transcriptional analysis of *Tsn1* showed that the expression of the gene was low under continuous dark conditions proving that light was an important regulatory factor of the interaction. Transcriptional analysis also showed that expression of *Tsn1* was downregulated in plants treated with ToxA, and it was speculated that the downregulation could have been due to the pathogen's ability to manipulate the photosynthetic pathway and not due to the direct impact of ToxA on *Tsn1* (Faris et al. 2010).

Molecular characterization of both ToxA and Tsn1 has revealed that mechanisms associated with ETI against biotrophs overlap with those of necrotrophic effector-triggered susceptibility for necrotrophic pathogens. Therefore, the SnTox1-Snn1 and SnToxA-Tsn1 interactions show how *P. nodorum* potentially manipulates both PTI and ETI systems to cause disease (Friesen et al. 2006; Faris et al. 2010; Liu et al. 2012; Shi et al. 2016a).

2.3 *SnTox2-Snn2*

The *SnTox2-Snn2* interaction was the third NE-susceptibility gene interaction to be characterized in the wheat-*P. nodorum* pathosystem. Friesen et al. (2007) extracted the culture filtrates of *P. nodorum* isolate Sn6 and infiltrated the recombinant inbred population derived from the cross BR34×Grandin (BG population). Sensitive lines produced a necrotic reaction within three days, whereas insensitive lines did not have a reaction. The population segregated 89 sensitive to 29 insensitive, which fit a 3:1 segregation ratio, suggesting that the population segregated for two sensitivity genes. PCR analysis using gene-specific primers showed that Sn6 contained a functional copy of *SnToxA*, suggesting that one of the NEs produced by Sn6 was SnToxA (Friesen et al. 2007). The second NE was designated SnTox2. SnTox2 was partially purified and separated from SnToxA using liquid chromatography. Initial characterization of SnTox2 showed that it was a protein between 3 and 10 kDa. The *SnTox2-Snn2* interaction was also shown to be light dependent, like SnToxA-Tsn1 and SnTox1-Snn1 (Friesen et al. 2007).

The BG wheat population was infiltrated with partially purified SnTox2 and the population segregated 1:1 for sensitivity to SnTox2, suggesting SnTox2 interacted with one sensitivity gene that segregated in this population. The sensitivity mapped as a single gene to the short arm of chromosome 2D, and it was designated *Snn2*. The BG population was inoculated with the SnToxA and SnTox2-producing isolate Sn6, and the *SnTox2-Snn2* interaction explained 47% of the variation in disease, whereas *SnToxA-Tsn1* explained 20% of the variation in disease (Table 1, Fig. 2) (Friesen et al. 2007). Furthermore, Friesen et al. (2007) showed that lines with both *Snn2* and *Tsn1* had an average higher disease reaction compared to the lines that harbored only one or the other sensitivity gene, suggesting that the *SnTox2-Snn2* and *SnToxA-Tsn1* interactions are largely additive (Friesen et al. 2007). This additive response is typically not seen in the systems that follow a gene-for-gene model where multiple effective resistance genes provide a similar level of resistance as does a single effective resistance gene. Therefore, this was the first study to show this difference between the inverse gene-for-gene and classical gene-for-gene models (Friesen and Faris 2010).

2.4 *SnTox3-Snn3-B1* and *SnTox3-Snn3-D1*

The *SnTox3-Snn3* interaction was the fourth interaction to be identified in the wheat-*P. nodorum* pathosystem. This interaction has proven unique since SnTox3 targeted two homoeologous host-susceptibility genes, *Snn3-B1* and *Snn3-D1*. The interaction was first reported by Friesen et al. (2008b) where infiltration of culture filtrates of *P. nodorum* isolate Sn1501, which does not harbor *SnToxA*, on the BG population resulted in a 3:1 sensitive:insensitive ratio suggesting two sensitivity genes segregated in the population. Subsequent genetic mapping revealed that one of the sensitivity genes was *Snn2* and the other mapped to a novel position on the short arm of chromosome 5B. The novel gene was designated *Snn3-B1*, and the NE that interacted with *Snn3-B1* was designated SnTox3 (Table 1) (Friesen et al. 2008b). Subsequent QTL analysis showed that *SnTox3-Snn3-B1* explained 13% of the variation in disease caused by Sn1501. It was also revealed that the disease reaction caused by the compatible *SnTox3-Snn3-B1* interaction was only observed when a compatible *SnTox2-Snn2* interaction was absent (Friesen et al. 2008b). Furthermore, QTL analysis conducted using SN15, which harbors *SnToxA*, and SN15KO18 which lacks *SnToxA*, showed that the *SnTox3-Snn3-B1* interaction was only significant in the absence of the *SnToxA-Tsn1* interaction (Friesen et al. 2008b).

Zhang et al. (2011) was the first to show *Aegilops tauschii*, the diploid D-genome donor of common wheat, was sensitive to SnTox3. Infiltration of *A. tauschii* accessions TA2377 and AL8/78 with culture filtrates of Sn4 resulted in a sensitive and insensitive reaction, respectively (Zhang et al. 2011). Infiltration of an F_2 population derived from a cross between TA2377 and AL8/78 with culture filtrates of Sn4 showed a 3:1 sensitive to insensitive segregation across the progeny. Therefore, a single host gene was responsible for the sensitivity to culture filtrates of Sn4. *P. nodorum* isolate Sn4 was known to produce multiple NEs including SnToxA, SnTox1, SnTox2 and SnTox3. Infiltration of culture filtrates of Sn79 + SnTox3A that only produced SnTox3 on two parental *A. tauschii* accessions matched the Sn4 infiltration reactions (Zhang et al. 2011). Bulked-segregate analysis performed using TA2377, AL8/78 and F_2 progeny mapped SnTox3 sensitivity to the distal end of the short arm of wheat chromosome 5D (Table 1) (Zhang et al. 2011).

This was the first time that two sensitivity loci were identified for a single necrotrophic effector. To compare the *SnTox3-Snn3-B1* and *SnTox3-Snn3-D1* interactions in the same genetic background, Zhang et al. (2011) developed an F_2 mapping population derived from the cross between the synthetic hexaploid wheat line, LDN-TA2377 (*snn3-B1snn3-B1/Snn3-D1Snn3-D1*) and the SnTox3 differential line BG220 (*Snn3-B1Snn3-B1/ snn3-D1snn3-D1*) and found that the *SnTox3-Snn3-D1* interaction contributed more to disease severity compared to the *SnTox3-Snn3-B1* interaction. Furthermore, Zhang et al. (2011) showed that the order of markers in the region of *Snn3* was well conserved in chromosome 5B and 5D. Zhang et al. (2011) speculated that both genes were derived from a common ancestor, and therefore, were homologous. Genetic evaluation of six hexaploid wheat lines showed that they only contained *Snn3-B1*, but not *Snn3-D1*. Zhang et al. (2011) suggested that the *A. tauschii* accession that was involved in the hybridization event that gave rise to hexaploid wheat did not contain *Snn3-D1* and therefore, sensitivity to SnTox3 in common wheat was derived from the B-genome progenitor. High-resolution maps have been developed for both *Snn3-B1* and *Snn3-D1*, and candidate regions have been delineated to 1.5 cM and 1.4 cM, respectively (Fig. 2) (Shi et al. 2016b; Zhang et al. 2011).

Both ultrafiltration and size exclusion chromatography suggested that SnTox3 was a protein with a size ranging from 10 to 30 kDa (Friesen et al. 2008b). Friesen et al. (2008b) also showed that SnTox3-Snn3-B1 was not light dependent. From the size exclusion chromatography and SDS-PAGE gel visualization, mass spectra for candidate proteins for SnTox3 were identified. PCR amplification of the corresponding candidate genes in avirulent isolates that did not produce SnTox3 and virulent isolates that produced SnTox3 revealed that *SNOG_08981* was the gene that encoded for SnTox3, and *SNOG_08981* was subsequently renamed as *SnTox3* (Liu et al. 2009). *SnTox3* was located at the distal end of chromosome 11(Table 2). The gene consists of one exon that spans 693 bp and 5′ and 3′ UTRs that span over 80 bp and 146 bp, respectively (Fig. 3) (Liu et al. 2009). *SnTox3* encodes for a protein with 230 amino acids and a mass of 25.85 kDa. Like SnToxA, SnTox3 is a pre-pro protein. A predicted pre-domain contains the signal peptide that consists of 20 amino acids. Very recently, Outram et al. (2020) characterized the crystal structure of SnTox3 and showed that the pro-domain consisting of 52 amino acids was necessary for protein folding and NE activity. In addition, Outram et al. (2020) showed that the protease Kex2 removed the pro-domain prior to secretion of mature SnTox3. Liu et al. (2009)

showed that the mature SnTox3 protein consisted of six cysteine residues that were predicted to produce disulfide bonds. These disulfide bonds were predicted to provide stability to the protein. Elucidation of the crystal structure of the mature SnTox3 protein showed that it consisted of a unique β-barrel formed from eight β-strands connected via three di-sulfide bonds formed between cysteine residues. In addition, SnTox3 also consisted of two other β-strands forming β-hairpin-like structures (Outram et al. 2020). Expression of *SnTox3* peaked at 72 h post inoculation, and maximum expression coincided with leaf colonization and the onset of the necrotic lesions Liu et al. (2009).

Liu et al. (2009) analyzed 923 *P. nodorum* isolates which were collected from eight different global geographical regions and found that 60.1% of the isolates had *SnTox3*. In addition, McDonald et al. (2013) analyzed a different global collection of *P. nodorum* that consisted of 1000 isolates and found that 61% of those contained *SnTox3*. Furthermore, it was suggested that frequency of *SnTox3* varied in different geographical regions, and the variation was determined by the presence of *Snn3* in the host population (Liu et al. 2009; McDonald et al. 2013; Richards et al. 2019). Liu et al. (2009) obtained *SnTox3* sequences from 245 isolates, identifying eleven haplotypes that encoded for four protein isoforms, all of which were biologically active. However, results did not show that *SnTox3* was undergoing positive selection.

Winterberg et al. (2014) carried out an Affymetrix microarray analysis of plants treated with SnTox3 and found that more than 5 700 genes were expressed differentially in the first 48 h. Out of these differentially expressed genes, a significant number of the downregulated genes were predicted to be involved in the downregulation of photosystem I and ATP synthase in light reactions of photosynthesis (Winterberg et al. 2014). Therefore, downregulation of photosynthesis by the SnTox3-Snn3 interaction was different from the downregulation of photosynthesis induced by the SnToxA-Tsn1 interaction that was predicted to regulate the genes impacting the Calvin cycle. Furthermore, upon the SnTox3-Snn3 interaction, genes typically involved in PTI pathways were upregulated. Upregulated genes included *TaMPK3* in the first 48 h after infiltration, suggesting that SnTox3 was also used to manipulate the host defense response (Winterberg et al. 2014). Upregulation of *TaMPK3* was also induced by SnTox1, suggesting that multiple NEs of *P. nodorum* manipulate the same gene in the MAPK pathway (Winterberg et al. 2014; Shi et al. 2016a). Winterberg et al. (2014) showed that SnTox3 also induced other components of host defense including PR proteins, proteins of the jasmonic acid pathway and the phenylpropanoid pathway.

A yeast two-hybrid study showed that SnTox3 directly interacts with multiple PR proteins that belong to the protein family of TaPR-1 (Breen et al. 2016). Out of these PR protein-encoding genes, the strongest interaction was observed with the protein TaPR-1-1, which was confirmed *in-planta* through co-immunoprecipitation. Breen et al. (2016) also showed that the PR-1-derived defense signaling peptide from the C-terminus, known as CAPE1, increased the infection caused by *P. nodorum* in wheat in the presence of the *SnTox3-Snn3-B1/D1* interactions.

2.5 *SnTox4-Snn4*

SnTox4 was the fifth NE to be identified in *P. nodorum* and it was partially purified from culture filtrates of the Swiss isolate Sn99CH 1A7a (Abeysekara et al. 2009). SnTox4 was a protein with an estimated size that ranged from 10 to 30 kDa (Abeysekara et al. 2009). Sensitivity to SnTox4 was conferred by a single dominant gene designated *Snn4* and was mapped to the short arm of chromosome 1A (Table 1). Infiltration of SnTox4 in leaves induced a mottled necrosis on wheat lines harboring *Snn4*. As with some other NE interactions, the compatible SnTox4-Snn4 interaction was also light dependent. Subsequent QTL analysis of the population derived from the cross between the Swiss cultivars 'Arina' and 'Forno' revealed that 41% of the variation in disease was explained by the *SnTox4-Snn4* interaction (Abeysekara et al. 2009). Evaluation of disease caused by Sn99CH 1A7a on another recombinant inbred line (RIL) population derived from the cross Salamouni × Katepwa showed that the *SnTox4-Snn4* interaction explained 23.5% of the variation in disease (Abeysekara et al. 2012). Abeysekara et al. (2009) delineated the *Snn4* candidate region to a 2.5 cM interval (Fig. 2).

2.6 *SnTox5-Snn5*

SnTox5-Snn5 was the seventh interaction to be discovered in the wheat-*P. nodorum* pathosystem. Friesen et al. (2012) evaluated a tetraploid doubled haploid population derived from the durum wheat cultivar 'Lebsock' and the *T. turgidum* ssp. *carthlicum* accession PI 94749 (LP749 population) for SNB disease caused by *P. nodorum* isolate Sn2000. Friesen et al. (2012) found that in addition to the *SnToxA-Tsn1* interaction, another interaction contributed significantly to SNB disease. QTL analysis mapped the novel susceptibility gene to the long arm of chromosome 4B, and it was designated *Snn5*. The NE targeting *Snn5* was designated SnTox5 (Table 1, Fig. 2).

Initial characterization showed that SnTox5 was a protein with a size ranging from 10 to 30 kDa, which is typical for a small secreted NE (Friesen et al. 2012). As seen in the *SnToxA-Tsn1*, *SnTox1-Snn1*, *SnTox2-Snn2* and *SnTox4-Snn4* interactions, the *SnTox5-Snn5* interaction was also light dependent. In addition, partially purified SnTox5 remained active even after 60 min of boiling, suggesting that SnTox5 was heat stable and may contain multiple cysteine residues forming disulfide bonds (Friesen et al. 2012).

SnTox5-Snn5 explained 37% of the variation in disease and in the absence of the *SnToxA-Tsn1* interaction it explained 63% of the variation in disease.

Friesen et al. (2012) showed that the two interactions together cause more disease compared to the presence of them individually. *Snn5* was mapped between microsatellite markers that spanned over 16.1 cM (Friesen et al. 2012). Both *Snn5* and *SnTox5* cloning efforts are currently underway.

2.7 *SnTox6-Snn6*

Gao et al. (2015) infiltrated culture filtrates of *P. nodorum* isolate Sn6 on the parental lines of the ITMI population. Parental line 'W-7984' showed an insensitive reaction, whereas the other parental line 'Opata85' showed a sensitive reaction. The ITMI population was infiltrated with the culture filtrate of Sn6 and subsequent regression analysis showed that two loci were involved with sensitivity. Out of the two loci, one locus was found on the distal end of the short arm of chromosome 5B and was identified as *Snn3B-1*, the known target of SnTox3, with the other locus mapping to the long arm of chromosome 6A (Fig. 2), which was a novel NE-sensitivity locus (Gao et al. 2015). Therefore, Gao et al. (2015) designated the novel NE and the corresponding susceptibility gene SnTox6 and *Snn6*, respectively. This was the eighth NE-susceptibility gene interaction to be identified in this system.

Initial characterization revealed that SnTox6 was a small secreted protein (Gao et al. 2015). Based on size exclusion chromatography, Gao et al. (2015) predicted that SnTox6 had a molecular mass equal to or less than 12 kDa. With the use of two-dimensional gel electrophoresis followed by mass spectroscopy, Gao et al. (2015) was able to derive a putative amino acid sequence which was used to screen predicted protein databases of *P. nodorum*. Three potential candidates were identified. As most of the interactions in the wheat-*P. nodorum* pathosysthem, *SnTox6-Snn6* was also light dependent.

The ITMI population was inoculated with Sn6 and the QTL analysis showed that the *SnTox6-Snn6* interaction accounted for 27% of the variation in disease (Gao et al. 2015). Sn6 also harbored both *SnTox1* and *SnTox3* and the ITMI population segregated for *Snn1* and *Snn3B-1*. Reverse transcriptase-PCR (RT-PCR) analysis showed that *SnTox1* expression was not detectable; however, *SnTox3* was expressed (Gao et al. 2015). Disease was not associated with either the *Snn1* or *Snn3B-1* loci in QTL analysis. The lack of disease association with *Snn1* could be explained by the lack of expression; however, the fact that the *SnTox3-Snn3B-1* interaction did not play a significant role in disease even though *SnTox3* was expressed, added to the complexity of this system.

2.8 *SnTox7-Snn7*

Culture filtrates of Sn6 were used to screen Chinese Spring, 'Timstein' (TM) and CS-Tm disomic chromosome substitution lines to reveal that Timstein, CS-Tm 2D and CS-Tm 5B substitution lines had sensitive reactions. The NE that induced necrosis on CS-Tm 2D was considered novel and designated SnTox7 and the corresponding wheat-sensitivity gene was named *Snn7* (Shi et al. 2015). This was the ninth and the latest NE-sensitivity gene interaction to be characterized in this system.

Partial characterization of SnTox7 revealed that it was a proteinaceous NE with a size range from 10 to 30 kDa. Shi et al. (2015) showed that disulfide bonds were required for necrosis-inducing activity and that the protein was heat stable.

Shi et al. (2015) developed an F_2 population by crossing Chinese Spring (CS) and CS-Tm2D, and mapped the *Snn7* to the long arm of chromosome 2D between microsatellite markers that spanned 2.7 cM (Fig. 2). The *SnTox7-Snn7* interaction explained 33% of the variation in disease segregating in the population. The infiltration of common wheat lines with Sn6 culture filtrates showed that Timstein is the only line sensitive to SnTox7 and suggested that the prevalence of *Snn7* is very low (Shi et al. 2015). However, understanding this interaction is important to the understanding of the wheat-*P. nodorum* pathosystem as a whole.

3 Genetic relationship between NE-sensitivity gene interactions

In a pathosystem that follows the classical gene-for-gene model, the resistance provided by the presence of a single resistance gene-avirulence effector gene interaction is often the same as that when multiple resistance gene-avirulent effector gene interactions are present (Friesen and Faris 2010; Peters-Haugrud et al. 2019). However, in host-pathogen interactions such as the wheat-*P. nodorum*, wheat-*P. tritici-repentis*, and wheat-*B. sorokiniana* systems, the inverse gene-for-gene model is largely based on the NE-sensitivity gene interactions and how they interact with each other. Therefore, in this instance susceptibility/resistance is often identified as quantitative unless the individual interactions are isolated (Faris and Friesen 2009; Friesen and Faris 2010; Kariyawasam et al. 2016; Liu et al. 2017).

Friesen et al. (2007) was the first to evaluate two NE-sensitivity gene interactions in the same host population. In this work, Friesen et al. (2007) showed that the disease response was governed by both the *SnTox2-Snn2* and the *SnToxA-Tsn1* interactions. The two reactions acted additively and together explained 65% of the variation in disease. Similarly, in the LP749 population, *SnTox5-Snn5* and *SnToxA-Tsn1* accounted for 37% and 31% of the variation in disease, respectively, showing that the two interactions act additively and independently to cause 60% of the variation in disease (Friesen et al. 2012). In both studies, it was shown that host genotypes that harbored multiple host-sensitivity genes had significantly higher disease reactions than genotypes with one or no sensitivity genes (Friesen et al. 2007; Friesen et al. 2012). However, Friesen et al. (2008b) reported that the *SnTox3-Snn3* interaction was only significant in the absence of the *SnTox2-Snn2* and *SnToxA-Tsn1* interactions. Later, it was also shown that the interactions *SnTox1-Snn1*, *SnTox5-Snn5* and *SnTox6-Snn6* had an epistatic effect on the *SnTox3-Snn3* interaction (Friesen et al. 2012; Gao et al. 2015; Phan et al. 2016). However, in a recent study Peters-Haugrud et al. (2019) showed that *SnTox3-Snn3* and *SnToxA-Tsn1* could also act additively but only in the absence of a compatible *SnTox1-Snn1* interaction. In addition, Peters-Haugrud et al. (2019) showed that the *SnTox3-Snn3* and *SnToxA-Tsn1* interactions together had an epistatic effect on the *SnTox1-Snn1* interaction, contradicting the results of Phan et al. (2016). Peters-Haugrud et al. (2019) suggested that the variation in results could be due to host allelic differences or variation in phenotyping between the two studies.

Faris et al. (2011) inoculated the BG population that segregated for *Snn2* and *Tsn1* with two *P. nodorum* isolates, Sn4 and Sn5, which both produced SnTox2 and SnToxA. Genetic analysis revealed that *SnTox2-Snn2* and *SnToxA-Tsn1* interactions played an equal role in disease caused by Sn4, explaining 26% and 25% of the variation in the phenotype, respectively. Interestingly, the *SnToxA-Tsn1* interaction explained 56% and *SnTox2-Snn2* explained only 6% of the variation in disease conferred by Sn5. Sequence analysis of the *SnToxA* locus in the two isolates showed 99.5% identity with no polymorphism in coding regions or the predicted promoter region. The highest expression of *SnToxA* for both isolates was observed at 26 h post inoculation, but the expression of *SnToxA* in Sn5 was more than two times that of Sn4. Therefore, this study revealed that different isolates of *P. nodorum* can express the same NE at different levels and that higher expression leads to a higher association with the corresponding host target gene. Furthermore, since the host genetic background was the same, this study also suggested that variation in *SnToxA* expression is governed by the fungal genetic factors involved in NE gene expression. Similarly, Virdi et al. (2016) inoculated a tetraploid population derived from durum wheat varieties 'Altar84' and 'Langdon' with *P. nodorum* isolate Sn2000 and *P. tritici-repentis* isolate 86-124. The population segregated for *Tsn1* and both isolates produced ToxA. However, the *ToxA-Tsn1* interaction was only significant in the disease caused by *P. nodorum* and not *P. tritici-repentis*. The level of *ToxA* expression was differential among the two pathogens where expression was undetectable in *P. tritici-repentis* (Virdi et al. 2016). Recently, Phan et al. (2016) also showed that transcription of *SnTox3* was elevated significantly in the *P. nodorum* isolates when *SnTox1* was inactivated. Therefore, Phan et al. (2016) speculated that epistatic action of *SnTox1-Snn1* over *SnTox3-Snn3* occurs at the pathogen level through transcriptional regulation. RNA-seq analysis between three *P. nodorum* isolates SN15, Sn4 and Sn5 showed that *SnTox1* was expressed significantly higher in SN15 at 48 h post inoculation suggesting expression of *SnTox1* changed based on the isolate (Peters-Haugrud et al. 2019). Furthermore, Peters-Haugrud et al. (2019) suggested that variation in expression of *SnTox1* among the isolates could be due to multiple polymorphisms in the promoter region of *SnTox1*. These changes could theoretically affect the function of transcription factors, altering the amount of expression of the *NE* gene. Therefore, these studies provide further proof that fungal transcription factors involved in NE gene expression likely play a role in the regulation of NE-sensitivity gene interactions.

One of the major genetic elements that regulate gene expression is transcription factors. IpCho et al. (2010) was the first to show that expression of *SnTox3* was significantly reduced in the deletion mutants of transcription factor *SnStuA*. However, inactivation of *SnStuA* leads to multiple deleterious effects in *P. nodorum*, suggesting the downregulation of *SnTox3* could be a pleiotropic effect rather than a direct regulation of expression. Rybak et al. (2017) showed that the GAL4-like Zn_2Cys_6 transcription factor PnPf2 regulated the expression of *SnTox3* and *SnToxA* in *P. nodorum*. Furthermore, deletion of an orthologue of *PnPf2* abolished the expression of *Ptr ToxA* in *P. tritici-repentis*. Therefore, Rybak et al. (2017) provided the first evidence of a functionally conserved genetic factor involved in regulation of the *ToxA-Tsn1* interaction in two different pathosystems. Recently, Lin et al. (2018) showed that the C2H2 zinc finger transcription factor PnCon7 was involved in direct regulation of *SnTox3*. Deletion of *PnCon7* resulted in the downregulation of *SnTox3*, suggesting that the transcription factor had a positive regulatory role in *SnTox3* expression. Lin et al. (2018) also showed that PnCon7 affected the transcription of *SnToxA* and *SnTox1* in a non-dose-dependent manner, suggesting that it regulated the expression of those two NEs indirectly. PnCon7 was also shown to regulate the expression of other putative effector genes. Collectively, this suggests that similar transcription factors play a major role in the regulation of multiple NE genes involved in NE-sensitivity gene interactions.

4 Importance of NE-sensitivity gene interactions in the field

Friesen et al. (2009) was the first to evaluate the role of NE-sensitivity gene interactions under field conditions. The BG population, which was used to evaluate the *SnToxA-Tsn1*, *SnTox2-Snn2* and *SnTox3-Snn3* interactions in the seedling stage under greenhouse conditions, was evaluated for the role of these interactions in the adult stage under field conditions using a local *P. nodorum* isolate. Trials were conducted in two locations for two years. Results showed that the *SnToxA-Tsn1* and *SnTox2-Snn2* interactions explained 18% and 15% of the variation in disease on the flag leaf, respectively (Friesen et al. 2009). As in the seedling stage, lines harboring both sensitivity genes *Tsn1* and *Snn2* showed significantly higher susceptibility compared to the lines that only harbored one or neither of the two susceptibility genes (Friesen et al. 2007; Friesen et al. 2009). Therefore, even under field condition, these two interactions acted additively (Friesen et al. 2009). In addition, even though the isolate used for inoculation produced SnTox3 and the population segregated for *Snn3*, the *SnTox3-Snn3* interaction did not play a significant role in adult plant susceptibility in the field (Friesen et al. 2009). Therefore, as in the seedling stage, the *SnToxA-Tsn1* interaction may have been epistatic to the *SnTox3-Snn3* interaction in the adult plant stage (Friesen et al. 2009). However, Lu and Lillemo (2014) showed that the *SnTox3-Snn3* interaction played a minor but significant role under some naturally infected field conditions in a bi-parental mapping population derived from the cross SHA3/CBRD×Naxos. Later, Ruud et al. (2017) saturated the genetic map of the same population with the Illumina 90K SNP chip and showed that the *SnTox3-Snn3* interaction explained 24% of the variation in disease under field conditions. This was the first report of the importance of the *SnTox3-Snn3* interaction under field conditions. In addition, it is important to note that none of the other known NE-sensitivity gene interactions were segregating in this population (Lu and Lillemo 2014; Ruud et al. 2017).

Phan et al. (2016) evaluated a double haploid population derived from 'Calingiri' and 'Wyalkatchem' for adult plant resistance for the disease caused by *P. nodorum* isolate SN15. In this population, the *SnTox1-Snn1* interaction accounted for 19% of the variation in disease under adult plant field conditions. This was the first report of the involvement of the *SnTox1-Snn1*interaction in adult plant susceptibility. In another study, Ruud et al. (2018) evaluated the adult plant sensitivity to SnToxA, SnTox1 and SnTox3 in 157 spring wheat genotypes and their reaction to natural infection of *P. nodorum* under field conditions. Trials were conducted for seven years in one location. Ruud et al. (2018) found that lines insensitive to SnToxA were highly resistant to *P. nodorum* compared to others regardless of the sensitivity to SnTox3. In this study, Ruud et al. (2018) found no significant association between sensitivity to SnTox3 and disease.

Variation in disease explained by the NE-sensitivity gene interactions in adult plants in the field was substantially lower than that of plants at the seedling stage (Friesen et al. 2009). This lower significance is likely due to the uncontrolled environmental factors that result in the inability to account for the true effects of these and other interactions (Friesen et al. 2009). Additionally, some of these studies were based on natural infection; therefore, the significance of the NE-sensitivity gene interaction is contingent on the presence of NE producing genes in the local population (Ruud et al. 2017). These results suggested that SNB in the field is at least partially based on NE-susceptibility gene interactions. Removing these interactions can disrupt the pathogen's life cycle and lead to genetic host resistance (Friesen and Faris 2010). Therefore, it is important to characterize more such NE-sensitivity gene interactions in the wheat-*P. nodorum* pathosystem to breed complete resistance against *P. nodorum*.

5 Additional QTL associated with susceptibility/resistance to *P. nodorum*

In addition to known host sensitivity gene loci, many other loci associated with SNB have been reported across 15 wheat chromosomes. These include, chromosomes 1A (Ruud et al. 2017), 1B (Friesen et al. 2007; Francki et al. 2011), 2A (Francki et al. 2011; Phan et al. 2016; Lin et al. 2020), 2B (Czembor et al. 2003), 2D (Aguilar et al. 2005; Shankar et al. 2008; Phan et al. 2016; Ruud et al. 2017; Lin et al. 2020), 3A (Phan et al. 2018), 3B (Czembor et al. 2003; Lu and Lillemo 2014), 4B (Liu et al. 2004b, Aguilar et al. 2005; Phan et al. 2018), 5A (Friesen et al. 2007, 2009; Shankar et al. 2008; Phan et al. 2018); 5B (Czembor et al. 2003; Shankar et al. 2008; Francki et al. 2011; Phan et al. 2018; Lin et al. 2020), 5D (Czembor et al. 2003), 6A (Arseniuk et al. 2004; Lin et al. 2020); 6B (Phan et al. 2016), 7A (Abeysekara et al. 2012; Phan et al. 2018), 7B (Aguilar et al. 2005; Ruud et al. 2017) and 7D (Phan et al. 2018; Lin et al. 2020). These QTL may represent additional host susceptibility genes for NEs that are yet to be identified or they may represent genes that are not targets of NEs (Friesen et al. 2007), but rather are involved in other forms of resistance/susceptibility. Further research is required to characterize the genes underlying each of these loci to fully understand this and other necrotrophic interactions.

6 Impact of genome sequencing in characterizing NE-sensitivity gene interactions

Advancement of genome sequencing has paved the way to characterize many traits at the molecular level. Genome sequences of both *P. nodorum* and wheat have significantly improved the potential of cloning NE genes and host susceptibility genes.

The reference quality genome for *P. nodorum* was first developed and published for the isolate SN15 and consisted of 38 Mb assembled into 107 nuclear scaffolds (Hane et al. 2007). Genome assembly consisted of at least 10 762 genes and the annotation was used to identify and clone *SnToxA*, *SnTox1* and *SnTox3* (Friesen et al. 2006; Liu et al. 2012; Liu et al. 2009). However, different *P. nodorum* isolates produce different combinations of NEs and the cloned effector genes have been shown to exhibit presence-absence variation among isolates (Stukenbrock and McDonald 2007; Liu et al. 2009; McDonald et al. 2013; Richards et al. 2019). Therefore, more isolates needed to be sequenced to identify novel effector-producing genes. Syme et al. (2013) re-sequenced *P. nodorum* isolate Sn4, which seemed to produce a different set of effectors as compared to SN15 and the avirulent isolate Sn79-1087. These genome sequences were developed using Illumina sequencing technology. Sn4 and Sn79-1087 consisted of 14 391 and 14 352 predicted protein coding genes, respectively. Syme et al. (2013) screened these gene models using criteria based on the characteristics of known NEs and other effectors. These criteria included size of the encoded protein, cysteine content, proximity to repeat content and the presence of a signal peptide. Resulting candidates were further screened based on their presence/absence on avirulent isolate Sn79-1087 and the evolution rate, and 159 gene models were identified as effector gene candidates. Four candidate effector-encoding genes that were unique to Sn4 were also identified. Later, Syme et al. (2016) significantly improved the SN15 reference assembly by correcting base calling errors, closing assembly gaps and improving gene annotation and functional characteristics of the gene models that were supported by new RNA sequencing data. New annotation identified new genes, some of which fit the effector gene criteria.

Recently, Richards et al. (2018) released reference quality-annotated assemblies for *P. nodorum* isolates Sn4, Sn2000 and Sn79-1087 using SMRT long read sequencing technology, resulting in telomere-to-telomere assemblies for nearly all chromosomes of each isolate. Genome assemblies of Sn4, Sn2000 and Sn79-1087 spanned approximately 37.7 Mb, 37.5 Mb and 34.9 Mb, respectively. Both Sn4 and Sn2000 consisted of 23 chromosomes compared to Sn79-1087, which consisted of only 22. Each assembly contained more than 13 000 annotated genes and the Sn4, Sn2000 and Sn79-1087 annotations contained 287, 281 and 237 genes that encoded for predicted effector genes, respectively (Richards et al. 2018).

Even with the availability of new high-quality *P. nodorum* genome assemblies, no new effector genes have been reported in close to a decade. Genome-wide association mapping (GWAS) is a powerful tool to identify genomic regions associated with traits in natural populations (Gibson and Muse 2009). Since sexual populations of *P. nodorum* are difficult to produce under laboratory conditions and variation in cloned NE-producing genes in natural population have been observed, GWAS is a great new approach to identify high-priority candidate genes encoding effectors (Gao et al. 2016). To validate GWAS as a tool to identify such genes, a global collection of 191 *P. nodorum* isolates which were used to identify 2983 SNP markers were phenotyped on the wheat cultivar 'Sumai3' (contains *Tsn1* and *Snn3*) and 'Alsen' (*Tsn1*). As expected, strong marker trait associations (MTAs) were observed between the *SnToxA* gene and the disease phenotype for inoculation on both lines. An MTA was also observed between disease on Sumai3 and *SnTox3* (Gao et al. 2016). However, higher marker saturation was shown to be necessary. In a subsequent study, Richards et al. (2019) further validated the MTA associated with *SnToxA* and *SnTox3* by using 322 613 SNPs/InDels and phenotypic data generated by inoculating 197 *P. nodorum* isolates on four different differential lines including 'Jerry' (*Tsn1*), 'TAM 105' (*Tsn1*), 'Massey' (*Snn3*), and 'F/G95195' (*Snn3*).

Therefore, availability of multiple reference genome sequences and a large marker set suitable for GWAS provides the greatest opportunity to identify and clone novel NE encoding genes. Our team has already cloned *SnTox2* and *SnTox5* using these tools, and further functional characterization of the genes is currently underway (Richards et al. unpublished; Kariyawasam et al. unpublished).

The completion of the wheat genome sequence and publication of the wheat genome annotation has helped to make sequence analysis, marker development and subsequent candidate gene identification in wheat less labor-intensive and time-consuming (Appels et al. 2018). The wheat genome consists of 16 billion base pairs that account for 107 891 high confident gene models. Therefore, cloning of wheat resistance/susceptibility genes is difficult and cloning of the NE genes can facilitate the process of identifying these genes. The cloning of susceptibility genes provides a valuable tool for removing these genes from breeding material. Removal of these susceptibility genes will result in additional resistance against SNB. Currently, our wheat genetics lab has been conducting intense wheat gene cloning projects to clone susceptibility genes that confer sensitivity to NEs secreted by *P. nodorum* (Pers. Comm. Justin D. Faris). With the advancement of genome sequencing and bioinformatic capabilities, molecular characterization of the NE-sensitivity gene interactions will continue to move faster in the coming years. Having more characterized genes and proteins in both *P. nodorum* and wheat will expand our understanding of this plant-pathogen interaction, allowing us to provide solutions to this and other necrotrophic diseases.

7 Where to look for further information

In the last two decades, major advances have been made in understanding how *P. nodorum* infects wheat and how wheat can defend itself against this necrotrophic pathogen. Future research will provide a greater understanding of how necrotrophic plant pathogens exploit their host. This research will involve the functional characterization of pathogen-produced effectors as well as the host targets for these effectors. Technological improvements in genomics, transcriptomics, proteomics, and metabolomics will increase the speed of discovery of the various components contributing to the *P. nodorum*-wheat interaction.

Several groups in the United states, Europe, and Australia are working on the *P. nodorum* - wheat patho-system. Relevant reviews describing the *P. nodorum* - wheat interaction include Friesen and Faris (2021, Annual Review of Phytopathology submitted) Faris and Friesen (2020), Oliver et al. (2012),

Friesen and Faris (2010), Friesen et al. (2008a), and Solomon et al. (2006). Additional information on this pathogen can be found at the National Center for Biotechnology Information (NCBI) (www.ncbi.nlm.nih.gov) using Bio Projects PRJNA686477, PRJNA13754 and PRJNA21049, which provides information on the original reference genome sequence of SN15 (Hane et al. 2007; Syme et al. 2016) as well as Bio Projects PRJNA398070, which provides reference quality genome sequences of three additional *P. nodorum* isolates (Richards et al. 2018). Resequencing of more than 200 *P. nodorum* isolates can be found under Bio Projects PRJNA476481 (Syme et al. 2018) and PRJNA398070 (Richards et al. 2019).

8 References

Abeysekara, N. S., Friesen, T. L., Keller, B. and Faris, J. D. (2009). Identification and characterization of a novel host-toxin interaction in the wheat-*Stagonospora nodorum* pathosystem, *Theor. Appl. Genet.* 120(1), 117–126.

Abeysekara, N. S., Faris, J. D., Chao, S., McClean, P. E. and Friesen, T. L. (2012). Whole-genome analysis of Stagonospora nodorum blotch resistance and validation of the SnTox4-*Snn4* interaction in hexaploid wheat, *Phytopathology* 102(1), 94–104.

Adhikari, T. B., Bai, J., Meinhardt, S. W., Gurung, S., Myrfield, M., Patel, J., Ali, S., Gudmestad, N. C. and Rasmussen, J. B. (2009). Tsn1-mediated host responses to ToxA from *Pyrenophora tritici-repentis*, *Mol. Plant Microbe Interact.* 22(9), 1056–1068.

Aguilar, V., Stamp, P., Winzeler, M., Winzeler, H., Schachermayr, G., Keller, B., Zanetti, S. and Messmer, M. M. (2005). Inheritance of field resistance to *Stagonospora nodorum* leaf and glume blotch and correlations with other morphological traits in hexaploid wheat (*Triticum aestivum* L.), *Theor. Appl. Genet.* 111(2), 325–336.

Andrew, M., Barua, R., Short, S. M. and Kohn, L. M. (2012). Evidence for a common toolbox based on necrotrophy in a fungal lineage spanning necrotrophs, biotrophs, endophytes, host generalists and specialists, *PLoS ONE* 7(1), e29943.

Appels, R., Eversole, K., Feuillet, C., Keller, B., Rogers, J., Stein, N., Pozniak, C. J., Choulet, F., Distelfeld, A., Poland, J. and Ronen, G. (2018). Shifting the limits in wheat research and breeding using a fully annotated reference genome, *Science* 361(6403), eaar7191.

Arseniuk, E., Czembor, P. C., Czaplicki, A., Song, Q., Cregan, P. B., Hoffman, D. L. and Ueng, P. P. (2004). QTL controlling partial resistance to *Stagonospora nodorum* leaf blotch in winter wheat cultivar Alba, *Euphytica* 137(2), 225–231.

Balance, G. M., Lamari, L. and Bernier, C. C. (1989). 'Purification and characterization of a host selective toxin from *Pyrenophora tritici-repentis*, *PMPP* 35(3), 203–213.

Balance, G. M., Lamari, L., Kowatsch, R. and Bernier, C. C. (1996). Cloning, expression and occurrence of the gene encoding the Ptr necrosis toxin from *Pyrenophora tritici-repentis*, *Mol. Plant Pathol.*, 1209.

Bergstrom, G. C. (2010). Stagonospora nodorum blotch and Stagonospora avenae blotch. In: *Compendium of Wheat Diseases and Pests* Bockus, W. W., Bowden, R. L., Hunger, R. M., Morrill, W. L., Murray, T. D. and Smiley, R. W. (Eds). American Phytopathological Society, St. Paul, MN, pp. 75–77.

Breen, S., Williams, S. J., Winterberg, B., Kobe, B. and Solomon, P. S. (2016). Wheat PR-1 proteins are targeted by necrotrophic pathogen effector proteins, *Plant J.* 88(1), 13–25.

Bousquet, J.-F. and Skajennikoff, M. (1974). Isolement et mode d'action d'une phytotoxine produite en culture par Septoria nodorum Berk, *Phytopathology. Z.* 80, 355–360.

Bousquet, J. F., de Franqueville, H. B., Kollmann, A. and Fritz, R. (1980). Action of septorin, a phytotoxin synthesized by Septoria nodorum, on oxidative phosphorylation in mitochondria isolated from wheat coleoptiles, *Can. J. Bot.* 58(24), 2575–2580.

Brenchley, R., Spannagl, M., Pfeifer, M., Barker, G. L., D'Amore, R., Allen, A. M., McKenzie, N., Kramer, M., Kerhornou, A., Bolser, D., Kay, S., Waite, D., Trick, M., Bancroft, I., Gu, Y., Huo, N., Luo, M. C., Sehgal, S., Gill, B., Kianian, S., Anderson, O., Kersey, P., Dvorak, J., McCombie, W. R., Hall, A., Mayer, K. F., Edwards, K. J., Bevan, M. W. and Hall, N. (2012). Analysis of the bread wheat genome using whole-genome shotgun sequencing, *Nature* 491(7426), 705–710.

Carlile, A. J., Bindschedler, L. V., Bailey, A. M., Bowyer, P., Clarkson, J. M. and Cooper, R. M. (2000). Characterization of SNP1, a cell wall-degrading trypsin, produced during infection by *Stagonospora nodorum*, *Mol. Plant Microbe Interact.* 13(5), 538–550.

Chu, C. G., Faris, J. D., Xu, S. S. and Friesen, T. L. (2010). Genetic analysis of disease susceptibility contributed by the compatible *Tsn1*-SnToxA and *Snn1*-SnTox1 interactions in the wheat-*Stagonospora nodorum* pathosystem, *Theor. Appl. Genet.* 120(7), 1451–1459.

Ciuffetti, L. M., Tuori, R. P. and Gaventa, J. M. (1997). A single gene encodes a selective toxin causal to the development of tan spot of wheat, *Plant Cell* 9(2), 135–144.

Ciuffetti, L. M., Francl, L. J., Ballance, G. M., Bockus, W. W., Lamari, L., Meinhardt, S. W. and Rasmussen, J. B. (1998). Standardization of toxin nomenclature in the *Pyrenophora tritici-repentis*/wheat interaction, *Can. J. Plant Pathol.* 20(4), 421–424.

Cockram, J., Scuderi, A., Barber, T., Furuki, E., Gardner, K. A., Gosman, N., Kowalczyk, R., Phan, H. P., Rose, G. A., Tan, K. C., Oliver, R. P. and Mackay, I. J. (2015). Fine-mapping the wheat Snn1 locus conferring sensitivity to the *Parastagonospora nodorum* necrotrophic effector SnTox1 using an eight founder multiparent advanced generation inter-cross population, *G3 (Bethesda)* 5(11), 2257–2266.

Cook, D. E., Mesarich, C. H. and Thomma, B. P. H. J. (2015). Understanding plant immunity as a surveillance system to detect invasion, *Annu. Rev. Phytopathol.* 53, 541–563.

Czembor, P. C., Arseniuk, E., Czaplicki, A., Song, Q., Cregan, P. B. and Ueng, P. P. (2003). QTL mapping of partial resistance in winter wheat to *Stagonospora nodorum* blotch, *Genome* 46(4), 546–554.

Devys, M., Bousquet, J. F., Kollmann, A. and Barbier, M. (1978). La septorine, nouvelle pyrazine substituée, isolée du milieu de culture de Septoria nodorum Berk., champignon phytopathogène, *C.R. Acad. Sci.* 286, 457–458.

Devys, M., Barbier, M., Bousquet, J. F. and Kollmann, A. (1994). Isolation of the (−)-(3R)-5-hydroxymellein from the fungus *Septoria nodorum*, *Phytochemistry* 35(3), 825–826.

Dodds, P. N. and Rathjen, J. P. (2010). Plant immunity: towards an integrated view of plant-pathogen interactions, *Nat. Rev. Genet.* 11(8), 539–548.

Eyal, Z., Scharen, A. L., Prescott, J. M. and van Ginkel, M. (1987). *The Septoria Diseases of Wheat: Concepts and Methods of Disease Management*, CIMMYT, Mexico.

Faris, J. D. and Friesen, T. L. (2009). Reevaluation of a tetraploid wheat population indicates that the *Tsn1*-ToxA interaction is the only factor governing Stagonospora nodorum blotch susceptibility, *Phytopathology* 99(8), 906-912.

Faris, J. D., Anderson, J. A., Francl, L. J. and Jordahl, J. G. (1996). Chromosomal location of a gene conditioning insensitivity in wheat to a necrosis-inducing culture filtrate from *Pyrenophora tritici-repentis*, *Phytopathology* 86(5), 459–463.

Faris, J. D. and Friesen, T. L. (2020). Plant genes hijacked by necrotrophic fungal pathogens. *Curr. Opin. Plant Biol.* 56, 74–80.

Faris, J. D., Zhang, Z., Lu, H., Lu, Z., Reddy, L., Cloutier, S., Fellers, J. P., Meinhardt, S. W., Rasmussen, J. B., Xu, S. S., Oliver, R. P., Simons, K. J. and Friesen, T. L. (2010). A unique wheat disease resistance-like gene governs effector-triggered susceptibility to necrotrophic pathogens, *Proc. Natl. Acad. Sci.* 107(30), 13544–13549.

Faris, J. D., Zhang, Z., Rasmussen, J. B. and Friesen, T. L. (2011). Variable expression of the *Stagonospora nodorum* effector SnToxA among isolates correlated with levels of disease in wheat, *Mol. Plant Microbe Interact.* 24(12), 1419–1426.

Faris, J. D., Liu, Z. and Xu, S. S. (2013). Genetics of tan spot resistance in wheat, *Theor. Appl. Genet.* 126(9), 2197–2217.

Figueroa, M., Hammond-Kosack, K. E. and Solomon, P. S. (2018). A review of wheat diseases–a field perspective, *Mol. Plant Pathol.* 19(6), 1523–1536.

Flor, H. H. (1956). Complementary genetic systems in flax and flax rust, *Adv. Genet.* 8, 29–54.

Francki, M. G., Shankar, M., Walker, E., Loughman, R., Golzar, H. and Ohm, H. (2011). New quantitative trait loci in wheat for flag leaf resistance to Stagonospora nodorum blotch, *Phytopathology* 101(11), 1278–1284. doi:10.1094/PHYTO-02-11-0054.

Friesen, T. L., Stukenbrock, E. H., Liu, Z., Meinhardt, S., Ling, H., Faris, J. D., Rasmussen, J. B., Solomon, P. S., McDonald, B. A. and Oliver, R. P. (2006). Emergence of a new disease as a result of interspecific virulence gene transfer, *Nat. Genet.* 38(8), 953–956.

Friesen, T. L., Meinhardt, S. W. and Faris, J. D. (2007). The *Stagonospora nodorum*-wheat pathosystem involves multiple proteinaceous host-selective toxins and corresponding host sensitivity genes that interact in an inverse gene-for-gene manner, *Plant J.* 51(4), 681–692.

Friesen, T. L., Faris, J. D., Solomon, P. S. and Oliver, R. P. (2008a). Host specific toxins: effectors of necrotrophic pathogenicity. *Cell Micro* 10, 1421–1428.

Friesen, T. L., Zhang, Z., Solomon, P. S., Oliver, R. P. and Faris, J. D. (2008b). Characterization of the interaction of a novel *Stagonospora nodorum* host-selective toxin with a wheat susceptibility gene, *Plant Physiol.* 146(2), 682–693.

Friesen, T. L., Chu, C. G., Liu, Z. H., Xu, S. S., Halley, S. and Faris, J. D. (2009). Host-selective toxins produced by *Stagonospora nodorum* confer disease susceptibility in adult wheat plants under field conditions, *Theor. Appl. Genet.* 118(8), 1489–1497.

Friesen, T. L. and Faris, J. D. (2010). Characterization of the wheat-*Stagonospora nodorum* system: what is the molecular basis of this quantitative necrotrophic disease interaction?, *Can. J.Plant Pathol.* 32(1), 20–28.

Friesen, T. L., Chu, C., Xu, S. S. and Faris, J. D. (2012). SnTox5-*Snn5*: a novel *Stagonospora nodorum* effector-wheat gene interaction and its relationship with the SnToxA-*Tsn1* and SnTox3-*Snn3-B1* interactions, *Mol. Plant Pathol.* 13(9), 1101–1109.

Friesen, T. L., Holmes, D. J., Bowden, R. L. and Faris, J. D. (2018). ToxA is present in the US Bipolaris sorokiniana population and is a significant virulence factor on wheat harboring Tsn1, *Plant Dis.* 102(12), 2446–2452.

Gao, Y., Faris, J. D., Liu, Z., Kim, Y. M., Syme, R. A., Oliver, R. P., Xu, S. S. and Friesen, T. L. (2015). Identification and characterization of the SnTox6-*Snn6* interaction in the

Parastagonospora nodorum-wheat pathosystem, *Mol. Plant Microbe Interact.* 28(5), 615–625.

Gao, Y., Liu, Z., Faris, J. D., Richards, J., Brueggeman, R. S., Li, X., Oliver, R. P., McDonald, B. A. and Friesen, T. L. (2016). Validation of genome-wide association studies as a tool to identify virulence factors in *Parastagonospora nodorum*, *Phytopathology* 106(10), 1177–1185.

Ghaderi, F., Sharifnabi, B., Javan-Nikkhah, M., Brunner, P. and McDonald, B. A. (2020). SnToxA, SnTox1 and SnTox3 originated in *Parastagonospora nodorum* in the Fertile Crescent, Plant Pathol. 69(8), 1482–1491.

Gibson, G. and Muse, S. V. (2009). *A Primer of Genome Science*, Sinauer Associates, Sunderland, MA.

Haen, K. M., Lu, H., Friesen, T. L. and Faris, J. D. (2004). Genomic targeting and high-resolution mapping of the Tsn1 gene in wheat, *Crop Sci.* 44(3), 951–962.

Hane, J. K., Lowe, R. G. T., Solomon, P. S., Tan, K. C., Schoch, C. L., Spatafora, J. W., Crous, P. W., Kodira, C., Birren, B. W., Galagan, J. E., Torriani, S. F. F., McDonald, B. A. and Oliver, R. P. (2007). Dothideomycete-plant interactions illuminated by genome sequencing and EST analysis of the wheat pathogen *Stagonospora nodorum*, *Plant Cell* 19(11), 3347–3368.

IpCho, S. V., Tan, K. C., Koh, G., Gummer, J., Oliver, R. P., Trengove, R. D. and Solomon, P. S. (2010). The transcription factor StuA regulates central carbon metabolism, mycotoxin production, and effector gene expression in the wheat pathogen Stagonospora nodorum, *Eukaryot. Cell* 9(7), 1100–1108.

Jones, J. D. G. and Dangl, J. L. (2006). The plant immune system, *Nature* 444(7117), 323–329.

Kariyawasam, G. K., Carter, A. H., Rasmussen, J. B., Faris, J. D., Xu, S. S., Mergoum, M. and Liu, Z. H. (2016). Genetic relationships between race-nonspecific and race-specific interactions in the wheat- *Pyrenophora tritici-repentis* pathosystem, *Theor. Appl. Genet.* 129(5), 897–908.

Keller, B., Winzeler, H., Winzeler, M. and Fried, P. M. (1994). Differential sensitivity of wheat embryos against extracts containing toxins of Septoria nodorum: first steps towards in vitro selection, *J. Phytopathol.* 141(3), 233–240.

Lalaoui, F., Halama, P., Dumortier, V. and Paul, B. (2000). Cell wall-degrading enzymes produced in vitro by isolates of *Phaeosphaeria nodorum* differing in aggressiveness, *Plant Pathol.* 49(6), 727–733.

Langridge, P., Lagudah, E. S., Holton, T. A., Appels, R., Sharp, P. J. and Chalmers, K. J. (2001). Trends in genetic and genome analyses in wheat: a review, *Aust. J. Agric. Res.* 52(12), 1043–1077.

Lehtinen, U. (1993). Plant cell wall degrading enzymes of Septoria nodorum, Physiol. Mol. Plant P. 43(2), 121–134.

Lin, S. Y., Chooi, Y. H. and Solomon, P. S. (2018). The global regulator of pathogenesis PnCon7 positively regulates Tox3 effector gene expression through direct interaction in the wheat pathogen *Parastagonospora nodorum*, *Mol. Microbiol.* 109(1), 78–90.

Lin, M., Corsi, B., Ficke, A., Tan, K. C., Cockram, J. and Lillemo, M. (2020). Genetic mapping using a wheat multi-founder population reveals a locus on chromosome 2a controlling resistance to both leaf and glume blotch caused by the necrotrophic fungal pathogen *Parastagonospora nodorum*, *Theor. Appl. Genet.* 133(3), 785–808.

Liu, Z. H., Faris, J. D., Meinhardt, S. W., Ali, S., Rasmussen, J. B. and Friesen, T. L. (2004a). Genetic and physical mapping of a gene conditioning sensitivity in wheat to a partially purified host-selective toxin produced by *Stagonospora nodorum*, *Phytopathology* 94(10), 1056–1060.

Liu, Z. H., Friesen, T. L., Rasmussen, J. B., Ali, S., Meinhardt, S. W. and Faris, J. D. (2004b). Quantitative trait loci analysis and mapping of seedling resistance to *Stagonospora nodorum* leaf blotch in wheat, *Phytopathology* 94(10), 1061–1067.

Liu, Z., Friesen, T. L., Ling, H., Meinhardt, S. W., Oliver, R. P., Rasmussen, J. B. and Faris, J. D. (2006). The *Tsn1*-ToxA interaction in the wheat-*Stagonospora nodorum* pathosystem parallels that of the wheat-tan spot system, *Genome* 49(10), 1265–1273.

Liu, Z., Faris, J. D., Oliver, R. P., Tan, K. C., Solomon, P. S., McDonald, M. C., McDonald, B. A., Nunez, A., Lu, S., Rasmussen, J. B. and Friesen, T. L. (2009). *SnTox3* acts in effector triggered susceptibility to induce disease on wheat carrying the Snn3 gene, *PLoS Pathog.* 5(9), e1000581.

Liu, Z., Zhang, Z., Faris, J. D., Oliver, R. P., Syme, R., McDonald, M. C., McDonald, B. A., Solomon, P. S., Lu, S., Shelver, W. L., Xu, S. and Friesen, T. L. (2012). The cysteine rich necrotrophic effector SnTox1 produced by *Stagonospora nodorum* triggers susceptibility of wheat lines harboring *Snn1*, *PLoS Pathog.* 8(1), e1002467.

Liu, Z., Gao, Y., Kim, Y. M., Faris, J. D., Shelver, W. L., de Wit, P. J. G. M., Xu, S. S. and Friesen, T. L. (2016). SnTox1, a *Parastagonospora nodorum* necrotrophic effector, is a dual-function protein that facilitates infection while protecting from wheat-produced chitinases, *New Phytol.* 211(3), 1052–1064, doi:10.1111/nph.13959.

Liu, Z., Zurn, J. D., Kariyawasam, G., Faris, J. D., Shi, G., Hansen, J., Rasmussen, J. B. and Acevedo, M. (2017). Inverse gene-for-gene interactions contribute additively to tan spot susceptibility in wheat, *Theor. Appl. Genet.* 130(6), 1267–1276.

Lorang, J. M., Sweat, T. A. and Wolpert, T. J. (2007). Plant disease susceptibility conferred by a "resistance" gene, *Proc. Natl. Acad. Sci.* 104(37), 14861–14866.

Lu, Q. and Lillemo, M. (2014). Molecular mapping of adult plant resistance to Parastagonospora nodorum leaf blotch in bread wheat lines 'Shanghai-3/Catbird' and 'Naxos', *Theor. Appl. Genet.* 127(12), 2635–2644.

Lu, H. J., Fellers, J. P., Friesen, T. L., Meinhardt, S. W. and Faris, J. D. (2006). Genomic analysis and marker development for the Tsn1 locus in wheat using bin-mapped ESTs and flanking BAC contigs, *Theor. Appl. Genet.* 112(6), 1132–1142.

Lu, S., Faris, J. D., Sherwood, R., Friesen, T. L. and Edwards, M. C. (2014). A dimeric PR-1-type pathogenesis-related protein interacts with ToxA and potentially mediates ToxA-induced necrosis in sensitive wheat, *Mol. Plant Pathol.* 15(7), 650–663.

Manning, V. A. and Ciuffetti, L. M. (2005). Localization of Ptr ToxA produced by *Pyrenophora tritici-repentis* reveals protein import into wheat mesophyll cells, *Plant Cell* 17(11), 3203–3212.

Manning, V. A., Hardison, L. K. and Ciufetti, L. M. (2007). Ptr ToxA interacts with a chloroplast-localized protein, *Mol. Plant Microbe Interact.* 20(2), 168–177.

Manning, V. A., Chu, A. L., Steeves, J. E., Wolpert, T. J. and Ciufetti, L. M. (2009). A host-selective toxin of *Pyrenophora* tritici-repentis, Ptr ToxA, induces photosystem changes and reactive oxygen species accumulation in sensitive wheat, *Mol. Plant Microbe Interact.* 22(6), 665–676.

Magro, P. (1984). Production of polysaccharide-degrading enzymes by Septoria nodorum in culture and during pathogenesis' ,*Plant Sci. Lett.* 37(1-2), 63–68.

McDonald, M. C., Oliver, R. P., Friesen, T. L., Brunner, P. C. and McDonald, B. A. (2013). Global diversity and distribution of three necrotrophic effectors in *Phaeosphaeria nodorum* and related species, *New Phytol.* 199(1), 241-251.

McDonald, M. C., Ahren, D., Simpfendorfer, S., Milgate, A. and Solomon, P. S. (2018). The discovery of the virulence gene *ToxA* in the wheat and barley pathogen *Bipolaris sorokiniana*, *Mol. Plant Pathol.* 19(2), 432–439.

McDonald, M. C., Taranto, A. P., Hill, E., Schwessinger, B., Liu, Z., Simpfendorfer, S., Milgate, A. and Solomon, P. S. (2019). Transposon-mediated horizontal transfer of the host-specific virulence protein ToxA between three fungal wheat pathogens, *mBio* 10(5), e01515-19.

Mehra, L. K., Adhikari, U., Ojiambo, P. S. and Cowger, C. (2019). Septoria nodorum blotch of wheat, *Plant Health Instr.* doi:10.1094/PHI-I-2019-0514-01.

Meinhardt, S. W., Cheng, W., Kwon, C. Y., Donohue, C. M. and Rasmussen, J. B. (2002). Role of the arginyl-glycyl-aspartic motif in the action of Ptr ToxA produced by Pyrenophora tritici-repentis, *Plant Physiol.* 130(3), 1545–1551.

Murray, G. M. and Brennan, J. P. (2009). Estimating disease losses to the Australian wheat industry, *Austral. Plant Pathol.* 38(6), 558–570.

Oliver, R. P., Friesen, T. L., Faris, J. D. and Solomon, P. S. (2012). *Stagonospora nodorum*: from pathology to genomics and host resistance, *Annu. Rev. Phytopathol.* 50, 23–43.

Outram, M. A., Sung, Y. C., Yu, D., Dagvadorj, B., Rima, S., Jones, D., Ericsson, D. J., Sperschneider, J., Solomon, P., Kobe, B. and Williams, S. J. (2020). The crystal structure of SnTox3 from the necrotrophic fungus Parastagonospora nodorum reveals a unique effector fold and insights into Kex2 protease processing of fungal effectors, *bioRxiv*, doi:10.1101/2020.05.27.120113.

Pandelova, I., Betts, M. F., Manning, V. A., Wilhelm, L. J., Mockler, T. C. and Ciuffetti, L. M. (2009). Analysis of transcriptome changes induced by Ptr ToxA in wheat provides insights into the mechanisms of plant susceptibility, *Mol. Plant* 2(5), 1067–1083.

Peters-Haugrud, A. R., Zhang, Z., Richards, J. K., Friesen, T. L. and Faris, J. D. (2019). Genetics of variable disease expression conferred by inverse gene-for-gene interactions in the wheat-*Parastagonospora nodorum* pathosystem, *Plant Physiol.* 180(1), 420–434.

Phan, H. T. T., Rybak, K., Furuki, E., Breen, S., Solomon, P. S., Oliver, R. P. and Tan, K. C. (2016). Differential effector gene expression underpins epistasis in a plant fungal disease, *Plant J.* 87(4), 343–354.

Phan, H. T. T., Rybak, K., Bertazzoni, S., Furuki, E., Dinglasan, E., Hickey, L. T., Oliver, R. P. and Tan, K. C. (2018). Novel sources of resistance to Septoria nodorum blotch in the Vavilov wheat collection identified by genome-wide association studies, *Theor. Appl. Genet.* 131(6), 1223–1238.

Quaedvlieg, W., Verkley, G. J. M., Shin, H. D., Barreto, R. W., Alfenas, A. C., Swart, W. J., Groenewald, J. Z. and Crous, P. W. (2013). Sizing up Septoria, *Stud. Mycol.* 75(1), 307–390.

Reddy, L., Friesen, T. L., Meinhardt, S. W., Chao, S. and Faris, J. D. (2008). Genomic analysis of the *Snn1* locus on wheat chromosome arm 1BS and the identification of candidate genes, *Plant Genome* 1(1), 55–66.

Richards, J. K., Wyatt, N. A., Liu, Z., Faris, J. D. and Friesen, T. L. (2018). 'Reference quality genome assemblies of three *Parastagonospora nodorum* isolates differing in virulence on wheat, *G3-Genes Genom. Genet.* 8(2), 393–399.

Richards, J. K., Stukenbrock, E. H., Carpenter, J., Liu, Z., Cowger, C., Faris, J. D. and Friesen, T. L. (2019). Local adaptation drives the diversification of effectors in the fungal wheat pathogen *Parastagonospora nodorum* in the United States, *PLoS Genet.* 15(10), e1008223.

Ruud, A. K., Windju, S., Belova, T., Friesen, T. L. and Lillemo, M. (2017). Mapping of SnTox3-*Snn3* as a major determinant of field susceptibility to Septoria nodorum blotch in the SHA3/CBRD × Navos population, *Theor. Appl. Genet.* 130(7), 1361–1374, doi:10.1007/s00122-017-2893-5.

Ruud, A. K., Dieseth, J. A. and Lillemo, M. (2018). Effects of three *Parastagonospora nodorum* necrotrophic effectors on spring wheat under Norwegian field conditions, *Crop Sci.* 58(1), 159–168.

Rybak, K., See, P. T., Phan, H. T. T., Syme, R. A., Moffat, C. S., Oliver, R. P. and Tan, K. C. (2017). A functionally conserved Zn_2Cys_6 binuclear cluster transcription factor class regulates necrotrophic effector gene expression and host-specific virulence of two major Pleopsporales fungal pathogens of wheat, *Mol. Plant Pathol.* 18(3), 420–434.

Sarma, G. N., Manning, V. A., Ciuffetti, L. M. and Karplus, P. A. (2005). Structure of Ptr ToxA: an RGD-containing host-selective toxin from *Pyrenophora tritici-repentis*, *Plant Cell* 17(11), 3190–3202.

Searchinger, T., Waite, R., Hanson, C., Ranganathan, J., Dumas, P., Matthews, E. and Klirs, C. (2019). Creating a sustainable food future: a menu of solutions to feed nearly 10 billion people by 2050. Final report.

Shankar, M., Walker, E., Golzar, H., Loughman, R., Wilson, R. E. and Francki, M. G. (2008). Quantitative trait loci for seedling and adult plant resistance to *Stagonospora nodorum* in Wheat, *Phytopathology* 98(8), 886–893.

Shi, G., Friesen, T. L., Saini, J., Xu, S. S., Rasmussen, J. B. and Faris, J. D. (2015). The wheat *Snn7* gene confers susceptibility on recognition of the *Parastagonospora nodorum* necrotrophic effector SnTox7, *Plant Genome* 8(2), eplantgenome2015.02.0007.

Shi, G., Zhang, Z., Friesen, T. L., Raats, D., Fahima, T., Brueggeman, R. S., Lu, S., Trick, H. N., Liu, Z., Chao, W., Frenkel, Z., Xu, S. S., Rasmussen, J. B. and Faris, J. D. (2016a). The hijacking of a receptor kinase-driven pathway by a wheat fungal pathogen leads to disease, *Sci. Adv.* 2(10), e1600822.

Shi, G., Zhang, Z., Friesen, T. L., Bansal, U., Cloutier, S., Wicker, T., Rasmussen, J. B. and Faris, J. D. (2016b). Marker development, saturation mapping, and high-resolution mapping of the Septoria nodorum blotch susceptibility gene *Snn3-B1* in wheat, *Mol. Genet. Genomics* 291(1), 107–119, doi:10.1007/s00438-015-1091-x.

Singh, R. P., Singh, P. K., Rutkoski, J., Hodson, D. P., He, X., Jørgensen, L. N., Hovmøller, M. S. and Huerta-Espino, J. (2016). Disease impact on wheat yield potential and prospects of genetic control, *Ann. Rev. Phytopathol.* 54, 303–322.

Solomon, P. S., Lowe, R. G. T., Tan, K. C., Waters, O. D. C. and Oliver, R. P. (2006). *Stagonospora nodorum*: cause of Stagonospora nodorum blotch of wheat, *Mol. Plant Pathol.* 7(3), 147–156.

Stukenbrock, E. H. and McDonald, B. A. (2007). Geographical variation and positive diversifying selection in the host-specific toxin SnToxA, *Mol. Plant Pathol.* 8(3), 321–332.

Syme, R. A., Hane, J. K., Friesen, T. L. and Oliver, R. P. (2013). Resequencing and comparative genomics of *Stagonospora nodorum*: sectional gene absence and effector discovery, *G3-Genes Genom. Genet.* 3(6), 959–969.

Syme, R. A., Tan, K. C., Hane, J. K., Dodhia, K., Stoll, T., Hastie, M., Furuki, E., Ellwood, S. R., Williams, A. H., Tan, Y. F., Testa, A. C., Gorman, J. J. and Oliver, R. P. (2016). Comprehensive annotation of the *Parastagonospora nodorum* reference genome using next-generation genomics, transcriptomics and proteogenomics, *PLoS ONE* 11(2), e0147221. doi:10.1371/journal.pone.0147221.

Syme, R. A., Tan, K. C., Rybak, K. Friesen, T. L., McDonald, B. A., Oliver, R. P. and Hane, J. K. (2018). Pan-Parastagonospora comparative genome analysis–effector prediction and genome evolution, *Genome Biol. Evol.* 10(9), 2443-2457.

Tai, Y. S., Bragg, J. and Meinhardt, S. W. (2007). Functional characterization of ToxA and molecular identification of its intracellular targeting protein in wheat, *Am. J. Plant Physiol.* 2(2), 76-89.

Teixeira, P. J. P., Colaianni, N. R., Fitzpatrick, C. R. and Dangl, J. L. (2019). Beyond pathogens: microbiota interactions with the plant immune system, *Curr. Opin. Microbiol.* 49, 7-17.

Tomás, A. and Bockus, W. W. (1987). Cultivar specific toxicity of culture filtrate of *Pyrenophora tritici-repentis*, *Phytopathology* 77(9), 1337-1366.

Tuori, R. P., Wolpert, T. J. and Ciuffetti, L. M. (2000). Heterologous expression of functional Ptr ToxA, *Mol. Plant Microbe Interact.* 13(4), 456-464.

van Kan, J. A. (2006). Licensed to kill: the lifestyle of a necrotrophic plant pathogen, *Trends Plant Sci.* 11(5), 247-253.

Virdi, S. K., Liu, Z., Overlander, M. E., Zhang, Z., Xu, S. S., Friesen, T. L., and Faris, J. D. (2016). New insights into the roles of host gene-necrotrophic effector interactions in governing susceptibility of durum wheat to tan spot and spetoria nodorum blotch, *G3-Genes Genet. Genom.*, 6(12), 4139-4150.

Wicki, W., Messmer, M., Winzeler, M., Stamp, P. and Schmid, J. E. (1999). In vitro screening for resistance against Septoria nodorum blotch in wheat, *Theor. Appl. Genet.* 99(7-8), 1273-1280.

Winterberg, B., Du Fall, L. A., Song, X., Pascovici, D., Care, N., Molloy, M., Ohms, S. and Solomon, P. S. (2014). The necrotrophic effector protein SnTox3 re-programs metabolism and elicits a strong defence response in susceptible wheat leaves, *BMC Plant Biol.* 14(1), 215.

Wolpert, T. J., Dunkle, L. D. and Ciuffetti, L. M. (2002). Host-selective toxins and avirulence determinants: what's in a name?, *Annu. Rev. Phytopathol.* 40(1), 251-285.

Zhang, Z., Friesen, T. L., Xu, S. S., Shi, G., Liu, Z., Rasmussen, J. B. and Faris, J. D. (2011). Two putatively homoeologous wheat genes mediate recognition of SnTox3 to confer effector-triggered susceptibility to *Stagnonospora nodorum*, *Plant J.* 65(1), 27-38.

Chapter 11

Advances in genetic mapping of Septoria nodorum blotch resistance in wheat and applications in resistance breeding

Min Lin and Morten Lillemo, Norwegian University of Life Sciences, Norway

1 Introduction

As one of the most important staple food sources, wheat (*Triticum* spp.) is a key component of global food security, with allohexaploid bread wheat (*Triticum aestivum* L., AABBDD) dominating wheat production as the most widely grown cereal species. However, wheat cultivation is limited by many stresses, including fungal diseases. Septoria nodorum blotch (SNB), caused by the fungal pathogen *Parastagonospora nodorum*, is notably a problem in wheat-growing areas with high rainfall such as North-Western, Central and Eastern Europe, Canada, the United States, South America and western Australia. For a general review of the disease we refer to Ruud and Lillemo (2018) for details. In this chapter, we aim to provide an in-depth overview of recent findings with a focus on the genetics of adult plant SNB resistance in wheat and its implications for resistance breeding.

http://dx.doi.org/10.19103/AS.2021.0092.15

2 Pathogen

Parastagonospora nodorum (syn. *Phaeosphaeria nodorum* (E. Müll.), syn. *Leptosphaeria nodorum* (E. Müll.), syn. *Stagonospora nodorum* (Berk.), syn. *Septoria nodorum* (Berk.)) is a typical necrotrophic fungal pathogen. It belongs to the Dothideomycetes class of the Ascomycota (Quaedvlieg et al. 2013). *P. nodorum* is mainly known as a pathogen of wheat, but can occasionally also infect barley (*Hordeum vulgare*) with less damage, reviewed by Cunfer (2000), as well as wild grasses (Williams and Jones 1973). The disease caused by *P. nodorum* is most commonly called Septoria nodorum blotch (SNB), but is also known as Stagonospora nodorum blotch. When *P. nodorum* infects glumes, the resulting disease is called wheat glume blotch (Oliver et al. 2016).

2.1 Symptoms

Symptoms of *P. nodorum* infection on wheat leaves start as oval brown necrotic lesions that are typically surrounded by chlorosis. As they develop, these lesions turn into irregular dark brown areas of necrotic tissue (Fig. 1a). In the field, the symptoms can easily be confused with those caused by two other important wheat leaf blotch fungal diseases: tan spot (TS) (caused by *Pyrenophora tritici-repentis*) and Septoria tritici blotch (STB) (caused by *Zymoseptoria tritici*) (Ficke et al. 2018a). Coinfection of these three diseases is common in field conditions (Blixt et al. 2010; Justesen et al. submitted). However, as both STB and TS are well-known as leaf diseases which seldom cause symptoms on wheat heads,

Figure 1 Symptoms of Septoria nodorum blotch in wheat. (a) Symptoms on leaves, photographed by Morten Lillemo. (b) Symptoms on head, photographed by Andrea Ficke.

Published by Burleigh Dodds Science Publishing Limited, 2022.

glume infection is an important indicator of SNB infection. Figure 1b shows the SNB symptoms on wheat heads, which start from brown to dark brown spots on the glumes.

2.2 Infection cycle and epidemics

P. nodorum is a heterothallic fungus which requires two mating-type idiomorphs in the population for sexual reproduction (Bennett et al. 2003). Wind-spread ascospores released from wheat debris serve as the major primary source of inoculum early in the season (Bathgate and Loughman 2001). As shown in Fig. 2, *P. nodorum* is also seed-transmitted (Sommerhalder et al. 2006). Infected seeds can also be a source of primary inoculum when seed treatments are poorly applied. After the establishment of the initial infection, the pathogen starts producing massive pycnidiospores which can be spread through rain splash to neighboring plants. The pathogen is polycyclic and can complete multiple cycles in one growing season, producing a considerable amount of pycnidiospores as secondary inoculum (Eyal et al. 1987; Sommerhalder et al. 2011). Epidemics of *P. nodorum* used to be common in all wheat-growing areas with suitable climatic condition for disease development across all six

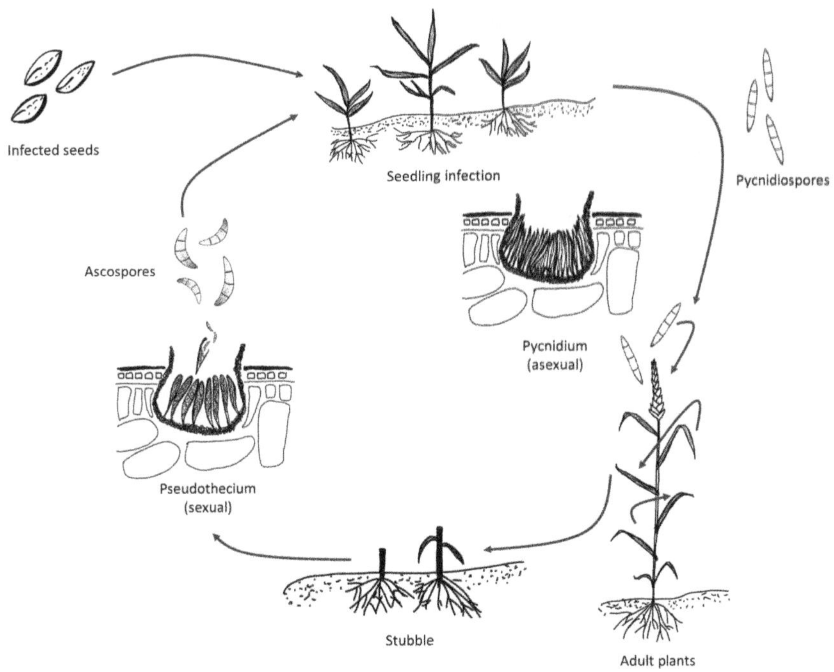

Figure 2 Infection cycle of *P. nodorum*. Drawing by Ling Su, adapted from Sommerhalder et al. (2011).

wheat-growing continents (Ficke et al. 2018a; Leath et al. 1993). The pathogen shares the same center of origin as its wheat host in the Fertile Crescent and probably spread during wheat germplasm exchange (McDonald et al. 2012; Ghaderi et al. 2020). As mentioned previously, *P. nodorum* and *Z. tritici* often cause coinfection on the host in the same field, because the asexual spores of both pathogens are spread by rain splash and prefer to grow in similar warm and humid conditions (Bearchell et al. 2005). *P. nodorum* was the dominating leaf blotch pathogen in Europe until the 1980s (Bearchell et al. 2005). Since then, it has in most of Europe been replaced by *Z. tritici*, and STB management is now responsible for large proportions of the fungicide applications in wheat (Bearchell et al. 2005; Shaw et al. 2008; Torriani et al. 2015). The reduction of *P. nodorum* epidemics in the UK has been correlated with the decrease in sulfur pollution (Shaw et al. 2008). However, the same sulfur theory could not explain the dominance of *P. nodorum* in western Australia (Oliver et al. 2012), or Norway, which is one of the remaining *P. nodorum* hot spots in Europe.

2.3 Population genetics studies of **P. nodorum**

P. nodorum undergoes frequent sexual reproduction, and high frequency of recombination results in pathogen populations with high genetic diversity (McDonald and Linde 2002). Studies on *P. nodorum* population structure have been carried out based on different molecular markers or sequence variations of selected genes (McDonald et al. 2012; Murphy et al. 2000; Stukenbrock et al. 2006). McDonald et al. (1994) investigated the genetic variability of two *P. nodorum* populations in the United States with eight restriction fragment length polymorphism (RFLP) markers and high level of genetic diversity was found even among isolates collected from the same lesion. A similar study investigated genetic variations of *P. nodorum* collections from Europe and the United States, while only small genetic differentiations were observed between populations (Keller et al. 1997). A western Australian *P. nodorum* population genetic study was undertaken using the same RFLP markers described above, and no evidence of population subdivision was observed due to high genetic variability within the population (Murphy et al. 2000). Similar results of a highly diverse pathogen population with no signs of population structure were recently also shown for the Norwegian *P. nodorum* population (Lin et al. 2020b); see Section 5 of this chapter. Stukenbrock et al. (2006) used 12 simple sequence repeat (SSR) markers to characterize the population structure of *P. nodorum* at a global scale, based on nine populations from five continents. As expected, high genetic diversity was observed within each population, but only moderate differentiation was detected between genetically divided populations (Stukenbrock et al. 2006). McDonald et al. (2013) compared the genetic diversity of three *P. nodorum* necrotrophic effector (NE) genes (*SnToxA*,

SnTox1 and *SnTox3*) and found significant differences in allele frequencies of the three genes among populations. However, *P. nodorum* populations with high *SnTox* gene sequence diversities were not correlated with high diversity at neutral loci, suggesting that the *SnTox* genes were under selection by local host cultivars (McDonald et al. 2013). A high frequency of the *SnToxA*, *SnTox1* and *SnTox3* genes and higher genetic diversity at neutral SSR loci of isolates collected in Iran compared to other geographic origins provided further evidence to the hypothesis that *P. nodorum* originated as a pathogen on wheat in the Fertile Crescent (Ghaderi et al. 2020). A recent study of 197 *P. nodorum* isolates originating from the United States found evidence of two populations that corresponded to the Upper Midwest and South-Eastern United States (Richards et al. 2019). Interestingly, isolates belonging to the South-Eastern populations lacked the *SnToxA* effector gene, which fitted well with the absence of the ToxA-sensitivity gene, *Tsn1*, in the widely grown winter wheat cultivars in the region. Overall, this suggests that effector genes in pathogen populations are under selection pressure for fitness on host genotypes.

3 Agricultural importance and management of Septoria nodorum blotch (SNB)

In modern agroecosystems, where high density of crops are grown, favorable environments are provided for multi-infections where different strains of the same pathogen can infect the same host, especially for pathogens with large population sizes combined with high genetic diversity (McDonald and Stukenbrock 2016; van Baalen and Sabelis 1995). It also promotes the development of virulence due to competition between strains from the same pathogen species and/or among pathogen species under coinfection conditions (McDonald and Stukenbrock 2016; van Baalen and Sabelis 1995). Under favorable climate conditions, *P. nodorum* can cause severe grain shriveling and reduce yield up to 30% (Bhathal et al. 2003) by destroying the photosynthetic capacity of the leaves and glumes.

Ascospores originating from wheat debris serve as the primary inoculum of SNB (Bathgate and Loughman 2001). Reduced tillage, which is advocated to reduce soil erosion, can promote SNB as it leaves higher amounts of infected wheat straw on the soil surface, which can then serve as primary inoculum (Ficke et al. 2018a). In addition, the pycnidiospores are spread by rain-splash. The high density of plants within the wheat fields makes it easier for pycnidiospores to spread to neighboring plants.

Disease management of SNB includes cultivar resistance, fungicide treatment and stubble management. Both seed treatment and fungicide application in the field are widely used chemical control methods for SNB management (Solomon et al. 2006). However, Blixt et al. (2009) reported that the majority of tested Swedish *P. nodorum* isolates showed reduced sensitivity to strobilurins, which was caused by amino acid substitutions in the *cytochrome b* gene. Pereira et al. (2017) reported variations in sensitivities to sterol demethylation inhibitors (DMIs) within a global *P. nodorum* collection, and the reduced sensitivity was due to non-synonymous mutations in the *CYP51* gene. Neither study included *P. nodorum* isolates collected in recent years (2011-present) for fungicide resistance testing. Therefore, higher frequency of fungicide insensitive isolates would be expected in the natural *P. nodorum* population due to the high selection pressure. However, field resistance to azoles or DMIs has not been reported so far.

As mentioned previously, reduced tillage increases the amount of crop residues that can serve as inoculum for residue-borne leaf blotch diseases (Ficke et al. 2018a; Shaner 1995). SNB disease severity in the field has shown significant correlations with the amount of residues, as can be expected (Mehra et al. 2015). Residue management (e.g. burial of residues and crop rotation) can effectively reduce the disease severity by decreasing the amount of primary inoculum when healthy seeds are used (Mehra et al. 2015).

4 Methods used in genetic studies of Septoria nodorum blotch (SNB)

Due to the large genome size of hexaploid bread wheat (approx. 16 Gb), genetic studies of wheat were among the most complicated of all cultivated plants (International Wheat Genome Sequencing 2014). Previous studies show that the inheritance of SNB resistance is quantitative (Bostwick et al. 1993; Wicki et al. 1999). However, neither the genetic location nor the molecular basis of SNB resistance genes was clear before the usage of molecular markers in the mapping of quantitative trait loci (QTL) (Czembor et al. 2003; Friesen and Faris 2010). Thanks to the rapid development of high-throughput genotyping technologies, genotyping by single nucleotide polymorphism (SNP) chips are nowadays widely used in wheat (Allen et al. 2017; Wang et al. 2014a), and have largely replaced the use of SSR or DArT (diversity arrays technology) markers which were the most commonly used marker types for wheat genetic studies less than a decade ago (Akbari et al. 2006; Langridge et al. 2001), resulting in an increasing numbers of published QTL mapping studies. In addition to QTL mapping using biparental populations, genome wide association studies (GWAS) serve as an alternative approach to genetic mapping of SNB resistance (Adhikari et al. 2011; Gurung et al. 2014; Jighly et al. 2016; Liu et al. 2015; Ruud et al. 2019). GWAS has two main advantages. First, the association mapping (AM) panel saves the cost and time to construct mapping populations (Bernardo 2016; Gupta et al. 2014). Second, the genetic diversity and map resolution are higher in an AM panel compared to biparental populations, since multiple historical recombination and multiple alleles per locus are available in diverse germplasm collections (Bernardo 2016; Gupta et al. 2014). However, genetic subpopulation structure in AM panels may result in false positive associations (Breseghello and Sorrells 2006; Gupta et al. 2014). In addition, GWAS analysis has its limitation for detecting QTL associated with rare alleles or rare variants; therefore, it can only be used for detecting QTL controlled by alleles with relatively high allele frequency in the panel (Bernardo 2016; Breseghello and Sorrells 2006; Gupta et al. 2014). Recently, multiparent advanced generation intercross (MAGIC) populations have also been used for SNB resistance studies (Cockram et al. 2015; Downie et al. 2018; Lin et al. 2020a; Lin et al. 2021). The design of MAGIC populations involves intercrossing of multiple parental lines (2^n), followed by several rounds of selfing to achieve a RIL population, which allows for increased allelic diversity and genetic recombination relative to comparatively sized biparental populations, while reducing the risk of false positive associations caused by the population structure commonly present in AM panels (Cavanagh et al. 2008; Mackay et al. 2014).

4.1 Genetics of SNB resistance and the inverse gene-for-gene model

Resistance to SNB is truly quantitative in the sense that no cultivar has been identified with complete resistance (Aguilar et al. 2005). In addition, due to the complexity of the pathogen population, seedling resistance to single isolates does not guarantee adult plant resistance in the field. Moreover, the fact that leaf blotch and glume blotch might be controlled by different genetic mechanisms (Aguilar et al. 2005; Wicki et al. 1999) adds further to the difficulties of SNB resistance breeding.

Most known SNB resistance mechanisms are due to lack of sensitivity genes (*Snn*) in wheat genotypes to corresponding *P. nodorum* necrotrophic effectors (inverse gene-for-gene model) (reviewed by Ruud and Lillemo 2018). The gene-for-gene model is well known for interactions between biotrophic pathogens, such as those causing powdery mildew and rust diseases, and their host plants. As biotrophic pathogens require living host tissues, when the host plant contains the resistance (R) gene able to recognize the product of a pathogen's avirulence (*Avr*) gene, a hypersensitive reaction (HR) will be induced resulting in programmed cell death (PCD), thus limiting the infection of the biotrophic pathogens (Fig. 3a). On the contrary, necrotrophic pathogens, such as *P. nodorum*, utilize nutrients from dead or dying host tissues and interact with host plants via an inverse gene-for gene model (Friesen et al. 2007). Via the production of NEs, which interact with host susceptibility genes (*Snn*), necrotrophic pathogens can trigger PCD to accelerate their infection (Fig. 3b).

(a) Biotrophic pathogen

	Avr	avr
R	Resistant	Susceptible
r	Susceptible	Susceptible

Gene for Gene Hypothesis

(b) Necrotrophic pathogen

	NE	ne
S	Susceptible	Resistant
s	Resistant	Resistant

Inverse Gene for Gene Hypothesis

Figure 3 Different plant-pathogen interaction mechanisms. (a) Gene-for-gene model, adapted from Flor (1971). (b) Inverse gene-for-gene model, adapted from Friesen et al. (2007). R: resistant gene, r: absence of resistant gene; Avr: avirulence gene, avr: absence of avirulence gene; NE: necrotrophic effector, ne: absence of necrotrophic effector; S: susceptibility gene, s: absence of susceptibility gene.

Up to now, nine NE-*Snn* interactions have been described; however, only three *P. nodorum* NE coding genes (*SnToxA*, *SnTox1* and *SnTox3*) and two host NE sensitivity genes (*Tsn1* and *Snn1*) have been cloned (reviewed by Peters Haugrud et al. (2019); Ruud and Lillemo (2018)).

ToxA was characterized as a 13-kDa polypeptide host selective toxin (now referred to as NE) produced by the wheat tan spot pathogen, *Pyrenophora tritici-repentis*, and which interacts with the sensitivity gene *Tsn1* on the long arm of wheat chromosome 5B (Faris et al. 1996; Tomas et al. 1990). Later, Liu et al. (2006) reported that *P. nodorum* also contains a ToxA coding gene, the product of which targets the same wheat *Tsn1* locus as tan spot. Through gene diversity analysis, it was shown that the ToxA coding gene *SnToxA* in *P. tritici-repentis* likely originated from *P. nodorum* through a recent horizontal gene transfer event (Friesen et al. 2006). The cloned ToxA sensitivity gene, *Tsn1*, has a typical R gene structure containing nucleotide binding site (NBS) and a leucine-rich repeat (LRR) domain, as well as a serine/threonine protein kinase (S/TPK) domain (Faris et al. 2010). However, the Tsn1 protein does not directly interact with ToxA, suggesting that *Tsn1* may mediate the signaling pathway of effector-triggered immunity (ETI) but is not the ToxA receptor (Faris et al. 2010). Recently, it was shown that another wheat and barley pathogen, *Bipolaris sorokiniana*, the cause of spot blotch, also possesses a *ToxA* gene that likely originated from *P. nodorum*, pointing to a selective advantage of carrying the virulence factor ToxA (Friesen et al. 2018; McDonald et al. 2018).

Tox1-Snn1 was the first reported NE-Snn interaction in the wheat-*P. nodorum* pathosystem, where Tox1 was characterized as a NE produced in *P. nodorum* culture filtrates interacting with the wheat sensitivity locus *Snn1* on chromosome 1B (Liu et al. 2004a). However, the cloning of *SnTox1* was not

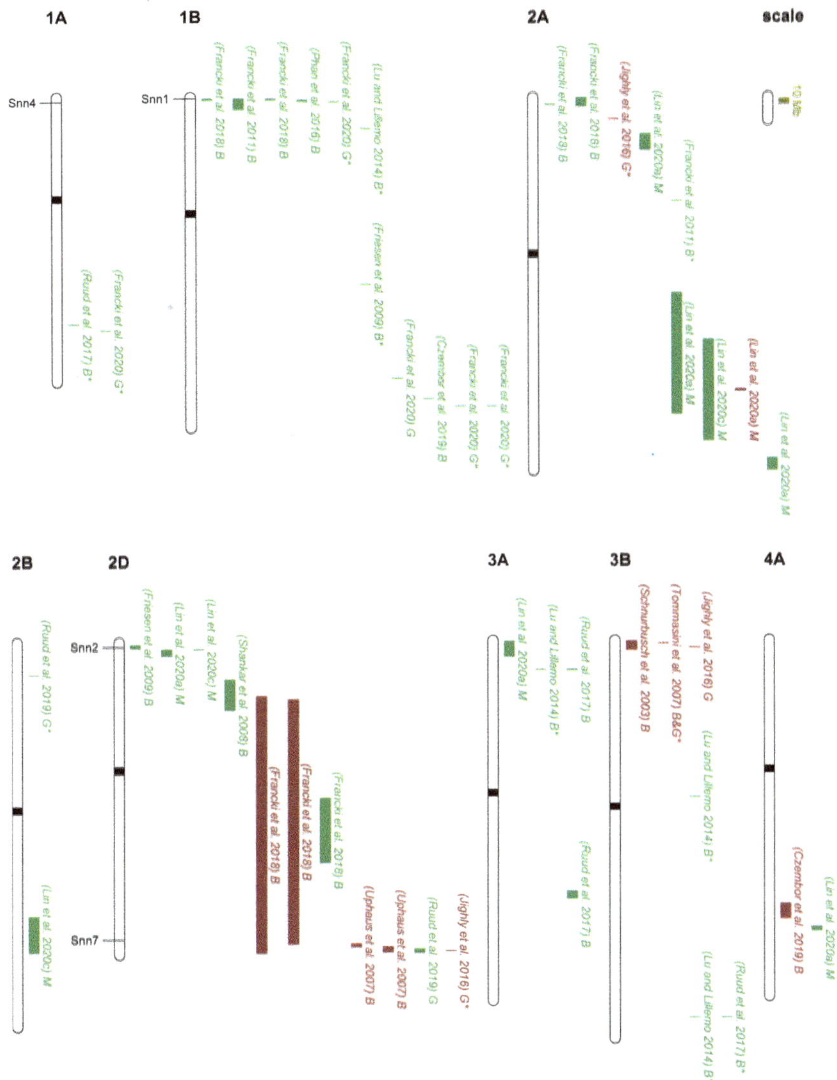

Figure 4 Projection of published field resistance QTL for SNB leaf blotch (green), glume blotch (dark red) onto the wheat reference genome assembly (RefSeq v1.0; IWGSC 2018). B: biparental population, G: GWAS panel, M: MAGIC population. If sequence of only single marker was available for the QTL, QTL were marked with a superscript asterisk.

achieved until eight years after the discovery of the Tox1-Snn1 interaction (Liu et al. 2012). *SnTox1* encodes a cysteine-rich protein with 117 amino acids which is light dependent and critical for fungal penetration (Liu et al. 2012). Further research on Tox1 showed that it serves as a dual function protein, which can bind the host chitinases to protect fungal infection as well as behaving like

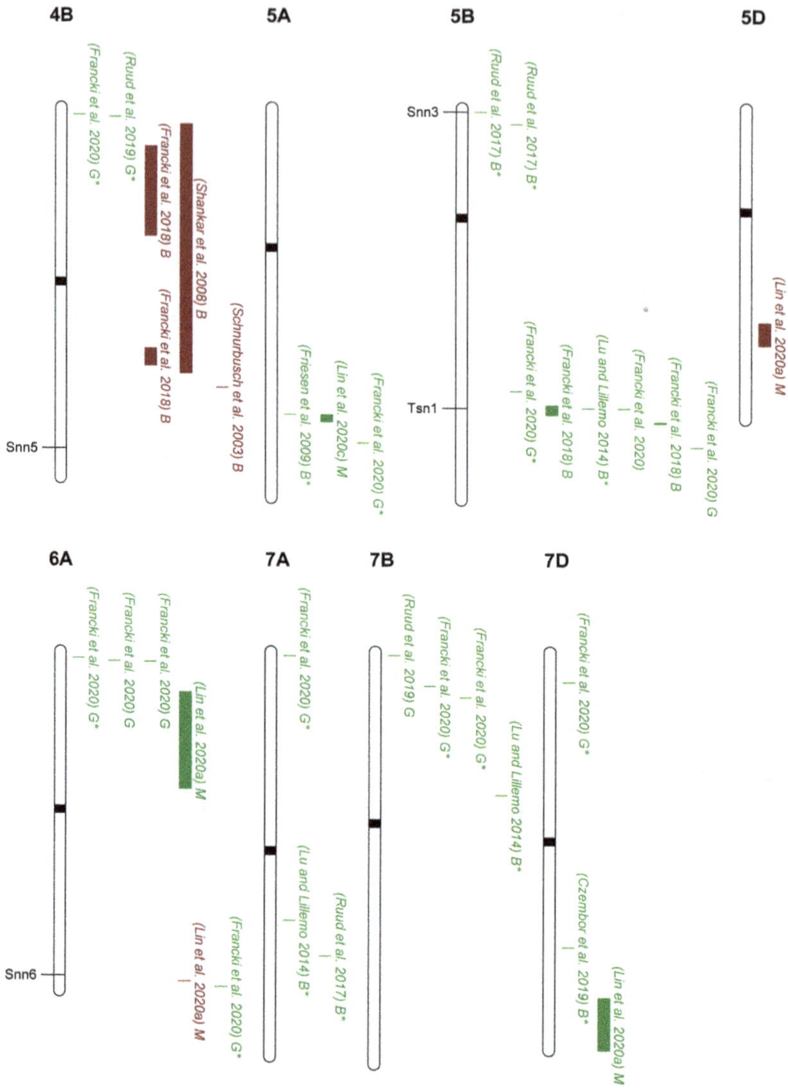

Figure 4 (*Continued*).

a virulent NE (Liu et al. 2016). In the same year, Shi et al. (2016) cloned the Tox1 wheat sensitivity gene *Snn1*, which encodes a wall-associated kinase (WAK). WAK proteins are pattern recognition receptors (PRRs) which can interact directly with pathogen-associated molecular patterns (PAMPs), such as oligogalacturonides (OGs), which trigger PCD and are involved in plant defense mechanisms against biotrophic pathogens (Brutus et al. 2010). In contrast to ToxA-Tsn1, which interacts in the ETI pathway, the cloning of *Snn1* revealed that *P. nodorum* could also hijack the PAMP-triggered immunity (PTI) pathway against biotrophic pathogens and lead to disease (Shi et al. 2016).

Tox3 was characterized after the discovery of ToxA, Tox1 and Tox2 by Friesen et al. (2008) as a novel *P. nodorum* NE, interacting with the wheat sensitivity locus designated as *Snn3* on the short arm of chromosome 5B. The Tox3 protein is around 26 kDa in size and the coding gene, which was cloned by Liu et al. (2009), showed little homology to the previously cloned *P. nodorum* NE gene *SnToxA*. In addition, sensitivity to Tox3 was also shown in the diploid wheat progenitor species *Aegilops tauschii* and the sensitivity locus was mapped to the short arm of chromosome 5D and is probably homoeologous to the *Snn3* locus on chromosome 5B (Zhang et al. 2011). Accordingly, the *Snn3* loci present in bread wheat and *Ae. tauschii* are now denoted as *Snn3-B1* and *Snn3-D1*, respectively. Though the remaining six NE-*Snn* interactions have been characterized and the sensitivity loci were mapped to wheat chromosomes (Fig. 4) (Abeysekara et al. 2012; Abeysekara et al. 2009; Friesen et al. 2012; Friesen et al. 2007; Gao et al. 2015; Shi et al. 2015), further studies are required to fine-map and elucidate the molecular mechanisms of these sensitivity loci.

4.2 Relationship between NE-Snn interactions and field SNB severity

All the described NE-*Snn* interactions were firstly characterized by seedling inoculations and infiltrations in greenhouse conditions, but there is evidence that some NE-*Snn* interactions also contribute to field SNB susceptibilities (Ruud and Lillemo 2018). Friesen et al. (2009) studied adult plant resistance in the field by spray inoculating a mapping population segregating for *Tsn1*, *Snn2* and *Snn3-B1* with an isolate producing both ToxA and Tox2. They found the *Tsn1* and *Snn2* loci to explain 18% and 15% of the phenotypic variation, respectively. In another study, a mapping population segregating for *Snn1* and *Snn3* was inoculated in the field with an isolate producing all three known NEs, showing that the *Snn1* locus could explain 19% of the phenotypic variation for adult plant disease severity (Phan et al. 2016). Furthermore, a field study conducted with natural *P. nodorum* inoculum showed that *Snn3-B1* was the major determinant of SNB susceptibility in the 'SHA3/CBRD × Naxos' population, explaining up to 24% of the phenotypic variation (Ruud et al. 2017). It is still unclear whether there are more NE-*Snn* interactions influencing the adult plant susceptibilities. A recent study by Lin et al. (2020a) reported that one robust adult plant susceptibility QTL co-located with a seedling sensitivity QTL to an unknown NE, which will be discussed in detail in the case study.

4.3 Overview of reported QTL for adult plant leaf blotch resistance

Since Aguilar et al. (2005) published the QTL study for both adult plant leaf blotch and glume blotch resistance, more than 60 QTL have been characterized for adult plant leaf blotch resistance on 15 different wheat chromosomes (1A, 1B, 2A, 2B, 2D, 3A, 3B, 4A, 4B, 5A, 5B, 6A, 7A, 7B and 7D) (Fig. 4). As shown in Fig. 4, some of the adult plant leaf blotch resistance/susceptibility QTL colocate with the wheat NE sensitivity loci, such as *Tsn1*, *Snn1*, *Snn2* and *Snn3* (Francki et al. 2018; Friesen et al. 2009; Lin et al. 2020a; Lu and Lillemo 2014; Phan et al. 2016; Ruud et al. 2017). By basic local alignment search tool (BLAST) analysis of the marker sequences of published QTL against the reference genome (International Wheat Genome Sequencing et al. 2018), more adult plant leaf blotch resistance QTL, which were detected using different mapping populations and different genetic markers, were projected into the similar physical map region and could potentially be the same QTL (Fig. 4). For example, overlapping QTL interval was found for the QTL reported by Friesen et al. (2009) and Lin et al. (2021) on the long arm of chromosome 5A (Fig. 4). However, most of the adult plant leaf blotch resistance QTL have not been fine-mapped and the molecular mechanisms behind these QTL are still unclear.

4.4 Relationship between seedling and adult plant resistance

Previous studies showed that SNB seedling resistance and adult plant resistance were not highly correlated (Francki 2013; Ruud and Lillemo 2018; Shankar et al. 2008b; Uphaus et al. 2007), which could partly be due to the fact that many of these studies used different isolates in the greenhouse experiments and the field testing (Ruud and Lillemo 2018; Ruud et al. 2017). It has also been shown that the correlations between seedling and adult plant leaf blotch resistance can be relatively high when the same isolate is used for both seedling tests in the greenhouse and adult plant tests in the field (Jönsson 1985). However, in general, the correlations between SNB resistance at these two growth stages are low. Since the natural *P. nodorum* populations are genetically diverse, it is difficult to identify representative isolates for greenhouse assays. In addition, even though Shankar et al. (2008b) used the same isolate mixture for inoculation in the greenhouse and in the field in the year 2003, they found only a weak correlation between seedling and flag leaf disease scores ($r = 0.31$), and a non-significant relationship between seedling and glume blotch severity ($r = 0.09$). However, results of field testing might still be influenced by the natural *P. nodorum* population even though specific isolate or isolates mixture are used as inoculum.

4.5 The genetics of glume blotch resistance

The first QTL mapping study for SNB glume blotch resistance was done by Schnurbusch et al. (2003) using a Swiss biparental winter wheat mapping population 'Arina × Forno', where two significant QTL were detected on chromosomes 3B and 4B, respectively. Up to date, 17 SNB glume blotch QTL have been characterized on chromosomes 2A, 2B, 2D, 3B, 4A, 4B and 5D (Fig. 4) (Czembor et al. 2019; Francki et al. 2018; Jighly et al. 2016; Lin et al. 2020a; Schnurbusch et al. 2003; Shankar et al. 2008a; Tommasini et al. 2007; Uphaus et al. 2007). Although leaf blotch and glume blotch are symptoms caused by the same pathogen on the same host, previous studies showed that the genetic mechanisms controlling resistance to SNB leaf blotch and glume blotch are different (Aguilar et al. 2005; Fried and Meister 1987; Wicki et al. 1999). Aguilar et al. (2005) carried out the first study to investigate leaf blotch and glume blotch resistance by assessing the disease on the same mapping population in the same field. They identified one common QTL for both leaf blotch and glume blotch on chromosome 2B; however, that QTL was associated to confounding morphological traits such as heading date and ear length (Aguilar et al. 2005). Therefore, they concluded that the resistance of leaf blotch and glume blotch were controlled by genetically independent mechanisms. A recent study showed that one glume blotch resistance QTL on chromosome 2A colocated with a leaf blotch resistance QTL, which was detected in the same study using a MAGIC population tested in the field (Lin et al. 2020a). However, the haplotype analysis revealed that the haplotype effects for the leaf and glume blotch QTL were different, which was congruent with previous studies that different mechanisms controlled the resistance to leaf and glume blotch (Lin et al. 2020a). In addition, to our knowledge, no glume blotch resistance QTL has been characterized at the sequence level nor being applied in resistance breeding by MAS.

5 Case study: Septoria nodorum blotch (SNB) resistance in wheat in Norway

5.1 Status of SNB knowledge in Norwegian wheat

SNB is the dominating leaf blotch disease on wheat in Norway, and there is a need to improve resistance in current cultivars. As reviewed by Ruud and Lillemo (2018), the first genetic studies of SNB resistance in Norwegian germplasm started in 2010. Based on the 'SHA3/CBRD × Naxos' population, SNB-resistant QTL were detected on wheat chromosomes 1B, 3A, 3B, 5B, 7A and 7B (Lu and Lillemo 2014). Ruud et al. (2017) used the same population for both greenhouse and field testing, and confirmed that the QTL on the short arm of chromosome

5B detected in the previous study (Lu and Lillemo 2014) was the Tox3 sensitivity locus *Snn3-B1*, which showed a major effect on wheat susceptibility at both seedling and adult plant stages. In addition, through NE infiltrations, Ruud et al. (2018) found that large proportions of lines in the Norwegian spring wheat collection were sensitive to ToxA and Tox3. Interestingly, *SnToxA* and *SnTox3* frequencies in their Norwegian *P. nodorum* isolate collections were also quite high, with *SnToxA* and *SnTox3* frequencies of 69% and 76%, respectively (Ruud et al. 2018). A recent study by Ruud et al. (2019) identified many adult plant resistant QTL by GWAS using a collection of 121 Nordic spring wheat cultivars and breeding lines. Among those, one QTL on chromosome 2D was robust in most of the tested years and significant correlations were found between field disease severity and sensitivity to ToxA (Ruud et al. 2019). However, although ToxA sensitivity is common in Norwegian spring wheat cultivars and showed positive correlation with SNB severity in the field (Ruud et al. 2018), the *Tsn1* locus was not significantly detected by this association study (Ruud et al. 2019).

With this as a background, we set out to do a more in-depth study of the pathogen population in Norwegian wheat fields and conduct genetic mapping of more germplasm using two winter wheat MAGIC populations and an expanded GWAS panel of both spring and winter wheat lines.

5.2 Background and main objectives of recent SNB studies in Norway

Cultivar resistance is usually considered as an effective and environmentally friendly disease management method. However, the durability of cultivar resistance is always being challenged due to the long time frame required for resistance breeding and the fast evolution of virulent pathogens. The pathogen population can adapt to new sources of host resistance quickly once a new cultivar is released to the market. Therefore, knowledge of the pathogen population is vital to optimize resistance breeding strategies and to help exploit the limited resistance resources in an effective way. In addition, apart from improving the usage of cultivar resistance, knowledge of the local pathogen population would also be beneficial to improve other disease management approaches such as chemical application and agronomic control methods. For example, a pathogen population that undergoes regular sexual reproduction and has a high mutation rate, high gene flow and a large population size is considered to have high evolutionary potential (McDonald and Linde 2002). Typically, the risk of breaking qualitative cultivar resistance is high for such populations. Additionally, fungicide resistance alleles may also spread quickly when pathogen populations are under high selection pressure. Moreover, as such diverse populations evolve rapidly, they will also adapt quickly to changing environments. Therefore, disease management should be adjusted accordingly. For instance, stacking different qualitative resistance genes or using quantitative resistance to breed resistant cultivar can make it more difficult for a highly adaptive pathogen population to overcome the host resistance.

In order to gain knowledge of the local Norwegian *P. nodorum* pathogen population, we collected and genotyped single spore isolates from spring and winter wheat fields across three seasons in the main wheat-growing regions in Norway. Different methods were used to analyze potential subdivision of the pathogen population due to sampling location, time or the cultivars the isolates were collected from. Since coinfection of *P. nodorum* and *Z. tritici* was more common on winter wheat but less on spring wheat (Justesen et al., submitted), we hypothesized to find differences between pathogen populations on different wheat types due to different competition pressure. Moreover, we analyzed the allele frequencies of *SnTox* genes and investigated whether local adaptation was evident due to these virulence factors.

Cultivar resistance is still insufficient and currently no cultivar shows complete resistance to SNB. One of the main objectives of our genetic studies has been to identify loci associated with SNB resistance in breeder-relevant germplasm in order to improve resistance breeding. Up to now, nine NE-*Snn* interactions have been characterized, with only a few of them showing important role in SNB susceptibility in the field as reviewed in section 4.2. However, since the *P. nodorum* populations are characterized by high genetic diversity, we hypothesized that additional NE-*Snn* interactions might be present, but not discovered yet. Three genetic studies of SNB resistance will be described here: (1) using one winter wheat MAGIC population (NIAB Elite) from the UK (Lin et al. 2020a), (2) a winter wheat MAGIC population (BMWpop) from Germany (Lin et al. 2021) and (3) a GWAS of field resistance using two Norwegian association mapping panels: one winter wheat panel and one spring wheat panel (Lin et al. in prep.).

The main objectives were to:

- investigate the genetic structure of the Norwegian *P. nodorum* pathogen population;
- evaluate NE-*Snn* interaction-related QTL at both seedling and adult plant stages using a MAGIC population of UK origin;
- assess the genetics of SNB disease severity of another MAGIC population of German origin and compare the QTL detected from the two MAGIC populations; and
- discover robust SNB resistance QTL in the field using two association mapping panels.

5.3 Main results and discussions of case study

5.3.1 Case study 1: Investigating the genetic structure of Norwegian *P. nodorum* populations

In the pathogen population study (Lin et al. 2020b), wheat leaf samples were collected from three major Norwegian wheat-growing areas including five counties, from 2015 to 2017. A total of 165 single spore isolates of *P. nodorum* were isolated from the leaf samples and recorded with information on sampling location, year and source cultivar. We genotyped the isolate collection together with nine foreign isolates using 20 SSR markers, three known *SnTox* genes (*SnToxA*, *SnTox1* and *SnTox3*) and two mating-type idiomorphs (Lin et al. 2020b). We used different methods to analyze whether the pathogen population could be subdivided by location, time or the cultivars the isolates were collected from. McDonald and Linde (2002) hypothesized that pathogen populations with combined sexual and asexual reproduction, high gene flow, high mutation rate and large population size have relatively high evolutionary potential. The *P. nodorum* population in Norway was found to perfectly fit this model. Both mating types (MAT1-1 and MAT1-2) were present in all tested locations and the ratio between the two mating types did not significantly ($p < 0.05$) deviate from 1:1 in any location, although the nationwide mating type ratio showed a slight deviation from 1:1 ($p < 0.05$). Overall, the Norwegian *P. nodorum* population exhibited the signature of sexual reproduction. Results from principal component analysis (PCA), STRUCTURE and 'snapclust' analyses revealed that there was no genetic population structure in the collected *P. nodorum* isolates. The analysis of molecular variance (AMOVA) confirmed that the genetic variations were larger within location, year, cultivar or wheat type rather than between these classifications, and no population subdivisions could be observed by location, year, cultivar or wheat types. Interestingly, we did not even find any substantial genetic differentiation between the Norwegian *P. nodorum* population and the nine foreign isolates included in the PCA analysis. This finding was supported by the observation that genetic distances estimated between individual Norwegian isolates were as large as, or sometimes even larger than, those between Norwegian isolates and foreign isolates. Consistent with previous *P. nodorum* population genetic studies (Keller et al. 1997; McDonald et al. 2012; Murphy et al. 2000; Stukenbrock et al. 2006), high genetic variability was found in the Norwegian *P. nodorum* population. Even though we found two isolates with the same SSR multilocus genotype, no clonal isolates were found in a single sampling location. This finding is also an indicator of large effective population size. As the mutation rate for a pathogen is usually fixed and generally low (McDonald and Linde 2002), large population size also means that a large number of mutants exist in the population.

Pathogens with high evolutionary potential usually have specific characteristics, such as a high potential of adaptation, high risk of breaking down qualitative host resistance, substantial advantage in competition with other pathogens and a high risk of developing fungicide resistance (McDonald and Linde 2002). The management control for *P. nodorum* should take all these characteristics into account. The results of the pathogen population study (Lin et al. 2020b) showed that Norwegian *P. nodorum* isolates have significantly higher frequency of the virulence gene, *SnToxA* (67.9 %), compared to a previously reported European population study (12%) (McDonald et al. 2013). We hypothesized that this large difference in *SnToxA* allele frequency was due to the local adaptation to the high frequency of the ToxA sensitivity allele *Tsn1* in Norwegian spring wheat cultivars; in Norway more spring wheat is grown than winter wheat, and ToxA sensitivity is common in Norwegian spring wheat cultivars (Ruud et al. 2018). Our 165 *P. nodorum* isolates were collected from 13 cultivars which covered 95% of the wheat market share from 2015 to 2017. Among these cultivars, only four are sensitive to ToxA and three of those are spring wheat cultivars. The only ToxA-sensitive winter wheat cultivar was the relatively old cultivar 'Magnifik' which had only 6–7% of the market share, while the widely cultivated recent spring wheat cultivars 'Mirakel', 'Krabat' and 'Demonstrant' are all sensitive to ToxA (Ruud et al. 2018), implying a high frequency of ToxA-sensitive alleles at *Tsn1* in current Norwegian spring wheat cultivars. The main results from the GWAS (Lin et al. in prep.) also supported this, showing that *Tsn1* was significantly detected in the field using the full set of the Norwegian spring wheat association panel. As mentioned in Section 4.1, ToxA is not only a virulence factor for *P. nodorum*, but also for the tan spot (*D. tritici-repentis*) and spot blotch (*B. sorokiniana*) pathogens. Removing the *Tsn1* susceptibility allele from current wheat cultivars may, therefore, be beneficial for multiple diseases.

However, eliminating a single susceptibility gene will not manage the disease completely. As discussed earlier, qualitative resistance is easy to break by a pathogen with high evolutionary potential. *P. nodorum* isolates likely carry more than one NE gene, and could regulate the expression level of effector genes based on the host sensitivity (Peters Haugrud et al. 2019). For example, *Tsn1* was not significantly associated with SNB in the winter wheat panel (Lin et al. in prep.), as most winter wheat lines did not carry this susceptibility gene. However, two other NE sensitivity loci, *Snn1* and *Snn3*, which were not significant in the Norwegian spring wheat association panel, were significantly associated with SNB resistance/susceptibility in the winter wheat panel. One could also hypothesize that high genetic diversity and frequent sexual recombination leads to more complicated effector profiles in the natural *P. nodorum* population. Therefore, stacking more resistance QTL is needed to decrease the SNB disease severity (Lin et al. in prep.).

As mentioned in Section 2.2, *Z. tritici* is the dominant pathogen in the leaf blotch disease complex in many other European countries, while *P. nodorum* is still the major leaf blotch pathogen in Norway (Ficke et al. 2018b; Justesen et al., submitted). One possible explanation could be that, since more spring wheat is grown in Norway, the limited growing season of spring wheat is too short for the longer latent period in *Z. tritici* development, which limits the expansion of its population size. In the meantime, *P. nodorum* could successfully maintain large natural populations on spring wheat due to its shorter latent period (Cunfer 1999) and rapid adaptation, which makes it more competitive in comparison to *Z. tritici* on the same host.

Resistances to different groups of fungicides have been reported in *P. nodorum* populations by different studies, as described in Section 3. As Norwegian *P. nodorum* has a large effective population size and has been treated with fungicides for decades, we would expect a large amount of fungicide-resistant mutations to exist in the population. When being consistently exposed to the same fungicide, the mutant allele may spread quickly in the population due to rapid sexual recombination and massive production of asexual pycnidiospores. Therefore, the large population size, frequent sexual reproduction and high genetic variability in Norwegian *P. nodorum* indicated a potentially high risk of fungicide resistance.

Thus, integrated pest management (IPM) is recommended to control SNB. Firstly, wheat debris should be removed before the next growing season in order to reduce the primary inoculum source, and consequently the pathogen population size. Alternatively, a two-year crop rotation appears to effectively reduce the risk of leaf blotch epidemics even under conducive environmental conditions (Pedersen and Hughes 1992). Secondly, cultivars insensitive to known effectors and possessing other resistance mechanisms should be preferred. Thirdly, use of healthy or fungicide-treated seeds should be recommended to decrease the spread of disease. Lastly, combining or rotating fungicides with different modes of action may effectively decrease the selection pressure of fungicide-resistant mutants in the pathogen population.

5.3.2 Case study 2: QTL mapping using NIAB MAGIC population

The NIAB Elite MAGIC population consists of more than 1000 RILs and was genotyped using an Illumina iSelect 90K SNP array (Mackay et al. 2014; Wang et al. 2014a). A subset of around 500 RILs were tested in Norway for four years using infected straw as inoculum and in the UK for two years of SNB resistance testing, inoculated by spore suspension. We also conducted seedling infiltration and inoculation assays using three *P. nodorum* isolates in order to compare QTL identified in the field with those identified under controlled conditions at the seedling stage (Lin et al. 2020a). The population segregates for sensitivity to

(a)

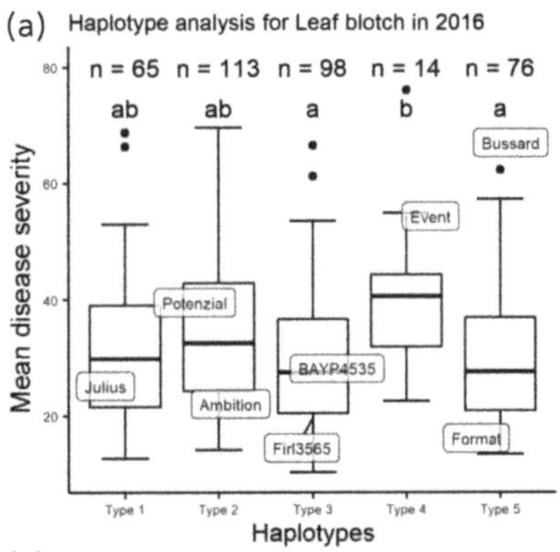

(b)

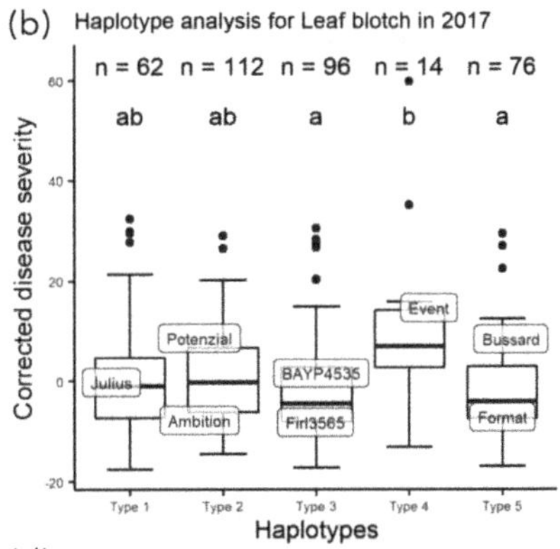

(c)

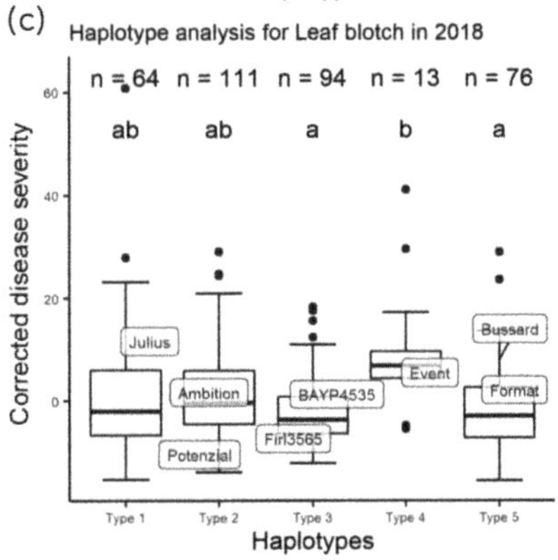

(d)

Haplotype	Parents	SNP1	SNP2	SNP3	SNP4	SNP5
Type1	Julius	0	0	0	0	2
Type2	Potenzial/ Ambition	0	0	2	2	0
Type3	BAYP4535/ Firl3565	0	2	0	0	0
Type4	Event	2	0	0	0	0
Type5	Bussard/ Format	2	0	0	0	2

Figure 5 Haplotype analysis for BMWpop leaf blotch QTL *QSnb.nmbu-2A.1/2018*. (a) Haplotype analysis of mean disease severity in field season 2016. (b–c) Haplotype analysis of corrected disease severity in field season 2017 and 2018, respectively, and the mean disease ratings for the eight founders are indicated. Haplotypes labeled with same letter represented no significant differences between haplotype disease severities as detected by Kruskalmc test ($p < 0.05$). (d) Genotype of each haplotype based on five SNP markers. SNP marker names are listed in order as below: *wsnp_CAP8_c2677_1394934*, *AX-95661975*, *RAC875_c38018_278*, *AX-94508462* and *BS00090569_51* (Lin et al. 2021).

all three well-characterized NEs (SnToxA, SnTox1 and SnTox3), and showed a broad genetic variation for both seedling and adult plant resistance. However, the correlation in resistance between seedling and adult plant resistance was low.

As described above, disease severities of both adult plant leaf blotch and glume blotch were scored in field trials in Norway, while only leaf blotch was scored in the UK. By extracting residuals of disease scores from a linear regression using plant height and days to heading as covariates, we tried to reduce the influences caused by these confounding traits in order to detect true SNB resistance QTL (Lin et al. 2020a). When assessing the disease severities of the MAGIC founders, rankings of disease severities were different for leaf blotch and glume blotch. Notably, the most susceptible founder Xi19 for leaf blotch showed moderate resistance to glume blotch, suggesting that the resistance mechanisms to leaf blotch and glume blotch might be different. Eight QTL on chromosomes 2A, 2D, 3A, 4A, 6A and 7D were identified for leaf blotch, while three QTL on chromosomes 2A, 5D and 6A were identified for glume blotch, and most of the QTL were different. Interestingly, we identified an overlapping QTL, *QSnb.niab-2A.3*, on chromosome 2A for both leaf blotch and glume blotch. Therefore, we considered that they represented a common QTL associated with both leaf blotch and glume blotch. Results of haplotype analysis showed that haplotype effects were significant for leaf blotch in many years and one year for glume blotch. However, the susceptible haplotype for leaf blotch showed the opposite effect for glume blotch, implying that leaf blotch and glume blotch resistance might indeed be controlled by different mechanisms and that the QTL identified might represent two closely located but independent genes or gene clusters, which is in agreement with the results from other studies (Aguilar et al. 2005; Francki 2013; Shankar et al. 2008b). Nevertheless, when comparing published QTL with QTL identified in our study, we found colocation of the glume blotch QTL with an NE sensitivity locus. *QSnb.niab-6A.2* detected for glume blotch in 2016 might colocate with the Tox6 sensitivity locus *Snn6* (Gao et al. 2015) indicating that NE-*Snn* interactions might potentially also play a role in glume blotch susceptibility. NE-*Snn* interactions were deeply investigated for leaf blotch at the seedling stage with only a few cases where seedling resistance QTL showed an effect on adult plant leaf resistance. However, whether NE-*Snn* interactions play a role in glume blotch resistance is still unknown. More research on glume blotch is needed to clarify the genetic mechanism of glume blotch resistance and whether NE-*Snn* interactions are involved in glume susceptibility.

5.3.3 Case study 3: QTL mapping using BMWpop MAGIC population

We validated the most robust QTL, *QSnb.niab-2A.3*, identified using the NIAB MAGIC population on an independent multi-founder population with different genetic background, the BMWpop MAGIC (Lin et al. 2021). The BMWpop consists of 394 $F_{6:8}$ RILs and was genotyped using the 20K Infinium iSelect array (Stadlmeier et al. 2018). Field testing of SNB resistance using BMWpop was conducted in Norway from 2016 to 2018 using the same methodology as for the NIAB MAGIC population (Lin et al. 2020a).

The QTL (*QSnb.nmbu-2A.1*) was significantly detected in two years out of a three-year study, and the haplotype effect of the QTL (Fig. 5, (Lin et al. 2021)) was significantly associated with field SNB susceptibility across all years tested,

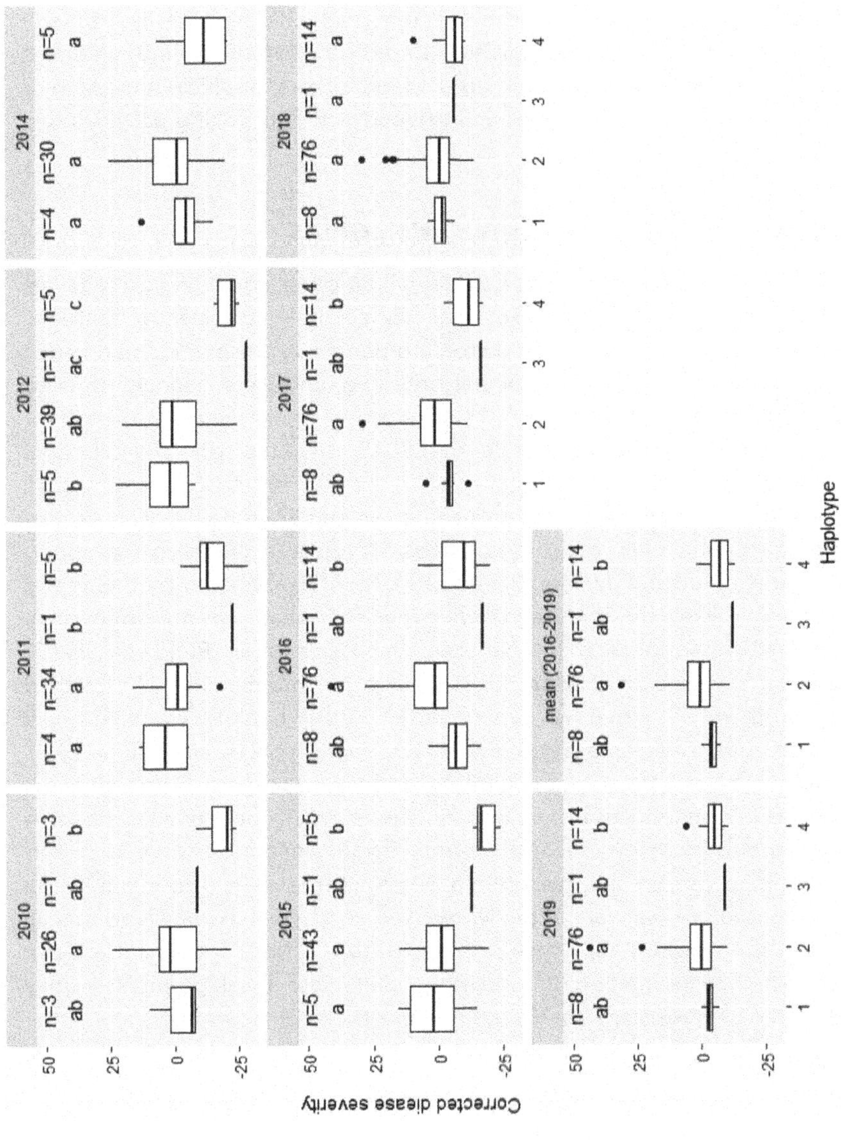

Figure 6 Haplotype analysis of the 2AS QTL in the winter wheat panel of nine years' field trial and mean of four years from 2016 to 2019 (Bottom right). Same letter on boxplots indicate no significant difference between haplotypes determined by Tukey's HSD test ($p < 0.05$).

indicating the robustness of the QTL and the potential value of applying it in MAS. In addition, another robust adult plant resistance QTL, *QSnb.nmbu-5A .1*, which was only found in this BMWpop MAGIC population, showed additive effect to *QSnb.nmbu-2A.1*. Therefore, it was concluded that using MAS to combine resistance alleles at these two loci could be a promising strategy for improving SNB resistance in wheat breeding. Moreover, the multi-locus haplotypes determined in the study provided markers for efficient tracking of these beneficial alleles in future wheat genetics and breeding activities (Lin et al. 2021).

5.3.4 Case study 4: QTL detected using GWAS

The GWAS was conducted using a winter wheat panel consisting of 103 lines and a spring wheat panel of 296 lines. Both panels consisted of cultivars, breeding lines and frequently used crossing parents representing the relevant genetic diversity for Norwegian wheat breeding, and were genotyped using the 35K Axiom array (Allen et al. 2017). The winter wheat panel was tested for SNB resistance in field trials from 2016 to 2019, while the spring wheat panel was tested from 2016 to 2018 (Lin et al. in prep.). The most consistent QTL across years were detected on chromosomes 2A and 5A using the winter wheat association panel, while on chromosomes 1A, 2B and 5B using the spring wheat panel. Interestingly, the significant QTL on 5B detected by the spring wheat panel is likely the ToxA sensitivity gene *Tsn1*, which was not significantly detected when the subset of this panel was used for analysis (Ruud et al. 2019). In addition, one robust QTL on the short arm of chromosome 2A detected by the winter wheat panel was of interest for further haplotype analysis, as it explained a relatively large proportion of the phenotypic variation (R^2 = 14%). By combining historical data, the haplotype effect was shown to be significant in seven out of nine tested years (Fig. 6). The resistant allele on chromosome 2A was carried by many lines with German origin while most Norwegian and Swedish lines carried the susceptible allele. Moreover, we found that the resistant haplotype was rare in the Norwegian spring wheat panel and that all lines with the resistant haplotype originated from CIMMYT. It was, therefore, concluded that integrating this resistant allele into local germplasm may improve the SNB resistance in Norwegian wheat (Lin et al., in prep.).

5.3.5 Can sensitivity to NEs explain differences in host SNB resistance in the field?

Since the discovery of the first NE-*Snn* interaction in the wheat *P. nodorum* pathogen system by Liu et al. (2004b), many studies have focused on these interactions and more NEs have been characterized (described in Section 4.1).

However, debates as to how much of the host resistance/susceptibility can be explained by NE-*Snn* interactions are still ongoing (Francki 2013; Ruud and Lillemo 2018). We also found interesting results from our own studies. In the GWAS (Lin et al., in prep.), we identified SNB field QTL to colocate with numerous NE sensitivity loci, such as *Tsn1*, *Snn1*, *Snn2* and *Snn3-B1*, even though some of them were not consistently detected across multiple years. However, our results highlighted the potential usefulness of screening NE sensitivities in breeding lines to reduce field SNB susceptibility. For the QTL mapping using the NIAB Elite MAGIC population, except for QTL *QSnb.niab-2A.3* and *QSnb.niab-3A*, which were detected by both crude CF infiltration and in the field, no other QTL were detected as significant in both greenhouse (i.e. culture filtrate, NE infiltration and seedling *P. nodorum* resistance) and field conditions (Lin et al. 2020a). Furthermore, the disease severities of seedling testing and field testing were not highly correlated. One possible explanation could be that the isolates selected in our greenhouse testing were not representative for the Norwegian *P. nodorum* population. Probably some uncharacterized NEs played an important role in the field but were not expressed or possessed by the selected isolates in our greenhouse study. Since naturally infected straw was used as inoculum for field experiments in Norway, multiple infections by different isolates were expected in the field. From the pathogen population study (Lin et al. 2020b) we knew that all of the three well-characterized NE genes (*SnToxA*, *SnTox1* and *SnTox3*) were common in the Norwegian *P. nodorum* population. NIAB Elite MAGIC segregates for all related sensitivity loci *Tsn1*, *Snn1* and *Snn3-B1*, but these QTL were not detected in the field in any of the tested years neither in Norway nor in the UK, while only the *Snn2* locus was identified in one year from the UK trial as a 'weak QTL' (Lin et al. 2020a). Similar results were shown for the BMWpop, which segregated for both *Snn1* and *Snn3-B1*; however, neither of these QTL were identified in the field using the BMWpop (Lin et al. 2021). Therefore, collectively our observations supported the hypothesis proposed by Peters Haugrud et al. (2019) that *P. nodorum* isolates may not simultaneously express all of the NE genes they harbor. Instead, depending on the host genetic background, the pathogen exploits a 'cost-effective' way to choose which NE to express when the host possesses many sensitivity loci (Peters Haugrud et al. 2019). Epistatic effects caused by host *Snn* genes may result in this phenomenon as well; however, more gene expression analyses on host *Snn* genes are required to disentangle this issue. As resistance/susceptibility to SNB is a quantitative trait, plant defense mechanisms other than NE-*Snn* interactions are also likely to be involved in the *P. nodorum*-wheat interaction in the field. For example, some resistant wheat cultivars minimize fungal penetration by producing lignified papillae (Bird and Ride 1981). Recently, Zhang et al. (2019) reported the genetic introgression of novel resistance genes to both tan spot and SNB from the diploid wheat species *Aegilops speltoides* to bread wheat by chromosome engineering, which provided new resistance resources and opportunities to investigate SNB resistance mechanisms other than the NE-*Snn* interactions. Other general resistance mechanisms are still unexplored.

5.3.6 Field inoculation methods

From an SNB resistance breeding point of view, our results illustrated the importance of field testing using natural pathogen populations as inoculum instead of arbitrarily selecting isolates for resistance screening. Spraying spore suspensions for SNB field inoculation is a standard method used by many studies (Fried and Meister 1987; Laubscher et al. 1966; Uphaus et al. 2007; Wicki et al. 1999), and it has its own specific advantages. For instance, the same isolates could be used in both greenhouse and field studies. Therefore, higher correlations between field and controlled environments would be expected compared to using natural inoculum. In addition, as inoculum is applied directly to either wheat leaves or heads, this method could reduce the influence of confounding traits such as plant height and days to heading, which was observed in the UK trial of the NIAB Elite MAGIC (Lin et al. 2020a). However, as discussed in Section 5.3.1, the high genetic diversity of the *P. nodorum* natural population made it difficult to select representative isolates. Such diversity is expected when the natural population has been long established and commonly undergoes sexual reproduction, and the resulting large effective population size increased the difficulty of choosing representative isolates for screening. Moreover, large proportions of field resistance could not be explained by known NE-*Snn* interactions alone. Therefore, NE screening under controlled greenhouse conditions and field testing with natural *P. nodorum* populations should be combined in order to breed for better resistance.

5.3.7 Influences of plant height and days to heading on SNB disease severity in the field

Under natural conditions, SNB develops from the lower leaves to the upper leaves. Thus, tall and late lines may avoid heavy infection due to morphological and phenological avoidance. Therefore, as previously discussed, confounding traits such as plant height and days to heading would interfere with the detection of the true SNB QTL. Correcting the effects caused by such traits were achieved in this study by extracting the residuals of disease scores from a linear regression using plant height and days to heading as covariates. Additionally, we compared QTL detected by the corrected disease data and uncorrected disease data for the NIAB Elite MAGIC population to further investigate the influences caused by these traits (Lin et al. 2020a). In general, all 'strong QTL'

Published by Burleigh Dodds Science Publishing Limited, 2022.

identified by uncorrected leaf blotch data were detected by corrected data, indicating that the methodology used was robust. In addition, less QTL were identified by the uncorrected dataset and common significant QTL detected by both datasets became less significant using uncorrected data, highlighting the influence of the confounding traits on the reliable detection of true SNB QTL. In addition, QTL analysis of uncorrected glume blotch data showed that only two 'strong QTL' were detected and both were common with plant height QTL on chromosomes 4B and 4D. 'Weak QTL' detected by uncorrected glume blotch data were all detected using corrected glume blotch data. And similar to the analysis of leaf blotch, those true SNB QTL detected using uncorrected glume blotch phenotypes were less significant than the ones detected using the corrected glume blotch phenotypes (Lin et al. 2020a).

5.4 Case study conclusions

- The Norwegian *P. nodorum* population has high evolutionary potential, high genetic diversity and no detectable population subdivision.
- Due to its ability of rapid local adaptation and risk of evolving fungicide resistance, the principles of integrated disease management should be used in order to control SNB in Norway. The overall disease pressure should be reduced by the joint efforts of good agronomic practice (stubble management, crop rotation), choice of resistant cultivars and application of fungicides when needed.
- ToxA is the major virulence factor in the Norwegian *P. nodorum* population, probably due to the local adaptation to Norwegian spring wheat cultivars. Eliminating the ToxA sensitivity allele *Tsn1* in Norwegian spring wheat cultivars may reduce future SNB infection.
- Three robust QTL were detected: two by QTL mapping in MAGIC (on the long arm of chromosome 2A and on 5A) and one by GWAS (on the short arm of 2A). All effects were confirmed by haplotype analysis and the corresponding markers can be used in MAS.
- Correlations between SNB seedling resistance and adult plant resistance are generally low and not all seedling-stage NE-*Snn* interactions detected in the greenhouse are of field relevance.
- In order to improve SNB resistance, field testing should be carried out using natural *P. nodorum* populations as inoculum and cannot be completely replaced by greenhouse assays.
- Resistance to SNB leaf blotch and glume blotch are controlled by different mechanisms.
- SNB resistance is quantitative and most of the SNB resistance-associated QTL each explain a low proportion of the phenotypic variations (<10%) under field conditions. However, stacking of resistant alleles in MAGIC

RILs and the GWAS panels showed significant effect on reducing disease severity.

6 Future trends

6.1 Applying CRISPR/Cas9 technology to improve SNB resistance

Classic plant breeding is mainly based on crossing and selection, which requires genetic recombination and allelic variability. However, sometimes allelic diversity is limited in domesticated crops or it is only some specific traits, such as disease resistance, that is the intended breeding target. In order to introduce beneficial traits from wild relatives to domesticated crops, it often takes 10–15 years to break the linkage drag with undesired traits (Steuernagel et al. 2016). While mutation breeding is a fast way to create genetic variation, mutations caused by either physical or chemical mutagens are usually unpredictable. Genome editing can precisely edit the plant genome and create predictable mutations in elite cultivars, thereby saving the time for backcrossing and overcoming linkage drag. Simultaneously, it also accelerates the selection process by decreasing the effect of random mutations (Li et al. 2012). Genome editing based on the recent CRISPR (clustered, regularly interspaced, short palindromic repeats)/Cas9 system (Zhang et al. 2014) has already been applied to multiple crops. CRISPR/Cas9 can also target multiple homoeologues simultaneously in polyploid crops, such as hexaploid bread wheat and tetraploid potato (Andersson et al. 2017; Feng et al. 2016; Liang et al. 2017; Wang et al. 2014b; Zong et al. 2017). In addition, CRISPR/Cas9 has been successfully used for editing susceptibility genes to key pathogens, with the knock-out mutants all showing enhanced resistance (Berg et al. 2017; Peng et al. 2017; Wang et al. 2014b). As discussed in Section 4.1 (this chapter), two NE sensitivity loci, *Tsn1* and *Snn1*, have been characterized at the gene level, and no yield penalty associated with NE insensitivity has been found so far (Oliver et al. 2014). Eliminating such NE sensitivity alleles by CRISPR/Cas9 from elite cultivars would reduce susceptibility to SNB, and at the same time keep all other desired traits. However, applying this technology requires sequence knowledge of the candidate gene, and is in many countries also subject to the same regulations as genetically modified organisms (Turnbull et al. 2021). At the moment, the other seven known *Snn* loci remain uncloned, therefore eliminating susceptibility at these loci currently still relies on MAS or NE screening assays.

6.2 Marker-assisted selection (MAS)

MAS has been widely used in wheat breeding for selecting major agronomic and quality traits, as well as resistances to different diseases (Bernardo 2008;

Dreisigacker et al. 2016; Miedaner and Korzun 2012; Toth et al. 2019). In comparison to traditional phenotypic selection, which usually requires selection in many generations, selections based on molecular markers can fix the allele of the desired trait in the early generations with high accuracy in the absence of phenotypic data (Dreisigacker et al. 2016; Toth et al. 2019). It is practical especially for breeding disease resistance traits, as MAS can be applied for stacking multiple resistance genes even when some of the diseases have not been present in the field environment (Toth et al. 2019). The selection accuracy of MAS also largely depends on the markers being used. Functional markers which locate within the target genes are diagnostic; however, such markers are not always available. Alternatively, markers linked to the target gene/QTL that to a large degree co-segregate with the desired allele can also be used in MAS. But the selection accuracy depends on the genetic distance between the marker and the gene/QTL, and sometimes may result in false positives in a different population. Therefore, linked markers should be validated in populations with different genetic background before being applied in MAS. As described in the previous sections, MAS can be used to eliminate the NE sensitivity loci in the breeding germplasm. For example, Zhang et al. (2009) developed SSR markers linked to *Tsn1* and *Snn2* which could be used for MAS. In recent years, mapping populations for SNB resistance are genotyped by SNP chips (Cockram et al. 2015; Francki et al. 2020; Lin et al. 2020a; Lin et al. 2021). Significant SNP markers can be simply converted into KASP genotyping system for selection (Cockram et al. 2015; Downie et al. 2018). Moreover, as *Tsn1* and *Snn1* genes have been cloned, diagnostic markers can be easily developed based on the gene sequence. Nevertheless, the resistance/susceptibility to SNB is a quantitative trait. There will be limitation to how much of the genetic variation can be captured in MAS even if diagnostic markers would become available for all the important sensitivity loci.

6.3 Genomic selection (GS)

Genomic prediction/selection uses large amounts of genetic markers covering all chromosomes and prior phenotypic and genotypic data of a training population to estimate the breeding values or predict the phenotypic performance of genotypes with unknown phenotypes (Meuwissen et al. 2001). Unlike conventional MAS which selects breeding germplasm using a relatively small number of genetic markers linked to previously detected major QTL/genes, GS includes genetic information of the whole genome which might give a more accurate prediction of an individual's breeding value (Bernardo and Yu 2007). As reviewed by Bernardo (2016), with the increase in marker density, size of the training population and heritability of the trait, the accuracy of genomic prediction will also increase. In addition, compared to treating all markers with equal random effects as is normally used in GS, fixing marker effects of major genes can also improve the prediction accuracy (Bernardo 2014). Genomic selection may well represent an efficient solution toward SNB resistance breeding as the SNB resistance is a complicated quantitative trait and only few diagnostic markers are available for MAS. Indeed, in our GWAS study only a small part (≤ 17%) of the genetic variation for the trait was explained by the QTL detected (Lin et al. in prep.). Besides, many markers were detected, which would make MAS unfeasible. Genomic selection using genome-wide markers will likely capture much more of the genetic variances. Juliana et al. (2017) conducted the first genomic prediction study on wheat leaf blotch resistances including SNB. However, the study was based on seedling SNB resistance which might not highly correlate with field resistance, as discussed previously in this chapter. Odilbekov et al. (2019) conducted a genomic prediction study on another wheat leaf blotch disease – STB – using a Nordic winter wheat panel, and found that by setting significant markers from GWAS as fixed effects, the prediction accuracy improved from 0.47 to 0.62. With the prior knowledge of the significant markers obtained from our GWAS study, genomic prediction based on field SNB resistance will probably be more relevant for breeders to use in practice.

7 Conclusion

Wheat resistance to *P. nodorum* is quantitatively inherited and is usually a result of complicated interactions. Interactions between NEs produced by *P. nodorum* and wheat sensitivity loci have been well characterized since Liu et al. (2004a). In addition to that, additive and epistatic effects were found between different NE-*Snn* interactions (Peters Haugrud et al. 2019; Phan et al. 2016; Shi et al. 2016). Moreover, due to the coevolutionary nature of the host and pathogens, *P. nodorum* populations may also shift virulence factors, for example, which NE to produce, by adapting to the NE sensitivities of the host cultivars (Lin et al. 2020b; Phan et al. 2019; Richards et al. 2019). Though leaf blotch and glume blotch are just different symptoms caused by the same pathogen on the same host, resistance to leaf blotch and glume are likely controlled by different genetic mechanisms (Aguilar et al. 2005; Francki 2013; Lin et al. 2020a; Shankar et al. 2008b), which makes it more difficult to breed for SNB resistance. It has been proved that some of the NE sensitivity loci play important roles in field susceptibility (Friesen et al. 2009; Phan et al. 2016; Ruud et al. 2017), for which either MAS or NE infiltration can be applied on germplasm screening to eliminate the sensitivity alleles. Despite the fact that many NE sensitivity loci have been characterized, functional markers are in most cases not available. Thanks to the rapid development of wheat genome studies, such as the Chinese spring reference genome (International Wheat Genome

Sequencing et al. 2018) and the recently published pan-genome of 15 diverse cultivars representing the global wheat diversity (Walkowiak et al. 2020), it is expected that the new bioinformatic tools will assist the cloning of more and more wheat resistance/susceptibility genes in the upcoming years. Thus, gene editing technology might also be applied to knock out the susceptibility genes in elite cultivars to reduce their SNB severity. Nevertheless, most of the NE-*Snn* interactions were based on seedling testing with selected isolates, which were quite different from the field conditions where plants were usually exposed to the highly diverse natural pathogen population. Therefore, MAS based on *Snn* genes might still not be able to significantly reduce the field susceptibility. Genomic selection might provide higher selection accuracy than MAS for SNB adult plant resistance, as it is suitable for predicting quantitative traits and taking genotype-by-environment interactions into account.

8 Where to look for further information

Below are some selected papers which are recommended for further reading. Friesen and Faris (2012) provide detailed descriptions of the methodologies for investigating the NE-*Snn* interactions. The short review by Oliver et al. (2016) studied the most common necrotrophic wheat diseases in Australia and wheat resistance mechanisms against those diseases. Francki (2013) reviewed the SNB resistance from a breeding point of view and proposed strategies for improving the SNB resistance in breeding programs. The study by Ruud and Lillemo (2018) generally reviewed SNB and provided details of published SNB-related QTL. Insights into breeding strategies based on the current knowledge of resistance mechanisms are also provided by Downie et al. (2020). A recent published research article by Peters Haugrud et al. (2019) investigated the interactions of the wheat-*P. nodorum* pathosystem, which illustrated the complicated nature of SNB resistance.

9 Acknowledgements

The case study in this chapter was conducted as part of the ERA-NET project 'EfectaWheat' with funding for both authors received from the Research Council of Norway (grant NFR 251894). We are grateful for the collaboration with consortium partners at NIAB, UK (James Cockram, Beatrice Corsi, Rowena Downie), Bavarian State Research Center for Agriculture, Germany (Lorenz Hartl, Volker Mohler, Melanie Stadlmeier), Aarhus University, Denmark (Lise Nistrup Jørgensen, Annemarie Fejer Justesen), Curtin University, Australia (Richard Oliver, Kar-Chun Tan) as well as the national collaborators at NIBIO (Andrea Ficke) and Graminor (Jon Arne Dieseth, Muath Alsheikh).

10 References

Abeysekara, N. S., Faris, J. D., Chao, S., McClean, P. E. and Friesen, T. L. (2012). Whole-genome QTL analysis of Stagonospora nodorum blotch resistance and validation of the SnTox4-Snn4 interaction in hexaploid wheat. *Phytopathology* 102(1):94–104 doi:10.1094/PHYTO-02-11-0040.

Abeysekara, N. S., Friesen, T. L., Keller, B. and Faris, J. D. (2009). Identification and characterization of a novel host-toxin interaction in the wheat-Stagonospora nodorum pathosystem. *Theor. Appl. Genet.* 120(1):117–126 doi:10.1007/s00122-009-1163-6.

Adhikari, T. B., Jackson, E. W., Gurung, S., Hansen, J. M. and Bonman, J. M. (2011). Association mapping of quantitative resistance to Phaeosphaeria nodorum in spring wheat landraces from the USDA National Small Grains Collection. *Phytopathology* 101(11):1301–1310 doi:10.1094/PHYTO-03-11-0076.

Aguilar, V., Stamp, P., Winzeler, M., Winzeler, H., Schachermayr, G., Keller, B., Zanetti, S. and Messmer, M. M. (2005). Inheritance of field resistance to Stagonospora nodorum leaf and glume blotch and correlations with other morphological traits in hexaploid wheat (Triticum aestivum L.). *Theor. Appl. Genet.* 111(2):325–336 doi:10.1007/s00122-005-2025-5.

Akbari, M., Wenzl, P., Caig, V., Carling, J., Xia, L., Yang, S., Uszynski, G., Mohler, V., Lehmensiek, A., Kuchel, H., Hayden, M. J., Howes, N., Sharp, P., Vaughan, P., Rathmell, B., Huttner, E. and Kilian, A. (2006). Diversity arrays technology (DArT) for high-throughput profiling of the hexaploid wheat genome. *Theor. Appl. Genet.* 113(8):1409–1420 doi:10.1007/s00122-006-0365-4.

Allen, A. M., Winfield, M. O., Burridge, A. J., Downie, R. C., Benbow, H. R., Barker, G. L., Wilkinson, P. A., Coghill, J., Waterfall, C., Davassi, A., Scopes, G., Pirani, A., Webster, T., Brew, F., Bloor, C., Griffiths, S., Bentley, A. R., Alda, M., Jack, P., Phillips, A. L. and Edwards, K. J. (2017). Characterization of a Wheat Breeders' Array suitable for high-throughput SNP genotyping of global accessions of hexaploid bread wheat (Triticum aestivum). *Plant Biotechnol. J.* 15(3):390–401 doi:10.1111/pbi.12635.

Andersson, M., Turesson, H., Nicolia, A., Falt, A. S., Samuelsson, M. and Hofvander, P. (2017). Efficient targeted multiallelic mutagenesis in tetraploid potato (Solanum tuberosum) by transient CRISPR-Cas9 expression in protoplasts. *Plant Cell Rep.* 36(1):117–128 doi:10.1007/s00299-016-2062-3.

Bathgate, J. A. and Loughman, R. (2001). Ascospores are a source of inoculum of Phaeosphaeria nodorum, P. avenaria f. sp. avenaria and Mycosphaerella graminicola in Western Australia. *Austral. Plant Pathol.* 30(4):317–322 doi:10.1071/AP01043.

Bearchell, S. J., Fraaije, B. A., Shaw, M. W. and Fitt, B. D. (2005). Wheat archive links long-term fungal pathogen population dynamics to air pollution. *Proc. Natl. Acad. Sci. U.S A.* 102(15):5438–5442 doi:10.1073/pnas.0501596102.

Bennett, R. S., Yun, S. H., Lee, T. Y., Turgeon, B. G., Arseniuk, E., Cunfer, B. M. and Bergstrom, G. C. (2003). Identity and conservation of mating type genes in geographically diverse isolates of Phaeosphaeria nodorum. *Fungal Genet. Biol.* 40(1):25–37 doi:10.1016/s1087-1845(03)00062-8.

Berg, J. A., Appiano, M., Bijsterbosch, G., Visser, R. G. F., Schouten, H. J. and Bai, Y. (2017). Functional characterization of cucumber (Cucumis sativus L.) Clade V MLO genes. *BMC Plant Biol.* 17(1):80 doi:10.1186/s12870-017-1029-z.

Bernardo, R. (2008). Molecular markers and selection for complex traits in plants: learning from the last 20 years. *Crop Sci.* 48(5):1649–1664 doi:10.2135/cropsci2008.03.0131.

Bernardo, R. (2014). Genomewide selection when major genes are known. *Crop Sci.* 54(1):68–75 doi:10.2135/cropsci2013.05.0315.

Bernardo, R. (2016). Bandwagons I, too, have known. *Theor. Appl. Genet.* 129(12):2323–2332 doi:10.1007/s00122-016-2772-5.

Bernardo, R. and Yu, J. M. (2007). Prospects for genome-wide selection for quantitative traits in maize. *Crop Sci.* 47(3):1082–1090 doi:10.2135/cropsci2006.11.0690.

Bhathal, J. S., Loughman, R. and Speijers, J. (2003). Yield reduction in wheat in relation to leaf disease from yellow (tan) spot and Septoria nodorum blotch. *Eur. J. Plant Pathol.* 109(5):435–443 doi:10.1023/A:1024277420773.

Bird, P. M. and Ride, J. P. (1981). The resistance of wheat to Septoria-nodorum - fungal development in relation to. *Physiol. Plant Pathol.* 19(3):289–299 doi:10.1016/S0048-4059(81)80063-X.

Blixt, E., Djurle, A., Yuen, J. and Olson, Å (2009). Fungicide sensitivity in Swedish isolates of Phaeosphaeria nodorum. *Plant Pathol.* 58(4):655–664 doi:10.1111/j.1365-3059.2009.02041.x.

Blixt, E., Olson, Å., Lindahl, B., Djurle, A. and Yuen, J. (2010). Spatiotemporal variation in the fungal community associated with wheat leaves showing symptoms similar to Stagonospora nodorum blotch. *Eur. J. Plant Pathol.* 126(3):373–386 doi:10.1007/s10658-009-9542-z.

Bostwick, D. E., Ohm, H. W. and Shaner, G. (1993). Inheritance of Septoria-glume blotch resistance in wheat. *Crop Sci.* 33(3):439–443 doi:10.2135/cropsci1993.0011183X003300030005x.

Breseghello, F. and Sorrells, M. E. (2006). Association analysis as a strategy for improvement of quantitative traits in plants. *Crop Sci.* 46(3):1323–1330 doi:10.2135/cropsci2005.09-0305.

Brutus, A., Sicilia, F., Macone, A., Cervone, F. and De Lorenzo, G. (2010). A domain swap approach reveals a role of the plant wall-associated kinase 1 (WAK1) as a receptor of oligogalacturonides. *Proc. Natl. Acad. Sci. U.S.A.* 107(20):9452–9457 doi:10.1073/pnas.1000675107.

Cavanagh, C., Morell, M., Mackay, I. and Powell, W. (2008). From mutations to MAGIC: resources for gene discovery, validation and delivery in crop plants. *Curr. Opin. Plant Biol.* 11(2):215–221 doi:10.1016/j.pbi.2008.01.002.

Cockram, J., Scuderi, A., Barber, T., Furuki, E., Gardner, K. A., Gosman, N., Kowalczyk, R., Phan, H. P., Rose, G. A., Tan, K. C., Oliver, R. P. and Mackay, I. J. (2015). Fine-mapping the wheat Snn1 locus conferring sensitivity to the Parastagonospora nodorum necrotrophic effector SnTox1 using an eight founder multiparent advanced generation inter-cross population. *G3 (Bethesda)* 5(11):2257–2266 doi:10.1534/g3.115.021584.

Cunfer, B. M. (1999). Stagonospora and septoria pathogens of cereals: the infection process. In van Ginkel, M., McNab, A. and Krupinsky, J. (Eds) Septoria and Stagonospora Diseases of Cereals: A Compilation of Global Research. CIMMYT, Mexico, pp 41–45.

Cunfer, B. M. (2000). Stagonospora and Septoria diseases of barley, oat and rye. *Can. J. Plant Pathol.* 22(4):332–348 doi:10.1080/07060660009500452.

Czembor, P. C., Arseniuk, E., Czaplicki, A., Song, Q., Cregan, P. B. and Ueng, P. P. (2003). QTL mapping of partial resistance in winter wheat to Stagonospora nodorum blotch. *Genome* 46(4):546–554 doi:10.1139/g03-036.

Czembor, P. C., Arseniuk, E., Radecka-Janusik, M., Piechota, U. and Słowacki, P. (2019). Quantitative trait loci analysis of adult plant resistance to Parastagonospora nodorum blotch in winter wheat cv. Liwilla (Triticum aestivum L.). *Eur. J. Plant Pathol.* 155(3):1001–1016 doi:10.1007/s10658-019-01829-5.

Downie, R. C., Bouvet, L., Furuki, E., Gosman, N., Gardner, K. A., Mackay, I. J., Campos Mantello, C., Mellers, G., Phan, H. T. T., Rose, G. A., Tan, K. C., Oliver, R. P. and Cockram, J. (2018). Assessing European wheat sensitivities to Parastagonospora nodorum necrotrophic effectors and fine-mapping the Snn3-B1 locus conferring sensitivity to the effector SnTox3. *Front. Plant Sci.* 9:881 doi:10.3389/fpls.2018.00881.

Downie, R. C., Lin, M., Corsi, B., Ficke, A., Lillemo, M., Oliver, R. P., Phan, H., Tan, K. C. and Cockram J. (2020). Septoria nodorum blotch of wheat: disease management and resistance breeding in the face of shifting disease dynamics and a changing environment. *(Early View) Phytopathology* 27. doi:10.1094/PHYTO-07-20-0280-RVW.

Dreisigacker, S., Sukumaran, S., Guzmán, C., He, X., Lan, C., Bonnett, D. and Crossa, J. (2016). Molecular marker-based selection tools in spring bread wheat improvement: CIMMYT experience and prospects. In: Rajpal V. RS, Raina S. (ed) Molecular Breeding for Sustainable Crop Improvement. *Sustainable Development and Biodiversity* 11:421–474 Switzerland. doi:10.1007/978-3-319-27090-6.

Eyal, Z., Scharen, A. L., Prescott, J. M. and Ginkel, M. V. (1987). *The Septoria Diseases of Wheat : Concepts and Methods of Disease Management*. CIMMYT, Mexico.

Faris, J. D., Anderson, J. A., Francl, L. J. and Jordahl, J. G. (1996). Chromosomal location of a gene conditioning insensitivity in wheat to a necrosis-inducing culture filtrate from Pyrenophora tritici-repentis. *Phytopathology* 86(5):459–463 doi:10.1094/Phyto-86-459.

Faris, J. D., Zhang, Z., Lu, H., Lu, S., Reddy, L., Cloutier, S., Fellers, J. P., Meinhardt, S. W., Rasmussen, J. B., Xu, S. S., Oliver, R. P., Simons, K. J. and Friesen, T. L. (2010). A unique wheat disease resistance-like gene governs effector-triggered susceptibility to necrotrophic pathogens. *Proc. Natl Acad. Sci. U.S.A.* 107(30):13544–13549 doi:10.1073/pnas.1004090107.

Feng, C., Yuan, J., Wang, R., Liu, Y., Birchler, J. A. and Han, F. (2016). Efficient targeted genome modification in maize using CRISPR/Cas9 system. *J. Genet. Genomics* 43(1):37–43 doi:10.1016/j.jgg.2015.10.002.

Ficke, A., Cowger, C., Bergstrom, G. and Brodal, G. (2018a). Understanding yield loss and pathogen biology to improve disease management: Septoria nodorum Blotch - A case study in wheat. *Plant Dis.* 102(4):696–707 doi:10.1094/PDIS-09-17-1375-FE.

Ficke, A., Dieseth, J. A., Kim, M. O. and Lillemo, M. (2018b). Bladsjukdommer i norsk hvete. Forekomst, betydning of tiltak. In: *Jord- og Plantekultur 2018/NIBIO BOK*. Norwegian Institute of Bioeconomy Research:108–112.

Flor, H. H. (1971). Current status of the gene-for-gene concept. *Annu. Rev. Phytopathol.* 9(1):275–296 doi:10.1146/annurev.py.09.090171.001423.

Francki, M. G. (2013). Improving Stagonospora nodorum resistance in wheat: a review. *Crop Sci.* 53(2):355–365 doi:10.2135/cropsci2012.06.0347.

Francki, M. G., Walker, E., Li, D. A. and Forrest, K. (2018). High-density SNP mapping reveals closely linked QTL for resistance to Stagonospora nodorum blotch (SNB) in flag leaf and glume of hexaploid wheat. *Genome* 61(2):145–149 doi:10.1139/gen-2017-0203.

Francki, M. G., Walker, E., McMullan, C. J. and Morris, W. G. (2020). Multi-location evaluation of global wheat lines reveal multiple QTL for adult plant resistance to Septoria nodorum Blotch (SNB) detected in specific environments and in response to different isolates. *Front. Plant Sci.* 11:771 doi:10.3389/fpls.2020.00771.

Fried, P. M. and Meister, E. (1987). Inheritance of leaf and head resistance of winter wheat to septoria nodorum in a diallel cross. *Phytopathology* 77(10):1371–1375 doi:10.1094/Phyto-77-1371.

Friesen, T. L., Chu, C., Xu, S. S. and Faris, J. D. (2012). SnTox5-Snn5: a novel Stagonospora nodorum effector-wheat gene interaction and its relationship with the SnToxA-Tsn1 and SnTox3-Snn3-B1 interactions. *Mol. Plant Pathol.* 13(9):1101–1109 doi:10.1111/j.1364-3703.2012.00819.x.

Friesen, T. L., Chu, C. G., Liu, Z. H., Xu, S. S., Halley, S. and Faris, J. D. (2009). Host-selective toxins produced by Stagonospora nodorum confer disease susceptibility in adult wheat plants under field conditions. *Theor. Appl. Genet.* 118(8):1489–1497 doi:10.1007/s00122-009-0997-2.

Friesen, T. L. and Faris, J. D. (2010). Characterization of the wheat-Stagonospora nodorum disease system: what is the molecular basis of this quantitative necrotrophic disease interaction? *Can. J. Plant Pathol.* 32(1):20–28 doi:10.1080/07060661003620896.

Friesen, T. L. and Faris, J. D. (2012). Characterization of plant-fungal interactions involving necrotrophic effector-producing plant pathogens. In: Bolton, M. D. and Thomma, B. P. H. J. (Eds) *Plant Fungal Pathogens: Methods and Protocols*. Humana Press, Totowa, NJ, pp 191–207.

Friesen, T. L., Holmes, D. J., Bowden, R. L. and Faris, J. D. (2018). ToxA is present in the U.S. Bipolaris sorokiniana population and is a significant virulence factor on wheat harboring Tsn1. *Plant Dis.* 102(12):2446–2452 doi:10.1094/PDIS-03-18-0521-RE.

Friesen, T. L., Meinhardt, S. W. and Faris, J. D. (2007). The Stagonospora nodorum-wheat pathosystem involves multiple proteinaceous host-selective toxins and corresponding host sensitivity genes that interact in an inverse gene-for-gene manner. *Plant J.* 51(4):681–692 doi:10.1111/j.1365-313X.2007.03166.x.

Friesen, T. L., Stukenbrock, E. H., Liu, Z., Meinhardt, S., Ling, H., Faris, J. D., Rasmussen, J. B., Solomon, P. S., McDonald, B. A. and Oliver, R. P. (2006). Emergence of a new disease as a result of interspecific virulence gene transfer. *Nat. Genet.* 38(8):953–956 doi:10.1038/ng1839.

Friesen, T. L., Zhang, Z., Solomon, P. S., Oliver, R. P. and Faris, J. D. (2008). Characterization of the interaction of a novel Stagonospora nodorum host-selective toxin with a wheat susceptibility gene. *Plant Physiol.* 146(2):682–693 doi:10.1104/pp.107.108761.

Gao, Y., Faris, J. D., Liu, Z., Kim, Y. M., Syme, R. A., Oliver, R. P., Xu, S. S. and Friesen, T. L. (2015). Identification and characterization of the SnTox6-Snn6 interaction in the Parastagonospora nodorum-wheat pathosystem. *Mol. Plant Microbe Interact.* 28(5):615–625 doi:10.1094/MPMI-12-14-0396-R.

Ghaderi, F., Sharifnabi, B., Javan-Nikkhah, M., Brunner, P. C. and McDonald, B. A. (2020). SnToxA ,SnTox1, and SnTox3 originated in Parastagonospora nodorum in the Fertile Crescent. *Plant Pathol.* 69(8):1482–1491 doi:10.1111/ppa.13233.

Gupta, P. K., Kulwal, P. L. and Jaiswal, V. (2014). Association mapping in crop plants: opportunities and challenges. *Adv. Genet.* 85:109–147 doi:10.1016/B978-0-12-800271-1.00002-0.

Gurung, S., Mamidi, S., Bonman, J. M., Xiong, M., Brown-Guedira, G. and Adhikari, T. B. (2014). Genome-wide association study reveals novel quantitative trait loci associated with resistance to multiple leaf spot diseases of spring wheat. *PLoS ONE* 9(9):e108179 doi:10.1371/journal.pone.0108179.

International Wheat Genome Sequencing Consortium (IWGSC) (2014). A chromosome-based draft sequence of the hexaploid bread wheat (Triticum aestivum) genome. *Science* 345(6194):1251788 doi:10.1126/science.1251788.

International Wheat Genome Sequencing Consortium (IWGSC), IWGSC RefSeq principal investigators: Appels, R., Eversole, K., Feuillet, C., Keller, B., Rogers, J., Stein, N., IWGSC whole-genome assembly principal investigators: Pozniak, C. J., Stein, N., Choulet, F., Distelfeld, A., Eversole, K., Poland, J., Rogers, J., Ronen, G., Sharpe, A. G., Whole-genome sequencing and assembly: Pozniak, C., Ronen, G., Stein, N., Barad, O., Baruch, K., Choulet, F., Keeble-Gagnère, G., Mascher, M., Sharpe, A. G., Ben-Zvi, G., Josselin, A. A., Hi-C data-based scaffolding: Stein, N., Mascher, M., Himmelbach, A., Whole-genome assembly quality control and analyses: Choulet, F., Keeble-Gagnère, G., Mascher, M., Rogers, J., Balfourier, F., Gutierrez-Gonzalez, J., Hayden, M., Josselin, A. A., Koh, C., Muehlbauer, G., Pasam, R. K., Paux, E., Pozniak, C. J., Rigault, P., Sharpe, A. G., Tibbits, J., Tiwari, V., Pseudomolecule assembly: Choulet, F., Keeble-Gagnère, G., Mascher, M., Josselin, A. A., Rogers, J., RefSeq genome structure and gene analyses: Spannagl, M., Choulet, F., Lang, D., Gundlach, H., Haberer, G., Keeble-Gagnère, G., Mayer, K. F. X., Ormanbekova, D., Paux, E., Prade, V., Šimková, H., Wicker, T., Automated annotation: Choulet, F., Spannagl, M., Swarbreck, D., Rimbert, H., Felder, M., Guilhot, N., Gundlach, H., Haberer, G., Kaithakottil, G., Keilwagen, J., Lang, D., Leroy, P., Lux, T., Mayer, K. F. X., Twardziok, S., Venturini, L., Manual gene curation: Appels, R., Rimbert, H., Choulet, F., Juhász, A., Keeble-Gagnère, G., Subgenome comparative analyses: Choulet, F., Spannagl, M., Lang, D., Abrouk, M., Haberer, G., Keeble-Gagnère, G., Mayer, K. F. X., Wicker, T., Transposable elements: Choulet, F., Wicker, T., Gundlach, H., Lang, D., Spannagl, M., Phylogenomic analyses: Lang, D., Spannagl, M., Appels, R., Fischer, I., Transcriptome analyses and RNA-seq data: Uauy, C., Borrill, P., Ramirez-Gonzalez, R. H., Appels, R., Arnaud, D., Chalabi, S., Chalhoub, B., Choulet, F., Cory, A., Datla, R., Davey, M. W., Hayden, M., Jacobs, J., Lang, D., Robinson, S. J., Spannagl, M., Steuernagel, B., Tibbits, J., Tiwari, V., van Ex, F., Wulff, B. B. H., Whole-genome methylome: Pozniak, C. J., Robinson, S. J., Sharpe, A. G., Cory, A., Histone mark analyses: Benhamed, M., Paux, E., Bendahmane, A., Concia, L., Latrasse, D., BAC chromosome MTP IWGSC-Bayer Whole-Genome Profiling (WGP) tags: Rogers, J., Jacobs, J., Alaux, M., Appels, R., Bartoš, J., Bellec, A., Berges, H., Doležel, J., Feuillet, C., Frenkel, Z., Gill, B., Korol, A., Letellier, T., Olsen, O. A., Šimková, H., Singh, K., Valárik, M., van der Vossen, E., Vautrin, S., Weining, S., Chromosome LTC mapping and physical mapping quality control: Korol, A., Frenkel, Z., Fahima, T., Glikson, V., Raats, D., Rogers, J., RH mapping: Tiwari, V., Gill, B., Paux, E., Poland, J., Optical mapping: Doležel, J., Číhalíková, J., Šimková, H., Toegelová, H., Vrána, J., Recombination analyses: Sourdille, P., Darrier, B., Gene family analyses: Appels, R., Spannagl, M., Lang, D., Fischer, I., Ormanbekova, D., Prade, V., CBF gene family: Barabaschi, D., Cattivelli, L., Dehydrin gene family: Hernandez, P., Galvez, S., Budak, H., NLR gene family: Steuernagel, B., Jones, J. D. G., Witek, K., Wulff, B. B. H., Yu, G., PPR gene family: Small, I., Melonek, J., Zhou, R., Prolamin gene family: Juhász, A., Belova, T., Appels, R., Olsen, O. A., WAK gene family: Kanyuka, K., King, R., Stem solidness (SSt1) QTL team: Nilsen, K., Walkowiak,

S., Pozniak, C. J., Cuthbert, R., Datla, R., Knox, R., Wiebe, K., Xiang, D., Flowering locus C (FLC) gene team: Rohde, A., Golds, T., Genome size analysis: Doležel, J., Čížková, J., Tibbits, J., MicroRNA and tRNA annotation: Budak, H., Akpinar, B. A., Biyiklioglu, S., Genetic maps and mapping: Muehlbauer, G., Poland, J., Gao, L., Gutierrez-Gonzalez, J., N'Daiye, A., BAC libraries and chromosome sorting: Doležel, J., Šimková, H., Číhalíková, J., Kubaláková, M., Šafář, J., Vrána, J., BAC pooling, BAC library repository, and access: Berges, H., Bellec, A., Vautrin, S., IWGSC sequence and data repository and access: Alaux, M., Alfama, F., Adam-Blondon, A. F., Flores, R., Guerche, C., Letellier, T., Loaec, M., Quesneville, H., Physical maps and BAC-based sequences: 1A BAC sequencing and assembly: Pozniak, C. J., Sharpe, A. G., Walkowiak, S., Budak, H., Condie, J., Ens, J., Koh, C., Maclachlan, R., Tan, Y., Wicker, T., 1B BAC sequencing and assembly: Choulet, F., Paux, E., Alberti, A., Aury, J. M., Balfourier, F., Barbe, V., Couloux, A., Cruaud, C., Labadie, K., Mangenot, S., Wincker, P., 1D, 4D, and 6D physical mapping: Gill, B., Kaur, G., Luo, M., Sehgal, S., 2AL physical mapping: Singh, K., Chhuneja, P., Gupta, O. P., Jindal, S., Kaur, P., Malik, P., Sharma, P., Yadav, B., 2AS physical mapping: Singh, N. K., Khurana, J., Chaudhary, C., Khurana, P., Kumar, V., Mahato, A., Mathur, S., Sevanthi, A., Sharma, N., Tomar, R. S., 2B, 2D, 4B, 5BL, and 5DL IWGSC-Bayer Whole-Genome Profiling (WGP) physical maps: Rogers, J., Jacobs, J., Alaux, M., Bellec, A., Berges, H., Doležel, J., Feuillet, C., Frenkel, Z., Gill, B., Korol, A., van der Vossen, E., Vautrin, S., 3AL physical mapping: Gill, B., Kaur, G., Luo, M., Sehgal, S., 3DS physical mapping and BAC sequencing and assembly: Bartoš, J., Holušová, K., Plíhal, O., 3DL BAC sequencing and assembly: Clark, M. D., Heavens, D., Kettleborough, G., Wright, J., 4A physical mapping, BAC sequencing, assembly, and annotation: Valárik, M., Abrouk, M., Balcárková, B., Holušová, K., Hu, Y., Luo, M., 5BS BAC sequencing and assembly: Salina, E., Ravin, N., Skryabin, K., Beletsky, A., Kadnikov, V., Mardanov, A., Nesterov, M., Rakitin, A., Sergeeva, E., 6B BAC sequencing and assembly: Handa, H., Kanamori, H., Katagiri, S., Kobayashi, F., Nasuda, S., Tanaka, T., Wu, J., 7A physical mapping and BAC sequencing: Appels, R., Hayden, M., Keeble-Gagnère, G., Rigault, P., Tibbits, J., 7B physical mapping, BAC sequencing, and assembly: Olsen, O. A., Belova, T., Cattonaro, F., Jiumeng, M., Kugler, K., Mayer, K. F. X., Pfeifer, M., Sandve, S., Xun, X., Zhan, B., 7DS BAC sequencing and assembly: Šimková, H., Abrouk, M., Batley, J., Bayer, P. E., Edwards, D., Hayashi, S., Toegelová, H., Tulpová, Z., Visendi, P., 7DL physical mapping and BAC sequencing: Weining, S., Cui, L., Du, X., Feng, K., Nie, X., Tong, W., Wang, L., Figures: Borrill, P., Gundlach, H., Galvez, S., Kaithakottil, G., Lang, D., Lux, T., Mascher, M., Ormanbekova, D., Prade, V., Ramirez-Gonzalez, R. H., Spannagl, M., Stein, N., Uauy, C., Venturini, L., Manuscript writing team: Stein, N., Appels, R., Eversole, K., Rogers, J., Borrill, P., Cattivelli, L., Choulet, F., Hernandez, P., Kanyuka, K., Lang, D., Mascher, M., Nilsen, K., Paux, E., Pozniak, C. J., Ramirez-Gonzalez, R. H., Šimková, H., Small, I., Spannagl, M., Swarbreck, D. and Uauy, C. (2018). Shifting the limits in wheat research and breeding using a fully annotated reference genome. *Science* 361(6403):eaar7191 doi:10.1126/science.aar7191.

Jighly, A., Alagu, M., Makdis, F., Singh, M., Singh, S., Emebiri, L. C. and Ogbonnaya, F. C. (2016). Genomic regions conferring resistance to multiple fungal pathogens in synthetic hexaploid wheat. *Mol. Breed.* 36(9) doi:10.1007/s11032-016-0541-4.

Jönsson, J. (1985). Evaluation of leaf resistance to Septoria nodorum in winter wheat at seedling and adult plant stage. *Agri Hortique Genet.* 43:52–68.

Juliana, P., Singh, R. P., Singh, P. K., Crossa, J., Rutkoski, J. E., Poland, J. A., Bergstrom, G. C. and Sorrells, M. E. (2017). Comparison of models and whole-genome profiling approaches for genomic-enabled prediction of Septoria tritici Blotch, Stagonospora nodorum blotch, and tan spot resistance in wheat. *Plant Genome* 10(2):1-16 doi:10.3835/plantgenome2016.08.0082.

Justesen, A. F, Corsi, B., Ficke, A., Hartl, L., Holdgate, S., Jørgensen, L. N., Lillemo, M., Lin, M., Mackay, I. J., Mohler, V., Stadlmeier, M., Tan, K. C., Turner, J., Oliver, R. P. and Cockram, J. (submitted). Hidden in plain sight: a molecular field survey of three wheat leaf blotch fungal diseases in north-western Europe shows co-infection is widespread. Submitted to *Eur. J. Plant Pathol.*

Keller, S. M., McDermott, J. M., Pettway, R. E., Wolfe, M. S. and McDonald, B. A. (1997). Gene flow and sexual reproduction in the wheat glume blotch pathogen Phaeosphaeria nodorum (anamorph Stagonospora nodorum). *Phytopathology* 87(3):353-358 doi:10.1094/PHYTO.1997.87.3.353.

Langridge, P., Lagudah, E. S., Holton, T. A., Appels, R., Sharp, P. J. and Chalmers, K. J. (2001). Trends in genetic and genome analyses in wheat: a review. *Aust. J. Agric. Res.* 52(12):1043-1077 doi:10.1071/AR01082.

Laubscher, F. X., Wechmar, B. and Schalkwyk, D. (1966). Heritable resistance of wheat varieties to glume blotch (Septoria nodorum Berk.). *J. Phytopathol.* 56(3):260-264 doi:10.1111/j.1439-0434.1966.tb02262.x.

Leath, S., Scharen, A. L., Dietzholmes, M. E. and Lund, R. E. (1993). Factors associated with global occurrences of Septoria-nodorum blotch and Septoria-tritici blotch of wheat. *Plant Dis.* 77(12):1266-1270 doi:10.1094/PD-77-1266.

Li, T., Liu, B., Spalding, M. H., Weeks, D. P. and Yang, B. (2012). High-efficiency TALEN-based gene editing produces disease-resistant rice. *Nat. Biotechnol.* 30(5):390-392 doi:10.1038/nbt.2199.

Liang, Z., Chen, K., Li, T., Zhang, Y., Wang, Y., Zhao, Q., Liu, J., Zhang, H., Liu, C., Ran, Y. and Gao, C. (2017). Efficient DNA-free genome editing of bread wheat using CRISPR/Cas9 ribonucleoprotein complexes. *Nat. Commun.* 8:14261 doi:10.1038/ncomms14261.

Lin, M., Corsi, B., Ficke, A., Tan, K. C., Cockram, J. and Lillemo, M. (2020a). Genetic mapping using a wheat multi-founder population reveals a locus on chromosome 2A controlling resistance to both leaf and glume blotch caused by the necrotrophic fungal pathogen Parastagonospora nodorum. *Theor. Appl. Genet.* 133(3):785-808 doi:10.1007/s00122-019-03507-w.

Lin, M., Ficke, A., Cockram, J. and Lillemo, M. (2020b). Genetic structure of the Norwegian Parastagonospora nodorum population. *Front. Microbiol.* 11:1280 doi:10.3389/fmicb.2020.01280.

Lin, M., Stadlmeier, M., Mohler, V., Tan, K. C., Ficke, A., Cockram, J. and Lillemo, M. (2021). Identification and cross-validation of genetic loci conferring resistance to Septoria nodorum blotch using a German multi-founder winter wheat population. *Theor. Appl. Genet.* 134(1):125-142 doi:10.1007/s00122-020-03686-x.

Liu, Z., El-Basyoni, I., Kariyawasam, G., Zhang, G., Fritz, A., Hansen, J., Marais, F., Friskop, A., Chao, S., Akhunov, E. and Baenziger, P. S. (2015). Evaluation and association mapping of resistance to tan spot and Stagonospora nodorum blotch in adapted winter wheat germplasm. *Plant Dis.* 99(10):1333-1341 doi:10.1094/PDIS-11-14-1131-RE.

Liu, Z., Faris, J. D., Oliver, R. P., Tan, K. C., Solomon, P. S., McDonald, M. C., McDonald, B. A., Nunez, A., Lu, S., Rasmussen, J. B. and Friesen, T. L. (2009). SnTox3 acts in effector triggered susceptibility to induce disease on wheat carrying the Snn3 gene. *PLoS Pathog.* 5(9):e1000581 doi:10.1371/journal.ppat.1000581.

Liu, Z., Friesen, T. L., Ling, H., Meinhardt, S. W., Oliver, R. P., Rasmussen, J. B. and Faris, J. D. (2006). The Tsn1-ToxA interaction in the wheat-Stagonospora nodorum pathosystem parallels that of the wheat-tan spot system. *Genome* 49(10):1265-1273 doi:10.1139/g06-088.

Liu, Z., Zhang, Z., Faris, J. D., Oliver, R. P., Syme, R., McDonald, M. C., McDonald, B. A., Solomon, P. S., Lu, S., Shelver, W. L., Xu, S. and Friesen, T. L. (2012). The cysteine rich necrotrophic effector SnTox1 produced by Stagonospora nodorum triggers susceptibility of wheat lines harboring Snn1. *PLoS Pathog.* 8(1):e1002467 doi:10.1371/journal.ppat.1002467.

Liu, Z. H., Faris, J. D., Meinhardt, S. W., Ali, S., Rasmussen, J. B. and Friesen, T. L. (2004). Genetic and physical mapping of a gene conditioning sensitivity in wheat to a partially purified host-selective toxin produced by stagonospora nodorum. *Phytopathology* 94(10):1056-1060 doi:10.1094/PHYTO.2004.94.10.1056.

Liu, Z. H., Gao, Y., Kim, Y. M., Faris, J. D., Shelver, W. L., de Wit, P. J., Xu, S. S. and Friesen, T. L. (2016). SnTox1, a Parastagonospora nodorum necrotrophic effector, is a dual-function protein that facilitates infection while protecting from wheat-produced chitinases. *New Phytol.* 211(3):1052-1064 doi:10.1111/nph.13959.

Lu, Q. and Lillemo, M. (2014). Molecular mapping of adult plant resistance to Parastagonospora nodorum leaf blotch in bread wheat lines 'Shanghai-3/Catbird' and 'Naxos'. *Theor. Appl. Genet.* 127(12):2635-2644 doi:10.1007/s00122-014-2404-x.

Mackay, I. J., Bansept-Basler, P., Barber, T., Bentley, A. R., Cockram, J., Gosman, N., Greenland, A. J., Horsnell, R., Howells, R., O'Sullivan, D. M., Rose, G. A. and Howell, P. J. (2014). An eight-parent multiparent advanced generation inter-cross population for winter-sown wheat: creation, properties, and validation. *G3 (Bethesda)* 4(9):1603-1610 doi:10.1534/g3.114.012963.

McDonald, B. A. and Linde, C. (2002). Pathogen population genetics, evolutionary potential, and durable resistance. *Annu. Rev. Phytopathol.* 40:349-379 doi:10.1146/annurev.phyto.40.120501.101443.

McDonald, B. A., Miles, J., Nelson, L. R. and Pettway, R. E. (1994). Genetic-variability in nuclear-DNA in-field populations of stagonospora-nodorum. *Phytopathology* 84(3):250-255 doi:10.1094/Phyto-84-250.

McDonald, B. A. and Stukenbrock, E. H. (2016). Rapid emergence of pathogens in agro-ecosystems: global threats to agricultural sustainability and food security. *Philos. Trans. R. Soc. Lond. B* 371(1709) doi:10.1098/rstb.2016.0026:28080995.

McDonald, M. C., Ahren, D., Simpfendorfer, S., Milgate, A. and Solomon, P. S. (2018). The discovery of the virulence gene ToxA in the wheat and barley pathogen Bipolaris sorokiniana. *Mol. Plant Pathol.* 19(2):432-439 doi:10.1111/mpp.12535.

McDonald, M. C., Oliver, R. P., Friesen, T. L., Brunner, P. C. and McDonald, B. A. (2013). Global diversity and distribution of three necrotrophic effectors in Phaeosphaeria nodorum and related species. *New Phytol.* 199(1):241-251 doi:10.1111/nph.12257.

McDonald, M. C., Razavi, M., Friesen, T. L., Brunner, P. C. and McDonald, B. A. (2012). Phylogenetic and population genetic analyses of Phaeosphaeria nodorum and its close relatives indicate cryptic species and an origin in the Fertile Crescent. *Fungal Genet. Biol.* 49(11):882–895 doi:10.1016/j.fgb.2012.08.001.

Mehra, L. K., Cowger, C., Weisz, R. and Ojiambo, P. S. (2015). Quantifying the effects of wheat residue on severity of Stagonospora nodorum blotch and yield in winter wheat. *Phytopathology* 105(11):1417–1426 doi:10.1094/PHYTO-03-15-0080-R.

Meuwissen, T. H. E., Hayes, B. J. and Goddard, M. E. (2001). Prediction of total genetic value using genome-wide dense marker maps. *Genetics* 157(4):1819–1829.

Miedaner, T. and Korzun, V. (2012). Marker-assisted selection for disease resistance in wheat and barley breeding. *Phytopathology* 102(6):560–566 doi:10.1094/PHYTO-05-11-0157.

Murphy, N. E., Loughman, R., Appels, R., Lagudah, E. S. and Jones, M. G. K. (2000). Genetic variability in a collection of Stagonospora nodorum isolates from Western Australia. *Aust. J. Agric. Res.* 51(6):679–684 doi:10.1071/AR99107.

Odilbekov, F., Armonienė, R., Koc, A., Svensson, J. and Chawade, A. (2019). GWAS-assisted genomic prediction to predict resistance to Septoria tritici Blotch in Nordic winter wheat at seedling stage. *Front. Genet.* 10:1224 doi:10.3389/fgene.2019.01224.

Oliver, R., Lichtenzveig, J., Tan, K.-C., Waters, O., Rybak, K., Lawrence, J., Friesen, T. and Burgess, P. (2014). Absence of detectable yield penalty associated with insensitivity to Pleosporales necrotrophic effectors in wheat grown in the West Australian wheat belt. *Plant Pathol.* 63(5):1027–1032 doi:10.1111/ppa.12191.

Oliver, R. P., Friesen, T. L., Faris, J. D. and Solomon, P. S. (2012). Stagonospora nodorum: from pathology to genomics and host resistance. *Annu. Rev. Phytopathol.* 50:23–43 doi:10.1146/annurev-phyto-081211-173019

Oliver, R. P., Tan, K.-C. and Moffat, C. S. (2016). *Necrotrophic Pathogens of Wheat. Encyclopedia of Food Grains* (2nd edn.). Academic Press, Elsevier, Oxford, Amsterdam.

Pedersen, E. A. and Hughes, G. R. (1992). The effect of crop rotation on development of the septoria disease complex on spring wheat in Saskatchewan. *Can. J. Plant Pathol.* 14:152–158.

Peng, A. H., Chen, S., Lei, T., Xu, L., He, Y., Wu, L., Yao, L. and Zou, X. (2017). Engineering canker-resistant plants through CRISPR/Cas9-targeted editing of the susceptibility gene CsLOB1 promoter in citrus. *Plant Biotechnol. J.* 15(12):1509–1519 doi:10.1111/pbi.12733.

Pereira, D. A., McDonald, B. A. and Brunner, P. C. (2017). Mutations in the CYP51 gene reduce DMI sensitivity in Parastagonospora nodorum populations in Europe and China. *Pest Manag. Sci.* 73(7):1503–1510 doi:10.1002/ps.4486.

Peters Haugrud, A. R., Zhang, Z., Richards, J. K., Friesen, T. L. and Faris, J. D. (2019). Genetics of variable disease expression conferred by inverse gene-for-gene interactions in the wheat-Parastagonospora nodorum pathosystem. *Plant Physiol.* 180(1):420–434 doi:10.1104/pp.19.00149.

Phan, H. T., Rybak, K., Furuki, E., Breen, S., Solomon, P. S., Oliver, R. P. and Tan, K. C. (2016). Differential effector gene expression underpins epistasis in a plant fungal disease. *Plant J.* 87(4):343–354 doi:10.1111/tpj.13203.

Phan, H. T. T., Jones, D. A. B., Rybak, K., Dodhia, K. N., Lopez-Ruiz, F. J., Valade, R., Gout, L., Lebrun, M. H., Brunner, P. C., Oliver, R. P. and Tan, K. C. (2019). Low amplitude boom-and-bust cycles define the Septoria nodorum blotch interaction. *Front. Plant Sci.* 10:1785 doi:10.3389/fpls.2019.01785.

Quaedvlieg, W., Verkley, G. J., Shin, H. D., Barreto, R. W., Alfenas, A. C., Swart, W. J., Groenewald, J. Z. and Crous, P. W. (2013). Sizing up Septoria. *Stud. Mycol.* 75(1):307-390 doi:10.3114/sim0017.

Richards, J. K., Stukenbrock, E. H., Carpenter, J., Liu, Z., Cowger, C., Faris, J. D. and Friesen, T. L. (2019). Local adaptation drives the diversification of effectors in the fungal wheat pathogen Parastagonospora nodorum in the United States. *PLOS Genet.* 15(10):e1008223 doi:10.1371/journal.pgen.1008223.

Ruud, A. K., Dieseth, J. A., Ficke, A., Furuki, E., Phan, H. T. T., Oliver, R. P., Tan, K. C. and Lillemo, M. (2019). Genome-wide association mapping of resistance to septoria nodorum leaf blotch in a Nordic spring wheat collection. *Plant Genome* 12(3):1-15 doi:10.3835/plantgenome2018.12.0105.

Ruud, A. K., Dieseth, J. A. and Lillemo, M. (2018). Effects of three Parastagonospora nodorum necrotrophic effectors on spring wheat under Norwegian field conditions. *Crop Sci.* 58(1):159-168 doi:10.2135/cropsci2017.05.0281.

Ruud, A. K. and Lillemo, M. (2018). Diseases affecting wheat: septoria nodorum blotch. In: *Burleigh Dodds Series in Agricultural Science.* Burleigh Dodds Science Publishing Limited, Cambridge, UK:109-144 doi:10.19103/AS.2018.0039.06.

Ruud, A. K., Windju, S., Belova, T., Friesen, T. L. and Lillemo, M. (2017). Mapping of SnTox3-Snn3 as a major determinant of field susceptibility to Septoria nodorum leaf blotch in the SHA3/CBRD x Naxos population. *Theor. Appl. Genet.* 130(7):1361-1374 doi:10.1007/s00122-017-2893-5.

Schnurbusch, T., Paillard, S., Fossati, D., Messmer, M., Schachermayr, G., Winzeler, M. and Keller, B. (2003). Detection of QTLs for Stagonospora glume blotch resistance in Swiss winter wheat. *Theor. Appl. Genet.* 107(7):1226-1234 doi:10.1007/s00122-003-1372-3.

Shaner, G. (1995). Epidemiology of leaf blotch of soft red winter wheat caused by septoria tritici and stagonospora nodorum. *Plant Dis.* 79(9):928-938 doi:10.1094/PD-79-0928.

Shankar, M., Walker, E., Golzar, H., Loughman, R., Wilson, R. E. and Francki, M. G. (2008). Quantitative trait loci for seedling and adult plant resistance to Stagonospora nodorum in wheat. *Phytopathology* 98(8):886-893 doi:10.1094/PHYTO-98-8-0886.

Shaw, M. W., Bearchell, S. J., Fitt, B. D. L. and Fraaije, B. A. (2008). Long-term relationships between environment and abundance in wheat of Phaeosphaeria nodorum and Mycosphaerella graminicola. *New Phytol.* 177(1):229-238 doi:10.1111/j.1469-8137.2007.02236.x.

Shi, G., Zhang, Z., Friesen, T. L., Raats, D., Fahima, T., Brueggeman, R. S., Lu, S., Trick, H. N., Liu, Z., Chao, W., Frenkel, Z., Xu, S. S., Rasmussen, J. B. and Faris, J. D. (2016). The hijacking of a receptor kinase-driven pathway by a wheat fungal pathogen leads to disease. *Sci. Adv.* 2(10):e1600822 doi:10.1126/sciadv.1600822.

Shi, G. J., Friesen, T. L., Saini, J., Xu, S. S., Rasmussen, J. B. and Faris, J. D. (2015). The wheat Snn7 gene confers susceptibility on recognition of the Parastagonospora nodorum necrotrophic effector SnTox7. *Plant Genome* 8(2):eplantgenome2015.02.0007 doi:10.3835/plantgenome2015.02.0007.

Solomon, P. S., Lowe, R. G., Tan, K. C., Waters, O. D. and Oliver, R. P. (2006). Stagonospora nodorum: cause of Stagonospora nodorum blotch of wheat. *Mol. Plant Pathol.* 7(3):147-156 doi:10.1111/j.1364-3703.2006.00326.x.

Sommerhalder, R. J., McDonald, B. A., Mascher, F. and Zhan, J. (2011). Effect of hosts on competition among clones and evidence of differential selection between pathogenic and saprophytic phases in experimental populations of the wheat pathogen Phaeosphaeria nodorum. *BMC Evol. Biol.* 11:188 doi:10.1186/1471-2148-11-188.

Sommerhalder, R. J., McDonald, B. A. and Zhan, J. (2006). The frequencies and spatial distribution of mating types in Stagonospora nodorum are consistent with recurring sexual reproduction. *Phytopathology* 96(3):234–239 doi:10.1094/PHYTO-96-0234.

Stadlmeier, M., Hartl, L. and Mohler, V. (2018) Usefulness of a multiparent advanced generation intercross population with a greatly reduced mating design for genetic studies in winter wheat. *Front Plant Sci.* 9:1825. https ://doi.org/10.3389/fpls.2018 .01825.

Steuernagel, B., Periyannan, S. K., Hernández-Pinzón, I., Witek, K., Rouse, M. N., Yu, G., Hatta, A., Ayliffe, M., Bariana, H., Jones, J. D., Lagudah, E. S. and Wulff, B. B. (2016). Rapid cloning of disease-resistance genes in plants using mutagenesis and sequence capture. *Nat. Biotechnol.* 34(6):652–655 doi:10.1038/nbt.3543.

Stukenbrock, E. H., Banke, S. and McDonald, B. A. (2006). Global migration patterns in the fungal wheat pathogen Phaeosphaeria nodorum. *Mol. Ecol.* 15(10):2895–2904 doi:10.1111/j.1365-294X.2006.02986.x.

Tomas, A., Feng, G. H., Reeck, G. R., Bockus, W. W. and Leach, J. E. (1990). Purification of a cultivar-specific toxin from pyrenophora-tritici-repentis, causal agent of tan spot of wheat. *Mol. Plant Microbe.* 3:221–224. doi:10.1094/Mpmi-3-221.

Tommasini, L., Schnurbusch, T., Fossati, D., Mascher, F. and Keller, B. (2007). Association mapping of Stagonospora nodorum blotch resistance in modern European winter wheat varieties. *Theor. Appl. Genet.* 115(5):697–708 doi:10.1007/s00122-007-0601-6.

Torriani, S. F., Melichar, J. P., Mills, C., Pain, N., Sierotzki, H. and Courbot, M. (2015). Zymoseptoria tritici: a major threat to wheat production, integrated approaches to control. *Fungal Genet. Biol.* 79:8–12 doi:10.1016/j.fgb.2015.04.010.

Toth, J., Pandurangan, S., Burt, A., Mitchell Fetch, J. and Kumar, S. (2019) Marker-assisted breeding of hexaploid spring wheat in the Canadian prairies. *Can. J. Plant Sci.* 99(2):111–127 doi:10.1139/cjps-2018-0183.

Turnbull, C., Lillemo, M. and Hvoslef-Eide, T. A. K. (2021) Global regulation of genetically modified crops amid the gene edited crop boom – a review. *Front. Plant Sci.* 12:630396. doi: 10.3389/fpls.2021.630396.

Uphaus, J., Walker, E., Shankar, M., Golzar, H., Loughman, R., Francki, M. and Ohm, H. (2007). Quantitative trait loci identified for resistance to Stagonospora glume blotch in wheat in the USA and Australia. *Crop Sci.* 47(5):1813–1822 doi:10.2135/cropsci2006.11.0732.

van Baalen, M. and Sabelis, M. W. (1995). The dynamics of multiple infection and the evolution of virulence. *Am. Nat.* 146(6):881–910 doi:10.1086/285830.

Walkowiak, S., Gao, L., Monat, C., Haberer, G., Kassa, M. T., Brinton, J., Ramirez-Gonzalez, R. H., Kolodziej, M. C., Delorean, E., Thambugala, D., Klymiuk, V., Byrns, B., Gundlach, H., Bandi, V., Siri, J. N., Nilsen, K., Aquino, C., Himmelbach, A., Copetti, D., Ban, T., Venturini, L., Bevan, M., Clavijo, B., Koo, D. H., Ens, J., Wiebe, K., N'Diaye, A., Fritz, A. K., Gutwin, C., Fiebig, A., Fosker, C., Fu, B. X., Accinelli, G. G., Gardner, K. A., Fradgley, N., Gutierrez-Gonzalez, J., Halstead-Nussloch, G., Hatakeyama, M., Koh, C. S., Deek, J., Costamagna, A. C., Fobert, P., Heavens, D., Kanamori, H., Kawaura, K., Kobayashi, F., Krasileva, K., Kuo, T., McKenzie, N., Murata, K., Nabeka, Y., Paape, T., Padmarasu, S., Percival-Alwyn, L., Kagale, S., Scholz, U., Sese, J., Juliana, P., Singh, R., Shimizu-Inatsugi, R., Swarbreck, D., Cockram, J., Budak, H., Tameshige, T., Tanaka, T., Tsuji, H., Wright, J., Wu, J., Steuernagel, B., Small, I., Cloutier, S., Keeble-Gagnère, G.,

Muehlbauer, G., Tibbets, J., Nasuda, S., Melonek, J., Hucl, P. J., Sharpe, A. G., Clark, M., Legg, E., Bharti, A., Langridge, P., Hall, A., Uauy, C., Mascher, M., Krattinger, S. G., Handa, H., Shimizu, K. K., Distelfeld, A., Chalmers, K., Keller, B., Mayer, K. F. X., Poland, J., Stein, N., McCartney, C. A., Spannagl, M., Wicker, T. and Pozniak, C. J. (2020). Multiple wheat genomes reveal global variation in modern breeding. *Nature* 588(7837):277-283 doi:10.1038/s41586-020-2961-x.

Wang, S., Wong, D., Forrest, K., Allen, A., Chao, S., Huang, B. E., Maccaferri, M., Salvi, S., Milner, S. G., Cattivelli, L., Mastrangelo, A. M., Whan, A., Stephen, S., Barker, G., Wieseke, R., Plieske, J., International Wheat Genome Sequencing Consortium, Lillemo, M., Mather, D., Appels, R., Dolferus, R., Brown-Guedira, G., Korol, A., Akhunova, A. R., Feuillet, C., Salse, J., Morgante, M., Pozniak, C., Luo, M. C., Dvorak, J., Morell, M., Dubcovsky, J., Ganal, M., Tuberosa, R., Lawley, C., Mikoulitch, I., Cavanagh, C., Edwards, K. J., Hayden, M. and Akhunov, E. (2014a). Characterization of polyploid wheat genomic diversity using a high-density 90,000 single nucleotide polymorphism array. *Plant Biotechnol. J.* 12(6):787-796 doi:10.1111/pbi.12183.

Wang, Y., Cheng, X., Shan, Q., Zhang, Y., Liu, J., Gao, C. and Qiu, J. L. (2014b). Simultaneous editing of three homoeoalleles in hexaploid bread wheat confers heritable resistance to powdery mildew. *Nat. Biotechnol.* 32(9):947-951 doi:10.1038/nbt.2969.

Wicki, W., Winzeler, M., Schmid, J. E., Stamp, P. and Messmer, M. (1999). Inheritance of resistance to leaf and glume blotch caused by Septoria nodorum Berk. in winter wheat. *Theor. Appl. Genet.* 99(7-8):1265-1272 doi:10.1007/s001220051332.

Williams, J. R. and Jones, D. G. (1973) Infection of grasses by Septoria nodorum and S. tritici. *Trans. Br. Mycol. Soc.* 60:355-358.

Zhang, F., Wen, Y. and Guo, X. (2014). CRISPR/Cas9 for genome editing: progress, implications and challenges. *Hum. Mol. Genet.* 23(R1):R40-R46 doi:10.1093/hmg/ddu125.

Zhang, W., Zhu, X., Zhang, M., Shi, G., Liu, Z. and Cai, X. (2019). Chromosome engineering-mediated introgression and molecular mapping of novel Aegilops speltoides-derived resistance genes for tan spot and Septoria nodorum blotch diseases in wheat. *Theor. Appl. Genet.* 132(9):2605-2614 doi:10.1007/s00122-019-03374-5.

Zhang, Z., Friesen, T. L., Xu, S. S., Shi, G., Liu, Z., Rasmussen, J. B. and Faris, J. D. (2011). Two putatively homoeologous wheat genes mediate recognition of SnTox3 to confer effector-triggered susceptibility to Stagonospora nodorum. *Plant J.* 65(1):27-38 doi:10.1111/j.1365-313X.2010.04407.x.

Zhang, Z. C., Friesen, T. L., Simons, K. J., Xu, S. S. and Faris, J. D. (2009). Development, identification, and validation of markers for marker-assisted selection against the Stagonospora nodorum toxin sensitivity genes Tsn1 and Snn2 in wheat. *Mol. Breed.* 23(1):35-49 doi:10.1007/s11032-008-9211-5.

Zong, Y., Wang, Y., Li, C., Zhang, R., Chen, K., Ran, Y., Qiu, J. L., Wang, D. and Gao, C. (2017). Precise base editing in rice, wheat and maize with a Cas9-cytidine deaminase fusion. *Nat. Biotechnol.* 35(5):438-440 doi:10.1038/nbt.3811.

Chapter 12

Advances in breeding techniques for durable resistance to spot blotch in cereals

Ramesh Chand, Institute of Agricultural Sciences, Banaras Hindu University, India; Sudhir Navathe, Agharkar Research Institute, India; and Sandeep Sharma, Institute of Agricultural Sciences, Banaras Hindu University, India

1 Introduction

Spot blotch, caused by the hemi-biotrophic fungus *Bipolaris sorokiniana*, is one of the production constraints for wheat crops grown in warm, humid tropics (Singh et al. 2015, Gupta et al. 2018). The disease is identified as one of the major biotic problems in the warm, humid tropics of India, Bangladesh, Nepal, Brazil, Bolivia, Argentina and Paraguay (Fig. 1; Gurung et al. 2014, Singh et al.

http://dx.doi.org/10.19103/AS.2021.0092.18

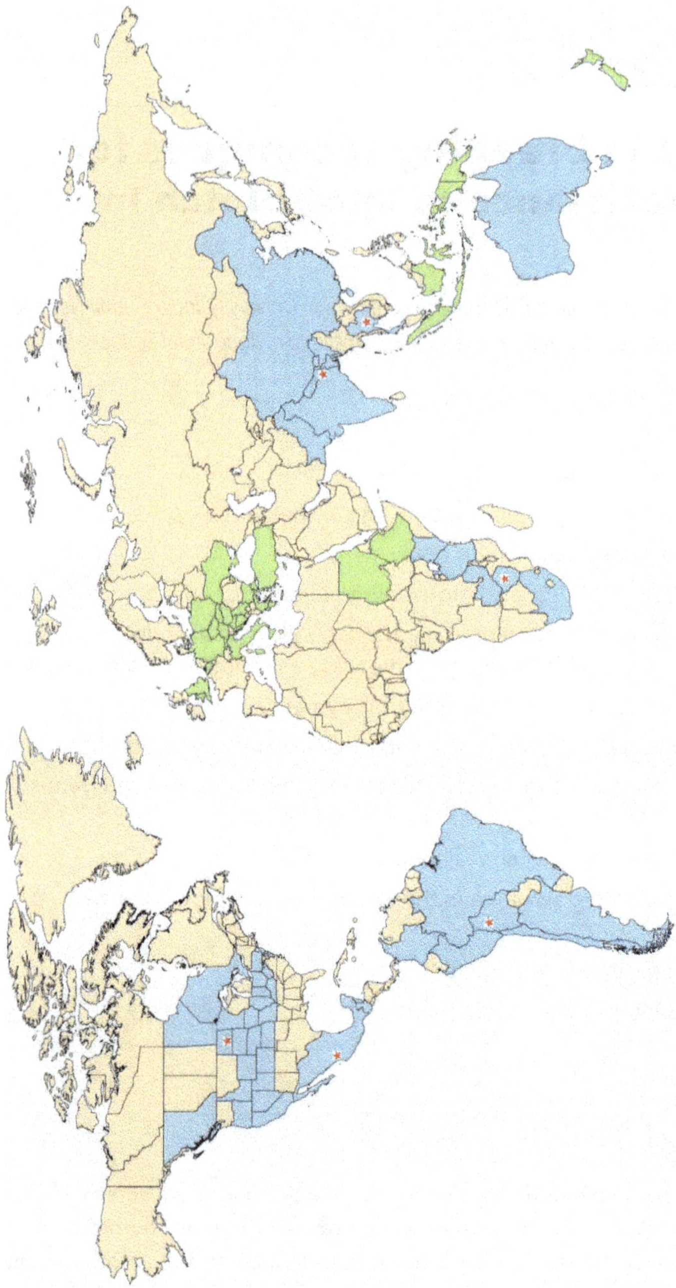

Figure 1 Global occurrence of the spot blotch disease on wheat and barley. The area highlighted in blue indicates the major occurrence while green indicates the area of minor importance. The asterisk (*) indicates the sites where screening for spot blotch resistance can be carried out under field conditions.

2015). Under favorable climate the disease spreads rapidly after anthesis, damaging the photosynthetic capacity of the plant during the grain-filling period, affecting the grain yield and quality (Joshi et al. 2007a, Singh et al. 2018). The disease is aggravated by intermittent rain with warm temperatures during February and calls for additional efforts to manage losses. Spot blotch resistance has been shown to be under polygenic control (Joshi et al. 2004, Kumar et al. 2014, 2016, Singh et al. 2018). Resistant cultivars exhibit fewer and smaller lesions than do susceptible ones. The reproduction of spores in these lesions has been observed to be delayed and reduced in volume (Eisa et al. 2013, Poudel et al. 2019). Breeding efforts have brought many improved wheat genotypes resistant to spot blotch (Singh et al. 2015). However, the desired levels of resistance are not yet achieved as none of the previously released wheat lines showed complete immunity against spot blotch infection (Sharma and Duveiller 2006, Joshi et al. 2007b). Delivery of new genotypes to farmers with a good level of resistance and a high yield would offer a cost-effective and environmentally friendly solution to the fight against spot blotch (Singh et al. 2015, Gupta et al. 2018).

2 Spread, economic importance and disease cycle

Foliar blight of wheat in South Asia is often referred to as Spot blotch or Helminthosporium leaf blight. The causal agent *B. sorokiniana* is a hemi-biotrophic fungus that is distributed worldwide. This non-specific pathogen is reported on several graminaceous hosts and a wide variety of other plants. Among cereals, it affects wheat and barley. The spot blotch is prevalent in 10 million hectares of wheat in South Asia, mainly in the East Gangetic Plains (EGP). The disease is responsible for significant yield losses ranging from 15.5 to 19.6 percent per year (Dubin and Ginkel 1991) and possibly up to 100 percent in some year during the epidemic (Singh et al. 2015). Globally, an estimated 25 mha of wheat is affected to some degree by spot blotch, which represents approximately 12 percent of the total wheat area (Duveiller et al. 2005). Spot blotch tends to flare up due to favorable weather such as warm air temperature, >12 h of leaf wetness and plant age (especially after ear emergence; Zadok's GS 55, Chaurasia et al. 1999, Joshi et al. 2007c). In addition, an increasing number of cloudy and foggy days from November to February favored the early establishment of the pathogen (Duveiller et al. 2005, Joshi et al. 2007c). *B. sorokiniana* is considered to be a relatively weak parasite; its success majorly relies on weakened host plants responding to various environmental stresses such as high temperature (Rosyara et al. 2007a, 2008), nutrient deficiency (Regmi et al. 2002) and water stress (Sharma and Duveiller 2006). These multiple stresses reduce the fitness of the plants that make *B. sorokiniana* more invasive and promote the development of spot blotch. The spot blotch is forecasted to become more serious in the future due to rising problems of climate change and nutritional and water deficiency (Sharma and Duveiller 2006). Spot blotch in combination with terminal heat stress results in loss of photosynthesis, premature leaf senescence, reduced grain filling, low kernel weight and severe reductions in grain yield. If the prediction that mean air temperature is likely to rise in the coming decades proves correct, the disease is expected to be a more serious constraint for wheat production (Gupta et al. 2018). The current procedure for spot blotch screening depends mostly on multi-location screening nurseries (Singh et al. 2015). Pathogenic variation in spot blotch fungus populations has also been reported that adopt the newly resistant cultivars after few years of selection (Maraite et al. 1998, Pandey et al. 2008, Sultana et al. 2018).

3 Diversity of the pathogen and physiological specialization

In contrast to other foliar diseases such as rust, spot blotch resistance in wheat cannot be characterized solely on the basis of lesion characteristics. A series of *B. sorokiniana* strains have been studied for aggressiveness, and symptom types using wheat genotypes with different resistance and continuous variation have been observed (Maraite et al. 2006, Chand et al. 2003, Pandey et al. 2008). Some strains produced very restricted lesions in the most susceptible genotypes, while others cause symptoms even in the most resistant one (Maraite et al. 1998, Chand et al. 2010, Navathe et al. 2020). More and less aggressive strains can be isolated simultaneously in one country or even in one region. Thus, *B. sorokiniana* strains do not show clear patterns and consist of a continuum of strains of varying aggressiveness (Maraite et al. 1998, Duveiller and Altamirano 2000). This is also supported by molecular and pathogenicity studies (Zhong and Steffenson 2001, Chand et al. 2003, Arabi and Jawhar 2004, Ghazvini and Tekauz 2007, Mahto et al. 2012, Sultana et al. 2018). The observed variability in pathogenicity between the strains confirms there is no physiological specialization or physiological race (Duveiller and Sharma 2009). Isolates may be differentiated in pathotypes based on virulence and aggressiveness (Valjavec-Gratian and Steffenson 1997, Poloni et al. 2009, Knight et al. 2010). Previously, in North Dakota, USA, three pathotypes (0, 1, and 2) of *B. sorokiniana* were identified on the basis of their virulence patterns in three differential barley lines (ND 5883, Bowman and ND B112; Valjavec-Gratian and Steffenson 1997). Recently a few other highly virulent pathotypes of *B. sorokiniana* have been identified on barley in Canada, North Dakota (Gyawali 2010, Leng et al. 2016) and Morocco (Gyawali et al. 2018). The regular progression of mean disease severity from less virulent to more virulent strains of different origins suggests variation in quantitative aggressiveness. The widespread occurrence of aggressive strains of pathogen is capable of overcoming promising sources of resistance under experimental conditions. This raises the question of the possible rapid adaptation of strains when resistant cultivars are released (Maraite et al. 1998, Chand et al. 2003, Sultana et al. 2018). A critical tool for the breeding of wheat varieties with sustainable genetic resistance is the collection of isolates representing virulence diversity. If this collection is global, the spread of virulent pathotypes to new regions can be combated by the distribution of germplasms that are resistant to specific pathotypes from all parts of the world. At present, as per the World Data Center for Microorganisms-Global Microorganism Catalog (WDCM-GCM, http:/gcm.wfcc.info/), there are a total of 154 isolates of *B. sorokiniana* available in different microbial repositories in different countries (see isolate details in Table 1). Systematic disease surveillance, global isolates collection with their genotyping data (often including genome assemblies as well as effector gene and fungicide resistance alleles) and, phenotypic data (virulence testing on wheat germplasm) could provide insights into possible pathogenicity changes in global populations of *B. sorokiniana*.

4 Identification of resistance sources

4.1 Screening of the genotypes under field conditions

Most of the studies done under controlled growth conditions do not mimic natural agro-climatic conditions of the field. This might be due to rapid disease progression and difficulties in assessing small phenotypic differences between genotypes under field environment. The resistance mechanism identified at the early seedling stage does not correlate with the resistance mechanisms of the adult plant stage. Thus, breeding and screening for spot blotch resistance must be carried out in the field where environment is favorable to the disease (See Fig. 1). Field screening for spot blotch resistance at many locations in South Asia is based on natural infection at hotspot sites such as the lowland (Terai) region of Nepal, the eastern Uttar Pradesh (India) or West Bengal (Joshi et al. 2007b). Other possible sites for the effective screening of genotype include Jessore (Bangladesh), Chiangmai (Thailand), Encarnacion (Paraguay), El Batan (Mexico), North Dakota (USA) and Zambia. The pathogen may be maintained *in vitro* in jars on sterilized sorghum grains and selection pressure may be increased by spreading these infested grains at different developmental stages like ear initiation or at the physiological Zadoks 55 (Zadoks et al. 1974, Chand et al. 2010). The technique is also followed at the CIMMYT (International Maize and Wheat Improvement Center) site in Agua Fria, Mexico, where screening for spot blotch is carried out on a large scale (Singh et al. 2015). Inoculation of spore suspension at anthesis is also followed commonly in South Asia.

Table 1 List *of Bipolaris sorokiniana* isolates in World Data Centre for Micro-organisms isGlobal catalogue of Microorganism (WDCM-GCM, http://gcm.wfcc.info/)

Sr.	Name of Culture collection	Country	Number of accessions	Accession Number	Link to repository/ Catalogue/Reference
1	NCCS-National Centre for Microbial Resources	India	70	MCC 1503, MCC 1533, MCC 1534, MCC 1535, MCC 1536, MCC 1537, MCC 1538, MCC 1539, MCC 1540, MCC 1541, MCC 1542, MCC 1543, MCC 1544, MCC 1545, MCC 1546, MCC 1547, MCC 1594, MCC 1592, MCC 1593, MCC 1570, MCC 1571, MCC 1572, MCC 1573, MCC 1574, MCC 1575, MCC 1576, MCC 1577, MCC 1578, MCC 1579, MCC 1580, MCC 1581, MCC 1582, MCC 1663, MCC 1664, MCC 1665, MCC 1782, MCC 1746, MCC 1747, MCC 1748, MCC 1673, MCC 1853, MCC 1537, MCC 1536, MCC 1535, MCC 1534, MCC 1533, MCC 1907, MCC 1908, MCC 1909, MCC 1910, MCC 1911, MCC 1912, MCC 1913, MCC 1914, MCC 1915, MCC 1916, MCC 1917, MCC 1918, MCC 1919, MCC 1920, MCC 1921, MCC 1922, MCC 1923, MCC 1924, MCC 1925, MCC 1926, MCC 1927, MCC 1928, MCC 1929	https://www.nccs.res.in/index.php/TeamsNCCS/Repositories
2	ATCC-American Type Culture collection	USA	15	ATCC 32093, ATCC 32094, ATCC 32095, ATCC 32096, ATCC 42133, ATCC 38793, ATCC 18949, ATCC 58474, ATCC 11404, ATCC 24324, ATCC 42957, ATCC 16092, ATCC 58315, ATCC 24082, ATCC 24081	https://www.atcc.org/Products/Collections/Microbiology_Collections/Fungi_and_Yeast.aspx
3	VKM-All-Russian Collection Of Microorganism	Russia	4	VKM F-1446, VKM F-3045, VKM F-3707, VKM F-4006	http://www.vkm.ru/Catalogue.htm
4	UAMH- The UAMH Centre For Global Microfungal Biodiversity	Canada	7	UAMH 517, UAMH 1216, UAMH 1379, UAMH 2677, UAMH 4194, UAMH 4673, UAMH 8282	https://www.uamh.ca/searchCatalogue.php
5	NAIMCC, India	India	2	NAIMCC-F-03005, NAIMCC-F-03006	http://nbaim.org.in/pages/services-culture-collectionnaimccculture-collectionnai
6	KACC- Korean Agricultural Culture Collection	Korea	4	KACC 40920, KACC 40921, KACC 41019, KACC 44884	http://genebank.rda.go.kr/eng/uat/uia/actionMain.do#LINK
7	JMRC: Jena Microbial Resource Collection	Germany	3	SF005612, SF005612, SF002706	https://www.leibniz-hki.de/en/jena-microbial-resource-collection.html

8	IRAN, Iran	Iran	3	IRAN 219C, IRAN 906C, IRAN 3102C	
9	ICMP: International Collection of Microorganisms from Plants	New Zealand	33	ICMP 11701, ICMP 13651, ICMP 13583, ICMP 13584, ICMP 6234, ICMP 6238, ICMP 6239, ICMP 6247, ICMP 6240, ICMP 6256, ICMP 6257, ICMP 6809, ICMP 10307, ICMP 9928, ICMP 9932,ICMP 10104, ICMP 6251, ICMP 10321, ICMP 10362, ICMP 6252, ICMP 6253, ICMP 6249, ICMP 6235, ICMP 6236, ICMP 6237, ICMP 6241, ICMP 6242, ICMP 6243, ICMP 6233, ICMP 6245, ICMP 6246, ICMP 6113, ICMP 10322	https://www.landcareresearch.co.nz/resources/collections/icmp
10	The Leibniz Institute DSMZ - German Collection of Microorganisms and Cell Cultures GmbH	Germany	1	DSM 62608	https://www.dsmz.de/collection/catalogue/microorganisms/catalogue
11	CBS-KNAW- Westerdijk Fungal Biodiversity Institute	Netherlands	24	CBS 110.14, CBS 120.24, CBS 138.29, CBS 139.29, CBS 140.31, CBS 143.32, CBS 144.32, CBS 145.32, CBS 304.50, CBS 147.56, CBS 148.56, CBS 149.56, CBS 150.56, CBS 311.64, CBS 312.64, CBS 510.65, CBS 511.65, CBS 325.68, CBS 537.72, CBS 538.72, CBS 234.77, CBS 127722, CBS 127723, CBS 127724	http://www.wi.knaw.nl/Collections/
12	BCCM/MUCL- IHEM, Belgian Co-ordinated Collections of Micro-organisms	Belgium	2	MUCL 46854, MUCL 46855, IHEM 2024	http://bccm.belspo.be/
13	Isolates reported in various studies but not in the catalogue of designated repositories	Bangladesh	169	-	Sultana et al. 2018, Plant Pathol. J. 34(2) : 93-103
		Brazil	60	-	Mann et al. 2014, Tropical Plant Pathology, 39(2):163-171
		Pakistan	87	-	Asad et al. 2009, Pak. J. Bot., 41(1): 301-308
		Nepal	96	-	Gurung et al. 2013, Plant Dis. 97:62-73
		USA	58	-	Zhong and Steffenson 2001, Phytopathology 91:469-476

4.2 Infection process and symptom types

Bipolaris sorokiniana is best referred to as hemibiotrophic pathogen. The spores adhere to the surface of the leaf, germinate and cause initial infection, forming penetration pegs. During the latent phase, the mycelium grows and invades the living epidermal host cells around the sites of the initial penetration, forming haustoria (biotrophic phase). During 3-7 days of asymptomatic latent growth, the fungus appears to be necrotrophic. Hyphae invade the mesophyll cells, resulting in cell collapse. This cell collapse is mainly due to toxin production because cells die without direct contact of fungal hyphae. The distinguishable necrotic symptoms can be induced by the infiltration of the toxins. During this phase considerable electrolytic leakage will also be observed (Kumar et al. 2002, Eisa et al. 2013, Poudel et al. 2019). The necrotrophic phase leads to various types of macroscopically visible spot blotch symptoms.

B. sorokiniana causes spot blotch on leaves, node canker, and rarely seedling blight. Early leaf lesions are characterized by small, dark brown, 1-2 mm in diameter, with no chlorotic margin (Fig. 2) In susceptible genotypes, these lesions spread several centimeters rapidly to form oval to elongated light to dark brown blotches that coalesce and result in leaf tissue necrosis (Duveiller et al. 2005, Chand et al. 2010). Abundant production of conidia can be observed in old lesions under humid weather. The chlorotic margin is sometimes seen as diffusing from the border of the lesion as a result of the phytotoxin or necrotrophic effector (*BsToxA*) produced by the pathogen (Navathe et al. 2020). Common root rot of wheat is usually found in cooler dry environments and is characterized by brown spotting on the inner crown and internodes. In the field, asexual anamorphic stage *B. sorokiniana* is found to be associated with root or leaf lesions. The spot blotch symptoms appear on the leaf, the sheath, the node and

Figure 2 Symptoms (a) Wheat line highly susceptible to the spot blotch. (b) various types of symptoms on the wheat leaves. (c) black point symptoms on wheat seeds.

the glume (Kumar et al. 2002, Chand et al. 2003). Generally, the disease begins with older leaves and gradually progresses upward. In most cases, the spikes are also affected, displaying black points on seeds (Chand et al. 2010, Bashyal et al. 2011a).

4.3 Scoring disease severity and quantifying disease progress

Spot blotch infection develops different types of lesions, for example, pin head size, dark brown to black, scattered, chlorotic, elongated brown with and without chlorotic halo in wheat (Chand et al. 2010); however, majority of these lesions have been identified in the wheat seedlings under controlled condition. Under the field, lesions progress rapidly under conducive environment and coalesce irrespective of genotype response (Vergnes et al. 2006, Chand et al. 2010, Bashyal et al. 2011b). Resistance cannot, therefore, be scored on the basis of the type of infection, such as rust. Several scales have been proposed for disease score over the years to assess resistance germplasm (Adlakha et al. 1984, Duveiller et al. 1998, Nagarajan and Kumar 1998, Fetch and Steffenson 1999). The scale-combining severity and apparent type of lesions are often neither easy to use nor accurate. The severity score of the disease based on the percentage of diseased leaf area is more reliable. Since the spot blotch is either seed or soil-transmitted, the field resistance assessment is based on a visual assessment of the upward disease progression from the lower canopy level. The double-digit scale (00-99) developed as a modification of the scale of severity of Saari and Prescott (Saari and Prescott 1975, Eyal et al. 1987) is most effective and widely adopted. The first digit (D1) of this scale indicates the vertical progression of the disease in the plant. The second digit (D2) indicates the severity of the disease based on the total leaf area. The disease score was converted to disease severity (DS) in percentage terms using the formula (Eyal et al. 1987, Duveiller et al. 1998).

$$\% \text{ Severity} = (D1/9) \times (D2/9) \times 100$$

Repeated observation is needed as the spot blotch progresses rapidly after anthesis; therefore, disease scores must be recorded at 3–7 days' intervals over a 3-4 week period between the anthesis and the dough stage after inoculation (more specifically growth stages include GS 63:beginning of anthesis to half-complete, GS 69:anthesis complete and GS 77:late milking) (Zadoks et al. 1974). This provides better information on the development of the disease and the resistance level in the germplasm.

The area under the disease progress curve measures the progress of a disease, which is generally used to evaluate the resistance in different genotypes. The AUDPC is calculated using the formula of Shaner and Finney (1977) and Madden et al. (2007).

$$\text{AUDPC} = \sum_{i=0}^{n-1}\left[\left\{\left(Y_i + Y_{(i+1)}\right)/2\right\} \times \left(t_{(i+1)} - t_i\right)\right]$$

Where, Y_i = % disease severity level at time t_i, $t_{(i+1)} - t_i$ = Time (days) between two disease scores, n = number of dates at which spot blotch was recorded.

Physiology and growth stages influence the disease severity in the crop. Growing seasons, sowing dates, and locations also add confounding effects to the data of disease severity. Most of these factors can be addressed by AUDPC/day (Duveiller and Sharma 2009). While selecting resistance in the field, it is not always easy to find a good comparison between disease severity, earliness and yield components, because these attributes are independent (Saxena et al. 2017). The selection index (SI) was developed by combining AUDPC, DHD (days to heading) and TKW (1000 kernel weight) and was found effective:

$$\text{SI} = \text{AUDPC rank in ascending order} + \text{DHD rank in ascending order} + \text{TKW rank in descending order.}$$

This index was calculated for each genotype, and the entry with the lowest SI was found promising (Sharma and Duveiller 2006). This approach can be followed for the effective selection of an improved offspring.

4.4 Image-based quantification of disease

Knowledge of the quantity of disease is particularly important to decision-makers in plant-breeding programs where various germplasms, varieties and/or cultivars need to be rated, and for disease management. Early estimation of disease severity included visual estimation, qualitatively or quantitatively, often with the aid of various assessment scales or keys or standard area diagrams. (Madden et al. 2007, Bock et al. 2010, Domiciano et al. 2014). Various methods for estimating or measuring disease severity include approaches such as visual assessment, digital photography and image analysis, and hyperspectral imagery (Bock et al. 2010, Mohanty et al. 2016). Of the methods available for the disease measurements, image-based analysis is an efficient technique to study the plant-pathogen interaction. This technique provides real time, precise and reproducible information of pathogen infection (Mohanty et al. 2016, Mutka et al. 2016). Quantitative resistance is expressed as a lower degree of damage to the host and measured as a percentage of the leaf area covered by the minute, restricted necrotic lesions. The type of disease response can be compared with susceptible controls and measured as a total necrotic area against the green leaf area. Understanding the infection cycle of spot blotch is important to distinguish and measure the most relevant aspects of quantitative resistance. *B. sorokiniana* spores germinate and penetrate the leaves through the stomata or directly through the pegs (Chand et al. 2003, Poudel et al. 2019). After penetration, the pathogen grows within the leaves for 3-5 days without any visible symptoms. In several studies, the importance of resistance that suppresses the reproduction of pathogens (i.e. the production of conidia) was recognized on the basis of quantitative observations of the coverage of lesions. Tools for image analysis such as ImageJ (http://imagej.nih.gov/ij/) and Adobe Photoshop (www.adobe.com/) can be easily accessed to analyze images taken from individual genotypes at different time points of disease progress (Bock et al. 2010, Stewart and McDonald 2014, Karisto et al. 2019). These tools are used mainly to measure leaf area and symptoms such as necrosis and lesions through the color threshold method. The percentage lesion area against the green leaf area can be easily converted to disease severity and then to disease progression curve (Bock et al. 2010). The genotype ranking for the spot blotch resistance based on image analysis can be more precise and accurate than the rankings based on traditional visual scoring. (Stewart and McDonald 2014, Karisto et al. 2018, 2019)

4.5 An account of national and international nurseries

The present methodology majorly relies on the screening of genotypes for spot blotch resistance under multi-location field trials. The concept of shuttle breeding and multi-location screening pursued by the CIMMYT breeding program has succeeded in accumulating a wide range of resistance/ tolerance in the breeding pool (Pask et al. 2014, Singh et al. 2015). Spot blotch resistance is additive in nature (Joshi et al. 2004), and therefore, germplasm from different stress nurseries, when tested against spot blotch, gives a higher chance of selecting stable and broad-based resistance. This strategy was successful in selecting most of the alleles that contributed to the resistance mechanisms against multiple stresses (Pask et al. 2013, Singh et al. 2015). One example of such special nurseries is the CIMMYT-CSISA-SB, which was established in 2009. Since then, CSISA-SB nurseries have been released every year, including the most promising spot blotch resistant lines developed at CIMMYT. The nursery was not only sent to South Asian countries, but also to Zambia, Brazil, Paraguay, Bolivia, and so on, where spot blotch is a major concern. The genotype identified in one environment may not necessarily be effective in other environments. Multi-year/multi-site screening is therefore justified for the identification of genotypes that are resistant across environments. Nurseries such as CSISA-SB use multiple evaluation systems to identify the stable resistant genotype that may be used in the breeding program or even may directly release as the varieties (Fig. 3).

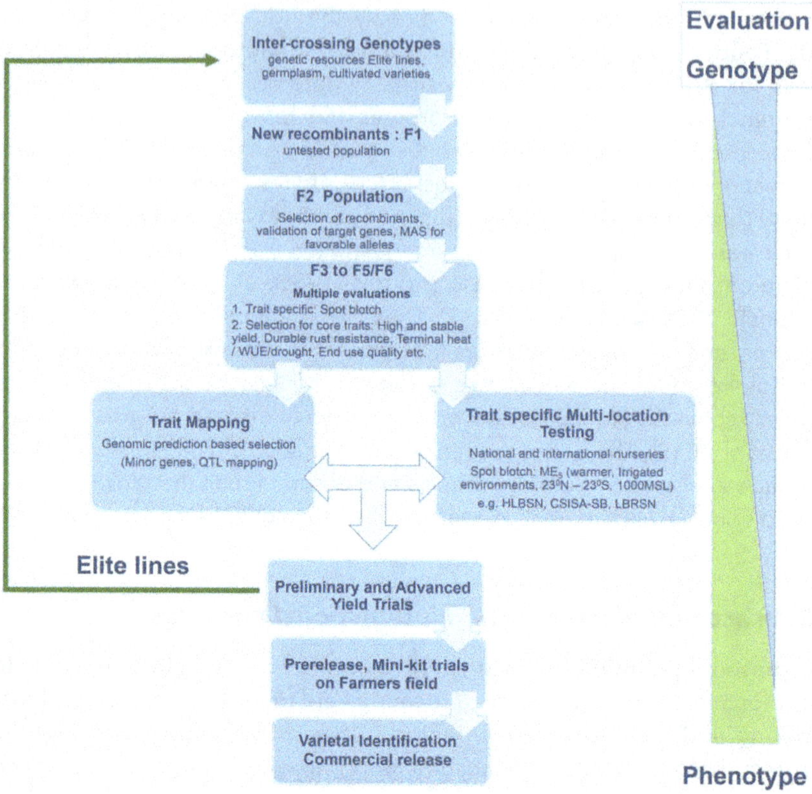

Figure 3 The breeding scheme followed for the selection of the spot blotch resistant genotypes. *ME5: Mega-environment 5, warm humid tropics where spot blotch becomes vulnerable.

5 Histological, biochemical and morphological components of resistance

The biochemical and histological components of disease resistance along with host morphology are significant as they could be used to strengthen the selection for resistance. This is more relevant for a disease such as spot blotch where immunity is not known and breeders are struggling to increase resistance beyond a certain level (Joshi et al. 2007b, Singh et al. 2015). Resistance mechanisms at biochemical and histochemical levels, along with the development of slow disease components, could be combined to provide a broad basis for the selection of long-lasting resistance to spot blotch. Effective resistance can be expressed as a reduction of epidemiological components and, therefore, a slow progression of the disease. The wheat defense model for *B. sorokiniana* was proposed based on a cell-wall-associated mechanism in which the fungus first overcomes epidermis-related defenses and then grows within mesophyll tissues (Ibeagha et al. 2005, Eisa et al. 2013). Many times, disease development is poor under field conditions, possibly due to dryness of the weather or less deposition of the dew on the leaf surface. This gives a biased interpretation about performance of the genotypes. In contrast, biochemical and histochemical parameters are less variable than the morphological parameters and, thus, may serve as more reliable markers to evaluate resistance of the wheat genotypes against spot blotch infection.

5.1 Phenotypic markers

The severity of spot blotch infection in wheat has been influenced by various phenotypic traits. The leaf angle as a feature of spot blotch resistance was studied by Joshi and Chand (2002). Genotypes with erect leaf posture showed less disease susceptibility than those with drooping leaves. Joshi et al. (2002) further established that resistance to spot blotch is independent of plant height and days of maturity. Therefore, it has been suggested that, in order to improve resistance, the breeding population should have susceptible controls of different maturity and height groups. Further, it was advised to combine disease severity score with the growth stage for effective selection of resistant genotypes in the segregating generations.

Leaf tip necrosis (Ltn) is another phenotypic marker associated with spot blotch resistance in wheat (Fig. 4a). *Ltn* is linked to three different loci that confer multiple disease resistance. These loci are *Lr34/Yr18/Pm38* on 7DS (Singh

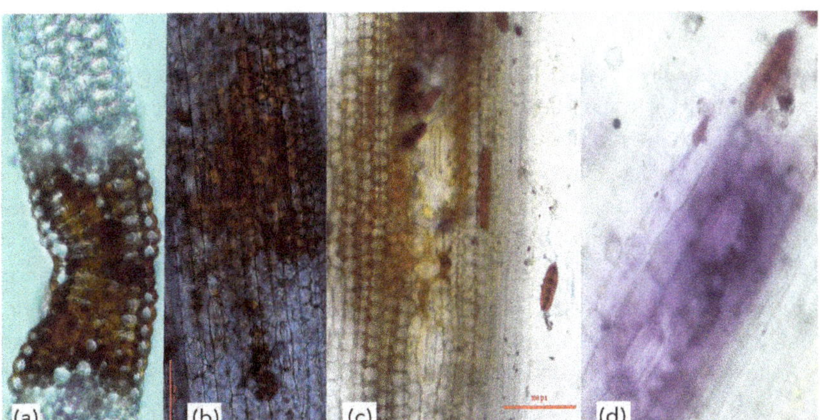

Figure 4 Histochemical staining for identification resistance in response to *B. sorokiniana*. (a-b) Phloroglucinol-HCl (Wiesner) test for lignin staining. (c) 3, 3′-diaminobenzidine (DAB) staining for localization of the H_2O_2 in during pathogenesis and (d) Nitro-blue tetrazolium (NBT) staining for the localization of superoxide anion (O_2^-).

1992), *Lr46/Yr29/Pm39* on 1BL (Rosewarne et al. 2006) and *Lr67/Yr46/Pm46* on 4DL (Herrera-Foessel et al. 2011). Wheat lines positive for Ltn perform better against spot blotch (Joshi et al. 2004). Although the presence of Ltn was associated with resistance to spot blotch, its absence was not always associated with the susceptibility. Further, a QTL identified for spot blotch resistance is found to be co-located with *Lr34* and *Lr46* genes (Lillemo et al. 2013). Use of Ltn as the phenotypic marker facilitates in selection of wheat genotypes with improved resistance to spot blotch infection.

The lesion-mimic (LM) phenotype in wheat expresses distinct whitish-brown lesions on the leaf blades, leaf sheaths and flowering tissues at the boot stage in the absence of any pathogen (Fig. 4c). Lesion mimics produce necrosis in the leaf tissue and promote the growth of necrotrophic/hemibiotrophic pathogens, emphasizing the importance of the lesion mimic locus. The early maturing wheat genotypes with higher expression of lesion mimic phenotypes showed a significant increase in spot blotch susceptibility (Pandey et al. 2016). Therefore, lesion mimic needs adequate attention during selection for early maturing spot blotch resistant genotypes.

Several other components of quantitative resistance, such as lesion number, frequency of infection, lesion size, rate of spore production, incubation period and latent period, that significantly influence the spreading of spot blotch are investigated (Bashyal et al. 2011a,b). Thus, these traits must be considered while selecting the resistant wheat genotypes. It appears that an optimum number of resistance components are required to have a low disease score and slow blotching. These components work together in an additive way and fortify the resistance (Eisa et al. 2013, Sahu et al. 2016, Poudel et al. 2019).

Stay green is a phenomenon where flag leaf of wheat remains photosynthetically active for a longer duration. This prolonged greening of the flag leaf during terminal heat stress and attack of various biotic pathogens was associated with the maintenance of the redox homeostasis. The genotypes with the stay green trait usually perform better against spot blotch. In support of this observation, a positive link was established between stay green and spot blotch resistance (Joshi et al. 2007a, Rosyara et al. 2007a). Additionally, Kumari et al. (2007) established an association of stay green (SG) traits with canopy temperature depression (CTD) and yield characteristics under terminal heat stress, implying that the stay green trait in combination with CTD may be used as effective selection criteria for heat stress tolerance as well as spot blotch. The role of various leaf anatomical characteristics, such as width of large vascular bundles, percentage of small vascular bundles with two girders and number of large veins, was studied for their association with spot blotch resistance in wheat (Rosyara et al. 2007b). However, the contribution of these traits in wheat resistance is not clear, and a large-scale analysis is needed to elucidate their functional roles in spot blotch resistance. A higher chlorophyll content and a low soil plant analysis development (SPAD) value with CTD during grain filling were associated with tolerance to spot blotch and terminal heat (Rosyara et al. 2007a,b). In comparison to ambient temperature, the temperature variation of the plant canopies was used in timely and late-seeding as a criterion for screening of spot blotch and heat stress.

5.2 NDVI-based monitoring and sensing the spot blotch disease

The normalized vegetation difference index (NDVI) is an easy and quick criterion for characterizing the genetic resources of wheat in response to both abiotic and biotic stresses. It is measured by handheld equipment called 'Green Seeker'. The rate of senescence/necrosis and residual 'greenness' can be calculated by sequential measurements of NDVI. Green Seeker is an effective tool in predicting grain yield, vegetation cover, senescence parameters such as stays green, and spot blotch resistance (Montazeaud et al. 2016, Kumar et al. 2016). Stay green has been visually assessed in association with spot blotch in the past (Joshi et al. 2007a). Later, use of soil plant analysis development (SPAD) chlorophyll measurements (Harris et al. 2007, Rosyara et al. 2010) and green seeker for NDVI (Lopes and Reynolds 2012) added precision to data recording. Both these methods have been updated continuously (Kipp et al. 2013, Christopher et al. 2014) to provide a more comprehensive view of the different components of stay green. Kumar et al. (2016) used an NDVI sensor for the first time to map the spot blotch resistance in 335 wheat lines derived from the cross of YS116-9 × Sonalika. A strong negative correlation between NDVI and AUDPC was reported, and spot blotch resistance locus *Sb2* was mapped. The study has successfully demonstrated the use of NDVI decay measurements in estimating the progression of spot blotch disease in wheat.

5.3 Biochemical markers

Biochemical parameters have received considerable attention due to their roles in resistance/disease susceptibility in plants (Bell 1981). Their early response to pathogen attack and significant changes in their quantity ascertained that biochemical parameters (total phenol, chlorophyll and phenylalanine ammonia-lyase (PAL)) could be used for the preliminary determination of resistance to different pathogens. It is reported that the formation of appressorium by the germ tube in fungi is influenced by various plant metabolites such as sucrose, phenols and plant exudates (Hoch and Staples 1991). A couple of recent studies have shown active roles of PAL and phenolic compounds in wheat resistant mechanisms against spot blotch (Sahu et al. 2016, Sharma et al. 2018, Poudel et al. 2019). Different levels of resistance in infected plants may depend on the rate of synthesis and/or the pre-infection level of phenol. *In vitro* studies showed growth inhibitory effect of certain phenols (e.g. syringic acid) and reactive oxygen species (ROS) on mycelial growth, appressoria formation, penetration and lesion size (Sharma et al. 2018, Poudel et al. 2019). The hypersensitive response is one of the earliest defense reactions observed in plants in response to pathogens, including the rapid production of large amounts of reactive oxygen species (Zurbriggen et al. 2010). Superoxide radicals (O_2^-), hydrogen peroxide (H_2O_2) and hydroxyl radicals (OH) are the major forms of ROS that accumulate in high amounts during plant-pathogen interactions (Fig. 5). Under control (non-stress) condition, ROS level is low in the chloroplasts and mitochondria of plant cells. However, levels increased significantly when plants are subjected to biotrophic and hemibiotrophic pathogens. This finally triggers

Figure 5 Phenotypic markers for the section of genotypes against spot blotch under field conditions. (a) Leaf tip necrosis (LTN) indicates the presence of *Lr34* and *Lr46* and spot blotch resistance loci *Sb1*. (b) Pseudo black Chaff indicates the presence of *Sr2*-multiple disease resistance locus. (c) Lesion mimic associated with susceptibility to spot blotch.

oxidative burst that leads to the activation of hypersensitive (HR) defense response (Zubriggen et al. 2010). Higher accumulation of ROS was observed in the early hours of spot blotch infection, which may directly inhibit the pathogen growth and may also induce downstream signaling to activate the plant defense response (Sahu et al. 2016, Poudel et al. 2019). Also, a fine-tuning between ROS and antioxidants protect wheat plants from oxidative damage. In agreement, the antioxidant level was observed to be higher in resistant genotypes than in susceptible genotypes (Sahu et al. 2016).

Malondialdehyde (MDA) is a product of lipid peroxidation and has been considered as a reliable maker to study the effect of various stresses, including spot blotch. Less MDA content in pathogen-treated wheat means the genotypes have less oxidative damage as compared to susceptible genotypes. The higher levels of antioxidant enzymes, balanced ROS levels and reduced MDA content were observed in the resistant genotypes (Yusuf et al. 2016, Sharma et al. 2018). Quantitative analysis of disease progression showed remarkable differences between resistant and susceptible genotypes, suggesting that plant response mechanisms are differentially regulated. This is supported by patterns shown by various defense signaling components, that is ROS, MDA, SOD, CAT, SA and syringic acid (Alscher et al. 2002, Sahu et al. 2016, Sharma et al. 2018). Further, strong negative correlations between defense components and spot blotch progression were observed, which indicates that these signaling pathways are important regulators of wheat resistance mechanisms against spot blotch infection. This has opened up possibilities for engineering these biochemical markers into susceptible genotypes for their long-lasting resistance.

6 Molecular approaches for the resistance breeding: an account of quantitative trait loci (QTL) and genome-wide association studies (GWAS)

The resistance to spot blotch disease is polygenic, quantitative and subjected to genotype-by-environment interaction. Although the resistance components are largely additive in their effect (Kumar et al. 2014, Singh et al. 2018), the combination of these components is expected to lead to stable and genetically broad resistance. Early research involved screening for new resistant germplasms from Zambia, Brazil, and the Yangtze River Valley in China (Singh et al. 2015). Genotypes with satisfactory resistance levels have been identified, widely used in CIMMYT wheat breeding programs and tested in international nurseries in epidemic regions (Dubin and Rajaram 1996, Mehta 1998, Dubin et al. 1998, van Ginkel and Rajaram 1998, Singh et al. 2015, 2018). Resistance sources were also sought in large cross-derivatives produced at CIMMYT, and a number of lines with good resistance were identified (Singh et al. 2018). On the basis of extensive evaluations for many years, several genotypes with

Table 2 Sources of spot blotch resistance reported in wheat.

Origin	Names of genotypes	References
Afghanistan	Pamir 94, Katia-1,OK82282//BOW/NKT/3/F4105, RENESANSA, VORONA/CUPE	Bainsla et al. (2020)
Bangladesh	BAW 969, BAW 1006, BAW 1008	Sharma et al. (2004), Siddique et al. (2006)
Brazil	BH 1146, CEP 14, CNT 1, Ocepar 7, Trigo BR 8	Mehta (1998), Sharma et al. (2004b, 2007b), Caierao et al. (2014)
China	Chuanmai 18, Fang 60, G162, Jinmai 4058, Longmai 10, Longmai 10370, Ning 8201, Ning 8319, Ning 9415 Quangfeng, Shanghai #4, Shanghai #158, Suzhoe #1-58, Suzhoe #8, Suzhoe #128-OY, Yangmai 6	Sharma et al. (1997, 2004), van Ginkel and Rajaram (1998), Joshi et al. (2004, 2007a,b), Ibeagha et al. (2005), Sharma and Duveiller (2007), Kumar et al. (2009, 2010), Bainsla et al. (2020)
CIMMYT	Attila = NL781 =PBW343, BOW 'S', M3, Chirya 1, Chirya 3, Chirya 7, Chukui#1, Cigm 90.455, FFN/VEE#5, HLB25, Kauz/Vee/Muna, Milan/Shanghai #7, SM-4-HSN24, Vayi#1, ALTAR84/AE.SQ//2*SERI/3/CHIR3, PSN/BOW//ROEK/3/MILAN, and CNDO/R143//ENTE/MEXI.2/3/AE.SQUARROSA(TAUS)/4/WEAVER	Chaurasia et al. (1999), Sharma et al. (2004a,b,c), Ragiba et al. (2004a), Duveiller et al. (2005), Ibeagha et al. (2005), Joshi et al. (2007a,b,c), Neupane et al. (2007), Sharma and Duveiller (2007), Kumar et al. (2009, 2010), Singh et al. (2015)
India	ACC 8226, BW 14999, CPAN 3003, CPAN 3048, CPAN 4006, CPAN 4007, CPAN 4011, CPAN 4042, CPAN 4065, CPAN 4070, HD 2662, HD 2819, HP 1729, HP 1808, HUW234, HUW206, HUW289, HUW302, HUW305, HUW323, HUW325, HW 2093, K 9107, M3109, PBW 343, PBW 486, RAJ 3702, Triveni, WH542, YS116 (Yangmai 6/Sonalika), 56 accessions in NBPGR	Chaurasia et al. (1999), Joshi and Chand (2002), Joshi et al. (2002, 2004), Sharma et al. (2004), Sharma and Duveiller (2007), Singh and Singh (2009), Khan and Chowdhury (2011), Kumar et al. (2015, 2016a, b)
Nepal	Achyut, Bhrikuti, BL1693, BL1724, BL1740, BL1813, BL1883, BL2069, BL2127, BL3704, BL4148, Gautam, Mayoor, NL835, NL868, NL872, WK 1204	Sharma et al. (2004a), Sharma and Duveiller (2006), Joshi et al. (2007b), Mahto et al. (2011)
Pakistan	Abadgar 93, Anmal 91, Auqab 2000, Bahawalpur 2000, Bahkhar 2002, Bakhtawar 92, Darawar 97, Faisalabad 85, Inqilab 91, Iqbal 2000, Kaghan 93, Kirin 95, Kohistan 97, Kohsar 95, Magalla 99, Mexi Pak, Moomal 2002, Nowshera 96, Parwaz 94, Pasban 90, Pirsabak 2005, Punjab 96, Saleem 2000, Sariab 92, SH 2002, Shafaq 2006, Shaheen 94, Shahkar 95, Soughat 90, Wafaq 01, Watan 94	Iftikhar et al. (2012)

good amount of resistance to spot blotch are now available for the breeding program. Resistant sources in wheat such as Ning 8201, Chirya 3, Yangmei 6, HD4502, Saar, BH1146, Chirya 1, Tksn1081/*A. squarrosa* and Mayoor (Kumar et al. 2009, 2010, 2016, Lillemo et al. 2013, Singh et al. 2014, 2018, Zhu et al. 2014) are being used in the spot blotch study (see Table 2).

The use of molecular markers in breeding practices could accelerate improvement resistance. However, the related research is limited, and no marker-assisted selection (MAS) has been performed for spot blotch resistance due to a lack of tightly linked markers or stable QTLs (Sharma et al. 1997, Das et al. 2002, Kumar et al. 2009, 2010, Makandar and Prabhu 2009). However, Lillemo et al. (2013) reported that two rust-resistant genes, *Lr34* and *Lr46*, also confer SB resistance to the CIMMYT breeding line Saar, indicating the potential value of these two genes along with their associated marker in the MAS for accelerating the selection.

Identification of fitness-related markers and spot blotch resistance is likely to address multiple screening problems. A bunch of interval mapping experiments have been performed where mapping populations like recombinant inbred lines (RIL) or double haploid lines (DH) were used to identify the QTLs associated with spot blotch resistance in wheat. By using different sources of resistance, several QTLs have been identified for spot blotch resistance in wheat. A few of these are on chromosomes 2A (*QSB.bhu-2A*), 2B (*QSB.bhu-2B*), 2D (*QSB.bhu-2D*), 3B (*QSB.bhu-3B*), 5B (*QSB.bhu-5B*), 6D (*QSB.bhu-6D*), 7B (*QSB.bhu-7B*), 7D (*QSB.bhu-7D*) and a few more (Kumar et al. 2009, 2010). The details of reported QTLs and markers are listed in Table 3. Although these studies provided critical information about the genetic basis of spot blotch resistance in wheat, majority of them used RILs or DH population that can only be identified a few at a time. However, considering the complexity of spot blotch resistance, a population with a broad genetic base was used in the study to uncover the mechanisms of spot blotch resistance (Adhikari et al. 2012, Gurung et al. 2014, Singh et al. 2018).

Backcross breeding has recently been used to pyramid number of disease resistance QTLs in single susceptible background to improve its resistance to spot blotch. Vasistha et al. (2016) introgressed two spot blotch resistance QTLs, *Qsb.bhu-2A* and *Qsb.bhu-5B*, in the susceptible wheat cultivar HUW 234 from two donor parents Chirya 3 and Ning 8201. The observation suggested that the technique may be used successfully to introduce multiple desirable alleles from different sources into a single genotype.

A genome-wide association study (GWAS) is an alternative approach to dissect the genetic basis of complex traits and overcome the limitations of QTL mapping. This approach also identifies the genetic architecture of complex multi-gene traits (Korte and Farlow 2013). Despite being advantageous over other approaches like QTL mapping, GWAS is prone to detecting false positives

Table 3 Summary of quantitative trait loci (QTL) and genome wide association studies (GWAS) in wheat*

Resistance genotypes and mapping strategy	Chromosome/arm/ QTL designation	Associated/flanking markers (interval in cM)	References
A set of 18 genotypes and F_2 population	5B	Xgwm67	Sharma et al. (2007)
	6A	Xgwm570	
	6D	Xgwm469	
Ning 8201, Chirya 3, Yangmai 6 and QTL mapping through RILs	QSb.bhu-2A QSb.bhu-2B	Xgwm425-Xbarc159 (15) Xgwm148-Xbarc91 (9)	Kumar et al. (2010)
	QSb.bhu-5B	Xgwm67-Xgwm213 (8)	
	QSb.bhu-7D	Xgwm111-Xgwm1168 (25)	
	QSb.bhu-2B	Xgwm148-Xgwm129 (15)	
	QSb.bhu-2D	Xgwm455-Xgwm815 (9)	
	QSb.bhu-3B	Xgwm533-Xgwm1037 (8)	
	QSb.bhu-7B	Xgwm263-Xgwm255 (5)	
	QSb.bhu-7D	Xgwm111-Xswm008 (25)	
	QSb.bhu-2A	Xbarc353-Xgwm445 (35.4)	
	QSb.bhu-2B	Xgwm148-Xgwm374 (15)	Kumar et al. (2009)
	QSb.bhu-5B	Xgwm067-Xgwm371 (13.2)	
	QSb.bhu-6D	Xbarc175-Xgwm732 (30.1)	
HD4502 and validation of RAPD markers using F_2 population		OPA 05, OPA 07, OPA 10, OPA 16, OPI 11, OPI 20, OPN 15	Patil and Hanchinal (2011)
526 wheat accessions and GWAS	1A	wPt-730148 (58)	Adhikari et al. (2012)
	1A	wPt-668214 (58)	
	3B	wPt-1159 (53.2)	
	3B	wPt-5769 (86.1)	
	7B	wPt-2838 (223.5)	
	7D	wPt-664459 (175.1)	

<table>
<tr><td>Saar and QTL mapping through RILs</td><td>1BL</td><td>hbe248-ncw1-V (120)</td><td>Lillemo et al. (2013)</td></tr>
<tr><td></td><td>7DS</td><td>gdm145-barc352 (59)</td><td></td></tr>
<tr><td></td><td>7DS</td><td>wPt-7654-gdm88 (42)</td><td></td></tr>
<tr><td>BH1146, Chirya 1 and screening in 2 RILs F_{10} RILs of two cross</td><td>2A
2B</td><td>Xgwm425
Xgwm148</td><td>Singh et al. (2014)</td></tr>
<tr><td></td><td>7D</td><td>Xgwm111</td><td></td></tr>
<tr><td>Tksn1081/A. squarrosa, Mayoor and QTL mapping in DH populations</td><td>QSb.cim-1B
QSb.cim-3B</td><td>Xwmc128-Xgwm374 (0.4)
990937|F|0-1123330|F| (2.7)</td><td>Zhu et al. (2014)</td></tr>
<tr><td></td><td>QSb.cim-5A</td><td>1086218|F|0-982608|F| (12.1)</td><td></td></tr>
<tr><td>528 spring wheat accessions and GWA Study</td><td>UnkD</td><td>wsnp_JD_c12281_12555386 (0)</td><td>Gurung et al. (2014)</td></tr>
<tr><td></td><td>Unk</td><td>wsnp_Ku_c44362_51657973 (0)</td><td></td></tr>
<tr><td></td><td>1B</td><td>wsnp_Ex_c24700_33953160 (37.1)</td><td></td></tr>
<tr><td></td><td>1B</td><td>wsnp_JD_c8926_9848514 (37.2)</td><td></td></tr>
<tr><td></td><td>5A</td><td>wsnp_Ex_c15342_23592740 (76.5)</td><td></td></tr>
<tr><td></td><td>5A</td><td>wsnp_Ku_c17951_27138894 (76.5)</td><td></td></tr>
<tr><td></td><td>5B</td><td>wsnp_Ex_rep_c70120_69069789 (109.5)</td><td></td></tr>
<tr><td></td><td>5B</td><td>wsnp_Ku_c50354_55979952 (146.8)</td><td></td></tr>
<tr><td></td><td>5B</td><td>wsnp_Ku_c20701_30355248 (147)</td><td></td></tr>
<tr><td></td><td>6B</td><td>wsnp_Ex_c15785_24157360 (90.3)</td><td></td></tr>
<tr><td></td><td>7B</td><td>wsnp_Ex_c52527_56097039 (56.8)</td><td></td></tr>
<tr><td>Yangmai 6 and fine mapping in RILs</td><td>Sb2-5B</td><td>Xgwm639-Xgwm1043 (0.28)</td><td>Kumar et al. (2016b)</td></tr>
<tr><td>621-18-3 and fine mapping in $F_{2:3}$</td><td>Sb3-3B</td><td>Xbarc133-Xbarc147 (0.15)</td><td>Lu et al. (2016)</td></tr>
<tr><td>H1146 and QTL mapping through RILs</td><td>QSb.iiwbr-7B</td><td>wmc758-wmc335 (8.6)</td><td>Singh et al. (2016)</td></tr>
<tr><td></td><td>QSb.iiwbr-7D</td><td>wmc653-barc121 (26.6)</td><td></td></tr>
<tr><td>SOKOLL//W15.92/WBLL1' x 'CIANO79'</td><td>5AL</td><td>2341646-Vrn-A1 (Position 44.2-48.6</td><td>He et al. (2020)</td></tr>
</table>

(Continued)

Table 3 *(Continued)*

Resistance genotypes and mapping strategy	Chromosome/arm/ QTL designation	Associated/flanking markers (interval in cM)	References
Hard winter wheat association mapping panel (HWWAMP) of 294 genotypes, and GWA study	QSb.sdsu-2D	SNP: Kukri_c31121_1460, R2 = 4%)	Ayana et al. (2018)
	QSb.sdsu-3A	SNP: Excalibur_c46082_440, R2 = 4%	
	QSb.sdsu-4A	SNP: IWA8475, R2 = 5.5%	
	QSb.sdsu-4B	SNP: Excalibur_rep_c79414_306, R2 = 4%),	
	QSb.sdsu-5A	SNP: Kukri_rep_c104877_2166, R2 = 6%	
	QSb.sdsu-7B	SNP: TA005844-0160, R2 = 6%	
BARTAI × CIANO T79 and QTL mapping using 232 $F_{2:7}$ SSD progenies	1B	995296-5410777 (0.4–2.8)	Singh et al. (2018)
		5324988-1110815 (209.4–218.6)	
	4B	1247021-1702494 (111.3–117.1)	
	5A	Vrn-A1-3064415 (175.9–179.4)	
WUYA × CIANO T79 and QTL mapping 232 $F_{2:7}$ SSD progenies	1B	1283166-2263671 (232.3–252.9)	
	4B	Rht-B1-1241652 (47.9–54.1)	
		1217659-997944 (99.2–107.4)	
	5A	1135154-2260918 (147.5–148.4)	
301 wheat genotypes from Afghanistan and GWA Study	1B	SNP: 1113207	Bainsla et al. (2020)
	5A	SNP:5411867, 998276	
	5B	SNP: 6036625	
GY17 × Zhongyu1211 and F5 fine mapping population	Sb4-4B	SNP: B6811-B6901 (7.14) SSR: YK12831-YK13104 (1.19)	Zhang et al. (2020)
139 advanced breeding lines of spring wheat and GWA Study	*Q.Sb.bisa*-1A	SNP: S1A_497201550, S1A_497201682	Tomar et al. (2021)
	Q.Sb.bisa-1D	SNP: S1D_89835681	
	Q.Sb.bisa-6D	SNP: S6D_6395796, S6D_6395796, S6D_7194112	
Various Population	*Tsn1-BsToxA* **NETS**	McDonald et al. (2018), Friesan et al. (2019), Navathe et al. (2020), Wu et al. (2020)	

* Updated after Gupta et al. (2018)

due to the confounding population structure or the effect of phenology. These limitations may be avoided by using a mixed linear model framework described with a fixed and random effect (Yu et al. 2006). Quantile plots (Q-Q plots) in multiple models for all phenological traits may be evaluated to select the best-fit models for the identification of markers trait association (MTAs). Multiple attempts have been made to study the genetics of the spot blotch resistance in wheat using GWAS (Gurung et al. 2014, Ayana et al. 2018, Ahirwar et al. 2018, Bainsla et al. 2020). The details of these studies and detected associations are listed in Table 3. An integrated map of all the QTL/genes discovered through interval mapping and GWAS indicates that the genomic regions on chromosome arms 7DS and 3BS certainly carry important genes (Gupta et al. 2018).

7 Resistance genes and their possible deployment

To date, only four spot blotch (Sb) resistance genes (*Sb1-Sb4*) have been mapped (Lillemo et al. 2013, Kumar et al. 2015, Lu et al. 2016, Zhang et al. 2020). *Sb1* was mapped on 7DS, which was also shown to be co-located with leaf rust resistance locus, *Lr34*, with pleiotropic effects on stripe rust (*Yr18*), stem rust (*Sr57*), powdery mildew (*Pm38*) resistance and leaf tip necrosis (Ltn1) (Lillemo et al. 2013). *Sb2* was delimited to 5BL in 0.62-cM genetic area between *Xgwm639* and *Xgwm1043* markers (Kumar et al. 2015). The third gene, *Sb3*, was located within a 0.15-cM genetic interval, spanning a 602-kb physical genomic region of the Chinese Spring chromosome 3BS (Lu et al. 2016). *Sb4* was recently mapped in a 1.19-cM genetic interval corresponding to a 1.34-Mb physical genomic region of chromosome 4BL using bulked RNA-Seq (BSR-Seq) technology (Zhang et al. 2020). The physical position of *Sb4* is found in the same chromosome area of the minor QTL at 4BL reported earlier by Singh et al. 2018.

In wheat, no source with all of the above four spot blotch resistance genes is currently available for use as a donor. The pyramiding of these four genes must be done by transferring them separately to the same host and then bringing them together through intercrossing. All four genes need to be assembled in a single donor and transferred to a high yield susceptible cultivar.

8 Low molecular weight toxins and their possible role in pathogenicity

The low molecular weight pathotoxins, for example prehelminthosporol, helminthosporol, helminthosporic acid sorokinianin and bipolenin K-J, are responsible for inducing more severe symptoms in adult plants (Carlson et al. 1991, Nakajima et al. 1998, Mercado Vergnes et al. 2006, Phan et al. 2019).

Genetic analysis of the isolate ND90Pr of *B. sorokiniana* revealed that a single locus (*VHv1*) is responsible for its virulence in the barley cultivar Bowman (Valjavec-Gratian and Steffenson 1997, Zhong et al. 2002). Further, genome sequencing and functional analysis identified two genes encoding non-ribosomal peptide synthetases (NRPSs) as essential for conferring the virulence ability to the ND90Pr isolate (Condon et al. 2013). The observation also indicated that the virulence factor of ND90Pr is probably a secondary host selective toxin (HST) metabolite. Additional studies on *CsLaeA*, *CsVeA*, *CsVelB*, *CsVelc* and *CsVosA* (global regulators in Velvet complex) have also shown that secondary metabolite factors play an important role in virulence (Wang et al. 2015, 2016).

9 Necrotrophic effector triggered susceptibility (NETS)-associated genes and breeding for the spot blotch resistance in wheat

Plants have developed two-tier immunity system, that is, the pattern triggered immunity (PTI) and the effector-triggered immunity (ETI), including the effector-triggered susceptibility (ETS), for the protection against the attack of various pathogens. The classic zig-zag model of plant immunity provides detailed insights into the molecular mechanisms of defense-related signaling in the presence of biotrophic fungi. This R-*Avr* mechanism follows a gene-for-gene pattern where the R gene in the host interacts with the pathogen-specific *Avr* gene, leading to a hypersensitive response, a characteristic feature of the ETI (Jones and Dangl 2006).

Intriguingly, necrotrophic pathogens required dead plant tissue to survive and they have evolved ways to trigger cell death responses in the host plants. An inappropriate elicitation of host defense responses by the action of necrotrophic pathogens activates the host dominant susceptibility factor that resulted in programmed cell death. This provides a source of necrotrophic nutrients, leading to a compatible reaction. At the molecular level, this interaction essentially reflects the "gene-for-gene" model, but the outcome shifts from resistance to susceptibility due to the phytopathogens' lifestyle. In addition, "gene-for-gene" interactions are generally qualitative but "inverse gene-for-gene" interactions in necrotrophic pathosystems often have an additive or quantitative effect. Characterization of the host susceptibility factors show the presence of homologous domains in the resistance as do the genes identified by Faris et al. (2010). The necrotrophic effector (ToxA) sensitive gene (*Tsn1*) in wheat has all the features of resistance, that is, S/TPK and NBS-LRR domains (Faris et al. 2010).

Necrotrophic effectors, generally seen as small, secreted proteins with high cysteine content, have evolved to manipulate the plant immune system for the benefit of the pathogen. Four economically important necrotrophic plant pathogens of wheat and barley follow "inverse gene-for-gene" and cause disease: *Pyrenophora tritici-repentis*, *P. teres*, *Parastagonospora nodorum* and *B. sorokiniana*. In the hypothetical case of a necrotrophic pathogen interacting with the host receptor, the diversifying selection may then be placed on the host receptor, attempting to eradicate the recognition.

The typical interactions occur in the inverse gene-for-gene model as demonstrated by the Tsn1-ToxA interaction. This interaction has recently been identified in the wheat-*B. sorokiniana* pathosystem (McDonald et al. 2018, Friesen et al. 2018, Navathe et al. 2020, Wu et al. 2020) from Australia, the USA and India. It has been shown that *Tsn1* and *ToxA* do not interact directly, but *Tsn1* is explicitly needed for the induction of host defense responses and the translocation of *ToxA* into a cell (Faris et al. 2010). *Tsn1* may be operating as a "guard" to the *ToxA* target. Upon modification of the target effector, *Tsn1* may recognize the alteration of the target protein and initiate cell-signaling pathways.

This has opened up a unique opportunity for genetic resistance in breeding programs by selecting null alleles. Traditionally, dominant R-genes are selected for biotrophic pathogens in particular. Although dominant resistance to necrotrophic pathogens has been identified, in most cases, resistance is not necessarily attributed to the dominant functional gene, but to the lack of a number of functional susceptibility factors requiring selection against functional genes or towards null alleles.

Necrotrophic specialists commonly select multiple effectors that target unique loci through the host genome with a wide range of pathogen populations (Friesen et al. 2008). In this situation several host loci are necessary to successfully eliminate all the effectors and achieve "resistance." Successful breeding for resistance to necrotrophic specialists including *B. sorokiniana* requires knowledge of the diversity of pathogen isolates in the region as well as the target for host susceptibility. In order to achieve effective breeding against NETS, it is imperative to understand and characterize the diversity of NEs in the pathogen.

10 Breeding for the spot blotch resistance in barley

In the last few years, great efforts have been made to identify and map QTLs and genes in barley germplasms for spot blotch resistance (Bilgic et al. 2005, 2006, Bovill et al. 2010, Grewal et al. 2012, Castro et al. 2012, Zhou and Steffenson 2013, Berger et al. 2013, Wang et al. 2017, 2019, Novakazi et al. 2020). Among these, *Rcs5* and *Rcs6* have been studied the most. *Rcs5* was originally mapped with short arm chromosome 7H for resistance to pathotype 1 (isolate ND85F) using a double-haploid (DH) cross arm population of barley derived from the cross of cultivars Steptoe and Morex (Steffenson et al. 1996, Bilgic et al. 2005). Further, fine mapping located the gene at ~240 kb region of 7H chromosome in genotype Morex (Drader 2011). *Rcs6* was originally mapped to the short arm of chromosome 1H for resistance to pathotype 2 (ND90Pr isolate) using DH population of the Calicuchima-sib/Bowman-BC cross (Bilgic et al. 2006). Further, fine mapping delimited *Rcs6* and its counterpart susceptibility allele *Scs6* at the same locus at a genomic interval of 125 kb at 1H Morex (Leng et al. 2018). Fine mapping of the susceptibility allele (*Scs6*) at the *Rcs6* locus in Bowman confined the gene to the *Mla* complex locus of chromosome 1H (Leng et al. 2018). Both *Rcs5* and *Rcs6* were found to be recessive alleles, and the corresponding dominant alleles are the functional ones responsible for susceptibility (Ameen et al. 2016, Leng et al. 2018). At least four QTLs for resistance on 1H, 2H, 4H and 5H in the advanced backcross mapping population derived from the PI 466423 (wild barley accession) and the recurrent parent 'Rasmussen' (the malting cultivar from Minnesota) have been reported by Haas et al. (2016). Through genome-wide-associated analysis of 1642 global barley accessions, the four QTLs on chromosomes of 1H, 2H, 3H and 6H were identified by Wang et al. (2017). Gyawali et al. (2018) validated the QTL on chromosome 3H resistant to the new pathotype, by association analysis of a spring barley collection in the ICARDA (International Center for Agricultural Research in Dry Areas).

Recently, Wang et al. (2019) have identified and mapped the first dominant spot blotch resistance locus '*Rbs7*' in barley to a new pathotype (isolate ND4008, pathotype 7) They mapped *Rbs7* in the genomic interval 304 kb on chromosome 6H using PI 356741 (highly susceptible to ND4008) and PI 235186 (highly resistant to ND4008). *Rbs7* mapping covered a distance of 304 kb genomic regions between two markers, M13.06 and M13.37, on Morex chromosome 6H. The function of *Rbs7* might differ from previously characterized *Rcs5* and *Rcs6/Scs6* genes. Eleven Mendelian loci (*Rcs1–Rcs6*, *Rbs7*, two unnamed genes and two loci associated with *gsh2* and *vrs3*) for spot blotch resistance have been identified in barley (for review see Bilgic et al. 2006). *Rcs1*, *Rcs2*, *Rcs3*, *Rcs5*, *Rcs6* and *Rbs7* were identified on chromosomes 2(2H), 5(1H), 7(5H), 1(7H), 5(1H) and 6H, respectively. The loci associated with the *gsh2* locus was identified on chromosome 3(3H) while *vrs3* locus on chromosome 5(1H).

In addition to the dominant resistance locus, barley *rcs5* QTL, which gives recessive resistance to spot blotch disease, was described by Ameen et al. (2020) in a recent study. The *rcs5* genetic interval was delimited to ~0.23 cM, representing a genomic region of ~234 kb containing four wall-associated kinase (WAK) genes in chromosome 7H. The genes are called *HvWak2*, *Sbs1*, *Sbs2* (Susceptibility to *B. sorokiniana* 1&2) and *HvWak5*, respectively. The gene/s underlying '*rcs5* QTL' constitutes an important goal for cloning and functional resistance determination.

Until now, plenty of genetic resources have been used to identify resistance genes and to develop user-friendly molecular markers that are useful for the development of spot-resistant cultivars in barley breeding programs.

11 Farmers participatory research in the release of spot blotch resistant variety

Participatory Variety Selection (PVS) can be used effectively to identify the varieties that are acceptable by the farmers. This may overcome the constraints of the farmers to grow old or obsolete varieties (Joshi and Witcombe 1996). In addition, participatory research increases the efficiency of scientists and knowledge of farmers, enabling them to effectively maintain the useful germplasm from year to year. PVS also helps to reduce the cost of research and adoption rates increased if farmers participate in the testing and selection of varieties (Ortiz-Ferrera et al. 2007). Moreover, production increases when farmers adopt new varieties identified in participatory research (Joshi et al. 2005).

Participatory Variety Selection (PVS) research was initiated in Bangladesh, Nepal and India in the early 21st century to involve farmers in selecting their preferred varieties according to their socio-economic needs and to disseminate them in a noble manner so that farmers can reap the benefits of new wheat varieties without delay. The PVS was implemented with a "Mother-Baby" trialing system that was originally adopted for maize breeding in Africa. In this trialing system the Mother trial consisted of 12–15 elite wheat lines and the most popular check variety. A large number of Baby trials are conducted on a single elite line from the Mother trial along with check in farmer's field. The mother and baby trials are assessed in a collaborative and consultative mode by the farmers and scientists. The feedback from the farmers in terms of yield and other adoptive traits were analyzed to ascertain the farmers' preference. The preferred genotype then becomes ready to be released as a variety in the broad sense.

Farmers in the hills of Nepal have selected and adopted three new wheat varieties – BL-1473, BL-1813 and WK-1204 – on the basis of their own selection criteria, which include, in particular, spot blotch disease resistance, high yield, bold seed, early maturity and reduced lodging.

In the period 2002–2005, farmers in Bangladesh identified at least three new varieties of wheat through PVS: Shatabdi, BAW-1006 and BAW-1008. In that period, the area under Shatabdi, the preferred farmer variety, increased from 0% to 36% in the participating villages. Farmers prefer the Shatabdi variety for their spot blotch resistance, higher yield of grain, early maturity and "chapatti making" quality (Pandit et al. 2007, Ortiz-Ferrera et al. 2007).

Varietal diversification and income generation have also increased in eastern India, particularly in the villages around Varanasi, eastern Uttar Pradesh and neighboring Bihar state. The activities of PVS at a few sites in the area began in 1997. Old wheat varieties, such as HUW-234, have now been replaced by farmer-preferred varieties (HUW-468, HUW-510, HUW-516 BHU-32 and others) identified as collaborating farmers in the project funds provided by the United Kingdom Department for International Development (DFID), the Indian

Council for Agricultural Research (ICAR), the Rice Wheat Consortium (RWC; Ortiz-Ferrera et al. 2007, Joshi et al. 2007a).

This indicated the potential for the integration of the 'Mother-Baby' test system and the technology for the conservation of resources used in the generation, distribution and seed multiplication for new varieties through PVS.

12 Challenges and future prospects

Defense signaling components interact and work synergistically to enhance the resistance and to improve the overall fitness of the host plants. The number of grains per spike, test weight and yield are important phenological components and are rated for cultivar development. Once the seed is packed with resistance and yield components, the technology is ready to deliver. As a result, a resistant, high-yield cultivar is very much necessary to empower the farmers with efficient, low-cost and environment-friendly technology for maximum benefit. Consequently, breeding for heat tolerance and spot resistance are two important objectives of the wheat improvement programs targeting the most tropical and subtropical plains growing wheat. Genetic diversity for heat tolerance and spot blotch in cultivated wheat is well explored separately, and resistance sources are identified. However, limited information is available for tolerance to the combined stress of spot blotch and terminal heat.

In any situation, tolerance to disease could be beneficial to reducing the risk of high yield losses. Limited information is available on the diversity of isolates, pathotypes, phytotoxins, necrotrophic effectors and the various secretive pathways that work together to impose susceptibility to the host. Understanding the potential molecular interactions that may occur between pathogen isolates and adapted germplasm will allow for the successful identification of the potentially susceptible loci from breeding materials and the effective introduction of resistance.

13 Where to look for further information

- The following articles provide a good overview of the subject that includes the detailed information of spot blotch resistance genes (*Sb1, Sb2, Sb3 and Sb4*) mapped so far.
  - Lillemo, M., Joshi, A. K., Prasad, R., Chand, R. and Singh, R. P. 2013. QTL for spot blotch resistance in bread wheat line Saar co-locate to the biotrophic disease resistance loci *Lr34* and *Lr46*. *Theoretical and Applied Genetics*, *126*(3), pp. 711–719.
  - Kumar, S., Röder, M. S., Tripathi, S. B., et al. 2015. Mendelization and fine mapping of a bread wheat spot blotch disease resistance QTL. *Molecular Breeding*, *35*(11), p. 218.

- Lu, P., Liang, Y., Li, D., et al. 2016. Fine genetic mapping of spot blotch resistance gene *Sb3* in wheat (*Triticum aestivum*). *Theoretical and Applied Genetics*, *129*(3), pp. 577–589.
- Zhang, P., Guo, G., Wu, Q. et al. 2020. Identification and fine mapping of spot blotch (*Bipolaris sorokiniana*) resistance gene *Sb4* in wheat. *Theoretical and Applied Genetics*. https://doi.org/10.1007/s00122-020-03610-3.

- The horizontal transfer of the necrotrophic effector gene 'ToxA' to the spot blotch Bipolaris sorokiniana has been detailed by: McDonald, M. C., Ahren, D., Simpfendorfer, S., Milgate, A. and Solomon, P. S. 2018. The discovery of the virulence gene ToxA in the wheat and barley pathogen Bipolaris sorokiniana. *Molecular Plant Pathology*, 19(2), pp. 432–439.
- Key research in wheat is globally driven through various research programs run by the International Maize and Wheat Improvement Center (www.cimmyt.org). The data of the various experiments and publications is available at publication repository: https://repository.cimmyt.org/; whereas Data repository: https://www.cimmyt.org/resources/data/.
- Various symposiums aim to bring together the entire research community-not only in academia, but also in research centers and institutes and across governments and organizations. The International Symposium on Cereal Leaf Blights (ISCLB) interest in a wide range of issues related to cereal leaf blights. The symposium is organized every year. More details can be found at: https://www.isclb2021.com/.

14 References

Adhikari, T. B., Gurung, S., Hansen, J. M., Jackson, E. W. and Bonman, J. M. (2012). Association mapping of quantitative trait loci in spring wheat landraces conferring resistance to bacterial leaf streak and spot blotch. *Plant Genome* 5(1), 1–16.

Adlakha, K. L., Wilcoxson, R. D. and Raychaudhuri, S. P. (1984). Resistance of wheat to leaf spot caused by *Bipolaris sorokiniana*. *Plant Dis.* 68(4), 320–321.

Ahirwar, R. N., Mishra, V. K., Chand, R., Budhlakoti, N., Mishra, D. C., Kumar, S., Singh, S. and Joshi, A. K. (2018). Genome-wide association mapping of spot blotch resistance in wheat association mapping initiative (WAMI) panel of spring wheat (*Triticum aestivum* L.). *PLoS ONE*, 13(12), e0208196.

Alscher, R. G., Erturk, N. and Heath, L. S. (2002). Role of superoxide dismutases (SODs) in controlling oxidative stress in plants. *J. Exp. Bot.* 53(372), 1331-1341.

Ameen, G., Drader, T., Sager, L., Steffenson, B. J., Kleinhofs, A. and Brueggeman, R. S. (2016). Rcs5 is a wall associated kinase gene that putatively functions as a dominant susceptibility factor in the barley–Cochliobolus sativus interaction. In: *The 12th International Barley Genetics Symposium*, 26-30 June 2016, Minneapolis.

Ameen, G., Solanki, S., Drader, T., Sager-Bittara, L., Steffenson, B., Kleinhofs, A., Vogiatzis, C. and Brueggeman, R. S. (2020). *rcs5*-mediated spot blotch resistance in barley is conferred by wall-associated kinases that resist pathogen manipulation. *bioRxiv*. https://doi.org/10.1101/2020.04.13.040238.

Arabi, M. I. E. and Jawhar, M. (2004). Identification of *Cochliobolus sativus* (spot blotch) isolates expressing differential virulence on barley genotypes in Syria. *J. Phytopathol.* 152(8-9), 461-464.

Ayana, G. T., Ali, S., Sidhu, J. S., Gonzalez-Hernandez, J. L., Turnipseed, B. and Sehgal, S. K. (2018). Genome-wide association study for spot blotch resistance in hard winter wheat. *Front. Plant Sci.* 9, 926.

Bainsla, N. K., Phuke, R. M., He, X., Gupta, V., Bishnoi, S. K., Sharma, R. K., Ataei, N., Dreisigacker, S., Juliana, P. and Singh, P. K. (2020). Genome-wide association study for spot blotch resistance in Afghan wheat germplasm. *Plant Pathol.* 69(6), 1161-1171 https://doi.org/10.1111/ppa.13191.

Bashyal, B. M., Chand, R., Kushwaha, C., Joshi, A. K. and Kumar, S. (2011a). *Bipolaris sorokiniana* of barley: infection behaviour in different members of Poaceae. *Indian Phytopath.* 64(1), 28-31.

Bashyal, B. M., Chand, R., Prasad, L. C. and Joshi, A. K. (2011b). Partial resistance components for the management of spot blotch pathogen *Bipolaris sorokiniana* of barley (*Hordeum vulgare* L.). *Acta Phytopathol. Entomol. Hung.* 46(1), 49-57.

Bell, A. A. (1981). Biochemical mechanisms of disease resistance. *Annu. Rev. Plant Physiol.* 32(1), 21-81.

Berger, G. L., Liu, S., Hall, M. D., Brooks, W. S., Chao, S., Muehlbauer, G. J., Baik, B. K., Steffenson, B. and Griffey, C. A. (2013). Marker-trait associations in Virginia Tech winter barley identified using genome-wide mapping. *Theor. Appl. Genet.* 126(3), 693-710.

Bilgic, H., Steffenson, B. J. and Hayes, P. M. (2005). Comprehensive genetic analyses reveal differential expression of spot blotch resistance in four populations of barley. *Theor. Appl. Genet.* 111(7), 1238-1250.

Bilgic, H., Steffenson, B. J. and Hayes, P. M. (2006). Molecular mapping of loci conferring resistance to different pathotypes of the spot blotch pathogen in barley. *Phytopathology* 96(7), 699-708.

Bock, C. H., Poole, G. H., Parker, P. E. and Gottwald, T. R. (2010). Plant Disease severity estimated visually, by digital photography and image analysis, and by hyperspectral imaging. *Crit. Rev. Plant Sci.* 29(2), 59-107.

Bovill, J., Lehmensiek, A., Sutherland, M. W., Platz, G. J., Usher, T., Franckowiak, J. and Mace, E. (2010). Mapping spot blotch resistance genes in four barley populations. *Mol. Breed.* 26(4), 653-666.

Caierao, E., Scheeren, P. L., Silva, M. S. and Castro, R. L. (2014). History of wheat cultivars released by Embrapa in forty years of research. *Crop Breed. Appl. Biotechnol.* 14, 216-223.

Carlson, H., Nilsson, P., Jansson, H. B. and Odham, G. (1991). Characterization and determination of prehelminthosporol, a toxin from the plant pathogenic fungus *Bipolaris sorokiniana*, using liquid chromatography/mass spectrometry. *J. Microbiol. Methods* 13(4), 259-269.

Castro, A. J., Gamba, F., German, S., Gonzalez, S., Hayes, P. M., Pereyra, S. and Perez, C. (2012). Quantitative trait locus analysis of spot blotch and leaf rust resistance in the BCD47 Baronesse barley mapping population. *Plant Breed.* 131(2), 258-266.

Chand, R., Pandey, S. P., Singh, H. V. and Joshi, A. K. (2003). Variability and its probable cause in the natural population of spot blotch pathogen *Bipolaris sorokiniana* of Wheat (*T. aestivum*). *J. Plant Dis. Prot.* 110, 27-35.

Chand, R., Pradhan, P. K., Prasad, L. C., Kumar, D., Verma, R. P. S., Singh, D. P. and Joshi, A. K. (2010). Diversity and association of isolates and symptoms of spot blotch caused by *Bipolaris sorokiniana* of barley (*Hordeum vulgare* L.). *Indian Phytopath.* 63(2), 154-157.

Chaurasia, S., Joshi, A. K., Dhari, R. and Chand, R. (1999). Resistance to foliar blight of wheat: a search. *Genet. Resour. Crop Evol.* 46(5), 469-475.

Christopher, J. T., Veyradier, M., Borrell, A. K., Harvey, G., Fletcher, S. and Chenu, K. (2014). Phenotyping novel stay-green traits to capture genetic variation in senescence dynamics. *Funct. Plant Biol.* 41(11), 1035-1048.

Condon, B. J., Leng, Y., Wu, D., Bushley, K. E., Ohm, R. A., Otillar, R., Martin, J., Schackwitz, W., Grimwood, J., MohdZainudin, N., Xue, C., Wang, R., Manning, V. A., Dhillon, B., Tu, Z. J., Steffenson, B. J., Salamov, A., Sun, H., Lowry, S., LaButti, K., Han, J., Copeland, A., Lindquist, E., Barry, K., Schmutz, J., Baker, S. E., Ciuffetti, L. M., Grigoriev, I. V., Zhong, S. and Turgeon, B. G. (2013). Comparative genome structure, secondary metabolite, and effector coding capacity across *Cochliobolus* pathogens. *PLoS Genet.* 9(1), e1003233. https://doi.org/10.1371/journal.pgen.1003233al.pgen.1003233.

Das, S., Aggarwal, R., Srivastava, K. D. and Tomar, S. M. S. (2002). Isozyme variations in some wheat genotypes in response to spot blotch infection. *Indian Phytopathol.* 55(3), 253-257.

Domiciano, G. P., Duarte, H. S. S., Moreira, E. N. and Rodrigues, F. A. (2014). Development and validation of a set of standard area diagrams to aid in estimation of spot blotch severity on wheat leaves. *Plant Pathol.* 63(4), 922-928.

Drader, T. B. (2011). Cloning of the seedling spot blotch resistance gene Rcs5. Ph.D. thesis. Washington State University.

Dubin, H. J., Arun, B., Begum, S. N., Bhatta, M., Dhari, R., Goel, L. B., Joshi, A. K., Khanna, B. M., Malaker, P. K., Pokhrel, D. R. and Rahman, M. M. (1998). Results of the South Asia Regional *Helminthosporium* leaf blight an yield experiment, 199-394. In: Duveiller, E., Dubin, H. J., Reeves, J. and McNab, A. (Eds), *Helminthosophism Diseases of Wheat: Spot Blotch and Tan Spot*, pp. 182-187. CIMMYT, Mexico.

Dubin, H. J. and Ginkel, M. (1991). The status of wheat disease and disease research in warmer areas. In: Sanders, D. A. (Ed), *Wheat for the Nontraditional Warm Areas. A Proceeding of the International Conference*, pp. 125-145. CIMMYT, Mexico.

Dubin, H. J. and Rajaram, S. (1996). Breeding disease resistant wheats for tropical highlands and lowlands. *Annu. Rev.Phytopathol.* 34, 503-526.

Duveiller, E. and Altamirano, G. I. (2000). Pathogenicity of *Bipolaris sorokiniana* isolates from wheat roots, leaves and seeds in Mexico. *Plant Pathol.* 49, 235-242.

Duveiller, E., Dubin, H. J., Reeves, J. and McNab, A. (Eds) (1998). Helminthosporium Blights of Wheat: Spot Blotch and Tan Spot. In: *Proceedings of the Helminthosporium Blights of Wheat Workshop, 1997*, pp. 376, El Batan, Mexico; CIMMYT, Mexico DF.

Duveiller, E., Kandel, Y. R., Sharma, R. C. and Shrestha, S. M. (2005). Epidemiology of foliar blights (spot blotch and tan spot) of wheat in the plains bordering the Himalayas. *Phytopathology* 95(3), 248-256.

Duveiller, E. M. and Sharma, R. C. (2009). Genetic improvement and crop management strategies to minimize yield losses in warm non-traditional wheat growing areas due to spot blotch pathogen Cochliobolus sativus. *J. Phytopath.* 157(9), 521-534.

Eisa, M., Chand, R. and Joshi, A. K. (2013). Biochemical and histochemical traits: a promising way to screen resistance against spot blotch (*Bipolaris sorokiniana*) of wheat. *Eur. J. Plant Pathol.* 137(4), 805-820.

Eyal, Z., Scharen, A. L., Prescott, J. M. and van Ginkel, M. (1987). *The Septoria Disease of Wheat: Concepts and Methods of Disease Management*. CIMMYT, Mexico D.F., Mexico.

Faris, J. D., Zhang, Z., Lu, H., Lu, S., Reddy, L., Cloutier, S., Fellers, J. P., Meinhardt, S. W., Rasmussen, J. B., Xu, S. S., Oliver, R. P., Simons, K. J. and Friesen, T. L. (2010). A unique wheat disease resistance-like gene governs effector-triggered susceptibility to necrotrophic pathogens. *PNAS* 107(30), 13544-13549.

Fetch, T. G. and Steffenson, B. J. (1999) Rating Scales for Assessing Infection Responses of Barley Infected with Cochliobolus sativus. *Plant Dis.* 83(3), 213-217.

Friesen, T. L., Faris, J. D., Solomon, P. S. and Oliver, R. P. (2008). Host-specific toxins: effectors of necrotrophic pathogenicity. *Cell. Microbiol.* 10(7), 1421 -1428.

Friesen, T. L., Holmes, D. J., Bowden, R. L. and Faris, J. D. (2018). ToxA is present in the US *Bipolaris sorokiniana* population and is a significant virulence factor on wheat harboring *Tsn1*. *Plant Dis.* 102(12), 2446-2452.

Ghazvini, H. and Tekauz, A. (2007). Virulence diversity in the population of *Bipolaris sorokiniana*. *Plant Dis.* 91(7), 814-821.

Grewal, T. S., Rossnagel, B. G. and Scoles, G. J. (2012). Mapping quantitative trait loci associated with spot blotch and net blotch resistance in a doubled-haploid barley population. *Mol. Breed.* 30(1), 267-279.

Gupta, P. K., Chand, R., Vasistha, N. K., Pandey, S. P., Kumar, U., Mishra, V. K. and Joshi, A. K. (2018). Spot blotch disease of wheat: the current status of research on genetics and breeding. *Plant Pathol.* 67(3), 508-531.

Gurung, S., Mahto, B. N., Gyawali, S. and Adhikari, T. B. (2013). Phenotypic and molecular diversity of *Cochliobolus sativus* populations from wheat. *Plant Dis.* 97(1), 62-73.

Gurung, S., Mamidi, S., Bonman, J. M., Xiong, M., Brown-Guedira, G. and Adhikari, T. B. (2014). Genome wide association study reveals novel quantitative trait loci associated with resistance to multiple leaf spot diseases of spring wheat. *PLoS ONE* 9(9), e108179.

Gyawali, S. (2010). Association mapping of resistance to common root rot and spot blotch in barley, and population genetics of Cochliobolus sativus. Ph.D. Dissertation. North Dakota State University, Fargo, ND.

Gyawali, S., Chao, S., Vaish, S. S., Singh, S. P., Rehman, S., Vishwakarma, S. R. and Verma, R. P. S. (2018). Genome wide association studies (GWAS) of spot blotch resistance at the seedling and the adult plant stages in a collection of spring barley. *Mol. Breed.* 38, 62. https ://doi.org/10.1007/s1103.

Haas, M., Menke, J., Chao, S. and Steffenson, B. J. (2016). Mapping quantitative trait loci conferring resistance to a widely virulent isolate of *Cochliobolus sativus* in wild barley accession PI 466423. *Theor. Appl. Genet.* 129(10), 1831-1842.

Harris, K., Subudhi, P. K., Borrell, A., Jordan, D., Rosenow, D., Nguyen, H., Klein, P., Klein, R. and Mullet, J. (2007). Sorghum stay-green QTL individually reduce post-flowering drought-induced leaf senescence. *J. Exp. Bot.* 58(2), 327-338.

He, X., Dreisigacker, S., Sansaloni, C., Duveiller, E., Singh, R. P. and Singh, P. K. (2020). Quantitative trait loci mapping for spot blotch resistance in two biparental mapping populations of bread wheat. *Phytopathology* 110(12), 1980-1987.

Herrera-Foessel, S. A., Lagudah, E. S., Huerta-Espino, J., Hayden, M. J., Bariana, H. S., Singh, D. and Singh, R. P. (2011). New slow-rusting leaf rust and stripe rust resistance genes Lr67 and Yr46 in wheat are pleiotropic or closely linked. *Theor. Appl. Genet.* 122(1), 239-249.

Hoch, H. C. and Staples, R. C. (1991). Signaling for infection structure formation in fungi. In: Cole, G. T. and Hoch, H. C. (Eds), *The Fungal Spore and Disease Initiation in Plants and Animals*, pp. 25-46. Springer, Boston, MA.

Ibeagha, A. E., Hückelhoven, R., Schäfer, P., Singh, D. P. and Kogel, K. H. (2005). Model wheat genotypes as tools to uncover effective defense mechanisms against the hemibiotrophic fungus *Bipolaris sorokiniana*. *Phytopathology* 95(5), 528-532.

Iftikhar, S., Asad, S., Rattu, A., Munir, A. and Fayyaz, M. (2012). Screening of commercial wheat varieties to spot blotch under controlled and field conditions. *Pak. J. Bot.* 44(1), 361-363.

Jones, J. D. and Dangl, J. L. (2006). The plant immune system. *Nature* 444(7117), 323-329.

Joshi, A. K. and Chand, R. (2002). Variation and inheritance of leaf angle, and its association with spot blotch (*Bipolaris sorokiniana*) severity in wheat (*Triticum aestivum*). *Euphytica* 124(3), 283-291.

Joshi, A. K., Chand, R. and Arun, B. (2002). Relationship of plant height and days to maturity with resistance to spot blotch in wheat. *Euphytica* 123(2), 221-228.

Joshi, A. K., Chand, R., Chandola, V. K., Tripathi, R., Arun, B., Prasad, L. C. and Ortiz Ferrara, G. (2005). Approaches to germplasm dissemination and adoption–Reaching farmers in the eastern Gangetic Plains. In: *7th International Wheat Conference, Mar del Plata (Argentina); 27 Nov-2 Dec 2005. Abstracts of oral and poster presentations C2005* (No. CIS-4632. CIMMYT.)..

Joshi, A. K., Kumar, S., Chand, R. and Ortiz-Ferrara, G. (2004). Inheritance of resistance to spot blotch caused by *Bipolaris sorokiniana* in spring wheat. *Plant Breed.* 123(3), 213-219.

Joshi, A. K., Kumari, M., Singh, V. P., Reddy, C. M., Kumar, S., Rane, J. and Chand, R. (2007a). Stay green trait: variation, inheritance and its association with spot blotch resistance in spring wheat (Triticum aestivum L.). *Euphytica* 153(1-2), 59-71.

Joshi, A. K., Mishra, B., Chatrath, R., Ortiz-Ferrara, G. and Singh, R. P. (2007b). Wheat improvement in India: present status, emerging challenges and future prospects. *Euphytica* 157(3), 431-446.

Joshi, A. K., Ortiz-Ferrara, G., Crossa, J., Singh, G., Alvarado, G., Bhatta, M. R., Duveiller, E., Sharma, R. C., Pandit, D. B., Siddique, A. B., Das, S. Y., Sharma, R. N. and Chand, R. (2007c). Associations of environments in South Asia based on spot blotch disease of wheat caused by *Cochliobolus sativus*. *Crop Sci.* 47(3), 1071-1081.

Joshi, A. and Witcombe, J. R. (1996). Farmer participatory crop improvement. II. Participatory varietal selection, a case study in India. *Exp. Agric.* 32(4), 461 -477.

Karisto, P., Dora, S. and Mikaberidze, A. (2019). Measurement of infection efficiency of a major wheat pathogen using time-resolved imaging of disease progress. *Plant Pathol.* 68(1), 163-172.

Karisto, P., Hund, A., Yu, K., Anderegg, J., Walter, A., Mascher, F., McDonald, B. A. and Mikaberidze, A. (2018). Ranking quantitative resistance to Septoria tritici blotch in elite wheat cultivars using automated image analysis. *Phytopathology* 108(5), 568-581.

Khan, H. and Chowdhury, S. (2011). Identification of resistance source in wheat germplasm against spot blotch disease caused by Bipolaris sorokiniana. *Arch. Phytopathol. Plant Protect.* 44(9), 840-844.

Kipp, S., Mistele, B. and Schmidhalter, U. (2013). Identification of stay -green and early senescence phenotypes in high yielding winter wheat: and their relationship to grain yield and grain protein concentration using high-throughput phenotyping techniques. *Funct. Plant Biol.* 41(3), 227-235.

Knight, N. L., Platz, G. J., Lehmensiek, A. and Sutherland, M. W. (2010). An investigation of genetic variation among Australian isolates of *Bipolaris sorokiniana* from different cereal tissues and comparison of their abilities to cause spot blotch on barley. *Austral. Plant Pathol.* 39(3), 207-216.

Korte, A. and Farlow, A. (2013). The advantages and limitations of trait analysis with GWAS: a review. *Plant Methods* 9(1), 29.

Kumar, J., Schäfer, P., Hückelhoven, R., Langen, G., Baltruschat, H., Stein, E., Nagarajan, S. and Kogel, K. H. (2002). *Bipolaris sorokiniana*, a cereal pathogen of global concern: cytological and molecular approaches towards better controldouble dagger. *Mol. Plant Pathol.* 3(4), 185-195.

Kumar, S., Kumar, U., Singh, A. K., Singh, R., Prasad, L. C. and Joshi, A. K. (2014). Study of inheritance and allelic relation of resistance to spot blotch (*Bipolaris sorokiniana*) of wheat. *Biotech Today* 3(2), 31-38.

Kumar, S., Röder, M. S., Singh, R. P., Kumar, S., Chand, R., Joshi, A. K. and Kumar, U. (2016). Mapping of spot blotch disease resistance using NDVI as a substitute to visual observation in wheat (*Triticum aestivum* L.). *Mol. Breed.* 36(7), 95.

Kumar, U., Joshi, A. K., Kumar, S., Chand, R. and Röder, M. S. (2009). Mapping of resistance to spot blotch disease caused by *Bipolaris sorokiniana* in spring wheat. *Theor. Appl. Genet.* 118(4), 783-792.

Kumar, U., Joshi, A. K., Kumar, S., Chand, R. and Röder, M. S. (2010). Quantitative trait loci for resistance to spot blotch caused by *Bipolaris sorokiniana* in wheat (T. aestivum L.) lines 'Ning 8201' and 'Chirya 3'. *Mol. Breed.* 26(3), 477-491.

Kumari, M., Singh, V. P., Tripathi, R. and Joshi, A. K. (2007). Variation for stay green trait and its association with canopy temperature depression and yield traits under terminal heat stress in wheat. In: Buck, H. T., Nisi, J. E. and Salomon, N. (Eds), *Wheat Production in Stressed Environments*, pp. 357-363. Springer, Netherlands.

Leng, Y., Wang, R., Ali, S., Zhao, M. and Zhong, S. (2016). Sources and genetics of spot blotch resistance to a new pathotype of *Cochliobolus sativus* in the USDA National small grains collection. *Plant Dis.* 100(10), 1988-1993.

Leng, Y., Zhao, M., Wang, R., Steffenson, B. J., Brueggeman, R. S. and Zhong, S. (2018). The gene conferring susceptibility to spot blotch caused by *Cochliobolus sativus* is located at the *Mla* locus in barley cultivar Bowman. *Theor. Appl. Genet.* 131(7), 1531-1539.

Lillemo, M., Joshi, A. K., Prasad, R., Chand, R. and Singh, R. P. (2013). QTL for spot blotch resistance in bread wheat line Saar co-locate to the biotrophic disease resistance loci *Lr34* and *Lr46*. *Theor. Appl. Genet.* 126(3), 711-719.

Lopes, M. S. and Reynolds, M. P. (2012). Stay-green in spring wheat can be determined by spectral reflectance measurements (normalized difference vegetation index) independently from phenology. *J. Exp. Bot.* 63(10), 3789-3798.

Lu, P., Liang, Y., Li, D., Wang, Z., Li, W., Wang, G., Wang, Y., Zhou, S., Wu, Q., Xie, J., Zhang, D., Chen, Y., Li, M., Zhang, Y., Sun, Q., Han, C. and Liu, Z. (2016). Fine genetic mapping of spot blotch resistance gene *Sb3* in wheat (*Triticum aestivum*). *Theor. Appl. Genet.* 129(3), 577-589.

Madden, L. V., Hughes, G. and Bosch, F. (2007). *The Study of Plant Disease Epidemics*. American Phytopathological Society (APS Press), St. Paul, MN.

Mahto, B. N., Gurung, S. and Adhikari, T. B. (2011). Assessing genetic resistance to spot blotch, Stagonospora nodorum blotch and tan spot in wheat from Nepal. *Eur. J. Plant Pathol.* 131(2), 249-260.

Mahto, B. N., Gurung, S., Nepal, A. and Adhikari, T. B. (2012). Morphological, pathological and genetic variations among isolates of *Cochliobolus sativus* from Nepal. *Eur. J. Plant Pathol.* 133(2), 405-417.

Makandar, R. and Prabhu, K. V. (2009). Inheritance of resistance to spot blotch disease in wheat. *Indian J. Genet.* 69(3), 178-182.

Maraite, H., Di Zinno, T., Longrée, H., Daumerie, V. and Duveiller, E. (1998). Fungi associated with foliar blight of wheat in warmer areas. In: Duveiller, E., Dubin, H. J., Reeves, J. and McNab, A. (Eds), *Proceedings of the International Workshop on Helminthosporium Blights of Wheat: Spot Blotch and Tan Spot*, pp. 293-300. El Batán, Mexico.

Maraite, H., Mercado Vergnes, D., Renard, M. E., Zhanarbekova, A. and Duveiller, E. (2006). Relevance of pathogen diversity in management of leaf spot and leaf blight diseases on wheat in Central Asia. *Agromeridian* 2, 105-114.

McDonald, M. C., Ahren, D., Simpfendorfer, S., Milgate, A. and Solomon, P. S. (2018). The discovery of the virulence gene *ToxA* in the wheat and barley pathogen *Bipolaris sorokiniana*. *Mol. Plant Pathol.* 19(2), 432-439.

Mehta, Y. R. (1998). Constraints on the integrated management of spot blotch of wheat. In: Duveiller, E., Dubin, H. J., Reeves, J. and McNab, A. (Eds), *Helminthosporium Blight of Wheat: Spot Blotch and Tan Spot*, CIMMYT, Mexico, pp. 18-27.

Mohanty, S. P., Hughes, D. P. and Salathé, M. (2016). Using deep learning for image-based plant disease detection. *Front. Plant Sci.* 7, 1419.

Montazeaud, G., Karatoğma, H., Özturk, I., Roumet, P., Ecarnot, M., Crossa, J., Özer, E., Özdemir, F. and Lopes, M. S. (2016). Predicting wheat maturity and stay-green parameters by modeling spectral reflectance measurements and their contribution to grain yield under rainfed conditions. *Field Crops Res.* 196, 191-198.

Mutka, A. M., Fentress, S. J., Sher, J. W., Berry, J. C., Pretz, C., Nusinow, D. A. and Bart, R. (2016). Quantitative, image-based phenotyping methods provide insight into spatial and temporal dimensions of plant disease. *Plant Physiol.* 172(2), 650-660.

Nagarajan, S. and Kumar, J. (1998). An overview of the increasing importance of research of foliar blights of wheat in India: germplasm improvement and future challenges towards a sustainable high yielding wheat production. In: Duveiller, E., Dubin, H. J., Reeves, J. and McNab, A. (Eds), *Proceedings of the Helminthosporium Blights of Wheat: Spot Blotch and Tan Spot Workshop*, pp. 52-58. CIMMYT, El Batan, Mexico DF.

Nakajima, H., Toratsu, Y., Fujii, Y., Ichinoe, M. and Hamasaki, T. (1998). Biosynthesis of sorokinianin a phytotoxin of *Bipolaris sorokiniana*: evidence of mixed origin from the sesquiterpene and TCA pathways. *Tetrahedron Lett.* 39(9), 1013-1016.

Neupane, R. B., Sharma, R. C., Duveiller, E., Ortiz-Ferrara, G., Ojha, B. R., Rosyara, U. R., Bhandari, D. and Bhatta, M. R. (2007). Major gene controls of field resistance to spot blotch in wheat genotypes 'Milan/Shanghai# 7'and 'Chirya. 3'. *Plant Dis.* 91(6), 692-697.

Navathe, S., Yadav, P. S., Vasishtha, N. K., Chand, R., Mishra, V. K., Meher, P. K., Joshi, A., Gupta, P. K. (2020). ToxA-Tsn1 interaction for spot blotch susceptibility in Indian wheat: An example of inverse gene-for-gene relationship. *Plant Dis.* 104, 71-81 https://doi.org/10.1094/PDIS-05-19-1066-RE.

Novakazi, F., Afanasenko, O., Lashina, N., Platz, G. J., Snowdon, R., Loskutov, I. and Ordon, F. (2020). Genome-wide association studies in a barley (Hordeum vulgare) diversity set reveal a limited number of loci for resistance to spot blotch (*Bipolaris sorokiniana*). *Plant Breed.* 139(3), 521-535.

Ortiz-Ferrara, G., Joshi, A. K., Chand, R., Bhatta, M. R., Mudwari, A., Thapa, D. B., Sufian, M. A., Saikia, T. P., Chatrath, R., Witcombe, J. R., Virk, D. S. and Sharma, R. C. (2007). Partnering with farmers to accelerate adoption of new technologies in South Asia to improve wheat productivity. *Euphytica* 157(3), 399-407.

Pandey, A., Ansari, S. Q., Navathe, S., Chand, R., Mishra, V. K. and Joshi, A. K. (2016). Association of lesion mimic trait with spot blotch resistance in wheat. *Trop. Plant Pathol.* 41(6), 406-414.

Pandey, S. P., Sharma, S., Chand, R., Sahi, P. and Joshi, A. K. (2008). Clonal variability in the spot blotch pathogen *Bipolaris sorokiniana* of wheat and its relevance in generation of new pathotypes. *Curr. Microbiol.* 56(1), 33-41.

Pandit, D. B., Islam, M. M., Harun-Ur-Rashid, M. and Sufian, M. A. (2007). Participatory variety selection in wheat and its impact on scaling-up seed dissemination and varietal diversity. *Bangladesh J. Agric. Res* 32(3), 473-486.

Pask, A., Joshi, A. K., Manès, Y., Sharma, I., Chatrath, R., Singh, G. P., Sohu, V. S., Mavi, G. S., Sakuru, V. S. P., Kalappanavar, I. K., Mishra, V. K., Arun, B., Mujahid, M. Y., Hussain, M., Gautam, N. R., Barma, N. C. D., Hakim, A., Hoppitt, W., Trethowan, R. and Reynolds, M. P. (2014). A wheat phenotyping network to incorporate physiological traits for climate change in South Asia. *Field Crops Res.* 168, 156-167.

Pask, A., Reynolds, M., Sharma, I., Chatrath, R., Singh, G. P., Sohu, V. S., Mavi, G. S., Sukaru, V. S. P., Kalappanavar, I. K., Mishra, V. K. and Balasubramaniam, A. (2013). The CSISA wheat phenotyping network. In: *International Workshop of the Wheat Yield Consortium, 3. Proceedings; CENEB, CIMMYT*. Cd. Obregon, Sonora (Mexico); 5-7 Mar 2013 pp.70-72.

Patil, L. C. and Hanchinal, R. R. (2011). Identification of RAPD markers for spot blotch resistance in wheat. *Karnataka J. Agr.Sci.* 24, 273-276.

Phan, C. S., Li, H., Kessler, S., Solomon, P. S., Piggott, A. M. and Chooi, Y. H. (2019). Bipolenins K-N: new sesquiterpenoids from the fungal plant pathogen *Bipolaris sorokiniana*. *Beilstein J. Org. Chem.* 15, 2020-2028. https://doi.org/10.3762/bjoc.15.198.

Poloni, A., Pessi, I. S., Frazzon, A. P. G. and van der Sand, S. T. (2009). Morphology, physiology, and virulence of *Cochliobolus sativus* isolates. *Curr. Microbiol.* 59(3), 267-273.

Poudel, A., Navathe, S., Chand, R., Mishra, V. K., Singh, P. K. and Joshi, A. K. (2019). Hydrogen peroxide prompted lignification affects pathogenicity of hemi-biotrophic pathogen *Bipolaris sorokiniana* to Wheat (Triticum aestivum L.) *Plant Pathol J* 35(4), 287-300.

Regmi, A. P., Ladha, J. K., Pasuquin, E. M., Pathak, H., Hobbs, P. R., Shrestha, L., Gharti, D. and Duveiller, E. (2002). The role of potassium in sustaining yields in a long-term rice-wheat experiment in the Indo-Gangetic plains of Nepal. *Biol. Fertil. Soils* 36(3), 240-247.

Rosewarne, G. M., Singh, R. P., Huerta-Espino, J., William, H. M., Bouchet, S., Cloutier, S., McFadden, H. and Lagudah, E. S. (2006). Leaf tip necrosis, molecular markers and b1-proteasome subunits associated with the slow rusting resistance genes Lr46/Yr29. *Theor. Appl. Genet.* 112(3), 500-508.

Rosyara, U. R., Khadka, K., Subedi, S., Sharma, R. C. and Duveiller, E. (2007a). Heritability of stay green traits and association with spot blotch resistance in three spring wheat populations. *J. Genet. Plant Breed.* 61, 140-148.

Rosyara, U. R., Pant, K., Duveiller, E. and Sharma, R. C. (2007b). Variation in chlorophyll content, anatomical traits and agronomic performance of wheat genotypes differing in spot blotch resistance under natural epiphytotic conditions. *Aust. Plant Pathol.* 36, 245–251.

Rosyara, U. R., Subedi, S., Duveiller, E. and Sharma, R. C. (2010). Photochemical efficiency and SPAD value as indirect selection criteria for combined selection of spot blotch and terminal heat stress in wheat. *J. Phytopath.* 158(11–12), 813–821.

Rosyara, U. R., Vromman, D. and Duveiller, E. (2008). Canopy temperature depression as an indication of correlative measure of spot blotch resistance and heat stress tolerance in spring wheat. *J. Plant Path.* 90, 103–107.

Saari, E. E. and Prescott, J. M. (1975). A scale for appraising the foliar intensity of wheat disease. *Plant Dis. Report.* 59, 377–380.

Sahu, R., Sharaff, M., Pradhan, M., Sethi, A., Bandyopadhyay, T., Mishra, V. K., Chand, R., Chowdhury, A. K., Joshi, A. K. and Pandey, S. P. (2016). Elucidation of defense-related signaling responses to spot blotch infection in bread wheat (*Triticum aestivum* L.). *Plant J.* 86(1), 35–49.

Saxena, R. R., Mishra, V. K., Chand, R., Chowdhury, A. K., Bhattacharya, P. M. and Joshi, A. K. (2017). Pooling together spot blotch resistance, high yield with earliness in wheat for eastern Gangetic Plains of South Asia. *Field Crops Res.* 214, 291–300.

Shaner, G. and Finney, R. E. (1977). The effect of nitrogen fertilization on the expression of slow-mildewing resistance in Knox wheat. *Phytopathology* 77(8), 1051–1056.

Sharma, R. C., Dubin, H. J., Bhatta, M. R. and Devkota, R. N. (1997). Selection for spot blotch resistance in four spring wheat populations. *Crop Sci.* 37(2), 432–435.

Sharma, R. C., Duveiller, E., Ahmed, F., Arun, B., Bhandari, D., Bhatta, M. R., Chand, R., Chaurasiya, P. C. P., Gharti, D. B., Hossain, M. H. and Joshi, A. K. (2004a). Helminthosporium leaf blight resistance and agronomic performance of wheat genotypes across warm regions of South Asia. *Plant Breed.* 123(6), 520–524.

Sharma, R. C., Duveiller, E., Gyawali, S., Shrestha, S. M., Chaudhary, N. K. and Bhatta, M. R. (2004b). Resistance to Helminthosporium leaf blight and agronomic performance of spring wheat genotypes of diverse origins. *Euphytica* 139(1), 33–44.

Sharma, R. C., Sah, S. N. and Duveiller, E. (2004c). Combining ability analysis of resistance to Helminthosporium leaf blight in spring wheat. *Euphytica* 136(3), 341–348.

Sharma, R. C. and Duveiller, E. (2006). Spot blotch continues to cause substantial grain yield reductions under resource limited farming conditions. *J. Phytopathol.* 154(7-8), 482–488.

Sharma, R. C. and Duveiller, E. (2007). Advancement toward new spot blotch resistant wheats in South Asia. *Crop Sci.* 47(3), 961–968.

Sharma, S., Sahu, R., Navathe, S., Mishra, V. K., Chand, R., Singh, P. K., Joshi, A. K. and Pandey, S. P. (2018). Natural variation in elicitation of defense-signaling associates to field resistance against the spot blotch disease in bread wheat (*Triticum aestivum* L.). *Front. Plant Sci.* 9, 636.

Siddique, A. B., Hossain, M. H., Duveiller, E. and Sharma, R. C. (2006). Progress in wheat resistance to spot blotch in Bangladesh. *J. Phytopathol.* 154(1), 16–22.

Singh, U. P. and Singh, R. (2009). Molecular basis of resistance in wheat varieties against spot blotch disease. *J. Appl. Nat. Sci.* 1(2), 191–195.

Singh, G., Sheoran, S., Chowdhury, A. K., Tyagi, B. S., Bhattacharya, P. M., Singh, V., Ojha, A., Turan, R. and Sharma, I. (2014). Phenotypic and marker aided identification of donors for spot blotch resistance in wheat. *J. Wheat Res.* 6, 98–100.

Singh, P. K., He, X., Sansaloni, C. P., Juliana, P., Dreisigacker, S., Duveiller, E., Kumar, U., Joshi, A. K. and Singh, R. P. (2018). Resistance to spot blotch in two mapping populations of common wheat is controlled by multiple QTL of minor effects. *Int. J. Mol. Sci.* 19(12), 4054.

Singh, P. K., Zhang, Y., He, X., Singh, R. P., Chand, R., Mishra, V. K., Malaker, P. K., Reza, M. A., Rahman, M. M., Islam, R., Chowdhury, A. K., Bhattacharya, P. M., Kalappanavar, I. K., Crossa, J. and Joshi, A. K. (2015). Development and characterization of the 4th CSISA-spot blotch nursery of bread wheat. *Eur. J. Plant Pathol.* 143(3), 595-605.

Singh, R. P. (1992). Association between gene Lr34 for leaf rust resistance and leaf tip necrosis in wheat. *Crop Sci.* 32(4), 874-878.

Singh, V., Singh, G., Chaudhury, A., Ojha, A., Tyagi, B. S., Chowdhary, A. K. and Sheoran, S. (2016). Phenotyping at hot spots and tagging of QTLs conferring spot blotch resistance in bread wheat. *Mol. Biol. Rep.* 43(11), 1293-1303.

Steffenson, B. J., Hayes, P. M. and Kleinhofs, A. (1996). Genetics of seedling and adult plant resistance to net blotch (*Pyrenophora teres* f. sp. *teres*) and spot blotch (*Cochliobolus sativus*) in barley. *Theor. Appl. Genet.* 92(5), 552-558.

Stewart, E. L. and McDonald, B. A. (2014). Measuring quantitative virulence in the wheat pathogen *Zymoseptoria tritici* using high-throughput automated image analysis. *Phytopathology* 104(9), 985-992.

Sultana, S., Adhikary, S. K., Islam, M. M. and Rahman, S. M. M. (2018). Evaluation of pathogenic variability based on leaf blotch disease development components of *Bipolaris sorokiniana* in *Triticum aestivum* and agroclimatic origin. *Plant Pathol. J.* 34(2), 93-103.

Tomar, V., Singh, D., Singh, R. P., Poland, J., Joshi, A. K., Dhillon, G. S., Singh, P. K., Kumar, S., Rahman, M. M., Tiwari, B. S. and Kumar, U. (2020). New QTLs for spot blotch disease resistance in wheat (*Triticum aestivum* L.) using genome-wide association mapping. *Front. Genet.* 11, 1740.

Valjavec-Gratian, M. and Steffenson, B. J. (1997). Pathotypes of *Cochliobolus sativus* on barley in North Dakota. *Plant Dis.* 81(11), 1275-1278.

van Ginkel, M. and Rajaram, S. (1998). Breeding for resistance to spot blotch in wheat: global perspective. In: Duveiller, E., Dubin, H. J., Reeves, J. and McNab, A. (Eds), *Helminthosporium Blights of Wheat: Spot Blotch and Tan Spot*. CIMMYT, Mexico, pp.162-170.

Vasistha, N. K., Balasubramaniam, A., Mishra, V. K., Chand, R., Srinivasa, J., Yadav, P. S. and Joshi, A. K. (2016). Enhancing spot blotch resistance in wheat by marker-aided backcross breeding. *Euphytica* 207(1), 119-133.

Vergnes, D. M., Renard, M. E., Duveiller, E. and Maraite, H. (2006). Effect of growth stage on host sensitivity to helminthosporol toxin and susceptibility to *Cochliobolus sativus* causing spot blotch on wheat. *Physiol. Mol. Plant Path.* 68(1-3), 14-21.

Wang, R., Leng, Y., Ali, S., Wang, M. and Zhong, S. (2017). Genome-wide association mapping of spot blotch resistance to three different pathotypes of *Cochliobolus sativus* in the USDA barley core collection. *Mol. Breed.* 37(4), 44.

Wang, R., Leng, Y. and Zhong, S. (2015). The regulatory gene *VosA* affects conidiogenesis and is involved in virulence of the fungal cereal pathogen *Cochliobolus sativus*. *Fungal Biol.* 119(10), 884-900.

Wang, R., Leng, Y., Shrestha, S. and Zhong, S. (2016). Coordinated and independent functions of velvet-complex genes in fungal development and virulence of the fungal cereal pathogen *Cochliobolus sativus*. *Fungal Biol.* 120(8), 948-960.

Wang, R., Leng, Y., Zhao, M. and Zhong, S. (2019). Fine mapping of a dominant gene conferring resistance to spot blotch caused by a new pathotype of *Bipolaris sorokiniana* in barley. *Theor. Appl. Genet.* 132(1), 41-51.

Wu, L., He, X., Lozano, N., Zhang, X. and Singh, P. K. (2020). ToxA, a significant virulence factor involved in wheat spot blotch disease, exists in the Mexican population of Bipolaris sorokiniana. *Trop. Plant Pathol.* 1-6.

Yu, J., Pressoir, G., Briggs, W. H., Vroh-Bi, I., Yamasaki, M., Doebley, J. F., McMullen, M. D., Gaut, B. S., Nielsen, D. M., Holland, J. B., Kresovich, S. and Buckler, E. S. (2006). A unified mixed-model method for association mapping that accounts for multiple levels of relatedness. *Nat. Genet.* 38(2), 203-208.

Yusuf, C. S., Chand, R., Mishra, V. K. and Joshi, A. K. (2016). The association between leaf malondialdehyde and lignin content and resistance to spot blotch in wheat. *J. Phytopath.* 164, 1-8.

Zadoks, J. C., Chang, T. T. and Konzak, C. F. (1974). A decimal code for the growth stages of cereals. *Weed Res.* 14(6), 415-421.

Zhang, P., Guo, G., Wu, Q., Chen, Y., Xie, J., Lu, P., Li, B., Dong, L., Li, M., Wang, R., Yuan, C., Zhang, H., Zhu, K., Li, W. and Liu, Z. (2020). Identification and fine mapping of spot blotch (*Bipolaris sorokiniana*) resistance gene *Sb4* in wheat. *Theor. Appl. Genet.* 133(8), 2451-2459 https://doi.org/10.1007/s00122-020-03610-3.

Zhong, S. and Steffenson, B. J. (2001). Virulence and molecular diversity in *Cochliobolus sativus*. *Phytopathology* 91(5), 469-476.

Zhong, S., Steffenson, B. J., Martinez, J. P. and Ciuffetti, L. M. (2002). A molecular genetic map and electrophoretic karyotype of the plant pathogenic fungus *Cochliobolus sativus*. *Mol. Plant Microbe Interact.* 15(5), 481-492.

Zhou, H. and Steffenson, B. (2013). Genome-wide association mapping reveals genetic architecture of durable spot blotch resistance in US barley breeding germplasm. *Mol. Breed.* 32(1), 139-154.

Zhu, Z., Bonnett, D., Ellis, M., Singh, P., Heslot, N., Dreisigacker, S., Gao, C. and Mujeeb-Kazi, A. (2014). Mapping resistance to spot blotch in a CIMMYT synthetic-derived bread wheat. *Mol. Breed.* 34(3), 1215-1228.

Zurbriggen, M. D., Carrillo, N. and Hajirezaei, M. R. (2010). ROS signaling in the hypersensitive response: when, where and what for? *Plant Signal. Behav.* 5(4), 393-396.

Part 5

Fungal diseases of cereals: net blotch

Chapter 13

Advances in understanding the epidemiology, molecular biology and control of net blotch and the net blotch barley interaction

Anke Martin[1], Barsha Poudel and Buddhika Amarasinghe Dahanayaka, Centre for Crop Health, University of Southern Queensland, Australia; Mark S. McLean, Agriculture Victoria, Victorian Department of Economic Development, Jobs, Tourism and Resources, Australia; Lisle Snyman, Queensland Department of Agriculture and Fisheries, Australia; and Francisco J. Lopez-Ruiz, Centre for Crop and Disease Management, Curtin University, Australia

1 Introduction

The net blotches are stubble-borne diseases in which primary infection is derived from barley crop stubble or residue. Infection can be derived from either air-borne conidia which are produced on conidiophores that grow on

1 All authors contributed equally.

http://dx.doi.org/10.19103/AS.2021.0092.31

the stubble surface or from ascospores derived from pseudothecia. The asexual morph consists of conidia of variable lengths, 25–300 μm long, 7–11 μm thick, straight, cylindrical in shape, smooth with rounded ends, inconspicuous hilum, and have 1–14 (usually 4–6) pseudosepta (Smedegård-Petersen, 1971). Conidia are dispersed over short distances with the majority wind dispersed within 100 cm and approximately 50% dispersed within 25 cm (Deadman and Cooke, 1989). The sexual morph is produced in pseudothecia which are 1–2 mm in diameter, globose to elongated and covered with dark septate setae at maturity (Steffenson, 1997). Pseudothecia contain asci with 3–8 ascospores which are 10–13 μm long and 20–23 μm thick (Steffenson, 1997; Van den Berg, 1988). Ascospores are air-borne and are able to travel greater distances to infect neighbouring barley fields (Steffenson, 1997). Net form net blotch can also be seed-borne. Laboratory and glasshouse tests have shown that mycelium in the caryopsis grows to infect the developing coleoptile that then penetrate the leaves. Spot form net blotch has not been shown to be seed-borne (Jordan, 1981).

Disease cycles of spot form and net form of net blotch of barley are illustrated in Figs 1 and 2. Primary infection occurs when conidia or ascospores land on the surface of the leaf, sheathes or floral bracts and cool wet weather conditions are present for a prolonged period. Infection takes place at temperatures ranging

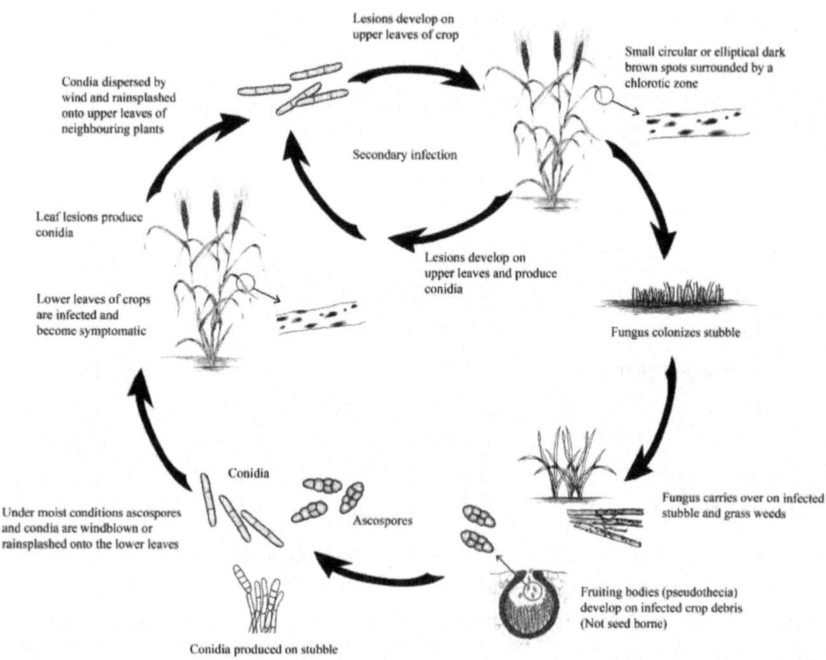

Figure 1 Disease cycle of spot form net blotch of barley.

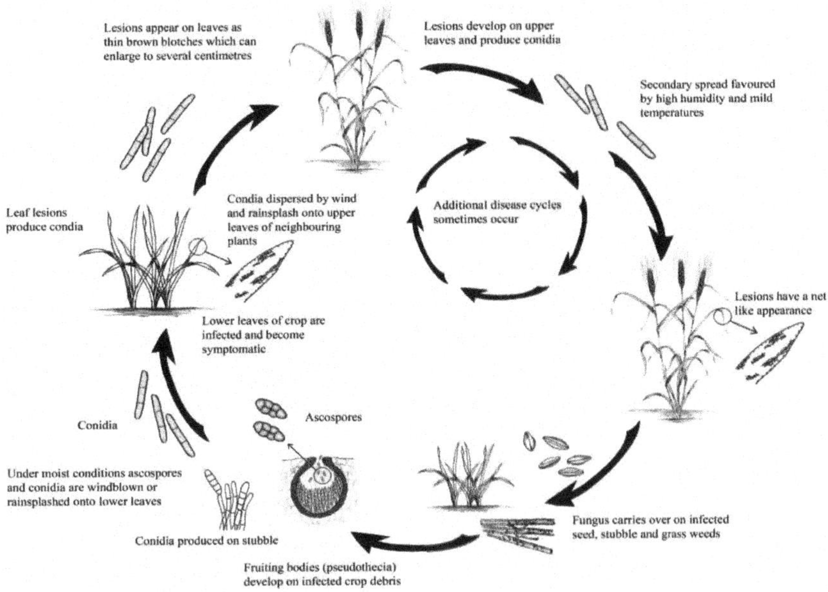

Figure 2 Disease cycle of net form net blotch of barley.

from 5°C to 25°C with optimum temperatures of 15°C–22°C (Hargreaves and Keon, 1983). The infection process is slower during cooler temperatures, taking 6 h at 10°C and is faster in warmer temperatures, taking as little as 1 h at 25°C (Hargreaves and Keon, 1983). The infection process consists of ascospores or conidia germinating and producing appressoria and penetration pegs that penetrate the epidermal cell wall, forming an intracellular infection vesicle (Hargreaves and Keon, 1983). After initial penetration, the fungus grows throughout the epidermal cell layer, colonising the apoplast of the mesophyll tissue (Hargreaves and Keon, 1983). Symptoms start to appear within 24–48 h as dark pin-point spots. These develop into lesions with varying amounts of necrosis and chlorosis depending on the climatic conditions and resistance/susceptibility of the host. On susceptible hosts, spot form net blotch develops as dark brown necrotic spots that increase in size to form elliptical or fusiform lesions measuring from 3 mm to 6 mm. These are surrounded by a chlorotic zone of varying width (Smedegård-Petersen, 1971). Symptoms of net form net blotch are initially very similar to those of spot form net blotch, but on susceptible varieties develop as fine dark brown lines extending across the leaf surface to produce a network pattern that may later coalesce and form irregular stripes (Ellis and Waller, 1973), which can extend and coalesce to destroy the entire leaf. Symptoms of spot form net blotch and net form net blotch on resistant varieties are small pin-point, dark brown necrotic lesions that do not increase in size but may develop a small chlorotic halo depending on the type of resistance present. The differences observed between net and spot form symptoms are due to differences in fungal growth. Spot form of net blotch initially grows as a biotroph forming intracellular vesicles within epidermal cells before switching to necrotrophic (intercellular) growth in the mesophyll. Net form of net blotch does not have the biotrophic stage and penetrates the mesophyll more quickly (Lightfoot and Able, 2010).

Secondary conidia development occurs from 14 days to 22 days after initial infection. Conidia develop readily from large lesions with severe necrosis and chlorosis and less frequently from small lesions, especially on resistant varieties. Conidia development occurs during the night and conidia are typically released following strong winds (Jordan, 1981). Secondary infection occurs as often as cool, wet conditions are repeated throughout the growing season. Infection can occur in the head and grain of plants if conditions are favourable late in the season.

Pyrenophora teres survives from one season to the next on crop stubble or residue. Once the crop has senesced, mycelial growth colonises the surface during wet conditions. Mycelium and pseudothecia are viable as long as barley stubble is present on the soil surface. The rate at which pseudothecia develop depends on the environment. They have been found to develop rapidly in the first or second season, with development reducing in the third season (Duczek et al., 1999).

Pyrenophora teres can also infect and survive on other cereals such as barley grass (*H. leporinum*), wheat (*Triticum aestivum*) and oat (*Avena sativa*) (Van den Berg, 1988; Uranga et al., 2020) and can infect a wide host range of other cereal grasses such as *Agropyron*, *Bromus*, *Elymus*, *Hordelymus*, *Stipa* and other *Hordeum* spp. (Sampson and Watson, 1985; Brown et al., 1993). However, these are minor hosts and generate little infection.

2 Hybrids

In vitro sexual recombination has been observed within and between the two forms of *P. teres*. McDonald (1963) was able to induce ascospores in culture when two compatible single-spore *P. teres* isolates were crossed. He indicated that *P. teres* are heterothallic in nature. In heterothallic species, sexual reproduction occurs between individuals of two different mating types. *Pyrenophora teres* has a single mating type locus (MAT) with two idiomorphs designated as MAT1-1 and MAT1-2.

Smedegård-Petersen (1971) produced viable hybrid progeny from crosses between *P. teres* f. *teres* and *P. teres* f. *maculata*. Hybrids inoculated on barley leaves produced symptoms similar to those of net or spot form net blotch or intermediate to those of the parents (Smedegård-Petersen, 1971; Smedegård-Petersen, 1976; Crous et al., 1995). They are genetically stable and can retain their virulence and fertility over generations (Campbell and Crous, 2003). Hybrids produced in a laboratory had reduced sensitivity to triazole fungicides as compared to the parents (Campbell et al., 1999). Different virulence patterns were observed when comparing the progeny with either of the parental isolates, with some hybrids being virulent on barley lines on which both parents were avirulent (Jalli, 2011).

Pyrenophora teres hybrids are relatively rarely found in the field and appear to be reproductively isolated (Rau et al., 2003; Serenius et al., 2007; Ellwood et al., 2012; Poudel et al., 2017, 2019a). The reproductive isolation could arise due to pre- or post-mating barriers and has been reviewed in a paper by Poudel et al. (2019a). The two most likely reasons suggested for the reproductive isolation include sexual preferences for the same form or reduced fitness compared to their parents. The finding of *P. teres* hybrids in nature, however, suggests that reproductive barriers under some conditions can be permissive, thereby allowing hybrids to be formed more readily. Seven hybrids have been identified under field conditions to date, one in South Africa (Campbell et al., 2002), two in the Czech Republic (Leisova et al., 2005b), one in Hungary (Dahanayaka et al., 2020), one in Japan (Dahanayaka et al., 2020) and two in Australia (McLean et al., 2014; Turo et al., 2021). A hybrid collected from fields in Western Australia was highly resistant to some Group 3 (azole or demethylase inhibitor) fungicides and was found to rapidly propagate by asexual reproduction (Turo et al., 2021). Although rare, *P. teres* hybrids are of concern as they can rapidly evolve acquiring virulence and fungicide resistance from both net blotch forms due to sexual recombination, thus producing more resistant strains. There is therefore a need to constantly monitor *P. teres* populations in fields for the presence of hybrids. Random amplified polymorphic DNA (RAPD) and amplified fragment length polymorphism (AFLP) markers (Campbell et al., 1999, 2002) have been used to identify hybrids, but these techniques are demanding and time-consuming. Recently, more efficient and accurate PCR-based markers have been developed to identify hybrids using six unique regions from each *P. teres* form (Poudel et al., 2017).

3 Molecular markers to accurately diagnose *P. teres* isolates

It is important to accurately characterise pathogens for disease management, plant breeding and epidemiological studies. As the spore morphology of the two forms of *P. teres* is very similar, symptom expression is the usual means of identification. It can, however, be difficult to distinguish between the two forms of *P. teres*, especially at early stages of symptom development. Spot form net blotch symptoms are also difficult to distinguish from spot blotch symptoms caused by *Bipolaris sorokiniana* Shoemaker 1959 (Williams et al., 2001; Lehmensiek et al., 2010). The most reliable characterisation of the blotches is thus by using molecular markers. A number of molecular assays have been developed, and the first of these was developed in 2001 by Williams et al. (2001).

Williams et al. (2001) cloned a random amplified polymorphic DNA (RAPD) marker that was present in *P. teres* f. *maculata* isolates and absent from *P. teres* f. *teres* isolates. The cloned segment was sequenced and primers designed which amplified the same size fragment in both forms. The amplicons from both forms were thus sequenced and primers designed specific to each form of net blotch (Table 1). Testing of the developed primer sets across 27 *P. teres* isolates collected in Australia, Canada, the USA and Germany confirmed that they were specific to the two forms. The markers did not amplify on DNA of *B. sorokiniana* but did amplify DNA from *P. graminea*. In European countries it is important to be able to differentiate between *P. teres* and *P. graminea* as the seed infection threshold for *P. graminea* is lower than that for *P. teres* and is applied for both species if the species has not been diagnosed (Justesen et al., 2008).

In 2005, Leisova et al. (2005a) developed form-specific markers that could distinguish the net blotches from *P. graminea* as well as distinguish between the two forms. They used sequenced amplified fragment length polymorphism (AFLP) markers to identify form-specific markers that were also distinct amongst different species, including *P. graminea*, *P. tritici-repentis* and *Helminthosporium sativum*. The form-specific markers were cloned and sequenced and primer sets were designed. Primers were validated across 66 *P. teres* isolates, five *P. graminea*, three *P. tritici-repentis*, one *P. flavispora* and four *B. sorokiniana* isolates. Two primer sets for each form were both species and form specific (Table 1).

A further set of diagnostic markers were developed by Kępiper et al. (2008). These nine microsatellite or simple sequence repeat (SSR) markers amplified loci in one form only, amplified loci with different allele size ranges in each form, or amplified different numbers of loci in each form. Of the nine markers, three were form specific, one to *P. teres* f. *maculata* and two to *P. teres* f. *teres* (Table 1). The primers were only tested across six *P. teres* isolates of each form and species specificity was not determined.

In 2010, mating type locus-specific markers were developed by Lu et al. (2010) which could be used to differentiate between the two forms of *P. teres*. The mating type loci were cloned and sequenced and form- and species-specific primers were designed across single nucleotide polymorphism differences. Two sets of primers for each mating type were developed (Table 1). Specificity of each primer set was validated across 54 *P. teres* isolates and DNA from *P. graminea*, *P. tritici-repentis* and other ascomycetes. Unfortunately, the specificity of the markers with *B. sorokiniana*, which produces disease symptoms very similar to those of *P. teres* f. *maculata*, was not tested. The advantage of using the mating type markers is that the pathogenicity and mating type can be simultaneously characterised whilst confirming the species and *P. teres* form.

Sequence-specific PCR markers were produced by Poudel et al. (2017) to identify the two forms of *P. teres* and progeny derived from recombination between the two forms (hybrids). Hybrids normally have similar lesion types to those of their parents and can therefore only be diagnosed using molecular markers. Previously developed diagnostic markers were not ideal for identifying hybrids and could therefore not be used to monitor the occurrence of hybrids in natural environments. Whole-genome assemblies and RNA-seq derived assembled and aligned transcripts were utilised to identify unique *P. teres* f. *teres* and *P. teres* f. *maculata* regions. Twelve form-specific expressed regions were identified and six sets of primers designed for each *P. teres* form (Table 1). The specificity of the primers sets was validated across 78 Australian and 46 South African *P. teres* isolates, seven *P. teres* isolates obtained from barley grass (*Hordeum leporinum*), six *B. sorokiniana* isolates and one isolate each of *Exserohilum rostratum* and *P. tritici-repentis*. The primer sets were specific to each form with all six markers present and no amplification on DNA of the other species tested. Hybrids had between two and eleven markers but always at least one marker specific to each form. Interestingly the *P. teres* isolates collected from barley grass only amplified four of the six *P. teres* f. *teres*-specific markers and none of the *P. teres* f. *maculata*-specific markers. AFLP analysis of both *P. teres* isolates collected from barley and barley grass revealed different fingerprints for the two groups of isolates (Poudel et al., 2017). Only 2–4 markers out of the 12 need to be amplified to confirm the species and form of *P. teres*. However, to detect hybrids all 12 markers need to be amplified. As the fragments produced by the markers are relatively small, they can be amplified and detected using a quantitative assay without needing to run an agarose gel (Dahanayaka et al., 2021).

In conclusion, a number of molecular markers have been developed to correctly diagnose *P. teres* isolates. Some markers have an added advantage over others such as the Lu et al. (2010) mating type markers which simultaneously characterise the mating type of the isolate and the Poudel et al. (2017) markers which can simultaneously identify hybrids. The choice of markers depends on the aim of the diagnosis.

4 Genetic variation and population genetics of *P. teres*

Genetic variation refers to the naturally occurring genetic differences between individuals of the same species that define population structures (Milgroom, 1996). The population structure of a pathogen can be used to predict how rapidly a pathogen can evolve and overcome host or fungicide resistance. Molecular markers such as RAPD, AFLP, SSR and Diversity Arrays Technology (DArT) have been applied effectively to understand genetic diversity in *P. teres* populations worldwide, including Canada, Germany, USA (Peever and Milgroom, 1994; Liu

Table 1 List of primers developed by different groups to distinguish the two forms of *P. teres* and identify hybrids

Species	Primer name (F = forward; R = reverse)	Sequence 5′-3′	Amplicon size (bp)	Reference
P. teres f. *teres*	PTT-F PTT-R	CTCTGGCGAACCGTTC ATGATGGAAAAGTAATTTGTA	378	(Williams et al., 2001)
P. teres f. *maculata*	PTM-F PTM-R	TGCTGAAGCGTAAGTTTC ATGATGGAAAAGTAATTTGTG	411	(Williams et al., 2001)
P. teres f. *teres*	DTT471h-F DTT471h-R	CCTGAGTAACTTGCCCCACC GAAAAGAGATGATGCGGACAC	91	(Leisova et al., 2005a)
P. teres f. *teres*	DTT339i-F DTT339i-R	TGATGCGCTGGAGTGAGACAC TGTACATACGCCGCATCACG	81	(Leisova et al., 2005a)
P. teres f. *maculata*	DTM494d-F DTM494d-R	TATTCTGCTAAGAGCTAGCATCCTA ACTGCGTACCAATTCTCTA CAACTA	161	(Leisova et al., 2005a)
P. teres f. *maculata*	DTM348j-F DTM348j-R	CTTGATGCGCTGGAGTGAGA TGCATTTCCACCTACTGGTATGTAC	66	(Leisova et al., 2005a)
P. teres f. *teres*	hSPT2_4agac-F hSPT2_4agac-R	CCTTGGTGGTTTCTGTGGTTTCT AGAGAGAGAGAGAGACACACAC	128–132	(Keiper et al., 2008)
P. teres f. *teres*	hSPT2_24tcac-F hSPT2_24tcac-R	ACTTCGCTGAGTGTTAGTTGCATC TCTCTCTCTCTCACACACAC	90–106	(Keiper et al., 2008)
P. teres f. *maculata*	hSPT2_24agac-F hSPT2_24agac-R	ATACTTGTGGTAGCCTACTTTGCA AGAGAGAGAGAGAGACACACAC	126–168	(Keiper et al., 2008)
P. teres f. *teres*	Ptt-MAT1-F Ptt-MAT1-R Ptt-MAT2-F Ptt-MAT2-R	ATGAGACGCTAGTTCAGAGTCT GATGCCCAGCCAAGGACAA TACGTTGATGCAGCTTTCTCAAT AACACCGTCCAAAGCACCT	1 143 1 421	(Lu et al., 2010)
P. teres f. *maculata*	Ptm-MAT1-F Ptm-MAT1-R Ptm-MAT2-F Ptm-MAT2-R	TGTTAGAGACCCCACCAGCGT CAGCTTTCTTGGCCTTCTGAA ACGCAAGGTACTCTGTACGCA GAC GTCGAGGGAGTCCATTT	149 939	(Lu et al., 2010)

P. teres f. *teres*	PttQ1-F PttQ1-R	GGATGATGACCTCGCCAGAT GCGATGGTATGTTCTGCGAA	70	(Poudel et al., 2017)
P. teres f. *teres*	PttQ2-F PttQ2-R	AACACTCTGAACGTGGTTGC TTCAGTTGTAAGCTGCGTGG	110	(Poudel et al., 2017)
P. teres f. *teres*	PttQ3-F PttQ3-R	CCTCGTCCTAAGTTGACTCGA TTACACGGGTTCCCTCCATC	130	(Poudel et al., 2017)
P. teres f. *teres*	PttQ4-F PttQ4-R	CGTCCCGCCGAAATTTTGTA CAAGGACTTACGCGCTCAAA	150	(Poudel et al., 2017)
P. teres f. *teres*	PttQ5-F PttQ5-R	GCATTGTCTAGCACTCGTCG CGCGGACTCAGAAGACATTG	173	(Poudel et al., 2017)
P. teres f. *teres*	PttQ6-F PttQ6-R	TCAGAATACTCCGCGGACTC GTCCGCATTGTCTAGCACTC	188	(Poudel et al., 2017)
P. teres f. *maculata*	PtmQ7-F PtmQ7-R	GTAGAGGCTGTAGGAGATGTGATT CATGGCAAATTGTTCGTAATCCTG	140	(Poudel et al., 2017)
P. teres f. *maculata*	PtmQ8-F PtmQ8-R	ACGCTAAGACCACCTCGTTT TCGACCAGAGAGAGCACAAA	161	(Poudel et al., 2017)
P. teres f. *maculata*	PtmQ9-F PtmQ9-R	AATGCTCAATTCTGGTGGCG TGTTCGAGTGCAAACTTGGG	201	(Poudel et al., 2017)
P. teres f. *maculata*	PtmQ10-F PtmQ10-R	TGCTGTGGACTTAGACGAGG TGGGGATCCTTGACCAACTC	220	(Poudel et al., 2017)
P. teres f. *maculata*	PtmQ11-F PtmQ11-R	GATTAGACCATTACCACACTAGCG ACCACCACATCTTTCCTACTAACT	260	(Poudel et al., 2017)
P. teres f. *maculata*	PtmQ12-F PtmQ12-R	CTAACCAAAGAACTTCACAGACGA CCTTATTAGCCAATTCCATGTCGA	279	(Poudel et al., 2017)

et al., 2012; Akhavan et al., 2015), Finland (Peltonen et al., 1996; Serenius et al., 2005, 2007), Sweden (Jonsson et al., 2000), Italy (Rau et al., 2003), the Czech and Slovak Republic (Leisova et al., 2005b; Leisova et al., 2014), South Africa (Campbell et al., 2002; Lehmensiek et al., 2010), Iceland (Stefansson et al., 2012) and Australia (Serenius et al., 2007; Bogacki et al., 2010; Lehmensiek et al., 2010; McLean et al., 2010b, 2014; Poudel et al., 2019b). As the genetic markers used in these studies were different, the outcomes are not directly comparable. Nevertheless, high population genetic diversity has always been reported for *P. teres* populations.

In one of the earliest such studies, using RAPD markers, five *P. teres* f. *teres* populations originating from Canada, Germany and the USA were examined (Peever and Milgroom, 1994). These populations shared a high number of alleles implying that *P. teres* was derived from a common source population; however, a significantly high degree of variability was detected within four of the populations. This implied that sexual reproduction occurred regularly in field populations and was consistent with most of the other population studies from a wide range of geographical locations, in which the genetic structure of *P. teres* f. *teres* and *P. teres* f. *maculata* were investigated (Akhavan et al., 2015; Campbell et al., 2002; Rau et al., 2003; Liu et al., 2012; Bogacki et al., 2010; McLean et al., 2010b; Jonsson et al., 2000; Serenius et al., 2005; Leisova et al., 2005b, 2014; Peltonen et al., 1996). The genetic variation within a population can occur within a field or a plant (Peltonen et al., 1996; Jonsson et al., 2000; Liu et al., 2012; Bogacki et al., 2010; Poudel et al., 2019b) and the population structure can change substantially each year (Liu et al., 2012; Poudel et al., 2019b). Diversity in *P. teres* populations is strongly associated with geographical distances. Distinct genetic lineages were detected among populations separated by wide geographical areas (such as states, countries and continents) due to limited gene flow (Serenius et al., 2007; Lehmensiek et al., 2010). Within a field or at closely located fields, high genetic variability was observed due to sexual reproduction and/or gene exchange through migration (Serenius et al., 2005; Bogacki et al., 2010; Liu et al., 2012; Poudel et al., 2019b). In contrast, some studies have indicated low genetic variability and occurrence of clonal populations within a field indicating that reproduction within *P. teres* populations occurs asexually (Jonsson et al., 2000; Lehmensiek et al., 2010; Rau et al., 2003; Serenius et al., 2007). The prevalence of mode of reproduction and genetic similarity between geographical regions can be highly variable depending on environmental conditions, specific barley varieties grown in the regions, agricultural practices such as stubble retention, close barley rotation and seed exchange (Rau et al., 2003; Serenius et al., 2007; Liu et al., 2012; Fowler et al., 2017).

Collectively, frequent sexual recombination within each form, generation of conidia by asexual reproduction and gene flow via air-borne spores and/

or infected seeds are important in epidemic development of the disease. The rapid change in *P. teres* populations due to sexual reproduction can lead to the rapid development of new virulences, which has the potential to overcome major host resistance genes and fungicides.

5 Pathogenic variation and changes in virulence

The net blotch pathogens pose considerable threats to the barley industry as they are able to frequently overcome the resistance in barley varieties (Linde and Smith, 2019). These virulence changes are particularly devastating if several barley varieties with common resistances are grown (Martin et al., 2020). It is widely accepted that widespread intensive cultivation and the development of genetically homogeneous crops lead to selective pressure on fungal pathogen populations (Ellwood et al., 2012). Many countries have reported an increased incidence of net form, probably caused by the practice of growing continuous barley crops in the same paddock (Arabi et al., 2003) and no-till farming practices (McLean et al., 2009). Douiyssi et al. (1998) indicated isolates collected from regions where barley is widely grown to be more virulent than isolates originating from regions with limited barley production.

Changes in resistance can be attributed to the appearance of new physiologic forms or races (Christensen and Graham, 1934). Fungi that reproduce both asexually and sexually are considered to pose a higher evolutionary risk and are predicted to overcome newly deployed resistances more rapidly than those reproducing only asexually or sexually (McDonald and Linde, 2002). During sexual reproduction, virulence alleles from two individuals combine into the same genetic background and these new gene combinations are rapidly propagated through asexual reproduction (Linde and Smith, 2019). Pathogens that exhibit mixed reproduction exhibit higher genotype diversity as a result of recombination and have greater potential for local adaptation to a changing environment (McDonald and Linde, 2002).

Most studies on host resistance have focussed on resistance identified at the seedling stage leading many to conclude that seedling tests can be confidently used to identify resistance and offering many advantages over field screening (Buchannon and McDonald, 1965). A numerical scale (Fig. 3) developed to classify reactions of barley to *P. teres* at the seedling stage is widely used (Tekauz, 1985). Resistance expressed at the seedling stage was reported to also express in adult plants (Arabi et al., 1990). It was however noted that several barley lines screened for resistance to *P. teres* f. *teres* showed no correlation between seedling tests in the glasshouse (one to two leaf stage), disease reactions on the fourth and flag leaves of plants grown in growth chambers and disease levels observed under field conditions, while for others, the correlation was significant (Jonsson et al., 1998).

The occurrence of the two forms of net blotch varies between geographical regions with one form usually dominating in an area (McLean et al., 2009). The net form was identified to be dominant in Canada (Tekauz, 1990), California (Steffenson and Webster, 1992a), Sweden (Jonsson et al., 1997) and Morocco (Jebbouj and El Yousfi, 2010), whereas the spot form is more prevalent in France (Arabi et al., 1992). Fewer and less in-depth studies on virulence profiles of *P. teres* f. *maculata* have been published (Liu et al., 2011).

Physiologic specialization in *P. teres* f. *teres* was first reported in 1949 with the identification of different races and differences in pathogenicity towards varieties (Pon, 1949). Plant pathologists and breeders can utilise knowledge about the variation in virulence of pathogen populations in the deployment of resistant sources that are likely to be effective against the spectrum of pathotypes in the area (Steffenson and Webster, 1992a).

In California, no distinct geographical differences in the distribution of pathotypes were identified (Steffenson and Webster, 1992a). Despite clusters of pathotypes being identified in some areas, others contained mixtures of pathotypes, identified even within a single crop. Sixteen pathotypes were identified from the 91 Californian isolates. A small number of isolates from Minnesota, Mexico and England included in this study were distinctly different from each other with only the Mexican pathotypes being similar to those from California in virulence to specific differential genotypes (Steffenson and Webster, 1992a). A study using 48 Algerian isolates screened across a set of 22 differentials indicated high levels of variation in the virulence of the pathogen and 12 pathotypes were identified. Similar to the Californian isolates, the most common isolate did not display a high level of virulence (Boungab et al., 2012). The *P. teres* f. *teres* population in Algeria was highly variable from region to region, possibly in response to locally gown varieties (Boungab et al., 2012).

A set of 18 differential genotypes were used to pathotype 25 Swedish and two Canadian isolates into 14 pathotypes. Host selection on the pathogen was evident with the same pathotype collected from the same variety (Jonsson et al., 1997); isolates collected from susceptible hosts were reported to be more virulent than those collected from less susceptible hosts (Robinson and Jalli, 1996). It was also concluded that *P. teres* f. *teres* dominates in Sweden as only one of the isolates collected were identified to be *P. teres* f. *maculata* (Jonsson et al., 1997).

Two hundred and nineteen net form isolates, collected in western Canada were characterised on a set of 12 differential barley genotypes, resulting in 82% identified to be *P. teres* f. *teres* and 18% *P. teres* f. *maculata*. For *P. teres* f. *teres*, 45 pathotypes were identified and 20 for *P. teres* f. *maculata*. Net form net blotch was generally distributed throughout the collection area; however, *P. teres* f. *maculata* was most prevalent in Saskatchewan (Tekauz, 1990).

Twenty-five barley genotypes were used to pathotype 23 *P. teres* f. *teres* and eight *P. teres* f. *maculata* isolates collected from 12 different barley-growing

Figure 3 The infection response eight days after inoculation, scores based on Tekauz (1985) scoring scale. (a) 10-point scale for net form net blotch and (b) 9-point scale for spot form net blotch.

regions of the world. Fifteen *P. teres* f. *teres* pathotypes and four *P. teres* f. *maculata* pathotypes were identified, with a broader spectrum and higher level of virulence observed in the *P. teres* f. *teres* compared to the *P. teres* f. *maculata* isolates (Wu et al., 2003).

High levels of pathogenic variability were identified in 15 *P. teres* f. *teres* isolates collected in Morocco and assessed at seedling and adult plant stage on a set of 38 differential genotypes, with none of the 15 isolates being identical (Douiyssi et al., 1998). A collection of 61 *P. teres* f. *teres* and 21 *P. teres* f. *maculata* isolates collected in the semi-arid regions of Morocco, were pathotyped on differential sets of 22 *P. teres* f. *teres* and 20 *P. teres* f. *maculata* genotypes, identifying ten pathotypes of *P. teres* f. *teres* and nine of *P. teres* f. *maculata* (Jebbouj and El Yousfi, 2010).

Twenty pathotypes were identified on 14 differential genotypes in a collection of 104 *P. teres* f. *teres* isolates collected in Tunisia and Syria (Bouajila et al., 2011). A successive study, using the same differential lines, identified 23 pathotypes in 85 *P. teres* f. *teres* isolates collected across Tunisia, with variability within groups higher than between groups observed (Bouajila et al., 2012). Significant variation in pathogenicity was also demonstrated in *P. teres* f. *teres* isolates collected in France and Syria with five clusters identified in 23 isolates screened on a set of 11 genotypes (Arabi et al., 2003).

A study in New Zealand identified 11 pathotypes in 29 isolates using 31 differential genotypes. Low levels of variation in the *P. teres* f. *teres* population was identified, with low levels of virulence observed for the most common collected pathotype, despite it being recorded throughout the surveyed area (Cromey and Parkes, 2003).

Pathotype variations of *P. teres* f. *teres* in Western Australia (WA) have been extensively studied (Khan and Boyd, 1969a; Khan, 1982; Gupta and Loughman, 2001; Fowler et al., 2017). The first evidence of physiologic specialization in *P. teres* f. *teres* isolates in Australia were reported in 1969 (Khan and Boyd, 1969a) with the identification of three physiologic races in WA. A major shift in the pathogen population was reported in WA in 1982, with a decline in the area grown to the susceptible variety Beecher. Dampier, also susceptible, remained a popular variety until 1976, whereas the resistant variety Clipper was introduced in the early 1970s becoming the major variety. Isolates collected post 1976 in WA identified a new group of isolates, avirulent on the variety Beecher, whereas Dampier was susceptible to all 52 isolates. It was concluded that the shift in the pathogen population was due to changes in commercially grown barley varieties (Khan, 1982). In the period 1995–1996, net form net blotch was found to be widely distributed across barley growing areas of WA, with *P. teres* f. *teres* frequently observed in surveyed crops (Gupta and Loughman, 2001). Seventy-nine *P. teres* f. *teres* isolates collected during this period were classified into two distinct groups, based on virulence to the variety Beecher, with the majority of isolates avirulent on Beecher. They concluded that virulence to *P. teres* f. *teres* in WA remained stable over a period of at least 19 years.

The pathogenic variation of *P. teres* f. *maculata* in Western Australia between 2001 and 2002 was explored by screening a collection of 99 isolates across 26 differential barley lines (Gupta et al., 2012). Seven isolate groups were identified and the 26 differential lines were classified into four line groups according to their differential responses to the spot form net blotch isolates.

The *P. teres* f. *teres* populations in South Australia have evolved over time in response to cultivation of barley varieties (Wallwork et al., 2016). Prior to 1993, the main barley varieties cultivated, Clipper and Schooner had good levels of adult plant resistance. After a period of low detection, *P. teres* f. *teres* re-emerged in South Australia in 1994 after the release of the susceptible variety Franklin. Increased levels of net form net blotch in crops resulted in the breakdown of seedling resistance in the variety Skiff, followed by virulence in Keel in 2007 and Maritime in 2009 (Wallwork et al., 2016). Virulence on Fleet was also detected in 2009 with virulence on the variety Oxford detected in 2012. It is unknown if the virulences evolved locally or were the result of pathogen migration from other Australian barley growing regions (Linde and Smith, 2019). Cluster analysis suggests that the virulence on varieties Maritime and Keel evolved in WA and was introduced to SA, possibly mediated by humans (Linde and Smith, 2019).

A more recent comprehensive study by Fowler et al. (2017) identified *P. teres* f. *teres* isolates to be different between eastern and Western Australia, suggesting regional evolution of pathotypes, dependent on varieties prevalent in each region. This study reported virulence of 123 *P. teres* f. *teres* isolates collected across Australia over a period of 27 years. Phenotype cluster analysis identified seven line groups, which clustered into four distinct groups, indicated by differential virulence to four key barley genotypes: Maritime, Prior, Skiff and Tallon. Distinct differences were observed between isolate groups in eastern Australia and WA, whereas all isolate groups were detected in southern Australia (Fowler et al., 2017). Results indicated *P. teres* f. *teres* populations of each state to be quite unique and reflect the cultivation of locally adapted varieties. Virulence to superseded varieties was detected in all states, indicating that virulences remain in the pathogen population, despite the varieties not being present (Fowler et al., 2017).

6 Differential sets

Virulences in *P. teres* f. *teres* populations have been studied extensively in many regions using sets of differential lines (Arabi et al., 2003; Cromey and Parkes, 2003; Douiyssi et al., 1998; Fowler et al., 2017; Gupta and Loughman, 2001; Gupta et al., 2003; Khan, 1982; Khan and Boyd, 1969a; Platz et al., 2000; Steffenson and Webster, 1992a; Tuohy et al., 2006; Wallwork et al., 2016). However, due to the lack of a universal set of genotypes, pathogen populations cannot be compared globally and this hampers monitoring of pathogen populations across regions and over time.

Afanasenko et al. (1995) identified a set of 12 differentials for characterising *P. teres* f. *teres* populations that could be used to compare results of population studies between countries and regions and extend knowledge of evolution in pathogen populations. Ongoing attempts to define the set for international use by screening 1000 isolates from Russia, Europe, Australia and Canada saw the proposed standard, international differential set reduced to nine genotypes (Afanasenko et al., 2009).

For a differential set to have local benefit, the inclusion of regional differential cultivars is required, enabling breeders and pathologists to detect local virulences not identified by the standard set (Cromey and Parkes, 2003). Due to the ever-changing virulence profile of *P. teres* f. *teres* populations, there will be an ongoing need to include barley genotypes representing new sources of resistance (Afanasenko et al., 2009).

Thirty-one barley genotypes were used by Fowler et al. (2017) to determine the diversity of 123 Australian *P. teres* f. *teres* pathotypes, consisting of a combination of genotypes used in international diversity studies, Australian diversity studies and six Australian varieties. Twenty of the differentials used by Fowler et al. (2017) were used to phenotype 188 Australian isolates for genome-wide association mapping to identify regions associated with virulence in the net form net blotch pathogen (Martin et al., 2020). These 20 genotypes, together with ten current commercial varieties (Bass, Compass, Explorer, Flinders, Navigator, Oxford, RGT Planet, Rosalind, Shepherd and Urambie) constitutes the current differential set used for pathotyping Australian *P. teres* f. *teres* isolates.

To determine the pathogenic variation in the worldwide *P. teres* f. *maculata* populations an initial differential set consisting of 19 barley lines was proposed by McLean et al. (2012). This set was later validated for reaction response to 60 *P. teres* f. *maculata* isolates collected throughout the Australian barley-growing region (McLean et al., 2014). The need for further studies to develop a consolidated spot form net blotch differential set was expressed.

Resistance expression, such as lesion size and the presence of necrosis and/or chlorosis, can be influenced by factors other than pathogen virulence and host resistance, including growth stage, temperature and light (Douiyssi et al., 1998; Khan and Boyd, 1969a; Liu et al., 2011), complicating the development of international differential sets.

7 The *P. teres* genome

The first genome sequencing and assembly of the *P. teres* f. *teres* isolate 0-1 provided a significant resource for understanding the necrotrophic lifestyle and pathogenicity of *P. teres (Ellwood et al.,* 2010). This genome was sequenced using short-read paired-end Illumina sequencing resulting in an assembly with 6412 contigs and a total size of 41.95 Mb (Table 2). Recently, using Pacific Biosciences (PacBio) single-molecule real-time sequencing, the genome sequence of isolate 0-1 was improved to generate a high-quality reference genome assembly and annotation (Wyatt et al., 2018). The updated genome of 0-1 has 86 contigs with a total genome size of 46.5 Mb (Table 2). The quality of the genome assembly was improved due to the use of long reads, which can assemble low complexity, repeat dense regions. Gene annotation was performed with an evidence-based approach that included assembled RNA-seq transcripts sequenced using an Illumina platform, multiple *ab initio* gene predictions, protein evidence from the closely related species *P. tritici-repentis* (Manning et al., 2013) and an available *P. teres* f. *teres* 0-1 annotation (Ellwood et al., 2010). Secreted proteins and potential candidate effectors were predicted which are likely to be involved in the barley–*P. teres* interaction.

A large-scale genomic comparison between *P. teres* f. *teres* and *P. teres* f. *maculata* was conducted to explore structural variations, co-linearity and orthology (Syme et al., 2018). Five *P. teres* f. *teres* (W1-1, Stir9-2, NB29, NB73 and NB85) and four *P. teres* f. *maculata* (SG1, Cad6-4, M2 and FGOB10Ptm-1) isolates were included in the study. To derive a comprehensive landscape of the *P. teres* genome structure, a combination of long and short DNA reads along with RNA reads and optical genetic mapping (isolate W1-1) were employed. Single molecule PacBio long reads and Illumina short reads were combined to assemble the reference genomes of the *P. teres* f. *teres* (W1-1) and *P. teres* f. *maculata* (SG1) isolates. Both *P. teres* forms comprised of 12 chromosomes with no chromosomal fusion or rearrangements between them. The genome size of *P. teres* f. *teres* isolates ranged from 46.31 Mbp to 51.76 Mbp and *P. teres* f. *maculata* isolates ranged from 39.27 Mbp to 41.28 Mbp (Table 2). The chromosomal-scale differences between *P. teres* forms were entirely due to expansion or insertion of repetitive repertoires, more particularly transposable elements (TEs) with Line-like Tad1 and DNA transposon Tc1/Mariner TEs mainly associated with the differentiation. Similar number of genes was predicted between *P. teres* forms and the majority of the genes within GC-rich syntenic regions were stable and conserved. Genic differences found between the forms were mainly located in gene-sparse regions near or within TE-rich regions and often harboured fungal effectors. Gene interruptions by TE insertions resulting in pseudogenes were also observed in *P. teres* f. *teres*, which is an effective mechanism for removing avirulence genes and producing new effector-like genes. Secondary metabolites, which are associated with adaptation to a particular niche and virulence, were compared between *P. teres* forms (Moolhuijzen et al., 2020). In all *P. teres* isolates compared, non-ribosomal peptides synthases (NRPS), NRPS-like, type I polyketide synthase (T1PKS), type III PKS (T3PKS), hybrid T1PKS-NRPS and terpene secondary metabolites were detected. *Pyrenophora teres* f. *teres* had almost twice the number of NRPS and polyketide synthase (PK) clusters compared to *P. teres* f. *maculata* (Table 2). The NRPS expanded regions

Table 2 Genome sequencing and assembly statistics of currently sequenced *P. teres* f. *teres* and *P. teres* f. *maculata* isolates

	P. teres f. *teres*		
Isolate name	0-1	W1-1	NB29
Collection site	Ontario, Canada	Western Australia	Western Australia
Collection date	Before 1998	2009	1985
Sequencing information			
Sequencing platform	Solexa; PacBio	PacBio; Illumina; HiSeq; Optical map	PacBio
RNA-Seq	Illumina Nextseq	Illumina; Hiseq	NA
Genetic maps	0-1x15A ; FGOH04 × BB25	NA	NB29 × NB85
Genome assembly statistics			
Genome size (Mb)	46.50	51.76	50.12
Total contigs	86	74	55
Gene number	11 541	11 245	11 214
GC% content	46.68	45.21	45.57
Repeats (%)	31.90	35.80	NA
Secondary Metabolites	72	82	38
Predicted effectors	282	NA	NA
Genome data source			
NCBI BioProject	PRJNA50389;PRJNA392275	PRJEB18107	PRJNA577527
NCBI GenBank accession	GCA_000166005.1; GCA_006112615.1	GCA_900232045.2	GCA_009728665.1
References	Ellwood et al., 2010. Wyatt et al., 2018, 2020; Moolhuijzen et al., 2020;	Syme et al., 2018; Moolhuijzen et al., 2020	Syme et al., 2018; Moolhuijzen et al., 2020

were located in repeat regions of chromosomes 1, 2 and 3 of *P. teres* f. *teres* and on chromosome 3 for *P. teres* f. *maculata*. These differences highlight the evolutionary divergence, host-pathogen interactions and adaption of *P. teres*.

The diversity in genomic architecture and gene content within *P. teres* f. *teres* isolates has been compared to identify core and isolate-specific effectors using a pan-genome approach (Wyatt et al., 2020). Using this pan-genome approach, the full gene repertoires including genes common to all individuals (core) and those unique to individuals (accessory) could be defined. When compared between *P. teres* f. *teres* isolates, the genomic synteny was mainly conserved except for one instance where fusion between chromosome 1 and 2 was discovered. The authors considered this fusion to be a recent event and

Table 2
(*Continued*)

		P. teres f. *teres*		
NB73	NB85	15A	6A	FGOH04Ptt-21
Queensland, Australia	Queensland, Australia	California, USA	California, USA	North Dakota, USA
1994	1995	1984-1986	Unknown	Before 2017
PacBio	PacBio	PacBio	PacBio	PacBio
NA	NA	Illumina; Nextseq	Illumina; Nextseq	Illumina; Nextseq
NA	NB29 × NB85	15Ax6A	15Ax6A	FGOH04 × BB25
48.03	49.03	45.3	48.60	49.7
43	47	163	52	42
11 199	11 193	12 183	11 551	11 557
45.94	45.71	NA	NA	NA
NA	NA	27.50	34.30	31.80
39	59	77	79	76
NA	NA	205	201	207
PRJNA577527	PRJNA577527	PRJNA434142	PRJNA434143	PRJNA434144
GCA_009728675.1	GCA_009728655.1	GCA_008086755.1	GCA_008086725.1	GCA_008086845.1
Syme et al., 2018; Moolhuijzen et al., 2020	Syme et al., 2018; Moolhuijzen et al., 2020	Wyatt et al., 2020; Moolhuijzen et al., 2020	Wyatt et al., 2020; Moolhuijzen et al., 2020	Wyatt et al., 2020; Moolhuijzen et al., 2020

not the result of ancestral inheritance or an interspecies hybridization event. Orthologous gene analysis identified multiple copies of orthologous groups representing paralogs of gene family expansions within five *P. teres* isolates, and several of these orthologous groups contained genes responsible for the production of secondary metabolites. Comparative analysis of effector proteins between five *P. teres* genomes identified isolate-specific effectors that were found to be expressed *in planta* and possibly responsible for isolate's differential response on different barley genotypes. In *P. teres* genomes, accessory regions were identified on the ends of chromosomes within sub-telomeric regions where breaks in synteny and synteny between non-homologous chromosomes occurred. These regions clustered nearest to TEs and contained higher number

Table 2
(*Continued*)

P. teres f. *teres*			*P. teres* f. *maculata*	
BB25	HRS09122	HRS09139	SG1	FGOB10Ptm-1
Denmark	New South Wales, Australia	South Australia	Western Australia	North Dakota, USA
Before 2017	2009	2009	1996	2010
PacBio	PacBio	PacBio	PacBio; Illumina; HiSeq	PacBio
Illumina ; Nextseq	NA	NA	Illumina; Hiseq	NA
FGOH04 × BB25		NA	FGOB10 × SG1	FGOB10 × SG1
50.60	47.98	50.89	41.28	39.27
115	44	91	47	46
11 986	10 555	10 579	11 165	11 080
NA	NA	NA	46.86	47.37
35.60	NA	NA	21.00	NA
75	61	36	47	45
201	NA	NA	NA	NA
PRJNA434145	PRJNA577527	PRJNA577527	PRJEB18107	PRJNA417860
GCA_008086785.1	GCA_009728645.1	GCA_009728635.1	GCA_900231935.2	NA
Wyatt et al., 2020; Moolhuijzen et al., 2020	Moolhuijzen et al., 2020;	Moolhuijzen et al., 2020	Syme et al., 2018; Moolhuijzen et al., 2020	Syme et al., 2018; Moolhuijzen et al., 2020

of accessory genes. Five *P. teres* f. *teres* isolates included in this study were used in at least one bi-parental mapping study (0-1 × 15A, 15A × 6A, FGOH04P. teres f. teres-21 × BB25) that reported a total of 15 virulence/avirulence quantitative trait loci (QTL) (Weiland et al., 1999; Lai et al., 2007; Shjerve et al., 2014; Koladia et al., 2017). Out of 15 unique QTL identified, 14 spanned accessory genomic regions and 10 were localised in sub-telomeric regions. This suggests that the sub-telomeric accessory genomic regions harbour most of the known virulence loci and can undergo rapid evolution.

8 Identification of genes associated with virulence/avirulence by QTL and association mapping

Bi-parental and genome-wide association mapping studies have been conducted to identify QTL/genes associated with virulence/avirulence in

P. teres (Beattie et al., 2007; Carlsen et al., 2017; Kinzer, 2015; Koladia et al., 2017; Lai et al., 2007; Martin et al., 2020; Shjerve et al., 2014; Weiland et al., 1999). Comparisons between QTL regions identified in the different studies has been difficult as different marker systems were used and chromosomes were not consistently numbered, with some studies using the *P. tritici-repentis* genome sequence as a reference to allocate numbers to *P. teres* chromosomes. Reference genomes for both *P. teres* f. *teres* and *P. teres* f. *maculata* have recently become available and enable us now to locate previously identified QTL regions on the *P. teres* genomes (Syme et al., 2018; Wyatt et al., 2018). We have summarised the QTL regions identified in previous studies and located them on the reference genomes for convenient comparison in future studies. The chromosomes of *P. teres* f. *teres* and *P. teres* f. *maculata* referred to in this section are classified according to the reference genomes W1-1 (Table 2) and SG1 (Table 2), respectively.

All published QTL regions have been summarised in Tables 3 and 4 and Figs 4 and 5. The first mapping study in *P. teres* was conducted by Weiland et al. (1999) who used 82 progeny from a cross between the two *P. teres* f. *teres* isolates 0-1 from Ontario, Canada and 15A from California, USA. The locus *AvrHar* was identified in 15A, conferring low virulence/avirulence on the barley line Harbin (Table 3, Fig. 4) and was closely linked to three co-localised RAPD markers on chromosome 5. As the barley line Harbin possesses a dominant resistance gene against net form net blotch (Mode and Schaller, 1958), it was suggested that *AvrHar* could be the respective dominant avirulence gene according to the gene-for-gene model (Lai et al., 2007). Another study implementing AFLP markers on 78 isolates from the same *P. teres* f. *teres* 0-1/15A cross and phenotyping these on Prato, Tifang, Canadian Lake Shore (CLS) and Ming reported a locus (*AvrPra2*) avirulent on CLS and Tifang. According to Lai et al. (2007) the identified locus *AvrPra2* co-segregated with the *AvrHar* locus on chromosome 5 described by Weiland et al. (1999). However, the avirulence of *AvrPra2* and *AvrHar* were conferred by parent 0-1 and 15A, respectively. Lai et al. (2007) suggested that the co-segregation might be due to different alleles of the same gene or different genes located close together. According to the QTL locations on the W1-1 reference genome (Table 3, Fig. 4), *AvrHar* and *AvrPra2* are situated approximately 1.2 Mb apart, confirming that these QTL are represented by different genes.

A unique avirulence gene, $Avr_{Heartland}$, was reported for the barley line Heartland using a cross between two Canadian isolates WRS1906 and WRS1607 with 67 progeny (Beattie et al., 2007). Unfortunately this gene could not be located on the reference genome. Another QTL mapping study reported four virulent loci *VK1, VK2, VR1* and *VR2*, with *VK1* and *VK2* virulent on barley Kombar and *VR1* and *VR2* virulent on Rika (Shjerve et al., 2014). A mapping population of 118 progeny derived from two USA *P. teres* f. *teres* isolates, 15A and 6A was

Table 3 Summary of QTL/genes reported for *P. teres*

Locus	Marker type	No. pro[a]	Cross	Vir/Avi[b]	Chro[c]	Position[d] Starting	Ending	Marker at peak of QTL	Genotype[e]	LOD score[f]	R^2 [g]	Parent contributing the QTL	Reference
QTL identified in *P. teres* f. *teres*													
AvrHar	RAPD	82	0-1/15A	Avi	5	4193688	-	-	Harbin	36	72	15A	Weiland et al. (1999)
AvrPra2	AFLP	78	0-1/15A	Avi	5	3008702	-	M11E13190-M12E11250	Tifang, Canadian Lake Shore	5.3	-	0-1	Lai et al. (2007)
AvrPra1	AFLP	78	0-1/15A	Avi	9	-	1256349	M15E20400-M12E11250	Prato	7.2	-	0-1	Lai et al. (2007)
$Avr_{Heartland}$	AFLP	67	WRS1607/WRS1906	Avi	1	-	-	GTTA285-CGAA1600	Heartland	-	-	WRS1906	Beattie et al. (2007)
VR1	SNP, SSR, AFLP	118	6A/15A	Vir	2	2066532	3939100	07628_18	Rika	5–10	35	6A	Shjerve et al. (2014)
VR2	SNP, SSR, AFLP	118	6A/15A	Vir	10	1516021	2300448	10177_27	Rika	10–15	20	6A	Shjerve et al. (2014)
VK1	SNP, SSR, AFLP	118	6A/15A	Vir	3	1041300	1650040	18850_67	Kombar	15–20	26	15A	Shjerve et al. (2014)
VK2	SNP, SSR, AFLP	118	6A/15A	Vir	2	442489	507296	03948_8	Kombar	10–15	19	15A	Shjerve et al. (2014)
PttTif1*	SNP	109	BB25/FGOH04Ptt-21	Vir	1	1519813	2279473	1579_4251	CI4822, Tifang, Manchurian	11,30,35	45,67,74	FGOH04Ptt-21	Koladia et al. (2017)
PttTif2*	SNP	109	BB25/FGOH04Ptt-21	Vir	8	-	593132	547_32651	Tifang	4.4	3	FGOH04Ptt-21	Koladia et al. (2017)
PttBee1*	SNP	109	BB25/FGOH04Ptt-21	Vir	1	-	2776486	1588_12100	Beecher	24.0	56	FGOH04Ptt-21	Koladia et al. (2017)

PttBee2*	SNP	109	BB25/ FGOH04Ptt-21	Vir	5	316575	525281	752_3220	Beecher	7.0	17	FGOH04Ptt-21	Koladia et al. (2017)
PttPin1*	SNP	109	BB25/ FGOH04Ptt-21	Vir	3	5804230	-	1667_1175	Pinnacle	14.0	49	BB25	Koladia et al. (2017)
PttPin2*	SNP	109	BB25/ FGOH04Ptt-21	Vir	12	1044631	1438885	2428_2378	Pinnacle	3.1	11	FGOH04Ptt-21	Koladia et al. (2017)
PttCel1*	SNP	109	BB25/ FGOH04Ptt-21	Vir	8	2843621	3029139	1454_3802	Tifang, Celebration	3.2,7.0	7,17	FGOH04Ptt-21	Koladia et al. (2017)
PttCel2*	SNP	109	BB25/ FGOH04Ptt-21	Vir	9	2562601	2806018	994_25330	Celebration	5.1	17	FGOH04Ptt-21	Koladia et al. (2017)
PttHec1*	SNP	109	BB25/ FGOH04Ptt-21	Vir	8	2652633	-	252_25719	Hector, Stellar	3.1,6.5	11,18	FGOH04Ptt-21	Koladia et al. (2017)
PttSki_3	DArTseq	78	NB29/ HRS09122	Vir	3	114611	-	28946459	Skiff	6.6	24	HRS09122	Martin et al. (2020)
PttBee_5	DArTseq	78	NB29/ HRS09122	Vir	5	5183980	5208563	28948016	Beecher	4.0	15	NB29	Martin et al. (2020)
PttSki_5	DArTseq	78	NB29/ HRS09122	Vir	5	3980200	4457075	28948170	Skiff	4.8	19	HRS09122	Martin et al. (2020)
PttBee_9	DArTseq	78	NB29/ HRS09122	Vir	9	1073073	1122817	28945299	Beecher	3.0	11	NB29	Martin et al. (2020)
PttBee_3	DArTseq	72	NB29/NB85	Vir	3	796216	970812	28945535	Beecher	12.0	36	NB29	Martin et al. (2020)
PttBee_7	DArTseq	72	NB29/NB85	Vir	7	2166986	2928737	28946946	Beecher	3.9	11	NB85	Martin et al. (2020)
PttPri_7	DArTseq	72	NB29/NB85	Vir	7	2166986	2928737	28949493	Prior	3.6	18	NB85	Martin et al. (2020)
PttBee_8	DArTseq	72	NB29/NB85	Vir	8	1883966	1972858	28949931	Beecher	3.0	7	NB29	Martin et al. (2020)

(*Continued*)

Table 3 (*Continued*)

Locus	Marker type	No. pro[a]	Cross	Vir/ Avi[b]	Chro[c]	Position[d] Starting	Position[d] Ending	Marker at peak of QTL	Genotype[e]	LOD score[f]	R^2 [g]	Parent contributing the QTL	Reference
QTL identified in *P. teres* f. *maculata*													
VQTL1A	SNP	105	SG1/ FGOB10Ptm-1	Vir	1	117776	117958	SNP_11381_87	TR326, Skiff	8.4,5.8	21,23	FGOB10Ptm-1	Carlsen et al. (2017)
VQTL1B						170676	170836	SNP_2207_88	81-82/033	5.5	21		
VQTL1C						316665	316851	SNP_16439_27	PI 392501	9.4	34		
VQTL2	SNP	105	SG1/ FGOB10Ptm-1	Vir	3	3154965	3155149	SNP_12879_149	Skiff	5.5	22	FGOB10Ptm-1	Carlsen et al. (2017)
VQTL3	SNP	105	SG1/ FGOB10Ptm-1	Vir	5	2709841	2710025	SNP_41831_15	Skiff	5.3	20	FGOB10Ptm-1	Carlsen et al. (2017)
VQTL4	SNP	105	SG1/ FGOB10Ptm-1	Vir	2	1108007	1108161	SNP_21264_143	81-82/033 PI 392501	8.0,11.0	30,37	FGOB10Ptm-1	Carlsen et al. (2017)
VQTL5	SNP	105	SG1/ FGOB10Ptm-1	Vir	3	2419962	2420099	SNP_2673_169	81-82/033, TR326, PI 392501	6.0,6.6, 9.3	33,26, 34	FGOB10Ptm-1	Carlsen et al. (2017)
VQTL6	SNP	105	SG1/ FGOB10Ptm-1	Vir	4	1618736	1618905	SNP_26064_6	PI 392501	5.0	20	FGOB10Ptm-1	Carlsen et al. (2017)

[a] Number of progeny, [b] virulence nature of the allele (Vir= virulent; Avi=avirulent), [c,d] chromosome location according to W1-1 reference genome, [e]name of the genotype that the allele is virulent/avirulent on, [f]logarithm of odds, [g]percentage of phenotypic variation explained by the QTL.

used with the genetic map consisting of AFLP, SSR and single nucleotide repeat (SNP) markers. Virulence loci *VK1* and *VK2* were located on chromosomes 3 and 2 with 26% and 19% of the phenotypic variance explained, respectively, while *VR1* and *VR2* were positioned on chromosomes 2 and 10 with 35% and 20% of the phenotypic variance explained, respectively.

A QTL mapping study conducted by Koladia et al. (2017) reported nine QTL associated with eight barley genotypes using a *P. teres* f. *teres* population consisting of 109 progeny developed by crossing Danish isolate, BB25, with the USA isolate FGOH04Ptt-21. Three QTL were reported to be major QTL accounting for more than 45% of the phenotypic variation. Two of these QTL were located on chromosome 1 and one on chromosome 3. One of the QTL on chromosome 1 conferred virulence on the commonly used differential cultivars Manchurian, Tifang and CI4922, while the other QTL on chromosome 1 was virulent on Beecher. The chromosome 3 QTL was associated with virulence on Pinnacle.

A genome-wide association mapping study of 188 Australian *P. teres* f. *teres* isolates collected from five states in Australia and assessed on 20 barley lines identified the presence of 14 different genomic regions associated with virulence, mainly on chromosome 3 and 5 with one each on chromosomes 1, 6 and 9 (Martin et al., 2020) (Table 4, Fig. 4). In order to confirm these identified genomic regions, QTL analysis of two bi-parental mapping populations, NB029/HRS09122 (78 progeny) and NB029/NB085 (72 progeny) was undertaken. Progeny were phenotyped on Beecher, Skiff and Prior. Four regions identified by GWAS were confirmed by bi-parental QTL mapping. The DArTseq marker system was implemented for both GWAS and QTL mapping. Martin et al. (2020) reported that QTL *VK1* on chromosome 3 which conferred virulence on Kombar (Shjerve et al., 2014), was in a similar location to PttBee_3 and QTL3, with a distance of 58 kb between their flanking markers. This is similar to the 70 kb distance observed on the W1-1 reference genome map (Fig. 4). The Skiff virulent QTL, PttSki_5 on chromosome 5 identified by Martin et al. (2020) in the bi-parental mapping population NB29/HRS09122 seems to be in the same location as the QTL *AvrHar*, which Weiland et al. (1999) identified to be avirulent on Harbin. Co-localising of these QTL suggests that this region may contain multiple closely linked genes associated with virulence/avirulence or that different alleles of the same gene confer virulence/avirulence. Both QTL associated with the Beecher virulence contributed by a USA isolate (Koladia et al., 2017) are in a different location to the Beecher virulent QTL identified by Martin et al. (2020) in isolates from Australia. This suggests that QTL associated with virulence on the same cultivar may not be conserved between isolates from different geographical regions.

Only one QTL mapping study in *P. teres* f. *maculata* has been reported (Carlsen et al., 2017). The *P. teres* f. *maculata* parent FGOB10Ptm-1 from the

Table 4 Summary of QTL reported for *P. teres* f. *teres* by genome-wide association mapping (Martin et al., 2020)

<table>
<tr><th rowspan="2">QTL ID[a]</th><th rowspan="2">Genotype[b]</th><th rowspan="2">Marker</th><th rowspan="2">Chro[c]</th><th colspan="2">Position[d]</th><th rowspan="2">LOD[e]</th><th rowspan="2">R^{2} [f]</th></tr>
<tr><th>Starting</th><th>Ending</th></tr>
<tr><td>QTL1</td><td>Beecher</td><td>28947393</td><td>1</td><td>679123</td><td></td><td>6.07</td><td>14.58</td></tr>
<tr><td>QTL2</td><td>Fleet, Skiff Herta</td><td>41806492|F|0-7:A>G-7:A>G</td><td>3</td><td>102867</td><td>114611</td><td>5.02-6.7</td><td>18.93-16.33</td></tr>
<tr><td>QTL3</td><td>Beecher, Kombar, Maritime, Beecher</td><td>28946135</td><td>3</td><td>788584</td><td>796281</td><td>5.26-9.77</td><td>12.07-23.58</td></tr>
<tr><td>QTL4</td><td>Beecher, Maritime</td><td>36352701</td><td>3</td><td>1028769</td><td>-</td><td>5.34-7.46</td><td>12.71-17.48</td></tr>
<tr><td>QTL5</td><td>CIho 11458</td><td>100133310</td><td>3</td><td>1480380</td><td>-</td><td>5.41</td><td>12.46</td></tr>
<tr><td>QTL6</td><td>Herta, Skiff</td><td>28945761</td><td>3</td><td>1905727</td><td>-</td><td>6.71-6.95</td><td>16.12-16.93</td></tr>
<tr><td>QTL7</td><td>Beecher</td><td>28949829|F|0-28:C>T-28:C>T</td><td>3</td><td>3985143</td><td>-</td><td>5.15</td><td>11.03</td></tr>
<tr><td>QTL8</td><td>Commander</td><td>28945998</td><td>3</td><td>5556417</td><td>-</td><td>5.05</td><td>12.69</td></tr>
<tr><td>QTL9</td><td>Fleet,Yerong</td><td>28948035</td><td>5</td><td>4110850</td><td>-</td><td>7.59-8.14</td><td>18.72-21.47</td></tr>
<tr><td>QTL10</td><td>Fleet,Yerong</td><td>36347132</td><td>5</td><td>4456299</td><td>-</td><td>7.27-8.89</td><td>17.55-22.98</td></tr>
<tr><td>QTL11</td><td>Harbin,Prior</td><td>28950148|F|0-5:A>C-5:A>C</td><td>5</td><td>5195933</td><td>5209123</td><td>5.11-8.55</td><td>13.5-21.66</td></tr>
<tr><td>QTL12</td><td>Corvette, Harbin, Orior</td><td>36352140|F|0-22:T>C-22:T>C</td><td>5</td><td>5243799</td><td>-</td><td>4.79 -8.34</td><td>14.19-21.61</td></tr>
<tr><td>QTL13</td><td>Tallon</td><td>100139818</td><td>6</td><td>410017</td><td>-</td><td>5.04</td><td>10.9</td></tr>
<tr><td>QTL14</td><td>Beecher</td><td>28949038</td><td>9</td><td>839959</td><td>-</td><td>4.82</td><td>10.19</td></tr>
</table>

[a] QTL identification, [b] marker at QTL, [c,d] chromosome location on W1-1 reference genome, [e] logarithm of odds, [f] percentage of phenotypic variation explained.

USA was crossed with SG1 from Western Australia and 105 progeny produced. Six QTL were detected and explained 20–37% of the phenotypic variance in barley lines Skiff, 81-82/033, TR326, and PI 392501 (Table 3, Fig. 5).

Multiple QTL with both major and minor effect have been identified within the *P. teres* genomes. Studies conducted using bi-parental mapping populations and GWAS suggest that *P. teres* f. *teres* chromosomes 3 and 5 harbour QTL/gene-rich regions (Fig. 4) associated with virulence/avirulence. With the exception of chromosomes 4 and 11, all other chromosomes of the *P. teres* f. *teres* genome had at least one virulence/avirulence QTL demonstrating the complex genetic basis of *P. teres* f. *teres* virulence. Virulences on different barley genotypes are governed by different genes and genes that confer virulence on one barley genotype may be avirulent on another. Genes which are responsible for virulence on one barley genotype (e.g. Beecher) in one geographical area (e.g. USA) may not be associated with the same virulence genes for the same barley genotype from another geographical area (e.g. Australia) (Koladia et al., 2017; Martin et al., 2020). Thus, in order to better understand this complex pathosystem, more studies are warranted as improving our understanding of the genetic basis of the *P. teres*-barley pathosystem is crucial for developing cultivars.

9 Managing the net blotches

Net blotches are economically important diseases and are regarded a major constraint to barley production in most barley-growing regions worldwide (Liu et al., 2011). Net form net blotch of barley is common throughout the major barley-growing regions of the world, including Morocco (Amezrou et al., 2018; Douiyssi et al., 1998), the USA (Buchannon and McDonald, 1965; Steffenson and Webster, 1992a), Australia (Cakir et al., 2003; Ellwood et al., 2019; Fowler et al., 2017; Khan, 1973; Lehmensiek et al., 2007; Shipton, 1966; Wallwork et al., 2016), South Africa (Campbell et al., 2002), Canada (Tekauz, 1990; Xi et al., 1999), Sweden (Jonsson et al., 1997), Finland (Robinson and Jalli, 1996), Syria and Tunisia (Bouajila et al., 2011), Russia and Belarus (Novakazi et al., 2019a), Sardinia (island west of Italy) (Rau et al., 2003), Germany (Vatter et al., 2017), Norway (Wonneberger et al., 2017), New Zealand (Cromey and Parkes, 2003), and France (Youcef-Benkada et al., 1994). Spot form net blotch is a prevalent disease of barley in many countries (McLean et al., 2009), including Canada (Tekauz, 1990), Denmark (Smedegård-Petersen, 1971), Finland (Mäkelä, 1975), France (Arabi et al., 1992), South Africa (Louw, 1996; Campbell et al., 1999), Turkey (Oğuz and Karakaya, 2017) and the USA (Liu and Friesen, 2010; Lartey et al., 2013). It was first reported in Western Australia in 1977 (Khan and Tekauz, 1982), however, was not detected in eastern Australia until 1990 (Wallwork et al., 1992; Wallwork, 1995) and has since spread through southern and eastern Australia (Wallwork, 1995; McLean et al., 2009).

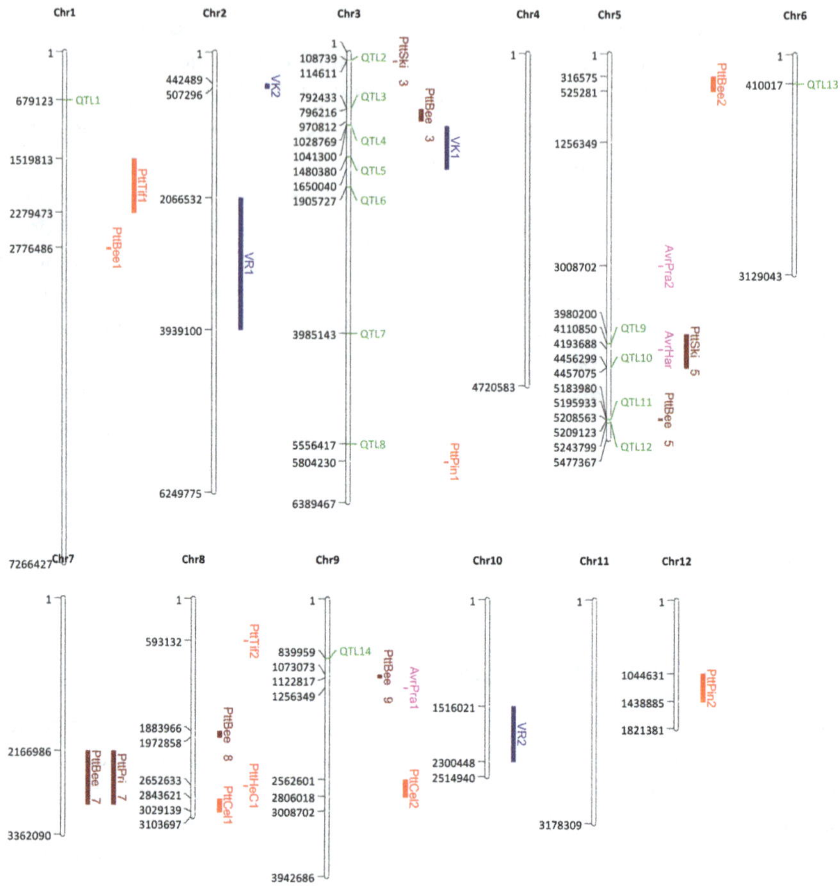

Figure 4 Map of the *Pyrenophora teres* f. *teres* reference genome W1-1 indicating genomic regions associated with virulence/avirulence identified through QTL mapping and GWAS (green). Map distance in bps is indicated on the left of the bar. Created using MapChart 2.32v (Voorrips, 2002).

Yield losses associated with net form net blotch typically range between 10% and 40% (Steffenson and Webster, 1992a). Under Australian conditions yield losses above 20% are common (Shipton, 1966; Khan, 1987) with losses in excess of 60% reported in Queensland (Qld) in crops of Gilbert (Poulsen et al., 1999) and up to 70% on the susceptible variety Maritime in South Australia (SA) under epidemic conditions (Wallwork, 2011). Yield loss reductions are often associated with significant reductions in seed weight and grain size (Khan, 1987; Shipton, 1966; Sutton and Steele, 1983; Poulsen et al., 1999; Rees et al., 1999). Yield loss response curve trials conducted over a three-year period (2014–2016) in Qld, Australia indicated increased yield loss with increased susceptibility to net form net blotch. Yield loss ranged between 9.5% and

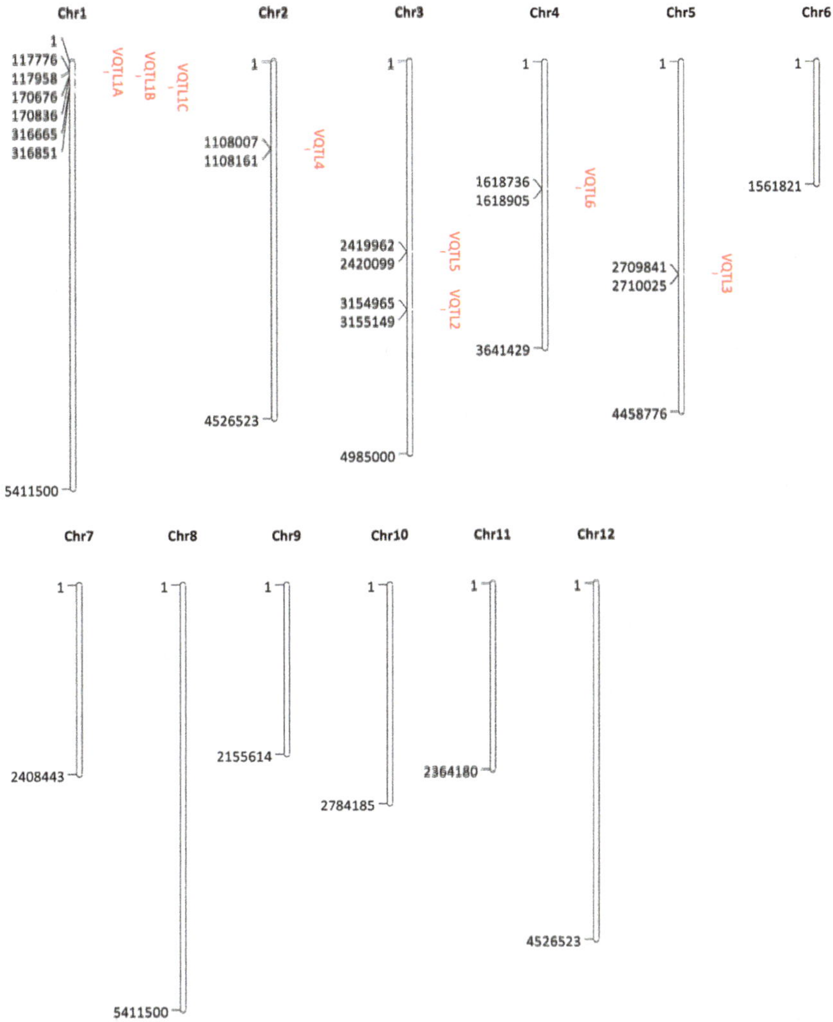

Figure 5 Map of the *P. teres* f. *maculata* SG1 reference genome indicating genomic regions associated with virulence/avirulence identified through QTL mapping. Map distance in bps is indicated on the left of the bar. Created using MapChart 2.32v (Voorrips, 2002).

18.6% in the variety Compass (moderately resistant to moderately susceptible), whereas yield loss in the very susceptible variety Charger were in excess of 40%. Disease levels did not have a big impact on grain quality, with biggest differences observed in seed retention levels (% of seed > 2.5 mm) (Snyman et al., 2017). Yield loss due to spot form net blotch is not well documented with most information reported for Australia (McLean et al., 2009). In Western Australia, yield losses up to 44% as a result of spot form net blotch infections were recorded, dependant on season, sowing date and variety (Khan, 1989; Jayasena et al., 2007).

Several methods are applied to manage *P. teres*, including fungicides, cultural practices and host resistance (Liu et al., 2011). Rotation to reduce or eliminate primary inoculum is an important factor in cultural management, as is chemical control using either seed dressing to reduce primary inoculum or foliar application to lower disease levels (Liu et al., 2011). As the net blotch pathogens persist on plant residue, the adoption of reduced- or zero-till practices has significantly increased the incidence in recent years (McLean et al., 2009). Crop rotation forms an integral part of the successful cultivation of barley and growers are advised not to plant barley on barley (Rees et al., 1999). Planting successive barley crops in the same paddock increases the incidence of the net blotch diseases and cultivation of the same variety will lead to an increase in the presence of pathotypes virulent on that particular variety and put increased pressure on effective resistance genes. Best practice includes crop rotation with non-host crops. During prolonged periods of drought, stubble breakdown occurs over a longer period of time and stubble from crops grown a few years earlier could serve as a source of inoculum, emphasising the need for crop rotation.

In Australia, all current barley varieties and varieties considered for release are rated for resistance to a suite of diseases and pathogens through the National Variety Trial disease screening process (https://www.nvtonline.com.au). They are categorised in nine resistance categories rating from resistant (R) to very susceptible (VS). These genotypes are screened annually in nationwide disease nurseries, with disease ratings assigned and reviewed on a yearly basis.

Growing a high-yielding well-adapted resistant variety provides the most economic and environmentally friendly means of disease control. A study conducted in Qld between 2014 and 2017 on yield loss associated with net form net blotch indicated that under disease pressure representative of that expected in commercial paddocks, the level of net form net blotch disease in varieties with some level of resistance did not reach the levels anticipated when screening under high disease pressures. This indicates that growing varieties with at least some level of resistance to net form net blotch can aid in limiting yield and quality loss (Forknall et al., 2019).

The spread of pathotypes depends on the importance of long distance dispersal of seed-borne inoculum or wind-borne ascospores. The relative fitness of pathotypes will depend on barley varieties grown, cultivation practices and chemical control measures (Jonsson et al., 1999). An increase in the cultivation of varieties susceptible to the net blotch diseases would see an increase in incidence and severity (McLean et al., 2010a; McLean, 2016).

Pyrenophora teres f. *teres* is a seed-borne pathogen that can spread with infected seed and could aid the spread of isolates resistant to fungicides. It is important that growers use an appropriate seed treatment, registered for net blotch control at the recommended rate and to ensure that seed treatment is effectively applied, reducing the effect of seed-borne inoculum (Rees et al., 1999). Foliar fungicides are applied routinely in most barley crops and should be aimed at protecting the top two leaf layers. To ensure that fungicides remain effective, it is important to limit fungicide application by spraying only when necessary, rotate fungicides with different modes of action and use fungicides at recommended rates. Fungicide applications are more effective if applied before the disease becomes established in the crop. This requires regular monitoring to ensure crops can be sprayed at the first sign of disease. When conditions are favourable for disease development, more frequent crop inspections will be needed and repeat fungicide applications may be necessary.

9.1 Fungicide resistance

Along with cultural practices, the application of fungicides and the use of resistant varieties are the main control strategies available for the management of fungal diseases. However, due to the lack of adequate genetic resistance, fungicides are frequently solely relied upon for the control of many fungal diseases. Despite a considerable body of research that has been developed over the last 50 years, fungicide resistance continues to be a major concern to agriculture worldwide mirroring a situation that clinics and animal production facilities are facing as a result of overuse of particular fungicide modes of action (MOA) (Leadbeater, 2014; Talbot et al., 2018; Buil et al., 2019). When a fungal disease develops resistance to one fungicide, often many other fungicides that share the same MOA are also at risk due to a phenomenon called cross-resistance (Brent, 1995). This frequently leads to scenarios where pathogens survive the application of multiple fungicides with different MOA (Rupp et al., 2017). Unfortunately, the rate at which new MOA are released to the market cannot match that of practical fungicide resistance development, resulting in a progressive reduction in the number of available effective fungicides. This arms race has exacerbated in recent years with the introduction of bans on specific agricultural fungicides (e.g. chlorothalonil) due to health risk concerns (Arena et al., 2018).

Fungicides used against the net blotch diseases are predominantly of the azole or demethylase inhibitor (DMI, group 3) class, although there are other groups like the quinone outside inhibitor (QoI, group 11) and, more recently, the succinate dehydrogenase inhibitor (SDHI, group 7), that also play an important role in the control of these two diseases (Sierotzki et al., 2007; Mair et al., 2016).

Despite their relatively widespread use since their introduction in 1996, QoI fungicides have traditionally provided good levels of control over the net blotch diseases and no occurrences of crop protection failure have been reported to date. The mutation G143A at the CYTb target site has been found in all pathogens isolated from crops where field resistance to QoIs was reported. This mutation is associated with cross-resistance to all QoI compounds and control failure is to be expected in crops where isolates carrying this mutation dominate the population (FRAC, 2020b). However, the presence of an intron that interrupts the codon responsible for this mutation in the *Cytb* gene of net blotch pathogens makes the evolution of this mutation quite unlikely, as codon 143 is part of the splice site (Sierotzki et al., 2007; Oliver and Hewitt, 2014; Grasso et al., 2006). This phenomenon, that seems to provide an apparent advantage of QoIs over other MOAs, needs to be considered carefully when designing adequate disease management strategies given that other lesser mutations could still develop. Since 2003, reduced sensitivity to QoIs has been detected in several European barley-growing regions and, in all cases, mutation F129L was found to be widespread, but not associated with significant reductions in the control of net blotches (Table 5) (Semar et al., 2007; Sierotzki et al., 2007; Rehfus, 2018).

Reduced sensitivity to SDHI fungicides in net blotch diseases was first reported in Europe in 2012 (Stammler et al., 2014). At that time, analysed isolates were found to carry mutation *SdhB*-H277Y. Since then, a plethora of mutations affecting subunits C (N75S, G79R, H134R, S135R) and D (D124N, D124E, H134R, D145G, E178K) in the Sdh complex have been described and cross resistance between the different SDHI members established (Rehfus et al., 2016; FRAC, 2020c). Mutations associated with varying levels of lower sensitivity to the majority of SDHIs are now widespread in regions of Germany, France and Australia (Rehfus et al., 2016). Prior to 2019, crop protection failure had not been observed and *in vivo* glasshouse experiments had shown that control of the mutant genotypes was always achieved when foliar SDHIs were applied at full rates (Stammler et al., 2014; Rehfus et al., 2016; FRAC, 2020c). However, a widespread outbreak of resistance to a fluxapyroxad seed dressing formulation in a barley-intensive growing region of South Australia in 2019 seems to challenge these results. The high frequency at which mutations *SdhC*-H134R and *SdhD*-D145G were found in South Australia, together with the SDHI treatment used, being a seed dressing formulation, might have contributed to the lack of control seen in several farms across this region.

For many years, DMI fungicides have been a key component in many spray programs aimed at controlling the net blotch diseases. Introduced in the 70s, this group of fungicides has been used against net blotches with only one significant fungicide resistance outbreak being reported in the twentieth century. In 1983, seed dressing application with DMI triadimenol failed to control net blotch in regions of the North Island in New Zealand (Sheridan et al., 1985). Interestingly, isolates collected from barley fields in 1984 seemed to have similar levels of resistance to triadimenol and nuarimol to that of isolates originating from UK and Denmark. These findings suggest that net blotch DMI resistant genotypes were already present in Europe prior to the findings reported in New Zealand.

In 2013, reduced sensitivity levels to several DMIs were found during the *in vitro* analysis of *P. teres* f. *teres* isolates collected across the central region of the Western Australian wheatbelt (Mair et al., 2016). The molecular characterisation of these isolates revealed an overexpressed copy of the *Cyp51A* gene carrying mutation F489L was the mechanism responsible for the recorded reduction in sensitivity (Table 5). Crop protection failure to different DMIs was reported in 2016 in barley crops affected with spot form net blotch in the southern region of the Western Australian wheatbelt (Mair et al., 2020). The *in vitro* analysis showed very high resistance levels to different DMIs in *P. teres* f. *maculata* isolates collected from these fields. Although the resistant isolates carried the same mutation discovered a few years earlier in *P. teres* f. *teres*, the high levels of resistance found, seemed to be due to the combination of F489L and the presence of a 134 bp insertion in the promoter of the *Cyp51A* gene (Table 5). Isolates carrying only the promoter insertion were also found, but in this case, the level of resistance was significantly lower and could not be correlated with crop protection failure (Mair et al., 2020).

P. teres spp. are classified as 'medium-risk' pathogens in terms of fungicide resistance development (FRAC, 2020a). However, the surprising capacity of these two pathogens to evolve resistance to several SDHI and DMI fungicides through different mechanisms is a reflection of how their highly plastic genomes can respond to adapt to the selection pressure applied by the use of fungicides. This probably challenges the validity of the current classification and calls for a revision of their placement within the existing fungicide resistance risk scale.

Selection pressure is the driving force responsible for the evolution of pathogen populations able to overcome management strategies. Disease management requires a balanced combination of multiple effective control methodologies in order to be sustainable and successful in the long run, and in the case of net blotch diseases this includes the strategic use of adequate genetic resistance and fungicides (Walters et al., 2012). The lack of a sufficiently diverse pool of strategies will inevitably result in some strategies being under higher selection pressure than others. The time required to breed a new resistant cultivar is often comparable to that of developing a new fungicide (Morton and Staub, 2008). This indicates how important the management of time is in this arms race scenario and how damaging it can be to loose these resources too quickly due to poor disease management practices.

Despite the availability of some barley varieties with resistance to the net blotches, their lack of suitability for all barley growing regions and market demand for susceptible varieties has led to growers developing a chronic over-reliance on chemical control. This has resulted in increased selection pressure and fungicide resistance development (Sierotzki et al., 2007; Tucker et al., 2015; Mair et al., 2016; Rehfus et al., 2016). With the three major fungicide MOA compromised in net blotch diseases, it is paramount that growers and advisers have up-to-date fungicide resistance information that can guide their variety choices to produce optimal combinations of these management tools suitable for particular crop environments. The deployment of monitoring programmes capable of providing information on the resistance status of regional and local net blotch pathogen populations, to determine best fungicide treatments according to frequency of resistant mutations, seem to be the most logical approach. To some extent, such programmes already exist in cereal-growing regions of Europe. An example is the EuroWheat project, which has been collecting information on fungicide-resistant mutations affecting DMI and SDHI fungicides in European populations of the wheat pathogen *Zymoseptoria tritici* over a number of years (EuroWheat, 2020). Although less current, similar information has been made available for the spread of mutations affecting SDHI performance in European populations of net blotches (Rehfus et al., 2016). Despite the enormous advantage of having these resources, no monitoring strategy has yet addressed the need to provide within-season fungicide resistance information to growers and advisers. Ultimately, the use of the most effective fungicides can contribute to slowing down the rate at which net blotch diseases overcome genetic resistance, which in turn will lead to a reduction in fungicide use and a delay of fungicide resistance development.

9.2 Breeding strategies

Breeding for resistance to multiple diseases has been an objective of many breeding programs and should remain a priority (Platz, 2005). High-risk pathogens, such as *P. teres*, require great effort to achieve durable resistance as virulence mutations can be recombined into many genetic backgrounds and can be dispersed over long distances. Breeding efforts should focus on quantitative resistance in an effort to achieve durability. If quantitative resistance cannot be achieved, major genes will provide protection, but should be managed aggressively (McDonald and Linde, 2002). Pyramiding major resistance genes will prolong the durability of each gene; however, higher levels of resistance could be attained by the accumulation of resistance genes with minor effects (Jonsson et al., 1999; Qamar et al., 2008; Grewal et al., 2008).

The *P. teres* pathogens are very adaptive, highlighting the need for barley breeders to use multiple sources of resistance to safeguard against mutations (Poudel et al., 2019b). Ideally, resistance should be effective to all pathotypes of *P. teres* present in all areas where a widely adapted barley variety is grown (Buchannon and McDonald, 1965). Resistance that is effective against a wide

Table 5 Summary of fungicide resistance mechanisms described in *P. teres* f. *teres* and *P. teres* f. *maculata* (Table modified from IRELAND, 2021)

Mechanism	Phenotypes by pathogen and affected MOAs		Examples of resistant mutations
Target-site mutation 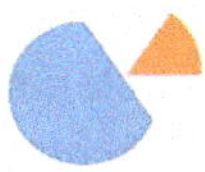	Reduced sensitivity	*P. teres* f. *teres* - DMI, Group 3	*Cyp51A*-F489L
		P. teres f. *teres* - SDHI, Group 7	*SdhD*-D145G
		P. teres f. *maculata* - Group 3	*Cyp51A*-F489L
		P. teres[a] - QoI, Group 11	*Cytb*-F129L
		P. teres[a] - Group 7	*SdhC*-N75S
	Resistant	*P. teres* f. *teres* - Group 7	*SdhC*-H134R
Multiple target-site mutations 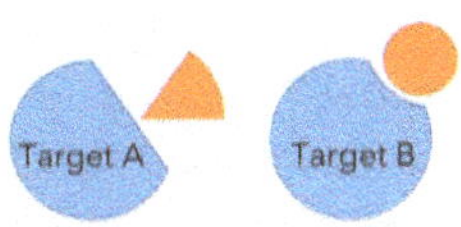	Reduced sensitivity	*P. teres* f. *teres* - Group 7	*SdhD*-D145G for Group 7
	Resistant	*P. teres* f. *teres* - Group 7	*SdhC*-H134R for Group 7
Target-site over-expression 	Reduced sensitivity	*P. teres* f. *maculata* - Group 3	134-bp insertion at different positions in the *Cyp51A* promoter
Target-site mutation + gene over-expression 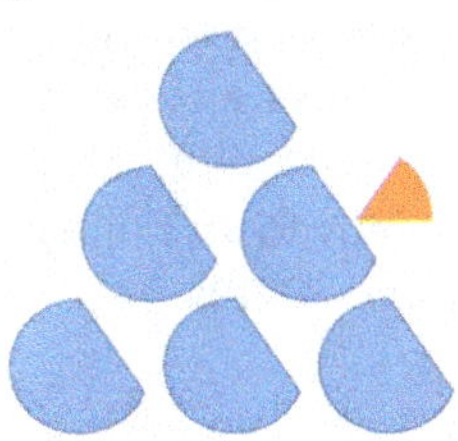	Resistant	*P. teres* f. *maculata* - Group 3	134-bp insertion in the *Cyp51* promoter + *Cyp51A*-F489L

[a] No distinction was made between *P. teres* f. *teres* and *P. teres* f. *maculata*.

spectrum of pathotypes may last longer than resistance exhibiting strong differential reactions to different genotypes of the pathogen (Steffenson and Webster, 1992b).

The *P. teres* populations are diverse and ever changing and will be influenced by current varieties, environmental conditions and cultivation practices favouring infection (Wallwork et al., 2016). The reported evolution of these pathogens emphasise the importance of screening breeding material with multiple isolates from different growing regions to account for varying virulences (McLean et al., 2014). Considerable variation observed between seedling glasshouse and field responses highlight the risk associated with the use of single isolates for glasshouse screening which may result in susceptibility in the field (Douiyssi et al., 1998). This is also supported by the development of adult plant virulent pathotypes, emphasising the need for multi-pathotype screening at adult plant stage under field conditions (Platz et al., 2009). Lines with resistance to a wide range of Australian isolates at both seedling and adult plant stage have been identified (Gupta et al., 2003; Wallwork et al., 2016) and could be used as diverse resistance sources.

Population genetic analysis to date suggests that sexual reproduction occurs within the worldwide *P. teres* population, but the contribution of sexual and asexual reproduction varies between regions, possibly based on environmental differences (Liu et al., 2011). Sexual recombination/hybridization between and within forms can lead to the formation of novel pathotypes. This in turn would increase genetic diversity in the population and increase challenges in the host plant to prevent disease (Syme et al., 2018).

Wild relatives and landraces represent valuable reservoirs of traits left behind as a consequence of domestication and may prove useful for crop improvement. Varying levels of resistance to various pathogens, including powdery mildew (*Blumeria graminis*), scald (*Rhynchosporium secalis*), leaf rust (*Puccinia hordei*) and net form net blotch (*P. teres* f. *teres*) have been identified in the Spanish barley core collection (Silvar et al., 2010) and resistance to *P. teres* f. *teres* and spot blotch (*Bipolaris sorokiniana*) in the Vavilov collection (Novakazi et al., 2019a,b). Non-host resistance in barley grass should be explored to combat the disease in cultivated barley (Linde and Smith, 2019). Breeding strategies, however, will need to be renewed regularly to stay ahead of ever changing pathogen populations (McDonald and Linde, 2002).

10 Conclusion and future trends

It is clear that there is a high variability in *P. teres* with respect to population genetics, phenotypic interactions with different barley lines and the genomic location and number of virulence/avirulence genes. The complex nature of the *P. teres*-barley pathosystem indicates that a number of different factors need to be taken into account when breeding for durable resistance against this pathogen. Phenotyping and genetic diversity and QTL mapping studies further confirm that there are clear differences between *P. teres* f. *teres* and *P. teres* f. *maculata* and that these pathogens need to be studied as separate entities. To date more studies have involved the form *teres* than the form *maculata*.

Although sexual reproduction between the two forms of *P. teres* have been reported to be rare, the recent identification of a fungicide-resistant hybrid in Western Australia that has spread very quickly via asexual conidia production and is virulent on the widely grown barley cultivar Oxford illustrates that sexual reproduction between the forms should not be ruled out and that there is a need to monitor hybrid occurrences.

A number of markers associated with virulence/avirulence in *P. teres* have been identified and the reference genomes for both forms are now available. This will aid in the characterization of the underlying mechanisms involved in the pathogen/host interactions and will be the next step towards breeding for resistance against this pathogen.

As with other pathogens, the monitoring of *P. teres* is a continual process as these pathogens are constantly evolving due to frequent sexual recombination within each form and possibly between forms. These changes can lead to the rapid development of new virulences with the potential to overcome major host resistance genes and fungicides.

11 Where to look for further information

11.1 Further reading

A thorough review of both forms of *P. teres* has been published by Liu et al. (2011).

A good review specifically on spot form of net blotch is McLean et al. (2009b). Resistance/susceptibility loci associated with the host and effector loci identified in *P. teres* have recently been reviewed by Clare et al. (2020).

11.2 Key conferences

The International Workshop on Barley Leaf Diseases is attended by barley foliar disease researchers from many different countries.

The International Barley Genetics Symposium is well attended by members of the barley community.

12 Acknowledgements

The authors would like to thank Dr Nathan Wyatt, Friesen Lab, USDA-ARS Cereal Crop Research Unit, who helped to find QTL locations on the W1-1 reference genome.

13 References

Afanasenko, O. S., Hartleb, H., Guseva, N. N., Minarikova, V. and Janosheva, M. 1995. A set of differentials to characterize populations of *Pyrenophora teres* Drechs for international use. *Journal of Phytopathology* 143(8), 501–507.

Afanasenko, O. S., Jalli, M., Pinnschmidt, H. O., Filatova, O. and Platz, G. J. 2009. Development of an international standard set of barley differential genotypes for Pyrenophora teres f. *teres*. *Plant Pathology* 58(4), 665–676.

Akhavan, A., Turkington, T. K., Kebede, B., Tekauz, A., Kutcher, H. R., Kirkham, C., Xi, K. Q., Kumar, K., Tucker, J. R. and Strelkov, S. E. 2015. Prevalence of mating type idiomorphs in *Pyrenophora teres* f. *teres* and P. *teres* f. *maculata* populations from the Canadian prairies. *Canadian Journal of Plant Pathology* 37(1), 52–60.

Amezrou, R., Verma, R. P. S., Chao, S., Brueggeman, R. S., Belqadi, L., Arbaoui, M., Rehman, S. and Gyawali, S. 2018. Genome-wide association studies of net form of net blotch resistance at seedling and adult plant stages in spring barley collection. *Molecular Breeding* 38(5), 58.

Arabi, M. I., Barrault, G., Sarrafi, A. and Albertini, L. 1992. Variation in the resistance of barley cultivars and in the pathogenicity of *Drechslera teres* f. sp. *maculata* and *D. teres* f. sp. *teres* isolates from France. *Plant Pathology* 41(2), 180–186.

Arabi, M. I., Sarrafi, A., Barrault, G. and Albertini, L. 1990. Inheritance of partial resistance to net blotch in Barley. *Plant Breeding* 105(2), 150–155.

Arabi, M. I. E., Al-Safadi, B. and Charbaji, T. 2003. Pathogenic variation among isolates of *Pyrenophora teres*, the causal agent of barley net blotch. *Journal of Phytopathology* 151(7–8), 376–382.

Arena, M., Auteri, D., Barmaz, S., Bellisai, G., Brancato, A., Brocca, D., Bura, L., Byers, H., Chiusolo, A., Court Marques, D., Crivellente, F., De Lentdecker, C., Egsmose, M., Erdos, Z., Fait, G., Ferreira, L., Goumenou, M., Greco, L., Ippolito, A., Istace, F., Jarrah, S., Kardassi, D., Leuschner, R., Lythgo, C., Magrans, J. O., Medina, P., Miron, I., Molnar, T., Nougadere, A., Padovani, L., Parra Morte, J. M., Pedersen, R., Reich, H., Sacchi, A., Santos, M., Serafimova, R., Sharp, R., Stanek, A., Streissl, F., Sturma, J., Szentes, C., Tarazona, J., Terron, A., Theobald, A., Vagenende, B., Verani, A. and Villamar-Bouza, L. 2018. Peer review of the pesticide risk assessment of the active substance chlorothalonil. *EFSA Journal European Food Safety Authority* 16(1), e05126.

Beattie, A. D., Scoles, G. J. and Rossnagel, B. G. 2007. Identification of molecular markers linked to a *Pyrenophora teres* avirulence gene. *Phytopathology* 97(7), 842–849.

Bogacki, P., Keiper, F. J. and Oldach, K. H. 2010. Genetic structure of South Australian *Pyrenophora teres* populations as revealed by microsatellite analyses. *Fungal Biology* 114(10), 834–841.

Bouajila, A., Zoghlami, N., Ahmed, M. A., Baum, M. and Nazari, K. 2012. Pathogenicity spectra and screening for resistance in barley against Tunisian *Pyrenophora teres* f. *teres*. *Plant Disease* 96(10), 1569–1575.

Bouajila, A., Zoghlami, N., Al Ahmed, M., Baum, M., Ghorbel, A. and Nazari, K. 2011. Comparative virulence of *Pyrenophora teres* f. *teres* from Syria and Tunisia and screening for resistance sources in barley: implications for breeding. *Letters in Applied Microbiology* 53(5), 489–502.

Boungab, K., Belabid, L., Fortas, Z. and Bayaa, B. 2012. Pathotype diversity among Algerian isolates of *Pyrenophora teres* f. *Teres*. *Phytopathologia Mediterranea* 51, 577–586.

Brent, K. J. 1995. Fungicide resistance in crop pathogens: how can it be managed? [Online]. Available at: https://ci.nii.ac.jp/naid/10007833541/en/ [Accessed].

Brown, M. P., Steffenson, B. and Webster, R. 1993. Host range of Pyrenophora teres f. *teres* isolates from California. *Plant Disease* 77(9), 942–947.

Buchannon, K. W. and McDonald, W. C. 1965. Sources of resistance in barley to *Pyrenophora teres*. *Canadian Journal of Plant Science* 45(2), 189–193.

Buil, J. B., Hare, R. K., Zwaan, B. J., Arendrup, M. C., Melchers, W. J. G. and Verweij, P. E. 2019. The fading boundaries between patient and environmental routes of triazole resistance selection in *Aspergillus fumigatus*. *PLoS Pathogens* 15(8), e1007858.

Cakir, M., Gupta, S., Platz, G. J., Ablett, G. A., Loughman, R., Emebiri, L. C., Poulsen, D., Li, C. D., Lance, R. C. M., Galwey, N. W., Jones, M. G. K. and Appelsb, R. 2003. Mapping and validation of the genes for resistance to *Pyrenophora teres* f. *teres* in barley (Hordeum vulgare L.). *Australian Journal of Agricultural Research* 54(12), 1369–1377.

Campbell, G. F. and Crous, P. W. 2003. Genetic stability of net x spot hybrid progeny of the barley pathogen *Pyrenophora teres*. *Australasian Plant Pathology* 32(2), 283–287.

Campbell, G. F., Crous, P. W. and Lucas, J. A. 1999. *Pyrenophora teres* f. *maculata*, the cause of Pyrenophora leaf spot of barley in South Africa. *Mycological Research* 103(3), 257–267.

Campbell, G. F., Lucas, J. A. and Crous, P. W. 2002. Evidence of recombination between net- and spot-type populations of *Pyrenophora teres* as determined by RAPD analysis. *Mycological Research* 106(5), 602–608.

Carlsen, S. A., Neupane, A., Wyatt, N. A., Richards, J. K., Faris, J. D., Xu, S. S., Brueggeman, R. S. and Friesen, T. L. 2017. Characterizing the *Pyrenophora teres* f. *maculata*-barley interaction using pathogen genetics. *G3: Genes, Genomes, Genetics*, 7(8), 2615–2626.

Christensen, J. J. and Graham, T. W. 1934. Physiologic specialization and variation in *Helminthosporium gramineum* Rab. *Technical Bulletin 95, University of Minnesota, Agricultural Experiment Station*. University Farm, St. Paul.

Clare, S. J., Wyatt, N. A., Brueggeman, R. S. and Friesen, T. L. 2020. Research advances in the *Pyrenophora teres*-barley interaction. *Molecular Plant Pathology* 21(2), 272–288.

Cromey, M. G. and Parkes, R. A. 2003. Pathogenic variation in *Drechslera teres* in New Zealand. *New Zealand Plant Protection* 56, 251–256.

Crous, P. W., Janse, B. J. H., Tunbridge, J. and Holz, G. 1995. DNA homology between *P. japonica* and *P. teres*. *Mycological Research* 99(9), 1098–1102.

Dahanayaka, B., Vaghefi, N., Knight, N. L., Bakonyi, J., Prins, R., Seress, D., Snyman, L. and Martin, A. 2021. Population structure of Pyrenophora teres f. teres barley pathogens from different continents. *Phytopathology*. https://doi.org/10.1094/PHYTO-09-20-0390-R

Deadman, M. L. and Cooke, B. M. 1989. An analysis of rain-mediated dispersal of *Drechslera teres* conidia in field plots of spring barley. *Annals of Applied Biology* 115(2), 209–214.

Douiyssi, A., Rasmusson, D. C. and Roelfs, A. P. 1998. Responses of barley cultivars and lines to isolates of *Pyrenophora teres*. *Plant Disease* 82(3), 316–321.

Duczek, L. J., Sutherland, K. A., Reed, S. L., Bailey, K. L. and Lanford, G. P. 1999. Survival of leaf spot pathogens on crop residues of wheat and barley in Saskatchewan. *Canadian Journal of Plant Pathology* 21(2), 165–173.

Ellis, M. B. and Waller, J. M. 1973. Pyrenophora teres. CMI Descriptions of Fungi and Bacteria. Set No.39. Bacteria No. 390. CAB International, Wallingford, UK.

Ellwood, S. R., Liu, Z., Syme, R. A., Lai, Z., Hane, J. K., Keiper, F., Moffat, C. S., Oliver, R. P. and Friesen, T. L. 2010. A first genome assembly of the barley fungal pathogen *Pyrenophora teres* f. *teres*. *Genome Biology* 11, R109.

Ellwood, S. R., Piscetek, V., Mair, W. J., Lawrence, J. A., Lopez-Ruiz, F. J. and Rawlinson, C. 2019. Genetic variation of Pyrenophora teres f. *teres* isolates in Western Australia and emergence of a Cyp51A fungicide resistance mutation. *Plant Pathology* 68(1), 135–142.

Ellwood, S. R., Syme, R. A., Moffat, C. S. and Oliver, R. P. 2012. Evolution of three *Pyrenophora* cereal pathogens: recent divergence, speciation and evolution of non-coding DNA. *Fungal Genetics and Biology: FG and B* 49(10), 825–829.

Eurowheat, P. 2020. Map with CYP51 mutation frequency in Europe [Online]. Available at: https://agro.au.dk/forskning/internationale-platforme/eurowhea.

Forknall, C. R., Snyman, L., Platz, G. and Kelly, A. 2019. Measuring the impact of environment on yield and quality losses from net form net blotch. 19th Australian Barley Technical Symposium. Perth, Western Australia.

Fowler, R. A., Platz, G. J., Bell, K. L., Fletcher, S. E. H., Franckowiak, J. D. and Hickey, L. T. 2017. Pathogenic variation of *Pyrenophora teres* f. *teres* in Australia. *Australasian Plant Pathology* 46(2), 115–128.

FRAC 2020a. Pathogen risk list [Online]. Available at: http://www.frac.info.

FRAC 2020b. QoI working group [Online]. Available at: http://www.frac.info.

FRAC 2020c. SDHI working group [Online]. Available at: http://www.frac.info.

Grasso, V., Palermo, S., Sierotzki, H., Garibaldi, A. and Gisi, U. 2006. Cytochrome b gene structure and consequences for resistance to Qo inhibitor fungicides in plant pathogens. *Pest Management Science* 62(6), 465–472.

Grewal, T. S., Rossnagel, B. G., Pozniak, C. J. and Scoles, G. J. 2008. Mapping quantitative trait loci associated with barley net blotch resistance. *TAG. Theoretical and Applied Genetics. Theoretische und Angewandte Genetik* 116(4), 529-539.

Gupta, S. and Loughman, R. 2001. Current virulence of *Pyrenophora teres* on barley in Western Australia. *Plant Disease* 85(9), 960–966.

Gupta, S., Loughman, R., D'antuono, M. and Bradley, J. 2012. Characterisation and diversity of *Pyrenophora teres* f. *maculata* isolates in Western Australia. *Australasian Plant Pathology* 41(1), 31–40.

Gupta, S., Loughman, R., Platz, G. J. and Lance, R. C. M. 2003. Resistance in cultivated barleys to *Pyrenophora teres* f. *teres* and prospects of its utilisation in marker identification and breeding. *Australian Journal of Agricultural Research* 54(12), 1379–1386.

Hargreaves, J. A. and Keon, J. P. 1983. The binding of isolated mesophyll cells from barley leaves of hyphae of *Pyrenophra teres*. *Plant Cell Reports* 2(5), 240–243.

Ireland, K. B., Beard, C., Cameron, J., Chang, S., Davidson, J., Dodhia, K. N., Garrard, T. A., Hills, A. L., Hollaway, G., Jayasena, K. W., Kiss, L., Mair, W., Marcroft, S. J., McLean, M. S., Milgate, A., Poole, N., Simpfendorfer, S., Snyman, L., Thomas, G. J., Wallwork, H., Van de Wouw, A. P., Zulak, K. G. and Lopez-Ruiz, F. J. 2021. Fungicide resistance management in Australian grain crops. Grains Research and Development Corporation, Australia.

Jalli, M. 2011. Sexual reproduction and soil tillage effects on virulence of *Pyrenophora teres* in Finland. *Annals of Applied Biology* 158(1), 95–105.

Jayasena, K. W., Burgel, A., Tanaka, K., Majewski, J. and Loughman, R. 2007. Yield reduction in barley in relation to spot-type net blotch. *Australasian Plant Pathology* 36(5), 429–433.

Jebbouj, R. and El Yousfi, B. 2010. An integrated multivariate approach to net blotch of barley: virulence quantification, pathotyping and a breeding strategy for disease resistance. *European Journal of Plant Pathology* 127(4), 521–544.

Jonsson, R., Bryngelsson, T. and Gustafsson, M. 1997. Virulence studies of Swedish net blotch isolates (Drechslera teres) and identification of resistant barley lines. *Euphytica* 94(2), 209–218.

Jonsson, R., Bryngelsson, T., Jalli, M. and Gustafsson, M. 1998. Effect of growth stage on resistance to *Drechslera teres* f. *teres* in Barley. *Journal of Phytopathology* 146(5–6), 261–265.

Jonsson, R., Sail, T. and Bryngelsson, T. 2000. Genetic diversity for random amplified polymorphic DNA (RAPD) markers in two Swedish populations of *Pyrenophora teres*. *Canadian Journal of Plant Pathology* 22(3), 258–264.

Jonsson, R., Säll, T. S., Kraft, T. and Gustafsson, M. 1999. Inheritance of resistance to *Pyrenophora teres* f. *teres* in spring barley. *Plant Breeding* 118(4), 313–317.

Jordan, V. W. L. 1981. Aetiology of barley net blotch casued by *Pyrenophora teres* and some effects on yield. *Plant Pathology* 30(2), 77–87.

Justesen, A. F., Hansen, H. J. and Pinnschmidt, H. O. 2008. Quantification of *Pyrenophora graminea* in barley seed using real-time PCR. *European Journal of Plant Pathology* 122(2), 253–263.

Keiper, F. J., Grcic, M., Capio, E. and Wallwork, H. 2008. Diagnostic microsatellite markers for the barley net blotch pathogens, *Pyrenohora teres* f. *Maculata* and *Pyrenophora teres* f. *Australasian Plant Pathology* 37(4), 428–430.

Khan, T. 1987. Relationship between net blotch (*Drechslera teres*) and losses in grain yield of barley in Western Australia. *Australian Journal of Agricultural Research* 38(4), 671–679.

Khan, T. 1989. Effect of spot-type net blotch (*Drechslera teres* (Sacc.) Shoem) infection on barley yield in short season environment of northern cereal belt of Western Australia. *Australian Journal of Agricultural Research* 40(4), 745–752.

Khan, T. and Boyd, W. 1969. Physiologic specialization in *Drechslera teres*. *Australian Journal of Biological Sciences* 22(5), 1229–1236.

Khan, T. N. 1973. Host specialization by Western Australian isolates causing net blotch symptoms on Hordeum. *Transactions of the British Mycological Society* 61(2), 215–220.

Khan, T. N. 1982. Changes in pathogenicity of *Drechslera teres* relating to changes in barley cultivars grown in Western Australia. *Plant Disease* 66(1), 655–656.

Khan, T. N. and Tekauz, A. 1982. Occurrence and pathogenicity of *Dreschlera teres* isolates causing spot-type symptoms on barley in Western Australia. *Plant Disease* 66(1), 423–425.

Kinzer, K. M. D. 2015. Characterizing *Pyrenophora teres* f. maculata in the northern United States and impact of spot form net blotch on yield of barley (PhD Dissertation). North Dakota State University, ND, USA.

Koladia, V. M., Richards, J. K., Wyatt, N. A., Faris, J. D., Brueggeman, R. S. and Friesen, T. L. 2017. Genetic analysis of virulence in the *Pyrenophora teres* f. *teres* population BB25 × FGOH04Ptt-21. *Fungal Genetics and Biology* 107, 12–19.

Lai, Z., Faris, J. D., Weiland, J. J., Steffenson, B. J. and Friesen, T. L. 2007. Genetic mapping of *Pyrenophora teres* f. *teres* genes conferring avirulence on barley. *Fungal Genetics and Biology: FG and B* 44(5), 323–329.

Lartey, R. T., Caesar-Tonthat, T. C., Caesar, A. J., Sainju, U. M. and Evans, R. G. 2013. First report of spot form net blotch caused by *Pyrenophora teres* f. *maculata* on barley in the Mon-Dak area of the United States. *Plant Disease* 97(1), 143–143.

Leadbeater, A. 2014. Plant health management: fungicides and antibiotics. In: Van Alfen, N. K. (Ed.) *Encyclopedia of Agriculture and Food Systems* (2nd edn.). Elsevier, Amsterdam, pp 408–424.

Lehmensiek, A., Bester-Van Der Merwe, A. E., Sutherland, M. W., Platz, G., Kriel, W. M., Potgieter, G. F. and Prins, R. 2010. Population structure of South African and Australian *Pyrenophora teres* isolates. *Plant Pathology* 59(3), 504–515.

Lehmensiek, A., Platz, G. J., Mace, E., Poulsen, D. and Sutherland, M. W. 2007. Mapping of adult plant resistance to net form of net blotch in three Australian barley populations. *Australian Journal of Agricultural Research*, 58(12), 1191–1197.

Leisova, L., Kucera, L., Minarikova, V. and Ovesna, J. 2005a. AFLP-based PCR markers that differentiate spot and net forms of *Pyrenophora teres*. *Plant Pathology* 54(1), 66–73.

Leisova, L., Minarikova, V., Kucera, L. and Ovesna, J. 2005b. Genetic diversity of *Pyrenophora teres* isolates as detected by AFLP analysis. *Journal of Phytopathology* 153(10), 569–578.

Leisova, L., Minarikova, V., Matusinsky, P., Hudcovicova, M., Ondreickova, K. and Gubis, J. 2014. Genetic structure of *Pyrenophora teres* net and spot populations as revealed by microsatellite analysis. *Fungal Biology* 118(2), 180–192.

Lightfoot, D. J. and Able, A. J. 2010. Growth of *Pyrenophora teres* in planta during barley net blotch disease. *Australasian Plant Pathology* 39(6), 499–507.

Linde, C. C. and Smith, L. M. 2019. Host specialisation and disparate evolution of Pyrenophora *teres* f. *teres* on barley and barley grass. *BMC Evolutionary Biology* 19(1), 139.

Liu, Z. H., Ellwood, S. R., Oliver, R. P. and Friesen, T. L. 2011. *Pyrenophora teres*: profile of an increasingly damaging barley pathogen. *Molecular Plant Pathology* 12(1), 1–19.

Liu, Z. H. and Friesen, T. L. 2010. Identification of *Pyrenophora teres* f. *maculata*, causal agent of spot type net blotch of barley in North Dakota. *Plant Disease* 94(4), 480–480.

Liu, Z. H., Zhong, S., Stasko, A. K., Edwards, M. C. and Friesen, T. L. 2012. Virulence profile and genetic structure of a North Dakota population of *Pyrenophora teres* f. *teres*, the causal agent of net form net blotch of barley. *Phytopathology* 102(5), 539–546.

Louw, J. 1996. *Relative importance of the barley net blotch pathogens Pyrenophora teres f. teres (net-type) and P. teres f. maculata*. *African Plant Protection* 2(2), 89–95.

Lu, S. W., Platz, G. J., Edwards, M. C. and Friesen, T. L. 2010. Mating type locus-specific polymerase chain reaction markers for differentiation of *Pyrenophora teres* f. *teres* and P. *teres* f. *maculata*, the causal agents of barley net blotch. *Phytopathology* 100(12), 1298–1306.

Mair, W. J., Deng, W. W., Mullins, J. G. L., West, S., Wang, P. H., Besharat, N., Ellwood, S. R., Oliver, R. P. and Lopez-Ruiz, F. J. 2016. Demethylase inhibitor fungicide resistance in *Pyrenophora teres* f. sp *teres* associated with target site modification and inducible overexpression of Cyp51. *Frontiers in Microbiology* 7, Article No. 1279.

Mair, W. J., Thomas, G. J., Dodhia, K., Hills, A. L., Jayasena, K. W., Ellwood, S. R., Oliver, R. P. and Lopez Ruiz, F. J. 2020. Parallel evolution of multiple mechanisms for demethylase inhibitor fungicide resistance in the barley pathogen *Pyrenophora teres* f. *Maculata*. *Fungal Genetics and Biology* 145. Available at: https://doi.org/10.1016/j.fgb.2020.103475.

Mäkelä, K. 1975. Occurrence of *Helminthosporium* species on cereals in Finland in 1971–1973. *Agricultural and Food Science* 47(3), 181–217.

Manning, V. A., Pandelova, I., Dhillon, B., Wilhelm, L. J., Goodwin, S. B., Berlin, A. M., Figueroa, M., Freitag, M., Hane, J. K., Henrissat, B., Holman, W. H., Kodira, C. D., Martin, J., Oliver, R. P., Robbertse, B., Schackwitz, W., Schwartz, D. C., Spatafora, J. W., Turgeon, B. G., Yandava, C., Young, S., Zhou, S., Zeng, Q., Grigoriev, I. V., Ma, L. J. and Ciuffetti, L. M. 2013. Comparative genomics of a plant-pathogenic fungus, *Pyrenophora tritici-repentis*, reveals transduplication and the impact of repeat elements on pathogenicity and population divergence. *G3: Genes, Genomes, Genetics*, 3(1), 41-63.

Martin, A., Moolhuijzen, P., Tao, Y., Mcilroy, J., Ellwood, S. R., Fowler, R. A., Platz, G. J., Kilian, A. and Snyman, L. 2020. Genomic regions associated with virulence in *Pyrenophora teres* f. *teres* Identified by Genome-Wide Association Analysis and Biparental Mapping. *Phytopathology* 110(4), 881-891.

McDonald, B. A. and Linde, C. 2002. Pathogen population genetics, evolutionary potential and durable resistance. *Annual Review of Phytopathology* 40, 349-379.

McDonald, W. C. 1963. Heterothallism in *Pyrenophora teres*. *Phytopathology* 53, 771-773.

McLean, M. 2016. Net blotches of barley. *Agricultural Notes*, Department of Primary Industries, Horsham, Victoria.

McLean, M. S., Howlett, B. J. and Hollaway, G. J. 2009. Epidemiology and control of spot form of net blotch (*Pyrenophora teres f. maculata*) of barley: a review. *Crop and Pasture Science* 60(4), 303-315.

McLean, M. S., Howlett, B. J. and Hollaway, G. J. 2010a. Spot form of net blotch, caused by *Pyrenophora teres* f. *maculata*, is the most prevalent foliar disease of barley in Victoria, Australia. *Australasian Plant Pathology* 39(1), 46-49.

McLean, M. S., Howlett, B. J., Turkington, T. K., Platz, G. J. and Hollaway, G. J. 2012. Spot form of net blotch resistance in a diverse set of barley lines in Australia and Canada. *Plant Disease* 96(4), 569-576.

McLean, M. S., Keiper, F. J. and Hollaway, G. J. 2010b. Genetic and pathogenic diversity in *Pyrenophora teres* f. *maculata* in barley crops of Victoria, Australia. *Australasian Plant Pathology* 39(4), 319-325.

McLean, M. S., Martin, A., Gupta, S., Sutherland, M. W., Hollaway, G. J. and Platz, G. J. 2014. Validation of a new spot form of net blotch differential set and evidence for hybridisation between the spot and net forms of net blotch in Australia. *Australasian Plant Pathology* 43(3), 223-233.

Milgroom, M. G. 1996. Recombination and the multilocus structure of fungal populations. *Annual Review of Phytopathology* 34, 457-477.

Mode, C. J. and Schaller, C. W. 1958. Two additional factors for host resistance to net blotch in barley 1. *Agronomy Journal* 50(1), 15-18.

Moolhuijzen, P. M., Muria-Gonzalez, M. J., Syme, R., Rawlinson, C., See, P. T., Moffat, C. S. and Ellwood, S. R. 2020. Expansion and conservation of biosynthetic gene clusters in pathogenic *Pyrenophora* spp. *Toxins* 12(4), 242.

Morton, V. and Staub, T. 2008. *A Short History of Fungicides* [Online]. *APSnet Features*.

Novakazi, F., Afanasenko, O., Anisimova, A., Platz, G. J., Snowdon, R., Kovaleva, O., Zubkovich, A. and Ordon, F. 2019a. Genetic analysis of a worldwide barley collection for resistance to net form of net blotch disease (*Pyrenophora teres* f. *teres*). *TAG. Theoretical and Applied Genetics. Theoretische und Angewandte Genetik* 132(9), 2633-2650.

Novakazi, F., Afanasenko, O., Lashina, N., Platz, G., Snowdon, R., Loskutov, I. and Ordon, F. 2019b. Genome-wide association studies in barley (*Hordeum vulgare*) diversity set reveal a limited number of loci for resistance to spot blotch (*Bipolaris sorokiniana*). *Plant Breeding* 00, 1–15.

Oğuz, A. Ç. and Karakaya, A. 2017. Pathotypes of *Pyrenophora teres* on barley in Turkey. *Phytopathologia Mediterranea* 56, 224–234.

Oliver, R. P. and Hewitt, H. G. 2014. Fungicides in crop protection. In: *Fungicide Resistance* (2nd edn.). CAB International, Wallingford, UK, pp. 123–149.

Peever, T. L. and Milgroom, M. G. 1994. Genetic structure of *Pyrenophora teres* populations determined with random amplified polymorphic DNA markers. *Canadian Journal of Botany* 72(7), 915–923.

Peltonen, S., Jalli, M., Kammiovirta, K. and Karjalainen, R. 1996. Genetic variation in *Drechslera teres* populations as indicated by RAPD markers. *Annals of Applied Biology* 128(3), 465–477.

Platz, G. 2005. Accelerated selection for resistance to multiple diseases of barley. 12th Australian Barley Technical Symposium. Hobart, Tasmania.

Platz, G., Bell, K. L., Rees, R. G. and Galea, V. J. 2000. Pathotype variation of the Australian net blotch populations. Proceedings of the 8th International Barley Genetics Symposium, Adelaide University, Glen Osmond, South Australia.

Platz, G., Wallwork, H., Butt, M. and Barsby, J. 2009. A new pathotype of net form net blotch in South Australia. 14th Australian Barley Technical Symposium. Sunshine Coast, Australia.

Pon, D. S. 1949. Physiologic specialization and variation in *Helminthosporium teres*. *Phytopatology* 39, 18.

Poudel, B., Ellwood, S. R., Testa, A. C., McLean, M., Sutherland, M. W. and Martin, A. 2017. Rare *Pyrenophora teres* hybridization events revealed by development of sequence-specific PCR markers. *Phytopathology* 107(7), 878–884.

Poudel, B., McLean, M. S., Platz, G. J., Mcilroy, J. A., Sutherland, M. W. and Martin, A. 2019a. Investigating hybridisation between the forms of *Pyrenophora teres* based on Australian barley field experiments and cultural collections. *European Journal of Plant Pathology* 153(2), 465–473.

Poudel, B., Vaghefi, N., McLean, M. S., Platt, G. J., Sutherland, M. W. and Martin, A. 2019b. Genetic structure of a *Pyrenophora teres* f. *teres* population over time in an Australian barley field as revealed by Diversity Arrays Technology markers. *Plant Pathology* 68(7), 1331–1336.

Poulsen, D. M. E., Johnston, R. P., Platz, G. J., Fox, G., Kelly, A., Sturgess, J. M., Fromm, R. L., Laufer, M. J., Inkerman, P. A. and Butler, D. 1999. Effects of foliar diseases on northern region grain production in the 1998 cropping season. 9th Australian Barley Technical Symposium. Melbourne, Australia.

Qamar, M., Liu, Z. H., Faris, J. D., Chao, S., Edwards, M. C., Lai, Z., Franckowiak, J. D. and Friesen, T. L. 2008. A region of barley chromosome 6h harbors multiple major genes associated with net type net blotch resistance. *Theoretical and Applied Genetics* 117, 1261–1270.

Rau, D., Brown, A. H. D., Brubaker, C. L., Attene, G., Balmas, V., Saba, E. and Papa, R. 2003. Population genetic structure of *Pyrenophora teres* Drechs. the causal agent of net blotch in Sardinian landraces of barley (*Hordeum vulgare* L.). *TAG. Theoretical and Applied Genetics. Theoretische und Angewandte Genetik* 106(5), 947–959.

Rees, R. G., Strong, W. M. and Neale, T. J. 1999. The effects of foliar diseases on production of wheat and barley in the northern region in 1998. Report to GRDC Northern Panel.

Rehfus, A. 2018. Analysis of the emerging situation of resistance to succinate dehydrogenase inhibitors in *Pyrenophora teres* and Zymoseptoria *tritici* in Europe. PhD, Universität Hohenheim.

Rehfus, A., Miessner, S., Achenbach, J., Strobel, D., Bryson, R. and Stammler, G. 2016. Emergence of succinate dehydrogenase inhibitor resistance of *Pyrenophora teres* in Europe. *Pest Management Science* 72(10), 1977–1988.

Robinson, J. and Jalli, M. 1996. Diversity among Finnish net blotch isolates and resistance in barley. *Euphytica* 92(1–2), 81–87.

Rupp, S., Weber, R. W., Rieger, D., Detzel, P. and Hahn, M. 2017. Spread of *Botrytis cinerea* strains with multiple fungicide resistance in German horticulture. *Frontiers in Microbiology* 7, 2075.

Sampson, M. G. and Watson, A. K. 1985. Host specificity of five leaf-spotting pathogens of agropyron repens. *Canadian Journal of Plant Pathology* 7(2), 161–164.

Semar, M., Strobel, D., Koch, A., Klappach, K. and Stammler, G. 2007. Field efficacy of pyraclostrobin against populations of *Pyrenophora teres* containing the F129L mutation in the cytochrome b gene. *Journal of Plant Diseases and Protection* 114(3), 117–119.

Serenius, M., Manninen, O., Wallwork, H. and Williams, K. 2007. Genetic differentiation in *Pyrenophora teres* populations measured with AFLP markers. *Mycological Research* 111(2), 213-223.

Serenius, M., Mironenko, N. and Manninen, O. 2005. Genetic variation, occurrence of mating types and different forms of *Pyrenophora teres* causing net blotch of barley in Finland. *Mycological Research* 109(7), 809–817.

Sheridan, J. E., Grbavac, N. and Sheridan, M. H. 1985. Triadimenol insensitivity in Pyrenophora *teres*. *Transactions of the British Mycological Society* 85(2), 338–341.

Shipton, W. 1966. Effect of net blotch infection of barley on grain yield and quality. *Australian Journal of Experimental Agriculture* 6(23), 437-440.

Shjerve, R. A., Faris, J. D., Brueggeman, R. S., Yan, C., Zhu, Y., Koladia, V. and Friesen, T. L. 2014. Evaluation of a *Pyrenophora teres* f. *teres* mapping population reveals multiple independent interactions with a region of barley chromosome 6H. *Fungal Genetics and Biology* 70, 104–112.

Shoemaker, R. A. 1959. Nomenclature of *Drechslera* and *Bipolaris*, grass parasites segregated from 'Helminthosporium'. *Canadian Journal of Botany* 37(5), 879-887.

Sierotzki, H., Frey, R., Wullschleger, J., Palermo, S., Karlin, S., Godwin, J. and Gisi, U. 2007. Cytochrome b gene sequence and structure of *Pyrenophora teres* and *P. tritici-repentis* and implications for QoI resistance. *Pest Management Science* 63(3), 225–233.

Silvar, C., Casas, A. M., Kopahnke, D., Habekuß, A., Schweizer, G., Gracia, M. P., Lasa, J. M., Ciudad, F. J., Molina-Cano, J. L., Igartua, E. and Ordon, F. 2010. Screening the Spanish Barley Core Collection for disease resistance. *Plant Breeding* 129(1), 45-52.

Smedegård-Petersen, V. 1971. *Pyrenophora teres* f. *maculata* f. (nov). and *Pyrenophora teres* f. *Teres* on barley in Denmark. Royal Veterinary and Agricultural University Yearbook, 124–144.

Smedegård-Petersen, V. 1976. Pathogenesis and genetics of net-spot blotch and leaf stripe of barley caused by *Pyrenophora teres* and *Pyrenophora graminea* (Doctoral Dissertation), Royal Veterinary and Agricultural University, Copenhagen.

Snyman, L., Platz, G. and Forknall, C. R. 2017. Is the resistance rating of a barley variety correlated to yield and quality loss as a result of disease? 18th Australian Barley Technical Symposium. Hobart, Tasmania.

Stammler, G., Rehfus, A., Prochnow, J., Bryson, R. and Strobel, D. 2014. New findings on the development of insensitive isolates of *Pyrenophora teres* towards SDHI fungicides. *Julius-Kühn-Archiv*, 447, 568.

Stefansson, T. S., Serenius, M. and Hallsson, J. H. 2012. The genetic diversity of Icelandic populations of two barley leaf pathogens, *Rhynchosporium commune* and *Pyrenophora teres*. *European Journal of Plant Pathology* 134(1), 167–180.

Steffenson, B. 1997. *Compendium of Barley Diseases*, The American Phytopathological Society, Bozeman.

Steffenson, B. J. and Webster, R. K. 1992a. Pathotype diversity of *Pyrenophora teres* f. *teres* on Barley. *Phytopathology* 82(2), 170–177.

Steffenson, B. J. and Webster, R. K. 1992b. Quantitative resistance to *Pyrenophora teres* f. *teres* in Barley. *Phytopathology* 82(4), 407–411.

Sutton, J. C. and Steele, P. 1983. Effects of seed and foliar fungicides on progress of net blotch and yield in barley. *Canadian Journal of Plant Science* 63(3), 631–639.

Syme, R. A., Martin, A., Wyatt, N. A., Lawrence, J. A., Muria-Gonzalez, M. J., Friesen, T. L. and Ellwood, S. R. 2018. Transposable element genomic fissuring in *Pyrenophora teres* is associated with genome expansion and dynamics of host-pathogen genetic interactions. *Frontiers in Genetics* 9, 130.

Talbot, J. J., Subedi, S., Halliday, C. L., Hibbs, D. E., Lai, F., Lopez-Ruiz, F. J., Harper, L., Park, R. F., Cuddy, W. S., Biswas, C., Cooley, L., Carter, D., Sorrell, T. C., Barrs, V. R. and Chen, S. C. A. 2018. Surveillance for azole resistance in clinical and environmental isolates of *Aspergillus fumigatus* in Australia and *cyp51A* homology modelling of azole-resistant isolates. *Journal of Antimicrobial Chemotherapy* 73(9), 2347–2351.

Tekauz, A. 1985. A numerical scale to classify reactions of barley to *Pyrenophora teres*. *Canadian Journal of Plant Pathology* 7(2), 181–183.

Tekauz, A. 1990. Characterization and distribution of pathogenic variation in *Pyrenophora teres* f. *teres* and P. *teres* f. *maculata* from western Canada. *Canadian Journal of Plant Pathology* 12(2), 141–148.

Tucker, M., Lopez-Ruiz, F., Jayasena, K. and Oliver, R. 2015. *Origin of Fungicide Resistant Barley Powdery Mildew in Western Australia: Lessons to Be Learned*, Springer, Japan.

Tuohy, J. M., Jalli, M., Cooke, B. M. and Sullivan, E. O. 2006. Pathogenic variation in populations of Drechslera teres f. *teres* and D. *teres* f. maculata and differences in host cultivar responses. *European Journal of Plant Pathology* 116, 177–185.

Turo, C., Mair, W. J., Martin, A., Ellwood, S. R., Oliver, R. P. and Lopez-Ruiz, F. J. 2021. Species hybridisation and clonal expansion as a new fungicide resistance evolutionary mechanism in Pyrenophora teres spp. *bioRxiv*. 454422; doi: https://doi.org/10.1101/2021.07.30.454422.

Uranga, J. P., Schierenbeck, M., Perello, A. E., Lohwasser, U., Boerner, A. and Simon, M. R. 2020. Localization of QTL for resistance to *Pyrenophora teres* f. *maculata*, a new wheat pathogen. *Euphytica* 216(4). Available at: https://link.springer.com/article/10.1007%2Fs10681-020-02593-y.

Van den Berg, J. 1988. Epidemiology of Pyrenophora teres and its effect on grain yield of Hordeum vulgare. PhD, University of Saskatchewan.

Vatter, T., Maurer, A., Kopahnke, D., Perovic, D., Ordon, F. and Pillen, K. 2017. A nested association mapping population identifies multiple small effect QTL conferring resistance against net blotch (Pyrenophora teres f. *teres*) in wild barley. *PLoS ONE* 12(10), e0186803.

Voorrips, R. E. 2002. MapChart: software for the graphical presentation of linkage maps and QTLs. *Journal of Heredity* 93(1), 77-78.

Wallwork, H. 1995. Barley leaf blights in Australia and New Zealand: historical perspective and current situation. *Rachis* 14, 75-81.

Wallwork, H. 2011. Barley pathogens in South Australia - current threats and future challenges. 15th Australian Barley technical Symposium. Adelaide, South Australia.

Wallwork, H., Butt, M. and Capio, E. 2016. Pathogen diversity and screening for minor gene resistance to *Pyrenophora teres* f. *teres* in barley and its use for plant breeding. *Australasian Plant Pathology* 45(5), 527-531.

Wallwork, H., Lichon, A. and Sivanesan, A. 1992. *Pyrenophora hordei* - a New Ascomycete with Drechslera anamorph Affecting Barley in Australia. *Mycological Research* 96(12), 1068-1070.

Walters, D. R., Avrova, A., Bingham, I. J., Burnett, F. J., Fountaine, J., Havis, N. D., Hoad, S. P., Hughes, G., Looseley, M., Oxley, S. J. P., Renwick, A., Topp, C. F. E. and Newton, A. C. 2012. Control of foliar diseases in barley: towards an integrated approach. *European Journal of Plant Pathology* 133(1), 33-73.

Weiland, J. J., Steffenson, B. J., Cartwright, R. D. and Webster, R. K. 1999. Identification of molecular genetic markers in *Pyrenophora teres* f. *teres* Associated with Low Virulence on 'Harbin' Barley. *Phytopathology* 89(2), 176-181.

Williams, K. J., Smyl, C., Lichon, A., Wong, K. Y. and Wallwork, H. 2001. Development and use of an assay based on the polymerase chain reaction that differentiates the pathogens causing spot form and net form of net blotch of barley. *Australasian Plant Pathology* 30(1), 37-44.

Wonneberger, R., Ficke, A. and Lillemo, M. 2017. Identification of quantitative trait loci associated with resistance to net form net blotch in a collection of Nordic barley germplasm. *TAG. Theoretical and Applied Genetics. Theoretische und Angewandte Genetik* 130(10), 2025-2043.

Wu, H.-L., Steffenson, B. J., Zhong, S., Li, Y. and Oleson, A. E. 2003. Genetic variation for virulence and RFLP markers in *Pyrenophora teres*. *Canadian Journal of Plant Pathology* 25(1), 82-90.

Wyatt, N. A., Richards, J. K., Brueggeman, R. S. and Friesen, T. L. 2018. Reference assembly and annotation of the *Pyrenophora teres* f. *teres* Isolate 0-1. *G3: Genes, Genomes, Genetics* 8(1), 1-8.

Wyatt, N. A., Richards, J. K., Brueggeman, R. S. and Friesen, T. L. 2020. A comparative genomic analysis of the barley pathogen *Pyrenophora teres* f. *teres* Identifies Subtelomeric Regions as Drivers of Virulence. *Molecular Plant-Microbe Interactions: MPMI* 33(2), 173-188.

Xi, K., Helm, J. H., Burnett, P. A., Turkington, T. K. and Tekauz, A. 1999. Determining quantitative resistance of barley cultivars at the seedling stage to net blotch caused by *Pyrenophora teres*. *Canadian Journal of Plant Pathology* 21(3), 284-290.

Youcef-Benkada, M., Bendahmane, B. S., Sy, A. A., Barrault, G. and Albertini, L. 1994. Effects of inoculation of barley inflorescences with *Drechslera teres* upon the location of seed-borne inoculum and its transmission to seedlings as modified by temperature and soil moisture. *Plant Pathology* 43(2), 350-355.

Chapter 14

Understanding plant-pathogen interactions in net blotch infection of cereals

Karl M. Effertz, Shaun J. Clare, Sarah M. Harkins and Robert S. Brueggeman, Washington State University, USA

1 Introduction

Fungal diseases are among the most important factors limiting the quality and yield of barley across the major growing regions around the world. An economically important disease that infects barley and causes significant yield and quality losses is net blotch, caused by the necrotrophic fungal pathogen *Pyrenophora teres*. To reduce the impact of net blotch on the supply of quality barley, the research community is engaged in basic and applied research to enhance genetic resistances, as this is the most economic and sustainable management strategy. Durable disease resistance against *Pyrenophora teres* will be a major achievement toward the goal of developing widely adapted barley varieties that have high yields and quality across dynamic environments. Due to the economic and social importance of malt barley for the brewing and distilling industries, it is essential that we achieve this goal to ensure adequate production of quality malt barley. However, global climate change is predicted to cause world barley shortages through shifting temperature and precipitation patterns. To meet production goals in the future, varieties will need to include

http://dx.doi.org/10.19103/AS.2021.0092.32

durable resistance to important fungal pathogens and stability across uncertain environments. Thus, breeders must have a thorough understanding of the *Pyrenophora teres* population structure and its virulence profiles, as well as the host-pathogen genetic interactions, to intelligently deploy disease resistance with gene stewardship in mind. This chapter focuses on a thorough review of the latest knowledge of both host resistance/susceptibility and pathogen virulence/avirulence in this important pathosystem, and the implications this knowledge will have on deploying sustainable resistances to this destructive pathogen of barley.

2 Host

Barley (*Hordeum vulgare* L.), being one of the first domesticated crops, has been sown by mankind since the dawn of agriculture. A hardy crop with a multitude of uses, barley is often grown on suboptimal cropland, leaving it exposed to an abundance of biotic and abiotic stresses. Barley has applications as food and animal feed, but perhaps its paramount use, which has had the biggest impact on civilization and human social interaction is as malt in the production of beer and distilled spirits. This robust cereal crop has played a major role in human society and the rise of civilization, particularly in social interactions dating back to the dawn of civilization up to modern times. Today, the multi-billion dollar-added value of barley due to the brewing and distilling industries makes it an important commodity across the globe.

Global climate change, as a result of man-made greenhouse emissions, has led to detrimental changes in the climate worldwide, which have been predicted to present challenges for the production of quality malting barley to meet the needs of the brewing and distilling industries (Xie et al., 2018). These anticipated world barley shortages are based on the predicted effects of abiotic stresses in regions of the globe experiencing elevated temperatures and drought. As temperatures rise and rainfall patterns change, barley production is predicted to suffer substantial yield losses, ranging from 3-17%, depending on the severity of conditions (Xie et al., 2018). However, as some regions of the globe experience precipitation deficits, others will experience excess precipitation and higher temperatures, which provide environmental conditions conducive to the development of plant disease epidemics caused by fungal pathogens. The shifting land use patterns accompanying the altered global climate could leave barley exposed to unexpected environmental conditions and pathogen populations. Modeling in Northern France estimated that early-season barley fields could see an additional four infection days per month by the year 2070, considering the current climate trends (Launay et al., 2014). Changes in the dynamics of this host-pathogen relationship have the potential to devastate future barley crops, with early-season infections creating additional opportunities for secondary infections and further losses to disease epidemics.

The second edition of the *Compendium of Barley Diseases* lists 77 different infectious agents capable of causing disease in cultivated barley (Mathre, 1997). Of these, 38 are fungal pathogens, with biotrophic, hemibiotrophic, and necrotrophic lifestyles. Several necrotrophic pathogens affect global barley production (see Table 1) (Brueggeman et al., 2020). The total yield losses to pathogens and pests average ~25% for all cereal grains, totaling ~$125 billion in annual economic losses from 2014 to 2016 (Savary et al., 2019; Knopff and Morell, 2019). A significant proportion of these losses is due to necrotrophic pathogens (Wang et al., 2014), with *Pyrenophora teres,* the causal agent of the diseases net form net blotch (NFNB) and spot form net blotch (SFNB), causing ~$309 million (Australian) of annual economic loss to the Australian barley industry alone (Murray and Brennan, 2010). Due to the lack of pathometry studies in other barley growing regions of the world, and the complexes of diseases present in regions conducive to fungal pathogens, it is difficult to even estimate the total annual worldwide losses due to *Pyrenophora teres*.

Barley is a hardy crop capable of growing on diverse geographies, exposed to a wide range of pathogens. This presents significant challenges to researchers and breeders tasked with identifying and incorporating multiple resistances while maintaining the desirable agronomic traits and producing varieties that are widely adapted. The uncertainty of a secure supply of quality malting barley, due in part to the shifting climate and unpredictable losses due to fungal diseases, highlights the importance of robust disease management systems to control multiple diseases including net blotch. While cultural practices and fungicide applications can limit the effects of *Pyrenophora teres*

Table 1 Destructive necrotrophic pathogens of Barley

Pathogen	Common Disease Name(s)	Synonyms
Bipolaris sorokiniana	Spot blotch, common root rot	*Cochliobolus sativus*
Ramularia collo-cygni	Ramularia leaf spot	*Ophiocladium hordei*
Septoria passerinii	Septoria speckled leaf blotch (SSLB)	*Mycosphaerella spp.*
Rhynchosporium commune	Scald	*Rhynchosporium secalis, Rhynchosporium graminicola, Marsonia secalis*
Several *Fusarium spp.*	Fusarium head blight (FHB), scab	*Fusarium graminearum, Gibberella zeae*
Pyrenophora teres f. *teres*, *Pyrenophora teres* f. *maculate*	Net form and spot form of net blotch	*Helminthosporium teres, Drechslera teres*

infections, the most desirable method for controlling this pathogen is robust and durable resistance.

2.1 Host diversity and breeding challenges

Barley spread from its domestication source in the Fertile Crescent and established five distinct subpopulations as it took hold in each center of diversification (Muñoz-Amatriaín et al., 2014). These five subpopulations spread across the globe to every habitable continent, with both artificial and natural selection allowing landraces to evolve into adapted varieties. Regional pathogen populations applied selective pressures on landraces, influencing farmers' artificial selection, thereby enriching these diverse lines with resistance. Moreover, the wild progenitor of domestic barley, *Hordeum vulgare* ssp. *spontaneum,* has a wide distribution with marked genomic diversity, both across its range (Fang et al., 2014) and within the specific microenvironments of a given region (Bedada et al., 2014). The diversity present within landraces and wild barley from world barley core collections represent a tremendous resource from which to identify novel resistances to aid in breeding efforts.

Although there is remarkable diversity within the barley primary germplasm pool, many barley-breeding programs have produced genetic bottlenecks through inbreeding within their programs. For example, genetic data revealed that the majority of US six-row barley had descended from the European six-row subpopulation four (Muñoz-Amatriaín et al., 2014), with all upper Midwest US six-row barley tracing back to only ten accessions (Horsley et al., 1995). Diversity was introduced from Manchurian lines. However, this example highlights the significant bottlenecks that occur, limiting the genetic diversity of modern barley cultivars, as breeding programs rely on inbreeding of the elite materials from their programs. As improving yield and malt quality have been the primary objectives of most US barley breeding programs, diseases that consistently affect these parameters are a priority. However, diseases that are less consistent or that have smaller effects on yield and quality are often a secondary concern and are not addressed by breeding programs. Moreover, these diseases persist, and evolve virulence shifts or environmental adaptability that allow them to explode onto the scene as a major problem. With climate change and uncertainty, crop producers may see more of these fringe pathogens rapidly shift to becoming major problems in the near future. When resistance genes are deployed in varieties without proper gene stewardship, the monocultural nature of modern agriculture applies selective pressure on pathogen populations, leading to the evolution of virulence shifts on these deployed resistances and the 'boom and bust' cycles of resistance genes (Brown, 2015). With rapidly evolving pathogens such as *Pyrenophora teres,* resistances can be overcome quickly. Thus, resistance must be intelligently deployed, with gene stewardship in mind (McDonald and Linde, 2002). This highlights the importance of constant monitoring of the pathogen population in each barley growing region, a task that traditionally relied on differential sets to determine the pathotype of regional isolates. The increased availability of affordable DNA sequencing and the advancements in barley and *Pyrenophora teres* genomic resources have enabled much more detailed examinations of *Pyrenophora teres* population dynamics (Clare et al., 2020).

3 Pathogen

3.1 Taxonomy and evolution

Originally classified as *Helminthosporium teres,* a later reexamination of the genus led to the anamorphic form of this species being reclassified as *Drechslera teres* (Shoemaker, 1959). The genus *Drechslera* is characterized as fungi that have 'cylindrical, not curved conidia, germinating from every cell and associated with *Pyrenophora*' (Alcorn, 1988). Two other *Pyrenophora* spp. commonly found in barley fields are *P. japonicum* and *P. graminea.* Both *P. japonicum* and *P. graminea* are closely related to *Pyrenophora teres* but can be distinguished from each other by their perithecia and conidia morphology as well as differential line susceptibility (Ito and Kuribayashi, 1931). In general, *P. graminea* is only found on the debris of previously infected plants, while *P. japonicum* and *Pyrenophora teres* spread to the stubble of uninfected plants and volunteer seedlings. The morphological similarities of *P. japonicum* and *Pyrenophora teres* and their ability to form genetically stable hybrids led to the two species being reclassified as interspecies of *Pyrenophora teres* (McDonald, 1967; Smedegård-Petersen, 1971). *Pyrenophora teres* was divided into *Pyrenophora teres* f. *teres* (*Ptt*) and *Pyrenophora teres* f. *maculata* (*Ptm*) based on their disease symptoms, causing net form net blotch and spot form net blotch, respectively (Smedegård-Petersen, 1971). Scott (1991) debated the genetic validity of two interspecies. However, subsequent studies showed that the two forms were not separate species but formed distinct genetic groups (Campbell et al., 1999, 2002; Crous et al., 1995; Lehmensiek et al., 2010; Leisova et al., 2005; Leišova et al., 2005; Louw et al., 1994, 1995; Rau et al., 2003, 2007; Serenius et al., 2005). Based on phylogenetic analysis, the wheat pathogen *P. tritici-repentis* (*Ptr*) is the closest ancestral relative (Bakonyi and Justesen, 2007; Leisova et al., 2005; Leišova et al., 2005; Rau et al., 2007). The two forms of *Pyrenophora teres* were calculated to have diverged from each other 519 kya (±30) and from *Ptr* 8.04 Mya (±138 kya) (Ellwood et al., 2012; Ohm et al., 2012; Syme et al., 2018).

3.2 Pyrenophora teres *infection of wheat*

The wheat pathogen *Ptr* is closely related to *Pyrenophora teres* and is the causal agent of the disease tan spot of wheat, a necrotic leaf spot disease similar to net blotch. Tan spot has the potential to reduce wheat yields by 50%, making it a disease of major concern (Wegulo, 2011). In recent years, both *Ptt* and *Ptm* have emerged as potential threats to wheat fields worldwide, with the first report of *Pyrenophora teres* being isolated from Canadian wheat in 1995 (see Table 2). From 1995–1997, ~1.2% of sampled Canadian wheat fields were found to harbor *Pyrenophora teres* seed infection (Turkington et al., 2002). Sampling of wheat fields in northwestern Russia and Hungary from 2005–2008 revealed *Pyrenophora teres* infections in ~20% of the wheat fields sampled (Mikhailova et al., 2010; Tóth et al., 2008). In Brazil, *Ptt* infection was identified on the wheat cv. Toruk in 2016 from lesions that resembled *Ptr* infection (Garozi et al., 2020). In addition, a recent report in Argentina found a 30%–80% incidence of *Ptm* in wheat fields, with symptoms again resembling those of *Ptr* infection (Perelló et al., 2019). The high levels of disease incidence and severity observed in Argentinian wheat fields led researchers to investigate genetic resistance to *Ptm* in wheat. Uranga et al. (2020) identified seven quantitative trait loci (QTLs) associated with *Ptm* resistance, contributing ~6–13% of the phenotypic variation. Together, these studies highlight the importance of monitoring both *Ptt* and *Ptm* as emerging pathogens affecting global wheat production, as the identification and control of novel emerging wheat pathogens is of paramount importance to worldwide food security.

Table 2 Host range of *P. teres*

Host(s)	Growing region	Conditions	Citation
Primary: *H. vulgare, H. Sponteneum*	Worldwide	Field, Greenhouse, Natural infection	N/A
H. hystrix, H. murinum, Bromus diandrus*	Australia	Field, Natural infection	Shipton, 1966, Linde and Smith 2019*
Genera: *Agropyron, Avena, Brachypodium, Bromus, Cyndon, Elymus, Hordelymus, Stipa, Triticum*	North America	Field, Greenhouse	Brown et al., 1993
Genera: *Dechampia, Festuca, Lolium, Phalaris*	North America	Field	Brown et al., 1993
Triticum aestivum (Common wheat - both spring and winter varieties)	Europe, North America, South America	Field, Natural infection	Turkington et al., 2002, Tóth et al., 2008, Mikhailova et al., 2010, Perello et al., 2019, Garozi et al., 2020

3.3 *Genetic structure and virulence diversity of* Pyrenophora teres

Given *Pyrenophora teres's* ability to rapidly overcome host resistances, it is paramount for researchers to continually monitor the population dynamics of both *Ptt* and *Ptm*. Wu et al. (2003) found that *Ptt* is more genetically diverse than *Ptm*, based on cluster analysis of restriction fragment length polymorphism markers. However, the sample size was small and skewed, with 23 *Ptt* isolates and eight *Ptm* isolates. A subsequent study to assess the genetic diversity of *Pyrenophora teres* utilized 128 *Ptt* isolates and 92 *Ptm* isolates using simple sequence repeat markers (Akhavan et al., 2016) found that *Ptt* isolates were more genetically diverse than *Ptm* isolates, supporting the claim from Wu et al. (2003). This contrasted with Rau et al. (2003), that found 65 *Ptm* isolates to be more diverse than 85 *Ptt* isolates using amplified fragment length polymorphism markers. The largest comparison used 167 *Ptt* isolates and 258 *Ptm* isolates, identifying 24 and 26 pathotypes, respectively (Çelík Oğuz and Karakaya, 2017), suggesting that *Ptm* is more diverse. However, this could be due to the skewed sampling. Wallwork et al. (2016) tested pathogen diversity using adult plant resistance (APR) and concluded that APR was more common than seedling resistance and that all of the 24 *Ptt* samples collected had a different virulence profile, suggesting a complex array of host–pathogen genetic interactions, and that the current ability to identify virulence profiles is limited by the number of differential barley lines.

Although artificial hybridization of *Ptt* and *Ptm* can be achieved under laboratory conditions (Campbell and Crous, 2003), they are considered genetically separate populations (Akhavan et al., 2016; Jalli, 2011; Rau et al., 2007) because natural hybrids are rare (Campbell et al., 2002; Leisova et al., 2005; McLean et al., 2014). Smedegård-Petersen (1976) was the first to report on the identification of *Ptt* × *Ptm* hybrids producing viable progeny. Smedegård-Petersen (1977a) crossed *Ptt* to *Ptm*, concluding that spot- and net-type lesions were controlled by two allelic independent genes, and not influenced by environmental factors. Four different classes of progeny isolates were observed when *Ptt* and *Ptm* were hybridized: spot form, net form, intermediate, and avirulent. Due to the haploid nature of *Pyrenophora*, progeny isolates segregated in a 1:1:1:1 manner for spot form, net form, intermediate, and avirulent phenotypes, or 1:1 when considered as parental:recombinant types. Additional support was provided by backcrossing intermediate forms to a respective net- or spot-parental type that resulted in progeny isolates that segregated 1:1 for a single parental phenotype or the intermediate type. Smedegård-Petersen (1977a) predicted that two genes would be present in *Pyrenophora teres*, one in each form, to result in either net- or spot- form. A natural barrier to fertility has been suggested to maintain separate *Pyrenophora teres* populations (Syme et al., 2018), as artificial hybrids are readily induced in a laboratory environment (Campbell and Crous, 2003; Smedegård-Petersen, 1977a). However, when over 1000 samples were analyzed from field experiments, they were all shown to be true-breeding forms (Poudel et al., 2019), although recently, a hybrid clonal population most closely related to *Ptm* isolates was detected in Western Australia (Lopez-Ruiz et al., 2020). Possible explanations for rare hybrid isolates are that some hybrid crosses are sterile (Minaříková and Polišenská, 1999), produce a reduced number of ascospores (Serenius et al., 2005) that are subsequently outcompeted, or we are simply not detecting hybrids genetically - since Akhavan et al. (2015) could not identify hybrids but identified intermediate phenotypes. These studies raise the question of whether current markers can adequately detect any residual level of hybridization.

4 *Pyrenophora teres*-barley pathosystem

4.1 *Genetics of barley reactions to* Pyrenophora teres *f.* teres

The first investigation of *Ptt* resistance in barley found that it was inherited as a single dominant gene (Geschele, 1928). Since that time, studies have identified dominant, recessive, and incomplete dominant resistances in this complex pathosystem. Schaller and Wiebe (1952) screened a globally diverse panel of 4526 barley lines to identify sources of *Ptt* resistance, finding a number of highly resistant lines of Manchurian origin, including Tifang, Harbin, and Canadian Lake Shore (Schaller and Wiebe, 1952). Further studies of a Tifang × Atlas population revealed a single resistance gene contributed by Tifang that was effective against naturally occurring Californian *Pyrenophora teres* isolates (Schaller, 1955). This locus was dubbed Pt_1, the first *Pyrenophora teres* resistance locus identified in barley. Mode and Schaller (1958) used a combination of crosses to identify Pt_2 and Pt_3, the two resistance loci harbored by barley lines CI4911 and Canadian Lake Shore. Additional studies using resistant lines CI5791, CI9825, and Canadian Lake Shore reported single dominant genes controlling *Pyrenophora teres* resistance, yet no loci were mapped (Gray, 1966; McDonald and Buchanon, 1962). Examination of the three established resistance loci using the Australian *Pyrenophora teres* isolate WA-2 revealed that Pt_1, Pt_2, and Pt_3 were, at least partially, isolate-specific resistances (Khan and Boyd, 1969b). Khan and Boyd (1969b) identified Pt_a as an independent locus conferring resistance to the WA-2 *Ptt* isolate. Additionally, this study reported that the genetic makeup of the resistant Manchurian lines differed, often making resistances unstable, leading the authors to favor resistance derived from the Ethiopian line CI5791 for net blotch resistance in Western Australia (Khan and Boyd, 1969b).

Khan and Boyd (1969b) also suggested that Pt_1 and Pt_2 were allelic, a claim that was given further support via trisomic analysis by Bockelman et al. (1977). This study used a susceptible trisomic line, Betzes, crossed by resistant lines CI9819 and CI7584, to investigate the heritability of known *Pyrenophora teres* resistance genes. This resulted in several shifts in resistance gene nomenclature. Pt_1 and Pt_2 were collapsed into the *Rpt1* locus on chromosome 3H (Fig. 1), with differing resistance alleles for Tifang (*Rpt1.a*) and CI9819 (*Rpt1.b*). The additional resistance loci identified in this study were *Rpt2.c* and *Rpt3.d*, contributed by CI9819 and CI7384 (Fig. 1), respectively (Bockelman et al., 1977). Later, additional loci *Pt,,a*, *Qrpts3L*, and *QNFNBSLR.AI/S-3H* were collapsed under *Rpt1* on the long arm of chromosome 3H (Graner et al., 1996; Lehmensiek et al., 2007; Raman et al., 2003). The *Rpt2* locus identified by Bockelman et al. (1977) was originally claimed to be located on chromosome

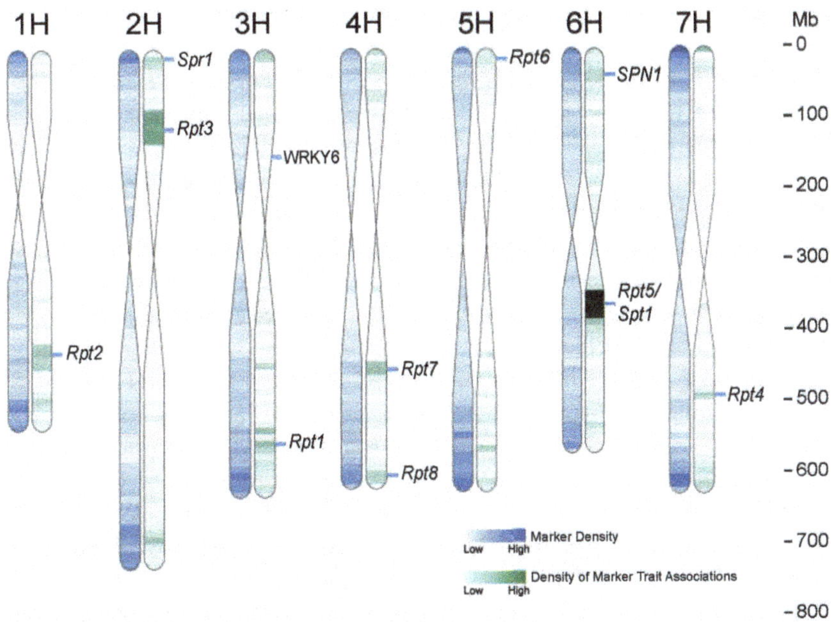

Figure 1 Chromosome karyotype map showing SNP marker density, marker trait associations, and designated resistance/susceptibility loci. The marker density (blue heatmap on left chromosome) depicts the Morex barley genome with marker density for the 50k iSelect SNP Array. The marker trait associations (green heatmap on right chromosome) show the number of studies that report marker trait association loci for either *Pyrenophora teres* f. *maculate* or *P. teres* f. *teres* resistance/susceptibility based on the absolute position of reported marker or collapsed into the locus, by Clare et al. (2020)], within 10 Mbp windows. Previously designated resistance/susceptibility loci are annotated to the right of each chromosome pair. The scale in Mbp is shown to the far right. The chromosome figures were constructed using the *RIdeogram 0.2.2* package in *R 3.6.3*.

5H; however, further studies revealed its actual location on chromosome 1H (Fig. 1), with the additional loci *Rpt-1H-5-6* and *QNFNBAPR.Ar/F-1H* being nested under *Rpt2* (Lehmensiek et al., 2007; Manninen et al., 2006; Yun et al., 2005). Additional loci that are considered synonymous with *Rpt3* include *Pt3*, *QRPtts2*, and several unnamed QTLs assigned to chromosome 2H (Fig. 1; Cakir et al., 2011; Emebiri et al., 2005; Grewal et al., 2012; Mode and Schaller, 1958; Raman et al., 2003; Steffenson et al., 1996).

The Ethiopian line CI9819 was found to harbor an additional resistance locus, *Rpt5*, first identified in a cross with susceptible line Rolfi (Manninen et al., 2006). The Pt_a locus identified by Khan and Boyd (1969b) has also been reclassified as *Rpt5*. This locus, located near the centromere of chromosome 6H (Fig. 1), has been reported in an abundance of studies and harbors multiple alleles, some of which confer broad resistance against globally diverse *Ptt* isolates (Abu Qamar et al., 2008; Adhikari et al., 2019; Cakir et al., 2011; Emebiri et al., 2005; Grewal et al., 2012; Gupta et al., 2011; Koladia et al., 2017a; Liu et al., 2015; Manninen et al., 2000, 2006; Novakazi et al., 2019; Raman et al., 2003; Richards et al., 2016; Steffenson et al., 1996). Published alleles for *Rpt5* include *Rpt5.f*, harbored by CI5791 and CI9819, which confers robust, broad resistance against globally diverse isolates of *Ptt* (Clare et al., 2020; Koladia et al., 2017a; Manninen et al., 2006). The resistance harbored by CI5791 and CI9819 at this locus appears to act through an early resistance mechanism and is effective against nearly all global *Ptt* isolates, highlighting the importance of characterizing this complex locus (Jalli and Robinson, 2000; Koladia et al., 2017a; Manninen et al., 2006). The other loci mapped to this region that may represent additional *Rpt5* alleles include *Pt,,d, QRpts6L, QRpt, QRpt6*, $Rpt_{CIho2291/Nomini}$, *QRPtm6-2, QPt.6H-1/2, NBP_QRPtt6-1, QRPtta-6H-49.79, QRPtt.6H-54-55, Qns-6H.5, Qnfnb-6H.1-4, Qsfnb-6H*, and *QRPtm-6H-55-64* (summarized in Table 3) (reviewed in Clare et al., 2020). The sheer number of loci mapped to this region demonstrates the importance and complexity of this locus in the barley–*Pyrenophora teres* pathosystem. In addition, Gyawali et al. (2020) screened the ICARDA AM-2014 Panel for resistance to two North Dakotan *Ptt* isolates in a variety of field conditions and plant stages; however, they did not map any loci. Five of the cultivars screened (~1.4%) were resistant to both isolates, including CI5791, which averaged a highly resistant score of 1.25. This again highlights the importance of this Ethiopian landrace as a source of robust resistance for barley breeders.

While multiple studies utilizing diverse populations and pathogen isolates have characterized dominant resistances mapping to the *Rpt5* locus, it has also been shown to harbor dominant susceptibility genes as well. Abu Qamar et al. (2008) took advantage of the reciprocal susceptibility observed with cultivars Rika and Kombar challenged with Californian *Ptt* isolates 6A and 15A to map the isolate-specific susceptibility factors in these two lines. These susceptibility loci

Table 3 Summary of current designated barley genes defined in the *Pyrenophora teres* f. *teres* and *P. teres* f. *maculata*–barley pathosystems

Designated locus (Ref)*	Current synonym (Ref)*	Previous synonyms (Ref)*	Known alleles	Location	Form
Rpt1 (Bockelman et al., 1977)		*Pt_1* (Mode and Schafer, 1958) *Pt_2* (Mode and Schafer, 1958) *Pt1a/Rpt1a* (Bockelman et al., 1977) *Pt1b/Rpt1b* (Bockelman et al., 1977) *Pt,,a* (Graner et al., 1996) *QRpts3L* (Raman et al., 2003) *QNFNBSLR.Al/S-3H* (Lehmensiek et al., 2007) *QNFNBSLR.W/Al-3H* (Lehmensiek et al., 2007) *QNFNBAPR.Al/S-3H* (Lehmensiek et al., 2007) *QNFNBAPR.W/Al-3H* (Lehmensiek et al., 2007) QRptts-3H-106.96 (Adhikari et al., 2020)	*Rpt1.a* *Rpt1.b*	3H	*P. teres* f. *teres*
Rpt2 (Bockelman et al. 1977)		*Pt2c/Rpt2c* (Bockelman et al., 1977) *Rpt-1H-5-6* # (Yun et al., 2005) *QNFNBAPR.Ar/F-1H* # (Lehmensiek et al., 2007)	*Rpt2.c*	1H	*P. teres* f. *teres*
Rpt3 (Bockelman et al., 1977)		*Pt_3* (Mode and Schaller, 1958) *Pt.d/Pt3d/Rpt3d* (Bockelman et al., 1977) *QRpts2S* (Raman et al., 2003) *QRpts2Sa* (Raman et al., 2003) *QRptts2* (Grewal et al., 2008) *QNFNBSLR.Al/S-2H* (Lehmensiek et al., 2007) *QNFNBSLR.W/Al-2H* (Lehmensiek et al., 2007) *QNFNBSLR.Ar/F-2Ha* (Lehmensiek et al., 2007) *QNFNBAPR.Ar/F-2H* (Lehmensiek et al., 2007) *QPt.2H-1/2* (Vatter et al., 2017)	*Rpt3.d*	2H	***P. teres* f. *teres*** *P. teres* f. *maculata*

(*Continued*)

Table 3 (*Continued*)

Designated locus (Ref)*	Current synonym (Ref)*	Previous synonyms (Ref)*	Known alleles	Location	Form
Rpt4 (Williams et al., 1999, 2003)		*QRpt7* (Grewal et al., 2008) *QNFNBAPR.Al/S-7Hb* # (Lehmensiek et al., 2007) *QNFNBAPR.W/Al-7Hb* # (Lehmensiek et al., 2007) *QRpt7* (Tamang et al., 2015) *QRptm7-3/4/5/6/7/8* (Wang et al., 2015) *NBP_QRPtt7-2* (Wonneberger et al., 2017b) *Qns-7H.2* (Daba et al., 2019) *QRptm-7H-119-137* (Tamang et al., 2019)	*Rpt4.e*	7H	*P. teres* f. *teres* ***P. teres* f. *maculata***
Rpt5 (Manninen et al., 2006)	*Spt1* (Richards et al., 2016)	Pt_a (Khan and Boyd, 1969a) *Pt,,d* (Graner et al., 1996) *QRpts6L* (Raman et al., 2003) *QRpt* (Emebiri et al., 2005) *rpt.r/rpt.r* (Abu Qamar et al., 2008) *QRpt6* (Grewal et al., 2008, 2012) $Rpt_{Ciho2291/Nomini}$ (O'Boyle et al., 2014) *QRptm6-2* (Wang et al., 2015) *QPt.6H-1/2* (Vatter et al., 2017) *NBP_QRPtt6-1* (Wonneberger et al., 2017b) *QRptta-6H-49.79* (Amezrou et al., 2018) *QRptt.6H-54-55* (Amezrou et al., 2018) *Qns-6H.5* (Daba et al., 2019) *Qnfnb-6H.1-4* (Daba et al., 2019) *Qsfnb-6H* (Daba et al., 2019) *QRptm-6H-55-64* (Tamang et al., 2019) *QRptts-6H-175.80* (Adhikari et al., 2019) QRptts_6H_57.64-60.21 (Adhikari et al., 2020)	*Rpt5.f* *Spt1.k* *Spt1.r*	6H	***P. teres* f. *teres*** *P. teres* f. *maculata*

Rpt6 (Manninen et al., 2006)	-	*Rpt6.g*	5H	*P. teres* f. *teres* ***P. teres* f. *maculata***
Rpt7 (Franckowiak and Platz, 2013)	*QRpts4* (Raman et al., 2003)† *Rpt-4H-5-7* (Yun et al., 2005) *QNFNBAPR.Al/S-4Ha* # (Lehmensiek et al., 2007) *AL_QRptt4-1* (Wonneberger et al., 2017a) *QPt.4H-3* # (Vatter et al., 2017) *QRptm-4H-58-64* (Tamang et al., 2019) QRptts_4H_53.87 (Adhikari et al. 2019)	*Rpt7.h*	4HL	***P. teres* f. *teres*** *P. teres* f. *maculata*
Rpt8 (Franckowiak and Platz, 2013)	QTL on 4H (Friesen et al., 2006a,b) *QRpts4* # (Grewal et al., 2008)† *Qns-4H.2* (Daba et al., 2019)	*Rpt8.j*	4HS	***P. teres* f. *maculata***

*First paper to use gene designation.
#Not confirmed.
†Multiple papers use this designation.
Bold indicates the predominant form the resistance is associated with.

were identified in a Rika × Kombar doubled haploid (DH) mapping population, that delimited the region to 5.9 cM near the centromere on chromosome 6H. The two loci were originally thought to be different tightly linked genes in repulsion (Abu Qamar et al., 2008) and were originally given the recessive resistance gene nomenclature resistance to *Pyrenophora teres* Rika and Kombar (*rpt.r* and *rpt.k*) because only homozygous individuals at the locus provided resistance. Additional markers were used to saturate the *rpt* region within the Rika x Kombar DH population, further delimiting the loci to 1.6 cM (Liu et al., 2010). This study took advantage of the synteny between barley chromosome 6H and chromosomes 2 and 3 of rice and *Brachypodium*, respectively. These syntenic relationships revealed two potential candidate genes within the delimited region, both LRR-kinases with a typical resistance gene structure. A later reexamination of the nature of necrotrophic pathogen dynamics led to the renaming of these loci *Spt1.r/Spt1.k*, as functional susceptibility genes required for the pathogen to cause disease (Liu et al., 2015; Richards et al., 2016). Richards et al. (2016) went on to fine-map this region, screening 2976 Rika x Kombar recombinant gametes to delimit the region to ~0.24 cM. *Brachypodium* synteny was again leveraged to construct a partial physical map of the 6H region in barley. This region corresponded to a ~466 kb region of the *Brachypodium* genome, containing 62 annotated genes, and approximately 9.5 MB of the barley genome, containing 39 annotated genes. Six receptor-like genes were present in the region of the barley genome; however, only one displayed polymorphism between the parental lines Rika and Kombar. This lead to the hypothesis that rather than two genes linked in repulsion, the effects of *Spt1.r/Spt1.k* were perhaps mediated by two alleles of the same gene (Richards et al., 2016).

An investigation utilizing the *Ptt* effector *PttNE1* also supported interactions between the *Ptt* and barley at chromosome 6H (Liu et al., 2015). The use of intracellular wash fluid infiltration revealed necrosis on susceptible parent Hector, while necrosis was absent on resistant parent NDB-112. Mapping the interaction using 118 Hector x NBD-112 recombinant inbred lines (RILs) leads to a strong peak over the 6H centromeric region, explaining 30% of the disease variability. This locus was dubbed *SPN1*, which has been reported as nested underneath the *Rpt5/Spt1* locus, although *SPN1* maps differently than the fine-mapping reported by Richards et al. (2016). This deviation exemplifies the complexity of the numerous *Ptt* resistance/susceptibility loci that map to overlapping genomic regions, making distinctions and nomenclature problematic (Clare et al., 2020). Analysis of F_2 progeny showed a 1:3 resistant-to-susceptible ratio, indicating a single gene operating in an inverse gene-for-gene manner where recognition of a necrotrophic effector results in susceptibility, which is common with necrotrophic pathogens (Faris and Friesen, 2020; Faris et al., 2010; Friesen et al., 2007; Liu et al., 2015). Early on, these loci were given recessive-resistance gene nomenclatures, but this trend is changing as we have begun to understand necrotrophic effector-triggered susceptibility and the inverse gene-for-gene model rather than viewing these loci as resistance loci, as was the dogma followed due to our earlier understanding of typical plant innate immune responses against biotrophic pathogens.

An additional resistance locus, *Rpt7*, was identified in the English malting barley Halcyon and has been mapped to the long arm of chromosome 4H (Fig. 1; Raman et al., 2003; Read et al., 2003). Several other resistance loci have been identified on chromosome 4H, including *Rpt-4H-5-7, QNFNBAPR. AVS-4Ha, AL_QRPtt4-1*, and *Qpt.4H-3*; however, it remains unclear whether these loci are part of the *Rpt7* locus (Clare et al., 2020; Lehmensiek et al., 2007; Vatter et al., 2017; Wonneberger et al., 2017a; Yun et al., 2005). Further high-resolution mapping efforts would be required to better understand the relationships between these loci.

While most early research into the genetic interplay between barley and *Ptt* was performed using biparental populations, there has been a recent increase in genome-wide associated studies (GWAS) aiming to clarify the important loci at play in this interaction. GWAS eliminates the need to generate populations and enables the use of globally diverse panels that capture more diversity than a typical biparental population. Since 2017, a total of eight GWAS manuscripts have been published in this pathosystem (Adhikari et al., 2019, 2020; Amezrou et al., 2018; Daba et al., 2019; Novakazi et al., 2019; Richards et al., 2017; Rozanova et al., 2019; Wonneberger et al., 2017a). Barley panels used to explore resistance to globally diverse *Ptt* isolates include the Barley Core Collection (Muñoz-Amatriaín et al., 2014; Richards et al., 2017), the Nordic Barley Panel (Wonneberger et al., 2017b), the ICARDA AM-2014 Panel (Amezrou et al., 2018), a combined Ethiopian, ICARDA, and North Dakota State University (NDSU) Barley Panel (Daba et al., 2019), the Siberian Barley Panel (Rozanova et al., 2019), the Ethiopian and Eritrean Barley Collection (Adhikari et al., 2019; Mamo and Steffenson, 2015), the Vavilov Research Institute Collection (Novakazi et al., 2019), and the Barley CAP Breeders Panel comprised of 3490 elite barley lines from ten United States breeding programs (Adhikari et al., 2020). The analysis of *Ptt* resistance in these diverse panels revealed between 7 and 31 independent significant markers for each GWAS performed. Each of these studies identified at least one significant marker for NFNB resistance near the *Rpt5/Spt1* locus (Fig. 1), with many having a majority of significant markers localized to this region.

The convergence of both GWAS and biparental studies on the 6H region harboring *Rpt5/Spt1* demonstrates the importance of this region and the resistances harbored by CI5791 and CI9819. While early analysis of this region had suggested a single potential candidate gene for *Spt1* (Richards et al., 2016), new annotations of the Morex reference assembly (Monat et al., 2019) reveal five polymorphic genes encoding protein kinases or receptor-like proteins (RLPs) within the delimited region that may confer isolate-specific susceptibility. Interestingly, recent high-resolution mapping efforts using 1920 Tifang x CI5791 recombinant gametes have delimited the region harboring *Rpt5* to the same 3 MB region harboring *Spt1* (Effertz et al. unpublished). This, combined with the region's enrichment for resistance-like genes (~14% compared to ~1.4% for the entire genome), highlights this region as an area where pathogen pressure has potentially driven host evolution (Andersen et al., 2016). Current validation efforts are underway to determine the causal genes underlying both *Rpt5* and *Spt1*, as well as the evolutionary mechanisms driving diversification at this unique and highly important locus (Brueggeman et al., unpublished).

A recent study demonstrated the importance of the transcription factor *HvWRKY6* in *Rpt5*-mediated *Ptt* resistance in the Ethiopian landrace CI5791 (Brueggeman et al., unpublished). The WRKY class of transcription factors is one of the largest in plants, and members have been implicated in both activation and repression of immune responses, with roles in PTI, ETI, and acquired systemic resistance. An ortholog of *HvWRKY6*, *Arabidopsis AtWRKY6*, has been shown to regulate defense responses to *Pseudomonas syringae* p.v. *tomato* (Robatzek et al., 2002). Additionally, wheat *TaWRKY70* contributes to fusarium head blight (FHB) resistance by regulating downstream resistance genes *TaACT*, *TaDGK*, and *TaGLI* (Kage et al., 2017). *HvWRKY6* was discovered by screening ~10 000 gamma irradiated M_2 CI5791 seedlings with *Ptt* isolate 0-1. Two independent mutants, CI5791-γ3 and CI5791-γ8, which had a single base pair deletion and a complete gene deletion of *HvWRKY6*, respectively, displayed a major shift toward susceptibility (disease scores = 6.5, 6.0) compared to wild-type (disease score = 1.5). The qPCR and RNA-seq data revealed temporal differences in *HvWRKY6* expression, with resistant plants showing an early upregulation and susceptible plants showing a later upregulation. Mapping of the mutant lines revealed that *HvWRKY6* lies directly underneath a resistance QTL on chromosome 3H, identified in a CI5791 x Tifang population (Koladia et al., 2017a), although allele analysis revealed no polymorphism between the two parents (Brueggeman et al., unpublished). Interestingly, these *HvWRKY6* mutants also have compromised resistance against *Ptm* isolates as well. While further research is necessary to explore these mutant lines, this study represents an important step in determining the molecular mechanisms underlying the robust *Ptt* resistance contributed by line CI5791.

4.2 Genetics of barley reactions to Pyrenophora teres f. maculate

The genetics of barley reactions to *Ptm* was initially thought to be simplistic in comparison to *Ptt*, containing only three major genes designated *Rpt4*,

Rpt6, and *Rpt8* (Fig. 1; Clare et al., 2020). However, due to the continuous nature of *Ptm* infection responses, the full genetic architecture was most likely not captured (Burlakoti et al., 2017). To date, 36 loci have been implicated in the barley response to *Ptm* infection, five of which are specific to *Ptm* and 31 showing potential overlap with *Ptt* loci (summarized in Table 3) (Clare et al., 2020), greatly exceeding the original three major loci.

The first resistance locus implicated in the *Ptm*-barley pathosystem was undesignated and contributed by Léger (Ho et al., 1996; Molnar et al., 2000), possibly representing *Rpt3;* however, this is unconfirmed (Clare et al., 2020). The first major designated locus, *Rpt4,* was identified shortly after by Williams et al. (1999) on chromosome 7H of Galleon (Fig. 1), followed by CI9214, Keel, Tilga, Chebec (Williams et al., 2003), PI67381, PI84314 (Tamang et al., 2019), possibly TR251 (Grewal et al., 2008), and Baudin (Cakir et al., 2011) and subsequent GWAS by Tamang et al. (2015), Wang et al. (2015), and Daba et al. (2019). The *Rpt4* locus has been used in Australian breeding programs. However, the lines had lower resistance than expected, leading to the discovery of adult plant resistance loci by Williams et al. (2003). The second major designated locus, *Rpt6*, was identified (Manninen et al., 2006) on chromosome 5H (Fig. 1) of CI9819 and one GWAS (Tamang et al., 2015). The third and last major designated locus, *Rpt8*, was identified by Friesen et al. (2006a) on chromosome 4H of Q21861 (Fig. 1; Franckowiak and Platz, 2013) and subsequent GWAS (Daba et al., 2019; Tamang et al., 2015).

Rpt1 (Burlakoti et al., 2017; Tamang et al., 2015), *Rpt5/Spt1* (Daba et al., 2019; Grewal et al., 2008; Tamang et al., 2015; Wang et al., 2015), and *Rpt7* (Grewal et al., 2008; Tamang et al., 2015) are predominantly associated with *Ptt* infection but have also been identified with *Ptm* isolates (Clare et al., 2020). *Rpt1* was potentially identified by Tamang et al. (2015) against the New Zealand isolate *Ptm* NZKF2 and the North Dakota isolate FGOB10*Ptm*-1, and by Burlakoti et al. (2017) against the Montana isolate SNFB-MT09. The *Rpt5/Spt1* locus has been identified in the *Ptm* interaction in one biparental mapping study (Grewal et al., 2008) and three GWAS (Daba et al., 2019; Tamang et al., 2015, 2019; Wang et al., 2015). *Rpt5/Spt1* was identified as *QRPtm-6H-55-64* by Tamang et al. (2019) against FGOB10*Ptm*-1, the Montana isolate PA14, the Danish isolate DEN2.6 and NZKF2, confirming the previous GWAS with the same isolates except PA14 (Tamang et al., 2015). *Rpt7* was potentially identified by Grewal et al. (2008) against the Canadian isolate WRS857 and Tamang et al. (2015) against isolates DEN2.6, NZKF2 and the Australian isolate SG1.

In addition, GWAS by multiple groups have revealed that the genetics involved in *Ptm* interaction are not as simple as anticipated (Burlakoti et al., 2017; Daba et al., 2019; Tamang et al., 2015; Wang et al., 2015). Over 29 further loci have been implicated in association with *Ptm* infection, in addition to the loci previously described above. However, this is likely an underestimate as not all loci could be anchored to the barley genome by Clare et al. (2020). Broad-range resistance has been identified in multiple lines, with fifteen lines from the barley core collection resistant to four *Ptm* isolates (Neupane et al., 2015), thirteen landraces resistant to three *Ptm* isolates (Çelik Oğuz et al., 2017), and twelve lines from the ICARDA AM-2014 panel providing stable resistance across hot spot environments to one isolate (Gyawali et al., 2019). Analysis to identify these loci and alleles will be important in assessing which genotypes will be the most useful for introgression in breeding programs.

As mentioned earlier, both *Ptt* and *Ptm* have emerged as potential threats to wheat production, with limited research showing that *Ptm* more readily infects wheat than *Ptt* (Ondrej, 2000; Leisova-Svobodova et al., 2010). Recent work to characterize *Ptm* resistance loci in wheat found partial overlap with *Ptr* resistance loci, leaving only three of the seven genomic regions in wheat as being novel (Uranga et al., 2020). The study did not conduct syntenic relationship comparisons of the wheat *Ptm* resistance loci to barley loci, but this would be an interesting comparison. Interestingly, the collapsed loci *SFNB-2H-8-10* designated by Burlakoti et al. (2017), but first identified by Tamang et al. (2015; Clare et al., 2020), is in the same region on barley chromosome 2H as the recently identified *Spr1* (Fig. 1), the first locus mapped in barley conferring dominant susceptibility to *Ptr* which lies in a syntenic position to the *Tsc2,* sensitivity to the ToxB effector of *Ptr,* region of wheat (Wei et al., 2020). Whether this is the same gene being targeted by *Ptm*, *Ptt*, and *Ptr* is still to be validated, but poses interesting evolutionary questions involving multiple closely related host and pathogen species.

4.3 Genetics of *Pyrenophora teres* effectors on barley

The first effectors identified in the *Ptt*-barley system were secondary metabolites and include four toxins (N-(2-amino-2-carboxyethyl)-aspartic acid [Toxin A], anhydroaspergillomarasmine A [Toxin B], aspergillomarasmine A [Toxin C], and aspergillomarasmine B [Toxin D]) of the marasmine class, which chelate iron ions (Bach et al., 1979; Friis et al., 1991; Sarpeleh et al., 2008, 2009; Smedegård-Petersen, 1977b). These toxins were initially used to screen for host resistance (Sharma, 1984; Weiergang et al., 2002) as they were shown to be associated with chlorosis (Keon and Hargreaves, 1983; Sarpeleh et al., 2007). However, these secondary metabolites do not appear to show isolate or host specificity (Sarpeleh et al., 2007). Recent studies to map biosynthetic gene clusters (BGCs) that are responsible for secondary metabolites have been undertaken in the *Pyrenophora* species (Moolhuijzen et al., 2020). *Ptt* was found to have higher than expected non-ribosomal peptide synthase-indole BGCs, postulated to be due to the expansion of transposable elements within *Ptt* (Moolhuijzen et al., 2020). *Ptm*, *Ptt*, and *Ptr* were all found to have a conserved type III polyketide synthase BGC, whereas *Ptm* and *Ptt* were both found to be lacking the triticone BGC (Moolhuijzen et al., 2020).

Although some secondary metabolites can cause necrosis, such as aspergillomarasmine A, necrosis has predominately been associated with proteinaceous effectors (Sarpeleh et al., 2007). The proteinaceous ToxA that is present in *Ptr, Parastagonospora nodorum and Biploris sorokiniana* (Friesen et al., 2006b, 2018; McDonald et al., 2018, 2019) may also be present in *Pyrenophora teres* (Leisova-Svobodova et al., 2010; Mikhailova et al., 2010), but these results need further confirmation before speculation can be made on the function of ToxA in *Pyrenophora teres*. ToxA may have been one of the proteins unknowingly identified by Sarpeleh et al. (2008), but this isolate would need to be sequenced for confirmation. The first confirmed proteinaceous effector within the *Ptt* interaction was *Ptt*NE1, isolated from Canadian isolate 0-1 that was associated with necrosis on the barley line Hector, interacting with the *SPN1* locus on chromosome 6H (Fig. 1; Liu et al., 2015). Additional proteins have been postulated to play a role in the *Ptt*-barley pathosystem including proteins involved in the transition from colonization to necrotrophy at approximately 48 hr post infection, based on differential gene expression analysis, but the functional analysis of these genes needs to be carried forward for validation (Ismail and Able, 2016, 2017; Ismail et al., 2014). Virulent *Ptt* isolates have also been shown to have increased protease production than less virulent isolates, which may allow faster establishment and nutrient acquisition (Dikilitas et al., 2018).

The first mapped effector QTL was the avirulence effector *AvrHar* using an isolate 15A × 0-1 mapping population, with the 15A allele conferring avirulence on the barley line Harbin (Weiland et al., 1999). In another 15A × 0–1 mapping population, the *AvrPra2* locus also mapped to the *AvrHar* locus, with isolate 0–1 conferring avirulence on barley line Prato (Lai et al., 2007). Lai et al. (2007) also identified *AvrPra* that conferred avirulence on Prato from isolate 0-1. The last avirulence locus detected was $Avr_{Heartland}$, identified in an isolate WRS 1906 × WRS 1607 population to map the avirulence on barley variety Heartland (Beattie et al., 2007). Additional avirulence loci and the added caveat of suppressor loci have been reported (Afanasenko et al., 2007; Mironenko et al., 2005) but have not been mapped; therefore it is not known if they correspond to other identified loci or are novel. As genetic analysis of bi-parental pathogen populations switched to the more robust restriction-site associated-digest–genotyping-by-sequencing (RAD-GBS) there was a coincident acceleration of the identification of virulence loci (Carlsen et al., 2017; Koladia et al., 2017b; Shjerve et al., 2014). Four virulence loci given the designation *VK1*, *VK2*, *VR1*, and *VR2* (Virulence on Kombar 1, Kombar 2, Rika 1 and Rika 2) were identified utilizing a bi-parental population developed from a cross between the isolates 15A and 6A. Interestingly, all four loci mapped to separate loci within the *Ptt* genome, yet all targeted the *Rpt5/Spt1* locus described earlier (Shjerve et al., 2014). Each isolate contributed two of the loci with reciprocal virulence, in that 15A possessed *VK1* and *VK2*, conferring virulence on Kombar and avirulence on Rika, whereas 6A possessed *VR1* and *VR2*, conferring virulence on Rika and avirulence on Kombar (Shjerve et al., 2014). Additional loci implicated in the *Ptt* interaction have included nine novel loci contributed by the North Dakotan *Ptt* isolate FGOH04*Ptt*-21 and one by the Danish isolate BB25 identified in an FGOH04*Ptt*-21 × BB25 mapping population (Koladia et al., 2017b). In the only *Ptm* effector mapping study to date, utilizing a bi-parental mapping population of the US isolate FGOB10*Ptm*-1 crossed with the Australian isolate SG1, FGOB10*Ptm*-1 contributed five virulence loci and SG1 contributed one locus (Carlsen et al., 2017).

These loci were recently summarized by Clare et al., (2020) to a total of twenty-three *Pyrenophora teres* effector loci, with seventeen loci specific to *Ptt* (Beattie et al., 2007; Koladia et al., 2017b; Lai et al., 2007; Shjerve et al., 2014; Weiland et al., 1999) and six *Ptm* specific loci (Carlsen et al., 2017). However, the first and only GWAS into the genetics of *Ptt* effectors identified fourteen unique genomic loci, with four confirmed with additional biparental mapping (Martin et al., 2020). Five QTLs were identified on Beecher, three on Fleet, two on Harbin, Herta, Maritime, Prior, Skiff and Yerong, and one on Clho11458, Commander, Corvette, Kombar, and Tallon (Martin et al., 2020). Seven and four QTLs were present on chromosomes 3 and 5, respectively (Martin et al., 2020). The number of predicated effectors based on EffectorP ranged from zero to three for a total of 20 effectors (Martin et al., 2020); however, EffectorP is known to be overly stringent, thus the number could be higher. In addition, two biosynthetic gene clusters were identified in close proximity to two of the QTLs on chromosome 3 (Martin et al., 2020). It is currently unknown which, if any of the loci currently identified, correspond to previous loci, for example, the virulence QTL against Beecher on chromosome 3 is approximately 58 kbp from the previously identified *VK1* locus (Martin et al., 2020; Shjerve et al., 2014). This brings the current total of *Pyrenophora teres* effector loci to 37 (Clare et al., 2020; Martin et al., 2020).

Mutations in fungi occur from the rapid reproductive cycle and large population size that select upon the fittest individual isolates (Möller and Stukenbrock, 2017; Rau et al., 2015). In addition, repeat-induced point (RIP) mutation leakage, where RIP Defective targets repetitive genomic regions and inadvertent adjacent regions allow the rapid accumulation of polymorphisms and potential beneficial effector recognitions or neo-functionalization (Oliver, 2012). There is an increased likelihood that a gene is affected by RIP the closer the gene is located to a repeat element (Ohm et al., 2012). These processes, coupled with sexual recombination, provide the pathogen with genome mutational mechanisms to generate diversity to rapidly overcome deployed resistances. Thus, host resistance or lack of susceptibility genes/loci must be intelligently deployed to maintain resistance.

5 Downstream responses of barley to *Pyrenophora teres* infection

The first reported physiological response of barley to *Pyrenophora teres* infection was increased respiration of susceptible plants (Smedegård-Petersen, 1977c). Downstream of initial recognition, multiple genes were found to be induced during *Ptt* infection of barley leaves that were also found to be induced by *Puccinia hordei* (causal agent of barley leaf rust) and *Erysiphe* (*Blumeria*) *graminis* (causal agent of barley powdery mildew) infection (Reiss and Bryngelsson, 1996). The genes found to be upregulated by *Ptt* included pathogenesis-related (PR) proteins such as PR1a and PR1b, 1,3-ß-glucanase (PR2), chitinases (PR3, 4, 8, 11), peroxidases (PR9) and several thaumatin-like (PR5) proteins (Reiss and Bryngelsson, 1996; Reiss and Horstmann, 2001). These upregulated genes were also identified in leaves that were treated with culture filtrate containing *Ptt* effectors (Reiss and Bryngelsson, 1996). In addition to the above PR proteins (PR1-5, 9), lipid transferases (PR14) and an oxalate oxidase-like protein (PR17) have also been identified (Hassett et al., 2020). The PR1, PR2, and PR5 protein families are all shown to be upregulated by the salicylic acid (SA) pathway during plant defense responses (Breen et al., 2017). Bogacki et al. (2008) found similar results as Reiss and Bryngelsson (1996), identifying 28 defense response genes that were differentially expressed in resistant reactions to *Ptt* and *Ptm*. The 28 differentially expressed genes related to carbohydrate metabolism, detoxification, signal transduction, calcium signal perception, and stress/defense response (including PR1 and chitinases), with some differentially regulated genes being of unknown function (Bogacki et al., 2008). The SA-dependent signaling pathway is often associated with the production of antimicrobial compounds, such as the previously mentioned pathogenesis-related proteins and phytoalexins, and additionally a hypersensitive response that can lead to an increase in reactive oxygen species (ROS) production and programmed cell death (PCD) (Pieterse et al., 2012). Therefore, the SA-dependent signaling pathway is associated with resistance to biotrophic pathogens and susceptibility to necrotrophic pathogens that hijack the typical disease resistance signaling pathway in an inverse gene-for-gene relationship (Pieterse et al., 2012). Alternatively, jasmonic acid and ethylene-dependent signaling pathways are typically thought to activate defense responses to necrotrophic pathogens (Pieterse et al., 2012).

One gene implicated in the downstream resistance response of barley to *Pyrenophora teres* is the *PR2* gene (Al-Daoude et al., 2018; Arabi et al., 2020). In both the susceptible line WI2291 and the resistant line Banteng, SA signaling was activated 24 hours post inoculation with *Ptt*. However, the SA levels in the resistant line Banteng increased three-fold in comparison to the susceptible line WI2291 (Al-Daoude et al., 2018). In addition, the *PR2* gene was significantly increased in response to SA, with a two-fold increase in the susceptible line and a six-fold increase in the resistant line, totaling a three-fold increase over the susceptible line by 96 hours post inoculation with *Ptt* (Al-Daoude et al., 2018). This is similar to a second study using Banteng that found a 6.4-fold increase in *PR2* expression (Arabi et al., 2020). The *PR2* gene encodes a 1,3-ß-glucanase that hydrolyses the ß-O-glycosidic bond of ß-glucan in the plant cell walls that allows the expansion and loosening of cells, resulting in cell wall leakage (Al-Daoude et al., 2018). Paradoxically, the induction of *PR2* in the resistant line Banteng should theoretically result in a susceptible response to *Ptt*, a necrotrophic pathogen that induces cell lysis. However, a sufficient early defense response to sequester the pathogen early in the infection process could be explained by the faster and higher constitutive expression of *PR2* in the resistant line Banteng (Al-Daoude et al., 2018). In addition, ß-glucan hydrolysis could facilitate cell wall remodeling and the release of antimicrobial compounds or endogenous substrates to further induce defenses that are harbored in the cell wall (Oide et al., 2013). It remains to be seen whether the early response resistance activated by *Rpt5/Spt1* in CI5791 is through the same mechanism of activating the SA-dependent pathway and *PR2* expression found in Banteng. A second gene, *PAL,* has also been shown to have higher constitutive and induced expression in the resistant line Banteng, compared to a susceptible line Furat 1, when infected with *Pyrenophora teres* (Arabi et al., 2020). Although the study did not state if this induced *PAL* expression was due to SA-signaling. *PAL* is a phenylalanine ammonia lyase, an enzyme involved in defense through the production of important secondary metabolites (Huang et al., 2010). The gene, *S40,* was found to be induced by both SA and by JA in leaf areas that were undergoing chlorosis and necrosis from *Pyrenophora teres* (Krupinska et al., 2002). The induction of *S40* from the JA-signaling pathway and the SA-signaling pathway occurred after 16 h and 40 h, respectively (Krupinska et al., 2002). Krupinska et al. (2002) concluded that the S40 protein is predicted to be 15.4 kD and to contain two nuclear localization sequences allowing predominant accumulation in the nucleus of mesophyll cells, and plays a key role in the breakdown of chloroplast during senescence. Krupinska et al. (2002) suggested that the impairment of the photosynthetic apparatus leads to production of ROS, which in turn activates JA-signaling to upregulate *S40*, which will breakdown chloroplasts and lead to chlorosis.

Reactive oxygen species (ROS) including singlet oxygen (O_2), hydroxyl radical (OH-), superoxide (O_2-/HO_2), and hydrogen peroxide (H_2O_2) are involved in a highly conserved signaling network for growth, development, and stress response, but become toxic at high levels (oxidative stress), which induces programmed cell death - the hypersensitive response (HR), that suppresses pathogen colonization (Baxter et al., 2014; Lightfoot et al., 2017). In contrast, during infection from a necrotrophic pathogen, the opposite occurs, in that the rapid ROS production (oxidative burst) initiated within seconds or minutes results in cell death that allows the necrotrophic pathogen to proliferate (Baxter et al., 2014). Able (2003) determined that ROS was produced in larger quantities in barley leaves that were susceptible to *Ptt*. However, the avirulent Australian isolate NB41 produced significantly more superoxide than the virulent isolate NB30, yet higher levels of superoxide dismutase (SOD) activity were detected in the resistant plants. Superoxide was found in the epidermal cells immediately adjacent to the phenolic browning, and hydrogen peroxide was found in the majority of mesophyll cells, suggesting different roles in plant signaling (Able, 2003). Able (2003) hypothesized that superoxide was produced by *Ptt* to signal the production of plant-derived superoxide from the epidermal cells, and in resistant reactions, SOD activity was also upregulated.

Lightfoot et al. (2017) identified a gene involved in barley responses to *Ptt* as *CSD1*, a gene encoding a SOD involved in catalyzing ROS to maintain redox homeostasis. CSD1, a cytosolic copper zinc SOD, uses the metal cofactors copper and zinc to catalyze superoxide into the less reactive ROS of singlet oxygen and hydrogen peroxide. In the resistant line CI9214, CSD1 and CuZnSOD are upregulated when infected with the *Ptt*, suggesting that increased SOD activity is required to maintain redox status by driving the conversion of superoxide to singlet oxygen and hydrogen peroxide (Lightfoot et al., 2017). Interestingly, knockdown of *CSD1* in the susceptible cultivar Golden Promise was less severe and delayed infection when compared to the wild type, indicating that a complex interaction between the fungal pathogen and host plant exists (Lightfoot et al., 2017). The reduction of *Ptt* disease severity in the CSD1 knockdown suggests that superoxide levels might impact the growth of *Ptt*, possibly due to the lack of nutrients delivered from PCD or induced plant defense responses (Lightfoot et al., 2017). Experiments showing that ROS increase during necrosis and before sporulation of *Pyrenophora teres* (Able, 2003); and the use of infiltrated ROS scavengers on infected barley lines suggest that *Pyrenophora teres* can cope with oxidative stress (Lightfoot et al., 2017). Lightfoot et al. (2017) concluded that CSD1 knockdown resistance may be a result of a different mechanism and that many factors may be important in the interaction such as the specific ROS involved, the stage of infection, and whether ROS production is modulated by fungal effectors. In addition, proteins involved in the photosynthetic capability and carbon metabolism of the cell have been shown to be downregulated when infected with *Ptt*, which may promote cell death (Hassett et al., 2020). Within the seed, the only protein to be found to be consistently upregulated was an oxalate oxidase, suggesting that genes mediating disease resistance are primarily present in leaves and stems (Kerr et al., 2019).

5.1 Genetic and genomic resources in barley and Pyrenophora teres

Barley is at the dawn of pangenomics (Monat et al., 2018), with a recently re-released reference genome for the cv. Morex (Monat et al., 2019) that improved upon the previous versions (Beier et al., 2017; IBGSC, 2012; Mascher et al., 2017). Additional assemblies have included Barke, Bowman (IBGSC, 2012), Haruna Nijo (Sato et al., 2016), Lasa Goumang (Zeng et al., 2015, 2020), Zangqing320 (Dai et al., 2018), and Golden Promise (Schreiber et al., 2020) plus the wild species *H. bulbosum* (Wendler et al., 2017) and *H. pubiflorum* (Mascher et al., 2013). Igri, Haruna Nijo, and a wild *H. spontaneum* species were only sequenced and aligned to the first Morex reference by IBGSC (2012). Sequencing of the barley mitochondrial genome revealed a significant number of fragments within the nuclear genome assembly that could aid future misassemblies and annotations (Hisano et al., 2016). Sequencing technology is beginning to overcome the challenges posed by the repetitive nature of the barley genome (Monat et al., 2018). One of the methods to reduce the amount of sequencing required was exome capture (Mascher et al., 2013). However, this was based on gene models at the time and would still inadvertently not capture all exons. As these resources become more robust when coupled with tools such as BRIDGE (König et al., 2020), we will be able to further consolidate genes and alleles.

Barley has also benefitted from a rapid development of universal genotyping platforms such as the Affymetrix Barley1 GeneChip (Close et al., 2004), two Illumina GoldenGate production barley oligonucleotide pool assays (BOPAs) (Close et al., 2009), and the Illumina Infinium 9k (Comadran et al., 2012) and 50k iSelect SNP arrays (Bayer et al., 2017). Current technology allows for the reduction or novel SNP targeting with the use of the Illumina BeadXpress platform (Moragues et al., 2010), restriction site-associated digest-genotyping by sequencing (Elshire et al., 2011), and polymerase chain reaction-genotyping by sequencing (Sharma Poudel et al., 2018), compared to the fixed SNP arrays. These rapid advancements in genotyping platforms have resolved the origins and dissemination of domesticated barley (Pankin et al., 2018; Pourkheirandish et al., 2018; Russell et al., 2016; Zeng et al., 2018) that will aid in identifying novel and durable resistances. There have also been three released targeting-induced local lesions in genomes (TILLING) populations in the cvs. Morex (TILLMore) (Sparla et al., 2014; Talamè et al., 2008), Sebastian (*Hor*TILLUS) (Szurman-Zubrzycka et al., 2018), and Golden Promise (Schreiber et al., 2019). In terms of specifics to the *Pyrenophora teres*–barley system, multiple barley mapping populations have been developed or harnessed (reviewed by Clare et al., 2020).

For example, the recent and or notable biparental mapping populations have included CI5791 × Tifang (Koladia et al., 2017a), Falcon × Azhul (Islamovic et al., 2017), OUH602 × Haruna Nijo (Hisano et al., 2017), Arve × Lavran (Wonneberger et al., 2017a), UVC8 × Erica (Martin et al., 2018) and Canadian Lake Shore × Harrington (Dinglasan et al., 2019) for *Ptt* and Tradition × PI67381, Pinnacle × PI67381 and Pinnacle × PI84314 for *Ptm* (Tamang et al., 2019). A total of twelve association panels have been harnessed that have included natural or artificial (breeding lines) collections or a combination of both since 2015 (Adhikari et al., 2019, 2020; Amezrou et al., 2018; Burlakoti et al., 2017; Daba et al., 2019; Novakazi et al., 2019; Richards et al., 2017; Rozanova et al., 2019; Tamang et al., 2015; Vatter et al., 2017; Wang et al., 2015; Wonneberger et al., 2017b).

Pyrenophora teres was initially thought to contain six chromosomes (Aragona et al., 2000). However, this was revised to nine chromosomes based on the first *Ptt* draft genome assembly (Ellwood et al., 2010) and twelve with the first reference quality genome (Wyatt et al., 2018). The Canadian *Ptt* isolate 0-1 was used for the draft and reference quality genome assemblies, increasing the genome size from 41.95 Mbp to 46.5 Mbp (Ellwood et al., 2010; Wyatt et al., 2018). A total of eight other *Ptt* genomes have been published, to varying degrees of quality, including the Australian isolates NB29, NB73, NB85, W1-1 (Syme et al., 2018), HRS09122 and HRS09139 (Moolhuijzen et al., 2020), the US isolates 15A, 6A, FGOH04*Ptt*-21, and the Danish isolate BB25 (Wyatt et al., 2020). The *Ptm* genome is smaller than the *Ptt* genome, at 39.27-41.28 Mbp, a similar size to the *Ptr* genome, with the FGOB10*Ptm*-1 and SG1 genomes currently published (Syme et al., 2018) and the FGOB10*Ptm*-1 serving as the *de facto* reference genome. Several biparental mapping populations exist on the pathogen side, including *Ptt* crosses: 0-1 × 15A (Lai et al., 2007; Weiland et al., 1999), WRS 1906 × WRS 1607 (Beattie et al., 2007), H22 × 92-178/9, 181-6 × A80, 181-6 × 8ax (Afanasenko et al., 2007; Mironenko et al., 2005), 15A × 6A (Shjerve et al., 2014), BB25 × FGOH04*Ptt*-21 (Koladia et al., 2017b), NB29 × NB85 (Syme et al., 2018), NB029 × HRS09122 (Martin et al., 2020), and *Ptm* cross: FGOB10*Ptm*-1 × SG1 (Carlsen et al., 2017). Beyond doubt, additional mapping populations and genome assemblies exist and are facilitated by the dramatic reduction in sequencing cost associated with either assembly or genetic map development that also drives up marker density. Leveraging these resources will facilitate the validation of *Pyrenophora teres* effectors, which will subsequently serve to identify and validate host effector targets for the continued improvement of barley cultivars.

6 Conclusion and future trends

Barley's roles as food, feed, and for malted beverages have become cemented in human society since its domestication. As a durable crop that can be grown on the fringes of arable crop lands in the modern agroecosystem, barley faces a wide variety of abiotic and biotic stresses. As a destructive and rapidly evolving necrotic disease, net blotch represents one of the major biotic factors requiring attention from barley breeders and researchers worldwide. Cereal crop breeders today are challenged with feeding a rapidly increasing human population, while lower value cereal crops are being increasingly pushed to sub-optimal farmland. This in turn leads to increased disease pressure, which will be exacerbated by the shifting weather patterns that are predicted to accompany global climate change. The rapid evolution of both *Ptt* and *Ptm*, their ability to form hybrids, and their emergence as a potential threat to global wheat production highlight the importance of continuous monitoring of pathogen populations and the identification of novel resistance genes to protect against future net blotch epidemics. The wide and diverse pool of wild and landrace barleys represents an excellent source of genetic diversity and potential resistances that must be leveraged by researchers and breeders, going forward.

The recent advancements in sequencing technology and molecular assays are enabling researchers to gather information at a faster rate than ever before. Intelligent deployment of these technologies must be considered when planning and executing future research. The barley-*Pyrenophora teres* research community is close to identifying the first barley *Ptt* resistance/susceptibility genes and pathogen virulence factors. As more host-pathogen genetic interaction information becomes available to barley breeding programs, fine-tuned decisions can be made to release barley varieties that are more precisely adapted to pathogen populations in each unique barley growing region. As climate change continues to unfold, it is imperative to practice proper gene stewardship while incorporating resistances into elite cultivars. Over the coming decades, barley breeders will face significant challenges in balancing durable disease resistance in the face of changing pathogen pressures, with the maintenance and expansion of total yield gains driven by centuries of breeding efforts. The recent advancements seen in molecular biology will enable the existing loci identified to be consolidated into their allelic series or truly novel genes underlying these loci. These advances will also help elucidate the specific molecular plant-pathogen interactions that occur on regional scales, and will prove vital for appropriate introgression, stacking, and deployment of the most suitable genes required for resistance in locally planted elite lines. The implementation of these new resources will assist researchers and breeders in the deployment and maintenance of robust, durable resistances to both forms of net blotch of barley worldwide.

7 Where to look for further information

Research efforts examining interactions between cereal grains and *Pyrenophora teres* require international collaboration within the barley research community.

For example, international collaborations have recently resulted in the publication of the barley pangenome (Jayakodi et al. 2020), while efforts to sequence and release a pangenome representing the diversity of wild barley germplasm are currently underway by the wild barley diversity panel sequencing consortium. These genomic resources will certainly aid in the discovery of new *Pyrenophora teres* resistance sources that breeders can deploy for robust resistance to mitigate the damage of net blotch worldwide. These resources and others available to the barley and *Pyrenophora teres* research communities will also aid in resistance and susceptibility gene discovery as well as the identification of virulence effectors from the pathogen. Gene discovery and functional analysis of the host-pathogen interactions underlying resistance or susceptibility will further our ability to intelligently deploy resistance and monitor the virulence of pathogen populations. Further details about this pathogen and the molecular interactions between barley and *Pyrenophora teres* can be found in reviews by Liu et al. 2011 & Backes et al. 2021, and Clare et al. 2020, respectively.

Other great resources for the newest information and research on these pathosystems are international meetings and conferences that bring the research community together to present and discuss the latest research surrounding issues to mitigate diseases of barley. These include the International Barley Genetics Symposium, International Workshop on Barley Leaf Diseases, the International Plant and Animal Genome Meeting, and the International Society for Molecular Plant-Microbe Interactions Congress. The meeting websites provide information about upcoming meetings, including abstracts and presentations.

8 Acknowledgements

The United States Department of Agriculture National Institute of Food and Agriculture award # 2018-67014-28491 and the National Science Foundation grant award # 1759030 supported some of the unpublished research presented in this book chapter. Any opinions, findings, conclusions, or recommendations expressed in this material are those of the author(s) and do not necessarily reflect the views of the United States Department of Agriculture, National Institute of Food and Agriculture or the National Science Foundation.

9 References

Able, A. J. (2003). Role of reactive oxygen species in the response of barley to necrotrophic pathogens. *Protoplasma* 221(1-2), 137-143.

Abu Qamar, M., Liu, Z. H., Faris, J. D., Chao, S., Edwards, M. C., Lai, Z., Franckowiak, J. D. and Friesen, T. L. (2008). A region of barley chromosome 6h harbors multiple major genes associated with net type net blotch resistance. *Theor. Appl. Genet.* 117(8), 1261-1270.

Adhikari, A., Steffenson, B. J., Smith, M. J. and Dill-Macky, R. (2019). Genome-wide association mapping of seedling net form net blotch resistance in an Ethiopian and Eritrean barley collection. *Crop Sci.* 59(4), 1625–1638.

Adhikari, A., Steffenson, B. J. J., Smith, K. P. P., Smith, M. and Dill-Macky, R. (2020). Identification of quantitative trait loci for net form net blotch resistance in contemporary barley breeding germplasm from the USA using genome-wide association mapping. *Theor. Appl. Genet.* 133(3), 1019–1037.

Afanasenko, O. S., Mironenko, N., Filatova, O., Kopahnke, D., Krämer, I. and Ordon, F. (2007). Genetics of host-pathogen interactions in the *Pyrenophora teres* f. *teres* (net form)–barley (Hordeum vulgare) pathosystem. *Eur. J. Plant Pathol.* 117(3), 267–280.

Akhavan, A., Turkington, T. K., Kebede, B., Tekauz, A., Kutcher, H. R., Kirkham, C., Xi, K., Kumar, K., Tucker, J. R. and Strelkov, S. E. (2015). Prevalence of mating type idiomorphs in *Pyrenophora teres* f. *teres* and P. *teres* f. *maculata* populations from the Canadian prairies. *Can. J. Plant Pathol.* 37(1), 52–60.

Akhavan, A., Turkington, T. K., Kebede, B., Xi, K., Kumar, K., Tekauz, A., Kutcher, H. R., Tucker, J. R. and Strelkov, S. E. (2016). Genetic structure of *Pyrenophora teres* f. *teres* and P. *teres* f. *maculata* populations from western Canada. *Eur. J. Plant Pathol.* 146(2), 325–335.

Al-Daoude, A., Jawhar, M., Al-Shehadah, E., Shoaib, A., Orfi, M. and Arabi, M. I. E. (2018). Changes in salicylic acid content and pathogenesis–related (PR2) gene expression during barley–*Pyrenophora teres* interaction. *Hell. Plant Prot. J.* 11(2), 71–77.

Alcorn, J. L. (1988). The taxonomy of "Helminthosporium" species. *Annu. Rev. Phytopathol.* 26(1), 37–56.

Amezrou, R., Verma, R. P. S., Chao, S., Brueggeman, R. S., Belqadi, L., Arbaoui, M., Rehman, S. and Gyawali, S. (2018). Genome-wide association studies of net form of net blotch resistance at seedling and adult plant stages in spring barley collection. *Mol. Breed.* 38(5), 58.

Andersen, E. J., Ali, S., Reese, R. N., Yen, Y., Neupane, S. and Nepal, M. P. (2016). Diversity and evolution of disease resistance genes in barley (*Hordeum vulgare* L.). *Evol. Bioinforma. Online* 12, 99–108, EBO S38085.

Arabi, M. I. E., Alek, H., Jawhar, M. and Al-Shehadah, E. (2020). Expression of PAL and PR2 pathogenesis related genes in barley plants challenged with closely related Pyrenophora species. *Cereal Res. Commun.* 48(2), 211–216.

Aragona, M., Montigiani, M. and Porta-Puglia, A. (2000). Electrophoretic karyotypes of the phytopathogenic Pyrenophora graminea and P. teres. *Mycol. Res.* 104(7), 853–857.

Bach, E., Christensen, S., Dalgaard, L., Larsen, P. O., Olsen, C. E. and Smedegård-Petersen, V. (1979). Structures, properties and relationship to the aspergillomarasmines of toxins produced by *Pyrenophora teres*. *Physiol. Plant Pathol.* 14(1), 41–46.

Backes, A., Guerriero, G., Ait Barka, E. and Jacquard, C. (2021). Pyrenophora teres: taxonomy, morphology, interaction with barley, and mode of control. *Frontiers in Plant Science* 0, 509. DOI: 10.3389/FPLS.2021.614951.

Bakonyi, J. and Justesen, A. F. (2007). Genetic relationship of Pyrenophora graminea, P. teres f. *maculata* and P. teres f. *teres* Assessed by RAPD analysis. *J. Phytopathol.* 155(2), 76–83.

Baxter, A., Mittler, R. and Suzuki, N. (2014). ROS as key players in plant stress signalling. *J. Exp. Bot.* 65(5), 1229–1240.

Bayer, M. M., Rapazote-Flores, P., Ganal, M., Hedley, P. E., Macaulay, M., Plieske, J., Ramsay, L., Russell, J., Shaw, P. D., Thomas, W. and Waugh, R. (2017). Development and Evaluation of a Barley 50k iSelect SNP Array. *Frontiers in plant science*, 8, 1792. https://doi.org/10.3389/fpls.2017.01792

Beattie, A. D., Scoles, G. J. and Rossnagel, B. G. (2007). Identification of molecular markers linked to a *Pyrenophora teres* avirulence gene. *Phytopathology* 97(7), 842–849.

Bedada, G., Westerbergh, A., Nevo, E., Korol, A. and Schmid, K. J. (2014). DNA sequence variation of wild barley Hordeum spontaneum (L.) across environmental gradients in Israel. *Heredity (Edinb)* 112(6), 646–655.

Beier, S., Himmelbach, A., Colmsee, C., Zhang, X. Q., Barrero, R. A., Zhang, Q., Li, L., Bayer, M., Bolser, D., Taudien, S., Groth, M., Felder, M., Hastie, A., Šimková, H., Staňková, H., Vrána, J., Chan, S., Muñoz-Amatriaín, M., Ounit, R., Wanamaker, S., Schmutzer, T., Aliyeva-Schnorr, L., Grasso, S., Tanskanen, J., Sampath, D., Heavens, D., Cao, S., Chapman, B., Dai, F., Han, Y., Li, H., Li, X., Lin, C., McCooke, J. K., Tan, C., Wang, S., Yin, S., Zhou, G., Poland, J. A., Bellgard, M. I., Houben, A., Doležel, J., Ayling, S., Lonardi, S., Langridge, P., Muehlbauer, G. J., Kersey, P., Clark, M. D., Caccamo, M., Schulman, A. H., Platzer, M., Close, T. J., Hansson, M., Zhang, G., Braumann, I., Li, C., Waugh, R., Scholz, U., Stein, N. and Mascher, M. (2017). Construction of a map-based reference genome sequence for barley, *Hordeum vulgare* L. *Sci. Data* 4, 170044.

Bockelman, H. E., Sharp, E. L. and Eslick, R. F. (1977). Trisomic analysis of genes for resistance to scald and net blotch in several barley cultivars. *Can. J. Bot.* 55(15), 2142–2148.

Bogacki, P., Oldach, K. H. and Williams, K. J. (2008). Expression profiling and mapping of defence response genes associated with the barley-*Pyrenophora teres* incompatible interaction. *Mol. Plant Pathol.* 9(5), 645–660.

Breen, S., Williams, S. J., Outram, M., Kobe, B. and Solomon, P. S. (2017). Emerging insights into the functions of pathogenesis-related protein 1. *Trends Plant Sci.* 22(10), 871–879.

Brown, J. K. M. (2015). Durable resistance of crops to disease: a Darwinian perspective. *Annu. Rev. Phytopathol.* 53, 513–539.

Brown, M. P., Steffenson, B. J. and Webster, R. K. (1993). Host range of *Pyrenophora teres* f. *teres* isolates from California. *Plant Dis.* 77(9), 942.

Brueggeman, R. S., Solanki, S., Ameen, G., Effertz, K., Poudel, R. S. and Karakaya, A. (2020). Fungal diseases affecting barley. In: *Achieving Sustainable Cultivation of Barley,* Fox, G. and Li, C. (Eds). Cambridge, UK: Burleigh Dodds Science Publishing Limited. pp. 1–58.

Burlakoti, R. R., Gyawali, S., Chao, S., Smith, K. P., Horsley, R. D., Cooper, B., Muehlbauer, G. J. and Neate, S. M. (2017). Genome-wide association study of spot form of net blotch resistance in the Upper Midwest barley breeding programs. *Phytopathology* 107(1), 100–108.

Cakir, M., Gupta, S., Li, C., Hayden, M., Mather, D. E., Ablett, G. A., Platz, G. J., Broughton, S., Chalmers, K. J., Loughman, R., Jones, M. G. K. and Lance, R. C. M. (2011). Genetic mapping and QTL analysis of disease resistance traits in the barley population Baudin × AC Metcalfe. *Crop Pasture Sci.* 62(2), 152–161.

Campbell, G. F. and Crous, P. W. (2003). Genetic stability of net × spot hybrid progeny of the barley pathogen *Pyrenophora teres*. *Australas. Plant Pathol.* 32(2), 283–287.

Campbell, G. F., Crous, P. W. and Lucas, J. A. (1999). *Pyrenophora teres* f. *maculata*, the cause of *Pyrenophora* leaf spot of barley in South Africa. *Mycol. Res.* 103(3), 257–267.

Campbell, G. F., Lucas, J. A. and Crous, P. W. (2002). Evidence of recombination between net- and spot-type populations of *Pyrenophora teres* as determined by RAPD analysis. *Mycol. Res.* 106(5), 602–608.

Carlsen, S. A., Neupane, A., Wyatt, N. A., Richards, J. K., Faris, J. D., Xu, S. S., Brueggeman, R. S. and Friesen, T. L. (2017). Characterizing the *Pyrenophora teres* f. *maculata* - barley interaction using pathogen genetics. *G3 (Bethesda)* 7(8), 2615–2626 LP-2626.

Çelík Oğuz, A. and Karakaya, A. (2017). Pathotypes of *Pyrenophora teres* on barley in Turkey. *Phytopathol. Mediterr.* 56, 224.

Çelik Oğuz, A., Karakaya, A., Ergün, N. and Sayim, İ. (2017). Turkish barley landraces resistant to net and spot forms of *Pyrenophora teres*. *Phytopathol. Mediterr.* 56, 217–223.

Clare, S. J., Wyatt, N. A., Brueggeman, R. S. and Friesen, T. L. (2020). Research advances in the *Pyrenophora teres*-barley interaction. *Mol. Plant Pathol.* 21(2), 272–288.

Close, T. J., Wanamaker, S. I., Caldo, R. A., Turner, S. M., Ashlock, D. A., Dickerson, J. A., Wing, R. A., Muehlbauer, G. J., Kleinhofs, A. and Wise, R. P. (2004). A new resource for cereal genomics: 22K barley GeneChip comes of age. *Plant Physiol.* 134(3), 960–968.

Close, T. J., Bhat, P. R., Lonardi, S., Wu, Y., Rostoks, N., Ramsay, L., Druka, A., Stein, N., Svensson, J. T., Wanamaker, S., Bozdag, S., Roose, M. L., Moscou, M. J., Chao, S., Varshney, R. K., Szucs, P., Sato, K., Hayes, P. M., Matthews, D. E., Kleinhofs, A., Muehlbauer, G. J., DeYoung, J., Marshall, D. F., Madishetty, K., Fenton, R. D., Condamine, P., Graner, A. and Waugh, R. (2009). Development and implementation of high-throughput SNP genotyping in barley. *BMC Genomics* 10, 582.

Comadran, J., Kilian, B., Russell, J., Ramsay, L., Stein, N., Ganal, M., Shaw, P., Bayer, M., Thomas, W., Marshall, D., Hedley, P., Tondelli, A., Pecchioni, N., Francia, E., Korzun, V., Walther, A. and Waugh, R. (2012). Natural variation in a homolog of Antirrhinum CENTRORADIALIS contributed to spring growth habit and environmental adaptation in cultivated barley. *Nat. Genet.* 44(12), 1388–1392.

Crous, P. W., Janse, B. J. H., Tunbridge, J. and Holz, G. (1995). DNA homology of *Pyrenophora japonica* and *P. teres*. *Mycol. Res.* 99(9), 1098–1102.

Daba, S. D., Horsley, R., Brueggeman, R., Chao, S. and Mohammadi, M. (2019). Genome-wide association studies and candidate gene identification for leaf scald and net blotch in barley (Hordeum vulgare L.). *Plant Dis.* 103(5), 880–889.

Dai, F., Wang, X., Zhang, X. Q., Chen, Z., Nevo, E., Jin, G., Wu, D., Li, C. and Zhang, G. (2018). Assembly and analysis of a qingke reference genome demonstrate its close genetic relation to modern cultivated barley. *Plant Biotechnol. J.* 16(3), 760–770.

Dikilitas, M., Çelík Oğuz, A. and Karakaya, A. (2018). Extracellular protease activity and glucose production in isolates of net blotch pathogens differing in virulence. *Zemdirbyste Agric.* 105(1), 89–94.

Dinglasan, E., Hickey, L., Ziems, L., Fowler, R., Anisimova, A., Baranova, O., Lashina, N. and Afanasenko, O. (2019). Genetic characterization of resistance to *Pyrenophora teres* f. *teres* in the International Barley Differential Canadian Lake Shore. *Front. Plant Sci.* 10, 326.

Ellwood, S. R., Liu, Z., Syme, R. A., Lai, Z., Hane, J. K., Keiper, F., Moffat, C. S., Oliver, R. P. and Friesen, T. L. (2010). A first genome assembly of the barley fungal pathogen *Pyrenophora teres* f. *teres*. *Genome Biol.* 11(11), R109.

Ellwood, S. R., Syme, R. A., Moffat, C. S. and Oliver, R. P. (2012). Evolution of three *Pyrenophora* cereal pathogens: recent divergence, speciation and evolution of non-coding DNA. *Fungal Genet. Biol.* 49(10), 825–829.

Elshire, R. J., Glaubitz, J. C., Sun, Q., Poland, J. A., Kawamoto, K., Buckler, E. S. and Mitchell, S. E. (2011). A robust, simple genotyping-by-sequencing (GBS) approach for high diversity species. *PLoS ONE* 6(5), e19379.

Emebiri, L. C., Platz, G. and Moody, D. B. (2005). Disease resistance genes in a doubled haploid population of two-rowed barley segregating for malting quality attributes. *Aust. J. Agric. Res.* 56(1), 49-56.

Fang, Z., Gonzales, A. M., Clegg, M. T., Smith, K. P., Muehlbauer, G. J., Steffenson, B. J. and Morrell, P. L. (2014). Two genomic regions contribute disproportionately to geographic differentiation in wild barley. *G3 (Bethesda)* 4(7), 1193–1203.

Faris, J. D. and Friesen, T. L. (2020). Plant genes hijacked by necrotrophic fungal pathogens. *Curr. Opin. Plant Biol.* 56, 74–80.

Faris, J. D., Zhang, Z., Lu, H., Lu, S., Reddy, L., Cloutier, S., Fellers, J. P., Meinhardt, S. W., Rasmussen, J. B., Xu, S. S., Oliver, R. P., Simons, K. J. and Friesen, T. L. (2010). A unique wheat disease resistance-like gene governs effector-triggered susceptibility to necrotrophic pathogens. *Proc. Natl. Acad. Sci. U. S. A.* 107(30), 13544–13549.

Franckowiak, J. D. and Platz, G. J. (2013). International database for barley genes and barley genetic stocks. *Barley Genet. Newsl.* 43, 48-223.

Friesen, T. L., Faris, J. D., Lai, Z. and Steffenson, B. J. (2006a). Identification and chromosomal location of major genes for resistance to *Pyrenophora teres* in a doubled-haploid barley population. *Genome* 49(7), 855–859.

Friesen, T. L., Stukenbrock, E. H., Liu, Z., Meinhardt, S., Ling, H., Faris, J. D., Rasmussen, J. B., Solomon, P. S., McDonald, B. A. and Oliver, R. P. (2006b). Emergence of a new disease as a result of interspecific virulence gene transfer. *Nat. Genet.* 38(8), 953–956.

Friesen, T. L., Meinhardt, S. W. and Faris, J. D. (2007). The *Stagonospora nodorum*-wheat pathosystem involves multiple proteinaceous host-selective toxins and corresponding host sensitivity genes that interact in an inverse gene-for-gene manner. *Plant J.* 51(4), 681–692.

Friesen, T. L., Holmes, D. J., Bowden, R. L. and Faris, J. D. (2018). ToxA is present in the U.S. Bipolaris sorokiniana population and is a significant virulence factor on wheat harboring Tsn1. *Plant Dis.* 102(12), 2446-2452.

Friis, P., Olsen, C. E. and Møller, B. L. (1991). Toxin production in *Pyrenophora teres*, the ascomycete causing the net-spot blotch disease of barley (*Hordeum vulgare* L.). *J. Biol. Chem.* 266(20), 13329-13335.

Garozi, R., Angelotti, P., Feksa, H. R. and Tessmann, D. J. (2020). First report of *Pyrenophora teres* f. *teres* causing leaf spot on wheat in Brazil. *Plant Dis.* 104(12), 3254.

Geschele, E. E. (1928). The response of barley to parasitic fungi *Helminthosporium teres* Sacc. *Bull. Appl. bot. Genet. Plant Breed.* 19(19), 371-384 (in Rev. Appl. Mycol. 8, 165).

Graner, A., Foroughi-Wehr, B. and Tekauz, A. (1996). RFLP mapping of a gene in barley conferring resistance to net blotch (*Pyrenophora teres*). *Euphytica* 91, 229–234.

Gray, G. G. (1966). *Genetic Systems in the Net Blotch Disease Complex of Barley*. Fargo, ND: North Dakota State University.

Grewal, T. S., Rossnagel, B. G., Pozniak, C. J. and Scoles, G. J. (2008). Mapping quantitative trait loci associated with barley net blotch resistance. *Theor. Appl. Genet.* 116(4), 529–539.

Grewal, T. S., Rossnagel, B. G. and Scoles, G. J. (2012). Mapping quantitative trait loci associated with spot blotch and net blotch resistance in a doubled-haploid barley population. *Mol. Breed.* 30(1), 267–279.

Gupta, S., Li, C., Loughman, R., Cakir, M., Westcott, S. and Lance, R. (2011). Identifying genetic complexity of 6H locus in barley conferring resistance to *Pyrenophora teres* f. *teres*. *Plant Breed.* 130(4), 423–429.

Gyawali, S., Amezrou, R., Verma, R. P. S., Brueggeman, R., Rehman, S., Belqadi, L., Arbaoui, M., Tamang, P. and Singh, M. (2019). Seedling and adult stage resistance to spot form of net blotch (SFNB) in spring barley and stability of adult stage resistance to SFNB in Morocco. *Eur. J. Plant Pathol.* 153(2), 475–487.

Gyawali, S., Amezrou, R., Verma, R. P. S., Brueggeman, R., Rehman, S., Belqadi, L., Arbaoui, M., Tamang, P. and Singh, M. (2020). Seedling and adult stage resistance to net form of net blotch (NFNB) in spring barley and stability of adult stage resistance to NFNB in Morocco. *J. Phytopathol.* 168(5), 254–266.

Hassett, K., Ellwood, S. R., Zulak, K. G. and Muria-Gonzalez, M. J. (2020). Analysis of apoplastic proteins expressed during net form net blotch of barley. *J. Plant Dis. Prot.* 127(5), 683–694.

Hisano, H., Tsujimura, M., Yoshida, H., Terachi, T. and Sato, K. (2016). Mitochondrial genome sequences from wild and cultivated barley (Hordeum vulgare). *BMC Genomics* 17(1), 824.

Hisano, H., Sakamoto, K., Takagi, H., Terauchi, R. and Sato, K. (2017). Exome QTL-seq maps monogenic locus and QTLs in barley. *BMC Genomics* 18(1), 125.

Ho, K. M., Choo, T. M., Tekauz, A. and Martin, R. A. (1996). Genetic studies on net blotch resistance in a barley cross. *Can. J. Plant Sci.* 76(4), 715–719.

Horsley, R. D., Schwarz, P. B. and Hammond, J. J. (1995). Genetic diversity in malt quality of North American six-rowed Spring barley. *Crop Sci.* 35(1), 113–118.

Huang, J., Gu, M., Lai, Z., Fan, B., Shi, K., Zhou, Y. H., Yu, J. Q. and Chen, Z. (2010). Functional analysis of the Arabidopsis PAL gene family in plant growth, development, and response to environmental stress. *Plant Physiol.* 153(4), 1526–1538.

IBGSC (2012). A physical, genetic and functional sequence assembly of the barley genome. *Nature* 491(7426), 711–716.

Islamovic, E., Bregitzer, P. and Friesen, T. L. (2017). Barley 4H QTL confers NFNB resistance to a global set of *P. teres* f. *teres* isolates. *Mol. Breed.* 37(3), 29.

Ismail, I. A. and Able, A. J. (2016). Secretome analysis of virulent *Pyrenophora teres* f. *teres* isolates. *Proteomics* 16(20), 2625–2636.

Ismail, I. A. and Able, A. J. (2017). Gene expression profiling of virulence-associated proteins in planta during net blotch disease of barley. *Physiol. Mol. Plant Pathol.* 98, 69–79.

Ismail, I. A., Godfrey, D. and Able, A. J. (2014). Fungal growth, proteinaceous toxins and virulence of *Pyrenophora teres* f. *teres* on barley. *Australas. Plant Pathol.* 43(5), 535–546.

Ito, S. and Kuribayashi, K. (1931). The ascigerous forms of some graminicolous species of Helminthosporium in Japan. *J. Fac. Agric.* 29, 85–125.

Jalli, M. (2011). Sexual reproduction and soil tillage effects on virulence of *Pyrenophora teres* in Finland. *Ann. Appl. Biol.* 158(1), 95–105.

Jalli, M. and Robinson, J. (2000). Stable resistance in barley to *Pyrenophora teres* f. *teres* isolates from the Nordic-Baltic region after increase on standard host genotypes. *Euphytica* 113(1), 71–77.

Jayakodi, M., Padmarasu, S., Haberer, G. et al. (2020). The barley pan-genome reveals the hidden legacy of mutation breeding. *Nature* 588, 284–289. https://doi.org/10.1038/s41586-020-2947-8.

Kage, U., Yogendra, K. N. and Kushalappa, A. C. (2017). TaWRKY70 transcription factor in wheat QTL-2DL regulates downstream metabolite biosynthetic genes to resist Fusarium graminearum infection spread within spike. *Sci. Rep.* 7, 42596.

Keon, J. P. R. and Hargreaves, J. A. (1983). A cytological study of the net blotch disease of barley caused by *Pyrenophora teres*. *Physiol. Plant Pathol.* 22(3), 321-IN14.

Kerr, E. D., Phung, T. K., Caboche, C. H., Fox, G. P., Platz, G. J. and Schulz, B. L. (2019). The intrinsic and regulated proteomes of barley seeds in response to fungal infection. *Anal. Biochem.* 580, 30–35.

Khan, T. N. and Boyd, W. J. R. (1969). Inheritance of resistance to net blotch in barley. II. Genes conditioning resistance against race W.A.-2. *Can. J. Genet. Cytol.* 11(3), 592–597.

Knopff, M. and Morell, M. (2019). The cereals imperative of future food systems. *CIMMYT* (online).

Koladia, V. M., Faris, J. D., Richards, J. K., Brueggeman, R. S., Chao, S. and Friesen, T. L. (2017a). Genetic analysis of net form net blotch resistance in barley lines CIho 5791 and Tifang against a global collection of P. teres f. *teres* isolates. *Theor. Appl. Genet.* 130(1), 163–173.

Koladia, V. M., Richards, J. K., Wyatt, N. A., Faris, J. D., Brueggeman, R. S. and Friesen, T. L. (2017b). Genetic analysis of virulence in the *Pyrenophora teres* f. *teres* population BB25×FGOH04Ptt-21. *Fungal Genet. Biol.* 107, 12–19.

König, P., Beier, S., Basterrechea, M., Schüler, D., Arend, D., Mascher, M., Stein, N., Scholz, U. and Lange, M. (2020). Bridge–a visual analytics web tool for barley GeneBank genomics. *Front. Plant Sci.* 11, 701.

Krupinska, K., Haussühl, K., Schäfer, A., van der Kooij, T. A. W., Leckband, G., Lörz, H. and Falk, J. (2002). A novel nucleus-targeted protein is expressed in barley leaves during senescence and pathogen infection. *Plant Physiol.* 130(3), 1172-1180 LP-1180.

Lai, Z., Faris, J. D., Weiland, J. J., Steffenson, B. J. and Friesen, T. L. (2007). Genetic mapping of *Pyrenophora teres* f. *teres* genes conferring avirulence on barley. *Fungal Genet. Biol.* 44(5), 323-329.

Launay, M., Caubel, J., Bourgeois, G., Huard, F., Garcia de Cortazar-Atauri, I., Bancal, M. O. and Brisson, N. (2014). Climatic indicators for crop infection risk: application to climate change impacts on five major foliar fungal diseases in Northern France. *Agric. Ecosyst. Environ.* 197, 147–158.

Lehmensiek, A., Platz, G. J., Mace, E., Poulsen, D. and Sutherland, M. W. (2007). Mapping of adult plant resistance to net form of net blotch in three Australian barley populations. *Aust. J. Agric. Res.* 58(12), 1191-1197.

Lehmensiek, A., Bester-van der Merwe, A. E., Sutherland, M. W., Platz, G., Kriel, W. M., Potgieter, G. F. and Prins, R. (2010). Population structure of South African and Australian *Pyrenophora teres* isolates. *Plant Pathol.* 59(3), 504–515.

Leisova, L., Kucera, L. and Minarikova, V. (2005). AFLP-based PCR markers that differentiate spot and net forms of *Pyrenophora teres*. *Plant Pathol.* 54(1), 66–73.

Leišova, L., Minarikova, V., Kucera, L. and Ovesna, J. (2005) Genetic diversity of *Pyrenophora teres* isolates as detected by AFLP analysis. *J. Phytopathol.* 153(10), 569–578.

Leisova-Svobodova, L., Hanzalova, A. and Kucera, L. (2010). Expansion and variability of the Ptr ToxA gene in populations of Pyrenophora tritici-repentis and *Pyrenophora teres*. *J. Plant Pathol.* 92, 729–735.

Lightfoot, D. J., McGrann, G. R. D. and Able, A. J. (2017). The role of a cytosolic superoxide dismutase in barley-pathogen interactions. *Mol. Plant Pathol.* 18(3), 323–335.

Linde, C. C. and Smith, L. M. (2019). Host specialisation and disparate evolution of *Pyrenophora teres* f. *teres* on barley and barley grass. *BMC Evol. Biol.* 19(1), 139.

Liu, Z., Ellwood, S. R., Oliver, R. P. and Friesen, T. L (2011). Pyrenophora teres: profile of an increasingly damaging barley pathogen. *Molecular plant pathology*, 12(1), 1–19. https://doi.org/10.1111/j.1364-3703.2010.00649.x.

Liu, Z., Faris, J. D., Edwards, M. C. and Friesen, T. L. (2010). Development of expressed sequence tag (EST)-based markers for genomic analysis of a barley 6H region harboring multiple net form net blotch resistance genes. *Plant Genome* 3(1), 41–52.

Liu, Z., Holmes, D. J., Faris, J. D., Chao, S., Brueggeman, R. S., Edwards, M. C. and Friesen, T. L. (2015). Necrotrophic effector-triggered susceptibility (NETS) underlies the barley-*Pyrenophora teres* f. *teres* interaction specific to chromosome 6H. *Mol. Plant Pathol.* 16(2), 188–200.

Lopez-Ruiz, F., Mair, W., Thomas, G., Jayasena, K., Hills, A. and Anke, M. (2020). Azole resistance in spot-form net blotch in Western Australia. Grain Research and Development Corporation Update Papers, Australia.

Louw, J. P. J., Crous, P. W. and Holz, G. (1994). Pyrenophora japonica occurring on barley in South Africa. *Plant Pathol.* 43(2), 420–423.

Louw, J. P. J., Victor, D., Crous, P. W., Holz, G. and Janse, B. J. H. (1995). Characterization of *Pyrenophora* isolates associated with spot and net type lesions on barley in South Africa. *J. Phytopathol.* 143(3), 129–134.

Mamo, B. E. and Steffenson, B. J. (2015). Genome-wide association mapping of Fusarium Head Blight resistance and agromorphological traits in barley landraces from Ethiopia and Eritrea. *Crop Sci.* 55(4), 1494–1512.

Manninen, O., Kalendar, R., Robinson, J. and Schulman, A. H. (2000). Application of BARE-1 retrotransposon markers to the mapping of a major resistance gene for net blotch in barley. *Mol. Gen. Genet.* 264(3), 325–334.

Manninen, O. M., Jalli, M., Kalendar, R., Schulman, A., Afanasenko, O. and Robinson, J. (2006). Mapping of major spot-type and net-type net-blotch resistance genes in the Ethiopian barley line CI 9819. *Genome* 49(12), 1564–1571.

Martin, A., Platz, G. J., de Klerk, D., Fowler, R. A., Smit, F., Potgieter, F. G. and Prins, R. (2018). Identification and mapping of net form of net blotch resistance in South African barley. *Mol. Breed.* 38(5), 53.

Martin, A., Moolhuijzen, P., Tao, Y., McIlroy, J., Ellwood, S. R., Fowler, R. A., Platz, G. J., Kilian, A. and Snyman, L. (2020). Genomic regions associated with virulence in *Pyrenophora teres* f. *teres* identified by genome-wide association analysis and biparental mapping. *Phytopathology®* 110, 881–891.

Mascher, M., Richmond, T. A., Gerhardt, D. J., Himmelbach, A., Clissold, L., Sampath, D., Ayling, S., Steuernagel, B., Pfeifer, M., D'Ascenzo, M., Akhunov, E. D., Hedley, P. E., Gonzales, A. M., Morrell, P. L., Kilian, B., Blattner, F. R., Scholz, U., Mayer, K. F., Flavell, A. J., Muehlbauer, G. J., Waugh, R., Jeddeloh, J. A. and Stein, N. (2013). Barley whole exome capture: a tool for genomic research in the genus Hordeum and beyond. *Plant J.* 76(3), 494–505.

Mascher, M., Gundlach, H., Himmelbach, A., Beier, S., Twardziok, S. O., Wicker, T., Radchuk, V., Dockter, C., Hedley, P. E., Russell, J., Bayer, M., Ramsay, L., Liu, H., Haberer, G., Zhang, X. Q., Zhang, Q., Barrero, R. A., Li, L., Taudien, S., Groth, M., Felder, M., Hastie, A., Šimková, H., Staňková, H., Vrána, J., Chan, S., Muñoz-Amatriaín, M., Ounit, R., Wanamaker, S., Bolser, D., Colmsee, C., Schmutzer, T., Aliyeva-Schnorr, L., Grasso, S., Tanskanen, J., Chailyan, A., Sampath, D., Heavens, D., Clissold, L., Cao, S., Chapman, B., Dai, F., Han, Y., Li, H., Li, X., Lin, C., McCooke, J. K., Tan, C., Wang, P., Wang, S., Yin, S., Zhou, G., Poland, J. A., Bellgard, M. I., Borisjuk, L., Houben, A., Doležel, J., Ayling, S., Lonardi, S., Kersey, P., Langridge, P., Muehlbauer, G. J., Clark, M. D., Caccamo, M., Schulman, A. H., Mayer, K. F. X., Platzer, M., Close, T. J., Scholz, U., Hansson, M., Zhang, G., Braumann, I., Spannagl, M., Li, C., Waugh, R. and Stein, N. (2017). A chromosome conformation capture ordered sequence of the barley genome. *Nature* 544(7651), 427-433.

Mathre, D. E. (1997). *Compendium of Barley Diseases*. St. Paul, MN: APS Press.

McDonald, B. A. and Linde, C. (2002). Pathogen population genetics, evolutionary potential, and durable resistance. *Annu. Rev. Phytopathol.* 40, 349-379.

McDonald, M. C., Ahren, D., Simpfendorfer, S., Milgate, A. and Solomon, P. S. (2018). The discovery of the virulence gene ToxA in the wheat and barley pathogen *Bipolaris sorokiniana*. *Mol. Plant Pathol.* 19(2), 432-439.

McDonald, M. C., Taranto, A. P., Hill, E., Schwessinger, B., Liu, Z., Simpfendorfer, S., Milgate, A. and Solomon, P. S. (2019). Transposon-mediated horizontal transfer of the host-specific virulence protein ToxA between three fungal wheat pathogens. *mBio* 10(5), e01515-19.

McDonald, W. C. (1967). Variability and inheritance of morphological mutants in *Pyrnophora teres*. *Phytopathology* 57, 747-755.

McDonald, W. C. and Buchanon, K. W. (1962). The inheritance of variability in *Pyrenophora teres*. *Barley. Newsl* 6:40.

McLean, M. S., Martin, A., Gupta, S., Sutherland, M. W., Hollaway, G. J. and Platz, G. J. (2014). Validation of a new spot form of net blotch differential set and evidence for hybridisation between the spot and net forms of net blotch in Australia. *Australas. Plant Pathol.* 43(3), 223-233.

Mikhailova, L. A., Ternyuk, I. G. and Mironenko, N. V. (2010). *Pyrenophora teres*, an agent causing wheat leaf spot. *Microbiology* 79(4), 561-565.

Minaříková, V. and Polišenská, I. (1999). Analysis of populations of *Pyrnophora teres* on Barley in the Czech Republic. *Plant Prot. Sci.* 35(4), 115-120.

Mironenko, N. V., Afanasenko, O. S., Filatova, O. A. and Kopahnke, D. (2005). Genetic control of virulence of *Pyrenophora teres* Drechs., the causative agent of net blotch in barley. *Genetika* 41(12), 1674-1680.

Mode, C. J. and Schaller, C. W. (1958). Two additional factors for host resistance to net blotch in barley 1. *Agron. J.* 50(1), 15-18.

Möller, M. and Stukenbrock, E. H. (2017). Evolution and genome architecture in fungal plant pathogens. *Nat. Rev. Microbiol.* 15(12), 756-771.

Molnar, S. J., James, L. E. and Kasha, K. J. (2000). Inheritance and RAPD tagging of multiple genes for resistance to net blotch in barley. *Genome* 43(2), 224-231.

Monat, C., Schreiber, M., Stein, N. and Mascher, M. (2018). Prospects of pan-genomics in barley. *Theor. Appl. Genet.* 132(3), 785-796.

Monat, C., Padmarasu, S., Lux, T., Wicker, T., Gundlach, H., Himmelbach, A., Ens, J., Li, C., Muehlbauer, G. J., Schulman, A. H., Waugh, R., Braumann, I., Pozniak, C., Scholz, U.,

Mayer, K. F. X., Spannagl, M., Stein, N. and Mascher, M. (2019). TRITEX: chromosome-scale sequence assembly of Triticeae genomes with open-source tools. *Genome Biol.* 20(1), 284.

Moolhuijzen, P. M., Muria-Gonzalez, M. J., Syme, R., Rawlinson, C., See, P. T., Moffat, C. S. and Ellwood, S. R. (2020). Expansion and conservation of biosynthetic gene clusters in pathogenic Pyrenophora spp. *Toxins (Basel)* 12(4), 242.

Moragues, M., Comadran, J., Waugh, R., Milne, I., Flavell, A. J. and Russell, J. R. (2010). Effects of ascertainment bias and marker number on estimations of barley diversity from high-throughput SNP genotype data. *Theor. Appl. Genet.* 120(8), 1525–1534.

Muñoz-Amatriaín, M., Cuesta-Marcos, A., Endelman, J. B., Comadran, J., Bonman, J. M., Bockelman, H. E., Chao, S., Russell, J., Waugh, R., Hayes, P. M. and Muehlbauer, G. J. (2014). The USDA barley core collection: genetic diversity, population structure, and potential for genome-wide association studies. *PLoS ONE* 9(4), e94688.

Murray, G. M. and Brennan, J. P. (2010). Estimating disease losses to the Australian barley industry. *Australas. Plant Pathol.* 39(1), 85–96.

Neupane, A., Tamang, P., Brueggeman, R. S. and Friesen, T. L. (2015). Evaluation of a barley core collection for spot form net blotch reaction reveals distinct genotype-specific pathogen virulence and host susceptibility. *Phytopathology* 105(4), 509–517.

Novakazi, F., Afanasenko, O., Anisimova, A., Platz, G. J. J., Snowdon, R., Kovaleva, O., Zubkovich, A. and Ordon, F. (2019). Genetic analysis of a worldwide barley collection for resistance to net form of net blotch disease (*Pyrenophora teres* f. *teres*). *Theor. Appl. Genet.* 132(9), 2633–2650.

O'Boyle, P., Barnett, W., Berger, M., Steffenson, G., Stromberg, B., Maroof, E., Liu, S. and Griffey, S. (2014). Mapping Net Blotch Resistance in 'Nomini' and CIho 2291 Barley. *Crop Science*. 54. 2596–2606. 10.2135/cropsci2013.08.0514.

Ohm, R. A., Feau, N., Henrissat, B., Schoch, C. L., Horwitz, B. A., Barry, K. W., Condon, B. J., Copeland, A. C., Dhillon, B., Glaser, F., Hesse, C. N., Kosti, I., LaButti, K., Lindquist, E. A., Lucas, S., Salamov, A. A., Bradshaw, R. E., Ciuffetti, L., Hamelin, R. C., Kema, G. H., Lawrence, C., Scott, J. A., Spatafora, J. W., Turgeon, B. G., de Wit, P. J., Zhong, S., Goodwin, S. B. and Grigoriev, I. V. (2012). Diverse lifestyles and strategies of plant pathogenesis encoded in the genomes of eighteen Dothideomycetes fungi. *PLOS Pathog.* 8(12), e1003037.

Oide, S., Bejai, S., Staal, J., Guan, N., Kaliff, M. and Dixelius, C. (2013). A novel role of PR2 in abscisic acid (ABA) mediated, pathogen-induced callose deposition in *Arabidopsis thaliana*. *New Phytol.* 200(4), 1187–1199.

Oliver, R. (2012). Genomic tillage and the harvest of fungal phytopathogens. *New Phytol.* 196(4), 1015–1023.

Pankin, A., Altmüller, J., Becker, C. and von Korff, M. (2018). Targeted resequencing reveals genomic signatures of barley domestication. *New Phytol.* 218(3), 1247–1259.

Perelló, A. E., Couretot, L., Curti, A., Uranga, J. P. and Consolo, V. F. (2019). First report of spot lesion of wheat caused by *Pyrenophora teres* f. sp maculata observed in Argentina. *Crop Prot.* 122, 19–22.

Pieterse, C. M. J., Van der Does, D., Zamioudis, C., Leon-Reyes, A. and Van Wees, S. C. M. (2012). Hormonal modulation of plant immunity. *Annu. Rev. Cell Dev. Biol.* 28, 489-521.

Poudel, B., McLean, M. S., Platz, G. J., McIlroy, J. A., Sutherland, M. W. and Martin, A. (2019). Investigating hybridisation between the forms of *Pyrenophora teres* based on Australian barley field experiments and cultural collections. *Eur. J. Plant Pathol.* 153(2), 465-473.

Pourkheirandish, M., Kanamori, H., Wu, J., Sakuma, S., Blattner, F. R. and Komatsuda, T. (2018). Elucidation of the origin of 'agriocrithon' based on domestication genes questions the hypothesis that Tibet is one of the centers of barley domestication. *Plant J.* 94(3), 525-534.

Raman, H., Platz, G. J., Chalmers, K. J., Raman, R., Read, B. J., Barr, A. R. and Moody, D. B. (2003). Mapping of genomic regions associated with net form of net blotch resistance in barley. *Aust. J. Agric. Res.* 54(12), 1359-1367.

Rau, D., Brown, A. H. D., Brubaker, C. L., Attene, G., Balmas, V., Saba, E. and Papa, R. (2003). Population genetic structure of *Pyrenophora teres*Drechs. the causal agent of net blotch in Sardinian landraces of barley (Hordeum vulgare L.). *Theor. Appl. Genet.* 106(5), 947-959.

Rau, D., Attene, G., Brown, A. H. D., Nanni, L., Maier, F. J., Balmas, V., Saba, E., Schäfer, W. and Papa, R. (2007). Phylogeny and evolution of mating-type genes from *Pyrenophora teres*, the causal agent of barley "net blotch" disease. *Curr. Genet.* 51(6), 377-392.

Rau, D., Rodriguez, M., Leonarda Murgia, M., Balmas, V., Bitocchi, E., Bellucci, E., Nanni, L., Attene, G. and Papa, R. (2015). Co-evolution in a landrace meta-population: two closely related pathogens interacting with the same host can lead to different adaptive outcomes. *Sci. Rep.* 5, 12834.

Read, B. J., Raman, H., McMichael, G., Chalmers, K. J., Ablett, G. A., Platz, G. J., Raman, R., Genger, R. K., Boyd, W. J. R., Li, C. D., Grime, C. R., Park, R. F., Wallwork, H., Prangnell, R. and Lance, R. C. M. (2003). Mapping and QTL analysis of the barley population Sloop × Halcyon. *Aust. J. Agric. Res.* 54(12), 1145-1153.

Reiss, E. and Bryngelsson, T. (1996). Pathogenesis-related proteins in barley leaves, induced by infection with *Drechslera teres* (Sacc.) Shoem. and by treatment with other biotic agents. *Physiol. Mol. Plant Pathol.* 49(5), 331-341.

Reiss, E. and Horstmann, C. (2001). *Drechslera teres*-infected barley (*Hordeum vulgare* L.) leaves accumulate eight isoforms of thaumatin-like proteins. *Physiol. Mol. Plant Pathol.* 58(4), 183-188.

Richards, J., Chao, S., Friesen, T. and Brueggeman, R. (2016). Fine mapping of the barley chromosome 6h net form net blotch susceptibility locus. *G3 (Bethesda)* 6(7), 1809-1818.

Richards, J. K. K., Friesen, T. L. L. and Brueggeman, R. S. S. (2017). Association mapping utilizing diverse barley lines reveals net form net blotch seedling resistance/susceptibility loci. *Theor. Appl. Genet.* 130(5), 915-927.

Robatzek, S. and Somssich, I. E. (2002). Targets of AtWRKY6 regulation during plant senescence and pathogen defense. *Genes Dev.* 16(9), 1139-1149.

Rozanova, I. V., Lashina, N. M., Mustafin, Z. S., Gorobets, S. A., Efimov, V. M., Afanasenko, O. S. and Khlestkina, E. K. (2019). SNPs associated with barley resistance to isolates of *Pyrenophora teres* f. *BMC Genomics* 20(Suppl 3), 292.

Russell, J., Mascher, M., Dawson, I. K., Kyriakidis, S., Calixto, C., Freund, F., Bayer, M., Milne, I., Marshall-Griffiths, T., Heinen, S., Hofstad, A., Sharma, R., Himmelbach, A., Knauft, M., van Zonneveld, M., Brown, J. W., Schmid, K., Kilian, B., Muehlbauer, G. J., Stein, N. and Waugh, R. (2016). Exome sequencing of geographically diverse barley landraces and wild relatives gives insights into environmental adaptation. *Nat. Genet.* 48(9), 1024-1030.

Sarpeleh, A., Wallwork, H., Catcheside, D. E. A., Tate, M. E. and Able, A. J. (2007). Proteinaceous metabolites from *Pyrenophora teres* contribute to symptom development of barley net blotch. *Phytopathology* 97(8), 907-915.

Sarpeleh, A., Wallwork, H., Tate, M. E., Catcheside, D. E. A. and Able, A. J. (2008). Initial characterisation of phytotoxic proteins isolated from *Pyrenophora teres*. *Physiol. Mol. Plant Pathol.* 72(1-3), 73-79.

Sarpeleh, A., Tate, M. E., Wallwork, H., Catcheside, D. and Able, A. J. (2009). Characterisation of low molecular weight phytotoxins isolated from *Pyrenophora teres*. *Physiol. Mol. Plant Pathol.* 73(6), 154-162.

Sato, K., Tanaka, T., Shigenobu, S., Motoi, Y., Wu, J. and Itoh, T. (2016). Improvement of barley genome annotations by deciphering the Haruna Nijo genome. *DNA Res.* 23(1), 21-28.

Savary, S., Willocquet, L., Pethybridge, S. J., Esker, P., McRoberts, N. and Nelson, A. (2019). The global burden of pathogens and pests on major food crops. *Nat. Ecol. Evol.* 3(3), 430-439.

Schaller, C. W. (1955). Inheritance of resistance to net blotch of barley. *Phytopathology* 45, 174-176.

Schaller, C. W. and Wiebe, G. A. (1952). Sources of resistance to net blotch of barley 1. *Agron. J.* 44(6), 334-336.

Schreiber, M., Barakate, A., Uzrek, N., Macaulay, M., Sourdille, A., Morris, J., Hedley, P. E., Ramsay, L. and Waugh, R. (2019). A highly mutagenised barley (cv. Golden Promise) TILLING population coupled with strategies for screening-by-sequencing. *Plant Methods* 15, 99.

Schreiber, M., Mascher, M., Wright, J., Padmarasu, S., Himmelbach, A., Heavens, D., Milne, L., Clavijo, B. J., Stein, N. and Waugh, R. (2020). A genome assembly of the barley 'transformation reference' cultivar golden Promise. *G3 (Bethesda)* 10(6), 1823-1827.

Scott, D. B. (1991). Identity of Pyrenophora isolates causing net-type and spot-type lesions on barley. *Mycopathologia* 116(1), 29-35.

Serenius, M., Mironenko, N. and Manninen, O. (2005). Genetic variation, occurrence of mating types and different forms of *Pyrenophora teres* causing net blotch of barley in Finland. *Mycol. Res.* 109, 809-817.

Sharma, H. S. S. (1984). Assessment of the reaction of some spring barley cultivars to *Pyrenophora teres* using whole plants, detached leaves and toxin bioassay. *Plant Pathol.* 33(3), 371-376.

Sharma Poudel, R., Al-Hashel, A. F., Gross, T., Gross, P. and Brueggeman, R. (2018). Pyramiding rpg4- and Rpg1-mediated stem rust resistance in barley requires the Rrr1 gene for both to function. *Front. Plant Sci.* 9, 1789.

Shipton, W. A. (1966). Net blotch of barley. *J. Dept. Agric., Western Australia, Ser. 4* 7(3), Article 8.

Shjerve, R. A., Faris, J. D., Brueggeman, R. S., Yan, C., Zhu, Y., Koladia, V. and Friesen, T. L. (2014). Evaluation of a *Pyrenophora teres* f. *teres* mapping population reveals multiple independent interactions with a region of barley chromosome 6H. *Fungal Genet. Biol.* 70, 104-112.

Shoemaker, R. A. (1959). Nomenclature of *Drechslera* and *Bipolaris*, grass parasites segregated from "*Helminthosporium*." *Can. J. Bot.* 37(5), 879-887.

Smedegård-Petersen, V. (1971). *Pyrenophora teres* f. *maculata* f. nov. and *Pyrenophora teres* f. *teres* on barley in Denmark. *Kgl Vet. Landbohojsk Arsskr*, 124-144.

Smedegård-Petersen, V. (1976). *Pathogenesis and Genetics of Netspot Blotch and Leaf Stripe of Barley Caused by Pyrenophora teres and Pyrenophora graminea*. Royal Veterinary and Agricultural University, DK-1871, Copenhagen V. Denmark.

Smedegård-Petersen, V. (1977a). Inheritance of genetic factors for symptoms and pathogenicity in hybrids of *Pyrenophora teres* and Pyrenophora graminea. *J. Phytopathol.* 89(3), 193-202.

Smedegård-Petersen, V. (1977b). Isolation of two toxins produced by *Pyrenophora teres* and their significance in disease development of net-spot blotch of barley. *Physiol. Plant Pathol.* 10(3), 203-211.

Smedegård-Petersen, V. (1977c). Respiratory changes of barley leaves infected with *Pyrenophora teres* or affected by isolated toxins of this fungus. *Physiol. Plant Pathol.* 10(3), 213-220.

Sparla, F., Falini, G., Botticella, E., Pirone, C., Talamè, V., Bovina, R., Salvi, S., Tuberosa, R., Sestili, F. and Trost, P. (2014). New starch phenotypes produced by TILLING in barley. *PLoS ONE* 9(10), e107779.

Steffenson, B. J., Hayes, P. M. and Kleinhofs, A. (1996). Genetics of seedling and adult plant resistance to net blotch (*Pyrenophora teres* f. *teres*) and spot blotch (Cochliobolus sativus) in barley. *Theor. Appl. Genet.* 92(5), 552-558.

Syme, R. A., Martin, A., Wyatt, N. A., Lawrence, J. A., Muria-Gonzalez, M. J., Friesen, T. L. and Ellwood, S. R. (2018). Transposable element genomic fissuring in *Pyrenophora teres* is associated with genome expansion and dynamics of host-pathogen genetic interactions. *Front. Genet.* 9, 130.

Szurman-Zubrzycka, M. E., Zbieszczyk, J., Marzec, M., Jelonek, J., Chmielewska, B., Kurowska, M. M., Krok, M., Daszkowska-Golec, A., Guzy-Wrobelska, J., Gruszka, D., Gajecka, M., Gajewska, P., Stolarek, M., Tylec, P., Sega, P., Lip, S., Kudełko, M., Lorek, M., Gorniak-Walas, M., Malolepszy, A., Podsiadlo, N., Szyrajew, K. P., Keisa, A., Mbambo, Z., Todorowska, E., Gaj, M., Nita, Z., Orlowska-Job, W., Maluszynski, M. and Szarejko, I. (2018). HorTILLUS–A rich and renewable source of induced mutations for forward/reverse genetics and pre-breeding programs in barley (Hordeum vulgare L.). *Front. Plant Sci.* 9, 216.

Talamè, V., Bovina, R., Sanguineti, M. C., Tuberosa, R., Lundqvist, U. and Salvi, S. (2008). TILLMore, a resource for the discovery of chemically induced mutants in barley. *Plant Biotechnol. J.* 6(5), 477-485.

Tamang, P., Neupane, A., Mamidi, S., Friesen, T. and Brueggeman, R. (2015). Association mapping of seedling resistance to spot form net blotch in a worldwide collection of barley. *Phytopathology* 105(4), 500-508.

Tamang, P., Richards, J. K., Alhashal, A., Sharma Poudel, R., Horsley, R. D., Friesen, T. L. and Brueggeman, R. S. (2019). Mapping of barley susceptibility/resistance QTL against spot form net blotch caused by Pyrenophora teres f. *maculata* using RIL populations. *Theor. Appl. Genet.* 132(7), 1953-1963.

Tóth, B., Csősz, M., Kopahnke, D. and Varga, J. (2008). First report on *Pyrenophora teres* causing lesions of wheat leaves in Hungary. *Plant Pathol.* 57(2), 385.

Turkington, T. K., Orr, D. D., Clear, R. M., Patrick, S. K., Burnett, P. A. and Xi, K. (2002). Fungal plant pathogens infecting barley and wheat seed from Alberta, 1995-1997. *Can. J. Plant Pathol.* 24(3), 302-308.

Uranga, J. P., Schierenbeck, M., Perelló, A. E., Lohwasser, U., Börner, A. and Simón, M. R. (2020). Localization of QTL for resistance to *Pyrenophora teres* f. *maculata*, a new wheat pathogen. *Euphytica* 216, 1-13.

Vatter, T., Maurer, A., Kopahnke, D., Perovic, D., Ordon, F. and Pillen, K. (2017). A nested association mapping population identifies multiple small effect QTL conferring resistance against net blotch (*Pyrenophora teres* f. *teres*) in wild barley. *Plos One* 12(10), e0186803.

Wallwork, H., Butt, M. and Capio, E. (2016). Pathogen diversity and screening for minor gene resistance to *Pyrenophora teres* f. *teres* in barley and its use for plant breeding. *Australas. Plant Pathol.* 45(5), 527–531.

Wang, X., Jiang, N., Liu, J., Liu, W. and Wang, G. L. (2014). The role of effectors and host immunity in plant-necrotrophic fungal interactions. *Virulence* 5(7), 722–732.

Wang, X., Mace, E. S., Platz, G. J., Hunt, C. H., Hickey, L. T., Franckowiak, J. D. and Jordan, D. R. (2015). Spot form of net blotch resistance in barley is under complex genetic control. *Theor. Appl. Genet.* 128(3), 489–499.

Wegulo, S. N. (2011). Tan spot of cereals. *Plant Heal. Instr.*, 1–8. DOI: 10.1094/PHI-I-2011-0426-01.

Wei, B., Moscou, M. J., Sato, K., Gourlie, R., Strelkov, S. and Aboukhaddour, R. (2020). Identification of a locus conferring dominant susceptibility to Pyrenophora tritici-repentis in barley. *Front. Plant Sci.* 11, 158.

Weiergang, I., Lyngs Jørgensen, H. J., Møller, I. M., Friis, P. and Smedegaard-petersen, V. (2002). Correlation between sensitivity of barley to *Pyrenophora teres* toxins and susceptibility to the fungus. *Physiol. Mol. Plant Pathol.* 60(3), 121–129.

Weiland, J. J., Steffenson, B. J., Cartwright, R. D. and Webster, R. K. (1999). Identification of molecular genetic markers in *Pyrenophora teres* f. *teres* associated with low virulence on 'Harbin' barley. *Phytopathology* 89(2), 176–181.

Wendler, N., Mascher, M., Himmelbach, A., Bini, F., Kumlehn, J. and Stein, N. (2017). A high-density, sequence-enriched genetic map of Hordeum bulbosum and its collinearity to H. vulgare. *Plant Genome* 10(3).

Williams, K. J., Lichon, A., Gianquitto, P., Kretschmer, J. M., Karakousis, A., Manning, S., Langridge, P. and Wallwork, H. (1999). Identification and mapping of a gene conferring resistance to the spot form of net blotch (*Pyrenophora teres* f. *maculata*. *Theor. Appl. Genet.* 99(1-2), 323–327.

Williams, K. J., Platz, G. J., Barr, A. R., Cheong, J., Willsmore, K., Cakir, M. and Wallwork, H. (2003). A comparison of the genetics of seedling and adult plant resistance to the spot form of net blotch (*Pyrenophora teres* f. *maculata*). *Aust. J. Agric. Res.* 54(12), 1387–1394.

Wonneberger, R., Ficke, A. and Lillemo, M. (2017a). Mapping of quantitative trait loci associated with resistance to net form net blotch (*Pyrenophora teres* f. *teres*) in a doubled haploid Norwegian barley population. *PLoS ONE* 12(4), e0175773.

Wonneberger, R., Ficke, A. and Lillemo, M. (2017b). Identification of quantitative trait loci associated with resistance to net form net blotch in a collection of Nordic barley germplasm. *Theor. Appl. Genet.* 130(10), 2025–2043.

Wu, H.-L., Steffenson, B. J., Zhong, S., Li, Y. and Oleson, A. E. (2003). Genetic variation for virulence and RFLP markers in *Pyrenophora teres*. *Can. J. Plant Pathol.* 25(1), 82–90.

Wyatt, N. A., Richards, J. K., Brueggeman, R. S. and Friesen, T. L. (2018). Reference assembly and annotation of the *Pyrenophora teres* f. *teres* isolate 0-1. *G3 Genes|Genomes|Genetics* 8, 1–8.

Wyatt, N. A., Richards, J. K., Brueggeman, R. S. and Friesen, T. L. (2020). A comparative genomic analysis of the barley pathogen *Pyrenophora teres* f. *teres* Identifies subtelomeric Regions as Drivers of Virulence. *Mol. Plant Microbe Interact.* 33(2), 173–188.

Xie, W., Xiong, W., Pan, J., Ali, T., Cui, Q., Guan, D., Meng, J., Mueller, N. D., Lin, E. and Davis, S. J. (2018). Decreases in global beer supply due to extreme drought and heat. *Nat. Plants* 4(11), 964–973.

Yun, S. J., Gyenis, L., Hayes, P. M., Matus, I., Smith, K. P., Steffenson, B. J. and Muehlbauer, G. J. (2005). Quantitative trait loci for multiple disease resistance in wild barley. *Crop Sci.* 45(6), 2563–2572.

Zeng, X., Long, H., Wang, Z., Zhao, S., Tang, Y., Huang, Z., Wang, Y., Xu, Q., Mao, L., Deng, G., Yao, X., Li, X., Bai, L., Yuan, H., Pan, Z., Liu, R., Chen, X., WangMu, Q., Chen, M., Yu, L., Liang, J., DunZhu, D., Zheng, Y., Yu, S., LuoBu, Z., Guang, X., Li, J., Deng, C., Hu, W., Chen, C., TaBa, X., Gao, L., Lv, X., Abu, Y. B., Fang, X., Nevo, E., Yu, M., Wang, J. and Tashi, N. (2015). The draft genome of Tibetan hulless barley reveals adaptive patterns to the high stressful Tibetan Plateau. *Proc. Natl. Acad. Sci. U. S. A.* 112(4), 1095–1100 LP-1100.

Zeng, X., Guo, Y., Xu, Q., Mascher, M., Guo, G., Li, S., Mao, L., Liu, Q., Xia, Z., Zhou, J., Yuan, H., Tai, S., Wang, Y., Wei, Z., Song, L., Zha, S., Li, S., Tang, Y., Bai, L., Zhuang, Z., He, W., Zhao, S., Fang, X., Gao, Q., Yin, Y., Wang, J., Yang, H., Zhang, J., Henry, R. J., Stein, N. and Tashi, N. (2018). Origin and evolution of qingke barley in Tibet. *Nat. Commun.* 9(1), 5433.

Zeng, X., Xu, T., Ling, Z., Wang, Y., Li, X., Xu, S., Xu, Q., Zha, S., Qimei, W., Basang, Y., Dunzhu, J., Yu, M., Yuan, H. and Nyima, T. (2020). An improved high-quality genome assembly and annotation of Tibetan hulless barley. *Sci. Data* 7(1), 139.

Chapter 15

Breeding barley for durable resistance to net and spot forms of net blotch

Jerome D. Franckowiak, University of Minnesota, USA; and Gregory J. Platz, Hermitage Research Facility, Agri-Science Queensland, Australia

1 Introduction

The invention of pureline breeding by humankind caused instability in the plant-pathogen arms race. Instead of genetically diverse seed lots, farmers sow only one or a few host genotypes. Both the pathogen and host must evolve to survive in this environment. Plant breeders impact further changes in the host-pathogen interactions by incorporation of one or a few simply inherited resistance genes into new cultivars. The preferred and least expensive means of achieving durable resistance is the utilization of genes that provide adequate resistance to most, if not all, pathotypes of a given pathogen, which are known to occur in the targeted production area. This chapter will focus on the management of host-pathogen interactions involving barley (*Hordeum vulgare* L. subsp. *vulgare*) and two of its diseases: net form net blotch (NFNB), caused by *Pyrenophora teres* Drechs. f. *teres* Smedeg. (*Ptt*), and spot form net blotch (SFNB), caused by *P. teres* Drechs. f. *maculata* Smedeg (*Ptm*).

Net blotch diseases of barley cause devastating yield and quality losses worldwide. The causal organisms are facultative pathogens for which survival and genetic diversity are not dependent on their being aggressive pathogens.

http://dx.doi.org/10.19103/AS.2021.0092.33

Farmers and plant breeders have chosen resistant barley cultivars for hundreds of years, but some cultivars become highly susceptible to NFNB or SFNB within a few years after distribution to growers. Recent examples include 'Maritime' and 'Oxford' in Australia and 'Pinnacle' (PI 643354) in the United States (Ellwood et al., 2019; Koladia et al., 2017b; Linde and Smith, 2019). Achieving durable resistance is difficult because many attributes besides host-pathogen interactions contribute to the development and release of new cultivars. The epidemiology of the pathogen, the biology of the host, and target environments all have roles to play in disease management. For example, recommended control strategies in or near centers of diversity differ from those for production areas where the pathogen rarely causes losses. Barley is an annual crop sown in the fall or spring in various production areas. *Pyrenophora* species are seed-borne pathogens that can over-season on stubble. Losses occur when seedling infection by seed- or stubble-borne inoculum is followed by a rapid buildup on a highly susceptible host. Both crop rotation and seed treatments can limit the inoculum load and reduce losses caused by the net blotches; however, to minimize losses over many years in large production areas, genetic control is preferred.

2 Net form net blotch caused by *Pyrenophora teres* f. *teres*

Clare et al. (2020) summarized genetic studies on the interactions between *H. vulgare* cultivars and *P. teres* isolates. They describe both Reaction to *Pyrenophora teres* (*Rpt*) genes and quantitative trait loci (QTL) that contribute to host-pathogen interactions. This history of screening for genetic resistance reveals that some resistance factors could be a primary focus of barley breeders. Based on the severe losses caused by the sporadic occurrence of NFNB in California, Schaller and Wiebe (1952) conducted an initial field screening of accessions from the United States Department of Agriculture's World Barley Collection. Schaller (1955) and Mode and Schaller (1958) identified dominant resistance factors in spring, six-rowed accessions from the Manchurian region of China. Based on these studies, net form net blotch resistant, six-rowed cultivars were released for California including 'Kombar' (CIho 15694, Minn. 64-98-8/2*Briggs) (Matchett and Cantu, 1976). The accessions identified by Schaller were not resistant to all *Ptt* isolates collected in Manitoba; thus, Buchannon and McDonald (1965) conducted a seedling screening of accessions from the same world barley collection. Forty of the 6,174 accessions tested were resistant to the seven North American isolates of *P. teres* used. Nearly half the resistant accessions were landraces from Ethiopia. The F2 progenies from crosses to these accessions produced a large number of cross and *Ptt* isolate specific inheritance patterns (Afanasenko et al., 2007). In Canada, resistance from CIho 5791, an Ethiopian landrace, led to the development of the six-rowed spring barley breeding line BT 201 (Tekauz and Mills, 1974), the six-rowed spring cultivar 'Heartland' (Therrien and Wolfe, 1985), and the two-rowed spring cultivar 'Norbert' (Metcalfe and Bendelow, 1981).

The main genetic factor controlling response to NFNB in the Manchurian accession Tifang (PI 69426) and the Ethiopian accession CIho 9819 (PI 195985) was associated with chromosome 3H by trisomic analysis using *Ptt* isolates from North Africa (Bockelman et al., 1977). Other resistance factors were located in chromosomes 1H of Tennessee Awnless (CIho 7584) and 2H of CIho 9819. Reaction to *Pyrenophora teres* 1 (*Rpt1*), Reaction to *Pyrenophora teres* 2 (*Rpt2*), and Reaction to *Pyrenophora teres* 3 (*Rpt3*) are the recommended names and codes for these loci (Bockelman et al., 1977). Using an Australian *Ptt* isolate, Khan and Boyd (1969b) identified another gene for resistance to NFNB in Manchurian and Ethiopian accessions, which apparently differed from *Rpt1* and *Rpt3*. Graner et al. (1996) mapped moderate resistance in the European, winter barley cultivar 'Igri' (PI 428488) to the proximal region of the long arm of chromosome 3H. Steffenson et al. (1996) mapped a major effect QTL for NFNB resistance in chromosome 6H of 'Morex' (CIho 15773). Manninen et al. (2006) mapped resistance to *Ptt* isolates in CIho 9819 to the centromeric region of the long arm of chromosome 6H and named the locus Reaction to *Pyrenophora teres* 5 (*Rpt5*).

Many studies have identified QTL associated with the development of NFNB symptoms in barley (Clare et al., 2020). Each study identified multiple loci, which are spread across all seven barley chromosomes. These results suggest a long co-evolutionary history for resistance genes in *H. vulgare* and virulence genes in *Ptt* isolates (Clare et al., 2020). In developing new cultivars with durable resistance, barley breeders have sufficient resources to target only a few resistance genes because many other traits are critical components of new cultivars. Thus, breeders prefer to focus on only a few resistance genes for each disease. Using progeny of biparental crosses in several mapping studies, reactions to *Ptt* isolates were associated with the *Rpt1* locus in 3HL (Bockelman et al., 1977; Dinglasan et al., 2019; Graner et al., 1996; Lehmensiek et al., 2007), the *Rpt5* locus in 6HL (Friesen et al., 2006; Steffenson et al., 1996; St. Pierre et al., 2010), or both. Using highly resistant accessions, Hisano et al. (2017), Koladia et al. (2017a), Liu et al. (2015), Manninen et al. (2006), Novakazi et al. (2019), and Rozanova et al. (2019) identified QTL for resistance in the centromeric regions of both chromosomes 3H and 6H in H602, ND B112 (CIho 11531), Tifang, CIho 9819, and other accessions. Abu Qamar et al. (2008) observed that the susceptibility response at a 6H locus was isolate specific and behaved as a dominant trait. These results suggest that breeding efforts for NFNB resistance should focus on alleles at the *Rpt1* and *rpt5* loci.

3 Molecular markers for reactions to net form net blotch

Molecular marker data could aid in understanding of inconsistent responses to the array of *Ptt* isolates and supplement the use of seedling and adult-plant evaluations of breeding materials. Inheritance studies have associated the *Rpt1* locus with chromosome 3H (Bockelman et al., 1977) and with the centromeric region of the long arm (Graner et al., 1996; Koladia et al., 2017a; Liu et al., 2011; Raman et al., 2003; Rozanova et al., 2019). Based on linkage drag, most breeding materials should retain molecular markers within 1 cM (centiMorgan) of the critical locus. The high marker densities available using DArTseq (http://www.diversityarrays.com) (Wang et al., 2015) or 50k iSelect SNP array (Bayer et al., 2017) should identify groups of SNP molecular markers closely linked to specific alleles. These alleles are associated with 10–20 adjacent markers that have an identical pattern of marker phases or an allele-specific haplotype. Breeding lines and cultivars having the same haplotype likely have the same allele at the target locus. For the development of breeding strategies, the cM positions of loci are important for the manipulation of desirable genes. Relative cM positions of molecular markers in barley were published by Muñoz-Amatriaín et al. (2014). For example, the *Rpt1* locus is near the centromere in 3HL and near 50k SNP marker JHI-Hv50k-2016-184503, located at about 499.0 Mb (megabase pairs) and mapped at approximately 62.97 cM. Tamang et al. (2021) mapped the *HvWRKY6* locus for NFNB resistance in this region of 3H. The *rpt5* locus is near the centromere of 6H at 384.4 Mb (Daba et al., 2019) and maps near the marker JHI-Hv-2016-401470 (385.8 Mb, 59.71 cM). Alleles at other *Rpt* loci or QTL can be associated with specific marker haplotypes, but their expression seems dependent on alleles at the *rpt5* locus (Adhikari et al., 2020).

The discovery history for the *Rpt1* locus provides some information about host-pathogen interactions. Six-rowed spring barley accessions from Manchuria (the Heilongjiang, Jilin, and Liaoning Provinces of China) showed adult-plant resistance (APR) to one *Ptt* isolate in California (Schaller and Wiebe, 1952). Resistance was determined to be simply inherited (Schaller, 1955) and associated with 3H (Bockelman et al., 1977). Environmental conditions used to access seedling responses in Australia were evaluated to observe that higher infection type (IT) readings were caused by lower pre- and post-inoculation temperatures, lower light intensities, younger seedlings, and higher conidia concentrations (Khan and Boyd, 1969a). The results suggest that the accuracy of determining the presence of *Rpt1* alleles could be low based on seedling screenings. Adult-plant responses such as those reported by Schaller and Wiebe (1952) might be more definitive of resistance conferred by the *Rpt1.a1* allele of Manchurian accessions. The 50k molecular marker haplotype associated with the *Rpt1.a1* allele occurs in most six-rowed spring barley cultivars developed for the Upper Midwest of the United States. Since net form net blotch epidemics have not occurred in this production area, these cultivars provide effective NFNB resistance. The *Rpt1.a1* associated marker haplotype can be traced back to 'Barbless' (CIho 5105) released in 1929 as a selection from the cross Oderbrucker/Lion. Since the original Oderbrucker (CIho 4666) bulk was introduced to Wisconsin in 1889 using an accession from Manchuria via Russia and Canada (Wiebe and Reid, 1958), this source of NFNB resistance has been effective for over 130 years. The *Rpt1.a1* associated haplotype is present in a number of other net blotch resistant accessions: CIho 9214 (PI 186125) from South Korea, CIho 6311 and CIho 9776 (PI 195896) from Morocco, and CIho 9647 (PI 193813) from Ethiopia. The list of accessions with the *Rpt1.a1* allele also includes historic cultivars from Australia and Europe. Since another haplotype pattern exists in the CIho 9819, the source of the 3HL QTL used by Manninen et al. (2006), the allele symbol *Rpt1.a2* is suggested for this allele.

Because all cultivars with the *Rpt1.a1* or *Rpt1.a2* genes do not exhibit low ITs to all *Ptt* isolates, Richards et al. (2016) suggested that necrotrophic effectors of specific *Ptt* isolates and *Rpt* genes at the second locus alter the expression of *Rpt1* alleles. Even though TR251 (TR229//AC Oxbow/ND7556) has the haplotype associated with the presence of the *Rpt1.a1* allele, Grewal et al. (2008, 2012) did not detect a QTL for *Ptt* response in 3HL based on seedling or adult-plant screenings. Instead, the resistant reactions were associated with a QTL at or near the *rpt5* locus in 6HL. Adhikari et al. (2020) failed to detect resistance at the *Rpt1* locus based on seedling tests of breeding materials from several US barley programs because either seedling tests failed to identify differential responses or there exists a very high frequency of resistant alleles at the *Rpt1* locus. Adhikari et al. (2020) observed a QTL for NFNB resistance in 6H near the *rpt5* locus in most of the six populations of breeding material evaluated. Wallwork et al. (2016) demonstrated that all cultivars with highly susceptible adult-plant IT scores to specific *Ptt* isolates had high seedlings IT readings to the same isolates. However, many other cultivars with high seedling readings had low adult-plant scores. Many cultivars listed by Wallwork et al. (2016) as having APR to *Ptt* isolates collected in South Australia have the haplotype associated with the presence of the *Rpt1.a1* allele.

The most frequently identified source of resistance to *Ptt* isolates involves alleles at the *rpt5* locus in 6HL (Clare et al., 2020). Abu Qamar et al. (2008) identified two recessive genes at this locus with the 'Rika' (PI 189762) allele (*rpt.r*) conferring resistance to California isolate 15A and the Kombar allele (*rpf.k*) conferring resistance to California isolate 6A. Gupta et al. (2011) and Fowler et al. (2017) reported a similar relationship between barley cultivars and *Ptt* isolates in Australia where 'Prior' (PI 235185) is susceptible to one group of isolates while 'Skiff' (PI 573727) is susceptible to another group. Liu et al. (2015) demonstrated that necrosis of leaf tissue around lesions is a necrotrophic effector-triggered susceptibility response and involved specific *rpt5* alleles. Richards et al. (2016) suggested renaming this locus as Susceptibility to *Pyrenophora teres* f. *teres* and using the locus symbol *Spt1*. The original designation by Manninen et al. (2006) for this locus is applied as a recessive symbol, *rpt5*, in this report with designations *rpt5.k* and *rpt5.r* for the 'Kombar' and 'Rika' alleles, respectively.

Worldwide, the most effective *rpt5* allele is *rpt5.f*, which was initially identified in Ethiopian accessions by Buchannon and McDonald (1965). Later, the presence of this allele was reported in accessions from many other countries (Amezrou et al., 2018; Çelik Oğuz et al., 2017; Daba et al., 2019; Fowler et al., 2017; Grewal et al., 2008, 2012; Gupta et al., 2011; Manninen et al., 2006; Martin et al., 2018; Richards et al., 2017; Rozanova et al., 2019). The molecular marker haplotype associated with the presence of the *rpt5.f* allele in CIho 5791, BT 201, 'Norbert', and TR251 is also found in the accessions CIho 3576 from Egypt; CIho 9819, CIho 9825, and Kober (PI 60694) from Ethiopia; CIho 9831 (PI 197102) from Japan; and c-8755 from Ethiopia via Russia. The same haplotype is present in Western Canadian cultivars and breeding lines, in Australian cultivars 'Scrabble' and 'Vlamingh,' in the Uruguayan cultivar 'AmBev 293' (ND17293, PI 643312), and in North Dakotan cultivars 'Conlon' (PI 597789) and 'ND Genesis' (PI 677345). Unfortunately, the *rpt5.f* gene is not associated with durable resistance in all barley production areas. CIho 5791, CIho 9819, and/or CIho 9820 are susceptible to isolates in Ethiopia (Bekele Hundie et al., 2011), in Tunisia and Syria (Bouajila et al., 2011), and in Turkey (Çelik Oğuz and Karakaya, 2017).

4 Population dynamics of barley-*P. teres* f. *teres* interactions

Development of barley breeding materials with durable resistance to NFNB depends in part on the understanding of host-pathogen interactions specific to the barley-*Ptt* pathosystem. Even though Abu Qamar et al. (2008) observed specific *Ptt* isolate-cultivar interactions, the pathosystem is much more complex than the gene-for-gene model developed by Flor (1955). Wallwork et al. (2016) demonstrated that the diversity among *Ptt* isolates changes rapidly over time, and it is at least partially independent on the cultivars grown locally. Fowler et al. (2017) identified more host-pathogen response patterns than can be explained by pathogen virulence genes or alleles at the *rpt5* locus. For example, the South Australian cultivars 'Clipper' (PI 349366), 'Keel,' and 'Franklin' have the haplotype associated with the 'Rika' allele (*rpt5.r*), but very different seedling reactions to a large collection of Australian *Ptt* isolates. Other examples of a large difference in cultivars responses are 'Buloke' and 'Commander' with *rpt5.r* and 'Maritime' and 'Skiff' with the Stiff allele (*rpt5.s*) (Fowler, 2018). Linde and Smith (2019) demonstrated that *Ptt* isolates from 'Maritime' have a mating-type ratio of 1:1 and contain considerable microsatellite diversity, but likely arose from one initial change in virulence. Primary transmission of *Ptt* as a seed-borne pathogen can explain the close relationships among *Ptt* isolates from 'Maritime,' but the evolution of genetic diversity and maintaining the mating-type ratio require frequent matings with non-pathogenic isolates. Thus, a highly susceptible host can influence the genetic diversity in *Ptt* populations, but virulence factors are likely not lost in the absence of a susceptible host. Çelik Oğuz and Karakaya (2017) reported that nearly 40% of the *Ptt* isolates collected in Turkey are avirulent on net blotch differentials while only a small portion is virulent on a large number of differentials. Shjerve et al. (2014) and Wyatt et al. (2018) reported that alleles at the *rpt5* locus interact with at least two virulence factors in compatible *Ptt* isolates.

If certain host genes and gene combinations improve pathogen resistance, they should accumulate in wild relatives and landraces. Many wild barley (*H. vulgare* spp. *spontaneum*) accessions have resistant seedling reactions to *Ptt* isolates from Canada (Jana and Bailey, 1995), from Japan (Sato and Takeda, 1997), and from North Dakota (Fetch et al., 2003). Using highly virulent *Ptt* isolates collected in Turkey, Çelik Oğuz et al. (2019) found that only a small portion of the Turkish wild barley accessions had low IT readings. With the same *Ptt* isolates, Çelik Oğuz et al. (2017a) reported a low frequency of resistant accessions among Turkish landraces. Adhikari et al. (2019) recorded only a small portion of the landraces from Ethiopia and Eritrea as having very low IT reactions to *Ptt* isolates collected in Minnesota and North Dakota. These studies indicate that no one highly resistant host gene such as *rpt5.f* accumulated in wild barley populations or landraces.

Host-specific toxins or necrotrophic effectors are small proteins that cause chlorosis and necrosis around NFNB lesions in barley (Liu et al., 2015). Effectors of many, but not all, *Ptt* isolates studied using recombinant inbred lines (RILs) from a cross between 'Hector' (CIho 15514) and ND B112 exhibited response QTL in the centromeric region of 6H (Liu et al., 2015). Richards et al. (2016) demonstrated that effectors caused the isolate specific responses in 'Rika' and 'Kombar' at the *rpt5* locus. The array of necrotrophic effectors might explain why Steffenson and Webster (1992) and Wu et al. (2003) found many *Ptt* isolates from California virulent on 'Prato' (CIho 15815) seedlings and few virulent on accessions from Manchuria while Liu et al. (2012) found most North Dakotan isolates avirulent on 'Prato' and virulent on Manchurian accessions. All isolates were avirulent on seedlings of CIho 5791 and 'Heartland,' which have the *rpt5.f* allele. Even though the *rpt5.f* allele is likely present in landraces from Ethiopia and Eritrea, Adhikari et al. (2019) identified a significant QTL for seedling disease scores in 6HL near the *rpt5* locus with only one of four *Ptt* isolates from Minnesota and North Dakota. Liu et al. (2015) and Richards et al. (2016) referred to the relationship between *Ptt* effectors and alleles at the *rpt5* locus as necrotrophic effector-triggered susceptibility.

Since alleles at the *rpt5* locus are largely restricted to interactions with necrotrophic effectors, genes at other loci must play a role in resistance to *Ptt* isolates. Biparental mapping studies have identified many genomic regions associated with resistance to NFNB, and genome-wide association studies (GWAS) have found many QTL involved in reactions between barley and *Ptt* isolates (reviewed by Clare et al., 2020). Adhikari et al. (2020) identified 12 trait-marker associations in breeding lines from ten US barley programs. In most six-rowed barley programs, they found a QTL near the *rpt5* locus in 6HL and occasionally a QTL in the terminal region of 3HS at about 3.7 Mb (6.46 cM). In the North Dakota two-rowed program, a QTL in the centromeric region of 4HL was located at about 70.4 Mb (53.87 cM), near a locus previously assigned the locus symbol *Rpt7* (Clare et al., 2020). Based on a GWAS of a core collection from the world barley collection held in the United States, Richards et al. (2017) identified 16 QTL for seedling responses to three *Ptt* isolates. Using three isolates in seedling tests and adult-plant trials at three locations, Novakazi et al. (2019) mapped QTL in 15 genomic regions in a GWAS of the Russian world barley collection. Both studies reported a large effect QTL in 6H, but this centromeric region covered 10 cM interval, likely caused by linkage drag and several *rpt5* alleles. König et al. (2014) found QTL for NFNB reactions in six genomic locations in winter barley, but not in 6HL. Rozanova et al. (2019) mapped four QTL besides the *rpt5* region in a Siberian collection of spring barleys. Koladia et al. (2017a) demonstrated that RILs with the *rpt5.k* allele from Tifang and different alleles at the *Rpt1* locus had differential responses to *Ptt* isolates. The differential response to *Ptt* isolates that have effectors for the *rpt5.k* allele might have involved the *Rpt1.a1* allele of Tifang and the *Rpt1.a2* allele of CIho 5791. These studies suggest that the groupings of interacting *Ptt* isolates and cultivars developed by Fowler et al. (2017) are based largely on *Ptt* effector genes and alleles at the *rpt5* locus. However, specific interactions within groups are probably caused by resistances at other loci.

5 Durable resistance to net form net blotch

Because regional differences in the virulence genes are present in *Ptt* populations, breeding strategies designed to limit losses should differ among barley production areas. For example, the *rpt5.f* gene could be a long-term solution in many production areas, but it has already been defeated in the West Asia and North Africa (WANA) region. Wallwork et al. (2016) reported that Vlamingh with the *rpt5.f* allele had seedling resistance to more *Ptt* isolates than 'Clipper' and other Australian cultivars with the *rpt5.r* allele, but both types had APR. Algerian (CIho 1179, PI 539104) is resistant to all Australian *Ptt* isolates (Fowler et al., 2017) and has a unique *rpt5* allele based on its marker haplotype. Sequencing of the *rpt5* locus could identify other alleles that offer long-term protection against NFNB.

The *rpt5.f* allele is promising a source of NFNB resistance in several barley production areas, but its utilization has some breeding issues. Because *rpt5* alleles protect only against the effectors of some *Ptt* isolates, other resistance genes and QTL are needed in the genetic background for effective resistance. The most important secondary resistances might be the *Rpt1* alleles from northern China (*Rpt1.a1*) and Ethiopia (*Rpt1.a2*). The combination of the *rpt5.k* and *Rpt1.a1* genes in Midwest six-rowed, spring barleys has been an effective field resistance for over 130 years after its introduction to the Upper Midwest. This germplasm now shows higher seedling IT responses because the effectiveness of *Rpt1* alleles is relatively weak (Khan, 1969) or the *Ptt* isolates have become more virulent over time (Adhikari et al., 2020; Koladia et al., 2017a). Marker haplotypes indicate that the *Rpt1.a1* and *Rpt1.a2* alleles have accumulated in Upper Midwest two- and six-rowed barley breeding materials.

These complex host-pathogen interactions predict that disease screenings with local *Ptt* isolates cannot characterize breeding lines for the presence of specific alleles at the *Rpt1* and *rpt5* loci. In field screenings where only one *Ptt* pathotype predominates, resistant plants might have one of several *rpt5* alleles. Inoculation of leaf blades or cut leaf segments with exotic *Ptt* isolates might overcome some quarantine requirements involved in evaluation (Afanasenko et al., 2007; El-Mor et al., 2018). Molecular markers or pedigree information could help identify the specific *Rpt1* and *rpt5* alleles present in breeding lines. Seedling and adult-plant testing with local *Ptt* isolates will still be necessary to avoid unexpected changes in host-pathogen relationships.

Breeding problems can be caused by the physical chromosomal position of the *rpt5* locus because loci controlling grain protein content (*gpc1*) (Falcon et al., 2019) and plant height, the erectoides-k (*ert-k*) semidwarf (Kristensen et al., 2016), are closely linked. Low grain protein is a desirable attribute of malting barleys bred for production areas where heat stress shortens the grain-fill period. The *ert-k.32* mutant reduces plant height and limits lodging. The marker haplotype for the *gpc1.a* allele includes marker SCRI_RS_119674 (61.2 Mb, 54.10 cM), about 5.6 cM proximal from the *rpt5* locus, while the *ert-k* locus is near marker JHI-Hv50k-2016-405592 (429.4 Mb, 65.28 cM), about 5.5 cM distal from the *rpt5* locus. Lüpken et al. (2014) demonstrated that reduced recombination occurs in this region of 6H even though the molecular distances are large. The breeding line ND11231 (PI 643261) likely has the *gpc1.a* and *rpt5.f* alleles in coupling based on reactions to NFNB (Cakir et al., 2003), grain protein levels (Emebiri, 2015), and molecular marker haplotypes. Breeding lines with the *gpc1.a*, *rpt5.f*, and *ert-k.32* alleles in coupling are rare. Yet with these alleles in coupling, it should be easier to detect desirable lines phenotypically based on spike type and grain protein levels. Another potential breeding problem is the presence of the loci for reaction to barley yellow dwarf virus, *Ryd2* (Collins et al., 1996), near marker JHI-Hv50k-2016-164264 (45.3 Mb,

50.66 cM), and *Ryd3* (Lüpken et al., 2014), near marker JHI-Hv50k-2016-387365 (60.8 Mb, 55.08 cM), very close to the *Rpt1* and *rpt5* loci, respectively.

Targeting other *Rpt* genes and QTL could also be an effective breeding strategy for durable resistance, but more information about such alternatives is needed. Bekele Hundie et al. (2011) reported that effective sources of NFNB resistance are present among differential lines in Ethiopia. Although 'Buloke' (Franklin/2*VB9104) has the marker haplotype denoting the presence of the *rpt5.r* allele, Fowler et al. (2017) observed that it did not exhibit a highly susceptible seedling response to any Australian *Ptt* isolate.

6 Spot form net blotch caused by *P. teres* f. *maculata*

The review by Liu et al. (2011) provides evidence for the separation of the pathogens *P. teres* f. *teres* and *P. teres* f. *maculata* into subspecies and some information about resistant cultivars. They point out that the history of breeding for resistance to SFBN is far shorter than that for NFNB and that complex host-pathogen interactions exist. Even though many barley production areas suffer losses caused by SFNB, physical isolation and environmental conditions have slowed the spread of more aggressive *Ptm* isolates into other production areas.

Several sources of genetic resistance limit losses in yield and grain quality in barley caused by SFNB (reviewed by Clare et al., 2020). Tekauz and Mills (1974) and Tekauz and Buchannon (1977) reported differential cultivar responses to *Ptm* and *Ptt* isolates in Western Canada. Williams et al. (1999) mapped seedling resistance to SFNB in the long arm of chromosome 7H in the Australian cultivar 'Galleon' (Clipper/Hiproly//3*Proctor/CIho 3576) and suggested that the locus symbol *Rpt4* be assigned. Later, Williams et al. (2003) mapped seedling resistance to SFNB in several Australian cultivars and CIho 9214 to the same region of 7HL; however, APR was mapped to other regions of the genome. Grewal et al. (2008) reported that the Canadian breeding line TR251 has a QTL for APR (*QRpt7*) in the same region of 7HL. Subsequent biparental mapping studies in other production areas identified other resistant cultivars and QTL for low IT scores in other genomic regions. Grewal et al. (2008) identified the QTL *Qpts4* (*Rpt8.j*) for seedling resistance to SFNB and NFNB in 4HS of TR251. In cross of TR251 to a different susceptible cultivar, Grewal et al. (2012) reported QTL for seedling resistance to the same SFNB isolate in chromosomes 4HS (*Qptms4*) and 6H (*Qptms6*), but not in 7H. Habachi-Houimli et al. (2019) attempted to identify genes for resistance to Tunisian isolates of *Ptm* in the North Dakota cultivar 'Bowman' (PI 483237). Both 'Bowman' and CIho 9819 had low IT readings using a *Ptm* isolate collected in Montana (Burlakoti et al., 2017).

GWAS studies of barley populations identified other QTL for SFNB responses in all seven barley chromosomes (reviewed by Clare et al., 2020).

Wang et al. (2015) found QTL for seeding resistance and APR in all seven chromosomes of Australian breeding lines with large effect ones in 7HL. In a GWAS study of barley breeding lines from the Upper Midwest of the United States, Burlakoti et al. (2017) found QTL for seedling resistance to SNFB in all barley chromosomes, but not in 7HL. In a GWAS study of barley accessions from a world collection, Neupane et al. (2015) and Tamang et al. (2015) identified numerous pathogen-host interactions and QTL for SFNB resistance in all seven chromosomes. Only 15 of 2062 accessions had reaction scores of < 2 to all four SFNB isolates used (Neupane et al., 2015). For individual isolates, 50 had scores < 2 for the North Dakota isolate, 79 for the Australian isolate, 182 for the Denmark isolate, and 244 for the New Zealand isolate. The *Ptm* isolates generated QTL arrays with significant associations in 7–12 genomic regions for each isolate (Tamang et al., 2015). Other QTL for resistance were likely not detected because loci with minor allele frequencies of less than 5% were excluded from the analysis.

The numerous QTL for SFNB reactions suggest that *Ptm* population structure and evolutionary patterns are important in breeding for durable resistance. Tekauz (2009) reported an increase over years in the pathogenic variation of *Ptm* isolates collected in Western Canada. Gupta et al. (2012) found considerable diversity in a collection of *Ptm* isolates made in Western Australia with the most resistant cultivars being those not frequently grown in the region. He reported also that the virulence patterns of local *Ptm* isolates were different from those observed by McLean et al. (2010) for *Ptm* isolates collected in Victoria. Lehmensiek et al. (2010) reported that genetic diversity of *Ptm* isolates from Australia differs from that of isolates from South Africa. Such reports indicate that pathogenic variation among *Ptm* isolates increases with exposure to more resistant barley cultivars and that diversity in virulence is specific to given production areas. Ideally for barley breeders, a few simply inherited genes that can overcome this pathogenic diversity should accumulate in some production areas; however, this has not happened. Neupane et al. (2015) reported that many of the most resistant accessions in a world barley collection originated from Central Asia. When Tamang et al. (2019) evaluated with six *Ptm* isolates using RILs from three crosses to two of the most resistant accessions, PI 67381 from Turkistan and PI 84314 from Uzbekistan, they observed 12 QTL for seedling reactions to SFNB with the largest responses in 2HS and 7HL. The large effect QTL in 7HL exhibited a seedling response identified as monogenic dominant susceptibility, which is typical of a necrotrophic effector response (Tamang et al., 2019).

7 Molecular markers for reactions to spot form net blotch

Numerous genes and QTL involve reactions to *Ptm* isolates, but few are reported consistently. Identifying a few loci frequently associated with SFNB responses would make their accumulation in elite breeding materials easier and facilitate breeding for durable resistance. Williams et al. (2003) demonstrated that selecting for a linked molecular marker in 7HL could accumulate resistance for SFNB. In Australian cultivars, the critical region of 7HL is at about 625.8 Mb (130.44 cM) (Williams et al., 2003; Wang et al., 2015). Williams et al. (2003) named the seedling resistance locus in this region *Rpt4*, and Wang et al. (2015) identified the QTL as *QRptm7-4*. Grewal et al. (2008) reported that TR251 had a QTL for seedling resistance to SFNB in this region of 7HL, but the same QTL was not observed in the progeny of a different cross to TR251 (Grewal et al., 2012). Seedlings in both studies were inoculated with the same *Ptm* isolate. Tamang et al. (2019) reported the presence of major effect QTL in 7HL with four of the six *Ptm* isolates screened using RIL populations from crosses of the resistant accession PI 67381 to susceptible cultivars 'Pinnacle' and 'Tradition' (PI 612442). In one F2 progeny, they observed a 3:1 segregation ratio for susceptibility. They suggested that the strong response to specific isolates is caused by necrotrophic effectors. This response pattern is similar to that reported for *rpt5* alleles and *Ptt* isolates (Richards et al., 2016). Interestingly, the two *Ptm* isolates not showing a QTL in 7HL were virulent on fewer entries of the world barley collection (Neupane et al., 2015). This suggests that recessive alleles at this locus protect plants against toxins produced by more virulent *Ptm* isolates, but other genes and QTL regulate other aspects of disease development. More protective *rpt4* alleles should show a recessive seedling response to highly virulent *Ptm* isolates.

Williams et al. (2003) pointed out that the *rpt4* gene was simply inherited, but it did not confer APR. Instead they found low IT readings for adult plants associated with a more distal region of 7HL. Australian cultivars resistant to SFNB have identical molecular marker profiles in a 15 cM region of 7HL with the *rpt4* locus near the proximal end and near marker JHI-Hv50k-2016-504023 (625.7 Mb, 130.44 cM). The locus involved in APR should be near the distal end of the segment near marker SCRI_RS_174159 (638.7 Mb, 145.68). At this position, the same marker haplotype is present in SFNB resistant Australian cultivars and a few Canadian cultivars. For quick reference to this locus, the locus name Reaction to *Pyrenophora teres* 9 (*Rpt9*) and the allele symbol *Rpt9.m1* for the allele in 'Galleon' are suggested. Other alleles at these two loci can confer SFNB resistance. For example, Williams et al. (2003) demonstrated that SFNB resistance in the Japanese landrace Dairokkaku (CIho 10954) is allelic to that of Australian cultivars, but Dairokkaku has different marker haplotypes near both the *rpt4* and *Rpt9* loci. A second haplotype pattern near the *Rpt9* locus is present in the resistance cultivar 'Canela' from the ICARDA barley program in Mexico. This haplotype, *Rpt9.m2*, likely originated from Ethiopian landraces. A second haplotype at the *rpt4* locus, *rpt4.e2*, is present in the recently released Canadian cultivars and in the Ethiopian landrace Jet. Based on the examination of haplotypes near the *rpt4* and *Rpt9* loci in a limited sample of barley accessions, at least 15 different ones are associated with each locus. Since the exact location of these loci is unknown, some of the diversity in haplotypes might be the result of recombination. Additional genetic resources exist because resistant accessions identified in world collections of landraces and wild barleys have not been characterized. In addition, more research is needed on the resistance conferred by combinations of *Rpt* genes.

8 Deployment of durable resistances

Because both NFNB and SFNB cause losses in barley crops worldwide, combining durable resistance to both is desirable. Even our rudimentary knowledge of the barley-*Ptt* and barley-*Ptm* pathosystems might be adequate to chart a path toward durable resistance. Knowledge of the necrotrophic effectors present in the local pathogen populations (Koladia et al., 2017b; Liu et al., 2015) could be employed to forecast the probable success of new cultivars. Since seed- and stubble-borne inoculum tends to sustain the vulnerability of given cultivars, preventing toxin-induced necrotic lesions in seedlings might limit rapid inoculum buildup and epidemics in highly susceptible cultivars. Seed treatments or specific combinations of alleles at the *Rpt1*, *rpt4*, *rpt5*, and *Rpt9* loci could limit in crop inoculum buildup. The seedling responses of Canadian cultivars such as 'ACC Synergy' to *Ptm* and *Ptt* isolates (Akhavan et al., 2016) can be accounted for by the presence of the *rpt4.e2*, *rpt5.f*, and *Rpt9.m1* genes. The Ethiopian landrace CIho 9819, which had higher IT scores with *Ptm* isolates (Akhavan et al., 2016), has the same haplotypes at the first two loci, but a different one near the *Rpt9* locus. This suggests that cultivars with a preferred allele at only some loci might provide durable crop protection in some production areas.

Backcrossing selected alleles into elite cultivars could be a quick solution for both of these leaf spot diseases. Application of speed breeding technology (Hickey et al., 2017) can produce resistant breeding lines in a few years. The technology requires growing three to five progeny generations each year, effective seedling tests, and appropriate molecular markers. Unfortunately, close linkages to genes controlling other desirable traits could make adding a few genes from unadapted sources expensive and time consuming. In addition to the *ert-k.32* and *gpc1.a* genes mentioned above, there is a less than 1 cM separation between the *Rpt9* locus, at about 639.8 Mb, and the semidwarf gene *sdw4.ba* (Franckowiak, 2016), near marker JHI-HV50k-2016-509425 (638.0 Mb, 143.14 cM). More complex breeding and screening procedures could facilitate the accumulation of the other previously reported genes and QTL for net blotch resistance.

As reviewed by Walters et al. (2012), both host responses to pathogen effectors and local barley production practices are involved in breeding for resistance and failure of disease development. Where production practices include the application of seed treatments or foliar fungicides, a high level of net blotch resistance might not be needed. The *rpt4.a1* and *rpt5.f* alleles might delay inoculum buildup so that fungicide applications for other pests can minimize losses caused by NFNB and SFNB. The *Rpt1.a1* and *Rpt9.m1* alleles for APR could be combined with seed treatments. These alternative breeding goals suggest that adding even one resistance gene can be a viable breeding objective. This alternative might include transferring or generating new alleles with genome editing.

Breeding new barley cultivars involves combining many simply inherited traits with essential quantitative traits such as high yield, yield stability over environments, lodging resistance, and possibly good malt quality. Smith et al. (2018) outlined breeding strategies using genomic tools to improve essential, but complexly inherited traits. Yet, genes for traits such as smooth awns (*raw1.a* and *raw2.b*) and the lack of lemma barbs (*gth1.a*) might influence consumer acceptance. Local cultivars, for instance, might require durable resistance to leaf rust (*Puccinia hordei* Otth) produced by a combination of the *Rph15.ad* and *Rph20.ai* alleles for seedling and adult-plant resistance (Martin et al., 2020). Marker haplotypes for these loci are located in 2HS near marker JHI-Hv50k-2016-78541 (52.5 Mb, 53.26 cM) and in 5HS near marker JHI-Hv50k-2016-280892 (9.3 Mb, 21.24 cM), respectively. The knowledge accumulated by many barley researchers and the genetic resources preserved in barley germplasm collections will aid barley breeders in managing the long list of desirable qualitative and quantitative traits.

9 Where to look for further information

Control of losses in barley caused by the net blotches has both local and global aspects. Local reports on the control of and losses caused by the net blotches can be found in national and state reports and handbooks on barley diseases and cultivars. Examples are available at:

- pdfcopyNDSUBarley.indd (msu.edu) and Barley | Diseases and Pests, Description, Uses, Propagation (psu.edu).

For a recent summary of global information on the most common forms of *Pyrenophora teres* pathogenic on barley, see:

- Backes, A., Guerriero, G., Ait Barka, E. and Jacquard, C. (2021). *Pyrenophora teres*: taxonomy, morphology, interaction with barley, and mode of control. *Front. Plant Sci.* 12:614951.

For a summary of research on genetic resistance in barley to net blotches, see:

- Clare, S. J., Wyatt, N. A., Brueggeman, R. S. and Friesen, T. L. (2020). Research advances in the *Pyrenophora teres*–barley interaction. *Mol. Plant Pathol.* 21(2):272–288.

Summaries of current information specific barley loci and genes are published at http://www.nordgen.org/bgs. Further information about the barley cultivars and accessions mentioned in this report can be obtained from the USDA-ARS GRIN system at Search Accessions GRIN-Global (ars-grin.gov) or from NordGen at http://nordic-baltic-genebanks.org. Updated barley locus descriptions are published in the Barley Genetics Newsletter at GrainGenes | A Database for Triticeae and Avena (usda.gov).

10 References

Abu Qamar, M., Liu, Z. H., Faris, J. D., Chao, S., Edwards, M. C., Lai, Z., Franckowiak, J. D. and Friesen, T. L. (2008). A region of barley chromosome 6H harbors multiple major genes associated with net type net blotch resistance. *Theor. Appl. Genet.* 117(8):1261-1270.

Adhikari, A., Steffenson, B. J., Smith, M. J. and Dill-Macky, R. (2019). Genome-wide association mapping of seedling net form net blotch resistance in an Ethiopian and Eritrean barley collection. *Crop Sci.* 59(4):1625–1638.

Adhikari, A., Steffenson, B. J., Smith, K. P., Smith, M. J. and Dill-Macky, R. (2020). Identification of quantitative trait loci for net form net blotch resistance in contemporary barley breeding germplasm from the USA using genome-wide association mapping. *Theor. Appl. Genet.* 133(3):1019–1037.

Afanasenko, O., Mironenko, N., Filatova, O., Kopahnke, D., Krämer, I. and Ordon, F. (2007). Genetics of host-pathogen interactions in the *Pyrenophora teres* f. *teres* (net form) – barley (*Hordeum vulgare*) pathosystem. *Eur. J. Plant Pathol.* 117(3):267-280.

Akhavan, A.,Turkington, T. K., Kebede, B., Xi, K. Q., Kumar, K., Tekauz, A., Kutcher, H. R., Tucker, J. R. and Strelkov, S. E. (2016). Genetic structure of *Pyrenophora teres* f. *teres* and *P. teres* f. *maculata* populations from western Canada. *Eur. J. Plant Pathol.* 146(2):325–335.

Amezrou, R., Verma, R. P. S., Chao, S., Brueggeman, R. S., Belqadi, L., Arbaoui, M., Rehman, S. and Gyawali, S. (2018). Genome-wide association studies of net form of net blotch resistance at seedling and adult plant stages in spring barley collection. *Mol. Breed.* 38(5):58.

Bayer, M. M., Rapazote-Flores, P., Ganal, M., Hedley, P. E., Macaulay, M., Plieske, J., Ramsay, L., Russell, J., Shaw, P. D., Thomas, W. and Waugh, R. (2017). Development and evaluation of a barley 50k iselect SNP-array. *Front. Plant Sci.* 8:1792.

Bekele Hundie, Asnakech Tekalign, Fekede Abebe, Melkamu Ayalew, Getaneh Woldeab, Yonas Worku, Eshetu Abebe, and Birhanu Bekele, B. (2011). Achievements of research on barley *Helminthosporium* diseases in Ethiopia. Pp. 257-272. In: Bayeh Mulatu, and Grando, S. (Eds). *Barley Research and Development in Ethiopia. Proceedings of the 2nd National Barley Research and Development Review Workshop*. 28–30 November 2006, HARC, Holetta, Ethiopia. ICARDA, Aleppo, Syria.

Bockelman, H. E., Sharp, E. L. and Eslick, R. F. (1977). Trisomic analysis of genes for resistance to scald and net blotch in several barley cultivars. *Can. J. Bot.* 55(15):2142–2148.

Bouajila, A., Zoghlami, N., Ahmed, M. Al., Baum, M., Ghorbel, A. and Nazari, K. (2011). Comparative virulence of *Pyrenophora teres* f. *teres* from Syria and Tunisia and screening for resistance sources in barley: implications for breeding. *Lett. Appl. Microbiol.* 53(5):489–502.

Buchannon, K. W. and McDonald, W. C. (1965). Sources of resistance in barley to *Pyrenophora teres*. *Can. J. Plant Sci.* 45(2):189–193.

Burlakoti, R. R., Gyawali, S., Chao, S., Smith, K. P., Horsley, R. D., Cooper, B., Muehlbauer, G. J. and Neate, S. M. (2017). Genome-wide association study of spot form of net blotch resistance in the Upper Midwest barley breeding programs. *Phytopathology* 107(1):100–108.

Cakir, M., Gupta, S., Platz, G. J., Ablett, G. A., Loughman, R., Emebiri, L. C., Poulsen, D., Li, C. D., Lance, R. C. M., Galwey, N. W., Jones, M. G. K. and Appels, R. (2003). Mapping and validation of the genes for resistance to *Pyrenophora teres* f. *teres* in barley (*Hordeum vulgare* L.). *Aust. J. Agric. Res.* 54(12):1369–1377.

Çelik Oğuz, A. and Karakaya, A. (2017). Pathotypes of *Pyrenophora teres* on barley in Turkey. *Phytopathol. Mediterr.* 56:224–234.

Çelik Oğuz, A., Karakaya, A., Ergün, N. and Sayim, I. (2017). Turkish barley landraces resistant to net and spot forms of *Pyrenophora teres*. *Phytopathol. Mediterr.* 56:217–223.

Çelik Oğuz, A., Karakaya, A., Murat Duran, R. and Özbek, K. (2019). Identification of *Hordeum spontaneum* genotypes resistant to net blotch disease. *Tarim Bilimleri Derg.* 25(1):115–122.

Clare, S. J., Wyatt, N. A., Brueggeman, R. S. and Friesen, T. L. (2020). Research advances in the *Pyrenophora teres*–barley interaction. *Mol. Plant Pathol.* 21(2):272–288.

Collins, N. C., Paltridge, N. G., Ford, C. M. and Symons, R. H. (1996). The *Yd2* gene for barley yellow dwarf virus resistance maps close to the centromere on the long arm of barley chromosome 3. *Theor. Appl. Genet.* 92(7):858–864.

Daba, S. D., Horsley, R., Brueggeman, R., Chao, S. and Mohammadi, M. (2019). Genome-wide association studies and candidate gene identification for leaf scald and net blotch in barley (*Hordeum vulgare* L.). *Plant Dis.* 103(5):880–889.

Dinglasan, E., Hickey, L., Ziems, L., Fowler, R., Anisimova, A., Baranova, O., Lashina, N. and Afanasenko, O. (2019). Genetic characterization of resistance to *Pyrenophora teres* f. *teres* in the International Barley Differential Canadian Lake Shore. *Front. Plant Sci.* 10:326.

Ellwood, S. R., Piscetek, V., Mair, W. J., Lawrence, J. A., Lopez-Ruiz, F. J. and Rawlinson, C. (2019). Genetic variation of *Pyrenophora teres* f. *teres* isolates in Western Australia and emergence of a Cyp51A fungicide resistance mutation. *Plant Pathol.* 68(1):135–142.

El-Mor, I. M., Fowler, R. A., Platz, G. J., Sutherland, M. W. and Martin, A. (2018). An improved detached-leaf assay for phenotyping net blotch of barley caused by *Pyrenophora teres*. *Plant Dis.* 102(4):760–763.

Emebiri, L. C. (2015). Breeding malting barley for consistently low grain protein to sustain production against predicted changes from global warming. *Mol. Breed.* 35(1):18.

Falcon, C. M., Horsley, R., Hu, G., Blake, T. and Smith, K. P. (2019). Mapping QTLs for grain protein concentration and agronomic traits under different nitrogen levels in barley. *Crop Sci.* 59(1):68–83.

Fetch, T. G. Jr., Steffenson, B. J. and Nevo, E. (2003). Diversity and sources of multiple disease resistance in *Hordeum spontaneum*. *Plant Dis.* 87(12):1439–1448.

Flor, H. H. (1955). Host-parasite interaction in flax rust–its genetics and other implications. *Phytopathology* 45:680–685.

Fowler, R. A. (2018). Pathogenic variation of *Pyrenophora teres* f. *teres* on *Hordeum vulgare* in Australia and identification of genomic regions for resistance and susceptibility to net form net blotch. Ph.D. Thesis, University Queensland, Australia.

Fowler, R. A., Platz, G. J., Bell, K. L., Fletcher, S. E. H., Franckowiak, J. D. and Hickey, L. T. (2017). Pathogenic variation of *Pyrenophora teres* f. *teres* in Australia. *Australas. Plant Pathol.* 46(2):115-128.

Franckowiak, J. D. (2016). BGS 045, Semidwarf 4, *sdw4*, revised. *Barley Genet. Newsl.* 46:51-52.

Friesen, T. L., Faris, J. D., Lai, Z. and Steffenson, B. J. (2006). Identification and chromosomal location of major genes for resistance to *Pyrenophora teres* in a barley doubled haploid population. *Genome* 49:855-859.

Graner, A., Foroughi-Wehr, B. and Tekauz, A. (1996). RFLP mapping of a gene in barley conferring resistance to net blotch (*Pyrenophora teres*). *Euphytica* 91:229-234.

Grewal, T. S., Rossnagel, B. G., Pozniak, C. J. and Scoles, G. J. (2008). Mapping quantitative trait loci associated with barley net blotch resistance. *Theor. Appl. Genet.* 116(4):529-539.

Grewal, T. S., Rossnagel, B. G. and Scoles, G. J. (2012). Mapping quantitative trait loci associated with spot blotch and net blotch resistance in a doubled-haploid barley population. *Mol. Breed* 30:267-279.

Gupta, S., Li, C., Loughman, R., Cakir, M., Westcott, S. and Lance, R. (2011). Identifying genetic complexity of 6H locus in barley conferring resistance to *Pyrenophora teres* f. *teres*. *Plant Breed.* 130(4):423-429.

Gupta, S., Loughman, R., D'Antuono, M. and Bradley, J. (2012). Characterisation and diversity of *Pyrenophora teres* f. *maculata* isolates in Western Australia. *Australas. Plant Pathol.* 41(1):31-40.

Habachi-Houimli, Y., Cherif, M., Gharsallah, C., Sébéi, A., Makni, M. and Bouktila, D. (2019). Expression analysis of *Pyrenophora teres* f. *maculata*-responsive loci in *Hordeum vulgare*. *Braz. Arch. Biol. Technol.* 62:e19180331.

Hickey, L. T., Germán, S. E., Pereyra, S. A., Diaz, J. E., Ziems, L. A., Fowler, R. A., Platz, G. J., Franckowiak, J. D. and Dieters, M. J. (2017). Speed breeding for multiple disease resistance in barley. *Euphytica* 213(3):64.

Hisano, H., Sakamoto, K., Takagi, H., Terauchi, R. and Sato, K. (2017). Exome QTL-seq maps monogenic locus and QTLs in barley. *BMC Genomics* 18(1):125.

Jana, S. and Bailey, K. L. (1995). Responses of wild and cultivated barley from West Asia to net blotch and spot blotch. *Crop Sci.* 35(1):242-246.

Khan, T. K. (1969). Inheritance of resistance to net blotch in barley. I. Factors affecting the penetrance and expressivity of gene(s) conditioning resistance. *Can. J. Genet. Cytol.* 11:587-591.

Khan, T. N. and Boyd, W. J. R. (1969a). Environmentally induced variability in the host reaction of barley to net blotch. *Aust. J. Biol. Sci.* 22(5):1237-1244.

Khan, T. N. and Boyd, W. J. R. (1969b). Inheritance of resistance to net blotch in barley. II. Genes conditioning resistance against race W.A.-2. *Can. J. Genet. Cytol.* 11(3):592-597.

Koladia, V. M., Faris, J. D., Richards, J. K., Brueggeman, R. S., Chao, S. and Friesen, T. L. (2017a). Genetic analysis of net form net blotch resistance in barley lines CIho 5791 and Tifang against a global collection of *P. teres* f. *teres* isolates. *Theor. Appl. Genet.* 130(1):163-173.

Koladia, V. M., Richards, J. K., Wyatt, N. A., Faris, J. D., Brueggeman, R. S. and Friesen, T. L. (2017b). Genetic analysis of virulence in the *Pyrenophora teres* f. *teres* population BB25×FGOH04Ptt-21. *Fungal Genet. Biol.* 107:12-19.

König, J., Perovic, D., Kopahnke, D. and Ordon, F. (2014). Mapping seedling resistance to net form of net blotch (Pyrenophora *teres*) in barley using detached leaf assa. *Plant Breed.* 133(3):356-365.

Kristensen, P. S., Dockter, C., Lundqvist, U., Lu, Q., Gregersen, P. L., Thordal-Christensen, H. and Hansson, M.. (2016). Genetic mapping of the barley lodging resistance locus *Erectoides-k*. *Plant Breed.* 135(4):420-428.

Lehmensiek, A., Bester, A. E., Sutherland, M. W., Platz, G., Kriel, W. M., Potgieter, G. F. and Prins, R. (2010). Population structure of South African and Australian *Pyrenophora teres* isolates. *Plant Pathol.* 59(3):504-515.

Lehmensiek, A., Platz, G. J., Mace, E., Poulsen, D. and Sutherland, M. W. (2007). Mapping of adult plant resistance to net form of net blotch in three Australian barley populations. *Aust. J. Agric. Res.* 58(12):1191- 1197.

Linde, C. C. and Smith, L. M. (2019). Host specialisation and disparate evolution of Pyrenophora *teres* f. *teres* on barley and barley grass. *BMC Evol. Biol.* 19(1):139.

Liu, Z., Ellwood, S. R., Oliver, R. P. and Friesen, T. L. (2011). *Pyrenophora teres* profile of an increasingly damaging barley pathogen. *Mol. Plant Pathol.* 12(1):1-19.

Liu, Z., Holmes, D. J., Faris, J. D., Chao, S., Brueggeman, R. S., Edwards, M. C. and Friesen, T. L. (2015). Necrotrophic effector-triggered susceptibility (NETS) underlies the barley-*Pyrenophora teres* f. *teres* interaction specific to chromosome 6H. *Mol. Plant Pathol.* 16(2):188-200.

Liu, Z. H., Zhong, S., Stasko, A. K., Edwards, M. C. and Friesen, T. L. (2012). Virulence profile and genetic structure of a North Dakota population of *Pyrenophora teres* f. *teres*, the causal agent of net form net blotch of barley. *Phytopathology* 102(5):539-546.

Lüpken, T., Stein, N., Perovic, D., Habekuß, A., Serfling, A., Krämer, I., Hähnel, U., Steuernagel, B., Scholz, U., Ariyadasa, R., Martis, M., Mayer, K., Niks, R. E., Collins, N. C., Friedt, W. and Ordon, F. (2014). High-resolution mapping of the barley *Ryd3* locus controlling tolerance to BYDV. *Mol. Breed.* 33(2):477-488.

Manninen, O. M., Jalli, M., Kalendar, R., Schulman, A., Afanasenko, O. and Robinson, J. (2006). Mapping of major spot-type and net-type net-blotch resistance genes in the Ethiopian barley line CI 9819. *Genome* 49(12):1564-1571.

Martin, A., Platz, G. J., de Klerk, D., Fowler, R. A., Smit, F., Potgieter, F. G. and Prins, R. (2018). Identification and mapping of net form of net blotch resistance in South African barley. *Mol. Breed.* 38(5):53.

Martin, M. J., Chicaiza, O., Caffarel, J. C., Sallam, A. H., Druka, A., Waugh, R., Ordon, F., Kopahnke, D., Keilwagen, J., Perovic, D., Fetch, T. G. Jr., Jin, Y., Franckowiak, J. D. and Steffenson, B. J. (2020). Development of barley introgression lines carrying the leaf rust resistance genes *Rph1* to *Rph15*. *Crop Sci.* 60(1):282-302.

Matchett, R. W. and Cantu, O. P. (1976). Registration of Kombar barley (Reg. No. 149). *Crop Sci.* 16(5):738.

McLean, M. S., Keiper, F. J. and Hollaway, G. J. (2010). Genetic and pathogenic diversity in *Pyrenophora teres* f. *maculata* in barley crops of Victoria, Australia. *Australas. Plant Pathol.* 39(4):319-325.

Metcalfe, D. R. and Bendelow, V. M. (1981). Norbert barley. *Can. J. Plant Sci.* 61(4):1005-1007.

Mode, C. J. and Schaller, C. W. (1958). Two additional factors for host resistance to net blotch in barley. *Agron. J.* 50(1):15-18.

Muñoz-Amatriaín, M., Cuesta-Marcos, A., Endelman, J. B., Comadran, J., Bonman, J. M., Bockelman, H. E., Chao, S., Russell, J., Waugh, R., Hayes, P. M. and Muehlbauer, G. J.

(2014). The USDA barley core collection: genetic diversity, population structure, and potential for genome-wide association studies. *PLoS ONE* 9(4):e94688.

Neupane, A., Tamang, P., Brueggeman, R. S. and Friesen, T. L. (2015). Evaluation of a barley core collection for spot form net blotch reaction reveals distinct genotype-specific pathogen virulence and host susceptibility. *Phytopathology* 105(4):509–517.

Novakazi, F., Afanasenko, O., Anisimova, A., Platz, G. J., Snowdon, R., Kovaleva, O., Zubkovich, A. and Ordon, F. (2019). Genetic analysis of a worldwide barley collection for resistance to net form of net blotch disease (*Pyrenophora teres* f. *teres*). *Theor. Appl. Genet.* 132(9):2633–2650.

Raman, H., Platz, G. J., Chalmers, K. J., Raman, R., Read, B. J., Barr, A. R. and Moody, D. B. (2003). Mapping of genomic regions associated with net form of net blotch resistance in barley. *Aust. J. Agric. Res.* 54(12):1359–1367.

Richards, J., Chao, S., Friesen, T. and Brueggeman, R. (2016). Fine mapping of the barley chromosome 6h net form net blotch susceptibility locus. *G3* 6(7):1809–1818.

Richards, J. K., Friesen, T. L. and Brueggeman, R. S. (2017). Association mapping utilizing diverse barley lines reveals net form net blotch seedling resistance/susceptibility loci. *Theor. Appl. Genet.* 130(5):915–927.

Rozanova, I. V., Lashina, N. M., Mustafin, Z. S., Gorobets, S. A., Efimov, V. M., Afanasenko, O. S. and Khlestkina, E. K. (2019). SNPs associated with barley resistance to isolates of Pyrenophora *teres*. *BMC Genomics* 20(Suppl 3):292.

Sato, K. and Takeda, K. (1997). Net blotch resistance in wild species of *Hordeum*. *Euphytica* 95(2):179–185.

Schaller, C. W. (1955). Inheritance of resistance to net blotch of barley. *Phytopathology* 45:174–176.

Schaller, C. W. and Wiebe, G. A. (1952). Sources of resistance to net blotch of barley 1. *Agron. J.* 44(6):334–336.

Shjerve, R. A., Faris, J. D., Brueggeman, R. S., Yan, C., Zhu, Y., Koladia, V. and Friesen, T. L. (2014). Evaluation of a *Pyrenophora teres* f. *teres* mapping population reveals multiple independent interactions with a region of barley chromosome 6H. *Fungal Genet. Biol.* 70:104–112.

Smith, K. P., Thomas, W., Gutierrez, L. and Bull, H. (2018). Genomics-based barley breeding. Pp. 287–315. In: Stein, N. and Muehlbauer, G. J. (Eds.) *The Barley Genome*. Springer Nature, Switzerland.

St. Pierre, S., Gustus, C., Steffenson, B., Dill-Macky, R. and Smith, K. P. (2010). Mapping net form net blotch and Septoria speckled leaf blotch resistance loci in barley. *Phytopathology* 100(1):80–84.

Steffenson, B. J. and Webster, R. K. (1992). Pathotype diversity of *Pyrenophora teres* f. *teres* on barley. *Phytopathology* 82(2):170–177.

Steffenson, B. J., Hayes, P. M. and Kleinhofs, A. (1996). Genetics of seedling and adult plant resistance to net blotch (*Pyrenophora teres* f. *teres*) and spot blotch (*Cochliobolus sativus*) in barley. *Theor. Appl. Genet.* 92(5):552–558.

Tamang, P., Neupane, A., Mamidi, S., Friesen, T. and Brueggeman, R. (2015). Association mapping of seedling resistance to spot form net blotch in a worldwide collection of barley. *Phytopathology* 105(4):500–508.

Tamang, P., Richards, J. K., Alhashal, A., Poudel, R. S., Horsley, R. D., Friesen, T. L. and Brueggeman, R. S. (2019). Mapping of barley susceptibility/ resistance QTL against spot form net blotch caused by *Pyrenophora teres* f. *maculata* using RIL populations. *Theor. Appl. Genet.* 132(7):1953–1963.

Tamang, P., Richards, J. K., Solanki, S., Ameen, G., Poudel, R. S., Deka, P., Effertz, K., Clare, S. J., Hegstad, J., Bezbaruah, A., Li, X., Horsley, R. D., Friesen, T. L. and Brueggeman, R. S. (2021). The barley *HvWRKY6* transcription factor is required for resistance against *Pyrenophora teres* f. *teres*. *Front. Genet.* 11:601500.

Tekauz, A. (2009).Characterization and distribution of pathogenic variation in *Pyrenophora teres* f. *teres and P. teres f. maculata* from Western Canada. *Can. J. Plant Pathol.* 1990:141–148.

Tekauz, A. and Buchannòn, K. W. (1977). Distribution of and sources of resistance to biotypes of *Pyrenophora teres* in Western Canada. *Can. J. Plant Sci.* 57:389–395.

Tekauz, A. and Mills, J. T. (1974). New types of virulence in *Pyrenophora teres* in Canada. *Can. J. Plant Sci.* 54(4):731–734.

Therrien, M. C. and Wolfe, R. I. (1985). Heartland barley. *Can. J. Plant Sci.* 65(2):445–446.

Wallwork, H., Butt, M. and Capio, E. (2016). Pathogen diversity and screening for minor gene resistance to *Pyrenophora teres* f. *teres* in barley and its use for plant breeding. *Australas. Plant Pathol.* 45(5):527–531.

Walters, D. R., Avrova, A., Bingham, I. J., Burnett, F. J., Fountaine, J., Havis, N. D., Hoad, S. P., Hughes, G., Looseley, M., Oxley, S. J. P., Renwick, A., Topp, C. F. E. and Newton, A. C. (2012). Control of foliar diseases in barley: towards an integrated approach. *Eur. J. Plant Pathol.* 133(1):33–73.

Wang, X., Mace, E. S., Platz, G. J., Hunt, C. H., Hickey, L. T., Franckowiak, J. D. and Jordan, D. R. (2015). Spot form of net blotch resistance in barley is under complex genetic control. *Theor. Appl. Genet.* 128(3):489–499.

Wiebe, G. A. and Reid, D. A. (1958). Classification of barley varieties grown in the United States. *U. S Dept. Agr. Tech. Bul.* 1224, 234 pp. Illus.

Williams, K. J., Lichon, A., Gianquitto, P., Kretschmer, J. M., Karakousis, A., Manning, S., Langridge, P. and Wallwork, H. (1999). Identification and mapping of a gene conferring resistance to the spot form of net blotch (*Pyrenophora teres* f. *maculata*. *Theor. Appl. Genet.* 99(1-2):323–327.

Williams, K. J., Platz, G. J., Barr, A. R., Cheong, J., Willsmore, K., Cakir, M. and Wallwork, H. (2003). A comparison of the genetics of seedling and adult plant resistance to the spot form of net blotch (*Pyrenophora teres* f. *maculata*). *Aust. J. Agric. Res.* 54(12):1387–1394.

Wu, H.-L., Steffenson, B. J., Zhong, S., Li, Y. and Oleson, A. E. (2003). Genetic variation for virulence and RFLP markers in *Pyrenophora teres*. *Can. J. Plant Pathol.* 25(1):82–90.

Wyatt, N. A., Richards, J. K., Brueggeman, R. S. and Friesen, T. L. (2018). Reference assembly and annotation of the *Pyrenophora teres* f. *teres* isolate 0–1. *G3,* 8:1–8.

Part 6

Fungal diseases of cereals: tan spot, blast and Ramularia

Chapter 16

Tan spot disease under the lenses of plant pathologists

Reem Aboukhaddour and Mohamed Hafez, Agriculture and Agri-Food Canada, Canada; Stephen E. Strelkov, University of Alberta, Canada; and Myriam R. Fernandez, Agriculture and Agri-Food Canada, Canada

1 Introduction

Tan spot is a relatively new yet one of the most destructive foliar diseases of wheat worldwide (Savary et al., 2019). The causal agent, *Pyrenophora tritici-repentis* (Died.) Drechs. (Ptr), is an ascomycete fungus first identified as a pathogen of wheat in the 1920s (Nisikado, 1928). Until about 1970, Ptr was reported from many parts of the world, causing no or mild symptoms and occasional severe but localized outbreaks. The damaging effects of tan spot, however, became more pronounced starting in the 1970s, especially in the temperate regions of Australia, the American Great Plains, and the Canadian Prairies. On susceptible wheat, a yield loss of 49% was reported under conditions conducive for disease development (Rees et al., 1982). The emergence of tan spot as a major disease in North America and Australia was explained mainly by the shift in agricultural practices to minimum tillage, as Ptr is stubble-borne and survives on crop residues. The widespread cultivation of susceptible wheat, and the horizontal gene transfer of *ToxA*, a necrosis-inducing effector, to Ptr from a related fungal pathogen, are now also recognized as essential factors in its sudden emergence (Friesen et al., 2006; McDonald et al., 2019).

Early research on tan spot and its causal agent was of fundamental value in understanding its classification, life cycle, host range, cytology of infection, physiological specialization, epidemiology, and in establishing the methodology

http://dx.doi.org/10.19103/AS.2021.0092.19
Published by Burleigh Dodds Science Publishing Limited, 2022.

and protocols for artificial infection and symptom rating (reviewed by De Wolf et al., 1998). Today, the Ptr–wheat interaction serves as a model system to understand necrotrophic effectors, their interactions with the host, and the evolution of virulence. At the time when tan spot was recognized as a new emerging disease, host specificity concepts explained by the inverse gene-for-gene interactions (toxin-model) were already established for many fungal necrotrophic pathogens (Pringle and Scheffer, 1964). Leading research in Japan and the USA on the sudden rise of several epidemics caused by species of *Alternaria* and *Cochliobolus* was very active and confirmed the involvement of host-selective toxins (Yoder, 1980; Turgeon and Baker, 2007; Tsuge et al., 2013; Lorang, 2019).

In Ptr, the purification of ToxA, the most important virulence factor known so far, and the cloning of its coding gene by different research groups working around the same time in Canada and the USA (reviewed by Ciuffetti et al., 2014) was a milestone in shaping subsequent research in the Ptr–host interaction and related species, such as *Parastagonospora nodorum* (septoria nodorum blotch), which encode the same *ToxA* gene (Friesen et al., 2006). Tomas and Bockus (1987) provided the first evidence for the presence of a potential necrosis-causing effector secreted in the culture filtrate of producing Ptr isolates, which was subsequently purified as the ToxA protein (Ballance et al., 1989; Tomas et al., 1990; Tuori et al., 1995; Zhang et al., 1997). Later, the ability of Ptr to cause chlorosis in susceptible genotypes was linked to the production of additional, chlorosis-inducing, necrotrophic effectors. These findings facilitated the purification of the second protein effector, ToxB, at the University of Manitoba, Canada (Strelkov et al., 1998, 1999).

On the wheat side, the identification of independently inherited and dominant susceptibility genes that condition sensitivity to each effector has also contributed to understanding the genetics of the Ptr–wheat interaction (reviewed by Faris et al., 2013). More recently, the identification of a dominant single locus controlling susceptibility to Ptr in barley revealed the ability of Ptr to interact in a specific manner with a secondary host (Aboukhaddour and Strelkov, 2016; Wei et al., 2020). The cloning of the ToxA-sensitivity gene in wheat, the *Tsn1* gene (Faris et al., 2010), showed that Ptr can 'hijack' the plant defense mechanism to cause necrosis, and this finding enabled a better understanding of the whole pathogenicity mechanism used by necrotrophic fungi.

In the past 15 years, with the advancement in genomic tools, and more recently with the release of full genome sequences and the cloning of *Tsn1* and several other genes corresponding to host sensitivity to a closely related species, *Parastagonospora nodorum* (reviewed by Faris and Friesen, 2020), the evolution and dynamics of Ptr and its effectors are under investigation. Ptr effectors are unique in that the coding genes or their homologs are found in several ascomycete species (Ciuffetti et al., 2014; Hafez et al., 2020). The *ToxA* gene presented the first proof of a horizontal gene transfer in the fungal kingdom (Friesen et al., 2006). Yet, despite the improved understanding of the Ptr-wheat interaction, challenges remain in mitigating the impact of tan spot, especially in regions of the world where susceptible wheat is still cultivated over large areas of land with fungicide application as the main disease management strategy. The accurate identification of the various pathogens in the leaf spot disease complex is an obstacle in many parts of the world. The genomic studies to date have been limited to a small number of races and isolates, and the identification of additional effectors are all issues the coming research will hopefully address. In this chapter, we will highlight the advances in Ptr research over the past 50 years from a plant pathology point of view.

2 *Pyrenophora tritici-repentis*: the fungus

The genus *Pyrenophora* represents a monophyletic group within the family Pleosporaceae, order Pleosporales, class Dothideomycetes (previously known as the Loculoascomycetes), which is the largest class within the largest fungal phylum, Ascomycota (Kirk et al., 2008). The Dothideomycetes include more than 19000 species of diverse life styles associated with a wide range of hosts/substrates, including numerous plant-pathogenic and non-plant-pathogenic species. In addition to parasites, foliar endophytes, ectophytes, epiphytes, and several other life style types are found within the Dothideomycetes (Haridas et al., 2020). Pleosporales is the largest order in this group, with a quarter of all species (Kirk et al., 2008), and includes several plant pathogens that are adapted to necrotrophy, including species attacking cereals in the genera *Pyrenophora*, *Cochliobolus,* and *Phaeosphaeria*, and plant pathogens that cluster closely to saprobes (Haridas et al., 2020).

Pathogenic species in the genus *Pyrenophora* are commonly observed in their anamorphic stage (*Drechslera*). In 1804, *Pyrenophora* was described as *Sphaeria* and was placed in the Xylariaceae, but then reclassified in the Pleosporales (Fries, 1849). Barr (1987) placed *Pyrenophora* in the Pyrenophoraceae, but Berbee (1996) reclassified it in Pleosporaceae, where it remains until now. Three species of *Pyrenophora* infect barley (*Pyrenophora graminea*, *Pyrenophora teres* f. sp. *teres* and *Pyrenophora teres* f. sp. *maculata*) and one species infects oat (*P. avenae*). Ptr on wheat, and the barley pathogens, were suggested to have diverged prior to the emergence of agriculture, and before the divergence of wheat and barley. Ptr diverged from both forms of *Pyrenophora teres* about 8 Mya, while *Pyrenophora teres* f. sp. *teres* and *Pyrenophora teres.* f. sp. *maculata* diverged from each other about 519 Kya (Ellwood et al., 2012). This speciation was explained by a possible isolation in glacial refuges during the glaciation era (Ellwood et al., 2012).

Ptr overwinters on crop residues as pseudothecia, and the development of the fruiting bodies is initiated purely by non-sexual vegetative means, although the asci inside are formed following some kind of fertilization event (Martin et al., 2017). Ascomycetes are named after the ascus (plural asci), which in Latin means a sac; it is inside these sac-shaped structures that the ascospores are formed, following meiosis in sexual reproduction. In *Pyrenophora*, the ascus is bitunicate, meaning it has two layers, a thicker outer layer (ectotunica) and a thinner inner layer (endotunica). There are several morphological characteristics to distinguish the various groups in the Ascomycota, based on 1) the type of ascus formed inside the ascoma (fruiting body, plural ascomata), and 2) the structure and organization of the centrum (inside the ascoma). The ascus is either unitunicate (with a single wall) or bitunicate (with a double wall), and the ascomata are classified as cleistothecia, perithecia, apothecia, and pseudothecia.

At maturity, and after the rupture of the outer layer of the ascus at its apex, the inner layer protrudes beyond the outer layer as in a 'jack-in-the-box', allowing the discharge of the ascospores through an opening in the ascoma (Berbee, 1996). In typical Pleosporales, asci are formed at the base of a cavity within a flask-like fruiting body, with the asci growing upward from the base and sterile hyphae (paraphyses) growing down from the top to the base, so that the mature fruiting body contains asci intermingled with these paraphyses (Berbee, 1996). Species within the *Pyrenophora* are defined morphologically by the absence of the pseudoparaphyses in their pseudothecia. Pseudoparaphyses, which are present in most pleosporalean fungi, are vertical hyphae arising at the top of the centrum and growing down to become anchored at the base, developing before asci formation (Luttrell, 1955).

The pseudothecia of Ptr range in size from 200–700 μm in diameter, and are spherical in shape and black in color (Pfender et al., 1988). Inside each ascus, eight hyaline ascospores develop, and the asci range in size from 47–65 × 20–26 μm (Shoemaker, 1962). The ascospores are dictyosporous (both transverse and longitudinal septa), have three transverse septa and one or sometimes two longitudinal septa in the central cell. The classification and nomenclature of Ptr and its anamorph have changed through time. Ptr and its anamorph, *Drechslera tritici-repentis* (Deid.), have been given various classifications since 1823. In older literature, *Pleospora trichostoma* was often used for the teleomorph, while *Helminthosporium* was used instead of *Drechslera* for the anamorph. This continued even after the correct classification and nomenclature as *Drechslera* was resolved in 1959 by Shoemaker (1961, 1962). The fungus was initially isolated and characterized from the grass species *Agropyron repens* in Germany, and was named *Pleosporales trichostigma* (Diedicke, 1902). On wheat, this pathogen was first isolated in Japan (Nisikado, 1928). As both *A. repens* and wheat are considered hosts, this was reflected in its hyphenated name and teleomorph nomenclature as *Pyrenophora tritici-repentis* (Shoemaker, 1961, 1962; Sone et al., 1994). The anamorph was initially named as *Helminthosporium tritici-vulgaris*. However, Shoemaker (1959) divided *Helminthosporium* based on conidial germination into three genera - *Drechslera*, *Bipolaris*, and *Helminthosporium*. A distinct characteristic of *Drechslera* is the ability of its conidia to germinate from all the cells, while in *Bipolaris,* germination occurs at the two apical cells only. *Drechslera tritici-repentis* has a conidium composed of 5-7 cylindrical cells, with a size range of 95-165×14-18 µm, and it is distinct from closely related fungi by the cone shape of its basal cell, known as 'snake-headed' at the bottom of the cell, while in other species the ends are more hemispherical (Shoemaker, 1962).

3 *Pyrenophora tritici-repentis*: the pathogen

Ptr is a necrotrophic pathogen, and is believed to have the widest host range among the *Pyrenophora* species (Shoemaker, 1962; Morrall and Howard, 1975). It has been isolated from over 26 grass species worldwide, including bromegrass, orchard grass, and numerous cereal crops such as rye, oat, and barley, in addition to wheat (reviewed by De Wolf et al., 1998). Recently, it was even isolated from a dicot host together with its related species *Parastagonospora nodorum*, and described as an endophyte of *Populus trichocarpa* (California poplar tree) (Newcombe et al., 2020). Ptr has a higher rate of occurrence on native perennials compared with non-native annual grasses (Kendig et al., 2021).

For the past 40 years, Ptr has been recognized to cause disease only on certain genotypes within the wheat species (Lamari et al., 2003), but recently, Ptr was found to interact specifically with barley (Aboukhaddour and Strelkov, 2016), and a single dominant locus, *Spr1*, on the short arm of chromosome 2H was identified to condition susceptibility to Ptr (Wei et al., 2020).

The ability of Ptr to cause damage to grasses is controversial among several research groups (reviewed by Ali and Langham, 2015), but recently, five grass species - smooth bromegrass, intermediate wheat grass, western wheat grass, crested wheatgrass, and green needlegrass - were found to be resistant to Ptr and insensitive to its effectors (Ali and Langham, 2015). In Canada, tan spot infection on the native prairies was low, despite favorable environmental conditions for disease development (Morrall and Howard, 1975). The isolates that were recovered from grasses or from wheat were reported to be pathogenic to both wheat and grasses in several studies, regardless of whether the isolates were derived from wheat or grassy species (Hosford, 1971; Morrall and Howard, 1975; Krupinsky, 1987, 1992). In a few studies, however, isolates collected from grasses were non-pathogenic when tested on wheat (Ali and

Francl, 2003; Wei et al., 2021). More recently, Ptr isolates from grasses were found to be pathogenic on durum wheat (Guo et al., 2020).

On wheat, Ptr infection is associated with the ability of the fungus to produce its necrotrophic effectors. If an isolate is non-pathogenic (produces no effectors), or if it is pathogenic and comes into contact with a resistant host genotype, then penetration of the epidermal layer occurs, but invasion and colonization of the mesophyll progress only for pathogenic isolates in susceptible wheat genotypes (Lamari and Strelkov, 2010; Aboukhaddour et al., 2012). Ptr has the ability to grow saprophytically, and under field conditions, it was recovered from leaf lesions regardless of the cultivar resistance prior to leaf senescence; soon after senescence, however, it could be isolated not just from lesions but from throughout the leaf tissue (reviewed by De Wolf et al., 1998). Prior to crop harvest at maturity, Ptr was even isolated from stem tissue after progressive colonization of the leaf sheath and upper stem internode (reviewed by De Wolf et al., 1998). The genetics behind the shift from being a necrotrophic pathogen to being a saprophyte or an endophyte on its various hosts is not yet understood.

On barley, ToxB-producing isolates were found to cause chlorosis (moderate in intensity compared with that on wheat) in a specific manner in certain barley genotypes. The purified ToxB, the chlorosis-inducing effector, also induced chlorosis in a genotype-specific manner when infiltrated into certain barley cultivars, albeit this chlorosis was weaker than that observed on wheat (Aboukhaddour and Strelkov, 2016). The cytology of infection on barley resembled that on wheat, as described in Aboukhaddour and Strelkov (2016) and illustrated in detail in Fig. 1. In brief, on wheat, the fungus penetrates the epidermal layer in both susceptible and resistant genotypes, and it progresses to colonize the mesophyll layer only in susceptible wheat genotypes, with the invading hyphae staying in the apoplast (intercellular).

In resistant wheat, Ptr is unable to colonize the mesophyll layer (Amaike et al., 2008; Aboukhaddour et al., 2012), but interestingly, in resistant barley, the fungus is able to colonize the mesophyll layer without causing any visible symptoms (asymptomatic) (Aboukhaddour and Strelkov, 2016). The ability of a necrotrophic pathogen to grow inside host tissue without causing any visible symptoms has been reported in several other species (reviewed by Ware, 2006; Stergiopoulos and Gordon, 2014). These pathogens may have the capacity to shift their life style from pathogenic to endophytic/mutualistic or vice versa (Freeman and Rodriguez, 1993; Stukenbrock and McDonald, 2008; Aboukhaddour and Strelkov, 2016). This suggests that, in the Ptr–barley interaction, host specificity, and pathogenicity may not be under the same genetic control (Aboukhaddour and Strelkov, 2016). Similarly, in *Colletotrichum magna*, a single mutation in a single locus has caused a shift in the lifestyle of this fungus from pathogenic to endophytic, without affecting other aggressiveness

Published by Burleigh Dodds Science Publishing Limited, 2022.

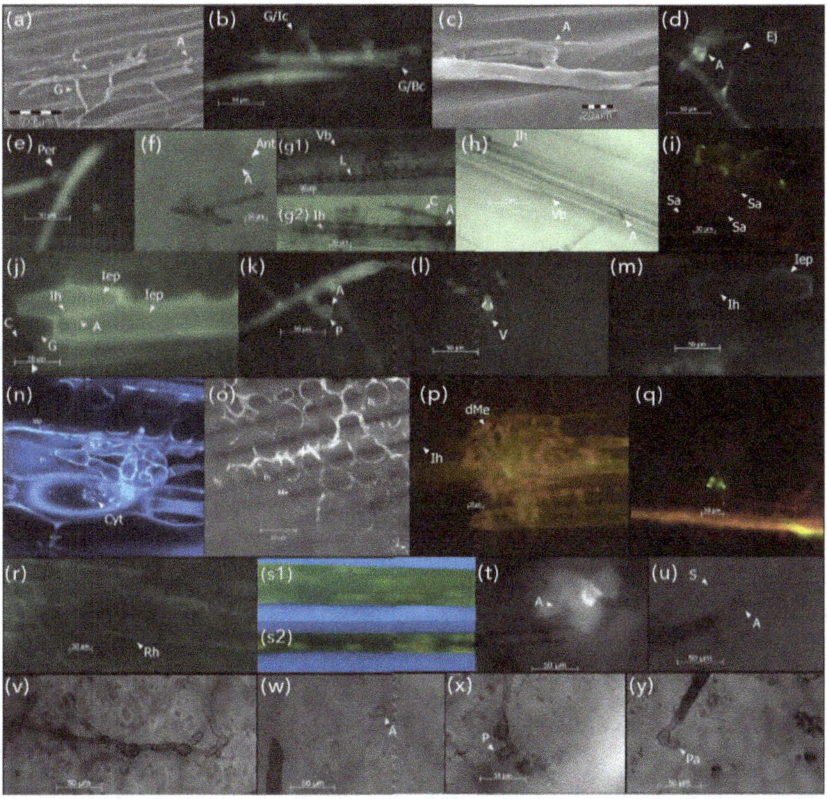

Figure 1 *Pyrenophora tritici-repentis* infection of barley and attempted infection of the non-host dicot canola. Published in Aboukhaddour and Strelkov (2016). Images a-r show the infection process on barley following inoculation with *P. tritici-repentis* race 5 isolate. Images (a), (b), (c), (d), (e), (f), (j), (k), (m), (n), (o), (p), (r) and (s1) were from a susceptible barley genotype, which developed visible chlorosis after inoculation; images (g1), (g2), (h), (i), (l) and (q) were from a resistant barley genotype. Images (s1) and (s2) show the lesion shape on susceptible barley and susceptible wheat genotypes, respectively, 5 days after inoculation. Attempted infection of a dicot non-host, canola, is shown in images t, u, v, w, x and y. (a) Scanning electron micrograph showing conidia germinating on the leaf surface 3 h after inoculation. (b) Fluorescence micrograph of a germ tube from basal or intercalary conidial cells. (c) Scanning electron micrograph of appressoria formed on the stomata 3 h after inoculation. (d) Fluorescence micrograph of appressoria formed at the junction of host epidermal cells. (e) Light micrograph of appressoria formed on the anticlinal wall of an epidermal cell, but with penetration having occurred over the periclinal wall. (f) Fluorescence micrograph of appressorium over the periclinal wall. (g1) Fluorescence micrograph of an infected long epidermal cell, running parallel to the vascular bundle, 12 h after infection. (g2) Fluorescence micrograph of appressorium on the long epidermal cell, with the intracellular hyphae growing inside. (h) Light micrograph of appressorium over the vascular bundle and invading a hypha growing inside the vascular bundle. (i) Fluorescence micrograph of infection at 24 h, showing secondary appressoria formed on invading mycelia. (j) Fluorescence micrograph of two neighboring epidermal cells infected by the same intracellular hypha and stained

attributes in sporulation levels, spore adhesion or appressoria formation, and infection (Freeman and Rodriguez, 1993).

Aside from wheat and barley, the cytology of Ptr on its other secondary hosts remains to be explored. The recent report of Ptr and *Parastagonospora nodorum* as endophytes on a tree species (Newcombe et al., 2020) is interesting and worthy of further exploration at the cytological level. The cytology of Ptr on a dicot crop, canola was explored in a single study by Aboukhaddour and Strelkov (2016). In brief, Ptr was never able to penetrate the epidermis or through the stomata, although the conidia germinated within 3 h and formed multiple appressoria (Fig. 1). However, no single penetration of the epidermal layer was observed, which suggested a lack of basic compatibility or specificity between Ptr and this dicot species.

4 Tan spot on the prairies: a case study

4.1 Ptr as one pathogen in a complex

Ptr is one species among several cereal pathogenic species on the prairies, all of which cause similar leaf spot symptoms and are commonly found in the same

Figure 1 (*Continued*).

deeply. (k) Fluorescence micrograph of a penetration peg formed from an appressorium. (l) Fluorescence micrograph showing an intracellular vesicle inside an epidermal cell. (m) Fluorescence micrograph of intracellular hypha growing inside an epidermal cell 12 h after inoculation, with hyphae starting to invade the mesophyll layer. (n) Confocal laser image of an infected epidermal cell 3 h after inoculation, showing aggregated cytoplasm. (o) Confocal laser image 6 h after inoculation, showing the invading hyphae growing intercellularly in the mesophyll layer. (p) Fluorescence micrograph of a lesion 24 h after inoculation, showing the advancing hyphae in green and the chlorotic mesophyll cells in red. (q) Fluorescence micrograph showing growth of hyphae in a resistant barley genotype, which did not develop chlorosis, 12 h after inoculation. (r) Fluorescence micrograph showing the hyphae running in a semi-straight fashion in infected leaf. (s1) Rectangular-shaped lesions on an infected barley leaf, compared with (s2) the oval shaped lesions in wheat. (t) Fluorescence micrograph of an appressorium formed on an epidermal cell of canola. (u) Light micrograph of an appressorium formed on the stomatal guard cell in canola. (v) Light micrograph of multiple appressoria formed on a thickened hypha trying to invade canola. (w) Light micrograph of the epidermal wall of a canola leaf pushed inside under an appressorium. (x) Light micrograph of a penetration peg formed from an appressorium. (y) Papilla formed under an appressorium in a canola epidermal cell. A, appressorium; Sa, secondary appressorium; C, conidia; G, germ tube; G/Bc, germ tube formed from a basal conidial cell; G/Ic, germ tube formed from an intracellular conidial cell; Ej, epidermal cell junction; Ant, anticlinal wall of host epidermal cell; Per, periclinal wall of host epidermal cell; Vb, vascular bundle; L, long epidermal cell; Ih, invading hypha; Iep, infected epidermal cell; P, penetration peg; V, vesicle; Cyt, aggregated cytoplasm; Me, mesophyll layer; dMe, damaged mesophyll cells; Rh, running hypha; S, stomata; Pa, papilla.

fields and in the same infected plants, and can sometimes even be recovered from the same lesion (Hafez et al., 2020). These species together are often referred to as the leaf spot complex. Ptr is the most prevalent of the pathogens in this complex on the Canadian prairies, followed by *Parastagonospora nodorum, Zymoseptoria tritici*, and *Cochliobolus sativus* (anamorph: *Bipolaris sorokiniana*) (blotch of wheat and barley) (Fernandez et al., 2016). In a 12-year survey (2001-2012) across Saskatchewan, the flag leaves of durum and bread crops were collected, and Ptr was the most prevalent and frequently isolated pathogen. Both Ptr and *B. sorokiniana* were recovered at a higher frequency from durum vs. bread wheat, while the reverse was the case for *Parastagonospora nodorum*. The same survey showed that, in various soil zones (brown, dark brown, black, or dark gray soil zones), *B. sorokiniana* was the only pathogen to be influenced by the soil type and was less common in the brown soil zone. Ptr was generally associated with warmer and drier conditions, but *B. sorokiniana* was prevalent at higher temperatures and precipitation. Recently, *Parastagonospora nodorum* and its sister species *Parastagonospora avenaria* f. sp. *tritici* (*Pat1*) were together the third most frequently recovered genus, after *Alternaria* and *Fusarium*, in grain and stem samples from a large number of field sites representing Alberta (south, central, north) and North-East Saskatchewan (Hafez et al., 2020).

Collectively, *Parastagonospora nodorum* and its sister species *Pat1* cause septoria blotch, and because *Pat1* is recovered only from grains, while *Parastagonospora nodorum* is isolated mainly from stem and leaf samples (McDonald et al., 2012; Hafez et al., 2020), it has been suggested that *Parastagonospora nodorum* primarily causes the leaf blotch, while *Pat1* causes the glume blotch (McDonald et al., 2012; Hafez et al., 2020).

The diagnosis and identification of leaf spot pathogens were initially based on the morphological characteristics of the spores, the sexual stage, and host specialization (Shaw, 1957a,b). While the ability to recognize these species by morphology is essential, a parallel molecular detection is powerful and can provide accuracy. These species are not easy to tell apart at times (McCartney et al., 2003). Several molecular diagnostic tools have been developed to detect and differentiate pathogens within the wheat leaf spot complex; these tools include species-specific primers that can be used in singleplex or multiplex PCR, in addition to other molecular tools like PCR restriction fragment length polymorphism (PCR-RFLP) and PCR-denaturing gradient gel electrophoresis (PCR-DGGE). Some of these important molecular tools and detailed information about the primers is included in Table 1

4.2 The emergence of tan spot and agronomic background

Tan spot is ranked among the top three leaf spot diseases in Canada, the USA, and Australia (Murray and Brennan, 2009; Wegulo et al., 2009; Fernandez

Table 1 PCR-based assays for the detection and quantification of leaf spot pathogens

Conventional PCR:

Target species	Primer	Sequence (5′ to 3′)	Target gene/region	Expected band length	Reference
Universal	ITS1-F	CTTGGTCATTTAGAGGAAGTAA	ITS	700 bp	Gardes and Bruns (1993)
	ITS4	TCCTCCG CTTATTGATATGC			White et al. (1990)
Universal	BMB-CR	GTACACACCGCCCGTCG	ITS	700 bp	White et al. (1990)
	ITS-4B	TTCCWCCGCTTATTGATATGC			
P. tritici-repentis	PtrUniqueF2	GGACTTTGGCTTTCTATT GTGC	Uncharacterized protein	490 bp	Antoni et al. (2010)
	PtrUniqueR2	CTTGGTGAATGGTGAAGATGG			
P. tritici-repentis	Pyr1	CCATATTCACCCATGTCTTT	ITS	300 bp	Mavragani et al. (2011)
	Pyr2	GWTTTTAAGGCGAGTCTYG*			
P. tritici-repentis	DTR1-F	ACCAATATGAAGCCGGACTG	ITS	382 bp	Hudcovicova et al. (2015)
	DTR1-R	CTCGGGAGAGAGACAAGACG			
P. nodorum	JB433	ACACTCAGTAGTTTACTACT	ITS	448 bp	Beck and Ligon (1995)
	JB434	TGTGCTGCGCTTCAATA			
B. sorokiniana	COSA-F	TCAAGCTGACCAAATCACCTTC	*Brn1*	519 bp	Matusinsky et al. (2010)
	COSA-R	CTTCTCACCAGCATCTGAATATATGA			

Real-time qPCR:

Target species	Primer/Probe	Sequence (5′ to 3′)	Target gene/region	Expected band length	Reference
P. tritici-repentis	Ptr-Forward	GTCTCCTCTGGTGGTATG	*ToxA* promotor	99 bp	Abdullah et al. (2018)
	Ptr-Reverse	GCTCTTAGTGAAGTTCAATC			
	Ptr-Probe	TACCTCTACTCGGTCGCCTATGG			
P. nodorum	Pn-Forward	ACCGCATTACCAAAGTTC	SNOG_01116_1	112 bp	Abdullah et al. (2018)
	Pn-Reverse	ACTGGAACTGGAACAATAAG			
	Pn-Probe	CCTGAATGCTCTTGACACTTGGTT			

PCR-RFLP:

Target species	Target gene/region	Restriction enzymes(s)	Restriction pattern (bp) for each species	Reference
P. nodorum *P. avenaria tritici*	*β-xylosidase*	*ScaI*	962 (*P. nodorum*) 695 + 267 (*P. avenaria tritici*)	McDonald et al. (2012)
P. nodorum *P. avenaria tritici* *P. tritici-repentis* *B. sorokiniana*	ITS	*HpaII* + *XapI*	80+210+10+120+280 (*P. nodorum*) 80+340+280 (*P. avenaria tritici*) 80+80+260+140+140 (*P. tritici-repentis*) 80+100+240+140+140 (*B. sorokiniana*)	Hafez et al. (2020)

et al., 2016). It emerged as a serious threat to wheat in these countries around the same time, in the early 1970s; its onset was largely attributed to the implementation of minimum-tillage practices (Rees and Platz, 1992; Lamari et al., 2005a; Oliver et al., 2008; Fernandez et al., 2009). Tillage promotes crop residue decomposition, reducing pathogen inoculum carried on it. Minimum-tillage practices allow Ptr inoculum to build up on the stubble, which remains on the soil surface, thereby contributing to increases in tan spot to epidemic levels (Rees and Platz, 1992). The pathogen itself was not new to Canada and, before the 1970s, Ptr fruiting bodies and conidia were identified in Saskatchewan in 1927 from the stubble of the wheat cultivar 'Marquis' (McCurry, 1929); ten years later, the fungus was identified from the leaves of durum wheat plants in Manitoba (Conners, 1937). Before that, Ptr was often isolated from local grasses, mainly *A. repens*, which prevail in the southern zone of the Canadian prairies and the northern plains of the USA.

The fungus may have been occurring on wheat in North America much earlier than first documented, as early reports described the presence of 'small brown spots with light borders' in southern Alberta in 1928 (reviewed by Lamari et al., 2005a). Reports of tan spot reaching epidemic levels in Manitoba were released in 1937 and 1939 (Conners, 1937, 1939), and they were the first records of major outbreaks of tan spot in wheat in North America (reviewed by Lamari et al., 2005a). Those early outbreaks were not as serious as those in the early 1970s, but were linked to the cropping practice in which stubble was not completely covered with soil, by using a one-way disc plough (Machacek cited by Conners, 1939). Overall, tan spot was sporadic in nature until the 1970s, with occurrences of the disease noted in 1941, 1944, and 1946 (Conners, 1941; Conners and Savile, 1944; Creelman, 1964). There was a gap of a few decades between the initial epidemics in the late 1930s and the serious ones in the 1970s, and the genetics of the dominant cultivars may have influenced the rise of tan spot in Canada (Lamari et al., 2005a).

The majority of wheat cultivars that have dominated the Canadian prairies over the past 100 years were susceptible to tan spot and sensitive to its effectors; of particular importance were ToxA and ToxC, the main effectors dominating North American Ptr populations (Lamari et al., 2005a; Tran et al., 2017; Hafez et al., 2020; Wei et al., 2021). The early cultivars 'Red Fife' followed by 'Marquis', which dominated Western Canada until 1938, were sensitive to ToxA but insensitive to ToxC, and both cultivars amplified *Tsn1*, the ToxA-sensitivity gene in wheat (Lamari et al., 2005a; Tran et al., 2017; Hafez et al., 2020). A new cultivar with stem rust resistance, 'Thatcher', replaced 'Marquis' and was predominant in Western Canada from 1936 to 1968 (McCallum and DePauw, 2008). 'Thatcher' tested insensitive to both ToxA and ToxC and lacked the *Tsn1* gene (Lamari et al., 2005a; Hafez et al., 2020). The cultivar 'Selkirk' replaced 'Thatcher' in the eastern prairies after the breakdown of stem rust resistance in the latter, due

Published by Burleigh Dodds Science Publishing Limited, 2022.

to the rise of the stem rust race group 15B (reviewed by Aboukhaddour et al., 2020). 'Selkirk' dominated the eastern prairies from 1956 to 1968, and was insensitive to ToxA (Lamari et al., 2005a). 'Neepawa', which dominated Western Canada from 1973 to 1985, was sensitive to all three Ptr effectors and carried the *Tsn1* gene (Lamari et al., 2005a; Tran et al., 2017; Hafez et al., 2020). The replacement of the ToxA-sensitive 'Marquis' with the ToxA-insensitive 'Thatcher' and 'Selkirk' could explain the gap in tan spot epidemics from the late 1930s until the 1970s, when another ToxA-sensitive cultivar, 'Neepawa', once again became predominant (reviewed by Lamari et al., 2005a). Sensitivity to ToxA in later and recent cultivars remains an issue in Canada today (Tran et al., 2017; Hafez et al., 2020; Wei et al., 2021).

Unlike the situation in Canada, it was not possible to find a clear correlation between variety selection and tan spot emergence in Australia, as both periods before and after the 1970s epidemics were dominated by ToxA-susceptible cultivars, and therefore tan spot emergence in Australia cannot be attributed solely to the host genetics (Oliver et al., 2008). In Western Europe, tan spot is not a major disease, with Septoria leaf blotch (*Z. tritici*) being the most important leaf spot. Susceptibility to ToxA-producing isolates, and the presence of the *Tsn1* gene, were reported at a lower frequency in European winter wheat genotypes than in the North American wheat (Wei et al., 2021).

Interestingly, cropping practices such as deep plowing and stubble burning were common during the 'Marquis' era, but were replaced by minimum or zero-tillage during the 'Neepawa' era (reviewed by Lamari et al., 2005a,b). This may also explain why the early epidemics during the 'Marquis' era were not as severe as the ones in the 1970s (Tekauz, 1976; Lamari et al., 2005a). Recently, the impact of cropping practices on the leaf spot complex in Canada was investigated (Fernandez et al., 2009). Tan spot infection in commercial bread wheat crops was prevalent in high-yielding and high-nitrogen systems, even with a previous non-cereal crop grown under zero tillage. In contrast, *Parastagonospora nodorum*, *Z. tritici*, and *Cochliobolus sativus* were associated with lower yielding crops and lower nitrogen fertilization with previous cereal crops under tillage (Fernandez et al., 2009); such conditions are likely similar to those in the early years before the emergence of tan spot.

4.3 Races and genetic diversity

Eight different races of Ptr have been identified on wheat worldwide, which differ in their ability to secrete different combinations of the three necrotrophic effectors - ToxA, ToxB, and ToxC (Lamari et al., 2003). Some races are described as 'basic' and produce only one effector each; these include races 2, 5, and 3, producing ToxA, ToxB, and ToxC, respectively. Three other races produce combinations of two effectors each, namely race 1 (ToxA, ToxC), race 6 (ToxB,

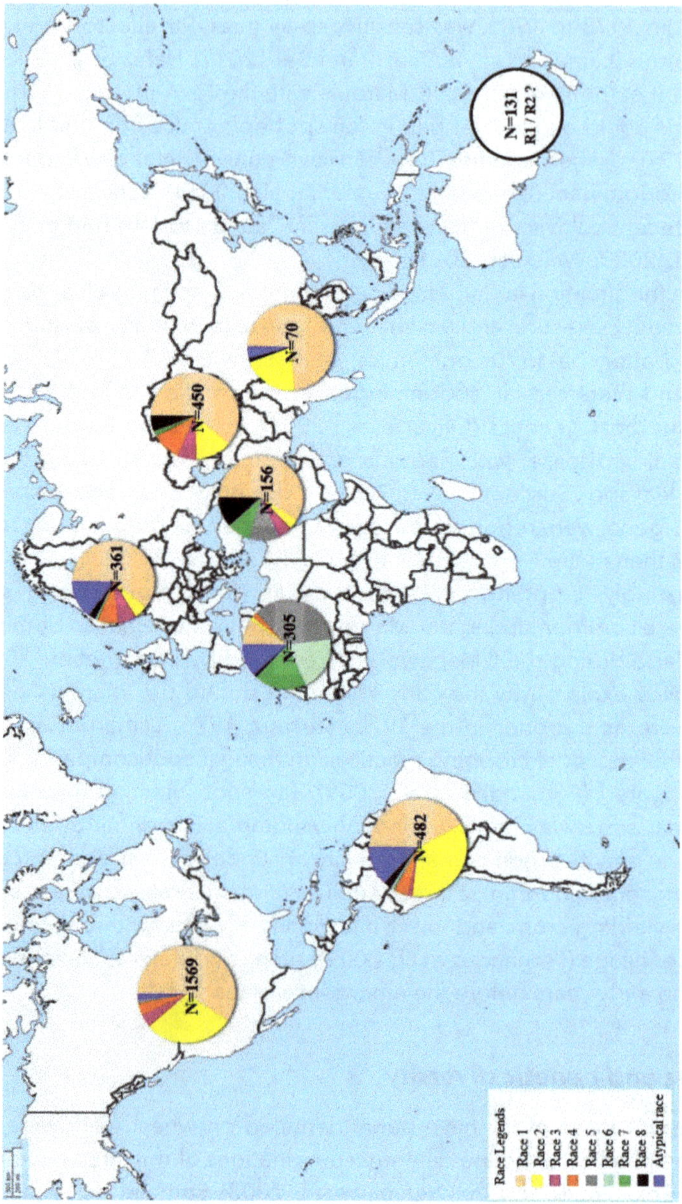

Figure 2 Distribution of Ptr races across geographical regions since 1989. As cited by Kamel et al., 2019.

ToxC), and race 7 (ToxA, ToxB). Race 8 produces all three effectors, while race 4 was until recently considered non-pathogenic and produces no known effectors (Lamari et al., 2003). Recently, isolates of race 4 were reported to cause extensive necrosis when tested on a number of durum wheat genotypes (Guo et al., 2020). Over 3500 Ptr isolates from across the world have been evaluated for their race classification since 1989 (Fig. 2) (reviewed by Kamel et al., 2019). About half of the tested isolates came from Canada and the USA (reviewed by Kamel et al., 2019).

As illustrated in Fig. 2, race 1 (ToxA, ToxC) is the most prevalent worldwide, found in North and South America, Europe, Asia, and Russia. Race 2 (ToxA) is the second most prevalent after race 1 in the Americas and in Asia, but race 2 was found at low frequencies in North Africa, the Caucasus, and Europe (Lamari et al., 2005b; Aboukhaddour et al., 2013; Kamel et al., 2019). While races 1 and 2 are predominant in the Americas, Europe, Russia, and Asia, the race structure in North Africa and part of the Caucasus region differs significantly, with ToxB-producing races being the most prevalent (reviewed by Kamel et al., 2019; Laribi et al., 2019). ToxB-producing isolates are absent in Australia, and are rare in the Americas (Lamari et al., 2005b; Antoni et al., 2010; Aboukhaddour et al., 2013). Races 3 and 4 have been reported from various regions but at lower frequencies, except on prairie grasses, where race 4 appears to predominate (Ali and Francl, 2003; Wei et al., 2020).

The limited capacity of the race system to accommodate all possible Ptr isolates, and the presence of isolates that do not fit the race system, was highlighted early on and by the same scientists who developed the race system (Lamari et al., 2003). There are several reports of the presence of atypical races, but in North America, this seems to be the exception (reviewed by Aboukhaddour et al., 2013). Nonetheless, the identification of 'atypical' isolates that do not fit the eight known races must be validated accurately at both the phenotypic and molecular level, especially when such findings can point to an unidentified new effector. For example, 44% of the Ptr isolates from Tunisia caused symptoms typical of race 7, where necrosis and chlorosis are visualized on 'Glenlea' and 6B662, respectively, while the remaining lines in the differential set showed a resistant reaction (Kamel et al., 2019). However, testing for the presence of *ToxA* in these isolates confirmed that they could not be designated as race 7, since they lacked the *ToxA* gene encoding ToxA that is usually responsible for necrosis on 'Glenlea'. These ToxA-lacking isolates, therefore, belong to an atypical virulence group, which may produce an additional necrosis-inducing effector(s) that awaits characterization (Kamel et al., 2019). The use of multiplex PCR, and the inclusion of a reference-positive fungal gene (chitin) that amplifies in all tested isolates, help to ensure the accuracy of such results. Similarly, repeating phenotyping experiments with a trusted seed source to confirm unexpected virulence patterns is also critical (Kamel et al., 2019).

North American and Australian Ptr populations exhibit the lowest diversity in terms of race structure, with races 1 and 2 usually found to be predominant in the routine surveys of these regions over the past 30 years. In contrast, up to five different races were recovered from a single wheat field in the regions encompassing the center of wheat diversity (Lamari et al., 2005b). Similarly, four to six races were identified in a small collection of isolates from North Africa (Benslimane et al., 2011; Gamba et al., 2017; Kamel et al., 2019; Laribi et al., 2019).

In some cases, significant variability in Ptr race composition was detected among countries that are near or adjacent to each other. For example, most races and Ptr effector combinations were identified in collections of the pathogen from Azerbaijan (Lamari et al., 2005b), while in the nearby countries of Uzbekistan, Kazakhstan, Kyrgyzstan, and Iran, only race 1 and/or race 2 were reported (Lamari et al., 2005b; Momeni et al., 2014). In North Africa, while the race composition in different countries was generally similar (reviewed by Kamel et al., 2019), some notable differences have been noted. Race 1 was not reported in Tunisia (Kamel et al., 2019; Laribi et al., 2019), while races 1 and 7 were predominant in Algeria, and races 5 and 6 were predominant in Morocco (Benslimane et al., 2011; Gamba et al., 2017).

The variation in race composition in nearby regions may be attributed to the host from which these isolates were recovered. In Azerbaijan, for instance, where more diversity in races and effectors was reported, local varieties and landraces of wheat are often planted. In the neighboring countries, cultivars developed by the International Maize and Wheat Improvement Center (CIMMYT) and the International Center for Agricultural Research in the Dry Areas (ICARDA), are more common (Lamari et al., 2005b; Momeni et al., 2014). The host ploidy level may also play a role in the variability in race structure. Indeed, ToxB and ToxC-producing isolates of Ptr have been obtained mainly from durum wheat, while isolates representing races 1 and 2, which lack the ability to produce ToxB, have predominated in hexaploid (bread) wheat in North America (Aboukhaddour et al., 2013; Wei et al., 2021). The non-pathogenic race 4 carrying the *toxb* homolog was recovered mainly from grass species in North America (Ali and Francl, 2003; Wei et al., 2021).

The variation in race structure by host is reflected at the genetic diversity level (Aung, 2001; Aboukhaddour et al., 2011). An analysis of the genetic diversity in a global collection of 80 isolates of Ptr with 31 polymorphic simple sequence repeat (SSR) markers spanning the Ptr genome and its various chromosomes indicated considerable differences between ToxA-producing isolates and ToxA-nonproducing isolates (Aboukhaddour et al., 2011). While most of the ToxA-nonproducing isolates were recovered from tetraploid (durum) wheat, ToxA-producers were collected mainly from hexaploid wheat. This suggested that the host-specificity imposed by the effectors on each ploidy level contributed to the differentiation between these races (Aboukhaddour et al., 2011). In contrast, no relationship between isolate race or geography was found in other studies (Friesen et al., 2005; Momeni et al., 2019), although these tested limited numbers of isolates or focused on collections from a restricted geographic origin. This could have represented constraints to their conclusions (reviewed by Ciuffetti et al., 2014).

There were also significant genetic differences between pathogenic races and race 4 (Aung, 2001; Lepoint et al., 2010; Aboukhaddour et al., 2011). These differences were confirmed at the chromosomal and genome levels, supporting the hypothesis that the non-pathogenic race 4, present mainly in grasses, is more diverged than the other races of Ptr (Lichter et al., 2002; Aboukhaddour et al., 2009; Manning et al., 2013).

4.4 ToxA as an important virulence factor

ToxA has been horizontally transferred between three fungal species, each known to cause a disease of wheat (tan spot, septoria nodorum, and spot blotch) (McDonald et al., 2019). *ToxA* is carried on a 14 Kb region typical of a type II DNA transposon (ToxhAT), and this ToxhAT resides on a larger transposon-rich region (140–250 Kb) only in the ToxA-producing isolates of all the three above-mentioned pathogenic species (McDonald et al., 2019). Likely, *ToxA* was horizontally transferred under two independent events facilitated by the transposon activity in the Ptr genome (McDonald et al., 2019). Ptr isolates from a global collection exhibited high plasticity in the structural organization of their genome and in chromosomal polymorphism with independent chromosomal locations of the effector genes (*ToxA* and *ToxB*) (Aboukhaddour et al., 2009). In Ptr and *B. sorkiniana* (spot blotch), there are events of interchromosomal transfer within each species, and the involvement of transposable elements is likely to play a role in such transfer and in the evolution of virulence in Ptr (Aboukhaddour et al., 2009; McDonald et al., 2019).

The *ToxA* gene has numerous haplotypes (Hafez et al., 2020). The diversity of these haplotypes provides insights into the different Pleosporales species and the evolution of virulence at a global level (Friesen et al., 2005; Stukenbrock and McDonald, 2007; McDonald et al., 2013; Ghaderi et al., 2020; Hafez et al., 2020). In the pathogens causing tan spot, septoria nodorum, and spot blotch, the *ToxA* gene sequence is very similar (Lu et al., 2015). However, a less similar homolog was reported in *Cochliobolus heterostrophus* (southern corn leaf blight) that codes for a protein, ChToxA, which is moderately similar (64%) to ToxA (Lu et al., 2015). A *ToxA* haplotype network is illustrated in Fig. 3, and represents the different *ToxA* haplotypes detected to date in four cereal leaf-spot pathogens occurring worldwide (Ptr, *Parastagonospora nodorum*, Pat1, and *B. sorokiniana*).

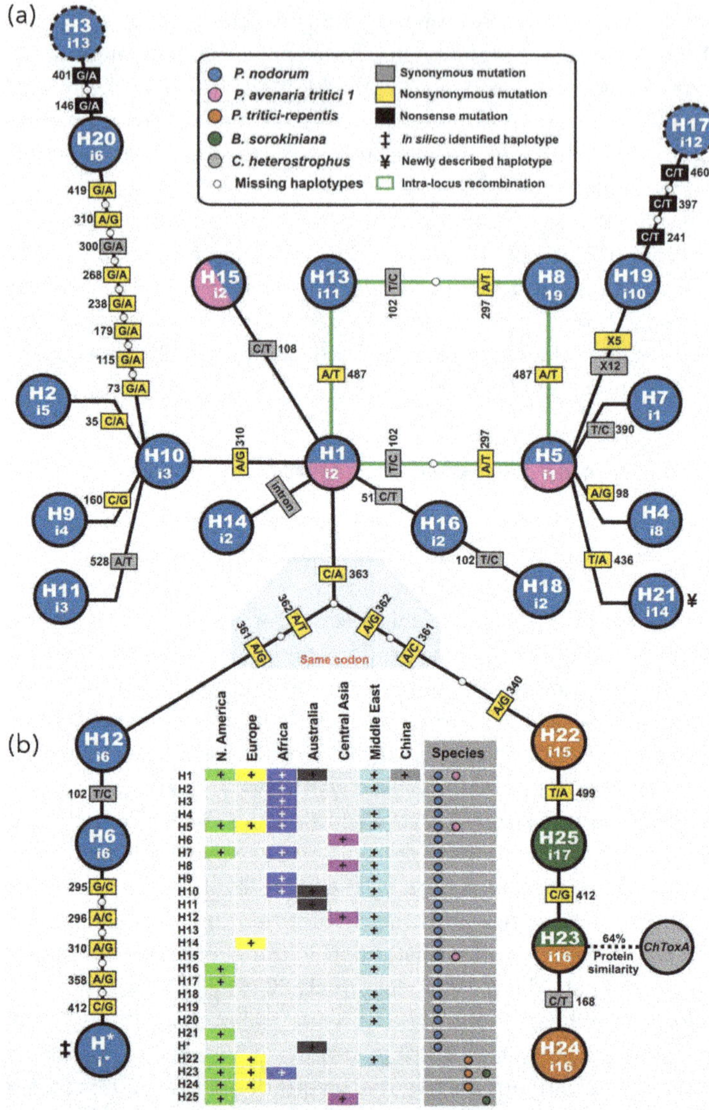

Figure 3 (a) *ToxA* haplotype network in *Parastagonospora nodorum* and other related species. Twenty-six *ToxA* haplotypes are described. Gray, yellow, and black boxes represent synonymous, non-synonymous and nonsense mutations, respectively, while the numbers beside each box represent the nucleotide position relative to the *ToxA* start codon. Three mutations (positions: 102, 297, and 487) in the loop between haplotypes H01, H05, H08, and H13 (green line) are shown twice due to intra-locus recombination events. (b) Geographic distribution of *ToxA* haplotypes in different geographic regions and the species found to contain each haplotype. A *ToxA*-like gene was identified in *Cochliobolus heterostrophus* (ChToxA) with 64% protein identity to Ptr-ToxA H23. This figure was published by Hafez et al. (2020).

In total, 26 different haplotypes were found, with only three haplotypes reported in Ptr (reviewed by Hafez et al., 2020). *Parastagonospora nodorum* has the highest diversity in terms of the *ToxA* sequence, with 22 different haplotypes, 19 of which are unique to *Parastagonospora nodorum*. The rest of the species have fewer haplotypes: Pat1 (3), Ptr (3), and *B. sorokiniana* (2). Among all of these haplotypes, H1 and H5 are the most frequent and widely distributed, with the former found in North America, Australia, Europe, Africa, central Asia, the Middle East and China, and the latter present in North America, Europe, Africa, central Asia and the Middle East (McDonald et al., 2013; Ghaderi et al., 2020; Hafez et al., 2020). In North America, four ToxA haplotypes were reported in *Parastagonospora nodorum*, H1, H5, H7, and H16, and three haplotypes were reported in Ptr, H22, H23, and H24 (Hafez et al., 2020). Haplotypes H1 and H23 represent the most abundant haplotypes in North America and worldwide for *Parastagonospora nodorum* and Ptr, respectively. It has been suggested that ancestral haplotypes could be identified based on their high frequency, their wide geographic distribution, and by their internal position in the haplotype network. Accordingly, H1 seems to be the ancestral haplotype, as it represents the most abundant and widely distributed haplotype, and is located at the center of the network (Stukenbrock and McDonald, 2007).

Unlike *ToxA*, the *ToxB* gene in Ptr is thought to be inherited vertically and may have been subjected to loss from the pathogen genome (Strelkov and Lamari, 2003; Aboukhaddour et al., 2009). *ToxB* is a multi-copy gene, and homologs, termed *toxb*, are present in non-producing isolates and other ascomycete species (Martinez et al., 2004; Strelkov et al., 2006; Andrie et al., 2007). Although *toxb* is transcriptionally active, isolates carrying this homolog failed to induce symptoms in ToxB-sensitive genotypes, and infiltration of heterologously expressed *toxb* in ToxB-sensitive lines caused no symptoms, indicating that toxb is not active (Kim and Strelkov, 2007; Amaike et al., 2008; Andrie and Ciuffetti, 2011). The variation in *ToxB* sequences from the wild-type resulted in a modification of the amino acid content. In certain races like race 3, this modification impacted the signal peptide regions leading to improper protein folding (Strelkov et al., 2006). In *Pyrenophora bromi*, a sister species to Ptr and the causal agent of brown leaf spot of smooth bromegrass, *ToxB* homologs (named *PbToxB*), when purified and infiltrated in *ToxB*-sensitive wheat genotypes, induced chlorosis but failed to induce symptoms on the grass host (Andrie and Ciuffetti, 2011). The presence of *ToxB* homologs in Dothideomycetes and Sordariomycetes suggested that Ptr likely inherited *ToxB* vertically from a common ancestor (Andrie et al., 2007).

Both *ToxA* and *ToxB* have independent chromosomal locations in Ptr and have not been found to co-locate on the same chromosome, suggesting an independent origin of the genes (Aboukhaddour et al., 2009). *ToxA* is present as a single-copy, highly conserved gene, and only in coding isolates, with no homologs reported in non-producing isolates, although its homolog chromosome remains present and essential in all isolates of the pathogen (Aboukhaddour et al., 2009). ToxC remains under-investigated, but the ability to manipulate the mating type genes in Ptr and to create outcrossing mapping populations may provide more insights into ToxC and its coding gene(s) (Ameen et al., 2017; Liu et al., 2017).

4.5 The development of the rating system and beyond

Having a sensitive and accurate bioassay to evaluate a plant's reaction to its pathogens is important, particularly so in the world of host-selective toxin discovery. With the exception of *ToxA* in *Parastagonospora nodorum* and in *Cochliobolus sativus* (Friesen et al., 2006; McDonald et al., 2018), the identification of necrotrophic effectors in fungal pathogens has usually been based on accurate and sensitive bioassays. Such bioassays have always included pathogenic isolates that produce their effectors in culture media, non-pathogenic isolates as controls, and semi-purified culture filtrates that can be infiltrated into plant tissues, to distinguish sensitive from insensitive genotypes. As simple as the bioassay criteria seem, there are many challenges and the process has not always been easy: here I mean it was not easy all the times (reviewed by Scheffer, 1989). Multiple factors can influence the sensitivity of the bioassay and the manifestation of symptoms, such as the temperature of incubation, the type of tissue being tested, the components of the culture media, the stability of the effector, and the ability to purify it (reviewed by Scheffer, 1989). For example, symptoms of Ptr necrosis or chlorosis failed to develop on susceptible wheat when the plants were maintained at a temperature above 27°C (Lamari and Bernier, 1994). Similarly, the Ptr–barley reaction shifted from compatible to incompatible when plants were kept at a temperature below 15°C or higher than 25°C, making evaluation difficult (Aboukhaddour and Strelkov, 2016).

The exposure to higher temperature has been associated with shifts in the host reaction to insensitivity in several pathosystems in addition to tan spot, including the PC toxin/milo disease of sorghum and AK toxin/black spot of Japanese pear (reviewed by Scheffer, 1989). In these examples, the effect of temperature was not the focus of the investigators initially, and was more of a fortunate accident of hot greenhouse obstacles at the time. In similar ways, the fortunate case of adding yeast extract to culture media facilitated the discovery of the HC toxin/maize leaf spot and ear mold, which otherwise would not have been secreted in sufficient quantity for the test (Scheffer and Ullstrup, 1965; Scheffer, 1989). It is not easy, however, to find all reports on such obstacles, as these are not often reported or even identified.

In Ptr, several researchers investigated variation in the ability of isolates to induce different types/severities of symptoms, and initially tan spot development was defined in quantitative terms. The evaluation of the reaction of wheat to tan spot in early research resulted in a large number of symptom-rating systems that ranged from semi-qualitative (Hosford, 1971; Gilchrist et al., 1984) to quantitative, including the use of parameters such as percentage of infection, lesion size, lesions per cm^2, and leaf position in a disease index (reviewed by De Wolf et al., 1998). The number of systems described within a short period reflects the challenges faced by plant pathologists dealing with tan spot disease in those years (Lamari et al., 1992). The quantitative approach captured the presence of variation in the virulence of Ptr on susceptible lines, but failed in identifying the specific host–pathogen interaction.

Early screening for resistance to tan spot in the USA started in the late 1960s and early 1970s, but the basis of genetic resistance was not fully understood, and was considered to depend on environmental factors, such as temperature and humidity, and how long the leaves stayed wet after inoculation (Hosford, 1971). Discrepancies in ratings among different research groups for the same cultivars/genotypes were often reported, but this was most likely due to the differences in inoculation methods and subjective ratings (Aboukhaddour et al., 2013). The development of improved inoculation methods, however, showed that lesion types were independent from leaf wetness period (Lamari et al., 1992).

A review of early protocols and inoculation methods indicates the lack of a standardized approach, with scientists still invested in optimizing the best methods to produce sufficient inoculum. There were difficulties in getting the fungus to sporulate *in vitro*, and older experiments described inoculation by dipping plants in a suspension of chopped mycelium, or collecting the conidia from the cutting bars of wheat-harvesting machines, described as 'green dust', which was then stored in freezers to maintain viability (Hosford, 1976). Later, optimizing the conditions to enhance conidia production in the lab helped to advance inoculation protocols.

In Canada, W.C. McDonald was able to produce conidia from the imperfect stage of the fungus by growing it on V8 juice agar with a 12-h photoperiod (Hosford, 1971). In Australia, the anamorphic pathogen was subjected to a cycle of light and dark, light for the conidiophore formation and dark for the conidia formation, similar to other diurnal sporulator *Drechslera* species (Khan, 1971). These findings were developed further to enable the production of a conidial suspension in the lab, which was essential for generating reliable inoculation results to determine virulence (Platt, 1979; Platt and Morrall, 1980; Odvody and Boosalis, 1982). The conidia produced in the lab caused more uniform lesions upon inoculation than the use of chopped mycelium (Lamari et al., 1992). Production of more conidia was achieved by culturing the fungus on potato dextrose agar (PDA), then transferring a disk from the actively growing colony onto a modified V8-agar (Platt et al., 1977). Raymond et al. (1985) simplified

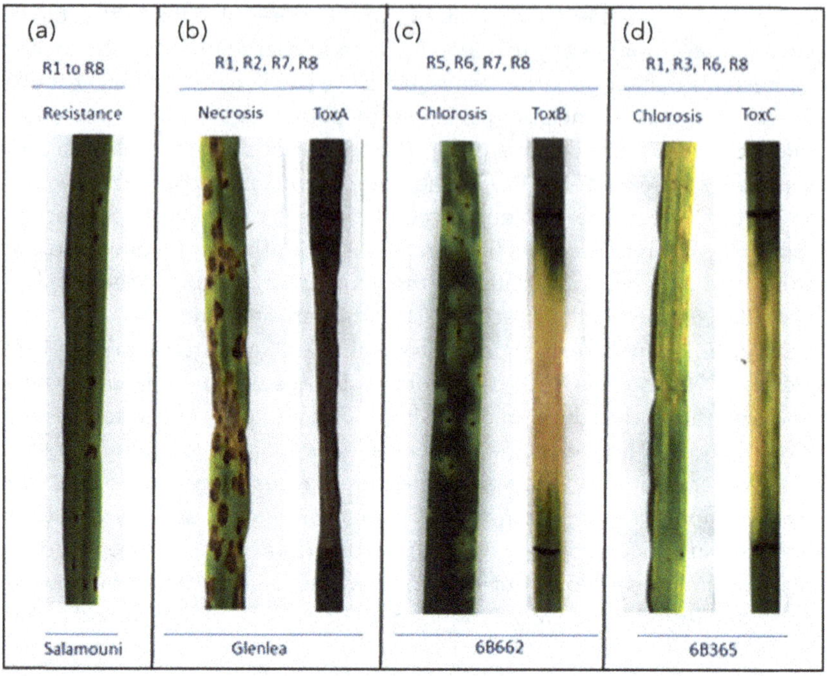

Figure 4 The symptoms caused by the races Ptr and their effectors. (a) Resistance on 'Salamouni', with only penetration sites observed but no necrosis or chlorosis. (b) Necrosis on 'Glenlea' caused by infection with ToxA-producing races (left) and infiltration with purified ToxA (right). (c) Chlorosis on 6B662 caused by infection with ToxB-producing races (left) and infiltration with purified ToxB. (d) Chlorosis on 6B365 caused by infection with ToxC-producing isolates (left) and infiltration with culture filtrates possessing putative ToxC activity.

this methodology further by including sectors of each medium on the same Petri plate. Lamari and Bernier (1989a) modified the protocols of Raymond et al. (1985) by combining both V8 and PDA into one culture medium (V8–PDA), and growing the culture under darkness instead of continuous light to prevent conidiophore formation during mycelial growth. Then, after they reached 4 cm diameter, the cultures were subjected to 18–24 h of light at room temperature followed by dark at 15°C to induce sporulation. The method of Lamari and Bernier (1989a) enabled the production of huge amounts of conidia *in vitro*, and has been adopted for use by the tan spot community to the present day, given the ease of its use and its efficiency over other methods.

Lamari and Bernier (1989b, 1991) also developed the symptom type-based tan spot rating system as a qualitative scheme, after finding that Ptr isolates could induce, in specific wheat genotypes, one of two symptoms, necrosis or chlorosis (Fig. 4). The isolates initially were grouped into four pathotypes based on their ability to induce these symptoms: pathotype 1 (both chlorosis and necrosis), pathotype 2 (necrosis only), pathotype 3 (chlorosis only), and pathotype 4 (neither symptom). The identification of other isolates (from Algeria) that were able to induce chlorosis on wheat genotypes that were resistant to all previously known isolates led to the development of the race-based classification system for Ptr (Lamari et al., 1995). Pathotypes 1 to 4 became races 1 to 4, while the new North African race became race 5 from which ToxB was later purified (Strelkov et al., 1999; Martinez et al., 2004). In subsequent years, races 6, 7, and 8 were identified from the Caucasus regions, the Fertile Crescent, and North Africa (Lamari et al., 2003; Lamari et al., 2005b). These symptoms are now known to reflect the specific interactions between Ptr effectors and sensitivity genes in wheat (Fig. 4).

The race-based system of classification utilized hexaploid genotypes, but on durum wheat, numerous isolates induced unexpected reactions such as necrosis instead of chlorosis by races 3 and 5 (Lamari and Bernier, 1989b; Liu et al., 2017; Wei et al., 2021). Recently, it has been found that the *Tsn1*-ToxA interaction does not condition the expected reaction, necrosis, in tetraploid wheat (Faris et al., 2020; Wei et al., 2021). This is likely explained by the presence of additional defense mechanisms in tetraploid wheat, which likely contribute to early resistance against pathogen invasion before the internalization of ToxA into the host cells (Faris et al., 2020).

Tsn1 lacks a transmembrane domain, and therefore is likely located inside the plant cell (Faris et al., 2010). The presence of a major dominant gene such as *Tsr7*, which acts in a non-specific manner, may inhibit Ptr penetration and proliferation at early infection stages, and may deprive the pathogen from delivering its ToxA inside the plant cell. As such, the host may still exhibit resistance despite the presence of *Tsn1* (Faris et al., 2020). This, however, does not explain the atypical necrosis induced by races 3 and 5 on durum wheat genotypes (Lamari and Bernier, 1989b; Virdi et al., 2016; Wei et al., 2021). Such necrosis may reflect either the presence of several sensitivity alleles for ToxC and ToxB in durum wheat, resulting in greater sensitivity leading to necrosis rather than chlorosis, or the presence of additional necrosis-inducing effectors specific to durum (Wei et al., 2021). The ToxB-*Tsc2* interaction appears to be more significant in durum than in hexaploid wheat (Virdi et al., 2016; Wei et al., 2021). It is worth noting that the *Tsn1*-ToxA interaction in *Parastagonospora nodorum* is significant in both tetraploid and hexaploid wheat (Virdi et al., 2016). In conclusion, the non-specific interactions between Ptr and its hosts need further exploration.

5 Conclusion and future trends

Over 90 years have passed since Ptr was recognized as a wheat pathogen (Nisikado, 1928), but it was not until 2006 that the link between the gain of a necrotrophic effector gene by horizontal gene transfer from another fungal species and the emergence of tan spot became evident (Friesen et al., 2006). In Australia and North America, where wheat is grown over large regions, selection pressure gave a limited number of fungal pathogens the ability to evolve and colonize this homogeneous crop. Ptr and related species that infect wheat and other cereals have evolved to share the same host, and therefore, the same effector genes, and exhibit the same mechanisms for interacting with the host plant.

We have achieved much progress in understanding these pathogens, the disease biology, and the associated mechanisms of virulence. Yet, in North America, we are still unable to implement an effective management without relying on fungicide treatments (Aboukhaddour et al., 2020). This can be attributed in part to the narrow genetic background of our wheat crop, which has been cultivated in one of the most expanded areas over the North American wheat belt and many other regions of the world. We need to protect our crops more than ever in the face of the challenges associated with climate change and an increased demand for food to satisfy the growing human population.

There is a need for a wider adoption of easy and reliable molecular diagnostic tools, coupled with conventional surveys to monitor any shifts in the leaf spot complex and its various effectors. This may ease the deployment of cultivars with durable resistance. Unlike breeding for resistance to obligate pathogens such as rust, which defeat resistance genes so often, tan spot resistance can be very durable (Faris et al., 2013). In wheat, major dominant genes confer resistance to rust, whereas with the leaf spots, resistance can be passive, with the dominant genes conferring susceptibility. Excluding or editing germplasm with known major susceptibility genes can be an easy first step.

The input of scientists cannot, however, be translated into action without the support of visionary, science-minded decision makers, and with more funds directed toward priorities playing an essential role. In Canada, despite the fact that leaf spots are one of the major diseases on the prairies, the leaf spot complex is not considered a top priority in the cultivar registration process, which recommends the developed cultivars to have a minimum level of resistance to priority #1 diseases (cereal rusts, common bunt, and fusarium head blight) (Aboukhaddour et al., 2020).

In Australia, it was reported that using ToxA as a tool for screening has reduced tan spot susceptibility considerably in a short time. According to Vleeshouwers and Oliver (2014), 'The area sown to ToxA-sensitive wheat in Western Australia fell from 30.4% in 2009–2010 to 16.9% in 2012–2013', preventing millions of dollars in losses. The Australian breeders adopted the use of effectors as a screening tool over the use of molecular markers. Testing with markers requires DNA extraction and lab analyses. The effector test is conducted in the field or in a greenhouse on seedlings and six-week-old adult plants. Thus, thousands of plants can be tested in a single day and any symptoms are visible and clear within a week. This gave Australian breeders the power to eliminate sensitive lines immediately, and save time and resources later on.

The use of the Ptr effectors ToxA and ToxB to screen wheat cultivars was first illustrated by Canadian scientists in 2005 (Lamari et al., 2005a) Yet, in Canada, screening for leaf spot resistance is still mainly conducted by testing the reaction of wheat lines to unidentified mixtures of several foliar pathogens under natural infection in the field. That explains why most wheat cultivars registered in Canada exhibit insufficient levels of resistance to leaf spot (Fernandez et al., 2016; Tran et al., 2017; Hafez et al., 2020; Wei et al., 2021). Given the presence of diverse species/races/effectors produced by varying pathogenic isolates, breeding strategies for enhanced resistance to leaf spot need to take into consideration the presence of such diversity, as well as the presence of dominant susceptibility genes in the host, rather than the often dominant resistance in other pathosystems. The presence of several pathogenic species with many races/effectors means that there is a need to identify the major loci susceptible to these specific necrotrophs.

Several reports of atypical virulent isolates that do not fit within the eight known races, and the potential involvement of additional, but not yet identified effectors, necessitate further exploration. The Ptr interaction with its various secondary hosts, and the genetics controlling such interactions, may provide insights into the evolution of Ptr as a pathogen of wheat. The increase in reports of tan spot in South America and North Africa in recent years underscores the threat that Ptr and related pathogens continue to pose to the sustainable production of wheat worldwide.

6 References

Abdullah, A. S., Turo, C., Moffat, C. S., Lopez-Ruiz, F. J., Gibberd, M. R., Hamblin, J. and Zerihun, A. (2018). Real-time PCR for diagnosing and quantifying co-infection by two globally distributed fungal pathogens of wheat. *Front. Plant Sci.* 9: 1086.

Aboukhaddour, R., Cloutier, S., Ballance, G. M. and Lamari, L. (2009). Genome characterization of Pyrenophora tritici-repentis isolates reveals high plasticity and independent chromosomal location of ToxA and ToxB. *Mol. Plant Pathol.* 10(2): 201-212.

Aboukhaddour, R., Cloutier, S., Lamari, L. and Strelkov, S. E. (2011). Simple sequence repeats and diversity of globally distributed populations of *Pyrenophora tritici-repentis*. *Can. J. Plant Pathol.* 33(3): 389-399.

Aboukhaddour, R., Fetch, T., McCallum, B. D., Harding, M. W., Beres, B. L. and Graf, R. J. (2020). Wheat diseases on the prairies: a Canadian story. *Plant Pathol.* 69(3): 418-432.

Aboukhaddour, R., Kim, Y. M. and Strelkov, S. E. (2012). RNA-mediated gene silencing of ToxB in *Pyrenophora tritici-repentis*. *Mol. Plant Pathol.* 13(3): 318-326.

Aboukhaddour, R. and Strelkov, S. E. (2016). Exploring de novo specificity: the *Pyrenophora tritici-repentis*-barley interaction. *Plant Pathol.* 65(8): 1347–1357.

Aboukhaddour, R., Turkington, T. K. and Strelkov, S. E. (2013). Race structure of *Pyrenophora triciti-repentis* (tan spot of wheat) in Alberta, Canada. *Can. J. Plant Pathol.* 35(2): 256–268.

Ali, S. and Francl, L. J. (2003). Population race structure of *Pyrenophora tritici-repentis* prevalent on wheat and noncereal grasses in the Great Plains. *Plant Dis.* 87(4): 418–422.

Ali, S. and Langham, M. A. (2015). Reaction of five non-cereal grasses to five races and two host selective toxins of *Pyrenophora tritici-repentis*. *Plant Pathol. J.* 31(3): 245–251.

Amaike, S., Ozga, J. A., Basu, U. and Strelkov, S. E. (2008). Quantification of ToxB gene expression and formation of appressoria by isolates of *Pyrenophora tritici-repentis* differing in pathogenicity. *Plant Pathol.* 57(4): 623–633.

Ameen, G., Kariyawasam, G., Shi, G., Friesen, T. L., Faris, J. D., Ali, S., Rasmussen, J. B. and Liu, Z. (2017). Molecular manipulation of the mating-type system and development of a new approach for characterizing pathogen virulence in *Pyrenophora tritici-repentis*. *Fungal Genet. Biol.* 109: 16–25.

Andrie, R. M. and Ciuffetti, L. M. (2011). *Pyrenophora bromi*, causal agent of brownspot of bromegrass, expresses a gene encoding a protein with homology and similar activity to Ptr ToxB, a host-selective toxin of wheat. *Mol. Plant Microbe Interact.* 24(3): 359–367.

Andrie, R. M., Pandelova, I. and Ciuffetti, L. M. (2007). A combination of phenotypic and genotypic characterization strengthens *Pyrenophora tritici-repentis* race identification. *Phytopathology* 97(6): 694–701.

Antoni, E. A., Rybak, K., Tucker, M. P., Hane, J. K., Solomon, P. S., Drenth, A., Shankar, M. and Oliver, R. P. (2010). Ubiquity of ToxA and absence of ToxB in Australian populations of *Pyrenophora tritici-repentis*. *Australas. Plant Pathol.* 39(1): 63–68.

Aung, T. S. T. (2001). Molecular polymorphism and virulence in *Pyrenophora tritici-repentis*. Master of Science. University of Manitoba, Winnipeg, Manitoba.

Ballance, G. M., Lamari, L. and Bernier, C. C. (1989). Purification and characterization of a host-selective necrosis toxin from *Pyrenophora tritici-repentis*. *Physiol. Mol. Plant Pathol.* 35(3): 203–213.

Barr, M. (1987). New taxa and combinations in the *Loculoascomycetes*. *Mycotaxon*: 29501–29505.

Barrus, M. (1942). Yellow-spot disease of wheat in New York State. *Plant Dis. Rep.* 26(11).

Beck, J. J. and Ligon, J. M. (1995). Polymerase chain reaction assays for the detection of Stagonospora nodorum and Septoria tritici in wheat. *Phytopathol* 85(3): 319–324.

Benslimane, H., Lamari, L., Benbelkacem, A., Sayoud, R. and Bouznad, Z. (2011). Distribution of races of *Pyrenophora tritici-repentis* in Algeria and identication of a new virulence type. *Phytopathol. Mediterr.* 50(2): 203–211.

Berbee, M. L. (1996). *Loculoascomycete* origins and evolution of filamentous ascomycete morphology based on 18S rRNA gene sequence data. *Mol. Biol. Evol.* 13(3): 462–470.

Ciuffetti, L. M., Manning, V. A., Pandelova, I., Faris, J. D., Friesen, T. L., Strelkov, S. E., Weber, G. L., Goodwin, S. B., Wolpert, T. J. and Figueroa, M. (2014). *Pyrenophora Tritici-Repentis: A Plant Pathogenic Fungus with Global Impact in Genomics of Plant-Associated Fungi: Monocot Pathogens*, 1–39. Springer.

Conners, I. L. (1937). Diseases in cereal crops. In: Conners, I. L. (Ed), *Seventeenth Annual Report of the Canadian Plant Disease Survey, 5*. Dominion of Canada Department of Agriculture Science Service.

Conners, I. L. (1939). Diseases in cereal crops. In: Conners, I. L. (Ed), *Nineteenth Annual Report of the Canadian Plant Disease Survey* (vols. 12–14). Dominion of Canada Department of Agriculture Science Service.

Conners, I. L. (1941). Diseases in cereal crops. In: Conners, I. L. (Ed), *Twenty-First Annual Report of the Canadian Plant Disease Survey, 6*. Dominion of Canada Department of Agriculture Science Service.

Conners, I. L. and Savile, D. B. O. (1944). Diseases in cereal crops in. In: Conners, I. L. and Savile, D. B. O. (Eds), *Twenty-Fourth Annual Report of the Canadian Plant Disease Survey, 5*. Dominion of Canada Department of Agriculture Science Service.

Creelman, D. (1964). Diseases of cereal crops in. In: Conners, I. L. and Savile, D. B. O. (Eds), *Fourty-Fourth Annual Report of the Canadian Plant Disease Survey*. Dominion of Canada Department of Agriculture Science Service, 15–21.

De Bary, A. (1886). Uber einige Sclerotinine und Sclerotienkrankheiten. *Bot. Zt* 44: 376–474.

De Wolf, E. D., Effertz, R. J., Ali, S. and Francl, L. J. (1998). Vistas of tan spot research. *Can. J. Plant Pathol.* 20(4): 349–370.

Diedicke, H. (1902). Ueber den Zusammenhang zwischen Pleospora-und Helminthosporium-Arten. *Centrablatt Bakteriol. Parasitenkunde Jena Abt.*: 9317–9329.

Ellwood, S. R., Syme, R. A., Moffat, C. S. and Oliver, R. P. (2012). Evolution of three *Pyrenophora* cereal pathogens: recent divergence, speciation and evolution of non-coding DNA. *Fungal Genet. Biol.* 49(10): 825–829.

Faris, J. D. and Friesen, T. L. (2020). Plant genes hijacked by necrotrophic fungal pathogens. *Curr. Opin. Plant Biol.* 56: 74–80.

Faris, J. D., Liu, Z. and Xu, S. S. (2013). Genetics of tan spot resistance in wheat. *Theor. Appl. Genet.* 126(9): 2197–2217.

Faris, J. D., Overlander, M. E., Kariyawasam, G. K., Carter, A., Xu, S. S. and Liu, Z. (2020) Identification of a major dominant gene for race-nonspecific tan spot resistance in wild emmer wheat. *Theor. Appl. Genet.* 133(3): 829–841.

Faris, J. D., Zhang, Z., Lu, H., Lu, S., Reddy, L., Cloutier, S., Fellers, J. P., Meinhardt, S. W., Rasmussen, J. B., Xu, S. S., Oliver, R. P., Simons, K. J. and Friesen, T. L. (2010). A unique wheat disease resistance-like gene governs effector-triggered susceptibility to necrotrophic pathogens. *Proc. Natl. Acad. Sci. U. S. A.* 107(30): 13544–13549.

Fernandez, M. R., Stevenson, C. F., Hodge, K., Dokken-Bouchard, F., Pearse, P. G., Waelchli, F., Brown, A. and Peluola, C. (2016). Assessing effects of climatic change, region and agronomic practices on leaf spotting of bread and durum wheat in the western Canadian prairies, from 2001 to 2012. *Agron. J.* 108(3): 1180–1195.

Fernandez, M. R., Pearse, P. G., Holzgang, G., Basnyat, P. and Zentner, R. P. (2009). Impacts of agronomic practices on the leaf spotting complex of common wheat in eastern Saskatchewan. *Can. J. Plant Sci.* 89(4): 717–730.

Freeman, S. and Rodriguez, R. J. (1993). Genetic conversion of a fungal plant pathogen to a nonpathogenic, endophytic mutualist. *Science* 260(5104): 75–78.

Fries, E. (1849). 1823: systema mycologicum 2: 275–621.

Friesen, T. L., Ali, S., Klein, K. K. and Rasmussen, J. B. (2005). Population genetic analysis of a global collection of *Pyrenophora tritici-repentis*, causal agent of tan spot of wheat. *Phytopathology* 95(10): 1144–1150.

Friesen, T. L., Stukenbrock, E. H., Liu, Z., Meinhardt, S., Ling, H., Faris, J. D., Rasmussen, J. B., Solomon, P. S., McDonald, B. A. and Oliver, R. P. (2006). Emergence of a new disease as a result of interspecific virulence gene transfer. *Nat. Genet.* 38(8): 953-956.

Gamba, F. M., Bassi, F. M. and Finckh, M. R. (2017). Race structure of *Pyrenophora tritici-repentis* in Morocco. *Phytopathol. Mediterr*: 119-126.

Gardes, M. and Bruns, T. D. (1993). ITS primers with enhanced specificity for basidiomycetes-application to the identification of mycorrhizae and rusts. *Mol. Ecol.* 2(2): 113-118.

Ghaderi, F., Sharifnabi, B., Javan-Nikkhah, M., Brunner, P. and McDonald, B. A. (2020). *SnToxA*, *SnTox1* and *SnTox3* originated in *Parastagonospora nodorum* in the Fertile Crescent. *bioRxiv*.

Gilchrist, L., Fuentes, S. and De Lourdes De La Isla De Bauer, M. (1984). Determinacion de fuentes de resistencia contra *Helminthosporium tritici-repentis* bajo condiciones de campo e invernadero. *Agrociencia*: 5695-5105.

Guo, J., Shi, G., Kalil, A., Friskop, A., Elias, E., Xu, S. S., Faris, J. D. and Liu, Z. (2020). *Pyrenophora tritici-repentis* race 4 isolates cause disease on tetraploid wheat. *Phytopathology* 110(11): 1781-1790.

Hafez, M., Gourlie, R., Despins, T., Turkington, T. K., Friesen, T. L. and Aboukhaddour, R. (2020). *Parastagonospora nodorum* and related species in western Canada: genetic variability and effector genes. *Phytopathology* 110(12): 1946-1958.

Haridas, S., Albert, R., Binder, M., Bloem, J., LaButti, K., Salamov, A., Andreopoulos, B., Baker, S. E., Barry, K., Bills, G., Bluhm, B. H., Cannon, C., Castanera, R., Culley, D. E., Daum, C., Ezra, D., González, J. B., Henrissat, B., Kuo, A., Liang, C., Lipzen, A., Lutzoni, F., Magnuson, J., Mondo, S. J., Nolan, M., Ohm, R. A., Pangilinan, J., Park, H. J., Ramírez, L., Alfaro, M., Sun, H., Tritt, A., Yoshinaga, Y., Zwiers, L. H., Turgeon, B. G., Goodwin, S. B., Spatafora, J. W., Crous, P. W. and Grigoriev, I. V. (2020). 101 *Dothideomycetes* genomes: a test case for predicting lifestyles and emergence of pathogens. *Stud. Mycol.* 96: 141-153.

Heath, M. C. (1987). Evolution of plant resistance and susceptibility to fungal invaders. *Can. J. Plant Pathol.* 9(4): 389-397.

Hiroe, I. and Aoe, S. (1954). Phytopathological studies on black spot disease of Japanese pears (I). *Sum. Fac. Agri.* 2(1): 1-26.

Hosford, R. M. (1971). A form of *Pyrenophora trichostoma* pathogenic to wheat and other grasses. *Phytopathology* 61(1): 28-32.

Hosford, R. M. (1976). Fungal leaf spot diseases of wheat in North Dakota. *North Dakota Agric. Exp. Stn. Bull.* 500: 12.

Hosford, R. M. (1982). *Tan Spot of Wheat and Related Diseases Workshop, July 14-15, 1981*. North Dakota State University, Fargo, ND. North Dakota Agricultural Experiment Station.

Hudcovicova, M., Matusinsky, P., Gubis, J., Leisova-Svobodova, L., Heinonen, U., Ondreickova, K., Mihalik, D., Gubisova, M., Majeska, M. and Jalli, M. (2015). DNA markers for identification of *Pyrenophora tritici-repentis* and detection of genetic diversity among its isolates. *Rom. Agric. Res.* 32: 263-272.

Kamel, S., Cherif, M., Hafez, M., Despins, T. and Aboukhaddour, R. (2019). *Pyrenophora tritici-repentis* in Tunisia: race structure and effector genes. *Front. Plant Sci.* 10: 1562.

Kendig, A. E., Spear, E. R., Daws, S. C., Flory, S. L. and Mordecai, E. A. (2021). Native perennial and non-native annual grasses shape pathogen community composition and disease severity in a California grassland. *J. Ecol.* 109(2): 900-912.

Khan, T. N. (1971). Effect of light on sporulation in *Drechslera tritici-repentis*. *Trans. Brit. Mycol. Soc.* 56(2): 309–311.

Kim, Y. M. and Strelkov, S. E. (2007). Heterologous expression and activity of Ptr ToxB from virulent and avirulent isolates of *Pyrenophora tritici-repentis*. *Can. J. Plant Pathol.* 29(3): 232–242.

Kirk, P., Cannon, P., Minter, D. and Stalpers, J. (2008). *Ainsworth & Bisby's Dictionary of the Fungi* (10th edn.). CABI Publishing, Wallingford.

Krupinsky, J. M. (1987). Pathogenicity on wheat of *Pyrenophora tritici-repentis* isolated from *Bromus inermis*. *Phytopathology* 77(5): 760–765.

Krupinsky, J. M. (1992). Aggressiveness of isolates of *Pyrenophora tritici-repentis* obtained from wheat in the Northern Great Plains. *Plant Dis.* 76(1): 87–91.

Lamari, L. and Bernier, C. C. (1989a). Evaluation of wheat lines and cultivars to tan spot [*Pyrenophora tritici-repentis*] based on lesion type. *Can. J. Plant Pathol.* 11(1): 49–56.

Lamari, L. and Bernier, C. C. (1989b). Virulence of isolates of *Pyrenophora tritici-repentis* on 11 wheat cultivars and cytology of the differential host reactions. *Can. J. Plant Pathol.* 11(3): 284–290.

Lamari, L. and Bernier, C. (1991). Genetics of tan necrosis and extensive chlorosis in tan spot of wheat caused by *Pyrenophora tritici-repentis*. *Phytopathology* 81(10): 1092–1095.

Lamari, L. and Bernier, C. C. (1994). Temperature-induced resistance to tan spot [*Pyrenophora tritici-repentis*] of wheat. *Can. J. Plant Pathol.* 16(4): 279–286.

Lamari, L., Bernier, C. and Ballance, G. (1992). The necrosis-chlorosis model in tan spot of wheat in. In: Francl, L. J., Kuprinsky, J. M. and McMullen, M. P. (Eds), *Advances in Tan Spot Research*. North Dakota Agriculture Experiment Station, Fargo, ND.

Lamari, L., McCallum, B. D. and Depauw, R. M. (2005a). Forensic pathology of Canadian bread wheat: the case of tan spot. *Phytopathology* 95(2): 144–152.

Lamari, L., Strelkov, S. E., Yahyaoui, A., Amedov, M., Saidov, M., Djunusova, M. and Koichibayev, M. (2005b). Virulence of *Pyrenophora tritici-repentis* in the countries of the Silk Road. *Can. J. Plant Pathol.* 27(3): 383–388.

Lamari, L., Sayoud, R., Boulif, M. and Bernier, C. C. (1995). Identification of a new race in *Pyrenophora tritici-repentis*: implications for the current pathotype classification system. *Can. J. Plant Pathol.* 17(4): 312–318.

Lamari, L., Strelkov, S. E., Yahyaoui, A., Orabi, J. and Smith, R. B. (2003). The identification of two new races of *Pyrenophora tritici-repentis* from the host center of diversity confirms a one-to-one relationship in tan spot of wheat. *Phytopathology* 93(4): 391–396.

Lamari, L. and Strelkov, S. E. (2010). The wheat/*Pyrenophora tritici-repentis* interaction: progress towards an understanding of tan spot disease. *Can. J. Plant Pathol.* 32(1): 4–10.

Laribi, M., Gamba, F. M., Hassine, M., Singh, P. K., Yahyaoui, A. and Sassi, K. (2019). Race structure and distribution of *Pyrenophora tritici-repentis* in Tunisia. *Phytopathol. Mediterr.* 58(3): 473–483.

Lepoint, P., Renard, M. E., Legrève, A., Duveiller, E. and Maraite, H. (2010). Genetic diversity of the mating type and toxin production genes in *Pyrenophora tritici-repentis*. *Phytopathology* 100(5): 474–483.

Lichter, A., Gaventa, J. M. and Ciuffetti, L. M. (2002). Chromosome-based molecular characterization of pathogenic and non-pathogenic wheat isolates of *Pyrenophora tritici-repentis*. *Fungal Genet. Biol.* 37(2): 180–189.

Liu, T., Wan, A., Liu, D. and Chen, X. (2017). Changes of races and virulence genes in *Puccinia striiformis* f. sp. *tritici*, the wheat stripe rust pathogen, in the United States from 1968 to 2009. *Plant Dis.* 101(8): 1522–1532.

Liu, Z., Ellwood, S. R., Oliver, R. P. and Friesen, T. L. (2011). *Pyrenophora teres*: profile of an increasingly damaging barley pathogen. *Mol. Plant Pathol.* 12(1): 1–19.

Lorang, J. (2019). Necrotrophic exploitation and subversion of plant defense: a lifestyle or just a phase, and implications in breeding resistance. *Phytopathology* 109(3): 332–346.

Lorang, J. M., Sweat, T. A. and Wolpert, T. J. (2007). Plant disease susceptibility conferred by a "resistance" gene. *Proc. Natl. Acad. Sci. U. S. A.* 104(37): 14861–14866.

Lu, S., Turgeon, B. G. and Edwards, M. C. (2015). A *ToxA*-like protein from *Cochliobolus heterostrophus* induces light-dependent leaf necrosis and acts as a virulence factor with host selectivity on maize. *Fungal Genet. Biol.* 81: 12–24.

Luttrell, E. S. (1955). The ascostromatic *Ascomycetes*. *Mycologia* 47(4): 511–532.

Manning, V. A., Pandelova, I., Dhillon, B., Wilhelm, L. J., Goodwin, S. B., Berlin, A. M., Figueroa, M., Freitag, M., Hane, J. K. and Henrissat, B. (2013). Comparative genomics of a plant-pathogenic fungus, *Pyrenophora tritici-repentis*, reveals transduplication and the impact of repeat elements on pathogenicity and population divergence *G3* 3(1): 41–63.

Martin, F., Spatafora, J. W., Aime, C. M., Grigorieva, I., Stajich, J. E. and Blackwell, M. (2017). The fungal tree of life: from molecular systematics to genome-scale phylogenies in *The Fungal Kingdom*. *American Society of Microbiology*, 3–34.

Martinez, J. P., Oesch, N. W. and Ciuffetti, L. M. (2004). Characterization of the multiple-copy host-selective toxin gene, ToxB, in pathogenic and nonpathogenic isolates of *Pyrenophora tritici-repentis*. *Mol. Plant Microbe Interact.* 17(5): 467–474.

Matusinsky, P., Frei, P., Mikolasova, R., Svacinova, I., Tvaruzek, L. and Spitzer, T. (2010). Species-specific detection of Bipolaris sorokiniana from wheat and barley tissues. *Crop Prot.* 29(11): 1325–1330.

Mavragani, D., Hamel, C. and Vujanovic, V. (2011). Species-specific PCR–DGGE markers to distinguish Pyrenophora species associated to cereal seeds. *Fungal Biol.* 115(2): 169–175.

McCallum, B. D. and DePauw, R. M. (2008). A review of wheat cultivars grown in the Canadian prairies. *Can. J. Plant Sci.* 88(4): 649–677.

McCartney, H. A., Foster, S. J., Fraaije, B. A. and Ward, E. (2003). Molecular diagnostics for fungal plant pathogens. *Pest Manag. Sci.* 59(2): 129–142.

McCurry, J. (1929). Report on the prevalence of plant diseases in the Dominion of Canada for the years 1927 and 1928. Canadian Department of Agriculture. Division of Botany, Dominion Experimental Farms.

McDonald, M. C., Oliver, R. P., Friesen, T. L., Brunner, P. C. and McDonald, B. A. (2013). Global diversity and distribution of three necrotrophic effectors in *Phaeosphaeria nodorum* and related species. *New Phytol.* 199(1): 241–251.

McDonald, M. C., Razavi, M., Friesen, T. L., Brunner, P. C. and McDonald, B. A. (2012). Phylogenetic and population genetic analyses of *Phaeosphaeria nodorum* and its close relatives indicate cryptic species and an origin in the Fertile Crescent. *Fungal Genet. Biol.* 49(11): 882–895.

McDonald, M. C., Ahren, D., Simpfendorfer, S., Milgate, A. and Solomon, P. S. (2018). The discovery of the virulence gene ToxA in the wheat and barley pathogen Bipolaris sorokiniana. *Mol. Plant Pathol.* 19(2): 432–439. https://doi.org/10.1111/mpp.12535.

McDonald, M. C., Taranto, A. P., Hill, E., Schwessinger, B., Liu, Z., Simpfendorfer, S., Milgate, A. and Solomon, P. S. (2019). Transposon-mediated horizontal transfer of the host-specific virulence protein ToxA between three fungal wheat pathogens. *mBio* 10(5).

Momeni, H., Aboukhaddour, R., Javan-Nikkhah, M., Razavi, M., Naghavi, M., Akhavan, A. and Strelkov, S. (2014). Race identification of *Pyrenophora tritici-repentis* in. *Iran. J. Plant Pathol.* 96(2): 287-294.

Momeni, H., Akhavan, A., Aboukhaddour, R., Javan-Nikkhah, M., Razavi, M., Naghavi, M. R. and Strelkov, S. E. (2019). Simple sequence repeat marker analysis reveals grouping of *Pyrenophora tritici-repentis* isolates based on geographic origin. *Can. J. Plant Pathol.* 41(2): 218-227.

Morrall, R. A. A. and Howard, R. J. (1975). The epidemiology of leaf spot disease in a native prairie. II. Airborne spore populations of *Pyrenophora tritici-repentis. Can. J. Bot.* 53(20): 2345-2353.

Murray, G. M. and Brennan, J. P. (2009). Estimating disease losses to the Australian wheat industry. *Australas. Plant Pathol.* 38(6): 558-570.

Nagy, E. D. and Bennetzen, J. L. (2008). Pathogen corruption and site-directed recombination at a plant disease resistance gene cluster. *Genome Res.* 18(12): 1918-1923.

Newcombe, G., Fraser, S. J., Ridout, M. and Busby, P. E. (2020). Leaf endophytes of *Populus trichocarpa* act as pathogens of neighboring plant species. *Front. Microbiol.* 11: 573056.

Nisikado, Y. (1928). Preliminary notes on yellow leaf spot diseases of wheat caused by *Helminthosporium tritici*-vulgaris Inst. *Agric. Biol.* 41: 03-109.

Odvody, G. and Boosalis, M. (1982). An efficient, synchronous conidial production technique for Pyrenophora trichostoma. In: Hosford, R. M. (Eds), *Tan Spot of Wheat and Related Diseases Workshop*, July 1981, 28-32. North Dakota State University, Fargo, ND.

Oliver, R. P., Lord, M., Rybak, K., Faris, J. D. and Solomon, P. S. (2008). Emergence of tan spot disease caused by toxigenic *Pyrenophora tritici-repentis* in Australia is not associated with increased deployment of toxin-sensitive cultivars. *Phytopathology* 98(5): 488-491.

Pfender, W. F., Pacey, C. and Zhang, W. (1988). Saprophytic growth and pseudothecia production by *Pyrenophora tritici-repentis* in plant tissue held at controlled water potentials. *Phytopathology* 78(9): 1205-1210.

Platt, H. W. (1979). Factors affecting spore production and liberation of *Pyrenophora tritici-repentis* (Died.) Drechs. PhD, Univserity of Saskatchewan, Saskatoon. Saskatchewan.

Platt, H. W. and Morrall, R. A. A. (1980). Effects of light intensity and relative humidity on conidiation in *Pyrenophora tritici-repentis. Can. J. Plant Pathol.* 2(2): 53-57.

Platt, H. W., Morrall, R. A. A. and Gruen, H. E. (1977). The effects of substrate, temperature, and photoperiod on conidiation of *Pyrenophora tritici-repentis. Can. J. Bot.* 55(3): 254-259.

Pringle, R. B. and Scheffer, R. P. (1964). Host-specific plant toxins. *Annu. Rev. Phytopathol.* 2(1): 133-156.

Raymond, P. J., Bockus, W. and Norman, B. (1985). Tan spot of winter wheat: procedures to determine host response. *Phytopathology* 75(6): 686-690.

Rees, R. and Platz, G. (1992). Tan spot and its control: some Australian experiences. In: Francl, L., Krupinsky, J. and McMullen, M. P. (Eds), *2nd International Tan Spot Workshop, 1992.* Agric. Exp. Stn. Fargo, ND.

Rees, R., Platz, G. and Mayer, R. (1982). Yield losses in wheat from yellow spot: comparison of estimates derived from single tillers and plots. *Aust. J. Agric. Res.* 33(6): 899-908.

Savary, S., Willocquet, L., Pethybridge, S. J., Esker, P., McRoberts, N. and Nelson, A. (2019). The global burden of pathogens and pests on major food crops. *Nat. Ecol. Evol.* 3(3): 430-439. https://doi.org/10.1038/s41559-018-0793-y.

Scheffer, R. and Ullstrup, A. J. (1965). A host-specific toxic metabolite from *Helminthosporium carbonum*. *Phytopathology* 55(9): 1037-1038.

Scheffer, R. P. (1989). Host-specific toxins in phytopathology origin and evolution of the concept. In Kohmoto, K. and Durbin, R. D. (Eds), Host-Specific Toxins: Recognition Specificity Factors in Plant Disease, 1-17. Tottori University, Tokyo, Japan.

Scheffer, R. P. (1991). Role of toxins in evolution and ecology of plant pathogenic fungi. *Experientia* 47(8): 804-811.

Scheffer, R. P. and Briggs, S. P. (1981). A perspective of toxin studies in plant pathology. In Durbin, R. (Eds), *Toxins in Plant Disease*, 1-20. Academic Press, London, UK.

Shaw, D. E. (1957a). Studies on *Leptosphaeria avenaria* f. sp. *avenaria*. *Can. J. Bot.* 35(1): 97-112.

Shaw, D. E. (1957b). Studies on *Leptosphaeria avenaria* f. sp. *triticea* on cereals and grasses. *Can. J. Bot.* 35(1): 113-118.

Shi, G., Zhang, Z., Friesen, T. L., Raats, D., Fahima, T., Brueggeman, R. S., Lu, S., Trick, H. N., Liu, Z., Chao, W., Frenkel, Z., Xu, S. S., Rasmussen, J. B. and Faris, J. D. (2016). The hijacking of a receptor kinase-driven pathway by a wheat fungal pathogen leads to disease. *Sci. Adv.* 2(10): e1600822.

Shoemaker, R. A. (1959). Nomenclature of *Drechslera* and *Bipolaris*, grass parasites segregated from *'Helminthosporium'*. *Can. J. Bot.* 37(5): 879-887.

Shoemaker, R. A. (1961). Pyrenophora phaeocomes (REB. Ex Fr.) FR. *Can. J. Bot.* 39(4): 901-908.

Shoemaker, R. A. (1962). Drechslera Ito. *Can. J. Bot.* 40(6): 809-836.

Sone, J., Bockus, W. and Claassen, M. (1994). Gradients of tan spot of winter wheat from a small-area source of *Pyrenophora tritici-repentis*. *Plant Dis.* 78(6): 622-627.

Stergiopoulos, I. and Gordon, T. R. (2014). Cryptic fungal infections: the hidden agenda of plant pathogens. *Front. Plant Sci.* 5: 506.

Strelkov, S. E., Kowatsch, R. F., Ballance, G. M. and Lamari, L. (2006). Characterization of the ToxB gene from North African and Canadian isolates of *Pyrenophora tritici-repentis*. *Physiol. Mol. Plant Pathol.* 67(3-5): 164-170.

Strelkov, S. E., Lamari, L. and Ballance, G. M. (1998). Induced chlorophyll degradation by a chlorosis toxin from *Pyrenophora tritici-repentis*. *Can. J. Plant Pathol.* 20(4): 428-435.

Strelkov, S. E. and Lamari, L. (2003). Host-parasite interactions in tan spot [*Pyrenophora tritici-repentis*] of wheat. *Can. J. Plant Pathol.* 25(4): 339-349.

Strelkov, S. E., Lamari, L. and Ballance, G. M. (1999). Characterization of a host-specific protein toxin (Ptr ToxB) from *Pyrenophora tritici-repentis*. *Mol. Plant Microbe Interact.* 12(8): 728-732.

Stukenbrock, E. H. and McDonald, B. A. (2007). Geographical variation and positive diversifying selection in the host-specific toxin *SnToxA*. *Mol. Plant Pathol.* 8(3): 321-332.

Stukenbrock, E. H. and McDonald, B. A. (2008). The origins of plant pathogens in agro-ecosystems. *Annu. Rev. Phytopathol.* 46: 75-100.

Tanaka, S. (1933). Studies on Black Spot Disease of the Japanese Pear Pirus serotina Rehd. Mem. Coll. Agr. Kyoto Imp. Univ. NO. 28.

Tekauz, A. (1976). Distribution, severity, and relative importance of leaf spot in western Canada in 1974. *Can. Plant Dis. Surv.* 56: 36-40.

Tomas, A. and Bockus, W. (1987). Cultivar-specific toxicity of culture filtrates of *Pyrenophora tritici-repentis*. *Phytopathology* 77(9): 1337–1340.

Tomas, A., Feng, G., Reeck, G., Bockus, W. and Leach, J. (1990). Purification of a cultivar-specific toxin from *Pyrenophora tritici-repentis*, causal agent of tan spot of wheat. *Mol. Plant Microbe Interact.* 3(4): 221-224.

Tran, V. A., Aboukhaddour, R., Strelkov, I. S., Bouras, N., Spaner, D. and Strelkov, S. E. (2017). The sensitivity of Canadian wheat genotypes to the necrotrophic effectors produced by *Pyrenophora tritici-repentis*. *Can. J. Plant Pathol.* 39(2): 149–162.

Tsuge, T., Harimoto, Y., Akimitsu, K., Ohtani, K., Kodama, M., Akagi, Y., Egusa, M., Yamamoto, M. and Otani, H. (2013). Host-selective toxins produced by the plant pathogenic fungus Alternaria alternata. *FEMS Microbiol. Rev.* 37(1): 44–66.

Tuori, R. P., Wolpert, T. J. and Ciuffetti, L. M. (1995). Purification and immunological characterization of toxic components from cultures of *Pyrenophora tritici-repentis*. *Mol. Plant. Microbe Interact.* 8(1): 41–48.

Turgeon, B. G. and Baker, S. E. (2007). Genetic and genomic dissection of the *Cochliobolus heterostrophus* Tox1 locus controlling biosynthesis of the polyketide virulence factor T-toxin. *Adv. Genet.* 57: 219–261.

Valder, P. and Shaw, D. E. (1952). Yellow spot disease of wheat in Australia. In *Proc. Linn. Soc. N.S.W.*, 1952.

Virdi, S. K., Liu, Z., Overlander, M. E., Zhang, Z., Xu, S. S., Friesen, T. L. and Faris, J. D. (2016), New insights into the roles of host gene-necrotrophic effector interactions in governing susceptibility of durum wheat to tan spot and Septoria nodorum blotch. *G3* 6(12): 4139–4150.

Vleeshouwers, V. G. and Oliver, R. P. (2014). Effectors as tools in disease resistance breeding against biotrophic, hemibiotrophic, and necrotrophic plant pathogens. *Mol. Plant Microbe Interact.* 27(3): 196–206.

Walton, J. D. (1996). Host-selective toxins: agents of compatibility. *Plant Cell* 8(10): 723.

Ware, S. B. (2006). Aspects of sexual reproduction in mycosphaerella species on wheat and barley: genetic studies on specificity, mapping, and fungicide resistance. PhD, University of Wageningen, Wageningen, Netherlands.

Wegulo, S. N., Breathnach, J. A. and Baenziger, P. S. (2009). Effect of growth stage on the relationship between tan spot and spot blotch severity and yield in winter wheat. *Crop Prot.* 28(8): 696–702.

Wei, B., Despins, T., Fernandez, M. R., Strelkov, S. E., Ruan, Y., Graf, R. and Aboukhaddour, R. (2021). Race distribution of *Pyrenophora tritici-repentis* in relation to ploidy level and susceptibility of durum and winter bread wheat. *Can. J. Plant Pathol.*

Wei, B., Moscou, M. J., Sato, K., Gourlie, R., Strelkov, S. and Aboukhaddour, R. (2020). Identification of a locus conferring dominant susceptibility to *Pyrenophora tritici-repentis* in barley. *Front. Plant Sci.* 11: 158.

White, T. J., Bruns, T., Lee, S. J. W. T. and Taylor, J. (1990). Amplification and direct sequencing of fungal ribosomal RNA genes for phylogenetics. PCR protocols: a guide to methods and applications 18(1): 315-322.

Yarwood, C. E. (1956). Obligate parasitism. *Annu. Rev. Plant Physiol.* 7(1): 115–142.

Yoder, O. C. (1980). Toxins in pathogenesis. *Annu. Rev. Phytopathol.* 18(1): 103–129.

Zhang, H. F., Francl, L. J., Jordahl, J. G. and Meinhardt, S. W. (1997). Structural and physical properties of a necrosis-inducing toxin from *Pyrenophora tritici-repentis*. *Phytopathology* 87(2): 154–160.

Chapter 17

Towards an early warning system for wheat blast: epidemiological basis and model development

J. M. Fernandes, Embrapa Trigo, Brazil; E. M. Del Ponte and J. P. Ascari, Universidade Federal de Viçosa, Brazil; T. J. Krupnik, International Maize and Wheat Improvement Center (CIMMYT), Bangladesh; W. Pavan, Universidade de Passo Fundo and SensorOn – Estrada do Trigo, Brazil; F. Vargas, SensorOn – Estrada do Trigo, Brazil; and T. Berton, Universidade de Passo Fundo, Brazil*

1 Introduction

Wheat blast is a relatively new disease to wheat (*Triticum aestivum* L.) caused by the fungus *Pyricularia oryzae Triticum* pathotype - PoT (syn. *Magnaporthe oryzae* B.C. Couch). The disease was first reported in the state of Paraná, south region of Brazil, in 1985 (Igarashi et al., 1986). During the following decades, wheat blast spread to almost all wheat regions in Brazil, and also to neighboring countries including Bolivia, Paraguay and northern Argentina (Kohli et al., 2011; Perelló et al., 2015). The impact of this disease to wheat production can be devastating; 100% yield losses have been reported multiple times in South America (Cruz et al., 2019; Gongora-Canul et al., 2020) and, more recently, large-scale outbreaks have occurred in Bangladesh (Islam et al., 2016; Malaker et al., 2016). However, significantly damaging wheat blast epidemics are sporadic

* International Fertilizer Development Center, Muscle Shoals, AL, U.S.A and Universidade de Passo Fundo - PPGCA, Rodovia, BR 285, Passo Fundo, RS, 99050-970, Brazil

http://dx.doi.org/10.19103/AS.2021.0092.22

and tend to be limited to situations when exceptionally wet and warm weather occurs prior and during the season, in regions where inoculum is not limiting, more typical of the tropics of Brazil and Bolivia (Coelho et al., 2016; Fernandez et al., 2007; Gongora-Canul et al., 2020; Goulart and Paiva, 2000). However, the frequency of severe outbreaks has increased in the Cerrado region, central-western of Brazil, where the disease is currently a major constraint to the expansion of wheat production in this important agroecosystem (Pasinato et al., 2018).

Until 2016, wheat blast was restricted to South America and of concern to wheat farmers in the Americas, as a risk analysis study highlighted the potential to its introduction into North America (Cruz et al., 2012). Unexpectedly, instead of the anticipated northward movement, wheat blast caused by PoT was found to suddenly devastate over 15000 ha, from February to March 2016 alone, in Bangladesh (Malaker et al., 2016). To date, the explosive outbreaks have not re-occurred in that country. Bangladesh's Department of Agricultural Extension (DAE) reported wheat blast-affected acreage ranging from 2 ha to 312 ha from 2016–2017 to 2019–2020 wheat growing seasons (DAE, Personal Communication). These numbers were backed by analyses conducted by the International Maize and Wheat Improvement Center (CIMMYT) and the Bangladesh Wheat and Maize Research Institute (BWMRI), in particular, by spore trapping efforts that confirmed the presence of airborne inoculum, though at low densities.

Another striking official report was the presence of the PoT fungus causing wheat blast in rain-fed wheat fields and experimental plots in Zambia, Africa (Tembo et al., 2020). Summer wheat production in Zambia is rain-fed, grown primarily by small-scale and resource-poor farmers who typically attain low yields given a range of limiting factors including foliar diseases such as spot blotch (*Bipolaris sativus*) (Tembo, 2019). The presence of wheat blast in Africa places current agricultural policies that have focused on increasing wheat acreage to reduce foreign currency expenditure on dependence on imports at risk (Tembo, 2019; Tembo et al., 2020).

The urgent need to deal with this relatively recent and poorly studied disease has mobilized scientists from all over the world after major epidemics were reported in Brazil and Bolivia in the early 2000s, but especially its appearance in Asia (Cruz and Valent, 2017). Original research and review articles have been published in various venues (Ceresini et al., 2018; Inoue et al., 2017; Juliana et al., 2020). Much has been learned in several aspects including genetics and genomics of the pathogen (Gladieux et al., 2018; Inoue et al., 2017), genetics of host resistance (Cruppe et al., 2020; Goddard et al., 2020; Gruner et al., 2020; Juliana et al., 2020), host-pathogen interactions at both physiological and molecular levels (Aucique-Pérez et al., 2020, 2017; Cruz et al., 2015b, 2016; Farman et al., 2017), as well as fungicide resistance (Castroagudín et al., 2015;

Dorigan et al., 2019; Oliveira et al., 2015) and biological control (Chakraborty et al., 2020).

Nevertheless, disease control is still a challenge for most farmers, and is complicated by two main factors, including (1) lack of genetic resistance in most wheat cultivars available commercially (Cruppe et al., 2020; Juliana et al., 2020) and (2) uncertainty in the profitability of fungicides that present inconsistent results with regards to efficacy and yield response (Cruz et al., 2019; Maciel, 2011). To counteract these challenges, shifting of sowing dates, particularly delaying planting in the Cerrado region, has been used as a disease avoidance measure (Coelho et al., 2016). Conversely, preliminary evidence in Bangladesh indicates that earlier sowing can help farmers to reduce disease risks by reducing the length of time that wheat experiences hotter temperatures favorable to infection later in the season (CSISA, 2020). Nonetheless, a close matching of disease-conducive factors over a large wheat production region is difficult to predict and can be devastating, such as the widespread severe epidemics of both leaf and head blast recorded in 2019 at the Cerrado region of Brazil (Fig. 1).

Much less research has, however, focused on crucially important and complex ecological interactions at the field, landscape or regional levels. Knowledge of disease epidemiology and ecology is a pillar of any integrated pest management (IPM) program, not only to refine management strategies, but also to develop and test models for risk analysis and prediction of the disease (Donatelli et al., 2017; Savary et al., 2018). During the season, fungicides are

Figure 1 Aerial view of the wheat fields (a) at the Cerrado region of Brazil, heavily affected on both (b) spikes and (c) leaves (with d) detail of lesions caused by wheat blast during the 2019 outbreak. Source: Embrapa (2019).

the last resort to minimize the impact of wheat blast, although the economic benefits of fungicides are questionable given the relatively low efficacy levels if improperly utilized (Cruz et al., 2019). Moreover, increase in production costs incurred with fungicide purchase limit the broad use of fungicides, especially for smallholder wheat farmers who may not be able to afford regular calendar-based applications. Nevertheless, a wheat blast warning system may be useful to indicate when and where the disease might occur, and thus preventative management advisories can be delivered to farmers several days in advance of potential disease outbreaks. This approach is aimed at rationalizing the use of fungicide, limiting it to locations and situations in which application is most likely to be necessary and economically rational (Shtienberg, 2013, 2000).

In this chapter, we review aspects of the epidemiology of wheat blast, mainly those related to inoculum and its role for the epidemics, which we believe are critical for more accurate predictions of wheat blast risk. We then describe the models that we have developed as well as the decision support system capable of operating at the global scale. Examples of the implementation of a warning system in Bangladesh and Brazil are illustrated.

2 Epidemiology and pathogen ecology

Epidemiology concerns the study of the change of the disease intensity in a population of host plants in space and time (Madden et al., 2007). In this context, epidemics are dynamic systems in which several different rates affect key processes that operate both at the 'infection chain' level (infection, colonization, sporulation) (plants) as well as at the landscape (fields) and regional (agroecosystems) levels. The latter two scales are affected by the spatial distribution and density of the main and the alternative hosts that harbor inoculum, seasonal weather patterns and source-sink relationships that influence the rate and severity of disease spread and infection (Plantegenest et al., 2007).

For wheat blast, the risk of severe attack and yield loss is increased when airborne inoculum is plentiful during the wheat heading period and the physical environment is wet and warm (Fig. 2). Therefore, knowledge of the presence and distribution of hosts of the wheat blast pathogen population, as well as factors that drive inoculum production and dispersal, are key not only to understand pathogen population genetics and evolution, but also and especially to predict current and future disease risks.

3 Inoculum sources and survival

Thus far, two sources of inoculum have been identified as important for wheat blast infection. These include (a) infected seeds (Gomes et al., 2018; Goulart and Paiva, 1990), which are responsible for the large distance (intercontinental)

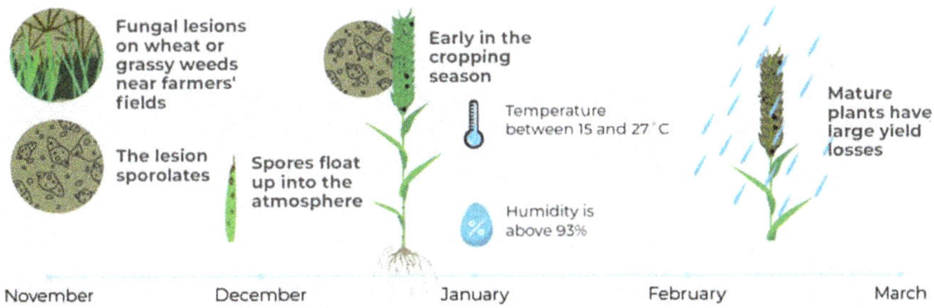

Figure 2 Generalization of the role of inoculum sources and key requirements of weather-based variables that lead to infection by *Pyricularia oryzae Triticum* (PoT) in wheat.

spread (Cruz and Valent, 2017) and (b) infected grasses present in the surroundings of wheat crops, before wheat sowing. These alternative hosts can harbor and spread the wheat blast pathogen inoculum early in the season.

In South America, the most abundant blast-susceptible grass is signal grass (*Urochloa* spp.), present in all tropical regions where wheat is grown in Brazil (Jank et al., 2014). Together with other grass hosts, signal grass likely serves as a major reservoir of primary inoculum for both leaf and head blast epidemics in commercial wheat fields (Urashima et al., 2017). Under experimental conditions, where infected crop residues were amended to field plots, the earliest symptoms of disease were recorded on newly emerged and youngest wheat leaves; these early infected leaves can further serve as a source of newly produced conidia (secondary inoculum) that move upwards, further infecting wheat heads (Cruz et al., 2015b; Gongora-Canul et al., 2020). In Brazil, a study showed that the blast fungus was capable of surviving for up to five months (Pizolotto et al., 2019). Therefore, it is more likely that residue-borne spores will be available to infect grasses during the off-season, and not directly infect wheat in the following season.

4 Production, release and aerial transport of inoculum

It has been known for a long time that *P. oryzae* conidia are probably forcibly discharged, but at a very short distance, by the active bursting of the minute stalk-cell by which it is attached to the conidiophore (Ingold, 1964). Environmental factors may also play an essential role in the spore-release process. High-humidity conditions may be accountable for turgor pressure buildup inside the stalk-cell, which explains the high availability of conidia under moistened conditions (Ingold, 1964; Meredith, 1962). The conidia release of *P. oryzae* can be triggered by either increasing levels (from 24% to 100%) or decreasing levels (from 100% to 24%) of relative humidity; maximum conidia release has been found when relative humidity levels increase from 76% to 100% (Danelli et al., 2019; Leach, 1980). Short rains (less than 5 mm) also seem to favor spore release, but intense rainfall events may also reduce *P. oryzae* conidia residence in sporulating lesions (Kim, 1994; Danelli et al., 2019). It is also plausible that under heavy rain, large raindrops vigorously damage conidiophores, releasing and washing off conidia, thereby reducing the amount of inoculum available for transport from plant to plant, field to field, and across regions. Silva and Prabhu (2005) also reported an inverse relationship between rainfall and trapped *P. oryzae* conidia. Another important factor for the release of *P. oryzae* conidia is light (Leach, 1980, 1962). Constant blue light and red light have been proven to induce higher conidia release than constant white light or darkness. The transition from a light to a dark period also induces the release of conidia (Lee et al., 2006).

Minimum wind speeds to induce spore release have been reported to vary among different fungi (Pasanen et al., 1991). However, there appears to be no evidence indicating that wind speed influences the conidia release of *P. oryzae*. Released spores can escape from the plant canopy (disease sources) and are dispersed or transported to different sites at various distances. Where they encounter a susceptible host, infections serve as a sink for the disease, in a pattern that is essential for the spread of the disease and perpetuation of an epidemic. Intermittent wind patterns also play a major role in spore removal and canopy escape, which in turn shape disease gradients for aerially dispersed fungal pathogens (Fitt et al., 2006; West and Kimber, 2015). One aerobiological study trapped *P. oryzae* conidia at distances as large as 1 km from a known source (Urashima et al., 2007).

A strong relationship between the number of conidia trapped in a volumetric sampler and rice blast disease progress was reported in a multiple-year study (Pinnschmidt et al., 1993). Increasing temperature may also be a factor leading to high airborne spore concentrations of *M. oryzae* during the rice-growing season in Southern Spain (Castejón-Muñoz, 2008). Daily spore release patterns for *P. oryzae* have been well described in the rice blast pathosystem. Among them is a nocturnal pattern of *M. oryzae* spore release that is observed in high concentrations at night (Leach, 1980; Meredith, 1962).

5 Infection and colonization

Detailed information on the effects of temperature and wetness duration period on infection by the wheat blast pathogen (PoT) is scarce (Cardoso et al., 2008). In their study conducted using PoT from Brazil, Cardoso and collaborators (2008) showed that a minimum of 10 hours was required for infection with an increase in infection efficiency positively correlated with an increase in wetness duration. Moreover, temperatures between 25°C and 30°C were more favorable for infection than 15°C (Cardoso et al., 2008). In the United States, temperature and leaf wetness duration models were fitted to data gathered in experiments that inoculated the gray leaf spot pathogen (*P. oryzae Lolium* pathotype, PoL) on perennial ryegrass (Uddin et al., 2003) or wheat (Mills et al., 2020). For the PoL-wheat pathosystem, the mean levels of disease intensity and rates of its increase were higher at 25°C and 30°C, with 24 and 48 hours of relative humidity >95%, when compared with 20°C and < 12 hours of high relative humidity (Mills et al., 2020). Overall the effects of temperature and moisture on infection are quite comparable among PoT (*Triticum* pathotype), PoO (*Oryzae* pathotype) and PoL (*Lolium* pathotype).

Apart from the weather influence, the infection site has shown to be important to define the intensity. In fact, massive wheat yield losses occur when *P. oryzae* infects the rachis at the base of the spike, thereby limiting grain development and filling in addition to destroying the spikes completely (Kohli et al., 2011). Cruz and collaborators (2015b) reported that within 72 hours after infection, the fungal hyphae reach the rachis and spread upward and downward along the rachis. Within rachis tissue, fungal hyphae massively colonized the epidermis, collenchyma, cortical and pith parenchyma and vascular bundle at just 96 hours after infection (Cruz et al., 2015b).

6 Modelling wheat blast epidemics

Models of plant disease epidemics are mathematical simplifications that, in most cases, allow the prediction of epidemic risks or future behavior, depending on their objective and complexity (De Wolf and Isard, 2007; Hardwick, 2006; Magarey et al., 2005). Model predictions are useful for aiding decisions during planning of risk mitigation procedures, establishing management strategies as well as controlling the disease in the field to minimize the risk of yield losses (Gent et al., 2013).

Depending on the specific conditions, the models vary in complexity and their development is not without challenges, depending on the goals (Cunniffe et al., 2015). As far as models to warn disease risk are concerned, traditionally they are based on requirements for infection (Magarey et al., 2005). However, this is valid only when the inoculum can be assumed as not limiting. This seems not be the case for wheat blast where grasses in the landscape serve as large reservoirs for the inoculum, but its production and amount available to infect commercial wheat is dependent on the seasonal weather (Danelli et al., 2019; Fernandes et al., 2017). Therefore, our group has employed efforts to develop weather-driven wheat blast models that take both inoculum production and infection risk into account. Thus far, our modelling efforts have resulted in two generations of wheat blast models that have been improved together with the decision support systems where the models are embedded and predictions deployed as risk maps at the country level.

6.1 First-generation model

The first dynamic wheat blast model that considered inoculum production on the primary sources (grasses) was parameterized a model to predict inoculum buildup and availability (Fernandes et al., 2017). The authors adapted the approach used in a previous study (Bregaglio and Donatelli, 2015) on the development of a simulation model for predicting the seasonal progress of rice blast (*Magnaporthe oryzae* Oryzae) based on relative humidity requirements for conidia production. The parameters in the wheat blast model were defined based on expert knowledge and observations of wheat blast outbreaks in Brazil. This model was purposely designed to be relatively simple in structure so as to be flexible and easy to use in practice. It was the first attempt to input variables which were limited to hourly temperature, relative humidity and precipitation observations or forecasts in the prediction of wheat blast.

The wheat blast model developed by Fernandes et al. (2017) has four components. The first component assumes that spores are present and estimates the rate of conidiophore development as a function of temperature and relative humidity, both of which are integrated to estimate blast inoculum potential by solving Equation 1 for the hourly sum of inoculum potential (IP) over the season:

$$IP = \begin{cases} 14.35 - 0.25 * Tif & 15C < T < 27C \text{ and } RH \geq 93\% \\ -8.5 - 0.59 * Tif & 27C < T < 35C \text{ and } RH \geq 93\% \\ 0 & \text{otherwise} \end{cases} \quad (1)$$

where T and RH are air temperature and relative humidity, respectively. Where RH is below the threshold in Equation 1, the model does not accumulate thermal time. The model also calculates the development of a spore cloud subject to assumptions of air current uptake, atmospheric diffusion and wind shear that affect spore longevity. Survival of spores while airborne may also be affected by temperature, solar and ultraviolet radiation, in addition to relative humidity. Spore cohorts were therefore assumed to have a half-life of three days within any seven-day window.

The model also determines the number and timing of days with weather favoring blast infection using a conditional ruleset. Days favoring infection were consequently declared following spore cloud development when the daily maximum temperature exceeded 23°C and temperature amplitude (calculated daily minimum temperature subtracted from daily maximum temperature) was > 13°C, with mean daily RH above 70%.

6.2 Second-generation model

The second generation was built on a previously developed generic plant disease model structure (Pavan et al., 2009). The model parameters were defined using values obtained from the literature or expert knowledge of wheat blast epidemics. In this model, hourly values of weather variables (such as temperature, relative humidity and precipitation) are the main drivers of wheat blast progress. The model assumes that alternative hosts (grasses) are presented at the surrounding of a wheat field. A submodule, parameterized for each region, predicts the host leaf area assumed to be a potential infection site. Spores originating from nearby sources that are deposited over the wheat canopy remain viable for a certain time that varies as a function of age and a cloudiness factor. Wheat crops are infected when a threshold value for airborne inoculum concentration is estimated, and lesions are formed and expand over time. At the end of the latency period, the lesion becomes infectious and newly produced spores produced are distributed over three spatial hierarchy layers, respectively, auto-deposition, allo-leaf-deposition and allo-plant-deposition (Willocquet and Savary, 2004).

The simulations are run for 200 days with a time step of one day. The model uses numerical weather model forecast outputs on a three-hourly timescale for precipitation, relative humidity and temperature as input. These represent environmental conditions for inoculum buildup expressed by continuous periods favoring sporulation (represented by a spore index) and the density of spores in the air (a spore cloud). The latter is simulated as the density of spores in a 1 m^3 area over the crop canopy, subject to assumptions of air current uptake, atmospheric diffusion and wind shear, all of which can affect and reduce spore longevity (Mousanejad et al., 2009). Importantly, this model also assumes the geographically uniform presence of PoT inoculum in the environment for which simulations are run. It therefore does not yet account for source-sink and spore dispersal mechanisms, but instead simulates the development of conidia and lesion cohorts from an assumed uniform base population.

7 Early warning system for wheat blast

Considering the strong dependence of wheat blast epidemics on seasonal weather, which varies across regions and seasons within a region, a prediction system that operates at large may be useful not only to anticipate the risk of outbreaks in the early season. In addition, such systems - when well-designed - can be used to map risk areas using historical data (Savary et al., 2012) or conduct risk analysis in regions where a pathogen has not been introduced (Hyatt-Twynam et al., 2017). Whatever the case, an early warning system (EWS) is a computerized framework that integrates several layers of geographic, climatic and biological information, which are processed and the results communicated via textual and graphical outputs to target users (Pavan et al., 2011).

The wheat blast early warning system (WB-EWS) uses weather data for simulating inoculum buildup during the vegetative growth stage of wheat which is key to estimate the risk of infection. The WB-EWS was developed by EMBRAPA and CIMMYT for Brazil and Bangladesh and is available at 'http://beattheblastews.net'. The current implementation of the WB-EWS integrates weather databases and scripts for specific submodules developed using different programming languages. For example, the spore index was implemented using the R programming language (R Core Team, 2020) based on previous work (Fernandes et al., 2017), and the spore cloud model was implemented using C++ based on a generic disease model (Pavan et al., 2009). All the submodules are integrated, and information on infection risks and crop management advisories are delivered to end users via e-mail, SMS and the web. The model diagram is presented in Fig. 3.

The WB-EWS has been made operational in Bangladesh and Brazil. In Bangladesh, to more accurately identify and map risk areas, three-hourly Weather Forecasting Research (WRF) model outputs are supplied by the Bangladesh Meteorological Department (BMD) at a 18-km^2 grid resolution for all wheat-growing areas of Bangladesh. WRF forecasts are generated on a three-hourly rolling basis for five days into the future, and combined with an automatic weather station network for observed data, advisories are generated at this spatial and temporal resolution. In Brazil, the Eta Model (Mesinger et al., 2012) output is provided by CPTEC/INPE 'https://www.cptec.inpe.br/' at 20-km^2 grid resolution for all wheat-growing areas. These are used to generate time- and location-specific advisories on a five-day forecast basis. The model adequately described the observed epidemic and non-epidemics years in Brazil and Bangladesh (Figs 4 and 5).

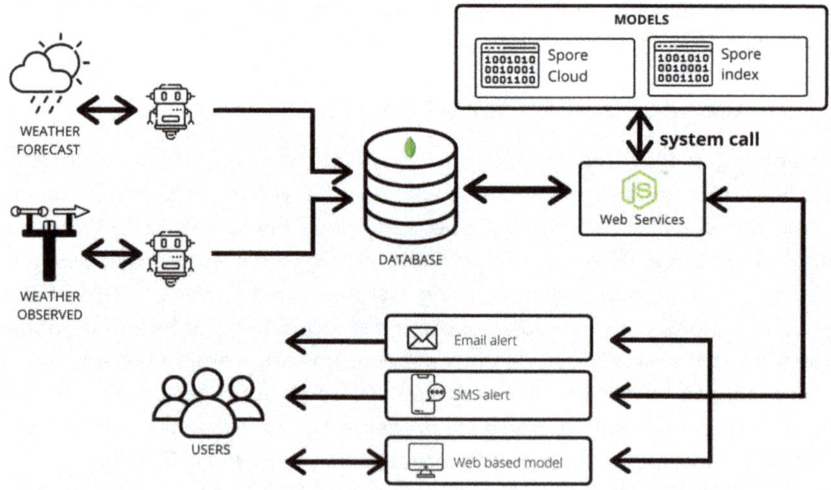

Figure 3 A relational diagram for the wheat blast-early warning system (WB-EWS) operational in Brazil and Bangladesh.

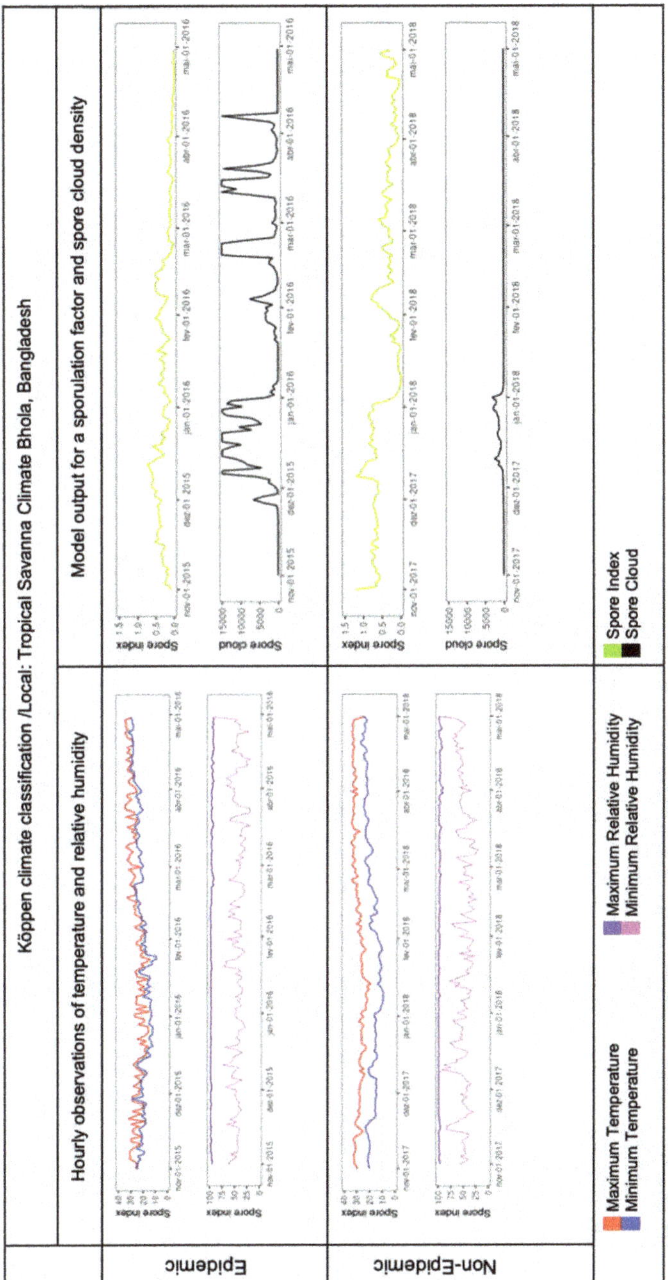

Weather data was kindly provided by the Bangladesh Meteorological Departament.

Figure 4 Procedural flow of information generating early warnings for wheat blast in Bangladesh and Brazil.

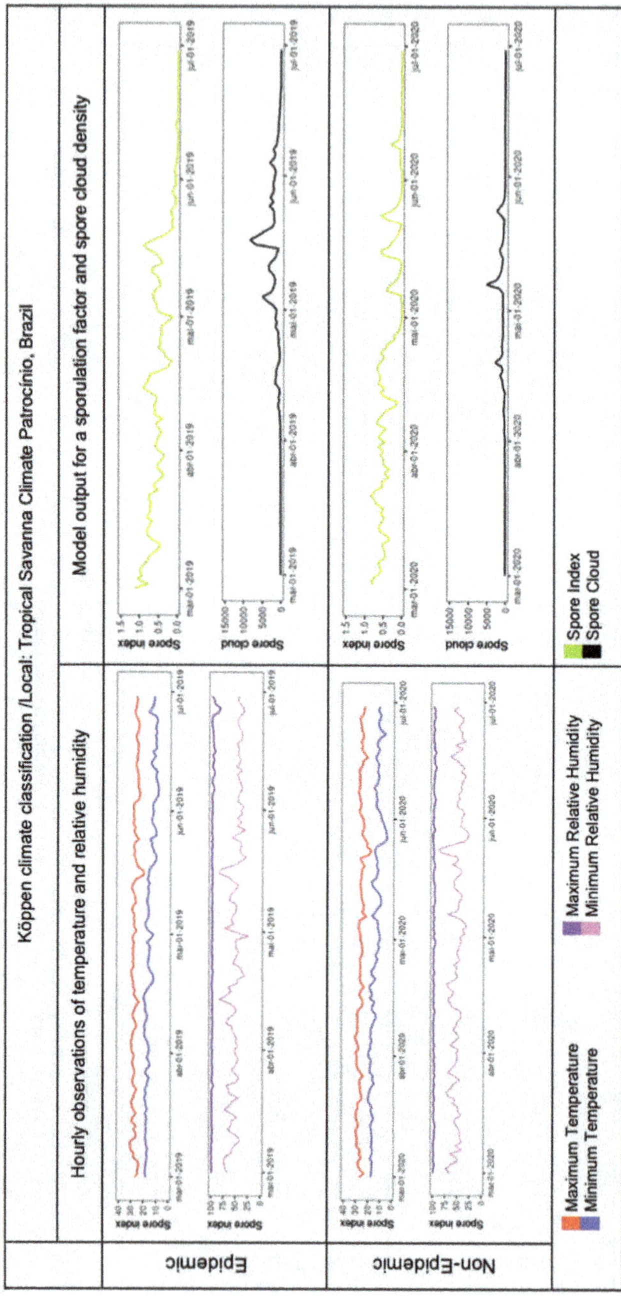

Weather data was kindly provided by the National Institute of Meteorology - INMET (Brazil).

Figure 5 Graphical outputs generated by the WB-EWS (http://beattheblastews.net) using historical weather data from epidemic and non-epidemic years in Bangladesh.

This system results in the triggering of an automated alert system deploying email and SMS warnings to extension agents and lead farmers that is operational during the corresponding typical phenological range from early heading to maturation for wheat in both countries (Fig. 5).

8 Model uses and future developments

In contrast to advisories that encourage calendar-based preventative sprays without field scouting, a data-driven approach can be used to adaptively advise farmers how to take effective and safe preventative action through the intelligent use of fungicides for disease control. The WB-EWS was approved for use in extension in Brazil in 2018 and in Bangladesh in 2019, and uses numerical weather model forecast outputs to map favorable areas and identify critical periods for wheat blast epidemics, which are currently being tested and improved with regard to accuracy. These are used to generate time- and location-specific advisories on a five-day forecast basis (Fig. 6).

We are also considering options for delivering blast warnings directly to farmers using interactive voice response (IVR) technologies for at-risk locations prior to potential infection. In contrast recommendation for standard sprays deployed as a function of the number of days after seeding and without field verification of disease, this data-driven approach can be used to adaptively advise farmers how to take effective and safe preventative action through the intelligent use of appropriate fungicides for disease control. In coupling a wheat growth model with the wheat blast model, our future objective is to provide

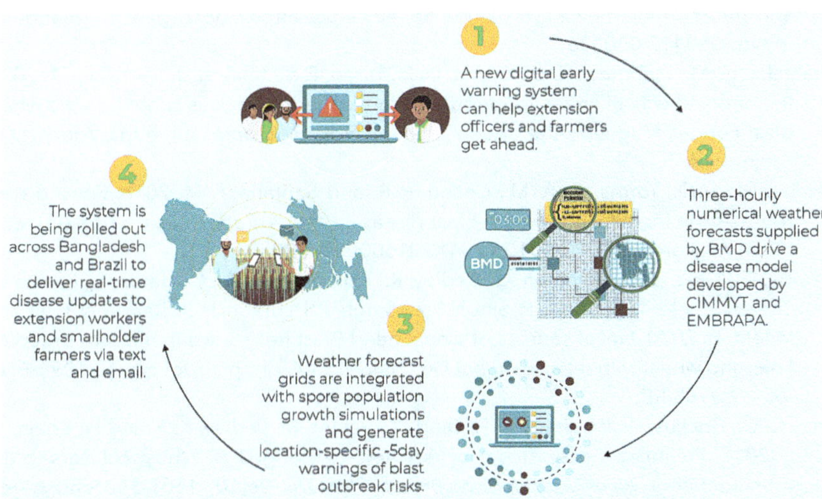

Figure 6 Graphical outputs generated by the WB-EWS (http://beattheblastews.net) using historical weather data from epidemic and non-epidemic years in Brazil.

farmers not only with forecast advisories, but also estimates of potential yield loss if they do not take preventative action to control blast.

9 References

Aucique-Pérez, C. E., de Menezes Silva, P. E., Moreira, W. R., DaMatta, F. M. and Rodrigues, F. Á. 2017. Photosynthesis impairments and excitation energy dissipation on wheat plants supplied with silicon and infected with *Pyricularia oryzae*. *Plant Physiol. Biochem.* 121, 196–205. https://doi.org/10.1016/j.plaphy.2017.10.023.

Aucique-Pérez, C. E., Resende, R. S., Martins, A. O., Silveira, P. R., Cavalcanti, J. H. F., Vieira, N. M., Fernie, A. R., Araújo, W. L., DaMatta, F. M. and Rodrigues, F. Á. 2020. How do wheat plants cope with Pyricularia oryzae infection? A physiological and metabolic approach. *Planta* 252(2), 24. https://doi.org/10/ghgh5z.

Bregaglio, S. and Donatelli, M. 2015. A set of software components for the simulation of plant airborne diseases. *Environ. Modell. Softw.* 72, 426–444. https://doi.org/10/f7tbwb.

Cardoso, C. AdA., Reis, E. M. and Moreira, E. N. 2008. Development of a warning system for wheat blast caused by Pyricularia grisea. *Summa Phytopathol.* 34(3), 216–221. https://doi.org/10.1590/S0100-54052008000300002.

Castejón-Muñoz, M. 2008. The effect of temperature and relative humidity on the airbone concentration of Pyricularia oryzae spores and the development of rice blast in southern Spain. *Span. J. Agric. Res.* 9. https://doi.org/10/ghjmnd.

Castroagudín, V. L., Ceresini, P. C., de Oliveira, S. C., Reges, J. T. A., Maciel, J. L. N., Bonato, A. L. V., Dorigan, A. F. and McDonald, B. A. 2015. Resistance to QoI fungicides is widespread in Brazilian populations of the wheat Blast Pathogen Magnaporthe oryzae. *Phytopathology* 105(3), 284–294. https://doi.org/10.1094/PHYTO-06-14-0184-R.

Ceresini, P. C., Castroagudín, V. L., Rodrigues, F. Á., Rios, J. A., Eduardo Aucique-Pérez, C., Moreira, S. I., Alves, E., Croll, D. and Maciel, J. L. N. 2018. Wheat blast: past, present, and future. *Annu. Rev. Phytopathol.* 56, 427–456. https://doi.org/10.1146/annurev-phyto-080417-050036.

Chakraborty, M., Mahmud, N. U., Gupta, D. R., Tareq, F. S., Shin, H. J. and Islam, T. 2020. Inhibitory effects of linear lipopeptides From a marine Bacillus subtilis on the wheat Blast Fungus Magnaporthe oryzae Triticum. *Front. Microbiol.* 11. https://doi.org/10/ghjmnh, 665.

Coelho, M. AdO., Torres, G. A. M., Cecon, P. R. and Santana, F. M. 2016. Sowing date reduces the incidence of wheat blast disease. *Pesq. Agropec. Bras.* 51(5), 631–637. https://doi.org/10.1590/S0100-204X2016000500025.

Cruppe, G., Cruz, C. D., Peterson, G., Pedley, K., Asif, M., Fritz, A., Calderon, L., Lemes da Silva, C., Todd, T., Kuhnem, P., Singh, P. K., Singh, R. P., Braun, H. J., Barma, N. C. D. and Valent, B. 2020. Novel sources of wheat head Blast Resistance in modern breeding lines and wheat wild relatives. *Plant Dis.* 104(1), 35–43. https://doi.org/10.1094/PDIS-05-19-0985-RE.

Cruz, C. D., Bockus, W. W., Stack, J. P., Tang, X., Valent, B., Pedley, K. F. and Peterson, G. L. 2012. Preliminary assessment of resistance Among U. S. Wheat cultivars to the triticum pathotype of Magnaporthe oryzae. *Plant Dis.* 96(10), 1501–1505. https://doi.org/10.1094/PDIS-11-11-0944-RE.

Cruz, C. D., Kiyuna, J., Bockus, W. W., Todd, T. C., Stack, J. P. and Valent, B. 2015a. Magnaporthe oryzae conidia on basal wheat leaves as a potential source of wheat blast inoculum. *Plant Pathol.* 64(6), 1491–1498. https://doi.org/10.1111/ppa.12414.

Cruz, C. D., Peterson, G. L., Bockus, W. W., Kankanala, P., Dubcovsky, J., Jordan, K. W., Akhunov, E., Chumley, F., Baldelomar, F. D. and Valent, B. 2016. The 2NS translocation from Aegilops ventricosa confers resistance to the triticum pathotype of Magnaporthe oryzae. *Crop Sci.* 56(3), 990–1000. https://doi.org/10.2135/cropsci2015.07.0410.

Cruz, C. D., Santana, F. M., Todd, T. C., Maciel, J. L. N., Kiyuna, J., Baldelomar, D. F., Cruz, A. P., Lau, D., Seixas, C. S., Goulart, A. C. P., Sussel, A. A., Schipanski, C. A., Chagas, D. F., Coelho, M., Montecelli, T. D. N., Utiamada, C., Custódio, A. P., Rivadeneira, M. G., Bockus, W. W. and Valent, B. 2019. Multi-environment assessment of fungicide performance for managing wheat head blast (WHB) in Brazil and Bolivia. *Trop. Plant Pathol.* 44(2), 183–191. https://doi.org/10.1007/s40858-018-0262-9.

Cruz, C. D. and Valent, B. 2017. Wheat blast disease: danger on the move. *Trop. Plant Pathol.* 42(3), 210–222. https://doi.org/10.1007/s40858-017-0159-z.

Cruz, M. F. Ad da, Silva, L. A. F., Rios, J. A., Debona, D. and Rodrigues, F. Á. 2015b. Microscopic aspects of the colonization of Pyricularia oryzae on the rachis of wheat plants supplied with silicon. *Bragantia* 74(2), 207–214. https://doi.org/10.1590/1678-4499.0023.

CSISA 2020. Annual report October 2019-September 2020. Building resilience in South Asia's cereal systems. Cereal systems initiative for South Asia (CSISA) Phase III. Available at: https://csisa.org/wp-content/uploads/sites/2/2020/11/201102-FINAL-CSISA-PIII-FY-2019-20-ANNUAL-REPORT_BD-NEPAL_web.pdf. Verified 18 Decembr 2020.

Cunniffe, N. J., Koskella, B., Metcalf, C. J., Parnell, S., Gottwald, T. R. and Gilligan, C. A. 2015. Thirteen challenges in modelling plant diseases. *Epidemics* 10, 6–10. https://doi.org/10.1016/j.epidem.2014.06.002.

Danelli, A. L. D., Fernandes, J. M. C., Maciel, J. L. N., Boaretto, C. and Forcelini, C. A. 2019. Monitoring Pyricularia sp. airborne inoculum in Passo Fundo, Rio Grande do Sul, Brazil. *Summa Phytopathol.* 45(4), 361–367. https://doi.org/10/ghjmnf.

De Wolf, E. D. and Isard, S. A. 2007. Disease cycle approach to plant disease prediction. *Annu. Rev. Phytopathol.* 45, 203–220. https://doi.org/10.1146/annurev.phyto.44.070505.143329.

Donatelli, M., Magarey, R. D., Bregaglio, S., Willocquet, L., Whish, J. P. M. and Savary, S. 2017. Modelling the impacts of pests and diseases on agricultural systems. *Agric. Syst.* 155, 213–224. https://doi.org/10.1016/j.agsy.2017.01.019.

Dorigan, A. F., Carvalho, G. D., Poloni, N. M., Negrisoli, M. M., Maciel, J. L. N. and Ceresini, P. C. 2019. Resistance to triazole fungicides in Pyricularia species associated with invasive plants from wheat fields in Brazil. *Acta Sci. Agron.* 41(1), 39332. https://doi.org/10.4025/actasciagron.v41i1.39332.

Embrapa 2019. *Brusone em lavouras de trigo no Brasil central - safra 2019 (Nota Técnica), Embrapa Trigo*. Embrapa, Passo Fundo, RS. Brasil.

Farman, M., Peterson, G., Chen, L., Starnes, J., Valent, B., Bachi, P., Murdock, L., Hershman, D., Pedley, K., Fernandes, J. M. and Bavaresco, J. 2017. The lolium pathotype of Magnaporthe oryzae recovered from a single blasted wheat plant in the United States. *Plant Dis.* 101(5), 684–692. https://doi.org/10.1094/PDIS-05-16-0700-RE.

Fernandes, J. M. C., Nicolau, M., Pavan, W., Hölbig, C. A., Karrei, M., de Vargas, F., Bavaresco, J. L. B., Lazzaretti, A. T. and Tsukahara, R. Y. 2017. A weather-based model for predicting early season inoculum build-up and spike infection by the wheat blast pathogen. *Trop. Plant Pathol.* 42(3), 230–237. https://doi.org/10.1007/s40858-017-0164-2.

Fernandez, M. R., Zentner, R. P., DePauw, R. M., Gehl, D. and Stevenson, F. C. 2007. Impacts of crop production factors on Fusarium Head Blight in barley in eastern Saskatchewan. *Crop Sci.* 47(4), 1574–1584. https://doi.org/10.2135/cropsci2006.09.0596.

Fitt, B. D. L., McCartney, H. A. and West, J. S. 2006. Dispersal of foliar plant pathogens: mechanisms, gradients and spatial patterns. In: Cooke, B. M., Jones, D. G. and Kaye, B. (Eds), *The Epidemiology of Plant Diseases*. Springer, Netherlands, Dordrecht, pp. 159–192. https://doi.org/10.1007/1-4020-4581-6_6.

Gent, D. H., Mahaffee, W. F., McRoberts, N. and Pfender, W. F. 2013. The use and role of predictive systems in disease management. *Annu. Rev. Phytopathol.* 51, 267–289. https://doi.org/10/ghjmpp.

Gladieux, P., Condon, B., Ravel, S., Soanes, D., Maciel, J. L. N., Nhani, A., Chen, L., Terauchi, R., Lebrun, M. H., Tharreau, D., Mitchell, T., Pedley, K. F., Valent, B., Talbot, N. J., Farman, M. and Fournier, E. 2018. Gene flow between divergent cereal- and grass-specific lineages of the rice Blast Fungus Magnaporthe oryzae. *mBio* 9(1). https://doi.org/10.1128/mBio.01219-17.

Goddard, R., Steed, A., Chinoy, C., Ferreira, J. R., Scheeren, P. L., Maciel, J. L. N., Caierão, E., Torres, G. A. M., Consoli, L., Santana, F. M., Fernandes, J. M. C., Simmonds, J., Uauy, C., Cockram, J. and Nicholson, P. 2020. Dissecting the genetic basis of wheat blast resistance in the Brazilian wheat cultivar BR 18-Terena. *BMC Plant Biol.* 20(1), 398. https://doi.org/10/ghjmnm.

Gomes, D. P., Rocha, V. S., Rocha, J. RdA. SdC., Pereira, O. L. and Souza, M. Ad 2018. Potential of transmission of Pyricularia graminis-tritici from plant to seed and from seed to seedling in wheat genotypes with different degrees of blast resistance. *J. Seed Sci.* 40(1), 16–24. https://doi.org/10.1590/2317-1545v40n1181833.

Gongora-Canul, C., Salgado, J. D., Singh, D., Cruz, A. P., Cotrozzi, L., Couture, J., Rivadeneira, M. G., Cruppe, G., Valent, B., Todd, T., Poland, J. and Cruz, C. D. 2020. Temporal dynamics of wheat Blast Epidemics and disease measurements using multispectral imagery. *Phytopathology* 110(2), 393–405. https://doi.org/10.1094/PHYTO-08-19-0297-R.

Goulart, A. C. P. and Paiva, F. de A. 2000. *Avaliação de Perdas no Rendimento de Grãos de Trigo Causadas por Pyricularia grisea, no Período de 1988 a 1992, em Mato Grosso do Sul* (7th edn.). Embrapa, Dourados/MS.

Goulart, A. C. P. and Paiva, F. A. 1990. Transmission of Pyricularia oryzae by wheat (Triticum aestivum) seeds. Fitopatol. Bras. *Fitopatol. Bras.* 15, 359–362.

Gruner, K., Esser, T., Acevedo-Garcia, J., Freh, M., Habig, M., Strugala, R., Stukenbrock, E., Schaffrath, U. and Panstruga, R. 2020. Evidence for Allele-specific levels of enhanced susceptibility of wheat mlo mutants to the hemibiotrophic fungal pathogen Magnaporthe Oryzae pv. Triticum. *Genes* 11(5), 517. https://doi.org/10/ghjmnk.

Hardwick, N. V. 2006. Disease forecasting. In: Cooke, B. M., Jones, D. G. and Kaye, B. (Eds), *The Epidemiology of Plant Diseases*. Springer, Netherlands, Dordrecht, pp. 239–267. https://doi.org/10.1007/1-4020-4581-6_9.

Hyatt-Twynam, S. R., Parnell, S., Stutt, R. O. J. H., Gottwald, T. R., Gilligan, C. A. and Cunniffe, N. J. 2017. Risk-based management of invading plant disease. *New Phytol.* 214(3), 1317–1329. https://doi.org/10.1111/nph.14488.

Igarashi, S., Utiamada, C. M., Igarashi, L. C., Kazuma, A. H. and Lopes, R. 1986. Ocurrence of Pyrcularia sp. in wheat (Triticum aestivum L.) in the State of Parana, Brazil. *Fitopatol. Bras.* 11, 351-352.

Ingold, C. T. 1964. Possible spore discharge mechanism in Pyricularia. *Br. Mycol. Soc.* 47(4), 573–575. https://doi.org/10/frdm4k.

Inoue, Y., Vy, T. T. P., Yoshida, K., Asano, H., Mitsuoka, C., Asuke, S., Anh, V. L., Cumagun, C. J. R., Chuma, I., Terauchi, R., Kato, K., Mitchell, T., Valent, B., Farman, M. and Tosa, Y. 2017. Evolution of the wheat blast fungus through functional losses in a host specificity determinant. *Science* 357(6346), 80–83. https://doi.org/10.1126/science.aam9654.

Islam, M. T., Croll, D., Gladieux, P., Soanes, D. M., Persoons, A., Bhattacharjee, P., Hossain, Md. S., Gupta, D. R., Rahman, Md. M., Salam, M. U., Mahboob, M. G., Cook, N., Surovy, M. Z., Sancho, V. B., Maciel, J. L. N., NhaniJúnior, A., Castroagudín, V. L., Reges, J. T. de A., Ceresini, P. C., Ravel, S., Kellner, R., Fournier, E., Tharreau, D., Lebrun, M.-H., McDonald, B. A., Stitt, T., Swan, D., Talbot, N. J., Saunders, D. G. O., Win, J. and Kamoun, S. 2016. Emergence of wheat blast in Bangladesh was caused by a South American lineage of Magnaporthe oryzae. *BMC Biol.* 14, 84. https://doi.org/10.1186/s12915-016-0309-7.

Jank, L., Barrios, S. C., Valle, C. B. do, Simeão, R. M. and Alves, G. F. 2014. The value of improved pastures to Brazilian beef production. *Crop Pasture Sci.* 65(11), 1132-1137. https://doi.org/10/f6qh8w.

Juliana, P., He, X., Kabir, M. R., Roy, K. K., Anwar, M. B., Marza, F., Poland, J., Shrestha, S., Singh, R. P. and Singh, P. K. 2020. Genome-wide association mapping for wheat blast resistance in CIMMYT's international screening nurseries evaluated in Bolivia and Bangladesh. *Sci. Rep.* 10(1), 15972. https://doi.org/10/ghgh5n.

Kim, C. K. 1994. Blast management in high input, high potential temperature rice ecosystem. In: Zeigler, R. S., Leong, S. A. and Teng, P. S. (Eds), *Rice Blast Disease*. CAB International, Wallingford, pp. 451-464.

Kohli, M. M., Mehta, Y. R., Guzman, E., De Viedma, L. and Cubilla, L. E. 2011. Pyricularia blast - a threat to wheat cultivation. *Czech J. Genet. Plant Breed.* 47(Special Issue), S130–S134. https://doi.org/10.17221/3267-CJGPB.

Leach, C. M. 1980. Influence of humidity, red-infrared radiation, and vibration on spore discharge by *Pyricularia* oryzae. *Phytopathology* 70(3), 201-205. https://doi.org/10/fmpctb.

Leach, C. M. 1962. Sporulation of diverse species of fungi under near-ultraviolet radiation. *Can. J. Bot.* 40(1), 151-161. https://doi.org/10/cspkq9.

Lee, K., Singh, P., Chung, W. C., Ash, J., Kim, T. S., Hang, L. and Park, S. 2006. Light regulation of asexual development in the rice blast fungus, Magnaporthe oryzae. *Fungal Genet. Biol.* 43(10), 694-706. https://doi.org/10/bh2cvp.

Maciel, J. 2011. Magnaporthe oryzae, the blast pathogen: current status and options for its control. *CAB Rev.* 6(50). https://doi.org/10.1079/PAVSNNR20116050.

Madden, L. V., Hughes, G. and van den Bosch, F. 2007. *The Study of Plant Disease Epidemics, Epidemiology*. APS Press, St. Paul. https://doi.org/10.1094/9780890545058.

Magarey, R. D., Sutton, T. B. and Thayer, C. L. 2005. A simple generic infection model for foliar fungal plant pathogens. *Phytopathology* 95, 92-100. https://doi.org/10/cznzc9.

Malaker, P. K., Barma, N. C. D., Tiwari, T. P., Collis, W. J., Duveiller, E., Singh, P. K., Joshi, A. K., Singh, R. P., Braun, H.-J., Peterson, G. L., Pedley, K. F., Farman, M. L. and Valent, B. 2016. First report of wheat blast caused by Magnaporthe oryzae pathotype triticum in Bangladesh. *Plant Dis.* 100(11), 2330-2330. https://doi.org/10.1094/PDIS-05-16-0666-PDN.

Meredith, D. S. 1962. Spore dispersal in Pyricularia grisea (Cooke) Sacc. *Nature* 195(4836), 92-93. https://doi.org/10/b7s2d7.

Mesinger, F., Chou, S. C., Gomes, J. L., Jovic, D., Bastos, P., Bustamante, J. F., Lazic, L., Lyra, A. A., Morelli, S., Ristic, I. and Veljovic, K. 2012. An upgraded version of the Eta model. *Meteorol. Atmos. Phys.* 116(3-4), 63-79. https://doi.org/10/f3xrnp.

Mills, K. B., Madden, L. V. and Paul, P. A. 2020. Quantifying the effects of temperature and relative humidity on the development of wheat Blast Incited by the lolium pathotype of Magnaporthe oryzae. *Plant Dis.* 104(10), 2622-2633. https://doi.org/10.1094/PDIS-12-19-2709-RE.

Mousanejad, S., Alizadeh, A. and Safaie, N. 2009. Effect of weather factors on spore population dynamics of rice blast fungus in Guilan province. *J. Plant Prot. Res.* 49(3), 319-329. https://doi.org/10.2478/v10045-009-0051-z.

Oliveira, S. Cd, Castroagudín, V. L., Maciel, J. L. N., Pereira, D. AdS. and Ceresini, P. C. 2015. Resistência cruzada aos fungicidas IQo azoxistrobina e piraclostrobina no patógeno da brusone do trigo Pyricularia oryzae no Brasil. *Summa Phytopathol.* 41(4), 298-304. https://doi.org/10.1590/0100-5405/2072.

Pasanen, A. -L., Pasanen, P., Jantunen, M. J. and Kalliokoski, P. 1991. Significance of air humidity and air velocity for fungal spore release into the air. *Atmos. Environ. A. Gen. Top.* 25(2), 459-462. https://doi.org/10/dgb5kn.

Pasinato, A., Cunha, G. R. da, Fontana, D. C., Monteiro, J. E. B. de A., Nakai, A. M., Oliveira, A. F. de, Pasinato, A., Cunha, G. R. da, Fontana, D. C., Monteiro, J. E. B. de A., Nakai, A. M. and de Oliveira, A. F. 2018. Potential area and limitations for the expansion of rainfed wheat in the Cerrado biome of Central Brazil. *Pesqui. Agropec. Bras.* 53, 779-790. https://doi.org/10/gfhg3g.

Pavan, W., Fernandes, J. M. C., Hölbig, C. A. and Fraisse, C. W. 2009. Padrões de projeto aplicados a modelos de simulação do crescimento e desenvolvimento de culturas. *Rev. Bras. Comput. Apl.* 1(1), 15-29. https://doi.org/10.5335/rbca.2009.003.

Pavan, W., Fraisse, C. W. and Peres, N. A. 2011. Development of a web-based disease forecasting system for strawberries. *Comput. Electron. Agric.* 75(1), 169-175. https://doi.org/10/cjncjn.

Perelló, A., Martinez, I. and Molina, M. 2015. First report of virulence and effects of Magnaporthe oryzae isolates causing wheat blast in Argentina. *Plant Dis.* 99(8), 1177. https://doi.org/10/ghjmnp.

Pinnschmidt, H. O., Klein-Gebbinck, H. W., Bonman, J. M. and Kranz, J. 1993. Comparison of aerial concentration, deposition, and infectiousness of conidia of Pyricularia grisea by spore-sampling techniques. *Phytopathology* 83(11), 1182-1189. https://doi.org/10/cktzbz.

Pizolotto, C. A., Maciel, J. L. N., Fernandes, J. M. C. and Boller, W. 2019. Saprotrophic survival of Magnaporthe oryzae in infested wheat residues. *Eur. J. Plant Pathol.* 153(2), 327-339. https://doi.org/10.1007/s10658-018-1578-5.

Plantegenest, M., Le May, C. and Fabre, F. 2007. Landscape epidemiology of plant diseases. *J. R. Soc. Interface* 4(16), 963-972. https://doi.org/10.1098/rsif.2007.1114.

R Core Team 2020. *R: A Language and Environment for Statistical Computing*. R Foundation for Statistical Computing, Vienna, Austria.

Savary, S., Nelson, A., Willocquet, L., Pangga, I. and Aunario, J. 2012. Modeling and mapping potential epidemics of rice diseases globally. *Crop Prot.* 34, 6–17. https://doi.org/10.1016/j.cropro.2011.11.009.

Savary, S., Nelson, A. D., Djurle, A., Esker, P. D., Sparks, A., Amorim, L., Bergamin Filho, A., Caffi, T., Castilla, N., Garrett, K., McRoberts, N., Rossi, V., Yuen, J. and Willocquet, L. 2018. Concepts, approaches, and avenues for modelling crop health and crop losses. *Eur. J. Agron.* 100, 4–18. https://doi.org/10.1016/j.eja.2018.04.003.

Shtienberg, D. 2013. Will decision-support systems be widely used for the management of plant diseases? *Annu. Rev. Phytopathol.* 51, 1–16. https://doi.org/10/ghjp84.

Shtienberg, D. 2000. Modelling: the basis for rational disease management. Crop Prot., XIVth International Plant Protection Congress 19. pp. 747–752. https://doi.org/10/dnfxvq.

Silva, G. Bd da and Prabhu, A. S. 2005. Quantificação de conídios de Pyricularia grisea no plantio direto e convencional de arroz de terras altas. *Fitopatol. Bras.* 30(6), 569–573. https://doi.org/10/fb5zbv.

Tembo, B. 2019. A review of rain-fed wheat production constraints in Zambia. *J. Agric. Crops.* 59, 158–161. https://doi.org/10/ghjmnn.

Tembo, B., Mulenga, R. M., Sichilima, S., M'siska, K. K., Mwale, M., Chikoti, P. C., Singh, P. K., He, X., Pedley, K. F., Peterson, G. L., Singh, R. P. and Braun, H. J. 2020. Detection and characterization of fungus (Magnaporthe oryzae pathotype Triticum) causing wheat blast disease on rain-fed grown wheat (Triticum aestivum L.) in Zambia. *PLOS ONE* 15(9), e0238724. https://doi.org/10/ghgh48.

Uddin, W., Serlemitsos, K. and Viji, G. 2003. A temperature and leaf wetness duration-based model for prediction of gray leaf spot of perennial ryegrass turf. *Phytopathology* 93, 336–343. https://doi.org/10/cckbnr.

Urashima, A. S., Alves, A. F., Silva, F. N., Oliveira, D. and Gazaffi, R. 2017. Host range, mating type and population structure of Magnaporthe sp. of a single barley field in São Paulo state, Brazil. *J. Phytopathol.* 165(7–8), 414–424. https://doi.org/10.1111/jph.12575.

Urashima, A. S., Leite, S. F. and Galbieri, R. 2007. Eficiência da disseminação aérea em Pyricularia grisea. *Summa Phytopathol.* 33(3), 275–279. https://doi.org/10.1590/S0100-54052007000300011.

West, J. S. and Kimber, R. B. E. 2015. Innovations in air sampling to detect plant pathogens. *Ann. Appl. Biol.* 166(1), 4–17. https://doi.org/10/f6x854.

Willocquet, L. and Savary, S. 2004. An epidemiological simulation model with three scales of spatial hierarchy. *Phytopathology* 94, 883–891. https://doi.org/10/dcskqm.

Chapter 18

Investigating the biology of rice blast disease and prospects for durable resistance

Vincent M. Were and Nicholas J. Talbot, The Sainsbury Laboratory, University of East Anglia, UK

1 Introduction

Rice blast disease is caused by the filamentous ascomycete fungus *Magnaporthe oryzae* (synonym of *Pyricularia oyzae*) (Zhang et al. 2016). The blast fungus has the ability to cause disease in almost 50 grass species, including many of economic importance, such as finger millet (*Eleusine coracana*), barley (*Hordeum vulgare*) and wheat (*Triticum aestivum*), as well as rice (*Oryza sativa*) (Talbot 2003; Ou 1980; Cruz and Valent 2017; Inoue et al. 2017). Despite efforts to control losses caused by the disease, the blast fungus still destroys 10–35% of the global harvest of rice and wheat. In 2015-16, for example, wheat losses caused by blast disease in Bangladesh affected approximately 15 000 ha across eight districts. This is equivalent to 3.4% of the total area of cultivated wheat in Bangladesh in that season (Islam et al.

http://dx.doi.org/10.19103/AS.2021.0092.23

2016). Efforts to control rice blast disease, including the use of fungicides, avoiding the excessive use of nitrogen-based fertilisers, and planting resistant rice varieties, have not been completely successful (Skamnioti and Gurr 2009) and blast disease remains a significant problem wherever rice is grown.

In addition to its economic significance, the biological processes involved in *M. oryzae* host colonisation have been extensively studied (Wilson and Talbot 2009). Importantly, the complete genomes of both the rice blast fungus and most of its hosts including rice, wheat and barley are available and offer powerful genomic-based analyses to be carried out on the interaction. In addition, an extensive molecular toolkit for studying the interaction has been developed including targeted gene replacement, CRISPR-Cas9 genome editing, advanced live-cell imaging procedures, proteomic and metabolomic resources (Dean et al. 2005; International Rice Genome Sequencing 2005; Foster et al. 2018; Parker et al. 2009; Kankanala et al. 2007; Giraldo et al. 2013).

There are important biological processes involved in rice blast disease that are now well-studied during the early events in plant infection which include:

- The cell biology of appressorium formation;
- The biology of invasive growth and effector secretion;
- The two distinct mechanisms of effector secretion;
- The nature of the plant-pathogen interface;
- PAMP-triggered immunity modulation by secreted effectors; and
- Effector-triggered immunity and blast resistance.

The devastating losses caused by the blast fungus have been documented in most grasses but this chapter will discuss:

- The use of major resistance genes to rice blast; and
- Wheat blast disease as an emerging threat to global food security.

This chapter also highlights an emerging approach to breed for durable resistance to plant pathogens using gene editing technologies with an example:

- CRISPR-Cas9 mutagenesis of dominant S-genes for disease control.

2 Cell biology of appressorium formation

Rice blast infection commences when a three-celled conidium lands on the leaf surface via wind or splash dispersal (Wilson and Talbot 2009). To enable attachment and anchorage to the waxy hydrophobic surface of the leaf, the conidium secretes an adhesive, called spore tip mucilage, from an apical compartment at the tip of the three-celled conidium (Hamer et al. 1988). Under favourable humid conditions, the spore germinates and produces a single polarised germ tube. The germ tube flattens at its tip after around 4 h, a process called hooking, that allows the conidium to recognise the rice leaf surface, prior to swelling at its tip leading to the development of a specialised dome-shaped, pressurised cell called the appressorium, as shown in Fig. 1a (Wilson and Talbot 2009; Talbot 1995; Bourett and Howard 1990b).

The process of appressorium-mediated plant infection by *M. oryzae* has been extensively studied. Turgor generated in the appressorium is necessary to facilitate physical rupture of the rice leaf cuticle (Ryder et al. 2019). Appressorium differentiation requires a hard, hydrophobic surface, the perception of which requires a series of membrane-localised sensor proteins, including stretch-activated ion channel proteins and the G-protein coupled receptor, Pth11 (Wilson and Talbot 2009; Mitchell and Dean 1995) (DeZwaan et al. 1999). The Pmk1 mitogen-activated protein kinase (MAPK) is critical to appressorium formation and mutants lacking the kinase are unable to form an appressorium, even when exposed to a hard, hydrophobic surface, or the rice leaf surface (Xu and Hamer 1996). Components of the Pmk1 MAPK pathway have been identified and confirmed to operate in this pathway, which is initiated by activation of a heterotrimeric G-protein complex and is also linked to downstream signalling via the cAMP-dependent protein kinase, a pathway that leads to appressorium maturation, turgor generation and melanin deposition in the cell wall of the incipient appressorium (Liu et al. 2007; Choi and Dean 1997; Xu 2000).

Development of the appressorium also requires two independent S-phase checkpoints of the cell cycle (Oses-Ruiz et al. 2017). Initial development of the appressorium formation requires an S-phase checkpoint acting through the DNA damage response (DDR) pathway, via the Cds1 kinase (Oses-Ruiz et al. 2017). This ensures that a single round of mitosis occurs in the three-celled conidium (Veneault-Fourrey et al. 2006). Incipient appressorium formation requires one of the daughter nuclei to move into the germ tube tip, while the remaining three nuclei are degraded by autophagy, resulting in programmed cell death of the conidium and transfer of all of its contents to the appressorium (Veneault-Fourrey et al. 2006) (Saunders et al. 2010a,b). Appressorium re-polarisation involves a DDR-independent S-phase checkpoint triggered by melanisation and turgor pressure generation as the infection structure matures (Oses-Ruiz et al. 2017).

The *M. oryzae* appressorium forms a highly differentiated cell wall rich in chitin and contains a thick layer of melanin on its inner surface (Bourett and Howard 1990a). The thick melanin layer provides an impermeable barrier to intracellular glycerol that accumulates inside the cell to generate high internal turgor pressure (Howard et al. 1991; Ryder and Talbot 2015). This generates

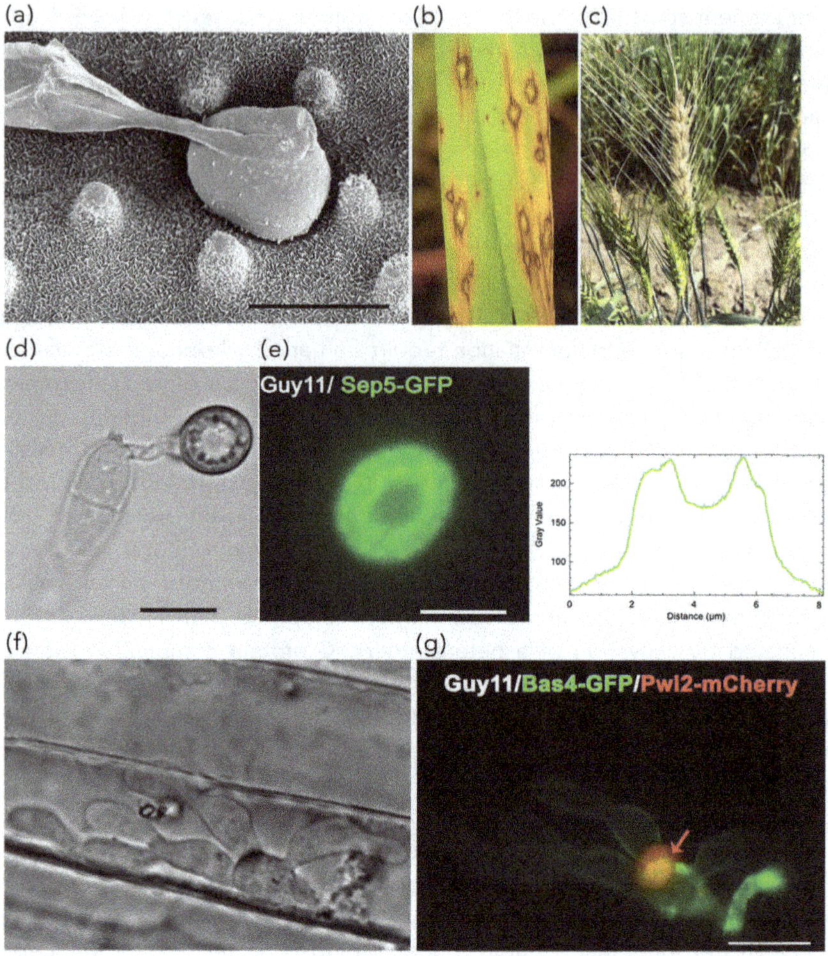

Figure 1 The biology of rice blast infection. (a) Scanning electron micrograph of a pressurised cell, appressorium on a rice leaf surface. Photograph, Kim Findlay, John Innes Centre and Lauren Ryder, The Sainsbury Laboratory. (b and c) Typical blast disease symptoms on rice and wheat caused by *M. oryzae* (b) a rice leaf characterised by a spreading brown lesion with a necrotic grey centre (c) wheat head showing blast symptoms. (d) Appressorium germination on hydrophobic glass coverslips. Bar = 10 μm. (e) Cellular localisation of septin ring protein, Sep5-GFP, during appressorium maturation. Bar = 5.0 μm (f) Live-cell imaging of a rice epidermal cell with a growing invasive hyphae. (g) Rice blast effectors present in two distinct localisation patterns. Pwl2:mRFP is visualised in the BIC (red), while Bas4:GFP is seen in the apoplast (green), the arrow shows position of the BIC. Bar = 10 μm.

hydrostatic turgor, which accumulates to a threshold necessary for leaf cuticle rupture to occur. Turgor is monitored by the Sln1 histidine-aspartate kinase, which acts to modulate both solute accumulation and melanin biosynthesis

(Ryder et al. 2019). Sln1 works in association with the protein kinase C-dependent cell integrity pathway to control turgor, and also to regulate appressorium re-polarisation (Ryder et al. 2019; Penn et al. 2015).

Re-polarisation of the appressorium requires guanosine nucleotide-binding proteins called septins to be recruited to the appressorium pore, where they form a large hetero-oligomeric complex as shown in Fig. 1e (Dagdas et al. 2012). Septin aggregation is regulated by the regulated synthesis of reactive oxygen species, via the Nox2-NoxR NADPH oxidase complex, which is necessary for appressorium function (Ryder et al. 2013; Egan et al. 2007). The septin ring structure bounds the appressorium pore allowing cellular asymmetry to be established at the base of the infection cell, re-modelling the F-actin cytoskeleton to the point of plant infection. The septin ring acts to rigidify the cortex of the appressorium, scaffold actin and form a lateral diffusion barrier that holds polarity determinants, such as Cdc42, Tea1 and Las17, in place at the point of penetration peg emergence (Dagdas et al. 2012; Gupta et al. 2015). The rigid penetration hypha then ruptures the rice cuticle, facilitating entry of the fungus to underlying leaf tissue.

3 The biology of invasive growth and effector secretion

Once inside the rice epidermis, the rice blast fungus is able to differentiate into bulbous, branched invasive hyphae that rapidly fill the initial cell colonised, as shown in Fig. 1f (Kankanala et al. 2007). The entry into rice cells coincides with production of a large suite of fungal effectors used by the invading fungus to suppress plant immunity responses and modulate the host physiology for its advantage (de Jonge et al. 2011). This enables the fungus to colonise the first invaded cell and move into adjacent rice cells, using pit field sites, where plasmodesmata accumulate. Plasmodesmata allow cytoplasmic continuity between plant cells and act as a conduit for transfer of proteins, metabolites and other signalling molecules. It is clear that plasmodesmatal conductance is manipulated by *M. oryzae* because effector proteins can freely move between plant cells during the early stages of infection (Mosquera et al. 2009; Khang et al. 2010) Furthermore, plasmodesmata are sites where a specific plant immunity response occurs (Faulkner et al. 2013).

The initial biotrophic phase of growth during *M. oryzae* invasion of the rice tissue is also characterised by the presence of a membrane-rich structure called the biotrophic interfacial complex (BIC), which is formed when the fungus is within the first epidermal cell. The BIC is thought to be the point from which a subset of effectors is secreted for delivery into host cells (de Jonge et al. 2011; Giraldo and Valent 2013; Khang et al. 2010). A large set of effector proteins are secreted directly into host cells, which is likely to involve the BIC and are called cytoplasmic, or symplastic effectors. Cytoplasmic effectors are involved in suppression of PAMP-triggered immunity, but can also be recognised by immune receptors in effector-triggered immunity (ETI), when they are termed avirulence proteins.

Apart from Ace1, an enzyme associated with synthesis of a secondary metabolite, all other known avirulence proteins, such as Pwl2, AVR-Piz-t, AVR-Pia, AVR-Pii, AVR-Pi9 AVR-Pi54 and AVR-Pita have been shown to localise to the BIC before translocating into the host cytoplasm as shown in Fig. 1g (Giraldo et al. 2013; Khang et al. 2010; Wu et al. 2015b; Giraldo and Valent 2013; Saitoh et al. 2012; Bohnert et al. 2004) where they are recognised by cognate immune receptors encoded by resistance (R) genes. Consequently, recognition of avirulence proteins can trigger a hypersensitive response (HR), stop proliferation of invasive hyphae, and lead to immunity. The only exception to this rule to date is Avr-CO39, which has been shown to localise in the apoplast. However, this observation needs more investigation because a constitutive ribosomal protein 27 (RP27) promoter was used to drive *AVR-CO39* instead of its native promoter. Other effectors that have not yet been shown to have avirulence functions include Bas1, Bas2 and Bas3, which also localise to the BIC and may play roles in suppression of PTI (Mosquera et al. 2009).

4 Two distinct mechanisms of effector secretion

To study how the rice blast fungus colonises rice tissue, strains with fluorescently labelled effector proteins can be observed after inoculation of translucent rice leaf sheath preparation in live-cell imaging experiments (Kankanala et al. 2007). Using fluorescently labelled BIC-localised effector proteins as markers, it has been shown that effector proteins accumulate preferentially in BICs before entry into rice cells (Giraldo et al. 2013; Mosquera et al. 2009). Two different approaches have been used to investigate whether the BIC is a portal for effector delivery in *M. oryzae*. First, a *M. oryzae* strain expressing a fluorescently labelled BIC-localised effector, Pwl2-RFP, was used to infect living rice epidermal strips, and invasive hyphae allowed to differentiate, before carrying out rice tissue plasmolysis. From this experiment, effector fluorescence was visualised in the host cell cytoplasm, suggesting their movement from invasive hyphae into the host cytoplasm (Khang et al. 2010). Furthermore, when a nuclear localisation sequence (NLS) was attached to fluorescently labelled Pwl2, the fluorescence signal concentrated in the nuclei of invaded rice cells, including those not yet occupied by the fungus (Giraldo et al. 2013; Khang et al. 2010). The best evidence, however, comes from a photobleaching experiment which provided evidence that Pwl2-RFP can move from the BIC-associated cell of the fungus into the BIC and then into plant tissue (Giraldo et al. 2013).

By use of live cell imaging, the BIC has been demonstrated to be a host plant-derived structure, located at an interface outside the fungal cell wall but bound to the invaginated plant plasma membrane (Giraldo et al. 2013). When, for example, a transgenic rice line expressing a fluorescently labelled plasma membrane marker, LTi6B, was infected using a *M. oryzae* strain transformed to express a BIC-localising effector, Pwl2-RFP, the LTi6B plasma membrane fluorescence outlined the BIC (Giraldo et al. 2013). Once an invasive hypha penetrates a new neighbouring living cell, the process of BIC formation and cytoplasmic effectors secretion is repeated, and this occurs at every new cell entered by the fungus. This is characterised by rupture of the plant plasma membrane that encloses the invasive hyphae (Jones et al. 2016).

A study by Giraldo et al. (2013) provided evidence that two distinct secretory pathways are necessary to secrete effector proteins by *M. oryzae*. Apoplastic effectors are delivered into the gap between the fungal cell wall and plant plasma membrane, or extra-invasive hyphal membrane (EIHM), while cytoplasmic effectors are delivered into rice cells through the BIC (Giraldo et al. 2013). Apoplastic effectors include a secreted LysM protein 1 called Slp1, the biotrophy-associated secreted proteins 4 (Bas4) and 113 (Bas113) and currently non-characterised effectors, such as MC69 and MoAa9 (Mentlak et al. 2012; Mosquera et al. 2009; Li et al. 2020). Giraldo and his co-workers (2013) showed that secretion of the apoplastic effector Bas-4 was inhibited by treatment with Brefeldin A (BFA), a chemical that inhibits secretion through the Golgi apparatus (Liu et al. 2013). However, treatment with BFA does not affect secretion of effector proteins that are delivered into the BIC, such as Bas1 or Pwl2 (Giraldo et al. 2013).

Cytoplasmic effectors instead require an exocyst-dependent, but Golgi-independent, unconventional secretion mechanism. In *M. oryzae, Δsec5* and *Δexo70* mutants expressing fluorescently labelled BIC-localising effectors, for example, showed partial retention of the effector proteins in BIC-associated hyphae, whereas these mutants are still able to secrete apoplastic effectors normally (Giraldo et al. 2013). A similar mechanism occurs in the un-related, oomycete, potato late blight pathogen, *Phytophthora infestans,* where an apoplastic effector EPIC1 has been observed to be conventionally secreted in a BFA-sensitive way compared to a cytoplasmic effector, Pi04314, which is BFA-insensitive and accumulates in the host nucleus where it suppresses immunity (Welsh et al. 2018; Wang et al. 2017c). The mechanism by which rice blast fungus effectors modulate plant immunity is discussed later in this chapter.

5 Understanding the plant-pathogen interface

Recent studies regarding the molecular dialogue between plants and pathogens during pathogen infection suggest that extracellular vesicles may be involved in the transport of small interfering RNA molecules from host plants into fungal cells, enabling suppression of the expression of fungal virulence-related genes (Rutter and Innes 2017; Baldrich et al. 2019). Conversely, fungal small RNAs from *Botrytis cinerea* that target plant genes have been isolated in *Arabidopsis thaliana* cells suggesting that cross-kingdom extra-vesicular (EV) transfer between a host and corresponding pathogen may occur in some instances (Cai et al. 2018). The study further suggests that *Botrytis cinerea* is able to deliver small RNAs into *A. thaliana* cells in order to perturb the host RNA interference defense response (Wang et al. 2017b). This phenomenon has not, however, been reported in *M. oryzae*, but extracellular vesicles may be present in BICs where they may be involved in effector delivery. Further experiments are necessary to investigate cargo associated with such vesicles and whether small RNA species are involved in this dialogue, as in other systems. In a recent review, Bozkurt and Kamoun compared the mechanisms of effector secretion in different pathogens and suggested that processes such as endocytosis may be involved in effector uptake (Bozkurt and Kamoun 2020). Effectors might, for example, be actively transported by translocons like those reported in human diseases, such as the malaria pathogen *Plasmodium falciparum* (Matthews et al. 2019), or by receptor-mediated endocytosis (Bozkurt and Kamoun 2020).

Plants have evolved two major lines of defence against pathogens attack (Jones and Dangl 2006; Eitas and Dangl 2010). First, plants possess plasma membrane-localised receptor proteins that can recognise conserved molecules from invading pathogens (Jones and Dangl 2006). These receptors have the ability to recognise conserved protein motifs called pathogen-associated molecular patterns (PAMPs). PAMPs might be shared in different plant pathogens and recognition by plant receptor kinases triggers the first line of plant immunity called PAMP-triggered immunity (PTI) (Jones and Dangl 2006; Sanchez-Vallet et al. 2015). The recognition of PAMPs by surface recognition receptors results in immune responses, for example, the production of reactive oxygen species, activation of both ion-channels and mitogen-activated protein (MAP) kinase signalling involved in plant immunity and changes in gene expression (Dodds and Rathjen 2010; Ngou et al. 2020). In rice, for example, a plasma membrane receptor called the chitin elicitor binding protein (CEBiP) that possesses a LysM domain is required to bind to and perceive chitin oligomers, and then initiate signalling (Liu et al. 2013; Dodds and Rathjen 2010; Jones and Dangl 2006). However, most pathogens have developed multiple mechanisms to circumnavigate this layer of host immunity and still cause disease.

To counter PTI suppression, plants have developed a second layer of defence known as effector-triggered immunity (ETI) that is directed towards strains carrying effectors that can be recognised by cognate immune receptors (Jones and Dangl 2006; Dodds and Rathjen 2010; Liu et al. 2013). Specific pathogenic isolates can be classified into physiological races, or pathotypes, defined by their repertoire of specific effectors (known as avirulence proteins) that they secrete, through direct or indirect recognition by plant resistance proteins (R). In a recent study using *Pseudomonas syringae* and the model plant *Arabidopsis thaliana,* it has been shown that ETI can, in some instances, work by restoring and magnifying PTI signalling (Ngou et al. 2020). This means that without the presence of PTI, ETI would be very rarely activated during natural infection (Ngou et al. 2020).

6 PAMP-triggered immunity modulation by secreted effectors

M. oryzae apoplastic effectors interfere with the accessibility of host receptors, or outcompete for their ability to bind PAMPs. For example, the LysM effector protein Slp1 acts by preventing the chitin elicitor binding protein (CEBiP) from perceiving and binding to chitin oligomers produced by the rice blast fungus during infection. The Slp1 possesses a higher chitin-binding affinity compared to the CEBiP host receptor, thereby suppressing the recognition of chitin oligomers by CEBiP and suppressing immunity (Mentlak et al. 2012). Similarly, another apoplastic secreted protein, MoAa91, appears to possess the ability to bind chitin and suppress PTI, although the protein has further, as yet uncharacterised enzymatic activities (Li et al. 2020). The tomato leaf mould fungus *Cladosporium fulvum* secretes an effector called Ecp6 that possesses a LysM domain protein like Slp1 and similarly competes for chitin oligomers resulting in supressed chitin-triggered immunity (Shimizu et al. 2010). In the corn smut fungus *Ustilago maydis*, an apoplastic effector, Pep1, has been shown to protect invasive hyphae from reactive oxygen species (ROS) by acting as a peroxidase inhibitor (Doehlemann et al. 2009). Another effector from the same pathogen, Cmu1, possesses a chorismite mutase activity and therein catalyses the conversion of chorismite to prephenate. By doing this, the effector is able to shift the shikimate pathway towards production of aromatic amino acids instead of the defense signalling hormone, salicylic acid (Djamei et al. 2011).

Effectors that translocate into different compartments of the host cell can also contribute to PTI suppression. AVR-Pizt, for example, can suppress chitin-induced PAMP-triggered immunity through binding and destabilising two rice RING E3 ubiquitin ligase proteins APIP6 and APIP10 (Park et al. 2012; Park et al. 2016). More studies have revealed that AVR-Pizt can also interact with a rice host cell death suppressor, APIP5, in rice cells lacking Piz-t. APIP5 is a bZIP-type transcription factor and inhibiting its transcriptional activity leads to effector-triggered necrosis, required at the necrotrophic stage of blast infection. Piz-t binds to and stabilises APIP5 inhibiting effector-triggered necrosis, while APIP5 is also needed to stabilise Piz-t (Wang et al. 2016; Park et al. 2012; Park et al. 2016). By use of genetic and phenotype analysis in a different study, AVR-Piz-t has also been characterised as having the ability to bind a K^+ channel protein,

OSAKT1, located at the rice cell plasma membrane leading to reduction of K^+ and an attenuated immune response (Shi et al. 2018).

The mechanism by which AVR-Piz-t is able to target multiple, structurally different and functionally distinct host proteins to suppress immunity is not understood and it is difficult to envisage this occurrence from an evolutionary perspective. At a structural level, AVR-Piz-t belongs to the MAX-fold type of effectors, a feature observed in many characterised *M. oryzae* avirulence proteins (Zhang et al. 2013). Despite lacking similarities at the nucleotide and amino acid level, MAX effectors contain six β-sheets in their secondary protein structure (de Guillen et al. 2015). However, unlike other MAX-fold type avirulence proteins, for example, AVR-Pia, AVR-CO39 and AVR-Pik alleles, which bind heavy metal-associated domain containing proteins, there remains no evidence for this activity in AVR-Piz-t. It is possible that AVR-Piz-t has an unusual ability to interact with a very wide range of proteins, through multiple, distinct interaction points, which may have been independently selected for, but this requires significant structural function studies to be carried out to verify such a characteristic.

Another avirulence protein, AVR-Pii, has also been shown to interact with different host proteins. First of all, AVR-Pii interacts with the host protein OsExo70-F3 during Pii-dependent resistance (Fujisaki et al. 2015). Secondly, AVR-Pii interacts with a rice NADP-malic enzyme (Os-NADP-ME2) in order to suppress ROS production to the benefit of the invading pathogen. Deletion of *Os-NADP-ME2* in rice leads to reduced chitin-induced ROS-burst compared to the wild type (Singh et al. 2016). Interaction with this alternative target, however, also needs to be investigated thoroughly in order to understand how such different specificities could have evolved in the same protein.

7 Effector-triggered immunity and blast resistance genes

As already described in this chapter the ETI response is directed towards pathogenic strains with cognate effectors (Jones and Dangl 2006; Dodds and Rathjen 2010; Liu et al. 2013). These effectors are therefore recognised when directly or indirectly bound by plant resistance (R) gene-encoded proteins. Most R genes encode proteins containing a centrally situated nucleotide-binding site (NB-ARC) domain and leucine-rich repeats (LRRs) in the C-terminal domain. These proteins are referred to as nucleotide-binding oligomerisation domain-like receptors, or in short, NOD-like receptors (NLRs). Many NLRs carry an additional coiled-coil (CC) or TOL1/interleukin receptor (TIR) domains in the N-terminus. These domains have been shown to be involved in the formation of an NLR homo complex, crucial for downstream signalling. Only effector proteins that translocate into the host cytoplasm can be recognised by the products of these R genes (Takken and Goverse 2012; Bernoux et al. 2011; Qi and Innes 2013; Maekawa et al. 2011).

Initially, it was thought that recognition of these effectors was only mediated by single NLRs, but there is increasing evidence of cases where more than one NLR can co-operate to recognise a pathogen and thereby initiate an immune response (Maqbool et al. 2015; Eitas and Dangl 2010; Wu et al. 2015a) (Bialas et al. 2017). Some R genes, such as the Cf-resistance proteins, in resistant tomato lines confer resistance to leaf mould disease caused by *C. fulvum*, and belong to a distinct group of R proteins referred to as receptor-like proteins (RLPs) (Stergiopoulos et al. 2010; Dixon et al. 1996). This type of R gene encodes for proteins that are membrane-spanning receptors with an extracellular domain designed to perceive apoplastic effectors, linked to a cytoplasmic domain that primarily contains LRRs (Stergiopoulos et al. 2010; Dixon et al. 1996).

8 Use of major resistance genes against rice blast

In plant genomes, some genes encoding NLRs can occur in paired, inverted tandem arrangements at the same locus. In a few reported cases, two NLRs may act to form a hetero-complex, where one NLR directly recognises the effector while the other plays a crucial role in downstream signalling (Eitas and Dangl 2010; Wu et al. 2015a; Bialas et al. 2017). Recently, several experiments have demonstrated that NLRs can also interact indirectly with effectors (Bialas et al. 2017; Park et al. 2016). In this case, they can either recognise a modified protein targeted by the effector, referred to as a 'guardee', or possess a modification that mimics the effector-targeted protein termed an 'integrated decoy' in one member of an NLR pair (Eitas and Dangl 2010; Cesari et al. 2014a; Williams et al. 2014).

More than 100 resistance genes against rice blast have been identified to date in plant breeding programmes, of which 36 have been molecularly cloned. Of these resistance genes, 34 encode NLRs, as shown in Table 1 (Dai et al. 2010; Fukuoka et al. 2014; Wang et al. 2017a; Zhao et al. 2018). Exceptions include the recessive resistance allele, *pi21*, which encodes for a proline-rich protein, required for susceptibility to rice blast (Fukuoka et al. 2009), and *Pi-d2,* which encodes a receptor-like kinase protein, that has an extracellular B lectin-like domain and a serine-threonine kinase domain (Chen et al. 2006). To date, approximately 24 avirulence genes have been identified in *M. oryzae* of which 12 have been cloned (Wang et al. 2017a), as shown in Table 2. With the exception of AVR-Pi9 and AVR-Pib, the interaction of the other 7 AVRs and their cognate R gene-encoded proteins has now been reported (Wang et al. 2017a). Interactions have either been reported to be direct, as in the case of AVR-Pita/Pita or indirect, as, for example, in the case of AVR-Piz-t that requires the presence of a 'guardee' host protein (Park et al. 2016).

Table 1 Cloned resistance genes against rice blast disease

R gene	Protein type	Donor line	Reference
Pita	NLR	Yashiro-mochi	(Bryan et al. 2000)
Piz-t	NLR	Zenith	(Zhou et al. 2006)
Pia	NLR	Sasanishiki	(Okuyama et al. 2011)
CO39	NLR	CO39	(Cesari et al. 2013)
Pii	NLR	Hitomebore	(Takagi et al. 2013)
Pik	NLR	Kusabue	(Zhai et al. 2014)
Pikm	NLR	Tsuyuake	(Ashikawa et al. 2008)
Pik-p	NLR	K60	(Zhai et al. 2014)
Pik-h	NLR	K3	(Zhai et al. 2014)
Pi9	NLR	75-1-127	(Qu et al. 2006)
Pib	NLR	Tohoku	(Wang et al. 1999)
Pi54	NLR	Tetep	(Sharma et al. 2005, 2010)
Pi37	NLR	St. No. 1	(Lin et al. 2007)
Pit	NLR	K59	(Hayashi and Yoshida 2009)
Pi35	NLR	Hokkai 188	(Fukuoka et al. 2014)
Pi64	NLR	Yangmaogu	(Ma et al. 2015)
Pish	NLR	Nipponbare	(Takahashi et al. 2010)
Pi21	Proline-rich metal binding protein	Owarihatamochi	(Fukuoka et al. 2009)
Pi63/ Pikahei-1(t)	NLR	Kahei	(Xu et al. 2014)
Pi-d2	B lectin receptor-like kinase	Digu	(Chen et al. 2006)
Pi-d3	NLR	Digu	(Shang et al. 2009)
Pi25	NLR	Gumei2	(Chen et al. 2011)
Pid3-A4	NLR	A4 (*Orya rufipogon*)	(Lv et al. 2013)
Pi2	NLR	Jefferson	(Zhou et al. 2006)
Pi50	NLR	NLR Er-Ba-zhan (EBZ)	(Zhu et al. 2012; Su et al. 2015)
Pi36	NLR	Kasalath	(Liu et al. 2007)
Pi5	NLR	RIL260	(Lee et al. 2009)
Pigm	NLR	Gumei4	(Deng et al. 2017)
Pi56	NLR	Sanhuangzhan	(Liu et al. 2013)
Pb1	NLR	Modan	(Hayashi et al. 2010)
Pi54rh	NLR	*Oryza rhizomatis* (nrcpb 002)	(Das et al. 2012)
Pike	NLR	Xiangzao143	(Chen et al. 2015)
Piks	NLR	Unknown	(Kanzaki et al. 2012)
Ptr	NLR	Katy	(Zhao et al. 2018)

8.1 Direct interaction between AVR proteins and unpaired NLRs

The interaction between AVR-Pita and Pita was the first direct avirulence and resistance protein interaction to be characterised in rice blast disease (Jia et al. 2000). *AVR-Pita* belongs to a gene family that includes *AVR-Pita-1*, *AVR-Pita-2* and *AVR-Pita-3*, of which *AVR-Pita-1* and *AVR-Pita-2* confer avirulence functions while *AVR-Pita-3* does not. *AVR-Pita* encodes a secreted neutral zinc metalloprotease protein, while *Pita* is a constitutively expressed resistance gene that encodes a classical cytoplasmic NLR (Bryan et al. 2000). The interaction between these proteins occurs at the leucine-rich domain (LRD) region and a mutation in the LRD region (LRD^{A918S}) leads to loss of recognition. Amino acid substitutions in the AVR-Pita-encoded protein ($avr\text{-}pita_{176}^{E177D}$ and $avr\text{-}pita_{176}^{M178W}$) that affect the protease activity of the protein also abolish recognition by Pita (Jia et al. 2000).

AVR-Pi54, like AVR-Pita, has a direct interaction with its cognate R protein Pi54. AVR-Pi54 belongs to a family of AVR genes called MAX effectors (de Guillen et al. 2015; Ray et al. 2016). After being identified, *Pi54* was initially named *Pi-k^h^* and encodes for an NLR with an additional zinc-finger domain in its LRD domain. Unlike *Pi-ta*, the expression of *Pi54* is induced upon pathogen infection. The interaction between AVR-Pi54 and Pi54 has been demonstrated in a yeast 2-hybrid interaction (Sharma et al. 2005, 2010; Devanna et al. 2014; Ray et al. 2016), but further studies are required to elucidate the biological function of AVR-Pi54.

8.2 Direct interaction between AVR proteins and paired NLRs

In paired NLRs, one of the proteins, known as a 'sensor NLR', acts by binding to a secreted effector from the pathogen, while the second protein, the 'helper NLR', activates defence signalling upon binding to its NLR pair protein (Wu et al. 2017). Well-studied and characterised resistance proteins in rice occur as paired NLRs and have been shown to cluster in a tandem organisation in the genome, as described in the Pii-1/Pii-2(Pi5-1/Pi5-2), Pik-1/Pik-2 and *Pi-CO39/Pia* (Rga4/Rga5) gene pairs (Cesari et al. 2014b). Recognition of AVR-CO39 requires a combination of two NLR proteins, Rga4 and Rga5, which work together. The two corresponding genes are situated adjacent to each other in the genome oriented in opposite directions. Rga5 possesses an additional integrated heavy metal-associated domain, involved in binding and perceiving the AVR-CO39 protein product. Rga4 acts by constitutively inducing disease resistance and cell death, while Rga5 acts as a repressor of this activity in the absence of the pathogen (Cesari et al. 2014b). The direct binding of Rga5 to AVR-CO39, or AVR-Pia, causes de-repression of Rga4, which results in the activation of immune response signalling and hypersensitive cell death (Cesari

Table 2 Characterised *M. oryzae* avirulence genes

Effector	ID number/MGG number	Function and location	Cognate R-gene
AVR-Pita	MGG_15370	Zinc metalloprotease	Pita
AVR-Piz-t	MGG_18041	Target rice U3 ubiquitin ligase	Piz-t
AVR-CO39	AF463528	BIC/Cytoplasmic	Pia/CO39
AVR-Pia	AB498873	BIC/Cytoplasmic	Pia/CO39
AVR-Pii	AB498874	Interacts with Exo70 or NADP-malic enzyme	Pii
AVR-Pik/km/kp	MGG_15972	BIC/Cytoplasmic	Pik/km/kp
AVR-Pi9	MGG_12655	BIC/Cytoplasmic	Pi9
AVR-Pib	KM887844	BIC/Cytoplasmic	Pib
PWL2	MGG_04301/MGG_13683	BIC/Cytoplasmic	-
ACE1	MMG_12447	Polyketide synthase	Pi33
AVR-Pi54	MGG_01947	BIC/Cytoplasmic	Pi54
AVR-Rmg8	LC223814	secreted protein	Rmg7/Rmg8
PWT3	MGG_14708	secreted protein	Rmg6 (Rtw3
PWT4	LC202655	secreted protein	Rmg1 (syn. Rwt4)

et al. 2014b). Recognition of AVR-Pia is similar to AVR-CO39 and mediated through the interaction of Rga4 and Rga5. Experiments carried out both in vitro and in planta suggest that Rga4 and Rga5 binding occurs at the CC domains of both proteins by forming either a hetero- or homo-oligomeric complex with the integrated HMA domain binding the cognate avirulence protein (Cesari et al. 2014b; Cesari et al. 2013).

AVR-Pik encodes a BIC-localised secreted protein. The corresponding R-gene, *Pik*, occurs on chromosome 11 at the same locus, as other alleles such as *Pi1*, *Pik*, *Pikm*, *Pikp*, *Piks Pikh* and *Pike* possess varied recognition to corresponding AVR-Pik alleles (Kanzaki et al. 2012). AVR-Pik/Pik interaction occurs when two NLR pairs Pik-1 and Pik-2, bind and identify AVR-Pik. Pikm-1, Pikp-1, Piks-1 and Pikh-1 are involved in recognising the different forms of AVR-Pik, while Pikm-2, Pikp-2, Piks-2 and Pikh-2, respectively, act as signal transducers that trigger resistance gene-specific immune responses. The rice NLR, Pik-1, has also been shown to contain an integrated HMA domain and to act together with the helper NLR, Pik-2, in the recognition of AVR-Pik as shown in Fig. 2 (Kanzaki et al. 2012). More cases of polymorphism have been reported in the coiled-coil region of *Pik-1*, in comparison to the low levels of polymorphism observed in the *Pik-2* region. This is explained by co-evolution of both the pathogen and host that has resulted in different alleles of *AVR-Pik* and *Pik* in natural populations, leading to different recognition specificities.

For example, rice plants with *Pikp* are resistant to isolates that carry *AVR-Pik-D*, but susceptible to those that harbour *AVR-Pik-E*, *A* and *C* (Kanzaki et al. 2012). Rice varieties possessing *Pikm* are resistant to strains that possess *AVR-Pik-D*, *E* and A, but produce a susceptible, compatible reaction when invaded by those that possess *AVR-Pik-C* (Kanzaki et al. 2012). Varieties with *Pik-k* are resistant to isolates that possess *AVR-Pik-D* and *AVR-Pik-E,* but susceptible to those that harbour *AVR-Pik-A* and *AVR-Pik-C* (Kanzaki et al. 2012). Resistance recognition specificities of rice varieties carrying *Piks* are identical to those with *Pikp*, while *Pikh* recognition is similar to *Pikm* (Kanzaki et al. 2012). Isolates that carry *AVR-Pik-D* possess less specificity and can produce incompatible reactions from almost all the *Pik* alleles including *Pikp*, *Pikm*, *Piks* and *Pikh* (Kanzaki et al. 2012). *AVR-Pik-E* isolates are recognised by *Pikm* and *Pikh* rice cultivars (Kanzaki et al. 2012). The crystal structure of AVR-PikD/Pikp-HMA interaction revealed that a mutation causing amino acid substitution in AVR-PikD (AVR-PikD^{His46Glu}, AVR-PikD^{Arg64Ala}, AVR-PikD^{Asp66Arg}) leads to loss of resistance due to lack of recognition by Pikp (Ashikawa et al. 2008; Bialas et al. 2017; Kanzaki et al. 2012; Maqbool et al. 2015).

The interaction between AVR-Pik and Pik has been demonstrated in the model plant *Nicotiana benthamiana* by testing for cell death responses following successful binding and recognition interactions. This assay can be

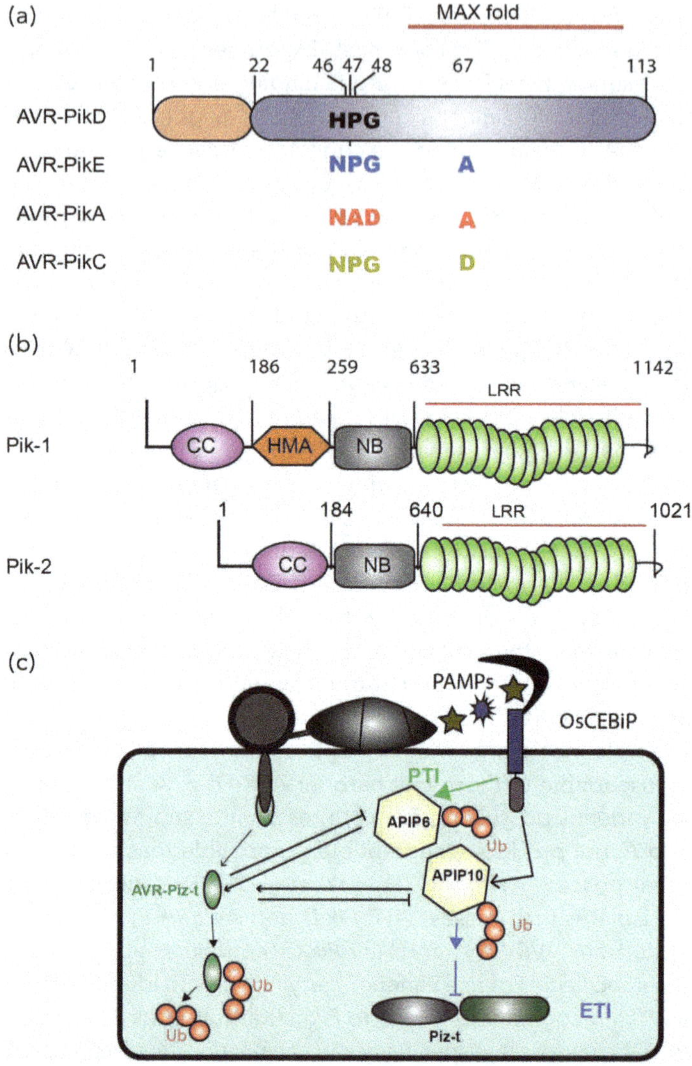

Figure 2 Direct and indirect interaction between AVR proteins and NLRs. (a) Schematic representation of the four distinct alleles of AVR-Pik, showing positions of polymorphic residues indicated as 46, 47, 48 and 67. (b) Rice Pik R protein pair. The 'sensor NLR' Pik-1 contains an integrated HMA domain situated between the coiled domain and nucleotide-binding site (NB-ARC). The 'helper NLR', Pik-2, activates defence signalling upon recognition of *M. oryzae* strains expressing AVR-Pik. (c) Recognition of AVR-Piz-t requires a 'guardee' APIP10, an E3 ligase protein (adapted from Kanzaki et al. 2012; Bialas et al. 2017; Park et al. 2016). Degradation of APIP10 by AVR-Piz-t suppresses immune response. APIP10 negatively regulates Piz-t, and thus the ubiquitination of APIP10 by AVR-Piz-t triggers ETI, due to accumulated Piz-t.

used to quickly study the effector-immune receptor interaction in numerous combinations (De la Concepcion et al. 2019). Such an approach has demonstrated that the Pikm recognition of AVR-PikE and AVR-PikA can be replicated in Pikp by altering two residues, Asn261Lys and Lys262Glu, both in vitro and in vivo. This strategy may be applied in future to genetically engineer broad spectrum R genes through increasing their binding ability to distinct avirulence protein (De la Concepcion et al. 2019).

8.3 Indirect interaction between AVR proteins and paired NLRs

Like most cloned AVR proteins, AVR-Pii has also been shown to be expressed and localise to the BIC during the biotrophic phase of infection (Singh et al. 2016). Unlike AVR-Pik and AVR-Pia/AVR-CO39, the interaction between AVR-Pii and Pii occurs indirectly and is dependent on a rice host protein. As previously discussed, AVR-Pii is able to interact with two rice proteins, Os-NADP-ME2 and OsExo70, to modulate host immunity in the absence of resistance protein, Pii. However, in the presence of the resistance protein, Pii, AVR-Pii binding to two exocytosis proteins, OsExo70-F2 and OsExo70-F3, results in a resistance response. OsExo70-F3, rather than OsExo70-F2, has been shown to play an important role in this interaction. This suggests that the rice Exo70 plays the role of 'helper' or 'decoy' during this interaction. (Fujisaki et al. 2015; Singh et al. 2016). Interestingly, a fusion of an NLR and Exo-70 domains has been identified in some wheat and barley cultivars, but not yet been reported in rice (Sarris et al. 2016). This suggests that the integration might be a relatively recent event in these lineages. A recent study has shown that resistance mediated by another NLR *Pi5* is similar to *Pii* and swapping *Pii-1* for *Pi5-1* and *Pii-2* for *Pi5-2* does not alter recognition of *M. oryzae* isolates carrying *AVR-Pii* (Vo et al. 2019). However, the recognition of AVR-Pii by Pi5 is independent of *Exo70F3* and knockout mutants of *OsExo70F3* in *Pi5* transgenic rice did not show effects in *Pi5*-mediated resistance. A different observation had been earlier made in *Exo70F3* RNAi loss of function mutants in *Pii* background (Vo et al. 2019; Fujisaki et al. 2015).

8.4 Indirect interation between AVR proteins and unpaired NLRs

The interaction between Piz-t and AVR-Piz-t occurs indirectly, providing a different mechanism in the evolution of recognition of effector and cognate immune receptors. AVR-Piz-t encodes a secreted protein that localises in the BIC before moving into rice host cell. The *Piz-t* gene occurs on chromosome 6 at the same locus as the cloned R genes, *Pi2/Pi9/Pigm and Pi50/*, and offer broad spectrum resistance (Zhou et al. 2006; Zhang et al. 2013; Zhu et al. 2012; Deng et al. 2017). Other R genes, for example, *Pi26/Pi40/Pi2-2* and *Piz,* have also been mapped to this locus, but are yet to be cloned (Jiang et al. 2012; Jeung et al. 2007; Inukai et al. 1994; Wu et al. 2005). The amino acid sequence of Piz-t is similar to Pi-2, but with an eight amino acid difference occurring in the LRR domain region, and this gives a difference in specificity (Zhou et al. 2006). In rice plants that contain Piz-t, an E3 ubiquitin ligase protein, APIP10, binds and stabilises Piz-t, preventing a trigger for cell death through ubiquitination, as shown in Fig. 2. However, when *M. oryzae* secretes AVR-Piz-t into these cells, binding of the effector causes degradation of APIP10, thereby removing suppression of Piz-t and resulting in ETI-induced cell death (Park et al. 2016).

8.5 Evolution and polymorphism of AVR genes

Using association genetics, Yoshida and colleagues identified three alleles of *AVR-Pik/km/kp* (Yoshida et al. 2009). When compared to the *M. oryzae* reference genome 70-15, the three alleles were located in an isolate-specific region of a *M. oryzae* field strain Ina168, suggesting high levels of polymorphism. Most recently, a combination of long-read sequencing, clamped homogenous electric field (CHEF) gel electrophoresis and analysis of genome synteny study revealed that *AVR-Pik* is located on a minichromosome in a rice infecting isolate, FR13 (Langner et al. 2020). This might explain the high levels of polymorphism observed for *AVR-Pik*. AVR-Pik exhibits polymorphism at amino acid 46–48 and 67, leading to varied recognition by cognate *Pik* alleles, as shown in Fig. 2. This suggests a high level of mutation of *AVR-Pik* being driven by selection from field deployment or emergence of new *Pik* alleles in natural populations (Yoshida et al. 2009).

Phenotypic and genetic analysis of field isolates of *M. oryzae* from five different geographic regions in China revealed sequence variation within another avirulence gene, *AVR-Pib* (Zhang et al. 2015). In this case, polymorphism was observed as either a complete gene deletion, segmental deletion, point mutation or transposable element (TE) insertion, suggesting positive selection (Zhang et al. 2015). An analysis of isolates within one region in China revealed different genetic and phenotypic structures, which led to a conclusion that this was due to selection pressure from the host rather than geography (Zhang et al. 2015). Isolates from the southern regions of China were more virulent compared to other regions, which can be explained by the fact that *Pib* was introduced into the region by the International Rice Research Institute (IRRI) in the early 1960s (Zhang et al. 2015). Similarly, *AVR-Pita* has been shown to be genetically unstable in field isolates with frequent occurrence of point mutations and partial deletions, both in the coding sequence and promoter region (Valent and Chumley 1991; Orbach et al. 2000; Chuma et al. 2011).

The polymorphism of an avirulence gene, *AVR-Pi9,* in relation to gain of virulence towards the *Pi9* resistance gene has also been reported (Wu et al.

2015b). Through whole genome sequencing of two closely related isolates from a sequential planting experiment, the study identified the *AVR-Pi9* gene in both isolates, but, in the pathogenic isolate, the *AVR-Pi9* locus was disrupted by a Mg-SINE element insertion (Wu et al. 2015b). This emphasised the important role played by transposable elements in the *M. oryzae* genome in relation to AVR gene evolution (Zhang et al. 2015; Wu et al. 2015b). Similarly, *AVR-CO39* has been inactivated in some rice-infecting isolates, through transposon insertions, leading to gain-of-virulence on rice (Tosa et al. 2005; Tosa et al. 2006). However, this gene has been retained in other non-*Oryza* infecting pathotypes, perhaps due to its virulence function, targeting small HMA proteins to suppress PTI (Tosa et al. 2006; Tosa et al. 2005).

The biological functions of AVR-CO39, AVR-Pik, AVR-Pi54, AVR-Pib and AVR-Pi9 during plant infection have not been fully elucidated, but the implication of these studies is that they contribute to pathogen fitness through their PTI-suppressing activities, or other functions that aid proliferation of *M. oryzae*. Understanding the biological role of these avirulence proteins can help to determine why genes thought to encode key virulence functions, for example, *AVR-Pi9,* have been widely maintained in the population while others have been more easily lost in the blast population.

8.6 Pathogen-guided R-gene deployment against rice blast disease

The presence of race-specific R genes has been shown to shape the genetic architecture of a *M. oryzae* population in a given geographic region (Zhang et al. 2015). Regional differences in rice blast pathotypes therefore pose a major challenge in deployment of single R-genes, because the deployed R gene will be quickly broken down after several growing seasons (McDonald and Linde 2002). The blast fungus is, furthermore, always evolving and generating new races through mutation of avirulence genes. To solve this challenge, an extensive analysis of the prevailing blast population followed by genome wide-analysis for the presence/absence of cloned and novel avirulence genes can assist in planning for effective rice blast control, providing information on which R genes to deploy (Mutiga et al. 2017). With this in mind, the best strategy may be to pyramid several R genes in locally adapted rice cultivars, after understanding the genetics and virulence spectrum of a given regional pathogen population of *M. oryzae* (Mutiga et al. 2017).

For example, extensive pathotype analysis has been carried out using monogenic rice lines carrying 24 rice blast resistance genes, *Pia, Pib, Pik, Pik-h, Pi-km, Pik-p, Pik-s, Pish, Pit, Pita, Pita-2, Piz, Pi-2, Piz-t, Pi1, Pii, Pi3, Pi5 (t), Pi7 (t), Pi9, Pi11 (t), Pi12 (t), Pi19 (t)* and *Pi20* in a large collection of rice blast isolates from sub-Saharan Africa (SSA). These showed a high proportion of non-compatible reactions on rice lines carrying *Pi9*, for example, suggesting that it may exclude a significant proportion of the prevailing population of the rice blast fungus in SSA (Mutiga et al. 2017). This means that *Pi9* is a suitable candidate to be used in combination with other R genes to breed for resistance in SSA (Mutiga et al. 2017). Major resistance genes, like *Pi9,* can be combined with recessive resistance alleles, such as *pi21* and partial resistance conditioned by QTLs, such as *Pi-A-35*, as a stack in order to offer long-term durable control of blast disease (Nguyen et al. 2006; Fukuoka et al. 2009). However, caution has to be taken when carrying out gene pyramiding in order to avoid introducing negative traits that may be closely linked to R genes from a donor line. For example, the partial resistance recessive allele *pi21* is tightly linked to a gene that affects grain quality in terms of poor chalky appearance and milling quality. These two traits have, therefore, to be successfully separated before introducing *pi21* into any recipient cultivar (Feng et al. 2019; Fukuoka et al. 2009).

9 Wheat blast as an emerging threat to global food security

Wheat blast disease is caused by *M. oryzae* isolates that originally derived from *Lolium*-infecting strains of the fungus and already causes one of the most devastating disease of wheat in South America. Wheat blast symptoms are most severe at the heading stages and are characterised by bleaching, similar to the symptoms observed in *Fusarium* head blight disease as shown in Fig. 1c (Cruz and Valent 2017). Wheat blast was first observed in Paraná state of Brazil in 1985 and has spread to wheat-producing regions across Brazil and neighbouring countries (Maciel et al. 2013). Deployment of wheat cultivars lacking *Rmg6* (syn. *Rtw3*) resulted in susceptibility to *Lolium*-infecting isolates of *M. oryzae* that express and avirulence gene called *PWT3* (Inoue et al. 2017). Furthermore, cultivation of *Rtw3* cultivars led to a further host jump through loss of function of *PWT3*. Two avirulence genes, *PWT3* and *PWT4*, that control infection of *Magnaporthe oryzae* isolates towards *Triticum aestivum* (Wheat) were therefore identified (Inoue et al. 2017). In 2016, wheat blast disease was detected for the first time outside South America, when Bangladesh reported cases affecting up to 16% of the total wheat-growing area (Islam et al. 2016). Comparative genomics studies and phylogenetic analysis revealed that wheat blast isolates from Bangladesh were genetically related to isolates from South America, and had resulted from pathogen movement through wheat grain importation (Islam et al. 2016). Use of fungicides has proven to be ineffective to control this disease and identification of durable resistance genes remains the most likely approach to wheat blast control (Cruz and Valent 2017).

To date, eight resistance genes (*Rmg1-8*) have been identified against wheat blast disease and each possesses different specificities against distinct pathotypes of *M. oryzae* (Ceresini et al. 2019; Takabayashi et al. 2002). *Rmg1* (syn. *Rwt4*), for example, is effective against oat (*Avena*)-infecting isolates, but less effective against *Triticum* isolates (Takabayashi et al. 2002; Anh et al. 2015). Two other resistance genes, *Rmg4* and *Rmg5*, showed resistance against *Digitaria*-infecting isolates, but not *Triticum* isolates. *Rmg6* (syn. *Rwt3*) conditions resistance against a *Lolium*-infecting isolate but not against *Triticum* isolates. However, a subset of R genes, *Rmg2, Rmg3, Rmg7* and *Rmg8*, have proven to provide resistance against a subgroup of *Triticum* isolates. In 2015, Anh and colleagues identified the avirulence gene/R gene interaction between *Rmg8* and the corresponding AVR *AVR-Rmg8*. Like most AVRs, *AVR-Rmg8* encodes a small secreted effector, with 109 amino acids. In 2017, they further showed that apart from being recognised by *Rmg8*, *AVR-Rmg8* was also recognised by *Rmg7* in tetraploid wheat (Anh et al. 2015).

Three R genes against wheat blast - *Rmg2*, *Rmg3* and *Rmg8* - occur in common wheat, while *Rmg7* occurs in emmer wheat (*T. dicoccum*), which is closely related to cultivated wheat (Anh et al. 2015; Tagle et al. 2015; Zhan et al. 2015). *Rmg2* and *Rmg3* are present in the common wheat variety, Thatcher, while *Rmg8* occurs in another line, S615 (Zhan et al. 2008). Three *T. dicoccum* lines, KU112 (st17), KU120 (st24) and KU122 (st25), meanwhile carry *Rmg7* (Tagle et al. 2015). In 2017, field trials were conducted in Brazil in wheat blast disease hotspots to test the effectiveness of resistance genes in two wheat varieties, Lagoa Vermelha and Thatcher (Ceresini et al. 2019). *Rmg2* and *Rmg3,* present in Thatcher, were not effective at high temperatures, compared to Lagoa Vermelha, which is a relative of Thatcher, but might contain other R genes against wheat blast that need more investigation to identify (Ceresini et al. 2019; Zhan et al. 2008). With the knowledge that *Rmg7* and *Rmg8* recognise the same avirulence gene, this limits the number of effective resistance specificities against *Triticum* isolates to only three known R-genes. Studies have, however, since shown that useful genes, including those that provide disease blast resistance, can be transferred from *Aegilops ventricosa and Ae. tauschiii* to wheat. Most of the disease resistances from *Ae. Ventricosa*, which include *Lr37, Yr17, Sr38, Rkn3* and Cre5, are a result of a 2NS-2AS chromosomal translocation (Helguera et al. 2003; Williamson et al. 2013; Jahier et al. 2001). The 2NS-2AS translocation includes a 25–38 cM segment of chromosome arm 2NS, from *Ae. ventricosa* to the furthest end of chromosome arm 2AS from wheat (Helguera et al. 2003; Maia 1967).

The 2NS/AS chromosomal translocation has been associated with resistance against wheat blast (Cruz et al. 2016). Field trials conducted in Bolivia in 2014 and 2015 showed that near-isogenic lines of both spring and winter wheat carrying 2NS/AS have reduced the severity to head blast resistance, especially under natural epidemic conditions, compared to those without the segment (Cruz and Valent 2017; Cruz et al. 2016). Additionally, a wheat variety called Milan, distributed by the International Maize and Wheat Improvement Centre (CIMMYT), that contains the 2NS segment, also showed relatively high levels of head blast resistance (Cruz and Valent 2017). Despite the fact that not all wheat varieties which contain the 2NS segment offer significant resistance to wheat blast, farmers have been advised to use these varieties to reduce the damage caused by wheat blast disease (Cruz et al. 2016).

After the first cases of wheat blast were reported in Bangladesh in 2016, the Bangladesh Agricultural Research Institute (BARI), in collaboration with CIMMYT, quickly developed a new variety called BARI-Gom 33 that is resistant to wheat blast disease (BARI 2017). Resistance of this variety to wheat blast is due to the 2NS-2AS translocation (Velu et al. 2019). Apart from being resistant to blast, BARI Gom 33 is a high-yielding variety, offering a 5–8% yield improvement compared to other varieties and is also zinc-fortified (BARI 2017; Mottaleb et al. 2019; Velu et al. 2019). This suggests that using locally adapted resistant varieties may be a strategy that can be used to alleviate losses caused by the spread of wheat blast disease in Bangladesh (Mottaleb et al. 2019).

10 CRISPR-Cas9 mutagenesis of dominant S-genes

Dominant R genes are widely deployed against pathogens to promote resistance. However, as reported and discussed in this chapter, race specificity of such resistance makes them prone to being easily broken down by mutation of avirulence genes in the pathogen population (McDonald and Linde 2002; Ballini et al. 2008). Outbreaks of such a new clone will therefore be devastating and spread very quickly in the field under selection. This means that new approaches are needed to replace or supplement the conventional approach to rice and wheat breeding (Ballini et al. 2008; Corwin and Kliebenstein 2017). Targeting genes that condition compatible interactions can be used as an alternative means to achieve durable resistance (Schie and Takken 2014; Pavan et al. 2009). This approach produces recessive resistance that may be more durable compared to dominant resistance genes (Mazier et al. 2011). This is because the dominant susceptibility function may be essential for the disease to occur, such that a pathogen cannot easily overcome the absence of such a compatibility factor in its host.

Caution should, however, be taken when selecting candidate genes to edit because some may be involved in important physiological processes in the host plant. For example, editing a lettuce (*Lactuca)*-susceptibility gene, *eIF4E*, resulted in resistance against potyvirus, but slightly impaired growth and fertility of the plant (Abdul-Razzak et al. 2009). Successful infection by the powdery mildew fungus, *Blumeria graminis,* requires the presence of a gene product encoded by *MLO* (mildew resistance locus O). *MLO* encodes a calmodulin-binding transmembrane protein involved in plant defence. Naturally occurring loss-of-function mutations in this gene leads to powdery mildew resistance by tomato (*Solanum lycopersicum*), pea (*Pisum sativum*) and barley (*Hordeum vulgare*) (Bai et al. 2008; Humphry et al. 2011; Jørgensen 1992). A knockdown by RNAi silencing of the *MLO* homolog in *Petunia hybrida* was recently found to increase resistance to powdery mildew (Jiang et al. 2016b).

To quickly generate resistant cereal crops with targeted modification to susceptibility genes, there is a need for faster and less laborious approaches (Das et al. 2019). Clustered regularly interspaced short palindromic repeats (CRISPR) associated with a DNA endonuclease, together with a guide RNA, offer a suitable approach to edit plant genomes to acquire desired traits (Das et al. 2019). For cereal crops, rice has been used as a model for genome editing, both for biological studies and agricultural improvement purposes (Weeks et al. 2016). In wheat, despite being a complicated system with a large and complex genome (as well as being hexaploid), gene editing using TALENs and CRISPR/Cas9 has been successfully carried out (Upadhyay et al. 2013; Wang et al. 2014; Shan et al. 2013). For example, transient expression has been used to carry out gene-specific modification of the wheat powdery mildew locus (*TaMLO*) using guide RNA-directed Cas9 (Shan et al. 2013). Additionally, this method has been successfully applied to edit *phytoene desaturase(pds)* and *inositol oxygenase* (*inox*) in wheat cell suspensions (Upadhyay et al. 2013). In cases where susceptibility genes occur in families of multiple genes in plant species or homoeoalleles for polyploid genomes like wheat, for example, *MLO* (the three homoeoalleles), CRISPR/Cas9 has also been used successfully (Wang et al. 2014; Jiang et al. 2016a). Furthermore, Nekrasov and colleagues used CRISPR/Cas9 gene editing to rapidly generate Tomelo, a non-transgenic tomato that is resistant to powdery mildew by introducing mutations in the *Mlo* locus. This study further confirmed that it is possible to generate mutations in a target gene, *Slmlo1,* without causing off-target effects (Nekrasov et al. 2017).

Interestingly, some proteins encoded by predicted susceptibility genes, for example, HMA domain-containing proteins, have been found to be integrated into plant NLRs (Sarris et al. 2016). It is thought that plant proteins with homology to those integrated in NLRs may be targets of pathogen-secreted effector proteins which promote susceptibility (Kroj et al. 2016; Sarris et al. 2016). The function of HMA domain-containing proteins in plants has not been discerned, but introduction of a mutation or silencing genes encoding HMA could be used to introduce resistance against blast fungus in cereals (Fukuoka et al. 2009; Sarris et al. 2016; Kroj et al. 2016). Additionally, understanding the function of potential effector host targets offers scientists potential gene targets for editing.

11 Conclusion and future trends

A combination of strategies is required to successfully manage blast disease, especially for economically important cereals such as rice, wheat, barley, finger millet and oat. Selecting and breeding for cultivars with durable race-specific resistance, together with non-race-specific resistance genes, may offer a solution for durable resistance in the future (Mutiga et al. 2017). Selection of resistance genes for breeding programmes should, however, also be guided by an understanding of the prevailing pathogen population (Mutiga et al. 2017). Such studies to forecast and map the diversity of a pathogen population can predict outbreaks due to pathogen evolution and the introduction of new pathogen races through seed importation, or the introduction of new varieties that provide specific selective pressure on the prevailing pathogen population. Above all, policymakers and governments should also consider lifting regulations against transgenic crops, as a way of achieving resistance against plant diseases, for example, as has been shown with rice-transgenic lines expressing the resistance gene *Pi-d2* (Chen et al. 2010). These measures must also be used in association with improved agricultural practices, for example, understanding the role played by non-crop cereals and weeds in the spread of blast disease (Asibi et al. 2019).

There is also a need for rapid comparative genomics efforts to identify new AVR-R gene interactions, which will be an important resource for pathologists and plant breeders. The number of already cloned AVRs is very low, for instance, compared to the number of cloned resistance genes (Wu et al. 2015b). Since the start of next-generation sequencing (NGS), there has been progress in comparative studies to characterise pathogens. The continuous development of these technologies, for example, the use of long-read sequencing offers a solution to study regions of pathogens that were initially impossible to study due to the abundance of repetitive sequences. These regions are thought to harbour virulence genes, secondary metabolite-encoding genes and effectors. Furthermore, this approach can further be coupled with other latest scaffolding technologies like Optical mapping, Chicago and Dovetails Hi-C to gain a better understanding of complicated genomic content of pathogens that drive evolution (Ghurye et al. 2019; Lieberman-Aiden et al. 2009; Putnam et al. 2016; Levy-Sakin and Ebenstein 2013).

It is also clear that the world's climate is changing and this will very likely affect food production and the prevalence of crop diseases, including rice blast (Bebber et al. 2014). It is clear that rice and wheat can, for example, be affected by abiotic stress, which can impair their ability to resist blast disease (Newton and Young 1996; Coakley et al. 1999; Luck et al. 2011). With climate change, it is expected that there will be an increase in the frequency of extreme weather conditions. Changes in the climate can also impact the distribution of the blast pathogen and its ability to move from one agroecological zone to another (Coakley et al. 1999; Asibi et al. 2019). This means that any measures taken to mitigate climate change can also impact how farmers and scientists combat plant disease in general (Asibi et al. 2019).

In summary, the rice blast pathogen is a significant, well-adapted micro-organism, which is difficult to control effectively. Its biology is specifically adapted to infect the tough outer layers of cereal host plants and it has the ability to cause diverse pathologies, affecting different tissues and growth stages of its host. This means that the yield can be affected in many different ways, from affecting the growth of seedlings, to the development of the inflorescence that carries rice grains, as well as seed set. Significant progress in understanding the fundamental biology of rice blast will provide new targets for both chemical and genetic intervention into blast disease. Understanding the interplay between effectors and their virulence targets, as well as the NLRs that have evolved to recognise them, also offers a very powerful means of developing predictive methods to protect the global rice harvest from this devastating disease.

12 Where to look for further information

12.1 Key journals and reviews

- Ceresini, P. C., V. L. Castroagudin, F. A. Rodrigues, J. A. Rios, C. E. Aucique-Perez, S. I. Moreira, D. Croll, E. Alves, G. de Carvalho, J. L. N. Maciel and B. A. McDonald (2019). "Wheat blast: from its origins in South America to its emergence as a global threat." *Molecular Plant Pathology* **20**(2): 155–172.
- Giraldo, M. C. and B. Valent (2013). "Filamentous plant pathogen effectors in action." *Nature Reviews Microbiology* **11**(11): 800–814.
- Wilson, R. A. and N. J. Talbot (2009). "Under pressure: investigating the biology of plant infection by Magnaporthe oryzae." *Nature Reviews Microbiology* **7**(3): 185–195.

13 References

Abdul-Razzak, A., Guiraud, T., Peypelut, M., Walter, J., Houvenaghel, M. C., Candresse, T., LE Gall, O. and German-Retana, S. 2009. Involvement of the cylindrical inclusion (CI) protein in the overcoming of an eIF4E-mediated resistance against Lettuce mosaic Potyvirus, *Molecular Plant Pathology* 10(1): 109-13.

Anh, V. L., Anh, N. T., Tagle, A. G., Vy, T. T., Inoue, Y., Takumi, S., Chuma, I. and Tosa, Y. 2015. Rmg8, a new gene for resistance to Triticum isolates of Pyricularia oryzae in hexaploid wheat, *Phytopathology* 105(12): 1568-72.

Ashikawa, I., Hayashi, N., Yamane, H., Kanamori, H., Wu, J., Matsumoto, T., Ono, K. and Yano, M. 2008. Two adjacent nucleotide-binding site-leucine-rich repeat class genes are required to confer Pikm-specific rice blast resistance, *Genetics* 180(4): 2267–76.

Asibi, A. E., Aziiba, Q. C. and Coulter, J. A. 2019. Rice blast: a disease with implications for global food security, *Agronomy* 9(8): 451–64.

Bai, Y., Pavan, S., Zheng, Z., Zappel, N. F., Reinstädler, A., Lotti, C., De Giovanni, C., Ricciardi, L., Lindhout, P., Visser, R., Theres, K. and Panstruga, R. 2008. Naturally occurring broad-spectrum powdery mildew resistance in a Central American tomato accession is caused by loss of mlo function, *Molecular Plant-Microbe Interactions: MPMI®* 21(1): 30–9.

Baldrich, P., Rutter, B. D., Karimi, H. Z., Podicheti, R., Meyers, B. C. and Innes, R. W. 2019. Plant extracellular vesicles contain diverse small RNA species and are enriched in 10- to 17-nucleotide "tiny" RNAs, *Plant Cell* 31(2): 315–24.

Ballini, E., Morel, J.-B., Droc, G., Price, A., Courtois, B., Notteghem, J.-L. and Tharreau, D. 2008. A genome-wide meta-analysis of rice blast resistance genes and quantitative trait loci provides new insights into partial and complete resistance, *Molecular Plant-Microbe Interactions: MPMI®* 21(7): 859–68.

BARI 2017. Proforma for obtaining approval of the National Seed Board of Bangladesh (NSB) for a new crop variety/cultivar. Bangladesh Agricultural Research Institute, BARI, and National Seed Board of Bangladesh (NATIONAL SCIENCE BOARD), Dhaka.

Bebber, D. P., Holmes, T. and Gurr, S. J. 2014. The global spread of crop pests and pathogens, *Global Ecology and Biogeography* 23(12): 1398–407.

Bernoux, M., Ve, T., Williams, S., Warren, C., Hatters, D., Valkov, E., Zhang, X., Ellis, J. G., Kobe, B. and Dodds, P. N. 2011. Structural and functional analysis of a plant resistance protein TIR domain reveals interfaces for self-association, signaling, and autoregulation, *Cell Host and Microbe* 9(3): 200–11.

Białas, A., Zess, E. K., De la Concepcion, J. C., Franceschetti, M., Pennington, H. G., Yoshida, K., Upson, J. L., Chanclud, E., Wu, C. H., Langner, T., Maqbool, A., Varden, F. A., Derevnina, L., Belhaj, K., Fujisaki, K., Saitoh, H., Terauchi, R., Banfield, M. J. and Kamoun, S. 2017. Lessons in effector and NLR biology of plant-microbe systems, *Molecular Plant-Microbe Interactions: MPMI* 31(1): 34–45.

Bohnert, H. U., Fudal, I., Dioh, W., Tharreau, D., Notteghem, J. L. and Lebrun, M. H. 2004. A putative polyketide synthase/peptide synthetase from Magnaporthe grisea signals pathogen attack to resistant rice, *Plant Cell* 16(9): 2499–513.

Bourett, T. M. and Howard, R. J. 1990a. Invitro development of penetration structures in the rice blast fungus Magnaporthe-grisea, *Canadian Journal of Botany-Revue Canadienne de Botanique* 68: 329–42.

Bourett, T. M. and Howard, R. J. 1990b. In vitro development of penetration structures in the rice blast fungus Magnaporthe grisea, *Canadian Journal of Botany* 68(2): 329–42.

Bozkurt, T. O. and Kamoun, S. 2020. The plant-pathogen haustorial interface at a glance, *Journal of Cell Science* 133(5): jcs237958.

Bryan, G. T., Wu, K. S., Farrall, L., Jia, Y., Hershey, H. P., McAdams, S. A., Faulk, K. N., Donaldson, G. K., Tarchini, R. and Valent, B. 2000. tA single amino acid difference distinguishes resistant and susceptible alleles of the rice blast resistance gene Pi-ta, *Plant Cell* 12(11): 2033–46.

Cai, Q., Qiao, L., Wang, M., He, B., Lin, F. M., Palmquist, J., Huang, S. D. and Jin, H. 2018. Plants send small RNAs in extracellular vesicles to fungal pathogen to silence virulence genes, *Science* 360(6393): 1126–29.

Ceresini, P. C., Castroagudin, V. L., Rodrigues, F. Á, Rios, J. A., Aucique-Perez, C. E., Moreira, S. I., Croll, D., Alves, E., de Carvalho, G., Maciel, J. L. N. and McDonald, B. A. 2019. Wheat blast: from its origins in South America to its emergence as a global threat, *Molecular Plant Pathology* 20(2): 155–72.

Cesari, S., Bernoux, M., Moncuquet, P., Kroj, T. and Dodds, P. N. 2014a. A novel conserved mechanism for plant NLR protein pairs: the "integrated decoy" hypothesis, *Frontiers in Plant Science* 5: 606.

Cesari, S., Kanzaki, H., Fujiwara, T., Bernoux, M., Chalvon, V., Kawano, Y., Shimamoto, K., Dodds, P., Terauchi, R. and Kroj, T. 2014b. The NB-LRR proteins RGA4 and RGA5 interact functionally and physically to confer disease resistance, *EMBO Journal* 33(17): 1941–59.

Cesari, S., Thilliez, G., Ribot, C., Chalvon, V., Michel, C., Jauneau, A., Rivas, S., Alaux, L., Kanzaki, H., Okuyama, Y., Morel, J. B., Fournier, E., Tharreau, D., Terauchi, R. and Kroj, T. 2013. The rice resistance protein pair RGA4/RGA5 recognizes the Magnaporthe oryzae effectors AVR-Pia and AVR1-CO39 by direct binding, *Plant Cell* 25(4): 1463–81.

Chen, D., Chen, X., Lei, C., Ma, B., Wang, Y. and Li, S. 2010 Rice blast resistance of transgenic rice plants with Pi-d2 gene, *Rice Science* 17(3): 179–84.

Chen, J., Peng, P., Tian, J., He, Y., Zhang, L., Liu, Z., Yin, D. and Zhang, Z. (2015). Pike, a rice blast resistance allele consisting of two adjacent NBS-LRR genes, was identified as a novel allele at the Pik locus. *Molecular Breeding* 35(5): 117.

Chen, J., Shi, Y., Liu, W., Chai, R., Fu, Y., Zhuang, J. and Wu, J. (2011). A Pid3 allele from rice cultivar Gumei2 confers resistance to Magnaporthe oryzae. *Journal of Genetics and Genomics* 38(5): 209–16.

Chen, X., Shang, J., Chen, D., Lei, C., Zou, Y., Zhai, W., Liu, G., Xu, J., Ling, Z., Cao, G., Ma, B., Wang, Y., Zhao, X., Li, S. and Zhu, L. 2006. A B-lectin receptor kinase gene conferring rice blast resistance, *Plant Journal: For Cell and Molecular Biology* 46(5): 794–804.

Choi, W. and Dean, R. A. 1997. The adenylate cyclase gene MAC1 of Magnaporthe grisea controls appressorium formation and other aspects of growth and development, *Plant Cell* 9(11): 1973–83.

Chuma, I., Isobe, C., Hotta, Y., Ibaragi, K., Futamata, N., Kusaba, M., Yoshida, K., Terauchi, R., Fujita, Y., Nakayashiki, H., Valent, B. and Tosa, Y. 2011. Multiple translocation of the AVR-Pita effector gene among chromosomes of the rice blast fungus magnaporthe oryzae and related species, *PLoS Pathogens* 7(7): e1002147.

Coakley, S. M., Scherm, H. and Chakraborty, S. 1999 Climate change and plant disease management, *Annual Review of Phytopathology* 37: 399–426.

Corwin, J. A. and Kliebenstein, D. J. 2017. Quantitative resistance: more than just perception of a pathogen, *The Plant Cell* 29(4): 655–65.

Cruz, C. D., Peterson, G. L., Bockus, W. W., Kankanala, P., Dubcovsky, J., Jordan, K. W., Akhunov, E., Chumley, F., Baldelomar, F. D. and Valent, B. 2016. The 2NS translocation from Aegilops ventricosa confers resistance to the triticum pathotype of magnaporthe oryzae, *Crop Science* 56(3): 990–1000.

Cruz, C. D. and Valent, B. 2017. Wheat blast disease: danger on the move, *Tropical Plant Pathology* 42(3): 210–22.

Dagdas, Y. F., Yoshino, K., Dagdas, G., Ryder, L. S., Bielska, E., Steinberg, G. and Talbot, N. J. 2012. Septin-mediated plant cell invasion by the rice blast fungus, Magnaporthe oryzae, *Science* 336(6088): 1590–5.

Dai, L., Wu, J., Li, X., Wang, X., Liu, X., Jantasuriyarat, C., Kudrna, D., Yu, Y., Wing, R. A., Han, B., Zhou, B. and Wang, G. L. 2010. Genomic structure and evolution of the Pi2/9 locus in wild rice species, *TAG. Theoretical and Applied Genetics. Theoretische und Angewandte Genetik* 121(2): 295–309.

Das, A., Sharma, N. and Prasad, M. 2019. CRISPR/Cas9: a novel weapon in the arsenal to combat plant diseases, *Frontiers in Plant Science* 9: 2008.

Das, A., Soubam, D., Singh, P. K., Thakur, S., Singh, N. K. and Sharma, T. R. (2012). A novel blast resistance gene, Pi54rh cloned from wild species of rice, Oryza rhizomatis confers broad spectrum resistance to Magnaporthe oryzae. *Functional & Integrative Genomics* 12(2): 215–28.

de Guillen, K., Ortiz-Vallejo, D., Gracy, J., Fournier, E., Kroj, T. and Padilla, A. 2015. Structure analysis uncovers a highly diverse but structurally conserved effector family in phytopathogenic fungi, *PLoS Pathogens* 11(10): e1005228.

de Jonge, R., Bolton, M. D. and Thomma, B. P. H. J. 2011. How filamentous pathogens co-opt plants: the ins and outs of fungal effectors, *Current Opinion in Plant Biology* 14(4): 400–6.

De la Concepcion, J. C., Franceschetti, M., MacLean, D., Terauchi, R., Kamoun, S. and Banfield, M. J. 2019. Protein engineering expands the effector recognition profile of a rice NLR immune receptor, *eLife* 8: e47713.

Dean, R. A., Talbot, N. J., Ebbole, D. J., Farman, M. L., Mitchell, T. K., Orbach, M. J., Thon, M., Kulkarni, R., Xu, J. R., Pan, H., Read, N. D., Lee, Y. H., Carbone, I., Brown, D., Oh, Y. Y., Donofrio, N., Jeong, J. S., Soanes, D. M., Djonovic, S., Kolomiets, E., Rehmeyer, C., Li, W., Harding, M., Kim, S., Lebrun, M. H., Bohnert, H., Coughlan, S., Butler, J., Calvo, S., Ma, L. J., Nicol, R., Purcell, S., Nusbaum, C., Galagan, J. E. and Birren, B. W. 2005. The genome sequence of the rice blast fungus Magnaporthe grisea, *Nature* 434(7036): 980–6.

Deng, Y., Zhai, K., Xie, Z., Yang, D., Zhu, X., Liu, J., Wang, X., Qin, P., Yang, Y., Zhang, G., Li, Q., Zhang, J., Wu, S., Milazzo, J., Mao, B., Wang, E., Xie, H., Tharreau, D. and He, Z. 2017. Epigenetic regulation of antagonistic receptors confers rice blast resistance with yield balance, *Science* 355(6328): 962–5.

Devanna, N. B., Vijayan, J. and Sharma, T. R. 2014. The blast resistance gene Pi54 of cloned from Oryza officinalis interacts with Avr-Pi54 through its novel non-LRR domains, *PLoS ONE* 9(8): e104840.

DeZwaan, T. M., Carroll, A. M., Valent, B. and Sweigard, J. A. 1999. Magnaporthe grisea pth11p is a novel plasma membrane protein that mediates appressorium differentiation in response to inductive substrate cues, *The Plant Cell* 11(10): 2013–30.

Dixon, M. S., Jones, D. A., Keddie, J. S., Thomas, C. M., Harrison, K. and Jones, J. D. 1996. The tomato Cf-2 disease resistance locus comprises two functional genes encoding leucine-rich repeat proteins, *Cell* 84(3): 451–9.

Djamei, A., Schipper, K., Rabe, F., Ghosh, A., Vincon, V., Kahnt, J., Osorio, S., Tohge, T., Fernie, A. R., Feussner, I., Feussner, K., Meinicke, P., Stierhof, Y. D., Schwarz, H., Macek, B., Mann, M. and Kahmann, R. 2011. Metabolic priming by a secreted fungal effector, *Nature* 478(7369): 395–8.

Dodds, P. N. and Rathjen, J. P. 2010. Plant immunity: towards an integrated view of plant-pathogen interactions, *Nature Reviews Genetics* 11(8): 539–48.

Doehlemann, G., van der Linde, K., Assmann, D., Schwammbach, D., Hof, A., Mohanty, A., Jackson, D. and Kahmann, R. 2009. Pep1, a secreted effector protein of Ustilago maydis, is required for successful invasion of plant cells, *PLoS Pathogens* 5(2): e1000290.

Egan, M. J., Wang, Z. Y., Jones, M. A., Smirnoff, N. and Talbot, N. J. 2007. Generation of reactive oxygen species by fungal NADPH oxidases is required for rice blast disease, *Proceedings of the National Academy of Sciences of the United States of America* 104(28): 11772–7.

Eitas, T. K. and Dangl, J. L. 2010. NB-LRR proteins: pairs, pieces, perception, partners, and pathways, *Current Opinion in Plant Biology* 13(4): 472–7.

Faulkner, C., Petutschnig, E., Benitez-Alfonso, Y., Beck, M., Robatzek, S., Lipka, V. and Maule, A. J. 2013. LYM2-dependent chitin perception limits molecular flux via plasmodesmata, *Proceedings of the National Academy of Sciences of the United States of America* 110(22): 9166–70.

Feng, X., Lin, K., Zhang, W., Nan, J., Zhang, X., Wang, C., Wang, R., Jiang, G., Yuan, Q. and Lin, S. 2019. Improving the blast resistance of the elite rice variety Kongyu-131 by updating the pi21 locus, *BMC Plant Biology* 19(1): 249.

Foster, A. J., Martin-Urdiroz, M., Yan, X., Wright, H. S., Soanes, D. M. and Talbot, N. J. 2018. CRISPR-Cas9 ribonucleoprotein-mediated co-editing and counterselection in the rice blast fungus, *Scientific Reports* 8(1): 14355.

Fujisaki, K., Abe, Y., Ito, A., Saitoh, H., Yoshida, K., Kanzaki, H., Kanzaki, E., Utsushi, H., Yamashita, T., Kamoun, S. and Terauchi, R. 2015. Rice Exo70 interacts with a fungal effector, AVR-Pii, and is required for AVR-Pii-triggered immunity, *Plant Journal: For Cell and Molecular Biology* 83(5): 875–87.

Fukuoka, S., Saka, N., Koga, H., Ono, K., Shimizu, T., Ebana, K., Hayashi, N., Takahashi, A., Hirochika, H., Okuno, K. and Yano, M. 2009. Loss of function of a proline-containing protein confers durable disease resistance in rice, *Science* 325(5943): 998–1001.

Fukuoka, S., Yamamoto, S.-I., Mizobuchi, R., Yamanouchi, U., Ono, K., Kitazawa, N., Yasuda, N., Fujita, Y., Nguyen, T. T. T., Koizumi, S., Sugimoto, K., Matsumoto, T. and Yano, M. 2014. Multiple functional polymorphisms in a single disease resistance gene in rice enhance durable resistance to blast, *Scientific Reports* 4(1): srep04550.

Ghurye, J., Rhie, A., Walenz, B. P., Schmitt, A., Selvaraj, S., Pop, M., Phillippy, A. M. and Koren, S. 2019. Integrating Hi-C links with assembly graphs for chromosome-scale assembly, *PLoS Computational Biology* 15(8): e1007273.

Giraldo, M. C., Dagdas, Y. F., Gupta, Y. K., Mentlak, T. A., Yi, M., Martinez-Rocha, A. L., Saitoh, H., Terauchi, R., Talbot, N. J. and Valent, B. 2013. Two distinct secretion systems facilitate tissue invasion by the rice blast fungus Magnaporthe oryzae, *Nature Communications* 4: 1996.

Giraldo, M. C. and Valent, B. 2013. Filamentous plant pathogen effectors in action, *Nature Reviews Microbiology* 11(11): 800–14.

Gupta, Y. K., Dagdas, Y. F., Martinez-Rocha, A.-L., Kershaw, M. J., Littlejohn, G. R., Ryder, L. S., Sklenar, J., Menke, F. and Talbot, N. J. 2015. Septin-dependent assembly of the exocyst is essential for plant infection by Magnaporthe oryzae, *The Plant Cell* 27(11): 3277–89.

Hamer, J. E., Howard, R. J., Chumley, F. G. and Valent, B. 1988. A mechanism for surface attachment in spores of a plant pathogenic fungus, *Science* 239(4837): 288–90.

Hayashi, K. and H. Yoshida (2009). Refunctionalization of the ancient rice blast disease resistance gene Pit by the recruitment of a retrotransposon as a promoter. *The Plant Journal* 57(3): 413–25.

Hayashi, N., Inoue, H., Kato, T., Funao, T., Shirota, M., Shimizu, T., Kanamori, H., Yamane, H., Hayano-Saito, Y., Matsumoto, T., Yano, M. and Takatsuji H. (2010). Durable panicle blast-resistance gene Pb1 encodes an atypical CC-NBS-LRR protein and was generated by acquiring a promoter through local genome duplication. *The Plant Journal* 64(3): 498–510.

Helguera, M., Khan, I. A., Kolmer, J., Lijavetzky, D., Zhong-qi, L. and Dubcovsky, J. 2003. PCR assays for the Lr37-Yr17-Sr38 cluster of rust resistance genes and their use to develop isogenic hard red spring wheat lines, *Crop Science* 43(5): 1839–47.

Howard, R. J., Ferrari, M. A., Roach, D. H. and Money, N. P. 1991. Penetration of hard substrates by a fungus employing enormous turgor pressures, *Proceedings of the National Academy of Sciences of the United States of America* 88(24): 11281–4.

Humphry, M., Reinstädler, A., Ivanov, S., Bisseling, T. and Panstruga, R. 2011. Durable broad-spectrum powdery mildew resistance in pea er1 plants is conferred by natural loss-of-function mutations in PsMLO1, *Molecular Plant Pathology* 12(9): 866–78.

Inoue, Y., Vy, T. T. P., Yoshida, K., Asano, H., Mitsuoka, C., Asuke, S., Anh, V. L., Cumagun, C. J. R., Chuma, I., Terauchi, R., Kato, K., Mitchell, T., Valent, B., Farman, M. and Tosa, Y. 2017. Evolution of the wheat blast fungus through functional losses in a host specificity determinant, *Science* 357(6346): 80–3.

International Rice Genome Sequencing, Project 2005. The map-based sequence of the rice genome, *Nature* 436(7052): 793–800.

Inukai, T., Nelson, R. J., Ziegler, R. S., Sarkarung, S., Mackill, D. J., Bonman, J. M., Takamure, I. and Kinoshita, T. 1994. Allelism of blast resistance genes in near-isogenic lines of rice, *Phytopathology* 84(11): 1278–83.

Islam, M. T., Croll, D., Gladieux, P., Soanes, D. M., Persoons, A., Bhattacharjee, P., Hossain, M. S., Gupta, D. R., Rahman, M. M., Mahboob, M. G., Cook, N., Salam, M. U., Surovy, M. Z., Sancho, V. B., Maciel, J. L., NhaniJunior, A., Castroagudin, V. L., Reges, J. T., Ceresini, P. C., Ravel, S., Kellner, R., Fournier, E., Tharreau, D., Lebrun, M. H., McDonald, B. A., Stitt, T., Swan, D., Talbot, N. J., Saunders, D. G., Win, J. and Kamoun, S. 2016. Emergence of wheat blast in Bangladesh was caused by a South American lineage of Magnaporthe oryzae, *BMC Biology* 14(1): 84.

Jahier, J., Abelard, P., Tanguy, M., Dedryver, F., Rivoal, R., Khatkar, S., Bariana, H. S. and Koebner, R. 2001. The Aegilops ventricosa segment on chromosome 2AS of the wheat cultivar 'VPM1' carries the cereal cyst nematode resistance gene Cre5, *Plant Breeding* 120(2): 125–8.

Jeung, J. U., Kim, B. R., Cho, Y. C., Han, S. S., Moon, H. P., Lee, Y. T. and Jena, K. K. 2007. A novel gene, Pi40(t), linked to the DNA markers derived from NBS-LRR motifs confers broad spectrum of blast resistance in rice, *TAG. Theoretical and Applied Genetics. Theoretische und Angewandte Genetik* 115(8): 1163–77.

Jia, Y., McAdams, S. A., Bryan, G. T., Hershey, H. P. and Valent, B. 2000. Direct interaction of resistance gene and avirulence gene products confers rice blast resistance, *EMBO Journal* 19(15): 4004–14.

Jiang, N., Li, Z., Wu, J., Wang, Y., Wu, L., Wang, S., Wang, D., Wen, T., Liang, Y., Sun, P., Liu, J., Dai, L., Wang, Z., Wang, C., Luo, M., Liu, X. and Wang, G. L. 2012. Molecular mapping of the Pi2/9 allelic gene Pi2-2 conferring broad-spectrum resistance to Magnaporthe oryzae in the rice cultivar Jefferson, *Rice* 5(1): 29.

Jiang, P., Chen, Y. and Dayton Wilde, H. D. 2016a. Identification and mutagenesis of disease susceptibility genes of Petunia hybrida, *Plant Cell, Tissue and Organ Culture (PCTOC)* 126(1): 117–25.

Jiang, P., Chen, Y. and Dayton Wilde, H. D. 2016b. Reduction of MLO1 expression in petunia increases resistance to powdery mildew, *Scientia Horticulturae* 201: 225-9.

Jones, J. D. and Dangl, J. L. 2006. The plant immune system, *Nature* 444(7117): 323-9.

Jones, K., Kim, D. W., Park, J. S. and Khang, C. H. 2016. Live-cell fluorescence imaging to investigate the dynamics of plant cell death during infection by the rice blast fungus Magnaporthe oryzae, *BMC Plant Biology* 16: 69.

Jørgensen, I. H. 1992. Discovery, characterization and exploitation of Mlo powdery mildew resistance in barley, *Euphytica* 63(1-2): 141–52.

Kankanala, P., Czymmek, K. and Valent, B. 2007. Roles for rice membrane dynamics and plasmodesmata during biotrophic invasion by the blast fungus, *Plant Cell* 19(2): 706–24.

Kanzaki, H., Yoshida, K., Saitoh, H., Fujisaki, K., Hirabuchi, A., Alaux, L., Fournier, E., Tharreau, D. and Terauchi, R. 2012. Arms race co-evolution of Magnaporthe oryzae AVR-Pik and rice Pik genes driven by their physical interactions, *Plant Journal: For Cell and Molecular Biology* 72(6): 894–907.

Khang, C. H., Berruyer, R., Giraldo, M. C., Kankanal, P., Park, S. Y., Czymmek, K., Kang, S. and Valent, B. 2010. Translocation of Magnaporthe oryzae effectors into rice cells and their subsequent cell-to-cell movement, *Plant Cell* 22(4): 1388–403.

Kroj, T., Chanclud, E., Michel-Romiti, C., Grand, X. and Morel, J. B. 2016. Integration of decoy domains derived from protein targets of pathogen effectors into plant immune receptors is widespread, *New Phytologist* 210(2): 618–26.

Langner, T., Harant, A., Gomez-Luciano, L. B., Shrestha, R. K., Win, J. and Kamoun, S. 2020. Genomic rearrangements generate hypervariable mini-chromosomes in host-specific lineages of the blast fungus, *bioRxiv*. https://doi.org/10.1101/2020.01.10.901983.

Lee, S.-K., Song, M.-Y., Seo, Y.-S., Kim, H.-K., Ko, S., Cao, P.-J., Suh, J.-P., Yi, G., Roh, J.-H., Lee, S., An, G., Hahn, T.-R., Wang, G.-L., Ronald, P. and Jeon, J.-S. (2009). Rice Pi5-mediated resistance to Magnaporthe oryzae requires the presence of two coiled-coil-nucleotide-binding-leucine-rich repeat genes. *Genetics* 181(4): 1627–38.

Levy-Sakin, M. and Ebenstein, Y. 2013. Beyond sequencing: optical mapping of DNA in the age of nanotechnology and nanoscopy, *Current Opinion in Biotechnology* 24(4): 690–8.

Li, Y., Liu, X., Liu, M., Wang, Y., Zou, Y., You, Y., Yang, L., Hu, J., Zhang, H., Zheng, X., Wang, P. and Zhang, Z. 2020. Magnaporthe oryzae auxiliary activity protein MoAa91 functions as chitin-binding protein to induce appressorium formation on artificial inductive surfaces and suppress plant immunity, *mBio* 11(2): e03304–19.

Lieberman-Aiden, E., van Berkum, N. L., Williams, L., Imakaev, M., Ragoczy, T., Telling, A., Amit, I., Lajoie, B. R., Sabo, P. J., Dorschner, M. O., Sandstrom, R., Bernstein, B., Bender, M. A., Groudine, M., Gnirke, A., Stamatoyannopoulos, J., Mirny, L. A., Lander, E. S. and Dekker, J. 2009. Comprehensive mapping of long-range interactions reveals folding principles of the human genome, *Science* 326(5950): 289–93.

Lin, F., Chen, S., Que, Z., Wang, L., Liu, X. and Pan, Q. (2007). The blast resistance gene Pi37 encodes a nucleotide binding site leucine-rich repeat protein and is a member of a resistance gene cluster on rice chromosome 1. *Genetics* 177(3): 1871–80.

Liu, H., Suresh, A., Willard, F. S., Siderovski, D. P., Lu, S. and Naqvi, N. I. 2007. Rgs1 regulates multiple Galpha subunits in Magnaporthe pathogenesis, asexual growth and thigmotropism, *EMBO Journal* 26(3): 690–700.

Liu, W., Liu, J., Ning, Y., Ding, B., Wang, X., Wang, Z. and Wang, G. L. 2013. Recent progress in understanding PAMP- and effector-triggered immunity against the rice blast fungus Magnaporthe oryzae, *Molecular Plant* 6(3): 605–20.

Liu, X., Lin, F., Wang, L. and Pan, Q. (2007). The in silico map-based cloning of Pi36, a rice coiled-coil nucleotide-binding site leucine-rich repeat gene that confers race-specific resistance to the blast fungus. *Genetics* 176(4): 2541-9.

Liu, Y., Liu, B., Zhu, X., Yang, J., Bordeos, A., Wang, G., Leach, J. E. and Leung, H. (2013). Fine-mapping and molecular marker development for Pi56(t), a NBS-LRR gene conferring broad-spectrum resistance to Magnaporthe oryzae in rice. *Theoretical and Applied Genetics* 126(4): 985–98.

Luck, J., Spackman, M., Freeman, A., Tre, P., Griffiths, W., Finlay, K. and Chakraborty, S. 2011 Climate change and diseases of food crops, *Plant Pathology* 60(1): 113-21.

Lv, Q., Xu, X., Shang, J., Jiang, G., Pang, Z., Zhou, Z., Wang, J., Liu, Y., Li, T., Li, X., Xu, J., Cheng, Z., Zhao, X., Li, S. and Zhu, L. (2013). Functional Analysis of Pid3-A4, an Ortholog of Rice Blast Resistance Gene Pid3 Revealed by Allele Mining in Common Wild Rice. *Phytopathology®* 103(6): 594–9.

Ma, J., Lei, C., Xu, X., Hao, K., Wang, J., Cheng, Z., Ma, X., Ma, J., Zhou, K., Zhang, X., Guo, X., Wu, F., Lin, Q., Wang, C., Zhai, H., Wang, H. and Wan, J. (2015). Pi64, Encoding a Novel CC-NBS-LRR Protein, Confers Resistance to Leaf and Neck Blast in Rice. *Molecular Plant-Microbe Interactions®* 28(5): 558–68.

Maciel, J. L. N., Ceresini, P. C., Castroagudin, V. L., Zala, M., Kema, G. H. J. and McDonald, B. A. 2013. Population structure and pathotype diversity of the wheat Blast Pathogen Magnaporthe oryzae 25 years After its emergence in brazil, *Phytopathology®* 104: 95–107.

Maekawa, T., Cheng, W., Spiridon, L. N., Toller, A., Lukasik, E., Saijo, Y., Liu, P., Shen, Q. H., Micluta, M. A., Somssich, I. E., Takken, F. L. W., Petrescu, A. J., Chai, J. and Schulze-Lefert, P. 2011. Coiled-coil domain-dependent homodimerization of intracellular barley immune receptors defines a minimal functional module for triggering cell death, *Cell Host and Microbe* 9(3): 187–99.

Maia, N. 1967. Obtention des bles tendres resistants Au pietin-verse par croisements interspecifiques bles x Aegilops, *Agriculturists Fr.* 53: 149–54.

Maqbool, A., Saitoh, H., Franceschetti, M., Stevenson, C. E., Uemura, A., Kanzaki, H., Kamoun, S., Terauchi, R. and Banfield, M. J. 2015. Structural basis of pathogen recognition by an integrated HMA domain in a plant NLR immune receptor, *eLife* 4: e08709.

Matthews, K. M., Pitman, E. L. and de Koning-Ward, T. F. 2019. Illuminating how malaria parasites export proteins into host erythrocytes, *Cellular Microbiology* 21(4): e13009.

Mazier, M., Flamain, F., Nicolaï, M., Sarnette, V. and Caranta, C. 2011. Knock-down of both eIF4E1 and eIF4E2 genes confers broad-spectrum resistance against potyviruses in tomato, *PLoS ONE* 6(12): e29595.

McDonald, B. A. and Linde, C. 2002. Pathogen population genetics, evolutionary potential, and durable resistance, *Annual Review of Phytopathology* 40: 349–79.

Mentlak, T. A., Kombrink, A., Shinya, T., Ryder, L. S., Otomo, I., Saitoh, H., Terauchi, R., Nishizawa, Y., Shibuya, N., Thomma, B. P. and Talbot, N. J. 2012. Effector-mediated suppression of chitin-triggered immunity by Magnaporthe oryzae is necessary for rice blast disease, *Plant Cell* 24(1): 322–35.

Mitchell, T. K. and Dean, R. A. 1995. The cAMP-dependent protein kinase catalytic subunit is required for appressorium formation and pathogenesis by the rice blast pathogen Magnaporthe grisea, *The Plant Cell* 7(11): 1869–78.

Mosquera, G., Giraldo, M. C., Khang, C. H., Coughlan, S. and Valenta, B. 2009. Interaction transcriptome analysis identifies Magnaporthe oryzae BAS1-4 as biotrophy-associated secreted proteins in rice Blast Disease, *Plant Cell* 21(4): 1273–90.

Mottaleb, K. A., Govindan, V., Singh, P. K., Sonder, K., He, X., Singh, R. P., Joshi, A. K., Barma, N. C. D., Kruseman, G. and Erenstein, O. 2019. Economic benefits of blast-resistant biofortified wheat in Bangladesh: the case of Bari Gom 33, *Crop Protection* 123: 45–58.

Mutiga, S. K., Rotich, F., Ganeshan, V. D., Mwongera, D. T., Mgonja, E. M., Were, V. M., Harvey, J. W., Zhou, B., Wasilwa, L., Feng, C., Ouedraogo, I., Wang, G. L., Mitchell, T. K., Talbot, N. J. and Correll, J. C. 2017. Assessment of the virulence spectrum and its association with genetic diversity in Magnaporthe oryzae populations from sub-Saharan Africa, *Phytopathology* 107(7): 852–63.

Nekrasov, V., Wang, Congmao, Win, J., Lanz, C., Weigel, D. and Kamoun, S. 2017. Rapid generation of a transgene-free powdery mildew resistant tomato by genome deletion, *Scientific Reports* 7(1): 482.

Newton, A. C. and Young, I. M. 1996. Temporary partial breakdown of Mlo-resistance in spring barley by the sudden relief of soil water stress, *Plant Pathology* 45(5): 973–7.

Ngou, B. P. M., Ahn, H. K., Ding, Pingtao and Jones, J. D. G. 2020. Mutual potentiation of plant immunity by cell-surface and intracellular receptors, *bioRxiv*. https://doi.org/10.1101/2020.04.10.034173.

Nguyen, T. T. T., Koizumi, S., La, T. N., Zenbayashi, K. S., Ashizawa, T., Yasuda, N., Imazaki, I. and Miyasaka, A. 2006. Pi35(t), a new gene conferring partial resistance to leaf blast in the rice cultivar Hokkai 188, *TAG. Theoretical and Applied Genetics. Theoretische und Angewandte Genetik* 113(4): 697–704.

Okuyama, Y., Kanzaki, H., Abe, A., Yoshida, K., Tamiru, M., Saitoh, H., Fujibe, T., Matsumura, H., Shenton, M., Galam, D. C., Undan, J., Ito, A., Sone, T. and Terauchi, R. (2011). A multifaceted genomics approach allows the isolation of the rice Pia-blast resistance gene consisting of two adjacent NBS-LRR protein genes. *The Plant Journal* 66(3): 467–79.

Orbach, M. J., Farrall, L., Sweigard, J. A., Chumley, F. G. and Valent, B. 2000. A telomeric avirulence gene determines efficacy for the rice blast resistance gene Pi-ta, *Plant Cell* 12(11): 2019–32.

Oses-Ruiz, M., Sakulkoo, W., Littlejohn, G. R., Martin-Urdiroz, M. and Talbot, N. J. 2017. Two independent S-phase checkpoints regulate appressorium-mediated plant infection by the rice blast fungus Magnaporthe oryzae, *Proceedings of the National Academy of Sciences of the United States of America* 114(2): E237–44.

Ou, S. H. 1980. Pathogen variability and host resistance in rice blast disease, *Annual Review of Phytopathology* 18(1): 167–87.

Park, C. H., Chen, S., Shirsekar, G., Zhou, B., Khang, C. H., Songkumarn, P., Afzal, A. J., Ning, Y., Wang, R., Bellizzi, M., Valent, B. and Wang, G. L. 2012. The Magnaporthe oryzae effector AvrPiz-t targets the RING E3 ubiquitin ligase APIP6 to suppress pathogen-associated molecular pattern-triggered immunity in rice, *Plant Cell* 24(11): 4748–62.

Park, C. H., Shirsekar, G., Bellizzi, M., Chen, S., Songkumarn, P., Xie, X., Shi, X., Ning, Y., Zhou, B., Suttiviriya, P., Wang, M., Umemura, K. and Wang, G. L. 2016. The E3 ligase APIP10 connects the effector AvrPiz-t to the NLR receptor Piz-t in rice, *PLOS Pathogens* 12(3): e1005529.

Parker, D., Beckmann, M., Zubair, H., Enot, D. P., Caracuel-Rios, Z., Overy, D. P., Snowdon, S., Talbot, N. J. and Draper, J. 2009. Metabolomic analysis reveals a common pattern of metabolic re-programming during invasion of three host plant species by Magnaporthe grisea, *The Plant Journal: For Cell and Molecular Biology* 59(5): 723–37.

Pavan, S., Jacobsen, E., Visser, R. G. F. and Bai, Y. 2009. Loss of susceptibility as a novel breeding strategy for durable and broad-spectrum resistance, *Molecular Breeding: New Strategies in Plant Improvement* 25(1): 1–12.

Penn, T. J., Wood, M. E., Soanes, D. M., Csukai, M., Corran, A. J. and Talbot, N. J. 2015. Protein kinase C is essential for viability of the rice blast fungus Magnaporthe oryzae, *Molecular Microbiology* 98(3): 403–19.

Putnam, N. H., O'Connell, B. L., Stites, J. C., Rice, B. J., Blanchette, M., Calef, R., Troll, C. J., Fields, A., Hartley, P. D., Sugnet, C. W., Haussler, D., Rokhsar, D. S. and Green, R. E. 2016. Chromosome-scale shotgun assembly using an in vitro method for long-range linkage, *Genome Research* 26(3): 342–50.

Qi, D. and Innes, R. W. 2013. Recent advances in plant NLR structure, function, localization, and signaling, *Frontiers in Immunology* 4: 348.

Qu, S., Liu, G., Zhou, B., Bellizzi, M., Zeng, L., Dai, L., Han, B. and Wang, G.-L. (2006). The broad-spectrum blast resistance gene Pi9 encodes a nucleotide-binding site-leucine-rich repeat protein and is a member of a multigene family in rice. Genetics 172(3): 1901–14.

Ray, S., Singh, P. K., Gupta, D. K., Mahato, A. K., Sarkar, C., Rathour, R., Singh, N. K. and Sharma, T. R. 2016. Analysis of Magnaporthe oryzae genome reveals a fungal effector, which is able to induce resistance response in transgenic rice line containing resistance gene, Pi54, *Frontiers in Plant Science* 7: 1140.

Rutter, B. D. and Innes, R. W. 2017. Extracellular vesicles isolated from the leaf apoplast carry stress-response proteins, *Plant Physiology* 173(1): 728–41.

Ryder, L. S., Dagdas, Y. F., Kershaw, M. J., Venkataraman, C., Madzvamuse, A., Yan, X., Cruz-Mireles, N., Soanes, D. M., Oses-Ruiz, M., Styles, V., Sklenar, J., Menke, F. L. H. and Talbot, N. J. 2019. A sensor kinase controls turgor-driven plant infection by the rice blast fungus, *Nature* 574(7778): 423–7.

Ryder, L. S. and Talbot, N. J. 2015. Regulation of appressorium development in pathogenic fungi, *Current Opinion in Plant Biology* 26: 8–13.

Ryder, L. S., Dagdas, Y. F., Mentlak, T. A., Kershaw, M. J., Thornton, C. R., Schuster, M., Chen, J., Wang, Z. and Talbot, N. J. 2013. NADPH oxidases regulate septin-mediated cytoskeletal remodeling during plant infection by the rice blast fungus, *Proceedings of the National Academy of Sciences of the United States of America* 110(8): 3179–84.

Saitoh, H., Fujisawa, S., Mitsuoka, C., Ito, A., Hirabuchi, A., Ikeda, K., Irieda, H., Yoshino, K., Yoshida, K., Matsumura, H., Tosa, Y., Win, J., Kamoun, S., Takano, Y. and Terauchi, R. 2012. Large-scale gene disruption in Magnaporthe oryzae identifies MC69, a secreted protein required for infection by monocot and dicot fungal pathogens, *PLOS Pathogens* 8(5): e1002711.

Sanchez-Vallet, A., Mesters, J. R. and Thomma, B. P. 2015. The battle for chitin recognition in plant-microbe interactions, *FEMS Microbiology Reviews* 39(2): 171–83.

Sarris, P. F., Cevik, V., Dagdas, G., Jones, J. D. and Krasileva, K. V. 2016. Comparative analysis of plant immune receptor architectures uncovers host proteins likely targeted by pathogens, *BMC Biology* 14: 8.

Saunders, D. G. O., Aves, S. J. and Talbot, N. J. 2010a Cell cycle-mediated regulation of plant infection by the rice blast fungus, *The Plant Cell* 22(2): 497–507.

Saunders, D. G. O., Dagdas, Y. F. and Talbot, N. J. 2010b. Spatial uncoupling of mitosis and cytokinesis during appressorium-mediated plant infection by the rice Blast Fungus Magnaporthe oryzae, *The Plant Cell* 22(7): 2417–28.

Schie, C. C. N. van and Takken, F. L. W. 2014. Susceptibility genes 101: how to be a good host, *Annual Review of Phytopathology* 52: 551–81.

Shan, Q., Wang, Y., Li, J., Zhang, Y., Chen, K., Liang, Z., Zhang, K., Liu, J., Xi, J. J., Qiu, J.-L. and Gao, C. 2013. Targeted genome modification of crop plants using a CRISPR-Cas system, *Nature Biotechnology* 31(8): 686–8.

Shang, J., Tao, Y., Chen, X., Zou, Y., Lei, C., Wang, J., Li, X., Zhao, X., Zhang, M., Lu, Z., Xu, J., Cheng, Z. Wan, J. and Zhu, L. (2009). Identification of a new rice blast resistance gene, Pid3, by genomewide comparison of paired nucleotide-binding site-leucine-rich repeat genes and their pseudogene alleles between the two sequenced rice genomes. *Genetics* 182(4): 1303–11.

Sharma, T. R., Madhav, M. S., Singh, B. K., Shanker, P., Jana, T. K., Dalal, V., Pandit, A., Singh, A., Gaikwad, K., Upreti, H. C. and Singh, N. K. 2005. High-resolution mapping, cloning and molecular characterization of the Pi-kh gene of rice, which confers resistance to Magnaporthe grisea, *Molecular Genetics and Genomics* 274(6): 569–78.

Sharma, T. R., Rai, A. K., Gupta, S. K. and Singh, N. K. 2010. Broadspectrum blast resistance gene Pi-kh cloned from rice line Tetep designated as Pi54, *Journal of Plant Biochemistry and Biotechnology* 19(1): 87–9.

Shi, X., Long, Y., He, F., Zhang, C., Wang, R., Zhang, T., Wu, W., Hao, Z., Wang, Y., Wang, G. L. and Ning, Y. 2018. The fungal pathogen Magnaporthe oryzae suppresses innate immunity by modulating a host potassium channel, *PLOS Pathogens* 14(1): e1006878.

Shimizu, T., Nakano, T., Takamizawa, D., Desaki, Y., Ishii-Minami, N., Nishizawa, Y., Minami, E., Okada, K., Yamane, H., Kaku, H. and Shibuya, N. 2010. Two LysM receptor molecules, CEBiP and OsCERK1, cooperatively regulate chitin elicitor signaling in rice, *Plant Journal: For Cell and Molecular Biology* 64(2): 204–14.

Singh, R., Dangol, S., Chen, Y., Choi, J., Cho, Y. S., Lee, J. E., Choi, M. O. and Jwa, N. S. 2016. Magnaporthe oryzae effector AVR-pii helps to establish compatibility by inhibition of the rice NADP-malic enzyme resulting in disruption of oxidative burst and host innate immunity, *Molecules and Cells* 39(5): 426–38.

Skamnioti, P. and Gurr, S. J. 2009. Against the grain: safeguarding rice from rice blast disease, *Trends in Biotechnology* 27(3): 141–50.

Stergiopoulos, I., van den Burg, H. A., Okmen, B., Beenen, H. G., van Liere, S., Kema, G. H. and de Wit, P. J. 2010. Tomato cf resistance proteins mediate recognition of cognate homologous effectors from fungi pathogenic on dicots and monocots, *Proceedings of the National Academy of Sciences of the United States of America* 107(16): 7610–5.

Su, J., Wang, W., Han, J., Chen, S., Wang, C., Zeng, L., Feng, A., Yang, J., Zhou, B. and Zhu, X. (2015). Functional divergence of duplicated genes results in a novel blast resistance gene Pi50 at the Pi2/9 locus. *Theoretical and Applied Genetics* 128(11): 2213–25.

Tagle, A. G., Chuma, I. and Tosa, Y. 2015. Rmg7, a new gene for resistance to Triticum isolates of Pyricularia oryzae identified in tetraploid wheat, *Phytopathology* 105(4): 495–9.

Takabayashi, N., Tosa, Y., Oh, H. S. and Mayama, S. 2002. A gene-for-gene relationship underlying the species-specific parasitism of Avena/Triticum isolates of Magnaporthe grisea on wheat cultivars, *Phytopathology* 92(11): 1182–8.

Takagi, H., Uemura, A., Yaegashi, H., Tamiru, M., Abe, A., Mitsuoka, C., Utsushi, H., Natsume, S., Kanzaki, H., Matsumura, H., Saitoh, H., Yoshida, K., Cano, L. M., Kamoun, S. and Terauchi, R. (2013). MutMap-Gap: whole-genome resequencing of mutant F2 progeny bulk combined with de novo assembly of gap regions identifies the rice blast resistance gene Pii. *New Phytologist* 200(1): 276–83.

Takahashi, A., Hayashi, N., Miyao, A. and Hirochika, H. (2010). Unique features of the rice blast resistance Pish locus revealed by large scale retrotransposon-tagging. *BMC Plant Biol* 10(175): 1–14.

Takken, F. L. W. and Goverse, A. 2012. How to build a pathogen detector: structural basis of NB-LRR function, *Current Opinion in Plant Biology* 15(4): 375–84.

Talbot, N. J. 1995. Having a blast: exploring the pathogenicity of Magnaporthe grisea, *Trends in Microbiology* 3(1): 9–16.

Talbot, N. J. 2003. On the trail of a cereal killer: exploring the biology of Magnaporthe grisea, *Annual Review of Microbiology* 57: 177–202.

Tosa, Y., Osue, J., Eto, Y., Oh, H. S., Nakayashiki, H., Mayama, S. and Leong, S. A. 2005. Evolution of an avirulence gene, AVR1-CO39, concomitant with the evolution and differentiation of Magnaporthe oryzae, *Molecular Plant-Microbe Interactions: MPMI* 18(11): 1148–60.

Tosa, Y., Tamba, H., Tanaka, K. and Mayama, S. 2006. Genetic analysis of host species specificity of Magnaporthe oryzae isolates from rice and wheat, *Phytopathology* 96(5): 480–4.

Upadhyay, S. K., Kumar, J., Alok, A. and Tuli, R. 2013. RNA-guided genome editing for target gene mutations in wheat, *G3* 3(12): 2233–8.

Valent, B. and Chumley, F. G. 1991. Molecular genetic analysis of the rice blast fungus, Magnaporthe grisea, *Annual Review of Phytopathology* 29: 443–67.

Velu, G., Crespo Herrera, L., Guzman, C., Huerta, J., Payne, T. and Singh, R. P. 2019. Assessing genetic diversity to breed competitive biofortified wheat With enhanced grain Zn and fe concentrations, *Frontiers in Plant Science* 9: 1971–71.

Veneault-Fourrey, C., Barooah, M., Egan, M., Wakley, G. and Talbot, N. J. 2006. Autophagic fungal cell death is necessary for infection by the rice blast fungus, *Science* 312(5773): 580–3.

Vo, K. T. X., Lee, S.-K., Halane, M. K., Song, M. Y., Hoang, T. V., Kim, C.-Y., Park, S.-Y., Jeon, J., Kim, S. T., Sohn, K. H. and Jeon, J.-S. 2019. Pi5 and pii paired NLRs are functionally exchangeable and confer similar disease resistance specificity, *Molecules and Cells* 42(9): 637–45.

Wang, B., Ebbole, D. J. and Wang, Z. 2017a. The arms race between Magnaporthe oryzae and rice: diversity and interaction of Avr and R genes, *Journal of Integrative Agriculture* 16(12): 2746–60.

Wang, M., Weiberg, A., Dellota, E., Jr., Yamane, D. and Jin, H. 2017b. Botrytis small RNA Bc-siR37 suppresses plant defense genes by cross-kingdom RNAi, *RNA Biology* 14(4): 421–8.

Wang, R., Ning, Y., Shi, X., He, F., Zhang, C., Fan, J., Jiang, N., Zhang, Y., Zhang, T., Hu, Y., Bellizzi, M. and Wang, G. L. 2016. Immunity to rice Blast Disease by suppression of effector-triggered necrosis, *Current Biology: CB* 26(18): 2399–411.

Wang, S., Boevink, P. C., Welsh, L., Zhang, R., Whisson, S. C. and Birch, P. R. J. 2017c. Delivery of cytoplasmic and apoplastic effectors from Phytophthora infestans haustoria by distinct secretion pathways, *New Phytologist* 216(1): 205–15.

Wang, Y., Cheng, X., Shan, Q., Zhang, Y., Liu, J., Gao, C. and Qiu, J.-L. 2014. Simultaneous editing of three homoeoalleles in hexaploid bread wheat confers heritable resistance to powdery mildew, *Nature Biotechnology* 32(9): 947–51.

Wang, Z. X., Yano, M., Yamanouchi, U., Iwamoto, M., Monna, L., Hayasaka, H., Katayose, Y. and Sasaki, T. (1999). The Pib gene for rice blast resistance belongs to the nucleotide binding and leucine-rich repeat class of plant disease resistance genes. *Plant J* 19(1): 55–64.

Weeks, D. P., Spalding, M. H. and Yang, B. 2016. Use of designer nucleases for targeted gene and genome editing in plants, *Plant Biotechnology Journal* 14(2): 483–95.

Welsh, L., Thorpe, P., Whisson, S. C., Boevink, P. C. and Birch, P. R. J. 2018. The phytophthora infestans haustorium is a site for secretion of diverse classes of infection-associated proteins, *mBio* 9: e01216-18.

Williams, S. J., Sohn, K. H., Wan, L., Bernoux, M., Sarris, P. F., Segonzac, C., Ve, T., Ma, Y., Saucet, S. B., Ericsson, D. J., Casey, L. W., Lonhienne, T., Winzor, D. J., Zhang, X., Coerdt, A., Parker, J. E., Dodds, P. N., Kobe, B. and Jones, J. D. 2014. Structural basis for assembly and function of a heterodimeric plant immune receptor, *Science* 344(6181): 299–303.

Williamson, V. M., Thomas, V., Ferris, H. and Dubcovsky, J. 2013. An Aegilops ventricosa translocation confers resistance against root-knot nematodes to common wheat, *Crop Science* 53(4): 1412–18.

Wilson, R. A. and Talbot, N. J. 2009. Under pressure: investigating the biology of plant infection by Magnaporthe oryzae, *Nature Reviews. Microbiology* 7(3): 185–95.

Wu, C. H., Abd-El-Haliem, A., Bozkurt, T. O., Belhaj, K., Terauchi, R., Vossen, J. H. and Kamoun, S. 2017. NLR network mediates immunity to diverse plant pathogens, *Proceedings of the National Academy of Sciences of the United States of America*, 114: 8113–18.

Wu, C. H., Krasileva, K. V., Banfield, M. J., Terauchi, R. and Kamoun, S. 2015a. The "sensor domains" of plant NLR proteins: more than decoys?, *Frontiers in Plant Science* 6: 134.

Wu, J., Kou, Y., Bao, J., Li, Y., Tang, M., Zhu, X., Ponaya, A., Xiao, G., Li, J., Li, C., Song, M. Y., Cumagun, C. J., Deng, Q., Lu, G., Jeon, J. S., Naqvi, N. I. and Zhou, B. 2015b. Comparative genomics identifies the Magnaporthe oryzae avirulence effector AvrPi9 that triggers Pi9-mediated blast resistance in rice, *New Phytologist* 206(4): 1463–75.

Wu, J. L., Fan, Y. Y., Li, D. B., Zheng, K. L., Leung, H. and Zhuang, J. Y. 2005. Genetic control of rice blast resistance in the durably resistant cultivar Gumei 2 against multiple isolates, *TAG. Theoretical and Applied Genetics. Theoretische und Angewandte Genetik* 111(1): 50–6.

Xu, J.-R. 2000. MAP kinases in fungal pathogens, *Fungal Genetics and Biology: FG and B* 31(3): 137–52.

Xu, J. R. and Hamer, J. E. 1996. MAP kinase and cAMP signaling regulate infection structure formation and pathogenic growth in the rice blast fungus Magnaporthe grisea, *Genes and Development* 10(21): 2696–706.

Xu, X., Hayashi, N., Wang, C.-T., Fukuoka, S., Kawasaki, S., Takatsuji, H. and Jiang, C.-J. (2014). Rice blast resistance gene Pikahei-1(t), a member of a resistance gene cluster on chromosome 4, encodes a nucleotide-binding site and leucine-rich repeat protein. *Molecular Breeding* 34(2): 691–700.

Yoshida, K., Saitoh, H., Fujisawa, S., Kanzaki, H., Matsumura, H., Yoshida, K., Tosa, Y., Chuma, I., Takano, Y., Win, J., Kamoun, S. and Terauchia, R. 2009. Association genetics reveals three novel avirulence genes from the rice Blast Fungal Pathogen Magnaporthe oryzae, *Plant Cell* 21(5): 1573–91.

Zhai, C., Zhang, Y. Yao, N. Lin, F. Liu, Z. Dong, Z. Wang, L. and Pan, Q. (2014). Function and interaction of the coupled genes responsible for Pik-h encoded rice blast resistance. *PLoS ONE* 9(6): e98067.

Zhan, J., Thrall, P. H., Papaïx, J., Xie, L. and Burdon, J. J. 2015. Playing on a pathogen's weakness: using evolution to guide sustainable plant disease control strategies, *Annual Review of Phytopathology* 53: 19–43.

Zhan, S. W., Mayama, S. and Tosa, Y. 2008. Identification of two genes for resistance to Triticum isolates of Magnaporthe oryzae in wheat, *Genome* 51(3): 216–21.

Zhang, H., Zheng, X. and Zhang, Z. 2016. The Magnaporthe grisea species complex and plant pathogenesis, *Molecular Plant Pathology* 17(6): 796–804.

Zhang, S., Wang, L., Wu, W., He, L., Yang, X. and Pan, Q. 2015. Function and evolution of Magnaporthe oryzae avirulence gene AvrPib responding to the rice blast resistance gene PIB, *Scientific Reports* 5: 11642.

Zhang, Z. M., Zhang, X., Zhou, Z. R., Hu, H. Y., Liu, M., Zhou, B. and Zhou, J. 2013. Solution structure of the Magnaporthe oryzae avirulence protein AvrPiz-t, *Journal of Biomolecular NMR* 55(2): 219-23.

Zhao, H., Wang, X., Jia, Y., Minkenberg, B., Wheatley, M., Fan, J., Jia, M. H., Famoso, A., Edwards, J. D., Wamishe, Yeshi, Valent, B., Wang, G. L. and Yang, Y. 2018. The rice blast resistance gene Ptr encodes an atypical protein required for broad-spectrum disease resistance, *Nature Communications* 9(1): 2039.

Zhou, B., Qu, S. H., Liu, G. F., Dolan, M., Sakai, H., Lu, G. D., Bellizzi, M. and Wang, G. L. 2006. The eight amino-acid differences within three leucine-rich repeats between Pi2 and Piz-t resistance proteins determine the resistance specificity to Magnaporthe grisea, *Molecular Plant-Microbe Interactions: MPMI* 19(11): 1216–28.

Zhu, X., Chen, S., Yang, J., Zhou, S., Zeng, L., Han, J., Su, J., Wang, L. and Pan, Q. 2012. The identification of Pi50(t), a new member of the rice blast resistance Pi2/Pi9 multigene family, *TAG. Theoretical and Applied Genetics. Theoretische und Angewandte Genetik* 124(7): 1295–304.

Chapter 19

Ramularia leaf spot in barley

Neil Havis, Scotland's Rural College (SRUC), UK

1 Introduction

One of the major issues challenging barley (*Hordeum vulgare*) growers in many countries is the control of Ramularia leaf spot (RLS), caused by the fungal pathogen *Ramularia collo-cygni*. This small, seemingly inconsequential fungus has moved from the realms of minor status in cereal-growing countries in the early to mid-20th century to a major economic barley disease across many continents, with only limited options in terms of effective control. Within this story lie emerging challenges and questions which are still in the process of being investigated by researchers to provide effective advice to farmers and growers.

2 Diagnosis of plant disease

Although the early reports of *R. collo-cygni* focussed on descriptions of the fungal structures (Sutton and Waller, 1988) the fungus was reported to produce small, elongated brown lesions with pinkish-white sporulating hyphae on the leaf underside of Triticale (*Triticum secalum*) (Sutton and Waller, 1988). A detailed description of symptoms appearing in barley crops was provided by Huss et al. (1987). Symptoms of necrosis were recorded on the upper leaves of the barley crop. Salamati and Reitan (2006) summarised disease development

http://dx.doi.org/10.19103/AS.2021.0092.28

in the following way. The appearance of small paper spots (2.5 × 0.5mm) on the upper canopy which spread and become abundant on leaves, sheaths and awns. The spots were described as having a yellow chlorotic halo. A guide to identifying RLS was produced in 2010 (BASF, 2010). More recently the distinguishing characteristics of RLS have been collated into a published guide (AHDB, 2018). The guide recommends that growers use the 5R's to successfully diagnose the symptoms of the disease: (1) Ring of chlorosis around the lesion, (2) Rectangular shape, (3) Restricted by the leaf veins, (4) Reddish-brown colouration and (5) Right through the leaf.

Perhaps the major challenge facing growers across many countries is successfully identifying the disease Ramularia leaf spot in barley. Physiological stresses are known to cause the production of necrotic areas on barley leaves, for example, damage due to excessive light levels (Wu and von Tiedemann, 2002) and lack of manganese in the crop (TEAGASC, 2020). In the case of abiotic stresses, many nutritional deficiencies produce symptoms which appear throughout the development of the crop, but physiological leaf spots (PLS) tend to appear on the upper canopy at the same time as RLS (Wu and von Tiedemann, 2002). A major difference between PLS and RLS is that the browning and necrosis is generally only seen on the upper leaf surface and symptoms are not mirrored on the underside of the leaf.

Barley (*Hordeum vulgare*) as a host is susceptible to many biotic agents as well as *R. collo-cygni*. Several of them can cause symptoms which may be confused with RLS, and during initial studies a degree of confusion occurred. In particular, the early symptoms of *Pyrenophora teres* (Sachs et al., 1998) caused problems for researchers. The spot form of the fungus *P. teres* f. *maculata* can form small rectangular lesions, which can be mistaken for RLS, but they do not have the chlorotic halo of RLS and tend to elongate along the length of the leaf, unlike RLS (BASF, 2010; AHDB, 2018).

Diagnosis of the RLS by visual means has been supplemented by additional tests. Sachs incubated leaves on acidified agar and the resultant reddish colour indicated the presence of the fungus. (Sachs, 2006). This simple lab test was soon surpassed by the development of molecular diagnostics assays (Havis et al., 2006; Frei et al., 2007; Taylor et al., 2010). These assays allowed the confirmation of important parts of the fungal life cycle and advances in epidemiology studies. (Walters et al., 2008; Havis et al., 2015).

3 Global distribution of disease

The initial description of Ramularia leaf spot symptoms was on barley in northern Italy (Cavara, 1893). The next description of the fungus was not available for a further 32 years, when the fungus was identified in Chile, and there were only sporadic reports in other countries. In the middle of the 1980s the fungus was identified on barley crops in the Lambach region of Austria by Herbert Huss (Huss et al., 1987). Since this initial report the disease had spread rapidly in Central Europe before moving into all of the major barley-growing regions of Europe at the start of the 21st century (Walters et al., 2008; Havis et al., 2015) see insert box in Fig. 1.

Although the fungus was described in South America in the early part of the twentieth century (Sutton and Waller, 1988) it was only reported as a serious economic issue at the start of this century. The disease was reported in the Buenos Aries province of Argentina in 2001 (Khier et al., 2002). Since that initial report, RLS has also been reported in other parts of Argentina and Uruguay (Carmona et al., 2013; Pereyra, 2013; Havis et al., 2014a).

In North America, the fungus was identified in many plant species in the 1970s (Sutton and Waller, 1988), and has also now been identified in barley crops in Canada (Kelly Turkington, personal communication).

The disease was identified in New Zealand at the end of the last century (Harvey, 2002), around the same time as its first appearance in the United Kingdom. Disease epidemics have been common in the last twenty years (Soonie Chng, personal communication). In recent years, after an initial report in Tasmania RLS has been identified in other regions of Australia (Tasmanian Government, 2020), *R. collo-cygni* has also been detected in seed samples from Tanzania and Israel (Michael Hess, personal communication) and is known to be present in barley crops in South Africa (Beukes et al., 2016). Major epidemics continue to be reported in New Zealand and across Europe, with 2017 being a particularly devastating year in the UK. The current distribution reflects the investigations so far carried out. More widespread testing may reveal that the fungus is present in other South American, North-African and possibly Asian countries.

4 Life cycle of the fungus

The fungus is known to infect grass and other hosts and exhibits a non-specific life cycle, which makes control a challenge. Host range is summarised in Table 1.

Infection of plants with *R. collo-cygni* occurs in a few ways. The fungus has been shown to be seed borne (Havis et al., 2006; Frei et al., 2007), and in barley the fungus has been shown to grow internally within the barley plant (Frei et al., 2007; Havis et al., 2014b) without causing any symptoms. Although fungal sporulation has been reported on many of the species in Table 1, the characteristic Ramularia leaf spot symptoms have only been observed on barley.

Several studies have sought to determine the exact location of *R. collo-cygni*. Detailed dissection of barley seed by Matusinsky indicated the

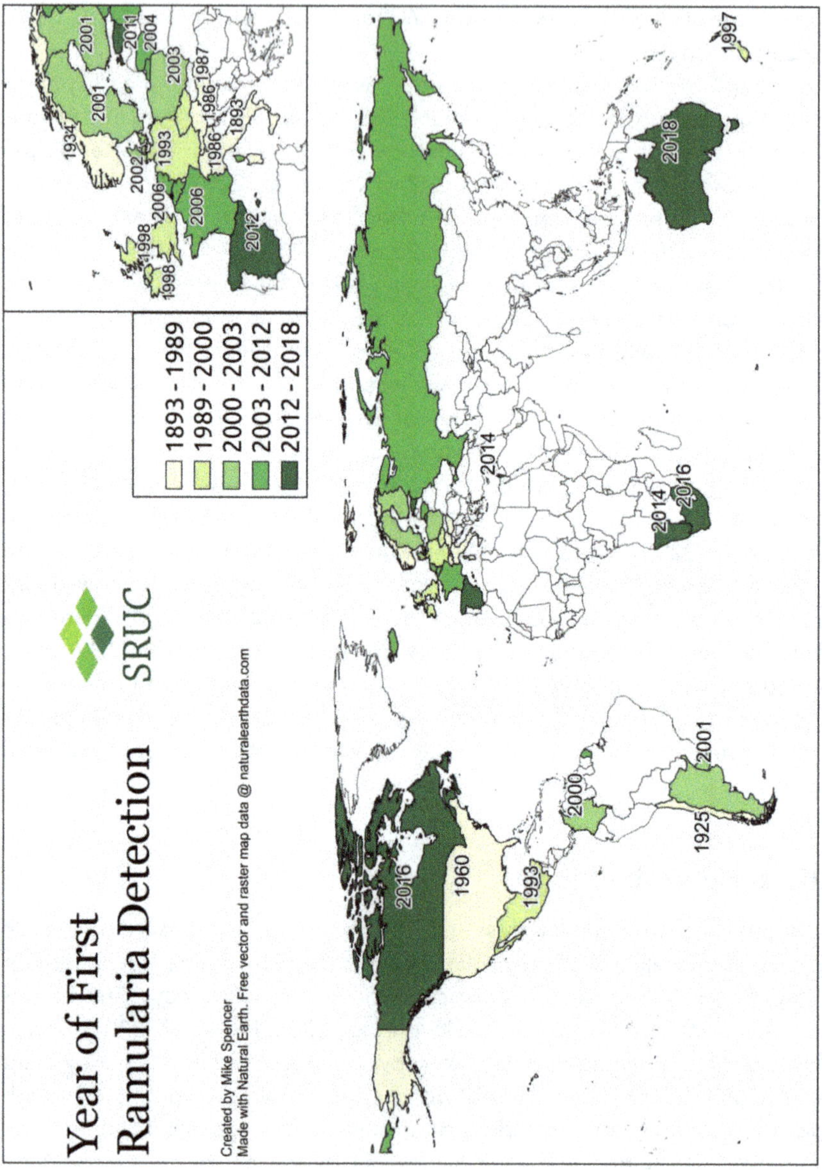

Figure 1 Global distribution of Ramularia leaf spot and the year of first report (Spencer et al., 2019).

Table 1 Reported host range for the fungus *Ramularia collo-cygni*

Host	Common name	Source
Hordeum vulgare	Barley	Cavara, 1893
Triticale secalum	Triticale	Sutton and Waller, 1988
Elytrygia repens	Scutch grass	Huss et al., 2004
Elymus caninus	Bearded couch grass	Huss et al., 2004
Echinochloa crus-galli	Annual grass	Huss, 2004
Avena sativa	Oat	Huss, 2004
Triticum aestivum	Wheat	Huss, 2004
Secale cereale	Rye	Huss, 2004
Quercus robur	Oak trees	Heuser and Zimmer, 2002
Triticum durum	Durum wheat	Frei and Gindrat, 2000
Poa pratensis	Common meadow grass	Frei and Gindrat, 2000
Lolium perenne	Perennial ryegrass	Frei and Gindrat, 2000
Agropyron repens	Couch grass	Frei and Gindrat, 2000
Brachypodium distachyon	Stiff brome grass	Peraldi et al., 2014
Hordeum murinum	Wall or false barley	Frei, 2004
Apera spica-venti	Silky bent grass	Frei, 2004
Agrostis spp	Bent grass species	Cromey et al., 2004
Bromus carthartICUS	Prairie grass	Cromey et al., 2004
Glyceria fluitans	Floating sweet grass	Cromey et al., 2004
Lollium multiflorum	Italian Ryegrass	Kaczmarek et al., 2016

presence of fungal DNA in the covering layers of the seed and the lemma, and in lower amounts in the pericarp and embryo (Matusinsky et al., 2011). In other studies, with higher fungal infection the fungus was identified in the endosperm (Havis et al., 2009a; Oxley and Havis, 2010; Havis et al., 2014a; Clemente et al., 2014; Hess et al., 2014). Indeed, later whole plant inoculation studies with a green fluorescing protein (GFP)-transformed isolate developed by Thirugnanasambandam et al. (2011) showed *R. collo-cygni* accumulation under the seed coat outside the aleurone layer and the GFP signal in all seed component parts, including the endosperm (Kaczmarek et al., 2013).

After the seed germinates and the plant develops the movement of the fungus has been tracked using a combination of microscopy and molecular diagnostics. The fungus has been shown to move in the plant asymptomatically (Havis et al., 2004; Salamati and Reitan, 2006). Huss reported that the fungus can survive as a saprophyte on senescing leaves in winter barley crops (Huss, 2004). The movement of the fungus can be tracked throughout the growing season as successive leaf layers emerge (Frei et al., 2007; Havis et al., 2014b; Kaczmarek et al., 2016). During the vegetative growth stage of the plant, the crop appears asymptomatic. The role of spore movement from senescent material or alternative hosts is still being studied and remains contentious. Although many studies have pointed to the importance of the seed-borne stage (Walters et al., 2008; Havis et al., 2015), a recent study in Estonia could not detect movement from seed (Mäe et al., 2018). The studies with the fluorescent isolates indicated that the fungus grows intercellularly, forming branched hyphae in the leaf mesophyll layer (Kaczmarek et al., 2013). A brick-like pattern is formed as hyphae surround the leaf cells but do not penetrate any of the cell walls. This study confirmed the initial growth pattern suggested by Sutton and Waller (1988).

Interestingly, there have been reports that in some southern hemisphere countries the pathogen displays symptoms earlier in the development of the crop, that is, pre-flowering (e.g. Harvey, 2002; Ignacio Erreguena, personal communication). The causes of this observation remain unclear.

After flowering in barley crops, disease symptoms begin to appear on the upper leaves in the canopy. The initial pepper spots formed on the leaf enlarge but remain bound by leaf veins. Detailed studies indicate that the hyphae always stop at leaf veins and make no attempt to penetrate a leaf vein (Kaczmarek et al., 2013). Neighbouring leaf spots eventually coalesce to form larger dark areas. As symptoms expand, the remainder of the leaf becomes chlorotic and then necrotic, usually starting at the leaf tip and margins (Huss, 2004). Fungal sporulation generally only occurs in necrotic tissue, with conidiophores emerging through stomata in caespituli of up to 15, from underlying stromata. The terminal part of each conidiophore is strongly curved (Sutton and Waller, 1988), with the apical region producing up to five conidia. *R. collo-cygni* grows preferably, although not exclusively, on abaxial leaf surfaces, and Huss (2004) estimates that a heavily infected leaf can produce up to 50 000 conidia.

The necrosis which appears in the leaf is due to the production of secondary metabolites by the fungus. Fungal mycelia were found to have a purple pigmentation, and Heiser et al. (2013) speculated that these could be a range of photoactive polycyclic aromatic toxins, similar to cercosporin, produced by several species of *Cercospora*. They identified some of the coloured metabolites, including an anthraquinoid, identified as rubellin D. When applied to barley leaves, rubellin D induced light- and concentration-dependent necrosis and, in a model system, was shown to exhibit photodynamic activity, triggering the light-dependent production of reactive oxygen species (ROS), and ultimately the peroxidation of α-linolenic acid (Heiser et al., 2003). Later work by Heiser et al. (2004) indicated that these toxins are not host specific. Several isomers of rubellin have been shown to be produced by the fungus and rapid conversion of types appears to take place during the expression of disease symptoms (Miethbauer et al., 2003; Heiser et al., 2004; Miethbauer et al., 2006). Rubellin B has been shown to be the dominant form found in leaves (Miethbauer et al., 2003). A role for rubellins in pathogenicity was suggested by Heiser et al. (2004).

However more recent work by Dussart et al. (2018a) suggested no correlation between susceptibility to rubellin D and disease susceptibility. At the same time research indicated a few gene clusters in the fungal genome relating to secondary metabolites (Dussart et al., 2018b). The interaction between the fungal secondary metabolism and the host remains under researched.

5 Environmental effects on disease expression

Understanding the interaction between the fungus, host and environment has presented researchers with significant challenges. The late season production of symptoms has challenged researchers attempting to define the factors which influence disease epidemics.

Case study 1 One of the early important studies was carried out by Huss et al. (2003) at the Lambach field station in Austria. In this study pots of two winter barley varieties (Virgo and Dido) were moved into a wind exposed polytunnel after first leaf emergence. One set of pots were placed into dry conditions overnight to remove the possibility of dew forming on the leaves. In the other set dew formed naturally. During the daytime, the plants were kept in identical conditions in a polytunnel. By the time the plants reached milk ripeness (approx. GS 75-80; Zadoks et al., 1974), the plants which has experienced dew on the leaves showed typical symptom formation and conidiophore formation on senescing leaves. The plants which had been kept dry stayed green and symptomless. This was the first experimental evidence of the importance of dew in symptom formation.

Studies following this identified relative humidity in crops at GS 31 as an indicator of RLS severity in Central Norway (Salamati and Reitan, 2006). Here, the authors investigated the links between weather conditions and disease severity. They found a correlation between long leaf wetness duration and days with precipitation at GS15-30 and final RLS severity. Havis et al. (2012) investigated the potential for leaf wetness for 14 days at GS30/31 to act as a risk warning for RLS in barley crops. Within season analysis of disease levels and meteorological data showed some correlation between leaf wetness and final disease levels, but analysis over multiple years showed that differences in disease levels could not be explained solely by one environmental parameter (Havis et al., 2018a).

Other environmental factors have been associated with the appearance of disease in barley crops. Heiser et al. (2003) reported on the unpublished work from E Sachs, which suggested light intensity was important in determining RLS symptom levels. However, subsequent work by Makepeace (2006) suggested light intensity prior to inoculation exerts a significant effect on RLS. Plants grown under low-light conditions before inoculation exhibited fewer RLS symptoms than those grown under high light. Increasing light intensity post inoculation led to fewer RLS symptoms, suggesting that although light is required for the toxins produced by *R. collo-cygni* (see Section 4) to exert their effects, excessively high light intensities might have a negative effect on pathogenicity (Makepeace, 2006).

Formayer et al. (2004) reported that high humidity levels were crucial for the outbreak of RLS epidemics in Austria, while radiation intensity was only of minor importance. Marik et al. (2011) found that stronger symptom expression was positively affected by a higher number of rainy days in the three weeks post heading. These authors also reported that higher temperatures and lower rainfall post flowering reduced disease levels in the Czech Republic. Havis et al. (2018a) found a correlation between RLS levels in spring barley and temperature and rainfall experienced by the crop from sowing up the ear emergence (GS59). However, this observation was not seen across all years of the study.

Understanding the impact of the environment and other biotic factors on RLS epidemics will require many more years of study.

6 Genetics of disease resistance

From the earliest reports of the appearance of RLS in barley crops it became apparent that the genetics of the host influenced the development of symptoms (Walters et al., 2008). The study from Harvey (2002) in New Zealand suggested that even low disease levels in the crop can exert a disproportionately negative effect on yield. Yield losses due the RLS were reported by both Huss et al. (1992) and Pinnschmidt and Hovmøller (2004), indicating a susceptibility of the plant to the disease. In Europe the average yield losses attributed to RLS in barley have been reported to be in the region of 0.4 t/ha (Hess et al., 2007; Walters et al., 2008). However varietal and regional effects have been reported on disease severity (Hess et al., 2011; Oxley and Havis, 2009). More recently epidemics in South American barley crops have produced yield losses as high as 70% in susceptible cultivars (Havis et al., 2015). In both Uruguay and Argentina, losses were associated with a significant reduction in grain size, with less than 10% of grains reaching 2.5mm in size (Pereyra, 2013; Clemente et al., 2014). In addition to reducing yield and grain size in crops, RLS has been implicated in a reduction in malt quality (Pinnschmidt and Jørgensen, 2009; Hess et al., 2011). Pinnschmidt and Jørgensen (2009) attributed the yield loss from RLS to a reduction in thousand grain weight in infected plots. The impact of the fungus on the plant during its long asymptomatic phase has recently garnered interest and a project was initiated in Scotland to examine the interaction between the fungus and photosynthesis in the plant prior heading. The interaction was examined using a combination of controlled conditions and field experiments (Burrell et al., 2020). The authors of this work observed no change in photosynthesis, measured as maximum quantum yield, in either system prior to symptom expression. They also noted that there was no difference in CO_2 uptake per unit leaf area in the same leaves and concluded that the yield response from *R. collo-cygni* was primarily related to the loss of photosynthetic area in the symptomatic phase (Burrell et al., 2020).

The effect of variety on RLS symptom severity has been studied for the last 20 years. Sheridan (2000) did one of the first studies in the Southern hemisphere to identify that differences in varietal susceptibility to RLS existed. The need for new varieties in the southern hemisphere was highlighted by Pereyra et al. (2014). The authors reported that of all the varieties used in Uruguay in 2013 the best options were still only moderately susceptible, and none were resistant.

Work in Denmark by Hans Pinnschmidt and colleagues on a large panel of spring and winter cultivars demonstrated a wide range of susceptibility but importantly also identified some moderately resistant lines (Pinnschmidt and Hovmøller, 2003, 2004; Pinnschmidt et al., 2006). This work suggested that varietal resistance could be utilised effectively to control RLS in Denmark. Improved resistance was also reported in Germany in the 6-row winter barley, Duet, by Hess et al. (2006). The authors also pointed out that the maturity of the varieties should be considered, when comparing disease scores.

This variation in varietal susceptibility has also been reported in a number of European countries, for example, Lithuania (Leistrumaitė and Liatukas, 2006), Norway (Reitan and Salamati, 2006), Slovakia (Gubiš et al., 2008) and also in the UK (Oxley et al., 2008). Significant differences in the RLS-susceptibility of winter barley cultivars was found in the Czech Republic (Mařík et al., 2011) but only limited variation between spring cultivars (Matušinsky et al., 2013). In all these cases, there was quantitative variation in resistance with no clear division between resistant and susceptible cultivars, suggesting that RLS-resistance is generally a quantitative trait.

An early project on breeding for RLS resistance identified a correlation between the presence of *mlo* mildew resistance and susceptibility of barley to RLS (Bistrich et al., 2006). The authors had examined a wide range of varieties, advanced breeding lines and Genebank accessions over several seasons and sites. The *Mlo* gene in barley is required to confer susceptibility to powdery mildew (*Blumeria graminis*); therefore non-functional (*mlo*) alleles confer recessively inherited resistance to this disease. Breeders have used the *mlo-11* allele especially, and *mlo-9* in spring barley breeding programmes across Europe (Jørgensen, 1992). Pinnschmidt et al. (2006) reported on trials of 75 barley cultivars in Denmark. Of the 13 most susceptible to RLS only one did not have *mlo* while all the 18 least susceptible lines lacked *mlo*. One of the difficulties of effectively trialling RLS in field experiments is the presence of other factors including disease. This was highlighted by an analysis of RLS scores from many naturally infected trials in Denmark. Here, two-thirds of variation between trial plots was found to be attributable to extraneous factors, including cultivars' susceptibility to other diseases. Only the remaining one-third of variation was related to cultivar susceptibility to RLS. However, removing the other factors revealed a strong association between the presence of *mlo* and high susceptibility to RLS. A series of inoculated trials in Denmark showed the same association (Pinnschmidt and Sindberg, 2009).

Several studies were carried out to investigate the relationship between *Mlo*. These experiments utilised seed lines, which varied in alleles of the *Mlo* gene. The results largely supported the conclusion of the field trials undertaken in Denmark but also raised other questions about the interaction between *mlo* mildew resistance and RLS. Makepeace et al. (2007) used near isogenic lines (NILs) to breed different *mlo* alleles, replacing *Mlo*$^+$ wild type allele, into a susceptible background. In field trials with low RLS disease levels *mlo* alleles were associated with a moderate decrease in RLS. These results contradicted the results of the Danish trials.

Case study 2 A separate experiment by Makepeace et al. (2008) had indicated that high light levels before inoculation could increase symptom formation in controlled conditions. In order to investigate the influence of light and *mlo* on RLS susceptibility, a further set of experiments was conducted on NILs with either *Mlo*$^+$ (mildew-susceptible wild-type) or *mlo-5* (mildew-resistant) in controlled environment chambers. When light levels were increased before inoculation RLS symptoms on both wild and modified *mlo* lines were increased but the disease levels were significantly greater in the *mlo* lines. Conversely lowering light levels before inoculation lowered RLDS symptoms but there were no significant differences between *Mlo*$^+$ and *mlo-5* lines (Brown and Makepeace, 2009). McGrann repeated this work and found that *R. collo-cygni* DNA levels were much higher in the *mlo-5* lines in field trials and controlled environment experiments (McGrann et al., 2014a).

The Coracle project was established in the UK to investigate the control of Ramularia in a changing climate. The final project report is available online (AHDB, 2016). This project examined in more detail the effect of *mlo* alleles on RLS by carrying out a series of field trials using a double haploid combination produced from a cross between a highly RLS susceptible *mlo-11* cultivar (Braemar) and a highly RLS susceptible *Mlo*$^+$ cultivar (Power). The field trials were carried out in a few countries to ameliorate the environmental effect on RLS development. The results of these trials were reported by McGrann et al. (2014a).

The authors reported that *mlo-11* was indeed associated with increased susceptibility to RLS but that this effect was much stronger in Bavarian trials rather than in other trial sites or controlled environment experiments. The authors also carried out a quantitative trial locus (QTL) analysis on disease expression in the lines and found the only significant QTL was the *Mlo* locus. At the site with the greatest differences between lines the QTL accounted for up to 37% of the genetic variation (McGrann et al., 2014a).

Interestingly, Hofer et al. (2015) reported no increase in *R. collo-cygni* DNA in grain from *mlo-5* varieties grown in Bavarian field trials over 4 seasons. The authors recommended a re-evaluation of the effect of *mlo* on the incidence and severity of necrotrophic pathogens.

However, the overall conclusion from the work on *mlo* and RLS expression in barley is that the mildew-resistance allele is very likely to contribute to the susceptibility of spring barley cultivars to RLS, but the strength of this effect varies with the environment, location and genetic background (Havis et al., 2015). This conclusion is based on the reports of significant variation in RLS-susceptibility between cultivars (Pinnschmidt et al., 2006) and other lines (McGrann et al., 2014a) with *mlo*. The future management of *mlo* in spring barley breeding programmes remains a challenge. Further work by McGrann et al. (2014a) showed that the use of *ror* mutants, which partially restore mildew susceptibility in *mlo* plants, could alleviate disease symptoms without affecting *R. collo-cygni* DNA levels in leaves. This suggests that resistance to fungal infection and colonisation and symptom formation may to some extent be under separate genetic control. In order to elucidate the role of plant genetic factors on disease expression. McGrann et al. (2014b) carried out a series of experiments with the stress responsive transcription factor (*SNAC1*). The NAC proteins are involved in the plant response to biotic and abiotic stress. In lines of the RLS and mildew susceptible variety, Golden Promise, over expression of this factor reduced RLS symptoms but had no effect on other disease levels. The transcription factor was found to have no effect on either stomatal conductance or reactive oxygen species induced cell death but was found to delay dark induced senescence (McGrann et al., 2014b). In an additional set of experiments the role of active oxygen species in the transition of the fungus from endophyte to necrotroph was investigated (McGrann and Brown, 2018). The authors reported that abiotic stress in seedling leaves prior to inoculation increased RLS symptoms in all susceptible and most partially resistant lines. This was attributed to accumulation of hydrogen peroxide in the leaves. This suggests that senescence in the leaf is not as important as changes in the plant reactive oxygen species.

The production of new tolerant/resistant varieties remains a challenge. One approach may be to use diverse germplasm as sources of partial resistance and select for lines with improved RLS, which still maintain broad spectrum resistance to other major economic pathogens and other desirable traits (Havis et al., 2015).

7 Pathogen variability

A major advance in our understanding of the host-fungus interactions has been the sequencing of the genome of the fungus and its evaluation alongside genetic information from the host. The first full *R. collo-cygni* genome sequence was published by McGrann et al. (2016). This has been followed by subsequent publication of new sequences (Stam et al., 2018). These publications gave the first insight into the genetic makeup of the fungus and offered some guidance into the likely interaction between fungus and host, for example, a limited number of genes coding for cell wall degrading enzymes but many genes associated with secondary metabolism and toxin production (McGrann et al., 2016). The publication of further sequence data from several isolates from barley and other hosts offered a glimpse into the evolution and global movement of the fungus (Stam et al., 2019). The authors noted that *R. collo-cygni* had diverged significantly from other dothidiomycte fungi and that the current global population was relatively uniform with no evidence of geographical clustering or host specialisation (Stam et al., 2019). They hypothesised that the global movement of the pathogen (described in Fig. 1) was due to manmade movement of seed.

Case study 3 A recent project has examined for the first time the interaction between the host and the fungal transcriptome during the infection process (Sjokvist et al., 2018). A susceptible barley variety was inoculated with an aggressive isolate of *R. collo-cygni*. The infection of the host and subsequent development of symptoms were related to changes in the transcriptome of both the fungus and the host. The infection was carried out using a mycelial suspension. After 3 days fungal growth was observed on the leaf surface and the barley cells were alive (Fig. 1). After 7 days the stomata were colonised, fungal growth was observed in mesophyll layer and the barley cells were alive (Fig. 2). After 12 days the fungal growth was extensive inside the leaf and the barley cells were chlorotic and collapsing (Fig. 3).

A few changes were observed in the plant and fungal transcriptome over the course of the experiment. In the fungus after only 3 days nearly 70% of the fungal genes were expressed early on. In particular, pathogenicity factors including genes coding for secreted effectors were detected. One of these potential effectors was a chitin-binding LysM domain protein. In response the host upregulated many of the classical chitin elicitor receptor kinases. As the fungus upregulated cell-wall degrading enzymes, the plant responded by upregulating chitinases and glucan-endo 1,3 β-glucosidases. The upregulation of pathogenesis-related proteins in barley suggests that the plant is detecting the pathogenesis-associated molecular patterns of *R. collo-cygni*. The induction of secondary metabolite transport in the plant also suggests host defence is based the production and secretion of plant

Figure 2 *Ramularia collo-cygni* lifecycle in barley.

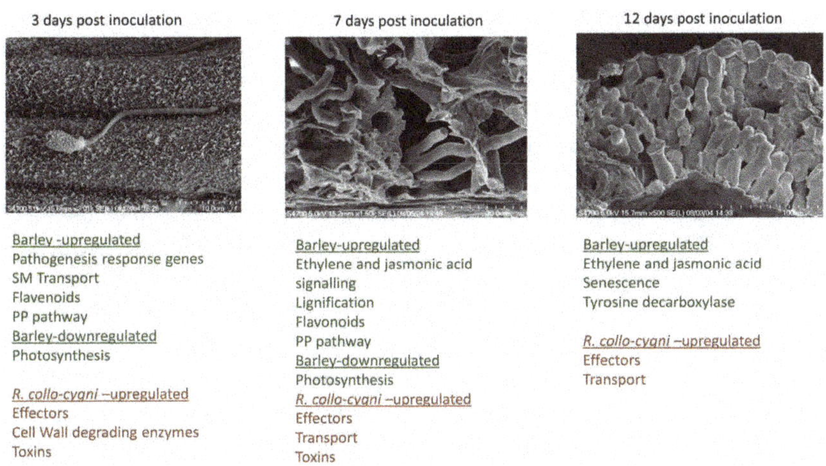

Figure 3 A pictorial representation of the infection process of barley by *R. collo-cygni* and the associated main molecular components identified to be induced or downregulated. SM = secondary metabolites, PP = phenylpropanoid.

phytotoxic compounds (Sjokvist et al., 2018). Photosynthesis in the host was strongly down regulated after 3 days with genes for light induced proteins affected. Photosynthesis was more subtly influenced through the rest of the colonisation experiment. After 7 days, with apoplastic infection taking place, the fungus reduced production of cell-wall degrading enzymes and glycosyl hydrolases. This could be a response to inhibitors, for example xylanase inhibitors, produced by the host.

The host also appears to be increasing the production of enzymes involved in lignifying the cell walls and boosts the production of flavonoids which are acting as phytoalexins. The colonisation of the apoplast appears to induce water surface tension on the hyphae and increase stress on the host plant. The strong increase in expression of genes involved in the uptake and metabolism of hexoses after 7days suggests that the fungus is deriving nutrition from the apoplast. In response to this the plant seems to be activating many of the calcium signalling pathways. The variety in the experiment had the wild type *Mlo* gene. It was found to be upregulated significantly after 7 days of post inoculation. The authors hypothesised that *Mlo* could mediate defence against *R. collo-cygni* but that the fungus produces effectors to supress *Mlo*. Plant defence responses are generally controlled by either the jasmonic/ethylene pathways or the salicyclic pathways, depending on the fungal lifestyle. The jasmonic/ethylene pathway is associated with necrotrophic pathogens and over the course of the experiment these pathways were observed to be upregulated. The salicylic acid pathway is associated with biotrophic pathogens and no changes in gene activity were recorded in this pathosystem.

After 12 days necrosis was appearing on the leaves and the apoplast was extensively colonised. The genes proposed to be involved in rubellin production in the fungus were not found to be upregulated after 12 days, but other genes involved in secondary metabolism were found to be upregulated (Sjokvist et al., 2018). The *R. collo-cygni* transcriptome appeared to be moving towards producing acidic conditions *in planta* which could facilitate toxin production. Genes encoding for senescence factors in the infected leaves were upregulated, suggesting that the natural senescence process is underway; however, the pathways involved in remobilisation of nutrients in the plant were not activated. The authors proposed that nutrients were being relocated into the fungus and contributed to lesion formation (Sjokvist et al., 2018). This interaction has been observed in the *Zymoseptoria tritici*/wheat pathosystem (Ma et al., 2018). This project is due shortly to reveal more details of changes within the transcriptome of the barley host and the fungus. The information will be ideally placed to identify the key molecular factors in the barley/*R. collo-cygni* interaction and will help in designing not only future breeding programmes but also potentially new control measures for the fungus (S. Radutiou, personal communication).

8 Control of Ramularia leaf spot by fungicides

Given the lack of effective genetic control of RLS, control of RLS has generally been based on the timely use of fungicides (Hess et al., 2007; Havis et al., 2015). At the time of the first reports of RLS in Europe the strobilurin-based fungicides (quinone outside inhibitors; QoI) were highly effective against the disease. The QoIs, which inhibit the cytochrome bc1 complex of the respiratory chain, are an important group of chemical fungicides, active against a broad spectrum of fungal diseases (Bartlett et al., 2002). These fungicides were used extensively to control cereal pathogens. Resistance to the major wheat pathogen *Zymoseptoria trictici* appeared in the early years of the century (Fraaije et al., 2005). At the same time, the strobilurin fungicides showed reduced efficacy against RLS in UK field trials. This decline in activity developed rapidly and spread to many countries (Oxley et al., 2006; McCabe, 2009). This rapid decline was attributed to the G143A point mutation in cytochrome b gene and was found to be prevalent in *R. collo-cygni* populations both in Scotland and in Denmark (Fountaine and Fraaije, 2009). When 302 isolates of *R.collo-cygni* collected in 12 locations in the Czech Republic in 2009 were tested for the mutation the resistant allele was detected in nearly 50% of the isolates (Matusinsky et al., 2010). In countries where strobilurin use was more regulated, the proportion of resistant:sensitive alleles remains closer to 50:50 (Jan-Eivind Kam Anderson, personal communication). The strobilurins are still used extensively in South America, suggesting they are still active. Field trials in Uruguay showed improved control of RLS when azoxystrobin and chlorothalonil were used together rather than chlorothalonil alone (Pereyra et al., 2014).

Control in more recent years has relied on the use of demethylation inhibitors (DMI), for example, prothioconazole and succinate dehydrogenase inhibitors (SDHI), for example, boscalid. The SDHI fungicides have been used for many years but the development of a second generation of SDHI compounds increased the control of RLS in many countries (Havis et al., 2015). The rapid development of resistance in *R. collo-cygni* to single site fungicides observed previously suggested careful stewarding of the DMI and SDHI fungicides had to be adopted. A study at the start of the decade suggested *R. collo-cygni* was medium to high risk for development of resistance to SDHI fungicides (Piotrowska et al., 2017). Many control programmes incorporated the highly effective multisite chlorothalonil to protect the single site chemicals. In some countries this multisite was not approved for use, for example, Denmark (Jørgensen and Christiansen, 2006). However, other multisites have been used in combination with other fungicides to control RLS in barley (e.g. folpet) (AHDB, 2020).

Despite the stewardship programmes in 2017 across Europe a rapid decline in activity of single site fungicides towards RLS was observed (Havis et al., 2018b). The first mutations conferring reduced efficacy to SDHI fungicides were detected in 2014 (Rehfus et al., 2019). Isolates from European countries were tested for the frequency of mutations from 2014 to 2017. The sdh mutations B-H266Y/R, B-T267I, B-I268V, C-N87S, C-H146R, C-H153R and some others) were detected with increasing frequencies since 2014. DMI adapted isolates of *R. collo-cygni* were also found at a high frequency in the countries under investigation. In total fifteen different *Cyp51* haplotypes were detected in the set of isolates from 2009 to 2017. The most frequent haplotype in 2017 was *C1* haplotype. In New Zealand 33 isolates were collected from both North and South islands (20 of them with either C-H142R, C-H149R, or C-N83S mutations in sdhi sub unit), so field efficacy is reducing but DMIs still remain effective (Soonie Chng, personal communication). Testing of isolates from South America is ongoing. The removal of chlorothalonil from the EU approved list in May 2020 presents a further challenge to barley growers (Dussart et al., 2020). New chemistry is slowly being released onto the global market, for example, mefentrifluconazole from BASF. However, the challenge of managing new single-site fungicides sustainably remains.

Timing of fungicide inputs has been studied extensively and found to have a significant effect on spray efficacy. Pereyra et al. (2014) showed that in Uruguay greatest RLS control could be achieved by a three-spray programme (applications at GS33, GS38 and GS47). This may not be practical or economic in many countries. Most studies indicate that the best effects from late treatments come from sprays at awns peeping (GS49). In Argentina, a protection window of barley against RLS by the application of an azoxystrobin plus isopyrazam mixture between barley stem elongation (GS30) and first awns visible (GS49) was determined by Erreguerena et al. (2014). The GS49 treatment has also been recommended for RLS control in Switzerland (Peter Frei, *personal communication*), Germany (Hess et al., 2007) and the UK (Havis et al., 2012). However, the choice and timing of fungicides in a growing season will depend on disease incidence and severity in the crop (Hess et al., 2014).

Alternatives to conventional fungicides are attracting increased attention. Plant defence elicitors which prime plants to provide broad spectrum disease control have been tested in field trials (Walters et al., 2008). These compounds are not recommended as solo treatments but did give significant reductions in RLS when applied at GS24 followed by reduced rate fungicides at GS31 and GS39 (Havis et al., 2009b). However, the use of elicitors needs to be managed carefully as other trials, which used a combination of defence inducing compounds, showed a negative effect on RLS (Walters et al., 2012). Therefore, more research is required before they can be used effectively (Havis et al., 2015).

The control of *R. collo-cygni* by seed treatments has been examined in many studies. Previously, good control of RLS was observed in field trials by fungicide seed treatments (Havis et al., 2010; Clemente et al., 2014). The control observed was not found to be related to restricting fungal movement in barley (Havis et al., 2010). However, the appearance of resistance to many of the fungicides has reduced the efficacy of these treatments. Non-chemical treatments have also been studied and hot water treatment has been shown to reduce fungal DNA levels (Havis et al., 2010; Zamani-Noor, 2011). The control from these treatments is not always consistent in reducing RLS in the barley crop. More studies are needed to evaluate the reasons behind this observation.

9 Future prospects

The global distribution of the fungus has been steadily expanding over the last twenty years and economic damage has been recorded on many continents. The future of Ramularia control presents many challenges. A sustainable approach will necessitate the adoption of Integrated Pest Management schemes. There are many factors to be considered in an IPM approach and some areas are under researched. A study from Croatia suggested that removing infected volunteer barley plants could be an effective cultural control technique (Koric et al., 2009) and work is ongoing to establish the possibility of transmission from barley stubble to following barley crops (Havis et al., unpublished). A successful IPM programme will need to incorporate more information on rotation, varietal tolerance, seed health, risk forecasting, host-pathogen interactions, alternative control measures and fungicide efficacy.

10 Where to look for further information

A number of useful reviews of research in Ramularia have been published:

- Walters, D. R., Havis, N. D. and Oxley, S. J. P. (2008). *Ramularia collo-cygni*: the biology of an emerging pathogen of barley. *FEMS Microbiology Letters* 279:1–7.
- Havis, N. D., Brown, J. K. M., Clemente, G., Frei, P., Jedryczka, M., Kaczmarek, J., Kaczmarekm, M., Matusinsky, P., McGrann, G. R. D., Pereyra, S., Piotrowska, M., Sghyer, H., Tellier, A. and Tess, M. (2015). *Ramularia collo-cygni* - an emerging pathogen of barley crops. *Phytopathology* 105:895–904.
- Dussart, F., Creissen, H. E. and Havis, N. D. (2020). *Ramularia collo-cygni* - an enemy in waiting. In: *eLS*. John Wiley & Sons, Ltd, Chichester. DOI:10.1002/9780470015902.a0028896.

The proceedings from a couple of meetings are available online and these can provide extra information and background.

- First European Ramularia workshop, 2006: http://webdoc.sub.gwdg.de/pub/mon/2007/koopmann.pdf.
- AHDB sponsored Ramularia workshop, 2018: https://projectblue.blob.core.windows.net/media/Default/Imported%20Publication%20Docs/AHDB%20Cereals%20&%20Oilseeds/Disease/Ramularia%20leaf%20spot/Ramularia%20workshop%202018%20(presentations,%20biographies%20and%20abstracts)%20041018.pdf.

Ramularia also featured in a number of barley meetings:

- Fourth international workshop on barley leaf blights, 2011: https://www.hutton.ac.uk/events/fourth-international-workshop-barley-leaf-blights.
- Second international barley leaf disease Rabat. Morocco, 2017: https://repo.mel.cgiar.org/handle/20.500.11766/8020.

11 References

AHDB (2016). Control of Ramularia in a Changing Climate. Final Project Report PR 533. Available online [https://projectblue.blob.core.windows.net/media/Default/Research%20Papers/Cereals%20and%20Oilseed/rd-2007-3441-final-1.pdf] Accessed 02/June/2020.

AHDB (2018). Ramularia leaf spot in barley. Available online [https://projectblue.blob.core.windows.net/media/Default/Imported%20Publication%20Docs/AHDB%20Cereals%20&%20Oilseeds/Disease/Ramularia%20guide%20(January%202019).pdf] Accessed 14/March/2020.

AHDB (2020). Fungicide programmes in barley. Available online [https://ahdb.org.uk/knowledge-library/fungicide-programmes-for-barley]. Accessed online 15/June/2020.

Bartlett, D. W., Clough, J. M., Godwin, J. R., Hall, A. A., Hamer, M. and Parr-Dobrzanski, B. (2002). The strobilurin fungicides. *Pest Management Science* 58(7):649–662.

BASF (2010). A guide to the recognition and understanding of Ramularia leaf spot and other leaf spots of barley. Available online [https://www.agro.basf.fi/Documents/migrated_files/fi_files/informationmaterial_files/basf_ramularia.pdf]. Accessed 14/March/2020.

Beukes, I., Jensen, T. and Coertze, S. (2016). First report of Ramularia collo-cygni infecting barley in South Africa. *Journal of Plant Pathology* 98(2):376.

Bistrich, H., Breun, J., Emmert, G., Fleck, A., Jaiser, H., Kempe, H. and Lemmens, M. (2006). Screening for leaf spot resistance–results and impact on practical breeding. *Proceedings of the 1st European Ramularia Workshop*, March 2006. Göttingen, Germany, pp. 83–84.

Brown, J. K. M. and Makepeace, J. C. (2009). The effect of genetic variation in barley on responses to *Ramularia collo-cygni*. *Aspects of Applied Biology* 92:43–47.

Burrell, C., Havis, N. D., Spoel, S. H. and Bingham, I. (2020). Quantifying the effects of asymptomatic and symptom expressing phases of *Ramularia collo-cygni* infection on photosynthesis and yield formation in spring barley. *Proceedings Crop Production in Northern Britain 2020*, pp. 97–102.

Carmona, M. A., Scandiani, M. M., Formento, A. N. and Luque, A. (2013). Epidemias de *Ramularia collo-cygni*, organismo causal del salpicado necrótico de la cebada. Campaña 2012-2013 Revista Cultivos Invernales en SD. de Aapresid: 44–47. Avaiable online [www.aapresid.org.ar Publicaciones]. Cultivos Invernales, AAPRESID. ISSN 1850-0633.

Cavara, F. (1893). Ueber einiger parasitische Pilze auf dem Getreide. *Zeitschrift fur Pfianzenkrankheiten* 3, 16–26.

Clemente, G., Quintana, S., Aguirre, N., Rosso, A., Cordi, N. and Havis, N. D. (2014). State of art of *Ramularia collo-cygni* (leaf spot of barley) in Argentina and detection and quantification of *R. collo-cygni* by Real Time PCR in barley plantlets and seeds treated with fungicide. *Proceedings 11th Conference of the European Foundation for Plant Pathology*, p. 100.

Cromey, M. C., Harvey, I. C., Sheridan, J. E. and Grbavac, N. (2004). Occurrence, importance and control of *Ramularia collo-cygni* in New Zealand. *Proceedings of the Second International Workshop on Barley Leaf Blights*, 7–11 April 2002. Aleppo, Syria: ICARDA, pp. 337–342.

Dussart, F., Douglas, R., Sjökvist, E., Hoebe, P. N., Spoel, S. H. and McGrann, G. R. D. (2018a). Genome-based discovery of polyketide-derived secondary metabolism pathways in the barley pathogen *Ramularia collo-cygni*. *Molecular Plant-Microbe Interactions: MPMI* 31(9):962–975.

Dussart, F., Hoebe, P. N., McGrann, G. R. D. and Spoel, S. H. (2018b). Determining the role of the phytotoxic secondary metabolite rubellin D in the pathology of *Ramularia collo-cygni*, the fungus responsible for Ramularia leaf spot disease of barley. In: Heilbronn, T. D. (ed), *Proceedings Crop Production in Northern Britain 2018*, pp. 97–102.

Dussart, F., Creissen, H. E. and Havis, N. D. (2020). *Ramularia collo-cygni*–an enemy in waiting. In: *eLS*, John Wiley & Sons, Ltd, Chichester. DOI:10.1002/9780470015902.a0028896.

Erreguerena, I. A., Quiroz, F. J., Montoya, M. R., Maringolo, A., Lazzaro, N. and Giménez, F. (2014). Ventana de protección para el control químico de *Ramularia collo-cygni* y *Rhynchosporium secalis* en cebada en el sudeste bonaerense. *Resúmenes III Congreso Argentino de Fitopatología*, p. 359.

Formayer, H., Huss, H. and Kromb-Kolb, H. (2004). Influence of climatic factors on the formation of symptoms of *Ramularia collo-cygni*. In: Yahyaoui, A. H., Brader, L., Tekauz, A., Wallwork, H. and Steffenson, B. (eds), *Proceedings of the Second International Workshop on Barley Leaf Blights*, pp. 329–330.

Fountaine, J. and Fraaije, B. A. (2009). Development of QoI resistant alleles in populations of *Ramularia collo-cygni*. *Aspects of Applied Biology* 92:123–126.

Fraaije, B. A., Cools, H. J., Fountaine, J., Lovell, D. J., Motteram, J., West, J. S. and Lucas, J. A. (2005). Role of ascospores in further spread of QoI-resistant cytochrome b alleles (G143A) in field populations of *Mycosphaerella graminicola*. *Phytopathology* 95(8):933–941.

Frei, P. and Gindrat, D. (2000). Le champignon Ramularia collo-cygni provoque une forme de grillures sur les feuilles d'orge d'automne et de gramine es adventices. *Revue Suisse d'Agriculture* 32:229–223.

Frei, P. (2004). *Ramularia collo-cygni*: cultivation, storage, and artificial infection of barley and weed grasses under controlled conditions. In: Yahyaou, A., Brader, L.,Tekauz, A., Wallwork, H. and Steffenson, B. (eds), *Proceedings of the Second International Workshop on Barley Leaf Blights*, 7–11 April 2002. Aleppo, Syria: ICARDA, pp. 351–354.

Frei, P., Gindro, K., Richter, H. and Schürch, S. (2007). Direct-PCR detection and epidemiology of *Ramularia collo-cygni* associated with barley necrotic leaf spots. *Journal of Phytopathology* 155(5):281–288.

Gubiš, J., Hudcovicova, M. and Klcova, L. (2008). First report of *Ramularia collo-cygni* in Slovakian. *Journal of Plant Pathology* 90:149.

Harvey, I. C. (2002). Epidemiology and control of leaf and awn spot of barley caused by *Ramularia collo-cygni*. *New Zealand Plant Protection* 55:331–335.

Havis, N. D., Piper, S. R., Oxley, S. J. P. and Langrell, S. R. H. (2004). Development of a PCR based identification and detection assay for *Ramularia collo-cygni* direct from barley leaf tissue. *Meeting the Challenges of Barley Blights*. In: Yahyaoui, A. H., Brader, L., Tekauz, A., Wallwork, H. and Steffenson, B. (eds), *Proceedings of the Second International Workshop on Barley Leaf Blights*, April 2002. Aleppo, Syria: ICARDA, pp. 343-350.

Havis, N. D., Oxley, S. J. P., Piper, S. R. and Langrell, S. R. H. (2006). Rapid nested PCR-based detection of *Ramularia collo-cygni* direct from barley. *FEMS Microbiology Letters* 256(2):217-223.

Havis, N. D., Taylor, J. M. G., Nyman, M. and Oxley, S. J. P. (2009a). Epidemiology of *Ramularia collo-cygni*. *Aspects of Applied Biology* 92:1-7.

Havis, N. D., Paterson, L., Taylor, J. M. G. and Walters, D. R. (2009b). Use of Resistance Elicitors to control *Ramularia collo-cygni* in spring barley. *Aspects of Applied Biology* 92:127-132.

Havis, N. D., Nyman, M. and Oxley, S. J. P. (2010). Potential of seed treatment to control *Ramularia- collo-cygni* in barley. *Proceedings of the Crop Protection in Northern Britain*, pp. 97-102.

Havis, N. D., Oxley, S. J. P., Burnett, F. and Hughes, G. (2012). Epidemiology of *Ramularia collo-cygni*. *Proceedings of the Crop Protection in Northern Britain*, pp. 119-124.

Havis, N. D., Gorniak, K., Carmona, M. A., Formento, A. N., Luque, A. G. and Scandiani, M. M. (2014a). First molecular detection of *Ramularia* leaf spot (*Ramularia collo-cygni*) in seeds and leaves of barley in Argentina. *Plant Disease* 98(2):277.

Havis, N. D., Nyman, M. and Oxley, S. J. P. (2014b). Evidence for seed transmission and symptomless growth of *Ramularia collo-cygni* in barley (*Hordeum vulgare*). *Plant Pathology* 63(4):929-936.

Havis, N. D., Brown, J. K. M., Clemente, G., Frei, P., Jedryczka, M., Kaczmarek, J., Kaczmarek, M., Matusinsky, P., McGrann, G. R. D., Pereyra, S., Piotrowska, M., Sghyer, H., Tellier, A. and Hess, M. (2015). *Ramularia collo-cygni*–an emerging pathogen of barley crops. *Phytopathology* 105(7):895-904.

Havis, N. D., Evans, N. and Hughes, G. (2018a). Development of UK wide risk forecast for Ramularia leaf spot in barley. AHDB Report PR600. Available at: [https://cereals.ahdb.org.uk/publications/2018/november/15/ramularia-leaf-spot-in-barley.aspx].

Havis, N. D., Gorniak, K., Taylor, J., Stanisz-Migal, M. and Burnett, F. J. (2018b). Controlling Ramularia leaf spot in barley crops. *Proceedings of the Crop Production in Northern Britain*, pp. 91-96.

Heiser, I., Sachs, E. and Liebermann, B. (2003). Photodynamic oxygen activation by rubellin D, a phytotoxin produced by *Ramularia collo-cygni* (Sutton etWaller). *Physiological and Molecular Plant Pathology* 62(1):29-36.

Heiser, I., Heß, M., Schmidtke, K.-U., Vogler, U., Miethbauer, S. and Liebermann, B. (2004). Fatty acid peroxidation by rubellin B, C and D, phytotoxins produced b *Ramularia collo-cygni* (Sutton & Waller). *Physiological and Molecular Plant Pathology* 64(3):135-143.

Hess, M., Habeker, R., Hausladen, H. and Heiser, I. (2006). Monitoring the epidemics of Ramularia collo-cygni: comparison of varieties, sites, years and methods. *Proceedings 1st European Ramularia Workshop*, March 2006. Göttingen, Germany, pp. 47-52.

Hess, M., Habecker, R., Kick, M., Martin, M. and Hausladen, H. (2007). Occurrence of the late leaf spot complex of barley and its consequences on optimized disease control. *Gesunde Pflanzen* 59:47-54.

Hess, M., Gastl, M., Weigand, S., Henkelmann, G. and Rychlik, M. (2011). Influence of crop health and fungal contamination of spring barley on mycotoxin content and malting quality. *Proceedings 33rd European Brewery Convention Congress*, pp. 460.

Hess, M., Sghyer, H., Hausladen, H. and Weigand, S. (2014). Studying the epidemiology of Ramularia collo-cygni for the improvement of an Integrated Pest Management system in a changing climate. *Proceedings of the 11th Conference European Foundation of Plant Pathology*, p. 99.

Heuser, T. and Zimmer, W. (2002). Quantitative analysis of phytopathogenic Ascomycota on leaves of pedunculate oaks (Quercus robur L.) by real time PCR. *FEMS Microbiology Letters* 209(2):295–299.

Hofer, K. M., Linkmeyer, A., Textor, K., Hückelhoven, R. and Hess, M. (2015). MILDEW LOCUS O Mutation does not affect resistance to grain infections with Fusarium spp. and *Ramularia collo-cygni*. *Phytopathology* 105(9):1214–1219.

Huss, H. (2004). The biology of *Ramularia collo-cygni*. Meeting the Challenges of Barley Blights. In: Yahyaoui, A. H., Brader, L., Tekauz, A., Wallwork, H. and Steffenson, B. (eds), *Proceedings of the Second International Workshop on Barley Leaf Blights*, April 2002. Aleppo, Syria: ICARDA, pp. 321-328.

Huss, H., Mayerhofer, H. and Wetschnig, W. (1987). Ophiocladium hordei CAV. (Fungi Imperfecti), ein für Österreich neuer parasitischer Pilz der Gerste. *Der Pflanzenarzt* 40:167–169.

Huss, H., Mayrhofer, H. and Ingolič, E. (1992). *Ramularia collo-cygni* Sutton & Waller (Fungi imperfecti), ein wirtschaftlich beeunter Parasit der Gerste in det Steiermark. *Mitt. Naturwiss. Ver. Steiermark* 122:87–95.

Huss, H., Formayer, H. and Stabentheimer, E. (2003). Die Sprenkelkrankheit –Weiter auf dem Vormarsch. *Pflanzenschutz* 5:8–11.

Huss, H., Formayer, H. and Stabentheimer, E. (2004). Wo Sie auf die Sprenkelkrankheit achten sollten. *Pflanzenshutz* 5:8–10.

Jørgensen, I. H. (1992). Discovery, characterization and exploitation of *mlo* powdery mildew resistance in barley. *Euphytica* 63(1–2):141-152.

Jørgensen, L. and Christiansen, A. (2006). Control of Ramularia in winter barley and spring barley using different fungicides–experience from Denmark. In Koopmann, B., Oxley, S. J. P., Schützendübel, A. and von Tiedemann, A. (eds), *Ramularia collo-cygni A new disease and challenge in barley production. Proceedings First European Ramularia Workshop*, 12–14 March 2006. Göttingen, Germany, pp. 113–119.

Kaczmarek, M., Fountaine, J. M., Newton, A. C., Read, N. D. and Havis, N. D. (2013). The life history of Ramularia collo-cygni. *27th Fungal Genetics Conference*, Asilomar, California, p. 564. Available online [www.fgsc.net/27thFGC/FungalProgramBook2013.pdf].

Kaczmarek, M., Piotrowska, M. J., Fountaine, J. M., Gorniak, K., McGrann, G. R. D., Armstrong, A., Wright, K. M., Newton, A. C. and Havis, N. D. (2016). Infection strategy of *Ramularia collo-cygni* and development of Ramularia leaf spot on barley and alternative graminaceous hosts. *Plant Pathology* 66(1):45–55.

Khier, M., Carmona, M., Sachs, E., Delhey, R., Frayssinet, S. and Barreto, D. (2002). Salpicado necrótico, nueva enfermedad de la cebada en Argentina causada por *Ramularia collo-cygni*. Resúmenes. *Xi Jornadas Fitosanitarias Argentinas* p. 47.

Koric, B., Tomic, Z., Simala, M. and Milek, M. (2009). *Ramularia* Leaf Spot on Barley in the Republic of Croatia. *Zbornik Predavanj in Referatov 9. slovenskega posvetovanja o varstvu rastlin z mednarodno udeležbo*, pp. 273–279.

Leistrumaite, A. and Liatukas, Z. (2006). Resistance of spring barley cultivars to the new disease *Ramularia* leaf spot, caused by *Ramularia collo-cygni*. *Agronomy Research* 4:251.

Ma, X., Keller, B., McDonald, B. A., Palma-Guerrero, J. and Wicker, T. (2018). Comparative transcriptomics reveals how wheat responds to infection by Zymoseptoria tritici. *Molecular Plant-Microbe Interactions: MPMI* 31(4):420–431.

Mäe, A., Sooväli, P. and Põllumaa, L. (2018). *Ramularia collo-cygni*: a new pathogen spreading in barley fields in Estonia. *Agricultural and Food Science* 27(2):138–145.

Makepeace, J. C. (2006). The effect of the mlo mildew resistance gene on spotting diseases of barley. PhD Thesis, University of East Anglia, Norwich, UK.

Makepeace, J. C., Oxley, S. J. P., Havis, N. D., Hackett, R., Burke, J. I. and Brown, J. K. M. (2007). Associations between fungal and abiotic leaf spotting and the presence of *mlo* alleles in barley. *Plant Pathology* 56(6):934–942.

Makepeace, J. C., Havis, N. D., Burke, J. I., Oxley, S. J. P. and Brown, J. K. M. (2008). A method of inoculating barley seedlings with *Ramularia collo-cygni*. *Plant Pathology* 57(6):991–999.

Mařík, P., Snejdar, Z. and Matusinsky, P. (2011). Expression of resistance to *Ramularia* leaf spot in winter barely cultivars grown in conditions of the Czech Republic. *Czech Journal of Genetics and Plant Breeding* 47(1):37–40.

Matusinsky, P., Leisova-Svobodova, L., Marik, P., Tvaruzek, L., Stemberkova, L., Hanusova, M., Minarikova, V., Vysohlidova, M. and Spitzer, T. (2010). Frequency of a mutant allele of cytochrome b conferring resistance to QoI fungicides in the Czech population of *Ramularia collo-cygni*. *Journal of Plant Disease and Protection* 117:248–252.

Matusinsky, P., Leisova-Svobodova, L., Gubis, J., Hudcovicova, M., Klcova, L., Gubisova, M., Marik, P., Tvaruzek, L. and Minarikova, V. (2011). Impact of the seed-borne stage of *Ramularia collo-cygni* in barley seed. *Journal of Plant Pathology* 93:679–689.

Matusinsky, P., Hanusová, M., Stemberková, L., Mařík, P., Minaříková, V., Tvarůžek, L., Langer, I. and Spitzer, T. (2013). Response of spring barley cultivars to *Ramularia* leaf spot in conditions of the Czech Republic. *Cereal Research Communications* 41(1):126–132.

McCabe, T. (2009). Fungicide strategies for the control of *Ramularia collo cygni* to optimise grain yield and quality on spring barley in Ireland. *Aspects of Applied Biology* 92:133–136.

McGrann, G. R. D., Stavrinides, A., Russell, J., Corbitt, M. M., Booth, A., Chartrain, L., Thomas, W. T. B. and Brown, J. K. M. (2014a). A trade-off between *mlo* resistance to powdery mildew and increased susceptibility of barley to a newly important disease, *Ramularia* leaf spot. *Journal of Experimental Botany* 65(4):1025–1037.

McGrann, G. R. D., Steed, A., Burt, C., Goddard, R., Lachaux, C., Bansal, A., Corbitt, M., Gorniak, K., Nicholson, P. and Brown, J. K. M. (2014b). Contribution of the drought tolerance-related Stress-responsive NAC1 transcription factor to resistance of barley to Ramularia leaf spot. *Molecular Plant Pathology* 16(2):201–209.

McGrann, G. R., Andongabo, A., Sjokvist, E., Trivedi, U., Dussart, F., Kaczmarek, M., Mackenzie, A., Fountaine, J. M., Taylor, J. M., Paterson, L. J., Gorniak, K., Burnett, F., Kanyuka, K., Hammond-Kosack, K. E., Rudd, J. J., Blaxter, M. and Havis, N. D. (2016). The genome of the emerging barley pathogen Ramularia collo-cygni. *BMC Genomics* 17:584.

McGrann, G. R. D. and Brown, J. K. M. (2018). The role of reactive oxygen in the development of Ramularia leaf spot disease in barley seedlings. *Annals of Botany* 121(3):415–430.

Miethbauer, S., Heiser, I. and Liebermann, B. (2003). The phytopathogenic fungus *Ramularia collo-cygni* produces biologically active rubellins on infected barley leaves. *Journal of Phytopathology* 151(11-12):665-668.

Miethbauer, S., Haase, S., Schmidtke, K. U., Günther, W., Heiser, I. and Liebermann, B. (2006). Biosynthesis of photodynamically active rubellins and structure elucidation of new anthraquinone derivatives produced by *Ramularia collo-cygni*. *Phytochemistry* 67(12):1206-1213.

Oxley, S. J. P. and Havis, N. D. (2009). Understanding *Ramularia collo-cygni* in the past, present and future. *Aspects of Applied Biology* 92:141-146.

Oxley, S. J. P., Havis, N. D. and Hackett, R. (2006). Impact of fungicides and varietal resistance on Ramularia collo-cygni in spring barley. In: Koopmann, B., Oxley, S. J. P., Schützendübel, A. and von Tiedemann, A. (eds), *Ramularia collo-cygni: A New Disease and Challenge in Barley Production. Proceedings First European Ramularia Workshop*, 12-14 March 2006. Göttingen, Germany, pp. 103-112.

Oxley, S. J. P., Havis, N. D., Brown, J. K. M., Makepeace, J. C. and Fountaine, J. (2008). Impact and interactions of *Ramularia collo-cygni* and oxidative stress in barley. Project report 431. Available online PR431 Final Project Report (windows.net).

Oxley, S. J. P. and Havis, N. D. (2010). Managing Ramularia collo-cygni through varietal resistance, seed health and forecasting. HGCA Project Report 463. Available online [www.hgca.com/media/267653/pr463.pdf].

Peraldi, A., Griffe, L. L., Burt, C., McGrann, G. R. D. and Nicholson, P. (2014). *Brachypodium distachyon* exhibits compatible interactions with *Oculimacula* spp. and *Ramularia collo-cygni*, providing the first pathosystem model to study eyespot and *Ramularia* leaf spot diseases. *Plant Pathology* 63(3):554-562.

Pereyra, S. (2013). Herramientas disponibles para el manejo de dos enfermedades relevantes de la pasada zafra: fusariosis de la espiga en trigo y *Ramularia* en cebada. *Actividades de Difusión INIA* 720:33-41.

Pereyra, S. A., Viera, J. P. and Havis, N. (2014). Managing Ramularia leaf spot of barley in Uruguay. *Proceedings of the APS-CPS Joint Meeting*, p. 274.

Piotrowska, M. J., Fountaine, J. M., Ennos, R. A., Kaczmarek, M. and Burnett, F. J. (2017). Characterisation of *Ramularia collo-cygni* laboratory mutants resistant to succinate dehydrogenase inhibitors. *Pest Management Science* 73(6):1187-1196.

Pinnschmidt, H. O. and Hovmøller, M. S. (2003). Ramularia, a new disease of Barley–a review of present knowledge. *DJF Rep* 89: 313-321.

Pinnschmidt, H. O. and Hovmøller, M. S. (2004). Ramularia-bladplet pa byg. (Ramularia leaf spots of barley). *Gron Viden, Markbrug* 290:6.

Pinnschmidt, H. O., Sindberg, S. A. and Willas, J. (2006). Expression of resistance of barley varieties to Ramularia leaf spot and the status of the disease in Denmark. *Proceedings of the First European Ramularia Workshop*, pp. 85-93.

Pinnschmidt, H. O. and Sindberg, S. A. (2009). Assessing *Ramularia* leaf spot resistance of spring barley cultivars in the presence of other diseases. *Aspects of Applied Biology* 92:71-80.

Pinnschmidt, H. O. and Jørgensen, L. N. (2009). Yield effects of *Ramularia* leaf spot on spring barley. *Aspects of Applied Biology* 92:57-66.

Rehfus, A., Matusinsky, P., Strobel, D., Bryson, R. and Stammler, G. (2019). Mutations in target genes of succinate dehydrogenase inhibitors and demethylation inhibitors in *Ramularia collo-cygni* in Europe. *Journal of Plant Diseases and Protection* 126(5):447-459.

Reitan, L. and Salamati, S. (2006). Field screening in Norway for resistance to Ramularia collo-cygni in old and new barley material. In: von Tiedemann, A., Schützendübel, A. and Koopman, B. (eds), Book of Abstracts of the First European Ramularia Workshop, March. Georg-August University Göttingen, Germany, p. 20.

Sachs, E. (2006). The history or research into *Ramularia collo-cygni*. *Proceedings of the 1st European Ramularia Workshop*, March 2006. Göttingen, Germany, pp. 9–15. Available online [*Ramularia collo-cygni* (gwdg.de)].

Sachs, E., Greil, P., Amelung, D. and Huss, H. (1998). Ramularia collo-cygni. A re-discovered Barley pathogen in Europe. *Mitteilungen aus der Biologischen Bundesanstalt fur Land- und Forstwirtschaft* 357:96–97.

Salamati, S. and Reitan, L. (2006). Ramularia collo-cygni on spring barley, an overview of its biology and epidemiology. In: von Tiedemann, A., Schutzendubel, A. and Koopman, B. (eds), *Book of Abstracts of the First European Ramularia Workshop*, March 2006. Georg-August University Gottingen, Germany, p. 13. Avaiavle online [http://wwwuser.gwdg.de/_rcc/index.htm] Accessed 12/March/2020.

Sheridan, J. E. (2000). Cereal Diseases 1999–2000 (Including Peas and Gooseberry Mildew). Disease Survey and Disease Control in the Waiarapa, New Zealand. Mycology and Plant Pathology Report No. 37, Botany Department, Victoria University of Wellington, Wellington, New Zealand.

Sjokvist, E., Lemcke, R., Kamble, M., Turner, F., Blaxter, M., Havis, N. D., Lyngkjaer, M. and Radutiou, S. (2018). Dissection of Ramularia leaf spot disease by integrated analysis of barley and *Ramularia collo-cygni* transcriptome responses. *Molecular Plant-Microbe Interactions* 38:176–193.

Spencer, M., Havis, N. D. and Creissen, H. E. (2019). Ramularia detection dates by country. *Zenodo*. DOI:10.5281/zenodo.3375540.

Stam, R., Münsterkötter, M., Pophaly, S. D., Fokkens, L., Sghyer, H., Güldener, U., Hückelhoven, R. and Hess, M. (2018). A new reference genome shows the one-speed genome structure of the barely pathogen Ramularia collo-cygni. *Genome Biology and Evolution* 10(12):3243–3249.

Stam, R., Sghyer, H., Tellier, A., Heß, M. and Hückelhoven, R. (2019). The current epidemic of the barley pathogen Ramularia collo-cygni derives from a recent population expansion and shows global admixture. *Phytopathology* 109(12):2161–2168.

Sutton, B. C. and Waller, J. M. (1988). Taxonomy of *Ophiocladium hordei* causing leaf lesions on Triticale and other Gramineae. *Transactions of the British Mycological Society* 90(1):55–61.

Tasmanian Government (2020). Biosecurity Tasmania-Ramularia leaf spot of barley. Available online [https://dpipwe.tas.gov.au/biosecurity-tasmania/plant-biosecurity/pests-and-diseases/ramularia-leaf-spot]. Accessed 10/May/2020.

Taylor, J. M. G., Paterson, L. and Havis, N. D. (2010). A quantitative real-time PCR assay for the detection of *Ramularia collo-cygni* from barley (*Hordeum vulgare*). *Letters in Applied Microbiology* 50:493–499.

TEAGASC (2020). Crops. Available at: https://www.teagasc.ie/crops/soil--soil-fertility/nutrient-deficiencies/manganese/. Accessed 17/May/2020.

Thirugnanasambandam, A., Wright, K. M., Havis, N. D., Whisson, S. C. and Newton, A. C. (2011). Agrobacterium-mediated transformation of the barley pathogen *Ramularia collo-cygni* with fluorescent marker tags and live tissue imaging of infection development. *Plant Pathology* 60(5):929–937.

Walters, D. R., Havis, N. D. and Oxley, S. J. P. (2008). *Ramularia collo-cygni*: the biology of an emerging pathogen of barley. *FEMS Microbiology Letters* 279(1):1-7.

Walters, D. R., Avrova, A., Bingham, I. J., Burnett, F. J., Fountaine, J., Havis, N. D., Hoad, S. P., Hughes, G., Looseley, M., Oxley, S. J. P., Renwick, A., Topp, C. F. E. and Newton, A. C. (2012). Control of foliar diseases in barley: towards an integrated approach. *European Journal of Plant Pathology* 133(1):33-73.

Wu, Y. X. and von Tiedemann, A. (2002). Evidence for oxidative stress involved in physiological leaf spot formation in winter and spring barley. *Phytopathology* 92(2):145-155.

Zadoks, J. C., Chang, T. T. and Konzak, C. F. (1974). A decimal code for the growth stages of cereals. *Weed Research* 14(6):415-421.

Zamani-Noor, N. (2011). Studies on Ramularia Leaf Spots on barley–resistance phenotyping, epidemiology and pathogenicity. PhD thesis, Georg-August-University Göttingen, Germany.

Part 7

Barley yellow dwarf virus

Chapter 20

Advances in understanding the biology and epidemiology of barley yellow dwarf virus (BYDV)

Douglas Lau, Embrapa Trigo, Brazil; Talita Bernardon Mar, National Council for Scientific and Technological Development Fellow (CNPq) (Embrapa-CNPq), Brazil; Carlos Diego Ribeiro dos Santos, Postgraduate Program in Plant Science, Faculty of Agronomy, Federal University of Rio Grande do Sul (UFRGS), Brazil; Eduardo Engel, Postgraduate Program in Entomology, University of São Paulo, Brazil; and Paulo Roberto do Valle da Silva Pereira, Embrapa Florestas, Brazil

1 Introduction

This introduction provides an overview of key components of the pathosystem involving cereals, the aphids that feed on them and transmitted viruses. It covers the following:

- Virus pathology,
- Virus transmission by aphid vectors, and
- Dynamics and management of aphid populations.

The overview provides a context for a more detailed discussion of pathosystem components and their interactions in the following sections.

http://dx.doi.org/10.19103/AS.2021.0092.34

1.1 Virus pathology

Barley/cereal yellow dwarf viruses (B/CYDVs, Luteoviridae) are a group of viruses that infect several plant species from the Poaceae family. They are distributed over almost all areas of the planet where there are grasses. First described in 1951 in the United States in barley (*Hordeum vulgare*), due to its wide distribution and economic damage caused to cereal crops, it is one of the most important groups of plant viruses. The disease caused by these viruses is commonly called 'yellow dwarf disease' (YDD). The reduction in the plant's stature and the yellowing of the leaves (always in the apex-base direction) are evident characteristics of this pathology. In fact, the color resulting from the virus infection depends on the host. Wheat and barley generally show yellowing, and oats respond to infection with a reddish, purplish or orange color and triticale with a more whitish color (Fig. 1).

Leaf symptoms result from infection of the virus in phloem cells. The degeneration of phloem cells hinders the circulation of the elaborated sap. As the photoassimilates do not circulate properly through the plant, there occurs an accumulation of these substances in source organs. This accumulation causes the degeneration of plastid systems. With the loss of chlorophyll, the leaves become yellowish, reddish or whitish. Deficiency in the transport of photoassimilates harms the sink tissues and organs. The root system shows reduced growth and, as a result, all growth of the aerial part is impaired (Fig. 2). Regarding the formation of reproductive organs, smaller ears are formed containing fewer grains. The grains are also smaller. Phloem degeneration can also affect leaf blade development, resulting in the characteristic symptom of serrated edges, often reminding the damage caused by chewing insects. Another effect of viral infection is the delay in plant development, increasing the duration of its different growth stages.

There is great variation in yield losses. Among the cultivated cereals, there are those that do not produce ears and, therefore, are not capable of producing grains. There are sources with some level of tolerance to infection. However, even the best tolerant materials have a productivity reduction of around 20%.

1.2 Virus transmission by aphid vectors

The epidemiology of B/CYDVs is intrinsically linked to its vectors, the aphids. Just as there are several species of BYDV and CYDV, there are also several species of aphids that host on grasses as a food source and are able to transmit these viruses. The transmission efficiency of virus species by aphid species is variable.

Among aphid species important vectors for B/CYDVs are *Rhopalosiphum padi* (Linnaeus, 1758), *Sitobion avenae* (Fabricius, 1775), *Metopolophium dirhodum* (Walker, 1849), *Schizaphis graminum* (Rondani, 1852) (Fig. 3). Also

Figure 1 Symptoms of yellow dwarf disease caused by barley yellow dwarf virus (BYDV) in the aerial part of different hosts: (a) wheat, (b) triticale, (c) oats and (d) barley.

reported are *Rhopalosiphum maidis* (Fitch, 1856), *Rhopalosiphum rufiadominalis* (Sasaki, 1899), *Diuraphis noxia* (Mordvilko), *Sipha maydis* (Passerini, 1860) and *Sipha flava* (Forbes, 1884).

The virus is acquired by the aphid during feeding. Upon ingesting the phloem sap, the aphid acquires the viral particles. These circulate through the digestive tract, penetrate hemocele and accumulate in the salivary gland. The passage of viral particles from the digestive tract to the hemocele depends on the action of aphid endosymbiont bacteria. Since the virus needs to circulate in the insect's body and accumulate at salivary gland to be transmitted, there is an acquisition period and a latency period until the insect is able to transmit the virus. The virus is not able to multiply in the insect's cells and is also not transmitted to the progeny.

Figure 2 Symptoms caused by barley yellow dwarf virus (BYDV) in different plant organs. (a) leaf blade with serrated edges, (b) reduced height and delay in the development cycle, (c) impaired ear development and (d) reduction in the number and weight of grains and deformed grains.

In a distant crop without viruses, the epidemic begins with the arrival of winged viruliferous aphids. In many regions, this is a long-distance movement. Temperate regions with severe winter, receive migrating populations of aphids that have multiplied in smaller latitudes. These winged insects, after trying the host plant, insert their stylets between the cells of the epidermis and parenchyma until reaching the vascular system. Along with the injection of saliva, which contains a cocktail of substances that establish the compatibility of the aphid and the host plant (for example, calcium chelating proteins that prevent the obstruction of phloem vessels), the viral particles are also injected. Once in the host plant, the virus starts the process of multiplication and movement in plant cells. Meanwhile, the aphid begins to reproduce. When feeding, nymphs generated on the infected plant will acquire the virus. The nymphs will grow and

Figure 3 Main aphid species vectors of barley yellow dwarf virus (BYDV). (a) *Rhopalosiphum padi*, (b) *Sitobion avenae*, (c) *Schizaphis graminum* and (d) *Metopolophium dirhodum*.

accumulate viral particles. At first, colonies are formed by apterous (wingless) individuals. With the growth of the colony, a movement begins in search of new places in the plants and the migration between plants. When moving to a new plant, the aphid will transmit the virus and restart the process. The virus and aphids multiply in the plant and spread to neighboring plants. When the populations of aphids on a plant become high (high population density) or when the host enters senescence, a signal for the formation of winged individuals is triggered. These individuals can then migrate over short distances to form secondary foci in the same field or nearby fields or initiate long-distance migrations (Fig. 4).

The occurrence of aphids and the rate of their multiplication depend on the availability of host plants, the favorability of ambient conditions and the ability of natural enemies to suppress aphid population growth.

As aphids need host plants to feed themselves, landscape composition is a determining factor in population dynamics. The vegetation is determined

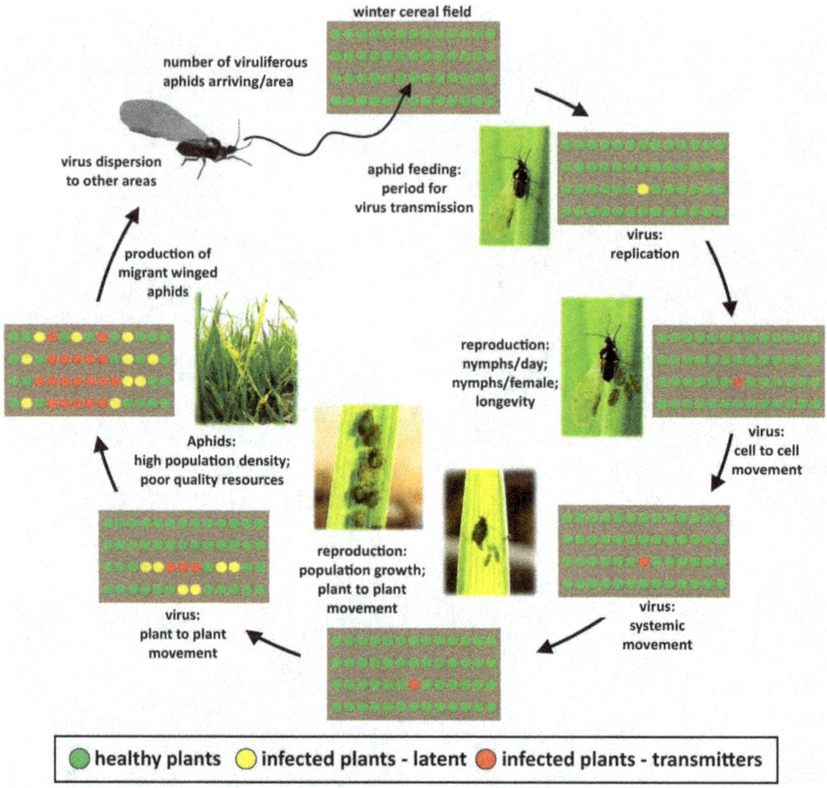

Figure 4 Steps in the epidemiology of BYDV.

by the region's climate. The annual thermal amplitude and water availability can determine greater or lesser stability in the supply of adequate vegetation to aphids. At high latitudes, the rigor of winter forces the disappearance of vegetation that could support aphids. In other regions, temperature may not be the limiting factor, but there are fluctuations between seasons with adequate water availability and periods of drought. Thus, depending on the stability of the climate, aphids can remain in the region throughout the year. In regions of greater contrasts, at each cycle occurs a new colonization by means of winged migrating aphids. Aphids can also use more resistant forms to overcome more difficult times. In regions of harsh winter, sexual reproduction occurs with the production of eggs that is maintained in the winter period. At lower latitudes, with less severe winters, aphids can be maintained through asexual (parthenogenetic) reproduction throughout the year.

In natural landscapes, aphid populations can lodge in native grasses. Agricultural landscapes impose new challenges. The annual crop cycles defined by human beings as a function of climatic viability, define the availability of hosts. Aphids can alternate between winter cereals (oats, wheat, barley, triticale, rye) and summer cereals (mainly corn and sorghum). Pastures also constitute important reservoirs of aphid populations. In some regions, these hosts are green bridges between which aphids can migrate and survive through the year. In addition, voluntary plants of cultivated species may remain between growing seasons and can serve as a refuge for aphids.

1.3 Dynamics and management of aphid populations

Against the background of host availability, all stages of aphid growth, development and reproduction are also regulated by environmental factors. The climate of each region determines its level of epidemiological risk. One of the main factors that determine aphid growth and multiplication rates is temperature. The temperature variation over the seasons determines those with the highest occurrence of aphids and virus transmission. Another key factor is rainfall. Intense rains can be harmful to aphid colonies, but in very dry regions rains are necessary to maintain the growth of host plants.

In sequence, with host and favorable climate, the next 'layer' that regulates aphid populations is composed by natural enemies which include predators and parasitoids. Predators are less specific. The relationships between aphids and parasitoids form complex trophic networks. The richness of species of aphids and parasitoids and their multiple interactions are key factors in understanding how they regulate aphid populations.

The management of YDD epidemics involves understanding all the factors mentioned above. Regarding the direct management of the virus, as there is no virus infection treatment, its management involves the adoption of plants resistant or tolerant to viral infection. In wheat, the sources of resistance may come from the species itself or from nearby species. There are also alternatives based on genetic engineering. In the case of the aphids, it is possible to use host plant resistance, chemical control, and also biological control.

Both chemical and biological control to be more successful require in-depth knowledge of epidemiology in the region. The monitoring of the aphid populations, the understanding of the annual population dynamics, of their oscillation patterns help in the decision making for adequate and efficient management.

In this sense, tools related to information technology, data storage and construction of forecasting models are essential. The current challenges that arise are ways to carry out monitoring in large areas in real time, associated with forecasting models for decision making. Such forecasting models must consider the agricultural landscape (extension of host plants available to viruses and aphids), climate favorability (temperature and precipitation) and natural enemies entomofauna. All this should be integrated with the understanding of the long-distance aphid migration patterns.

2 Components of the pathosystem: host plants, viruses and aphid vectors

There are three primary components in the pathosystem:

- host plants,
- viruses, and
- aphid vectors.

All these components are under the influence of the environment. The interaction between these components determines the progress of the disease in the field (Robert, 2001; Jones, 2014).

2.1 Host - grasses

Plants can be firstly classified as host and non-host for virus and its vectors. In this system, grasses (Poaceae) are primary producers and have great species variability. Spread across the globe, grasses are the species that enable the occupation of areas that have some kind of restriction on the establishment of more complex biomes. Savannahs in Africa; the cerrado in central Brazil; the pampas in southern Brazil, Uruguay and Argentina; and North American grasslands are some examples of large areas inhabited by these plants.

YDD viruses infects over 150 species of Poaceae and reduces the grain yield of crops such as wheat (*Triticum aestivum* L.), barley (*Hordeum vulgare* L.), corn (*Zea mays* L.), oats (*Avena sativa* L.), rye (*Secale cereale* L.) and rice (*Oryza sativa* L.) (D'arcy and Burnett, 1995). Several wild plant species that are also hosts can play an important role as virus reservoirs during off season are discussed later. Therefore, both crops and non-cultivated hosts are important from an epidemiologic point of view (Robert, 2001).

2.2 Pathogen -virus

YDD is caused by a group of viruses in the family Luteoviridae that can be transmitted by several cereal aphid species (Hemiptera, Aphididae) in a highly specific circulative, nonpropagative manner (Domier, 2012; Wang et al., 2015). The origin of the Luteoviridae species occurred approximately between 4000 and 9000 years ago, so its species origin may be related with the expansion of agricultural systems (Fargette et al., 2008; Pagan and Holmes, 2010).

Acquisition and transmission of luteovirids by aphids is highly specific. These insects have developed close interspecific relationships with viruses that cause YDD (Power and Gray, 1995). Initially, Rochow (1969) classified these viruses into strains based on their aphid transmission specificities. Each strain was named according to the first letter of the name of the main aphid vector species,for example, MAV is transmitted by *S. avenae* (=*Macrosiphum avenae*); SGV is transmitted by *S. graminum* ; PAV is transmitted by *R. padi* and *S. avenae*; RPV is transmitted by *R. padi*; and RMV is transmitted by *R. maidis* (Sadeghi et al., 1997; Gray et al., 1998; Lucio-Zavaleta et al., 2001).

Subsequently, genome sequence analysis has enabled taxonomy elucidation and reports of new species (Zhang et al., 2009; Krueger et al., 2013), with YDD viruses comprising the species BYDV-PAV, BYDV-MAV, BYDV-PAS, BYDV-kerII and BYDV-kerIII in the genus Luteovirus; cereal yellow dwarf virus (CYDV)-RPS, CYDV-RPV, maize yellow mosaic virus (MaYMV) and maize yellow dwarf virus (MYDV)-RMV in the genus Polerovirus; and BYDV-GPV and BYDV-SGV, which are not yet assigned to any genus (Lefkowitz et al., 2018) (Fig. 5).

There are a few differences between the genome organization of Luteovirus and Polerovirus. Both encode an RNA-dependent RNA polymerase (RdRp) translated by a ribosomal frameshift of open reading frame (ORF)-1 and ORF2. ORF3 encodes the coat protein (CP) and completely contains ORF4. ORF4 encodes the movement protein (MP) required for cell-to-cell movement and systemic infection. Translation of ORF3a initiates at a non-AUG start codon. This protein is required for long-distance movement of the virus in the plant. ORF5 is expressed as a translational fusion with the product of ORF3. The readthrough domain (RTD) is implicated in aphid transmission and virus stability. The translational readthrough of ORF3 and ORF5 leads to the translation of the CP-RTD, which is necessary for aphid transmission and acts to facilitate long-distance virus movement in phloem tissue (Liu et al., 2012; Smirnova et al., 2015; Lefkowitz et al., 2018). Some luteovirus and polerovirus genomes contain an ORF 6, predicted to encode a small (≤6 kDa) protein, but no functions have been assigned to the proteins. ORF 6 is neither conserved nor required for virus infectivity (Chalhoub et al., 1994; Domier et al., 2002; Miller et al., 2002). However, there is evidence of RNA-silencing suppressor activity of the BYDV-GAV P6 protein (Liu et al., 2012).

Recombination plays an important role in shaping genetic diversity in some plant RNA viruses (Chare and Holmes, 2006). Intra- and interspecific recombination are frequently reported between species of the Luteoviridae family (Smith et al., 2000; Pagan and Holmes, 2010; Boulila, 2011; Wu et al., 2011). The evidence of recombination breakpoints between RdRp and CP gene boundaries suggests that these genes may have undergone different evolutionary histories (Chare and Holmes, 2006; Pagan and Holmes, 2010). With few exceptions, the structural proteins and MPs are conserved in all Luteoviridae family members.

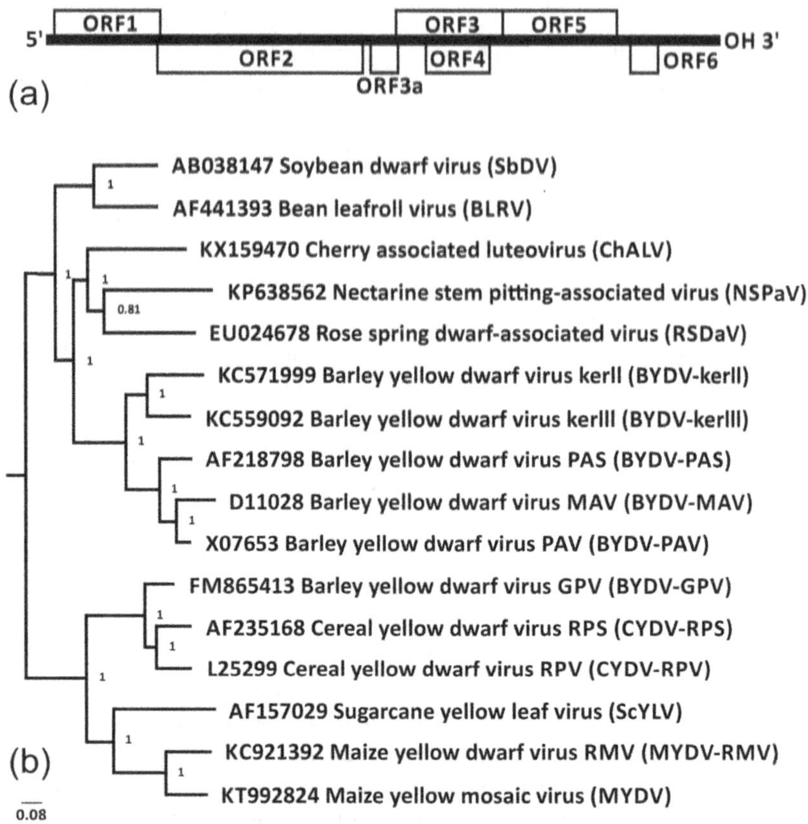

Figure 5 Yellow dwarf viruses genome and phylogenetics relationships. (a) Linearized representation of the BYDV-PAV genoma, individual open reading frames (ORFs) are shown as open boxes: ORF1 and ORF2 encode replication-related proteins; ORF3 and ORF5 encode the major coat protein and readthrough domain (RTD); ORF3a encodes a protein required for long-distance movement; ORF4 encodes a protein required for virus cell-to-cell movement and ORF6 - present in some luteovirus and polerovirus genomes there is evidence of RNA-silencing suppressor activity; (b) midpoint-rooted bayesian inference tree based on full-length nucleotide sequences of species members of the family Luteoviridae, the scale bars indicate nucleotide substitutions per site. The tree was built by the authors based on sequences deposited at GenBank.

In contrast, the replication machinery of Polerovirus and Luteovirus is different. Polerovirus RdRps are more closely related to the Sobemovirus genus, and genes involved in replication and translational control signals of Luteovirus are more closely related to those of the Tombusviridae family (Miller et al., 2002).

The speciation events in the family Luteoviridae tend to occur within the same plant host species and country of origin (Pagan and Holmes, 2010). Among the causal agents of YDD, BYDV-PAV is widely spread throughout the world (Griesbach et al., 1990; Moriones et al., 1993; McElhany et al., 1995; Makkouk et al., 2001; Kapooria and Ndunguru, 2004; Pakdel et al., 2010; Siddiqui et al., 2012; Hall and Little, 2013; Yu et al., 2013; Najar et al., 2017). At the intraspecific level, BYDV-PAV variability can be explained by local geographic adaptation as well as by host-driven adaptation. Therefore, isolates from different countries tend to show divergence in both genomic sequences and virulence (Bisnieks et al., 2006; Wu et al., 2011; Najar et al., 2017; Nancarrow et al., 2018; Hodge et al., 2020). The genomes evolve at a substitution rate of 3.158×10^{-4} nucleotide substitutions/site/year and under purifying selection in the ORFs. BYDV-PAV exhibits a high frequency of homologous recombination, and recombination breakpoints appear distributed along the entire genome (Wu et al., 2011).

2.3 Vectors -aphids

Cereal aphids are among the greatest entomological concerns for world agriculture. With a wide geographical distribution, this group is cosmopolitan. Most of aphid species considered to be of economic importance come from the Paleartic region, corresponding to Europe and Asia (Harrington and Clark, 2010).

The main species of winter cereal aphids belong to the Aphididae family, especially in the Aphidini, Macrosiphini and Siphini tribes. In the Aphidini tribe, *R. padi*, *R. rufiabdominalis*, *R. maidis* and *S. graminum* can be highlighted as cereal pests. In the Macrosiphini: *S. avenae*, *M. dirhodum* and *D. noxia*. In the Siphini, *S. flava* and *S. maydis*.

The establishment and spatio-temporal distribution of aphids is determined by its adaptability in the face of environmental conditions, such as host distribution, adequate climatic conditions and biotic factors such as potential pressure from natural enemies and trophic compatibility (Corrales et al., 2007; Harrington et al., 2007; Harrington and Clark, 2010). With global warming, the distribution of aphids tends to be boosted to areas where previously there were no host plants or minimal weather conditions for their establishment (Finlay and Luck, 2011).

Among the factors that help us to understand the seasonal fluctuation of aphids in conjunction with the agricultural landscape, are the meteorological variables, mainly air temperature and rainfall (Harrington et al., 2007; Davis et al., 2014). The air temperature together with the host plant acts directly on the biology of these insects, influencing the duration of the life cycle and the rate of population growth (Auad et al., 2009). Air temperature is one of the most reliable variables for predicting dynamics of cereal aphids and its increase is positively related to the population accumulation (Zhang et al., 2019).

The limiting and optimal temperatures for the development and reproduction of aphids vary according to the species or group. Dean's classic study (1974) demonstrated the response of cereal aphids to temperature for the species *M. dirhodum*, *S. avenae* and *R. padi*. These species have a shorter development time between 15°C and 25°C, increasing the number of generations and resulting in faster population growth (Fig. 6a, b and c). *R. padi* was the species with the highest optimum temperature, 25°C. For *S. graminum*, Tofangsazi et al. (2010) found a shorter development time range between 15°C and 26°C, with 26°C considered as the optimal temperature for this species (Fig. 6d). Both studies used barley as a host plant.

Populations of each species can also adapt to local conditions. For example, Asin and Pons (2001) determined that the most suitable temperature for population development was ≈28.5°C for *R. padi*, 26.5°C for *S. avenae* and 24.5°C for *M. dirhodum*. These temperatures are higher than those established by Dean (1974). Since Dean studied insects from northern Europe (cold region) and Asin and Pons (2001) populations from southern Europe (warm region), this suggests the ability of aphid populations to adapt to high temperatures in the Iberian Peninsula.

The temperature range to which these species of aphids are adapted allows them to establish themselves in different regions of the planet where cereals are grown. Another example, referring to the potential of world aphid

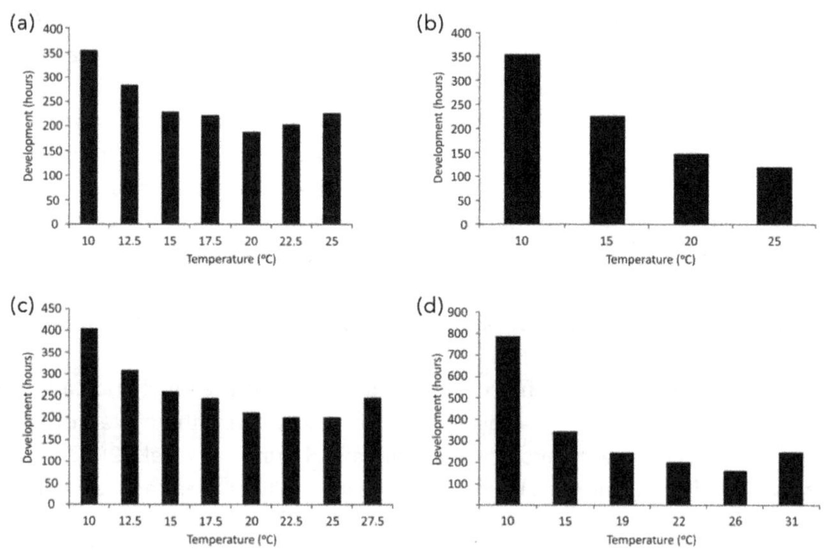

Figure 6 Development time (hours) of cereal aphids in barley as a function of temperature. (a) *Metopolophium dirhodum*, (b) *Sitobion avenae*, (c) *Rhopalosiphum padi*. Data adapted from Dean (1974) and (d) *Schizaphis graminum*. Data adapted from Tofangsazi et al. (2010).

distribution, considering temperature, presence of primary host and forms of reproduction, is demonstrated by Macfadyen and Kriticos (2012) for *R. padi* species in scenarios with natural precipitation and irrigation.

Despite being widely distributed, some species are still in the process of spreading across the globe. Two examples of the introduction and recent adaptation of aphids in South America are the species *D. noxia* and *S. maydis*. The aphid *D. noxia*, found in Chile in 1987 and, later, in Argentina in 1991, has a high damage capacity, mainly in cereals such as barley and wheat (Reed and Kindler, 1994; Brewer et al., 2019). Despite the proximity, there are still no records of its occurrence in the Brazilian cereal fields, indicating that factors such as the climate may be preventing its geographical expansion. Conversely, *S. maydis* is one of the recurrent species in cereals in southern Brazil. Since its arrival in the South American continent in 2002, it has had a rapid expansion, with records in Brazil, Argentina, Chile, Paraguay, Bolivia, Peru, Ecuador and Colombia (CABI, 2020b). When comparing the distribution of both species in Argentina, it was found that *S. maydis* reached the same level of geographical distribution as *D. noxia* in just 4 years, whereas *D. noxia* took 10 years to reach its current distribution in this country (Corrales et al., 2007).

On the other way, *S. flava* that is a Nearctic species, widely distributed throughout North, Central (including the Caribbean Islands and Puerto Rico) and South Americas (Colombia in 1982, Venezuela, 1988, Argentina in 1996, Chile in 2002, Brazil in 2003), it has also been more recently introduced in North Africa (Morocco in 2008), South Europe (Spain in 2014), and Austral Africa (South Africa in 2014) (CABI, 2020a).

3 Virus-aphid interactions

Aphid vectors carry B/CYDVs actively from plant to plant, crop to crop and region to region. On a crop scale, the dissemination of the virus occurs in two stages. Primary spread, that is the virus introduction in a field, and secondary spread, that is the transmission of the virus infection from primarily infected plants to healthy ones (Robert, 2001). B/CYDVs are transmitted in a highly specific circulative, nonpropagative manner, and consequently, the virus remains infectious within the body of the aphid for a long time. The ability of the vector to maintain the virus for a long period of time enables them to carry the virus over distances (Power et al., 1991). Long-distance transfers are achieved each year by aphids allowing reinfection of new areas where vectors would not normally survive harsh winter conditions. Even long distance spread of B/CYDVs through maritime trade between the United States and Australia has been previously proposed after analyzing old gramineae samples conserved at herbaria (Malmstrom et al., 2007).

The movement of the aphid vectors in temperate climate regions is described in detail by Van den Eynde et al. (2020). Briefly, the primary transmission of B/CYDVs to winter cereals occurs during the autumn. Winged aphid vectors flies and colonize newly sown winter cereal grains infecting seedling. The primary transmission is a determinant of virus infection and subsequently yield loss. The secondary spread occurs as the aphids reproduce by parthenogenesis. This fast population growth allows YDD spreading within the crop. During the drop of temperature, as winter approaches, aphids develop wings and migrate to their primary host. Each species of aphid lay eggs in a specific host where the sexual reproduction takes place. The eggs survive over the cold temperatures of the winter and hatch during the spring. After a few generations, they develop wings and recolonize small grains field. In places where the winters are warmer, aphids can overwinter anholocyclically (without eggs) on small grains. During the summer, in the off-season, winged aphid vectors colonize summer hosts, such as maize, grasses and volunteer small grain plants that act as the green bridge between seasons.

Different species of aphids can transmit viruses with different efficiency of transmission (ET) (Robert, 2001). Isolates of BYDV-PAV from South of Brazil are transmitted by *R. padi* with ET of 94.4%, while the ET associated with *S. avenae* and *M. dirhodum* are equal to 76.1% and 63.4%, respectively (Parizoto et al., 2013). Transmission assays point out that the correspondence proposed by Rochow (1969) is not absolute. Viral strains from southwestern Idaho are transmitted by other species as well its described vectors. In an assay, MYDV-RMV was also transmitted by other species than *R. maidis*; BYDV-MAV was also transmitted by *M. dirhodum* and *Myzus persicae* as well as by *S. avenae* (vector originally described); BYDV-SGV was transmitted by *R. padi* and *S. avenae* as well as its described vector *S. graminum*; CYDV-RPV was transmitted by *S. graminum* and *Rhopalosiphum insertum* (Walker, 1849) as well as by *R. padi* (original vector) (Halbert et al., 1992).

The ET also can be significantly different between aphid clones of the same species. Different clones of *R. padi* and *S. avenae* can transmit BYDV-PAV, BYDV-MAV and CYDV-RPV with significant different efficiencies. The difference in the ET is probably determined by the ability to prevent virions passage from haemocoel to salivary duct and/or different capacities for the retention (Jing-Quan et al., 1997). The virions passage from haemocoel to salivary duct is affected by bacterial endosymbionts, consequently influencing the ET within the same aphid species or clones (Power and Gray, 1995). Bacteria from the genus Buchnera have an intracellular symbiotic relationship with aphid vectors of luteoviruses and produce large quantities of a protein homologue to the chaperonin GroEL. These proteins are abundant in the hemolymph of aphids, and it is essential for the stability of the luteovirus particles over the movement through the aphid vector. It is known that chaperonin expressed by symbionts of *S. avenae* and *R. padi* bind to the RTD domain of BYDV species (Plumb, 2002; Reavy and Mayo, 2002).

Species of virus that share vector species are usually found aggregated spatially, indicating the importance of the insect vectors for predicting the spatial distribution of coinfection risk by B/CYDVs (Kendig et al., 2017). Some vector groups have developed close interspecific relationships with selected strains of viruses. Once the ET of each B/CYDVs species varies for different aphid vector combination, we can consider that groups of vectors have different abilities to disperse and thus spread the viruses into a landscape. This can result in virus strains becoming dominant in areas where their specific vectors prevail. It is reasonable to estimate that by collecting data about the prevailing aphid species in a certain region, period or crop it is possible to predict the YDD severity and the main species of virus associated with the disease (Robert, 2001; Van den Eynde et al., 2020).

Among the causal agents of YDD, BYDV-PAV is widely spread throughout the world. BYDV-PAV prevalence is probably related to the high transmission efficiency by its vector *R. padi* which has broad host range and wide geographic distribution (Griesbach et al., 1990; Moriones et al., 1993; McElhany et al., 1995; Kapooria and Ndunguru, 2004; Pakdel et al., 2010; Siddiqui et al., 2012; Hall and Little, 2013; Yu et al., 2013; Najar et al., 2017). BYDV-PAS, a species very similar to BYDV-PAV, was proposed by Mayo (2002). This species is distinguished from BYDV-PAV by approximately 10% and 22% nucleotide divergence in their CP and replication-related genes, respectively. These two species are indistinguishable by serology, and only molecular methods can differentiate them (Bencharki et al., 1999; Hall, 2006). Also, they share the same vector species, *R. padi* and *S. avenae*. Besides the prevalence of BYDV-PAV in most regions, there are those that are the exception, such as the Czech Republic and Alaska, where molecular diagnostic methods revealed BYDV-PAS as the most frequently occurring species (Robertson and French, 2007; Jarošová et al., 2013). In Czech Republic, BYDV-PAS was the most prevalent species over 3 years of monitoring YDD viruses in cereals and grasses. BYDV-PAS was detected in 66% of the samples, followed by BYDV-MAV being detected in 18% and BYDV-PAV being detected in 17%. The prevalence of BYDV-PAS in the Czech Republic may be related to the ET of each aphid species, because for that population all four species, *R. padi*, *S. avenae*, *R. maidis* and *M. dirhodum* were able to transmit BYDV-PAS. While BYDV-PAV was transmitted only by *R. padi* and *S. avenae*, *S. avenae* was the only aphid species able to transmit BYDV-MAV (Jarošová et al., 2013).

For other populations, some studies aimed to explain the prevalence of BYDV-PAV over BYDV-PAS based on the ET. In a greenhouse experiment, BYDV-PAV transmission by a single *R. padi* aphid was significantly greater than BYDV-PAS independent of viral title. The selection of BYDV-PAV by the predominant vector *R. padi* may help to explain its higher incidence. Also, in the same study, BYDV-PAV fails to cause infection on wild Italian ryegrass, while BYDV-PAS was able to infect the host. The infection of wild Italian ryegrass suggests a mechanism by which BYDV-PAS is not competitively excluded from the population by BYDV-PAV, and provides evidence that host diversity facilitates the persistence of diversity in the virus population (Hall and Little, 2013).

4 Virus-host interactions

Plant species plays a strong role in determining virus prevalence, and the identification of of B/CYDVs species is a key factor to understand the disease epidemiology. Besides the vector specificities, on a crop scale, each species can induce symptoms with different severity in the same host, and mixed infections can aggravate the symptoms. Cultivars of oats, barley and wheat react differently when infected with the CYDV-RPV or BYDV-PAV isolates. Mixed infection by CYDV-RPV and BYDV-PAV produced significantly more severe effects on plant height, number of tillers per plant and grain yield (Baltenberger et al., 1987). Experiments using isolates from Morocco demonstrated different biological properties when tested in a set of eight barley cultivars and an oat cultivar. Isolates of BYDV-PAS (North American PAV-129 isolate) produced more severe symptoms and stunting than did those closely related to the BYDV-PAV isolates from Australia, North America and Japan (Bencharki et al., 1999).

Some studies also pointed for a correlation between B/CYDVs with a specific host, indicating host-driven adaptation. In a survey covering wheat-grown areas in Australia, during 1989 to 1991, BYDV-PAV was the most common specie found in wheat samples, while CYDV-RPV was found in pasture grasses (Henry et al., 1992). The association of CYDV-RPV species with perennial grass leys has been reported in several countries worldwide (Doodson, 1967; Fargette et al., 1982; Paliwal, 1982; Lister et al., 1984; Guy et al., 1987; Henry and Dedryver, 1991; Henry et al., 1993). In a survey in western France, during 1984–1986, BYDV-PAV, BYDV-MAV and CYDV-RPV were detected in pastures grasses, but BYDV-MAV and CYDV-RPV were only detected from ryegrass and brome grass, respectively (Henry and Dedryver, 1991). Also, in western France and south-west England, during 1987 and 1988, CYDV-RPV was strongly associated with perennial grass, while BYDV-MAV was associated with cereal crops (Henry et al., 1993).

Kendall et al. (1996) described grasses with respect to their susceptibility to B/CYDVs species. Three groups were identified as grasses predominantly infected by CYDV-RPV and BYDV-MAV (*Festuca rubra*, *Dactylis glomerata*, *Lolium perenne*, *Bromus ramosus* and *Bromus* spp), grasses infected equally by all isolates BYDV-PAV, CYDV-RPV and BYDV-MAV (*Lolium multiflorum* and

Holcus lanatus), and grasses infected mainly by BYDV-PAV (*Poa annua*). The variation in the incidence was consistent with the aphid preference.

From 1989 to 1992, a survey in Scotland found a correlation between B/CYDVs species occurring in barley and *P. annua* through years and location. In July and September, during the senescence of crop plants, the aphid species *S. avenae*, *M. dirhodum* and *R. padi* infested *P. annua* increasing the sources of viruliferous aphids (Masterman et al., 1994). Comparisons between four frequently occurring grass species demonstrated that *Poa pratensis* and *Paspalum dilatatum* were more frequently infected than either *Schedonorus arundinaceus* or *Eleusine indica*. *P. pratensis* and *P. dilatatum* were more likely to be infected with CYDV-RPV and BYDV-MAV than with BYDV-PAV. Those differences may be related to the specificities of the relationship among the host plant and B/CYDVs or the host plant and B/CYDVs aphid vectors species (Rúa et al., 2014).

The impacts of the disease in the crop are affected by the phenological stage at which a host gets infected. Younger plants are more susceptible to both vector and virus infection. Also, the expression of symptoms, such as yellowing, can attract aphids to infected plants during the disease cycle (Robert, 2001).

B/CYDVs do not multiply in the vector and acquisition and transmission periods are usually long lasting (several hours) (Power et al., 1991). As a consequence, the population dynamics of B/CYDVs species usually correlates with the plant community preferred by the aphid vector, for example, the occurrence and spread of BYDV-PAV and BYDV-MAV in the field is related to that of their main vector species (Leclercq-Le Quillec et al., 2000). The plant genetic diversity affects virus dispersal by influencing aphid feeding behavior. The incidence of B/CYDVs were consistently lower in genetically diverse oat populations. The reduction is probably related to change in the movement pattern of the aphid vectors. In a genetically diverse oat population, *S. avenae* increased their movement rates among plants and reduced the plant feeding time. Since B/CYDVs require several hours of feeding for effective transmission, its reduction affected the efficiency of virus transmission (Power, 1991).

B/CYDVs species that is prevalent usually correlates with the plant community. Host diversity is an important component of YDD pathosystem, as B/CYDVs species usually correlates with the plant community based on the preference of vectors that transmit the species with different specificity (Rochow, 1979). Besides BYDV-PAV is the most reported species occurring in cereal fields (Lanzarini et al., 2007; Mar et al., 2013; Parizoto et al., 2013); landscape level has demonstrated an important contribution to determine the predominant B/CYDVs species in an area. Differences in the infection prevalence among host species could be a result of the plant-vector interaction that determines the host's ability to support vector and virus population. Those differences can lead to effects of plant community composition on virus prevalence (Borer et al., 2010; Delmiglio et al., 2010).

Surveys in grass species demonstrated that other species than BYDV-PAV could be predominant in non-managed grasslands (Dempster and Holmes, 1995; Garrett et al., 2004). In a survey on biomass crops, the grass species *Miscanthus sinensis*, *Echinochloa crusgalli* and *Phalaris arundinacea* exposed to infection during a growing season were infected only by BYDV-MAV. But in a transmission assay using *R. padi* or *S. avenae* as vectors, *E. crusgalli* demonstrated to be a host of BYDV-PAV, BYDV-MAV and BYDV-RPV, while *P. arundinacea* was infected only by BYDV-MAV and BYDV-RPV, and *M. sinensis* only by BYDV-MAV (Lamptey et al., 2003). The prevalence of the virus that infects *M. sinensis* could be related with the effects on vector preference. *R. padi* was unable to complete its development on *M. sinensis*, probably because of its inability to penetrate the phloem tubes of C4 plants. While it is a suitable host for *R. maidis*, enabling CYDV-RPV transmission (Huggett et al., 1999).

5 Natural enemies of aphids

Several natural enemies act on aphid populations. Considering the biotic potential of aphids, the effect of these agents in halting population growth is significant. Among the natural enemies of aphids, there are the micro-biological agents (entomopathogenic fungi) and macro-biological agents (predators and parasitoids).

Entomopathogenic fungi occupy a prominent place to control aphids. The Entomophthorales are considered as entomopathogenic with non-cellular mycelium, currently categorized in the class Zygomycetes. The species are widely distributed and classified as one of the most important micro-bioagent to control aphid colonies. The success of this bioagent depends on climatic conditions, and high relative humidity is essential for these fungi to accomplish the two key stages of their life cycle: sporulation and infection (Brodeur et al., 2017; Dedryver, 2018).

Predators are a very wide range of organisms. Those most commonly seen in wheat crops are larvae of hoverflies, green lacewings, larvae and adults of lady beetle and brown lacewings, arachnids and earwigs (Fig. 7). Specialist predators (coccinellid and syrphids) are dependent on the cereal aphid population, whereas generalist predators (neuropterans, dermapterans, spiders, harvestmen, carabid and staphylinid beetles, cockroaches, ants and one species of bird) can feed on other insect species or food sources as well, which makes the generalist population more stable in an environment where the cereal aphid populations are changing (Toft, 2005; Van den Eynde et al., 2020).

Figure 7 Natural enemies of aphids. (a) entomopathogenic fungi. (b) coccinellid larvae. (c) adult coccinellid. (d) green-lacewing eggs. (e) adult green-lacewing. (f) brown-lacewing larvae. (g) adult brown-lacewing. (h) coccinellid larvae. (i) syrphid larvae. (j) syrphid ovipositing on aphid colony. (k–l) aphid mummies. (m, n, o) adult parasitoid.

The two predator groups thus play very different roles in aphid control and exert their effect in different phases of the aphid population cycle: the generalists may work as a preventive force against a prospective outbreak, whereas the specialists may suppress an ongoing outbreak (Toft, 2005).

Parasitoids are the most efficient natural enemies of aphids and are used in biological control programs for these agricultural pests (Carver, 1989; Schmidt et al., 2003). It is a group of several species belonging mainly to the Braconidae and Aphelinidae families, and these organisms are widely studied worldwide. The parasitoid cycle comprises the location of host aphids by adult winged parasitoids. Then, parasitoids oviposit in the aphids' body. Larvae hatch from the egg and consume their hosts, leaving only mummies. So, mummy is the cocoon formed after the parasitism of micro-hymenoptera in aphids, represented by their dead bodies.

In the cereal crops there are interactions between assemblages of herbivores and natural enemies, that in large agricultural areas are overlapped and complex (Ortiz-Martínez and Lavandero, 2018; Ortiz-Martínez et al., 2019, 2020, Sturza et al., 2012). Knowledge is especially lacking on the tri-trophic interactions and intra-guild competitions in biological pest control. Also, in this pathosystem there is the direct influence of landscape that operates on epidemiological risk of crops, as well, the spontaneous grasses could act as a reservoir of B/CYDVs and green bridges to maintain aphid populations during non-cereal crop seasons (Gilabert et al., 2017; Spiesman et al., 2020).

6 Case study: biological control of cereal aphids and BYDV management in Brazil

In Brazil, the first report of symptoms similar to YDD was in wheat in 1929. The association of yellowing or reddening leaves and dwarfing symptoms with BYDV dates to 1968 (Caetano, 1968). At that time, areas cultivated with wheat were rapidly increasing; populations of aphids, like *M. dirhodum* and *S. avenae*, were high; and its aphid-transmitted virus species were predominant (Caetano, 1972; Caetano, 1982). In 1978, a biological control program for wheat aphids was successfully introduced (Salvadori and Salles, 2002). In the 1980s, the adoption of a no-till system changed the agricultural landscape, and crops, such as black oats, were grown in large areas. Nowadays, *R. padi* and *S. avenae* are the most abundant aphid species found associated with cereal crops in Brazil (Rebonatto et al., 2015), transmitting mainly BYDV-PAV (Parizoto et al., 2013). Other species associated with YDD in Brazil are CYDV-RPV, BYDV-MAV and MYDV-RMV which are detected at a low frequency (Lanzarini et al., 2007; Mar et al., 2013; Gonçalves et al., 2017).

The Brazilian agricultural landscape is diverse. Wheat is grown in different biomes and climates. The Brazilian wheat regions are divided into four (Fig. 8).

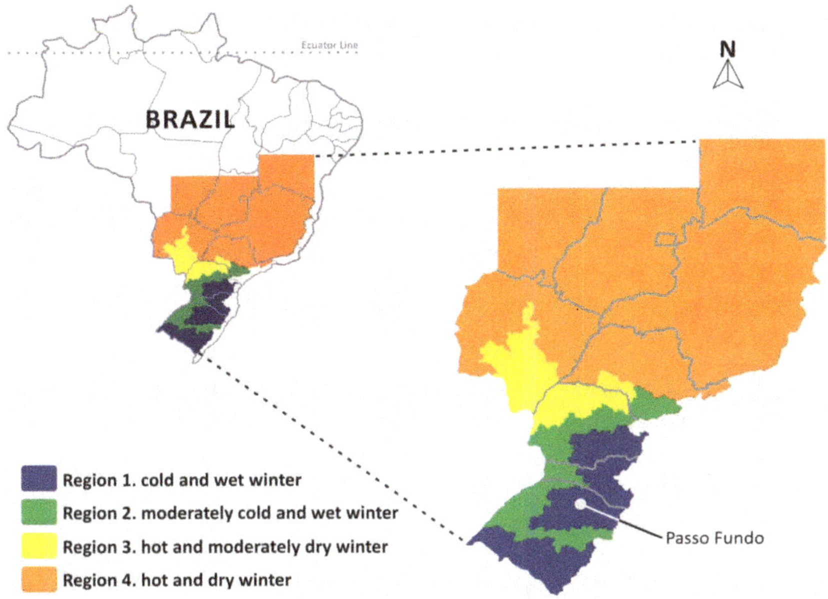

Figure 8 Brazilian wheat regions. Region 1 - cold and wet winter; Region 2 - moderately cold and wet winter; Region 3 - hot and moderately dry winter and; Region 4 - hot and dry winter. Regions 1 and 2 are located in southern Brazil. Region 3 is a transition region located in the south, southeast and central west. Region 4 is mainly located in central Brazil. Passo Fundo city is the headquarters of Embrapa Trigo, where aphids and their parasitoids have been monitored. In this place, parasitoids were created for the biological control program of wheat aphids.

Three of them are distributed in southern Brazil. Most of the following examples are from this region that concentrates 90% of the wheat production in Brazil, and due to the succession of host plants during crop seasons, there are well-established conditions for YDD epidemics.

In central Brazil (region 4), wheat is grown under a hot and dry winter. The original biome in most of these areas is called 'cerrado'. It can be considered the Brazilian equivalent to African savannas. The climate is characterized by a lower annual temperature variation when compared to the south of the country. However, it has well-defined rainy and dry seasons. In the autumn and winter period (April to September), rainfall is scarce. At this time, wheat crops are feasible only under irrigation. The sowing of wheat in the central region of Brazil, according to the Agricultural Climate Risk Zoning (ZARC), occurs from February to May. There are two cultivation systems: dryland cultivation when wheat is sown from February to March or irrigated cultivation when wheat is sown from April to May. The aphid species that can be easily found in these areas are *R. padi*, *S. avenae* and *S. graminum*. Also, several of their natural enemies are found. Yellow dwarf viruses also occur, but its incidence does not appear to be very significant. In summer, most areas in this region are cultivated with soybean. Corn and sorghum are also cultivated on a smaller scale.

Corn can be grown as a second summer crop, and may overlap with the wheat-growing period. Corn can act as a green bridge for aphids and B/CYDVs. The aphids more frequently found in corn are of the genus *Rhopalosiphum* (*R. padi* and *R. maidis*). In sorghum, *S. graminum* is commonly reported (Waquil et al., 2015). In addition to corn, there are also often pastures in this region, with the presence of *Brachiaria*, where *R. padi* can be detected (Auad et al., 2009). However, it is possible that these summer hosts are not sufficient to maintain high populations of B/CYDVs and viruliferous aphids. As the wheat crop areas are small in relation to the large areas in central Brazil, a multiplicative phase of the virus in wheat and migration of viruliferous aphids between crops is not sufficient to create conditions for a high incidence of the disease, which is more common in crop borders.

In the south of Brazil, formed by the states of Paraná (north), Santa Catarina and Rio Grande do Sul (south), B/CYDVs occurs frequently in the three wheat regions. Region 3, further north of the region, has a moderately hot and dry climate in winter. Wheat sowing is carried out in late summer and autumn (March to April), after soybean harvest. In this region, corn and wheat compete for area during autumn and winter. As the climate is becoming warmer due to global warming, the risk of frost has been reduced and the area of corn in this region has increased year after year. Aphids are common and *R. padi*, *S. avenae* and *S. graminum* being the most frequent. *M. dirhodum* is uncommon. B/CYDVs can be observed frequently. The virus and aphids can also be found in other winter crops such as black and white oats.

Regions 1 and 2 are the two southernmost regions. Region 2 has average temperatures slightly higher than those in Region 1, but both have a good distribution of rainfall throughout the wheat crop season. In region 2, wheat is sown earlier than in region 1. The agricultural landscape of these two regions is similar, and after summer crops (mainly soybeans and corn), oat and wheat are sown. In the state of Paraná (the northernmost of the three states in the southern region), second-season corn is cultivated, but in the southernmost states, the risk of frost is high and prohibitive. Autumn oat areas are particularly prone to aphids (mainly *R. padi* and *S. graminum*) and B/CYDVs. As in average, only 20% of the area has been cultivated with wheat (although the percentage that varies according to the micro-region), the oat areas occupying most of the landscape. From the north to south and east to west direction there is a gradient of the sowing dates of wheat. So, unlike the central region, in which there is greater distance and isolation between crops, in the southern region this gradient and proximity of crops allows successive cycles of aphids to migrate from crop to crop as the host's phenological stages continue to succeed.

The increase in wheat crop area in Brazil began at the southern region and occurred mainly after the 1960s. With the great expansion of the host area, the pest population and the incidence of diseases increased significantly. The problem with cereal aphids intensified in the late 1960s and 1970s. Several management practices needed to be introduced to reduce the damage and maintain the economic viability of the crops. At that time aphids control was carried out exclusively by chemical insecticides (Caetano, 1973; Zúñiga-Salinas, 1982), which resulted in the constant elimination of natural enemies.

In southern Brazil, between 1968 and 1976, natural parasitism in cereal aphids was associated with the species: *Aphidius colemani* Vierek, *Diaeretiella rapae* (M'Intosh) and *Lysiphlebus testaceipes* (Cresson) (Silva et al., (1968); Pimenta and Smith, 1976). However, the biological control performed by these species was considered inefficient (Pimenta and Smith, 1976).

Deal with this situation and the success achieved with the program for biological control of aphids in Chile, Embrapa Trigo, in 1978, together with the Food and Agriculture Organization of the United Nations (FAO) introduced the program for the biological control of wheat aphids (BCPWA) in Brazil (Zúñiga-Salinas, 1982). In this program, 12 species of hymenopterans were imported from the regions of origin of the aphids, massively produced and released in wheat fields, mainly in Rio Grande do Sul and Paraná States (Gassen and Tambasco, 1983; Salvadori and Tonet, 2001, Pereira and Salvadori, 2005). Also, two species of Coccinellidae were introduced from United State of America and Israel, but due to the low production of ladybugs in the laboratory, predator releases were discontinued in the first year of the program (Embrapa, 1982).

After the introduction of parasitoids the rates of aphid parasitism in the wheat crop fields exceeded the expected target of 10–15% of parasitism, sometimes in cereals the parasitism rate achieved 60% (Zúñiga-Salinas, 1982). After these releases, four Braconidae species had been established in Rio Grande do Sul: *Aphidius ervi* Haliday, *Aphidius uzbekistanicus* Luzhetzki, *Aphidius rhopalosiphi* De Stefani, and *Praon volucre* (Haliday). The program was completed in the early 1990s (Salvadori and Salles, 2002).

The impact of biological control associated with other changes in the agricultural landscape on the populations of aphids and transmission of B/CYDVs was significant. Right after the program starts, there was an increase in parasitism levels and a reduction in aphid populations. The populations of *M. dirhodum*, the main aphid of the 1970s declined significantly. In the 1970s, this species corresponded to about 90% of the annual aphid population collected in traps, with well-defined peaks between August and October (Fig. 9) (Caetano and Caetano, 1978). Currently, this species corresponds to less than 1% of winged aphids collected in traps (Rebonatto et al., 2015). Some species of parasitoids that were introduced, such as *A. rhopalosiphi* and *A.*

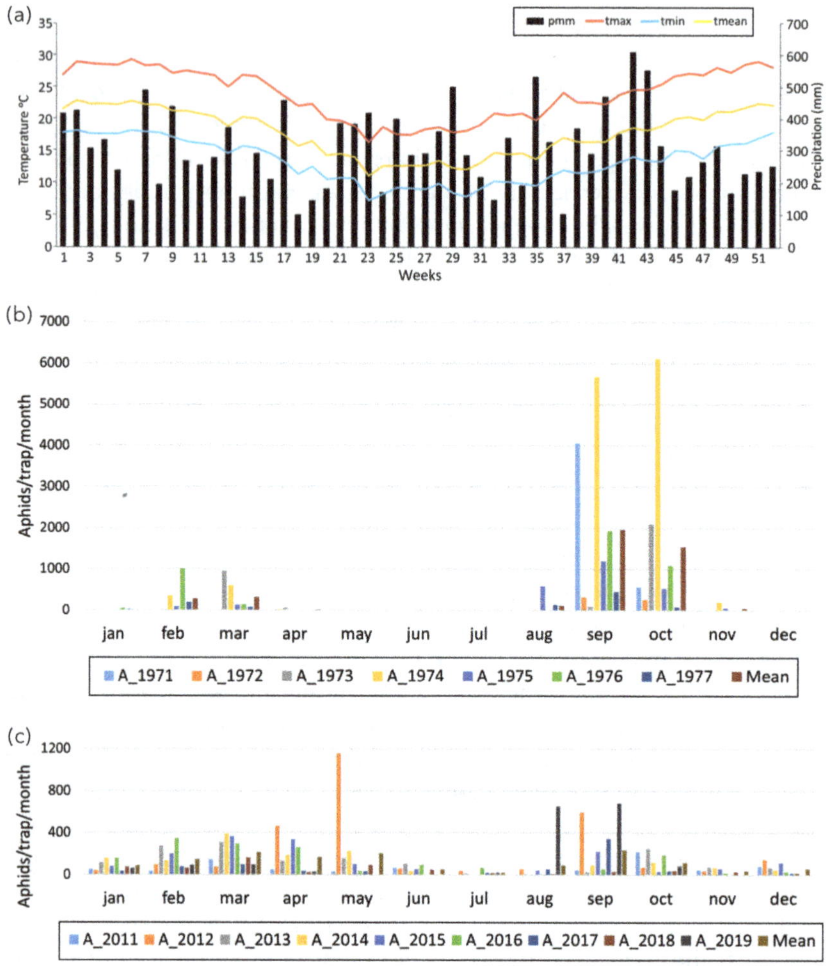

Figure 9 Seasonal variation on aphid populations and meteorological variables in southern Brazil (Passo Fundo, Rio Grande do Sul State). (a) Mean of minimum, mean, maximum air temperature (tmin, tmean and tmax) and accumulated precipitation (pmm) over the seven days preceding the sampling of aphids. (b) Mean population fluctuation of cereal aphids before the introduction of biological control program - series 1971-1977 (Caetano and Caetano, 1978). (c) Mean population fluctuation of cereal aphids - series 2011 to 2019 (Trapsystem, 2011 to 2019).

uzbezkistanicus, that efficiently parasitize aphids of the Macrosiphini tribe, may be the key elements for the reduction of *M. dirhodum* populations.

Another turning point was the implementation of the no-till system in Brazil. With the implementation of this system, some species of grasses began to be cultivated as a soil cover in annual rotation systems with wheat. One of the main species that began to be cultivated was black oats (*Avena strigosa* Schreb). This species is sown before of wheat and occupies large areas. In fact, currently in southern Brazil, wheat occupies less than 20% of the arable area, with most areas filled with oats. The sowing of oats in the beginning of autumn may have favored the populations of *R. padi* (which is also known as oat aphid), and *S. graminum,* both being the main aphids found in this culture. In this scenario, in autumn there is a large increase in areas with hosts favorable to the aphid multiplication (especially the two species of the Aphidini tribe). In traps, an autumnal peak of winged aphids is observed (Fig. 9). In this period also occurs a multiplicative phase of B/CYDVs. In addition to the increase in winged aphids, there is also an increase in the proportion of viruliferous aphids (Parizoto et al., 2013). Next, wheat sowing takes place. It is assumed that most of the aphid and virus inoculum in wheat come from oats. In relation to natural enemies, an increase in the parasitoid population can also be observed from winter onwards. The main species detected is *Aphidius platensis* (Santos et al., 2019). High populations of *A. platensis* multiply in the autumn on aphid populations of the Aphidini tribe. In addition, the multiplication of the parasitoid species *A. rhopalosiphi* and *A. uzbekistanicus* also occurs, to a lesser extent. The winter period, especially the months of June and July, is marked by a reduction in aphid populations. This significant reduction, however is not total. Unlike regions with high latitudes (temperate regions), winter is not as severe. Aphids are able to maintain asexual reproduction throughout this period. Species like *R. padi* are able to withstand frosts, remaining hidden close to the ground. From August, with the days getting longer and the temperature increasing, there is a new aphid population growth. This period is marked by a greater diversity in the population of aphids and parasitoids. Depending on the weather conditions, the aphid population curve grows faster, increasing the spread of the virus. At that time, aphids of the Aphidini tribe and also of the Macrosiphini tribe are found, mainly *S. avenae* and in a smaller number *M. dirhodum*. Parasitoid populations follow the growth of aphid populations with a slight delay (typical of a prey-predator curve). The presence of mummies is remarkable, and the percentage of parasitism reaches 60%. Then, with the senescence of winter cereals, aphid populations begin to decline significantly. Aphids can survive the summer in crops such as corn, sorghum and many are housed in voluntary wheat and oat plants that grow in the midst of corn and soybeans (the main summer crop in the region). Summer represents a strong bottleneck effect for aphids, B/CYDVs and also for parasitoids. In fact, one of the few species of parasitoids found in this period is *L. testaceipes*. The percentages of parasitism are usually low. The new increase of aphid populations will take place only from next March. The maximum temperatures in the region are not a limiting factor to the population growth of aphids, at least of the main current species (*R. padi*). Studies offering hosts (wheat) throughout the year demonstrate that summer is the most favorable period for the population growth of this species.

As temperatures drop, multiplication decreases, reaching its minimum in winter and increasing progressively in spring. So, the agricultural landscape and increased availability of hosts is one of the key factors for increasing and reducing aphid populations. Another factor that cannot be ignored is the profound effect of natural enemies. Studies with restricted action of natural enemies demonstrate the powerful capacity of natural enemies in reducing aphid populations. In summer, the action of predators is very significant and during autumn, winter and spring the action of parasitoids is important.

Another key factor is precipitation. In southern Brazil, precipitation is evenly distributed throughout the year (Fig. 9). There are no significant water restrictions that affect, for example, the development of hosts. However, there may be more or less dry years. Dry years are those with the highest populations of aphids and YDD epidemics. For example, for the 1970s, 1974 was a very dry year and had very large populations of aphids. In contrast, 1972 was the rainiest year in the series and aphid populations were very small. For the current years, the years 2012 and 2019 were the least rainy years and the populations of aphids were very high.

The yield loss caused by B/CYDVs in southern Brazil is estimated at around 20%. This value was obtained from a series of more than 10 years comparing plots with and without insecticide treatment. This value is a consequence of the susceptibility and intolerance of the cultivars and the moment of infection of the plants. Assays composed by cultivars recommended for the South region are conducted annually to evaluate the reaction to BYDV-PAV. Thirty cultivars representing more than 90% of the wheat-cultivated area are evaluated. According to the results, on average, commercial wheat cultivars, if inoculated at the beginning of their cycle, lose 40% of their productive potential. The most tolerant cultivars lose around 20% and there are those that lose above 80% (Fig. 10).

The second factor that determines the losses is the incidence of the virus in the field. This depends on the conditions of the year and the sowing time. If the sazonal curve of aphid populations presented above is considered, it is possible to infer that if sowing occurs earlier, even in autumn, the probability of infection by the virus is greater. The closer to winter, the less chance of the plant being infected at the beginning of the cycle. The amount of aphids and the incidence of B/CYDVs will depend on the temperature and rainfall. During warmer and drier years populations of aphids could be larger and the transmission of the virus high. Plot monitoring has already indicated infestation above 80% in extremely dry years. Therefore, 20% corresponds to an average, but the combination of intolerant cultivars and high incidence can cause losses between 60% and 80%.

The management of YDD in this region is based primarily on biological control. As pointed out, the successful biological control program brought the

Figure 10 Tolerance to BYDV and yield loss. (a) Intolerant cultivar. (b) Tolerant cultivar. (c) Frequency distribution of yield loss (%) caused by BYDV in Brazilian wheat cultivars infected at the beginning of the development cycle. In figures (a) and (b), in each figure the pots on the left contain plants that have been inoculated with BYDV and the pots on the right contain healthy plants. Yield loss caused by BYDV on grain production is estimated by comparing the 'Inoculated Plants' treatment (I) with the 'Non-Inoculated Plants' treatment (NI), according to the following formula: Yield loss% = (NI - I) / NI * 100, where: NI = yield/plant for the treatment of non-inoculated plants, and I = yield / plant for the treatment of inoculated plants. (N = 30 cultivars × 9 years = 270).

population levels of aphids to a level of 20% average damage in the absence of other control measures. Before this program, infestations of 100% of plants with viruses were a reality (Caetano, 1972). The second most used measure is chemical control in seed treatment and aerial application (Stoetzer, 2013; Stoetzer et al., 2014). Seed treatment is carried out with neonicotinoids (McKirdy and Jones 1996). This could be a problem in the near future, as in some regions of the world, like Europe, this type of active ingredient has already been banned. Success of seed treatment depends on the region and time of sowing. When the effect of seed treatment is finished, insecticides are applied on aerial part. The action threshold currently adopted in Brazil is 10% of plants with aphids. Products with varied active ingredients are applied (predominantly mixtures of pyrethroids with neonicotinoids). There are no reports of the existence of aphid populations resistant to active ingredients. But in fact, this monitoring is not carried out.

The great challenge to decision making for aerial application of insecticides is the monitoring of aphids. As technical assistance needs to cover large areas, in many cases the easy solution is application based on calendar system. In order to circumvent this problem, a network monitoring system using traps has been adopted. The idea is to better understand the patterns of aphid populations by region, relating them to damage levels. In this way, it is necessary to create tools to facilitate and expand monitoring by automation.

Recently, computer vision techniques have been developed to facilitate the counting of aphids in traps (Lins et al., 2020), aiming to make the process less dependent on human labor. The success of such techniques could expand the monitoring networks and improve the understanding of the aphid population patterns, increasing the precision of decision making for management.

7 Conclusions and future trends

The grass-B/CYDVs-aphid system is constantly adapting and changing. As much as this system is already dispersed across the globe, some species of grass-viruses and vectors can be introduced in areas that were not yet present, altering the original balance. With global warming the patterns of aphid populations have also been changing. Therefore, the risk for each region needs to be recalculated based on these new scenarios.

Surveys with specialists point out that there are many diagnostic difficulties, with confusion of symptoms. Consequently, there is an inaccuracy in the estimates of the impact that the disease causes in several regions. Proper management of this pathosystem requires correct diagnosis, damage estimates, knowledge of epidemiological factors and the effectiveness of available management tools.

The first step is the biological balance between aphid populations and natural enemies. This balance brings epidemic to more reasonable levels where other tools can be used. In the host field, resistance to the vector and the virus should be sought. Especially now, that some widely used insecticides, such as neonicotinoids, are being banned.

In the field of vector monitoring platforms with alert systems and technologies that facilitate monitoring as computer vision should be pursued. For future perspectives of bioagent managements within YDD epidemiological risk, the monitoring and modeling should be performed considering the local and landscape scales of trophic interactions, accounting the ecosystem services in temporal dimensions, and the viruliferous aphid densities oscillations.

8 Where to look for further information

For more information we recommend the following reviews and books:

Aphids as Crop Pests (2017) Edited by: Helmut van Emden, University of Reading, UK, Richard Harrington, Rothamsted Research, Harpenden, UK. This book contemplates several aspects and advances in knowledge about aphids.

Barley Yellow Dwarf: 40 Years of Progress (1995) Edited by Cleora J. D'Arcy, Peter A. Burnett APS Press, 1995. This book contains an important historical synthesis of knowledge about BYDV.

Viruses and Virus Diseases of Poaceae (Gramineae) (2004) Edited by: Hervé Lapierre, Pierre-A. Signoret Science Publishers, U.S. Recommended for those who would like an overview of grass viruses.

Truol, G. (2009). Enfermedades virales asociadas al cultivo de trigo en Argentina: reconocimiento, importancia, formas de transmisión y manejo (77 pp). Córdoba: Biglia Press. Important book for those who would like to have information about viruses and their vectors in wheat in South America (Argentina).

A very interesting example of a management tool for B/CYDVs can be found at https://ahdb.org.uk/bydv. At https://ahdb.org.uk/aphid-news you can also find updated newsletters on populations in the United Kingdom. Unfortunately, monitoring networks like this are not yet a reality in many countries suffering from YDD epidemics.

9 References

Asin, L. and Pons, X. (2001). Effect of high temperature on the growth and reproduction of corn aphids (Homoptera: Aphididae) and implications for their population dynamics on the northeastern Iberian Peninsula, *Environmental Entomology* 30(6), 1127-1134. doi: 10.1603/0046-225X-30.6.1127.

Auad, A. M., Alves, S. O., Carvalho, C. A., Silva, D. M., Resende, T. T. and Veríssimo, B. A. (2009). The impact of temperature on biological aspects and life table of *Rhopalosiphum padi* (Hemiptera: Aphididae) fed with signal grass, *Florida Entomologist* 92(4), 569-577. doi: 10.1653/024.092.0406.

Baltenberger, D. E., Ohm, H. W. and Foster, J. E. (1987). Reactions of oat, barley, and wheat to infection with barley yellow dwarf virus Isolates1, *Crop Science* 27(2), 195-198.

Bencharki, B., Mutterer, J., El Yamani, M. E., Ziegler-Graff, V., Zaoui, D. and Jonard, G. (1999). Severity of infection of Moroccan barley yellow dwarf virus PAV isolates correlates with variability in their coat protein sequences, *The Annals of Applied Biology* 134(1), 89–99.

Bisnieks, M., Kvarnheden, A., Turka, I. and Sigvald, R. (2006). Occurrence of barley yellow dwarf virus and cereal yellow dwarf virus in pasture grasses and spring cereals in Latvia, *Acta Agriculturae Scandinavica, Section B - Plant Soil Science* 56(3), 171–178.

Borer, E. T., Seabloom, E. W., Mitchell, C. E. and Power, A. G. (2010). Local context drives infection of grasses by vector-borne generalist viruses, *Ecology Letters* 13(7), 810–818.

Boulila, M. (2011). Selective constraints, molecular recombination structure and phylogenetic reconstruction of isometric plant RNA viruses of the families Luteoviridae and Tymoviridae, *Biochimie* 93(2), 242–253.

Brewer, M. J., Peairs, F. B. and Elliott, N. C. (2019). Invasive cereal aphids of North America: ecology and pest management, *Annual Review of Entomology* 64, 73–93. doi: 10.1146/annurev-ento-011118-111838.

Brodeur, J., Hajek, A. E., Heimpel, G. E., Sloggett, J. J., Mackauer, M., Pell, J. K. and Völkl, W. (2017). Predators, parasitoids and pathogens. In: van Emden, H. F. and Harrington, R. (Eds), *Aphids as Crop Pests* (2nd edn.), pp. 225–261. Wallingford, UK: CABI.

CABI (2020a). *Sipha flava* (Forbes). In: Invasive Species Compendium. Wallingford, UK: CAB International. www.cabi.org/isc (accessed 01 July 2020).

CABI (2020b). *Sipha maydis* (Passerini). In: Invasive Species Compendium. Wallingford, UK: CAB International. www.cabi.org/isc (accessed 10 June 2020).

Caetano, V. da R. (1968). Nota prévia sobre a ocorrência de uma virose em cereais de inverno no Rio Grande do Sul, *Revista da Sociedade Brasileira de Fitopatologia* 2, 53–66.

Caetano, V. da R. (1972). Estudo sobre o vírus do nanismo amarelo da cevada, em trigo, no Rio Grande do Sul. Tese (Doutorado), Universidade de São Paulo-ESALQ, Piracicaba.

Caetano, V. da R. (1973). Estudos sobre os afídeos vetores do vírus do nanismo amarelo da cevada, em especial de Acyrthosiphon dirhodum, em trigo no Sul do Brasil. Tese (Doutorado), Programa de Pós-Graduação em Biologia, Universidade Estadual de Campinas, Campinas.

Caetano, V. da R. (1982). Viroses. In: Eduardo A. Osório (Ed), *Trigo no Brasil*, vol. 2, pp. 545–579. Campinas: Fundação Cargill.

Caetano, V. da R.and Caetano, V. da R. (1978). Variações sazonais de pulgões de trigo. In: EMBRAPA. Centro Nacional de Pesquisa de Trigo. Solos e técnicas culturais, economia e sanidade. Passo Fundo, vol. 2, pp. 102–114. X Reunião Anual Conjunta de Pesquisa de Trigo, Porto Alegre.

Carver, M. (1989). Biological control of aphids. In: Minks, A. K. and Harrewijn, P. (Eds), *Aphids: Their Biology, Natural Enemies and Control*, pp. 141–165. Amsterdam: Elsevier.

Chalhoub, B. A., Kelly, L., Robaglia, C. and Lapierre, H. D. (1994). Sequence variability in the genome-3′-terminal region of BYDV for 10 geographically distinct PAV-like isolates of barley yellow dwarf virus: analysis of the ORF6 variation, *Archives of Virology* 139(3–4), 403–416.

Chare, E. R. and Holmes, E. C. (2006). A phylogenetic survey of recombination frequency in plant RNA viruses, *Archives of Virology* 151(5), 933–946.

Corrales, C. E., Castro, A. M., Ricci, M. and Dixon, A. F. G. (2007). *Sipha maydis*: distribution and Host Range of a New Aphid Pest of Winter Cereals in Argentina, *Journal of Economic Entomology* 100(6), 1781-1788. doi: 10.1603/0022-0493(2007)100[1781:smdahr]2.0.co;2.

D'Arcy, C. J. and Burnett, P. A. (1995). *Barley Yellow Dwarf–40 Years of Progress*. USA: APS Press.

Davis, T. S., Abatzoglou, J. T., Bosque-Pérez, N. A., Halbert, S. E., Pike, K. and Eigenbrode, S. D. (2014). Differing contributions of density dependence and climate to the population dynamics of three eruptive herbivores, *Ecological Entomology* 39(5), 566-577. doi: 10.1111/een.12134.

Dean, G. J. (1974). Effect of temperature on the cereal aphids *Metopolophium dirhodum* (Wlk.), *Rhopalosiphum padi* (L.) and *Macrosiphum avenue* (F.) (Hem., Aphididae), *Bulletin of Entomological Research* 63(3), 401-409.

Dedryver, C. A. (2018). Encyclop'Aphid, Aphids and their environment, Antagonists, Entomophthorales. https://www6.inrae.fr/encyclopedie-pucerons_eng/Aphids-and-their-environment/Antagonists/Entomorphorales (accessed 15 June 2020).

Delmiglio, C., Pearson, M. N., Lister, R. A. and Guy, P. L. (2010). Incidence of cereal and pasture viruses in New Zealand's native grasses, *Annals of Applied Biology* 157(1), 25-36.

Dempster, L. C. and Holmes, S. J. I. (1995). The incidence of strains of barley yellow dwarf virus in perennial ryegrass crops in south-west and central Scotland, *Plant Pathology* 44(4), 710-717.

Domier, L. L. (2012). *Family Luteoviridae*. Oxford: Elsevier/Academic Press.

Domier, L. L., McCoppin, N. K., Larsen, R. C. and D'Arcy, C. J. (2002). Nucleotide sequence shows that Bean leafroll virus has a luteovirus-like genome organization, *Journal of General Virology* 83(7), 1791-1798.

Doodson, J. K. (1967). A survey of barley yellow dwarf virus inS.24 perennial ryegrass in England and Wales , *Plant Pathology* 16(1), 42-45.

Embrapa (1982). Relatório técnico anual do centro de pesquisa do trigo, 1979-1980. Passo-Fundo, Empresa Brasileira de Pesquisa Agropecuária, CNPT, 176p.

Fargette, D., Lister, R. and Hood, E. (1982). Grasses as a reservoir of barley yellow dwarf virus in Indiana, *Plant Disease* 66(1), 1041-1045.

Fargette, D., Pinel-Galzi, A., Sereme, D., Lacombe, S., Hebrard, E., Traoré, O. and Konaté, G. (2008). Diversification of rice yellow mottle virus and related viruses spans the history of agriculture from the Neolithic to the present, *PLoS Pathogens* 4(8), e1000125.

Finlay, K. J. and Luck, J. E. (2011). Response of the bird cherry-oat aphid (*Rhopalosiphum padi*) to climate change in relation to its pest status, vectoring potential and function in a crop-vector-virus pathosystem, *Agriculture, Ecosystems and Environment* 144(1), 405-421. doi: 10.1016/j.agee.2011.08.011.

Garrett, K. A., Dendy, S. P., Power, A. G., Blaisdell, G. K., Alexander, H. M. and McCarron, J. K. (2004). Barley yellow dwarf disease in natural populations of dominant tallgrass prairie species in Kansas, *Plant Disease* 88(5), 574-574.

Gassen, D. N. and Tambasco, F. J. (1983). Controle biológico dos pulgões do trigo no Brasil. Informe Agropecuário. *Belo Horizonte* 9(104), p. 49-51.

Gilabert, A., Gauffre, B., Parisey, N., Le Gallic, J., Lhomme, P., Bretagnolle, V., Dedryver, C., Baudry, J. and Plantegenest, M. (2017). Influence of the surrounding landscape on the colonization rate of cereal aphids and phytovirus transmission in autumn, *Journal of Pest Science* 90(2), 447-457.

Gonçalves, M. C., Godinho, M., Alves-Freitas, D. M. T., Varsani, A. and Ribeiro, S. G. (2017). First report of maize yellow mosaic virus infecting maize in Brazil, *Plant Disease* 101(12), 2156.

Gray, S. M., Chapin, J. W., Smith, D. M., Banerjee, N. and Thomas, J. S. (1998). Barley yellow dwarf luteoviruses and their predominant aphid vectors in winter wheat grown in South Carolina, *Plant Disease* 82(12), 1328–1333.

Griesbach, J. A., Steffenson, B. J., Brown, M. P., Falk, B. W. and Webster, R. K. (1990). Infection of grasses by barley yellow dwarf viruses in California, *Crop Science* 30(6), 1173–1177.

Guy, P., Johnstone, G. and Morris, D. (1987). Barley yellow dwarf viruses in, and aphids on, grasses (including cereals) in Tasmania, *Australian Journal of Agricultural Research* 38(1), 139–152.

Halbert, S. E., Connelly, B. J., Lister, R. M., Klein, R. E. and Bishop, G. W. (1992). Vector specificity of barley yellow dwarf virus serotypes and variants in southwestern Idaho, *Annals of Applied Biology* 121(1), 123–132.

Hall, G. (2006). Selective constraint and genetic differentiation in geographically distant barley yellow dwarf virus populations, *Journal of General Virology* 87(10), 3067–3075.

Hall, G. S. and Little, D. P. (2013). Within-host competition between barley yellow dwarf-PAV and -PAS, *Virus Research* 174(1–2), 148–151.

Harrington, R. and Clark, S. (2010). Trends in the timings of the start and end of annual flight periods. In: Kindlmann, P., Dixon, A. and Michaud, J. (Eds), *Aphid Biodiversity under Environmental Change*. Dordrecht: Springer. doi: 10.1007/978-90-481-8601-3_3.

Harrington, R., Clark, S. J., Welham, S. J., Verrier, P. J., Denholm, C. H., Hullé, M., Maurice, D., Rounsevell, M. D. and Cocu, N. (2007). Environmental change and the phenology of European aphids, *Global Change Biology* 13(8), 1550–1564. doi: 10.1111/j.1365-2486.2007.01394.x.

Henry, M. and Dedryver, C. A. (1991). Occurrence of barley yellow dwarf virus in pastures of western France, *Plant Pathology* 40(1), 93–99.

Henry, M., Francki, R. I. B. and Wallwork, H. (1992). Occurrence of barley yellow dwarf virus in cereals and grasses of the low-rainfall wheatbelt of South Australia, *Plant Pathology* 41(6), 713–721.

Henry, M., George, S., Arnold, G. M., Dedryver, C. A., Kendall, D. A., Robert, Y. and Smith, B. D. (1993). Occurrence of barley yellow dwarf virus (BYDV) isolates in different farmland habitats in western France and south-west England, *Annals of Applied Biology* 123(2), 315–329.

Hodge, B. A., Paul, P. A. and Stewart, L. R. (2020). Occurrence and high throughput sequencing of viruses in Ohio wheat, *Plant Disease* 104(6), 1789–1800.

Huggett, D. A. J., Leather, S. R. and Walters, K. F. A. (1999). Suitability of the biomass crop Miscanthus sinensis as a host for the aphids Rhopalosiphum padi (L.) and Rhopalosiphum maidis (f.), and its susceptibility to the plant Luteovirus Barley Yellow Dwarf Virus, *Agricultural and Forest Entomology* 1(2), 143–149.

Jarošová, J., Chrpová, J., Šíp, V. and Kundu, J. K. (2013). A comparative study of the Barley yellow dwarf virus species PAV and PAS: distribution, accumulation and host resistance, *Plant Pathology* 62(2), 436–443.

Jing-Quan, G., Lapierre, H. and Moreau, J. P. (1997). Vectoring ability of aphid clones of Rhopalosiphum padi (L.) and Sitobion avenae (Fabr.) and their capacity to retain barley yellow dwarf virus, *Annals of Applied Biology* 131(2), 179–188.

Jones, R. A. C. (2014). Plant virus ecology and epidemiology: historical perspectives, recent progress and future prospects, *Annals of Applied Biology* 164(3), 320-347.

Kapooria, R. G. and Ndunguru, J. (2004). Occurrence of viruses in irrigated wheat in Zambia, EPPO Bulletin, *EPPO Bulletin* 34(3), 413-419.

Kendall, D. A., George, S. and Smith, B. D. (1996). Occurrence of barley yellow dwarf viruses in some common grasses (Gramineae) in south west England, *Plant Pathology* 45(1), 29-37.

Kendig, A. E., Borer, E. T., Mitchell, C. E., Power, A. G. and Seabloom, E. W. (2017). Characteristics and drivers of plant virus community spatial patterns in US west coast grasslands, *Oikos* 126(9), 1281-1290.

Krueger, E. N., Beckett, R. J., Gray, S. M. and Miller, W. A. (2013). The complete nucleotide sequence of the genome of Barley yellow dwarf virus-RMV reveals it to be a new Polerovirus distantly related to other yellow dwarf viruses, *Frontiers in Microbiology* 4, 205.

Lamptey, J. N. L., Plumb, R. T. and Shaw, M. W. (2003). Interactions between the grasses Phalaris arundinacea, Miscanthus sinensis and Echinochloa crus-galli, and barley and cereal yellow dwarf viruses, *Journal of Phytopathology* 151(7-8), 463-468.

Lanzarini, A. C., Schons, J., Salvadori, J. R., Nienow, A. A., Nicolini-Teixeira, F., Binotto-Missiura, F. and Deuner, E. (2007). Avaliação de danos causados pelo Barley yellow dwarf virus: PAV em cultivares de trigo no Brasil, *Fitopatologia Brasileira* 32(6), 512-517.

Leclercq-Le Quillec, F., Plantegenest, M., Riault, G. and Dedryver, C. A. (2000). Analyzing and modeling temporal disease progress of Barley yellow dwarf virus serotypes in barley fields, *Phytopathology* 90(8), 860-866.

Lefkowitz, E. J., Dempsey, D. M., Hendrickson, R. C., Orton, R. J., Siddell, S. G. and Smith, D. B. (2018). Virus taxonomy: the database of the International Committee on Taxonomy of Viruses (ICTV), *Nucleic Acids Research* 46(D1), D708-D717.

Lins, E. A., Rodriguez, J. P. M., Scoloski, S. I., Pivato, J., Lima, M. B., Fernandes, J. M. C., Pereira, P. R. V. da S., Lau, D. and Rieder, R. (2020). A method for counting and classifying aphids using computer vision, *Computers and Electronics in Agriculture* 169, 1-11. ISSN 0168-1699, https://doi.org/10.1016/j.compag.2019.105200.

Lister, R., Clement, D., Skaria, M. and Foster, J. (1984). Biological differences between barley yellow dwarf viruses in relation to their epidemiology and host reactions. *Barley Yellow Dwarf Workshop*, Mexico, 16-25.

Liu, Y., Zhai, H., Zhao, K., Wu, B. and Wang, X. (2012). Two suppressors of RNA silencing encoded by cereal-infecting members of the family Luteoviridae, *Journal of General Virology* 93(8), 1825-1830.

Lucio-Zavaleta, E., Smith, D. M. and Gray, S. M. (2001). Variation in transmission efficiency among Barley yellow dwarf virus-RMV isolates and clones of the normally inefficient aphid vector, Rhopalosiphum padi, *Phytopathology* 91(8), 792-796.

Macfadyen, S. and Kriticos, D. J. (2012). Modelling the geographical range of a species with variable life-history, *PLoS ONE* 7(7), e40313. doi: 10.1371/journal.pone .0040313.

Makkouk, K. M., Najar, A. and Kumari, S. G. (2001). First record of Barley yellow dwarf virus and Cereal yellow dwarf virus in Tunisia, *Plant Pathology* 50(6), 806.

Malmstrom, C. M., Shu, R., Linton, E. W., Newton, L. A. and Cook, M. A. (2007). Barley yellow dwarf viruses (BYDVs) preserved in herbarium specimens illuminate historical disease ecology of invasive and native grasses, *Journal of Ecology* 95(6), 1153-1166.

Mar, T. B., Lau, D., Schons, J., Yamazaki-Lau, E. and Nhani Jr., A. (2013). Molecular identification based on coat protein sequences of the Barley yellow dwarf virus from Brazil, *Scientia Agricola* 70(6), 428–434.

Masterman, A. J., Holmes, S. J. and Foster, G. N. (1994). The role of Poa annua in the epidemiology of barley yellow dwarf virus in autumn-sown cereals, *Plant Pathology* 43(4), 621–626.

Mayo, M. A. (2002). Virology division news: ICTV at the Paris ICV: results of the plenary session and the binomial ballot, *Archives of Virology* 147(11), 2254–2260.

McElhany, P., Real, L. A. and Power, A. G. (1995). Vector preference and disease dynamics: a study of barley yellow dwarf virus, *Ecology* 76(2), 444–457.

McKirdy, S. J. and Jones, R. A. C. (1996). Use of Imidacloprid and newer generation synthetic pyrethroids to control the spread of barley yellow dwarf Luteovirus in cereals, *Plant Disease* 80(8), 895–901.

Miller, W. A., Liu, S. and Beckett, R. (2002). Barley yellow dwarf virus: Luteoviridae or Tombusviridae?, *Molecular Plant Pathology* 3(4), 177–183.

Moriones, E., Ortego, F., Ruiz-Tapiador, M., Gutiérrez, C., Castañera, P. and García-Arenal, F. (1993). Epidemiology of RPV- and PAV-like barley yellow dwarf viruses on winter barley in central Spain, *Crop Protection* 12(3), 224–228.

Najar, A., Hamdi, I. and Varsani, A. (2017). Barley yellow dwarf virus in barley crops in Tunisia: prevalence and molecular characterization, *Phytopathologia Mediterranea* 56(1), 111–118.

Nancarrow, N., Aftab, M., Freeman, A., Rodoni, B., Hollaway, G. and Trębicki, P. (2018). Prevalence and incidence of yellow dwarf viruses Across a climatic gradient: a four-year field study in Southeastern Australia, *Plant Disease* 102(12), 2465–2472.

Ortiz-Martínez, S. A. and Lavandero, B. (2018). The effect of landscape context on the biological control of Sitobion avenae: temporal partitioning response of natural enemy guilds. *Journal of Pest Science* 91, 41–53. https://doi.org/10.1007/s10340-017-0855-y.

Ortiz-Martínez, S., Pierre, J. S., van Baaren, J., Le Lann, C., Zepeda-Paulo, F. and Lavandero, B. (2019). Interspecific competition among aphid parasitoids: molecular approaches reveal preferential exploitation of parasitized hosts, *Scientific Reports* 9(1), 19641. doi: 10.1038/s41598-019-56187-3.

Ortiz-Martínez, S., Staudacher, K., Baumgartner, V., Traugott, M. and Lavandero, B. (2020). Intraguild predation is independent of landscape context and does not affect the temporal dynamics of aphids in cereal fields, *Journal of Pest Science* 93(1), 235–249. doi: 10.1007/s10340-019-01142-4.

Pagan, I. and Holmes, E. C. (2010). Long-term evolution of the Luteoviridae: time scale and mode of virus speciation, *Journal of Virology* 84(12), 6177–6187.

Pakdel, A., Afsharifar, A., Niazi, A., Almasi, R. and Izadpanah, K. (2010). Distribution of cereal luteoviruses and molecular diversity of BYDV-PAV isolates in central and Southern Iran: proposal of a new species in the genus Luteovirus, *Journal of Phytopathology* 158(5), 357–364.

Paliwal, Y. C. (1982). Role of perennial grasses, winter wheat, and aphid vectors in the disease cycle and epidemiology of barley yellow dwarf virus, *Canadian Journal of Plant Pathology* 4(4), 367–374.

Parizoto, G., Rebonatto, A., Schons, J. and Lau, D. (2013). Barley yellow dwarf virus-PAV in Brazil: seasonal fluctuation and biological characteristics, *Tropical Plant Pathology* 38(1), 11–19.

Pereira, P. R. da S. and Salvadori, J. R. (2005). *Identificação dos principais parasitoides (Hymenoptera: Aphelinidae e Braconidae, Aphidinae) envolvidos no controle biológico de pulgões (Hemiptera: Aphididae) em trigo no sul do Brasil*, 8 p., Passo Fundo: Embrapa Trigo. il. color. (Embrapa Trigo. Comunicado técnico, 22). http://ainfo.cnptia.embrapa.br/digital/bitstream/item/84170/1/CNPT-COM.-TEC.-22-05.pdf (accessed 13 July 2020).

Pimenta, H. R. and Smith, J. G. (1976). *Afídeos, seus danos e inimigos naturais em plantações de trigo (Triticum sp.) no Estado do Paraná*. Curitiba: OCEPAR.

Plumb, R. T. (2002). Viruses of Poaceae: a case history in plant pathology, *Plant Pathology* 51(6), 673–682.

Power, A. and Gray, S. (1995). Aphid transmission of barley yellow dwarf viruses: interactions between viruses, vectors, and host plants. In: D'arcy, C. J. and Burnett, P. A. (Eds), *Barley Yellow Dwarf–40 Years of Progress*, pp. 259–289, USA: APS Press.

Power, A. G. (1991). Virus spread and vector dynamics in genetically diverse plant populations, *Ecology* 72(1), 232–241.

Power, A. G., Seaman, A. J. and Gray, S. M. (1991). Aphid transmission of barley yellow dwarf virus: inoculation access periods and epidemiological implications, *Phytopathology* 81(5), 545–548.

Reavy, B. and Mayo, M. A. (2002). Persistent transmission of luteoviruses by aphids, *Advances in Botanical Research* 36, 21–46.

Rebonatto, A., Salvadori, J. R. and Lau, D. (2015). Temporal changes in cereal aphids (Hemiptera: Aphididae) populations in northern Rio Grande do Sul, Brazil, *Journal of Agricultural Science* 7(10), 71–78. doi: 10.5539/jas.v7n10p71.

Reed, D. and Kindler, S. (1994). Report of trip to Argentina and Chile. *Proceedings 6th Russian Wheat Aphid Workshop*, 163–168.

Reunião da comissão Brasileira de pesquisa de trigo e triticale (2018). *Passo Fundo, R. S. Informações Técnicas Para Trigo e Triticale–Safra 2019*. Brasília, DF: Embrapa.

Robert, Y. (2001). Epidemiology of plant virus diseases. In eLS (Ed). doi: 10.1038/npg.els.0000759.

Robertson, N. L. and French, R. (2007). Genetic structure in natural populations of barley/cereal yellow dwarf virus isolates from Alaska, *Archives of Virology* 152(5), 891–902.

Rochow, W. F. (1969). Biological properties of four isolates of barley yellow dwarf virus, *Phytopathology* 59(11), 1580–1589.

Rochow, W. F. (1979). Field variants of barley yellow dwarf virus: detection and fluctuation during twenty years, *Phytopathology* 69(6), 655–660.

Rúa, M. A., McCulley, R. L. and Mitchell, C. E. (2014). Climate drivers, host identity and fungal endophyte infection determine virus prevalence in a grassland ecosystem, *Journal of Ecology* 102(3), 690–699.

Sadeghi, E., Dedryver, C. A. and Gauthier, J. P. (1997). Role of acquisition and inoculation time in the expression of clonal variation for BYDV-PAV transmission in the aphid species Rhopalosiphum padi, *Plant Pathology* 46(4), 502–508.

Salvadori, J. R. and Salles, L. A. B. (2002). Controle biológico dos pulgões do trigo. In: Parra, J. R. P. et al. (Eds), *Controle Biológico no Brasil: parasitóides e predadores* (vol. 26), pp. 427–447. São Paulo: Manole.

Salvadori, J. R. and Tonet, G. E. L. (2001). *Manejo integrado dos pulgões de trigo*. Passo Fundo: Embrapa Trigo, 52 p. (Embrapa trigo. Documentos, 34). http://ainfo.cnptia.

embrapa.br/digital/bitstream/item/36617/1/Manejo-integrado-dos-pulgoes-de-trigo.pdf (accessed 13 July 2020)..

Santos, C. D. R. D., Sampaio, M. V., Lau, D., Redaelli, L. R., Jahnke, S. M., Pivato, J. and Carvalho, F. J. (2019). Taxonomic Status and Population Oscillations of Aphidius colemani Species Group (Hymenoptera: Braconidae) in Southern Brazil, *Neotropical Entomology* 48(6), 983–991. doi: 10.1007/s13744-019-00716-2.

Schmidt, M. H., Lauer, A., Purtauf, T., Thies, C., Schaefer, M. and Tscharntke, T. (2003). Relative importance of predators and parasitoids for cereal aphid control, *Proceedings. Biological Sciences* 270(1527), 1905–1909.

Siddiqui, N. N., Ilyas, M., Mansoor, S., Azhar, A. and Saeed, M. (2012). Cloning and phylogenetic Analysis of Coat Protein of Barley yellow dwarf virus Isolates from Different Regions of Pakistan, *Journal of Phytopathology* 160(1), 13–18.

Silva, A. G. d'A. e, Gonçalves, C. R., Galvão, D. M., Gonçalves, A. J. L., Gomes, J., Silva, M., do N. and Simoni, L. de. (1968). Quarto catálogo dos insetos que vivem nas plantas do Brasil—Seus parasitas e predadores. *Rio de Janeiro: Minist. Agric.* (1) 622.

Smirnova, E., Firth, A. E., Miller, W. A., Scheidecker, D., Brault, V., Reinbold, C., Rakotondrafara, A. M., Chung, B. Y. and Ziegler-Graff, V. (2015). Discovery of a small non-AUG-initiated ORF in poleroviruses and luteoviruses that is required for long-distance movement, *PLOS Pathogens* 11(5), e1004868.

Smith, G. R., Borg, Z., Lockhart, B. E. L., Braithwaite, K. S. and Gibbs, M. J. (2000). Sugarcane yellow leaf virus: a novel member of the Luteoviridae that probably arose by inter-species recombination, *Journal of General Virology* 81(7), 1865–1869.

Spiesman, B. J., Iuliano, B. G. and Gratton, C. (2020). Temporal resource continuity increases predator abundance in a metapopulation model: insights for conservation and biocontrol. bioRxiv 2020.03.25.006882; doi: https://doi.org/10.1101/2020.03.25.006882.

Stoetzer, A. (2013). Afídeos vetores de vírus em trigo e cevada em Guarapuava-PR: Monitoramento, manejo de inseticidas e análise econômica associada ao controle químico. Dissertação (mestrado em Agronomia). Universidade Estadual do Centro-oeste, Guarapuava.

Stoetzer, A., Kawakami, J., Marsaro Júnior, A. L., Lau, D., Pereira, P. R. V. da S. and Antoniazzi, N. (2014). Protective effect and economic impact of insecticide application methods on barley, *Pesquisa Agropecuária Brasileira* 49(3), 153–162. doi: 10.1590/S0100-204X2014000300001.

Sturza, V. S., Poncio, S., dos Santos, A. Bd and Lopes-da-Silva, M. (2012). Infestation and natural parasitism of aphids in single and mixed pastures of black oats and ryegrass, *Revista Brasileira de Entomologia* 56(3), 363–367. doi: 10.1590/S0085-56262012005000052.

Tofangsazi, N., Kheradmand, K., Shahrokhi, S. and Talebi, A. A. (2010). Temperature-dependent life history of *Schizaphis graminum* on barley, *Bulletin of Insectology* 63, 79–84.

Toft, S. (2005). The quality of aphids as food for generalist predators: implications for natural control of aphids, *European Journal of Entomology* 102(3), 371–383.

Van den Eynde, R., Van Leeuwen, T. and Haesaert, G. (2020). Identifying drivers of spatio-temporal dynamics in barley yellow dwarf virus epidemiology as a critical factor in disease control, *Pest Management Science* 76(8), 2548–2556.

Waquil, J. M., Lopes-da-Silva, M., Salvadori, J. R., Lazzarotto, C. M., Pereira, P. R. V. da S. and Lau, D. (2015). Pulgão-Verde-dos-cereais, Schizaphis graminum (Rondani). In: Vilela,

E. F. and Zucchi, R. A. (Eds), *Pragas introduzidas no Brasil: insetos e ácaros* (vol. 38), pp. 653–681. Piracicaba: FEALQ.

Wang, H., Wu, K., Liu, Y., Wu, Y. and Wang, X. (2015). Integrative proteomics to understand the transmission mechanism of Barley yellow dwarf virus-GPV by its insect vector Rhopalosiphum padi, *Scientific Reports* 5(1), 10971.

Wu, B., Blanchard-Letort, A., Liu, Y., Zhou, G., Wang, X. and Elena, S. F. (2011). Dynamics of Molecular Evolution and Phylogeography of Barley yellow dwarf virus-PAV, *PLoS ONE* 6(2), e16896.

Yu, W., Xu, Z., Francis, F., Liu, Y., Cheng, D., Bragard, C. and Chen, J. (2013). Variation in the transmission of barley yellow dwarf virus-PAV by different Sitobion avenae clones in China, *Journal of Virological Methods* 194(1–2), 1–6.

Zhang, W., Cheng, Z., Xu, L., Wu, M., Waterhouse, P., Zhou, G. and Li, S. (2009). The complete nucleotide sequence of the barley yellow dwarf GPV isolate from China shows that it is a new member of the genus Polerovirus, *Archives of Virology* 154(7), 1125–1128.

Zhang, K., Pan, Q., Yu, D., Wang, L., Liu, Z., Li, X. and Liu, X. (2019). Systemically modeling the relationship between climate change and wheat aphid abundance, *Science of the Total Environment* 674, 392–400. doi: 10.1016/j.scitotenv.2019.04.143.

Zuñiga-Salinas, E. S. (1982). Controle biológico dos afídeos do trigo (Homoptera: Aphididae) por meio de parasitoides no planalto médio do Rio Grande do Sul, Brasil. Tese (Doutorado em Entomologia). Programa de Pós-Graduação em Ciências, Universidade Federal do Paraná, Curitiba.

Chapter 21

Resistance breeding in barley against Barley yellow dwarf virus (BYDV): avoiding negative impacts on anatomy and physiology

Torsten Will, Frank Ordon and Dragan Perovic, Julius Kühn-Institute (JKI), Federal Research Centre for Cultivated Plants, Germany

1 Introduction

Barley yellow dwarf (BYD) is one of the most widespread and damaging viral diseases of grasses and cereal crops (D'Arcy, 1995). The virus is phloem restricted and is transmitted in a persistent manner by many aphid species (Fig. 1a), for example, *Rhopalosiphum padi* and *Sitobion avenae* (Slykhuis, 1967; Jensen, 1969). Visible symptoms of a Barley yellow dwarf virus (BYDV) infection of barley are stunted growth and yellow discoloration of leaves. As a main consequence, BYDV infection leads to a collapse of sieve elements accompanied by an accumulation of 'wound gum' resulting in necrosis (Esau, 1957), most likely leading to a breakdown of nutrition transport inside plants at an early developmental stage. In case that infected plants produce ears, BYDV negatively affects the number and size of kernels per ear (Edwards et al., 2001). The nutritional value, that is, protein content, in general, increases, but the disease has a negative effect on malting quality, leading to, for example, increased malting losses and haze formation in beer (Burger and LaBerge, 1985).

BYDV, the type member of the genus *Luteovirus* (Mayo and Pringle, 1998), is distributed worldwide (Lister and Ranieri, 1995). In the late 1960s and early 1970s, Rochow (1969) and Rochow and Muller (1971) discovered

http://dx.doi.org/10.19103/AS.2021.0092.36

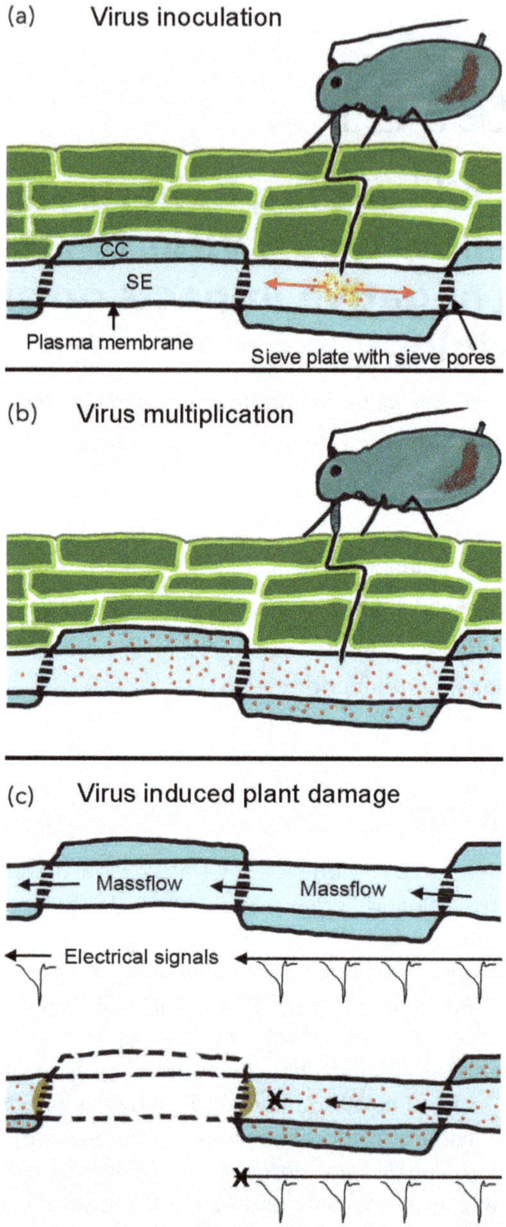

Figure 1 Graphical illustration of BYDV infection and how it affects phloem functions. (a) Together with watery saliva (orange dots) aphids transmit BYDV (red dots) into sieve elements (SE) of barley. Longitudinal ends of SEs form sieve plates that are permeated by sieve pores (SP) and connect adjacent SEs to sieve tubes. There BYDV most likely travels from the site of injection along with and against the massflow (red arrows), as, for example, observed for fluorescent dyes that were applied into sieve elements via a

that BYDV consists of different strains. These strains are characterized by their most important vector(s): MAV (*Sitobion (Macrosiphon) avenae*), PAV (*R. padi, S. avenae* and others), RMV (*R. maidis*), RPV (*R. padi*) and SGV (*Schizaphis graminum*) (Table 1). In 2002, the International Committee on Taxonomy of Viruses (ICTV) reclassified some strains that previously belonged to BYDV-PAV to the new strain BYDV-PAS, differing in its amino acid sequence of ORFs 1 and 2 (<90% homology) and symptom severity, representing now a distinct species within the group BYDV (Mayo, 2002). In the same year, sequencing revealed great differences between BYDV-PAV and RPV. The latter was given a new name, Cereal yellow dwarf virus-RPV (CYDV-RPV), and it was grouped in the genus *Polerovirus*. In 2013, Krueger et al. (2013) sequenced the genome of BYDV-RMV and concluded that RMV, due to a higher sequence identity with members of the genus *Polerovirus* than with the genus *Luteovirus*, should be assigned to the genus *Polerovirus* as well and suggested to rename it Maize yellow dwarf virus-RMV (MYDV-RMV). Sequencing data also suggest that recombination events occur between members of different BYDV strains as demonstrated for BYDV-PAV-CN, a Chinese strain, showing sequence similarities to BYDV-PAS and BYDV-GAV (Liu et al., 2007). Thus, the group of viruses causing BYD disease is diverse and consists of different virus strains belonging to different genera. The introduction of further strains is most likely continuing, as the taxonomy of viruses is still dynamic and is reworked based upon recent technical advances. Recent advances of comparative genomics and metagenomics uncover an increasing number of diverse viruses (Koonin et al., 2019). Koonin et al. (2019) concluded that these discoveries not only improved our understanding of the evolutionary relationships within the virosphere, but also emphasized that the existing taxonomic framework is inadequate to depict the relationships within the virosphere.

BYDV infection in general causes significant economic losses in different cereal crops. Out of the different viruses and strains mentioned, BYDV-PAV is probably the most important strain due to its severe symptoms (Chay et al., 1996) and the range of different aphid species vectoring BYDV-PAV (Rochow,

Figure 1 (*Continued*)
plant hopper stylet (Fujimaki et al., 2000). (b) Via pore plasmodesma units (PPUs) BYDV gets from SEs into companion cells (CC) where it multiplies due to the absence of a transcription machinery in SEs (Van Bel, 2003). The BYDV movement protein regulates translocation towards and through PPUs (SE-CC) and plasmodesmata (CC-PPC) as well as sieve pores (SE-SE). (c) Nutrients are transported along intact sieve tubes, driven by a mass flow (black arrow). Furthermore, long-distance signals, chemical as well as physical (e.g. electrical signals), are distributed along sieve tubes. In case of BYDV, induced necrosis of mass flow and signal transduction are interrupted leading to nutrition shortages in sink tissues as well as problems in systemic defence responses and, for example, growth regulation.

Table 1 Members of the BYDV group

Current		Old		
Virus name (including strain)	Genus	Virus name (including strain)	Genus	Main vector
Barley yellow dwarf virus-MAV	*Luteovirus*			*Sitobion (Macrosiphon) avenae*
Barley yellow dwarf virus-PAV	*Luteovirus*			*Rhopalosiphum padi, Sitobion avenae* and others
Barley yellow dwarf virus-PAS	*Luteovirus*			
Barley yellow dwarf virus-RMV	*Polerovirus*		*Luteovirus*	*Rhopalosiphum maidis*
Cereal yellow dwarf virus-RPV	*Polerovirus*	*Barley yellow dwarf virus*-RPV	*Luteovirus*	*Rhopalosiphum padi*
Barley yellow dwarf virus-SGV	*Luteovirus*			*Schizaphis graminum*

1969). In different regions, yield losses were estimated during the last decades in different years for wheat, oat and barley with huge variations between regions and years (Table 2).

In contrast to the indication of yield losses in percentages, McKirdy et al. (2002) suggested instead to present yield losses in t/ha due to an easy conversion into monetary losses. An exemplary calculation for BYDV impact on spring barley, based upon a mean yield of 5.12 t/ha and prices of, for example, 160 €/t with a yield reduction of 0.36–1.1 t/ha (Kennedy and Connery, 2005) would mean a financial loss between 57.6 €/ha (7%) and 176 €/ha (21.5%). Thus, the experimentally determined yield losses from individual infected fields are high and economically relevant. A recent study that is based upon an expert assessment estimates worldwide BYD-related yield losses in wheat on an overall level of 0.96% (7.25 million tons for 2016 (FAO Stat)) with a focus on NW-Europe (3.26%) and W-Asia/NW-Africa (2.9%) (Savary et al., 2019). Based upon such calculations/estimations an economic benefit of the growing of BYDV-tolerant/resistant cultivars is obvious. Virus resistance reduces multiplication or spreading from the infection site in the plant (Cooper and Jones, 1983) while in the case of tolerance, the plant does not affect pathogen multiplication but copes with them by less yield and quality losses when compared to susceptible cultivars (Schafer, 1971).

Table 2 Economic losses due to BYDV infection in different cereal crops

Crop	Region	Virus type	Time	Yield loss (%)	Yield loss (t/ha)	Reference
Wheat	W-Australia	PAV	1992–1994	n.d.	0.013-0.025	McKirdy et al., 2002
Wheat	The United States of America	PAV	1993 1994	23.8 33	n.d.	Hoffman and Kolb, 1998 (mean of eight cultivars)
Wheat	Australia	PAV	2017 2018	18–49 (early infection) 6–31 (late infection	n.d.	Choudhury et al., 2019a
Barley	The United States of America	PAV	1989 1990	32.5–38.0 8.5–19.8	n.d.	Edwards et al., 2001
Barley	Ireland	MAV	1990–1993 1996–2001	3.4–23.1 3.2–18.5	0.15–1.21 0.19–1.51	Kennedy and Connery, 2005
Oat	W-Australia	PAV	1993–1994	n.d.	0.044–0.055	McKirdy et al., 2002

n.d. means 'not determined'.

Several authors suggest that with regard to climate change, damage caused by BYDV will increase in the future in wheat and other cereal crops (e.g. reviewed in Trębicki et al., 2015; Trębicki et al., 2017; Walls et al., 2019). Climate change includes an increasing temperature and an increase of the CO_2 level, both of which are linked with each other, as well as changes in rainfall (IPCC, 2014). It appears likely, that the main regions for barley ((1) Europe, (2) Asia (FAOSTAT, 2018)) and wheat production ((1) Asia, (2) Europe (FAOSTAT, 2018)) will be affected by an increase of mean temperature within a range of 1–2°C until the year 2100 if efforts for the reduction of the emission of greenhouse gases are successful (RCP2.6 scenario; IPCC, 2014). However, future temperature increases will show drastic local differences with, for example, an assumed maximum increase in some Australian regions of up to 5°C. Such changes are expected for areas of wheat production, causing in comparison with current temperatures a rapid increase of BYDV virus titre in infected plants with a strong increase of visible symptoms (Nancarrow et al., 2014). The expected increase of the atmospheric CO_2 level under the prediction that air temperature will not increase above 1.5°C will be approximately up to 650 ppm, leading to an increase of BYDV virus titre as well (Trębicki et al., 2015). Trębicki et al. (2015) point out that a high virus titre may not only lead to stronger effects of the virus on its host, but could furthermore increase the probability of virus transmission by its vector.

Besides direct effects of climate change on BYDV infection, the vector is affected as well on multiple levels. *R. padi* populations reproduce holocyclic, including sexual reproduction and egg production on a primary winter host and asexual reproduction (parthenogenesis) on a secondary summer host (Blackman and Eastop, 2000). Under mild winter conditions and a constant decrease of the number of cold winter days as observed in the last decade in several parts of the world, the *R. padi* populations increase is exclusively based upon parthenogenetic reproduction as reported by Dixon (1971) as well as Strażyński and Ruszkowska (2015). Under these conditions, winged aphids colonize new fields earlier in spring and later in autumn, easily switching between cereal crops and wild grasses. Furthermore, an increase in temperatures leads to a shorter developmental time and an earlier start of reproduction (reviewed by Finlay and Luck, 2011).

Concerning reports about decreasing abundance and biodiversity of terrestrial insects in the field (e.g. Hallmann et al., 2017; Seibold et al., 2019) one may ask how aphids are affected by this development. Against the trend of decreasing biodiversity and abundance for many insect species, the aphids appear to do well in our agricultural environment. Recent studies indicate that the number of aphids in an agricultural environment is stable or increases in sampled regions of the United States (Crossley et al., 2020). Recently published data of Ziesche et al. (2020) shows furthermore that the number of aphids is continuously increasing since almost 30 years in a sampled region in Germany.

It can be expected that problems with cereal aphids and therefore with BYDV will increase in the future. This development is accompanied by a decreasing acceptance of using chemical insecticides in the public and by political decision-makers. In 2013, the European Union (EU) decided to ban three neonicotinoid seed-coating insecticides (clothianidin, imidacloprid and thiamethoxam) used to protect seedlings/young plants from early infestation by harmful insects, for example, aphids acting as virus vectors in maize and other cereals. Exceptions were possible between 2013 and 2018 and allowed for wheat and barley, seed and soil treatments if sowing takes place between July and December (Implementing regulation (EU) No 485/2013). In 2018 the ban was finally approved. In addition, from the beginning of 2021, the EU chose not to renew authorization of Thiacloprid, a commonly used insecticide of the same group. Europe does not play an isolated role in this context because discussions about a ban of neonicotinoids exist, for example, in the United States where the US Environmental Protection Agency adopted a contrary decision, deciding to keep five insecticides of that group on the market (Acetamiprid EPA-HQ-OPP-2012-0329, Clothianidin EPA-HQ-OPP-2011-0865, Dinotefuran EPA-HQ-OPP-2011-0920, Imidacloprid EPA-HQ-OPP-2008-0844 and Thiamethoxam EPA-HQ-OPP-2011-0581). The future consideration of factors such as, for example, landscape simplification as well as a low crop heterogeneity, supporting the occurrence of aphids and other pests, can help to reduce pressure on cereal crops in the field (Van der Eynde et al., 2020). The meaning of tolerance/resistance of cereal crops such as barley against BYDV and other viruses is gaining more importance in such a scenario. One form of tolerance of BYDV in barley is mediated by the gene *Ryd2* (Schaller *et al.*, 1964). The *Ryd2*-based tolerance is used in, for example, the winter barley cultivar 'Vixen' since 1986 (Parry and Habgood, 1986), but until now no information is available about the functional background of this tolerance. Further genes, such as *Ryd3* (Niks *et al.*, 2004) and *Ryd4*^{Hb} (Scholz *et al.*, 2009) and additional quantitative trait loci (Scheurer *et al.*, 2001) were detected.

2 The impact of Barley yellow dwarf virus (BYDV) on host and vector and the benefit of resistance

2.1 The tritrophic interaction of virus, vector and plant

The question arises as to why BYDV-PAV is actually the most important virus strain with the largest economic impact. DeFilippis and Villarreal (2000) have examined direct and indirect viral interactions that may contribute to the dominance of BYDV-PAV. They suggest that BYDV-PAV may be winning an evolutionary battle because of its interactions with aphid vector species that are favouring higher transmission rates. In contrast to BYDV-MAV, which is only vectored by *S. avenae*, BYDV-PAV has the ability to be transmitted by, for example, *S. avenae* and *R. padi* as mentioned above. Furthermore, observations by Gray et al. (1991) and Power et al. (1991) indicate that *R. padi* is a more efficient vector than *S. avenae*. However, clones within a species, as has been shown for *R. padi* and *S. avenae,* differ significantly in their efficiency to transmit BYDV (Habekuss et al., 1999; Bencharki et al., 2000). Unfortunately, no systematic monitoring and analysis of clones regarding their BYDV transmission efficiency was conducted so far. Therefore, it cannot be determined how important this trait of aphids is for the transmission of BYDV in the field. An induction of morphological changes of the offspring of an adult aphid from an unwinged (Apterae) to a winged stage (Alatae) as a result of BYDV-PAV infection of the host plant may enhance its spread in addition. Gildow (1980) observed an increased number of winged aphids of the species *R. padi* as well as *S. avenae* when feeding on BYDV-PAV-infected plants. A weaker effect occurred when aphids fed on BYDV-MAV-infected plants (Gildow, 1980). DeFilippis and Villarreal (2000) describe that this fact may lead to a transmission advantage of BYDV-PAV over BYDV-MAV. However, when assigning vectors to certain virus strains, one should not exclude the possibility of transmission by other aphid species. For BYDV-RMV, Lucio-Zavaleta et al. (2001) reported that under certain environmental conditions not only *R. maidis* but also *R. padi* might be an important vector. Regarding environmental effects, the authors describe that an increased temperature has a positive effect on BYDV-RMV inoculation by *R. padi* and transmission to wheat. The characteristics of inoculation and transmission efficiency varied significantly depending on the *R. padi* clone and virus isolate.

Aphids use plant volatile organic compounds (VOCs) to find their host plants (Powell et al., 2006). Changes of VOCs in a way that individual compounds increase or in other cases ratios between different compounds change may influence host plant identification by aphids (Bruce et al., 2005; Döring, 2014). After successful infection, virus infection changes the physiology of its host plant affecting the interaction with insect vectors. Positive vector responses such as attraction to virus-infected plants may be advantageous for the virus by promoting its spread as has been shown for the interaction of different aphid species with plants infected by persistently transmitted and phloem-located viruses (Dáder et al., 2017). For example, *R. padi* prefers susceptible and BYDV-infected wheat plants in comparison to non-infected plants (Ingwell et al., 2012). Aphids appear to benefit from this choice by shorter development times (Jiménez-Martínez et al., 2004a). Furthermore, Jiminéz-Martinez et al. (2004b) and Medina-Ortega et al. (2009) showed that *R. padi* carrying BYDV prefer the odour of susceptible BYDV-infected wheat cultivars. A BYDV-tolerant transgenic wheat line was not preferred, most likely due to the absence of symptoms. Recently Kern et al. (2021) conducted host-choice experiments for *R. padi* clones on two barley cultivars with and without BYDV infection. One of the two genotypes contained a *Ryd2*-based tolerance (Schaller et al., 1964) against BYDV. They observed that preferences to the BYDV-infected susceptible cultivar might be associated with the ability of an *R. padi* clone to transmit BYDV. Furthermore, they demonstrated that no preference occurs when BYDV-infected tolerant plants were offered as hosts. Kern et al. (2021) conclude that BYDV tolerance will not only provide a benefit for yield but may reduce the spread of BYDV due to the absence of an increased attraction of aphid vectors.

2.2 Anatomical and physiological aspects of the phloem

To understand the consequences of BYDV infection on plant anatomy and thus on physiology one must have a closer look on the ontology, the cellular composition and the function of the phloem. The phloem is a specialized tissue of higher plants and is composed out of sieve elements (SEs) and companion cells (CCs) that are closely connected by pore plasmodesma units (PPUs), forming a functional unit (Van Bel, 2003), and phloem parenchyma cells (PPCs). SEs are elongated living cells that contain only a reduced intracellular structure including a few mitochondria, a parietal located smooth endoplasmic reticulum, SE plastids and an extensive set of phloem-specific proteins (Knoblauch and van Bel, 1998; Gaupels et al., 2008). Due to this reason, BYDV multiplication most likely takes place in CCs from which BYDV spreads along sieve tubes (Fig. 1b). Located at the terminal ends of SEs, sieve plates with sieve pores connect adjacent SEs forming long sieve tubes. Mass flow through sieve tubes that distributes photoassimilates, amino acids, proteins and vitamins as well as signalling molecules such as phytohormones throughout the whole plant is driven by an osmotic pressure gradient between source (mature leaves) and sink (e.g. storage organs) tissues (Münch, 1930; Knoblauch *et al.,* 2016). Depending on the mechanism of how carbohydrates get from PPCs into the CC-SE complex, a distinction is made between different loading types. In barley, the CC-SE complex appears to have only a low number of plasmodesmata connecting CCs to surrounding parenchyma cells, indicating that CCs are mostly isolated from the PPCs. Thus, carbohydrates must change their way of transportation from their production site towards the sieve tubes from a symplastic to an apoplastic way (Evert et al., 1996). Besides the transport of nutrients and signal molecules, the phloem is also involved in long-distance communication by electropotential waves (EPWs) (e.g. Furch *et al.,* 2007, 2009). EPWs are complex electrical signals and include features of action potentials and variation potentials (Hafke *et al.,* 2009). They have been recorded in response to different mechanical and physical stimuli (Fromm and Spanswick,

1993; Rhodes et al., 1996; Mancuso, 1999; Furch et al., 2007) but also in response to biotic stimuli such as feeding of caterpillars (Salvador-Recatalà et al., 2014; Zimmermann et al., 2016). Furthermore, EPWs are involved in, for example, growth regulation (Trębacz et al., 2006).

Recent observations by Paulmann et al. (2018) for barley show that the phloem and the xylem are both negatively affected by a BYDV infection, which could be explained by necrotic obliteration of sieve elements and xylem vessels, already occurring in differentiating vascular bundles in highly susceptible barley cultivars (Esau, 1957). Pathological degeneration and necrosis of the phloem, involving sieve elements, companion cells and parenchyma cells, affect old and young vascular bundles in main and marginal veins. The fact that PPCs undergo necrosis as well indicates that phloem loading/unloading may be reduced inside source/sink areas because loading and unloading in barley leaves is probably based upon an apoplastic mechanism involving PPCs and CCs (Evert et al., 1996). Even if a previously observed low number of plasmodesmal connections between CCs and PPCs (Evert et al., 1996) might be an evidence for symplastic phloem unloading in sink leaves of barley (Haupt et al., 2001), necrosis of PPCs would most likely lead to a reduced phloem unloading. It can be proposed that reduced phloem loading/unloading leads to a reduction of mass flow rate from source to sink because it is driven by a photoassimilate gradient between source and sink (Münch, 1930). Although this is an assumption, it could be an explanation for the reduced growth in BYDV-infected susceptible plants. Direct effects of BYDV on the metabolism of meristematic cells can most likely be excluded because viruses cannot enter these due to the symplasmatic isolation, that is, the absence of plasmodesmata connections to surrounding tissues. The absence of viruses of the family Luteoviridae (including BYDV) was demonstrated for sugarcane yellow leaf virus in sugarcane by using meristem tip culture for the development of virus-free plants (Fitch et al., 2001). It can be suggested that a change of mass flow is the most relevant parameter that is responsible for changes in plant growth and respective yield losses. As outlined by Van Bel and Hafke (2005), a reduction of sieve tube radius to 50% of the original value will increase the flow resistance 6-fold according to the Hagen-Poiseuille equation. This would reduce the mass flow inside sieve tubes and constrain the nutrient transport (e.g. photoassimilates) inside leaves. This is supported by Erion and Riedell (2012) who reported that root dry weight and leaf area are reduced (each by 36%) when barley plants are infected by BYDV and also by Hoffman and Kolb (1997) who demonstrated that a smaller proportion of total assimilate is transported to the root system in BYDV-infected wheat. This reduction is most likely caused by limited nutrient transport in the phloem towards meristematic tissues as indicated by results of Esau (1957) who reported a reduced meristematic activity in leaves of BYDV-infected plants. Thus, BYDV-infected plants appear to suffer from nutrient deficiency caused by a reduced transport capacity, which might help to explain the observed reduction of plant/leaf sizes in cv. Rubina plants in comparison to cv. Vixen plants that lack a reduction of sieve tube area as observed for cv. Rubina (Paulmann et al., 2018).

In addition to nutrition transport, BYDV infection affects the physiological functionality of sieve elements regarding their propagation of EPWs (Fig. 1c). EPWs as long-distance signals are characterized by a decreasing propagation velocity along sieve tubes with increasing distance from the trigger site as observed, for example, in barley (Paulmann et al., 2018). Sieve tubes can be compared with cables in this context (Hedrich et al., 2016). As pointed out, BYDV infection leads to a reduced radius of single sieve elements (single cable) as well as a reduced total radius of vascular bundles (several parallel cables) (Paulmann et al., 2018), combined with 'cable breaks' induced by necrosis of single sieve elements, reduced sieve-element area and the loss of sieve-element membrane integrity (Esau, 1957). This may cause changes in the longitudinal (intracellular) resistance for electrophysiological signals according to the cable theory model for the calculation of the electrophysiological current along passive neurites (Taylor, 2013) and sieve tubes in plants (Hedrich et al., 2016) showing that a reduced cell radius will increase the longitudinal (intracellular) resistance for current flow. As a further effect, the lateral resistance (affecting the flow of electrical current through the membrane) will decrease with decreasing cell diameter due to a relative increase of the membrane surface with regard to the cell volume. Together, both parameters may negatively affect the propagation of electrophysiological current along a cable, leading to the observed loss of EPWs near their trigger site. It can be suggested that the suppression of EPW propagation may have a negative effect on, for example, plant defence (Trębacz et al., 2006). Observations by Paulmann et al. (2018) indicate that *Ryd2*-based tolerance prevents negative effects on phloem structures, and by doing so secures nutrient transport as well as defence-associated long-distance signalling.

3 Barley yellow dwarf virus control

Since the control of BYDV is not possible by chemical measures and the application of insecticides against BYDV-transmitting aphids has to be reduced due to the public and political demand, the use of virus-resistant or tolerant cultivars, respectively, is an effective solution to combat BYD. The pursuit to identify aphid and BYDV resistance genes, like in case of fungal pathogens, is nowadays enhanced by the emergence of new molecular techniques in plant science (Perovic et al., 2018). In this section we will present methods for screening of BYDV resistance and give an overview of currently known resistance genes as well as on molecular markers for their employment in barley breeding programmes. In addition, we will show how the availability of a high-standard reference genome of barley has ushered in a new era of genomics and the decoded blueprint opened up new avenues in genome sequence research for both applied and basic research of BYDV resistance (Perovic et al., 2020a).

3.1 Screening methods for the detection of BYDV tolerance/resistance

BYD represents the most important viral disease in cereals worldwide (Lister and Ranieri, 1995). Because of the need for tolerance/resistance in plants, there is a high demand for methods allowing screening of germplasm of cereals for the identification of resistant or tolerant genotypes. Attempts regarding exploitation of landraces and collections of wild relatives by plant breeders have been hampered by the limited availability of phenotypic and genotypic information as the identification of new resistances from these sources requires various steps that usually combine interdisciplinary efforts. The first step in this process is the development of methods for reliable screening for resistance under controlled conditions. For reliable selection of BYDV-tolerant/resistant cultivars, uniform inoculation is essential for which the natural infection in the field is usually too unreliable. Controlled application of viruliferous aphids enables the assessment of particular virus isolates and the inoculation at various developmental stages (Burnett and Gill, 1976). Therefore, to introduce, for example, BYDV resistance into elite cultivars that are usually susceptible to BYDV it is important to develop methods that enable a rapid and reliable phenotyping (Ordon et al., 2004). Choudhury et al. (2018) describe a reliable method to screen cereal crops for BYDV resistance in glasshouse experiments. The method entailed the evaluation of traits associated with infection such as leaf discolouration, reduction in growth, biomass and yield traits. The measurement of a relative virus titre is important to determine the percentage of virus-infected plants. This is done by using enzyme-linked immunosorbent assay (ELISA) and tissue blot immunoassay (TBIA). For barley and oat, visual evaluation of symptoms was considered as adequate for assessing BYDV resistance, but at the same time, for wheat Choudhury et al. (2018) highlighted the importance to evaluate BYDV resistance by ELISA tests, plant biomass (at an early stage of infection), grain number as well as yield (at a late infection stage) measurements. By using their phenotyping method, Choudhury et al. (2019b) identified multiple novel QTL for BYD resistance and some resistant wheat genotypes.

3.2 Resistance to aphids

According to Hogenhout et al. (2008), among 697 described plant viruses, aphids are vectors for 197 different virus species. Regarding this, breeding for aphid resistance may be an interesting and ecological-friendly, less harmful pest management strategy (Åhman and Ninkovic, 2010). Furthermore, resistance against a vector would target, under optimal conditions, multiple viruses transmitted by one vector. In one of the first reports about barley resistance (antibiosis and antixenosis) and tolerance against *Rhopalosiphum padi* L., Hus and Robinson (1962) stated that among 136 tested varieties no complete resistance was detected; rather a certain level of tolerance to infestation of the aphid could be identified in a set of 49 lines. Among tolerant lines in this study was the cultivar 'Rojo' as a donor of *ryd*1 resistance to BYDV (Suneson, 1955), indicating some correlation between vector resistance and virus resistance in barley in this genotype. More recently, Åhman and Bengtsson (2019) reported a successful transfer of resistance to *Rhopalosiphum padi* L. from *Hordeum vulgare* ssp. *spontaneum* to cultivated barley *H. vulgare* ssp. *vulgare*. In this study, individual aphid growth in the laboratory was used as a parameter for plant resistance. By using the Infinium iSelect 50 K SNP array, Åhman and Bengtsson (2019) mapped a QTL for aphid resistance on chromosome 2HS. Macaulay et al. (2020) confirmed a locus for aphid resistance on the distal part of chromosome 2HS and furthermore identified one or more QTL for gramine concentration, an indole alkaloid which is toxic to animals, on chromosome 3H.

According to recently published papers there is a high potential in identification of aphid resistance by combining various screening methods and molecular marker technologies. Nevertheless, phenotyping for aphid resistance remains labour-intensive and time-consuming (cf. Åhman and Bengtsson, 2019).

3.3 Genes and QTL conferring tolerance/resistance to BYDV

Breeding for BYDV resistance is the most feasible means of BYDV control. Up to now, five resistance genes and several quantitative trait loci (QTL) to BYDV in barley are known. Regarding BYDV-tolerance, *Ryd*1 was first detected in the spring barley cultivar 'Rojo' (Suneson, 1955) but was not used in barley breeding due to its low efficiency. After that, *Ryd*2 and *Ryd*3 with similar effects against BYDV-PAV and BYDV-MAV were detected in Ethiopian landraces (Schaller et al., 1964; Niks et al., 2004). Besides this, several QTL for BYDV tolerance have been identified on chromosomes 1H, 2H, 4H and 7H (Toojinda et al., 2000; Scheurer et al., 2001; Kraakman et al., 2006; Perovic et al., 2013) and various molecular markers for their use in breeding programmes have been developed.

The *Ryd*2 gene, which was identified in a number of Ethiopian landraces (Rasmusson and Schaller, 1959), is putatively multiallelic (Catherall et al., 1970; Chalhoub et al., 1995) and maps close to the centromere on the long arm of chromosome 3H (Collins et al., 1996). Several alleles in barley cultivars are known to confer tolerance/resistance against BYDV-PAV and BYDV-MAV (Skaria et al., 1985; Banks et al., 1992); however, Ryd2 effectiveness may depend on the genetic background (Jones and Catherall, 1970; Schaller, 1984). Collins et al. (1996) have also showed that the effect of the semi-dominant *Ryd*2 gene depends on the genetic background, the environmental conditions and the virus isolate. *Ryd2* has been reported to reduce the virus titre in young plants (Chalhoub et al., 1995; Sip et al., 2006), but no differences could be observed in older plants (Skaria et al., 1985; Scheurer et al., 2000). However, this does not seem to affect the economic relevance of *Ryd*2 because the infection of young plants has a strong impact on yield while there is only a minor or no effect when BYDV infects older plants (Collins et al., 1996).

The *Ryd*3 gene confers tolerance to BYDV-PAV and BYDV-MAV and the locus is located in barley chromosome 6H, near the centromere (Niks et al., 2004). In a population of 3 210 F_2 plants, Lüpken et al. (2014) identified 14 tightly linked markers of which 10 co-segregated with the *Ryd*3 gene. Furthermore, the study of Lüpken et al. (2014) showed that the centromeric region, where *Ryd*3 is located, suffers from suppressed recombination suggesting potential problems in achieving positional isolation of *Ryd*3 and marker-assisted selection (MAS) of the resistance in breeding programmes without the introduction of undesirable traits due to linkage drag.

In contrast to *Ryd*2 and *Ryd*3, a gene called *Ryd*4^{Hb}, conferring complete resistance to BYDV-PAV, has been transferred from *Hordeum bulbosum* L., the sole member of the secondary genepool of *H. vulgare* (von Bothmer et al., 1995), into cultivated barley. Already, *H. bulbosum* has proven to be a valuable source of genetic diversity for barley improvement. Despite crossing difficulties, successful introgressions of agronomical useful genes, for example, resistance to powdery mildew (Pohler and Szigat, 1982; Xu and Kasha, 1992; Pickering et al., 1995), leaf rust (Szigat et al., 1997; Pickering et al., 1998), the BaMMV/BaYMV complex (Ruge et al., 2003; Ruge-Wehling et al., 2006) and BYDV (Scholz et al., 2009) have been achieved. In addition, during mapping of *Ryd2* in the mapping population 'Post' × 'Vixen', Scheurer et al. (2001) detected a QTL for BYDV tolerance on the long arm of chromosome 2HL with the positive allele derived from 'Post'.

During screenings of barley lines carrying an introgression derived from *H. bulbosum*, Johnston et al. (2009) identified a line highly tolerant to BYDV carrying an introgression on chromosome 2HL. Later on, corresponding *H. bulbosum* introgression line called 203S11 was backcrossed twice to the spring barley cultivar 'Emir' and the BC2F1 plants were used for DH-line production and mapping (Perović et al., 2012). Within a population of 221 DH lines, Perović et al. (2012) identified 27 DH lines carrying a recombination event in the *H. bulbosum* fragment. Genotyping of these lines with the 9KiSelect bead chip resulted in the identification of about 50 polymorphic SNP markers allowing fine mapping of the *H. bulbosum* introgression conferring tolerance to BYDV, which is preliminary, called $Ryd_{203S11}{}^{Hb}$.

It is interesting to note that Schliephake et al. (2013) observed that *H. bulbosum* does not only appear to be resistant against BYDV, but furthermore shows resistance against the most important BYDV vector, that is, *R. padi*. On different *H. bulbosum* clones *R. padi* showed an increased developmental time combined with reduced reproduction. When observing feeding behaviour of *R. padi*, it turned out that aphids spend more than half of the observation time on the phloem (58.5%) nearly all of which was ingestion (98.5%) on the susceptible barley cultivar 'Rubina' while phloem activity (including ingestion) on the resistant *H. bulbosum* clone was only 0.3%. The mode of action that underlies aphid resistance in *H. bulbosum* is unknown so far but Schliephake et al. (2013) suggested a presence of secondary metabolites, for example, DIBOA that was observed to act in other wild relatives of *H. vulgare* against *Diuraphis noxia* and *Schizaphis graminum* (Gianoli and Niemeyer, 1998). Thus, Schliephake et al. (2013) point out that rather a vector than a virus resistance could explain BYDV resistance in some *H. bulbosum* clones.

In total, two BYDV resistance sources originating from *H. bulbosum* ($Ryd4^{Hb}$ and $Ryd_{2003S11}^{Hb}$) and two genes from *H. vulgare* (*Ryd*2, *Ryd*3), as well as fife minor QTLs on 1H, 2H, 4H and 7H (Scheurer et al., 2001; Toojinda et al., 2000), have been assigned to the barley chromosomes by mapping. Consequently, these genes and QTLs could be used in marker-assisted breeding schemes (Table 3). Corresponding markers flanking these resistance genes and QTLs can be efficiently used for pre-selection of BYDV-tolerant genotypes. These markers together with the doubled haploid technique facilitate a reliable selection for virus resistance in barley, for example, doubled haploid populations can be

Table 3 BYD resistance genes originating from *Hordeum bulbosum* and from *Hordeum vulgare* that have been mapped and could be used in breeding schemes

Gene/QTL	Chromosomal location	Type of resistance	Reference
*Ryd*1	--	Tolerance	Suneson, 1955
Ryd2	3HL	Tolerance	Collins et al., 1996, Paltridge et al., 1998, Ford et al., 1998
Ryd3	6H	Tolerance	Niks et al., 2004, Lüpken et al., 2014
$Ryd4^{Hb}$	3HL	Resistance	Scholz et al., 2009
Ryd_{203S11}^{Hb}	2HL	Tolerance	Perović et al., 2012
Ryd_{QTL}^{2HL}	2HL	Tolerance	Scheurer et al., 2001, Kraakman et al., 2006
Ryd_{QTL}^{1H}	1H	Tolerance	Toojinda et al., 2000
Ryd_{QTL}^{4H}	4H	Tolerance	Toojinda et al., 2000
Ryd_{QTL}^{7H}	7HL	Tolerance	Toojinda et al., 2000

screened by respective markers directly in vitro and only those plantlets carrying the resistance-encoding allele have to be transferred to the greenhouse. On a longer turn, these markers serve as starting points for the isolation of relevant genes and analysis of their function (Lüpken et al., 2014).

3.4 New genomic tools for marker development and marker-based breeding for resistance

Barley is among the oldest model species for genetic analysis. Initial molecular maps of barley comprised a limited number of RFLP markers (Graner et al., 1991; Kleinhofs et al., 1993) and were subsequently improved by PCR-based markers, that is, RAPDs (Williams et al., 1990), AFLPs (Vos et al., 1995) and SSRs (Ramsay et al., 2000; Varshney et al., 2007) and most recently by SNPs (Comadran et al., 2012; Bayer et al., 2017). After mapping of resistance genes, closely linked markers are developed as an efficient tool for improving resistance in barley by applying MAS procedures, that is, facilitating the selection of resistant plants without phenotypic analysis (Choudhury et al., 2018). This holds especially true for resistance to BYDV, which is difficult to phenotype (Habekuss et al., 2009). Furthermore, respective markers facilitate efficient marker-based backcrossing procedures (Ordon and Perovic, 2013) and pyramiding of respective genes leading to improved resistance (Riedel et al., 2011) or an extended use of partly overcome genes (Werner et al., 2005). An overview on BYDV resistance genes for which molecular markers are available in barley is given in Table 3.

While marker development was time-consuming and laborious in the past, today efficient high throughput genotyping systems are available (Comadran et al., 2012; Mascher et al., 2013; Bayer et al., 2017) together with a reference sequence of barley (Mascher et al., 2017), facilitating a fast marker development followed by the use in barley breeding programmes. This has ushered in a new era of barley genomics and the decoded genomic blueprint opens new avenues in exploring genome sequences for both applied and basic research (Perovic et al., 2020a). Recent advances in barley genomics enable researchers to go one-step further in the shortening and explicit demarcation on the barley genome regions conferring resistance to diseases as well as in the identification of candidate genes.

The public availability of a physical map and respective sequence assemblies (IBGS, 2012; Mascher et al., 2017) facilitated plenty of downstream applications, that is, the development of high throughput genotyping platforms, for example, the Illumina 9K iSelect array (Comadran et al., 2012) and the Illumina 50K iSelect chip (Bayer et al., 2017) and the exome capture-based sequencing (Mascher et al., 2013). One of the most important applications of the reference barley genome (Mascher et al., 2017) is the deciphering of the natural variation in barley and related species. After sequencing of additional barley genotypes, as a result of the barley PanGenome project (N. Stein, personal communication), and wild relatives (*H. spontaneum*, *H. bulbosum*) in the future, it will be possible to directly target resistance gene variants and to introduce these into new cultivars, by applying SNP, InDels and copy number variation (CNVs) markers, developed within these analyses. In addition exome sequencing (Mascher et al., 2013, 2014), RNA sequencing (RNASeq) (Wang et al., 2009), genotyping by sequencing (GBS) (Poland et al., 2012), resistance gene enrichment sequencing (RenSeq) (Andolfo et al., 2014) and whole genome sequencing (WGS) are available to develop markers and isolate resistance genes. More recently, MutRenSeq (Steuernagel et al., 2016), that combines mutational genetics and NGS-based sequencing, allows the rapid targeting and isolation of all types of genes for which a candidate gene approach is difficult. Most recently, cultivar-specific long-range chromosome assembly (Thind et al., 2017) allowed the rapid cloning of genes in hexaploid wheat, which will be even easier applicable in barley. However, the most powerful way for the detection of novel natural variants will be WGS (Beier et al., 2016) of resistant and tolerant genotypes. Identified polymorphisms of all above-mentioned methods in combination with the barley reference genome sequence allow the fast development of high-throughput PCR-based assays such as are pyrosequencing (Silvar et al., 2011) or competitive allele-specific polymerase chain reaction (KASP) (Allen et al., 2011) which are the base for MAS procedures in barley breeding. Furthermore, genome sequences of Triticeae-related species, as Brachypodium, wheat and wheat relatives will promote comparative genomics and will provide insights in the evolution of gene families that are responsible for resistance (Perovic et al., 2020b). Whole-genome sequencing of barley will in the near future enable efficient breeding for many traits at the allele level/for many allele-level traits, thus contributing to a broader genetic basis of BYDV breeding in barley. In this regard, comparative genomics will be extremely important, and it will play a special role in the isolation of important genes in cereals with large genomes near centromeres (Perovic et al., 2020b). Once the causal gene is known, efficient allele mining may be conducted (Yang et al., 2016; Hofinger et al., 2011) facilitating a directed use of genetic resources as well as efficient allele editing, for example, by CRISPS/Cas9.

Up to now, the pedigree selection scheme, marker-assisted backcrossing (MABC) and MAS are still basic methods of resistance breeding to BYDV. In this regard, Riedel et al. (2011) tested how pyramiding of two resistance genes, *Ryd*2 and *Ryd*3 and the QTL for tolerance on chromosome 2HL affects resistance to BYDV. The results of this experiment revealed that the QTL on chromosome 2H had only a small effect on the level of tolerance in those lines carrying only *Ryd*2, or *Ryd*3 or a combination of both, but the effect in comparison to lines carrying no tolerance allele was significant. In DH-lines carrying the combination *Ryd2* and *Ryd3*, a significant reduction of the virus titre was detected as compared to lines carrying only one of these genes. In summary, results by Riedel et al. (2011) showed that the combination of *Ryd2* and *Ryd3* in winter barley leads to quantitative resistance against BYDV-PAV instead of tolerance.

In future, molecular markers may be used for the incorporation of known and novel QTL or genes for aphid resistance in combination with resistant genes and QTL for BYD resistance. However, BYDV and aphid resistance is often identified in un-adapted germplasm. Therefore, backcrossing procedures are needed to combine both types of resistance with superior agronomic performance. This is more obvious for resistances originating from *H. bulbosum*, which belongs to the secondary pool of barley.

4 Conclusion and future trends

Recently, several breeding companies presented new barley varieties with a *Ryd2*-based tolerance against BYDV. Due to the genomic organization of the virus as single-stranded RNA and its low genome size, the viral genome is subjected to a high genomic mutation rate within a range of 10^{-6} and 10^{-4} substitutions per nucleotide (nt) site per cell infection (Peck and Lauring, 2018). The nucleotide sub rate was found to be $3.158*10^{-4}$ sites/year for BYDV isolates that were collected from the field (Wu et al., 2011). With a genome size of approximately 5600 nt for different BYDV strains (Jin et al., 2004; Liu et al., 2007) this means that 1.77 nucleotides change permanently within a BYDV population each year. This high mutation rate drives, for example, adaptation to environmental changes, which might lead to a change of virulence and a loss of function of major resistance genes, especially *Ryd4^{Hb}*, which confers complete resistance to BYDV. In addition, the loss of function of plant tolerance of *Cucurbita pepo* was reported as a consequence of a point mutation within the *Zucchini yellow mosaic virus* (Desbiez et al., 2007). Pyramiding of tolerance genes with different functional backgrounds might be a promising approach (e.g. reviewed in Fuchs, 2017), making it more difficult for the virus to adapt. To compensate the functional loss of single tolerance/resistance gens, it is necessary to continuously identify and characterise further resistance genes as well as QTLs and closely linked molecular markers to follow this aim. Economic success and therefore the establishment of the trait BYDV tolerance/resistance for future barley varieties is threatened by another pathogen. Symptoms in the field that are caused by Wheat dwarf virus (WDV) infection of barley look similar to those caused by BYDV, possibly resulting in a misinterpretation by farmers that BYDV resistance does not work successfully in the field. In addition, WDV causes significant yield losses as reported for the Czech Republic (Ripl et al., 2020). Since WDV spreads together with its vector *Psammotettix alienus* across Europe (Lindblad and Areno, 2002; Manurung et al., 2004; Abt et al., 2018), WDV resistance must not be neglected and research activities have to be intensified because no resistance was identified in barley so far. Based on these considerations, there might be a need to include WDV resistance in future barley varieties as well. A recent study demonstrated the relevance of eIF4E for the interaction of different viruses with wheat, in which silencing caused in parallel qualitative and quantitative resistance against several viruses e.g. BYDV (Rupp et al., 2019). The authors conclude that targeted gene editing using CRISPS/Cas can provide the genetic variation required for multiple virus resistance.

5 Where to look for further information

For BYDV taxonomy: International Committee on Taxonomy of Viruses (ICTV) - https://talk.ictvonline.org/taxonomy.

The Frontiers in Plant Science Research topic 'Advances in Plant-Hemipteran Interactions' gives a broad view about different aspects like the meaning of salivary effectors and the development of several aspects of plant resistance and tolerance against Hemiptera. A more detailed focus on the phloem is presented by the research topic 'Phloem: the integrative avenue for resource distribution, signalling and defence' in the same journal.

For an overview on molecular breeding for virus resistance, cf. Friedt and Ordon, 2007; Lange and Ordon, 2010; Ordon and Perovic, 2013; Ordon and Kühne, 2014). In the research highlighted 'Use of Barley and Wheat Reference Sequences: Downstream Applications in Breeding, Gene Isolation, GWAS and Evolution' at Frontiers in Plant Science, a collection of 19 articles sheds how knowledge of whole barley genome sequence promotes applied breeding (genome wide association study (GWAS), genomic selection (GS)), basic biological research (mapping of major genes and quantitative trait loci (QTLs)), accelerated isolation of novel genes, novel methods in sequence analysis and rapid detection of natural variation. Furthermore, the presented articles revealed efficient use of genetics and breeding methods in harnessing genetic resources of barley and wheat that promote the rapid improvement of cultivars.

6 References

Abt, I., Derlink, M., Mabon, R., Virant-Doberlet, M. and Jacquot, E. (2018). Integrating multiple criteria for the characterization of Psammotettix populations in European cereal fields. *Bull. Ent. Res.* 108(2), 185-202.

Åhman, I. and Bengtsson, T. (2019). Introgression of resistance to *Rhopalosiphum padi* L. from wild barley into cultivated barley facilitated by doubled haploid and molecular marker techniques. *Theor. Appl. Genet.* 132(5), 1397-1408.

Åhman, I. and Ninkovic, V. (2010). Volatile interactions between undamaged plants: effects and potential for breeding resistance to aphids. In: Baluška, F. and Ninkovic, V. (Eds), *Plant Communication from an Ecological Perspective*. Berlin: Springer, pp. 113-125.

Allen, A. M., Barker, G. L. A., Berry, S. T., Coghill, J. A., Gwilliam, R., Kirby, S., Robinson, P., Brenchley, R. C., D'Amore, R., McKenzie, N., Waite, D., Hall, A., Bevan, M., Hall, N. and Edwards, K. J. (2011). Transcript-specific, single-nucleotide polymorphism discovery and linkage analysis in hexaploid bread wheat (*Triticum aestivum* L.). *Plant Biotech. J.* 9(9), 1086-1099.

Andolfo, G., Jupe, F., Witek, K., Etherington, G. J., Ercolano, M. R. and Jones, J. D. (2014). Defining the full tomato NB-LRR resistance gene repertoire using genomic and cDNA RenSeq. *BMC Plant Biol.* 14, 120.

Banks, P. M., Waterhouse, P. M. and Larkin, P. J. (1992). Pathogenicity of three RPV isolates of *Barley Yellow Dwarf Virus* on barley, wheat and wheat alien addition lines. *Ann. Appl. Biol.* 121(2), 305-314.

Bayer, M. M., Rapazote-Flores, P., Ganal, M., Hedley, P. E., Macaulay, M., Plieske, J., Ramsay, L., Russell, J., Shaw, P. D., Thomas, W. and Waugh, R. (2017). Development and Evaluation of a Barley 50k iSelect SNP Array. *Front. Plant Sci.* 8, 1792.

Beier, S., Himmelbach, A., Schmutzer, T., Felder, M., Taudien, S., Mayer, K. F., Platzer, M., Stein, N., Scholz, U. and Mascher, M. (2016). Multiplex sequencing of bacterial artificial chromosomes for assembling complex plant genomes. *Plant Biotech. J.* 14(7), 1511-1522.

Bencharki, B., El Yamani, M. and Zaoui, D. (2000). Assessment of transmission ability of *Barley yellow dwarf virus*-PAV isolates by different populations of *Rhopalosiphum padi* and *Sitobion avenae*. *Europ. J. Plant Prot.* 106(5), 455-464.

Blackman, R. L. and Eastop, V. F. (2000). *Aphids on the World's Crops* (2nd edn.). Chichester: Wiley.

Bruce, T. J. A., Wadhams, L. J. and Woodcock, C. M. (2005). Insect host location: a volatile situation. *Trends Plant Sci.* 10(6), 269-274.

Burger, W. C. and LaBerge, D. E. (1985). Malting and brewing quality. In: Rasmusson, D. C. (Ed), *Agronomy Monographs*. Madison, WI: American Society of Agronomy, pp. 367-401.

Burnett, P. A. and Gill, C. (1976). The response of cereals to increased dosage with *Barley yellow dwarf virus*. *Phytopathology* 66(5), 646-651.

Catherall, P. L., Jones, A. T. and Hayes, J. D. (1970). Inheritance and effectiveness of genes in barley that condition tolerance to *Barley yellow dwarf virus*. *Ann. Appl. Bio.* 65(1), 153-161.

Chalhoub, B. A., Sarrafi, A. and Lapierre, H. D. (1995). Partial resistance in the barley (*Hordeum vulgare* L.) cultivar 'Chikurin-Ibaraki-l' to two PAV-like isolates of barley yellow-dwarf virus: allelic variability at the Yd2 gene locus. *Plant. Breed.* 114(4), 303-307.

Chay, C. A., Smith, D. M., Vaughan, R. and Gray, S. M. (1996). Diversity among isolates within the PAV serotype of barley yellow dwarf virus. *Phytopathology* 86(4), 370-377.

Choudhury, S., Al-Shammari, D., Hu, H., Meinke, H., Westmore, G., Birchall, C., Larkin, P. and Zhou, M. (2018). A screening method to detect BYDV-PAV resistance in cereals under glasshouse conditions. *Plant. Pathol.* 67(9), 1987-1996.

Choudhury, S., Larkin, P., Meinke, H., Hasanuzzaman, M. D., Johnson, P. and Zhou, M. (2019a). Barley yellow dwarf virus infection affects physiology, morphology, grain yield and flour pasting properties of wheat. *Crop Pasture Sci.* 70(1), 16-25.

Choudhury, S., Larkin, P., Xu, R., Hayden, M., Forrest, K., Meinke, H., Hu, H., Zhou, M. and Fan, Y. (2019b). Genome wide association study reveals novel QTL for barley yellow dwarf virus resistance in wheat. *BMC Genomics* 20(1), 891.

Collins, N. C., Paltridge, N. G., Ford, C. M. and Symons, R. H. (1996). The *Yd2* gene for barley yellow dwarf virus resistance maps close to the centromere on the long arm of barley chromosome 3. *Theor. Appl. Genet.* 92(7), 858-864.

Comadran, J., Kilian, B., Russell, J., Ramsay, L., Stein, N., Ganal, M., Shaw, P., Bayer, M., Thomas, W., Marshall, D., Hedley, P., Tondelli, A., Pecchioni, N., Francia, E., Korzun, V., Walther, A. and Waugh, R. (2012). Natural variation in a homolog of Antirrhinum CENTRORADIALIS contributed to spring growth habit and environmental adaptation in cultivated barley. *Nat. Genet.* 44(12), 1388-1392.

Cooper, J. I. and Jones, A. T. (1983). Responses of plants to viruses: proposals for use of terms. *Phytopathology* 73(2), 127-128.

Crossley, M. S., Meier, A. R., Baldwin, E. M., Berry, L. L., Crenshaw, L. C., Hartman, G. L., Lagos-Kutz, D., Nichols, D. H., Patel, K., Varriano, S., Snyder, W. E. and Moran, M. D. (2020). No net insect abundance and diversity declines across US long term ecological research sites. *Nat. Ecol. Evol.* 4(10), 1368-1376 doi: 10.1038/s41559-020-1269-4.

D'Arcy, C. J. (1995). Symptomatology and host range of barley yellow dwarf. In: D'Arcy, C. J. and Burnett, P. A. (Eds), *Barley Yellow Dwarf. 40 Years of Progress*. St Paul: APS Press, pp. 9-28.

Dáder, B., Then, C., Berthelot, E., Ducousso, M., Ng, J. C. K. and Drucker, M. (2017). Insect transmission of plant viruses: multilayered interactions optimize viral propagation. *Insect Sci.* 24(6), 929-946.

DeFilippis, V. R. and Villarreal, L. P. (2000). An introduction to the evolutionary ecology of viruses. In: Hurst, C. (Ed), *Viral Ecology*. San Diego: Academic Press, pp. 125-208.

Desbiez, C., Gal-On, A., Girard, M., Wipf-Scheibel, C. and Lecoq, H. (2007). Increase in *Zucchini yellow mosaic virus* symptom severity in tolerant zucchini cultivars is related to a point mutation in P3 protein and is associated with a loss of relative fitness on susceptible plants. *Virology* 93(12), 1478-1484.

Dixon, A. F. G. (1971). The life-cycle and host preferences of the bird cherry-oat aphid, *Rhopalosiphum padi* L., and their bearing on the theories of host alternation in aphids. *Ann. App. Biol.* 68(2), 135-147.

Döring, T. F. (2014). How aphids find their host plants, and how they don't. *Ann. App. Biol.* 165(1), 3-26.

Edwards, M. C., Fetch, T. G., Schwarz, P. B. and Steffenson, B. J. (2001) Effect of Barley yellow dwarf virus Infection on yield and malting quality of Barley. *Plant Dis.* 85(2), 202-207.

Erion, G. G. and Riedell, W. E. (2012). Barley yellow dwarf virus effects on cereal plant growth and transpiration. *Crop Sci.* 52(6), 2794-2799.

Esau, K. (1957). Phloem degeneration in Gramineae affected by the barley yellow-dwarf virus. *Am. J. Bot.* 44(3), 245-251.

Evert, R. F., Russin, W. A. and Botha, C. E. J. (1996). Distribution and frequency of plasmodesmata in relation to photoassimilate pathways and phloem loading in the barley leaf. *Planta* 198(4), 572-579.

Finlay, K. J. and Luck, J. E. (2011). Response of the bird cherry-oat aphid (*Rhopalosiphum padi*) to climate change in relation to its pest status, vectoring potential and function in a crop-vector-virus pathosystem. *Agric. Ecosyst. Environ.* 144(1), 405-421.

Fitch, M. M. M., Lehrer, A. T., Komor, E. and Moore, P. H. (2001). Elimination of *Sugarcane yellow leaf virus* from infected sugarcane plants by meristem tip culture visualized by tissue blot immunoassay. *Plant Pathol.* 50(6), 676–680.

Food and Agriculture Organization of the United Nations (FAOSTAT) (2018). Available at: http://www.fao.org/faostat/en/#data. Accessed 24.09.2020.

Ford, C. M., Paltridge, N. G., Rathjen, J. P., Moritz, R. L., Simpson, R. J. and Symons, R. H. (1998). Rapid and informative assays for *Yd2*, the barley yellow dwarf virus resistance gene, based on the nucleotide sequence of a closely linked gene. *Mol. Breed.* 4(1), 23–31.

Friedt, W. and Ordon, F. (2007). Molecular markers for gene pyramiding and resistance breeding in barley. In: Varshney, R. and Tuberosa, R. (Eds), *Genomics-Assisted Crop Improvement, vo12, Genomic Applications in Crops*. Heidelberg: Springer, pp. 81–101.

Fromm, J. and Spanswick, R. (1993). Characteristics of action potential in willow (*Salix viminalis* L). *J. Exp. Bot.* 44(7), 1119–1125.

Fuchs, M. (2017). Pyramiding resistance-conferring gene sequences in crops. *Curr. Opin. Virol.* 26, 36–42.

Fujimaki, S., Fujiwara, T. and Hayashi, H. (2000). A new method for direct introduction of chemicals into a single sieve tube of intact rice plants. *Plant Cell Physiol.* 41(1), 124–128.

Furch, A. C. U., Hafke, J. B., Schulz, A. and van Bel, A. J. E. (2007). Ca2+-mediated remote control of reversible sieve tube occlusion in *Vicia faba*. *J. Exp. Bot.* 58(11), 2827–2838.

Furch, A. C. U., van Bel, A. J. E., Fricker, M. D., Felle, H. H., Fuchs, M. and Hafke, J. B. (2009). Sieve element Ca^{2+} channels as relay stations between remote stimuli and sieve tube occlusion in *Vicia faba*. *Plant Cell* 21(7), 2118–2132.

Gaupels, F., Knauer, T. and van Bel, A. J. E. (2008). A combinatory approach for analysis of protein sets in barley sieve-tube samples using EDTA-facilitated exudation and aphid stylectomy. *J. Plant Physiol.* 165(1), 95–103.

Gianoli, E. and Niemeyer, H. M. (1998). DIBOA in wild Poaceae: sources of resistance to the Russian wheat aphid (*Diuraphis noxia*) and the Greenburg (*Schizaphis graminum*). *Euphytica* 102(3), 317–321.

Gildow, F. E. (1980). Increased production of alatae by aphids reared on oats infected with *Barley yellow dwarf virus*. *Ann. Entomol. Soc. Am.* 73(3), 343–347.

Graner, A., Jahoor, A., Schondelmaier, J., Siedler, H., Pillen, K., Fischbeck, G., Wenzel, G. and Herrmann, R. G. (1991). Construction of an RFLP map of barley. *Theor. Appl. Genet.* 83(2), 250–256.

Gray, S. M., Power, A. G., Smith, D. M., Seaman, A. J. and Altman, N. S. (1991). Aphid transmission of *Barley yellow dwarf virus*: acquisition access periods and virus concentration requirements. *Phytopathology* 81(5), 539–545.

Habekuss, A., Leistner, H. U. and Schliephake, E. (1999). Characterization of *Rhopalosiphum padi* genotypes differing in the geographical origin by transmission efficiency of *Barley yellow dwarf viruses* and molecular markers. *Z. Pflanzenkr. Pflanzenschutz* 106(4), 437–443.

Habekuss, A., Riedel, C., Schliephake, E. and Ordon, F. (2009). Breeding for resistance to insect-transmitted viruses in barley - an emerging challenge due to global warming. *J. Kulturpflanz.* 61(2), 53–61.

Hafke, J. B., Furch, A. C. U., Fricker, M. D. and van Bel, A. J. E. (2009). Forisome dispersion in *Vicia faba* is triggered by Ca2+ hotspots created by concerted action of diverse Ca2+ channels in sieve elements. *Plant Signal. Behav.* 4(10), 968–972.

Hallmann, C. A., Sorg, M., Jongejans, E., Siepel, H., Hofland, N., Schwan, H., Stenmans, W., Müller, A., Sumser, H., Hörren, T., Goulson, D. and de Kroon, H. (2017). More than 75 percent decline over 27 years in total flying insect biomass in protected areas. *PLoS ONE* 12(10), e0185809. doi: 10.1371/journal.pone.0185809.

Haupt, S., Duncan, G. H., Holzberg, S. and Oparka, K. J. (2001). Evidence for symplastic phloem unloading in sink leaves of barley. *Plant Physiol.* 125(1), 209–218.

Hedrich, R., Salvador-Recatala, V. and Dreyer, I. (2016). Electrical wiring and long-distance plant communication. *Trends Plant Sci.* 21(5), 376–387.

Hoffman, T. K. and Kolb, F. L. (1997). Effects of *barley yellow dwarf virus* on root and shoot growth of winter wheat seedlings grown in hydroponic culture. *Plant Dis.* 81(5), 497–500.

Hoffman, T. K. and Kolb, F. L. (1998). Effects of barley yellow dwarf virus on yield and yield components of drilled winter wheat. *Plant Dis.* 82(6), 620–624.

Hofinger, B. J., Russel, J. R., Bass, C. G., Baldwin, T., Dos Reis, M., Hedley, P. E., Li, Y., Macaulay, M., Waugh, R., Hammond-Kosack, K. E. and Kanyuka, K. (2011). An exceptional high nucleotide and haplotype diversity and a signature of positive selection for the eIF4E resistance gene in barley are revealed by allele mining and phylogenetic analyses of natural populations. *Mol. Ecol.* 20(17), 3653–3668.

Hogenhout, S. A., Ammar, E.-D., Whitfield, A. E. and Redinbaugh, M. G. (2008). Insect vector interactions with persistently transmitted viruses. *Ann. Rev. Phytopathol.* 46, 327–359.

Hus, S. J. and Robinson, A. G. (1962). Resistance of barley varieties to the aphid *Rhopalosiphum padi* (L.). *Can. J. Plant Sci.* 42(2), 247–251.

Ingwell, L. L., Eigenbrode, S. D. and Bosque-Pérez, N. A. (2012). Plant viruses alter insect behavior to enhance their spread. *Sci. Rep.* 2, 578.

International Barley Genome Sequencing Consortium (2012). A physical, genetic and functional sequence assembly of the barley genome. *Nature* 491(7426), 711–716.

The Intergovernmental Panel on Climate Change (IPCC) (2014). *Climate Change 2014: Synthesis Report. Contribution of Working Groups I, II and III to the Fifth Assessment Report of the Intergovernmental Panel on Climate Change* [Core Writing Team, Pachauri, R. K. and Meyer, L. A. (Eds)]. Geneva: IPCC, p. 151.

Jensen, S. G. (1969). Occurrence of virus particles in the phloem tissue of BYDV-infected barley. *Virology* 38(1), 83–91.

Jiménez-Martínez, E. S., Bosque-Pérez, N. A., Berger, P. H. and Zemetra, R. S. (2004a). Life history of the bird cherry-oat aphid, *Rhopalosiphum padi* (Homoptera: Aphididae), on transgenic and untransformed wheat challenged with *Barley yellow dwarf virus*. *J. Econ. Entomol.* 97(2), 203–212.

Jiménez-Martínez, E. S., Bosque-Pérez, N. A., Berger, P. H., Zemetra, R. S., Ding, H. and Eigenbrode, S. D. (2004b). Volatile cues influence the response of *Rhopalosiphum padi* (Homoptera: Aphididae) to *Barley yellow dwarf virus*-infected transgenic and untransformed wheat. *Environ. Entomol.* 33(5), 1207–1216.

Jin, Z., Wang, X., Chang, S. and Zhou, G. (2004). The complete nucleotide sequence and its organization of the genome of Barley yellow dwarf virus-GAV. *Sci. China C* 47(2), 175–182.

Johnston, P. A., Timmerman-Vaughan, G. M., Farnden, K. J. and Pickering, R. (2009). Marker development and characterization of *Hordeum bulbosum* introgression lines: a resource for barley improvement. *Theor. Appl. Genet.* 118(8), 1429–1437.

Jones, A. T. and Catherall, P. L. (1970). The relationship between growth rate and the expression of tolerance to barley yellow dwarf virus in barley. *Ann. Appl. Biol.* 65(1), 137–145.

Kennedy, T. F. and Connery, J. (2005). Grain yield reductions in spring barley due to barley yellow dwarf virus and aphid feeding. *Ir. J. Agric. Food Res.* 44, 111–128.

Kern, M., Meiners, T., Schliephake, E., Habekuss, A., Ordon, F., and Will, T. (2021). Effects of *Barley yellow dwarf virus* infection on the interaction of virus susceptible and tolerant barley genotypes with different clones of the aphid species *Rhopalosiphum padi*. Forthcoming.

Kleinhofs, A., Kilian, A., Saghai Maroof, M. A., Biyashev, R. M., Hayes, P., Chen, F. Q., Lapitan, N., Fenwick, A., Blake, T. K., Kanazin, V., Ananiev, E., Dahleen, L., Kudrna, D., Bollinger, J., Knapp, S. J., Liu, B., Sorrells, M., Heun, M., Franckowiak, J. D., Hoffman, D., Skadsen, R. and Steffenson, B. J. (1993). A molecular, isozyme and morphological map of the barley (*Hordeum vulgare*) genome. *Theor. Appl. Genet.* 86(6), 705–712.

Knoblauch, M., Knoblauch, J., Mullendore, D. L., Savage, J. A., Babst, B. A., Beecher, S. D., Dodgen, A. C., Jensen, K. H. and Holbrook, N. M. (2016). Testing the Münch hypothesis of long distance transport in plants. *eLife* 5, e15341. doi: 10.7554/eLife.15341.

Knoblauch, M. and van Bel, A. J. E. (1998). Sieve tubes in action. *Plant Cell* 10(1), 35–50.

Koonin, E. V., Dolja, V. V., Krupovic, M., Varsani, A., Wolf, Y. I., Yutin, N., Zerbini, M. and Kuhn J. H. (2019). Create a megataxonomic framework, filling all principal taxonomic ranks, for DNA viruses encoding vertical jelly roll-type major capsid proteins. Proposal (Taxoprop) No. 2019.003G. International Committee for Taxonomy of Viruses.

Kraakman, A. T. W., Martinez, F., Mussiraliev, B., van Eeuwijk, F. A. and Niks, R. E. (2006). Linkage disequilibrium mapping of morphological, resistance, and other agronomically relevant traits in modern spring barley cultivars. *Mol. Breed.* 17(1), 41–58.

Krueger, E. N., Beckett, R. J., Gray, S. M. and Miller, W. A. (2013). The complete nucleotide sequence of the genome of Barley yellow dwarf virus-RMV reveals it to be a new Polerovirus distantly related to other yellow dwarf viruses. *Front. Microbiol.* 4, 205.

Lindblad, M. and Areno, P. (2002). Temporal and spatial population dynamics of *Psammotettix alienus*, a vector of wheat dwarf virus. *Int. J. Pest Manag.* 48(3), 233–238.

Lister, R. M. and Ranieri, R. (1995). Distribution and economic importance of barley yellow dwarf. In: D'Arcy, C. J. and Burnett, P. A. (Eds), *Barley Yellow Dwarf. 40 Years of Progress*. St Paul: APS Press, pp. 29–53.

Liu, F., Wang, X., Liu, Y., Xie, J., Gray, S. M., Zhou, G. and Gao, B. (2007). A Chinese isolate of barley yellow dwarf virus-PAV represents a third distinct species within the PAV serotype. *Arch. Virol.* 152(7), 1365–1373.

Lucio-Zavaleta, E., Smith, D. M. and Gray, S. M. (2001). Variation in transmission efficiency among barley yellow dwarf virus-RMV isolates and clones of the normally inefficient aphid vector, *Rhopalosiphum padi*. *Phytopathology* 91(8), 792–796.

Lüpken, T., Stein, N., Perovic, D., Habekuß, A., Serfling, A., Krämer, I., Hähnel, U., Steuernagel, B., Scholz, U., Ariyadasa, R., Martis, M., Mayer, K., Niks, R. E., Collins, N. C., Friedt, W. and Ordon, F. (2014). High-resolution mapping of the barley Ryd3 locus controlling tolerance to BYDV. *Mol. Breed.* 33(2), 477–488.

Macaulay, M., Ramsay, L. and Åhman, I. (2020). Quantitative trait locus for resistance to the aphid *Rhopalosiphum padi* L. in barley (*Hordeum vulgare* L.) is not linked with a genomic region for gramine concentration. *Arthropod Plant Interact.* 14(1), 57–65.

Mancuso, S. (1999). Hydraulic and electrical transmission of wound induced signals in *Vitis vinifera*. *Functional Plant Biol.* 26(1), 55–61.

Manurung, B., Witsack, W., Mehner, S., Grüntzig, M. and Fuchs, E. (2004). The epidemiology of *Wheat dwarf virus* in relation to occurrence of the leafhopper *Psammotettix alienus* in Middle-Germany. *Virus Res.* 100(1), 109–113.

Mascher, M., Gundlach, H., Himmelbach, A., Beier, S., Twardziok, S. O., Wicker, T., Radchuk, V., Dockter, C., Hedley, P. E., Russell, J., Bayer, M., Ramsay, L., Liu, H., Haberer, G., Zhang, X. Q., Zhang, Q., Barrero, R. A., Li, L., Taudien, S., Groth, M., Felder, M., Hastie, A., Šimková, H., Staňková, H., Vrána, J., Chan, S., Muñoz-Amatriaín, M., Ounit, R., Wanamaker, S., Bolser, D., Colmsee, C., Schmutzer, T., Aliyeva-Schnorr, L., Grasso, S., Tanskanen, J., Chailyan, A., Sampath, D., Heavens, D., Clissold, L., Cao, S., Chapman, B., Dai, F., Han, Y., Li, H., Li, X., Lin, C., McCooke, J. K., Tan, C., Wang, P., Wang, S., Yin, S., Zhou, G., Poland, J. A., Bellgard, M. I., Borisjuk, L., Houben, A., Doležel, J., Ayling, S., Lonardi, S., Kersey, P., Langridge, P., Muehlbauer, G. J., Clark, M. D., Caccamo, M., Schulman, A. H., Mayer, K. F. X., Platzer, M., Close, T. J., Scholz, U., Hansson, M., Zhang, G., Braumann, I., Spannagl, M., Li, C., Waugh, R. and Stein, N. (2017). A chromosome conformation capture ordered sequence of the barley genome. *Nature* 544(7651), 427–433.

Mascher, M., Jost, M., Kuon, J. E., Himmelbach, A., Aßfalg, A., Beier, S., Scholz, U., Graner, A. and Stein, N. (2014). Mapping-by-sequencing accelerates forward genetics in barley. *Genome Biol.* 15(6), R78.

Mascher, M., Richmond, T. A., Gerhardt, D. J., Himmelbach, A., Clissold, L., Sampath, D., Ayling, S., Steuernagel, B., Pfeifer, M., D'Ascenzo, M., Akhunov, E. D., Hedley, P. E., Gonzales, A. M., Morrell, P. L., Kilian, B., Blattner, F. R., Scholz, U., Mayer, K. F., Flavell, A. J., Muehlbauer, G. J., Waugh, R., Jeddeloh, J. A. and Stein, N. (2013). Barley whole exome capture: a tool for genomic research in the genus *Hordeum* and beyond. *Plant J.* 76(3), 494–505.

Mayo, M. A. (2002). Virology division news: ICTV at the Paris ICV: results of the plenary session and the binomial ballot. *Arch.Virol.* 147(11), 2254–2260.

Mayo, M. A. and Pringle, C. R. (1998). Virus taxonomy-1997. *J. Gen.Virol.* 79(4), 649–657.

McKirdy, S. J., Jones, R. A. C. and Nutter, F. W. Jr. (2002). Quantification of yield losses caused by *Barley yellow dwarf virus* in wheat and oats. *Plant Dis.* 86(7), 769–773.

Medina-Ortega, K. J., Bosque-Pérez, N. A., Ngumbi, E., Jiménez-Martínez, E. S. and Eigenbrode, S. D. (2009). *Rhopalosiphum padi* (Hemiptera: Aphididae) responses to volatile cues from *Barley yellow dwarf virus*-infected wheat. *Environ. Entomol.* 38(3), 836–845.

Münch, E. (1930). *Stoffbewegungen in der Pflanze*. Jena: Gustav Fischer.

Nancarrow, N., Constable, F. E., Finlay, K. J., Freeman, A. J., Rodoni, B. C., Trebicki, P., Vassiliadis, S., Yen, A. L. and Luck, J. E. (2014). The effect of elevated temperature on Barley yellow dwarf virus-PAV in wheat. *Virus Res.* 186, 97–103.

Niks, R. E., Habekuss, A., Bekele, B. and Ordon, F. (2004). A novel major gene on chromosome 6h for resistance of barley against the *Barley yellow dwarf virus*. *Theor. Appl. Genet.* 109(7), 1536–1543.

Lange, T. and Ordon, F. (2010). Coordinator's report: disease and pest resistance genes. *Barley Genet. Newsl.* 39, 34–35.

Ordon, F., Friedt, W., Scheurer, K., Pellio, B., Werner, K., Neuhaus, G., Huth, W., Habekuss, A. and Graner, A. (2004). Molecular markers in breeding for virus resistance in barley. *J. Appl. Genet.* 45(2), 145–159.

Ordon, F. and Kühne, T. (2014). Response to viral pathogens. In: Kumlehn, K. and Stein, S. (Eds), *Biotechnology in Agriculture and Forestry*. Heidelberg: Springer, pp. 181-196.

Ordon, F. and Perovic, D. (2013). Virus resistance in barley. In: Varshney, R. K. and Tuberosa, R. (Eds), *Translational Genomics for Crop Breeding, Volume I: Biotic Stresses* (1st edn.). Hoboken: John Wiley & Sons, pp. 63-75.

Paltridge, N. G., Collins, N. C., Bendahmane, A. and Symons, R. H. (1998). Development of YLM, a codominant PCR marker closely linked to the *Yd2* gene for resistance to barley yellow dwarf disease. *Theor. Appl. Genet.* 96(8), 1170-1177.

Parry, A. L. and Habgood, R. M. (1986). Field assessment of the effectiveness of a *Barley yellow dwarf virus* resistance gene following its transference from spring to winter barley. *Ann. Appl. Biol.* 108(2), 395-401.

Paulmann, M. K., Kunert, G., Zimmermann, M. R., Theis, N., Ludwig, A., Meichsner, D., Oelmüller, R., Gershenzon, J., Habekuss, A., Ordon, F., Furch, A. C. U. and Will, T. (2018). Barley yellow dwarf virus Infection leads to higher chemical defense signals and lower electrophysiological reactions in susceptible compared to tolerant barley genotypes. *Front. Plant Sci.* 9, 145.

Peck, K. M. and Lauring, A. S. (2018). Complexities of viral mutation rates. *J. Virol.* 92(14), e01031-17. doi: 10.1128/JVI.01031-17.

Perović, D., Budak, H., Sato, K. and Sourdille, P. (2020a), Editorial: use of barley and wheat reference sequences: downstream applications in breeding, gene isolation, GWAS, and evolution. *Front. Plant Sci.* 11, 1017.

Perović, D., Dodig, D., Perović, J., Šurlan Momirović, G., Ordon, F. and Novo, P. (2020b). Comparative genomics of cereals in an era of new sequencing technologies. In: Grujić, R., Janjić, V. and Trkulja, R. (Eds), *Primary Agricultural Production*. Banja Luka: Academy of Sciences and Arts of Republika of Srpska, p. 18.

Perovic, D., Kopahnke, D., Habekuss, A., Ordon, F. and Serfling, A. (2018). Marker based harnessing of genetic diversity to improve resistance of barley to fungal and viral diseases. In: Miedaner, T. and Korzun, V. (Eds), *Applications of Genetic and Genomic Research in Cereals. Woodhead Publishing Series in Food Science, Technology and Nutrition*. Amsterdam: Elsevier, pp. 137-151.

Perovic, D., Kopahnke, D., Steffenson, B. J., Förster, J., König, J., Kilian, B., Plieske, J., Durstewitz, G., Korzun, V., Krämer, I., Habekuß, A., Johnston, P., Pickering, R. and Ordon, F. (2013). Genetic fine mapping of a novel leaf rust resistance gene and a Barley yellow dwarf virus tolerance (BYDV) introgressed from Hordeum bulbosum by the use of the 9K iSelect chip. In: Zhang, G., Li, C. and Liu, X. (Eds), *Advance in Barley Sciences. Proceedings of the 11th International Barley Genetics Symposium*. Dordrecht: Springer and Zhejiang University Press, pp. 269-284.

Pickering, R. A., Hill, A. M., Michel, M. and Timmermann-Vaughan, G. M. (1995). The transfer of a powdery mildew resistance gene from *Hordeum bulbosum* L. to barley (*H. vulgare* L.) chromosome 2 (2L). *Theor. Appl. Genet.* 91(8), 1288-1292.

Pickering, R. A., Steffenson, B. J., Hill, A. M. and Borovkova, I. (1998). Association of leaf rust and powdery mildew resistance in a recombinant derived from a *Hordeum vulgare* x *Hordeum bulbosum* hybrid. *Plant Breed.* 117(1), 83-84.

Pohler, W. and Szigat, G. (1982). Versuche zur rekombinativen Genu¨bertragung von der Wildgerste Hordeum bulbosum auf die Kulturgerste H. vulgare. 1. Mitt. Die Rückkreuzung VV 9 BBVV. *Arch Züchtungsforsch Berlin* 12(8), 7-100.

Poland, J. A., Brown, P. J., Sorrells, M. E. and Jannink, J. L. (2012). Development of high-density genetic maps for barley and wheat using a novel two-enzyme genotyping by-sequencing approach. *PLoS ONE* 7(2), e32253.

Powell, G., Tosh, C. R. and Hardie, J. (2006). Host plant selection by aphids: behavioral, evolutionary, and applied perspectives. *Ann. Rev. Entomol.* 51, 309–330.

Power, A. G., Seaman, A. J. and Gray, S. M. (1991). Aphid transmission of *Barley yellow dwarf virus*: inoculation access periods and epidemiological implications. *Phytopathology* 81(5), 545–548.

Ramsay, L., Macaulay, M., degli Ivanissevich, S., MacLean, K., Cardle, L., Fuller, J., Edwards, K. J., Tuvesson, S., Morgante, M., Massari, A., Maestri, E., Marmiroli, N., Sjakste, T., Ganal, M., Powell, W. and Waugh, R. (2000). A simple sequence repeat-based linkage map of barley. *Genetics* 156(4), 1997–2005.

Rasmusson, D. C. and Schaller, C. W. (1959). The inheritance of resistance in barley to the yellow dwarf virus. *Agron. J.* 51(11), 661–664.

Rhodes, J. D., Thain, J. F. and Wilson, D. C. (1996). The pathway for systemic electrical signal conduction in the wounded tomato plant. *Planta* 200(1), 50–57.

Riedel, C., Habekuß, A., Schliephake, E., Niks, R., Broer, I. and Ordon, F. (2011). Pyramiding of *Ryd2* and *Ryd3* conferring tolerance to a German isolate of *Barley yellow dwarf virus*-PAV (BYDV-PAV-ASL-1) leads to quantitative resistance against this isolate. *Theor. Appl. Genet.* 123(1), 69–76.

Ripl, J., Drab, T., Gadiou, S. and Kundu, J. K. (2020). Differences in responses to Wheat dwarf virus infection in contrasting wheat cultivars Ludwig and Svitava. *Plant Prot. Sci.* 56(2), 67–73.

Rochow, W. F. (1969). Biological properties of four isolates of *Barley yellow dwarf virus*. *Phytopathology* 59(11), 1580–1589.

Rochow, W. F. and Muller, I. (1971). A fifth variant of barley yellow dwarf in New York. *Plant Dis.* 55(10), 874–877.

Ruge, B., Linz, A., Pickering, R., Proeseler, G., Greif, P. and Wehling, P. (2003). Mapping of *Rym*14Hb, a gene introgressed from *Hordeum bulbosum* and confering resistance to BaMMV and BaYMV in barley. *Theor. Appl. Genet.* 107(6), 965–971.

Ruge-Wehling, B., Linz, A., Habekuß, A. and Wehling, P. (2006). Mapping of *Rym16*^{Hb}, the second soil-borne virus-resistance gene introgressed from *Hordeum bulbosum*. *Theor. Appl. Genet.* 113(5), 867–873.

Rupp, J. S., Cruz, L., Trick, H. N. and Fellers, J. P. (2019). RNAi-mediated silencing of endogenous wheat genes eIF(Iso)4E-2 and eIF4G induce resistance to multiple RNA viruses in transgenic wheat. *Crop Sci.* 59(6), 2642–2651.

Salvador-Recatalà, V., Tjallingii, W. F. and Farmer, E. E. (2014). Real-time, in vivo intracellular recordings of caterpillar-induced depolarization waves in sieve elements using aphid electrodes. *New Phytol.* 203(2), 674–684.

Savary, S., Willoquet, L., Pethybridge, S. J., Esker, P., McRoberts, N. and Nelson, A. (2019). The global burden of pathogens and pests on major food crops. *Nat. Ecol. Evol.* 3(3), 430–439.

Schafer, J. F. (1971). Tolerance to plant disease. *Annu. Rev. Phytopathol.* 9(1), 235–252.

Schaller, C. W. (1984). The genetics of resistance to barley yellow dwarf virus in barley. In: Burnett, P. A. (Ed), *Barley Yellow Dwarf. Proceedings of the a Workshop*. Mexico, DF: CIMMYT, p. 93.

Schaller, C. W., Qualset, C. O. and Rutger, J. N. (1964). Inheritance and linkage of the Yd2 gene conditioning resistance to the barley yellow dwarf virus disease in barley 1. *Crop Sci.* 4(5), 544–548.

Scheurer, K. S., Friedt, W., Huth, W., Waugh, R. and Ordon, F. (2001). QTL analysis of tolerance to a German strain of BYDV-PAV in barley (*Hordeum vulgare* L.). *Theor. Appl. Genet.* 103(6–7), 1074–1083.

Scheurer, K. S., Huth, W., Friedt, W. and Ordon, F. (2000). First results on BYDV-tolerance in barley estimated in pot experiments. *J. Plant Dis. Prot.* 107(4), 427–432.

Schliephake, E., Habekuss, A., Scholz, M. and Ordon, F. (2013). Barley yellow dwarf virus transmission and feeding behaviour of *Ropalosiphum padi* on *Hordeum bulbosum* clones. *Entomol. Exp. Appl.* 146(3), 347–356.

Scholz, M., Ruge-Wehling, B., Habekuss, A., Schrader, O., Pendinen, G., Fischer, K. and Wehling, P. (2009). Ryd4Hb: a novel resistance gene introgressed from *Hordeum bulbosum* into barley and conferring complete and dominant resistance to the barley yellow dwarf virus. *Theor. Appl. Genet.* 119(5), 837–849.

Seibold, S., Gossner, M. M., Simons, N. K., Blüthgen, N., Müller, J., Ambarlı, D., Ammer, C., Bauhus, J., Fischer, M., Habel, J. C., Linsenmair, K. E., Nauss, T., Penone, C., Prati, D., Schall, P., Schulze, E. D., Vogt, J., Wöllauer, S. and Weisser, W. W. (2019). Arthropod decline in grasslands and forests is associated with landscape-level drivers. *Nature* 574(7780), 671–674.

Silvar, C., Perovic, D., Casas, A. M., Igartua, E. and Ordon, F. (2011). Development of a cost-effective Pyrosequencing approach for SNP genotyping in barley. *Plant Breed.* 130(3), 394 -397.

Sip, V., Sirlova, L. and Chrpova, J. (2006). Screening for barley yellow dwarf virus-resistant barley genotypes by assessment of virus content in inoculated seedlings. *J. Phytopathol.* 154(6), 336–342.

Skaria, M., Lister, R. M., Foster, J. E. and Shaner, G. (1985). Virus content as an index of symptomatic resistance to barley yellow dwarf virus in cereals. *Phytopathology* 75(2), 212–216.

Slykhuis, J. T. (1967). Methods for experimenting with mite transmission of plant viruses. *Methods Virol.* 1, 347–368.

Steuernagel, B., Periyannan, S. K., Hernández-Pinzón, I., Witek, K., Rouse, M. N., Yu, G., Hatta, A., Ayliffe, M., Bariana, H., Jones, J. D., Lagudah, E. S. and Wulff, B. B. (2016). Rapid cloning of disease-resistance genes in plants using mutagenesis and sequence capture. *Nat. Biotechnol.* 34(6), 652–655.

Strażyński, P. and Ruszkowska, M. (2015). The life cycle functional response of *Rhopalosiphum padi* (L.) to higher temperature: territorial expansion of permanent parthenogenetic development as a result of warmer weather conditions. *J. Plant Prot. Res.* 55(2), 162–165.

Suneson, C. A. (1955). Breeding for resistance to yellow dwarf virus in barley. *Agron. J.* 47(6), 283.

Szigat, G., Herrmann, M. and Rapke, H. (1997). Integration von Bastardpflanzen mit der Wildgerste *Hordeum bulbosum* in den Zuchtprozeß von Wintergerste. In: Begemann, F. (Ed), *Züchterische Nutzung Pflanzengenetischer Ressourcen. Ergebnisse und Forschungsbedarf. Schriften zu Genetischen Ressourcen 8*. Bonn: Informationszentrum für Genetische Ressourcen (IGR)Zentralstelle für Agrardokumentation und -information, pp. 267–270.

Taylor, R. E. (2013). Cable theory. *Phys. Tech. Biol. Res.* 6, 219–262.

Thind, A. K., Wicker, T., Šimková, H., Fossati, D., Moullet, O., Brabant, C., Vrána, J., Doležel, J. and Krattinger, S. G. (2017). Rapid cloning of genes in hexaploid wheat using cultivar-specific long-range chromosome assembly. *Nat. Biotechnol.* 35(8), 793-796.

Toojinda, T., Broers, L. H., Chen, X. M., Hayes, P. M., Kleinhofs, A., Korte, J., Kudrna, D., Leung, H., Line, R. F., Powell, W., Ramsay, L., Vivar, H. and Waugh, R. (2000). Mapping quantitative and qualitative disease resistance genes in a doubled haploid population of barley (*Hordeum vulgare*). *Theor. Appl. Genet.* 101(4), 580-589.

Trębacz, K., Dziubińska, H. and Król, E. (2006). Electrical signals in long-distance communication in plants. In: Baluska, F., Mancuso, S. and Volkmann, D. (Eds), *Communication in Plants*. Berlin-Heidelberg: Springer-Verlag, pp. 277-290.

Trębicki, P., Dáder, B., Vassiliadis, S. and Fereres, A. (2017). Insect-plant-pathogen interactions as shaped by future climate: effects on biology, distribution, and implications for agriculture. *Insect Sci.* 24(6), 975-989.

Trębicki, P., Nancarrow, N., Cole, E., Bosque-Pérez, N. A., Constable, F. E., Freeman, A. J., Rodoni, B., Yen, A. L., Luck, J. E. and Fitzgerald, G. J. (2015). Virus disease in wheat predicted to increase with a changing climate. *Glob. Chang. Biol.* 21(9), 3511-3519.

Van Bel, A. J. E. (2003). The phloem, a miracle of ingenuity. *Plant Cell Environ.* 26(1), 125-149.

Van Bel, A. J. E. and Hafke, J. B. (2005). Physiochemical determinants of phloem transport. In: Holbrook, N. M. and Zwieniecki, M. (Eds), *Vascular Transport in Plants*. Amsterdam: Elsevier, pp. 19-44.

Van der Eynde, R., van Leeuwen, T. and Haesaert, G. (2020). Identifying drivers of spatio-temporal dynamics in *Barley yellow dwarf virus*. *Pest Manag. Sci.* 76(8), 2548-2556. doi: 10.1002/ps.5851.

Varshney, R. K., Marcel, T. C., Ramsay, L., Russell, J., Röder, M. S., Stein, N., Waugh, R., Langridge, P., Niks, R. E. and Graner, A. (2007). A high density barley microsatellite consensus map with 775 SSR loci. *Theor. Appl. Genet.* 114(6), 1091-1103.

von Bothmer, R., Jacobsen, N., Baden, C., Jorgensen, R. and Linde-Laursen, I. (1995). *An Ecogeographical Study of the Genus Hordeum* (2nd edn.). Rome: International Plant Genetic Resources Institute.

Vos, P., Hogers, R., Bleeker, M., Reijans, M., Van der Lee, T., Hornes, M., Frijters, A., Pot, J., Peleman, J. and Kuiper, M. (1995). AFLP: a new technique for DNA fingerprinting. *Nucleic Acids Res.* 23(21), 4407-4414.

Walls, J. III, Rajotte, E. and Rosa, C. (2019). The past, present, and future of Barley yellow dwarf management. *Agriculture* 9(1), 23.

Wang, Z., Gerstein, M. and Snyder, M. (2009). RNA-Seq: a revolutionary tool for transcriptomics. *Nat. Rev. Genet.* 10(1), 57-63.

Werner, K., Friedt, W. and Ordon, F. (2005). Strategies for pyramiding resistance genes against the *Barley yellow mosaic virus* complex. *Mol. Breed.* 16(1), 45-55.

Williams, J. G. K., Kubelik, A. R., Livak, K. J., Rafalski, J. A. and Tingey, S. V. (1990). DNA polymorphisms amplified by arbitrary primers are useful as genetic markers. *Nucleic Acids Res.* 18(22), 6531-6535.

Wu, B., Blanchard-Letort, A., Liu, Y., Zhou, G., Wang, X. and Elena, S. F. (2011). Dynamics of molecular evolution and phylogeography of *Barley yellow dwarf virus*-PAV. *PLoS ONE* 6(2), e16896.

Xu, J. and Kasha, K. J. (1992). Transfer of a dominant gene for powdery mildew resistance and DNA from *Hordeum bulbosum* into cultivated barley (*H. vulgare*). *Theor. Appl. Genet.* 84(7-8), 771-777.

Yang, J., Worley, E., Ma, Q., Li, J., Torres-Jerez, I., Li, G., Zhao, P. X., Xu, Y., Tang, Y. and Udvardi, M. (2016). Nitrogen remobilization and conservation, and underlying senescence-associated gene expression in the perennial switchgrass *Panicum virgatum*. *New Phytol.* 211(1), 75-89.

Ziesche, T. M., Bell, J., Ordon, F., Schliephake, E. and Will, T. (2020). Long-term monitoring of insects in agricultural landscapes. *Mitteilungen der DGaaE* 22, 101-106.

Zimmermann, M. R., Mithöfer, A., Will, T., Felle, H. H. and Furch, A. C. U. (2016). Herbivore-triggered electrophysiological reactions: candidates for systemic signals in higher plants and the challenge of their identification. *Plant Physiol.* 170(4), 2407-2419.

Part 8

Fungal diseases of cereals: Regional strategies

Chapter 22

Key challenges in breeding durable disease-resistant cereals: North America

Christina Cowger, USDA-ARS, USA

1 Introduction

In this chapter, we first describe the challenges of diverse climates, diseases, and market classes that are faced by North American small-grain cereal breeders and producers. We touch briefly on challenges inherent in the complex systems of cereal breeding on the continent, and the changing resistance priorities brought about by shifting pathogen races and production practices. The remainder of the chapter is devoted (in rough order of priority) to the status and prospects for durable resistance to the main pests currently confronting the continent: Fusarium head blight, rusts, powdery mildew, leaf (and glume) blotches, viruses, Hessian fly, and bacterial leaf streak.

http://dx.doi.org/10.19103/AS.2021.0092.37
Published by Burleigh Dodds Science Publishing Limited, 2022.

2 Key challenge 1: diversity of production settings

North America is a globally important producer of cereals, with approximately 25 million ha of wheat and 4 million ha of barley harvested annually (USDA Foreign Agricultural Service, or FAS). Canada is usually the world's third largest barley producer, with roughly 10 million metric tons per year (FAS) (Fig. 1), and is among the world's top durum wheat producers (Lin et al. 2018; Statistics Canada, https://www150.statcan.gc.ca/t1/tbl1/en/tv.action?pid=3210035901) (Fig. 2). Ninety percent of Canada's wheat production occurs in the Prairie Provinces of Alberta, Saskatchewan, and Manitoba (Laroche et al. 2019). Mexico harvests roughly 600 000 ha of wheat annually, with about 60% of production (in metric tons) being durum wheat (Fig. 3), and ~350 000 ha of barley (FAS) (Fig. 4). The United States is generally the world's fifth largest wheat producer, with about 50–60 million metric tons annually (FAS), and harvests around 800 000 ha of barley annually (Fig. 5).

Chief among the challenges to breeding durably disease-resistant cereals in North America is the great diversity of climates and thus of specific disease and pest problems. Wheat is grown across vast areas of the United States and Canada, and thus a wide array of environmental conditions govern its production. Production areas include the cool Canadian maritime provinces

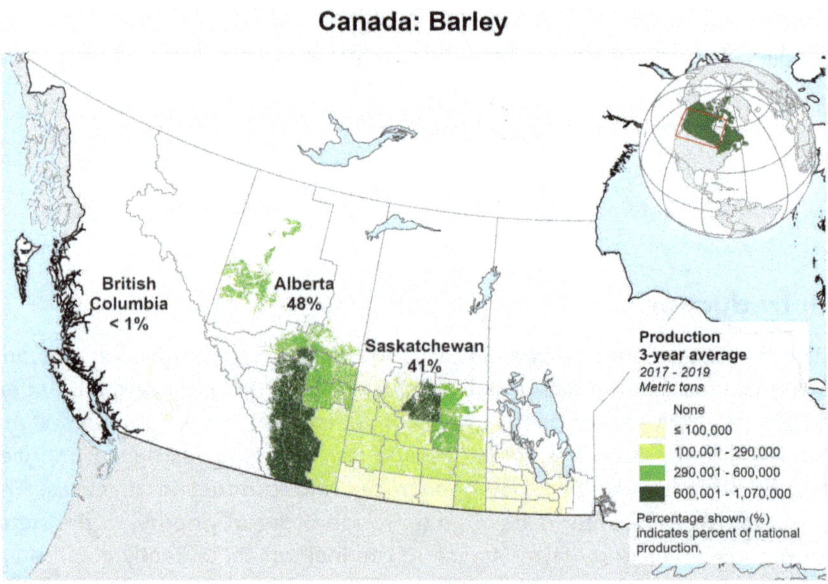

Figure 1 Canada is usually the world's third largest barley producer, with roughly 10 million metric tons per year (Map by U.S. Department of Agriculture Foreign Agricultural Service with data from Statistics Canada and AgCanada).

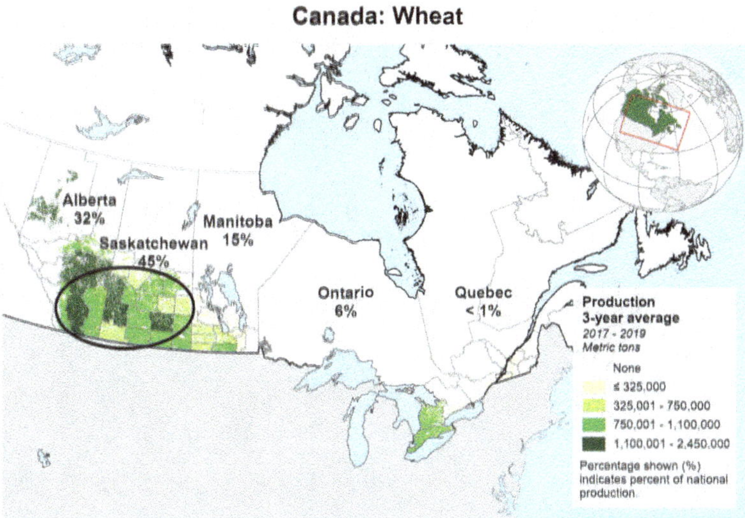

Figure 2 Total wheat production in Canada. Hard red spring wheat and durum wheat (the latter shown by oval) are mainly produced in the Plains provinces of Alberta, Saskatchewan, and Manitoba, while Ontario produces soft red winter wheat (Map by U.S. Department of Agriculture Foreign Agricultural Service with data from Statistics Canada, AgCanada and Canadian Grain Commission).

and U.S. northeast, the dry plains of Canada and the United States, the humid mid-Atlantic and southeastern United States, the arid intermountain landscapes of the western United States, the cool and wet conditions west of the Cascade Mountains in the Pacific Northwest, the deserts of Sonora and Baja California, and the cool Mexican Highlands.

These climatically diverse regions also possess a wide array of soil types. Moreover, the adoption of conservation tillage has been particularly robust in North America compared to Europe, Africa, and most of Asia (Huggins and Reganold 2008). In 2004, about 41% of all planted U.S. cropland was under conservation tillage, as compared to 26% in 1990. Most of the increase was due to the expanded practice of no-till agriculture, which by 2004 was employed on 22% of U.S. farmland. Reduced tillage has many important benefits such as avoiding erosion, improving soil quality, sequestering carbon, and facilitating biological diversity. However, in the case of small-grain cereals, it may also contribute to an increase in diseases harbored in debris, such as Fusarium head blight (FHB) and leaf blotches (Aboukhaddour et al. 2020) or Rhizoctonia root rot (Schroeder and Paulitz 2006).

Within its large U.S. eco-geographic range, wheat is produced in five major market classes: hard red spring, hard red winter, soft red winter, soft white winter, and durum wheat (Fig. 6). Barley of both spring and winter habits is

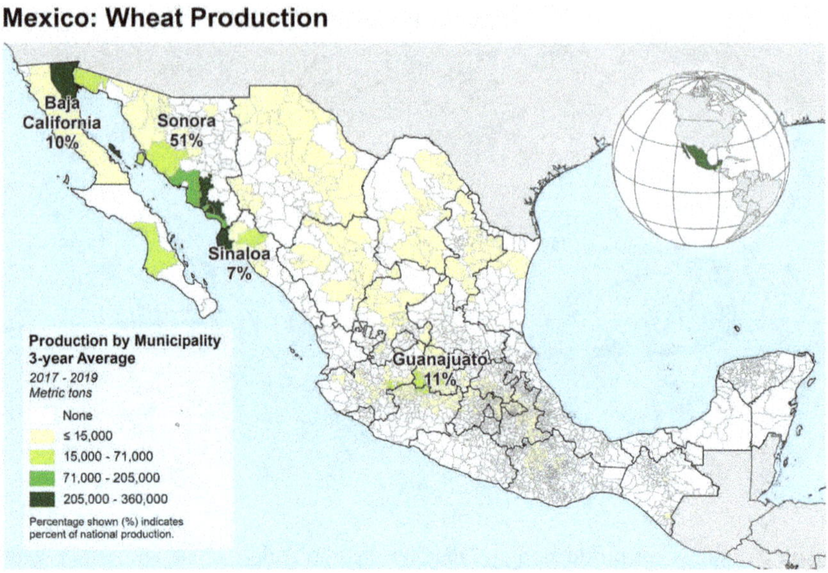

Figure 3 Mexican wheat production is concentrated in the northwestern states of Sonora, Baja California, and Sinaloa, with additional cultivation in Guanajuato and adjacent highland states (Map by U.S. Department of Agriculture Foreign Agricultural Service with data from INEGI-SIAP).

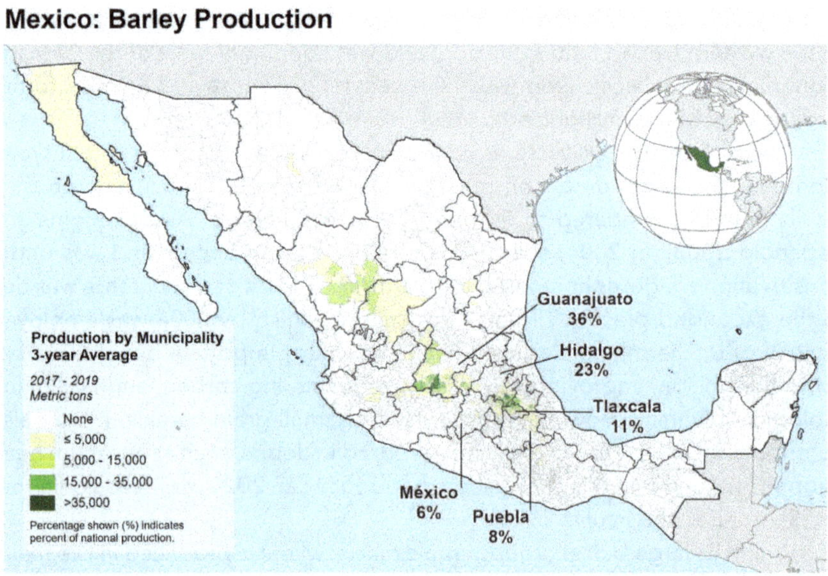

Figure 4 In Mexico, barley is produced mainly in the interior highlands (Map by U.S. Department of Agriculture Foreign Agricultural Service with data from INEGI-SIAP).

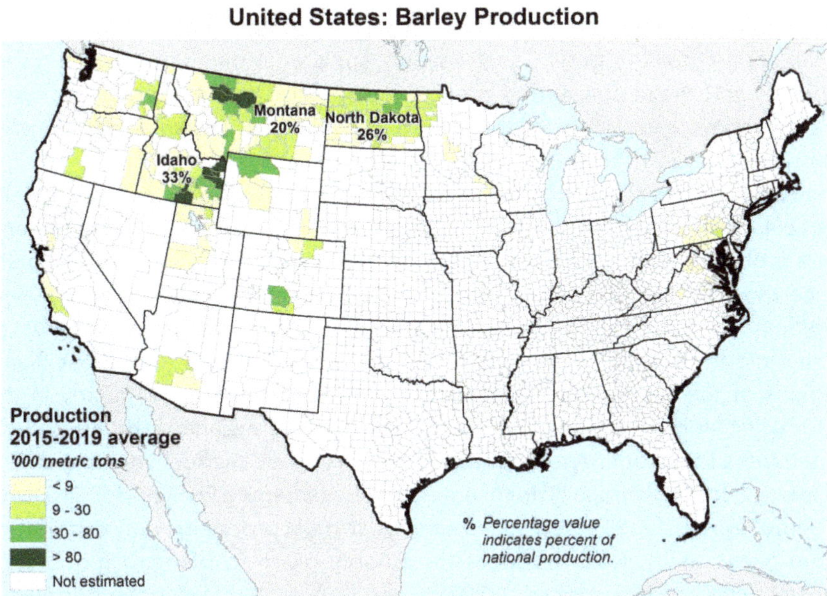

Figure 5 The majority of US barley production occurs in the northern states of Idaho, Montana, and North Dakota, although malting barley is grown on a smaller scale in many other states (Map by U.S. Department of Agriculture Foreign Agricultural Service with data from the USDA National Agricultural Statistics Service).

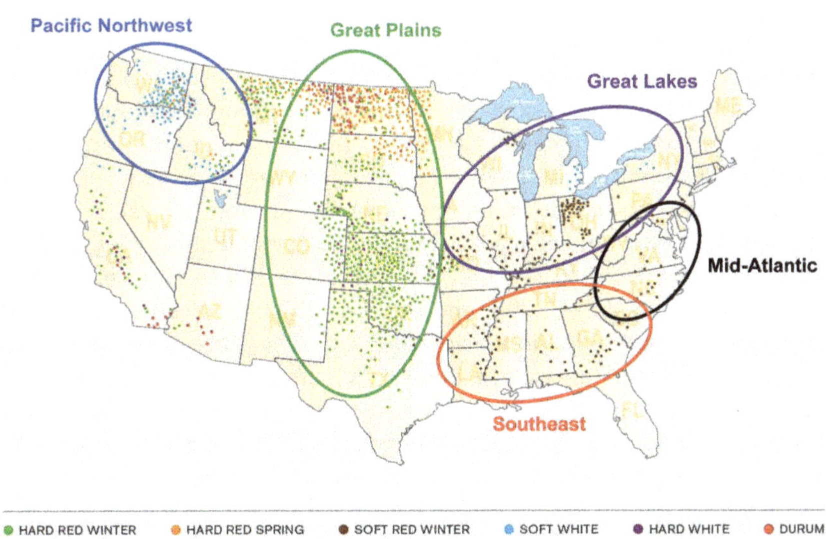

Figure 6 US wheat production by region; dots indicate where the six main wheat market classes are produced (Source: National Association of Wheat Growers).

produced for two end uses, feed and malting. In addition, a handful of specialty small-grain crops are grown, such as club, spelt, and emmer wheat; heirloom varieties and those adapted to organic production (Kucek et al. 2015); food-grade barley; and wheat adapted to particular local markets and climates (Brouwer et al. 2015).

Naturally, then, each North American market class of each crop has its particular set of climate conditions, rotational partner crops, and disease and pest problems (Table 1). While Fusarium head blight has become a problem since the 1990s in almost all production areas, many diseases are relatively regional. For example, powdery mildew and wheat streak mosaic virus as economic threats are usually confined to the mid-Atlantic and Great Plains regions of the United States, respectively. Stripe rust occurs regularly in the U.S. Pacific Northwest and Great Plains, and in western Canada, but irregularly in the rest of the continent. Even where they are most destructive, many cereal diseases occur episodically: for example, while damaging FHB epidemics occur in some region of the United States every year, most producers only experience them occasionally, which reduces the priority given to planting moderately resistant cultivars (Cowger et al. 2020a) and thus probably also the priority for that trait in cultivar release decisions.

3 Key challenge 2: structure of North American cereal breeding

Adding to the complexity of the germplasm challenges are the diversity of breeding programs and the strained resources for public breeding, pre-breeding and breeder training. In the United States, for example, there are over 40 public and private small-grain breeding programs engaged directly in variety development, and many additional laboratories conducting research on cereal genetics, gene discovery, germplasm enhancement, and other breeding-related fields. However, it was observed over 10 years ago that in the previous two decades, a decline in U.S. investment in plant breeding had weakened university plant breeding programs and reduced the number of public plant breeders (Stuber and Hancock 2008). In recent times, private companies have become more involved in U.S. wheat breeding, despite the decline in U.S. wheat area since the 1990s, although this investment ebbs and flows (Baenziger 2019, National Association of Wheat Growers 2014). Development of wheat and barley varieties is still principally a public-sector activity in Canada, although private companies have been expanding their role (Skerritt 2016).

In tandem with these changes, there has been a tightening of intellectual property constraints, although the effect on diversity of disease-resistance sources and resistance durability is not clearly documented. Historically, in

Table 1 Principal North American market classes of wheat and barley and the main disease resistance targets in breeding programs for each

Country	Crop	Market class	Principal end uses	Main production region(s)	Current top disease resistance breeding targets
The United States	Wheat	Soft white winter	Noodles	Pacific Northwest (PNW)	Stripe rust, Cephalosporium stripe, strawbreaker foot rot, Fusarium crown rot, Septoria tritici blotch (STB)
	Wheat	Hard red spring	Milling and baking (bread)	Upper Midwest, Montana	Fusarium head blight (FHB), bacterial streak and black chaff, leaf rust, stripe rust, tan spot
	Wheat	Hard red winter	Milling and baking (bread)	Great Plains, PNW	Leaf rust, stripe rust, FHB, barley yellow dwarf virus, Hessian fly
	Wheat	Soft red winter	Pastries, flat breads, animal feed	Six-state, Mid-Atlantic, Southeast	FHB, leaf rust, Hessian fly, powdery mildew, stripe rust
	Wheat	Durum	Semolina for pasta, couscous	Upper Midwest	FHB, tan spot, leaf rust
	Barley	Malting (two-row, six-row)	Malting	North Dakota, Montana, Idaho	FHB, net blotch, spot blotch
	Barley	Six-row	Animal feed	Upper Midwest, eastern United States (scattered)	FHB
Canada	Wheat	Hard red spring	Milling and baking (bread)	Alberta, Saskatchewan, Manitoba	Stripe rust, FHB, tan spot
	Wheat	Soft red winter wheat	Cakes, pastries, cereal, crackers	Ontario	FHB, powdery mildew, STB, stripe rust
	Wheat	Durum	Semolina for pasta, couscous	Alberta, Saskatchewan	FHB, tan spot, leaf rust
	Barley	Malting (2-row, 6-row)	Malting	Alberta, Saskatchewan	FHB, spot blotch, stem rust
Mexico	Wheat	Hard red spring	Milling and baking (bread)	Northwest (Baja California, Sonora)	Stripe rust, leaf rust, stem rust, karnal bunt
	Wheat	Hard red spring	Milling and baking (bread)	Highlands (Tlaxcala, Puebla and Mexico states)	FHB, STB, other leaf blotches
	Wheat	Spring durum	Semolina, couscous	Northwest (Baja California, Sonora) and Guanajuato	Leaf rust, stripe rust, leaf blotches, stem rust, FHB
	Barley	Malting (6-row)	Malting	South-Central (Guanajuato, Hidalgo, Tlaxcala, Mexico, Puebla)	Stripe rust, net and spot blotch, leaf rust, scald, Alternaria leaf blight

Sources: USDA-FAS, Canadian Grain Commission, Schwarz and Horsley 1996, Aboukhaddour et al. 2020, P. K. Singh (CIMMYT).

wheat breeding, U.S. public researchers worked together, sharing germplasm through organizations such as Triticeae CAP (Coordinated Agricultural Project) and the International Wheat Genome Sequencing Consortium (National Association of Wheat Growers 2014). North American wheat germplasm used to be shared freely according to the Wheat Workers Code of Ethics (https://www.ars.usda.gov/ARSUserFiles/30421000/wheatcode.html), but more recently such exchange has come under the aegis of material transfer agreements (MTAs). Other changes include the ability of public universities to own cultivars and the application of private contract law to ownership of genetic resources (Baenziger 2019, Luby et al. 2015). Restricted access to germplasm could limit the within-species diversity available for the development of new disease-resistant cultivars (Luby et al. 2015). Another factor that could operate to reduce available diversity of disease-resistance genes is the centralization of end-use standards (Brouwer et al. 2015). It is unclear to what extent these changes have impeded the exchange of wheat or barley germplasm among breeders or reduced the diversity of disease-resistance and other QTL entering and moving through breeding pipelines.

Particularly in the eastern U.S. wheat market, a challenge to the selection and prioritization of durable resistance is the ephemeral nature of commercial soft wheat cultivars, which turn over on average about every 3–4 years. Few of these cultivars attain a large market share during their brief time on the market. Their short lifespan can make it difficult to determine which cultivars possess durable resistance, especially in the case of FHB, where rankings from multiple environments are essential. Also, many soft wheat growers do not prioritize the selection of disease-resistant cultivars, as wheat can be an 'afterthought' crop when the price is low, and private-sector varieties often have names that consist of similar letters and numbers (Cowger, personal observation; Cowger et al. 2020a).

4 Key challenge 3: evolution of principal disease threats to small-grains cereals

Another challenge is the shifting disease and pest 'target list'. Over relatively short time spans, factors such as the evolution of cropping practices have altered which diseases are most damaging to North American wheat and barley, and consequently which breeding targets are a priority. For example, the growing use of conservation tillage has increased the importance of residue-borne diseases, especially Fusarium head blight, tan spot in Canada (Aboukhaddour et al. 2020), and possibly Hessian fly (Ratcliffe et al. 2000, but see Del Conte et al. 2005).

In western Canada, the top five breeding targets for wheat are rusts, FHB and common bunt, and intermediate-or-better resistance to those diseases is required for cultivar registration (Aboukhaddour et al. 2020, Brar et al. 2019b,c). More specifically, to the west of Saskatoon, the most important of those are stripe rust and common bunt, while east of Saskatoon the priorities are stem rust, leaf rust, FHB, and common bunt (Laroche et al. 2019). However, stem rust, leaf rust, and common bunt have become better managed in the Canadian prairies due to cultivar resistance and fungicides, and the top three diseases causing economic damage on the Canadian prairies are currently stripe rust, FHB and tan spot (Aboukhaddour et al. 2020).

On a continental scale, while leaf rust has diminished in importance in western Canada, it continues to be a problem in the central and eastern United States. Stripe rust, which in the twentieth century was mainly confined to the U.S. Pacific Northwest, has generally increased in importance across the continent since the emergence of new races in about the year 2000 (Brar et al. 2019b). Stem rust continues to be rare in most of North America, but it represents a potent future threat should Ug99 races arrive.

In another example of shifting disease priorities, wheat soil-borne mosaic (WSBM), caused by soil-borne wheat mosaic virus (SBWMV), was formerly a common problem in the mid-Atlantic and southeastern United States, but decreased to insignificance in that region starting in about 2011 (C. Cowger, personal observations). The reasons are not fully known; one factor could be that the resistance gene *Sbm1* is estimated to occur in at least three-quarters of eastern US soft red winter wheat varieties, based on marker data from elite experimental lines (USDA https://www.ars.usda.gov/southeast-area/raleigh-nc/plant-science-research/). At the same time, WSBM was first discovered in the Pacific Northwest's dryland wheat area in 2008, and within a decade had increased its range (Kroese et al. 2020).

Conversely, Pythium root rot was historically a problem in dryland wheat production in Washington state (Paulitz and Adams 2003), especially in spring wheat and under direct-seeded (no-till) conditions (Mavrodi et al. 2012). Unknown previously in the mid-Atlantic states, Pythium root rot emerged in 2016 as a widespread problem in central and eastern North Carolina, where wheat is grown under conventional tillage (Lookabaugh et al. 2017) in rotation with corn, soybeans, and tobacco. It has continued to do damage in North Carolina, perhaps owing to longer periods of winter soil waterlogging, with three particular species of *Pythium* being especially frequent in stunted wheat (*P. irregulare, P. spinosum, and P. vanterpoolii*; Reeves et al. 2021).

Thus, North America's principal small-grain disease problems vary by region, leading to a complex mosaic of resistance breeding and disease management priorities (Table 1). Following is a summary of the status and prospects of durable resistance to the continent's main cereal diseases.

Most of the major North American small-grain diseases of importance are caused by fungi, as the following sections show, starting with Fusarium head blight. A handful of second-tier challenges are arthropod-related or bacterial, and those are discussed later.

5 Specific disease challenges: Fusarium head blight

Globally and in North America, Fusarium head blight (FHB) is caused by a variety of *Fusarium* species, with the main causal species being *F. graminearum*. As a result, deoxynivalenol (DON) is the principal *Fusarium* mycotoxin of concern in small-grain cereals. However, in Canada, Mexico, and parts of the United States, FHB-causing species have been detected in small-grain crops that belong to the *Fusarium tricinctum* species complex and produce 'emerging mycotoxins' such as moniliformin, rather than DON (Bec et al. 2015, Cerón-Bustamante et al. 2018, Cowger et al. 2020c, Gräfenhan et al. 2013).

Continentally, FHB is now at or near the top of the list of breeding challenges due to the increased conduciveness of large parts of the North American agricultural landscape, a situation that appears likely to persist indefinitely. The disease emerged as one of 'devastating impact' in the U.S. Upper Midwest and Canada in the early 1990s (McMullen et al. 1997), and also around the same time in the North-Central and Highland regions of Mexico (Cerón-Bustamante et al. 2018). FHB proceeded to become a threat in much of the United States, wherever small grains are grown. In some areas of the US South, FHB is a major factor in the reduced production of wheat (Castro Aviles et al. 2020). The widening of the FHB threat in the United States is due to expanded maize cultivation, broader use of reduced tillage, and climatic changes (Cowger and Sutton 2005, Dong et al. 2018, McMullen et al. 2012). Together, these factors leave increased amounts of maize and other cereal debris on the soil surface, where they decompose more slowly. As infected residues have become more widely distributed in the North American landscape, the degree to which tillage can reduce disease pressure within a given field appears to have declined, as airborne inoculum may be dispersed regionally (Bergstrom et al. 2010, 2013, Del Ponte et al. 2003, Dill-Macky and Jones 2000). FHB is the most damaging wheat disease in Canada, occurring both in the prairie states (Aboukhaddour et al. 2020, Haile et al. 2019) and in Ontario (Crippin et al. 2020).

As fungicides are at best only partially effective in controlling FHB and its mycotoxins, cultivar resistance is an essential tool for small grain producers to hedge against potential losses due to low yield, test weight dockage, inability to sell grain for human consumption, and even load rejection. In organic production, resistance is key: after weed competitive ability, FHB resistance was regarded as the second most important trait in wheat cultivars by organic farmers in the northeast United States (Kucek et al. 2015). Resistance to FHB is

Published by Burleigh Dodds Science Publishing Limited, 2022.

quantitatively inherited and mainly additive (Arruda et al. 2015). On the positive side, quantitative resistance (QR) to FHB is expected to be relatively durable: selection pressure for erosion of efficacy is low, as the disease is monocyclic and no genes confer complete immunity. On the negative side, due to the high level of economic damage caused by the disease, the levels of QR currently available in commercial cultivars are frequently inadequate, and there is still a shortage of highly effective QTL.

Fhb1, a large-effect gene originally found in Chinese cultivars such as Sumai-3 and Wangshuibai, has gradually assumed greater prominence in North American FHB resistance breeding. The gene is primarily associated with Type II resistance, or resistance to spread within the spike after initial infection (Buerstmayr et al. 2002). Cloned by three research groups (Li et al. 2019, Rawat et al. 2016, Su et al. 2019), *Fhb1* has been incorporated into hard red spring cultivars accounting for substantial shares of the commercial market in the U.S. Upper Midwest and Canada (Hao et al. 2020, Zhu et al. 2019), as well as a hard white spring germplasm line (Cao et al. 2016). Indeed, in the United States, moderately FHB-resistant cultivars occupy a substantially larger proportion of planted area in the hard red spring wheat market class than in any other class (Cowger et al. 2020a). In the past decade, the use of *Fhb1* has also expanded in soft winter wheat breeding. This is likely due to the breaking of the linkage drag previously associated with this locus (Bakhsh et al. 2013, Brar et al. 2019c).

In North America, while *Fhb1* has increased in frequency in both spring and winter commercial wheat varieties, the same does not appear to be true for other Mendelized Fusarium head blight resistance genes (*Fhb2-Fhb7*). In some cases, these genes confer relatively high levels of FHB resistance, and their introgression from wild relatives into wheat germplasm has been a long-term project. For example, as part of a decades-long 'evolutionary breeding' effort, *Fhb3* was introgressed on a compensating translocation from the allotetraploid wild rye *Leymus racemosus* into spring wheat germplasm, and specific markers were identified (Qi et al. 2008). *Fhb5,* originating from cvs. Sumai-3 and Wangshuibai, was found to confer resistance equivalent to that provided by *Fhb1* in one Canadian hard spring wheat bi-parental NIL population, but not in another (Brar et al. 2019a). Also, the Type II gene *Fhb6* was introgressed from the Asian perennial grass species, *Elymus tsukushiensis,* into hexaploid winter wheat backgrounds, and molecularly tagged for marker-assisted breeding (Cainong et al. 2015). However, these genes do not as yet appear to be in widespread use in breeding programs, perhaps because of complex, undesirable interactions that are specific to genotypic backgrounds (Brar et al. 2019c).

By contrast, 'native resistance', that is, resistance not borrowed from Sumai-3 or other Asian spring wheats, has dominated in soft winter wheat FHB-resistance breeding for two decades. Many native QTL have been identified in the eastern

United States wheat gene pool (e.g., Benson et al. 2012, Carpenter et al. 2020?, Castro Aviles et al. 2020, Islam et al. 2016, Liu et al. 2013, Petersen et al. 2016, Petersen et al. 2017, Zhu et al. 2019). Genomic selection and genome-wide association studies have become practical for incorporating this type of FHB resistance, which is usually due to additive effects of multiple loci (Arruda et al. 2015, Larkin et al. 2020, Rutkoski et al. 2012, Tessmann et al. 2019). Explorations of native FHB resistance have also begun to occur in Canadian spring wheat, with numerous QTL identified in bi-parental mapping populations (McCartney et al. 2016, Thambugala et al. 2020).

While the lack of effective FHB resistance QTL is starting to be overcome in hexaploid wheat, it is still a major hurdle in durum wheat. Variation for FHB resistance in durum is relatively narrow, resistance is not as well characterized as in hexaploid wheat, and no gene as effective as *Fhb1* has been identified (Haile et al. 2019, Zhang et al. 2014b). However, FHB QTL with promise for North American durum breeding have been detected in North Dakotan and Canadian durum germplasm, domesticated emmer (*T. turgidum* subsp. *dicoccum*), Persian wheat (*T. turgidum* subsp. *carthlicum*), an exotic tetraploid landrace of unknown origin, and Tunisian durum lines (Ghavami et al. 2011, Huhn et al. 2012, Ruan et al. 2012, Sari et al. 2018, Zhang et al. 2014b, Zhu et al. 2016). A major QTL has also been introgressed into durum from a hexaploid wheat line (Zhao et al. 2018).

In barley, most efforts to identify FHB resistance have occurred in the spring barley region. Genomic selection has potential for barley, as for wheat (Lorenz et al. 2012). For winter barleys, differences in resistance have been observed, although distinguishing levels of resistance to DON accumulation is hampered by high variability in DON rankings among environments and a lack of genes providing high levels of partial resistance (Berger et al. 2014, Cowger et al. 2019, Khatibi et al. 2011). Winter barley is relatively sensitive to cold temperatures and wet soil, which results in unevenness in heading within FHB phenotyping experiments. This leads to high error variances and compounds the difficulty of assigning precise ranks to genotypes based on symptoms and DON resistance.

6 Specific disease challenges: stripe rust, leaf rust and stem rust

6.1 Stripe rust

The race structure of the U.S. *P. striiformis* f. sp. *tritici* (*Pst*) population, which causes stripe rust (also known as yellow rust) of wheat, is extremely complex (Liu et al. 2017, Wan and Chen 2014). Since 2000, North America has seen a considerable expansion of *Pst* races and concomitantly the area where damaging epidemics occur (Chen 2007). The new races are characterized by virulence to *Yr8* and *Yr9*, as well as greater aggressiveness and adaptation to warmer temperatures (Milus et al. 2009). These races have widened the range where the disease occurs regularly from the U.S. Pacific Northwest (PNW) to the Great Plains and the Canadian Prairies (Brar and Kutcher 2016, Brar et al. 2018).

While it occurs annually in the western and central United States and Canada, and is frequently severe there, wheat stripe rust is episodic in much of the eastern United States, being more common in the south and less common in the north (Milus et al. 2015b, Wan et al. 2016). Most Canadian durum cultivars are resistant to stripe rust, but new races virulent to genes used in durum wheat such as *Yr24* were recently identified in Mexico and could pose a challenge to more northerly durum crops (Huerta-Espino and Singh 2017).

Prioritization of stripe rust resistance by wheat breeding programs follows the occurrence patterns described above. Thus, resistance to stripe rust is a high priority for wheat breeding programs in the PNW (Chen 2013, Chen 2014, Naruoka et al. 2015). It is also a breeding priority in the U.S. Great Plains and Canadian prairie provinces, but less so in the eastern United States. In an informal survey of central and eastern U.S. wheat breeders in 2013, the majority of hard winter wheat breeders considered stripe rust resistance a high priority, while the majority of soft winter wheat breeders considered it a moderate to low priority (Milus et al. 2015b). Most of the hard wheat breeders relied upon cultivars with unknown sources of resistance, while the soft wheat breeders relied to a greater extent on specific genes or markers. When selecting for stripe rust resistance, most breeders in both groups relied on evaluating adult plants in the field.

Many major genes for stripe rust resistance have been named, but only a few are actually deployed commercially in the United States (Liu et al. 2020b, Mu et al. 2020). Of more than 80 identified *Yr* genes, only a handful, including *Yr5* and *Yr15*, remained effective against all detected races of the pathogen in the United States (Wang and Chen 2017). *Yr5* and *Yr15* have been incorporated into several spring wheat cultivars in combination, but have not been introduced into U.S. winter wheat (Liu et al. 2020b, Mu et al. 2020). This has renewed interest among breeders in partial resistance as a more durable approach.

High-temperature adult-plant (HTAP) resistance to *P. striiformis* is generally considered to be relatively durable. Most HTAP QTL have small effects (Zhou et al. 2014), and some have been effective for over 50 years (Chen 2013, Wang and Chen 2017). For example, the HTAP resistance sources in U.S. PNW cultivars Madsen and Stephens have been effective for over 30 and 40 years, respectively, and are attributed to combinations of different QTL (Liu et al. 2018, Santra et al. 2008, Vazquez et al. 2012). *Yr36* confers HTAP resistance that is believed to be race-nonspecific, post-haustorial, and effective at a wide range of temperatures, but its durability has yet to be demonstrated, as the gene is originally from *Triticum turgidum* ssp. *dicoccoides* and has been incorporated into a few U.S. common wheat cultivars only within the last 15 years (Bryant et al. 2014, Fu et al. 2009, Liu et al. 2020b, Lowe et al. 2011, Wang and Chen 2017). To overcome the tight linkage between *Yr36* and *Gpc-B1*, a gene for high grain protein that is often undesirable for soft wheat, tetraploid and hexaploid germplasm lines were released that carry *Yr36* and a non-functional *Gpc-B1* allele (Hale et al. 2013).

HTAP resistance to stripe rust has been used extensively in the U.S. PNW. More than 40 HTAP genes have been introgressed into wheat cultivars in the PNW, where most wheat cultivars possess HTAP resistance (Chen 2013, Liu et al. 2020b, Mu et al. 2020). Examples of recently identified HTAP genes introgressed into PNW spring wheat genotypes from international sources are *Yr52* from India (Ren et al. 2012), *Yr59* from Iraq (Zhou et al. 2014), *Yr62* from Portugal (Lu et al. 2014), and *Yr79* from Pakistan (Feng et al. 2018). These genes have been used in U.S. wheat breeding programs (X. M. Chen, personal communication). The lines carrying these genes were found in the National Small Grains Collection (NSGC) (http://www.ars-grin.gov/cgi-bin/npgs/acc/search.pl?accid=PI+183527).

Hard winter wheat with HTAP QTL has also been studied in the PNW (Chen 2013), where an association-mapping approach has yielded markers associated with both seedling and adult-plant QTL in a large panel of winter wheat genotypes in several regional market classes (Naruoka et al. 2015). The genome-wide association study (GWAS) approach was recently used to investigate stripe rust resistance in cultivars, breeding lines, and genetic stocks of both spring and winter wheats developed or used in various regions throughout the United States (Liu et al. 2020b, Mu et al. 2020). These studies identified a large number of stripe rust resistance genes or QTL utilized with varying frequencies in different geographic regions.

In Canada, although *Lr34/Yr18* appeared in an increasing share of wheat cultivars since the 1970s, breeding for stripe rust resistance only became a priority after the arrival of new races in 2000 (Brar et al. 2019b). Numerous major and possibly novel QTL were mapped in Canadian hexaploid spring wheats, some overlapping with known genes such as *Yr18* and *Yr36*, but some not (Bokore et al. 2017, Singh et al. 2014). However, due to concern over linkage drag with new genes, breeding programs still rely mainly on *Yr18*, and to a lesser degree on *Yr36* and *Yr29* (Brar et al. 2019b). In Canadian durum wheat, QTL for stripe rust resistance were identified on chromosomes 7B and 5B in a biparental mapping population, with additional support for the 7B locus found in a global durum breeding panel (Lin et al. 2018). An increasing number of wheat breeding lines from Canada, especially from Alberta, have been resistant to stripe rust in the last 10 years (X. M. Chen, personal communication).

Beyond the PNW and Canadian prairies, it is difficult to get a comprehensive overview of breeding progress toward durable stripe rust resistance, but there are scattered insights. In the U.S. Great Plains, a stripe rust resistance QTL was identified in the cultivar TAM 111 that was environmentally dependent and probably race specific (Basnet et al. 2014). TAM 111 was popular in the south-central part of the region and resistant to stripe rust until it became moderately susceptible in 2012. *Yr17* has been used by many south-central U.S. breeding programs, but was overcome by a new race of *P. striiformis* around that time; however, the new cultivar TAM 114 showed resistance in the field (Milus et al. 2015a, Rudd et al. 2018). The effects of *Yr17* are clearest in field-grown adult plants, while in seedlings they are dependent on temperature, relative humidity, and light (Milus et al. 2015a). The broad use of *Yr17* in U.S. winter wheat breeding programs, with more than 60% of Great Plains nursery entries positive for the gene, is attributable to its linkage with genes for resistance to several other diseases and to the availability of useful markers (Mu et al. 2020). *Yr17* has been used less frequent in spring wheat breeding programs, but still is present in nearly 15% of spring wheat cultivars and breeding lines in the PNW. As virulence to *Yr17* appeared in the United States in the early 1970s and has become more frequent throughout the United States since 2000 (Liu et al. 2017, Wan et al. 2016), the continued strong performance of cultivars carrying *Yr17* is attributed to a linked QTL (Liu et al. 2018, 2020b, Mu et al. 2020).

In the eastern U.S. soft wheat region, race-specific adult-plant resistance (APR) has been held to be the only effective type of resistance (Milus et al. 2015b), while another study detected more than 30 genes or QTL among eastern soft wheat lines, including both race-specific all-stage resistance (ASR) and non-race-specific HTAP resistance (Mu et al. 2020). A resistance source termed *Yr_4BL* was mapped in cultivars VA96W270 and USG3555 and was present in an average of 51% of Uniform Eastern (UE) entries and 67% of Uniform Southern (US) nursery entries in 2018 and 2019 (USDA Eastern Regional Small Grains Genotyping Laboratory, or ERSGGL). *Yr17* occurred in an average of 34% of UE nursery entries and 49% of US entries in 2017–2019. *Yr5* was absent from U.S. Uniform Southern and Uniform Eastern nursery entries in 2017 (Mu et al. 2020) and 2019 (USDA ERSGGL).

6.2 Leaf rust

Leaf rust (*Puccinia triticina*) is a perennial problem throughout the central and eastern U.S. and Canadian wheat region. It is the most common disease of wheat in the north-central U.S. Great Plains and the adjoining Canadian prairie provinces (Kolmer et al. 2020). Frequencies of *P. triticina* races have been monitored regularly in the United States and Canada (e.g. Kolmer 2019, McCallum et al. 2020).

Most named leaf rust resistance genes (*Lr* genes) are seedling genes that confer all-stage resistance (ASR). Some leaf rust ASR genes have lost efficacy soon after introduction, such as *Lr10* and *Lr16* in Canadian spring wheat (McCallum et al. 2016). Certain leaf rust ASR genes have remained durable over long periods of time; for example:

- *Lr18* from *T. timopheevii* retains good partial efficacy in the southeastern U.S. (author's personal observations; Green et al. 2014), even though *Lr18*-virulent races are found at a low frequency in the region (J. Kolmer, personal communication). The gene is present in about 20% of the USDA's Uniform Eastern Nursery entries and over 50% of Uniform Southern Nursery entries (USDA Eastern Regional Small Grains Genotyping Laboratory).
- *Lr21* was introgressed into hexaploid wheat from *Ae. tauschii* in the 1970s. Virulence to *Lr21* did not appear until 2010–2011 in the northern United States and Canada; the gene is now no longer effective (J. Kolmer, personal communication; McCallum et al. 2016).
- *Lr37* (=*Yr17*), introgressed from *Triticum ventricosum,* is present in a third to a half of recent Uniform Eastern and Uniform Southern Nursery entries (USDA Eastern Regional Small Grains Genotyping Laboratory) and retains moderate efficacy (C. Cowger, personal observation). However, in the Great Plains, hard red winter wheats such as cultivar Jagger that have only *Lr37* are very susceptible (J. Kolmer, personal observation).

Some genes for leaf rust ASR are effective but not in widespread use in North America. For example, *Lr9*, introgressed from *Aegilops umbellulata*, is generally effective in the eastern United States (Kolmer 2019) but hardly utilized (USDA Eastern Regional Small Grains Genotyping Laboratory), possibly because of a linked yield reduction (Ortelli et al. 1996).

At present, there are at least 15 leaf rust APR genes with an *Lr* gene designation, about half race-specific and the other half race-non-specific (Pinto da Silva et al. 2018). Genes conferring APR to leaf rust have long been used in North American hexaploid wheat. First identified in cultivar Frontana (Dyck et al. 1966), *Lr13* is most effective in adult plants but is highly race specific; it was identified in hard red spring cultivars Manitou and Chris (released in 1965 and 1966, respectively), and still remains effective in Canada, although not in Mexico (Pinto da Silva et al. 2018). In the northern spring wheat region where it is relatively common, APR gene *Lr34* has proven to be effective over time, especially when combined with other genes, where it has an enhancing effect (McCallum et al. 2016). *Lr34* is also present at a low frequency in hard red winter cultivars (J. Kolmer, personal communication). In the southeastern United States, *Lr34* confers only a modest degree of resistance and is not widely used in cultivar development: in the 2017–2019 period, only one of the approximately 200 advanced experimental lines tested in the Uniform Southern and Uniform Eastern winter wheat nurseries possessed *Lr34* (= *Sr57*) (USDA Eastern Regional Small Grains Genotyping Laboratory).

Genome-wide association studies on large international sets of spring wheat accessions have identified numerous potentially novel leaf rust resistance QTL, with the top 10 QTL each for ASR and APR collectively explaining roughly 30% of the total variation in the respective types of resistance (Gao et al. 2016, Turner et al. 2017). Overall, at least 250 leaf rust resistance QTL have been identified, and a study of meta-QTL has identified 35 regions in hexaploid wheat having two or more closely linked leaf rust QTL (Pinto da Silva et al. 2018). Association mapping studies have been carried out on 96 winter wheat genotypes, 3640 spring wheat genotypes, and 295 mixed types. The B genome carries the largest number of leaf rust resistance QTL, and the A genome the fewest (Pinto da Silva et al. 2018).

Gene combinations constitute a major strategy for durable resistance to leaf rust. Generally, a combination of several APR QTL is necessary to provide adequate levels of leaf rust resistance (Pinto da Silva et al. 2018). In particular, combinations involving *Lr13* and *Lr34* remain effective even when the individual genes have been defeated (Kolmer et al. 2020).

Further, several old wheat varieties in various market classes possess very durable QR to leaf rust, the genetic basis of which is unknown or still under investigation.

- In soft red winter wheat, one such variety is Caldwell, which was released in Indiana in 1981 and widely grown for four years. Its APR remains moderately to highly effective in the eastern United States (author's personal observation), may be present in other soft winter wheat cultivars, and was recently traced to a major QTL on chromosome 3BS (Kolmer et al. 2018a).
- In hard red spring wheat, an example is AC Taber, a Canadian hard red spring wheat variety released in 1991 with durable leaf rust resistance conferred by *Lr13* on chromosome 2BS and another QTL on 3BS near adult plant resistance gene *Lr74* (Kolmer et al. 2020). Cultivars with long-lasting resistance have possessed combinations of ASR genes *Lr16* and *Lr23* with APR genes *Lr13* and *Lr34*, for example, Norm and Alsen (Oelke and Kolmer 2005).
- Among hard red winter wheats, durable resistance is possessed by the old varieties Sturdy (*Lr34*), Santa Fe (*Lr37*, *Lr77*), and Duster (*Lr34*, *Lr46*, and *Lr77*) (Kolmer 2017, Kolmer et al. 2018b, 2019).

Leaf rust epidemics on durum wheat in Mexico in the 2001–2009 period caused significant economic damage, while the Mexican bread wheat crop remained less affected due to APR (Huerta-Espino et al. 2011). While durum was historically considered more stably resistant to leaf rust than hexaploid wheat, new races emerged and overcome resistance of widely adapted durum cultivars in northwestern Mexico (Herrera-Foessel et al. 2014, Singh et al. 2013), and isolates of those races found in the United States have raised concerns for the U.S. and Canadian durum crops (Aoun et al. 2016, Kolmer 2015). Efforts are underway to identify and utilize APR that will be durable in North American durum cultivars confronting the new *P. triticina* races (Kthiri et al. 2019, Lan et al. 2017, Lan et al. 2019, Li et al. 2020).

6.3 Stem rust

Wheat stem rust, caused by *Puccinia graminis* f. sp. *tritici*, has not been economically important in North America since the late 1970s (Zhang et al. 2014a). However, stem rust resistance has excited increased attention in this continent since the emergence of the east African race Ug99, with its broad virulence to North American wheats (Jin and Singh 2006), and the evolution of additional virulences in its descendants (Newcomb et al. 2016).

Seedling or all-stage resistance (ASR) genes can be rapidly overcome, and there is a paucity of identified durable adult-plant resistance (APR) genes for stem rust (Singh et al. 2011, Zhang et al. 2014a). *Sr2* is the only APR gene so far shown to be durable, and the resistance it confers is modest (Singh et al. 2015, Zhang et al. 2014a). The *Sr2* gene complex plays an important role in CIMMYT germplasm (Rutkoski et al. 2014). However, *Sr2* is either closely linked to or pleiotropic for pseudo-black chaff, a discoloration of wheat glumes that can be expressed weakly or strongly, depending on the wheat genotype and environment, and may be a deterrent to producer acceptance (Juliana et al. 2015). Other genes that confer APR to stem rust are *Sr55*, *Sr57*, and *Sr58* - all pleiotropically encoding resistance to leaf rust, stripe rust and powdery mildew - and *Sr56* (Singh et al. 2015). Gene *Sr12* was shown to confer adult plant resistance to stem rust in Kenya when combined with other QTL (Rouse et al. 2014).

A major challenge to the identification of additional sources of durable APR to Ug99 races in U.S. wheat experimental lines is that the relevant field phenotyping of adult plants cannot take place in North America. For that reason, and to identify new sources of ASR and APR, U.S. and Canadian wheat lines are screened annually along with other nations' materials at the field stem rust nursery in Njoro, Kenya, established by the Kenyan Agricultural and Livestock Research Organization in conjunction with CIMMYT (Jin et al. 2007). This program allows researchers to capture adult-plant responses to the Ug99 races that cannot be observed in U.S. field nurseries (e.g. Anderson et al. 2018) because as living isolates, those races may only be handled within containment facilities.

The combination of Kenyan field data on adult plants with controlled-environment data on seedlings has allowed the screening of CIMMYT and U.S. spring wheat lines and the identification of accessions with either ASR or APR to the Ug99 race group (Rouse et al. 2011b). Using seedling data and Kenyan field data, a Montenegrin spring wheat accession with Ug99 resistance was found to have a combination of resistance QTL resembling that in the broadly stem rust-resistant Canadian spring wheat cultivar Thatcher, from which it may be descended (Zurn et al. 2018). Kenyan field data on APR have been used for a GWAS on North American spring wheats that identified several likely novel genomic regions useful in cultivar development, even though only about 6% of the wheat lines showed seedling resistance to Ug99 races (Bajgain et al. 2015). Also, Kenyan field data have shown enhanced efficacy when stem rust APR from Thatcher (Kolmer et al. 2011, Rouse et al. 2014) or seedling resistance (*SrCad*) from cvs. AC Cadillac and Peace (Hiebert et al. 2011) is combined with *Lr34*.

At present, known stem rust APR genes with efficacy against the Ug99 race group do not appear to be widely present in U.S. wheat. For example, only one of the approximately 200 advanced experimental lines tested in the Uniform Southern and Uniform Eastern winter wheat nurseries in the 2017–2019 period possessed *Sr57* = *Lr34* (USDA Genotyping Lab). However, some U.S. wheat germplasm displays moderate-to-complete resistance to stem rust in Kenya with the source of that resistance remaining unknown. In current U.S. spring wheat breeding lines, ASR genes effective in Kenya are infrequent (Bajgain et al. 2015). However, several lines do possess APR genes including cultivars that have been recently released such as 'UI Platinum' (*Sr2*) and 'Linkert' (*Sr12* and *Sr57*) (Anderson et al. 2018, Edae et al. 2018). Genomic selection using genome-wide marker coverage offers the possibility of improving APR to stem rust that is conferred by combining multiple small-effect genes to achieve a high level of resistance (reviewed in Rutkoski et al. 2011).

Gene pyramids, particularly those involving *Sr2*, have contributed to managing stem rust in twentieth-century North American spring wheat cultivation, regions where CIMMYT germplasm is heavily used, and Australia (Mundt 2018). Pyramids of genes effective against Ug99, such as *SrCad*, *Sr33*, and *Sr57* = *Lr34*, have been developed in spring wheat germplasm (Zhang et al. 2019). *Sr21* from *Triticum monococcum* is also proposed as a pyramid component; it confers partial resistance to Ug99 races and its efficacy increases at higher temperatures (Chen et al. 2018). *Sr* gene pyramids are also relatively common in hard red winter wheat (Y. Jin, personal communication). Novel QTL for what is likely ASR to stem rust were identified in an accession of cultivated emmer (*T. turgidum* ssp. *dicoccum*) that is broadly resistant to Ug99 races, and could be used in gene pyramids in either durum or common wheat (Sharma et al. 2019). Additional genes for Ug99 resistance that could be utilized in pyramids have been found in triticale (X *Triticosecale*) (Olivera et al. 2013) and *Aegilops* species (Olivera et al. 2018, Rouse et al. 2011a).

7 Specific disease challenges: wheat powdery mildew

In North America, wheat powdery mildew (caused by *Blumeria graminis* f. sp. *tritici*, or *Bgt*) is most frequently a problem in the eastern soft wheat region, especially the Mid-Atlantic US states. However, the disease can flare up in unusual locales (e.g. the states of Georgia, Montana, and Oklahoma) if susceptible cultivars are planted in a large area (Cowger et al. 2018). Powdery mildew is adequately managed by wheat cultivars having a high level of quantitative or adult-plant resistance (APR). The *Bgt* population possesses qualities that furnish it with very high evolutionary potential: approximately annual sexual recombination, enormous effective population sizes, and the build-up of huge quantities of asexual spores via polycyclic reproduction during epidemics. Thus while major resistance (*Pm*) genes are important to provide resistance during severe epidemics, the U.S. genes such as *Pm3a*, *Pm4a*, and *Pm17* have been overcome quickly (Cowger et al. 2009). The best strategy to deploy still effective *Pm* genes is in QR backgrounds and/or in pyramids.

A large set of novel *Pm* genes were introgressed from wild wheat relatives into the well-adapted soft winter wheat background of cv. Saluda for purposes of breeding in the eastern United States (Cowger et al. 2012, Maxwell et al. 2009, 2010, 2012, Miranda et al. 2006, 2007a,b, Murphy et al. 2009, Perugini et al. 2008, Petersen et al. 2015). Pyramids of two, three, and four of the genes were composed in the background of the soft red winter wheat cultivar Saluda, which is adapted to the mid-Atlantic United States, and the germplasm is publicly available.

APR to wheat powdery mildew, or 'slow-mildewing', was studied in the United States in the 1970s–1990s, but less so in recent years. Several soft red wheat cultivars have exhibited APR to powdery mildew that remained durable over many years, included Knox and its derivatives Knox 62 and Massey, as well as Genesee, Redcoat, Houser, and Maris Huntsman (Cowger et al. 2012, Liu et al. 2001, 2012). Two cultivars that have exhibited good levels of partial resistance to powdery mildew since they were released in 2000 are AGS2000 and P26R61, resistance QTL were identified in both (Hao et al. 2015). In recent years, the frequency of heavy wheat powdery mildew epidemics appears to have declined along the mid-Atlantic seaboard, with possible reasons including improved average cultivar resistance, more scattered crops due to reduced wheat planting, and the warming climate.

8 Specific disease challenges: leaf blotch diseases

The three most important wheat leaf blotch diseases in North America (see Fig. 1, Cowger et al. 2020b for distribution) are:

1 Septoria nodorum blotch (SNB), also known as Stagonospora nodorum blotch, caused by the necrotrophic ascomycete *Parastagonospora nodorum*. SNB occurs widely across the United States and is the only important wheat leaf blotch disease east of the Appalachian Mountains. In wetter areas, it also causes glume blotch, which in addition to disfiguring the crop can reduce test weight and result in seed transmission.
2 Septoria tritici blotch (caused by *Zymoseptoria tritici*), which occurs together with SNB in the rainier parts of the Pacific Northwest, where STB is the greater economic problem, and in parts of the U.S. Midwest and Plains.
3 Tan spot (caused by *Pyrenophora tritici-repentis*) can be a significant problem in the Canadian prairies and Great Plains states of the United States and is found occasionally in a wider area.

The rise of reduced tillage in North America has increased inoculum availability for these diseases in various regions. Moreover, it can be difficult to distinguish the lesions of the three leaf blotches in field locations where they co-occur, potentially confounding the phenotypic ratings that are necessary for selection of wheat genotypes with quantitative resistance to a specific blotch disease. For example, an impediment to STB resistance breeding is the fact that in most of its U.S. range, the disease co-occurs with SNB and in some cases tan spot (Cowger et al. 2020b). While resistance to the three diseases is often under separate genetic control, wheat genotypes have been identified with multiple blotch resistance in controlled environment studies of seedlings (e.g. Gurung et al. 2012).

8.1 Septoria nodorum blotch

Breeding for durable resistance to SNB is challenging for several reasons. SNB epidemics are patchy, and most breeders are unable to apply sufficiently intense and uniform disease pressure to adequately distinguish levels of resistance (Cowger and Murphy 2007). Screening under uniformly inoculated and irrigated conditions is essential to distinguish different levels of resistance; irrigation is often vital to insure that inoculum is splashed up to the spikes irrespective of cultivar height. Knowledgeable, accurate rating is critical so that SNB (Mehra et al. 2019), black chaff, and 'false black chaff' caused by the stem rust resistance gene *Sr2* (Kota et al. 2006) are properly distinguished.

In North Dakota, a series of necrotrophic effectors (NE or SnTox genes) and corresponding wheat sensitivity (Snn) genes that interact in a one-to-one, specific manner to produce susceptibility have been identified using biparental mapping populations (summarized in Richards et al. 2019). Also, QTL not thought to be associated with NE-Snn interactions were identified in U.S. winter wheat germplasm as conferring adult-plant foliar resistance (Francki et al. 2011) or glume resistance (Uphaus et al. 2007). SNB glume resistance gene *QSng.sfr-3BS* from the Swiss winter wheat cv. Arina was mapped between *Fhb1* and *Sr2* on chromosome 3B, as a step toward developing a line with all three tightly linked genes in coupling as a breeder-friendly linkage block (Thapa et al. 2016).

In general, SNB resistance is controlled by many loci with small effects, so marker-assisted breeding is of limited utility. To date, there have been few association mapping studies of SNB resistance in the United States (but see Adhikari et al. 2011b for a spring wheat landrace AM study). A workable strategy for SNB resistance breeding in the eastern United States is to screen against high foliar or glume susceptibility with a combination of field phenotyping under inoculated, irrigated conditions (Cowger et al. 2020b), and genotyping for *Tsn1*, the *Snn* gene that corresponds to NE *SnToxA*. Screening against *Tsn1* is relatively straightforward because a dominant marker is internal to the gene (Bertucci et al. 2014, Faris et al. 2010). *Tsn1* is not widely present in eastern U.S. soft winter wheat germplasm (Bertucci et al. 2014, Cowger et al. 2020b) but is more frequent in hard winter wheats from the U.S. Great Plains (Richards et al. 2019). *Tsn1* marker data are routinely provided to breeding programs by the eastern and central USDA-ARS genotyping laboratories.

8.2 Septoria tritici blotch

In the United States, research on this disease has focused principally on population biology and epidemiology and on the molecular basis of *Stb* gene resistance (the latter summarized in Goodwin 2012). Relatively little has been published on advances in breeding sustainably STB-resistant wheat cultivars in the United States. As an exception, high-throughput automated image analysis using a flat-bed scanner was developed and refined to facilitate the rating of quantitative resistance to STB (Stewart et al. 2016).

Most of the U.S. research on population biology and epidemiology of STB has occurred in western Oregon, in the high-rainfall part of the Pacific Northwest, in part because conditions there are favorable for epidemic development and the rapid evolution of the pathogen (summarized in McDonald and Mundt 2016). Both the defeat of qualitative host resistance (Cowger et al. 2000) and the more gradual decline of moderate resistance (Krenz et al. 2008, Mundt et al. 2002) have been observed there. As qualitative resistance is readily overcome by *Z. tritici*, and insensitivity to both major classes of fungicides has begun to appear (Estep et al. 2015), quantitative STB resistance remains the main viable strategy for prolonging the life of agronomically desirable cultivars in that region.

8.3 Tan spot

Tan spot is Canada's most damaging wheat leaf-spotting disease (Zou et al. 2017), and it can also be destructive in the northern U.S. Great Plains. Considerable effort has been dedicated to identifying QTL for tan spot resistance in the relevant wheat market classes.

In the spring wheat region, varieties and advanced experimental lines in the hard red spring, white, and durum market classes were screened as seedlings for resistance to isolates of *P. tritici-repentis* races 2, 3, and 5, and via culture filtrate infiltration of toxins Ptr ToxA and ToxB (Singh et al. 2006); four hard red spring varieties and lines were resistant to all races and toxins. Of the 11 durum varieties, all were susceptible, but a tetraploid *T. turgidum* accession was identified that had a high level of resistance. Sources of resistance were identified by screening spring wheat accessions from CIMMYT (Ali et al. 2008), as well as accessions belonging to *T. turgidum* L. subspecies *T. carthlicum* , *T. polonicum* , *T. turgidum* , *T. dicoccum* , and *T. turanicum* (Chu et al. 2008). More recently, a QTL for resistance to tan spot and other diseases were identified in a mapping population derived from crossing the Canadian spring wheat cultivars Attila and CDC Go (Zou et al. 2017).

Farther south, winter wheat accessions were screened for tan spot resistance from nine major hard wheat-growing states and 16 major soft wheat-growing states in the United States (Noriel et al. 2011). About 60% of 379 accessions were insensitive to Ptr ToxA, and QR lines were identified. Six hard red winter wheat lines had high levels of QR to both tan spot and SNB, indicating they were not sensitive to ToxA (Liu et al. 2015).

9 Specific disease challenges: insect- and mite-related diseases

9.1 Viruses

In a nine-state survey of the U.S. Great Plains, the three most prevalent wheat viruses were all vectored by the wheat curl mite *Aceria tosichella*: *Wheat streak mosaic virus* (WSMV), *Wheat mosaic virus* (WMoV), and *Triticum mosaic virus* (TriMV) (Burrows et al. 2016). The survey took advantage of over 2400 samples of symptomatic cereals, the large majority from winter wheat, submitted in a three-year period to diagnostic clinics; thus, the sampling was not random. Of these three viruses, more breeding effort has been focused on WSMV; resistance to WSMV from *Thinopyrum intermedium* and other sources has been incorporated into hard red winter wheat (e.g. Graybosch et al. 2009, Seifers et al. 2007).

However, at least historically, the most serious North American cereal virus disease has been yellow dwarf (YD), caused by barley yellow dwarf virus (BYDV) and cereal yellow dwarf virus (CYDV), and spread by several aphid vectors (mainly *Rhopalosiphum padi*, *Sitobion avenae*, *Rhopalosiphum madis,* and *Schizaphis graminum*). Losses in Midwestern U.S. states were estimated at 10–20% in 2002 (Balaji et al. 2003). Measured by yield reduction, YD is the fourth most important wheat disease in Kansas, the top wheat-producing state in the United States (Gaunce and Bockus 2015). YD risk may be higher in reduced-tillage production systems, where unincorporated surface crop residues may attract or harbor aphids (Cowger et al. 2010).

In wheat, YD tolerance is polygenic and complex; a gene associated with tolerance is *Bdv1* (Ayala et al. 2002). The *Bdv2* gene from wheatgrass (*Thinopyrum intermedium*), which has been available in breeding lines since the 1990s, confers actual resistance by allowing the host to induce PAMP-triggered immunity, thereby reducing virus titer (Ayala-Navarrete et al. 2013, Balaji et al. 2003, Walls III et al. 2019). However, *Bdv2* is also linked to undesirable traits such as yellow flour color. In recent years, only low percentages (3–11%) of lines in the elite USDA-ARS Uniform Eastern and Uniform Southern wheat nurseries have carried *Bdv2* or *Bdv3*, for which the same marker is used in screening (USDA-ARS Genotyping Nursery, Raleigh, North Carolina). It is unclear why these genes are little used in U.S. wheat breeding, particularly in regions such as the eastern United States where a large proportion of wheat serves as animal feed, and consequently flour color is of lesser importance.

In hard red winter wheat, strides toward YD resistance have been made in Kansas (Bockus et al. 2016). The most prominent example is the cultivar Everest (PI 659807), which had moderate YD resistance and was widely planted in Kansas for many years. Everest reduced yield losses to YD by 73% compared to susceptible cultivars (Gaunce and Bockus 2015).

In barley, cereal yellow dwarf virus causes significant losses in the U.S. Resistance to CYDV is an important trait in California malting barley breeding, and CYDV tolerance in the CIMMYT malting barley breeding line Madre Selva was localized to two major QTL with markers to facilitate selection (del Blanco et al. 2014).

In oat, an association mapping panel of 428 spring lines assembled by the Collaborative Oat Research Enterprise was evaluated for BYDV tolerance in an aphid-inoculated field nursery in Illinois (Foresman et al. 2016). Two QTL for tolerance were identified and a large number of the lines, most from breeding programs in Canada and the midwestern United States, had favorable alleles at both loci.

Overall, it appears that research on YD resistance in North American cereals has declined in recent years and is currently at a low level, despite the economic importance of the disease and the difficulty of achieving adequate chemical control. This dearth of research attention is likely due to the difficulties of maintaining and propagating aphid populations as vectors to ensure even levels of inoculation, as well as the complexity of testing for viral presence by ELISA or RT-PCR (Ayala-Navarrete et al. 2013).

9.2 Hessian fly

An obligate parasite of wheat and related grasses, Hessian fly (HF; *Mayetiola destructor* [Say]) was introduced to North America, likely on multiple occasions, and is now present in most U.S. wheat-growing regions (Morton and Schemerhorn 2013, Schmid et al. 2018, Shukle et al. 2012). The region that has seen the most frequent and damaging HF outbreaks is the U.S. Southeast, due to its relatively mild winters (Hao et al. 2013). The concept of the 'fly-free' planting date (Schmid et al. 2018) generally does not apply in the Southeast, where no feasible date for planting winter wheat crops is late enough for cold temperatures to reliably protect seedlings from Hessian fly (HF) attack.

While the insect was extensively reported as a damaging pest of wheat in the eastern United States during the latter half of the twentieth century, its emergence as a serious problem in Pacific Northwest (PNW) wheat was more recent, with economic damage first reported in southeast Washington and northern Idaho in the 1980s (Ratcliffe et al. 2000). In the PNW, HF may be a problem in winter wheat, but it causes greater damage in spring wheat due to the timing of stand establishment in relation to the fly's life cycle (Ando et al. 2018). Reduced tillage has inconsistent effects on HF abundance, at least in the PNW (Del Conte et al. 2005). In the northern Great Plains, warming winters and the increase in conservation tillage may convert HF from an occasional to a more frequent economic problem as more winter wheat creates a green bridge with the traditional spring wheat crop (Anderson and Harris 2019).

HF presents a particular challenge to breeding for resistance durability due to the high adaptive potential of the insect population. At least 35 resistance genes (*H1-H34* and *Hdic*) have been officially named, but only some have been deployed commercially (Liu et al. 2020a). These dominant resistance genes, termed H genes, unleash plant defenses in response to saliva from actively feeding larvae or trigger protein production that targets the larval midgut and eventually results in larval death (Schmid et al. 2018). H genes interact on a gene-for-gene basis with HF, where a susceptible insect larva is generally killed via antibiosis triggered by an effective H gene. In addition, so-called Hfr genes operate specifically against first instar larval feeding; an example is *HfrDrd*, linked to a SNP marker (Subramanyam et al. 2013, Tan et al. 2015). Thus, there is considerable selective pressure on the fly population to evolve to insensitivity against deployed H genes, and 18 fly biotypes have been identified based on their ability to defeat particular H genes or gene combinations.

Indeed, H genes are typically defeated within 6–8 years if deployed singly in cultivars that are widely planted (Schmid et al. 2018). Many H genes have been overcome in the southeastern United States. For example, in a 2008 survey of HF-infested wheat from 20 locations in Alabama, Georgia, North Carolina, South Carolina, and Louisiana, only 5 out of 21 tested H genes were found to provide effective protection against HF in that region; those genes were *H12*, *H18*, *H24*, *H25*, and *H26* (Cambron et al. 2010). In a 2014 follow-up survey, collections were made from five locations in North Carolina, Georgia, and Alabama and tested on 16 genes; *H13* and *H33* were also found to be widely effective, although local fly populations that had previously shown insensitivity to *H13* were not included (Shukle et al. 2015). In both surveys, *H9* was only protective against some HF populations from the Southeast. In an adjacent area of the south-central U.S., another survey of HF-damaged regions in Texas, Louisiana, and Oklahoma found that *H12*, *H13*, *H17*, *H18*, *H22*, *H25*, *H26*, and *Hdic* were highly effective against all tested HF populations from that region (Garcés-Carrera et al. 2014).

Due to the rapidity with which the few commonly used H genes may be defeated, there have been efforts to usher new genes into the breeding pipeline. *Aegilops tauschii*, the D-genome donor of *H13, H22, H23, H24, H26,* and *H32*, is one promising source. For example, resistance to the prevalent local fly biotype GP was observed among synthetic wheat lines derived from crosses of *Ae. tauschii* and the durum wheat cultivar Langdon in North Dakota (Friesen et al. 2008). The North Carolina germplasm line NC09MDD14 (PI 656395) had two *Ae. tauschii* accessions in its pedigree, and its monogenic resistance to HF biotype L mapped very close to and may be allelic with *H13*, but is evidently novel, as it was effective against the HF biotype vH13 that could defeat *H13* (Miranda et al. 2010). In addition, *Ae. tauschii* was proposed as a useful system for functional analysis and characterization of H genes effective in hexaploid wheat (Nemacheck et al. 2019).

Some new H genes are being identified in existing *T. aestivum* and *T. durum* gene pools. For example, while many H genes originating from durum wheat are no longer effective in the southeastern United States, *H33,* a gene derived from a Portuguese durum accession, was fully effective against all tested southeastern HF populations and has been introgressed into hexaploid wheat (McDonald et al. 2014). As another example, the resistance in a popular PNW soft white spring cultivar, Louise, is believed to be different from but linked to

H3 and *H5* in an H gene-rich region on wheat chromosome 1AS, and molecular markers can facilitate pyramiding (Carter et al. 2014).

As for durable HF protection, given the vulnerability of H genes when singly deployed, gene rotation and pyramiding are commonly promoted strategies (Carter et al. 2014, Liu et al. 2020a, Schmid et al. 2018, Shukle et al. 2015). An interesting proposed strategy involves pyramiding trans-genes that encode antinutrient and toxic proteins with highly effective and un-deployed H genes to prolong the life of the latter (Shukle et al. 2012). At present, however, it is difficult to ascertain whether any particular H gene or gene pyramid is present in most North American wheat cultivars (Schmid et al. 2018). It appears likely that the majority of commercial wheat cultivars contain no currently effective H genes (but see Hao et al. 2013). For example, over the years 2017-2019, 11% of winter wheat experimental lines in the elite Uniform Eastern Nursery possessed *H9*, and 16% of entries in the Uniform Southern Nursery had *H13* (USDA-ARS Genotyping Lab, Raleigh NC). In both nurseries, fewer than 3% of entries contained the other one of these two widely effective genes in that time period.

As yet, there are only a few published studies of genes conferring tolerance or incomplete resistance to HF. Among them, three QTL were identified in the parents of a P26R61xAGS2000 biparental mapping population that conferred tolerance to *H13*-virulent flies (Hao et al. 2013). In another report, tolerance conferred by the recessive gene *h4* from an old wheat cultivar, Java, is reported to enable better tillering after HF damage, making the gene useful for pyramiding (Niu et al. 2020). Interestingly, southeastern U.S. wheat breeders observe quantitative resistance and/or tolerance in the field that is not identified by screening seedlings in controlled environments (P. Murphy, personal communication).

10 Specific disease challenges: bacterial leaf streak and black chaff

After last receiving substantial research attention in the United States in the late 1980s and early 1990s, bacterial leaf streak (BLS) and black chaff have re-emerged in recent years as challenges to small-grain cereal production in the northern Great Plains region of the United States (Curland et al. 2018, Wen et al. 2018). These diseases are caused by *Xanthomonas translucens* pv. *translucens* in barley, and *X. translucens* pv. *undulosa* (*Xtu*) in wheat, barley, triticale, rye, and oat (Curland et al. 2020, Curland et al. 2018, Sapkota et al. 2018).

Sources of resistance have been sought in both spring and winter wheat germplasm. In the early 1990s, a set of 5000 winter wheat accessions from the USDA National Small Grain Collection (NSGC) was evaluated and from it, 26 quantitatively BLS-resistant accessions were identified as useful for breeding in the southern United States (Tillman et al. 1996). In another set of over 600 NSGC winter wheat lines, there was significantly more resistance in accessions from North America than in those originating in Europe or south-central Asia (Adhikari et al. 2011a). A hard winter wheat association mapping panel of 299 lines from across the U.S. Great Plains region yielded several useful QTL for BLS resistance (Ramakrishnan et al. 2019). In hard red spring wheat, varieties were identified with quantitative BLS resistance superior to that in susceptible varieties (Kandel et al. 2012, Kandel et al. 2015). Using association mapping, several QTL were identified in an NGSC collection of 566 spring wheat landraces that conferred resistance to *Xtu* or *Bipolaris sorokiniana*, cause of spot blotch, and in some cases to both pathogens (Adhikari et al. 2012).

In wheat, only partial resistance to BLS has been detected, and it is influenced by host phenology and environmental conditions (Curland et al. 2018, Tillman et al. 1996, Wen et al. 2018). In triticale, however, more complete resistance has been observed (Sapkota et al. 2018). Recently, a major gene (or tightly linked genes) for BLS resistance was localized to rye chromosome 5R in the triticale cultivar Siskiyou and designated *Xct1* (Wen et al. 2018). This dominant gene could be transferred to wheat via molecular marker-mediated chromosome engineering. It is unknown whether *Xct1*-mediated resistance is triggered by a type III virulence effector in *Xtu*, and the potential durability of this resistance is also unclear.

11 Conclusion

Common to breeding efforts on most major cereal diseases in North America is the recognition that partial resistance, conditioned by multiple or even numerous loci with small effects, is the foundation of breeding durably resistant cultivars. Major or all-stage resistance genes may be introgressed or pyramided, but having a quantitatively resistant background is crucial. Across the spectrum of diseases, certain older cultivars have maintained useful levels of incomplete resistance over decades of exposure, for example, Stephens (stripe rust), Caldwell and Duster (leaf rust), and Knox (powdery mildew). Using such sources as parents, breeders were able to select durable resistance without the advantages of newer breeding tools such as genomic selection and association mapping.

Now that these techniques are widely utilized, many challenges still lie ahead. In cases such as the rusts, considerable research has been done to locate and characterize QTL in regionally adapted breeding materials, while with other diseases such as powdery mildew or Septoria tritici blotch, that work largely remains to be done. In the case of Fusarium head blight and the leaf blotch diseases, where genes for immunity do not exist, there is no alternative but to accumulate many small-effect loci - perhaps in addition to *Fhb1* - if high levels of resistance are to be attained.

It is increasingly imaginable to take a strategic approach to combining QTL with different modes or timings of action, thanks to our ability to clone partial resistance genes and investigate gene interaction networks (Lowe et al. 2011). For example, the goal may be to combine large numbers of minor genes that confer resistance to multiple fungi at varying points in the infection process with genes encoding tolerance to Hessian fly and barley yellow dwarf virus. Such strategies may help to slow pathogen adaptation and extend the useful commercial life of small-grain cultivars. Robust public investment in plant breeding would help to support complex, regionally specific pre-breeding programs and the human and technical resources to develop high-yielding cultivars from them.

12 Where to look for further information

For more information about efforts to breed durably disease-resistant wheat cultivars in North America, readers may contact these organizations: Agriculture and Agri-Food Canada, US Department of Agriculture-Agricultural Research Service, CIMMYT, and US Wheat & Barley Scab Initiative.

13 Acknowledgements

I greatly appreciate knowledgeable reviews and comments by Xianming Chen, Yue Jin, James Kolmer, Zhaohui Liu, Matthew Rouse, Paul Murphy, and Pawan Kumar Singh. They significantly strengthened the chapter.

14 References

Aboukhaddour, R., Fetch, T., McCallum, B. D., Harding, M. W., Beres, B. L. and Graf, R. J. 2020. Wheat diseases on the prairies: a Canadian story. *Plant Pathol.* 69(3):418–432.

Adhikari, T. B., Gurung, S., Hansen, J. M., Jackson, E. W. and Bonman, J. M. 2012. Association mapping of quantitative trait loci in spring wheat landraces conferring resistance to bacterial leaf streak and spot blotch. *Plant Genome* 5(1):1–16.

Adhikari, T. B., Hansen, J. M., Gurung, S. and Bonman, J. M. 2011a. Identification of new sources of resistance in winter wheat to multiple strains of *Xanthomonas translucens* pv. *undulosa*. *Plant Dis.* 95(5):582–588.

Adhikari, T. B., Jackson, E. W., Gurung, S., Hansen, J. M. and Bonman, J. M. 2011b. Association mapping of quantitative resistance to *Phaeosphaeria nodorum* in spring wheat landraces from the USDA National Small Grains Collection. *Phytopathology* 101(11):1301–1310.

Ali, S., Singh, P. K., McMullen, M. P., Mergoum, M. and Adhikari, T. B. 2008. Resistance to multiple leaf spot diseases in wheat. *Euphytica* 159(1-2):167–179.

Anderson, J. A., Wiersma, J. J., Linkert, G. L., Reynolds, S. K., Kolmer, J. A., Jin, Y., Rouse, M., Dill-Macky, R., Hareland, G. A. and Ohm, J.-B. 2018. Registration of 'Linkert' spring wheat with good straw strength and adult plant resistance to the Ug99 family of stem rust races. *J. Plant Reg.* 12(2):208–214.

Anderson, K. M. and Harris, M. O. 2019. Susceptibility of North Dakota Hessian fly (Diptera: Cecidomyiidae) to 31 *H* genes mediating wheat resistance. *J. Econ. Entomol.* 112(5):2398–2406.

Ando, K., Rynearson, S., Muleta, K. T., Gedamu, J., Girma, B., Bosque-Pérez, N. A., Chen, M. S. and Pumphrey, M. O. 2018. Genome-wide associations for multiple pest resistances in a Northwestern United States elite spring wheat panel. *PLoS ONE* 13(2):e0191305.

Aoun, M., Breiland, M., Kathryn Turner, M., Loladze, A., Chao, S., Xu, S. S., Ammar, K., Anderson, J. A., Kolmer, J. A. and Acevedo, M. 2016. Genome-wide association mapping of leaf rust response in a durum wheat worldwide germplasm collection. *Plant Genome* 9(3):eplantgenome2016.01.0008.

Arruda, M. P., Brown, P. J., Lipka, A. E., Krill, A. M., Thurber, C. and Kolb, F. L. 2015. Genomic selection for predicting Fusarium head blight resistance in a wheat breeding program. *Plant Genome* 8(3):eplantgenome2015.01.0003.

Ayala, L., Henry, M., van Ginkel, M., Singh, R., Keller, B. and Khairallah, M. 2002. Identification of QTLs for BYDV tolerance in bread wheat. *Euphytica* 128(2): 249–259.

Ayala-Navarrete, L. I., Mechanicos, A. A., Gibson, J. M., Singh, D., Bariana, H. S., Fletcher, J., Shorter, S. and Larkin, P. J. 2013. The Pontin series of recombinant alien translocations in bread wheat: single translocations integrating combinations of *Bdv2*, *Lr19* and *Sr25* disease-resistance genes from *Thinopyrum intermedium* and *Th. Ponticum*. *Theor. Appl. Genet.* 126(10):2467–2475.

Baenziger, P. S. 2019. Modern plant breeding: a perspective from the public sector in the United States. Pages 148–161. In: *Sustaining Global Food Security: The Nexus of Science and Policy* Zeigler, R. (Ed). CSIRO Publishing, Clayton South, Australia.

Bajgain, P., Rouse, M. N., Bulli, P., Bhavani, S., Gordon, T., Wanyera, R., Njau, P. N., Legesse, W., Anderson, J. A. and Pumphrey, M. O. 2015. Association mapping of North American spring wheat breeding germplasm reveals loci conferring resistance to Ug99 and other African stem rust races. *BMC Plant Biol.* 15:249.

Bakhsh, A., Mengistu, N., Baenziger, P. S., Dweikat, I., Wegulo, S. N., Rose, D. J., Bai, G. and Eskridge, K. M. 2013. Effect of Fusarium head blight resistance gene *Fhb1* on agronomic and end-use quality traits of hard red winter wheat. *Crop Sci.* 53(3):793–801.

Balaji, B., Bucholtz, D. B. and Anderson, J. M. 2003. Barley yellow dwarf virus and Cereal yellow dwarf virus quantification by real-time polymerase chain reaction in resistant and susceptible plants. *Phytopathology* 93(11):1386–1392.

Basnet, B. R., Ibrahim, A. M. H., Chen, X., Singh, R. P., Mason, E. R., Bowden, R. L., Liu, S., Hays, D. B., Devkota, R. N., Subramanian, N. K. and Rudd, J. C. 2014. Molecular mapping of stripe rust resistance in hard red winter wheat TAM 111 adapted to the U.S. High Plains. *Crop Sci.* 54(4):1361–1373.

Bec, S., Ward, T., Farman, M., O'Donnell, K., Hershman, D. E., Van Sanford, D. A. and Vaillancourt, L. J. 2015. Characterization of *Fusarium* strains recovered from wheat with symptoms of head blight in Kentucky. *Plant Dis.* 99(11):1622–1632.

Benson, J., Brown-Guedira, G., Paul Murphy, J. and Sneller, C. 2012. Population structure, linkage disequilibrium, and genetic diversity in soft winter wheat enriched for Fusarium head blight resistance. *Plant Genome* 5(2):71–80.

Berger, G., Green, A., Khatibi, P., Brooks, W., Rosso, L., Liu, S., Chao, S., Griffey, C. and Schmale, D. 2014. Characterization of Fusarium Head blight resistance and deoxynivalenol accumulation in hulled and hulless winter barley. *Plant Dis.* 98(5):599-606.

Bergstrom, G. C., Cummings, J. A., Waxman, K. D., Bradley, C. A., Wegulo, S. N., Hazelrigg, A. L., Hershman, D. E., Nagelkirk, M. and Sweets, L. E. 2013. Effects of local corn debris management on FHB and DON levels in seventeen U.S. wheat environments in 2011-2013. Pages 92-93. In: *2013 National Fusarium Head Blight Forum* Canty, S., Clark, A., Salat, Y. and Van Sanford, D. (Eds). U.S. Wheat and Barley Scab Initiative. East Lansing, Michigan / Lexington, Kentucky.

Bergstrom, G. C., Waxman, K. D., Schmale III, D. G., Bradley, C. A., Sweets, L. E., Wegulo, S. N. and Keller, M. D. 2010. Effects of within-field corn debris in microplots on FHB and DON in eleven U.S. wheat environments in 2010. Pages 69-70. In: *National Fusarium Head Blight Forum* (vol. 2010) Canty, S., Clark, A., Anderson-Scully, A., Sanford, D. E. and Van Sanford, D. (Eds). U.S. Wheat and Barley Scab Initiative, Lexington, Kentucky.

Bertucci, M., Brown-Guedira, G., Murphy, J. P. and Cowger, C. 2014. Genes conferring sensitivity to *Stagonospora nodorum* necrotrophic effectors in Stagonospora nodorum blotch-susceptible U.S. wheat cultivars. *Plant Dis.* 98(6):746-753.

Bockus, W. W., De Wolf, E. D. and Todd, T. C. 2016. Management strategies for barley yellow dwarf on winter wheat in Kansas. *Plant Health Prog.* 17(2):122-127.

Bokore, F. E., Cuthbert, R. D., Knox, R. E., Randhawa, H. S., Hiebert, C. W., DePauw, R. M., Singh, A. K., Singh, A., Sharpe, A. G., N'Diaye, A., Pozniak, C. J., McCartney, C., Ruan, Y., Berraies, S., Meyer, B., Munro, C., Hay, A., Ammar, K., Huerta-Espino, J. and Bhavani, S. 2017. Quantitative trait loci for resistance to stripe rust of wheat revealed using global field nurseries and opportunities for stacking resistance genes. *Theor. Appl. Genet.* 130(12):2617-2635.

Brar, G. S. and Kutcher, H. R. 2016. Race characterization of *Puccinia striiformis* f. sp. *tritici*, the cause of wheat stripe rust, in Saskatchewan and Southern Alberta, Canada and virulence comparison with races from the United States. *Plant Dis.* 100(8):1744-1753.

Brar, G. S., Brûlé-Babel, A. L., Ruan, Y., Henriquez, M. A., Pozniak, C. J., Kutcher, H. R. and Hucl, P. J. 2019a. Genetic factors affecting Fusarium head blight resistance improvement from introgression of exotic Sumai 3 alleles (including *Fhb1*, *Fhb2*, and *Fhb5*) in hard red spring wheat. *BMC Plant Biol.* 19(1):179.

Brar, G. S., Dhariwal, R. and Randhawa, H. S. 2018. Resistance evaluation of differentials and commercial wheat cultivars to stripe rust (*Puccinia striiformis*) infection in hot spot regions of Canada. *Eur. J. Plant Pathol.* 152(2):493-502.

Brar, G. S., Fetch, T., McCallum, B. D., Hucl, P. J. and Kutcher, H. R. 2019b. Virulence dynamics and breeding for resistance to stripe, stem, and leaf rust in Canada since 2000. *Plant Dis.* 103(12):2981-2995.

Brar, G. S., Pozniak, C. J., Kutcher, H. R. and Hucl, P. J. 2019c. Evaluation of Fusarium head blight resistance genes *Fhb1*, *Fhb2*, and *Fhb5* introgressed into elite Canadian hard red spring wheats: effect on agronomic and end-use quality traits and implications for breeding. *Mol. Breed.* 39(3):44.

Brouwer, B. O., Murphy, K. M. and Jones, S. S. 2015. Plant breeding for local food systems: a contextual review of end-use selection for small grains and dry beans in Western Washington. *Renew. Agr. Food Syst.* 31:172-184.

Bryant, R. R. M., McGrann, G. R. D., Mitchell, A. R., Schoonbeek, H. J., Boyd, L. A., Uauy, C., Dorling, S. and Ridout, C. J. 2014. A change in temperature modulates defence to yellow (stripe) rust in wheat line UC1041 independently of resistance gene *Yr36*. *BMC Plant Biol.* 14:10.

Buerstmayr, H., Lemmens, M., Hartl, L., Doldi, L., Steiner, B., Stierschneider, M. and Ruckenbauer, P. 2002. Molecular mapping of QTL for Fusarium head blight resistance in spring wheat. Resistance to fungal spread (type II resistance). *Theor. Appl. Genet.* 104(1):84–91.

Burrows, M., Thomas, C., McRoberts, N., Bostock, R. M., Coop, L. and Stack, J. 2016. Coordination of diagnostic efforts in the Great Plains: wheat virus survey and modeling of disease onset. *Plant Dis.* 100(6):1037–1045.

Cainong, J. C., Bockus, W. W., Feng, Y., Chen, P., Qi, L., Sehgal, S. K., Danilova, T. V., Koo, D. H., Friebe, B. and Gill, B. S. 2015. Chromosome engineering, mapping, and transferring of resistance to Fusarium head blight disease from *Elymus tsukushiensis* into wheat. *Theor. Appl. Genet.* 128(6):1019–1027.

Cambron, S. E., David Buntin, G. D., Weisz, R., Holland, J. D., Flanders, K. L., Schemerhorn, B. J. and Shukle, R. H. 2010. Virulence in Hessian fly (Diptera: Cecidomyiidae) field collections from the southeastern United States to 21 resistance genes in wheat. *J. Econ. Entomol.* 103(6):2229–2235.

Cao, W., Fedak, G., Voldeng, H., Somers, D., Xue, A. G., Fregeau-Reid, J., McCartney, C. and Humphreys, G. 2016. Hard white spring wheat line WS175 with a high level of resistance to Fusarium head blight. *Can. J. Plant Sci.* 96:1–5.

Carpenter, N. R., Wright, E., Malla, S., Singh, L., Van Sanford, D., Clark, A., Harrison, S., Murphy, J. P., Costa, J., Chao, S., Brown-Guedira, G. L., McMaster, N., Schmale III, D. G., Griffey, C. A. and Rawat, N. 2020. Identification and validation of FHB resistance QTL in the US soft red winter wheat variety Jamestown. *Crop Sci.* 60(6):2919–2930.

Carter, A. H., Cambron, S. E., Ohm, H. W., Bosque-Pérez, N. A. and Kidwell, K. K. 2014. Hessian fly (*Mayetiola destructor* [Say]) resistance identified on chromosome 1AS in the spring wheat (*Triticum aestivum* L.) cultivar 'Louise'. *Crop Sci.* 54(3):971–981.

Castro Aviles, A., Alan Harrison, S., Joseph Arceneaux, K., Brown-Guidera, G., Esten Mason, R. and Baisakh, N. 2020. Identification of QTLs for resistance to Fusarium head blight using a doubled haploid population derived from southeastern United States soft red winter wheat varieties AGS 2060 and AGS 2035. *Genes* 11(6):699.

Cerón-Bustamante, M., Ward, T. J., Kelly, A., Vaughan, M. M., McCormick, S. P., Cowger, C., Leyva-Mir, S. G., Villaseñor-Mir, H. E., Ayala-Escobar, V. and Nava-Díaz, C. 2018. Regional differences in the composition of Fusarium Head Blight pathogens and mycotoxins associated with wheat in Mexico. *Inter. J. Food Microbiol.* 273:11–19.

Chen, S., Zhang, W., Bolus, S., Rouse, M. N. and Dubcovsky, J. 2018. Identification and characterization of wheat stem rust resistance gene *Sr21* effective against the Ug99 race group at high temperature. *PLoS Genet.* 14(4):e1007287.

Chen, X. M. 2007. Challenges and solutions for stripe rust control in the United States. *Aust. J. Agric. Res.* 58(6):648–655.

Chen, X. M. 2013. Review article: high-temperature adult-plant resistance, key for sustainable control of stripe rust. *Am. J. Plant Sci.* 04(3):608–627.

Chen, X. M. 2014. Integration of cultivar resistance and fungicide application for control of wheat stripe rust. *Can. J. Plant Pathol.* 36(3):311–326.

Chu, C. G., Friesen, T. L., Faris, J. D. and Xu, S. S. 2008. Evaluation of seedling resistance to tan spot and Stagonospora nodorum blotch in tetraploid wheat. *Crop Sci.* 48(3):1107–1116.

Cowger, C. and Murphy, J. P. 2007. Artificial inoculation of wheat for selecting resistance to Stagonospora nodorum blotch. *Plant Dis.* 91(5):539–545.

Cowger, C. and Sutton, A. L. 2005. The southeastern U.S. Fusarium head blight epidemic of 2003. *Plant Health Prog.* 6(1) doi:10.1094/PHP-2005-1026-01-RS.

Cowger, C., Arellano, C., Marshall, D. and Fitzgerald, J. 2019. Managing Fusarium head blight in winter barley with cultivar resistance and fungicide. *Plant Dis.* 103(8):1858–1864.

Cowger, C., Hoffer, M. E. and Mundt, C. C. 2000. Specific adaptation by *Mycosphaerella graminicola* to a resistant wheat cultivar. *Plant Pathol.* 49(4):445–451.

Cowger, C., Mehra, L., Arellano, C., Meyers, E. and Murphy, J. P. 2018. Virulence differences in *Blumeria graminis* f. sp. *tritici* from the central and eastern United States. *Phytopathology* 108(3):402–411.

Cowger, C., Miranda, L., Griffey, C., Hall, M., Murphy, P. and Maxwell, J. 2012. Wheat powdery mildew. Pages 84–119. In: *Disease Resistance in Wheat* Sharma, I. (Ed). CABI, London.

Cowger, C., Parks, R. and Marshall, D. 2009. Appearance of powdery mildew of wheat caused by *Blumeria graminis* f. sp. *tritici* on *Pm17*-bearing cultivars in North Carolina. *Plant Dis.* 93(11):1219.

Cowger, C., Smith, J., Boos, D., Bradley, C. A., Ransom, J. and Bergstrom, G. C. 2020a. Managing a destructive, episodic crop disease: a national survey of wheat and barley growers' experience with Fusarium head blight. *Plant Dis.* 104(3): 634–648.

Cowger, C., Ward, B., Brown-Guedira, G. and Brown, J. K. M. 2020b. Role of effector-sensitivity gene interactions and durability of quantitative resistance to Septoria nodorum blotch in eastern U.S. wheat. *Front. Plant Sci.* 11:155.

Cowger, C., Ward, T. J., Nilsson, K., Arellano, C., McCormick, S. P. and Busman, M. 2020c. Regional and field-specific differences in *Fusarium* species and mycotoxins associated with blighted North Carolina wheat. *Inter. J. Food Microbiol.* 323:108594.

Cowger, C., Weisz, R., Anderson, J. M. and Horton, J. R. 2010. Maize debris increases Barley yellow dwarf virus severity in North Carolina winter wheat. *Agron. J.* 102(2):688–695.

Crippin, T., Limay-Rios, V., Renaud, J. B., Schaafsma, A. W., Sumarah, M. W. and Miller, J. D. 2020. *Fusarium graminearum* populations from maize and wheat in Ontario, Canada. *World Mycotoxin J.* 13(3):355–366.

Curland, R. D., Gao, L., Bull, C. T., Vinatzer, B. A., Dill-Macky, R., Van Eck, L. and Ishimaru, C. A. 2018. Genetic diversity and virulence of wheat and barley strains of *Xanthomonas translucens* from the Upper Midwestern United States. *Phytopathology* 108(4):443–453.

Curland, R. D., Gao, L., Hirsch, C. D. and Ishimaru, C. A. 2020. Localized genetic and phenotypic diversity of *Xanthomonas translucens* associated with bacterial leaf streak on wheat and barley in Minnesota. *Phytopathology* 110(2):257–266.

del Blanco, I. A., Hegarty, J., Gallagher, L. W., Falk, B. W., Brown-Guedira, G., Pellerin, E. and Dubcovsky, J. 2014. Mapping of QTL for tolerance to cereal yellow dwarf virus in two-rowed spring barley. *Crop Sci.* 54(4):1468–1475.

Del Conte, S. C. C., Bosque-Pérez, N. A., Schotzko, D. J. and Guy, S. O. 2005. Impact of tillage practices on Hessian fly-susceptible and resistant spring wheat cultivars. *J. Econ. Entomol.* 98(3):805–813.

Del Ponte, E. M., Shah, D. A. and Bergstrom, G. C. 2003. Spatial patterns of Fusarium head blight in New York wheat fields suggest role of airborne inoculum. *Plant Health Prog.* 4(1):6.

Dill-Macky, R. and Jones, R. K. 2000. The effect of previous crop residues and tillage on Fusarium head blight of wheat. *Plant Dis.* 84(1):71–76.

Dong, H., Wang, R., Yuan, Y., Anderson, J., Pumphrey, M., Zhang, Z. and Chen, J. 2018. Evaluation of the potential for genomic selection to improve spring wheat resistance to Fusarium head blight in the Pacific Northwest. *Front. Plant Sci.* 9:911.

Dyck, P. L., Samborshi, D. J. and Anderson, R. G. 1966. Inheritance of adult-plant leaf rust resistance derived from the common wheat varieties Exchange and Frontana. *Can. J. Genet. Cytol.* 8(4):665–671.

Edae, E. A., Pumphrey, M. O. and Rouse, M. N. 2018. A genome-wide association study of field and seedling response to individual stem rust pathogen races reveals combinations of race-specific genes in North American spring wheat. *Front. Plant Sci.* 9:52.

Estep, L. K., Torriani, S. F. F., Zala, M., Anderson, N. P., Flowers, M. D., McDonald, B. A., Mundt, C. C. and Brunner, P. C. 2015. Emergence and early evolution of fungicide resistance in North American populations of *Zymoseptoria tritici*. *Plant Pathol.* 64(4):961–971.

Faris, J. D., Zhang, Z., Lu, H., Lu, S., Reddy, L., Cloutier, S., Fellers, J. P., Meinhardt, S. W., Rasmussen, J. B., Xu, S. S., Oliver, R. P., Simons, K. J. and Friesen, T. L. 2010. A unique wheat disease resistance-like gene governs effector-triggered susceptibility to necrotrophic pathogens. *Proc. Natl. Acad. Sci. U. S. A.* 107(30):13544–13549.

Feng, J., Wang, M., See, D. R., Chao, S., Zheng, Y. and Chen, X. 2018. Characterization of novel gene *Yr79* and four additional quantitative trait loci for all-stage and high-temperature adult-plant resistance to stripe rust in spring wheat PI 182103. *Phytopathology* 108(6):737–747.

Foresman, B. J., Oliver, R. E., Jackson, E. W., Chao, S., Arruda, M. P. and Kolb, F. L. 2016. Genome-wide association mapping of barley yellow dwarf virus tolerance in spring oat (*Avena sativa* L.). *PLoS ONE* 11(5):e0155376.

Francki, M. G., Shankar, M., Walker, E., Loughman, R., Golzar, H. and Ohm, H. 2011. New quantitative trait loci in wheat for flag leaf resistance to Stagonospora nodorum blotch. *Phytopathology* 101(11):1278–1284.

Friesen, T. L., Xu, S. S. and Harris, M. O. 2008. Stem rust, tan spot, Stagonospora nodorum blotch, and Hessian fly resistance in Langdon durum–*Aegilops tauschii* synthetic hexaploid wheat lines. *Crop Sci.* 48(3):1062–1070.

Fu, D., Uauy, C., Distelfeld, A., Blechl, A., Epstein, L., Chen, X., Sela, H., Fahima, T. and Dubcovsky, J. 2009. A kinase-START gene confers temperature-dependent resistance to wheat stripe rust. *Science* 323(5919):1357–1360.

Gao, L., Turner, M. K., Chao, S., Kolmer, J. and Anderson, J. A. 2016. Genome wide association study of seedling and adult plant leaf rust resistance in elite spring wheat breeding lines. *PLoS ONE* 11(2):e0148671.

Garcés-Carrera, S., Knutson, A., Wang, H., Giles, K. L., Huang, F., Whitworth, R. J., Smith, C. M. and Chen, M. S. 2014. Virulence and biotype analyses of Hessian fly (Diptera:

Cecidomyiidae) populations from Texas, Louisiana, and Oklahoma. *J. Econ. Entomol.* 107(1):417–423.

Gaunce, G. M. and Bockus, W. W. 2015. Estimating yield losses due to barley yellow dwarf on winter wheat in Kansas using disease phenotypic data. *Plant Health Prog.* 16(1):1–6.

Ghavami, F., Elias, E. M., Mamidi, S., Ansari, O., Sargolzaei, M., Adhikari, T., Mergoum, M. and Kianian, S. F. 2011. Mixed model association mapping for Fusarium head blight resistance in Tunisian-derived durum wheat populations. *G3-Genes Genom. Genet.* 1:209–218.

Goodwin, S. B. 2012. Resistance in wheat to Septoria diseases caused by *Mycosphaerella graminicola* (*Septoria tritici*) and *Phaeosphaeria* (*Stagonospora*) *nodorum*. Pages 151–159. In: *Disease Resistance in Wheat* Sharma, I. (Ed). CABI, Wallingford, UK.

Gräfenhan, T., Patrick, S. K., Roscoe, M., Trelka, R., Gaba, D., Chan, J. M., McKendry, T., Clear, R. M. and Tittlemier, S. A. 2013. Fusarium damage in cereal grains from western Canada. 1. Phylogenetic analysis of moniliformin-producing *Fusarium* species and their natural occurrence in mycotoxin-contaminated wheat, oats, and rye. *J. Agric. Food Chem.* 61(23):5425–5437.

Graybosch, R. A., Peterson, C. J., Baenziger, P. S., Baltensperger, D. D., Nelson, L. A., Jin, Y., Kolmer, J., Seabourn, B., French, R., Hein, G., Martin, T. J., Beecher, B., Schwarzacher, T. and Heslop-Harrison, P. 2009. Registration of 'Mace' hard red winter wheat. *J. Plant Reg.* 3(1):51–56.

Green, A. J., Berger, G., Griffey, C. A., Pitman, R., Thomason, W. and Balota, M. 2014. Genetic resistance to and effect of leaf rust and powdery mildew on yield and its components in 50 soft red winter wheat cultivars. *Crop Prot.* 64:177–186.

Gurung, S., Hansen, J. M., Bonman, J. M., Gironella, A. I. N. and Adhikari, T. B. 2012. Multiple disease resistance to four leaf spot diseases in winter wheat accessions from the USDA National Small Grains Collection. *Crop Sci.* 52(4):1640–1650.

Haile, J. K., N'Diaye, A., Walkowiak, S., Nilsen, K. T., Clarke, J. M., Kutcher, H. R., Steiner, B., Buerstmayr, H. and Pozniak, C. J. 2019. Fusarium head blight in durum wheat: recent status, breeding directions, and future research prospects. *Phytopathology* 109(10):1664–1675.

Hale, I., Zhang, X., Fu, D. and Dubcovsky, J. 2013. Registration of wheat lines carrying the partial stripe rust resistance gene *Yr36* without the *Gpc-B1* allele for high grain protein content. *J. Plant Reg.* 7(1):108–112.

Hao, Y., Cambron, S. E., Chen, Z., Wang, Y., Bland, D. E., Buntin, G. D. and Johnson, J. W. 2013. Characterization of new loci for Hessian fly resistance in common wheat. *Theor. Appl. Genet.* 126(4):1067–1076.

Hao, Y., Parks, R., Cowger, C., Chen, Z., Wang, Y., Bland, D., Murphy, J. P., Guedira, M., Brown-Guedira, G. and Johnson, J. 2015. Molecular characterization of a new powdery mildew resistance gene *Pm54* in soft red winter wheat. *Theor. Appl. Genet.* 128(3):465–476.

Hao, Y., Rasheed, A., Zhu, Z., Wulff, B. B. H. and He, Z. 2020. Harnessing wheat *Fhb1* for Fusarium resistance. *Trends Plant Sci.* 25(1):1–3.

Herrera-Foessel, S. A., Huerta-Espino, J., Calvo-Salazar, V., Lan, C. X. and Singh, R. P. 2014. *Lr72* confers resistance to leaf rust in durum wheat cultivar Atil C2000. *Plant Dis.* 98(5):631–635.

Hiebert, C. W., Fetch, T. G., Zegeye, T., Thomas, J. B., Somers, D. J., Humphreys, D. G., McCallum, B. D., Cloutier, S., Singh, D. and Knott, D. R. 2011. Genetics and mapping of seedling resistance to Ug99 stem rust in Canadian wheat cultivars 'Peace' and 'AC Cadillac'. *Theor. Appl. Genet.* 122(1):143–149.

Huerta-Espino, J. and Singh, R. P. 2017. First detection of virulence in *Puccinia striiformis* f. sp. *tritici* to wheat resistance genes *Yr10* and *Yr24* (=*Yr26*) in Mexico. *Plant Dis.* 101(9):1676–1676.

Huerta-Espino, J., Singh, R. P., Germán, S., McCallum, B. D., Park, R. F., Chen, W. Q., Bhardwaj, S. C. and Goyeau, H. 2011. Global status of wheat leaf rust caused by *Puccinia triticina*. *Euphytica* 179(1):143–160.

Huggins, D. R. and Reganold, J. P. 2008. No-till: the quiet revolution. *Sci. Am.* 299(1):70–77.

Huhn, M. R., Elias, E. M., Ghavami, F., Kianian, S. F., Chao, S., Zhong, S., Alamri, M. S., Yahyaoui, A. and Mergoum, M. 2012. Tetraploid Tunisian wheat germplasm as a new source of Fusarium head blight resistance. *Crop Sci.* 52(1):136–145.

Islam, M. S., Brown-Guedira, G., Van Sanford, D., Ohm, H., Dong, Y. and McKendry, A. L. 2016. Novel QTL associated with the Fusarium head blight resistance in Truman soft red winter wheat. *Euphytica* 207(3):571–592.

Jin, Y. and Singh, R. P. 2006. Resistance in U.S. wheat to recent eastern African isolates of *Puccinia graminis* f. sp. *tritici* with virulence to resistance gene *Sr31*. *Plant Dis.* 90(4):476–480.

Jin, Y., Singh, R. P., Ward, R. W., Wanyera, R., Kinyua, M., Njau, P., Fetch, T., Pretorius, Z. A. and Yahyaoui, A. 2007Characterization of seedling infection types and adult plant infection responses of monogenic Sr gene lines to race TTKS of Puccinia graminis f. sp. tritici. *Plant Dis.* 91(9):1096–1099.

Juliana, P., Rutkoski, J. E., Poland, J. A., Singh, R. P., Murugasamy, S., Natesan, S., Barbier, H. and Sorrells, M. E. 2015. Genome-wide association mapping for leaf tip necrosis and pseudo-black chaff in relation to durable rust resistance in wheat. *Plant Genome* 8(2):eplantgenome2015.01.0002.

Kandel, Y. R., Glover, K. D., Osborne, L. E. and Gonzalez-Hernandez, J. L. 2015. Mapping quantitative resistance loci for bacterial leaf streak disease in hard red spring wheat using an identity by descent mapping approach. *Euphytica* 201(1):53–65.

Kandel, Y. R., Glover, K. D., Tande, C. A. and Osborne, L. E. 2012. Evaluation of spring wheat germplasm for resistance to bacterial leaf streak caused by *Xanthomonas campestris* pv. *translucens*. *Plant Dis.* 96(12):1743–1748.

Khatibi, P. A., Berger, G., Liu, S., Brooks, W. S., Griffey, C. A. and Schmale, D. G. 2011. Resistance to Fusarium head blight and deoxynivalenol accumulation in Virginia barley. *Plant Dis.* 96(2):279–284.

Kolmer, J., Turner, M. K., Rouse, M. N. and Anderson, J. A. 2020. Adult plant leaf rust resistance in AC Taber wheat maps to chromosomes 2BS and 3BS. *Phytopathology* doi:10.1094/PHYTO-03-20-0074-R.

Kolmer, J. A. 2015. First report of a wheat leaf rust (*Puccinia triticina*) phenotype with high virulence to durum wheat in the Great Plains region of the United States. *Plant Dis.* 99(1):156–156.

Kolmer, J. A. 2017. Genetics of leaf rust resistance in the hard red winter wheat cultivars Santa Fe and Duster. *Crop Sci.* 57(5):2500–2505.

Kolmer, J. A. 2019. Virulence of *Puccinia triticina*, the wheat leaf rust fungus, in the United States in 2017. *Plant Dis.* 103(8):2113–2120.

Kolmer, J. A., Chao, S., Brown-Guedira, G., Bansal, U. and Bariana, H. 2018a. Adult plant leaf rust resistance derived from the soft red winter wheat cultivar 'Caldwell' maps to chromosome 3BS. *Crop Sci.* 58(1):152–158.

Kolmer, J. A., Garvin, D. F. and Jin, Y. 2011. Expression of a Thatcher wheat adult plant stem rust resistance QTL on chromosome arm 2BL is enhanced by *Lr34*. *Crop Sci.* 51(2):526–533.

Kolmer, J. A., Su, Z., Bernardo, A., Bai, G. and Chao, S. 2018b. Mapping and characterization of the new adult plant leaf rust resistance gene *Lr77* derived from Santa Fe winter wheat. *Theor. Appl. Genet.* 131(7):1553–1560.

Kolmer, J. A., Su, Z., Bernardo, A., Bai, G. and Chao, S. 2019. A backcross line of Thatcher wheat with adult plant leaf rust resistance derived from Duster wheat has *Lr46* and *Lr77*. *Phytopathology* 109(1):127–132.

Kota, R., Spielmeyer, W., McIntosh, R. A. and Lagudah, E. S. 2006. Fine genetic mapping fails to dissociate durable stem rust resistance gene *Sr2* from pseudo-black chaff in common wheat (*Triticum aestivum* L.). *Theor. Appl. Genet.* 112(3):492–499.

Krenz, J. E., Sackett, K. E. and Mundt, C. C. 2008. Specificity of incomplete resistance to *Mycosphaerella graminicola* in wheat. *Phytopathology* 98(5):555–561.

Kroese, D. R., Schonneker, L., Bag, S., Frost, K., Cating, R. and Hagerty, C. H. 2020. Wheat soil-borne mosaic: yield loss and distribution in the US Pacific Northwest. *Crop Prot.* 132:105102.

Kthiri, D., Loladze, A., N'Diaye, A., Nilsen, K. T., Walkowiak, S., Dreisigacker, S., Ammar, K. and Pozniak, C. J. 2019. Mapping of genetic loci conferring resistance to leaf rust from three globally resistant durum wheat sources. *Front. Plant Sci.* 10:1247.

Kucek, L. K., Darby, H., Mallory, E., Dawson, J., Davis, M., Dyck, E., Lazor, J., O'Donnell, S., Mudge, S., Kimball, M., Molloy, T., Benscher, D., Tanaka, J., Cummings, E., and Sorrells, M. E. 2015. Participatory breeding of wheat for organic production. Proceedings of the Organic Agriculture Research Symposium. LaCrosse, WI, February 25–26, 2015.

Lan, C., Basnet, B. R., Singh, R. P., Huerta-Espino, J., Herrera-Foessel, S. A., Ren, Y. and Randhawa, M. S. 2017. Genetic analysis and mapping of adult plant resistance loci to leaf rust in durum wheat cultivar Bairds. *Theor. Appl. Genet.* 130(3):609–619.

Lan, C., Li, Z., Herrera-Foessel, S. A., Huerta-Espino, J., Basnet, B. R., Dreisigacker, S., Ren, Y., Lagudah, E. and Singh, R. P. 2019. Identification and mapping of two adult plant leaf rust resistance genes in durum wheat. *Mol. Breed.* 39(8):118.

Larkin, D. L., Holder, A. L., Mason, R. E., Moon, D. E., Brown-Guedira, G., Price, P. P., Harrison, S. A. and Dong, Y. 2020. Genome-wide analysis and prediction of Fusarium head blight resistance in soft red winter wheat. *Crop Sci.* 60(6):2882–2900.

Laroche, A., Frick, M., Graf, R. J., Larsen, J. and Laurie, J. D. 2019. Pyramiding disease resistance genes in elite winter wheat germplasm for Western Canada. *Crop J.* 7(6):739–749.

Li, G., Zhou, J., Jia, H., Gao, Z., Fan, M., Luo, Y., Zhao, P., Xue, S., Li, N., Yuan, Y., Ma, S., Kong, Z., Jia, L., An, X., Jiang, G., Liu, W., Cao, W., Zhang, R., Fan, J., Xu, X., Liu, Y., Kong, Q., Zheng, S., Wang, Y., Qin, B., Cao, S., Ding, Y., Shi, J., Yan, H., Wang, X., Ran, C. and Ma, Z. 2019. Mutation of a histidine-rich calcium-binding-protein gene in wheat confers resistance to Fusarium head blight. *Nat. Genet.* 51(7):1106–1112.

Li, Z., Yuan, C., Herrera-Foessel, S. A., Randhawa, M. S., Huerta-Espino, J., Liu, D., Dreisigacker, S., Singh, R. P. and Lan, C. 2020. Four consistent loci confer adult plant resistance to leaf rust in the durum wheat lines Heller#1 and Dunkler. *Phytopathology* 110(4):892–899.

Lin, X., N'Diaye, A., Walkowiak, S., Nilsen, K. T., Cory, A. T., Haile, J., Kutcher, H. R., Ammar, K., Loladze, A., Huerta-Espino, J., Clarke, J. M., Ruan, Y., Knox, R., Fobert, P., Sharpe,

A. G. and Pozniak, C. J. 2018. Genetic analysis of resistance to stripe rust in durum wheat (*Triticum turgidum* L. var. *durum*). *PLoS ONE* 13(9):e0203283.

Liu, G., Liu, X., Xu, Y., Bernardo, A., Chen, M., Li, Y., Niu, F., Zhao, L. and Bai, G. 2020a. Reassigning Hessian fly resistance genes *H7* and *H8* to chromosomes 6A and 2B of the wheat cultivar 'Seneca' using genotyping-by-sequencing. *Crop Sci.* 60(3):1488–1498.

Liu, L., Wang, M., Zhang, Z., See, D. R. and Chen, X. 2020b. Identification of stripe rust resistance loci in U.S. spring wheat cultivars and breeding lines using genome-wide association mapping and *Yr* gene markers. *Plant Dis.* 104(8):2181–2192.

Liu, L., Wang, M. N., Feng, J. Y., See, D. R., Chao, S. M. and Chen, X. M. 2018. Combination of all-stage and high-temperature adult-plant resistance QTL confers high-level, durable resistance to stripe rust in winter wheat cultivar Madsen. *Theor. Appl. Genet.* 131(9):1835–1849.

Liu, S., Griffey, C. A., Hall, M. D., Chen, J., Liu, S., Tucker, D. and Brooks, W. S. 2012. Registration of 'Becker'/'Massey' wheat recombinant inbred line mapping population. *J. Plant Reg.* 6(3):358–362.

Liu, S., Griffey, C. A., Hall, M. D., McKendry, A. L., Chen, J., Brooks, W. S., Brown-Guedira, G., Van Sanford, D. and Schmale, D. G. 2013. Molecular characterization of field resistance to Fusarium head blight in two US soft red winter wheat cultivars. *Theor. Appl. Genet.* 126(10):2485–2498.

Liu, S., Griffey, C. A. and Saghai Maroof, M. A. S. 2001. Identification of molecular markers associated with adult plant resistance to powdery mildew in common wheat cultivar Massey. *Crop Sci.* 41(4):1268–1275.

Liu, T., Wan, A., Liu, D. and Chen, X. 2017. Changes of races and virulence genes in *Puccinia striiformis* f. sp. *tritici*, the wheat stripe rust pathogen, in the United States from 1968 to 2009. *Plant Dis.* 101(8):1522–1532.

Liu, Z., El-Basyoni, I., Kariyawasam, G., Zhang, G., Fritz, A., Hansen, J., Marais, F., Friskop, A., Chao, S., Akhunov, E. and Baenziger, P. S. 2015. Evaluation and association mapping of resistance to tan spot and Stagonospora nodorum blotch in adapted winter wheat germplasm. *Plant Dis.* 99(10):1333–1341.

Lookabaugh, E. C., Shew, B. B. and Cowger, C. 2017. Three *Pythium* species isolated from severely stunted wheat during an outbreak in North Carolina. *Plant Health Prog.* 18(3):169–173.

Lorenz, A. J., Smith, K. P. and Jannink, J. -L. 2012. Potential and optimization of genomic selection for Fusarium head blight resistance in six-row barley. *Crop Sci.* 52(4):1609–1621.

Lowe, I., Cantu, D. and Dubcovsky, J. 2011. Durable resistance to the wheat rusts: integrating systems biology and traditional phenotype-based research methods to guide the deployment of resistance genes. *Euphytica* 179(1):69–79.

Lu, Y., Wang, M., Chen, X., See, D., Chao, S. and Jing, J. 2014. Mapping of *Yr62* and a small-effect QTL for high-temperature adult-plant resistance to stripe rust in spring wheat PI 192252. *Theor. Appl. Genet.* 127(6):1449–1459.

Luby, C. H., Kloppenburg, J., Michaels, T. E. and Goldman, I. L. 2015. Enhancing freedom to operate for plant breeders and farmers through open source plant breeding. *Crop Sci.* 55(6):2481–2488.

Mavrodi, O. V., Walter, N., Elateek, S., Taylor, C. G. and Okubara, P. A. 2012. Suppression of Rhizoctonia and Pythium root rot of wheat by new strains of *Pseudomonas*. *Biol. Control* 62(2):93–102.

Maxwell, J. J., Lyerly, J. H., Cowger, C., Marshall, D., Brown-Guedira, G. and Murphy, J. P. 2009. MlAG12: a *Triticum timopheevii*-derived powdery mildew resistance gene in common wheat on chromosome 7AL. *Theor. Appl. Genet.* 119(8):1489–1495.

Maxwell, J. J., Lyerly, J. H., Srnic, G., Murphy, J. P., Cowger, C., Parks, R., Marshall, D., Brown-Guedira, G. and Miranda, L. 2012. *MlNCD1*: A novel *Aegilops tauschii*-derived powdery mildew resistance gene identified in common wheat. *Crop Sci.* 52(3):1162–1170.

Maxwell, J. J., Lyerly, J. H., Srnić, G., Parks, R., Cowger, C., Marshall, D., Brown-Guedira, G. and Murphy, J. P. 2010. *MlAB10*: a *Triticum turgidum* subsp. *dicoccoides* derived powdery mildew resistance gene identified in common wheat. *Crop Sci.* 50(6):2261–2267.

McCallum, B. D., Hiebert, C. W., Cloutier, S., Bakkeren, G., Rosa, S. B., Humphreys, D. G., Marais, G. F., McCartney, C. A., Panwar, V., Rampitsch, C., Saville, B. J. and Wang, X. 2016. A review of wheat leaf rust research and the development of resistant cultivars in Canada. *Can. J. Plant Pathol.* 38(1):1–18.

McCallum, B. D., Seto-Goh, P., Reimer, E., Foster, A. and Xue, A. 2020. Physiological specialization of *Puccinia triticina*, the causal agent of wheat leaf rust, in Canada in 2013. *Can. J. Plant Pathol.* 42(2):243–249.

McCartney, C. A., Brûlé-Babel, A. L., Fedak, G., Martin, R. A., McCallum, B. D., Gilbert, J., Hiebert, C. W. and Pozniak, C. J. 2016. Fusarium head blight resistance QTL in the spring wheat cross Kenyon/86ISMN 2137. *Front. Microbiol.* 7:1542.

McDonald, B. A. and Mundt, C. C. 2016. How knowledge of pathogen population biology informs management of Septoria tritici blotch. *Phytopathology* 106(9):948–955.

McDonald, M. J., Ohm, H. W., Rinehart, K. D., Giovanini, M. P. and Williams, C. E. 2014. *H33*: a wheat gene providing Hessian fly resistance for the southeastern United States. *Crop Sci.* 54(5):2045–2053.

McMullen, M., Bergstrom, G., DeWolf, E. D., Dill-Macky, R., Hershman, D., Shaner, G. and Van Sanford, D. 2012. A unified effort to fight an enemy of wheat and barley: Fusarium head blight. *Plant Dis.* 96(12):1712–1728.

McMullen, M., Jones, R. and Gallenberg, D. 1997. Scab of wheat and barley: A re-emerging disease of devastating impact. *Plant Dis.* 81(12):1340–1348.

Mehra, L. K., Adhikari, U., Ojiambo, P. S. and Cowger, C. 2019. Septoria nodorum blotch of wheat. The Plant Health Instructor. American Phytopathological Society. https://www.apsnet.org/edcenter/disandpath/fungalasco/pdlessons/Pages/SeptoriaNodorum.aspx. https://doi.org/10.1094/PHI-I-2019-0514-01.

Milus, E. A., Kristensen, K. and Hovmøller, M. S. 2009. Evidence for increased aggressiveness in a recent widespread strain of *Puccinia striiformis* f. sp. *tritici* causing stripe rust of wheat. *Phytopathology* 99(1):89–94.

Milus, E. A., Lee, K. D. and Brown-Guedira, G. 2015a. Characterization of stripe rust resistance in wheat lines with resistance gene *Yr17* and implications for evaluating resistance and virulence. *Phytopathology* 105(8):1123–1130.

Milus, E. A., Moon, D. E., Lee, K. D. and Mason, R. E. 2015b. Race-specific adult-plant resistance in winter wheat to stripe rust and characterization of pathogen virulence patterns. *Phytopathology* 105(8):1114–1122.

Miranda, L. M., Bland, D. E., Cambron, S. E., Lyerly, J. H., Johnson, J., Buntin, G. D. and Murphy, J. P. 2010. Genetic mapping of an *Aegilops tauschii*-derived Hessian fly resistance gene in common wheat. *Crop Sci.* 50(2):612–616.

Miranda, L. M., Murphy, J. P., Marshall, D., Cowger, C. and Leath, S. 2007a. Chromosomal location of *Pm35*, a novel *Aegilops tauschii* derived powdery mildew resistance gene introgressed into common wheat (*Triticum aestivum* L.). *Theor. Appl. Genet.* 114(8):1451–1456.

Miranda, L. M., Murphy, J. P., Marshall, D. and Leath, S. 2006. *Pm34*: a new powdery mildew resistance gene transferred from *Aegilops tauschii* Coss. to common wheat (*Triticum aestivum* L.). *Theor. Appl. Genet.* 113(8):1497–1504.

Miranda, L. M., Perugini, L., Srnić, G., Brown-Guedira, G., Marshall, D., Leath, S. and Murphy, J. P. 2007b. Genetic mapping of a *Triticum monococcum*-derived powdery mildew resistance gene in common wheat. *Crop Sci.* 47(6):2323–2329.

Morton, P. K. and Schemerhorn, B. J. 2013. Population structure and the colonization route of one of the oldest North American invasive insects: stories from the worn road of the Hessian fly, *Mayetiola destructor* (Say). *PLoS ONE* 8(3): e59833.

Mu, J., Liu, L., Liu, Y., Wang, M., See, D. R., Han, D. and Chen, X. 2020. Genome-wide association study and gene specific markers identified 51 genes or QTL for resistance to stripe rust in U.S. winter wheat cultivars and breeding lines. *Front. Plant Sci.* 11:998.

Mundt, C. C. 2018. Pyramiding for resistance durability: theory and practice. *Phytopathology* 108(7):792–802.

Mundt, C. C., Cowger, C. and Garrett, K. A. 2002. Relevance of integrated disease management to resistance durability. *Euphytica* 124(2):245–252.

Murphy, J. P., Maxwell, J. J., Miranda, L. M., Lyerly, J. H., Parks, W. R., Srnic, G., Perugini, L., Cowger, C., Marshall, D., Van Esbroeck, G., Brown-Guedira, G., Jarrett, S. G. and Kolmer, J. A. 2009. Qualitative powdery mildew resistance mapping update. Pages 28-30. In: *2009 Eastern Wheat Workers/Southern Small Grain Workers–NCERA184 Conference*. EWW/SSGW, Baltimore, Maryland.

Naruoka, Y., Garland-Campbell, K. A. and Carter, A. H. 2015. Genome-wide association mapping for stripe rust (*Puccinia striiformis* f. sp. *tritici*) in US Pacific Northwest winter wheat (*Triticum aestivum* L.). *Theor. Appl. Genet.* 128(6):1083–1101.

National Association of Wheat Growers 2014. Farmer leadership creates a new 21st century opportunity for wheat. National Wheat Foundation, Washington, DC. Available at: https://wheatfoundation.org/wp-content/uploads/2017/01/NWF-Wheat-Research-whitepaper-FLM-20141009.pdf.

Nemacheck, J. A., Schemerhorn, B. J., Scofield, S. R. and Subramanyam, S. 2019. Phenotypic and molecular characterization of Hessian fly resistance in diploid wheat, *Aegilops tauschii*. *BMC Plant Biol.* 19(1):439.

Newcomb, M., Olivera, P. D., Rouse, M. N., Szabo, L. J., Johnson, J., Gale, S., Luster, D. G., Wanyera, R., Macharia, G., Bhavani, S., Hodson, D., Patpour, M., Hovmøller, M. S., Fetch, T. G. and Jin, Y. 2016 Kenyan isolates of Puccinia graminis f. sp. tritici from 2008 to 2014: virulence to SrTmp in the Ug99 race group and implications for breeding programs. *Phytopathology* 106(7):729–736.

Niu, F., Xu, Y., Liu, X., Zhao, L., Bernardo, A., Li, Y., Liu, G., Chen, M. S., Cao, L., Hu, Z., Xu, X. and Bai, G. 2020. The Hessian fly recessive resistance gene *h4* mapped to chromosome 1a of the wheat cultivar 'Java' using genotyping-by-sequencing. *Theor. Appl. Genet.* 133(10):2927–2935.

Noriel, A. J., Sun, X., Bockus, W. and Bai, G. 2011. Resistance to tan spot and insensitivity to Ptr ToxA in wheat. *Crop Sci.* 51(3):1059–1067.

Oelke, L. M. and Kolmer, J. A. 2005. Genetics of leaf rust resistance in spring wheat cultivars Alsen and Norm. *Phytopathology* 95(7):773–778.

Olivera, P. D., Pretorius, Z. A., Badebo, A. and Jin, Y. 2013. Identification of resistance to races of *Puccinia graminis* f. sp. *tritici* with broad virulence in triticale (×*Triticosecale*). *Plant Dis.* 97(4):479–484.

Olivera, P. D., Rouse, M. N. and Jin, Y. 2018. Identification of new sources of resistance to wheat stem rust in *Aegilops* spp. in the tertiary genepool of wheat. *Front. Plant Sci.* 9:1719.

Ortelli, S., Winzeler, H., Winzeler, M., Fried, P. M. and Nösberger, J. 1996. Leaf rust resistance gene *Lr9* and winter wheat yield reduction: I. yield and yield components. *Crop Sci.* 36(6):1590–1595.

Paulitz, T. C. and Adams, K. 2003. Composition and distribution of *Pythium* communities in wheat fields in eastern Washington state. *Phytopathology* 93(7):867–873.

Perugini, L. D., Murphy, J. P., Marshall, D. and Brown-Guedira, G. 2008. *Pm37*, a new broadly effective powdery mildew resistance gene from *Triticum timopheevii*. *Theor. Appl. Genet.* 116(3):417–425.

Petersen, S., Lyerly, J. H., Maloney, P. V., Brown-Guedira, G., Cowger, C., Costa, J. M., Dong, Y. and Murphy, J. P. 2016. Mapping of Fusarium head blight resistance quantitative trait loci in winter wheat cultivar NC-Neuse. *Crop Sci.* 56(4):1473–1483.

Petersen, S., Lyerly, J. H., McKendry, A. L., Islam, M. S., Brown-Guedira, G., Cowger, C., Dong, Y. and Murphy, J. P. 2017. Validation of Fusarium head blight resistance QTL in US winter wheat. *Crop Sci.* 57(1):1–12.

Petersen, S., Lyerly, J. H., Worthington, M. L., Parks, W. R., Cowger, C., Marshall, D. S., Brown-Guedira, G. and Murphy, J. P. 2015. Mapping of powdery mildew resistance gene *Pm53* introgressed from *Aegilops speltoides* into soft red winter wheat. *Theor. Appl. Genet.* 128(2):303–312.

Pinto da Silva, G. B., Zanella, C. M., Martinelli, J. A., Chaves, M. S., Hiebert, C. W., McCallum, B. D. and Boyd, L. A. 2018. Quantitative trait loci conferring leaf rust resistance in hexaploid wheat. *Phytopathology* 108(12):1344–1354.

Qi, L. L., Pumphrey, M. O., Friebe, B., Chen, P. D. and Gill, B. S. 2008. Molecular cytogenetic characterization of alien introgressions with gene *Fhb3* for resistance to Fusarium head blight disease of wheat. *Theor. Appl. Genet.* 117(7):1155–1166.

Ramakrishnan, S. M., Sidhu, J. S., Ali, S., Kaur, N., Wu, J. and Sehgal, S. K. 2019. Molecular characterization of bacterial leaf streak resistance in hard winter wheat. *PeerJ* 7:e7276.

Ratcliffe, R. H., Cambron, S. E., Flanders, K. L., Bosque-Perez, N. A., Clement, S. L. and Ohm, H. W. 2000. Biotype composition of Hessian fly (Diptera: Cecidomyiidae) populations from the southeastern, midwestern, and northwestern United States and virulence to resistance genes in wheat . *J. Econ. Entomol.* 93(4):1319–1328.

Rawat, N., Pumphrey, M. O., Liu, S., Zhang, X., Tiwari, V. K., Ando, K., Trick, H. N., Bockus, W. W., Akhunov, E., Anderson, J. A. and Gill, B. S. 2016. Wheat *Fhb1* encodes a chimeric lectin with agglutinin domains and a pore-forming toxin-like domain conferring resistance to Fusarium head blight. *Nat. Genet.* 48(12):1576–1580.

Reeves, E. R., Kerns, J. P., Cowger, C. and Shew, B. B. 2021. Characterization of *Pythium* spp. associated with stunting and root rot of winter wheat in North Carolina. *Plant Dis.*, in press, doi.org/10.1094/PDIS-09-20-2022-RE.

Ren, R. S., Wang, M. N., Chen, X. M. and Zhang, Z. J. 2012. Characterization and molecular mapping of *Yr52* for high-temperature adult-plant resistance to stripe rust in spring wheat germplasm PI 183527. *Theor. Appl. Genet.* 125(5):847–857.

Richards, J. K., Stukenbrock, E. H., Carpenter, J., Liu, Z., Cowger, C., Faris, J. D. and Friesen, T. L. 2019. Local adaptation drives the diversification of effectors in the fungal wheat pathogen *Parastagonospora nodorum* in the United States. *PLOS Genet.* 15(10):e1008223.

Rouse, M. N., Olson, E. L., Gill, B. S., Pumphrey, M. O. and Jin, Y. 2011a. Stem rust resistance in *Aegilops tauschii* germplasm. *Crop Sci.* 51(5):2074–2078.

Rouse, M. N., Talbert, L. E., Singh, D. and Sherman, J. D. 2014. Complementary epistasis involving *Sr12* explains adult plant resistance to stem rust in Thatcher wheat (*Triticum aestivum* L.). *Theor. Appl. Genet.* 127(7):1549–1559.

Rouse, M. N., Wanyera, R., Njau, P. and Jin, Y. 2011b. Sources of resistance to stem rust race Ug99 in spring wheat germplasm. *Plant Dis.* 95(6):762–766.

Ruan, Y., Comeau, A., Langevin, F., Hucl, P., Clarke, J. M., Brule-Babel, A. and Pozniak, C. J. 2012. Identification of novel QTL for resistance to Fusarium head blight in a tetraploid wheat population. *Genome* 55(12):853–864.

Rudd, J. C., Devkota, R. N., Ibrahim, A. M., Baker, J. A., Baker, S., Lazar, M. D., Sutton, R., Simoneaux, B., Opena, G., Rooney, L. W., Awika, J. M., Liu, S., Xue, Q., Bean, B., Duncan, R. W., Seabourn, B. W., Bowden, R. L., Jin, Y., Chen, M.-S. and Graybosch, R. A. 2018. 'TAM 114' wheat, excellent bread-making quality hard red winter wheat cultivar adapted to the southern High Plains. *J. Plant Reg.* 12(3):367–372.

Rutkoski, J., Benson, J., Jia, Y., Brown-Guedira, G., Jannink, J.-L. and Sorrells, M. 2012. Evaluation of genomic prediction methods for Fusarium head blight resistance in wheat. *Plant Genome* 5(2):51–61.

Rutkoski, J. E., Heffner, E. L. and Sorrells, M. E. 2011. Genomic selection for durable stem rust resistance in wheat. *Euphytica* 179(1):161–173.

Rutkoski, J. E., Poland, J. A., Singh, R. P., Huerta-Espino, J., Bhavani, S., Barbier, H., Rouse, M. N., Jannink, J.-L. and Sorrells, M. E. 2014. Genomic selection for quantitative adult plant stem rust resistance in wheat. *Plant Genome* 7:eplantgenome2014.02 .0006.

Santra, D. K., Chen, X. M., Santra, M., Campbell, K. G. and Kidwell, K. K. 2008. Identification and mapping QTL for high-temperature adult-plant resistance to stripe rust in winter wheat (*Triticum aestivum* L.) cultivar 'Stephens'. *Theor. Appl. Genet.* 117(5):793–802.

Sapkota, S., Zhang, Q., Chittem, K., Mergoum, M., Xu, S. S. and Liu, Z. 2018. Evaluation of triticale accessions for resistance to wheat bacterial leaf streak caused by *Xanthomonas translucens* pv. *undulosa*. *Plant Pathol.* 67(3):595–602.

Sari, E., Berraies, S., Knox, R. E., Singh, A. K., Ruan, Y., Cuthbert, R. D., Pozniak, C. J., Henriquez, M. A., Kumar, S., Burt, A. J., N'Diaye, A., Konkin, D. J., Cabral, A. L., Campbell, H. L., Wiebe, K., Condie, J., Lokuruge, P., Meyer, B., Fedak, G., Clarke, F. R., Clarke, J. M., Somers, D. J. and Fobert, P. R. 2018. High density genetic mapping of Fusarium head blight resistance QTL in tetraploid wheat. *PLoS ONE* 13(10):e0204362.

Schmid, R. B., Knutson, A., Giles, K. L. and McCornack, B. P. 2018. Hessian fly (Diptera: Cecidomyiidae) biology and management in wheat. *J. Integr. Pest Manag.* 9(1) 9(1):14; 1–12.

Schroeder, K. L. and Paulitz, T. C. 2006. Root diseases of wheat and barley during the transition from conventional tillage to direct seeding. *Plant Dis.* 90(9):1247–1253.

Schwarz, P. and Horsley, R. D. 1996. A comparison of North American two-row and six-row malting barley. *Brewing Technol.* 4:48–55.

Seifers, D. L., Martin, T. J., Harvey, T. L. and Haber, S. 2007. Temperature-sensitive wheat streak mosaic virus resistance identified in KS03HW12 wheat. *Plant Dis.* 91(8):1029–1033.

Sharma, J. S., Zhang, Q., Rouse, M. N., Klindworth, D. L., Friesen, T. L., Long, Y., Olivera, P. D., Jin, Y., McClean, P. E., Xu, S. S. and Faris, J. D. 2019. Mapping and characterization of two stem rust resistance genes derived from cultivated emmer wheat accession PI 193883. *Theor. Appl. Genet.* 132(11):3177–3189.

Shukle, R. H., Cambron, S. E., Moniem, H. A., Schemerhorn, B. J., Redding, J., David Buntin, G., Flanders, K. L., Reisig, D. D. and Mohammadi, M. 2015. Effectiveness of genes for Hessian fly (Diptera: Cecidomyiidae) resistance in the southeastern United States. *J. Econ. Entomol.* 109(1):399–405.

Shukle, R. H., Subramanyam, S. and Williams, C. E. 2012. Effects of antinutrient proteins on Hessian fly (Diptera: Cecidomyiidae) larvae. *J. Insect Physiol.* 58(1):41–48.

Singh, A., Knox, R. E., DePauw, R. M., Singh, A. K., Cuthbert, R. D., Campbell, H. L., Shorter, S. and Bhavani, S. 2014. Stripe rust and leaf rust resistance QTL mapping, epistatic interactions, and co-localization with stem rust resistance loci in spring wheat evaluated over three continents. *Theor. Appl. Genet.* 127(11):2465–2477.

Singh, A., Pandey, M. P., Singh, A. K., Knox, R. E., Ammar, K., Clarke, J. M., Clarke, F. R., Singh, R. P., Pozniak, C. J., Depauw, R. M., McCallum, B. D., Cuthbert, R. D., Randhawa, H. S. and Fetch, T. G. 2013 Identification and mapping of leaf, stem and stripe rust resistance quantitative trait loci and their interactions in durum wheat. *Mol. Breed.* 31(2):405–418.

Singh, P. K., Mergoum, M., Ali, S., Adhikari, T. B., Elias, E. M., Anderson, J. A., Glover, K. D. and Berzonsky, W. A. 2006. Evaluation of elite wheat germ plasm for resistance to tan spot. *Plant Dis.* 90(10):1320–1325.

Singh, R. P., Hodson, D. P., Huerta-Espino, J., Jin, Y., Bhavani, S., Njau, P., Herrera-Foessel, S., Singh, P. K., Singh, S. and Govindan, V. 2011. The emergence of Ug99 races of the stem rust fungus is a threat to world wheat production. *Annu. Rev. Phytopathol.* 49:465–481.

Singh, R. P., Hodson, D. P., Jin, Y., Lagudah, E. S., Ayliffe, M. A., Bhavani, S., Rouse, M. N., Pretorius, Z. A., Szabo, L. J., Huerta-Espino, J., Basnet, B. R., Lan, C. and Hovmøller, M. S. 2015. Emergence and spread of new races of wheat stem rust fungus: continued threat to food security and prospects of genetic control. *Phytopathology* 105(7):872–884.

Skerritt, J. H. 2016. Research organizations of the world: Europe and North America. Pages 436–443. In: *Encyclopedia of Food Grains: The World of Food* Grains Wrigley, C. W., Corke, H., Seetharaman, K. and Faubion, J. (Eds). Elsevier Ltd., Oxford, UK.

Stewart, E. L., Hagerty, C. H., Mikaberidze, A., Mundt, C. C., Zhong, Z. and McDonald, B. A. 2016. An improved method for measuring quantitative resistance to the wheat pathogen *Zymoseptoria tritici* using high-throughput automated image analysis. *Phytopathology* 106(7):782–788.

Stuber, C. W. and Hancock, J. 2008. Sustaining plant breeding-national workshop. *Crop Sci.* 48(1):25–29.

Su, Z., Bernardo, A., Tian, B., Chen, H., Wang, S., Ma, H., Cai, S., Liu, D., Zhang, D., Li, T., Trick, H., St. Amand, P., Yu, J., Zhang, Z. and Bai, G. 2019. A deletion mutation in *TaHRC* confers *Fhb1* resistance to Fusarium head blight in wheat. *Nat. Genet.* 51(7):1099–1105.

Subramanyam, S., Zheng, C., Shukle, J. T. and Williams, C. E. 2013. Hessian fly larval attack triggers elevated expression of disease resistance dirigent-like protein-encoding gene, HfrDrd, in resistant wheat. *Arthropod Plant Interact* 7(4): 389–402.

Tan, M.-K., El-Bouhssini, M., Emebiri, L., Wildman, O., Tadesse, W. and Ogbonnaya, F. C. 2015. A SNP marker for the selection of *HfrDrd*, a Hessian fly-response gene in wheat. *Mol. Breed.* 35(11):216.

Tessmann, E. W., Dong, Y. and Van Sanford, D. A. 2019. GWAS for Fusarium head blight traits in a soft red winter wheat mapping panel. *Crop Sci.* 59(5):1823–1837.

Thambugala, D., Brûlé-Babel, A. L., Blackwell, B. A., Fedak, G., Foster, A. J., MacEachern, D., Gilbert, J., Henriquez, M. A., Martin, R. A., McCallum, B. D., Spaner, D., Iqbal, M., Pozniak, C. J., N'Diaye, A. and McCartney, C. A. 2020. Genetic analyses of native Fusarium head blight resistance in two spring wheat populations identifies QTL near the *B1 Ppd-D1*, *Rht-1*, *Vrn-1*, *Fhb1*, *Fhb2* and *Fhb5* loci. *Theor. Appl. Genet.* 133(10):2775–2796.

Thapa, R., Brown-Guedira, G., Ohm, H. W., Mateos-Hernandez, M., Wise, K. A. and Goodwin, S. B. 2016. Determining the order of resistance genes against Stagonospora nodorum blotch, Fusarium head blight and stem rust on wheat chromosome arm 3BS. *BMC Res. Notes* 9:58.

Tillman, B. L., Harrison, S. A., Clark, C. A., Milus, E. A. and Russin, J. S. 1996. Evaluation of bread wheat germplasm for resistance to bacterial streak. *Crop Sci.* 36(4):1063–1068.

Turner, M. K., Kolmer, J. A., Pumphrey, M. O., Bulli, P., Chao, S. and Anderson, J. A. 2017. Association mapping of leaf rust resistance loci in a spring wheat core collection. *Theor. Appl. Genet.* 130(2):345–361.

Uphaus, J., Walker, E., Shankar, M., Golzar, H., Loughman, R., Francki, M. and Ohm, H. 2007. Quantitative trait loci identified for resistance to Stagonospora glume blotch in wheat in the USA and Australia. *Crop Sci.* 47(5):1813–1822.

Vazquez, M. D., James Peterson, C., Riera-Lizarazu, O., Chen, X., Heesacker, A., Ammar, K., Crossa, J. and Mundt, C. C. 2012. Genetic analysis of adult plant, quantitative resistance to stripe rust in wheat cultivar 'Stephens' in multi-environment trials. *Theor. Appl. Genet.* 124(1):1–11.

Walls III, J., Rajotte, E. and Rosa, C. 2019. The past, present and future of barley yellow dwarf management. *Agriculture* 9(1):23.

Wan, A. and Chen, X. 2014. Virulence characterization of *Puccinia striiformis* f. sp. *tritici* using a new set of *Yr* single-gene line differentials in the United States in 2010. *Plant Dis.* 98(11):1534–1542.

Wan, A., Chen, X. and Yuen, J. 2016. Races of *Puccinia striiformis* f. sp. *tritici* in the United States in 2011 and 2012 and comparison with races in 2010. *Plant Dis.* 100(5):966–975.

Wang, M. N. and Chen, X. M. 2017. Stripe rust resistance. In: *Stripe Rust*. Chen, X. M., and Kang, Z. S. (Eds). Springer, Dordrecht, 353–558.

Wen, A., Jayawardana, M., Fiedler, J., Sapkota, S., Shi, G., Peng, Z., Liu, S., White, F. F., Bogdanove, A. J., Li, X. and Liu, Z. 2018. Genetic mapping of a major gene in triticale conferring resistance to bacterial leaf streak. *Theor. Appl. Genet.* 131(3):649–658.

Zhang, B., Chi, D., Hiebert, C., Fetch, T., McCallum, B., Xue, A., Cao, W., Depauw, R. and Fedak, G. 2019. Pyramiding stem rust resistance genes to race TTKSK (Ug99) in wheat. *Can. J. Plant Pathol.* 41(3):443–449.

Zhang, D., Bowden, R. L., Yu, J., Carver, B. F. and Bai, G. 2014a. Association analysis of stem rust resistance in U.S. winter wheat. *PLoS ONE* 9(7):e103747.

Zhang, Q., Axtman, J. E., Faris, J. D., Chao, S., Zhang, Z., Friesen, T. L., Zhong, S., Cai, X., Elias, E. M. and Xu, S. S. 2014b. Identification and molecular mapping of quantitative trait loci for Fusarium head blight resistance in emmer and durum wheat using a single nucleotide polymorphism-based linkage map. *Mol. Breed.* 34(4):1677–1687.

Zhao, M., Leng, Y., Chao, S., Xu, S. S. and Zhong, S. 2018. Molecular mapping of QTL for Fusarium head blight resistance introgressed into durum wheat. *Theor. Appl. Genet.* 131(9):1939–1951.

Zhou, X. L., Wang, M. N., Chen, X. M., Lu, Y., Kang, Z. S. and Jing, J. X. 2014. Identification of *Yr59* conferring high-temperature adult-plant resistance to stripe rust in wheat germplasm PI 178759. *Theor. Appl. Genet.* 127(4):935–945.

Zhu, X., Zhong, S., Chao, S., Gu, Y. Q., Kianian, S. F., Elias, E. and Cai, X. 2016. Toward a better understanding of the genomic region harboring Fusarium head blight resistance QTL *Qfhs.ndsu-3AS* in durum wheat. *Theor. Appl. Genet.* 129(1):31–43.

Zhu, Z., Hao, Y., Mergoum, M., Bai, G., Humphreys, G., Cloutier, S., Xia, X. and He, Z. 2019. Breeding wheat for resistance to Fusarium head blight in the Global North: China, USA, and Canada. *Crop J.* 7(6):730–738.

Zou, J., Semagn, K., Chen, H., Iqbal, M., Asif, M., N'Diaye, A., Navabi, A., Perez-Lara, E., Pozniak, C., Yang, R.-C., Graf, R. J., Randhawa, H. and Spaner, D. 2017. Mapping of QTLs associated with resistance to common bunt, tan spot, leaf rust, and stripe rust in a spring wheat population. *Mol. Breed.* 37(12):144.

Zurn, J. D., Rouse, M. N., Chao, S., Aoun, M., Macharia, G., Hiebert, C. W., Pretorius, Z. A., Bonman, J. M. and Acevedo, M. 2018. Dissection of the multigenic wheat stem rust resistance present in the Montenegrin spring wheat accession PI 362698. *BMC Genomics* 19(1):67.

Chapter 23

Achievements in breeding cereals with durable disease resistance in Northwest Europe

James K. M. Brown, John Innes Centre, UK

1 Progress in breeding for effective, durable resistance

Resistant varieties make an increasingly important contribution to control of diseases and pests of wheat, barley and oats in northwest Europe. Resistance is increasingly durable and the era of 'boom-and-bust', in which new resistance genes were rapidly rendered ineffective by the emergence of virulent pathogens, is for the most part long gone. In some cases, resistance is now so effective, widespread and durable that the diseases themselves have become much less significant, while a growing proportion of varieties are

http://dx.doi.org/10.19103/AS.2021.0092.39

not very susceptible to any of the most important diseases. The 'arms race' between breeders and pathogens is not yet over, but we are increasingly in an era when most disease resistance is stable and further advances can be foreseen.

As cereal breeding in this region is done almost entirely by the private sector, breeding for resistance cannot be done in isolation; rather, to produce varieties which are popular in a highly competitive market, breeders have combined durable resistance to multiple diseases and pests with the many other traits required by farmers and end-users which contribute to high yield, marketable grain quality and desirable agronomic properties. In the chapter, I describe the achievements of the cereal breeding industry in northwest Europe in developing commercially successful varieties with durable resistance and reflect upon the reasons for this success.

1.1 Cereal production in Northwest Europe

In northwest Europe, I include the United Kingdom (UK), Ireland, Scandinavia, the Benelux countries, France, Germany, Switzerland and Austria. This is a region in which arable farming is intensive with higher use of fertiliser and pesticides than in most other cereal-growing areas and in which crop yields are correspondingly high. The drive to produce high yields of cereals poses a significant challenge for disease control, including breeding for resistance, because many fungal diseases are much more severe when greater amounts of nitrogen fertiliser are supplied.

Wheat and barley are the main cereals in this region with the three highest producers of both crops being France, Germany and the UK (Table 1). The three Scandinavian countries, Denmark, Norway and Sweden, together rank fourth for both species, with a total not far behind the UK. Among these countries, the highest yields of wheat are in the UK while France, Germany and the UK have similar average yields of barley. The highest yields of all are, in fact, in Ireland, with 9.4 t ha^{-1} wheat and 8.2 t ha^{-1} barley. Rye, triticale and oats are minor crops in northwest Europe as a whole but are locally important. The majority of rye is grown in northeast Germany while most triticale is produced in Germany and France, often on marginal land. Oats is an important crop in northern areas including the UK and Scandinavia; again, the highest yields of all are in Ireland with 8.1 t ha^{-1}.

Winter wheat, sown in early autumn, is much more important than spring wheat, sown in late winter in almost all areas of northwest Europe. In both barley and oats, winter and spring crops are equally important. In the UK, in 2018 and 2019, there was an average of 1.8 million hectares of wheat, almost all of it winter wheat, 1.2 million hectares of barley with just over 60% being spring barley, and 180 000 hectares of oats (DEFRA, 2020).

Table 1 Production areas and grain yields per unit area of small-grain cereals in northwest Europe. Data are from 2019.

	Growing area in northwest Europe in 2019 (Mha)					
Crop[a]	Total	France	Germany	UK	Scandinavia[b]	Others[c]
Wheat	12.07	5.24	3.12	1.82	1.12	0.77
Barley	6.25	1.94	1.71	1.16	1.00	0.43
Rye	0.93	0.03	0.64	0.03	0.19	0.05
Triticale	0.79	0.31	0.36	0.01	0.04	0.08
Oats	0.69	0.09	0.13	0.18	0.25	0.05
	Mean yield in 2019 (t ha^{-1})					
Wheat	7.8	7.7	7.4	8.9	7.6	7.6
Barley	6.7	7.0	6.8	6.9	5.7	7.4
Rye	5.2	4.7	5.1	2.4	6.1	4.6
Triticale	5.8	5.4	6.1	4.5	6.3	5.6
Oats	5.0	4.6	4.1	5.9	4.7	6.0

[a] Crop production data from FAOSTAT (http://www.fao.org/faostat).
[b] Denmark, Norway and Sweden.
[c] Austria, Belgium, Republic of Ireland, Luxembourg, The Netherlands and Switzerland.

2 Sources of information about breeding for resistance

Cereal breeding in northwest Europe is done largely by companies rather than universities or public-sector research institutes. Although the leading breeders have diverse corporate structures, an important feature they share is that most information about their activities is confidential. Even so, it is possible to infer many of the industry's priorities from the performance of the varieties they produce. Inevitably some of the information in this chapter is personal knowledge, often gleaned from conversations with breeders.

Additional information about breeding for resistance can be found in academic research papers. These are usually most informative when the authors have worked closely with the industry and thus have a realistic view of the significance of diseases in crop production and are well informed about plant breeders' requirements.

2.1 The AHDB Recommended List

In this chapter, I focus largely on the United Kingdom, partly because it is the country with which I am most familiar but also because the UK's Recommended List (RL), published annually, is a valuable source of long-running, comparative information about cereal varieties. It evaluates varieties for traits which are important to farmers and end-users, including disease and pest resistance, and is thus an important guide to breeders in prioritising the traits which most need to be advanced.

The RL was issued by the National Institute of Agricultural Botany from 1948 to 2001 and has been produced by the Agriculture and Horticulture Development Board (AHDB) since 2002. As AHDB is a levy body, the RL is largely funded by farmers with significant contributions in kind by breeders and other relevant industries. I make frequent references to the RL in this chapter, which can be found at https://ahdb.org.uk/rl for the current year and at https://ahdb.org.uk/rlarchive for 2004 onwards. Note the dating system used for the RL; from 2004 onwards, the RL for 2021/22, for example, recommended winter varieties for sowing in 2021 and spring varieties to be sown in 2022. Until 2003, the RL was published as a printed booklet with a different dating system. The RL for 2003, for example, recommends winter varieties to sow in 2003 and spring varieties for 2004. Other countries in northwest Europe operate systems similar to the UK's RL, although the details vary considerably. Priorities for breeders in the UK are generally similar to those elsewhere in the region because most of the leading companies are multinationals, with subsidiaries or affiliates across the continent. Differences in emphasis in different countries largely reflect the local significance of various diseases.

3 Release and recommendation of cereal varieties

3.1 Plant breeder's rights

The desire for confidentiality stems from the system of protecting intellectual property in cereal varieties in Europe. This is governed by Plant Breeder's Rights (PBR), also known as Plant Variety Rights, not by patents. PBR originated with the UPOV Convention (*Union Internationale pour la Protection des Obtentions Végétales*) in 1961 and were implemented in the UK by the Plant Varieties and Seeds Act 1964, amended by the Plant Varieties Act 1997 (DEFRA, 2011). In the European Union (EU), the corresponding legislation is a set of 12 Directives which regulate plant reproductive material (European Council, 2020).

An important advantage of this system of intellectual property is that once a variety has been granted PBR, the company that owns the PBR has the right to market the variety and to earn royalties from seed sales, but any other company is free to use it as a parental line in its own breeding programme. This enables advances made by one company to be spread widely throughout the industry, accelerating the spread of new germplasm for the benefit of growers and ultimately consumers. This has been highly beneficial in breeding cereals for resistance to pests and diseases.

Published by Burleigh Dodds Science Publishing Limited, 2022.

3.2 National List and Recommended List trials

Release of a new cereal variety in the UK is a two-step process. The first is National List (NL) testing, in which the variety is trialled over two years for distinctness, uniformity and stability (DUS) and value for cultivation and use (VCU), including resistance to important pests and diseases. A variety which passes NL trials can be awarded PBR. In the EU, varieties with PBR can be entered on the Common Catalogue of varieties approved for sale in all member countries.

It is not obligatory for new varieties to be entered into RL trials, which are more extensive and detailed than NL trials, but in practice, the RL is an important marketing tool. The RL is compiled by the Wheat Crop Committee and the Barley, Oats and Other Cereals Crop Committee of AHDB. Varieties may be provisionally recommended if they are successful in the first year of RL trials and fully recommended after a further two years of RL trials, making five years of official trials in total for the NL and RL. The RL for a particular cropping year is based on trials in years up to and including the previous year, so the 2021/22 RL, which was released in January 2021, is based on trial results in farming years up to and including 2019/20.

3.3 Disease ratings

Disease resistances are rated on a scale from 1 to 9, where 1 is very susceptible and 9 is largely immune. The scale is logarithmic, so differences in disease severity between integer ratings are much greater at the bottom of the scale than at the top. This is because when a variety has poor resistance to a disease, it does not matter very much just how poor it is, whereas smaller differences between good and very good resistance may influence farmers' choice of varieties.

The calculation of disease ratings is explained here: https://ahdb.org.uk/rl-disease-ratings. Disease severity in RL trials is scored on a percent scale, using a visual disease assessment key to promote consistency of disease scores at different trial sites. The use of percent data is more precise than the scoring systems used in most cases by breeders (see Section 5.3). Disease scores are transformed to logarithms and means for varieties are calculated across sites and across several years, recognising that disease severity is inherently variable. Linear interpolation is then used to convert the means of log-transformed disease scores to a 1–9 scale, with a very low score corresponding to 9 and a very high score to 1.

A complication in interpreting RL disease ratings is that they require the use of established varieties as controls to fix points on the 1–9 scale, usually 3 (susceptible) and 7 or 8 (resistant). If a control variety is significantly more or less susceptible than expected, the ratings of the varieties on the RL may change. This explains the drop of about 1 point in the mildew ratings of spring oat varieties in 2010 (Fig. 5a) because the variety used to fix the low end of the scale had been less diseased than expected in several trials. Rarely, a rating scale may be recalibrated to reflect changes in the importance of a disease, as was done for wheat yellow rust in the 2021/22 RL.

The traits scored in RL trials reflect the requirements of farmers and end-users. When considering a variety for recommendation, the Crop Committee balances several traits by their importance, according to a set of weightings which the Committee has agreed in advance (AHDB, 2020). Four categories of traits are assessed: yield in various situations, grain quality, agronomic properties such as height, maturity date and lodging (the tendency to fall over in wind and rain), and resistance to pests and diseases. AHDB's Crop Committees each include several farmers and agronomists in addition to breeders, specialists in end-use sectors, a grain trader and a plant pathologist. This ensures that the UK's RL system focusses on the requirements of farmers, not the opinions of academic experts.

A few diseases are especially significant for farmers, and thus, for breeding. In winter wheat, resistance to Septoria tritici blotch ('Septoria') is one of only three traits with very high importance on the RL while resistances to Rhynchosporium leaf blotch ('Rhynchosporium') and net blotch are the only traits with very high importance in winter barley grown for feed; in winter malting barley, they are outweighed, rather obviously, by malting quality. For comparison, it is essential for yield to exceed a target value but so long as it does so, yield is treated as a character of high importance in wheat, barley and oats, to be balanced with other traits (AHDB, 2020).

Some varieties are given a regional or specific recommendation if either the breeder or the Crop Committee believes they will be commercially successful only in part of the UK or only for a specific purpose. This can include an unusual resistance to a pest or disease. Wolverine winter wheat was included in the 2021/22 RL with a specific recommendation for resistance to Barley yellow dwarf virus although its yield was lower than most other recommended varieties in its quality class, Hard Group 4.

4 Demand for disease resistance in cereals

Breeders can only be commercially successful if they produce varieties which are improvements on those currently available. Disease resistance is never sufficient on its own; a successful variety must combine resistance with high yield (inevitably the most important requirement), marketable quality and appropriate agronomic properties (Summers and Brown, 2013).

4.1 Why do we need resistant varieties?

There is demand for disease-resistant cereal varieties in northwest Europe for at least five reasons. First, farmers need to make a profit by minimising the cost of inputs, including fungicides and insecticides. Varieties with at least moderate resistance to all pathogens usually need fewer pesticide applications. Second, the farming industry as a whole benefits from suppressing disease. By growing resistant varieties, a farmer avoids spreading pathogens to her neighbours' fields (no doubt hoping her neighbours are equally considerate!) Third, the EU and all countries in northwest Europe have a policy of reducing the use of pesticides, so farmers increasingly need resistant varieties to control diseases because fewer effective chemicals are available. Fourth, when crops are damaged by pests and diseases, the resources used to grow the fraction of the crop that has been lost are wasted. Disease control thus contributes to reducing greenhouse gas emissions from agriculture (Berry et al., 2010). Last, resistance to fungicides has evolved in several important cereal pathogens, especially powdery mildew and Septoria of wheat, and powdery mildew and Ramularia leaf spot ('Ramularia') of barley (FRAG-UK, 2020). Economic crop production now depends on having varieties with at least moderate resistance to those diseases.

4.2 Minimum standards

The most important outcome of resistance breeding is to avoid producing 'disease suckers': varieties which are so susceptible that they spread disease to other varieties nearby with otherwise acceptable resistance. To be included in the RL, a new variety must achieve at least the minimum standard for resistance to several diseases, so that growing it does not pose an unacceptable risk for other farmers locally (AHDB, 2020). In effect, the minimum standard is a level of susceptibility where acceptable disease control can be achieved with an effective spray programme, using currently available pesticides. This is 3 for most diseases of cereals but 4 for Septoria of wheat. There is currently no minimum standard for resistance to invertebrate pests or viruses. The minimum standard takes account of the availability of resistance in breeding programmes, the timescale for improving resistance and the likely durability of resistance. The evidence from diseases such as powdery mildew of winter wheat and spring oats, brown rust of winter barley and Septoria nodorum blotch ('nodorum') of wheat is that having moderate resistance in most varieties is an effective route to achieving disease control at the national level.

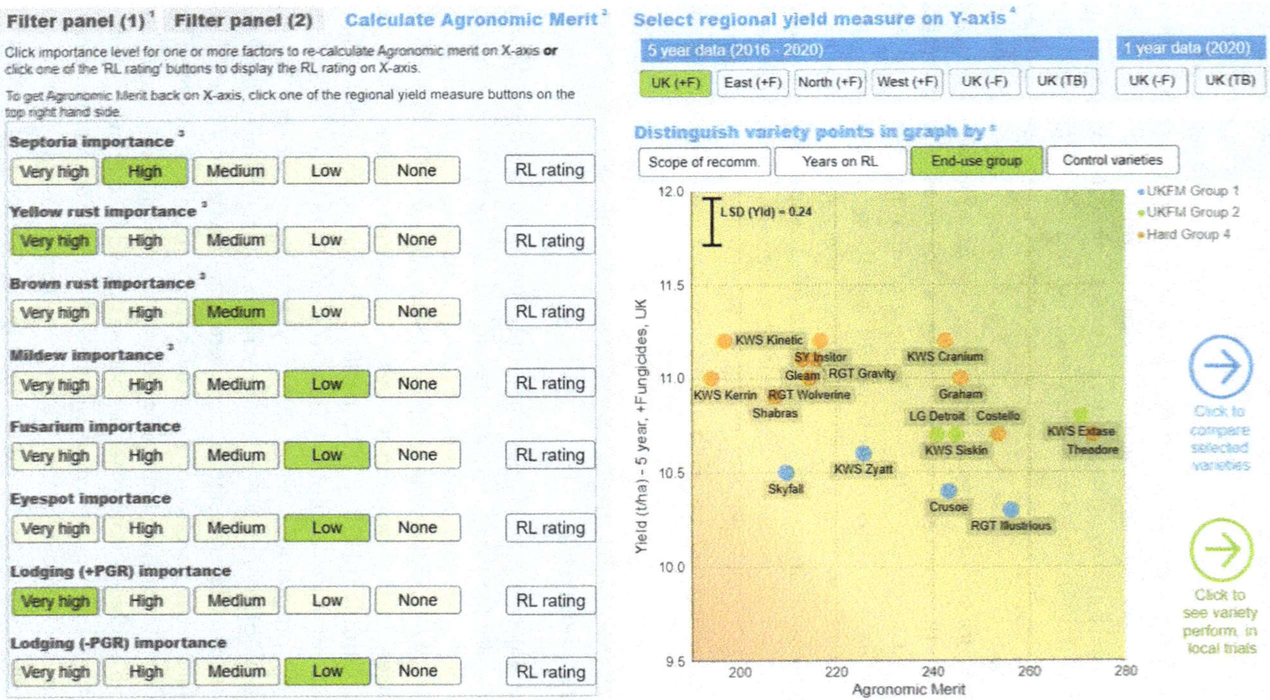

Figure 1 A screenshot of the AHDB Variety Selection Tool for wheat, plotting yield in fungicide-treated trials as a measure of yield potential against agronomic merit, including disease resistance. The default weightings of the importance of different traits, as in the AHDB Crop Committee Handbook (AHDB, 2020), have been adjusted to suit the preferences of a hypothetical farmer in the east of England who wishes to grow a hard endosperm wheat, following another crop but not maize, with a plant growth regulator to reduce lodging. The choice of varieties is based on their average yields in trials across the whole of the UK. Reproduced by courtesy of the AHDB. © AHDB.

4.3 Variety Selection Tool: choosing a disease-resistant, high-yielding variety

The AHDB has recently introduced a Variety Selection Tool to help growers choose varieties that meet their requirements (https://ahdb.org.uk/vst). Most farmers wish to balance high yield against the risk of losses if diseases and lodging are not controlled, but some aim for higher yields while accepting greater risks while others wish to minimise inputs even if that means foregoing some yield. Moreover, farmers may have their own experience of the local significance of different diseases. The Tool is thus a flexible way for farmers to apply RL data to their own requirements.

The Variety Selection Tool plots yield in fungicide-treated trials as a measure of yield potential if diseases are well controlled, against agronomic merit calculated either from RL importance weightings or according to the farmer's own requirements. Figure 1 shows this for a hypothetical farmer in the east of England who wishes to grow hard endosperm wheat, following another crop but not maize, with a plant growth regulator (PGR) to reduce lodging. She believes yellow rust is more significant on her farm than Septoria while mildew, Fusarium and eyespot are negligible. She bases her choice on the full UK-wide dataset of trial yields. Among Hard Group 4 and Group 2 varieties, KWS Cranium has one of the highest yields with greater agronomic merit than its competitors, but if she wishes to minimise risk, she could consider Theodore or KWS Extase instead. Among Group 1 varieties, KWS Zyatt has the highest potential yield while RGT Illustrious has the greatest agronomic merit.

5 Breeding elite cereal varieties with resistance to multiple diseases

Plant breeding is a greatly accelerated form of evolution by natural selection (Brown, 2015). In nature, when there is *variation* in a phenotype, the phenotype is *inherited* and the variation affects *fitness*, natural selection follows as an inevitable consequence in a large population. Similarly, a successful plant breeding programme relies on identifying parental varieties with useful traits, an efficient system for crossing lines and developing progeny generations, and effective trialling to identify the best lines. In the words of the famous Soviet geneticist N.I. Vavilov, plant breeding is 'evolution directed by the will of man'.

5.1 Variation in many traits, including disease resistance

The need to balance many traits is reflected in the very large population sizes with which cereal breeders work. In winter wheat, a typical company produces many hundred thousand or even more than a million F2 plants each year, which are reduced by stages to a few lines, often in single figures, submitted to NL trials each year. Of even this tiny subset of lines, only a minority become profitable as commercial varieties. These successful lines are exceptional because they are in the extreme of a distribution defined by many traits, controlled by rare combinations of genes. Part of the skill of a breeder is to balance the intensity of selection for all these features. This limits the potential for making rapid improvements in disease resistance in one generation or over very few generations because strong selection for a trait of moderate importance runs the risk of eliminating unusual combinations of genes which improve a more important trait (Summers and Brown, 2013). Nevertheless, the approach of making gradual improvements in the whole range of traits across successive generations has led to remarkable advances in durable resistance to most diseases of cereals in the UK and much of northwest Europe.

European breeders work with a range of sources of resistance. Crosses between leading local varieties have the highest chance of producing elite progeny lines with the traits required for current local conditions, including disease resistance, but do not introduce new genes. Crosses between varieties from different countries within northwest Europe can help to increase resistance above the level that can be achieved with local varieties alone. An example was the use of German wheat varieties to improve resistance to Septoria in the UK when it first became recognised as a major disease in the early 1980s (Arraiano and Brown, 2017). As the genetic distance between the source of resistance and the local population increases, however, the chance of introducing undesirable genes also increases. Arraiano and Brown (2017) found evidence that in the mid-1950s, when Septoria was barely known as a pathogen of wheat, genes increasing susceptibility to Septoria were inadvertently transferred into UK winter wheat from a German spring variety used as a source of increased yield and resistance to yellow rust. In more distant crosses, genes for resistance to mildew, rusts, eyespot and viruses have been introduced into wheat or barley from non-European landraces or wild grasses related to cultivated cereals. A challenge to using such exotic material routinely is that breeders must select lines which not only carry the desired resistance but also match the best varieties for yield, quality, agronomy and resistance to other diseases and pests. This requires topcrossing to the best current varieties for several generations to transfer the new resistance into a commercially competitive variety. The need to play catch-up limits the potential for breeders to use germplasm which has worthwhile resistance but is poorly adapted to local conditions; for example, it has restricted the use of resistance to Soil-borne cereal mosaic virus in European winter wheat (Summers and Brown, 2013).

It is challenging enough for breeders to decide how to balance the priority of many traits at any one time. Given that it takes nine years or more to bring a winter cereal variety from the initial cross to being fully recommended, however, it is an even greater challenge to predict what farmers and consumers will require a decade after the cross is made. Some demands are quite easy to foresee - many breeders currently aim to produce varieties with yields which will be more robust in a less predictable climate - but others are harder, not least when they involve changes in pathogen genotypes or species. Until 2010, the wheat yellow rust fungus had limited genetic diversity in northern Europe. The huge increase in diversity from 2011 onwards (Hubbard et al., 2015; Hovmøller et al., 2016) came as a great surprise and has presented new challenges for resistance breeding.

Changes in policy for regulation of food production and the environment affect farmers' decisions, including their choice of crop varieties. A foreseeable development to which breeders have been able to respond to some extent has been the removal of several fungicides from the market by EU regulations, first announced in 2008 and implemented in stages since then (Punja, 2018). It was perhaps harder to predict the ban, announced in 2018, on applying neonicotinoid insecticides to crops which are not insect-pollinated (European Commission, n.d.). This has caused an urgent demand for resistance to insect pests and insect-borne viruses. Given the timescale on which cereal breeding operates, sudden changes in policy do not help breeders to produce the varieties that meet the requirements of farmers, consumers or society in general.

5.2 Pedigrees and breeding systems

A critical feature of successful breeding is to choose parents which have desirable values of important features but are genetically diverse. This generates transgressive segregation with phenotypes which exceed the range of the parental lines, so breeders can select progeny with new combinations of genes which improve the most important traits. This is critical to improving disease resistance and any other character. The choice of parents, therefore, tends to be treated as confidential. Although the NL requires breeders to state varieties' parentage, it has become common over the last 20 years for one or both parents to be named with serial numbers which mean nothing to people outside the breeding company. This creates a minor obstacle to studying the recent history of cereal breeding although high-throughput marker systems are now capable of substantially reconstructing pedigrees (Fradgley et al., 2019).

Each company has its own preferred system for incorporating sources of disease resistance into breeding programmes to combine resistance with yield, quality and agronomic properties. They also have their own preferences for methods of advancing lines through the generations. Important considerations are to ensure that varieties are genetically uniform, to carry out selection on multiple traits efficiently, and to test enough lines to have a high chance of finding the very few which are so exceptional that they may become profitable.

The pedigree system is capable of testing very many lines for different traits while advancing them by selfing through the generations (Koebner and Summers, 2003) but a considerable amount of 'pure stocks' work is needed to minimise segregation of traits even in advanced generations, so that candidate varieties pass DUS testing. Doubled-haploid methods produce uniform lines and allow exactly the same genotype to be tested for diverse traits, but do not have the capacity of the pedigree system to produce huge numbers of progeny from which rare, exceptional lines can be selected (Weyen, 2008). Single-seed descent (Snape and Riggs, 1975), including its modern development as 'speed breeding' (Watson et al., 2018), is a widely-used and effective compromise. Lines are advanced by selfing in conditions which produce small amounts of seed quickly, to produce F4 or sometimes F5 plants which are homozygous at most loci. This system can produce more lines each year than the doubled-haploid method and the lines themselves are more uniform than in the pedigree system.

5.3 Selecting for disease resistance

Methods of scoring vary according to the disease, the purpose of the trial and the breeder's preference. There is a rather pervasive view among academics that quantitative resistance is difficult to select (to pick one example from the many papers that make such a statement, see Pilet-Nayal et al., 2017) but that is incorrect for many diseases of cereals. It may or may not be difficult to quantify accurately the amount of disease on a plant, but that is not the purpose of scoring disease in breeding trials. The main goal is to remove lines which are too susceptible to be advanced to the next generation or to official trials. At one extreme, great progress in breeding wheat for mildew resistance was made in the 1970s and 1980s by scoring plants in early generations on a two-point scale: acceptable or too susceptible. At the other extreme, semi-quantitative scores like the RL's 1-9 scale are sometimes used, but ordinal scales with a few categories such as resistant, intermediate and susceptible (sometimes with very resistant and very susceptible) are more common. Whatever the system, it must identify unacceptably susceptible varieties with some confidence and a breeder must be able to use it to score very many lines rapidly.

Most selection of resistance is based on field trialling and in the end, what really matters is varieties' performance on farms. The method which reflects farming practice most accurately is to choose sites where natural infection is predictably high so that more resistant and more susceptible varieties can be differentiated. A practical limitation is that this may require breeders to travel to several distant sites in early summer. For some diseases, infection is promoted by inoculation with the pathogen and for rusts, most breeders inoculate trials with pathogen races of particular interest. Field trials are supplemented by laboratory and greenhouse tests for some diseases. DNA markers are used by all companies to select important genes (Koebner and Summers, 2003) while the use of genomic selection is increasing (Poland and Rutkoski, 2016).

The choice of trial site is crucial for field selection of disease resistance phenotypes. For some diseases, varieties' susceptibility at a single well-chosen site gives a reasonably good estimate of their likely performance on farms. RL disease trials are run at geographically diverse sites to observe varieties' susceptibility in a range of conditions in case there is significant genotype-by-environment interaction (GxE), and the RL rating is calculated from the mean score across sites. It is not usually economic for breeders to run multiple trials for any but the most important diseases. GxE effects arise from local variation in pathogen virulence, especially in rusts and mildew, or from physiological variation in the expression of resistance, which is especially troublesome in Ramularia leaf spot of barley (Havis et al., 2015; Hoheneder et al., 2021).

A limitation on field phenotyping is the time that breeders can afford to score disease trials. This requires knowledge of the symptoms and sometimes the ability to distinguish different diseases, so there is great pressure on specialists within breeding companies to score many trials in a short period of time. There has been increasing interest in using image analysis technology, often combined with artificial intelligence, for scoring diseases in disease nurseries (e.g. Mi et al., 2020). This has the potential to increase the amount of disease data obtained at a lower cost.

6 Durable and non-durable resistance

Disease resistance which is durable is especially desirable. For farmers, the requirement is not so much for durable resistance over many years as for predictable resistance over a growing season. They do not want a previously resistant variety to become unexpectedly susceptible because this incurs costs of additional pesticide applications or substantial losses from a disease which cannot be controlled. For breeders, durable resistance protects their investment in germplasm. When a variety's resistance 'breaks down' because it has been overcome through the evolution of virulent pathogen genotypes, the breeding company may lose its investment not only in developing and marketing that variety but also other, related varieties with the same resistance gene.

6.1 Race-specific resistance

There are broadly two types of genetic system controlling disease resistance in cereals, but there is a great imbalance between how well they are understood scientifically and how useful they are in breeding. Almost all academic research focusses on major genes which have strong effects and are amenable to genetic and molecular analysis. Many of these genes, described as race-specific genes, encode receptors which induce defences when they recognise the presence of specific avirulent pathogen genotypes. The simple fact that they are overcome by pathogen mutations means that race-specific resistance is rarely durable against any disease of any crop and diseases of cereals in northwest Europe are no exception. Nevertheless, most breeders in northwest Europe are willing to use new race-specific resistances but accept that they are likely to break down sooner rather than later when virulent pathogen genotypes emerge. The situation that should be avoided is when resistance breeding relies heavily on repeated introductions of new race-specific genes, as happened in mildew of spring barley for several decades (Wolfe, 1984; Brown, 1994, 2015).

Toxin receptor genes are similar to race-specific resistance genes in biological function but not in coevolution with the pathogen. Necrotrophic toxins produced by a range of non-biotrophic fungi induce disease (rather than defence) through their interaction with a specific receptor (Faris and Friesen, 2020). Although the evolutionary biology of these interactions is less well understood than in race-specific resistance, mutations in a pathogen toxin may prevent it from interacting with the host's receptor and thus reduce pathogen virulence. Hence, resistance based on the absence of toxin receptors may be durable.

The length of time a pathogen takes to overcome a race-specific resistance depends on its mode of dispersal (Brown and Hovmøller, 2002; Wingen et al., 2013). Broadly, virulent mutants of wind-borne pathogens, including rusts and powdery mildews, can spread very rapidly over large areas while soil-borne pathogens disperse much more slowly. In the latter case, single genes for resistance to the soil-borne Barley yellow mosaic virus have been overcome by the pathogen, but the rate of dispersal of resistance-breaking viruses has been so low that breeders have had time to introduce new resistance genes (Ordon et al., 2009).

There has been significant interest among researchers in 'pyramiding' race-specific resistances, combining several such genes. This can be effective, at least temporarily, if the target pathogen has evolved virulence to each gene separately but genotypes which overcome the combination of all the resistances do not yet exist. The effectiveness of this strategy depends substantially on the pathogen's mode of reproduction. When it is asexual, breeders can introduce new race-specific resistances sequentially, keeping ahead of changes in the pathogen population. When it is conducted rigorously, this can be a durable strategy for deploying inherently non-durable resistances. It depends on having good survey data on the frequencies of pathogen virulences and races (Bayles et al., 1997; https://ahdb.org.uk/ukcpvs). Gene pyramiding is much less effective against sexual pathogens, however, because recombinant virulent genotypes can overcome the combination of race-specific resistances (Brown et al., 1993).

Unfortunately, while research on the molecular biology of resistance has accelerated, genetic analysis and mapping of resistance to most diseases of cereals in northwest Europe have all but stopped in the last 25 years. It is clear from varieties' performance that new race-specific resistances to rusts, mildew, Septoria and other diseases have been introduced into breeding programmes (Figs 3–6) but most genes have not been identified individually and have not been mapped. This makes it difficult to know whether varieties have the same resistance, how widespread a particular resistance is in breeding programmes, and if a resistant variety has a new gene or a new combination of existing genes.

6.2 Durable resistance: mostly quantitative, mostly polygenic and mostly race-non-specific

Much more important in European cereal breeding is resistance which is polygenic and quantitative, effective against all genotypes of a pathogen species, and controlled by dispersed genes with minor or sometimes moderate effects. This type of resistance has many descriptions which largely overlap. Here, I call it quantitative resistance; other common names are partial, background and minor gene resistance. It is sometimes called race-non-specific resistance because it is effective against all genotypes of a pathogen species. This lack of genotype-specificity means that quantitative resistance is usually durable. In quantitative trait locus (QTL) mapping, it is usual for the statistical analysis to identify some genes which control moderately large, significant fractions of quantitative resistance but not to account for a substantial fraction of genetic variation. Presumably, this unexplained resistance is controlled by genes with effects too small to be detected individually but one can infer that such minor genes must be present.

There is a striking contrast between the intense interest of academic scientists in race-specific resistance, which has limited value in breeding arable crops, and their comparative neglect of quantitative resistance, which has much greater practical significance but is less tractable for molecular biologists. This could be described as the Cappuccino Model of plant disease resistance: an appealing surface layer (major genes or the froth on the coffee) covering the much larger part that does the effective work (durable host defences or the body of the drink with caffeine). Despite its importance in breeding, the physiological basis of quantitative resistance is poorly understood but it may largely concern variation in downstream defence processes rather than pathogen recognition (Corwin and Kliebenstein, 2017). It is common for academic research papers to say that knowing more about the underlying mechanisms will help breeders select for disease resistance. This may involve microscopy of infection, modelling epidemiological parameters of infection processes or gene analysis expression associated with resistance, depending on the researchers' preference. The important point, however, is the simple one that breeders have successfully increased durable resistance to several diseases of cereals without such knowledge of defence mechanisms and often with little information even about the genetics of resistance. Having said that, the increasing power and rapidly decreasing costs of 'omics' technologies give academics the opportunity to restore some balance to the study of plant disease resistance.

'There is no single model for the genetic or phenotypic basis of durable resistance' (Johnson, 1993) and there are some important exceptions to the conflation of race-specific resistance with major genes, and quantitative, race-non-specific resistance with minor genes. A small minority of major genes control host defence rather than pathogen recognition, much the best known being *mlo* in spring barley, which gives near-complete control of powdery mildew (Kusch and Panstruga, 2017). It is not only genetically tractable but it confers durable resistance and is thus of great interest to both academics and breeders. Conversely, while almost all well-studied race-specific recognition genes have large effects against avirulent pathogens, there is wide variation in the effectiveness of such genes. This is apparent in many cereal diseases, including Septoria tritici blotch of wheat, where *Stb6* and *Stb15* have stronger effects on avirulent fungal isolates than other *Stb* genes (Brown et al., 2015). In powdery mildew of barley, some race-specific genes which have been named and mapped confer incomplete resistance or a high incompatible infection type (Jørgensen, 1994). There are similar examples in all the cereal rusts. Cowger and Brown (2019) suggested that the somewhat misleading view that all race-specific resistances are highly effective arose because breeders have chosen to use genes with the strongest effects, and that this has been reinforced by academics choosing to study the same genes because they are the most tractable.

Reviewing the durability of durable resistance, Cowger and Brown (2019) concluded that the rate of pathogen evolution to virulence on race-specific resistances with minor or moderate effects is slow enough for breeders to keep up with changes in the pathogen population. In practical terms, therefore, minor genes with race-specific effects can be treated as durable over a decade or more.

Race-specific recognition genes with strong effects are epistatic to variation in minor-gene, quantitative resistance. In some diseases, most strikingly powdery mildew of spring barley in Europe from the 1950s to the 1980s, reliance on such major genes has masked variation in quantitative 'background' resistance or susceptibility. This has led to spectacular breakdowns of resistance, when virulent pathogen genotypes have overcome a major gene in a variety with little background resistance. It is therefore unwise (and usually unprofitable) to rely excessively on major race-specific resistances.

By contrast, when used effectively, race-specific resistance and quantitative resistance can work together in breeding. A strong major gene can give a variety a high rating on the RL and thus a boost in the early stages of its commercial life. When that resistance breaks down, as is almost inevitable, high quantitative resistance means the disease can still be controlled and the variety can still be profitable (note the brown rust resistance of NFC Tipple spring barley in Fig. 4, the crown rust resistance of Mascani winter oats in Fig. 5 and the brown rust resistance of Gladiator winter wheat in Fig. 6). Genomic selection offers a solution to the long-standing problem of knowing when strong race-specific resistance is combined with high quantitative background resistance. The race-specific resistance gene can be identified by its effectiveness against diverse pathogen genotypes (or possibly by molecular sequence data) while genomic selection can predict the quantitative resistance that is likely to remain when the major gene has broken down (Poland and Rutkoski, 2016).

6.3 Standing on the shoulders of giants

It is important to appreciate that disease resistance is not achieved in a single step. As with yield and many other traits, progress in breeding for durable quantitative resistance has come from applying effective selection in diverse populations over several generations. Nor is breeding a matter of chance: it is a misleading caricature to say that breeders 'cross the best with the best and hope for the best' (e.g. Podevin et al., 2013 – again, citing just one of many papers that make such statement). Of course, the actual F2 genotypes present in a population have random reassortments of the parental genes, but the choice of parents, selection methods and breeding system allow continual advances to be achieved so long as population sizes are large enough for a few really outstanding lines to be identified and developed as commercial varieties, and to be used as the next generation of parents. A more accurate metaphor is that breeders build on their own past successes and those of their predecessors: 'standing on the shoulders of giants'.

7 Resistance to biotrophic fungal diseases

Cereal breeders in northwest Europe have been successful in producing varieties with resistance to multiple diseases. In winter wheat and winter barley, breeders in the UK have increasingly produced varieties which are not susceptible to any of the main foliar fungal pathogens. Currently, on the 2021/22 RL, more than 70% of varieties have ratings of 5 or higher for all these diseases.

In wheat, this reflects improvements in Septoria resistance in particular. In every year since the mid-1980s, the RL for both crops has included at least one variety with at least moderate resistance to all the main foliar diseases, indicated by a rating of 6 or higher (Fig. 2).

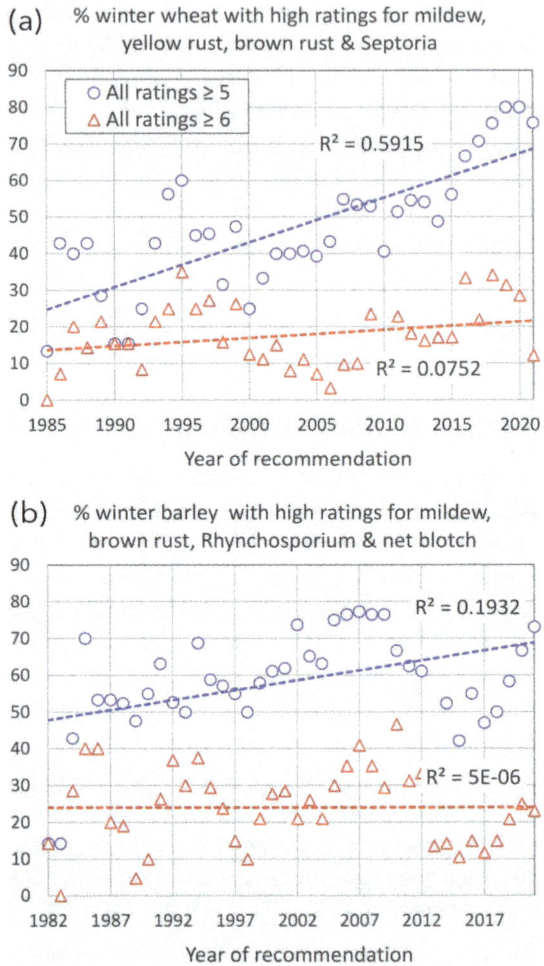

Figure 2 The proportion of varieties of (a) winter wheat and (b) winter barley on the AHDB Recommended List which are rated as not susceptible to any of the main foliar diseases of each crop with a rating of 5 or higher (blue circles) or as resistant to all these diseases with a rating of 6 or higher (red triangles). The graphs start in the years when Septoria tritici blotch resistance was first rated in wheat (1985) and net blotch resistance was first rated in barley (1982). The large decrease in the number of winter wheat lines with broad-spectrum resistance between 2020/21 and 2021/22 is the result of a major recalibration of wheat yellow rust ratings, not a sudden drop in varieties' resistance.

7.1 Powdery mildew of spring barley: 'From the ashes of disaster...'

Powdery mildew, caused by the ascomycete fungus *Blumeria graminis*, is a potentially significant disease of barley, wheat and oats, especially on lighter land in cool, damp environments and in crops given large amounts of inorganic nitrogen fertiliser (https://ahdb.org.uk/powderymildew). It used to be one of the most important diseases of cereals in northwest Europe but has been well controlled by breeding, especially in spring barley, winter wheat and spring oats. Powdery mildew fungi are notorious for the speed with which they evolve insensitivity to fungicides so breeding for resistance has been essential for success in controlling this formerly troublesome disease. Mildew of the major small-grain cereal crops is caused by different forms (*formae speciales*) which are highly specialised on their host species, with *B. graminis* f.sp. *avenae* on oats, f.sp. *hordei* on barley and f.sp. *tritici* on wheat. There is similar specialisation in several rust fungi.

The history of mildew resistance in spring barley illustrates vividly the failures and successes of different breeding strategies. Until about 1990, mildew was much the most important disease of barley in northwest Europe and indeed one of the most significant of all crop diseases. The source of the problem was that breeders relied heavily on major recognition genes, especially in spring barley. There was a substantial research effort to discover new *Ml* genes for mildew resistance in several countries, notably Denmark and the former East Germany, and these genes were introduced into European varieties at an average rate of one every three years between the mid-1950s and the mid-1980s. They included several alleles of the complex *Mla* gene as well as other genes, while *Mlh* and *Mlra* were largely used in winter barley. Most of these genes were introduced from landraces, mainly from outside Europe, although *MlLa* came from a wild barley species, *Hordeum laevigatum* (Hilbers et al., 1992). Several genes were introgressed more than once from different sources (Jørgensen, 1994). All these genes were rapidly overcome by mutation of the pathogen to virulence, sometimes in as little as two years after the release of the first varieties carrying them. Breeders increasingly used combinations of these genes ('gene pyramiding') but they proved no more successful. During this period, mildew resistance in spring barley showed the characteristics of a textbook boom-and-bust process, over and over again (Wolfe, 1984; Brown, 1994; Fig. 1a in Brown, 2015). In short, major recognition genes have made no contribution to durable control of barley mildew.

7.2 Powdery mildew of spring barley: '...grow the roses of success'

The situation changed radically from the early 1980s onwards, for two reasons. First, mildew resistance in spring barley provides one of the very few instances in plant pathology of a highly effective, durable, major gene. Recessive alleles of *Mlo* discovered as artificial mutations in the early 1940s confer almost complete resistance to mildew (Kusch and Panstruga, 2017). Initially, resistance was associated with more or less severe necrotic flecking which reduced yield. Although lines with less necrosis and higher yield could be selected, most mutant *mlo* alleles were unsuitable for use in breeding. Subsequently, a natural mildew-resistance allele in an Ethiopian landrace, *mlo11*, was introgressed successfully into spring barley breeding throughout northern Europe (Jørgensen, 1992). It is likely that *mlo11* is currently present in at least half of all varieties, covering at least half the spring barley growing area. In addition, a few varieties with an artificial mutation, *mlo9*, have been released.

Although the MLO protein is required for susceptibility of barley to powdery mildew, its biological function is not yet known. It is not a member of either of the main classes of recognition gene, the nucleotide-binding leucine-rich repeat protein (NLR) or receptor-like kinase (RLK) families. MLO is required for the fungus to successfully penetrate the plant's epidermal cells so it provides an example of resistance achieved by mutating a gene required for host susceptibility (Kusch and Panstruga, 2017). Pathogen genotypes with some adaptation to *mlo* resistance have been detected but the extent of adaptation has been minor and has not prevented *mlo* from being highly effective (Schwarzbach et al., 2002).

About the same time as *mlo11* and *mlo9* were first used on a large scale in the 1980s, interest also grew in selecting varieties with partial resistance to mildew, in the expectation that this would enable barley breeders to escape the endless treadmill of discovering, using and losing major recognition genes. Most current spring barley varieties which have a functional, mildew-susceptibility allele of *Mlo* have very high quantitative resistance. This advance may have been facilitated by fewer major race-specific genes being used, which has made it possible to detect variation in quantitative 'background' resistance. By one route or other, the importance of mildew on spring barley has been greatly diminished. It is given a weighting of high importance in the RL system because few fungicides effective against barley mildew are now available, so it is essential for this level of resistance to be maintained.

7.3 Powdery mildew of wheat, oats and winter barley

Breeding for resistance to powdery mildew of wheat (*B. graminis* f.sp. *tritici*) has been outstandingly successful in the UK and most other countries in northern Europe. Until the early 1980s, mildew was one of the most serious foliar diseases of wheat and it was still troublesome until the mid-1990s. Breeding for resistance has reduced its importance so much that it can be hard to find in crops in the dryer parts of the UK. Wheat mildew is now resistant to most fungicides so, like barley mildew, it is important for this generally high level of resistance to be maintained.

Some major genes were used in breeding both winter and spring wheat for mildew resistance between the 1950s and the 1980s (Bennett, 1984). Since the mid-1980s, several other, unnamed race-specific resistances have entered UK wheat breeding programmes, whether introduced intentionally or otherwise, but they have not contributed significantly to long-term control of mildew, and there has been little investigation of their genetics. Some of these, such as the mildew resistance genes in the wheat varieties Shamrock and Robigus came from a large set of lines derived from *Triticum dicoccoides* in the Netherlands, which were first used by Dutch breeding companies and then more widely (Lange and Jochemsen, 1992). The same *dicoccoides* material has contributed genes for resistance to yellow rust, brown rust and Septoria but so far as I am aware, none of them have been durable.

Quantitative resistance has been much more important than major genes in breeding for mildew resistance. It has been achieved by phenotypic selection over many years despite very limited knowledge of genes controlling quantitative resistance. In the UK, where breeding for mildew resistance in winter wheat and spring oats has been especially successful, no single quantitative resistance gene has been identified in either crop. With only a few exceptions, breeding for quantitative resistance to mildew in winter wheat has become a 'virtuous circle' because most parental lines have at least moderate resistance, so it is easy to select progeny with similar resistance. Most current varieties have RL ratings of 6 or 7 and many varieties from the mid-1980s onwards have had a rating of 8. Only two varieties recommended in the last 40 years have had a rating of 3. This degree of durable resistance protects varieties against severe epidemics when a major gene breaks down, as in Robigus, which had a new, unidentified mildew resistance gene, and its daughter variety Glasgow (Fig. 1b in Brown, 2015). Most spring wheat is grown in the south and east of England where severe mildew epidemics are now almost unknown and all currently recommended varieties have moderate to high mildew resistance with ratings from 6 to 8. A challenge for breeders now is to find trial sites where mildew epidemics occur regularly so that selection for resistance can be conducted reliably.

As the resistance allele is recessive, natural *mlo* mildew resistance is unknown in hexaploid wheat because the functional, wild-type, mildew-susceptibility allele in one genome complements *mlo* mildew resistance mutations the other genomes. Several workers have mutated or inactivated all three *Mlo* genes in bread wheat to generate complete resistance to mildew (Wang et al., 2014; Acevedo-Garcia et al., 2017) and both genes in durum wheat (Ingvardsen et al., 2019). This is undoubtedly an impressive technical feat although its practical benefit is perhaps less clear, at least in northwest Europe.

Despite its prevalence in spring barley, *mlo* has not been used successfully in winter barley breeding despite several attempts to introduce it, which suggests there may be a more significant penalty of *mlo* mildew resistance in winter than spring cereals. Moreover, quantitative resistance to mildew in wheat, which is no longer a major disease in many areas, has been one of the leading success stories of plant breeding in northern Europe. This does not mean that multiple

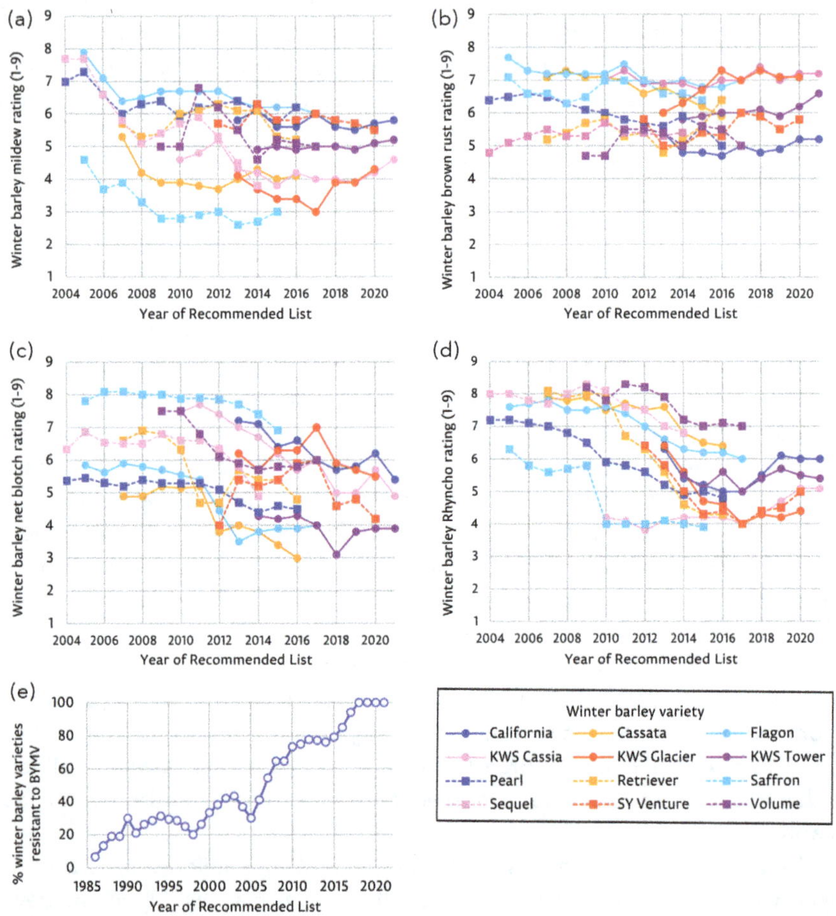

Figure 3 Ratings of winter barley varieties on the AHDB Recommended List for resistance to (a) powdery mildew, (b) brown rust, (c) net blotch and (d) Rhynchosporium over the period that the RL has been produced in its present form from 2004/05 to 2021/22. Varieties which were on the RL for at least seven years during this time are included. (e) The proportion of winter barley varieties on the RL with resistance to Barley yellow mosaic virus from the introduction of Sonate, the first resistant variety. Resistance is presumably conferred by *rym4*.

knockouts of the wild-type *Mlo* gene may not be useful elsewhere in the world, possibly in spring wheat by analogy with the use of *mlo* in spring barley.

Mildew resistance in winter barley also relies on polygenic, quantitative resistance although not at successfully as in wheat (Fig. 3a). Current varieties have ratings between 4 and 7 with only LG Flynn having 3. While it may be

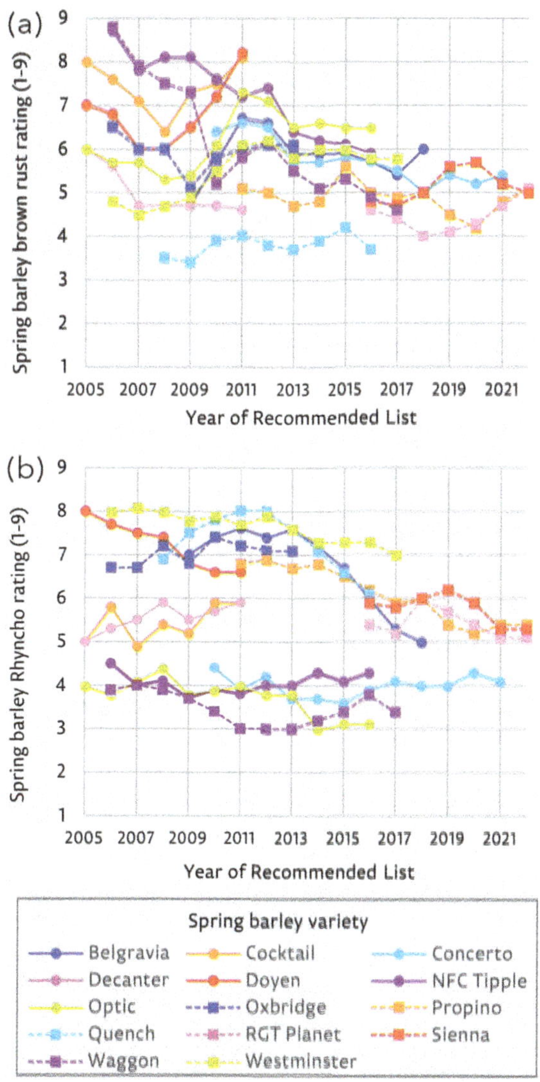

Figure 4 Ratings of spring barley varieties on the AHDB Recommended List for resistance to (a) brown rust and (b) net blotch between the 2004/05 and 2021/22 RL. Varieties which were on the RL for at least seven years during this time are included.

desirable to increase the average level of mildew resistance, breeders currently focus on Rhynchosporium and net blotch. *mlo* incurs significant fitness penalties (reviewed by Brown and Rant, 2013; also see McGrann et al., 2014; Gruner et al., 2020; Jacott et al., 2020) so one can speculate that in winter barley, the costs of this resistance may outweigh its benefits.

Mildew resistance in oats has moderate importance and has depended almost entirely on quantitative resistance. This has been especially successful in spring oats (Fig. 5a), which are on a par with winter wheat, as most popular varieties have moderate to high quantitative, durable resistance. Mildew resistance has been weaker in winter oats (Fig. 5b).

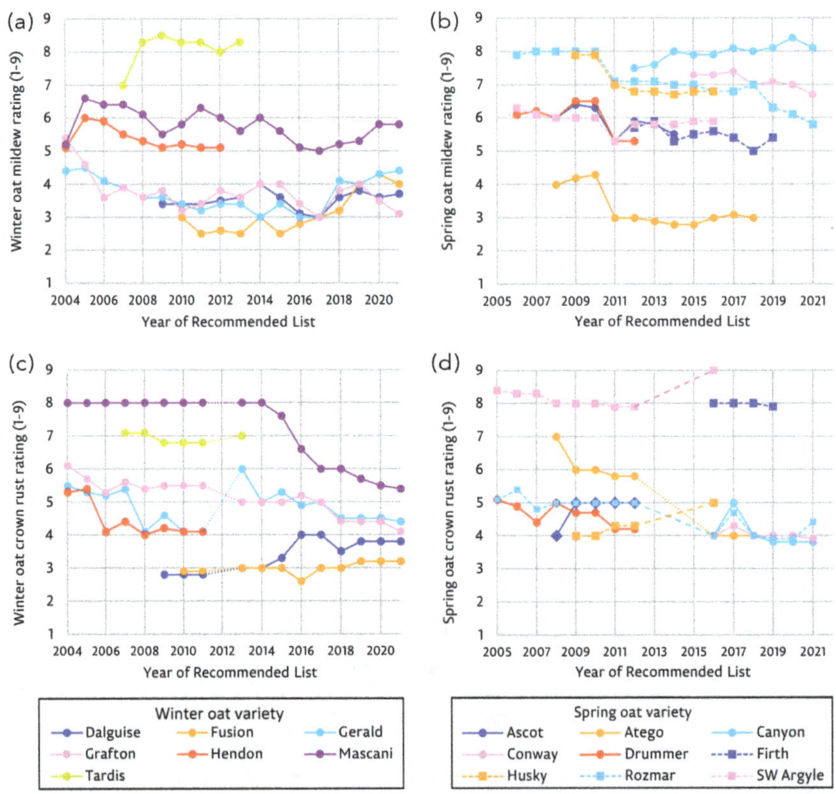

Figure 5 Ratings of oat varieties on the AHDB Recommended list for the two main diseases between 2004/05 and 2021/22. (a) Resistance to powdery mildew of winter oats, (b) powdery mildew of spring oats, (c) crown rust of winter oats and (d) crown rust of spring oats. Varieties which were on the RL for at least seven years during this time are included. Low disease levels meant that crown rust ratings were not produced in 2012/13 for winter oats or in 2012/13, 2013/14 and 2014/15 for spring oats.

7.4 Yellow rust and brown rust of wheat: 'The times they are a-changing'

Wheat in northwest Europe is currently affected by two rust diseases: yellow rust, also known as stripe rust, caused by *Puccinia striiformis* f.sp. *tritici* (*Pst*), and brown or leaf rust (*Puccinia triticina*). Yellow rust was historically more severe in cooler northerly areas of Europe but recent changes in the climate and the pathogen population have increased its importance more widely. Brown rust continues to be the dominant rust disease in warmer, southerly areas within northwest Europe but climate change is leading to more severe epidemics further north.

Breeding for wheat yellow rust resistance has continued in England for nearly 120 years, longer than any other disease of cereals. It currently has high importance for variety recommendation. Although yellow rust can cause yield losses comparable to Septoria, it can be controlled by fungicides on varieties with moderate resistance. All breeders run nurseries to select yellow rust resistance, either exposed to natural infection or inoculated with specific *Pst* isolates. Eastern England from Cambridge to York is the prime hot-spot for wheat yellow rust in the UK and indeed Europe, and is thus a favoured location for breeders' trials.

The history of breeding wheat in northwest Europe for yellow rust resistance divides into two phases. Until 2010, the *Pst* population in this region was entirely asexual (Hovmøller et al., 2002). Major genes were widely used to control yellow rust, including *Yr1*, *Yr2*, *Yr3*, *Yr4*, *Yr6*, *Yr9*, *Yr17* and *Yr32* while *Yr7* and *Yr8* were regionally significant. Several other major gene resistances were used in important varieties but were not identified or mapped. In several countries, there have been surveys of the *Pst* population to monitor changes in pathogen virulence to major genes in current varieties. The UK Cereal Pathogen Virulence Survey has been run continuously since 1967, following the spectacular breakdown of yellow rust resistance in Rothwell Perdix (Johnson, 1992). Up to 2010, the clonal population of *Pst* evolved in a step-wise manner so pathogen survey data could indicate if new varieties were at risk of attack from the limited range of races of *Pst* in the current population (Bayles et al., 1997).

The situation changed dramatically in 2011 when a new population of *Pst* with much greater genetic diversity infected wheat crops throughout western Europe. It was later discovered that these genotypes originated in eastern Asia, where *Pst* is a sexual organism with *Berberis* species (barberry and mahonia) as alternate hosts (Hovmøller et al., 2016). The new fungal population rapidly overcame major resistances in Warrior and Torch, and there was great concern that much of the yellow rust resistance in existing European varieties could become ineffective. There were further significant changes in genetics and virulence of the *Pst* population in 2014 and 2016 but it is not known if the new genotypes were selected from the population which arrived shortly before 2011 or from fresh incursions. The geographic range over which yellow rust is significant has expanded and the severity of yellow rust on established varieties has increased. This might have resulted from wider environmental adaptation among diverse *Pst* genotypes and from natural selection for greater aggressiveness (Pope de Vallavieille et al., 2018). Such adaptive variation has also been reported in *Pst* within France (Enjalbert et al., 2005). Although the new *Pst* population originates from a sexual population which has much greater capacity for telium production than the pre-2011 clonal races in northern Europe (Ali et al., 2010), there is as yet no evidence that it can reproduce sexually on *Berberis* in Europe.

Although the new *Pst* population overcame some major genes, perhaps the most remarkable feature of the new situation is that the quantitative resistance of many varieties continues to be effective while some varieties, notably Crusoe and Revelation, have very strong resistance which has been durable through the large genetic shifts in the *Pst* population in 2014 and 2016 (Fig. 1a in Cowger and Brown, 2019). In addition, Gladiator, which was on the RL until 2010, continues to be important as a source of durable resistance to yellow rust. Varieties with apparently durable resistance to yellow rust have also been identified in France, including Soissons and Scipion (Perronne et al., 2017). Evidently, durable resistance to the old clonal population was largely effective against the new, diverse *Pst* population. In retrospect, perhaps this should not have been surprising: if quantitative resistance does indeed largely involve downstream defences rather than pathogen recognition, it may be effective against all genotypes of a pathogen species and thus can control a new *Pst* genotype from East Asia just as well as an indigenous race from northwest Europe. The greater severity of epidemics caused by new, more aggressive genotypes of *Pst* means that recent advances in quantitative resistance to yellow rust (Fig. 1d in Cowger and Brown, 2019) need to be sustained.

An interesting feature of the new type of yellow rust is that some varieties have become more resistant during one or other of the three major recent changes in the *Pst* population, for example, JB Diego in 2011, KWS Santiago in 2014 and Viscount in 2016 (Fig. 1a in Cowger and Brown, 2019). Presumably, these varieties had recognition genes which were moderately effective against new *Pst* populations or there is some variation in the effectiveness of varieties' quantitative resistance to different fungal genotypes.

Breeding wheat for brown rust resistance has followed a similar pattern to yellow rust in the United Kingdom, with breeding for quantitative resistance punctuated by occasional breakdowns of major genes (Fig. 6). The average level of resistance to brown rust in the United Kingdom has not been as high as to yellow rust but it is a less important disease in most parts of the country. Almost all varieties on the 2021/22 RL have ratings between 5 and 8 except for Crusoe with 3 and LG Astronomer with 9; it is not known whether LG Astronomer has a major gene and thus may be at risk of breakdown or has outstanding quantitative resistance.

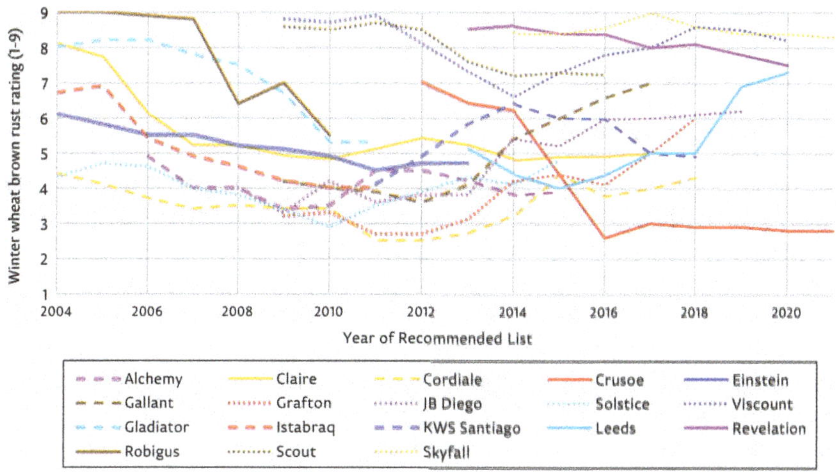

Figure 6 Ratings of winter wheat varieties on the AHDB Recommended List for resistance to brown rust between 2004/05 and 2021/22. Varieties which were on the RL for at least eight years during this time are included, as well as Robigus as an example of a strong resistance 'breaking down' owing to the evolution of virulent pathogen genotypes.

Less is known about the European *P. triticina* population than about *Pst* but it is being studied in a project funded by the EU, RustWatch, from 2018 to 2022 (https://cordis.europa.eu/project/id/773311). It is striking that as with yellow rust, the brown rust resistance ratings of some wheat varieties have risen over time. There was a rapid rise in the rating of Leeds between 2018 and 2019 and more gradual increases in the ratings of KWS Santiago, Grafton, Gallant and Viscount over several years in the 2010s. One would not normally expect resistance ratings of widely-grown varieties to rise significantly during their commercial lifetime, so these changes suggest that there may have been incursions of new populations of *P. triticina* into northwest Europe with less virulence or lower aggressiveness to some varieties, just as in *Pst*.

Surveys of variation in rust virulence are as important now as they were in the era of the older, simpler *Pst* population (Bayles et al., 1997; https://ahdb.org.uk/ukcpvs). One of the most important activities of the UKCPVS is to make a representative selection of *Pst* and *P. triticina* isolates for breeders to use in rust trials. This selection is updated each year and trials with diverse, relevant isolates have made a critical contribution to breeding for durable resistance to both yellow rust and brown rust.

A challenge for releasing resistant varieties is that quantitative resistance is sometimes not strongly expressed in seedlings. It is understandable that farmers who have suffered at the hands of yellow rust epidemics on Torch, Warrior and other varieties and brown rust on Stigg or less dramatically Crusoe (Fig. 6) may not completely trust the assurances of agronomists that durable, adult-plant resistance will protect their wheat fields. Hopefully, continuing success in advancing durable rust resistance will increase farmers' confidence in new varieties.

7.5 Brown rust of barley

The most important rust disease of barley in northwest Europe is brown rust (also known as leaf rust), caused by *Puccinia hordei*. Even so, it only has moderate importance for both winter and spring varieties on the UK's RL. Like yellow and brown rusts of wheat, it can be controlled quite easily with fungicides. Breeding winter barley for durable resistance to brown rust has been remarkably successful. Almost all recent varieties have had moderate to high quantitative resistance and ratings have been largely stable (Fig. 3b). The most significant exception was Valerie, which fell from 8.5 to 6.0 between the 2020/21 and 2021/22 RL. Quantitative resistance has not been as strong in spring barley as in winter barley in recent years, although varieties' ratings have been largely stable (Fig. 4a).

7.6 Yellow rust of barley

Where has all the yellow rust of barley gone? Until the 1980s, it was a significant disease although less so than yellow rust of wheat. It is still found occasionally in areas where wheat yellow rust is also important but at very low levels. It is an important disease of barley elsewhere in the world, while wheat yellow rust has become more significant than ever. Might this be an unheralded achievement of cereal breeding: the near-elimination of a formerly major disease by durable resistance?

7.7 Crown rust of oats

In contrast to mildew, popular winter oat varieties are slightly more resistant to crown rust (*Puccinia coronata*) on average than spring oats although most varieties are intermediate to moderately susceptible (Fig. 5c and d). Resistance is largely quantitative and durable. Mascani may have a major gene which broke down between 2014 and 2016 but it has been protected by moderate quantitative resistance. International research on crown rust resistance has largely focussed on major genes, and it is only recently that academics have paid significant attention to durable, quantitative resistance to crown rust, especially in Australia (Nazareno et al., 2018). This is surely an area for future development.

8 Resistance to non-biotrophic fungal diseases of wheat

In northwest Europe, the most significant diseases of wheat caused by non-biotrophic fungi are Septoria (*Zymoseptoria tritici*, formerly *Mycosphaerella graminicola*), nodorum blotch (*Parastagonospora nodorum*), tan spot (*Pyrenophora tritici-repentis*), eyespot (*Oculimacula yallundae* and *Oculimacula acuformis*) and Fusarium head blight (*Fusarium graminearum*, *Fusarium culmorum* and other *Fusarium* species; Xu and Nicholson, 2009). Of these, Septoria is the most economically important disease of wheat in northwest Europe as in many other parts of the world.

8.1 Septoria tritici blotch

Septoria is most significant in regions with mild winters and warm, damp springs, including much of northwest Europe. Several factors have combined to make Septoria a disease of very high importance in winter wheat. First, climate change has reduced the number of nights with sub-zero temperatures in November and December, the critical period for establishment of Septoria epidemics by *Z. tritici* ascospores (Hunter et al., 1999). Second, *Z. tritici* is highly sensitive to sulphite in rainwater (Chandramohan and Shaw, 2013), which has declined as less coal has been burnt for power generation. Third, in the mid-1950s, before Septoria was even recognised as a disease of wheat in Europe, genes which increase susceptibility or reduce resistance to Septoria were inadvertently introduced in a breeding programme which subsequently fed into all wheat pedigrees in the UK (Arraiano and Brown 2017). Lastly, Septoria is the main target for fungicide applications on wheat and as a result, *Z. tritici* has become increasingly resistant to most fungicides (FRAG-UK, 2020). In the UK, Septoria is most severe in southwest England, where winters are mild and spring rainfall is plentiful. It is especially destructive in southeast Ireland, which is thus a favoured location for Septoria trials. Spring wheat is much less affected by Septoria because it avoids infection by ascospores in late autumn.

Breeding for Septoria resistance has relied almost completely on selection for quantitative resistance, which has been durable (Fig. 1d in Brown, 2015) . The average level of resistance has increased greatly in the UK over the last 35 years. In 1985, all fully recommended winter wheat varieties had ratings or 3 or 4 but the last variety with a 3 rating entered the RL in 1989. Only two varieties on the 2021/22 RL have a rating of 4, while 13 of the 33 varieties have ratings of 7 or 8. Even so, breeders will need to produce varieties with ratings between 6 and 8 consistently to achieve good Septoria control without having to rely routinely on fungicides.

Selection for Septoria resistance uses trial sites in mild, humid regions although a significant new challenge is the spread of yellow rust into areas where it was not formerly important. Not only is it inherently difficult to select resistance to one disease when another pathogen is present at high levels but also, it is surprisingly easy to confuse Septoria and yellow rust at certain stages of the two diseases' life cycles. There are many QTL for Septoria resistance in

European wheat (Brown et al., 2015; also see Riaz et al., 2020) but as none of them are especially important, it is unlikely that breeders use marker-assisted selection for specific genes. Knowledge of mechanisms of host defence against Septoria is limited (O'Driscoll et al., 2014), but that is not very important in the context of selecting for quantitative advances in varieties' resistance. Septoria resistance, as a highly important trait with complex genetic control, is very much the type of character which can be advanced by genomic selection. Even so, phenotypic selection for quantitative resistance may continue to be the mainstay of improvements in Septoria resistance but breeders will need to maintain and even expand the genetic diversity in their plant material to continue forming new, more effective combinations of minor genes to improve Septoria control.

Several major genes were present in UK wheat pedigrees up to 2000, notably *Stb15*, *Stb6* and, in spring wheat, *Stb9*, but they played little if any role in field resistance (Arraiano et al., 2009; Arraiano and Brown, 2017). Since then, some major genes have been used but probably only *Stb16q* has been introduced deliberately in the last decade (Saintenac et al., 2021). As with major genes for rust and mildew resistance, it was overcome within just a few years (Kildea et al., 2020). An unmapped gene was introduced in Cougar, possibly from a line derived from *T. dicoccoides*, but virulent isolates were detected in the UK in 2015 (Caiazzo et al., 2019). Higher levels of Septoria than expected from RL ratings were detected on varieties bred from Cougar in Ireland in 2020 (Kildea et al., 2021) and there was a similar report from the UK in 2021 (AHDB, 2021). Cougar-virulent isolates were recovered from such varieties in Ireland (Kildea et al., 2021). As ever with major genes, the Cougar resistance may have selected virulent genotypes of *Z. tritici* and is thus unlikely to be durable. No doubt such major genes will continue to be introduced from time to time and, like the mildew resistance of Robigus wheat or the brown rust resistance of NFC Tipple barley, they may make some contribution to disease control if they are combined with durable, polygenic, quantitative resistance.

Disease escape is important in the susceptibility of wheat varieties to Septoria. As a splash-borne pathogen, *Z. tritici* spreads upwards between leaf layers in the crop as conidia are carried by heavy raindrops. Shorter plants with closer-spaced leaves therefore tend to be more susceptible to Septoria while varieties with earlier leaf emergence also tend to be more susceptible as their upper leaves, which are critical for yield formation, can suffer more extensive symptoms (van Beuningen and Kohli, 1990; Arraiano et al., 2009).

8.2 Septoria nodorum blotch

Nodorum or glume blotch used to be an important disease of leaves and glumes of wheat. It is another disease which has been successfully controlled by breeding in northwest Europe and is now only significant in Norway (Cowger et al., 2020). In the UK, it has become so insignificant that since 2021/22, varieties have not been rated for resistance to it (Fig. 1c in Brown, 2015). Resistant varieties were bred by phenotypic selection in the absence of detailed knowledge of genetics and information about the presence of toxin-sensitivity genes in UK varieties has only recently become available (Downie et al., 2018). In Norway, the toxin-sensitivity gene *Tsn1* was present in 45% of varieties and the matching SnToxA toxin in 69% of fungal isolates. There was some association between the presence of *Tsn1* and field susceptibility to nodorum, suggesting that resistance of Norwegian wheat could be raised by selecting against *Tsn1* (Ruud et al., 2018; Lin et al., 2020).

8.3 Tan spot

Tan spot is a significant fungal disease in regions of northwest Europe with warm summers, including southern Scandinavia and parts of Germany and France but it is less severe there than in eastern Europe or central Asia. Resistance is, as usual, polygenic with identified QTL having moderate, additive effects. As in nodorum blotch, toxin receptor genes promote susceptibility to tan spot (Kollers et al., 2014). In the UK, tan spot is quite common but it rarely causes serious crop losses and can be controlled by agronomic methods. As it is transmitted from stubble and crop debris, tan spot may have been stimulated by the trend to cultivate by minimum tillage rather than ploughing (https://ahdb.org.uk/tanspot).

8.4 Eyespot

Eyespot is a disease of moderate importance which attacks the stem base of cereal plants, causing lodging. Although it affects all cereals, it is most significant on wheat. As it is transmitted on crop debris, it is more time-consuming to run field trials for eyespot resistance than for wind-borne or splash-borne diseases. Two genes for eyespot resistance are present in UK winter wheat, of which the more important is *Pch1* on a chromosome segment introgressed from *Aegilops ventricosa* onto chromosome 7D of wheat (Pasquariello et al., 2017). Varieties with *Pch1* on the 2021/22 winter wheat RL have a mean rating 2.3 points higher than those without and breeders are able to test large populations for molecular markers closely linked to *Pch1*. A less potent gene, *Pch2* on chromosome 7A, was introduced into UK germplasm in Cappelle Desprez, the dominant variety in the early 1960s. *Pch1* and *Pch2* aside, resistance to eyespot is, as usual, controlled by polygenes dispersed throughout much of the genome (Lewien et al., 2018). The difficulty of trialling large populations and thus accumulating polygenes for resistance may explain why progress with selecting eyespot resistance has been slower than for most other wheat diseases. It might seem surprising that *Pch1* has not been used more widely, as it is only present in three of the 35 varieties on the 2021/22 RL. This reflects in part the moderate importance of eyespot, so breeders have higher priorities but in addition, the *Ae. ventricosa* segment bearing *Pch1* also carries genes which reduce yield. Battalion, recommended in 2007/08, was probably the first commercially successful variety to have *Pch1* without the yield penalty (Summers and Brown, 2013) but this has not resulted in a surge of similar varieties.

8.5 Fusarium head blight

Fusarium head blight (FHB) is less important in northwest Europe than in areas of central and eastern Europe with hotter summers but it can be significant in wheat following a maize crop. More important than the effect of FHB on yield, however, is the requirement to reduce the level of the fungal toxin deoxynivalenol (DON) in harvested grain to as close to zero as possible. In the UK, FHB is of medium importance but that does not reflect its significance to breeding companies with operations throughout Europe. Breeders run field trials for resistance in areas where FHB epidemics are predictable, such as central Germany. This is combined with selection for anther extrusion, a trait strongly associated with reduced severity of FHB, while marker-assisted selection for QTL of moderate importance is economically worthwhile, given the high cost of running field trials for this disease. Breeders use the gene *Fhb1* because when pathogen pressure is low to moderate, it prevents the fungus spreading into the ear and thus limits DON accumulation (Buerstmayr et al., 2020). There may be scope for improving FHB resistance in future by selecting mutations in genes which promote susceptibility (Hales et al., 2020).

9 Resistance to non-biotrophic fungal diseases of barley

9.1 Net blotch and Rhynchosporium

Barley is affected by several non-biotrophic leaf spotting diseases, with symptoms that can be difficult to distinguish, even for an experienced pathologist. Net blotch resistance is one of the most important traits in winter barley, with high importance in malting barley and very high importance in feed barley, but is not considered important in spring barley. It is caused by two forms of one fungal species, *Pyrenophora teres* f. *teres* and f. *maculata*, which produce the netting and spotting forms of net blotch respectively. There has been more research on the more widespread netting form, although the spotting form is also common. Resistance to net blotch is polygenic and as usual, is characterised by several QTL with effects large enough to be identified as single genes, in addition to a presumably large number of dispersed genes with small effects which have not been detected individually. Some resistance genes with larger effects have specific interactions with pathogen isolates, reflecting a possibly complex set of interactions between pathogen toxins and host receptors (reviewed by Clare et al., 2020).

Both winter and spring barley are attacked by *Rhynchosporium commune*, an ascomycete fungus which causes a disease known variously as leaf blotch, scald or most commonly, simply Rhynchosporium. It has very high importance in winter feed barley and high importance in winter malting barley and spring barley. Like net blotch and many other plant diseases, resistance involves polygenic quantitative resistance, including mapped QTLs as well as genotype-specific interactions, some of which are well characterised. Breeding for Rhynchosporium resistance has involved introgression of genes from wild barley as well as selection within cultivated germplasm (reviewed by Zhang et al., 2020).

The complex control of partial resistance is apparent in RL ratings for both these diseases (net blotch: Fig. 3c; Rhynchosporium: Figs 3d and 4b) as there has been a tendency for resistance ratings of most but not all varieties to decline over the years. This is consistent with evolution of pathogen virulence involving specific interactions with host resistance genes. Nevertheless, most current varieties have ratings which indicate moderate resistance to both diseases. Most winter barleys on the 2021/22 RL have ratings of 6 or 7 for net blotch and most winter and spring varieties have 5 or 6 for Rhynchosporium. The general pattern of resistance is consistent with a situation in which quantitative resistance is controlled largely by genes which provide durable resistance while some have effects which are moderately or weakly specific to certain fungal genotypes.

Despite quantitative resistance to net blotch and Rhynchosporium being less durable than mildew and brown rust resistance, some varieties are potentially reliable sources for use in barley breeding. Considering those that were on the RL for at least five years, Amarena, Boot, Camion and Colibri maintained ratings above 7.0 for net blotch resistance throughout their period of cultivation while several other varieties maintained ratings above 6 with little decrease in resistance. For Rhynchosporium in winter barley, Amarena, Pelican, Pict and Volume stand out as resistant over their commercial lifetime, as do Publican, Rebecca and Westminster in spring barley.

9.2 Ramularia leaf spot of barley: limits to breeding?

Ramularia leaf spot has been recognised as a disease of barley for nearly 40 years but it is only 20 years since it was identified as a significant threat. It is a late season disease which primarily affects the top two leaf layers and can cause substantial losses to grain yield and quality. When it first became commercially important, it was quickly realised that all the most susceptible varieties had *mlo* mildew resistance (Havis et al., 2015). In genetic analysis of a biparental cross, the mildew-resistance allele *mlo11* increased susceptibility to Ramularia at sites in Bavaria, Scotland and Norway, although the effect was somewhat variable (McGrann et al., 2014). There were advances in selecting for Ramularia resistance until the early 2010s but since then, progress has stalled.

A brief summary of the current situation is that, first, modern spring barley varieties are not as susceptible as many *mlo* varieties in the early 2000s so it seems that, as in other cereal diseases, breeders have successfully eliminated very susceptible genotypes. Ramularia resistance was rated on the barley RL in 2017/18 and 2018/19 but in trials across the UK, there was great variation in the relative susceptibility of varieties at different locations, even at sites which were close to one another. The average ratings in the RL therefore did not give a reliable prediction of varieties' responses to Ramularia on farms. The size of the GxE effect, which is much larger than variation between varieties' average susceptibility, explains why breeders have not been able to make further significant advances since culling the highly susceptible *mlo* lines.

The fungal pathogen, *Ramularia collo-cygni*, has been associated with barley since at least the mid-nineteenth century (Fountaine and Fraaije, 2009). The key point may not be to develop varieties which are resistant to Ramularia in current farming conditions so much as to understand why, after decades or centuries as an apparently harmless endophyte, *R. collo-cygni* became an aggressive parasite in the mid-1990s. The important questions may be agronomic, not genetic: what features of agriculture changed around that time, did they provoke *R. collo-cygni* to become a pathogen, and can they be reversed?

10 Resistance to viral diseases

Two types of viral disease are significant in cereals in northwest Europe, soil-borne mosaic viruses and the aphid-transmitted Barley yellow dwarf virus (BYDV).

10.1 Soil-borne mosaic viruses

Greatest progress has been made in breeding for resistance to Barley yellow mosaic virus (BYMV) and Barley mild mosaic virus, Bymoviruses which are transmitted by the slime mould *Polymyxa graminis*. Phenotypic selection for resistance requires a trial site infested with the virus, so marker-assisted selection for the resistance gene is preferred. All winter barley varieties released recently in the UK have resistance to both viruses, probably conferred by the recessive gene *rym4* (Fig. 3e). Although this resistance has been effective to date, resistance-breaking strains have been found in fields of *rym4* varieties (Rolland et al., 2017). Other *rym* genes, particularly *rym5* at the same locus as *rym4*, are being used in barley breeding. The slow dispersal of soil-borne mosaic viruses means it should be feasible to control BYMV with major genes because when a resistance-breaking strain appears, breeders have time to select other resistance genes (Ordon et al., 2009).

The most significant soil-borne virus of wheat is Soil-borne cereal mosaic virus (SBCMV). Resistance is provided by *Sbm1* on chromosome 5D and *Sbm2* on 2B, which are both present in the resistant variety Cadenza (Bayles et al., 2007). Cultivars had similar resistance at sites in France, Italy and the UK, indicating that SBCMV resistance should be useful in breeding across Europe (Budge et al., 2008).

10.2 Barley yellow dwarf virus

The aphid-borne pathogen BYDV has recently and quite suddenly become a cause for serious concern on all cereals, following the EU-wide ban on using neonicotinoid insecticides for crop protection in 2018 (European Commission, n.d.). This has generated strong demand for BYDV-resistant varieties. The first resistant wheat variety in the UK, RGT Wolverine, was recommended in 2021. Field selection for a resistant phenotype will only be reliable if viruliferous aphids become more common than they are at present, so marker-assisted selection for known genes is essential. RGT Wolverine has the *Bdv2* resistance gene, introgressed into wheat chromosome 7D from a wild grass, *Thinopyrum intermedium* (Zhang et al., 2009).

11 Resistance to insects

Aside from their role as vectors of pathogenic viruses, some insects are pests of cereals in their own right. In recent years, the most significant has been *Sitodiplosis mosellana*, the orange wheat blossom midge (OWBM), a gall midge. The yellow or lemon wheat blossom midge, *Contarinia tritici*, is also found in the UK but is rarely damaging (https://ahdb.org.uk/knowledge-library/how-to-identify-wheat-blossom-midges). OBM has a long-lived pupal phase so its numbers can be reduced by crop rotation and cultivation, but the most effective way of controlling it is by growing resistant varieties. *Sm1* on chromosome 2B of wheat provides strong resistance to OBM (Kassa et al., 2016) and has been used increasingly in European breeding. It is currently present in 23 of 33 winter wheat varieties on the 2021/22 RL, including all those in the soft endosperm Groups 3 and Soft 4, and in three of the eight spring varieties, all of which have hard endosperm (Fig. 7). Resistance is associated with greater induction of phenolic acids such as ferulic acid and *p*-coumaric acid during early grain development (Ding et al., 2000). All breeders select *Sm1* by marker-assisted selection.

As well as being vectors of BYDV, aphids used to be important pests of wheat in their own right before the introduction of modern insecticides. Resistance of the most important aphid, the grain aphid *Sitobion avenae*, to pyrethroid insecticides is widespread in England (Foster et al., 2012) and Ireland (Walsh et al., 2020). This involves a significant fitness cost but nonetheless will threaten aphid control if pyrethroids are used heavily instead of neonicotinoids. Research in the 1970s detected significant variation in resistance of wheat and barley to *S. avenae*, with the popular wheat variety Maris Huntsman being particularly susceptible (Lowe, 1980). In principle, therefore, it will be possible to breed cereal varieties with greater resistance to aphids but a significant challenge is to scale up the tests so that enough plants can be scored even to select resistant parents, let alone improved progeny lines.

12 Mitigating trade-offs of disease resistance

Disease resistance which involves induction of plant defences may be energetically costly. Fitness costs of resistance and their relevance to plant breeding were reviewed by Brown and Rant (2013). While one might intuitively suppose that such costs could limit the ability to increase quantitative resistance, the examples and data presented here show that if there are such costs in European cereal varieties, they have not been large enough to prevent

Figure 7 The proportions of winter wheat varieties with resistance to orange wheat blossom midge from 2006, when this trait was first included in the Recommended List. Resistant varieties are those that breeders claim to have *Sm1* based on DNA sequence data. Soft endosperm varieties are in Group 3 and Soft Group 4; hard endosperm varieties are in Group 1, Group 2 and Hard Group 4.

breeders from increasing yield and maintaining quality while simultaneously developing durable resistance to most diseases.

12.1 The best alleles in the best genomes

mlo mildew resistance in spring barley gives an insight into how fitness costs may affect cereal breeding (reviewed by Brown and Rant, 2013; also see McGrann et al., 2014; Gruner et al., 2020; Jacott et al., 2020). *mlo* mutations incur large yield penalties, increase susceptibility to some non-biotrophic diseases and decrease mycorrhization. The penalties of most artificial alleles are greater, however, than those of the natural allele *mlo11* and one of the artificial alleles, *mlo9*, which have been widely used in breeding. The first point, therefore, is that the process of accelerated natural selection that operates in plant breeding programmes enables breeders to select the alleles which have the largest benefits compared to any costs. In a segregating population of a cross of an *mlo5* line with one carrying wild-type *Mlo* (*Mlo*⁺), it was possible to select lines which had *mlo5* as well as grain yield close to that of the *Mlo*⁺ parent (Kjær et al., 1990). The second relevant point, therefore, is that breeders can select lines in which reassortment of background genes modifies the effect of *mlo* on traits other than mildew resistance. The rest of the genome is thus adapted to the presence of the *mlo* allele, reducing the penalties of the strong mildew resistance. Yields of the best *mlo11* lines are now not much lower than those of *Mlo*⁺ varieties but I understand from several barley breeders that crosses between the two genotype classes are rarely productive.

One can readily imagine that the same process applies to breeding for disease resistance in general. When breeders select an optimal phenotype, they are selecting alleles of disease resistance genes which have the greatest benefit in relation to any costs, and they are also reassorting the genome to minimise any remaining costs. In this way, if there are costs of durable resistance, they do not matter. At least, they have not prevented successful breeding for resistance to most diseases of cereals.

12.2 Disease escape

Disease escape results from the effects of crop morphology or development on disease progress. The spread of splash-borne diseases such as Septoria tritici blotch and Septoria nodorum blotch can be promoted by shorter plant height, earlier flowering, larger leaves and leaf layers overlapping so that spores are more easily transmitted from one leaf to the other (Scott et al., 1982; van Beuningen and Kohli, 1990; Lovell et al., 2004; Arraiano et al., 2009; Judge, 2015). Disease escape traits can be seen as costly because the ideal plant ideotype from the point of view of yield formation may be one which promotes disease: earlier leaf emergence, longer-lived leaves and larger leaves allow photosynthesis over a longer time or larger area, while shorter plants divert more photosynthate to the ear instead of vegetative tissue. Genes on chromosome 6A are associated with larger, longer-lived leaves and thus greater yield but this also results in greater severity of Septoria (Judge, 2015; Arraiano and Brown, 2017). Breeders need to balance maximising yield against minimising the cost of disease control. Selecting stronger disease resistance will reduce the relevance of disease escape and thus give breeders greater flexibility to modify developmental and morphological traits in such a way as to increase yield.

13 Future threats

13.1 Stem rust: a known unknown

In the epistemiological system of the late Donald H. Rumsfeld, the examples above are the 'known knowns' of cereal diseases: those we know can cause trouble. The outstanding disease among 'known unknowns' - those that are absent or not currently severe but may become more serious - is black or stem rust of wheat (*Puccinia graminis*). This has been largely absent from northwest Europe since the 1950s but has been found occasionally in this region in recent years (Lewis et al., 2018; Hovmøller et al., 2021). The most significant factor suppressing stem rust over the last 60 years may have been earlier harvesting of wheat, made possible by selection for earlier maturity (another example of disease escape driven by crop development). In future, the warmer summer weather that will result from climate change may make the northwest European environment more conducive for stem rust (Lewis et al., 2018).

UK wheat varieties are susceptible to stem rust (Lewis et al., 2018) and given the close relationship between UK varieties and those elsewhere in Europe, the same is probably true of other countries in this region. If stem rust becomes more common in northwest Europe, stronger, durable resistance will be needed. Breeders can take advantage of the globalisation of their industry to run trials for resistance to a disease in locations where it is common, aiming to select resistant lines before the pathogen becomes troublesome in northwest Europe. In pre-emptive breeding trials, European varieties can be tested in places where there are reliable epidemics of stem rust caused by genetically diverse populations of *P. graminis* f.sp. *tritici*. Lines which are susceptible to as many pathogen genotypes as possible with a high infection type as seedlings, but have low stem rust severity in field conditions, are likely to have race-non-specific, quantitative, probably durable resistance (for similar work on mildew of wheat in central China, see Yu et al., 2001). If such trials are run regularly over the next 20-30 years and if breeders are prepared to discard varieties which are excessively susceptible to stem rust, there is a good chance of building up an acceptable level of durable resistance to the disease before it re-emerges as a serious problem in northern Europe.

Oat stem rust (*P. graminis* f.sp. *avenae*) has re-emerged in Sweden following the repeal of a law requiring eradication of barberry in 1994 (Berlin et al., 2013). This has allowed the pathogen to complete its sexual cycle in Sweden and has resulted in outbreaks of oat stem rust being more frequent and occasionally severe. Despite the value of barberry to wildlife, those benefits should be balanced against the risks to production of cereals, including wheat and barley as well as oats. Perhaps eradicating barberry from hedgerows around arable fields would be an acceptable compromise.

13.2 Unknown unknowns

By contrast, 'unknown unknowns' are organisms which are not known to be major threats to crop production, perhaps because their potential for causing damage in northwest Europe is not appreciated or because they are not favoured by the current environment or agronomy. New, previously unknown diseases must be combatted by a broad range of methods because it takes several years to breed for resistance to them. It is not possible to breed for resistance to a pathogen the existence of which we do not yet know. When Ramularia leaf spot of barley, the most prominent new disease of cereals in recent decades, first became serious around 2000, it was quickly found that broad-spectrum fungicides which were active against many other ascomycete fungi also controlled *R. collo-cygni*. This gave breeders time to eliminate the most susceptible varieties. Breeders need to work closely with agronomists to control new diseases (as they did in the case of Ramularia), but agronomists need to have access to a reasonably wide range of pesticides to buy time for breeders (Brown, 2008). The fundamentals of breeding for resistance to a new pathogen are the same as those for established diseases: genetically diverse but well-adapted germplasm, an effective trialling system including locations where the disease is prevalent, and an efficient process for improving resistance to the new pathogen in germplasm with high yield, good quality, suitable agronomic properties and low susceptibility to other pests and diseases.

14 Acknowledgements

I thank several breeders and staff of AHDB for advice and discussions but the responsibility for all opinions expressed in this chapter is mine alone. I have no competing interests to declare. The writing of this chapter was supported by BBSRC through the Plant Health Institute Strategic Programme and the EU Horizon 2020 RustWatch project.

15 Where to look for further information

AHDB has a large Knowledge Library with information about almost all pests and diseases relevant to arable farming in the UK. Much of it is also relevant to other countries in northwest Europe. Go to: https://ahdb.org.uk/knowledge-library and filter first by Cereals & Oilseeds then by Crops & Grassland and a topic such as Diseases, Pests or Varieties. Or search by the name of a disease as a keyword.

Recommended Lists are available at: https://ahdb.org.uk/rl for the current year and https://ahdb.org.uk/rlarchive for previous years back to 2004/05.

The AHDB's Variety Selection Tool is at: https://ahdb.org.uk/vst.

Information about the plant breeding industry in the UK, which is closely integrated with companies elsewhere in Europe, is produced by the British Society of Plant Breeders at: https://bspb.co.uk/.

16 References

Acevedo-Garcia, J., Spencer, D., Thieron, H., Reinstädler, A., Hammond-Kosack, K., Phillips, A. L. and Panstruga, R. (2017). mlo-based powdery mildew resistance in hexaploid bread wheat generated by a non-transgenic TILLING approach. *Plant Biotechnology Journal* 15(3), 367–378. doi: 10.1111/pbi.12631.

AHDB (Agriculture and Horticulture Development Board) (2020). Recommended Lists Crop Committee Handbook. Available at: https://bit.ly/rlhandbook2020 (accessed on 29th June 2021).

AHDB (2021). Septoria tritici disease resistance in winter wheat. Available at: https://bit.ly/cougar2021 (accessed on 2nd July 2021).

Ali, S., Leconte, M., Walker, A. S., Enjalbert, J. and de Vallavieille-Pope, C. (2010). Reduction in the sex ability of worldwide clonal populations of *Puccinia striiformis* f.sp. *tritici*. *Fungal Genetics and Biology* 47(10), 828–838. doi: 10.1016/j.fgb.2010.07.002.

Arraiano, L. S. and Brown, J. K. M. (2017). Sources of resistance and susceptibility to Septoria tritici blotch of wheat. *Molecular Plant Pathology* 18(2), 276–292. doi: 10.1111/mpp.12482.

Arraiano, L. S., Balaam, N., Fenwick, P. M., Chapman, C., Feuerhelm, D., Howell, P., Smith, S. J., Widdowson, J. P. and Brown, J. K. M. (2009). Contributions of disease resistance and escape to the control of Septoria tritici blotch of wheat. *Plant Pathology* 58(5), 910–922. doi: 10.1111/j.1365-3059.2009.02118.x.

Bayles, R. A., Clarkson, J. D. S. and Slater, S. E. (1997). The UK Cereal Pathogen Virulence Survey. In: Crute, I. R., Holub, E. B. and Burdon, J. J. (Eds) *The Gene-for-Gene Relationship in Plant-Parasite Interactions*. CAB International, Wallingford, pp. 103–117.

Bayles, R., O'Sullivan, D. and Lea, V. (2007). Controlling soil-borne cereal mosaic virus in the UK by developing resistant wheat cultivars. AHDB Project Report No. 418. Available at: https://bit.ly/sbcmv2007 (accessed on 29th June 2021).

Bennett, F. G. A. (1984). Resistance to powdery mildew in wheat: a review of its use in agriculture and breeding programmes. *Plant Pathology* 33(3), 279–300. doi: 10.1111/j.1365-3059.1984.tb01324.x.

Berlin, A., Samils, B., Djurle, A., Wirsén, H., Szabo, L. and Yuen, J. (2013). Disease development and genotypic diversity of *Puccinia graminis* f.sp. *avenae* in Swedish oat fields. *Plant Pathology* 62(1), 32–40. doi: 10.1111/j.1365-3059.2012.02609.x.

Berry, P. M., Kindred, D. R., Olesen, J. E., Jorgensen, L. N. and Paveley, N. D. (2010). Quantifying the effect of interactions between disease control, nitrogen supply and land use change on the greenhouse gas emissions associated with wheat production. *Plant Pathology* 59(4), 753–763. doi: 10.1111/j.1365-3059.2010.02276.x.

Brown, J. K. M. (1994). Chance and selection in the evolution of barley mildew. *Trends in Microbiology* 2(12), 470–475. doi: 10.1016/0966-842X(94)90650-5.

Brown, J. K. M. (2008). Breeding for disease resistance as an integral component of crop protection. *Proceedings, ENDURE International Conference 2008, 'Diversifying Crop Protection'*. Available at: https://bit.ly/breeding-icm (accessed on 29th June 2021).

Brown, J. K. M. (2015). Durable resistance of crops to disease: a Darwinian perspective. *Annual Review of Phytopathology* 53, 513–539. doi: 10.1146/annurev-phyto-102313-045914.

Brown, J. K. M. and Hovmøller, M. S. (2002). Aerial dispersal of pathogens on the global and continental scales and its impact on plant disease. *Science* 297(5581), 537–541. doi: 10.1126/science.1072678.

Brown, J. K. M. and Rant, J. C. (2013). Fitness costs and trade-offs of disease resistance and their consequences for breeding arable crops. *Plant Pathology* 62(Suppl.1), 83–95. doi: 10.1111/ppa.12163.

Brown, J. K. M., Simpson, C. G. and Wolfe, M. S. (1993). Adaptation of barley powdery mildew populations in England to varieties with two resistance genes. *Plant Pathology* 42(1), 108–115. doi: 10.1111/j.1365-3059.1993.tb01476.x.

Brown, J. K. M., Chartrain, L., Lasserre-Zuber, P. and Saintenac, C. (2015). Genetics of resistance to *Zymoseptoria tritici* and applications to wheat breeding. *Fungal Genetics and Biology: FG and B* 79, 33–41. doi: 10.1016/j.fgb.2015.04.017.

Budge, G. E., Ratti, C., Rubies-Autonell, C., Lockley, D., Bonnefoy, M., Vallega, V., Pietravalle, S. and Henry, C. M. (2008). Response of UK winter wheat cultivars to Soil-borne cereal mosaic and Wheat spindle streak mosaic viruses across Europe. *European Journal of Plant Pathology* 120(3), 259–272. doi: 10.1007/s10658-007-9214-9.

Buerstmayr, M., Steiner, B. and Buerstmayr, H. (2020). Breeding for Fusarium head blight resistance in wheat - progress and challenges. *Plant Breeding* 139(3), 429–454. doi: 10.1111/pbr.12797.

Caiazzo, R., Corsi, B., Di Lenarda, S., Naska, E., Webb, A., Elliott, D., Brown, J., Fraaije, B., Wood, T. and Holdgate, S. (2019). Investigating a potential new variant of *Zymoseptoria tritici*, causal agent of Septoria leaf blotch, and implications for UK winter wheat varieties. AHDB Project Report No. 604. Available at: https://bit.ly/cougar2019 (accessed on 29th June 2021).

Chandramohan, P. and Shaw, M. W. (2013). Sulphate and sulphurous acid alter the relative susceptibility of wheat to *Phaeosphaeria nodorum* and *Mycosphaerella graminicola*. *Plant Pathology* 62(6), 1342–1349. doi: 10.1111/ppa.12052.

Clare, S. J., Wyatt, N. A., Brueggeman, R. S. and Friesen, T. L. (2020). Research advances in the *Pyrenophora teres*-barley interaction. *Molecular Plant Pathology* 21(2), 272–288. doi: 10.1111/mpp.12896.

Corwin, J. A. and Kliebenstein, D. J. (2017). Quantitative resistance: more than just perception of a pathogen. *Plant Cell* 29(4), 655–665. doi: 10.1105/tpc.16.00915.

Cowger, C. and Brown, J. K. M. (2019). Durability of quantitative resistance in crops: greater than we know? *Annual Review of Phytopathology* 57, 253–277. doi: 10.1146/annurev-phyto-082718-100016.

Cowger, C., Ward, B., Brown-Guedira, G. and Brown, J. K. M. (2020). Role of effector-sensitivity gene interactions and durability of quantitative resistance to Septoria nodorum blotch in eastern U.S. wheat. *Frontiers in Plant Science* 11, 155. doi: 10.3389/fpls.2020.00155.

DEFRA [Department for Environment, Food and Rural Affairs] (2011). Plant Varieties Act. Available at: https://bit.ly/pva1997 (accessed on 29th June 2021).

DEFRA (2020). Farming statistics – provisional arable crop areas, yields and livestock populations at 1 June 2020 United Kingdom. Available at: https://bit.ly/cropareas2020 (accessed on 29th June 2021).

Ding, H., Lamb, R. J. and Ames, N. (2000). Inducible production of phenolic acids in wheat and antibiotic resistance to *Sitodiplosis mosellana*. *Journal of Chemical Ecology* 26(4), 969–985. doi: 10.1023/A:1005412309735.

Downie, R. C., Bouvet, L., Furuki, E., Gosman, N., Gardner, K. A., Mackay, I. J., Campos Mantello, C., Mellers, G., Phan, H. T. T., Rose, G. A., Tan, K. C., Oliver, R. P. and Cockram, J. (2018). Assessing European wheat sensitivities to *Parastagonospora nodorum* necrotrophic effectors and fine-mapping the *Snn3-B1* locus conferring sensitivity to the effector SnTox3. *Frontiers in Plant Science* 9, 881. doi: 10.3389/fpls.2018.00881.

Enjalbert, J., Duan, X., Leconte, M., Hovmøller, M. S. and de Vallavieille-Pope, C. (2005). Genetic evidence of local adaptation of wheat yellow rust (*Puccinia striiformis* f. sp. *tritici*) within France. *Molecular Ecology* 14(7), 2065–2073. doi: 10.1111/j.1365-294X.2005.02566.x.

European Commission (n.d.). Neonicotinoids. Available at: https://bit.ly/neonicseu (accessed on 29th June 2021).

European Council (2020). Plant reproductive material. Available at: https://bit.ly/seedregseu (accessed on 29th June 2021).

Faris, J. D. and Friesen, T. L. (2020). Plant genes hijacked by necrotrophic fungal pathogens. *Current Opinion in Plant Biology* 56, 74–80. doi: 10.1016/j.pbi.2020.04.003.

Fountaine, J. M. and Fraaije, B. A. (2009). Development of QoI resistant alleles in populations of *Ramularia collo-cygni*. *Aspects of Applied Biology* 92, 123–126.

Foster, S. P., Paliwal, D., Martin, J. and Williamson, M. S. (2012). Evolution of knock-down resistance to pyrethroids in grain aphids (*Sitobion avenae*) in the UK. *Aspects of Applied Biology* 117, 95–96.

Fradgley, N., Gardner, K. A., Cockram, J., Elderfield, J., Hickey, J. M., Howell, P., Jackson, R. and Mackay, I. J. (2019). A large-scale pedigree resource of wheat reveals evidence for adaptation and selection by breeders. *PLoS Biology* 17(2), e3000071. doi: 10.1371/journal.pbio.3000071.

FRAG-UK (Fungicide Resistance Action Group UK) (2020). Fungicide resistance management in cereals. Available at: https://bit.ly/fraguk2020 (accessed on 29th June 2021).

Gruner, K., Esser, T., Acevedo-Garcia, J., Freh, M., Habig, M., Strugala, R., Stukenbrock, E., Schaffrath, U. and Panstruga, R. (2020). Evidence for allele-specific levels of enhanced susceptibility of wheat mlo mutants to the hemibiotrophic fungal pathogen *Magnaporthe oryzae* pv. *Triticum*. *Genes* 11(5), 517. doi: 10.3390/genes11050517.

Hales, B., Steed, A., Giovannelli, V., Burt, C., Lemmens, M., Molnár-Láng, M. and Nicholson, P. (2020). Type II Fusarium head blight susceptibility conferred by a region on wheat chromosome 4D. *Journal of Experimental Botany* 71(16), 4703–4714. doi: 10.1093/jxb/eraa226.

Havis, N. D., Brown, J. K. M., Clemente, G., Frei, P., Jedryczka, M., Kaczmarek, J., Kaczmarek, M., Matusinsky, P., McGrann, G. R., Pereyra, S., Piotrowska, M., Sghyer, H., Tellier, A. and Hess, M. (2015). *Ramularia collo-cygni* – an emerging pathogen of barley crops. *Phytopathology* 105(7), 895–904. doi: 10.1094/PHYTO-11-14-0337-FI.

Hilbers, S., Fischbeck, G. and Jahoor, A. (1992). Localization of the *Laevigatum* resistance gene *MlLa* against powdery mildew in the barley genome by the use of RFLP markers. *Plant Breeding* 109(4), 335–338. doi: 10.1111/j.1439-0523.1992.tb00193.x.

Hoheneder, F., Hofer, K., Groth, J., Herz, M., Heß, M. and Hückelhoven, R. (2021). Ramularia leaf spot disease of barley is highly host genotype-dependent and suppressed by continuous drought stress in the field. *Journal of Plant Diseases and Protection* 128(3), 749–767. doi: 10.1007/s41348-020-00420-z.

Hovmøller, M. S., Justesen, A. F. and Brown, J. K. M. (2002). Clonality and long-distance migration of *Puccinia striiformis* f.sp. *tritici* in north-west Europe. *Plant Pathology* 51(1), 24–32. doi: 10.1046/j.1365-3059.2002.00652.x.

Hovmøller, M. S., Walter, S., Bayles, R. A., Hubbard, A., Flath, K., Sommerfeldt, N., Leconte, M., Czembor, P., Rodriguez-Algaba, J., Thach, T., Hansen, J. G., Lassen, P., Justesen, A. F., Ali, S. and de Vallavieille-Pope, C. (2016). Replacement of the European wheat yellow rust population by new races from the centre of diversity in the near-Himalayan region. *Plant Pathology* 65(3), 402–411. doi: 10.1111/ppa.12433.

Hovmøller, M. S., Patpour, M., Rodriguez-Algaba, J., Thach, T., Justesen, A. F. and Hansen, J. G. (2021). GRRC report of yellow and stem rust genotyping and race analyses 2020. Available at: https://bit.ly/grrc2020 (accessed on 29th June 2021).

Hubbard, A., Lewis, C. M., Yoshida, K., Ramirez-Gonzalez, R. H., de Vallavieille-Pope, C., Thomas, J., Kamoun, S., Bayles, R., Uauy, C. and Saunders, D. G. (2015). Field pathogenomics reveals the emergence of a diverse wheat yellow rust population. *Genome Biology* 16, 23. doi: 10.1186/s13059-015-0590-8.

Hunter, T., Coker, R. R. and Royle, D. J. (1999). The teleomorph stage, *Mycosphaerella graminicola*, in epidemics of Septoria tritici blotch on winter wheat in the UK. *Plant Pathology* 48(1), 51–57. doi: 10.1046/j.1365-3059.1999.00310.x.

Ingvardsen, C. R., Massange-Sanchez, J. A., Borum, F., Uauy, C. and Gregersen, P. L. (2019). Development of mlo-based resistance in tetraploid wheat against wheat powdery mildew. *TAG. Theoretical and Applied Genetics. Theoretische und Angewandte Genetik* 132(11), 3009–3022. doi: 10.1007/s00122-019-03402-4.

Jacott, C. N., Charpentier, M., Murray, J. D. and Ridout, C. J. (2020). Mildew locus o facilitates colonization by arbuscular mycorrhizal fungi in angiosperms. *New Phytologist* 227(2), 343–351. doi: 10.1111/nph.16465.

Johnson, R. (1992). Past, present and future opportunities in breeding for disease resistance, with examples from wheat. *Euphytica* 63(1–2), 3–22. doi: 10.1007/BF00023908.

Johnson, R. (1993). Durability of disease resistance in crops: some closing remarks about the topic and the symposium. In: Jacobs, T. and Parlevliet, J. E. (Eds) *Durability of Disease Resistance*. Springer, Dordrecht, pp. 283–300. doi: 10.1007/978-94-011-2004-3_30.

Jørgensen, I. H. (1992). Discovery, characterization and exploitation of Mlo powdery mildew resistance in barley. *Euphytica* 63(1–2), 141–152. doi: 10.1007/BF00023919.

Jørgensen, J. H. (1994). Genetics of powdery mildew resistance in barley. *Critical Reviews in Plant Sciences* 13(1), 97-119. doi: 10.1080/07352689409701910.

Judge, C. (2015). Disease escape in relation to a trade off between Septoria tritici blotch and yield of wheat. Ph.D. Thesis. University of East Anglia. Available at: https://bit.ly/chrisjudge2015 (accessed on 29th June 2021)

Kassa, M. T., Haas, S., Schliephake, E., Lewis, C., You, F. M., Pozniak, C. J., Krämer, I., Perovic, D., Sharpe, A. G., Fobert, P. R., Koch, M., Wise, I. L., Fenwick, P., Berry, S., Simmonds, J., Hourcade, D., Senellart, P., Duchalais, L., Robert, O., Förster, J., Thomas, J. B., Friedt, W., Ordon, F., Uauy, C. and McCartney, C. A. (2016). A saturated SNP linkage map for the orange wheat blossom midge resistance gene *Sm1*. *TAG. Theoretical and Applied Genetics. Theoretische und Angewandte Genetik* 129(8), 1507-1517. doi: 10.1007/s00122-016-2720-4.

Kildea, S., Byrne, J. J., Cucak, M. and Hutton, F. (2020). First report of virulence to the Septoria tritici blotch resistance gene *Stb16q* in the Irish *Zymoseptoria tritici* population. *New Disease Reports* 41(1), 13. doi: 10.5197/j.2044-0588.2020.041.013.

Kildea, S., Sheppard, L., Cucak, M. and Hutton, F. (2021). Detection of virulence to *Septoria tritici* blotch (STB) resistance conferred by the winter wheat cultivar Cougar in the Irish *Zymoseptoria tritici* population and potential implications for STB control. *Plant Pathology*, in press. doi: 10.1111/ppa.13432.

Kjær, B., Jensen, H. P., Jensen, J. and Jorgensen, J. H. (1990). Associations between 3 *ml-o* powdery mildew resistance genes and agronomic traits in barley. *Euphytica* 46, 185-193. doi: 10.1007/BF00027217.

Koebner, R. M. D. and Summers, R. W. (2003). 21st century wheat breeding: plot selection or plate detection? *Trends in Biotechnology* 21(2), 59-63. doi: 10.1016/S0167-7799(02)00036-7.

Kollers, S., Rodemann, B., Ling, J., Korzun, V., Ebmeyer, E., Argillier, O., Hinze, M., Plieske, J., Kulosa, D., Ganal, M. W. and Röder, M. S. (2014). Genome-wide association mapping of tan spot resistance (*Pyrenophora tritici-repentis*) in European winter wheat. *Molecular Breeding* 34(2), 363-371. doi: 10.1007/s11032-014-0039-x.

Kusch, S. and Panstruga, R. (2017). *mlo*-based resistance: an apparently universal "weapon" to defeat powdery mildew disease. *Molecular Plant-Microbe Interactions* 30(3), 179-189. doi: 10.1094/MPMI-12-16-0255-CR.

Lange, W. and Jochemsen, G. (1992). Use of the gene pools of *Triticum turgidum* ssp. *dicoccoides* and *Aegilops squarrosa* for the breeding of common wheat (*T. aestivum*), through chromosome-doubled hybrids. I. Two strategies for the production of the amphiploids. *Euphytica* 59(2-3), 197-212. doi: 10.1007/BF00041273.

Lewien, M. J., Murray, T. D., Jernigan, K. L., Garland-Campbell, K. A. and Carter, A. H. (2018). Genome-wide association mapping for eyespot disease in US Pacific Northwest winter wheat. *PLoS ONE* 13(4), e0194698. doi: 10.1371/journal.pone.0194698.

Lewis, C. M., Persoons, A., Bebber, D. P., Kigathi, R. N., Maintz, J., Findlay, K., Bueno-Sancho, V., Corredor-Moreno, P., Harrington, S. A., Kangara, N., Berlin, A., García, R., Germán, S. E., Hanzalová, A., Hodson, D. P., Hovmøller, M. S., Huerta-Espino, J., Imtiaz, M., Mirza, J. I., Justesen, A. F., Niks, R. E., Omrani, A., Patpour, M., Pretorius, Z. A., Roohparvar, R., Sela, H., Singh, R. P., Steffenson, B., Visser, B., Fenwick, P. M., Thomas, J., Wulff, B. B. H. and Saunders, D. G. O. (2018). Potential for re-emergence of wheat stem rust in the United Kingdom. *Communications Biology* 1, 13. doi: 10.1038/s42003-018-0013-y.

Lin, M., Ficker, A., Cockram, J. and Lillemo, M. (2020). Genetic structure of the Norwegian *Parastagonospora nodorum* population. *Frontiers in Microbiology* 11, 1280. doi: 10.3389/fmicb.2020.01280.

Lovell, D. J., Parker, S. R., Hunter, T., Welham, S. J. and Nichols, A. R. (2004). Position of inoculum in the canopy affects the risk of Septoria tritici blotch epidemics in winter wheat. *Plant Pathology* 53(1), 11–21. doi: 10.1046/j.1365-3059.2003.00939.x.

Lowe, H. J. B. (1980). Resistance to aphids in immature wheat and barley. *Annals of Applied Biology* 95(1), 129–135. doi: 10.1111/j.1744-7348.1980.tb03978.x.

McGrann, G. R. D., Stavrinides, A., Russell, J., Corbitt, M. M., Booth, A., Chartrain, L., Thomas, W. T. B. and Brown, J. K. M. (2014). A trade off between *mlo* resistance to powdery mildew and increased susceptibility of barley to a newly important disease, Ramularia leaf spot. *Journal of Experimental Botany* 65(4), 1025–1037. doi: 10.1093/jxb/ert452.

Mi, Z., Zhang, X., Su, J., Han, D. and Su, B. (2020). Wheat stripe rust grading by deep learning with attention mechanism and images from mobile devices. *Frontiers in Plant Science* 11, 558126. doi: 10.3389/fpls.2020.558126.

Nazareno, E. S., Li, F., Smith, M., Park, R. F., Kianian, S. F. and Figueroa, M. (2018). *Puccinia coronata* f. sp. *avenae*: a threat to global oat production. *Molecular Plant Pathology* 19(5), 1047–1060. doi: 10.1111/mpp.12608.

O'Driscoll, A., Kildea, S., Doohan, F., Spink, J. and Mullins, E. (2014). The wheat-Septoria conflict: a new front opening up? *Trends in Plant Science* 19(9), 602–610. doi: 10.1016/j.tplants.2014.04.011.

Ordon, F., Habekuss, A., Kastirr, U., Rabenstein, F. and Kühne, T. (2009). Virus resistance in cereals: sources of resistance, genetics and breeding. *Journal of Phytopathology* 157(9), 535–545. doi: 10.1111/j.1439-0434.2009.01540.x.

Pasquariello, M., Ham, J., Burt, C., Jahier, J., Paillard, S., Uauy, C. and Nicholson, P. (2017). The eyespot resistance genes *Pch1* and *Pch2* of wheat are not homoeoloci. *TAG. Theoretical and Applied Genetics. Theoretische und Angewandte Genetik* 130(1), 91–107. doi: 10.1007/s00122-016-2796-x.

Perronne, R., Diguet, S., Pope de Vallavieille, Cd, Leconte, M. and Enjalbert, J. (2017). A framework to characterize the commercial life cycle of crop varieties: application to the case study of the influence of yellow rust epidemics on French bread wheat varieties. *Field Crops Research* 209, 159–167. doi: 10.1016/j.fcr.2017.05.008.

Pilet-Nayel, M. L., Moury, B., Caffier, V., Montarry, J., Kerlan, M. C., Fournet, S., Durel, C. E. and Delourme, R. (2017). Quantitative resistance to plant pathogens in pyramiding strategies for durable crop protection. *Frontiers in Plant Science* 8, 1838. doi: 10.3389/fpls.2017.01838.

Podevin, N., Davies, H. V., Hartung, F., Nogué, F. and Casacuberta, J. M. (2013). Site-directed nucleases: a paradigm shift in predictable, knowledge-based plant breeding. *Trends in Biotechnology* 31(6), 375–383. doi: 10.1016/j.tibtech.2013.03.004.

Poland, J. and Rutkoski, J. (2016). Advances and challenges in genomic selection for disease resistance. *Annual Review of Phytopathology* 54, 79–98. doi: 10.1146/annurev-phyto-080615-100056.

Pope de Vallavieille, C., Bahri, B., Leconte, M., Zurfluh, O., Belaid, Y., Maghrebi, E., Huard, F., Huber, L., Launay, M. and Bancal, M. O. (2018). Thermal generalist behaviour of invasive *Puccinia striiformis* f.sp. *tritici* strains under current and future climate conditions. *Plant Pathology* 67(6), 1307–1320. doi: 10.1111/ppa.12840.

Punja, N. (2018). The impact of Regulation (EC) no 1107/2009 on innovation and development of alternatives and new plant protection products. Available at: https://bit.ly/reg1107-2009 (accessed on 29th June 2021).

Riaz, A., Kock Appelgren, P., Hehir, J. G., Kang, J., Meade, F., Cockram, J., Milbourne, D., Spink, J., Mullins, E. and Byrne, S. (2020). Genetic analysis using a multi-parent wheat population identifies novel sources of Septoria tritici blotch resistance. *Genes* 11(8), 887. doi: 10.3390/genes11080887.

Rolland, M., Villemot, J., Marais, A., Theil, S., Faure, C., Cadot, V., Valade, R., Vitry, C., Rabenstein, F. and Candresse, T. (2017). Classical and next generation sequencing approaches unravel Bymovirus diversity in barley crops in France. *PLoS ONE* 12(11), e0188495. doi: 10.1371/journal.pone.0188495.

Ruud, A. K., Dieseth, J. A. and Lillemo, M. (2018). Effects of three *Parastagonospora nodorum* necrotrophic effectors on spring wheat under Norwegian field conditions. *Crop Science* 58(1), 159–168. doi: 10.2135/cropsci2017.05.0281.

Saintenac, C., Cambon, F., Aouini, L., Verstappen, E., Ghaffary, S. M. T., Poucet, T., Marande, W., Berges, H., Xu, S., Jaouannet, M., Favery, B., Alassimone, J., Sánchez-Vallet, A., Faris, J., Kema, G., Robert, O. and Langin, T. (2021). A wheat cysteine-rich receptor-like kinase confers broad-spectrum resistance against Septoria tritici blotch. *Nature Communications* 12(1), 433. doi: 10.1038/s41467-020-20685-0.

Scott, P. R., Benedikz, P. W. and Cox, C. J. (1982). A genetic study of the relationship between height, time of ear emergence and resistance to Septoria nodorum in wheat. *Plant Pathology* 31(1), 45–60. doi: 10.1111/j.1365-3059.1982.tb02811.x.

Schwarzbach, E., Slater, S. E. and Clarkson, J. D. S. (2002). Occurrence of partially mlo-virulent isolates of barley powdery mildew in agricultural environments in Europe. *Cereal Rusts and Powdery Mildews Bulletin* 2002, 0208schwarzbach. Available at: https://bit.ly/schwarzbach2002 (accessed on 29th June 2021).

Snape, J. W. and Riggs, T. J. (1975). Genetical consequences of single seed descent in the breeding of self-pollinating crops. *Heredity* 35(2), 211–219. doi: 10.1038/hdy.1975.85.

Summers, R. W. and Brown, J. K. M. (2013). Constraints on breeding for disease resistance in commercially competitive wheat cultivars. *Plant Pathology* 62(Suppl.1), 115–121. doi: 10.1111/ppa.12165.

van Beuningen, L. T. and Kohli, M. M. (1990). Deviation from the regression of infection on heading and height as a measure of resistance to Septoria tritici blotch in wheat. *Plant Disease* 74(7), 488–493. doi: 10.1094/PD-74-0488.

Walsh, L. E., Schmidt, O., Williamson, M. S. and Gaffney, M. T. (2020). In-field prevalence of resistant grain aphid *Sitobion avenae* (Fabricius). *Biology and Environment: Proceedings of the Royal Irish Academy* 120B, 29–38. doi: 10.3318/bioe.2020.05.

Wang, Y., Cheng, X., Shan, Q., Zhang, Y., Liu, J., Gao, C. and Qiu, J. L. (2014). Simultaneous editing of three homoalleles in hexaploid bread wheat confers heritable resistance to powdery mildew. *Nature Biotechnology* 32(9), 947–951. doi: 10.1038/nbt.2969.

Watson, A., Ghosh, S., Williams, M. J., Cuddy, W. S., Simmonds, J., Rey, M. D., Asyraf Md Hatta, M., Hinchliffe, A., Steed, A., Reynolds, D., Adamski, N. M., Breakspear, A., Korolev, A., Rayner, T., Dixon, L. E., Riaz, A., Martin, W., Ryan, M., Edwards, D., Batley, J., Raman, H., Carter, J., Rogers, C., Domoney, C., Moore, G., Harwood, W., Nicholson, P., Dieters, M. J., DeLacy, I. H., Zhou, J., Uauy, C., Boden, S. A., Park, R. F., Wulff, B. B. H. and Hickey, L. T. (2018). Speed breeding is a powerful tool to accelerate crop research and breeding. *Nature Plants* 4(1), 23–29. doi: 10.1038/s41477-017-0083-8.

Weyen, J. (2008). Barley and wheat doubled haploids in breeding. In: Touraev, A., Forster, B. P. and Jain, S. M. (Eds) *Advances in Haploid Production in Higher Plants*. Springer, Dordrecht, pp 179–187. doi: 10.1007/978-1-4020-8854-4_15.

Wingen, L. U., Shaw, M. W. and Brown, J. K. M. (2013). Long-distance dispersal and its influence on adaptation to host resistance in a heterogeneous landscape. *Plant Pathology* 62(1), 9–20. doi: 10.1111/j.1365-3059.2012.02621.x.

Wolfe, M. S. (1984). Trying to understand and control powdery mildew. *Plant Pathology* 33(4), 451–466. doi: 10.1111/j.1365-3059.1984.tb02868.x.

Xu, X. and Nicholson, P. (2009). Community ecology of fungal pathogens causing wheat head blight. *Annual Review of Phytopathology* 47, 83–103. doi: 10.1146/annurev-phyto-080508-081737.

Yu, D. Z., Yang, X. J., Yang, L. J., Jeger, M. J. and Brown, J. K. M. (2001). Assessment of partial resistance to powdery mildew in Chinese wheat varieties. *Plant Breeding* 120(4), 279–284. doi: 10.1046/j.1439-0523.2001.00592.x.

Zhang, X., Ovenden, B. and Milgate, A. (2020). Recent insights into barley and *Rhynchosporium commune* interactions. *Molecular Plant Pathology* 21(8), 1111–1128. doi: 10.1111/mpp.12945.

Zhang, Z., Lin, Z. and Xin, Z. (2009). Research progress in BYDV resistance genes derived from wheat and its wild relatives. *Journal of Genetics and Genomics* 36(9), 567–573. doi: 10.1016/S1673-8527(08)60148-4.

Chapter 24

Key challenges in breeding durable disease-resistant cereals: North Africa and West Asia

Sarrah Ben M'Barek, Regional Field Crops Research Center of Béja and CRP Wheat Septoria Phenotyping Platform, Tunisia; and Seyed Mahmoud Tabib Ghaffary, Safiabad Agricultural and Natural Resources Research and Education Center (AREEO), Iran

1 Introduction

The West Asia and Northern Africa (WANA) region embraces two continents: Africa and Asia and comprises currently 29 countries (Ibrahim and Lal, 2013; Sivakumar et al., 2013) (Fig. 1a). The WANA region is a vast geographic area characterized by large variations in agro-ecology, farming systems, soil types, cultural practices and climatic conditions. It is divided into five sub-regions:

- Sub-region 1 known as the Maghreb includes Mauritania, Morocco, Algeria, Tunisia, Libya and Malta.
- Sub-region 2 is the Nile Valley and Red Sea Sub-region which includes Egypt, Sudan, Eritrea, Ethiopia, Somalia and Yemen.
- Sub-region 3 is designated as the Mashreq, which includes Turkey, Cyprus, Lebanon, Palestine, Israel, Jordan, Syria and Iraq.
- Sub-region 4 is the Arabian Peninsula that is represented by Kuwait, Bahrain, Qatar, Saudi Arabia, the United Arab Emirates and Oman.

http://dx.doi.org/10.19103/AS.2021.0092.40

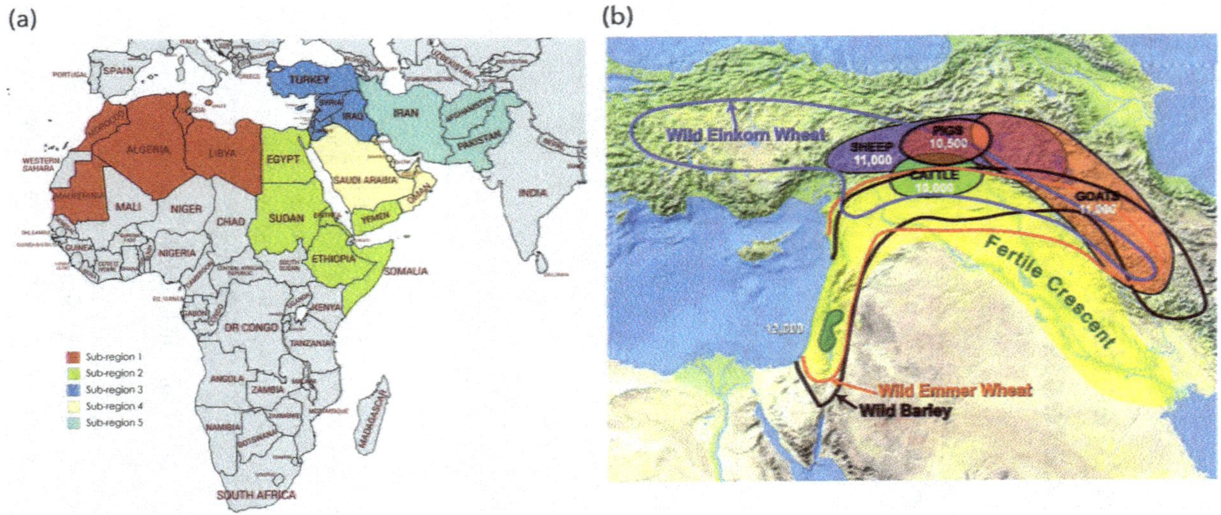

Figure 1 (a) WANA Region countries (colors indicate different sub-regions). (b) Map of the Near East indicating the Fertile Crescent. Colored lines enclose the wild ranges of Einkorn wheat, emmer wheat, and barley. Green-shaded area in southern Levant indicates the region where all three grains were first domesticated 12 000 years BP (Driscoll et al. 2009).

- Sub-region 5 is the west and central Asia sub-region which includes Iran, Afghanistan and Pakistan.

Almost 950 million people, representing 12.4% of the global population, live in the WANA region. The population increased by a factor of 5.4 between 1950 and 2018 (FAOSTAT-Data-Population, 2013) and is expected to increase further by 2050 together with the demand for foods such as cereals.

Most of the countries in the region have a Mediterranean climate, with hot and dry summers, cold winters and an erratic distribution of rainfall, and are expected to be hotter and drier in the future as a consequence of climate change (Cervigni and Dobardzic, 2012). The WANA region in general is vulnerable to drought where available renewable fresh water resources range between 10 mm/year and 50 mm/year and rarely up to 200 mm/year in some locations. The shortage of water restricts rice production, for example, to specific zones like the valley of lower Nile River in Egypt, the Caspian Sea coast in Iran and the Sindh and Punjab regions in Pakistan. Sub-region 5 has the richest water resources (Table 1) that makes it the main area of wheat production. According to the International Maize and Wheat Improvement Center (CIMMYT) world wheat survey, the WANA region is the second major rainfed wheat production zone in the developing world next to South Asia (Byerlee and Moya, 1993).

The region is known for its relative richness in indigenous plant genetic resources and in plant diversity; it has the world's three centers of origin of agricultural crops and three Vavilovian centers of crop diversity. Two of these centers (The Mediterranean Region and the Near Eastern Region) are considered the centers of origin of more than 150 cultivated plant species. Of particular importance, the Fertile Crescent which is a crescent-shaped region in the Middle East (Fig. 1b) is considered as the cradle of early agriculture and a rich niche of biodiversity, particularly for cereal crops (Gepts, 2004; Gepts and Papa, 2003; Gupta et al., 2010; Zeder, 2008). Early human settlers in the region likely harvested the wild forms of wheat and other cereals from natural stands before domestication of tetraploid and hexaploid wheat and the development of agriculture (Salamini et al., 2002) from around 7000 BC (Harlan and Zohary, 1966; Newton et al., 2010). The start of agriculture was enabled by a set of domesticated primitive landraces of cereals, many of which are still in use in various parts of the world (Faris, 2014; Gupta et al., 2010; Salamini et al., 2002). The domestication and adoption of the wild cereal species as potential staple crops was a crucial step for the onset of agriculture that particularly influenced the durum wheat-based diet in the Mediterranean basin, including pasta, couscous and bulgur (Elias, 1995). About 10% of the world's wheat area is covered by durum wheat of which 90% or approximately 11 million ha is cultivated in the drier areas of the Mediterranean region (Nachit, 1998), which also includes most of the WANA region.

Table 1 Renewable freshwater resources (FAOAQUASTAT)

		Long-term average annual precipitation(mm/year)	Total renewable water resources (10^9 m^3/year)
Sub-region 1	Mauritania	92	11.4
	Morocco	346	29
	Algeria	89	11.67
	Tunisia	207	4.615
	Libya	56	0.7
	Malta	560	0.0505
	Total		57.44
Sub-region 2	Egypt	18.1	57.5
	Eritrea	384	7.315
	Ethiopia	848	122
	Sudan	250	37.8
	Somalia	282	14.7
	Yemen	167	2.1
	Total		241.42
Sub-region 3	Turkey	593	211.6
	Cyprus	498	0.78
	Lebanon	661	4.503
	Palestine	402	0.837
	Israel	435	1.78
	Jordan	111	0.937
	Syria	252	16.8
	Iraq	216	89.86
	Total		327.1
Sub-region 4	Kuwait	121	0.02
	Saudi Arabia	59	2.4
	Qatar	74	0.058
	Bahrain	83	0.116
	United Arab Emirates	78	0.15
	Oman	125	1.4
	Total		4.14
Sub-region 5	Iran	228	137
	Afghanistan	327	65.33
	Pakistan	494	246.8
	Total		449.13

2 Cereal production in the West Asia and Northern Africa (WANA) region

Approximately 183–186 million metric tons of cereals were produced annually between 2013 and 2018 in the region, with consumption of approximately 140 million metric tons. The region has seen more than 35% improvement in daily calorie supply per capita between 1961 and 2013 (4% higher than the global average) with the exception of Afghanistan (Figs 2 and 3). According to

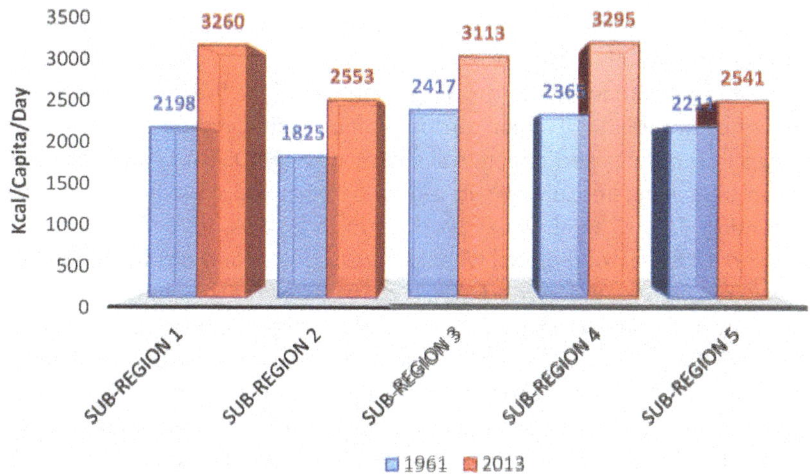

Figure 2 Average of Food supply for each sub-region in earliest and latest accessible FAOSTAT-Data-Crop, 2018.

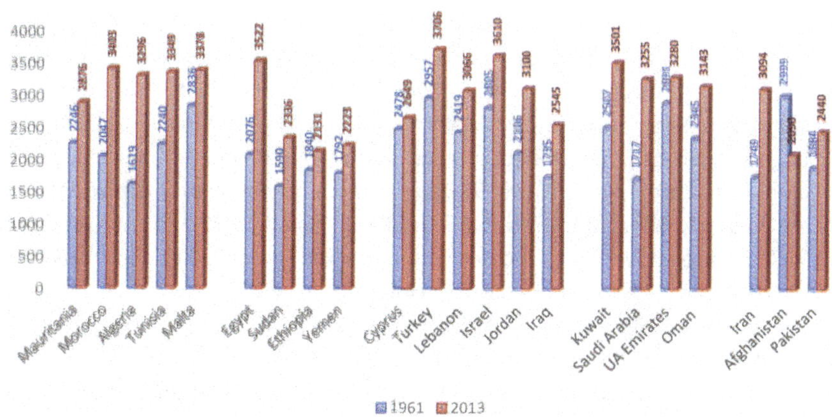

Figure 3 Available data supplied Daily Calorie for each country. Base data of Oman is existing from 1990 instead of 1961. For Libya, Somalia, Eritrea, Syria, Palestine, Bahrain and Qatar the data were not available in FAOSTAT-Data-Food supply.

the FAOSTAT database, wheat, rice, maize and barley are the major crops that support food security in the WANA region. Wheat is the principal staple food in most countries as it is an important source of proteins and carbohydrates and accessible to most rural communities. More than 40% of dietary calories are provided by wheat in Morocco, Algeria, Tunisia, Yemen, Iraq, Iran and Afghanistan. For 15 countries in the region, cereals, including wheat, rice, maize, barley, oat and rye, support 47–75% of daily energy supply per capita. After wheat, rice is the second major commodity in WANA, particularly in sub-region 4 as well as in Mauritania, Egypt and Iraq where rice contributes 12–18% of the diet. Other cereals such as maize, barley, oat and rye play an important role in Morocco, Egypt and Ethiopia and represent 16–33% of daily energy supply per capita (Table 2).

The most important wheat-growing countries of the WANA region in terms of area of cultivation, total annual production in decreasing order and yield are indicated in Table 3 based on latest available data (FAOSTAT-Data-Crop, 2018). With the best water resources, sub-region 5 produced 43 million metric tons representing 46% of total wheat production in WANA in 2018, representing 16.4% and 12.8% of total global wheat harvested area and production in 2018, respectively. During the period 1961–2018, the area of wheat production increased from 26.9 million ha to 35.1 million ha, while total production increased from 22 to 94 million tons with Pakistan, Turkey, Iran, Egypt and Morocco contributing 81% of total production in the region (Fig. 4). Average yields increased from 0.935 tons/ha to 2.67 tons/ha. More than 140 million metric tons (~75%) of total cereal production in the region is produced annually in Pakistan, Turkey, Ethiopia, Egypt and Iran (FAOSTAT-Data-Crop, 2018). Currently, Turkey (sub-region 3), as well as Pakistan and Iran (sub-region 5), are the main wheat growers and producers in the WANA region. Egypt (sub-region 2) is the fourth wheat producer with the highest average of wheat yield in the WANA region.

The increase in production is mainly due to the adoption of improved wheat varieties originating from CIMMYT and the International Center for Agricultural Research in the Dry Areas (ICARDA), more intensive utilization of inputs, better agronomic practices, increased area of production and favorable policies (Tadesse et al., 2017). In 1990 the area planted with modern varieties in WANA was only 42% compared to 88% in south Asia and 82% in Latin America. In 1997, the area covered by modern varieties had increased to 66% in WANA, 86% in Asia and 90% in Latin America (Heisey et al., 2003).

Given the small area of arable land compared with the total population in the region, national cereal production has never been sufficient for most of the countries in the region. Consequently, many countries in the region are major cereal importers. For example, imports of cereals in the region were 7.6, 34, 82 and 97 Million metric tons/year in 1961, 1991, 2008 and 2013, respectively. The region imported 6.03, 22.6, 44.8 and 48.4 million metric tons of wheat

Table 2 Contribution of wheat, rice, maize, barley, oat and rye food supply of the WANA region

		Kcal/capita/day						Contribution in food supply (%)		
		Total	Wheat	Rice	Maize	Barley	Oat & Rye	Wheat	Others	Wheat and Others
Sub-region 1	Mauritania	2876	818	481	28	3	4	28	18	46
	Morocco	3403	1392	11	273	281	1	41	17	58
	Algeria	3296	1424	30	129	94	0	43	8	51
	Tunisia	3349	1576	11	0	42	7	47	2	49
	Malta	3378	985	49	40	0	16	29	3	32
Sub-region 2	Egypt	3522	1175	417	596	1	0	33	29	62
	Ethiopia	2336	284	22	398	125	258	12	34	46
	Sudan	2131	197	14	4	0	12	9	1	10
	Yemen	2223	884	177	118	0	0	40	13	53
Sub-region 3	Cyprus	2649	671	50	28	0	7	25	3	28
	Turkey	3706	1285	97	194	0	25	35	9	44
	Israel	3066	863	165	99	4	10	28	9	37
	Lebanon	3610	983	110	4	10	1	27	3	30
	Jordan	3100	1149	200	23	3	1	37	7	44
	Iraq	2545	1141	408	3	41	0	45	18	63
Sub-region 4	Kuwait	3501	823	433	54	0	20	24	14	38
	Saudi Arabia	3255	797	385	186	0	22	24	18	42
	United Arab Emirates	3280	816	555	8	0	0	25	17	42
	Oman	3143	489	558	100	0	38	16	22	38
Sub-region 5	Iran	3094	1242	287	21	4	0	40	10	50
	Afghanistan	2090	1369	141	21	26	0	66	9	75
	Pakistan	2440	903	125	129	1	0	37	10	47

Table 3 Wheat-growing countries of the WANA region in decreasing order (FAO-Data-Crop-2018)

Country	Area (ha)	Country	Production metric ton	Country	Yield(kg/ha)
Pakistan	8797227	Pakistan	25076149	Egypt	6690
Turkey	7288622	Turkey	20000000	Saudi Arabia	6533
Iran	6700000	Iran	14500000	Malta	5080
Morocco	2842748	Egypt	8800000	Oman	3853
Algeria	2087003	Morocco	7320620	Lebanon	3367
Ethiopia	1748972	Ethiopia	4238836	Kuwait	3000
Afghanistan	1635000	Algeria	3981219	Pakistan	2851
Egypt	1315490	Afghanistan	3613300	Turkey	2744
Iraq	1180443	Iraq	3211382	Iraq	2721
Tunisia	846939	Tunisia	1500000	Sudan	2633
Sudan	226000	Sudan	595000	Morocco	2575
Libya	179558	Saudi Arabia	586413	Qatar	2500
Saudi Arabia	89756	Libya	138770	Ethiopia	2424
Yemen	64530	Lebanon	135411	Palestine	2217
Israel	52900	Yemen	105503	Afghanistan	2210
Lebanon	40216	Israel	70500	Iran	2164
Jordan	18049	Palestine	34167	Mauritania	2157
Palestine	15413	Jordan	20668	Cyprus	2128
Cyprus	8485	Cyprus	18054	Algeria	1908
Mauritania	3408	Malta	12208	Tunisia	1771
Somalia	2586	Mauritania	7349	Yemen	1635
Malta	2403	Oman	3579	Israel	1333
Oman	929	Somalia	1033	Jordan	1145
Qatar	6	Qatar	15	Libya	773
Kuwait	3	Kuwait	9	Somalia	399
Bahrain	No data	Bahrain	No data	Bahrain	No data
Eritrea	No data	Eritrea	No data	Eritrea	No data
Syria	No data	Syria	No data	Syria	No data
UAE	No data	UAE	No data	UAE	No data

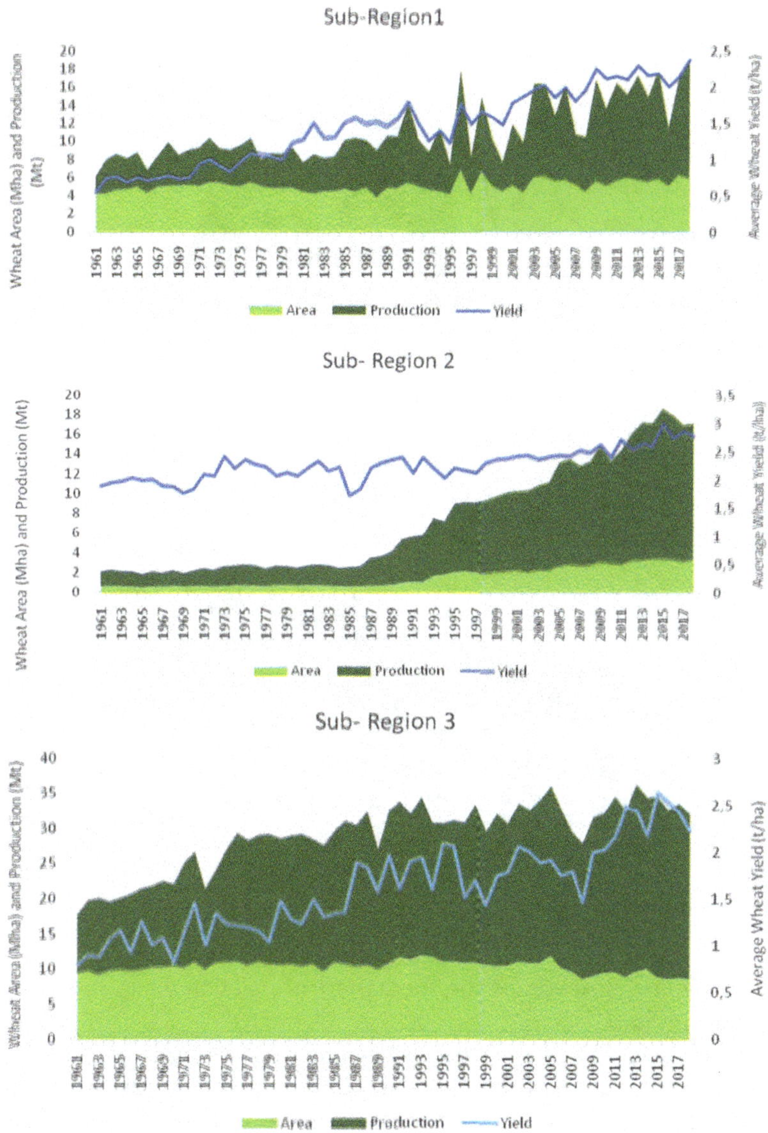

Figure 4 Wheat area, production and yield from 1961 to 2018 in WANA sub-regions (FAOSTAT-Data-Crop).

alone in 1961, 1991, 2008 and 2013, respectively. Among North African countries, Egypt is the largest importer with 10.3 million tons of imported wheat every year. According to Shiferaw et al. (2013), the demand for wheat is growing at 2.2%/year in North Africa, calling therefore for more research and development efforts to increase wheat productivity at regional and global

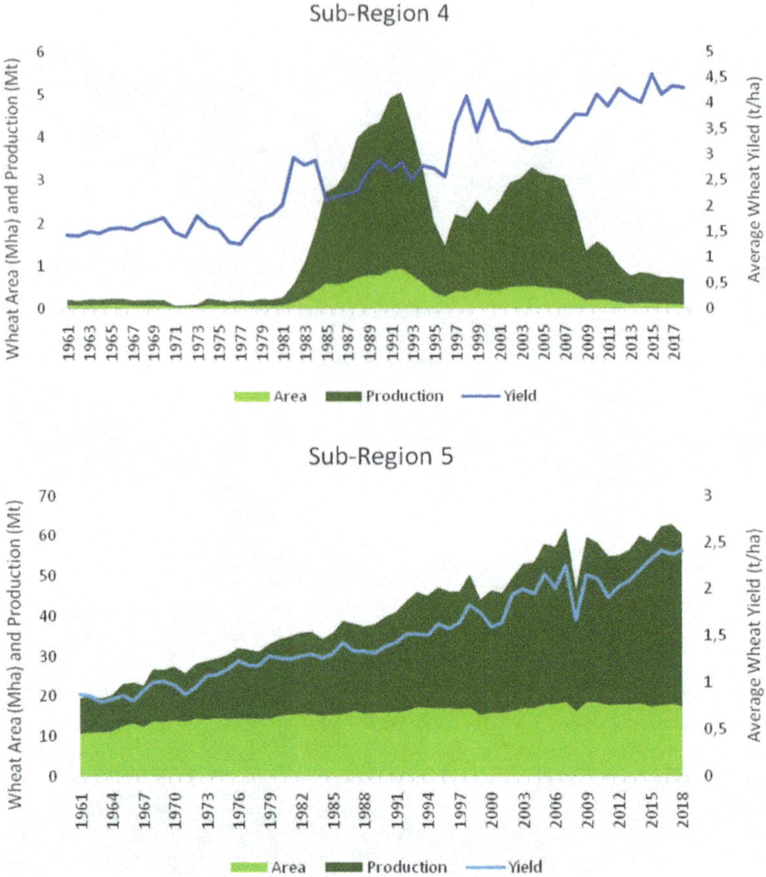

Figure 4 (*Continued*).

levels to meet the needs of the increasing population while reducing the reliance on imports.

Agricultural productivity in the WANA region suffers losses because of high temperature, drought, soil degradation and from pests and diseases, putting the food security of many countries under threat (Lal and Stewart, 2012). A recent survey considered the region as a food security hotspot, defined as a critical source and/or sink in the global food system (Savary et al., 2019). Biotic factors are important limiting factors for wheat production in this region that hosts some of the most damaging and virulent races of diseases and pests. Fungal pathogens in the Fertile Crescent co-evolved with wheat domestication 8000–9000 BC (Stukenbrock et al., 2007; Stukenbrock and McDonald, 2008). Potential crop losses due to pests and diseases are expected to increase with climate change (Chakraborty and Newton, 2011), human-mediated invasions by alien pests/diseases and disease re-emergence (Anderson et al., 2004). Yield loss estimates for each disease vary from country to country and season to season but in general, yield losses generated by pests and diseases in the WANA region are estimated to be around 10.1% (Savary et al., 2019) but can easily reach up to 40% during favorable conditions, causing major economic losses (Yahyaoui et al., 2000).

In the following sections, we will only focus on wheat and especially durum wheat as one of the most widely cultivated crops in the region. Wheat is severely affected by a number of diseases such as leaf, stem and stripe rust diseases, Septoria tritici blotch (STB; *Zymoseptoria tritici*), tan spot (*Pyrenophora tritici-repentis*), Fusarium (*Fusarium* spp.), powdery mildew, the bunts and root rots, barley yellow dwarf virus (Savary et al., 2019). Insect pests, which are often vectors for disease, include Hessian fly, Russian wheat aphid, wheat stem sawfly and sunn pest (Mamluk and Van Slageren, 1993; Tadesse et al., 2017; Yahyaoui et al., 2000).

3 Breeding for durable disease resistance: key challenges

Improving crop resistance to pathogens through breeding is an environmentally sound method for minimizing yield losses. However, introducing new resistance genes into varieties to produce varieties with effective, stable, broad-spectrum and durable resistance is both costly and challenging. Durability of resistance is dependent on many factors, including the biology, genetics and evolution of the target pathogen (Johnson, 1983;Brown et al., 2015). The highest risk pathogens that break down resistance genes are those with a mixed reproduction system, a high potential for genotype flow, large effective population sizes and high mutation rates, while the lowest risk are the pathogens with strict asexual reproduction, low potential for gene flow, small effective population sizes and low mutation rates (McDonald and Linde, 2002). Durability also depends on external factors (environmental conditions and agronomic practices) affecting the fitness of the pathogens (Brown et al., 2015; REX Consortium, 2016) (Fig. 5).

An example of the challenges is the disease STB (*Z. tritici*) on durum wheat. Because of the dynamic nature of wheat-*Z. tritici* interactions and of the very high level of genetic diversity within pathogen populations (Ghaffary, 2011; Marcel et al., 2017), it is a challenge to enhance the resistance levels of wheat cultivars which can take into account the continuous co-evolutionary battle between durum wheat and *Z. tritici* (Ghaffary, 2011; Marcel et al., 2017; Stukenbrock and Croll, 2014; Stukenbrock, 2014; Stukenbrock et al., 2011; Stukenbrock and McDonald, 2008). An effective durum wheat breeding for resistance to *Z. tritici* is also particularly challenging given the potential effects of climate change on disease dynamics (Xynias et al., 2020). It is therefore important to study the genetics of resistance to *Z. tritici* in durum wheat (Soriano et al., 2016; Tadesse

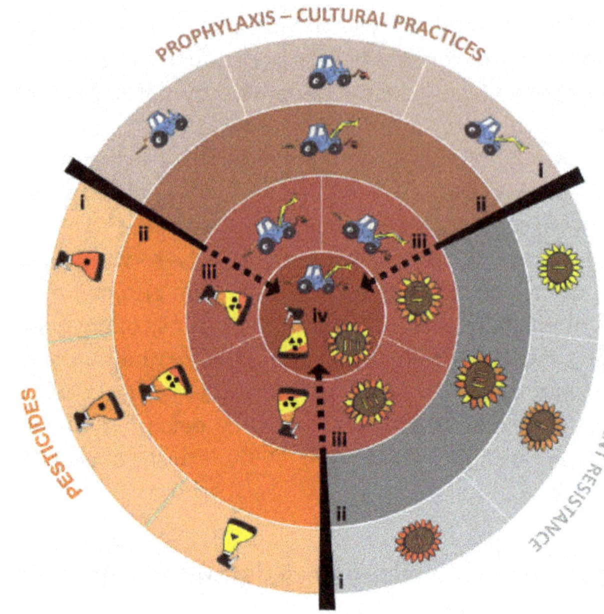

Figure 5 How to combine the various methods available to control plant pathogens to maximize durability. The three main categories of methods are presented by different symbols: a plant for plant resistance genes, a sprayer for pesticides, a tractor for farming methods. Strategies are obtained by combining these methods of pathogen management. The figure is constructed as a target, whose center maximizes the 'Degree of treatment heterogeneity' (i) Only one method (either a plant resistance gene, an antifungal mode of action or a prophylaxis method) is used to control a plant pathogen species. (ii) Several methods of the same type (several plant resistance genes, or several antifungal modes of actions, or several prophylaxis methods) are combined. (iii) Two methods of different types are combined. (iv) All possible methods are combined. This strategy maximizes DTH, and maximizes the durability of plant resistance genes and antifungals (REX Consortium, 2016).

et al., 2016). Efficient use of biodiversity in breeding programs is imperative for improving STB resistance as gene introgression from wild progenitors and landraces can greatly enhance the fitness of contemporary cultivars to *Z. tritici* and maintain their high yield potential (Merchuk-Ovnat et al., 2017; Sheikh et al., 2017; Valkoun, 2001; Xie and Nevo, 2008). Gene pyramiding and the use of diverse sources of resistance is an effective strategy for ensuring more durable resistance (Aouini, 2018; McDonald and Mundt, 2016; Mundt, 2014; Singh et al., 2011, 2014, 2015, 2016, 2017).

Achievement of durable resistance in bread wheat against STB has been reported either by employing a wide range of single major *Stb* resistance genes or *Stb* genes stacking (Chartrain et al., 2005; Ghaffary, 2011; Ghaffary et al., 2011). To examine the single gene route, seedlings of a differential set of 15 cultivars/lines, carrying 18 resistant *Stb* genes (Table 4) along with a susceptible variety (Taichung 29) were inoculated with 20 *Z. tritici* isolates that were collected from different geographical locations across the WANA region. In total, 20 isolates were tested including18 isolates originated from bread wheat and two isolates encoded 86022-TUR and 95052-ALG that originated from durum wheat and collected in Turkey and Algeria, respectively (Table 5). Durum isolates, in general, cannot infect bread wheat cultivars (including Taichung 29) and vice versa (Ghaffary, 2011; Kema et al., 1996). Hence, non-host resistance in wheat-*Z. tritici* interaction can result in a durable resistance by planting durum wheat in the area where bread wheat isolates are dominant and vice versa.

The bread wheat cultivars Arina, a *Stb15+Stb6* carrier, and M3 that carries *Stb16* and *Stb17*, are resistant to eight *Z. tritici* bread wheat strains originating from Algeria (ALG), Ethiopia (ETH), Syria (SYR), Turkey (TUR) and Iran (IRN). In contrast, Shafir (*Stb6*), Estanzuela Federal (*Stb7*), W7984 (*Stb8*) and Balance (*Stb18*) were the least effective genes in the WANA region in resistance to *Z. tritici*. Arina carries both *Stb15* and *Stb6* but, as *Stb6* carried by Shafir is not effective, it was concluded that *Stb15* plays a major role in Arina resistance. M3 is the most durable resistant line among the differential set not only in the WANA region but also against 10 other isolates originated from Canada (CAN), United States (USA), Mexico (MEX), Uruguay (URY), Argentina (ARG), The Netherlands (NL) and Portugal (PT). TE9111, carrying *Stb6*, *Stb7*and *Stb11*, is another effective resistant line to most applied isolates. *Stb11* is known as another efficient gene in WANA as well as globally (Ghaffary, 2011).

The STB-wheat pathosystem follows the gene-for-gene (GFG) concept, where specific host resistance genes interact with their corresponding avirulent pathogen genes. Therefore, another idea for achieving durable host resistance would be the use of a multi-layer defense strategy based on gene pyramiding in the host (Ghaffary et al., 2011). Detection of multiple QTLs in the Apache/Balance DH population related to four *Stb* genes support the gene pyramiding concept. Four QTLs were detected on chromosomes 3AS, 1BS, 6DS and 7DS representing *Stb6* and *Stb11*, *Stb18* and *Stb4*, respectively. Combinations of susceptible (S) and resistant (R) alleles of each related QTL to each *Stb* gene were classified into five groups where SSSS represents DH individuals carrying four susceptible alleles and RRRR demonstrates DH individuals having four resistance alleles (Fig. 6). Stacking of *Stb* genes can be achieved using markers-assisted breeding and has been shown to minimize STB development in commercial breeding programs (Ghaffary et al., 2011).

Rust diseases also present a challenge for durable disease resistance. Favorable conditions for rust, including optimum temperature and high relative humidity in the growing season, occur frequently in most of the years in high-input irrigated and rainfed fields. Inoculum movement between cool highlands and plains, and generation of aggressive races through sexual recombination,

Table 4 Differential set of 15 lines/cultivars carrying 18 *Stb* genes

Stb genes	Cultivars source	*Chromosomal position*	Closest (Flanking‡) marker	Marker type	Reference
Stb1	Bulgaria 88*	5BL	*Xbarc74*	SSR	(Adhikari et al., 2004b)
Stb2	Veranopolis*	3Bs	*Xgwm389*	SSR	(Adhikari et al., 2004a)
Stb3	Israel 493*	7As	*Xwmc83*	SSR	(Goodwin et al., 2015)
Stb4	Tadinia*	7Ds	*Xgwm111*	SSR	(Adhikari et al., 2004c)
Stb5	Cs Synthetic 6X (7D)*	7Ds	*Xgwm44*	SSR	(Arraiano et al., 2001)
Stb6	Shafir	3As	*Xgwm369*	SSR	(Brading et al., 2002)
Stb7	Estanzuela Federal	4AL	*Xwmc313*	SSR	(McCartney et al., 2003)
Stb8	W7984‡	7BL	*Xgwm146* *Xgwm577*	SSR	(Adhikari et al., 2003)
Stb9	Courtot‡	2B	*XksuF1* *Xfbb226*	RFLP	(Chartrain et al., 2009)
Stb10	KK4500†	1D	*Xgwm848*	SSR	(Chartrain et al., 2005)
Stb11	TE9111†	1Bs	*Xbarc008*	SSR	(Chartrain et al., 2005)
Stb12	KK4500†	4AL	*Xwmc313; Xwmc219*	SSR	(Chartrain et al., 2005)
Stb13	Salamouni	7BL	*Xwmc396*	SSR	(Brown et al., 2015)
Stb14	Salamouni	3Bs	*Xwmc500*	SSR	(Brown et al., 2015)
Stb15	Arina*‡	6As	*Xpsr563a* *Xpsr904*	RFLP	(Arraiano et al., 2007)
Stb16	M3	3DL	*Xwmc494* *Xbarc125*	SSR	(Ghaffary et al., 2012)
Stb17	M3	5AL	*Xhbg247*	SSR	(Ghaffary et al., 2012)
Stb18	Balance*‡	6Ds	*Xgpw5176* *Xgpw3087*	SSR	(Ghaffary et al., 2011)

* These lines are also carrying *Stb6*.
† These lines are also carrying *Stb6* and *Stb7*.

Table 5 Results of inoculation experiments with 20 *Zymoseptoria tritici* isolates on 15 differential cultivars that carry 18 *Stb* genes and a susceptible check. Figures represent P data. Colors indicate resistance (not significantly different from 0P, green boxes), intermediate significantly different from 0P as well as maxP, yellow boxes) and susceptible (not significantly different from maxP, red boxes)

	North and Central America				South America				Europe		WANA									
	94218-CAN	00003-USA	00005-USA	90006-MEX	87016-URY	90015-PRU	99015-ARG	86068-ARG	89011-NLD	92004-PRT	92034-ALG	95054-ALG	88004-ETH	88018-ETH	95036-SYR	86013-TUR	02159-IRN	02166-IRN	86022-TUR	95052-ALG
Bulgaria	0	6	0	0	1	0	0	8	4	0	1	1	3	6	2	1	4	1	0	0
Veranopolis	22	1	2	0	4	3	1	1	3	1	0	0	5	3	3	3	0	3	5	1
Israel 493	1	1	1	0	3	0	0	1	5	3	2	1	0	1	6	1	1	4	0	0
Tadinia	0	1	1	0	3	1	1	2	1	1	1	6	1	3	1	1	6	1	0	0
Cs Synthetic 6X (7D	25	3	1	0	2	1	1	1	0	1	6	2	4	1	1	1	1	1	2	2
Shafir	24	21	13	45	6	4	2	4	9	1	4	1	2	2	2	2	1	1	0	0
Estanzuela Federal	4	31	3	42	1	3	3	1	3	1	2	1	4	3	6	2	1	4	0	0
W7984	26	20	7	8	1	6	1	8	6	1	3	2	1	3	5	3	7	2	0	0
Courtot	1	52	1	30	5	4	5	3	1	1	4	1	4	1	1	1	0	1	0	0
KK4500	1	11	0	23	0	2	2	2	6	8	2	0	1	2	1	0	0	0	0	0
TE9111	0	10	0	0	0	0	0	1	1	4	4	0	1	1	0	2	1	0	0	0
Salamouni	0	1	1	1	3	4	7	8	7	1	2	2	2	1	5	3	1	2	0	0
Arina	0	6	0	0	2	5	0	2	6	3	1	0	0	1	1	0	0	1	0	0
M3	0	1	0	0	0	0	0	1	0	0	0	0	0	0	0	0	0	0	0	0
Balance	4	5	1	2	5	7	2	2	1	1	2	4	1	1	3	2	2	1	0	0
Taichung	42	63	15	45	5	3	4	1	4	2	4	3	3	2	6	3	1	4	1	0

mutation or somatic hybridization is also increasing the risk of yellow, leaf and stem rust epidemics in the WANA region (Singh et al., 2004). The three types of rusts are airborne pathogens that can migrate from one region to another. Figure 7 demonstrates the migration airborne route of a novel *Puccinia striiformis* (yellow rust) race virulent on Yr9 and the *Puccinia graminis* (stem rust) race known as Ug99. Both races migrated from the Eastern African highlands to South Asia, through route A (Yemen, Saudi Arabia, Iraq, Iran and Afghanistan) or route B directly moving via Yemen (Meyer et al., 2017; Singh et al., 2006; Singh et al., 2004). The WANA is a target region that could benefit from breeding of durable resistance to rust (Singh et al., 2004).

Application of marker-assisted selection of *Sr2*, a durable resistance gene for the stem rust Ug99 race, has been suggested using SSR marker *Xgwm533* (Vishwakarma et al., 2019). Several complex genes conferring resistance to rusts and other diseases have been reported. An example is the putative ABC transporter gene, *Lr34/Yr18/Pm38*, that is effective for resistance to leaf rust (Lr), yellow rust (Yr) and powdery mildew (Pm). Moreover, *Lr67/Yr46/Sr55/Pm* and *Lr46/Yr29* are two other suggested complexes to convey durable resistance to wheat rust diseases (Herrera-Foessel et al., 2014; Krattinger et al., 2009;

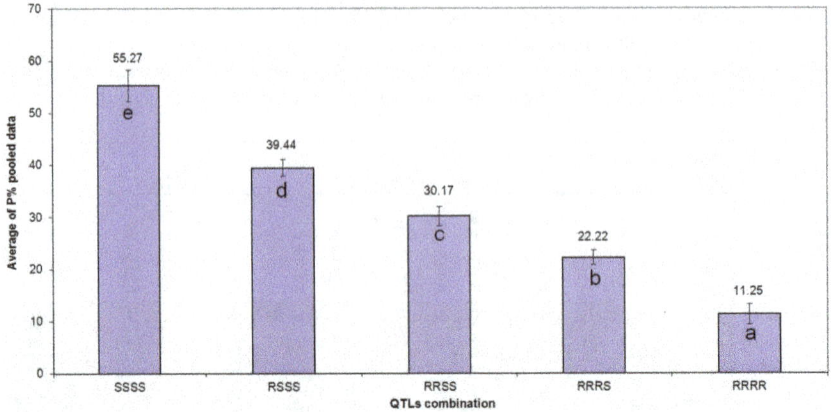

Figure 6 Stacking effect of individual QTLs that were detected in the Apache/Balance DH mapping population. Overall resistance to the five employed *Z. tritici* isolates (*P*) significantly (*P* = 0.01) increased with the number of QTLs in DH lines. SSSS and RRRR; Representative of individual DH lines merely carrying markers of susceptibility or resistance alleles associated with QTLs on chromosomes 3AS, 6DS,7DS and 1BS, respectively. RSSS and RRRS: indicate individual DH lines with combinations of one resistance allele linked to a QTL and three susceptibility alleles or vice versa. RRSS: indicates individual DH lines carrying two QTLs associated with resistance and two QTLs associated with susceptibility (Ghaffary et al., 2011).

Rosewarne et al., 2006). The transfer of complex durable resistance genes to commercial cultivars in national/regional collaborative breeding programs, the monitoring of emerging new aggressive races and the extension of as many as possible cultivars to improve genetic variation with different resistance genes can minimize the threat of rusts in the WANA region.

4 Challenges in breeding durable resistance in wheat: rust diseases in Iran

With a total land area of 1 628 760 km^2, Iran is the 16th biggest country in the world. Iran is located on the eastern tail of *Fertile Crescent* where wheat and barley domestication occurred 9000–12 000 years ago (Childe, 1953; Harlan and Zohary, 1966; Moore et al., 2000; Zohary and Hopf, 2000). Only about 9% of the land (~15 million hectares) can support arable cultivation (Fig. 8) (Anonymous 2005). Figure 9 indicates different climate zone classification where almost 84–88% of the country has been categorized as arid and semi-arid regions (Khalili and Rahimi, 2018; Vaghefi et al., 2019). Food security mainly relies on wheat and rice that solely provide 40% and 9.3% of daily calorie/capita, respectively (Table 2). The population of Iran has already reached 83 million people (MP), needing about 13 million tons (MT) wheat and 2.5 MT rice annually, increasing to an estimated 16 and 3.2 million tons when the

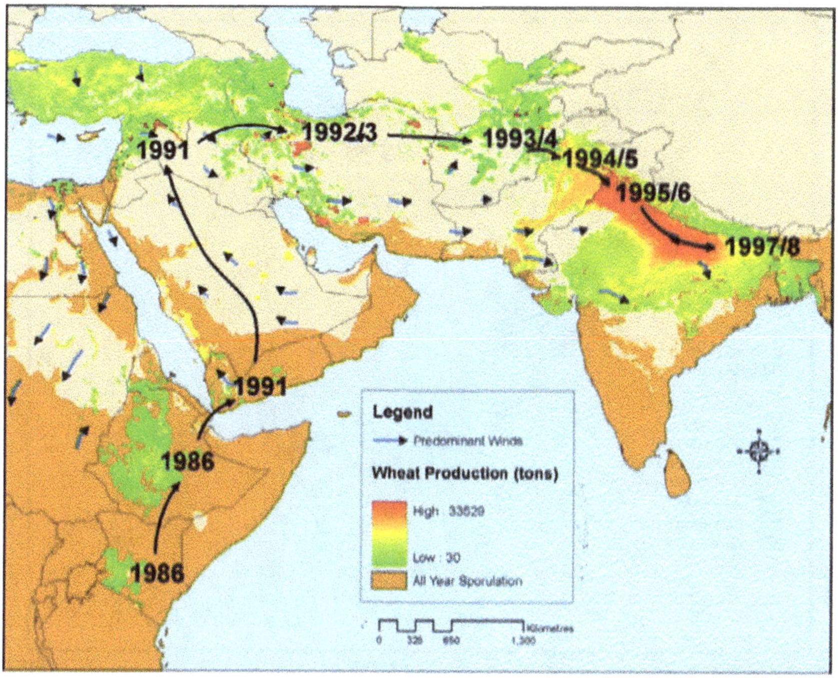

Figure 7 Year-round favorable climatic conditions, prevailing winds and historical, year-by-year movement of a Yr9-virulent race of the yellow rust pathogen that evolved in East Africa and migrated to Asia (Singh et al., 2006).

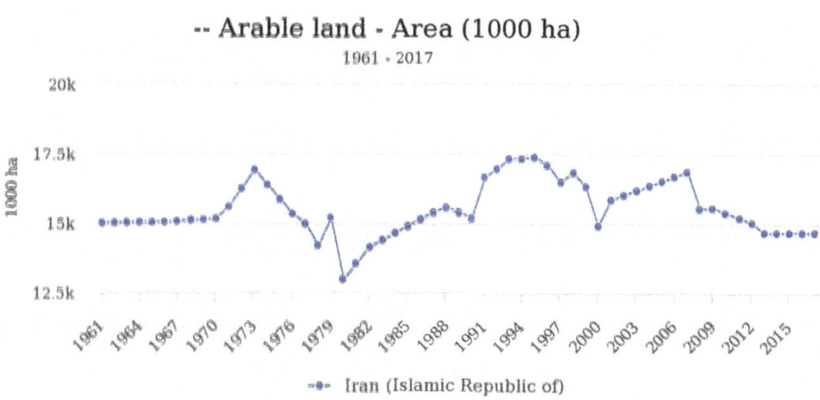

Figure 8 Iran's arable lands (FAOSTAT-land Use).

population reachesan estimated 105 million in 2061. Annual wheat production increased by a factor of 6.4 approximately from 1963 to 2007 and, since 2016, wheat production has not dropped below 14 MT, stopping reliance on wheat imports (Fig. 10).

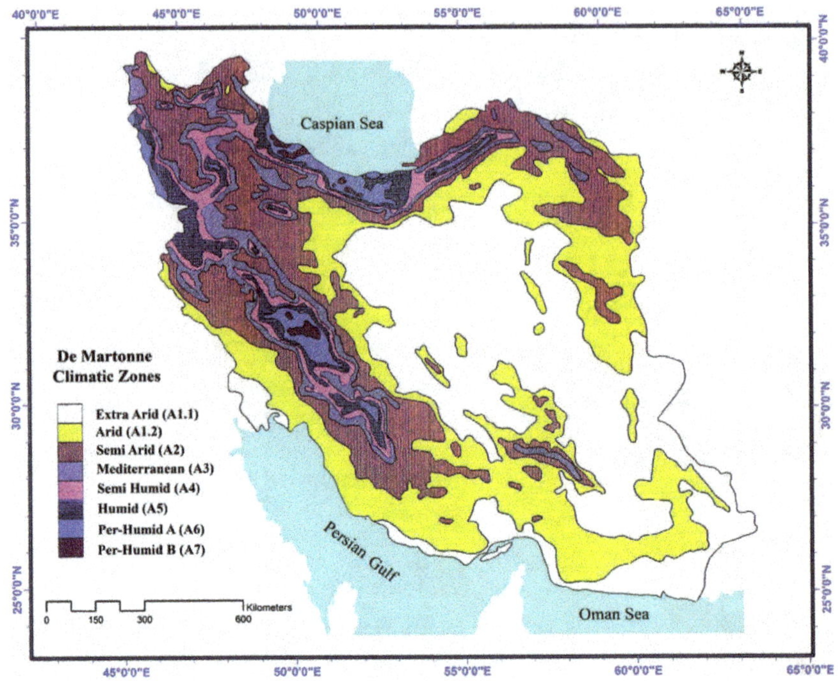

Figure 9 Iran's climate zone based on De Martonne.

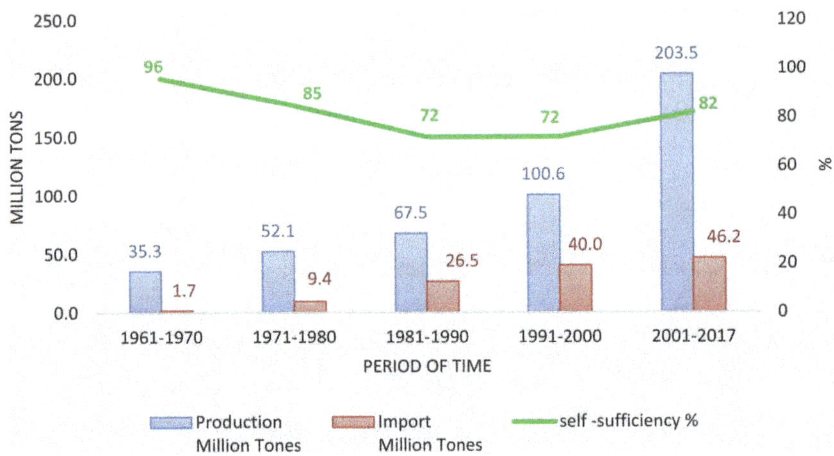

Figure 10 Iran's wheat production, import and self-sufficiency over time.

Cereal breeding in Iran began in 1930 by collecting and evaluating wheat local landraces. Six years later, cereal breeding programs officially started with the evaluation of 5100 wheat and barley pure lines and the release of two wheat cultivars 'Shahpasand', 'Ataee' as well as a barley cultivar 'California'. In 1948, Iran's ministry of agriculture became responsible for the cereal national breeding program collaborating with the FAO from 1952 to 1968. Since 1961, Iranian breeders have been trained by CIMMYT and ICARDA to develop a breeding program for bread and durum wheat, barley, triticale and rye that resulted in the release of 212 cultivars (Fig. 11). Currently, the Seed and Plan Improvement Institute (SPII-Cereal Department) and Dry land Agricultural Research Institute (DARI) manage breeding activities as well as releasing seeds in collaboration with private companies approved by the Seed and Plant Certification Research Institute.

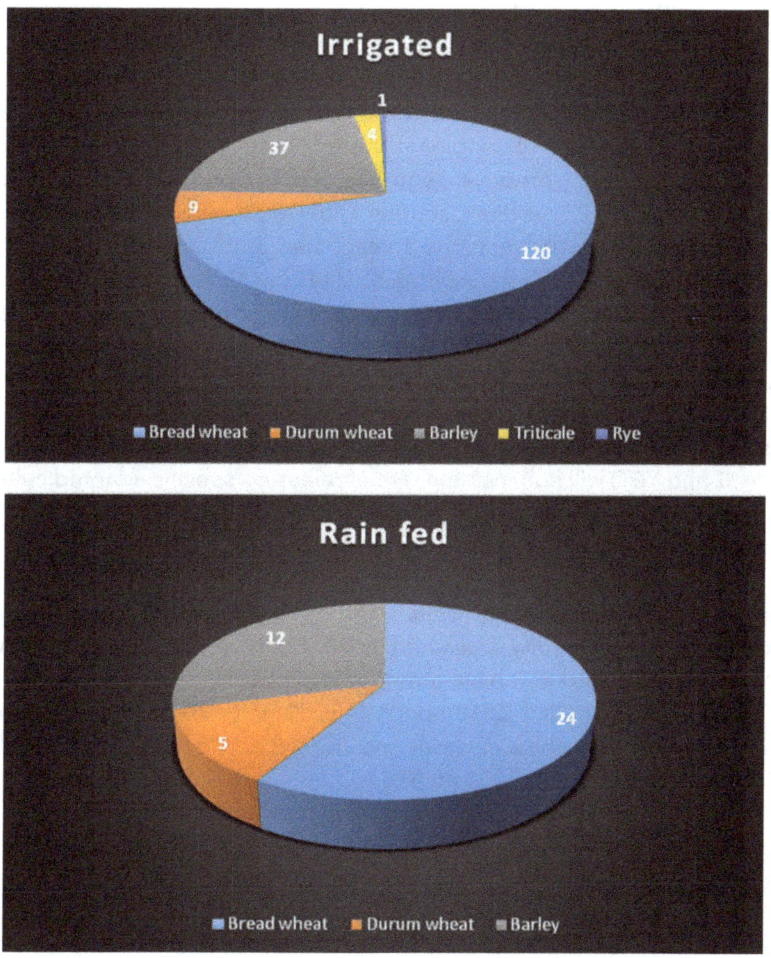

Figure 11 Released cultivars in Iran at irrigated and rainfed area.

The impact of wheat breeding in Iran was compared to China between 1961 and 2012, based on data from FAOSTAT and wheat atlas-released varieties. In order to segregate breeding impact on yield from other elements, a linear regression was conducted between the number of released varieties in each decade (1961–1970, 1971–1980, 1981–1990, 1991–2000 and 2001–2012) and yield in the last year of each time series. The highest explanation of total yield variation (R^2) was 0.90 for China, with an almost ninefold wheat yield increase from 559 kg/ha in 1961 to 4995 kg/ha in 2012. In comparison, the total yield variation by released varieties for Iran was 0.65 resulting in a 2.5 fold yield improvement. By the end of 2018 wheat yield had increased 2.7 times compared to 1961 with R^2 explaining 72% of total yield variation from 1961 to 2018.

The main objective of Iranian breeders is to increase the number of wheat genotypes with a high level of environmental adaptation and wide resistance to diseases in order to cover sub-climates within different climatic zones in Iran using 33 research field stations. Wheat germplasm management (Fig. 12) starts with a crossing program for each zone independently followed by segregating populations to produce recombinant inbred lines in the F6 generation. To reduce the period to produce segregating generations, shuttle breeding (Rajaram et al., 2002) between southern lowlands and northern highlands of Iran, managing two generations a year, is an approach that already have resulted in releasing cultivars such as Khalil (Bow"s"/Vee"s"//1-60-3/3/Cocoraque75/4/Chamran). Moreover, speed breeding (Ghosh et al., 2018; Watson et al., 2018) has been recently established at both SPII and DARI to minimize post cross recombinant inbred line production. Creating doubled haploid lines started in 2015, aiming at fast genetic purification of wheat germplasm after crossing. These techniques in combination with field evaluation and yield trials (PYT, AYT and YET) will enhance the rate of releasing specific adapted cultivars for a wide range of wheat-growing areas.

Abiotic stresses are major limiting factors in Iran, although biotic stresses including plant diseases are significant in wheat- and barley-growing areas where rainfall occurs during winter and coincides with disease optimal temperature and humidity. Fusarium head blight (FBH), powdery mildew (PM), tan spot (TS) and STB have emerged as important wheat foliar diseases compared to rust. In 2017–2018 about 445 000 ha (~8% of total wheat area) were sprayed against wheat diseases, including 210 000 ha for FHB, 83 000 ha on PM, 80 000 on TS and 50 000 on STB.

Yellow rust caused by *Puccinia striiformis* is particularly damaging in Iran. Yellow rust can survive on the wheat relatives or wild grasses where the temperature is moderate during summer and provides initial inoculum in the wheat-growing season (Torabi, 2012). In 1993, yellow rust destroyed almost 1.5 million tons representing about 14% of total national wheat production, in the west and North West of Iran where Falat and Qods (Ghuds) were the

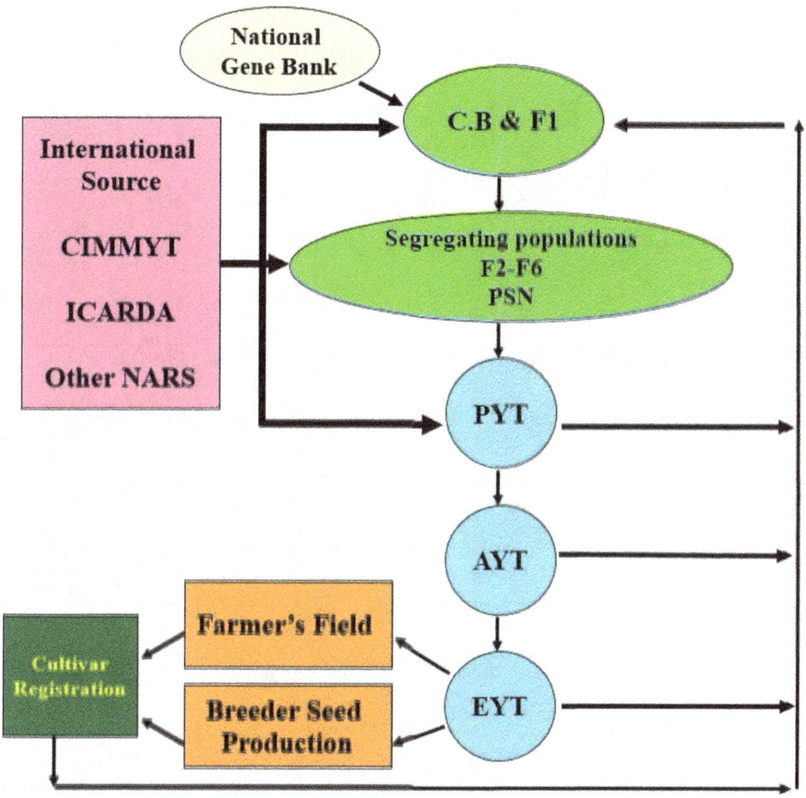

Figure 12 Breeding flowchart and wheat germplasm management in Iran (CB: crossing block, F1-F6: segregation generations, PSN: Local Preliminary Screen Nursery, PYT: Zone wide Preliminary Yield Trial, AYT: Advanced Yield Trial, EYT: Elite Yield Trial).

main cultivars (Torabi et al., 1995). Falat is a national registration name of Seri82 (Kavkaz/(SIB)Buha//Kalyansona/Bluebird) released in 1990 and carrying *Yr2*, *Yr7* and *Yr9* resistance genes. Qods (Red-Hussar/5/Willet/4/Norin-10/2*K-54// Frontana/3/Protor/6/Omid//Kalyasona/Bluebird) was also released in 1989 but with no reports of genotyping data.

After a number of yellow rust pandemics, trap nurseries were distributed annually over 40 different geographical locations in Iran to evaluate *Yr* genes efficiency using YR differential sets, Cobbity Near Isogenic Lines (NILs), foreign durable resistant varieties as well asa range of Iranian old and new wheat varieties. Table 6 indicates the evaluation of 33 *Yr* genes in the period of 1997–1999, 2001–2003 and 2011–2013 (Afshari et al., 2006; Nazari and Torabi, 2000; Safavi, 2015). The collaboration of wheat breeders and pathologists resulted in replacing several cultivars, including Chamran sourced from CIMMYTs' nurseries known as Attila (NordDesprez/VG9144//Kayansona/bluebird/3/(SI

Table 6 *Yr* resistance genes responses to yellow rust races over time and geographical locations in Iran

Gene	1997–1999			2001–2003				2011–2013
	North East	North	North West	North	West	Central	East	Ardabil North West
Yr1	S	R	R	R	R	R	R	No data
Yr2	S	S	S	S	S R (Kalyansona	S	R	S
Yr3	S	R	R	R	R	R	R	No data
Yr4	S	S	S	R	R	R	R	No data
Yr5	S	S	S	R	R	R	R	No data
Yr6	S	S	S	S	S	S	S	S
Yr7	S	S	S	S	S	S	S(*Yr7*/Avocat"s") R (Lee)	S
						R (Richersberg 42)		
Yr8	R	S	S	R	R	R	R	No data
Yr9	S	S	S	S	S	S	S	S
Yr10	S	S	S	R	R	R	R	No data
Yr11	S	R	S			No data		No data
Yr12	S	S	S			No data		No data
Yr13	R	R	S			No data		No data
Yr14	S	S	S			No data		No data
Yr15	S	S	S	R	R	R	R	No data
Yr16	R	R	S			No data		No data
Yr17	S	S	S	R	R	S	S	No data
Yr18	S	S	S	R	R	R	R	No data
Yr21		No data				No data		S
Yr22	S	S	S			No data		S
Yr23	S	S	S			No data		S
Yr24	S	S	S	R	R	R	R	S
Yr25	S	S	S	S	S	S	S	S
Yr26		No data				No data		S
Yr27	S	R	R	R	R	R	R	S
Yr31		No data				No data		S
YrA	S	S	S	S	S	S	S	S
YrC	S	S	S	R	R	R	R	No data
YrN		No data		R	R	R	R	No data
YrSD	S	S	S	R	R	R	R	No data
YrSP	S	S	S	R	R	R	R	No data
YrSU	R	S	S	R	R	S	R	S

B)Yaco/4/(SIB)Veery-5), in 1997. Chamran became quickly established as a high yielding resistant cultivar grown in 850 000 ha in 2011–2012, representing almost 40% of irrigated wheat harvested area (Wheat Atlas-Released Variety).

As a cultivar Chamran remained resistant for almost 16 years. In 2010 and later on in 2013, favorable climatic conditions spread rust disease in the WANA region (Fig. 13). In many countries, cool and wet conditions occurred and early yellow rust infections were observed. Very early yellow rust infections were observed in January 2013 in the south-west of Iran (South Khuzestan). The outbreak quickly developed and spread on Chamran cultivars, forcing farmers to spray up to 100 000 ha of infected field with fungicides in the south-west and west of Iran. Yellow rust on susceptible wheat cultivars was also reported in Ilam and Kermanshah on the west, Gorgan on the Caspian Sea region (North), Kerman and Jiroft on the south-east, south of Fars province and Ardabil on the North West in 2013. The virulent races circumvented *Yr27* resistance gene,

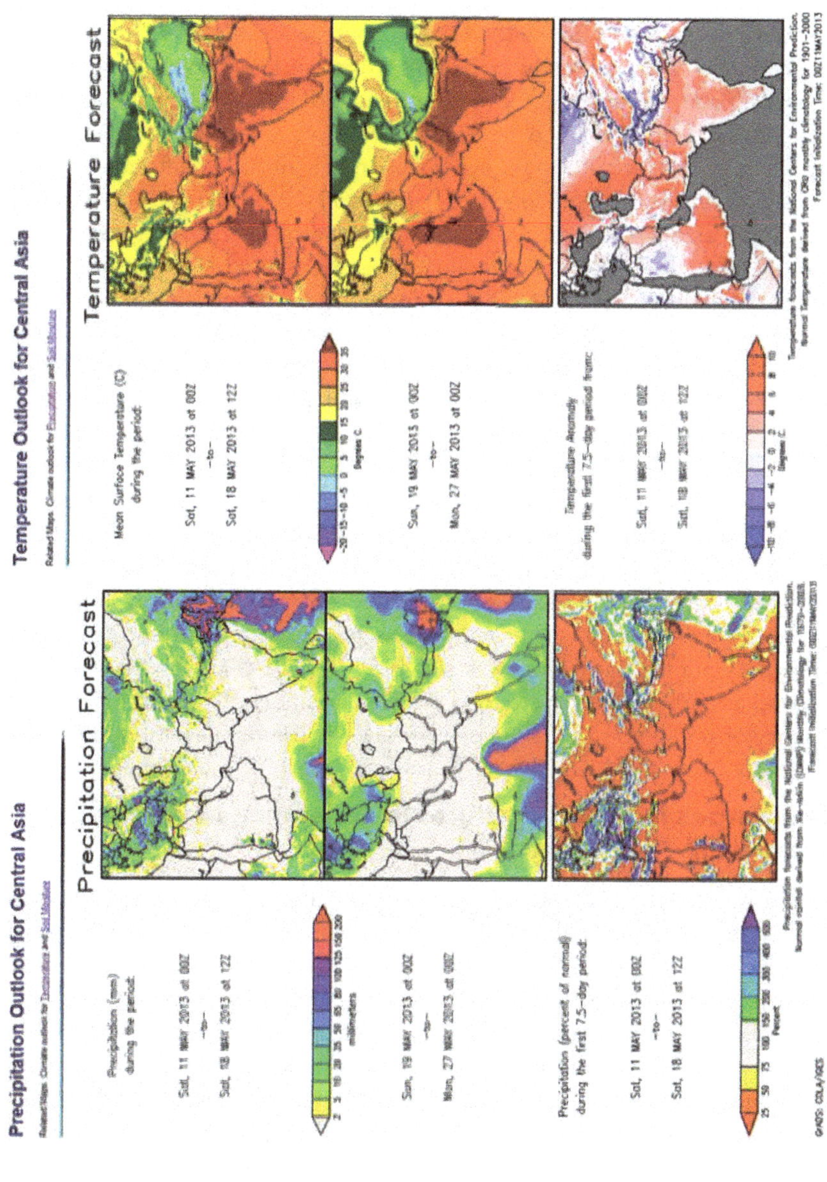

Figure 13 Precipitation and temperature over central Asia in 2013 (Source: COLA/IGES (http://wxmaps.org/).

but newly released cultivars including Chamran #2, Mehrgan, Parsi, Sivand, Sirvan, Gonbad, Morvareid, Zaerh, Uroum, Mihan, Dena, Arya, Kharkhah and Behranghave been classified as resistant cultivars (Hodson 2013).

Evaluations of wheat germplasm at the Safiabad Agricultural and Natural Resources Research and Education Center, south west of Iran, resulted in the identification of 17 completely or partially resistant lines within hundreds of genotypes. Pedigree analyses of resistant lines identified Batavia, Genaro81, Opata, Pastor, Trap and Yaco as potential donors of resistance. Genotype information of these cultivars is available on the http://wheatpedigree.net/ database, revealing the presence of *Yr33*, *Yr30+Yr18* and *Yr31* in Batavia, Opata and Pastor, respectively. Moreover, Genaro81, Trap and Yaco were known as the source of *Yr18* durable resistance gene. The progenies of selected plants in F_3 generation were genotyped for associated markers with *Yr18* (*Xgwm295-7D*), *Yr30* (flanking markers *Xgwm533.1-Xgwm493-3B*), *Yr31* (*Xgwm630/Xgwm374-2B* (*Lr13/Lr23*)) and *Yr33* (flanking markers *Xgwm111-Xgwm437-7D*). Currently, the lines are in yield trials and genotyping is in progress. Screening of wheat genotypes is conducted annually in hot spots where diseases occur (Table 7).

Genotyping and phenotyping wheat germplasm have enabled breeders to plan crossing programs rather than rely on random crossing. Since 2017–2018, the 'Increasing the Productivity of Wheat and Wheat-Based System in Iran' project has focused on wheat yield improvement and managing limitations including biotic stresses. A set of bread wheat germplasm with genotyping and phenotypic data for each entry was provided in 2016–2017 by CIMMYT and employed as parental lines in crossing block. A summary of genotyping and the frequency of single/combined genes carried by entries is presented in Table 8.

5 Challenges in breeding durable resistance in wheat: *Septoria* in North Africa

Northern Africa (NA) has been the cradle of wheat production since historical times. Wheat arrived from the Fertile Crescent and reached NA ca. 7000 years BC (Feldman et al., 2011). Later, Phoenicians have traded wheat along the Mediterranean shores and in 146 BC, Romans settled in Carthage (Tunisia), after which NA became the granary of the Empire (Hilali, 2013; Oliveira et al., 2012). During these times, cereal cultivation was also established by the Romans in the plateau of Sétif in Algeria and Numidia (Gautier, 1930). While durum wheat was well settled in NA at the end of the Roman Empire and its expansion further continued during the Early Arab empire (Martínez-Moreno et al. 2020), cultivation of bread wheat as a true crop started relatively recently after the French colonization in Tunisia, Algeria and Morocco (Boeuf, 1926; Ducellier, 1921; Miège, 1950).

Table 7 Hot spot locations to screen wheat genotypes against diseases in Iran

		North West			North				North East		Central	South	West				
Disease	Station	Miandoab	Moghan	Ardebil	Sari	Gorgan	Kalardasht	Gharakhil	Mashhad	Shirvan	Karaj	Zarghan	Kermanshah	Hamedan	Brojerd	Dezful	Ahvaz
Yellow Rust		✓	✓	✓	✓	✓			✓		✓	✓	✓	✓	✓	✓	
Leaf Rust			✓	✓	✓	✓	✓			✓	✓				✓	✓	✓
Stem Rust				✓			✓										
Powdery mildew						✓		✓									
Septoria			✓			✓		✓			✓					✓	
FHB			✓			✓					✓						

Table 8 Genotypic data over 531 bread wheat entries (399 Lines of Crossing Block Bread Wheat (CBBW) 2016/17 and 132 Lines of PPT Nursery) provided by CIMMYT Molecular laboratory. Each number on the table represents the frequency of genotypes carrying the single/combined resistance genes

Single/combined resistance genes	CBBW-2016/17	PPT
LR34 + (VPM) Lr37/Yr17/Sr38 + Lr46 + Lr68	1	0
LR34 + (VPM) Lr37/Yr17/Sr38 + Lr46	9	0
LR34 + (VPM) Lr37/Yr17/Sr38	1	0
LR34 + Lr46 + Lr68	3	0
LR34 + Lr46 + Sr35	1	0
LR34 + Lr46 + Lr68 Het	1	0
LR34 + Lr46	3	0
LR34 + Sr2	1	0
LR34	1	0
Lr34 Het + (VPM) Lr37/Yr17/Sr38 + Lr46	1	0
Lr34 Het + Lr46 + Lr68	1	0
Lr34 Het + Lr46	1	0
Lr34 Het + Lr46 + Lr68 Het	1	0
Lr46 + Lr68 Het	1	0
Lr46 + Sr2	1	0
Lr46 + Sr2 Het	2	0
Lr46 + Lr68	3	0
(VPM) Lr37/Yr17/Sr38 + Lr46 + Lr68 Het	3	3
(VPM) Lr37/Yr17/Sr38 + Lr46 + Lr68 + Sr2	3	2
(VPM) Lr37/Yr17/Sr38 + Lr46 + Lr68 + Sr2 Het	1	0
(VPM) Lr37/Yr17/Sr38 + Lr46 + Lr68	78	50
(VPM) Lr37/Yr17/Sr38 + Lr46 + Lr68 + Lr19/Sr25	4	0
(VPM) Lr37/Yr17/Sr38 + Lr46 + Lr19/Sr25	3	2
(VPM) Lr37/Yr17/Sr38 + Lr68	1	0
(VPM) Lr37/Yr17/Sr38 + Lr46	146	52
(VPM) Lr37/Yr17/Sr38 + Lr46 Het	0	1
(VPM) Lr37/Yr17/Sr38 + Lr46 + Sr2 Het	1	0
(VPM) Lr37/Yr17/Sr38 + Lr46 + Sr2	20	3
(VPM) Lr37/Yr17/Sr38 + Lr46 + Sr22	3	2
(VPM) Lr37/Yr17/Sr38	2	1
Lr46 + Lr68 + Sr2	2	1
Lr46 + Lr68	37	8
Lr46 + Lr68 Het	1	0
Lr46 + Sr2 + Yr15	2	0
Lr46 + Sr2	1	1

(Continued)

Table 8 (*Continued*)

Single/combined resistance genes	CBBW-2016/17	PPT
Lr46 + Sr2 Het	0	5
Lr46 + Lr19/Sr25	5	0
Lr46	50	1
Lr46 Het + Lr19/Sr25	1	0
Sr2	1	0
No data	2	0
Total	399	132

Wheat and barley are the main components of annual rainfed cropping systems in North Africa region (Algeria, Morocco, and Tunisia) with barley (3.5 million ha) being second in importance, of which 60% is grown in Morocco (Ullrich, 2010). Wheat is widely grown in marginal agricultural environments prone to drought due to a high degree of aridity and extreme temperatures (Tadesse et al., 2017). Durum wheat is widely grown because of its heat and drought tolerance and good performance in poor soils (Mastrangelo et al., 2005). Durum wheat is grown particularly for pasta and couscous (Tidiane Sall et al., 2019). In 2018, the largest wheat-producing country was Morocco with 7.3 million tons, followed by Algeria with approximately 3.9 million tons and Tunisia with 1.5 million tons (FAO-Data-Production, 2018).

In Morocco, wheat is grown on 2 million ha, while durum wheat is grown on 1 million ha. Yields have improved markedly, from 1.43 tons per ha at the end of 2007 to 1.92 tons per ha at the end of 2015 (Fig. 14). As part of the Green Morocco Plan the cereal sectors aims to reduce cereal acreage, increase production and reduce imports by 20% with the main aim to convert cereals into high value-added crops. In Algeria, cereals are grown on 3–3.5 million ha, representing nearly 40% of Algeria's total agricultural land (Benbelkacem, 2014). The agricultural sector essentially depends on rainfed agriculture (Cline, 2007; Schilling et al., 2012) with 50% of this area occupied by cereals, mostly by durum wheat which accounts for 2.2 million tons annually (Bensemane et al., 2011). In Tunisia, durum wheat is the most important cereal, grown on 0.54 million ha, representing about 85% of the area devoted to wheat which, in turn, accounted for more than half of the cereal cultivated area (Hammami and Gharbi, 2018) (Fig. 15).

From the beginning of the twentieth century breeding programs have generated a large number of varieties and lines, through either crossing local cultivars or native parents and foreign material. As a result, the grain yield of recently released varieties is three to four times than that of old varieties before the introduction of semi-dwarf wheat (Hammami and Gharbi 2018). Nevertheless, yields vary greatly from year to year and remain much lower than

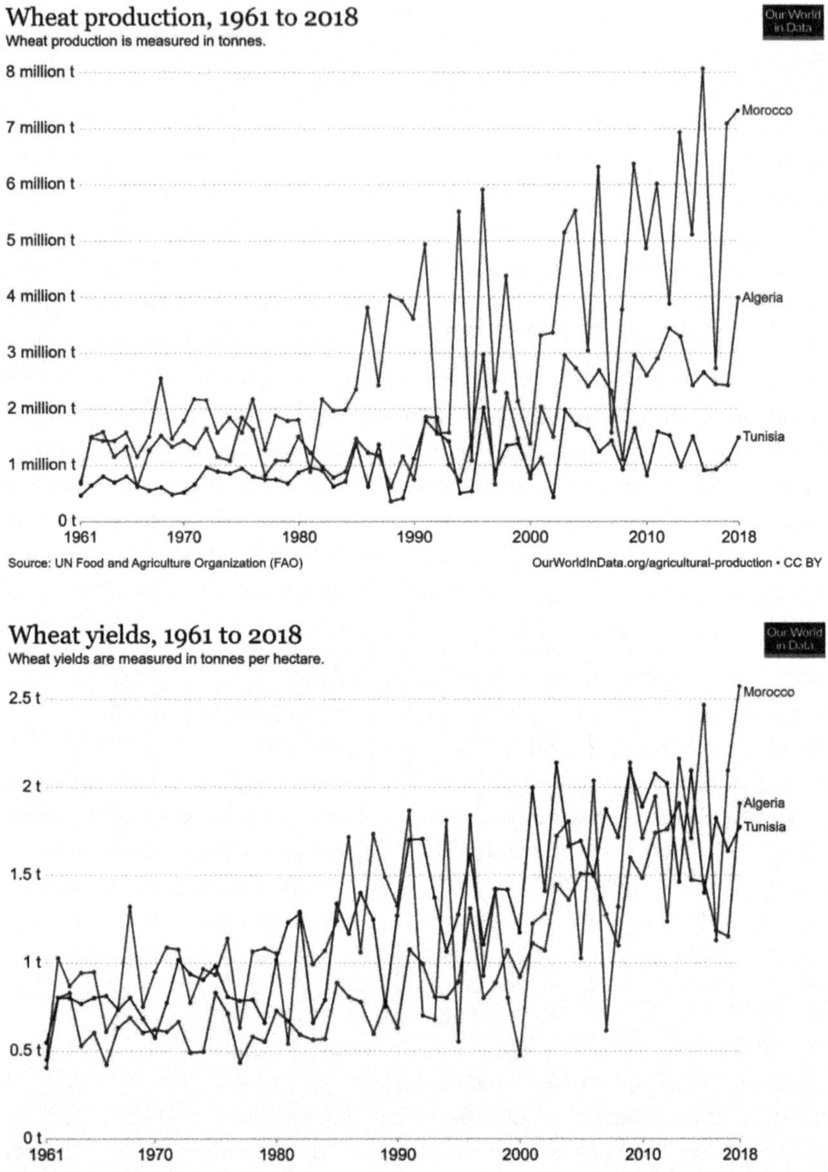

Figure 14 Wheat production and yields in Algeria, Morocco and Tunisia from 1961 to 2018 (FAOSTAT-Data-Crop, 2018).

in other regions of the world with similar climatic characteristics (Latiri et al., 2010) with an average yield of 1.6 tons/ha between 2000 and 2012 (Nefzaoui et al., 2012; Rastoin and Benabderrazek,, 2014) (Fig. 15).

Current durum wheat germplasm in Tunisia includes about 45 varieties (Medini et al., 2005) grouped in:

- traditional varieties selected from Tunisian landraces (developed up to the 1930s),
- old cultivars produced by crossing Tunisian traditional varieties and foreign materials (developed up to the 1970s), and
- modern cultivars obtained in national and international breeding centers (developed up to the present).

Traditional varieties such as 'Biskri', 'Bidi', 'Mahmoudi', 'Chili' and 'Jenah Khottifa' are still grown under traditional farming systems or where this germplasm is known to be tolerant to certain stresses (FAO, 2019). Although these landraces are grown on a reduced acreage since the adoption of modern cultivars, they are still highly appreciated as ingredients for local dishes such as frikeh, burghul and couscous. Tunisia is considered as a secondary center of diversity for durum wheat (Robbana et al., 2019). The diversification of the durum wheat genome and the development of a large collection of local populations such as in Algeria and Tunisia (Boeuf, 1932) is mainly due to the contrasting environmental conditions of NA but also due to the region's many influxes of peoples (Séguéla, 1946). This vast diversity also extends to the array of homemade foods derived from durum grains (Mackey, 2005).

Wheat production is far below its potential in most of these countries and they still rely on importing wheat from abroad. With the exception of Europe, North Africa (Algeria, Morocco, Tunisia, and Libya) is the largest import market for durum wheat (50% of the world market) (Bonjean and Mathijs, 2016). Wheat grain imports vary from 37% of the total need in Morocco to 66% in Algeria and 58% in Tunisia (FAOSTAT-Data-Imports). An FAO 2017 regional overview of food security in NA has highlighted the region's vulnerability (http://www.fao.org/3/I8336EN/i8336en.pdf).

In NA, landraces have been replaced by modern wheat cultivars (Martínez-Moreno et al., 2020; Soriano et al., 2016). The adoption of the improved germplasm from an alliance of NARS, CIMMYT, ICARDA and other advanced research institutes has led to major gains in wheat production and productivity (Payne, 2004; Reynolds and Borlaug, 2006; Trethowan and Crossa, 2007; Trethowan et al., 2002). In recent decades, the national breeding program conducted by the National Agricultural Research Institute of Tunisia in collaboration with CIMMYT has played a major role in improving wheat productivity (Ammar et al., 2011; El Felah and Gharbi, 2014; Hammami and

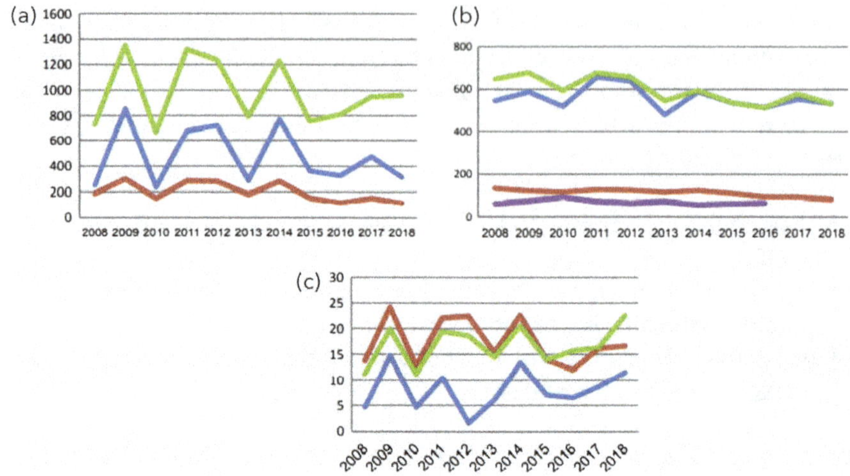

Figure 15 (a) Cereal production (1000 tons), (b) cultivated area of cereals (1000 ha), (c) yield q/ha (% cultivated area) in Tunisia (ONAGRI, 2018) (blue line: barley, green line: Durum wheat and red line: bread wheat; purple line: all cereals cultivated in irrigated area).

Gharbi, 2018). According to Ammar et al. (2011), wheat yield recorded during the period 2000–2010 is three times higher than it was in the 1960s.

However, introduction of more productive semi-dwarf cultivars has resulted in the loss of genetically diverse, locally well-adapted but unimproved landraces, and the loss of on-farm genetic variability. This so-called 'genetic erosion' (Maccaferri et al., 2005; Soleimani et al., 2002) has also resulted in a higher vulnerability of durum wheat germplasm to disease and particularly to fungal diseases such as STB, rusts and Fusarium head blight (Prat et al., 2016). STB caused by the foliar fungal pathogen *Z. tritici* is one of the most important biotic threats in Tunisia, Morocco and Algeria. The region suffered a severe outbreak from this fungus in 1970s that coincided with the introduction of novel semi-dwarf cultivars and the intensive use of fertilizers (Saari and Wilcoxson, 1974). In Tunisia, yield losses can reach up to 40% (Berraies et al., 2014) particularly on durum wheat that shows a higher level of susceptibility to *Z. tritici* compared to bread wheat, suggesting an adaptation of *Z. tritici* isolates to durum rather than to bread wheat (Bel Hadj Chedli et al., 2018; Ben Mohamed et al., 2000; Gharbi et al., 2000; Yahyaoui et al., 2000).

Currently, disease management relies on the use of fungicides and/or resistant cultivars. However, both fungicides and resistant cultivars are likely to rapidly lose efficacy against *Z. tritici* since it has a high evolutionary potential (McDonald and Linde, 2002). High levels of genetic diversity within the Moroccan and Tunisian *Z. tritici* populations have been reported (Bel Hadj Chedli et al., 2020; Siah et al., 2014). Furthermore, the occurrence of the teleomorph stage of *Z. tritici* has been confirmed (Ben Hassine and Hamada, 2014; Elbekali et al., 2012; Meamiche Neddaf et al., 2017) in the three countries and regular sexual reproduction cycles in Algeria and Tunisia occur (Ayad et al., 2014; Boukef et al., 2012; Hassine et al., 2019). The reported dichotomy and speciation in the *Z. tritici* populations (Ayad et al., 2014; Boukef et al., 2012; Hassine et al., 2019; Kema et al., 1996; Kema and van Silfhout, 1997; Ware, 2006) has prevented the use of well-defined *Z. tritici* isolates, further limiting resistance breeding in durum wheat. In addition, cultivation is dominated by the use of genetically uniform wheat cultivars, particularly the Karim cultivar that is used in more than 60% of durum wheat cultivation in Tunisia (Ammar et al., 2011; Ben Mohamed et al., 2000). This cultivar is highly susceptible to STB.

As a result, the demand for new *Septoria*-resistant wheat varieties by farmers has risen sharply. New varieties in Tunisia such as Nasr, Maali, Salim registered in 2004, 2007, 2009, respectively) and recently INRAT100 and Dhahbi (registered in 2017) seem to be more productive and resistant to *Septoria* than cultivar Karim (Hammami and Gharbi, 2018), though Salim and Nasr have now become susceptible to STB. With the high genetic variability of the pathogen (Linde et al., 2002; Morais et al., 2018; Zhan et al., 2003), it seems likely that most resistance genes will not last long, so there will be a continual need to identify new strategies for deployment of durable resistance. Recent studies by Ben M'Barek et al. (2020) showed that the use of variety mixtures in cultivation can improve disease control of STB.

Interestingly, the opposite situation exists in Morocco, where bread wheat is the major crop affected by STB in temperate rainfall bread wheat-growing regions and the yield losses can exceed 50% under favorable conditions (Mazouz et al., 1995; Zahri et al., 2014). It has been reported that isolates of *Z. tritici* exhibit both cultivar specificity (ability to infect only some cultivars of either durum or bread-wheat) and host species specificity (ability to only infect one or the other wheat (Kema et al., 1996; Kema and van Silfhout, 1997; Yahyaoui et al., 2000). On the other hand, STB affects both durum and bread wheat on coastal and sub-littoral regions in Algeria (Ayad et al., 2014; Benbelkacem, 2014).

Research to dissect STB resistance in contemporary tetraploid wheat and durum landraces showed an abundance of partial resistance in modern cultivars (Gharbi et al., 2000; Gharbi et al., 2008; Kidane et al., 2017). Studies using recombinant inbred lines from the cross between the susceptible high-yielding cv. Karim and the moderately resistant cv. Salim identified loci for quantitative STB partial resistance (Berraies et al. 2014). In addition, durum wheat landraces remain a reservoir of genetic diversity and thus a powerful tool for the introgression of novel resistance sources to STB in commercial breeding programs. However, efforts to explore diversity for resistance in a suite of

Tunisian landraces though promising, have begun recently (Aouini, 2018). Several studies conducted in Tunisia with regards to landraces have shown that these latter harbor sources of resistance to *Z. tritici* (Aouini, 2018; Ferjaoui et al., 2015; Ferjaoui et al., 2011; Ouaja et al., 2020). Aouini (2018) demonstrated that the broad efficacy of the observed STB resistance in the Tunisian landrace 'Agili39' is due to several stacked QTLs with partial and major effects at the seedling and adult stages (Ferjaoui et al., 2015; Ferjaoui et al., 2011).

However, it has been found that the high yielding Tunisian cv. Khiar, used in the cross with the landrace 'Agili39', shares a QTL on chromosome 2A that increased its susceptibility to *Z. tritici* in the derived RILs. This cultivar was registered in Tunisia in 1992 to increase yield potential in Tunisian durum wheats. This example shows breeders should be careful when introducing alleles for better yield that they do not also introduce unfavorable traits as was also observed among European bread wheats where susceptibility genes for *Z. tritici* were unconsciously introduced from CIMMYT germplasm (Arraiano and Brown, 2017). Trade-offs should be considered in plant breeding when several traits are important. It is a significant challenge for breeders to release varieties that combine good resistance and higher yields (Ammar et al., 2011) particularly when the R gene is linked to an important trait such as yield (Aouini, 2018).

The important sources of resistance to STB in landraces represent a good basis for further investigation and deployment in breeding programs (Aouini, 2018; Ferjaoui et al., 2015; Ouaja et al., 2020; Slim et al., 2019). The Tunisian genebank (NGBT) holds a total of 4000 accessions of durum wheat collected from the predominant remaining landraces, with the aim of fostering the cultivation of traditional Tunisian durum germplasm by on farm conservation (Slim et al., 2019). Diversity array technology has been used to investigate the genetic diversity and population structure of 196 durum wheat lines so that they can be compared with landraces from surrounding countries (Robbana et al., 2019).

To improve *Septoria* resistance and identify new resistance sources, while minimizing undesirable side-effects such as reductions in yield, intensive phenotyping under field conditions, plays a very important role. High-throughput field phenotyping that is likely to facilitate breeding for complex resistance by enabling many aspects of the plant resistance response to be screened simultaneously (Sankaran et al., 2015). The International Maize and Wheat Improvement Center (CIMMYT) established the *Septoria* Precision Phenotyping Platform in collaboration with the ICARDA and the Institution of Agricultural Research and Higher Education of Tunisia (IRESA) in Tunisia in 2015. The platform is part of a global network of precision field-based wheat phenotyping centers (Saint Pierre et al., 2019), where selected locations at key environments host platforms that generate data on prioritized traits, fostering global germplasm and data exchange. The platform aims to accelerate the transfer of STB resistant genes to durum wheat lines from national and international breeding programs, particularly CIMMYT and ICARDA breeding programs. An impressive diversity of durum wheat lines for resistance to the disease from research institutes across Tunisia, Morocco, Algeria, Mexico, France, Italy, the UK, USA and Canada were tested and evaluations of around 30 000 accessions are conducted at two major hotspots for the disease (Kodia at the National Institute of Field Crops-Irrigated & Beja at the Regional Field Crop Center of Beja-Rainfed) using artificial inoculation at seedling and adult growth stages using two distinctives STB-populations representatives of major durum wheat-growing areas. Several sources of resistance were identified, both in the elite germplasm panel and within the group of landraces (Yahyaoui et al., 2019).

Conventional selection methods for STB resistance can be improved by using genetic markers linked with STB resistance genes. These latter are very frequently used to measure the extent of genetic diversity and to identify and determine the genetic fingerprint of varieties. This will help not only in the identification of genotypes carrying desirable genes but also in the conservation of local and foreign genetic material and especially for the protection of new breeding lines. In addition, molecular marker-assisted selection accelerates the traditional selection method through genetic labeling and simplifies the management and evaluation of genetic pools. With the progress made in sequencing the wheat genome, it is becoming possible to introduce new high-speed sequencing technologies for genotyping and phenotyping and modeling in order to identify (by association genetics) new regions responsible for the control of target traits such as tolerance to biotic stresses (Hammami and Gharbi, 2018).

In Morocco, research for cultivars with better disease tolerance resulted in the release of cv. Nesma in 1976 that was replaced by cv. Amal in 1993 which was more tolerant (Jlibene and Elalaoui, 2002). There have been studies in Morocco on the potential of synthetic hexaploids as sources of resistance to *Z. tritici* in breeding programs, and to identify which of the known genes are present and effective in RILs populations (Ghizlan et al., 2014). There have been investigations into sources of resistance in a set of 377 advance breeding lines (ABLs) of bread wheat in the ICARDA breeding program to identify loci conferring resistance to STB both at the seedling as well as at adult plant stages. Genome-wide association studies (GWAS) revealed a total of 33 QTL at the seedling stage and 7 QTL at the adult plant stage. Identification of STB-resistant spring bread wheat germplasm, in combination with QTL found at both stages, will serve as an important resource in STB resistance breeding efforts (Louriki et al., 2019).

As in Tunisia and Morocco, numerous new cultivars from CIMMYT were introduced and cultivated over large areas by farmers in Algeria from 1968 to 1971, replacing old traditional varieties and resulting in increased productivity and some resistance to diseases (mainly to all rusts). The number of newly registered cultivated varieties increased from 6 in 1975 to 61 in 2012. The existing wheat collection comprising landraces, modern and new varieties have relatively low resistance rates (Bensemane et al., 2011). Genetic uniformity caused by mega-varieties, and overlapping and/or continued cultivation of susceptible varieties, are major constraints to wheat production in Algeria (Benbelkacem et al., 2016). Much effort has been devoted to screening wheat collections for leaf blight diseases (Meamiche Neddaf et al., 2017).

Algeria has invested in early warning systems for farmers about disease outbreaks and selection of highly resistant varieties used in crosses to improve the resistance of local varieties which are subsequently released and distributed to farmers to replace their susceptible varieties. This strategy was used for yellow rust where severe epidemics occurred in Algeria in 1977, 2004 and 2009 causing 80% yield losses. Newly released resistant varieties were deployed in different cereal growing areas providing a natural barrier between eastern to western Algeria to intercept the spread of the disease (Bentounsi, 2018).

Recent advances in genomic technologies, including marker-assisted selection, genetic transformation, new sequencing technologies and genomic editing, have opened up new possibilities for breeders (Gupta et al., 2010; Lusser et al., 2012). Genomic selection has strongly facilitated the combination of new resistance genes/assembling polygenic resistance into a single variety (Heffner et al., 2009; Tester and Langridge, 2010).

6 Conclusion and future trends

Whilst the Green Revolution significantly improved yields, it led to monocultures based on a relatively small number of varieties. This narrow genetic base has increased the vulnerability to diseases. More sustainable crop production calls for improved management practices (e.g. replacement of less sustainable monocultures by use of rotations), minimizing abiotic stresses by providing sufficient water and fertilizer, as well as effective control of disease, pests and weeds (Senapati et al., 2019). Crop diversification, for example, can disrupt the life cycle of common crop diseases such as *Fusarium* (Kazan and Gardiner, 2018; Kelly and Ward, 2018; Suproniene et al., 2019). Good management is essential since no cultivar is able to resist all types of stress. A targeted approach to improving disease resistance also involves mapping risks and targeting resources on the most damaging diseases, whether breeding programs or improved management, as is the case with wheat rust diseases in Iran.

Research is continually advancing our understanding of the resistance mechanisms, with a particular emphasis on R genes. However, this remains limited for durum wheat where genetic studies have only begun recently.

More needs to be done to use wheat landrace collections that contain a wide genetic diversity. Alternative breeding methods such as decentralized participatory plant breeding are also being used to develop more diverse varieties better adapted to organic and low-input systems in particular locations. This approach allows farmers to be involved at an early stage of the breeding process by selecting varieties that are adapted to their needs (van Frank et al., 2020). As these approaches suggest, we should focus more on specific adaptation rather than general adaptation by applying modern high-throughput genetic, breeding, and plant pathology, phenotyping techniques as well as satellite-based meteorology and geography to classify micro-target environments and release cultivars that are specific for each target. This approach can maximize genetic variation and make a multi-layer defense strategy using multi-resistance genes against a wide range of diseases.

7 Where to look for further information

- Martínez-Moreno, F., Solís, I., Noguero, D. et al. (2020). Durum wheat in the Mediterranean Rim: historical evolution and genetic resources. *Genet Resour Crop Evol*. https://doi.org/10.1007/s10722-020-00913-8.
- Mrabet, R. (2011). *No-Tillage Agriculture in West Asia and North Africa Rainfed Farming Systems* (pp. 1015-1042): Springer.
- Pala, M., Oweis, T., Benli, B., De Pauw, E., El Mourid, M., Karrou, M., Jamal, M. and Zencirci, N. (2011). *Assessment of Wheat Yield Gap in the Mediterranean: Case Studies from Morocco, Syria and Turkey*. International Center for Agricultural Research in the Dry Areas (ICARDA), Aleppo, Syria. iv, 963-921.
- Tadesse, W., Amri, A., Sanchez-Garcia, M., El Bouhssini, M., Karrou, M., Patil, S., Bassi, F., Baum, M. and Oweis, T. (2017). Improving wheat production in the Central and West Asia and North Africa (CWANA) region. In: Achieving sustainable cultivation of wheat Volume 1: Breeding, quality traits, pests and diseases. Burleigh Dodds Science Publishing Limited.
- William, A., Alain, B., Angus, W. J. and Van Ginkel, M. (2016). *The World Wheat Book: A History of Wheat Breeding*, Vol 3. Lavoisier.
- The International Maize and Wheat Improvement Center (CIMMYT) (www.cimmyt.org) (which focuses on breeding new varieties).

8 References

Adhikari, T. B., Anderson, J. M. and Goodwin, S. B. (2003). Identification and molecular mapping of a gene in wheat conferring resistance to *Mycosphaerella graminicola*. *Phytopathology* 93(9), 1158-1164.

Adhikari, T. B., Cavaletto, J. R., Dubovsky, J., Gieco, J. O., Schlatter, A. R. and Goodwin, S. B. (2004c). Molecular mapping of the *Stb4* gene for resistance to Septoria tritici blotch in wheat. *Phytopathology* 94(11), 1198–1206.

Adhikari, T. B., Wallwork, H. and Goodwin, S. B. (2004a). Microsatellite markers linked to the *Stb2* and *Stb3* for resistance to Septoria tritici blotch in wheat. *Crop Science* 44(4), 1403–1411.

Adhikari, T. B., Yang, X., Cavaletto, J. R., Hu, X., Buechley, G., Ohm, H. W., Shaner, G. and Goodwin, S. B. (2004b). Molecular mapping of *Stb1*, a potentially durable gene for resistance to Septoria tritici blotch in wheat. *TAG. Theoretical and Applied Genetics. Theoretische und Angewandte Genetik* 109(5), 944– 953.

Afshari, F., Torabi, M., Kia, S., Dadrezaei, S., Safavi, S., Chaichi, M., Karbalaei Khiavi, H., Zakeri, A., Nasrollahi, M., Patpour, M. and Ebrahimnezhad, S. (2006). Monitoring of virulence factors of Puccinia triticina Eriksson, the causal agent of wheat leaf rust in Iran during 2002-2004. *Seed Plant* 21, 485–496.

Ammar, K., Gharbi, M. S. andDeghais, M. (2011). Wheat in Tunisia. In: Bonjean, A. P. and Angus, W. J. (Eds) *The World Wheat Book: A History of Wheat Breeding*. Paris: Lavoisier, 443–465.

Anderson, R. M., Fraser, C., Ghani, A. C., Donnelly, C. A., Riley, S., Ferguson, N. M., Leung, G. M., Lam, T. H. and Hedley, A. J. (2004). Epidemiology, transmission dynamics and control of SARS: the 2002-2003 epidemic. *Philosophical Transactions of the Royal Society of London. Series B, Biological Sciences* 359(1447), 1091–1105.

Anonymous (2005). Fertilizer use by crop in the Islamic Republic of Iran. Contents published by FAO, Rome. Available at: http://www.fao.org/3/a0037e00.htm#.

Aouini, L. (2018). Durum wheat and Septoria tritici blotch: genes and prospects for breeding. PhD thesis, Wageningen University, The Netherlands.

Arraiano, L. S. and Brown, J. K. (2017). Sources of resistance and susceptibility to Septoria tritici blotch of wheat. *Molecular Plant Pathology* 18(2), 276–292.

Arraiano, L. S., Chartrain, L., Bossolini, E., Slatter, H. N., Keller, B. and Brown, J. K. M. (2007). A gene in European wheat cultivars for resistance to an African isolate of *Mycosphaerella graminicola*. *Plant Pathology* 56(1), 73– 78.

Arraiano, L. S., Worland, A. J., Ellerbrook, C. and Brown, J. K. M. (2001). Chromosomal location of a gene for resistance to Septoria tritici blotch (*Mycosphaerella graminicola*) in the hexaploid wheat 'synthetic 6x'. *Theoretical and Applied Genetics* 103(5), 758– 764.

Ayad, D., Sayoud, R., Benbelkacem, K. and Bouznad, Z. (2014). La tâche septorienne du blé: Première signalisation de la présence en Algérie des deux Mating types du téleomorphe Mycosphaerella graminicola (Fuckel) Schröter, (anamorphe: Septoria tritici Rob. ex Desm.) et diversité phénotypique de l'agent pathogène. Nature & Technology 11, 34.

Bel Hadj Chedli, R., Ben M'Barek, S., Souissi, A., Yahyaoui, A., Rezgui, S. and Chaabane, H. (2020). Screening for resistance of Tunisian, Moroccan and Algerian wheat cultivars to *Zymoseptoria tritici* in Northern Tunisia. *Journal of Plant Pathology* 102, 1–11.

Bel Hadj Chedli, R., Ben M'Barek, S., Yahyaoui, A., Kehel, Z. and Rezgui, S. (2018). Occurrence of Septoria tritici blotch (*Zymoseptoria tritici*) disease on durum wheat, triticale, and bread wheat in Northern Tunisia. *Chilean Journal of Agricultural Research* 78(4), 559–568.

Ben Hassine, A. H. and Hamada, W. (2014). First isolation of the *Mycosphaerella graminicola* teleomorph stage causing Septoria leaf blotch on wheat in Tunisia. *New Disease Reports* 29, 18.

Ben M'Barek, S., Karisto, P., Abdedayem, W., Laribi, M., Fakhfakh, M., Kouki, H., Mikaberidze, A. and Yahyaoui, A. (2020). Improved control of Septoria tritici blotch in durum wheat using cultivar mixtures. *Plant pathology* 00:1-11. DOI: 10.1111/ppa.13247.

Ben Mohamed, L., Rouaissi, M., Sebei, A., Hamza, S. and Harrabi, M. (2000). Effet du génotype, de la date de semis, de la fertilisation azotée et potassique et des fongicides sur le développement de Septoria tritici. In: Royo, C., Nachit, M., Di Fronzo, N. and Araus, J. L. (Eds) Durum *Wheat Improvement* in the Mediterranean *Region*: New *Challenges*. CHIEHAM- *Options Méditerranéenne*: Série A. Séminaires Méditerranéen(vol. 40). Zaragoza: CIHEAM, 349-356.

Benbelkacem, A. (2014). The history of wheat breeding in Algeria. In : Porceddu, E., Damania, A. B., Qualset, C. O. (Eds) *Proceedings of the International Symposium on Genetics and Breeding of Durum Wheat*. Bari: CIHEAM, pp. 363-370 (Options Méditerranéennes : Série A. Séminaires Méditerranéens; n. 110).

Benbelkacem, A., Jenadi, C. and Meamiche, H. (2016). Mitigation of the global threat of Septoria leaf blotch of cereals in Algeria. *International Journal of Research Studies in Agricultural Sciences* 2, 28-35.

Bensemane, L., Bouzerzour, H., Benmahammed, A. and Mimouni, H. (2011). Assessment of the phenotypic variation within two- and six rowedbarley (*Hordeum vulgare* L.) breeding lines under semiaridcondition. *Advancesin Environmental Biology* 5, 1454-1460.

Bentounsi, A. (2018). Occurrence of wheat rusts in Algeria and strategies to reduce crop losses. Paper presented at the BGRI, Marrakech, Morocco.

Berraies, S., Ammar, K., Gharbi, M. S., Yahyaoui, A. and Rezgui, S. (2014). Quantitative inheritance of resistance to Septoria tritici blotch in durum wheat in Tunisia. *Chilean Journal of Agricultural Research* 74(1), 35-40.

Boeuf, F. (1926). Amélioration de la culture du Blé en Tunisie (Suite et fin). Revue de botanique appliquée et d'agriculture coloniale,6n)blotch in durum w 64, 757-765.

Boeuf, F. (1932). Le blé en Tunisie: La plante. *Ann. Serv. Botanique et Agronomique Tunisie. Tunis* 1, 1-454.

Bonjean, I. and Mathijs, E. (2016). How transaction costs shape market power: conceptualization and policy implications. In: Conference Proceedings of the 12th European IFSA Symposium, 12-15 July 2016 at Harper Adams University, UK.

Boukef, S., McDonald, B. A., Yahyaoui, A., Rezgui, S. and Brunner, P. C. (2012). Frequency of mutations associated with fungicide resistance and population structure of *Mycosphaerella graminicola* in Tunisia. *European Journal of Plant Pathology* 132(1), 111-122.

Brading, P. A., Verstappen, E. C. P., Kema, G. H. J. and Brown, J. K. M. (2002). A gene-for-gene relationship between wheat and *Mycosphaerella graminicola*, the Septoria tritici blotch pathogen. *Phytopathology* 92(4), 439-445.

Brown, J. K. M., Chartrain, L., Lasserre-Zuber, P. and Saintenac, C. (2015). Genetics of resistance to *Zymoseptoria tritici* and applications to wheat breeding. *Fungal Genetics and Biology: FG and B* 79, 33-41.

Byerlee, D. and Moya, P. (1993). *Impacts of International Wheat Breeding Research in the Developing World, 1966-1990*. Mexico: CIMMYT.

Cervigni, R. and Dobardzic, S. (2012). Adaptation to climate change in the Middle East and North Africa region. Middle East and North Africa. Available at: http://web.worldbank.org/WBSITE/EXTERNAL/COUNTRIES/MENAEXT/0,,contentMDK:21596766~pagePK:146736~piPK:146830~theSitePK:256299,00.htm.

Chakraborty, S. and Newton, A. C. (2011). Climate change, plant diseases and food security: an overview. *Plant Pathology* 60(1), 2–14.

Chartrain, L., Berry, S. T. and Brown, J. K. (2005). Resistance of wheat line Kavkaz-K4500 L.6.a.4 to septoria tritici blotch controlled by isolate-specific resistance genes. *Phytopathology* 95(6), 664–671.

Chartrain, L., Sourdille, P., Bernard, M. and Brown, J. K. M. (2009). Identification and location of Stb9, a gene for resistance to Septoria tritici blotch in wheat cultivars Courtot and Tonic. *Plant Pathology* 58(3), 547–555.

Childe, V. G. (1953). *New Light on the Most Ancient East*. New York: Frederick A. Praeger.

CIMMYT.org. Available at: https://www.cimmyt.org/news/what-is-conservation-agriculture/.

Cline, W. (2007). *Global Warming and Agriculture*. Washington, DC: Peterson Institute for International Economics.

DARI. Available at: http://dari.areeo.ac.ir/HomePage.aspx?TabID=10535&Site=dari.areeo.ac&Lang=en-US.

Driscoll, C. A., Macdonald, D. W. and O'Brien, S. J. (2009). From wild animals to domestic pets, an evolutionary view of domestication. *Proceedings of the National Academy of Sciencesof the United States of America* 106(Suppl.1), 9971–9978. doi: 10.1073/pnas.0901586106.

Ducellier, L. (1921). Contribution à l'étude des espèces du genre Triticum cultivées dans le Nord de l'Afrique. *Bulletinde la société d'Histoire Naturellede l'Afrique du Nord*, pp. 66–68.

El Felah, M. and Gharbi, M. S. (2014). Les céréales en Tunisie: historiques, contraintes de développement et perspective. *Actes de la journée nationale sur la valorisation des résultats de la recherche dans le domaine des grandes cultures, Tunis* 110, 1–7.

Elbekali, A. Y., Siah, A., Ramdani, A., El Chartouni, L., Tisserant, B., Deweer, C., Hafidi, M., Reignault, Ph. andHalama, P. (2012). First report of the Mating type Mat1-1 and Mat1-2 of *Mycosphaerella graminicola* in Morocco. *Journal of Plant Pathology* 94(1), 243–247.

Elias, E. (1995). Durum wheat products. Durum wheat improvement in the Mediterranean region: new challenges. *Serie a: Séminaires Méditerranéennes* 40, 23–31.

FAO (2017). Available at: http://www.fao.org/3/I8336EN/i8336en.pdf

FAO (2019). Rural transformation-Key for sustainable development in the near East and North Africa. Overview of Food Security and Nutrition 2018. Cairo. 80pp. Licence: CCBY-NC-SA 3.0 IGO.

FAOAQUASTAT. Available at: http://www.fao.org/nr/water/aquastat/data/query/results.html

FAOSTAT-Data-Crop (2018). Available at: http://www.fao.org/faostat/en/#data/QC

FAOSTAT-Data-Food Supply. Available at: http://www.fao.org/faostat/en/#data/CC

FAOSTAT-Data-Imports. Available at: http://www.fao.org/faostat/en/#rankings/commodities_by_country_imports.

FAOSTAT-Data-Land Use. Available at: http://www.fao.org/faostat/en/#data/RL/visualize.

FAOSTAT-Data-Population (2013). Available at: http://www.fao.org/faostat/en/#data/.

FAO-Data-Production (2018). Available at: http://www.fao.org/faostat/en/#data/QC.

Faris, J. D. (2014). Wheat domestication: key to agricultural revolutions past and future. In: Tuberosa, R., Graner, A. and Frison, E. (Eds) *Genomics of Plant Genetic Resources*. Dordrecht: Springer, 439–464.

Feldman, M., Bonjean, A. P. and Angus, W. J. (2011). Origin of cultivated wheat. In Bonjean, A. P. and Angus, W. J. (Eds.) *The Worlds Wheat Book: A History of Wheat Breeding*, Paris, France: Lavoisier Publishing 3-56.

Ferjaoui, S., Ben M'Barek, S., Bahri, B., Slimane, R. B. and Hamza, S. (2015). Identification of resistance sources to Septoria tritici blotch in old Tunisian durum wheat germplasm applied for the analysis of the *Zymoseptoria tritici*-durum wheat interaction. *Journal of Plant Pathology* 97(3), 471–481.

Ferjaoui, S., Sbei, A., Aouadi, A. and Hamza, S. (2011). Monogenic inheritance of resistance to Septoria tritici blotch in durum wheat 'Agili'. *International Journal of Plant Breeding* 5(1), 17-20.

Gautier, M. E. F. (1930). L'Evolution de l'Algérie de 1830 à 1930. In: Comité National Métropolitain du Centenaire de l'Algérie(Ed)*Les 12 cahiers du Centenaire de l'Algérie*(vol 3). Orléans: A. Pigelet &Cie, 95 pp.

Gepts, P. (2004). Crop domestication as a long-term selection experiment. *Plant Breeding Reviews* 24(2), 1-44.

Gepts, P. and Papa, R. (2003). Evolution during domestication. *eLS*. doi:10.1038/npg.els.0003071.

Ghaffary, S. M. T. (2011). Efficacy and mapping of resistance to Mycosphaerella graminicola in wheat. PhD thesis, Wageningen University, The Netherlands, 233 pp.

Ghaffary, S. M. T., Faris, J. D., Friesen, T. L., Visser, R. G. F., van der Lee, T. A. J., Robert, O. and Kema, G. H. J. (2012). New broad-spectrum resistance to Septoria tritici blotch derived from synthetic hexaploid wheat. *Theoretical and Applied Genetics* 124, 125-142.

Ghaffary, S. M. T., Robert, O., Laurent, V., Lonnet, P., Margalé, E., van der Lee, T. A., Visser, R. G. and Kema, G. H. (2011). Genetic analysis of resistance to Septoria tritici blotch in the French winter wheat cultivars Balance and Apache. *TAG. Theoretical and Applied Genetics. Theoretische und Angewandte Genetik* 123(5), 741–754.

Gharbi, M. S., Deghais, M. and Ben Amar, F. (2000). Breeding for resistance to Septoria tritici in durum wheat. Durum wheat improvement in the mediterranean region: new challenges. CIHEAM, Zaragoza, Spain.(Options Méditerranéennes). *Série A: Séminaires Méditerranéens* 40, 397-401.

Gharbi, M. S., Fakhar, S. B., Ammar, K., Hajlaoui, M. R., Yahyaoui, A. and Trifi, M. (2008). Breeding durum wheat with resistance to Septoria leaf blotch for sustainable production under Mediterranean conditions. Paper presented at the Proceeding of International Durum Wheat Symposium from Seed to Pasta: The Durum Wheat Chain, Bologna, Italy.

Ghizlan, D., Rabha, A., Benchaba, M., Gaboun, F., Arhbal, A. and Ibriz, M. (2014). Marker assisted evaluation for resistance to Septoria tritici blotch in Moroccan wheat recombinant inbred lines (RILs). *International Journal of Agriculture Science and Research* 3, 213-221.

Ghosh, S., Watson, A., Gonzalez-Navarro, O. E., Ramirez-Gonzalez, R. H., Yanes, L., Mendoza-Suárez, M., Simmonds, J., Wells, R., Rayner, T., Green, P., Hafeez, A., Hayta, S., Melton, R. E., Steed, A., Sarkar, A., Carter, J., Perkins, L., Lord, J., Tester, M., Osbourn, A., Moscou, M. J., Nicholson, P., Harwood, W., Martin, C., Domoney, C., Uauy, C., Hazard, B., Wulff, B. B. H. and Hickey, L. T. (2018). Speed breeding in growth chambers and glasshouses for crop breeding and model plant research. *Nature Protocols* 13(12), 2944–2963.

Goodwin, S. B., Cavaletto, J. R., Hale, I. L., Thompson, I., Xu, S. X., Adhikari, T. B. and Dubcovsky, J. (2015). A new map location of gene *Stb3* for resistance to septoria tritici blotch in wheat. *Crop Science* 55(1), 35–43. https://doi.org/10.2135/cropsci2013.11.0766.

Gupta, P. K., Langridge, P. and Mir, R. R. (2010). Marker-assisted wheat breeding: present status and future possibilities. *Molecular Breeding* 26(2), 145–161.

Hammami, R. and Gharbi, M. S. (2018). Contribution of breeding program to the sustainability of durum wheat production. *Journal of New Sciences* 59, 3799–3805.

Harlan, J. R. and Zohary, D. (1966). Distribution of wild emmer wheat and barley. *Science* 153(3740), 1074–1080.

Hassine, M., Siah, A., Hellin, P., Cadalen, T., Halama, P., Hilbert, J. L., Hamada, W., Baraket, M., Yahyaoui, A., Legrève, A. and Duvivier, M. (2019). Sexual reproduction of *Zymoseptoria tritici* on durum wheat in Tunisia revealed by presence of airborne inoculum, fruiting bodies and high levels of genetic diversity. *Fungal Biology* 123(10), 763–772.

Heffner, E. L., Sorrells, M. E. and Jannink, J.-L. (2009). Genomic selection for crop improvement. *Crop Science* 49(1), 1–12.

Heisey, P., Lantican, M. and Dubin, H. (2003). Wheat 4. In: Evenson, R. (Ed.) *Crop Variety Improvement and Its Effect on Productivity*. USA: Yale University. Available at: 47.fao.org.

Herrera-Foessel, S. A., Singh, R. P., Lillemo, M., Huerta-Espino, J., Bhavani, S., Singh, S., Lan, C., Calvo-Salazar, V. and Lagudah, E. S. (2014). Lr67/Yr46 confers adult plant resistance to stem rust and powdery mildew in wheat. *TAG. Theoretical and Applied Genetics. Theoretische und Angewandte Genetik* 127(4), 781–789.

Hilali, A. (2013). Rome and agriculture in Africa proconsularis: land and hydraulic development. *Revue belge de philologie et d'histoire*91(1), 113–125.

Hodson, D. (2013). Caution–potentially serious outbreaks of yellow (stripe) rust in cwana region. Vigilance needed in East Africa. Available at: https://rusttracker.cimmyt.org/?p=5101RustTracker.org.

Ibrahim, M. and Lal, R. (2013). Climate change and land use in the WANA region with a specific reference to morocco. In: Sivakumar, M. V. K., Lal, R., Selvaraju, R. and Hamda, I. (Eds) *Climate Change and Food Security in West Asia and North Africa*. Dordrecht, Netherlands: Springer, 89–113.

Jlibene, M. and Elalaoui, A. (2002). Variétés de blé adaptées au Nord-Ouest du Maroc. *Bulletin Transfert de Technologie en Agriculture*88, 1–4.

Johnson, R. (1983). Genetic background of durable resistance. In: Lamberti, F., Waller, J. M., Van der Graaff, N. A. (Eds) *Durable Resistance in Crops. NATO Advanced Science Institutes Series (Series A: Life Sciences)*, vol 55. Boston, MA: Springer. https://doi.org/10.1007/978-1-4615-9305-8_2.

Kazan, K. and Gardiner, D. M. (2018). Transcriptomics of cereal–Fusarium graminearum interactions: what we have learned so far. *Molecular Plant Pathology* 19(3), 764–778.

Kelly, A. C. and Ward, T. J. (2018). Population genomics of Fusarium graminearum reveals signatures of divergent evolution within a major cereal pathogen. *PLoS ONE* 13(3), e0194616.

Kema, G., Annone, J., Sayoud, R. and Van Silfhout, C. (1996). Genetic variation for virulence and resistance in the wheat-Mycosphaerella graminicola pathosystem. I. Interactions between pathogen isolates and host cultivars. *Phytopathology*86(2), 200-212.

Kema, G. H. J. and van Silfhout, C. H. (1997). Genetic variation for virulence and resistance in the wheat-*Mycosphaerella graminicola* pathosystem III. Comparative seedling and adult plant experiments. *Phytopathology* 87(3), 266–272.

Khalili, A. and Rahimi, J., Siadat, H. and Farshad, A. (2018). *The Soils of Iran*. Cham: Springer.

Kidane, Y. G., Hailemariam, B. N., Mengistu, D. K., Fadda, C., Pè, M. E. and Dell'Acqua, M. (2017)Genome-Wide Association Study of Septoria tritici Blotch Resistance in Ethiopian Durum Wheat Landraces. *Frontiers in Plant Science* 8, 1586.

Krattinger, S. G., Lagudah, E. S., Spielmeyer, W., Singh, R. P., Huerta-Espino, J., McFadden, H., Bossolini, E., Selter, L. L. and Keller, B. (2009). A putative ABC transporter confers durable resistance to multiple fungal pathogens in wheat. *Science* 323(5919), 1360–1363.

Lal, R. and Stewart, B. A. (2012). *Soil Water and Agronomic Productivity*. Boca Raton, FL: CRC Press.

Latiri, K., Lhomme, J. P., Annabi, M. and Setter, T. L. (2010). Wheat production in Tunisia: progress, inter-annual variability and relation to rainfall. *European Journal of Agronomy* 33(1), 33–42.

Linde, C. C., Zhan, J. and McDonald, B. A. (2002). Population structure of *Mycosphaerella graminicola*: from lesions to continents. *Phytopathology* 92(9), 946–955.

Lourlki, S., Rehman, S., El Hanafi, A., Amri, A., Douira, A. and Tadesse, W. (2019, May 22–24). Identification of resistance sources and Genome wide association study of Septoria tritici blotch at seedling and adult plant stages in spring bread wheat *germplasm of ICARDA*. Paper presented at the International Symposium on Cereal Leaf Blights, Dublin, Ireland.

Lusser, M., Parisi, C., Plan, D. and Rodriguez-Cerezo, E. (2012). Deployment of new biotechnologies in plant breeding. *Nature Biotechnology* 30(3), 231–239.

Maccaferri, M., Sanguineti, M. C., Donini, P., Porcedu, E. and Tuberosa, R. (2005). *A Retrospective Analysis of Genetic Diversity in Durum Wheat Elite Germplasm Based on Microsatellite Analysis: A Case Study. Durum Wheat Breeding: Current Approaches and Future Strategies*. New York: Food Products Press, 99–142.

Mackey, J. (2005). Wheat: its concept, evolution and taxonomy. In: Royo, C., Nachit, M., Di Fonzo, N., Araus, J. L., Pfeiffer, W. H. and Slafer, G. A. (Eds) *Durum Wheat Breeding: Current Approaches and Future Strategies*. New York: Food Products Press, 3–61.

Mamluk, O. F. and Van Slageren, M. W. (1993). Resistance to common bunt, yellow rust, leaf rust and Septoria tritici blotch in wild einkorn and wild emmer wheat. *Phytopathologia Mediterranea* 32, 14–19.

Marcel, T., Ducasse, A., Compain, J., Lapalu, N., Chen, Z. C. Z., Hartmann, F. E., Confais, J., Goyeau, H., Palma-Guerrero, J., Kema, G. H. J., Amselem, J. and Croll, D. (2017). A first glimpse into the genetic architecture of pathogenicity of *Zymoseptoria tritici* on bread wheat. *29. Fungal Genetics Conference Asilomar 17*, Mar 2017, Pacific Grove, CA. p. 246. ⟨hal-01530825⟩.

Martínez-Moreno, F., Solís, I., Noguero, D., Blanco, A., Özberk, İ, Nsarellah, N., Elias, E., Mylonas, I. and Soriano, J. M. (2020). Durum wheat in the Mediterranean Rim: historical evolution and genetic resources. *Genetic Resources and Crop Evolution*67(6), 1415–1436.

Mastrangelo, A. M., Belloni, S., Barilli, S., Ruperti, B., Di Fonzo, N., Stanca, A. M. and Cattivelli, L. (2005). Low temperature promotes intron retention in two e-cor genes of durum wheat. *Planta* 221(5), 705–715.

Mazouz, H., Jlibene, M., Saadoui, E. M. and Bouami, F. (1995). Variabilité du pouvoir pathogène chez *Mycosphaerella graminocola* Fuckel (Schroeter) (Septoria tritici Rob. ex Desm) au Maroc. *Al Awamia* 88, 67–78.

McCartney, C. A., Brûlé-Babel, A. L., Lamari, L. and Somers, D. J. (2003). Chromosomal location of a race-specific resistance gene to *Mycosphaerella graminicola* in the spring wheat ST6. *TAG. Theoretical and Applied Genetics. Theoretische und Angewandte Genetik* 107(7), 1181-1186. doi: 10.1007/s00122-003-1359-0. Epub 2003 Jul 26. PMID: 12898022.

McDonald, B. A. and Linde, C. (2002). Pathogen population genetics, evolutionary potential, and durable resistance. *Annual Review of Phytopathology* 40(1), 349-379.

McDonald, B. A. and Mundt, C. C. (2016). How knowledge of pathogen population biology informs management of Septoria tritici blotch. *Phytopathology* 106(9), 948-955.

Meamiche Neddaf, H., Aouini, L., Bouznad, Z. and Kema, G. H. J. (2017). Equal distribution of mating type alleles and the presence of strobilurin resistance in Algerian *Zymoseptoria tritici* field populations. *Plant Disease* 101(4), 544-549.

Medini, M., Hamza, S., Rebai, A. and Baum, M. (2005). Analysis of genetic diversity in Tunisian durum wheat cultivars and related wild species by SSR and AFLP markers. *Genetic Resources and Crop Evolution* 52(1), 21-31.

Merchuk-Ovnat, L., Fahima, T., Ephrath, J. E., Krugman, T. and Saranga, Y. (2017). Ancestral QTL alleles from wild emmer wheat enhance root development under drought in modern wheat. *Frontiers in Plant Science* 8, 703.

Meyer, M., Cox, J. A., Hitchings, M. D. T., Burgin, L., Hort, M. C., Hodson, D. P. and Gilligan, C. A. (2017). Quantifying airborne dispersal routes of pathogens over continents to safeguard global wheat supply. *Nature Plants* 3(10), 780-786.

Miège, E. (1950). Les principales espèces et variétés de blé cultivées en Afrique du Nord (suite et fin). *Revuede Botanique Appliquéeet d'Agriculture Tropicale* 329-330, 203-215.

Moore, A. M. T., Hillman, G. C. and Anthony, J. (2000). *Village on the Euphrates: From Foraging to Farming at Abu Hureyra*. Oxford: Oxford University Press.

Morais, D., Duplaix, C., Sache, I., Laval, V., Suffert, F. and Walker, A.-S. (2018). Overall stability in the genetic structure of a *Zymoseptoria tritici* population from epidemic to interepidemic stages at a small spatial scale. *European Journal of Plant Pathology* 154(2), 423-436.

Mundt, C. C. (2014). Durable resistance: a key to sustainable management of pathogens and pests. *Infection, Genetics and Evolution: Journal of Molecular Epidemiology and Evolutionary Genetics in Infectious Diseases* 27, 446-455.

Nachit, M. (1998). Durum breeding research to improve dry-land productivity in the Mediterranean region. In: Nachit, M. M., et al. (Eds) *SEWANA Durum Research Network*. ICARDA Editions, 1-15.

Nazari, K. and Torabi, M. (2000). Distribution of yellow rust (Yr) resistance genes in Iran. *Acta Phytopathologica et Entomologica Hungarica* 35(1/4), 121-131.

Nefzaoui, A., Ketata, H. and El Mourid, M. (2012). Agricultural technological and institutional innovations for enhanced adaptation to environmental change in North Africa. In: *International Perspectives on Global Environmental Change*. Intech.

Newton, A. C., Akar, T., Baresel, J. P., Bebeli, P. J., Bettencourt, E., Bladenopoulos, K. V., Czembor, J. H., Fasoula, D. A., Katsiotis, A., Koutis, K., Koutsika-Sotiriou, M., Kovacs, G., Larsson, H., Pinheiro de Carvalho, M. A. A., Rubiales, D., Russell, J., dos Santos, T. M. M. and Vaz Patto, M. C. (2010). Cereal landraces for sustainable agriculture.

A review. *Agronomy for Sustainable Development* 30, 237–269. doi:10.1051/agro/2009032.
Oliveira, H. R., Campana, M. G., Jones, H., Hunt, H. V., Leigh, F., Redhouse, D. I., Lister, D. L. and Jones, M. K. (2012). Tetraploid wheat landraces in the Mediterranean basin: taxonomy, evolution and genetic diversity. *PLOS ONE* 7(5), e37063.
ONAGRI (2018). Available at: www.onagri.tn/uploads/filieres/cereales%20/tb-CORRIGe-cereales.pdf.
Ouaja, M., Aouini, L., Bahri, B., Ferjaoui, S., Medini, M., Marcel, T. C. and Hamza, S. (2020). Identification of valuable sources of resistance to *Zymoseptoria tritici* in the Tunisian durum wheat landraces. *European Journal of Plant Pathology*156(2), 647–661.
Payne, T. (2004). *The International Wheat Improvement Network (IWIN)*. Mexico: CIMMYT.
Prat, N., Guilbert, C., Prah, U., Wachter, E., Steiner, B., Langin, T., Robert, O. and Buerstmayr, H. (2016). QTL mapping of Fusarium head blight resistance in three related durum wheat populations. *Theoretical and Applied Genetics* 130(1), 13–27.
Rajaram, S., Borlaug, N. and Van Ginkel, M. (2002). CIMMYT international wheat breeding. In: Curtis, B. C., Rajaram, S. and Gomez, H. (Eds) Bread Wheat Improvement and Production. Rome: FAO, 103–117.
Rastoin, J. L. and Benabderrazek, H. (2014). Céréales et oléoprotéagineux au Maghreb: pour un codéveloppement de filières territorialisées. ⟨hal-02799980⟩.
REX Consortium (2016). Combining selective pressures to enhance the durability of disease resistance genes. *Frontiers in Plant Science* 7, 1916. doi:10.3389/fpls.2016.01916.
Reynolds, M. P. and Borlaug, N. E. (2006). Impacts of breeding on international collaborative wheat improvement. *Journalof Agricultural Science* 144(1), 3–17.
Robbana, C., Kehel, Z., Ben Naceur, M., Sansaloni, C., Bassi, F. andAmri, A.(2019). Genome-wide genetic diversity and population structure of Tunisian durum wheat landraces based on DArTseq technology. *International Journal of Molecular Sciences* 20(6), 1352.
Rosewarne, G. M., Singh, R. P., Huerta-Espino, J., William, H. M., Bouchet, S., Cloutier, S., McFadden, H. and Lagudah, E. S. (2006). Leaf tip necrosis, molecular markers and β1-proteasome subunits associated with the slow rusting resistance genes Lr46/Yr29. *TAG. Theoretical and Applied Genetics. Theoretische und Angewandte Genetik* 112(3), 500–508.
Saari, E. E. and Wilcoxson., R. D. (1974). Plant disease situation of high-yielding dwarf wheats in Asia and Africa. *Annual Reviewof Phytopathology* 12(1), 49–68.
Safavi, S. A. (2015). Effects of yellow rust on yield of race-specific and slow rusting resistant wheat genotypes. *Journal of Crop Protection* 4(3), 395–408.
Saint Pierre, C., Ben M'Barek, S., Azzimonti, G., Pereyra, S., Germán, S., Marza, F., Zucca, C., Nazari, K., Yahyaoui, A., Singh, P., Baum, M. and Braun, H. (2019). Global network of precision field-based wheat phenotyping platforms. 1st International Wheat Congress, Saskatoon-Canada, 21–26 July 2019.
Salamini, F., Özkan, H., Brandolini, A., Schäfer-Pregl, R. and Martin, W. (2002). Genetics and geography of wild cereal domestication in the near east. *Nature Reviews. Genetics* 3(6), 429–441.
Sankaran, S., Khot, L. R., Espinoza, C. Z., Jarolmasjed, S., Sathuvalli, V. R., Vandemark, G. J., Miklas, P. N., Carter, A. H., Pumphrey, M. O., Knowles, N. R. and Pavek, M. J. (2015). Low-altitude, high-resolution aerial imaging systems for row and field crop phenotyping: a review. *European Journal of Agronomy* 70, 112–123.

Savary, S., Willocquet, L., Pethybridge, S. J., Esker, P., McRoberts, N. and Nelson, A. (2019). The global burden of pathogens and pests on major food crops. *Nature Ecology and Evolution* 3(3), 430–439.

Schilling, J., Freier, K. P., Hertig, E. and Scheffran, J. (2012). Climate change, vulnerability and adaptation in North Africa with focus on Morocco. *Agriculture, Ecosystems and Environment* 156, 12–26.

Séguéla, J. (1946). Les étapes de l'amélioration génétique du blé dur en Tunisie. *Annales du Service Botanique et Agronomique de Tunisie* 19, 1–23.

Senapati, N., Brown, H. E. and Semenov, M. A. (2019). Raising genetic yield potential in high productive countries: designing wheat ideotypes under climate change. *Agricultural and Forest Meteorology* 271, 33–45.

Sheikh, F. A., Wani, S. H., Aziz, M. A. and Razvi, S. M. (2017). Role of wild relatives in imparting disease resistance to wheat rust. *International Journal of Pure and Applied Bioscience* 5(1), 645–659.

Shiferaw, B., Smale, M., Braun, H.-J., Duveiller, E., Reynolds, M. and Muricho, G. (2013). Crops that feed the world 10. Past successes and future challenges to the role played by wheat in global food security. *Food Security* 5(3), 291–317.

Siah, A., Elbekali, A. Y., Ramdani, A., Reignault, P., Torriani, S. F. F., Brunner, P. C. and Halama, P. (2014). QoI resistance and mitochondrial genetic structure of *Zymoseptoria tritici* in Morocco. *Plant Disease* 98(8), 1138–1144.

Singh, M., Mallick, N., Chand, S., Kumari, P., Sharma, J. B., Sivasamy, M., Jayaprakash, P., Prabhu, K. V., Jha, S. K. and Vinod (2017). Marker-assisted pyramiding of Thinopyrum-derived leaf rust resistance genes Lr19 and Lr24 in bread wheat variety HD2733. *Journal of Genetics* 96(6), 951–957.

Singh, R. P., Herrera-Foessel, S., Huerta-Espino, J., Singh, S., Bhavani, S., Lan, C. and Basnet, B. R. (2014). Progress towards genetics and breeding for minor genes based resistance to Ug99 and other rusts in CIMMYT high-yielding spring wheat. *Journal of Integrative Agriculture* 13(2), 255–261.

Singh, R. P., Hodson, D. P., Huerta-Espino, J., Jin, Y., Bhavani, S., Njau, P., Herrera-Foessel, S., Singh, P. K., Singh, S. and Govindan, V. (2011). The emergence of Ug99 races of the stem rust fungus is a threat to world wheat production. *Annual Review of Phytopathology* 49, 465–481.

Singh, R. P., Singh, P. K., Rutkoski, J., Hodson, D. P., He, X., Jørgensen, L. N., Hovmøller, M. S. and Huerta-Espino, J. (2016). Disease impact on wheat yield potential and prospects of genetic control. *Annual Review of Phytopathology* 54, 303–322.

Singh, R. P., Hodson, D. P., Jin, Y., Huerta-Espino, J., Kinyua, M. G., Wanyera, R., Singh, R. and Ward, R. W. (2006). Current status, likely migration and strategies to mitigate the threat to wheat production from race Ug99 (TTKS) of stem rust pathogen. *CAB Reviews: Perspectives in Agriculture, Veterinary Science, Nutrition and Natural Resources* 1(54), 1–13.

Singh, R. P., Hodson, D. P., Jin, Y., Lagudah, E. S., Ayliffe, M. A., Bhavani, S., Rouse, M. N., Pretorius, Z. A., Szabo, L. J., Huerta-Espino, J., Basnet, B. R., Lan, C. and Hovmøller, M. S. (2015). Emergence and spread of new races of wheat stem rust fungus: continued threat to food security and prospects of genetic control. *Phytopathology* 105(7), 872–884.

Singh, R. P., William, H. M., Huerta-Espino, J. and Rosewarne, G. (2004). Wheat rust in Asia: meeting the challenges with old and new technologies. Paper presented at the Proceedings of the 4th International Crop Science Congress.

Sivakumar, M. V., Lal, R., Selvaraju, R. and Hamdan, I. (2013). *Climate Change and Food Security in West Asia and North Africa*. The Netherlands: Springer.

Slim, A., Piarulli, L., Chennaoui Kourda, H., Rouaissi, M., Robbana, C., Chaabane, R., Pignone, D., Montemurro, C. and Mangini, G. (2019). Genetic structure analysis of a collection of Tunisian durum wheat germplasm. *International Journal of Molecular Sciences* 20(13), 3362.

Soleimani, V. D., Baum, B. R. and Johnson, D. A. (2002). AFLP and pedigree-based genetic diversity estimates in modern cultivars of durum wheat [Triticum turgidum L. subsp. durum (Desf.) Husn.]. *Theoretical and Applied Genetics* 104, 350-357.

Soriano, J. M., Villegas, D., Aranzana, M. J., del Moral, L. F. G. and Royo, C. (2016). Genetic structure of modern durum wheat cultivars and Mediterranean landraces matches with their agronomic performance. *PLoS ONE* 11(8), e0160983.

SPII-Cereal Department. Available at: http://www.spii.ir/en-US/DouranPortal/4954/page/Cereals-Research-Department-(CRD).

Stukenbrock, E. H. and Croll, D. (2014). The evolving fungal genome. *Fungal Biology Reviews* 28(1), 1-12.

Stukenbrock, E. H. (2014). Martin, F. (Ed.) *Advances in Botanical Research: Genomics of Plant-Related Fungi, Speciation Genomics of Fungal Plant Pathogens*. Elsevier, pp. 397-423.

Stukenbrock, E. H., Banke, S., Javan-Nikkhah, M. and McDonald, B. A. (2007). Origin and domestication of the fungal wheat pathogen Mycosphaerella graminicola via sympatric speciation. *Molecular Biology and Evolution* 24(2), 398-411.

Stukenbrock, E. H., Bataillon, T., Dutheil, J. Y., Hansen, T. T., Li, R., Zala, M., McDonald, B. A., Wang, J. and Schierup, M. H. (2011). The making of a new pathogen: insights from comparative population genomics of the domesticated wheat pathogen *Mycosphaerella graminicola* and its wild sister species. *Genome Research* 21(12), 2157-2166.

Stukenbrock, E. H. and McDonald, B. A. (2008). The origins of plant pathogens in agro-ecosystems. *Annual Review of Phytopathology* 46, 75-100. doi: 10.1146/annurev.phyto.010708.154114.

Suproniene, S., Kadziene, G., Irzykowski, W., Sneideris, D., Ivanauskas, A., Sakalauskas, S., Serbiak, P., Svegzda, P., Auskalniene, O. and Jedryczka, M. (2019). Weed species within cereal crop rotations can serve as alternative hosts for Fusarium graminearum causing Fusarium head blight of wheat. *Fungal Ecology* 37, 30-37.

Tadesse, W., Amri, A., Sanchez-Garcia, M., El Bouhssini, M., Karrou, M., Patil, S., Bassi, F. and Baum, M. (2017). *Improving Wheat Production in the Central and West Asia and North Africa (CWANA) Region*. Cambridge, UK: Burleigh Dodds Science Publishing.

Tadesse, W., Rajaram, S., Ogbonnaya, F., Sanchez-Garcia, M., Sohail, Q and Baum, M. (2016). Wheat. In: Singh, M. and Kumar, S. (Eds) *Broadening the Genetic Base of Grain Cereals*. India: Springer, 9-26.

Tester, M. and Langridge, P. (2010). Breeding technologies to increase crop production in a changing world. *Science* 327(5967), 818-822.

Tidiane Sall, A. C., Chiari, T., Legesse, W., Seid-Ahmed, K., Ortiz, R., van Ginkel, M. and Bassi, F. M. (2019). Durum wheat (*Triticum durum* Desf.): origin, cultivation and potential expansion in sub-Saharan Africa. *Agronomy* 9(5), 263.

Torabi, M. (2012). Epidemiology of wheat yellow rust *(Puccinia striiformis f. sp. tritici)* in Iran. Paper presented at the Meeting the Challenge of Yellow Rust in Cereal Crops.

Torabi, M., Mardoukhi, V., Nazari, K., Afshari, F., Forootan, A., Ramai, M., Golzar, H. and Kashani, A. (1995). Effectiveness of wheat yellow rust resistance genes in different parts of Iran. *Cereal Rusts and Powdery Mildews Bulletin* 23(1), 9-12.

Trethowan, R. and Crossa, J. (2007). Lessons learnt from forty years of international spring bread wheat trials. *Euphytica* 157(3), 385-390.

Trethowan, R. M., Ginkel, M. and Rajaram, S. (2002). Progress in breeding wheat for yield and adaptation in global drought affected environments. *Crop Science* 42(5), 1441-1446.

Ullrich, S. E. (2010). Barley: production, improvement, and uses. *World Agricultural Series*. Wiley-Blackwell, 656 pp.

Vaghefi, S. A., Keykhai, M., Jahanbakhshi, F., Sheikholeslami, J., Ahmadi, A., Yang, H. and Abbaspour, K. C. (2019). The future of extreme climate in Iran. *Scientific Reports* 9(1), 1464.

Valkoun, J. J. (2001). Wheat pre-breeding using wild progenitors. *Euphytica* 119(1-2), 17-23.

van Frank, G., Rivière, P., Pin, S ., Baltassat, R ., Berthellot, J.-F ., Caizergues, F ., Dalmasso, C., Gascuel, J.-S ., Hyacinthe, A ., Mercier, F ., Montaz, H ., Ronot, B. and Goldringer, I. (2020). Genetic diversity and stability of performance of wheat population varieties developed by participatory breeding. *Sustainability* 12(1), 384.

Vishwakarma, G., Sanyal, R. P., Shitre, A., Gadekar, D. A., Saini, A. and Das, B. K. (2019). Validation and marker-assisted selection of stem rust resistance gene Sr2 in Indian wheat using gel-based and gel-free methods. *Journal of Crop Science and Biotechnology* 22(4), 309-315.

Ware, S. B. (2006). Aspects of sexual reproduction in Mycosphaerella species on wheat and barley: genetic studies on specificity, mapping, and fungicide resistance. PhD thesis, Wageningen University, The Netherlands.

Watson, A., Ghosh, S., Williams, M. J., Cuddy, W. S., Simmonds, J., Rey, M. D., Asyraf Md Hatta, M., Hinchliffe, A., Steed, A., Reynolds, D., Adamski, N. M., Breakspear, A., Korolev, A., Rayner, T., Dixon, L. E., Riaz, A., Martin, W., Ryan, M., Edwards, D., Batley, J., Raman, H., Carter, J., Rogers, C., Domoney, C., Moore, G., Harwood, W., Nicholson, P., Dieters, M. J., DeLacy, I. H., Zhou, J., Uauy, C., Boden, S. A., Park, R. F., Wulff, B. B. H. and Hickey, L. T. (2018). Speed breeding is a powerful tool to accelerate crop research and breeding. *Nature Plants* 4(1), 23-29.

Wheat Atlas-Released Variety. Available at: http://wheatatlas.org/varieties.

Xie, W. and Nevo, E. (2008). Wild emmer: genetic resources, gene mapping and potential for wheat improvement. *Euphytica* 164(3), 603-614.

Xynias, I. N., Mylonas, I., Korpetis, E. G., Ninou, E., Tsaballa, A., Avdikos, I. D. and Mavromatis, A. G. (2020). Durum wheat breeding in the Mediterranean region: current status and future prospects. *Agronomy* 10(3), 432.

Yahyaoui, A. H., Ben M'Barek, S., Bockelman, H., Saint Pierre, C., Laribi, M., Madariaga, R., Castillo, D., Nefzaoui, M. and Araar, C. (2019). Novel sources of resistance to Zymoseptoria tritici in Mediterranean durum wheat landraces. 1st International Wheat Congress, Saskatoon-Canada, 21-26 July 2019.

Yahyaoui, A., Hakim, S., Al-Naimi, M. and Nachit, M. M. (2000). Multiple disease resistance in durum wheat (*Triticum turgidum* L. var. durum). In: Royo, C., Nachit, M., Di Fon zo, N., Araus, J. L. (Eds) *Durum Wheat Improvement in the Mediterranean Region: New Challenges*. Zaragoza: CIHEAM, 387-392.

Zahri, S., Farih, A. and Douira, A. (2014). Statut des principales maladies cryptogamiques foliaires du blé au Maroc en 2013. *Journal of Applied Biosciences* 77(1), 6543–6549.

Zeder, M. A. (2008). Domestication and early agriculture in the Mediterranean Basin: origins, diffusion, and impact. *Proceedings of the National Academy of Sciences of the United States of America* 105(33), 11597–11604.

Zhan, J., Pettway, R. E. and McDonald, B. A. (2003). The global genetic structure of the wheat pathogen *Mycosphaerella graminicola* is characterized by high nuclear diversity, low mitochondrial diversity, regular recombination, and gene flow. *Fungal Genetics and Biology: FG and B* 38(3), 286–297.

Zohary, D. and Hopf, M. (2000). *Domestication of Plants in the Old World: The Origin and Spread of Cultivated Plants in West Asia, Europe and the Nile Valley*. Oxford: Oxford University Press.

Index

http://dx.doi.org/10.19103/AS.2021.0092.10